Engineering Mechanics:
Dynamics

Gary L. Gray
Department of Engineering Science and Mechanics
Penn State

Francesco Costanzo
Department of Engineering Science and Mechanics
Penn State

Michael E. Plesha
Department of Engineering Physics
University of Wisconsin–Madison

Boston   Burr Ridge, IL   Dubuque, IA   New York   San Francisco   St. Louis
Bangkok   Bogotá   Caracas   Kuala Lumpur   Lisbon   London   Madrid   Mexico City
Milan   Montreal   New Delhi   Santiago   Seoul   Singapore   Sydney   Taipei   Toronto

The McGraw-Hill Companies

McGraw Hill

Higher Education

ENGINEERING MECHANICS: DYNAMICS

1 2 3 4 5 6 7 8 9 0 DOW/DOW 0 9

ISBN 978–0–07–282871–9
MHID 0–07–282871–4

Global Publisher: *Raghothaman Srinivasan*
Senior Sponsoring Editor: *Bill Stenquist*
Vice-President New Product Launches: *Michael Lange*
Developmental Editor: *Darlene M. Schueller*
Senior Marketing Manager: *Curt Reynolds*
Senior Project Manager: *Sheila M. Frank*
Senior Production Supervisor: *Sherry L. Kane*
Lead Media Project Manager: *Stacy A. Patch*
Digital Product Manager: *Daniel Wallace*
Senior Designer: *David W. Hash*
Cover/Interior Designer: *Greg Nettles/Squarecrow Design*
(USE) Cover Image: *SpaceShip Two ©Virgin Galactic*
Lead Photo Research Coordinator: *Carrie K. Burger*
Photo Research: *Sabina Dowell*
Compositor: *Aptara®, Inc.*
Typeface: *10/12 Times Roman*
Printer: *R. R. Donnelley Willard, OH*

Library of Congress has catalogued the main title as follows:

Costanzo, Francesco, 1964-
    Engineering mechanics : statics & dynamics / Francesco Costanzo, Michael E. Plesha, Gary L. Gray. -- 1st ed.
        p. cm.
    Includes index.
    Summary: This is a full version; do not confuse with 2 vol. set version (Statics 9780072828658 and Dynamics 9780072828719) which LC will not retain.
  ISBN 978–0–07–313412–3 — ISBN 0–07–313412–0 (hard copy : alk. paper) 1. Mechanics, Applied— Textbooks. I. Plesha, Michael E. II. Gray, Gary L. III. Title.
    TA350.C79 2010
    620.1--dc22

                                                                                                2008054286

www.mhhe.com

U.S. Customary and SI unit systems.

| Base Dimension | System of Units | |
	U.S. Customary	SI
force	pound (lb)	newton[a](N) $\equiv$ kg·m/s^2
mass	slug[a] $\equiv$ lb·s^2/ft	kilogram (kg)
length	foot (ft)	meter (m)
time	second (s)	second (s)

[a] Derived unit.

Conversion factors between U.S. Customary and SI unit systems.

	U.S. Customary		SI
length	1 in.	=	0.0254 m (2.54 cm, 25.4 mm)[a]
	1 ft (12 in.)	=	0.3048 m[a]
	1 mi (5280 ft)	=	1.609 km
force	1 lb	=	4.448 N
	1 kip (1000 lb)	=	4.448 kN
mass	1 slug (1 lb·s^2/ft)	=	14.59 kg

[a] Exact.

Common prefixes used in the SI unit systems.

Multiplication Factor		Prefix	Symbol
1 000 000 000 000 000 000 000 000	10^{24}	yotta	Y
1 000 000 000 000 000 000 000	10^{21}	zetta	Z
1 000 000 000 000 000 000	10^{18}	exa	E
1 000 000 000 000 000	10^{15}	peta	P
1 000 000 000 000	10^{12}	tera	T
1 000 000 000	10^{9}	giga	G
1 000 000	10^{6}	mega	M
1 000	10^{3}	kilo	k
100	10^{2}	hecto	h
10	10^{1}	deka	da
0.1	10^{-1}	deci	d
0.01	10^{-2}	centi	c
0.001	10^{-3}	milli	m
0.000 001	10^{-6}	micro	μ
0.000 000 001	10^{-9}	nano	n
0.000 000 000 001	10^{-12}	pico	p
0.000 000 000 000 001	10^{-15}	femto	f
0.000 000 000 000 000 001	10^{-18}	atto	a
0.000 000 000 000 000 000 001	10^{-21}	zepto	z
0.000 000 000 000 000 000 000 001	10^{-24}	yocto	y

Gary L. Gray is an Associate Professor of Engineering Science and Mechanics in the Department of Engineering Science and Mechanics at Penn State in University Park, PA. He received a B.S. in Mechanical Engineering (cum laude) from Washington University in St. Louis, MO, an S.M. in Engineering Science from Harvard University, and M.S. and Ph.D. degrees in Engineering Mechanics from the University of Wisconsin-Madison. His primary research interests are in dynamical systems, dynamics of mechanical systems, mechanics education, and multi-scale methods for predicting continuum-level properties of materials from molecular calculations. For his contributions to mechanics education, he has been awarded the Outstanding and Premier Teaching Awards from the Penn State Engineering Society, the Outstanding New Mechanics Educator Award from the American Society for Engineering Education, the Learning Excellence Award from General Electric, and the Collaborative and Curricular Innovations Special Recognition Award from the Provost of Penn State. In addition to dynamics, he also teaches mechanics of materials, mechanical vibrations, numerical methods, advanced dynamics, and engineering mathematics.

Francesco Costanzo is an Associate Professor of Engineering Science and Mechanics in the Engineering Science and Mechanics Department at Penn State. He received the Laurea in Ingegneria Aeronautica from the Politecnico di Milano, Milan, Italy. After coming to the U.S. as a Fulbright scholar he received his Ph.D. in aerospace engineering from Texas A&M University. His primary research interest is the mathematical and numerical modeling of material behavior. He has focused on the theoretical and numerical characterization of dynamic fracture in materials subject to thermo-mechanical loading via the use of cohesive zone models and various finite element methods, including space-time formulations. His research has also focused on the development of multi-scale methods for predicting continuum-level material properties from molecular calculations, including the development of molecular dynamics methods for the determination of the stress-strain response of nonlinear elastic systems. In addition to scientific research, he has contributed to various projects for the advancement of mechanics education under the sponsorship of several organizations, including the National Science Foundation. For his contributions, he has received various awards, including the 1998 and the 2003 GE Learning Excellence Awards, and the 1999 ASEE Outstanding New Mechanics Educator Award. In addition to teaching dynamics, he also teaches statics, mechanics of materials, continuum mechanics, and mathematical theory of elasticity.

Michael E. Plesha is a Professor of Engineering Mechanics in the Department of Engineering Physics at the University of Wisconsin-Madison. Professor Plesha received his B.S. from the University of Illinois-Chicago in structural engineering and materials, and his M.S. and Ph.D. from Northwestern University in structural engineering and applied mechanics. His primary research areas are computational mechanics, focusing on the development of finite element and discrete element methods for solving static and dynamic nonlinear problems, and the development of constitutive models for characterizing behavior of materials. Much of his work focuses on problems featuring contact, friction, and material interfaces. Applications include nanotribology, high temperature rheology of ceramic composite materials, modeling geomaterials including rock and soil, penetration mechanics, and modeling crack growth in structures. He is co-author of the book *Concepts and Applications of Finite Element Analysis* (with R. D. Cook, D. S. Malkus, and R. J. Witt). He teaches courses in statics, basic and advanced mechanics of materials, mechanical vibrations, and finite element methods.

The authors thank their families for their patience, understanding, and, most importantly, encouragement during the long years it took to bring these books to completion. Without their support, none of this would have been possible.

Brief descriptions of the chapter-opening photos and how they relate to dynamics are given below.

Chapter 1: Raphael's *School of Athens* depicts ancient Greek philosophers such as Aristotle, Plato, Euclid, and Pythagoras. This fresco celebrates the kinship that the renaissance humanists felt with the great minds from antiquity as they explored new ways of thinking about the arts, sciences, and engineering.

Chapter 2: Jim Wooding of the United States performing a long jump during the 1984 Olympic Games in Los Angeles, California. Projectile motion can be used to model the long jump.

Chapter 3: An aerobatic maneuver performed by the Red Arrows, the aerobatic team of the British Royal Air Force. Using Newton's second law we can relate the acceleration of these airplanes to the forces acting on them.

Chapter 4: Blanka Vlašić performing a high jump at the 2008 Indoor World Championship in Valencia, Spain. Using the work-energy principle we can see how speed is converted to height.

Chapter 5: The Space Shuttle Atlantis taking off from launch pad 39A at the Kennedy Space Center. The acceleration of the Shuttle due to the thrust of its rocket engines can be computed starting with the impulse-momentum principle.

Chapter 6: The Falkirk Wheel in central Scotland. This is a rotating lift system that connects the Union Canal with the Forth and Clyde Canal. Fixed axis rotations are an important special case of rigid body motions.

Chapter 7: Casey Stoner of the Ducati MotoGP Team, performing a turn with his racing motorcycle. By modeling the motorcycle as a rigid body we can determine the maximum acceleration of the motorcycle for which it does not slip or tip.

Chapter 8: Astronaut Dave Williams working at the installation of a new control moment gyroscope on the International Space Station. The concept of angular momentum is essential to understand the working principles of gyroscopes.

Chapter 9: The Thelonious Monk Quartet in 1957 at the Five Spot Café in New York City (from left to right, John Coltrane, Shadow Wilson, Thelonious Monk, and Ahmed Abdul-Malik). The music we hear is the result of vibrations induced in air by musical instruments.

Appendix A: The International Space Station as it existed in June 2008. Accurately determining its mass moments and products of inertia requires sophisticated computer codes, but the principles used in those codes are no different than what we study in this book.

Appendix B: Hurricane Isabelle on September 15, 2003 as seen from the International Space Station. Even though the air particles near the eye of a hurricane have a small moment arm relative to the eye, they contribute significantly to the total angular momentum because they have the highest speed.

BRIEF CONTENTS

Dynamics

TABLE OF CONTENTS

Dynamics is the science that relates motion to the forces that cause and are caused by that motion. Consequently, dynamics is at the heart of any branch of engineering dealing with the design and analysis of mechanical systems whose operating principles rely on motion or are meant to control motion. The engineering applications of dynamics are many and varied. Traditional applications include the design of mechanisms, engines, turbines, and airplanes. Other (perhaps less known) applications include the kinesiology of the human body, the analysis of cell motion, and the design of some micro- and nano-size devices including both sensors and actuators. All of these applications stem from the combination of kinematics, which describes the geometry of motion, with a few basic principles anchored in Newton's laws of motion, such as the work-energy and the impulse-momentum principles.

With this book we hope to provide a teaching and learning experience that is not only effective but also motivates the study and application of dynamics. We have structured the book to achieve four main objectives. First, we provide a rigorous introduction to the fundamental principles of particle and rigid body dynamics. In a constantly changing technological landscape, it is by relying on fundamentals that we can find new ways of applying what we know. Second, we motivate learning by a *problem-centered approach,* that is, by introducing the subject matter via concrete and relevant problems. Third, we incorporate those pedagogical principles that recent research in math, science, and engineering education has identified as essential for improving student learning. While it is commonly accepted that a good conceptual understanding is important to improve problem-solving skills, it has been discovered that problem-solving skills and concepts need to be taught in different ways. Fourth, have made *modeling* the underlying theme of our approach to problem solving. We believe that modeling, understood as the making of sensible assumptions to reduce a real complex problem to a simpler but solvable problem, is also something that must be taught and discussed alongside the basic principles. The four objectives that animate this textbook have been incorporated in a series of clearly identifiable features that are used consistently throughout the book. We believe these features make the book new and unique, and we hope that they will improve both the teaching and the learning experience.

This book is the second volume of a new Statics and Dynamics series. Let's see in detail what make these books different.

Why Another Statics and Dynamics Series?

These books provide thorough coverage of all the pertinent topics traditionally associated with statics and dynamics. Indeed, many of the currently available texts also provide this. However, the new books by Gray/Costanzo/Plesha offer several major innovations that enhance the learning objectives and outcomes in these subjects.

What Then Are the Major Differences between Gray/Costanzo/ Plesha and Other Engineering Mechanics Texts?

• A Consistent Approach to Problem Solving

The example problems in Gray/Costanzo/Plesha follow a structured four-step problem-solving methodology that will help you develop your problem-solving skills not only in statics and dynamics, but also in almost all other mechanics subjects that follow. This structured problem-solving approach consists of the following steps: Road Map & Modeling, Governing Equations, Computation, and Discussion & Verification. The Road Map provides some of the general objectives of the problem and develops a strategy for how the solution will be developed. Modeling is next, where a real-life problem is idealized by a model. This step results in the creation of a free body diagram and the selection of the balance laws needed to solve the problem. The Governing Equations step is devoted to writing all the equations needed to solve the problem. These equations typically include the Equilibrium Equations, and, depending upon the particular problem, Force Laws (e.g. spring laws or friction laws) and Kinematic Equations. In the Computation step, the governing equations are solved. In the final step, Discussion & Verification, the solution is interrogated to ensure that it is meaningful and accurate. This four-step problem-solving methodology is followed for all examples that involve a balance principle such as Newton's second law or the work-energy principle. Some problems (e.g., kinematics problems) do not involve balance principles, and for these the Modeling step is not needed.

• Contemporary Examples, Problems, and Applications

The examples, problem sets, and design problems were carefully constructed to help show you how the various topics of statics and dynamics are used in engineering practice. Statics and dynamics are immensely important subjects in modern engineering and science, and one of our goals is to excite you about these subjects and the career that lies ahead of you.

• A Focus on Design

A major difference between Gray/Costanzo/Plesha and other books is the systematic incorporation of design and modeling of real-life problems throughout. In statics, topics include important discussions on design, ethics, and professional responsibility. In dynamics, the emphasis is on parametric analysis and motion over ranges of time and space. These books show you that meaningful engineering design is possible using the concepts of statics and dynamics. Not only is the ability to develop a design very satisfying, but it also helps you develop a greater understanding of basic concepts and helps sharpen your ability to apply these concepts. Because the main focus of statics and dynamics textbooks should be the establishment of a firm understanding of basic concepts and correct problem-solving techniques, design topics do not have an overbearing presence in the books. Rather, design topics are included where they are most appropriate. While some of the discussions on design could be described as "common sense," such a characterization trivializes the importance and necessity for discussing pertinent issues such as safety, uncertainty in determining loads, the designer's responsibility to anticipate uses, even unintended uses, communications, ethics, and uncertainty in workmanship. Perhaps the most important feature of our inclusion of design and modeling topics is that you get a glimpse of what engineering is about and where your career in engineering is headed. The book is structured so that design topics and design problems are offered in a variety of places, and it is possible to pick when and where the coverage of design is most effective.

• Problem-Based Introductions

Many topics are presented using a problem-based introduction. By this approach, we hope to pique your interest and curiosity with a problem that has real-life significance and/or offers physical insight into the phenomena to be discussed. Using an interesting problem as a springboard, the necessary theory and/or tools needed to address the problem are developed. Problem-based introductions are used where they are especially effective, namely, topics that are challenging to visualize or understand.

• Computational Tools

Some examples and problems are appropriate for solution using computer software. The use of computers extends the types of problems that can be solved while alleviating the burden of solving equations. Such examples and problems give you insight into the power of computer tools and further insight into how statics and dynamics are used in engineering practice.

• Modern Pedagogy

Numerous modern pedagogical elements have been included. These elements are designed to reinforce concepts and they provide additional information to help you make meaningful connections with real-world applications. Marginal notes (i.e., Helpful Information, Common Pitfalls, Interesting facts, and Concept Alerts) help you place topics, ideas, and examples in a larger context. These notes will help you study (e.g., Helpful Information and Common Pitfalls), will provide real-world examples of how different aspects of statics and dynamics are used (e.g., Interesting Facts), and will drive home important concepts or dispel misconceptions (e.g., Concepts Alerts and Common Pitfalls). Mini Examples are used throughout the text to immediately and quickly illustrate a point or concept without having to wait for the worked-out examples at the end of the section. Occasionally in dynamics, the opportunity arises for in-depth discussions that are typically left for higher-level courses. Instead of shying away from these discussions we offer them in a feature called Advanced Topic, which we hope will help make connections with later courses and will motivate you in going beyond a first course in dynamics.

• Answers to Problems

Answers to most even-numbered problems are posted as a freely downloadable PDF file at www.mhhe.com/pgc. Providing answers in this manner allows for the inclusion of more complex information than would otherwise be possible. In addition to final numerical and/or symbolic answers, selected problems have more extensive information such as free body diagrams and/or plots for Computer Problems. This feature not only provides more complete answers in selected circumstances, but also provides a kick start that might help you on some homework problems. Furthermore, the free body diagrams that are provided for some problems as part of the model for that problem will give you the opportunity to practice constructing these on your own for extra problems. Appendix C gives an example of the additional information provided for a particular problem.

McGraw-Hill's 360° Development

McGraw-Hill's 360° Development Process is a continuous, market-oriented approach to building accurate and innovative print and digital products. It is dedicated to continual improvement driven by multiple customer feedback loops and checkpoints. This is initiated during the early planning stages of our new products and intensifies during the development and production stages, then begins again upon publication, in anticipation of the next edition.

This process is designed to provide a broad, comprehensive spectrum of feedback for refinement and innovation of our learning tools for both student and instructor. The 360° Development Process includes market research, content reviews, faculty and student focus groups, course- and product-specific symposia, accuracy checks, art reviews, and a Board of Advisors.

Here is a brief overview of the initiatives included in the 360° Development Process of the new statics and dynamics books.

Board of Advisors   A hand-picked group of trusted instructors active in teaching engineering mechanics courses served as chief advisors and consultants to the authors and editorial team during manuscript development. The Board of Advisors reviewed parts of the manuscript; served as a sounding board for pedagogical, media, and design concerns; and consulted on organizational changes.

Manuscript Review Panels   Numerous instructors reviewed the various drafts of the manuscript to give feedback on content, design, pedagogy, and organization. This feedback was summarized by the book team and used to guide the direction of the text.

Symposia   McGraw-Hill conducted several engineering mechanics symposia attended by instructors from across the country. These events are an opportunity for McGraw-Hill editors and authors to gather information about the needs and challenges of instructors teaching these courses. They also offered a forum for the attendees to exchange ideas and experiences with colleagues they might not have otherwise met.

Focus Group   In addition to the symposia, McGraw-Hill held a focus group with the authors and selected engineering mechanics professors. These engineering mechanics professors provided ideas on improvements and suggestions for fine tuning the content, pedagogy, and art.

Accuracy Check   A select group of engineering mechanics instructors reviewed the entire final manuscript for accuracy and clarity of the text and solutions.

Class Tests   Over a number of years, both books have been class tested by thousands of students. Dynamics has been carefully tested at Penn State. The students involved in the class testing have provided invaluable feedback via surveys on each individual chapter. In addition, some of these students have given more in-depth feedback through individual interviews and via the participation in focus groups.

Student Focus Groups   Student focus groups provided the editorial team with an understanding of how content and the design of a textbook impacts students' homework and study habits in the engineering mechanics course area.

Manuscript Preparation   The authors developed the manuscripts on Apple Macintosh laptop computers using LaTeX and Adobe Illustrator. This approach is novel to the publishing industry. The code generated by the authors was used to typeset the final manuscript. This approach eliminates the usual source of errors where the original authors' manuscript is rekeyed by the publisher to obtain the final manuscript.

The following individuals have been instrumental in ensuring the highest standard of content and accuracy. We are deeply indebted to them for their tireless efforts.

Board of Advisors

Janet Brelin-Fornari
Kettering University

Manoj Chopra
University of Central Florida

Pasquale Cinnella
Mississippi State University

Ralph E. Flori
Missouri University of Science and Technology

Christine B. Masters
Penn State

Mark Nagurka
Marquette University

David W. Parish
North Carolina State University

Gordon R. Pennock
Purdue University

Michael T. Shelton
California State Polytechnic University-Pomona

Joseph C. Slater
Wright State University

Arun R. Srinivasa
Texas A&M University

Carl R. Vilmann
Michigan Technological University

Ronald W. Welch
The University of Texas at Tyler

Robert J. Witt
University of Wisconsin-Madison

Reviewers

Makola M. Abdullah
Florida Agricultural and Mechanical University

Murad Abu-Farsakh
Louisiana State University

George G. Adams
Northeastern University

Farid Amirouche
University of Illinois at Chicago

Stephen Bechtel
Ohio State University

Kenneth Belanus
Oklahoma State University

Glenn Beltz
University of California-Santa Barbara

Haym Benaroya
Rutgers University

Sherrill B. Biggers
Clemson University

James Blanchard
University of Wisconsin-Madison

Janet Brelin-Fornari
Kettering University

Pasquale Cinnella
Mississippi State University

Ted A. Conway
University of Central Florida

Joseph Cusumano
Penn State

Bogdan I. Epureanu
University of Michigan

Ralph E. Flori
Missouri University of Science and Technology

Barry Goodno
Georgia Institute of Technology

Kurt Gramoll
University of Oklahoma

Hartley T. Grandin, Jr.
Professor Emeritus, Worcester Polytechnic Institute

Roy J. Hartfield, Jr.
Auburn University

Paul R. Heyliger
Colorado State University

James D. Jones
Purdue University

Yohannes Ketema
University of Minnesota

Carl R. Knospe
University of Virginia

Sang-Joon John Lee
San Jose State University

Jia Lu
The University of Iowa

Ron McClendon
University of Georgia

Paul Mitiguy
Consulting Professor, Stanford University

William R. Murray
California Polytechnic State University, San Luis Obispo

Mark Nagurka
Marquette University

Robert G. Oakberg
Montana State University

James J. Olsen
Wright State University

Chris Passerello
Michigan Technological University

Gary A. Pertmer
University of Maryland

David Richardson
University of Cincinnati

William C. Schneider
Texas A&M University

Sorin Siegler
Drexel University

Joseph C. Slater
Wright State University

Ahmad Sleiti
University of Central Florida

Arun R. Srinivasa
Texas A&M University

Josef S. Torok
Rochester Institute of Technology

John J. Uicker
Professor Emeritus, University of Wisconsin-Madison

David G. Ullman
Professor Emeritus, Oregon State University

Carl R. Vilmann
Michigan Technological University

Claudia M. D. Wilson
Florida State University

C. Ray Wimberly
University of Texas at Arlington

Robert J. Witt
University of Wisconsin-Madison

T. W. Wu
University of Kentucky

X. J. Xin
Kansas State University

Henry Xue
California State Polytechnic University, Pomona

Joseph R. Zaworski
Oregon State University

M. A. Zikry
North Carolina State University

Symposium Attendees

Farid Amirouche
University of Illinois at Chicago

Subhash C. Anand
Clemson University

Manohar L. Arora
Colorado School of Mines

Stephen Bechtel
Ohio State University

Sherrill B. Biggers
Clemson University

J. A. M. Boulet
University of Tennessee

Janet Brelin-Fornari
Kettering University

Louis M. Brock
University of Kentucky

Amir Chaghajerdi
Colorado School of Mines

Manoj Chopra
University of Central Florida

Pasquale Cinnella
Mississippi State University

Adel ElSafty
University of North Florida

Ralph E. Flori
Missouri University of Science and Technology

Walter Haisler
Texas A&M University

Kimberly Hill
University of Minnesota

James D. Jones
Purdue University

Yohannes Ketema
University of Minnesota

Charles Krousgrill
Purdue University

Jia Lu
The University of Iowa

Mohammad Mahinfalah
Milwaukee School of Engineering

Tom Mase
California Polytechnic State University, San Luis Obispo

Christine B. Masters
Penn State

Daniel A. Mendelsohn
The Ohio State University

Faissal A. Moslehy
University of Central Florida

LTC Mark Orwat
United States Military Academy at West Point

David W. Parish
North Carolina State University

Arthur E. Peterson
Professor Emeritus, University of Alberta

W. Tad Pfeffer
University of Colorado at Boulder

David G. Pollock
Washington State University

Robert L. Rankin
Professor Emeritus, Arizona State University

Mario Rivera-Borrero
University of Puerto Rico at Mayaguez

Hani Salim
University of Missouri

Brian P. Self
California Polytechnic State University, San Luis Obispo

Michael T. Shelton
California State Polytechnic University-Pomona

Lorenz Sigurdson
University of Alberta

Larry Silverberg
North Carolina State University

Joseph C. Slater
Wright State University

Arun R. Srinivasa
Texas A&M University

David G. Ullman
Professor Emeritus, Oregon State University

Carl R. Vilmann
Michigan Technological University

Anthony J. Vizzini
Mississippi State University

Andrew J. Walters
Mississippi State University

Ronald W. Welch
The University of Texas at Tyler

Robert J. Witt
University of Wisconsin-Madison

T. W. Wu
University of Kentucky

Musharraf Zaman
University of Oklahoma-Norman

Joseph R. Zaworski
Oregon State University

Focus Group Attendees

Janet Brelin-Fornari
Kettering University

Yohannes Ketema
University of Minnesota

Mark Nagurka
Marquette University

C. Ray Wimberly
University of Texas at Arlington

M. A. Zikry
North Carolina State University

Accuracy Checkers

Walter Haisler
Texas A&M University

Richard McNitt
Penn State

Mark Nagurka
Marquette University

ACKNOWLEDGMENTS

The authors would like to thank Jonathan Plant, former editor at McGraw-Hill, for his guidance in the early years of this project.

We are grateful to Chris Punshon for thoroughly proofreading the manuscript and for making suggestions to improve it. In addition, we would like to thank Chris Punshon, Andrew Miller, Chris O'Brien, Chandan Kumar, and Joseph Wyne for their substantial contributions to the solutions manual.

Most of all, we would like to thank Andrew Miller for infrastructure he created to keep the authors, manuscript, and solutions manual in sync. His knowledge of programming, scripting, subversion, and many other computer technologies made a gargantuan task feel just a little more manageable.

Chapter Introduction

Each chapter begins with an introductory section setting the purpose and goals of the chapter.

---

3 | *Force and Acceleration Methods for Particles*

In this chapter we show how Newton's second law, $\vec{F} = m\vec{a}$, is applied to study the motion of bodies that are modeled as particles. Throughout this chapter, Newton's second law is viewed as an *axiom*, i.e., a statement we treat as true and not derivable from other principles. As such, there is little to explain about the law itself. The chapter will therefore emphasize how $\vec{F} = m\vec{a}$ is applied, and it begins the study of *kinetics*, namely, the study of the forces that cause and are caused by motion. By the time we complete this chapter, we will be able to use the kinematics (discussed in Chapter 2) along with Newton's second law to either (1) predict the motion of a particle system caused by given forces or (2) determine the forces needed for a particle system to move in a prescribed way.

3.1 Rectilinear Motion

A chameleon capturing an insect

Suppose that a chameleon is propelling its long tongue out to snatch an insect for a meal (see Fig. 3.1). This process occurs so fast that a high-speed video camera needs to be used to capture the event. Since it is the "stickiness" of the chameleon's tongue that allows it to latch onto the insect, the question is, Can we use the video to estimate how much stickiness, i.e., force, is required to get the insect where the chameleon would like it to be?

We begin by describing the insect's motion based on the available video data, which tells us that it takes 0.15 s for the chameleon to completely retrieve the insect. If the insect is not initially moving, then the initial speed of the insect is zero. The final speed of the insect must be zero since it ends up in the chameleon's mouth for ingestion. You may recall the problem we studied in Section 2.2 in which a car travels between two stop signs—the motion of the insect is not unlike that of the car. As for the car, there are three considerations we need to take into account in generating the velocity versus time profile:

1. The time it takes for the chameleon to retrieve the insect.

2. The velocity, which must be zero at the start and end of the time interval.

3. The distance traveled by the insect from start to finish.

Figure 3.1
A chameleon capturing an insect.

🔑 **A Closer Look** This example demonstrates several features of the Coulomb friction model. First, Eq. (9) tells us that the crate will never move if $\cos\theta \leq \mu_s \sin\theta$ (since the denominator becomes negative). Also, the closer $\cos\theta$ is to $\mu_s \sin\theta$, the closer the denominator in Eq. (9) is to zero and the longer it takes to get the crate moving. 🖥️ ➡ An interesting representation of Eq. (9) can be found in Fig. 4, which shows t_s as a function of θ and μ_s. The black curve at the edge of the red region represents the θ and μ_s values for which the denominator of Eq. (9) is zero. Since the crate only moves for positive values of t_s, no matter how long we wait or how hard we push, the crate will never move for any θ and μ_s values above and to the right of the black curve. As we approach the black curve from below, it takes longer and longer to get the crate moving. For example, as θ approaches 90°, it becomes impossible to move the crate no matter how small μ_s is, unless $\mu_s = 0$. ⬅ 🖥️ Since friction is the key ingredient in this problem, it is illustrative to plot the friction force F as a function of time. Until the crate starts to move, Eq. (1) gives F as

$$F = P_0 t \cos\theta, \quad \text{for } t < t_s. \tag{11}$$

Computer Solutions

We make use of computer solutions in some problems and it is important that you be able to easily identify when this is the case. Therefore, anywhere a computer is used for a solution, you will see the symbol 🖥️. If this occurs within part of an example problem or within the discussion, then the part requiring the use of a computer will be enclosed in the following symbols 🖥️ ➡ ⬅ 🖥️. If one of the exercises requires a computer for its solution, then the computer symbol and its mirror image will appear on either side of the problem heading.

Mini-Examples

Mini-examples are used throughout the text to immediately and quickly illustrate a point or concept without having to wait for the worked-out examples at the end of the section. Mini-examples begin with the text ■ **Mini-Example.** and end with the symbol ———■.

■ **Mini-Example.** Using the planets Jupiter and Neptune (whose astronomical symbols are ♃ and ♆, respectively) as an example, the force on Jupiter due to the gravitational attraction of Neptune, $\vec{F}_{JN}$, is given by (see Fig. 1.3)

$$\vec{F}_{JN} = \frac{Gm_J m_N}{r^2}\,\hat{u}, \tag{1.7}$$

where r is the distance between the two bodies, m_J is the mass of Jupiter, m_N is the mass of Neptune, and $\hat{u}$ is a unit vector pointing from the center of Jupiter to the center of Neptune. The mass of Jupiter is 1.9×10^{27} kg and that of Neptune is 1.02×10^{26} kg. Since the mean radius of Jupiter's orbit is 778,300,000 km and that of Neptune is 4,505,000,000 km, one could assume that their closest approach to one another is approximately 3,727,000,000 km. Thus, at their closest approach, the magnitude of the force between these two huge planets is

$$|\vec{F}_{JN}| = \left(6.674 \times 10^{-11}\,\frac{\text{m}^3}{\text{kg}\cdot\text{s}^2}\right)\frac{(1.9 \times 10^{27}\,\text{kg})(1.02 \times 10^{26}\,\text{kg})}{(3.727 \times 10^{12}\,\text{m})^2} \tag{1.8}$$

$$= 9.31 \times 10^{17}\,\text{N}.$$

It is interesting to compare this force with the force of gravitation between Jupiter and the Sun. The Sun's mass is 1.989×10^{30} kg, and we have already stated that the mean radius of Jupiter's orbit is 778,300,000 km. Applying Eq. (1.7) between Jupiter and the Sun gives 4.16×10^{23} N, which is almost 450,000 times larger. ———■

E X A M P L E   3.4   *Projectile Motion with Drag*

Figure 1
Tiger Woods swinging an iron.

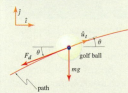

Figure 2
FBD of a golf ball in flight, modeled as a particle and subject to gravity and the aerodynamic drag force F_d. The unit vector $\hat{u}_t$ is tangent to the ball's path and indicates the direction of the ball's velocity.

Figure 3
Components of the velocity vector of the golf ball.

The projectile motion model presented in Section 2.3, in which a projectile is subject only to constant gravity and the trajectory is a parabola, is not adequate for studying the trajectory of objects like golf balls. In general, the trajectory of a golf ball is not a parabola and is significantly affected by many factors such as the dimple pattern on the ball, the spin imparted to the ball, the relative humidity of the atmosphere, the local density of the air, etc. Accounting for all of these factors is beyond the scope of this book. Here we can consider a simple improvement on the model in Section 2.3 by lumping the effects just mentioned into an aerodynamic drag, which we assume to be proportional to the square of the ball's speed and directed opposite to the ball's velocity. Use this model to derive the equations of motion of the golf ball.

SOLUTION ————

Road Map & Modeling   Referring to Fig. 2, we have modeled the golf ball as a particle subject to its own weight mg and a drag force F_d. We chose a Cartesian component system with the y axis parallel to gravity because it simplifies the representation of the weight force and it allows for an easy comparison between the current model and that in Section 2.3. The direction of the drag force is opposite to the ball's velocity, which is in the $\hat{u}_t$ direction. Although the orientation of $\hat{u}_t$ is currently unknown, for convenience we have oriented it via the time-dependent angle θ. As we saw in Section 3.1, the ball's equations of motion are derived by combining Newton's second law with the force laws and the kinematic equations.

Governing Equations

Balance Principles   Using the FBD in Fig. 2, Newton's second law, in component form, gives

$$\sum F_x: \qquad -F_d \cos\theta = ma_x, \tag{1}$$

$$\sum F_y: \quad -F_d \sin\theta - mg = ma_y. \tag{2}$$

Force Laws   Because F_d is proportional to the square of the speed, we have

$$F_d = C_d v^2, \tag{3}$$

where C_d is a *drag coefficient*[*] and v is the ball's speed.

Kinematic Equations   In the chosen component system, we have

$$a_x = \ddot{x} \quad \text{and} \quad a_y = \ddot{y}. \tag{4}$$

In addition, because the ball's velocity can be written as $\vec{v} = \dot{x}\,\hat{\imath} + \dot{y}\,\hat{\jmath}$ and the speed is the magnitude of $\vec{v}$, we have

$$v = \sqrt{\dot{x}^2 + \dot{y}^2}. \tag{5}$$

Finally, given that θ is the orientation of $\vec{v}$ relative to the x direction, we have

$$\cos\theta = \frac{\dot{x}}{v} = \frac{\dot{x}}{\sqrt{\dot{x}^2 + \dot{y}^2}} \quad \text{and} \quad \sin\theta = \frac{\dot{y}}{v} = \frac{\dot{y}}{\sqrt{\dot{x}^2 + \dot{y}^2}}, \tag{6}$$

where the components of $\vec{v}$ are depicted in Fig. 3.

[*] The coefficient C_d in Eq. (3) has dimensions of mass over length and should not be confused with the nondimensional drag coefficient normally used in aerodynamics.

Examples

Consistent Problem-Solving Methodology

Every problem in the text employs a carefully defined problem-solving methodology to encourage systematic problem formulation while reinforcing the steps needed to arrive at correct and realistic solutions.

Each example problem contains these four steps:

- **Road Map & Modeling**
- **Governing Equations**
- **Computation**
- **Discussion & Verification**

Some examples include a Closer Look (noted with a magnifying glass icon 🔎) that offers additional information about the example.

Concept Alerts and Concept Problems

Two additional features are the Concept Alert and the Concept Problems. These have been included because research has shown (and it has been our experience) that even though you may do quite well in a science or engineering course, your conceptual understanding may be lacking. **Concept Alerts** are marginal notes and are used to drive home important concepts (or dispel misconceptions) that are related to the material being developed at that point in the text. **Concept Problems** are mixed in with the problems that appear at the end of each section. These are questions designed to get you thinking about the application of a concept or idea presented within that section. They should never require calculation and should require answers of no more than a few sentences.

Concept Alert

Direction of velocity vectors. One of the most important concepts in kinematics is that the velocity of a particle is always tangent to the particle's path.

💡 **Problem 3.85** 💡

A car is being pulled to the right in the two ways shown. Neglecting the inertia of the pulleys and rope as well as any friction in the pulleys, if the car is allowed to roll freely, will the acceleration of the car in (a) be smaller, equal to, or larger than the acceleration of the car in (b)?
Note: Concept problems are about *explanations*, not computations.

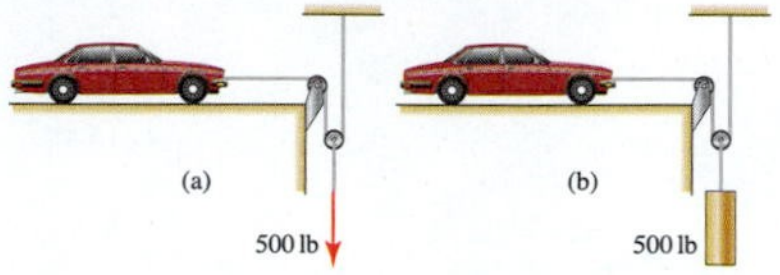

(a)   (b)

500 lb   500 lb

Figure P3.85

🎓 Advanced Topic 🎓
When is a force conservative?

Is there a way to tell whether or not a force is conservative? That is, can we determine whether or not an associated potential energy exists? If we establish that a potential energy exists, can we find it? Also, given a potential energy, can we find the associated force? Let's try to answer these questions.

Equation (4.31) implies that the infinitesimal work of a conservative force $\vec{F}$ can be written as

$$\vec{F} \cdot d\vec{r} = -dV, \qquad (4.40)$$

where we recall that a conservative force is only a function of position. In terms of Cartesian components, we can write the differential of V as

$$dV = \frac{\partial V}{\partial x} dx + \frac{\partial V}{\partial y} dy + \frac{\partial V}{\partial z} dz \qquad (4.41)$$

and $\vec{F} \cdot d\vec{r}$ as

$$\vec{F} \cdot d\vec{r} = (F_x\,\hat{\imath} + F_y\,\hat{\jmath} + F_z\,\hat{k}) \cdot (dx\,\hat{\imath} + dy\,\hat{\jmath} + dz\,\hat{k}) = F_x\,dx + F_y\,dy + F_z\,dz. \quad (4.42)$$

Inserting Eqs. (4.41) and (4.42) into Eq. (4.40), we see that

$$F_x\,dx + F_y\,dy + F_z\,dz = -\frac{\partial V}{\partial x} dx - \frac{\partial V}{\partial y} dy - \frac{\partial V}{\partial z} dz. \qquad (4.43)$$

Since dx, dy, and dz are independent of each other, for Eq. (4.43) to be satisfied, it must be true that

$$F_x = -\frac{\partial V}{\partial x}, \quad F_y = -\frac{\partial V}{\partial y}, \quad \text{and} \quad F_z = -\frac{\partial V}{\partial z}. \qquad (4.44)$$

Advanced Topic Feature

Our textbook is intended for use in introductory dynamics courses. However, occasionally the material presented leads naturally to questions and topics that are typically tackled in more advanced classes. While the answers to some of these deeper questions are indeed outside the scope of an introductory-level textbook, in some cases we felt that deferring discussion until a later course misses an opportunity for deeper learning, which can in turn help motivate students to take more advanced courses. Therefore, when compatible with the mathematical background of the intended audience, we have chosen to offer some brief departures into more challenging material in a feature called *Advanced Topic*. Since these discussions are meant to enrich the students' learning, this feature is set apart from the main body of the text by placing the words "Advanced Topic" between a pair of graduation caps, and by changing both the size and color of the font.

Marginal Notes

Marginal notes have been implemented that will help place topics, ideas, and examples in a larger context. This feature will help students study (using **Helpful Information** and **Common Pitfalls**) and will provide real-world examples of how different aspects of statics are used (using **Interesting Facts**).

Common Pitfall

About friction. A common misconception regarding the Coulomb friction model is that $F = \mu_s N$ when the body is not moving and $F = \mu_k N$ when the body is moving. When a wheel rolls without slip, it is moving, but the friction force F and the normal force N between the w... on which it rolls are su... In addition, the only t... $F = \mu_k N$ is when an... the surface on which... object and the surface... ities. When a body is... the surface on which it... say is that $|F| \le \mu_s|$... unless we know that... which case we can sa...

Interesting Fact

Engineers and trebuchets. Trebuchets were often called *engines* in Europe (from the Latin *ingenium*, or "an ingenious contrivance"). The people who designed, made, and used trebuchets were called *in-geniators*, and it is from this that we derive ...ngineer and *engineer*-...vedden, L. Eigenbrod, ...edel, "The Trebuchet," **273**(1), pp. 66–71, ...cle about trebuchets.

🧭 **Helpful Information**

Choosing a frame of reference. An important criterion for choosing a frame is *convenience*. Reference frames and (as we will see later in this chapter) coordinate systems profoundly impact how simply and directly the solution to a problem is obtained. Hence, a useful skill to cultivate is selection of the frame of reference and coordinate system that leads to the solution with greater ease.

Sections and End of Section Summary

Each chapter is organized into several sections. There is a wealth of information and features within each section, including examples, problems, marginal notes, and other pedagogical aids. Each section concludes with an end of section summary that succinctly summarizes the section. In many cases, cross-referenced important equations are presented again for review and reinforcement before the student proceeds to the examples and homework problems.

2.3   Projectile Motion

In this section we present a simple model to study the motion of projectiles: we model projectile motion as a constant acceleration motion.

Trajectory of a pumpkin used as a projectile

A pumpkin has been launched from a trebuchet (see Fig. 2.18) at the annual World Championship "Punkin Chunkin" competition in Millsboro, Delaware.[*] The pumpkin, shown in Fig. 2.19, is assumed to have been released at O with an initial speed v_0 and an elevation angle β. What is the trajectory of the pumpkin after it leaves the trebuchet? To answer this question, let's consider

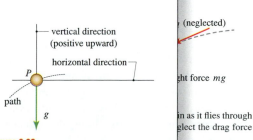

Figure 2.18
A working trebuchet (pronounced *treb-yoo-shay*).

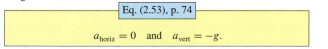

Figure 2.19. A pumpkin launched by a trebuchet showing the quantities described in the text. See Fig. 2.20 for some details on trebuchets.

the forces acting on the pumpkin while airborne. Referring to Fig. 2.21, these forces are the pumpkin's weight mg and the drag force $\vec{F}_d$ due to air resistance.

Figure 2.20
A trebuchet. The arm AB pivots clockwise about C due to the large counterweight D. As the arm reaches the vertical position, it launches the payload E to the right. The wheels at F and ~~out that when the~~ ~~rebuchet is allowed~~ ~~e is increased.~~

End of Section Summary

We defined *projectile motion* to be the motion of a particle in free flight, neglecting the forces due to air drag and neglecting changes in gravitational attraction with changes in height. In this case, referring to Fig. 2.23, the only force on the particle is the *constant* gravitational force, and the equations describing the motion are

Eq. (2.53), p. 74

$$a_{\text{horiz}} = 0 \quad \text{and} \quad a_{\text{vert}} = -g.$$

Since both the number 0 and the acceleration due to gravity g are constants, that is, since the horizontal and vertical components of the acceleration are constant, the constant acceleration equations developed in Section 2.2 can be applied both in the vertical and in the horizontal directions.

In projectile motion the trajectory is a parabola.   We showed that the trajectory of a projectile is a *parabola*. The mathematical form of the trajectory is of the type $y = C_0 + C_1 x + C_3 x^2$ if the motion is described using a Cartesian coordinate system with the y axis parallel to the direction of gravity.

vertical direction (positive upward)

horizontal direction

path

g

Figure 2.23
Figure 2.22 repeated. Acceleration of a point P in projectile motion.

PROBLEMS

Note: In all problems, all reference frames are stationary.

Problem 2.1

The position of a car traveling between two stop signs along a straight city block is given by $r = [9t - (45/2)\sin(2t/5)]$ m, where t denotes time and $0\,\text{s} \le t \le 17.7\,\text{s}$. Compute the displacement of the car between 2.1 and 3.7 s as well as between 11.1 and 12.7 s. For each of these time intervals compute the average velocity.

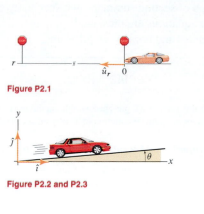

Figure P2.1

Problems 2.2 and 2.3

The position of the car relative to the coordinate system shown is

$$\vec{r}(t) = \left[(5.98t^2 + 0.139t^3 - 0.0149t^4)\,\hat{\imath} + (0.523t^2 + 0.0122t^3 - 0.00131t^4)\,\hat{\jmath} \right] \text{ft}.$$

Problem 2.2   Determine the velocity and acceleration of the car at $t = 15\,\text{s}$. In addition, again at $t = 15\,\text{s}$, determine the slope θ of the car's path relative to the coordinate system shown as well as the angle ϕ between velocity and acceleration.

Problem 2.3   Find the difference between the average velocity over the time interval $0\,\text{s} \le t \le 2\,\text{s}$ and the true velocity computed at the midpoint of the interval, i.e., at $t = 1\,\text{s}$. Repeat the calculation for the time interval $8\,\text{s} \le t \le 10\,\text{s}$. What do the results suggest about the approximation of the true velocity by the average velocity over different time intervals?

Figure P2.2 and P2.3

Problem 2.4

If $\vec{v}_{\text{avg}}$ is the average velocity of a point nitude of the average velocity, equal to t question?

Note: Concept problems are about *expl*

Problem 2.117

The end B of a robot arm is moving vertically down with a constant speed $v_0 = 2\,\text{m/s}$. Letting $d = 1.5\,\text{m}$, apply Eq. (2.62) to determine the rate at which r and θ are changing when $\theta = 37°$.

Problem 2.118

The end B of a robot arm is moving vertically down with a constant speed $v_0 = 6\,\text{ft/s}$. Letting $d = 4\,\text{ft}$, use Eqs. (2.62) and (2.64) to determine $\dot{r}$, $\dot{\theta}$, $\ddot{r}$, and $\ddot{\theta}$ when $\theta = 0°$.

Problem 2.119

A micro spiral pump consists of a spiral channel attached to a stationary plate. This plate has two ports, one for fluid inlet and the other for outlet, the outlet being farther from the center of the plate than the inlet. The system is capped by a rotating disk. The fluid trapped between the rotating disk and stationary plate is put in motion by the rotation of the top disk, which pulls the fluid through the spiral channel. With this in mind, consider a channel with geometry given by the equation $r = \eta\theta + r_0$, where $\eta = 12\,\mu\text{m}$

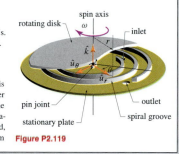

Figure P2.119

Modern Problems

Problems of varying difficulty follow each section. These problems allow students to develop their ability to apply concepts of statics and dynamics on their own. The most common question asked by students is "How do I set this problem up?" What is really meant by this question is "How do I develop a good mathematical model for this problem?" The only way to develop this ability is by practicing numerous problems. Answers to most even-numbered problems are posted as a freely downloadable PDF file at www.mhhe.com/pgc. Providing answers in this manner allows for more complex information than would otherwise be possible. In addition to final numerical or symbolic answers, selected problems have more extensive information such as free body diagrams and plots over intervals in space and/or time. Not only does this feature provide more complete answers in selected circumstances, but it also provides the kick start needed to get students started on some homework problems. Furthermore, the free body diagrams that are provided for some problems as part of the model for that problem will give students ample opportunity to practice constructing FBDs on their own for extra problems. Appendix C gives an example of the extensive information provided for a particular problem. Each problem in the book is accompanied by a thermometer icon that indicates the approximate level of difficulty. Those considered to be "introductory" are indicated with the symbol. Problems considered to be "representative" are indicated with the symbol, and problems that are considered to be "challenging" are indicated with the symbol.

Engineering Design and Design Problems

Several design problems are presented where appropriate throughout the book. These problems can be tackled with the knowledge and skill set that are typical of introductory-level courses, although the use of mathematical software is strongly recommended. These problems are open ended and their solution requires the definition of a parameter space in which the dynamics of the system must be analyzed. In this textbook we have chosen to emphasize the role played by *parametric analyses* in the overall design process, as opposed to cost-benefit analyses or the choice of specific materials and/or components.

DESIGN PROBLEMS

Design Problem 4.2

The plunger (the rod and rectangular block attached to the end of it) of a pinball machine has weight $W_p = 5$ oz. The weight of the ball is $W_b = 2.85$ oz, and the machine is inclined at $\theta = 8°$ with the horizontal. Design the springs k_1 and k_2 (i.e., their stiffnesses and unstretched lengths) so that the ball separates from the plunger with a speed $v = 15$ ft/s after pulling back the plunger 2 in. from its rest position and so that the plunger comes no closer than 0.5 in. from the stop. Note that as part of your design, you will also need to specify reasonable values for the dimensions ℓ_1 and ℓ_2.

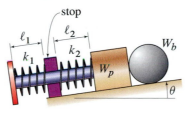

Figure DP4.2

2.9   Chapter Review

In this chapter we have presented some basic definitions needed to study the motion of objects. We have also developed some basic tools for the analysis of motion both in two and three dimensions. We now present a concise summary of the material covered in this chapter.

Position, velocity, acceleration

Position.   The position of a point is a *vector* going from the origin of the chosen frame of reference to the point in question. The position vectors of a point measured by different reference frames are different from one another.

Trajectory.   The trajectory of a moving point is the line traced by the point during its motion. Another name for trajectory is *path*.

Displacement.   The displacement between positions A and B is the vector going from A to B. In general, the magnitude of the displacement between two positions is not the distance traveled along the path between these positions.

Velocity.   The velocity vector is the time rate of change of the position vector. The velocity vector is the same with respect to any two reference frames that do not move relative to one another. The velocity is *always* tangent to the path.

Speed.   The speed is the magnitude of the velocity and is a nonnegative scalar quantity.

Acceleration.   The acceleration vector is the time rate of change of the velocity vector. As with the velocity, the acceleration is the same with respect to any two reference frames that are not moving relative to one another. Contrary to what happens for the velocity, the acceleration vector is, in general, not tangent to the trajectory.

Cartesian coordinates.   The Cartesian coordinates of a particle P moving along some path are shown in Fig. 2.63. The position vector is given by

> Eq. (2.13), p. 34
>
> $$\vec{r}(t) = x(t)\,\hat{\imath} + y(t)\,\hat{\jmath}.$$

In Cartesian components, the velocity and acceleration vectors are given by

> Eqs. (2.16) and (2.17), p. 34
>
> $$\vec{v}(t) = \dot{x}(t)\,\hat{\imath} + \dot{y}(t)\,\hat{\jmath} = v_x(t)\,\hat{\imath} + v_y(t)\,\hat{\jmath},$$
> $$\vec{a}(t) = \ddot{x}(t)\,\hat{\imath} + \ddot{y}(t)\,\hat{\jmath} = \dot{v}_x(t)\,\hat{\imath} + \dot{v}_y(t)\,\hat{\jmath} = a_x(t)\,\hat{\imath} + a_y(t)\,\hat{\jmath}.$$

Elementary motions and projectile motion

In applications, the acceleration of a point can be found as a function of time, position, and sometimes velocity.

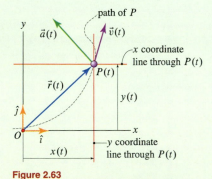

Figure 2.63
Figure 2.10 repeated. The position vector $\vec{r}(t)$ of the point P in Cartesian coordinates.

End-of-Chapter Review and Problems

Every chapter concludes with a succinct, yet comprehensive chapter review and a wealth of review problems.

REVIEW PROBLEMS

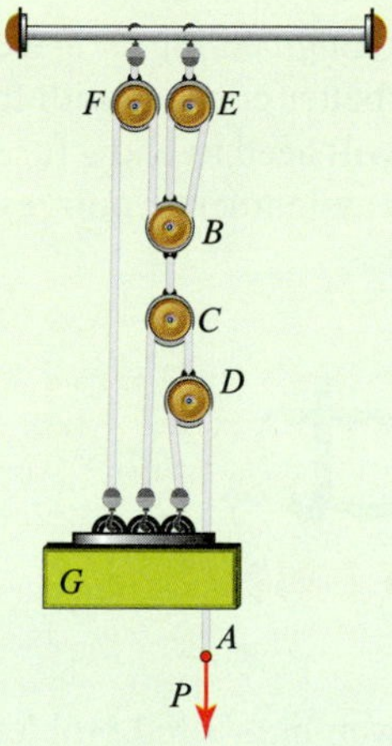

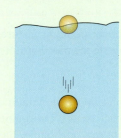

Figure P3.115 and P3.116

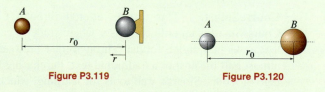

Figure P3.117 and P3.118

Problem 3.115

A constant force P is applied at A to the rope running behind the load G, which has a mass of 300 kg. Assuming that any source of friction and the inertia of the pulleys can be neglected, determine P such that G has an upward acceleration of $1\,\text{m/s}^2$.

Problem 3.116

A constant force $P = 300\,\text{lb}$ is applied at A to the rope running behind the load G, which weighs 1000 lb. If each of the pulleys weighs 7 lb, and assuming that any source of friction and the rotational inertia of the pulleys can be neglected, determine the acceleration of G and the tension in the rope connecting pulleys B and C.

Problem 3.117

A metal ball weighing 0.2 lb is dropped from rest in a fluid. If the magnitude of the resistance due to the fluid is given by $C_d v$, where $C_d = 0.5\,\text{lb·s/ft}$ is a drag coefficient and v is the ball's speed, determine the depth at which the ball will have sunk when the ball achieves a speed of 0.3 ft/s.

Problem 3.118

A metal ball weighing 0.2 lb is dropped from rest in a fluid. It is observed that after falling 1 ft, the ball has a speed of 2.25 ft/s. If the magnitude of the resistance due to the fluid is given by $C_d v$, where C_d is a drag coefficient and v is the ball's speed, determine the value of C_d.

Problem 3.119

Two particles A and B, with masses m_A and m_B, respectively, are a distance r_0 apart. Particle B is fixed in space, and A is initially at rest. Using Eq. (1.6) on p. 5 and assuming that the diameters of the masses are negligible, determine the time it takes for the two particles to come into contact if $m_A = 1\,\text{kg}$, $m_B = 2\,\text{kg}$, and $r_0 = 1\,\text{m}$. Assume that the two masses are infinitely far from any other mass.

Figure P3.119   **Figure P3.120**

Problem 3.120

The centers of two spheres A and B, with weights $W_A = 3\,\text{lb}$ and $W_B = 7\,\text{lb}$, respectively, are a distance $r_0 = 5\,\text{ft}$ apart when they are released from rest. Using Eq. (1.6) on p. 5, determine the speed with which they collide if the diameters of spheres A and B are $d_A = 2.5\,\text{in.}$ and $d_B = 4\,\text{in.}$, respectively. Assume that the two masses are infinitely far from any other mass.

What Resources Support This Textbook?

McGraw-Hill offers various tools and technology products to support *Engineering Mechanics: Statics* and *Engineering Mechanics: Dynamics.* Instructors can obtain teaching aids by calling the McGraw-Hill Customer Service Department at 1-800-338-3987, visiting our online catalog at www.mhhe.com, or contacting their local McGraw-Hill sales representative.

McGraw-Hill Connect Engineering. McGraw-Hill Connect Engineering is a web-based assignment and assessment platform that gives students the means to better connect with their coursework, with their instructors, and with the important concepts that they will need to know for success now and in the future. With Connect Engineering, instructors can deliver assignments, quizzes and tests easily online. Connect Engineering is available at www.mhhe.com/pgc.

Problem Answers. For the use of students, answers to most even-numbered problems are posted as a freely downloadable PDF file at www.mhhe.com/pgc. In addition to final numerical and/or symbolic answers, selected problems have more extensive information such as free body diagrams and/or plots for Computer Problems.

Solutions Manual. The Solutions Manual that accompanies the first edition features typeset solutions to the homework problems. The Solutions Manual is available on the password-protected instructor's website at www.mhhe.com/pgc.

VitalSource. VitalSource is a downloadable eBook. Students that choose the VitalSource eBook can save up to 45% off the cost of the print book, reduce their impact on the environment, and access powerful digital learning tools. Students can share notes with others, customize the layout of the eBook, and quickly search their entire eBook library for key concepts. Students can also print sections of the book for maximum portability. To learn more about VitalSource options, contact your sales representative or visit www.vitalsource.com.

CourseSmart. This text is offered through CourseSmart for both instructors and students. CourseSmart is an online browser where students can purchase access to this and other McGraw-Hill textbooks in a digital format. Through their browser, students can access the complete text online at almost half the cost of a traditional text. Purchasing the eTextbook also allows students to take advantage of CourseSmart's web tools for learning, which include full text search, notes and highlighting, and e-mail tools for sharing notes among classmates. To learn more about CourseSmart options, contact your sales representative or visit www.coursesmart.com.

Hands-on Mechanics.   Hands-on Mechanics is a website designed for instructors who are interested in incorporating three dimensional, hands-on teaching aids into their lectures. Developed through a partnership between the McGraw-Hill Engineering Team and the Department of Civil and Mechanical Engineering at the United States Military Academy at West Point, this website not only provides detailed instructions on how to build 3-D teaching tools using materials found in any lab or local hardware store but also provides a community where educators can share ideas, trade best practices, and submit their own demonstrations for posting on the site. Visit www.handsonmechanics.com.

Custom Publishing. Did you know that you can design your own text using any McGraw-Hill text and your personal materials to create a custom product that correlates specifically to your syllabus and course goals? Contact your McGraw-Hill sales representative to learn more about this option.

Engineering Mechanics:
Dynamics

Setting the Stage for the Study of Dynamics

The *dynamics* we study in this book is the part of mechanics concerned with the motion of bodies, the forces causing their motion, and/or the forces caused by their motion. Since the middle of the 20th century, dynamics has also included the study and analysis of any time-varying process, be it mechanical, electrical, chemical, biological, or some other kind. While we will focus on mechanical processes, much of what we will study is also applicable to other time-varying phenomena. Our goal is to provide an introduction to the science, skill, and art involved in modeling mechanical systems to predict their motion. We begin the study of dynamics with an overview of the part of its history that is relevant to this book, namely, that concerned with the motion of particles and rigid bodies. In Section 1.2, we review those elements of physics and vector algebra needed to develop the material in the remainder of the book. In Section 1.3, we conclude the chapter by touching upon the role of dynamics in engineering design.

1.1   A Brief History of Dynamics*

Early scientists and engineers were commonly called *philosophers*, and their noble undertaking was to use thoughtful reasoning to explain natural phenomena. Much of their focus was on understanding and describing motion of the Sun, Moon, planets, and stars. With few exceptions, their studies had to yield results that were intrinsically beautiful and/or compatible with the dominant religion of the time and location. What follows is a short historical survey of the major figures who profoundly influenced the development of dynamics.

Aristotle (384–322 B.C.) wrote about science, politics, and economics, and he proposed what is often called a "physics of common sense." He classified objects as either light or heavy, and he said that light objects fall slower than heavy objects. He recognized that objects can move other than up or down, that such motion is contrary to the body's natural motion, and that some force

---

*This history is culled from the excellent works of C. Truesdell, *Essays in the History of Mechanics*, Springer-Verlag, Berlin, 1968; I. Bernard Cohen, *The Birth of a New Physics*, revised and updated edition, W. W. Norton & Company, New York, 1985; and James H. Williams, Jr., *Fundamentals of Applied Dynamics*, John Wiley & Sons, New York, 1996.

must continuously act on the body for it to move this way. Most importantly, he said that the natural state of objects is for them to be at rest.

Claudius Ptolemæus (87–150*), who is commonly known as Ptolemy, authored a series of volumes now known as the *Almagest*. The *Almagest* presented the mathematical theory of the motions of the observable objects within our solar system. Ptolemy believed in a geocentric (Earth-centered) solar system, and he created a complex system of epicycles to explain the observed motion of the planets. Ptolemy's version of the solar system presented in the *Almagest* was not superseded until 1400 years later when the Polish astronomer Nicolaus Copernicus (1473–1543) presented his heliocentric (Sun-centered) theory in *De Revolutionibus* in 1543.

Johannes Kepler (1571–1630) believed in a heliocentric solar system. Using the huge amount of data on planetary motion collected by Tycho Brahe (1546–1601), he concluded that the assumption of circular planetary orbits with the Sun at their center did not match Brahe's data. The focus of Kepler's efforts was on the orbit of Mars, which has the largest orbital eccentricity (i.e., its orbit is the least circular) of the planets for which Kepler had good data. Kepler was able to explain the data of Brahe by making the planetary orbits *elliptical*, with the Sun at one focus of the ellipse, instead of circular. Over a period of almost 20 years, Kepler was able to formulate his three laws of planetary motion, *based entirely on observational data*.

1. The orbits of the planets are ellipses, with the Sun at one focus of the ellipse.

2. The line joining a planet to the Sun sweeps out equal areas in equal times as the planet travels around the Sun.

3. The ratio of the squares of the orbital periods for two planets is equal to the ratio of the cubes of their semimajor axes

$$\frac{P_1^2}{P_2^2} = \frac{r_1^3}{r_2^3}, \tag{1.1}$$

where P is the period, r is the length of the semimajor axis of the ellipse, and the subscripts 1 and 2 denote two different planets.

Galileo Galilei (1564–1642) had a strong interest in astronomy, and among his numerous discoveries was that Jupiter has moons. But for our purposes his most important contribution was his thought experiment in which he concluded that a body in its natural state of motion has *constant velocity*. Galileo also discovered the correct law for freely falling bodies; that is, the distance of the body's travel is proportional to the square of time. He also concluded that two bodies of different weight fall at the same rate and that any differences are due to air resistance.

---

* These dates are approximate as the exact years of Ptolemy's birth and death are unknown.

Isaac Newton (1643*–1727)

Newton (Fig. 1.1) is generally considered one of the greatest scientists of all time. He made important contributions to optics, astronomy, mathematics, and mechanics. His collection of three books entitled *Philosophiæ Naturalis Principia Mathematica*, or *Principia* as they are generally known, which were published in 1687, is considered by many to be the greatest collection of scientific books ever written.

In the *Principia*, Newton analyzed the motion of bodies, and he applied his results to orbital mechanics, projectiles, pendula, and objects in free fall near the Earth. By comparing his "law of centrifugal force" with Kepler's third law of planetary motion, Newton further demonstrated that the planets were attracted to the Sun by a force varying as the inverse square of the distance, and he generalized that all heavenly bodies mutually attract one another in the same way. In the first book of the *Principia* Newton developed his three laws of motion; in the second book he developed some concepts in fluid mechanics, waves, and other areas of physics; and in the third book he presented his law of universal gravitation. His contributions in the first and third books are especially significant to dynamics.

Newton's *Principia* was the final brick in the foundation of the laws that govern the motion of bodies. We say *foundation* because it took the work of Johann Bernoulli (1667–1748), Daniel Bernoulli (1700–1782), Jean le Rond d'Alembert (1717–1783), Joseph-Louis Lagrange (1736–1813), and Leonhard Euler (1707–1783) to clarify, refine, and advance mechanics into the form used today. Euler's contributions are especially notable since he used Newton's work to develop the theory for rigid body dynamics.

It has been said that "The principles of Newton suffice by themselves, without the introduction of any new laws, to explore thoroughly every mechanical phenomenon practically occurring."[†] While Newton's contribution is no doubt *foundational*, it is not the last word as this statement implies (and as many works in the history of science seem to indicate). Newton's work does have limitations, such as when dealing with very small objects (at the atomic scale) or with objects moving at close to the speed of light. The former is the realm of quantum mechanics, which is due to P. A. M. Dirac (1902–1984) and others, and the latter is the realm of relativity theory, which is due to Albert Einstein (1879–1955). In addition, Newton's concept of force was rather vague, and his notion of "bodies" sometimes refers to what we call point masses and sometimes refers to what we will call rigid bodies. Newton also showed no evidence of being able to derive differential equations of motion for mechanical systems, which is the cornerstone of modern dynamical analysis. It took the genius of Euler to show that Newton's laws applied to any point mass and that Newton's second law was not enough to describe the behavior of all mechanical systems. Finally, it was Augustin Cauchy (1789–1857) who developed the foundational theories of deformable bodies, thus beginning the area of study now known as continuum mechanics.

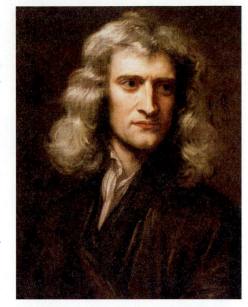

Figure 1.1
A portrait of Newton painted in 1689 by Sir Godfrey Kneller. It shows Newton before he went to London to take charge of the Royal Mint, when he was at his scientific peak.

Interesting Fact

Who invented calculus? Newton's works engendered considerable controversy during his lifetime (see, e.g., J. Gleick, *Isaac Newton*, Pantheon Books, 2003). His most famous argument was probably with Gottfried Wilhelm Leibniz (1646–1716) over which of the two had invented calculus (Newton called it the *method of fluxions*). While Newton was the first to discover calculus (about 10 years before Leibniz), Leibniz was the first to publish his own discovery (about 15 years before Newton) and to propose the notation we use today.

---

* This birth date is according to the Gregorian, or modern, calendar. According to the older Julian calendar, which was is use in England at that time, Newton's birth was in 1642 (this date is quoted by some sources). The difference between these calendars is due to how they treat leap years. At the time of Newton's birth, the Gregorian calendar was 10 days ahead of the Julian, and so Newton was born on December 25, 1642, by the Julian calendar and on January 4, 1643, by the Gregorian.

† This statement is attributed to Ernst Mach (1838–1916). See C. Truesdell, *Essays in the History of Mechanics*, Springer-Verlag, Berlin, 1968.

Newton's laws of motion

Newton's three laws of motion, stated in contemporary English, are as follows

First law *A particle remains at rest, or moves in a straight line with a constant speed, as long as the total force acting on the particle is zero.*

Second law *The time rate of change of momentum of a particle is equal to the resultant force acting on that particle.*

Third law *The forces of action and reaction between interacting particles are equal in magnitude, opposite in direction, and collinear.*

The second law, stated in modern mathematical notation, is

$$\vec{F} = \frac{d\vec{p}}{dt} = \frac{d(m\vec{v})}{dt}, \tag{1.2}$$

where $\vec{F}$ is the net force acting on the particle, $\vec{p}$ is the momentum of the particle, m is the mass of the particle, and $\vec{v}$ is the velocity of the particle. We have used the definition of momentum, which is $\vec{p} = m\vec{v}$. Throughout this book, we will denote vectors by using a superposed arrow ($\vec{\ }$). You are probably more familiar with seeing Newton's second law written as

$$\vec{F} = m\vec{a}, \tag{1.3}$$

which explicitly accounts for the fact that a particle is generally understood to have constant mass.* We shall learn in Chapter 3 that the first law is simply a special case of the second. The second and third laws, along with the ideas developed by Euler for rigid body dynamics, are all that is needed to solve a broad spectrum of problems involving particles and rigid bodies.

The third law, stated in modern mathematical notation, is

$$\vec{F}_{ij} = -\vec{F}_{ji}, \tag{1.4}$$

$$\vec{F}_{ij} \times (\vec{r}_i - \vec{r}_j) = \vec{0}, \tag{1.5}$$

where, referring to Fig. 1.2 for any interacting particles i and j, $\vec{F}_{ij}$ is the force on particle i due to particle j and $\vec{r}_i$ is the position of the ith particle. Some people refer to Newton's third law as just Eq. (1.4) while others require both Eqs. (1.4) and (1.5). Requiring both equations is sometimes referred to as the *strong form of Newton's third law*.[†]

Newton's universal law of gravitation

Newton used the laws postulated by Kepler, along with his laws of dynamics, to deduce the *law of universal gravitation*, which describes the force of attraction between two bodies. The gravitational force on a mass m_1 due to a mass

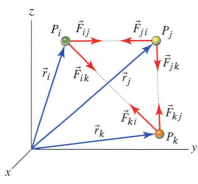

Figure 1.2
A system of particles interacting with one another.

---

* The application of Eq. (1.2) to variable mass systems will be considered in Section 5.5, and surprisingly, it does *not* take the form $\vec{F} = (dm/dt)\vec{v} + m(d\vec{v}/dt)$.

[†] Modern mechanics generally discards Newton's third law and replaces it with a much more general result called the *angular impulse-momentum principle*. In addition, it has been proposed since the 1950s that an even more general notion called *the principle of material frame indifference* could be used to replace Newton's third law. This latter principle states that the properties of materials and the actions of bodies on one another are the same for all observers.

m_2 a distance r away from m_1 is

$$\vec{F}_{12} = \frac{Gm_1m_2}{r^2}\,\hat{u}, \qquad (1.6)$$

where $\hat{u}$ is a unit vector pointing from m_1 to m_2 and G is the *universal gravitational constant** (sometimes called the *constant of gravitation* or *constant of universal gravitation*). The following example demonstrates the application of this law.

■ **Mini-Example.** Using the planets Jupiter and Neptune (whose astronomical symbols are ♃ and ♆, respectively) as an example, the force on Jupiter due to the gravitational attraction of Neptune, $\vec{F}_{JN}$, is given by (see Fig. 1.3)

$$\vec{F}_{JN} = \frac{Gm_Jm_N}{r^2}\,\hat{u}, \qquad (1.7)$$

where r is the distance between the two bodies, m_J is the mass of Jupiter, m_N is the mass of Neptune, and $\hat{u}$ is a unit vector pointing from the center of Jupiter to the center of Neptune. The mass of Jupiter is 1.9×10^{27} kg and that of Neptune is 1.02×10^{26} kg. Since the mean radius of Jupiter's orbit is 778,300,000 km and that of Neptune is 4,505,000,000 km, one could assume that their closest approach to one another is approximately 3,727,000,000 km. Thus, at their closest approach, the magnitude of the force between these two huge planets is

$$|\vec{F}_{JN}| = \left(6.674\times10^{-11}\,\frac{\text{m}^3}{\text{kg}\cdot\text{s}^2}\right)\frac{(1.9\times10^{27}\,\text{kg})(1.02\times10^{26}\,\text{kg})}{(3.727\times10^{12}\,\text{m})^2} \qquad (1.8)$$
$$= 9.31\times10^{17}\,\text{N}.$$

It is interesting to compare this force with the force of gravitation between Jupiter and the Sun. The Sun's mass is 1.989×10^{30} kg, and we have already stated that the mean radius of Jupiter's orbit is 778,300,000 km. Applying Eq. (1.7) between Jupiter and the Sun gives 4.16×10^{23} N, which is almost 450,000 times larger. ————————————————————————————■

Acceleration due to gravity. Equation (1.6) allows us to determine the force of Earth's gravity on an object of mass m on the surface of the Earth. This is done by noting that the radius of the Earth is 6371.0 km (see the marginal note) and the mass of the Earth is 5.9736×10^{24} kg and then applying Eq. (1.6):

$$F_s = \left(6.674\times10^{-11}\,\frac{\text{m}^3}{\text{kg}\cdot\text{s}^2}\right)\frac{(5.9736\times10^{24}\,\text{kg})m}{(6371.0\times10^3\,\text{m})^2} \qquad (1.9)$$
$$= (9.8222\,\text{m/s}^2)m.$$

This result[†] tells us that the force of gravity (in N) on a object on the Earth's surface is about 9.8 times the object's mass (in kg). This factor of 9.8 is so

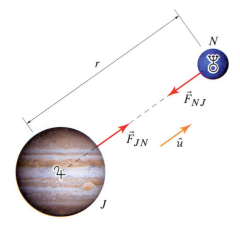

Figure 1.3
The gravitational force between the planets Jupiter (whose astronomical symbol is ♃) and Neptune (whose astronomical symbol is ♆). The relative sizes of the planets are accurate, but their separation distance is not.

Interesting Fact

The radius of the Earth. The Earth is not a perfect sphere. Therefore, there are different notions of "radius of the Earth." The given value of 6371.0 km is the *volumetric radius* when rounded to 5 significant digits. The Earth's volumetric radius is the radius of a perfect sphere with volume equal to that of the Earth. Other measures of the Earth's radius, rounded to 5 significant digits, are the *quadratic mean radius*, the *authalic mean radius*, and the *meridional Earth radius*, which are equal to 6372.8, 6371.0, and 6367.4 km, respectively.

---

* Henry Cavendish (1731–1810) was the first to measure G and did so in 1798. The generally accepted value is $G = 6.674\times10^{-11}$ m^3/(kg·s^2) $= 3.439\times10^{-8}$ ft^3/(slug·s^2).

† We will normally round the result of a calculation to 4 significant digits and round final results to 3 significant digits. Here we are using 5 significant figures because the data used in this particular calculation is known to such a degree of accuracy.

prevalent in engineering that it is given the label g, and it is called the *acceleration due to gravity* because it has units of acceleration and its value is the acceleration of objects in free fall near the surface of the Earth. We will take the value of g to be $9.81\,\text{m/s}^2$ in SI units and $32.2\,\text{ft/s}^2$ in U.S. Customary units. Notice that the value of g obtained in Eq. (1.9) is slightly greater than the $9.81\,\text{m/s}^2$ that we will use in this book. The difference between these values has several sources, including that the Earth is not perfectly spherical, does not have uniform mass distribution, and is rotating. Because of these sources, the actual acceleration due to gravity is about 0.27% lower at the equator, and 0.26% higher at the poles, relative to the standard value of $g = 9.81\,\text{m/s}^2$, which is for a north or south latitude of $45°$ at sea level. In addition, there may be small local variations in gravity due to geological formations. Nonetheless, throughout this book we will use the standard value of g stated above.

Change in acceleration due to altitude.   There is a formula that allows us to find how the acceleration due to gravity changes with altitude. To find it, we begin by equating Eqs. (1.3) and (1.6) to determine the acceleration a at a height h above the surface of the Earth

$$a = \frac{Gm_e}{(r_e + h)^2}, \tag{1.10}$$

where r_e is the radius of the Earth, m_e is the mass of the Earth, and we have canceled the mass of the object on both sides of the equation. Now, at the surface of the Earth, we know that $a = g$ and $h = 0$ so that Eq. (1.10) becomes

$$g = Gm_e/r_e^2 \quad \Rightarrow \quad Gm_e = gr_e^2. \tag{1.11}$$

Substituting Eq. (1.11) into Eq. (1.10), we see that a is given by

$$a = g\frac{r_e^2}{(r_e + h)^2}, \tag{1.12}$$

where g is the acceleration due to gravity at the surface of the Earth. Equation (1.12) is very handy because it requires knowledge of only the radius of the Earth to get the acceleration due to gravity rather than both the radius of the Earth *and* the universal gravitational constant G.

Leonhard Euler (1707–1783)

Leonhard Euler* (Fig. 1.4) spent a large part of his life taking the ideas of Newton, adding a huge number of his own ideas, and then developing the framework so that real problems could be solved. In fact, the dynamics we will present in this text is largely due to Euler. Euler realized that Newton's work was generally correct only when applied to particles (concentrated mass points). In addition, he was the first to recognize and employ acceleration as a kinematical quantity, and he was the first to use the concept of vector for velocity, acceleration, and other quantities (instead of just for forces). In 1744, Euler was the first to derive the exact differential equations of motion for a system of n bodies. This is the earliest example of what we now call the Newtonian method. It was Euler who, in 1752, first recognized that the "principle

Figure 1.4
Portrait of Leonhard Euler in 1753.

---

* Euler is pronounced as one would pronounce "oiler," not "you-ler."

of linear momentum" (i.e., Newton's second law) applied to mechanical systems of all kinds, whether discrete or continuous. In this work, he was the first to publish the "Newtonian" equations

$$F_x = ma_x, \qquad F_y = ma_y, \qquad F_z = ma_z, \tag{1.13}$$

where m is the mass. As we now know, Eqs. (1.13) are not enough to describe the behavior of *all* mechanical systems — we also need rotational equations of motion for finite-size bodies. To obtain these additional equations, Euler applied Eqs. (1.13) to the infinitesimal elements of mass that comprise a rigid body. In doing so, he wrote expressions for the acceleration of each element in terms of the *angular velocity vector* of the body. These expressions were reported in the same publication in which Eqs. (1.13) appeared, and this is the first time the angular velocity vector appears in mechanics. By taking the moments about the mass center of a body, he was able to derive the rotational equations of motion of a rigid body as well as to define the mass moments of inertia of a body. This was the first time the distinction was made between inertia and mass. The rotational equations of motion due to Euler are discussed in Chapter 7.

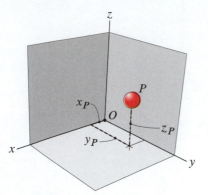

Figure 1.5
A point in a three-dimensional space.

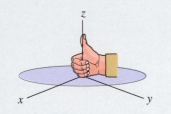

Helpful Information

The right-hand rule. In three dimensions a Cartesian coordinate system uses three orthogonal reference directions. These are the x, y, and z directions shown below.

Proper interpretation of many vector operations, such as the cross product, requires that the x, y, and z directions be arranged in a consistent manner. The convention in mechanics and vector mathematics in general is that if the axes are arranged as shown, then, according to the *right-hand rule*, rotating the x direction into the y direction yields the z direction. The result is called a *right-handed coordinate system*.

1.2 Fundamental Concepts

Space and time

Giving a precise meaning to the concepts of space and time is challenging.* Here we provide only the basic ideas about space and time used in this book.

Space

We view *space* as the environment in which objects move, and we consider it to be a collection of locations or *points*. The position of a point in space is indicated by specifying the point's coordinates relative to a chosen coordinate system. Figure 1.5 shows a three-dimensional *Cartesian coordinate system* with *origin* at O and mutually orthogonal axes x, y, and z. The *Cartesian coordinates* of the point P are x_P, y_P, and z_P, which are scalars obtained by measuring the distance between O and the perpendicular projections of the point P onto the axes x, y, and z, respectively. Note that x_P, y_P, and z_P have a positive or negative sign depending on whether, in going from O to the projections of P along each axis, one moves in the positive or negative direction of these axes. In Chapter 2 we will introduce two additional coordinate systems.

Time

We view *time* as a scalar variable that allows us to specify when an event occurs and to order a sequence of events. In classical mechanics and in this book, the most important assumption regarding time is that time is *absolute*. Specifically, we assume that the duration of an event is independent of the motion of the observer making time measurements and that the same clock can be used by all observers. Einstein's theory of relativity rejects this assumption.

Force, mass, and inertia

Force

The *force* acting on an object is the interaction between that object and its environment. A more precise description of this interaction requires that we know something about the interaction in question. For example, if two objects collide or slide against one another, we say that they interact via *contact* forces. Regardless of the type, a force has two essential characteristics: (1) magnitude and (2) direction. Therefore, we use *vectors* to mathematically represent forces.

Dynamics focuses on the description of the motion of a body under the influence of forces of any kind—what is important here is the motion-force relation and its mathematical description, which is rooted in Newton's laws of motion. Since these laws are stated in very general terms, we need to "specialize" their mathematical statement depending on the particular object we intend to study. In this book, we study only two types of bodies: particles and rigid bodies. Both of these are abstractions of real physical bodies and will be described later in this section.

---

* It is by a careful review of how we make measurements of space and time that Einstein's *theory of relativity* came to be (see, for example, P. G. Bergmann, *Introduction to the Theory of Relativity*, Dover Publications, Inc., 1976).

Mass

The *mass* of an object is a measure of the amount of matter in the object. Along with the concept of force, the concept of mass has been recognized as a *primitive concept*, i.e., not explainable via more elementary ideas.

Newton's second law postulates that the force acting on a body is *proportional* to the body's acceleration—the constant of proportionality is the *mass* of the body. It is important to understand that Newton's second law relates the concepts of mass and force, but it does not define either of them.

Inertia

Inertia is commonly understood as a body's resistance to changing its state of motion in response to the application of a force system. In this book, we use *inertia* as an umbrella term encompassing both the idea of mass and that of mass distribution over a region of space. We call *inertia properties* of an object the object's mass and a quantitative description of the mass distribution.

Particle and rigid body

Particle

In classical mechanics, a *particle* is an object whose mass is concentrated at a point and therefore is also called a point mass. The inertia properties of a particle consist only of the particle's mass. A particle is generally understood to have zero volume. It is meaningless to talk about the rotation of a particle whose position is held fixed, although we do say that a particle can "rotate about a point," meaning that a particle can move along a path around a point. Regardless of its volume, when we choose to model a real object as a particle, we choose to neglect the possibility that the object might "rotate" in the sense of "change its orientation" relative to some chosen reference object.

Rigid body

A rigid body is the other *model* of real physical objects that we consider in this book. A *rigid body* is an object whose mass is (1) distributed over a region of space and (2) such that the distance between any two points on it never changes. Since its mass is not concentrated at a point, the *rigid body* is the simplest model for the study of motions that include the possibility of rotation, i.e., a change of orientation relative to a chosen reference object. We model objects as rigid bodies when we want to account for the possibility of rotation while neglecting the effects of deformation. Finally, the mass distribution of a rigid body does not change relative to an observer moving with the body. This fact makes it possible to describe the inertia properties of a three-dimensional rigid body via seven pieces of information consisting of the body's mass and six mass moments of inertia.*

---

* For a definition of the mass moments of inertia of a rigid body, see Section 10.3 on p. 553 of M. E. Plesha, G. L. Gray, and F. Costanzo, *Engineering Mechanics: Statics*, McGraw-Hill, Dubuque, IA, 2010.

> **Interesting Fact**
>
> **History of mass.** The concept of mass was developed from the concept of weight (see I. B. Cohen, *The Birth of a New Physics*, W. W. Norton & Company, New York, 1985), the latter being the *force* Earth exerts on objects on its surface. In the theory of relativity the mass of an object is a function of the object's speed. For our purposes, the mass of an object is a constant independent of speed.

Vectors and their Cartesian representation

Notation

Scalars.   By *scalar* we mean a *real number*. Scalars will be denoted by italic roman characters (e.g., a, h, or W) or by Greek letters (e.g., α, ω, or δ).

Vectors.   In the text, we will *always* denote vectors by placing arrows over letters, such as $\vec{r}$ or $\vec{F}$.* The conventions we use to depict vectors in figures are shown in Fig. 1.6. The color scheme used in the figure is defined in the caption. Depending on what we want or need to emphasize in a figure, a vector will be labeled with a letter that has an arrow placed above it (e.g., $\vec{a}$ or $\vec{\omega}$) or with just a letter (e.g., a or ω). Specifically, we will use the following conventions:

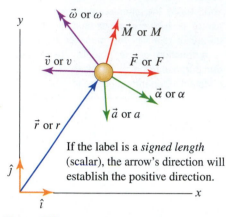

Figure 1.6
Notation and colors for commonly used vectors. Position vectors will always be blue ■ ($\vec{r}$), velocity (linear and angular) vectors purple ■ ($\vec{v}$ and $\vec{\omega}$), and acceleration (linear and angular) vectors green ■ ($\vec{a}$ and $\vec{\alpha}$). Forces and moments will always be red ■ ($\vec{F}$ and $\vec{M}$) and unit vectors orange ■ ($\hat{\imath}$ and $\hat{\jmath}$). Vectors with no particular physical significance will be black ■, magenta ■, or gray ■.

- In figures, a vector will be labeled with arrows over letters when it is important to emphasize the arbitrary directional nature of the vector (e.g., a velocity) or the vectorial nature of the quantity (e.g., a unit vector).

- Base vectors in Cartesian components will always be designated using the unit vectors $\hat{\imath}$, $\hat{\jmath}$, and $\hat{k}$. A *unit vector* is a vector with magnitude equal to 1. In any other context, e.g., in other component systems, unit vectors will be designated using a caret (sometimes called a "hat") over the letter u, that is, $\hat{u}$, often accompanied by a subscript indicating the direction of the vector, such as $\hat{u}_\theta$.

- In figures, the label of a vector with known direction will generally not be a letter with an arrow over it. A vector with known direction will most frequently be labeled as a *signed length* (i.e., a scalar component) whose positive direction is that of the arrow in the figure.

- Double-headed arrows will designate vectors associated with "rotational" quantities, i.e., moments, angular velocities, and angular accelerations.[†]

Cartesian vector representation

Here we review those aspects of vectors that are most important for our applications.[‡]

Figure 1.7 shows a two-dimensional description of the position of a point P with respect to the origin O of a rectangular coordinate system. The position of P is represented by the arrow that starts at O and ends at P, which we call the vector $\vec{r}_{P/O}$. The subscript "P/O" is to be read as "P relative to O," or "P as seen by O," or "P with respect to O." The vector $\vec{r}_{P/O}$ informs an observer of (1) the distance between P and O (magnitude) and (2) the orientation of segment $\overline{OP}$, where the bar over the letters O and P designates the line segment connecting the points O and P, with respect to a chosen reference direction.

The *Cartesian representation* of $\vec{r}$ is as follows:

$$\vec{r} = r_x\,\hat{\imath} + r_y\,\hat{\jmath}, \tag{1.14}$$

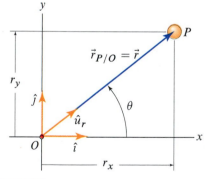

Figure 1.7
Description of the position of a particle P. The curved arrows indicating an angle with a single arrowhead designate an angle's positive direction.

---

* The vector notation used in the various engineering fields is diverse. Vectors may also be denoted by boldface letters, such as **v**, or by bars placed above or below a letter, such as $\bar{v}$ or $\underline{v}$.
† Angular velocities and accelerations will be discussed in Chapter 2.
‡ The discussion here is in two dimensions, but it is easily extendible to three dimensions.

where $\hat{i}$ and $\hat{j}$ are vectors of unit length in the x and y directions, respectively. The quantities r_x and r_y are the *(scalar) Cartesian components* of $\vec{r}$. Using trigonometry, we have

$$r_x = |\vec{r}|\cos\theta \quad \text{and} \quad r_y = |\vec{r}|\sin\theta, \tag{1.15}$$

where θ is the orientation of the segment $\overline{OP}$ relative to the x axis and $|\vec{r}|$, called the *magnitude of $\vec{r}$* or *length of $\vec{r}$*, is the length of $\overline{OP}$. Generalizing what we said about $\vec{r}_{P/O}$ and referring to Fig. 1.8, given points A and B with coordinates (x_A, y_A) and (x_B, y_B), respectively, the vector

$$\vec{r}_{A/B} = (x_A - x_B)\hat{i} + (y_A - y_B)\hat{j} \tag{1.16}$$

will be called *the position of A with respect to B*, or position of A relative to B.

Equation (1.14) could be written as $\vec{r} = \vec{r}_x + \vec{r}_y$, where the vectors $\vec{r}_x = r_x\hat{i}$ and $\vec{r}_y = r_y\hat{j}$ are called the x and y *vector* components of $\vec{r}$, respectively. In this book, *component* will always mean *scalar component*. When talking about *vector components*, we will explicitly say *vector components*.

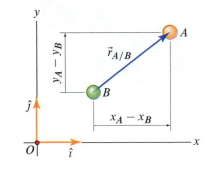

Figure 1.8
Vector representation of the position of A relative to B.

Vector operations

We now summarize the various vector operations we will use in this book.

1. A vector $\vec{r}$ can be multiplied by a scalar a in the following way:

$$a\vec{r} = ar_x\hat{i} + ar_y\hat{j}. \tag{1.17}$$

This *scales* the magnitude of $\vec{r}$ by the factor $|a|$. The object $a\vec{r}$ is a *vector* (not a scalar) with the same line of action as $\vec{r}$; the direction of $a\vec{r}$ is the same as that of $\vec{r}$ if $a > 0$, whereas it is opposite to $\vec{r}$ if $a < 0$.

2. Two vectors can be summed to obtain another vector as follows:

$$\vec{r} + \vec{w} = (r_x + w_x)\hat{i} + (r_y + w_y)\hat{j}, \tag{1.18}$$

which conforms to the triangle law of *vector addition* (see Fig. 1.9).

3. The operation of summing a scalar with a vector is not defined since it would be like "mixing apples and oranges."

4. Referring to Fig. 1.10, the *dot* or *scalar product* of two vectors $\vec{r}$ and $\vec{w}$ is denoted by $\vec{r} \cdot \vec{w}$ and yields the following *scalar* quantity:

$$\vec{r} \cdot \vec{w} = |\vec{r}||\vec{w}|\cos\theta. \tag{1.19}$$

5. Referring to Fig. 1.10, the *cross product* of two vectors $\vec{r}$ and $\vec{w}$ is the *vector* denoted by $\vec{r} \times \vec{w}$ with

   (a) magnitude
   $$|\vec{r} \times \vec{w}| = |\vec{r}||\vec{w}|\sin\theta, \tag{1.20}$$

   (b) line of action perpendicular to the plane containing $\vec{r}$ and $\vec{w}$, and direction determined by the right-hand rule.

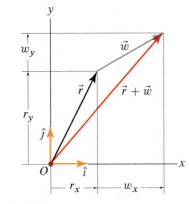

Figure 1.9
Graphical representation of the vector addition of $\vec{r}$ and $\vec{w}$ showing the "triangle law."

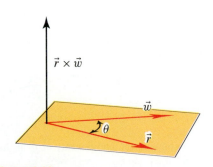

Figure 1.10
Graphical representation of the vector cross product of $\vec{r}$ and $\vec{w}$.

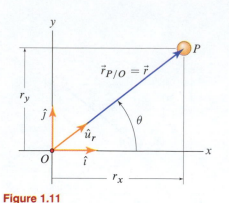

Figure 1.11

Figure 1.7 repeated. Description of the position of a particle P.

In Eqs. (1.19) and (1.20), θ is the smallest angle that will rotate one of the vectors into the other. For the cross product, such a choice of θ ensures that Eq. (1.20) will always yield a value greater than or equal to zero. For the dot product, since $\cos\theta = \cos(2\pi - \theta)$, θ can be replaced by $2\pi - \theta$. Finally, the definition of cross product implies that the cross product is *anticommutative*, that is,

$$\vec{r} \times \vec{w} = -\vec{w} \times \vec{r}. \tag{1.21}$$

Referring to Fig. 1.11, we recall that $\vec{r}$ represents the length and orientation (relative to some reference direction) of the segment $\overline{OP}$. Can we extract these pieces of information from r_x and r_y? From Fig. 1.11, using the Pythagorean theorem and trigonometry and choosing the x axis as reference direction, we see that

$$\text{length of } \vec{r} = |\vec{r}| = \sqrt{r_x^2 + r_y^2}, \tag{1.22}$$

and

$$\text{direction of } \vec{r} = \theta = \tan^{-1}\left(\frac{r_y}{r_x}\right). \tag{1.23}$$

Finally, observe that Eqs. (1.14) and (1.15) allow us to rewrite $\vec{r}$ as

$$\vec{r} = |\vec{r}|\,\hat{u}_r, \quad \text{where} \quad \hat{u}_r = \cos\theta\,\hat{\imath} + \sin\theta\,\hat{\jmath}. \tag{1.24}$$

Since $\hat{u}_r$ is a unit vector in the direction of $\vec{r}$, Eq. (1.24) implies that

> *The information carried by any vector can be written as the product of its magnitude and a unit vector pointing in the direction of that vector.*

Useful vector "tips and tricks"

Components of a vector

We now review how to find the components of a vector since this operation occurs so frequently in dynamics.

Figure 1.12 shows two perpendicular and oriented lines* ℓ_1 and ℓ_2. The lines are oriented using the unit vector $\hat{u}_1$ for ℓ_1 and $\hat{u}_2$ for ℓ_2. In addition, we have a vector $\vec{q}$ oriented arbitrarily relative to ℓ_1 and ℓ_2. Our goal is to find the scalar components of $\vec{q}$ along ℓ_1 and ℓ_2.

If we apply Eq. (1.19) to Fig. 1.12 and let $\vec{r}$ be $\vec{q}$, $\vec{w}$ be $\hat{u}_1$, and θ be θ_1, we see that the dot product gives us q_1 directly, that is,

$$q_1 = \vec{q} \cdot \hat{u}_1 = |\vec{q}||\hat{u}_1|\cos\theta_1 = |\vec{q}|\cos\theta_1. \tag{1.25}$$

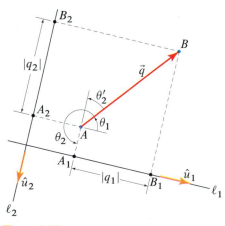

Figure 1.12

Diagram showing the components of $\vec{q}$ in the directions of $\hat{u}_1$ and $\hat{u}_2$.

The quantity q_1 in Eq. (1.25) is what we were looking for because, according to the definition of scalar component of a vector and Fig. 1.12,

1. $|q_1|$ is the distance between A_1 and B_1.

2. The sign of $\vec{q} \cdot \hat{u}_1$ is determined by the sign of $\cos\theta_1$, which is positive if $0° \le \theta_1 < 90°$ and negative if $90° < \theta_1 \le 180°$ (if $\theta_1 = 90°$, A_1 and B_1 coincide so that $q_1 = 0$).

---

* By *oriented*, we mean that they have a positive and a negative direction.

In summary,

$$\boxed{\text{component of } \vec{q} \text{ along } \ell_1 = q_1 = \vec{q} \cdot \hat{u}_1.} \qquad (1.26)$$

Repeating the foregoing discussion in the case of q_2 we have that the

$$\text{component of } \vec{q} \text{ along } \ell_2 = q_2 = \vec{q} \cdot \hat{u}_2 = |\vec{q}| \cos \theta_2 = -|\vec{q}| \cos \theta_2', \quad (1.27)$$

since $\theta_2 = \theta_2' + \pi$ and $\cos(\theta_2' + \pi) = -\cos \theta_2'$.

In dynamics we are often confronted with the situation depicted in Fig. 1.13, in which we need to calculate the Cartesian components of two mutually orthogonal vectors $\vec{q}$ and $\vec{r}$. If the angle θ defining the orientation of $\vec{q}$ relative to the y axis is given, to express $\vec{q}$ and $\vec{r}$ in components, we can write $\vec{q}$ as

$$\begin{aligned}
\vec{q} = (\vec{q} \cdot \hat{\imath}) \, \hat{\imath} + (\vec{q} \cdot \hat{\jmath}) \, \hat{\jmath} &= |\vec{q}| \cos(\theta + \tfrac{\pi}{2}) \, \hat{\imath} + |\vec{q}| \cos \theta \, \hat{\jmath} \\
&= -|\vec{q}| \sin \theta \, \hat{\imath} + |\vec{q}| \cos \theta \, \hat{\jmath} \\
&= |\vec{q}| \underbrace{(- \sin \theta \, \hat{\imath} + \cos \theta \, \hat{\jmath})}_{\text{unit vector}}, \qquad (1.28)
\end{aligned}$$

and then we can write $\vec{r}$ as

$$\begin{aligned}
\vec{r} = (\vec{r} \cdot \hat{\imath}) \, \hat{\imath} + (\vec{r} \cdot \hat{\jmath}) \, \hat{\jmath} &= |\vec{r}| \cos(\pi - \theta) \, \hat{\imath} + |\vec{r}| \cos(\theta + \tfrac{\pi}{2}) \, \hat{\jmath} \\
&= -|\vec{r}| \cos \theta \, \hat{\imath} - |\vec{r}| \sin \theta \, \hat{\jmath} \\
&= |\vec{r}| \underbrace{(- \cos \theta \, \hat{\imath} - \sin \theta \, \hat{\jmath})}_{\text{unit vector}}. \qquad (1.29)
\end{aligned}$$

Equations (1.28) and (1.29) demonstrate a very important and useful characteristic; that is, the two equations have the following structure:

- Each vector is equal to its magnitude times a unit vector with one sine and one cosine term.

- The argument of the sine and cosine terms is the angle orienting one of the vectors with respect to one of the component directions.

- In the two equations there will *always* be three positive terms and one negative term or three negative terms and one positive term. That is, three of the four sine and cosine terms will *always* be of one sign and only one of the terms will be of the opposite sign.

If, in planar problems, we take the components of two orthogonal vectors in two orthogonal directions and get anything other than this structure, then we should immediately recognize that we have made a mistake.

Cross products

Since we will often encounter cross products in dynamics, here is a useful procedure to help us evaluate them. Consider three unit vectors $\hat{\imath}$, $\hat{\jmath}$, and $\hat{k}$ such that, according to the right-hand rule, $\hat{\imath} \times \hat{\jmath} = \hat{k}$. Arrange the three vectors as shown in Fig. 1.14. To calculate the product of, say, $\hat{\jmath} \times \hat{k}$, just move around the circle, starting from $\hat{\jmath}$ and going toward $\hat{k}$. Now notice that (1) the next vector on the circle is $\hat{\imath}$ and (2) in going from $\hat{\jmath}$ to $\hat{k}$ we move with the arrow (counterclockwise). Hence, $\hat{\jmath} \times \hat{k} = +\hat{\imath}$. Now consider $\hat{k} \times \hat{\jmath}$ and notice that in going from $\hat{k}$ toward $\hat{\jmath}$, the next vector along the circle is $\hat{\imath}$, and we move opposite to the arrow. Therefore, the result is negative, and we have $\hat{k} \times \hat{\jmath} = -\hat{\imath}$.

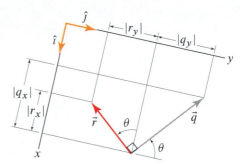

Figure 1.13
Diagram showing the Cartesian components of the vector $\vec{q}$ as well as the vector $\vec{r}$ that is orthogonal to $\vec{q}$.

Figure 1.14
A little "trick" to help remember the cross products between Cartesian unit vectors

Helpful Information

Cross products using determinants. You may be familiar with the following "determinant" method of evaluating the cross product of two vectors:

$$\vec{a} \times \vec{b} = \begin{vmatrix} \hat{\imath} & \hat{\jmath} & \hat{k} \\ a_x & a_y & a_z \\ b_x & b_y & b_z \end{vmatrix}.$$

For vectors in 3D, this method provides a very efficient evaluation. As an alternative, the cross product may be evaluated on a term-by-term basis by expanding the following product:

$$\begin{aligned}
\vec{a} \times \vec{b} = &\left(a_x \, \hat{\imath} + a_y \, \hat{\jmath} + a_z \, \hat{k} \right) \\
&\times \left(b_x \, \hat{\imath} + b_y \, \hat{\jmath} + b_z \, \hat{k} \right).
\end{aligned}$$

When expanded, nine terms such as $a_x \, \hat{\imath} \times b_x \, \hat{\imath}$ and $a_x \, \hat{\imath} \times b_y \, \hat{\jmath}$ must be evaluated, and this is accomplished quickly using Fig. 1.14. In this book, we primarily do cross products of vectors in 2D, and we will likely find the term-by-term evaluation to be quicker.

Units

Units are an essential part of any quantifiable measure. Newton's second law in scalar form, $F = ma$, provides for the formulation of a consistent and unambiguous system of units. We will employ both U.S. Customary units and SI units (International System*) as shown in Table 1.1. Each system has three

Table 1.1. U.S. Customary and SI unit systems

	System of units	
Base dimension	*U.S. Customary*	*SI*
force	pound (lb)	newton[a] (N) $\equiv$ kg·m/s^2
mass	slug[a] $\equiv$ lb·s^2/ft	kilogram (kg)
length	foot (ft)	meter (m)
time	second (s)	second (s)

[a] derived unit

base dimensions and a fourth *derived dimension*. In the U.S. Customary system, the base dimensions are force, length, and time, whose corresponding *base units* are lb (pounds), ft (feet), and s (seconds), respectively. The corresponding derived dimension is mass, which is obtained from the equation $m = F/a$. This gives the mass unit as lb·s^2/ft. This unit of mass is called the *slug*.

In the SI system, the base dimensions are mass, length, and time, whose corresponding base units are kg (kilogram), m (meter), and s (second), respectively. The corresponding derived dimension is force, the unit of which is obtained from the equation $F = ma$, which gives the force unit as kg·m/s^2. This unit of force is referred to as a *newton*, and its abbreviation is N.

Because of the difference in base dimensions between the U.S. Customary system and the SI system, when using the U.S. Customary system, we normally specify the weight of an object (typically in lb) instead of its mass; and, conversely, when using the SI system, we normally specify the mass of an object (typically in kg) instead of its weight.

For both systems, we may occasionally use different, but consistent, units for some dimensions. For example, we may use minutes rather than seconds, inches instead of feet, grams instead of kilograms. Nonetheless, the definitions of one newton and one slug are always as shown in Table 1.1.

Finally, note that the derived unit of angular measure in both the SI and U.S. Customary systems is the radian, which is dimensionless.

Dimensional homogeneity and unit conversions

When we use the symbol =, what is on the left-hand side of the symbol must be the same as what is on the right-hand side. Normally, this means that the left- and right-hand sides have the same numerical value, the same dimensions, and the same units.[†] Our strong recommendation is that appropriate

---

> **Common Pitfall**
>
> **Weight and mass.** Unfortunately, it is common to refer to weight using mass units. For example, the person who says, "I weigh 70 kg" really means "My mass is 70 kg". In science and engineering it is essential that accurate nomenclature be used. Specifically, weights and forces must be reported using appropriate force units, and masses must be reported using appropriate mass units.

---

* SI has been adopted as the abbreviation for the French *Le Système International d'Unités*.

† A simple example of an exception to this is the equation 12 in. = 1 ft. Such equations play a key role in performing unit conversions.

units always be used in all equations during a calculation to make sure that the results are dimensionally correct. Such practice helps avoid catastrophic blunders and provides a useful check on a solution, for if an equation is found to be dimensionally inconsistent, then an error has certainly been made. In September 1999, NASA (National Aeronautics and Space Administration) lost a $125 million Mars orbiter because the climate orbiter spacecraft team at the contractor who built the spacecraft used U.S. Customary units when computing rocket thrust while the mission navigation team at NASA used metric units for this key spacecraft operation. This units error, which came into play when the spacecraft was to be inserted into orbit around Mars, caused the spacecraft to approach Mars at too low an altitude, thus causing it to burn up in Mars' atmosphere.

Unit conversions are frequently needed and are easily accomplished using conversion factors, such as those shown in Table 1.2, and rules of algebra. The basic idea is to multiply either or both sides of an equation by dimensionless factors of unity, where each factor of unity embodies an appropriate unit conversion. This procedure is illustrated in the examples at the end of this section.

Prefixes

Prefixes are a useful alternative to scientific notation for representing numbers that are very large or very small. Common prefixes and a summary of rules for their use are given in Table 1.3.

Table 1.2

Conversion factors between U.S. Customary and SI unit systems.

	U.S. Customary	SI
length	1 in.	0.0254 m
		(25.4 mm)
	1 ft (12 in.)	0.3048 m
	1 mi (5280 ft)	1.609 km
force	1 lb	4.448 N
	1 kip (1000 lb)	4.448 kN
mass	1 slug (1 lb·s²/ft)	14.59 kg

Helpful Information

Abbreviation of inch. The abbreviation of the unit *inch* is "in.," as opposed to "in" — the period at the end of the abbreviation "in." helps avoid confusion in a sentence such as "Express your answer in in. and report it to three significant digits," where "in" means *using* whereas "in." is the required unit of length. If "in." occurs at the end of a sentence, then the period following "in" also indicates the end of the sentence.

Table 1.3. Common prefixes used in the SI unit systems.

Multiplication factor		Prefix	Symbol
1 000 000 000 000 000 000 000 000	10^{24}	yotta	Y
1 000 000 000 000 000 000 000	10^{21}	zetta	Z
1 000 000 000 000 000 000	10^{18}	exa	E
1 000 000 000 000 000	10^{15}	peta	P
1 000 000 000 000	10^{12}	tera	T
1 000 000 000	10^{9}	giga	G
1 000 000	10^{6}	mega	M
1 000	10^{3}	kilo	k
100	10^{2}	hecto	h
10	10^{1}	deka	da
0.1	10^{-1}	deci	d
0.01	10^{-2}	centi	c
0.001	10^{-3}	milli	m
0.000 001	10^{-6}	micro	μ
0.000 000 001	10^{-9}	nano	n
0.000 000 000 001	10^{-12}	pico	p
0.000 000 000 000 001	10^{-15}	femto	f
0.000 000 000 000 000 001	10^{-18}	atto	a
0.000 000 000 000 000 000 001	10^{-21}	zepto	z
0.000 000 000 000 000 000 000 001	10^{-24}	yocto	y

Helpful Information

Accuracy of numbers in calculations. Throughout this book, we will generally assume that the data given for problems is accurate to 3 significant digits. When calculations are performed, such as in example problems, all intermediate results are stored in the memory of a calculator or computer, using the full precision these machines offer. However, when these intermediate results are reported in this book, they are rounded to 4 significant digits. Final answers are usually reported with 3 significant digits. If you verify the calculations described in this book using the rounded numbers that are reported, you may occasionally calculate results that are slightly different from those shown.

Here is a list of common rules for correct prefix use:

1. With few exceptions, use prefixes only in the numerator of unit combinations; e.g., use the unit km/s (kilometer per second) and avoid the unit m/ms (meter per millisecond). One common exception to this rule is kg, which may appear in numerator or denominator; e.g., use the unit kW/kg (kilowatt per kilogram) and avoid the unit W/g (watt per gram).

2. Double prefixes must be avoided; e.g., use the unit GHz (gigahertz) and avoid the unit kMHz (kilo-megahertz).

3. Use a center dot or dash to denote multiplication of units, e.g., N·m or N-m. In this book, we denote multiplication of units by a dot, e.g., N·m.

4. Exponentiation applies to both the unit and prefix, e.g., mm² = (mm)².

5. When the number of digits on either side of a decimal point exceeds 4, it is common to group the digits into groups of 3, with the groups separated by commas or thin spaces. Since many countries use a comma to represent a decimal point, the thin space is sometimes preferable; e.g., 1234.0 could be written as is, but by contrast, 12345.0 should be written as 12,345.0 or as 12 345.0.

While prefixes can often be incorporated in an expression by inspection, the rules for accomplishing this are identical to those for performing unit transformations.

End of Section Summary

Review of vector operations. Referring to Fig. 1.15, the Cartesian representation of a two-dimensional vector $\vec{r}$ takes the form

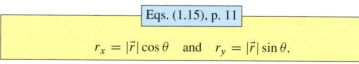

$$\vec{r} = r_x\,\hat{\imath} + r_y\,\hat{\jmath},$$

Eq. (1.14), p. 10

where $\hat{\imath}$ and $\hat{\jmath}$ are unit vectors in the positive x and y directions, respectively, and where r_x and r_y are the x and y (*scalar*) *components* of $\vec{r}$, respectively. Using trigonometry, r_x and r_y are given by

Eqs. (1.15), p. 11

$$r_x = |\vec{r}|\cos\theta \quad \text{and} \quad r_y = |\vec{r}|\sin\theta,$$

where θ is the orientation of the segment $\overline{OP}$ relative to the x axis and $|\vec{r}|$, called the *magnitude of* $\vec{r}$ or *length of* $\vec{r}$, is the length of $\overline{OP}$.

The *dot* or *scalar product* of the vectors $\vec{r}$ and $\vec{w}$ gives the *scalar*

Eq. (1.19), p. 11

$$\vec{r} \cdot \vec{w} = |\vec{r}||\vec{w}|\cos\theta,$$

with θ being the smallest angle that will rotate one of the vectors into the other.

Referring to Fig. 1.16, the *cross product* of two vectors $\vec{r}$ and $\vec{w}$ is denoted by $\vec{r} \times \vec{w}$ and gives a *vector* with

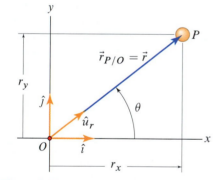

Figure 1.15

Figure 1.7 repeated. Description of the position of a particle P.

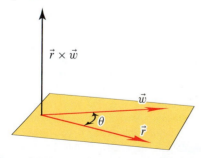

Figure 1.16

Figure 1.10 repeated. Graphical representation of the vector cross product of $\vec{r}$ and $\vec{w}$.

1. Magnitude equal to

Eq. (1.20), p. 11

$$|\vec{r} \times \vec{w}| = |\vec{r}||\vec{w}| \sin \theta,$$

where θ is the smallest angle that will rotate one of the vectors into the other (to ensure that $|\vec{r}||\vec{w}| \sin \theta \geq 0$).

2. Whose line of action is perpendicular to the plane containing $\vec{r}$ and $\vec{w}$ and whose direction is determined by the right-hand rule.

3. For which the cross product is anticommutative; i.e., $\vec{r} \times \vec{w} = -\vec{w} \times \vec{r}$.

Referring to Fig 1.15, given r_x and r_y, we compute $|\vec{r}|$ and θ as follows:

Eqs. (1.22) and (1.23), p. 12

$$\text{length of } \vec{r} = |\vec{r}| = \sqrt{r_x^2 + r_y^2},$$

$$\text{direction of } \vec{r} = \theta = \tan^{-1}\left(\frac{r_y}{r_x}\right).$$

Another useful representation of a vector $\vec{r}$ is as follows:

Eq. (1.24), p. 12

$$\vec{r} = |\vec{r}|\,\hat{u}_r, \quad \text{where} \quad \hat{u}_r = \cos\theta\,\hat{\imath} + \sin\theta\,\hat{\jmath},$$

where $\hat{u}_r$ is the unit vector in the direction of $\vec{r}$.

Useful vector "tips and tricks." Referring to Fig. 1.17, the component of a vector in a given direction can be computed using the dot-product. For example,

Eq. (1.25), p. 12

$$q_1 = \vec{q} \cdot \hat{u}_1 = |\vec{q}||\hat{u}_1| \cos\theta_1 = |\vec{q}| \cos\theta_1,$$

that is,

Eq. (1.26), p. 13

$$\text{component of } \vec{q} \text{ along } \ell_1 = q_1 = \vec{q} \cdot \hat{u}_1.$$

Referring to Fig. 1.18, for two mutually orthogonal vectors $\vec{r}$ and $\vec{q}$ the following relations hold:

Eqs. (1.28) and (1.29), p. 13

$$\vec{q} = (\vec{q} \cdot \hat{\imath})\,\hat{\imath} + (\vec{q} \cdot \hat{\jmath})\,\hat{\jmath} = |\vec{q}| \cos(\theta + \tfrac{\pi}{2})\,\hat{\imath} + |\vec{q}| \cos\theta\,\hat{\jmath}$$
$$= -|\vec{q}| \sin\theta\,\hat{\imath} + |\vec{q}| \cos\theta\,\hat{\jmath}$$
$$= |\vec{q}| (-\sin\theta\,\hat{\imath} + \cos\theta\,\hat{\jmath}),$$
$$\vec{r} = (\vec{r} \cdot \hat{\imath})\,\hat{\imath} + (\vec{r} \cdot \hat{\jmath})\,\hat{\jmath} = |\vec{r}| \cos(\pi - \theta)\,\hat{\imath} + |\vec{r}| \cos(\theta + \tfrac{\pi}{2})\,\hat{\jmath}$$
$$= -|\vec{r}| \cos\theta\,\hat{\imath} - |\vec{r}| \sin\theta\,\hat{\jmath}$$
$$= |\vec{r}| (-\cos\theta\,\hat{\imath} - \sin\theta\,\hat{\jmath}).$$

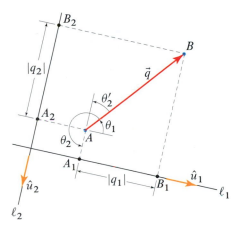

Figure 1.17
Figure 1.12 repeated. Diagram showing the components of $\vec{q}$ in the directions of $\hat{u}_1$ and $\hat{u}_2$.

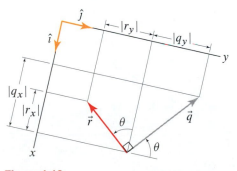

Figure 1.18
Figure 1.13 repeated. Diagram showing the Cartesian components of the vector $\vec{q}$ as well as the vector $\vec{r}$ that is orthogonal to $\vec{q}$.

E X A M P L E   1.1   *Position Vectors, Relative Position Vectors, and Components*

y, north

Figure 1

A map of the state of Minnesota with the locations of four of its cities. An east-north or xy coordinate frame with origin at O is also defined.

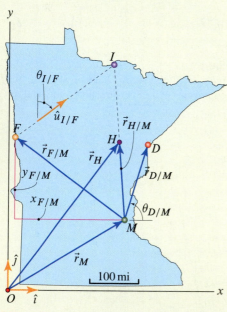

Figure 2

Vectors, projections, and angles needed to compute the quantities of interest.

The map of the state of Minnesota in Fig. 1 shows four of its cities and defines a coordinate system whose origin is at O. The coordinates of the four cities relative to O are given in Table 1. Assuming the Earth is flat and ignoring errors due to the map projection used, the x and y directions can be considered to be east and north, respectively. Using the information in Table 1, determine

(a) The position of Duluth (D) relative to Minneapolis/St. Paul (M), $\vec{r}_{D/M}$.

(b) The orientation, relative to north, of the position of International Falls (I) relative to Fargo/Moorhead (F).

(c) The east and north (scalar) components of the position of Fargo/Moorhead relative to Minneapolis/St. Paul.

(d) The position of the point H halfway between Minneapolis/St. Paul and International Falls.

Table 1. Coordinates of the four cities in the state of Minnesota shown in Fig. 1. All coordinates are relative to the origin O.

City	x, east (mi)	y, north (mi)
Minneapolis/St. Paul (M)	216	130
Duluth (D)	259	267
Fargo/Moorhead (F)	12	278
International Falls (I)	195	413

SOLUTION

―――――――――――――――――――― **Part (a)** ――――――――――――――――――――

Road Map   Since the coordinates of points D and M are available, the components of $\vec{r}_{D/M}$ can be found by using Eq. (1.16) on p. 11 to compute the difference between the coordinates of D and M.

Computation   Referring to Fig. 2 and Table 1 and then taking the difference between $\vec{r}_D$ and $\vec{r}_M$, we obtain

$$\begin{aligned} \vec{r}_{D/M} &= \vec{r}_D - \vec{r}_M = (x_D - x_M)\,\hat{\imath} + (y_D - y_M)\,\hat{\jmath} \\ &= [(259 - 216)\,\hat{\imath} + (267 - 130)\,\hat{\jmath}]\,\text{mi} \\ &= (43\,\hat{\imath} + 137\,\hat{\jmath})\,\text{mi} = 144\,\text{mi} \, @ \, 72.6°\measuredangle, \end{aligned} \tag{1}$$

where $\theta_{D/M} = 72.6°$.

Discussion & Verification   Referring to Fig. 2 and taking advantage of the scale indicated on the map, we can graphically verify that our answers are correct. We note that in this particular problem, the answer given as a length (144 mi) and a direction ($\theta_{D/M} = 72.6°$) is probably more straightforward to verify than the answer given in terms of Cartesian components, since we could directly measure the distance from Minneapolis/St. Paul to Duluth on the map itself.

―――――――――――――――――――― **Part (b)** ――――――――――――――――――――

Road Map   To determine the orientation of I relative to F, finding either the unit vector $\hat{u}_{I/F}$ or the angle $\theta_{I/F}$ will suffice. To find $\hat{u}_{I/F}$, we can find $\vec{r}_{I/F}$ and then divide by its magnitude (Eq. (1.22) on page 12).

Computation Again referring to Fig. 2 and Table 1 and using Eq. (1.16), the vector $\hat{u}_{I/F}$ is given by

$$\hat{u}_{I/F} = \frac{\vec{r}_{I/F}}{|\vec{r}_{I/F}|} = \frac{(195-12)\,\hat{\imath} + (413-278)\,\hat{\jmath}}{\sqrt{(195-12)^2 + (413-278)^2}} = 0.805\,\hat{\imath} + 0.594\,\hat{\jmath}. \quad (2)$$

Now, we can find the angle $\theta_{I/F}$ by applying Eq. (1.23), which gives

$$\theta_{I/F} = \tan^{-1}\left(\frac{x_{I/F}}{y_{I/F}}\right) = \tan^{-1}\left(\frac{183}{135}\right) = 53.6°, \quad (3)$$

which would be approximately northeast.

Discussion & Verification As with Part (a), we have expressed the answer in two different ways, and the version given in terms of an angle (i.e., Eq. (3)) probably allows for an easier verification that the solution is reasonable.

──────────────── **Part (c)** ────────────────

Road Map To find the east (x) and north (y) components of $\vec{r}_{F/M}$, we can use Eq. (1.26) on p. 13.

Computation Referring to Fig. 2 and using Eq. (1.26), we find that

$$x_{F/M} = \vec{r}_{F/M}\cdot\hat{\imath} = [(12-216)\,\hat{\imath} + (278-130)\,\hat{\jmath}]\cdot\hat{\imath}\text{ mi} = -204\text{ mi}, \quad (4)$$
$$y_{F/M} = \vec{r}_{F/M}\cdot\hat{\jmath} = [(12-216)\,\hat{\imath} + (278-130)\,\hat{\jmath}]\cdot\hat{\jmath}\text{ mi} = 148\text{ mi}, \quad (5)$$

where we have used the fact that $\hat{\imath}\cdot\hat{\imath} = \hat{\jmath}\cdot\hat{\jmath} = 1$ and $\hat{\imath}\cdot\hat{\jmath} = \hat{\jmath}\cdot\hat{\imath} = 0$.

Discussion & Verification We calculated that Fargo/Moorhead is 148 mi north of Minneapolis/St. Paul and it is 204 mi *west* (i.e., -204 mi east), which certainly seem reasonable given the figure.

──────────────── **Part (d)** ────────────────

Road Map As we can see from Fig. 2, the position of the point H, which is halfway between Minneapolis/St. Paul and International Falls, is given by the vector $\vec{r}_H$. The key to the solution is then seeing that we can write this position as $\vec{r}_H = \vec{r}_M + \vec{r}_{H/M}$.

Computation We begin with the decomposition of $\vec{r}_H$, which is given by

$$\vec{r}_H = \vec{r}_M + \vec{r}_{H/M}. \quad (6)$$

We can write $\vec{r}_{H/M}$ as

$$\vec{r}_{H/M} = \tfrac{1}{2}\vec{r}_{I/M} = \tfrac{1}{2}[(195-216)\,\hat{\imath} + (413-130)\,\hat{\jmath}]\text{ mi} = (-10.5\,\hat{\imath} + 141.5\,\hat{\jmath})\text{ mi}, \quad (7)$$

which, when substituted into Eq. (6), gives

$$\vec{r}_H = [(216\,\hat{\imath} + 130\,\hat{\jmath}) + (-10.5\,\hat{\imath} + 141.5\,\hat{\jmath})]\text{ mi} = (206\,\hat{\imath} + 272\,\hat{\jmath})\text{ mi}, \quad (8)$$

where we have used $\vec{r}_M = (216\,\hat{\imath} + 130\,\hat{\jmath})\text{ mi}$ from Table 1.

Discussion & Verification Again referring to Fig. 2, the result in Eq. (8) looks reasonable. In addition, the power of vectors starts to come into focus in this part of the example. That is, once we knew the position of International Falls relative to Minneapolis/St. Paul, computing the position at any fraction of the distance in between was trivial. Once that was computed, vector addition allowed us to easily find the position of the halfway point relative to the origin at O.

E X A M P L E 1.2   *Dimensional Analysis and Unit Usage*

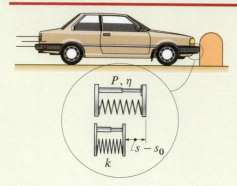

Figure 1
A car about to collide with a concrete block.

In a collision, the bumper of a car can undergo both elastic (i.e., reversible) as well as permanent deformation. A model for the force F transmitted by the bumper to the car is given by $F = P + k(s - s_0) + \eta\, ds/dt$, where $s - s_0$ is the compression experienced by the bumper and has the dimension of length, s_0 is a constant with dimension of length, and ds/dt is the time rate of change of s, which has the dimension of length over time. The quantity P is the force needed to permanently deform the bumper, k is the stiffness of the system, and η is a constant that relates the overall force to the speed of deformation. Determine

(a) The dimensions of P, k, and η, and

(b) The units that these quantities would have in the SI and the U.S. Customary systems.

SOLUTION

-- **Part (a)** --

Road Map   The first step in dimensional analysis is the identification of a basic relation, such as a law of nature, containing the quantities to analyze and for which the dimensions are known. Since we are dealing with the expression of a force, we can use Eq. (1.3) on p. 4 as the basic relation. Notice that the given force law consists of the sum of three terms. For this sum to be meaningful, each of the terms in question must have dimensions of force. We call this property *dimensional homogeneity*, and it is the key to solving the given problem.

Computation   Let L, M, and T denote length, mass, and time, respectively. Writing "[*something*]" to mean the "dimensions of *something*," for Eq. (1.3), we have

$$[F] = [ma] = [m][a] = M\frac{L}{T^2}. \tag{1}$$

Considering the given force law, we have

$$[F] = \left[P + k(s - s_0) + \eta\frac{ds}{dt} \right] = [P] + [k(s - s_0)] + \left[\eta\frac{ds}{dt} \right]. \tag{2}$$

> ## ⓘ Concept Alert
>
> **Dimensional homogeneity.** In writing Eq. (2) we have used a property stating that "[something + something else] = [something] + [something else]." This property expresses the requirement that the sum of two physical quantities makes sense only when these quantities have the same dimensions. Using a more formal language, we say that the quantities in question must satisfy *dimensional homogeneity* or, equivalently, must be *dimensionally homogeneous*.

Comparing Eq. (1) with Eq. (2) and enforcing dimensional homogeneity between them, we see that $[P]$, $[k(s-s_0)]$, and $[\eta\, ds/dt]$ must each be ML/T^2. Therefore the quantity P must have the dimensions of force. For the term k, we have

$$[k(s - s_0)] = [k][s - s_0] = [k]L = M\frac{L}{T^2}, \tag{3}$$

where we have used the fact that the dimension of s and s_0 is L. Simplifying the last equality in Eq. (3), we have

$$[k] = \frac{M}{T^2}. \tag{4}$$

Next, considering the term with η in Eq. (2), we have

$$\left[\eta\frac{ds}{dt} \right] = [\eta]\left[\frac{ds}{dt} \right] = [\eta]\frac{L}{T} = M\frac{L}{T^2}, \tag{5}$$

where we have used the fact that the dimensions of ds/dt are L/T. Simplifying the last equality in Eq. (5), we have

$$[\eta] = \frac{M}{T}. \tag{6}$$

Discussion & Verification   The correctness of our dimensional analysis in the case of P is apparent, since P must have the dimensions of a force. In the case of k and η, we can verify the correctness of our solution by substituting these quantities back into the expression for F. Doing so confirms that the dimensions of k and η are correct.

——————————————————— **Part (b)** ———————————————————

Road Map   To solve Part (b) of the problem, we need to match the dimensions obtained in Part (a) with their corresponding units according to the conventions established by the SI and U.S. Customary systems. To do this, we need only look at Table 1.1 on p. 14.

Computation   Since P has the same dimensions as a force, its SI units can be simply taken to be N (newtons) or, using the SI system base units, $\mathrm{kg\cdot m/s^2}$. In the U.S. Customary system, P is measured in lb (pounds).

Since the dimensions of k are mass over time squared, the corresponding units are $\mathrm{kg/s^2}$ in the SI system and lb/ft in the U.S. Customary system. Notice that in the SI system, the units of k can also be expressed as N/m.

Recalling that η has dimensions of mass over time, the units of η are kg/s in the SI system and slug/s in the U.S. Customary system. Using the base units of the U.S. Customary system, the units of η are lb·s/ft.

All of these results are summarized in Table 1.

Table 1. Summary of the solution to the second part of the problem.

Quantity	SI units	U.S. Customary units
P	N or $\mathrm{kg\cdot m/s^2}$	lb
k	$\mathrm{kg/s^2}$ or N/m	lb/ft
η	kg/s	lb·s/ft

Discussion & Verification   The correctness of our results can be verified by replacing the units with the dimensions they correspond to. For example, for P we have that the dimensions corresponding to the unit of pound are those of a force, i.e., ML/T^2, which are the dimensions of P obtained in Part (a) of the problem solution. Repeating this process for the other quantities and for both the SI system as well as the U.S. Customary system, we see that our results are indeed correct.

EXAMPLE 1.3 *Dimensional Analysis and Unit Conversion*

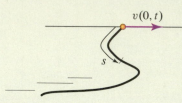

$v(0,t)$

s

Figure 1
A moving string with one end being dragged.

In studying the motion of a string, it is determined that the speed of the various points along the string is given by the function

$$v(s,t) = \alpha + \beta t^2 - \gamma s + \delta \frac{s}{t}, \tag{1}$$

where s is the coordinate of points along the string, t is time, and α, β, γ, and δ are constants.

(a) What are the dimensions of α, β, γ, and δ?

(b) If α, β, γ, and δ are all equal to 1 in SI units, what are they in U.S. Customary units?

SOLUTION

Part (a)

Road Map Since the dimensions of speed are L/T (with corresponding units of m/s in SI and ft/s in U.S. Customary), the dimensions of every term on the right-hand side of Eq. (1) must also be L/T.

Computation Begin with $[\alpha]$, which must have the same dimensions as those of the speed v, so they are simply L/T. As for $[\beta]$, we know that

$$\left[\beta t^2\right] = [\beta]T^2 = L/T \quad \Rightarrow \quad [\beta] = L/T^3. \tag{2}$$

To get $[\gamma]$, we proceed as in Eq. (2) to obtain

$$[\gamma s] = [\gamma]L = L/T \quad \Rightarrow \quad [\gamma] = 1/T. \tag{3}$$

Finally, $[\delta]$ is obtained similarly as

$$\left[\delta \frac{s}{t}\right] = [\delta]L/T = L/T \quad \Rightarrow \quad [\delta] \text{ is dimensionless.} \tag{4}$$

Discussion & Verification The verification of the results for α is immediate since no calculations were performed to obtain them. In the case of β, γ, and δ, substituting the results in Eqs. (2)–(4), we see that our results are correct.

Part (b)

Road Map After expressing α, δ, and γ via SI units, we need to convert them into U.S. Customary units. Note that no conversion is needed for δ since it is dimensionless.

Computation Based on Part (a), the SI units of α, β, and γ are m/s, m/s³, and s⁻¹, respectively. Now we need to convert $\alpha = 1\,\text{m/s}$, $\beta = 1\,\text{m/s}^3$, and $\gamma = 1\,\text{s}^{-1}$ to U.S. Customary units. Beginning with α, we have

$$\alpha = 1\,\text{m/s} = 1\frac{\text{m}}{\text{s}}\left(\frac{\text{ft}}{0.3048\,\text{m}}\right) = 3.28\,\text{ft/s}, \tag{5}$$

for β we have

$$\beta = 1\,\text{m/s}^3 = 1\frac{\text{m}}{\text{s}^3}\left(\frac{\text{ft}}{0.3048\,\text{m}}\right) = 3.28\,\text{ft/s}^3, \tag{6}$$

and finally, for γ we see that no conversion is needed since $1\,\text{s}^{-1}$ is the same in either SI or U.S. Customary units.

Discussion & Verification The only results to verify are those for α and β, which have dimensions of L/T and L/T^3, respectively. Therefore, since the base unit for time is the same in both the SI and U.S. Customary systems, the unit conversion was expected to affect only the L dimension and yield the same value (namely, 3.28) for both α and β. Finally, the value in question was expected to be close to 3 since a meter is a little over 3 ft. Consequently, our results appear to be correct.

PROBLEMS

Problem 1.1

Determine $(r_{B/A})_x$ and $(r_{B/A})_y$, the x and y components of the vector $\vec{r}_{B/A}$, so as to be able to write $\vec{r}_{B/A} = (r_{B/A})_x\,\hat{\imath} + (r_{B/A})_y\,\hat{\jmath}$.

Problem 1.2

If the positive direction of line ℓ is from D to C, find the component of the vector $\vec{r}_{B/A}$ along ℓ.

Problem 1.3

Find the components of $\vec{r}_{B/A}$ along the p and q axes.

Problem 1.4

Determine expressions for the vector $\vec{r}_{B/A}$ using both the xy and the pq coordinate systems. Next, determine $|\vec{r}_{B/A}|$, the magnitude of $\vec{r}_{B/A}$, using both the xy and the pq representations and establish whether or not the two values for $|\vec{r}_{B/A}|$ are equal to each other.

Problem 1.5

Suppose that you were to compute the quantities $\left.|\vec{r}_{B/A}|\right._{xy}$ and $\left.|\vec{r}_{B/A}|\right._{pq}$, that is, the magnitude of the vector $\vec{r}_{B/A}$ computed using the xy and pq frames, respectively. Do you expect these two scalar values to be the same or different? Why?
Note: Concept problems are about *explanations*, not computations.

Problem 1.6

The measure of angles in radians is defined according to the following relation: $r\theta = s_{AB}$, where r is the radius of the circle and s_{AB} denotes the length of the circular arc. Determine the dimensions of the angle θ.

Problem 1.7

A simple oscillator consists of a linear spring fixed at one end and a mass attached at the other end, which is free to move. Suppose that the periodic motion of a simple oscillator is described by the relation $y = Y_0\,\sin\,(2\pi\omega_0 t)$, where y has units of length and denotes the vertical position of the oscillator, Y_0 is called the oscillation amplitude, ω_0 is the oscillation frequency, and t is time. Recalling that the argument of a trigonometric function is an angle, determine the dimensions of Y_0 and ω_0, as well as their units in both the SI and the U.S. Customary systems.

Problem 1.8

To study the motion of a space station, the station can be modeled as a rigid body and the equations describing its motion can be chosen to be Euler's equations, which read

$$M_x = I_{xx}\,\alpha_x - \left(I_{yy} - I_{zz}\right)\omega_y\,\omega_z,$$
$$M_y = I_{yy}\,\alpha_y - \left(I_{zz} - I_{xx}\right)\omega_x\,\omega_z,$$
$$M_z = I_{zz}\,\alpha_z - \left(I_{xx} - I_{yy}\right)\omega_x\,\omega_y.$$

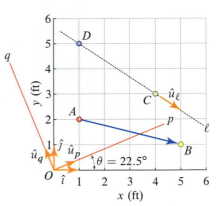

Figure P1.1–P1.5

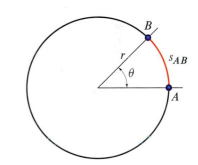

Figure P1.6

Figure P1.7

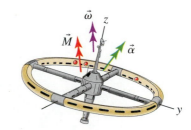

Figure P1.8

In the previous equations, M_x, M_y, and M_z denote the x, y, and z components of the moment applied to the body;* ω_x, ω_y, and ω_z denote the corresponding components of the angular velocity of the body, where angular velocity is defined as the time rate of change of an angle; α_x, α_y, and α_z denote the corresponding components of the angular acceleration of the body, where angular acceleration is defined as the time rate of change of an angular velocity. The quantities I_{xx}, I_{yy}, and I_{zz} are called the *principal mass moments of inertia* of the body. Determine the dimensions of I_{xx}, I_{yy}, and I_{zz} and determine their units in SI as well as in the U.S. Customary system.

Problem 1.9

The lift force F_L generated by the airflow moving over a wing is often expressed as follows:

$$F_L = \tfrac{1}{2}\rho v^2 C_L(\theta) A, \tag{1}$$

where ρ, v, and A denote the mass density of air, the airspeed (relative to the wing), and the wing's nominal surface area, respectively. The quantity C_L is called the *lift coefficient*, and it is a function of the wing's angle of attack θ. Find the dimensions of C_L and determine its units in the SI system.

Problem 1.10

Are the words *units* and *dimensions* synonyms?
Note: Concept problems are about *explanations*, not computations.

F_L

leading edge

trailing edge

Figure P1.9

---

* The xyz frame used in Euler's equations is a special frame that moves with the body, but for the purpose of the problem solution, it will suffice to say that it consists of three mutually orthogonal axes.

1.3　Dynamics and Engineering Design*

Design is the fundamental goal of all engineering endeavors. While there is no single accepted definition of engineering design, for our purposes, it can be thought of as the process that culminates in the specification of how a *system* is to be produced so that the *needs* and *requirements* that were identified in the design process are met. The "system" to which we refer might be as simple as a bolt or as complex as the Apollo missions. The need might be to hold two steel plates together or to put a human on the moon. The requirements to which we refer can be technical, environmental, political, financial, among others.

One of the hard things about engineering design is that neither the process used to reach the final design nor the final result of the design is unique (this is in contrast to the typical homework or exam problem you have been given where there is only *one* correct answer). There are generally many (sometimes infinitely many) designs that will satisfy the requirements of most design specifications.

In the design process, mechanicians[†] are generally concerned with the determination of a variety of mechanical responses such as deflection, stress, strain, and temperature in machines or structures due to imposed forces. Mechanicians are also concerned with relating these responses to the prediction of a structure's or machine's fatigue life, durability, and safety, as well as to the choice of materials and fabrication procedures. Dynamics plays an important role in this determination since accelerations are an important cause of forces in mechanical systems (via the equations developed by Newton and Euler).

There are several methods to structure the design process. Regardless of the method, the design process should be done iteratively through identifying needs, prioritizing, making value decisions, and exploiting physical laws to develop a sound solution that optimizes the objectives while satisfying the constraints that have been identified. As you progress through your education in your chosen field of study, you will learn about structured and standard procedures for design, including information about performance standards, safety standards, and design codes. For now, we will focus on that part of the design process that can be introduced based on your knowledge of calculus, statics, and dynamics.

Figure 1.19
Dynamic loading due to the interaction of the structure with the wind played a crucial role in the collapse of the Tacoma Narrows bridge on November 7, 1940.

Objectives of design

At a minimum, a mechanical design product must

1. Accomplish the design goals.

2. Not fail during normal use.

3. Minimize hazards.

4. Attempt to anticipate and account for all foreseeable uses.

5. Be thoroughly documented and archived.

---

* See D. G. Ullman, *The Mechanical Design Process*, 3rd edition, McGraw-Hill, 2003, for a thorough treatment of design.

[†] *Mechanicians* are people who study mechanics. *Mechanics* are people who fix cars or other machinery.

In addition, a design should also take into consideration

- Cost of manufacture, purchase, and ownership.

- Ease of manufacture and maintenance.

- Energy efficiency in manufacture and use.

- Impact on the environment in its manufacture, use, and retirement.

The goal of this book is not to teach you engineering design, but we will introduce you to some aspects of it. The aspects that we will focus on most heavily are *system modeling* and the associated *parametric studies*. In fact, it can be argued that dynamics is about modeling a mechanical system and then studying its behavior as parameters of the system are varied.

System modeling

As far as dynamics is concerned, we will consider *modeling* to be the process of translating "real life" into mathematical equations for the purpose of making predictions regarding the behavior of the model. The good news is that once you learn to model mechanical systems, you will have a very powerful tool at your disposal. The bad news is that the modeling process does not come easily for most students, and it is only mastered through *a lot* of practice.

When creating models, we have to remember that models are *not* exact representations of reality, but give (hopefully) enough information to tell us something meaningful about real physical systems. For example, a commercial airliner as a whole, and the wings in particular, undergo significant bending and torsion during flight. However, in building a model of an airplane's performance, we typically start by assuming that an airplane is a perfectly rigid object. To assume that an airplane is a rigid object allows us to simplify the equations of our model and more directly make predictions about the airplane's behavior for a large range of flight conditions. Based on these predictions, we can then refine the model to study the airplane's behavior in those circumstances where accounting for the airplane's deformation is imperative. The bottom line is any model we create will depend on the amount and type of information we wish to garner from it.

In general, once a model is created, you should compare the predictions of that model with data obtained from the real system. If the behavior of the system predicted by the model agrees with the real behavior of the system, then, within the assumptions used to create the model, you have some well-founded confidence to use that model to make further predictions of the system's behavior. If they do not agree, then you must go back to your model and determine what needs to be improved.

It is not always possible to compare the results of a model with the results of a laboratory test. For example, when engineers are creating a new type of aircraft, they cannot, for financial reasons and/or time constraints, build a prototype for every design they create to test it so that the results can be compared with the predictions of the model. They must use experience garnered from previous models and designs and extrapolate, using their experience and knowledge of engineering and physics to create a new system. In fact, when new commercial or military aircraft are designed, the first prototypes built are generally those that are flight-tested—there are no intermediate steps.

Helpful Information

Modeling and differential equations. Beginning in Chapter 3, you will see that models for mechanical systems generally consist of systems of algebraic equations or of systems of ordinary differential equations. In fact, Newton's second law is a second-order ordinary differential equation, so it is not hard to imagine that anytime we apply Newton's second law, we anticipate getting a differential equation that needs to be solved.

Particle and rigid body as models

Physics tells us that objects are comprised of an astonishing multitude of elementary particles. For example, a quick calculation indicates that 1 mol of iron, which has a mass of 55.844 g and fits very comfortably within a cube with 2 cm long sides, has 6.022×10^{23} atoms (6.022×10^{23} atoms/mol is Avogadro's number). Each atom, in turn, consists of a number of entities (electrons, protons, and neutrons), which, in turn, consist of other parts (e.g., quarks). This realization leads to the following question: Is it necessary to account for the existence of all of these particles to describe the motion of an object the size of a car or even the size of a human hair? The answer depends on how accurate we need to be and, even more importantly, on what information we need from our model. Fortunately, in most engineering applications, even for systems operating at the nanoscale (1 nm is a rough measure of a three-atom chain length), a satisfactory degree of accuracy can be achieved without resorting to the advanced models of atomic or subatomic physics. In fact, very good answers can be obtained across an extremely large range of length/time scales by using two very simple models: (1) the *particle* (defined in Section 1.2 on p. 9) and (2) the *rigid body* (defined in Section 1.2 on p. 9). This entire textbook is devoted to the use of these two fundamental models of mechanical systems for predicting forces and motion as well as their relation. In other courses, such as mechanics of materials, you will learn about new models in which bodies are assumed to be deformable and so they are neither a particle nor a rigid body.

Helpful Information

How big can a particle be? A particle is often defined to have zero volume and hence to occupy only one point in space. However, we often idealize real life objects, such as trucks, airplanes, and planets, to be particles! The accuracy of this modeling choice depends on whether or not the object's rotation, understood as a change in the object's orientation relative to some chosen reference, can be neglected. When rotations can be neglected, modeling an object as a particle, even if the object is very large, is legitimate. This will be seen in detail in Chapters 2, 3, 7, and 10.

Interesting Fact

Do rigid bodies really exist? All objects in nature are deformable, and thus there are no true rigid bodies. However, many objects are sufficiently stiff or maintain their shape so that a rigid body idealization is meaningful. In addition, even when dealing with a flexible object, such as an airplane wing, it is useful, as a first level of approximation, to model the system as rigid for the purpose of estimating loads.

2

Particle Kinematics

Kinematics studies the *geometry of motion* without reference to the causes of motion. It is rooted in vector algebra and calculus, and to many "it looks and feels like math." Kinematics is essential for the application of Newton's second law, $\vec{F} = m\vec{a}$, since it allows us to describe the $\vec{a}$ in $\vec{F} = m\vec{a}$. Unfortunately, just writing $\vec{F} = m\vec{a}$ doesn't tell us how something moves; it only describes the relationship between force and acceleration. In general, we also want to know a particle's *motion*, where by *motion* we mean "all the positions occupied by an object over time." It is kinematics that allows us to translate $\vec{a}$ into the *motion* of a point by using calculus. This is why in this chapter we study the concepts of position, velocity, and acceleration, and how these quantities relate to one another. In addition, we will learn how to write position, velocity, and acceleration in the various component systems most commonly used in dynamics.

2.1 Position, Velocity, Acceleration, and Cartesian Coordinates

Motion tracking along a racetrack

A motion tracking system like those used by TV networks during sporting events for the motion analysis of a football or a hockey puck can be used to analyze the motion of race cars along tracks such as that in Fig. 2.1. Two of the

Figure 2.1. A car racetrack.

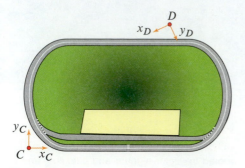

Figure 2.2

Tracking cameras C and D. The arrows at a camera indicate the directions of the Cartesian component system used by that camera.

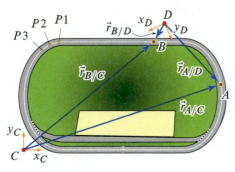

Figure 2.3

Position *vectors* of a car at two distinct times. Vectors $\vec{r}_{A/C}$ and $\vec{r}_{B/C}$ describe the position of a car at times $t_1 = 42.53$ s and $t_2 = 45.76$ s, respectively, as seen by camera C (see Table 2.1). Similarly, $\vec{r}_{A/D}$ and $\vec{r}_{B/D}$ are the positions of the *same* car and times relative to camera D.

cameras, C and D, that are part of the tracking system are shown in Fig. 2.2. The cameras can follow a car and measure the distance to it. The tracking information is then fed to a motion analyzer, which records it as in Table 2.1, i.e., as coordinates taken at regular time intervals. These coordinates are relative to the axes indicated in Fig. 2.2. How can we use the given data to learn how fast a car is moving and what acceleration the car is subject to?

Table 2.1. Cartesian coordinates of a car during a test run (only camera C and D data is reported). The system is able to measure time to within 0.01 s and distance to 0.1 ft.

Position	Time (s)	(x_C, y_C) (ft)	(x_D, y_D) (ft)
A	42.53	(1098.4, 360.6)	(−143.9, 437.2)
B	45.76	(723.2, 594.4)	(97.3, 66.7)
$P1$	48.87	(195.6, 593.0)	(576.1, −155.0)
$P2$	49.26	(145.6, 577.7)	(627.9, −162.3)
$P3$	49.66	(102.5, 548.8)	(679.2, −154.3)

Let's begin by observing that the locations of a car can be represented by arrows going from the camera that has recorded them to the locations in question (see Fig. 2.3). This suggests that we can use *vectors* to describe the positions of points. As can be seen in Fig. 2.3, the magnitude and direction of these vectors depend on the reference point (i.e., camera) that defines them.

The idea of velocity stems from computing the ratio between a change in position and the time interval spanned by the position change. Therefore we could measure a car's velocity by computing the difference between the coordinate pairs from two consecutive rows in Table 2.1 and dividing by the corresponding time difference. This operation raises various questions: If we take differences between two positions and then divide by the corresponding time interval, are we calculating the velocity at a *specific location* or are we calculating the velocity over some *range of positions?* Can we ever talk about "the velocity of a car at some specific location"? If position is described by a vector, is velocity also described by a vector? Should we expect the two velocity vectors measured by two different cameras to be different since position vectors relative to these cameras are different?

The questions raised are at the core of *kinematics* and tell us that to deal with any engineering application concerning the description of motion, we need rigorous definitions of the concepts of position, velocity, and acceleration. In the remainder of this section we will see that vectors play a fundamental role in how we define position, velocity, and acceleration and how we can correctly interpret information provided from multiple observation reference points. As we formally define position, velocity, and acceleration, we will come back to the car tracking problem to provide concrete applications to our definitions.

A notation for time derivatives

In studying kinematics, we will be writing derivatives with respect to time so often that it is convenient to have a shorthand notation. If $f(t)$ is a function of time, we write a dot over it to mean $df(t)/dt$. Furthermore, the *number* of dots over a quantity indicates the order of the derivative; that is,

$$\dot{f}(t) = \frac{df(t)}{dt}; \quad \ddot{f}(t) = \frac{d^2f(t)}{dt^2}; \quad \dddot{f}(t) = \frac{d^3f(t)}{dt^3}; \quad \text{etc.} \quad (2.1)$$

Position vector

Figure 2.4 shows a moving point P and a coordinate system with origin at O. We define the *position vector* of P at time t (relative to the origin O) as the *vector* $\vec{r}_P(t)$ going from O to P at time t. In two dimensions (Section 2.8 on p. 154 covers the three-dimensional case) and using Cartesian components, we have

$$\vec{r}_P(t) = x_P(t)\,\hat{\imath} + y_P(t)\,\hat{\jmath}, \quad (2.2)$$

where, as discussed in Section 1.2, an arrow over a letter denotes a vector and a "hat" over a letter denotes a unit vector.

The magnitude of $\vec{r}_P(t)$ is the distance between P and O, which we can compute as

$$\left|\vec{r}_P(t)\right| = \sqrt{x_P^2(t) + y_P^2(t)}. \quad (2.3)$$

The direction of the vector $\vec{r}_P(t)$ is that of the oriented line going from O to P.

The notion of position is meaningless without the specification of a reference point relative to which position is measured. In general, it is understood that position is measured relative to the origin of a coordinate system. If the reference point is not the origin of a coordinate system or if several coordinate systems are used concurrently, then we use the notation introduced in Section 1.2 (see Eq. (1.16) on p. 11) to indicate the position of a point *relative to* another. For example, if we go back to the motion tracking problem and refer to Fig. 2.3 as well as Table 2.1, at $t = 42.53\,\text{s}$, the position of the car at A *relative to* cameras C and D is given by the two vectors

$$\vec{r}_{A/C}(42.53\,\text{s}) = (1098.4\,\hat{\imath}_C + 360.6\,\hat{\jmath}_C)\,\text{ft}, \quad (2.4)$$

$$\vec{r}_{A/D}(42.53\,\text{s}) = (-143.9\,\hat{\imath}_D + 437.2\,\hat{\jmath}_D)\,\text{ft}. \quad (2.5)$$

While the vectors $\vec{r}_{A/C}$ and $\vec{r}_{A/D}$ are different from each other, they both tell us where the car is at the specified instant in time.

Trajectory

The *trajectory* of a moving point is the line traced through space by the point during its motion. Synonyms for *trajectory* are *path* and *space curve*.

Referring to Fig. 2.3, the trajectory of a car going around the track is, roughly speaking, the track itself. Let's observe that a car can keep running around the track while taking different amounts of time for different laps. This means that an object's trajectory by itself cannot tell us anything about how fast an object moves.

Velocity vector and speed

Displacement vector.   Vectors $\vec{r}(t_i)$ and $\vec{r}(t_j)$ in Fig. 2.5 are the positions of a point P at times t_i and t_j, respectively, with $t_j > t_i$. We define the *displacement (or position change)* of P between t_i and t_j as the vector $\Delta\vec{r}(t_i,t_j) =$

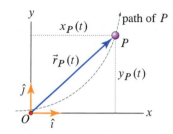

Figure 2.4
A moving point P and its trajectory (or path).

Concept Alert

Position is a *vector*. The position of a point P relative to a chosen reference point defines a vector with (1) magnitude equal to the distance between P and the reference point and (2) direction defined by the oriented line going from the reference point to the point P.

Concept Alert

Trajectory and time. Although the trajectory is the line traced by a point while moving, it does not tell us anything about the amount of time it took to go a specific distance along the trajectory — that depends on the *speed*.

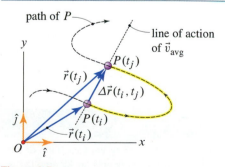

Figure 2.5
Displacement of a point between times t_i and t_j, with $t_j > t_i$. The distance traveled by P between t_i and t_j is highlighted in yellow.

$\vec{r}(t_j) - \vec{r}(t_i)$. In general, the length of $\Delta\vec{r}(t_i, t_j)$ *does not* measure the distance traveled by P between t_i and t_j (highlighted in yellow in Fig. 2.5). This is so because the distance traveled between t_i and t_j depends on the geometry of the path of P whereas $\Delta\vec{r}(t_i, t_j)$ depends only on $\vec{r}(t_i)$ to $\vec{r}(t_j)$ without reference to *how* P moved from one position to the other!

Average velocity vector.   We define the *average velocity vector* of P over the time interval (t_i, t_j) as

$$\vec{v}_{\text{avg}}(t_i, t_j) = \underbrace{\frac{1}{t_j - t_i}}_{\text{scalar}} \underbrace{\left[\vec{r}(t_j) - \vec{r}(t_i)\right]}_{\text{vector}} = \frac{\Delta\vec{r}(t_i, t_j)}{t_j - t_i}. \tag{2.6}$$

Let's observe that the vectors $\vec{v}_{\text{avg}}(t_i, t_j)$ and $\Delta\vec{r}(t_i, t_j)$ have the same direction since the term $1/(t_j - t_i)$ in Eq. (2.6) is a positive scalar.

Velocity vector.   Considering the average velocity over a time interval $(t, t + \Delta t)$, we define the *velocity vector* at time t as

$$\vec{v}(t) = \lim_{\Delta t \to 0} \vec{v}_{\text{avg}}(t, t + \Delta t) = \lim_{\Delta t \to 0} \frac{\Delta\vec{r}(t, t + \Delta t)}{(t + \Delta t) - t}. \tag{2.7}$$

Calculus tells us that the second limit in Eq. (2.7) is the time derivative of $\vec{r}(t)$, so that the velocity vector is normally written as

$$\boxed{\vec{v}(t) = \frac{d\vec{r}(t)}{dt} = \dot{\vec{r}}(t),} \tag{2.8}$$

that is, *The velocity vector is the time rate of change of the position vector.*

Speed.   The *speed* of a point is defined as the *magnitude of its velocity*:

$$\boxed{v(t) = \left|\vec{v}(t)\right|.} \tag{2.9}$$

Therefore, by definition, the *speed is a scalar quantity that is never negative*.

The velocity vector is *always* tangent to the path.   The velocity vector has an important property: *The velocity vector at a point along the trajectory is tangent to the trajectory at that point!* To see this, recall that after Eq. (2.6) we remarked that the average velocity between two time instants, say, t and $t + \Delta t$, has the same direction as the displacement vector $\Delta\vec{r}(t, t + \Delta t)$. Figure 2.6 illustrates that as $\Delta t \to 0$, the vector $\Delta\vec{r}(t, t + \Delta t)$ becomes *tangent* to the trajectory, thus causing the velocity vector to become tangent to the trajectory at the position $\vec{r}(t)$.

Additional properties of the velocity vector.   We can now go back and answer some of the questions raised by the car tracking problem. Specifically, if, using Table 2.1, we compute the ratio between a position change and the corresponding time interval, then (1) we now know that what we are actually computing is the car's *average velocity* and (2) the average velocity cannot be said to pertain to any particular position within the time interval considered. We can also answer the question about whether or not the data collected from two different cameras yields different velocity vectors. Observe that the vector

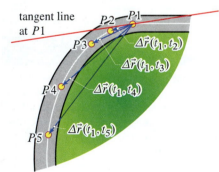

Figure 2.6
Displacement vectors $\Delta\vec{r}(t_i, t_j)$ between time position $P1$ at time t_1 and subsequent positions and times Pi at t_i, $i = 2, 3, 4, 5$.

$\Delta \vec{r}(t_A, t_B)$ in Fig. 2.7, between positions A and B, i.e., between times $t_A = 42.53\,\text{s}$ and $t_B = 45.76\,\text{s}$, respectively, is the same whether measured from camera C or camera D. Because cameras C and D do not move relative to one another, we can then conclude that a car's velocity (average or not) is independent of the frame of reference used to measure it. It is the *components* of the velocity vector that change in going from one frame to another, not the velocity vector itself. For example, using Table 2.1 and Eq. (2.6), we can write

$$\vec{v}_{\text{avg}}(t_A, t_B) = \underbrace{(-116.2\,\hat{\imath}_C + 72.4\,\hat{\jmath}_C)\,\tfrac{\text{ft}}{\text{s}}}_{\text{camera } C} = \underbrace{(74.7\,\hat{\imath}_D - 114.7\,\hat{\jmath}_D)\,\tfrac{\text{ft}}{\text{s}}}_{\text{camera } D}. \quad (2.10)$$

Clearly, the components of $\vec{v}_{\text{avg}}(t_A, t_B)$ relative to cameras C and D are different, but the vector $\vec{v}_{\text{avg}}(t_A, t_B)$ must be the same for the two cameras.

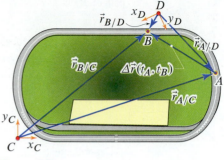

Figure 2.7
The displacement vector $\Delta \vec{r}(t_A, t_B)$ is the same for both cameras.

Acceleration vector

The *acceleration vector* is the time rate of change of the velocity vector, i.e.,

$$\vec{a}(t) = \frac{d\vec{v}(t)}{dt} = \frac{d^2\vec{r}(t)}{dt^2} = \dot{\vec{v}}(t) = \ddot{\vec{r}}(t), \quad (2.11)$$

all of which are equivalent.

The acceleration is generally not tangent to the path. Contrary to what we discovered about the velocity, the acceleration vector is generally not tangent to the path. To see why, consider a point P moving at constant speed as shown in Fig. 2.8. Although the speed is constant, the acceleration $\vec{a}(t)$ of P, which measures the changes in both the magnitude and the *direction* of $\vec{v}(t)$, is not zero because $\vec{v}(t)$ changes direction so as to remain tangent to the *curved* path. Note that the velocity change, i.e., $\vec{a}(t)$, is more pronounced between points A and B than it is between B and C since the path's curvature is larger between A and B than between B and C.* Furthermore, even if the speed $v(t)$ is constant, $\vec{a}(t)$ depends on the speed $v(t)$ because $v(t)$ determines how quickly P moves along its path and therefore how fast $\vec{v}(t)$ changes direction. These observations indicate that the acceleration vector along a curved path depends on both the speed and the path's curvature. Finally, to get a sense of the *direction* of $\vec{a}(t)$, consider the velocity between time t and time $t + \Delta t$, as shown in Fig. 2.9. Since, by definition, we have

$$\vec{a}(t) = \dot{\vec{v}}(t) = \lim_{\Delta t \to 0} \frac{\vec{v}(t + \Delta t) - \vec{v}(t)}{\Delta t}, \quad (2.12)$$

the direction of $\vec{a}(t)\,\Delta t$, which is the same as that of $\vec{a}(t)$, can be approximated by the direction of $\vec{v}(t + \Delta t) - \vec{v}(t)$ if Δt is small enough. Referring to Fig. 2.9, observe that the vector $\vec{v}(t + \Delta t) - \vec{v}(t)$, and therefore $\vec{a}(t)$, points toward the concave side of the trajectory instead of being tangent to it! While the arguments just provided are qualitative, in Section 2.5 we will quantitatively show that, for a curved path, not only does the acceleration have a component pointing toward the concave side of the trajectory, but also this component is proportional to the square of the speed and inversely proportional to the path's radius of curvature.

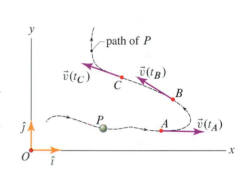

Figure 2.8
Velocity vectors of a particle P at three instants in time t_A, t_B, and t_C. The particle is moving with constant speed.

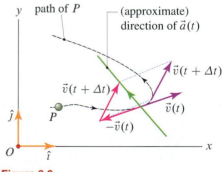

Figure 2.9
Velocity of P at times t and $t + \Delta t$ demonstrating the approximate direction of $\vec{a}(t)$.

---

* We will mathematically define *curvature* in Section 2.5.

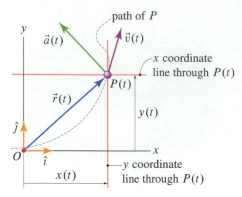

Figure 2.10
Position, velocity, and acceleration of a moving point P along with a Cartesian coordinate system.

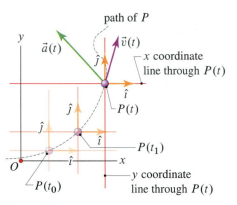

Figure 2.11
A moving point P at various times with its corresponding companion base vectors.

Cartesian coordinates

We now discuss the relation between *coordinates* in a Cartesian coordinate system and the *components* of the position, velocity, and acceleration vectors. Here we focus on motions viewed by a single observer situated at the origin of the coordinate system. The case of multiple observers moving relative to each other is discussed first in Section 2.7 and then in Section 6.4.

Figure 2.10 shows a moving point P with coordinates $(x(t), y(t))$ (for the three-dimensional case see Section 2.8) relative to a Cartesian coordinate system with origin at O and axes x and y. Cartesian coordinate systems have the distinctive property that the coordinates of a point are also the components of the position vector of this point relative to the system's origin.* Therefore the *position vector $\vec{r}(t)$ in a Cartesian component system* takes on the form

$$\vec{r}(t) = x(t)\,\hat{\imath} + y(t)\,\hat{\jmath}. \tag{2.13}$$

Observe that if x changes while y remains constant, the tip of $\vec{r}$ will trace a line through P parallel to the x axis. Similarly, if y changes while x remains constant, the tip of $\vec{r}$ will trace a line through P parallel to the y axis. These lines are called the *coordinate lines* through P.

The velocity $\vec{v}(t)$ and the acceleration $\vec{a}(t)$ of P are vectors based at $P(t)$, as shown in Fig. 2.10. Thus, we view the components of $\vec{v}(t)$ and $\vec{a}(t)$ as being relative to the *vectors $\hat{\imath}$ and $\hat{\jmath}$ that are also based at $P(t)$ and are tangent to the coordinate lines*, as shown in Fig. 2.11. This choice has two consequences. First, different sets of base vectors $\hat{\imath}$ and $\hat{\jmath}$ are used as P moves, as can be seen in Fig. 2.11, which shows the sets of base vectors at $P(t_0)$, $P(t_1)$, and $P(t)$. Hence, as P moves through space, its companion base vectors need to be regarded as (implicit) functions of time. Second, since the *unit* base vectors at $P(t_0)$, $P(t_1)$, and $P(t)$ must remain tangent to the coordinate lines, we have that in a Cartesian coordinate system the base vectors do not change in magnitude or direction as P moves. The first observation implies that when we differentiate Eq. (2.13) with respect to time to compute $\vec{v}(t)$, we must apply the product rule and write

$$\vec{v} = \dot{x}(t)\,\hat{\imath} + x(t)\,\frac{d\hat{\imath}}{dt} + \dot{y}(t)\,\hat{\jmath} + y(t)\,\frac{d\hat{\jmath}}{dt}. \tag{2.14}$$

However, the second observation implies that $\hat{\imath}$ and $\hat{\jmath}$ are constants so that

$$d\hat{\imath}/dt = \vec{0} \quad \text{and} \quad d\hat{\jmath}/dt = \vec{0}, \tag{2.15}$$

and the *velocity vector $\vec{v}(t)$ in a Cartesian component system* becomes

$$\vec{v} = \dot{x}(t)\,\hat{\imath} + \dot{y}(t)\,\hat{\jmath} = v_x(t)\,\hat{\imath} + v_y(t)\,\hat{\jmath}. \tag{2.16}$$

Similarly, the *acceleration vector $\vec{a}(t)$ in a Cartesian component system* is

$$\vec{a} = \ddot{x}(t)\,\hat{\imath} + \ddot{y}(t)\,\hat{\jmath} = \dot{v}_x(t)\,\hat{\imath} + \dot{v}_y(t)\,\hat{\jmath} = a_x(t)\,\hat{\imath} + a_y(t)\,\hat{\jmath}. \tag{2.17}$$

In later sections we will discover that in other component systems the *components* of the velocity and acceleration of a moving point are *not* obtained by the direct time differentiation of the *coordinates* of the point.

---

* In curvilinear coordinate systems, *coordinates* cannot be used *directly* as *components* of the position vector. For example, in polar coordinates (see Section 2.6 on p. 118) the coordinates of a point P are (r, θ), whereas the position vector of P is $r\hat{u}_r$, in which the *coordinate* θ does not appear as a *component*.

End of Section Summary

Position. The position of a point is a *vector* going from the origin of the chosen frame of reference to the point in question. The position vectors of a point measured by different reference frames are different from one another.

Trajectory. The trajectory of a moving point is the line traced by the point during its motion. Another name for trajectory is *path*.

Displacement. The displacement between positions A and B is the vector going from A to B. In general, the magnitude of the displacement between two positions is not the distance traveled along the path between these positions.

Velocity. The velocity vector is the time rate of change of the position vector. The velocity vector is the same with respect to any two reference frames that do not move relative to one another. The velocity is *always* tangent to the path.

Speed. The speed is the magnitude of the velocity and is a nonnegative scalar quantity.

Acceleration. The acceleration vector is the time rate of change of the velocity vector. As with the velocity, the acceleration is the same with respect to any two reference frames that are not moving relative to one another. Contrary to what happens for the velocity, the acceleration vector is, in general, not tangent to the trajectory.

Cartesian coordinates. The Cartesian coordinates of a particle P moving along some path are shown in Fig. 2.12. The position vector is given by

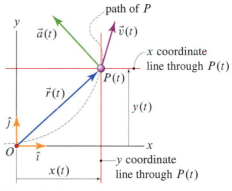

Figure 2.12
Figure 2.10 repeated. The position vector $\vec{r}(t)$ of the point P in Cartesian coordinates.

> Eq. (2.13), p. 34
>
> $$\vec{r}(t) = x(t)\,\hat{\imath} + y(t)\,\hat{\jmath}.$$

In Cartesian components, the velocity and acceleration vectors are given by

> Eqs. (2.16) and (2.17), p. 34
>
> $$\vec{v}(t) = \dot{x}(t)\,\hat{\imath} + \dot{y}(t)\,\hat{\jmath} = v_x(t)\,\hat{\imath} + v_y(t)\,\hat{\jmath},$$
> $$\vec{a}(t) = \ddot{x}(t)\,\hat{\imath} + \ddot{y}(t)\,\hat{\jmath} = \dot{v}_x(t)\,\hat{\imath} + \dot{v}_y(t)\,\hat{\jmath} = a_x(t)\,\hat{\imath} + a_y(t)\,\hat{\jmath}.$$

An important note regarding example problems. In Chapter 2, all examples will begin to employ *part* of the problem-solving framework that is formally introduced in Chapter 3. That is, each example problem will include a *Road Map* step, a *Computation* step, and a *Discussion & Verification* step.* The Road Map step will lay out the path to the solution by identifying given information and unknowns and then proposing a problem-solving strategy. The Computation step will set up the appropriate equations and will solve them. Finally, the Discussion & Verification step will check whether or not the solution is reasonable.

---

* The problem-solving framework introduced in Chapter 3 includes additional steps since it is applied to kinetics problems, which are generally more involved than kinematics problems.

EXAMPLE 2.1 *How Do You Get to Carnegie Hall? ...Practice!*

Figure 1
Cab route.

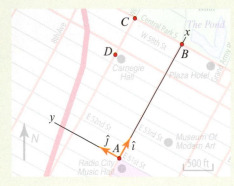

Figure 2
Cartesian coordinate system with origin at the pickup point A. The city grid is such that the line through B and C is parallel to the y axis and the line through C and D is parallel to the x axis.

Table 1
Coordinates of the points defining the cab's route.

Point	x (ft)	y (ft)
A	0	0
B	2200	0
C	2200	906
D	1500	906

A cab picks up a passenger outside Radio City Music Hall on the corner of the Avenue of the Americas and E 51st St. (point A) and drops her off in front of Carnegie Hall on 7th Ave. (point D) after 5 min, following the route shown. Find the cab's displacement, average velocity, distance traveled, and average speed in going from A to D. Note that the distance from A to B is 2200 ft, from B to C is 906 ft, and from C to D is 700 ft.

SOLUTION

Road Map To solve the problem, we need to set up a coordinate system and identify the coordinates of points A, B, C, and D that define the cab's path. Then the problem's questions can be answered by applying the definitions of displacement, average velocity, distanced traveled, and speed.

Computation Referring to Fig. 2, we select a Cartesian coordinate system with origin at A and aligned with the city grid. Using the given information, the coordinates of A, B, C, and D are given in Table 1. Since the displacement from A to D is the difference between the position vectors of points A and D, we have

$$\Delta \vec{r}(t_A, t_D) = \vec{r}(t_D) - \vec{r}(t_A) = (1500\,\hat{\imath} + 906\,\hat{\jmath})\,\text{ft}, \tag{1}$$

where t_A and t_D are times at which the cab is at A and D, respectively. Applying the definition of average velocity in Eq. (2.6), for the time interval (t_A, t_D) we have

$$\vec{v}_{avg}(t_A, t_D) = \frac{\Delta \vec{r}(t_A, t_D)}{t_D - t_A} = \frac{\vec{r}(t_D) - \vec{r}(t_A)}{t_D - t_A} = (5\,\hat{\imath} + 3.02\,\hat{\jmath})\,\text{ft/s}, \tag{2}$$

where $t_D - t_A = 5\,\text{min} = 300\,\text{s}$. Next, the distance traveled by the cab, which we will denote by d, is given by the sum of the lengths of the segments $\overline{AB}$, $\overline{BC}$, and $\overline{CD}$, i.e,

$$d = (x_B - x_A) + (y_C - y_B) + (x_C - x_D) = 3810\,\text{ft}. \tag{3}$$

Since the speed is the magnitude of the velocity, the *average speed* must be computed as the *average of the magnitude of the velocity*, i.e.,

$$v_{avg} = \frac{1}{t_D - t_A} \int_{t_A}^{t_D} |\vec{v}|\,dt = \frac{1}{t_D - t_A} \left(\int_{t_A}^{t_B} v_x\,dt + \int_{t_B}^{t_C} v_y\,dt - \int_{t_C}^{t_D} v_x\,dt \right), \tag{4}$$

since v is equal to v_x, v_y, and $-v_x$ during the time intervals (t_A, t_B), (t_B, t_C), and (t_C, t_D), respectively. Now notice that the last three integrals in Eq. (4) measure the distance traveled by the cab in each of the corresponding time intervals. Considering the integral, say, over (t_A, t_B), we can write

$$\int_{t_A}^{t_B} v_x\,dt = \int_{t_A}^{t_B} \frac{dx}{dt}\,dt = \int_{x_A}^{x_B} dx = x_B - x_A. \tag{5}$$

Proceeding similarly for the other two integrals, we find the average speed to be

$$v_{avg} = \frac{(x_B - x_A) + (y_C - y_B) - (x_D - x_C)}{t_D - t_A} = \frac{d}{t_D - t_A} = 12.7\,\text{ft/s}. \tag{6}$$

Discussion & Verification The results obtained are dimensionally correct. Since 12.7 ft/s corresponds to 8.66 mph, we can consider the result acceptable since such a speed is typical of city traffic such as can be found in midtown Manhattan.

🔎 **A Closer Look** Observing that $|\Delta \vec{r}(t_A, t_D)| = 1750\,\text{ft}$, and that $|\vec{v}_{avg}(t_A, t_D)| = 5.84\,\text{ft/s}$, we see this example reinforces the idea that we should never confuse distance traveled for displacement or average velocity for average speed.

EXAMPLE 2.2   *Trajectory, Velocity, and Acceleration*

Two stationary ships A and B track an object P launched from A and flying low and parallel to the water. Relative to the Cartesian frames A and B in Fig. 1 and for the first few seconds of flight, the recorded motion of P is

$$\vec{r}_{P/A}(t) = 2.35t^3\,\hat{\imath}_A \text{ m}, \tag{1}$$

$$\vec{r}_{P/B}(t) = \left[(225 + 2.13t^3)\,\hat{\imath}_B + (225 + 0.993t^3)\,\hat{\jmath}_B\right] \text{m}, \tag{2}$$

where t is in seconds.

(a) Determine the path of P as viewed by frame A and frame B.

(b) Using *both* $\vec{r}_{P/A}$ and $\vec{r}_{P/B}$, determine the velocity and the speed of P.

(c) Find the acceleration and its orientation in relation to the path in both frames A and B.

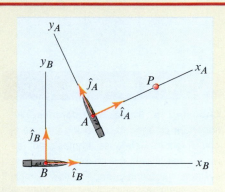

Figure 1
Two Cartesian frames A and B.

SOLUTION

Part (a): Path of P

Road Map   We have the motion of P as a function of time. Since the path of P is the line traced by P in *space*, we can find the path by eliminating time from the motion.

Computation   Starting with the motion in component form for frame B, we have

$$x_{P/B}(t) = (225 + 2.13t^3)\,\text{m} \quad \text{and} \quad y_{P/B}(t) = (225 + 0.993t^3)\,\text{m}. \tag{3}$$

To eliminate the time variable, we solve the first of Eqs. (3) for t^3 and then substitute the result in the second of Eqs. (3). This yields

$$\boxed{y_{P/B} = \left[225 + \frac{0.993}{2.13}\left(x_{P/B} - 225\right)\right] \text{m} = (120 + 0.466x_{P/B})\,\text{m}.} \tag{4}$$

Observe that in frame A, $y_{P/A} = 0$ at all times so that no calculation is needed to eliminate time from the y component of the motion of P. Therefore, the path of P in frame A is described by $y_{P/A} = 0$, which is the equation of the x_A axis.

Discussion & Verification   The path in frame B is given in Eq. (4), which is the equation of a straight line. This agrees with the calculation of the path in frame A, in which the path lies on the x_A axis, which is also a straight line. The trajectory as seen by the two frames is shown in Fig. 2. Note that Fig. 2 also indicates the location of P at $t = 0$ and the direction of motion.

Part (b): Velocity and Speed of P

Road Map   We are given the position in Cartesian components, so we can compute the velocity vector, using Eq. (2.16). The speed is then found by applying Eq. (2.9), i.e., by computing the magnitude of the velocity vector.

Computation   Using Eq. (2.16), the velocity of P in each of the two frames is

$$\boxed{\vec{v}_{P/A}(t) = \dot{\vec{r}}_{P/A}(t) = 7.05t^2\,\hat{\imath}_A \text{ m/s},} \tag{5}$$

$$\boxed{\vec{v}_{P/B}(t) = \dot{\vec{r}}_{P/B}(t) = (6.39t^2\,\hat{\imath}_B + 2.98t^2\,\hat{\jmath}_B)\,\text{m/s}.} \tag{6}$$

As far as the speed is concerned, applying Eq. (2.9) to Eqs. (5) and (6), we obtain

$$\boxed{v_{P/A}(t) = 7.05t^2\,\text{m/s},} \tag{7}$$

$$\boxed{v_{P/B}(t) = \sqrt{(6.39t^2)^2 + (2.98t^2)^2}\,\text{m/s} = 7.05t^2\,\text{m/s}.} \tag{8}$$

Helpful Information

Trajectory and time. The trajectory (or path) is what the motion looks like once time is removed from the motion's description.

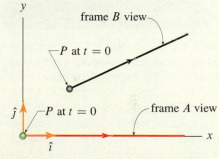

Figure 2
Path of P as seen by frames A and B.

Helpful Information

Velocity revisited. Since the frames we are using do not move relative to one another, the velocity vector is the same no matter what frame is used. This is why it is correct to say that $\dot{\vec{r}}_{P/A} = \dot{\vec{r}}_{P/B}$ in Eqs. (5) and (6).

Discussion & Verification   Since frames A and B do not move relative to each other, we expect the vectors in Eqs. (5) and (6) to be equal to each other, and therefore we expect to find the same value of speed in either frame. While it is not immediately obvious that Eqs. (5) and (6) are describing the same velocity vector, Eqs. (7) and (8) do confirm our expectation. In Example 2.3 we will show that $\vec{v}_{P/A} = \vec{v}_{P/B}$.

──────── **Part (c): Acceleration of P & Its Orientation** ────────

Road Map   Using the velocity results in Eqs. (5) and (6), we can find the acceleration by using Eq. (2.17). We can then find the orientation of the acceleration vector relative to the trajectory by finding the angle between the tangent to the trajectory and the x axis and comparing that with the angle between the acceleration vector and the x axis. Since the velocity is tangent to the trajectory, we can find the angle between the trajectory and the x axis by finding the angle between the velocity and the x axis.

Computation   Applying Eq. (2.17), the acceleration of P is given by

$$\vec{a}_{P/A}(t) = \dot{\vec{v}}_{P/A}(t) = 14.1t\,\hat{\imath}_A\,\text{m/s}^2, \tag{9}$$

$$\vec{a}_{P/B}(t) = \dot{\vec{v}}_{P/B}(t) = (12.8t\,\hat{\imath}_B + 5.96t\,\hat{\jmath}_B)\,\text{m/s}^2. \tag{10}$$

Determining the orientation of $\vec{a}$ relative to the trajectory is not difficult in frame A since both $\vec{a}_{P/A}$ and $\vec{v}_{P/A}$ are always in the positive x_A direction. Therefore, in frame A, $\vec{a}$ is always tangent to the trajectory. In frame B, the angle between the trajectory and the x_B axis is given by

$$\theta_v = \tan^{-1}\left(\frac{v_y}{v_x}\right) = \tan^{-1}\left(\frac{2.98t^2}{6.39t^2}\right) = 25.0°. \tag{11}$$

The angle between the acceleration and the x_B axis is

$$\theta_a = \tan^{-1}\left(\frac{a_y}{a_x}\right) = \tan^{-1}\left(\frac{5.96t}{12.8t}\right) = 25.0°. \tag{12}$$

Since the angles in Eqs. (11) and (12) are equal, we conclude that the acceleration is tangent to the path, even when using frame B data.

Discussion & Verification   Now that we found the acceleration in frames A and B in Eqs. (9) and (10), respectively, it is not obvious that these two vectors are the same. As with the velocity, we will show that this is so in Example 2.3. Regarding the orientation of $\vec{a}$ relative to the trajectory, we saw in frame A that $\vec{a}$ is always tangent to the trajectory. In frame B, Eqs. (11) and (12) tell us that the angle between $\vec{a}$ and the x_2 axis is always 25° and the angle between $\vec{v}$ and the x_2 axis is also always 25°. Therefore, we see that $\vec{a}$ is also always tangent to the trajectory in frame B. The fact that the accelerations in both frames are tangent to their respective trajectories gives us some confidence that the two accelerations we computed are the same and are correct.

Helpful Information

Choosing a frame of reference. An important criterion for choosing a frame is *convenience*. Reference frames and (as we will see later in this chapter) coordinate systems profoundly impact how simply and directly the solution to a problem is obtained. Hence, a useful skill to cultivate is selection of the frame of reference and coordinate system that leads to the solution with greater ease.

Common Pitfall

Does Eq. (12) contradict what we said earlier about the acceleration not being tangent to the trajectory? Earlier in the section we stated that *in general* the acceleration is not tangent to the path. We also indicated that if the trajectory is curved, then we must expect the acceleration not to be parallel to the path. Therefore we can conclude that (1) there can be points along a path at which the acceleration is tangent to the path and that (2) at these points the path's curvature must be equal to zero. This is consistent with what we found in our example since the path in this example is a straight line, that is, *a line with no curvature.*

EXAMPLE 2.3   *Acceleration Vector Measured by Two Stationary Observers*

In Example 2.2 we stated that the velocities and accelerations of point P are the same no matter what frame is used to study the point's motion, provided that the frames do not move relative to one another. Referring to Fig. 1 and using the relations for finding the components of a vector given in Section 1.2, show that $\vec{v}_{P/A} = \vec{v}_{P/B}$ and $\vec{a}_{P/A} = \vec{a}_{P/B}$.

SOLUTION

Road Map   We need to take the component representations of the velocity and acceleration of point P in one frame and turn them into the corresponding representations in the other frame. Since the motion has a particularly simple representation in frame A, we will turn this representation into that relative to frame B. For the velocity, we will start from Eq. (5) in Example 2.2 and show that we can transform it into Eq. (6). For the acceleration, we will start from Eq. (9) in Example 2.2 and show that we can transform it into Eq. (10).

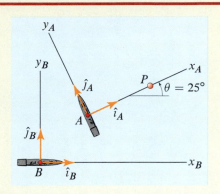

Figure 1
Frames A and B as well as the angle θ defining their relative orientation.

Computation   Recall that the velocity and acceleration of P in frame A are, respectively,

$$\vec{v}_{P/A} = 7.05t^2\,\hat{\imath}_A\,\text{m/s} \quad \text{and} \quad \vec{a}_{P/A} = 14.10t\,\hat{\imath}_A\,\text{m/s}^2. \tag{1}$$

Next, we express the unit vector $\hat{\imath}_A$, using the unit vectors $\hat{\imath}_B$ and $\hat{\jmath}_B$. Using the method demonstrated in Section 1.2, i.e., applying Eq. (1.28) on p. 13, we obtain

$$\begin{aligned}
\hat{\imath}_A &= (\hat{\imath}_A \cdot \hat{\imath}_B)\,\hat{\imath}_B + (\hat{\imath}_A \cdot \hat{\jmath}_B)\,\hat{\jmath}_B \\
&= \cos\theta\,\hat{\imath}_B + \sin\theta\,\hat{\jmath}_B \\
&= 0.906\,\hat{\imath}_B + 0.423\,\hat{\jmath}_B,
\end{aligned} \tag{2}$$

where, as can be seen in Fig. 1, θ is the angle formed between the directions of $\hat{\imath}_A$ and $\hat{\imath}_B$. Substituting Eq. (2) into the expression for $\vec{v}_{P/A}$, we obtain

$$\begin{aligned}
\vec{v}_{P/A} &= 7.05t^2\,\hat{\imath}_A\,\text{m/s} \\
&= 7.05t^2(0.906\,\hat{\imath}_B + 0.423\,\hat{\jmath}_B)\,\text{m/s} \\
&= \left(6.39t^2\,\hat{\imath}_B + 2.98t^2\,\hat{\jmath}_B\right)\text{m/s},
\end{aligned} \tag{3}$$

where we have expressed the final result to three significant digits. Repeating this process for the acceleration, we have

$$\vec{a}_{P/A} = 14.1t\,\hat{\imath}_A\,\text{m/s}^2 = (12.8t\,\hat{\imath}_B + 5.96t\,\hat{\jmath}_B)\,\text{m/s}^2. \tag{4}$$

Discussion & Verification   Comparing Eq. (3) with Eq. (6) of Example 2.2, we can verify that the expression for $\vec{v}_{P/A}$ given here coincides with that of $\vec{v}_{P/B}$ in Example 2.2. Similarly, comparing Eq. (4) with Eq. (10) of Example 2.2, we can verify that the expression for $\vec{a}_{P/A}$ given here coincides with that of $\vec{a}_{P/B}$ in Example 2.2.

EXAMPLE 2.4   *Determining Position and Acceleration from Velocity*

The velocity vector of a particle is $\vec{v}(t) = (3t^2\,\hat{\imath} - 4t\,\hat{\jmath})$ in./s. If the particle is at the origin when $t = 0$ s, determine the coordinates of the particle when $t = 4$ s, the equation of the particle's path, the acceleration of the particle, and the angle between the tangent to the path and the acceleration vector at $t = 4$ s.

SOLUTION

Road Map We are given the velocity vector of the particle, so we need to integrate the components of the velocity to get the components of the position. Then we will be able to find the path $y(x)$ by eliminating time from the position components. The acceleration will be obtained by differentiating the velocity with respect to time. Finally, since the velocity is always tangent to the path, the angle between the tangent to the path and the acceleration can be found by finding the angle between velocity and acceleration. In turn, this is accomplished by computing the dot product of the two vectors.

Computation Equation (2.16) tells us that $v_x = \dot{x} = dx/dt$ and $v_y = \dot{y} = dy/dt$. Therefore, the position components x and y are obtained as follows

$$dx = v_x\,dt \quad \Rightarrow \quad x(t) = \int v_x\,dt + C_1 = \int 3t^2\,dt + C_1 = t^3 + C_1, \qquad (1)$$

$$dy = v_y\,dt \quad \Rightarrow \quad y(t) = \int v_y\,dt + C_2 = \int -4t\,dt + C_2 = -2t^2 + C_2, \qquad (2)$$

where C_1 and C_2 are constants of integration. The constants are found by noting that x and y are both zero at $t = 0$, which gives $C_1 = C_2 = 0$. Thus the particle's positions in Cartesian coordinates at an arbitrary time t and at $t = 4$ s, respectively, are given by

$$\boxed{\vec{r}(t) = (t^3\,\hat{\imath} - 2t^2\,\hat{\jmath})\text{ in.} \quad \text{and} \quad \vec{r}(4\,\text{s}) = (64\,\hat{\imath} - 32\,\hat{\jmath})\text{ in.}} \qquad (3)$$

To get the particle's path, we take $x = t^3$ and $y = -2t^2$ and eliminate t. Solving the x equation for t and then substituting that into the y equation, we obtain

$$\boxed{y = -2x^{2/3}\text{ in.,}} \qquad (4)$$

whose plot, corresponding to the time interval $0\,\text{s} \le t \le 4\,\text{s}$, is shown in Fig. 1.

Differentiating the velocity with respect to time, we have that the acceleration is

$$\boxed{\vec{a}(t) = (6t\,\hat{\imath} - 4\,\hat{\jmath})\text{ in./s}^2.} \qquad (5)$$

Finally, by applying Eq. (1.19) on p. 11, the angle θ between the velocity and the acceleration vectors for any time t is

$$\cos\theta(t) = \frac{\vec{v}\cdot\vec{a}}{|\vec{v}||\vec{a}|} = \frac{18t^3 + 16t}{2t\sqrt{(4 + 9t^2)(16 + 9t^2)}}, \qquad (6)$$

which, when evaluated at $t = 4$ s, gives

$$\boxed{\cos\theta(4\,\text{s}) = 0.9878 \quad \Rightarrow \quad \theta(4\,\text{s}) = 8.97°.} \qquad (7)$$

A plot of $\theta(t)$ for $0\,\text{s} \le t \le 4\,\text{s}$ in given in Fig. 2. As argued in the Road Map, the angle θ between $\vec{v}$ and $\vec{a}$ is also the angle between the tangent to the path and the acceleration.

Discussion & Verification The path shown in Fig. 1 is not a straight line. Therefore the acceleration is not tangent to the path. Since the velocity is always tangent to the path, this means that the angle between velocity and acceleration is expected to be different from zero, and our result is consistent with this expectation.

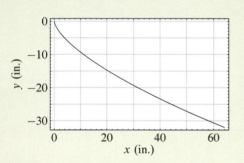

Figure 1
The path of the particle for $0 \le t \le 4$ s. Note that Eq. (3) tells us that when $t = 4$ s, $x = 64$ in.

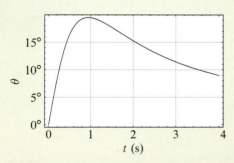

Figure 2
The angle between the velocity vector and the acceleration vector for $0 \le t \le 4$ s.

EXAMPLE 2.5   *Relating the Path and the Speed to Velocity and Acceleration*

A particle P moves with constant speed v_0 along the parabola $y^2 = 4ax$ with $\dot{y} > 0$. Determine the

(a) Cartesian components of the velocity vector as a function of y.

(b) Cartesian components of the acceleration vector as a function of y.

(c) Angle between the velocity and acceleration vectors as a function of y.

In addition, plot the Cartesian components of the velocity and acceleration vectors as a function y for $a = 0.2\,\text{m}$ and $v_0 = 3\,\text{m/s}$.

SOLUTION

Road Map   We are given the path of the particle and its speed along the path, but we do not know how x and y vary with time so we cannot immediately apply Eqs. (2.16) and (2.17). On the other hand, we can differentiate the equation of the curve with respect to time and make use of the fact that the speed is given by $\sqrt{\dot{x}^2 + \dot{y}^2}$.

Computation   We start by differentiating the particle's path with respect to time to obtain

$$\frac{d}{dt}(y^2 = 4ax) \quad \Rightarrow \quad \frac{d(y^2)}{dy}\frac{dy}{dt} = \frac{d(4ax)}{dt} \quad \Rightarrow \quad 2y\dot{y} = 4a\dot{x}, \tag{1}$$

where, in differentiating with respect to time, we have used the chain rule. Solving Eq. (1) for $\dot{y}$, we obtain

$$\dot{y} = \frac{2a\dot{x}}{y}. \tag{2}$$

Now, we also know that the speed is related to the components of the velocity via

$$v_0^2 = \dot{x}^2 + \dot{y}^2 \quad \Rightarrow \quad v_0^2 = \dot{x}^2 + \left(\frac{2a\dot{x}}{y}\right)^2, \tag{3}$$

where we have substituted in Eq. (2). We can now solve Eq. (3) for $\dot{x}$ to obtain

$$\boxed{\dot{x} = \frac{v_0 y}{\sqrt{y^2 + 4a^2}},} \tag{4}$$

where we have chosen the plus sign when taking the square root to be consistent with Eq. (2), which, given that $\dot{y} > 0$, tells us that for $y < 0$ we must have $\dot{x} < 0$ and for $y > 0$ we must have $\dot{x} > 0$. Now we can find $\dot{y}(y)$ by substituting Eq. (4) into Eq. (2), which gives

$$\boxed{\dot{y} = \frac{2v_0 a}{\sqrt{y^2 + 4a^2}},} \tag{5}$$

where $\dot{y} > 0$ as expected. Figure 2 shows $\dot{x}$ and $\dot{y}$ as functions of y.

There are several ways to obtain the x and y components of the acceleration. We will differentiate the last of Eq. (1) with respect to time, and after simplifying, we obtain

$$\dot{y}^2 + y\ddot{y} = 2a\ddot{x} \quad \Rightarrow \quad \ddot{x} = \frac{1}{2a}(\dot{y}^2 + y\ddot{y}). \tag{6}$$

We see that we need $\ddot{y}$ to know $\ddot{x}$. Hence, differentiating Eq. (5) with respect to time, we obtain

$$\ddot{y} = 2v_0 a\left[\frac{-y\dot{y}}{(y^2 + 4a^2)^{3/2}}\right], \tag{7}$$

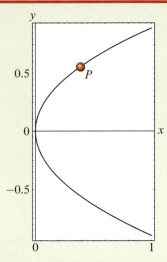

Figure 1
Parabolic path of the particle P for $a = 0.2\,\text{m}$.

Helpful Information

Why find components as a function of y instead of x? We chose to find the velocity and acceleration components as a function of y rather than x because of the fact that for any given value of y, there is one value of x. The converse is not true; that is, for a given value of x, there are two values of y, and this makes the analysis more complicated.

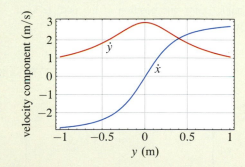

Figure 2
The x and y components of the velocity as a function of y.

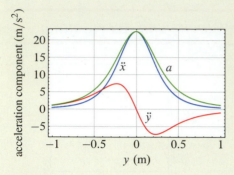

Figure 3
The x component (blue), y component (red), and the total acceleration (green) of the particle as a function of y.

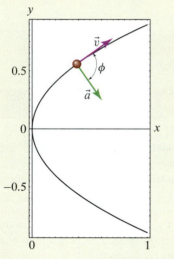

Figure 4
The velocity vector, acceleration vector, and the angle ϕ for one position of the particle as it moves along the parabola.

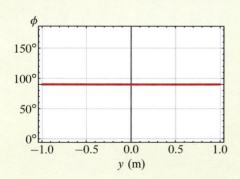

Figure 5
The heavy red line shows the angle $\phi(y)$ between the velocity vector and the acceleration vector.

which after plugging in $\dot{y}$ from Eq. (5) becomes

$$\ddot{y} = \frac{-4v_0^2 a^2 y}{(y^2 + 4a^2)^2}. \tag{8}$$

We can now get the final version of $\ddot{x}$ by substituting Eqs. (5) and (8) into the second of Eq. (6) — doing this gives

$$\ddot{x} = \frac{1}{2a}\left[\frac{4v_0^2 a^2}{y^2 + 4a^2} - y\,\frac{4v_0^2 a^2 y}{(y^2 + 4a^2)^2}\right] \quad\Rightarrow\quad \boxed{\ddot{x} = \frac{8v_0^2 a^3}{(y^2 + 4a^2)^2}.} \tag{9}$$

Figure 3 shows $\ddot{x}$, $\ddot{y}$, and $a = \sqrt{\ddot{x}^2 + \ddot{y}^2}$ as a function of y.

Finally, to find the angle between the velocity and acceleration vectors, we first need to form those vectors as

$$\vec{v} = \dot{x}\,\hat{\imath} + \dot{y}\,\hat{\jmath} \quad\text{and}\quad \vec{a} = \ddot{x}\,\hat{\imath} + \ddot{y}\,\hat{\jmath}. \tag{10}$$

Then, using Eq. (1.19) on p. 11, the angle ϕ between these two vectors is

$$\cos\phi = \frac{\vec{v}\cdot\vec{a}}{|\vec{v}||\vec{a}|} = \frac{(\dot{x}\,\hat{\imath} + \dot{y}\,\hat{\jmath})\cdot(\ddot{x}\,\hat{\imath} + \ddot{y}\,\hat{\jmath})}{\sqrt{\dot{x}^2 + \dot{y}^2}\,\sqrt{\ddot{x}^2 + \ddot{y}^2}} = \frac{\dot{x}\ddot{x} + \dot{y}\ddot{y}}{\sqrt{\dot{x}^2 + \dot{y}^2}\,\sqrt{\ddot{x}^2 + \ddot{y}^2}}. \tag{11}$$

Although it is tedious, we can substitute Eqs. (4), (5), (8), and (9) into the numerator of the last expression in Eq. (11) to find

$$\dot{x}\ddot{x} + \dot{y}\ddot{y} = 0 \quad\Rightarrow\quad \cos\phi = 0 \quad\Rightarrow\quad \boxed{\phi = 90°.} \tag{12}$$

Since $\phi = 90°$ for all values of y, then $\vec{v}$ is *always* perpendicular to $\vec{a}$.

Discussion & Verification   Recall that $\dot{y} > 0$. This agrees with the plot in Fig. 2, as it should, since we chose the plus sign when taking the square root to get Eq. (5). We also see that $\dot{y}$ is largest at $y = 0$. This also makes sense due to the fact that the particle has constant speed. Recalling that the velocity is always tangent to the path, note that at $y = 0$, the velocity must be completely in the y direction. Therefore, at $y = 0$, $\dot{x}$ must be zero and so $\dot{y}$ achieves a maximum equal to $v_0 = 3\,\text{m/s}$, as can be seen in Fig. 2. As for the x component of the velocity, it is negative for $y < 0$ and positive for $y > 0$. This is what we should expect, given that x is decreasing for $y < 0$ and is increasing for $y > 0$ (this is why we chose the $+$ sign in Eq. (4)). In addition, notice that $\dot{x} = 0$ when $y = 0$, which is also to be expected from the parabola shown in Fig. 1.

To verify our acceleration results, recall that the particle moves at a constant speed along a *parabola*. Since the particle's path is curved, we expect the components of the acceleration vector to be different from zero, as can be seen in Fig. 3. In addition, we expect $\ddot{y} = 0$ at $y = 0$ because, as argued earlier, $\dot{y}$ is maximum at $y = 0$. In addition, note that $\ddot{x}$ is largest at $y = 0$ even though $\dot{x} = 0$ there. While perhaps counterintuitive, this result is consistent with the fact that part of the acceleration is proportional to curvature of the path on which the particle moves. Since the curvature of the parabola is largest at $y = 0$, the magnitude of the acceleration, given by $a = \sqrt{\ddot{x}^2 + \ddot{y}^2}$, achieves a maximum at $y = 0$.*

As for the angle ϕ between $\vec{v}$ and $\vec{a}$, we discovered that it is always equal to $90°$. Since the speed is constant, we know that $\vec{v}$ *cannot* be changing along its line of action, and so any change in $\vec{v}$ (that is, any $\vec{a}$) must be perpendicular to $\vec{v}$. We will discuss this topic in detail in Sections 2.4 and 2.5.

_____

* This topic is discussed in detail in Section 2.5.

EXAMPLE 2.6   *From Position Data to Acceleration*

An experiment is performed in which a basketball is tossed in the air (Fig. 1). Once

Figure 1. A person tossing a basketball in the air.

it is airborne, the basketball's motion is captured by a camera. The basketball will be modeled as a point. The movie of the basketball's motion is digitized to extract a single set of coordinates per movie frame. The data is given in Table 1. It is claimed that the experiment was conducted on the surface of the Earth and that the y axis used in the experiment was perpendicular to the surface of the Earth and directed upward. Verify this claim by computing the trajectory, velocity, and acceleration of the basketball as a function of time.

SOLUTION

Road Map   The acceleration of gravity on the surface of the Earth is constant and equal to $9.81\,\text{m/s}^2$. We can therefore solve this problem by determining the acceleration and then checking whether or not it is as expected. Assuming gravity is the only force acting on the basketball while in flight, the x component of the acceleration should equal 0 and the y component of the acceleration should be approximately $-9.81\,\text{m/s}^2$.*

Computation   We start by using the data in Table 1 to plot the point's trajectory by plotting the point's y coordinate vs. its x coordinate. The trajectory, shown in Fig. 2, *seems* to be consistent with that of an object tossed in the air since it looks like a parabola. Let's keep checking.

Since we are *not* given the position as a function of time for which a derivative can be computed analytically, we *resign* ourselves to find the average velocity of the body over every time increment. In particular, if $\vec{r}_i$ and $\vec{r}_{i+1}$ are the position vectors of the object in frames[†] i and $i+1$, respectively, using the definition in Eq. (2.6), the average velocity between these two frames will be

$$\vec{v}_{\text{avg}}(t_i, t_{i+1}) = \frac{\vec{r}_{i+1} - \vec{r}_i}{t_{i+1} - t_i}, \qquad (1)$$

---

* We are implicitly using a model based on Newton's second law that neglects air resistance and other factors. We will see how this is done when we cover projectile motion in Section 2.3.

† Here *frame* refers to a frame in the digitized movie.

Table 1
Experimentally obtained positions vs. time.

Frame	t (s)	x (m)	y (m)
1	0.000	0.000	1.50
2	0.033	0.029	1.66
3	0.067	0.069	1.81
4	0.100	0.105	1.96
5	0.133	0.135	2.08
6	0.167	0.167	2.20
7	0.200	0.197	2.30
8	0.233	0.231	2.40
9	0.267	0.270	2.49
10	0.300	0.305	2.56
11	0.333	0.335	2.62
12	0.367	0.369	2.68
13	0.400	0.401	2.72
14	0.433	0.435	2.74
15	0.467	0.471	2.77
16	0.500	0.502	2.77
17	0.533	0.531	2.77
18	0.567	0.563	2.76
19	0.600	0.596	2.74
20	0.633	0.637	2.70
21	0.667	0.664	2.65
22	0.700	0.704	2.59
23	0.733	0.735	2.53
24	0.767	0.763	2.45
25	0.800	0.795	2.36
26	0.833	0.833	2.26
27	0.867	0.863	2.15
28	0.900	0.903	2.03
29	0.933	0.937	1.89
30	0.967	0.968	1.75
31	1.000	1.000	1.59
32	1.030	1.040	1.43

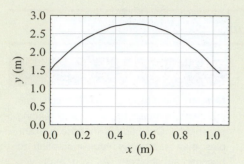

Figure 2

Trajectory of the object generated using the experimental data provided.

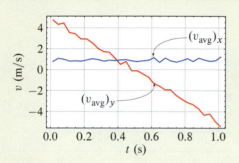

Figure 3

Plot of the components of the average velocity.

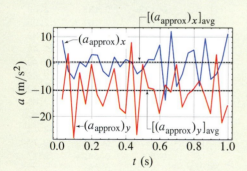

Figure 4

Plot of the components of the approximate acceleration.

which is defined neither at $t = t_i$ nor at $t = t_{i+1}$. Although it is *not* the only approach possible, we decide that $\vec{v}_{avg}(t_i, t_{i+1})$ corresponds to the time value midway between $t = t_i$ and $t = t_{i+1}$, namely, $t_{avg}(i, i+1) = (t_i + t_{i+1})/2$. Noting that this causes us to have one value of average velocity less than there are position values, we see the result plotted in Fig. 3.

We now turn to the acceleration, which we cannot compute as the time derivative of the velocity since we do not have an analytical expression for the velocity. To overcome this difficulty, observe that if we had *true* velocity values at the recorded times, the *average* acceleration between $t = t_i$ and $t = t_{i+1}$ would be

$$\vec{a}_{avg}(t_i, t_{i+1}) = \frac{\vec{v}(t_{i+1}) - \vec{v}(t_i)}{t_{i+1} - t_i}. \tag{2}$$

However, we have only $\vec{v}_{avg}(t_i, t_{i+1})$ values associated to corresponding $t_{avg}(i, i+1)$ time values. Therefore, we *resign* ourselves to *approximate* the acceleration via Eq. (2), replacing the *true* velocity and time values we need by the *average* velocity and time values we actually have, which yields the following formula:

$$\vec{a}_{approx}\big(t_{avg}(i, i+1), t_{avg}(i+1, i+2)\big) = \frac{\vec{v}_{avg}(t_{i+1}, t_{i+2}) - \vec{v}_{avg}(t_i, t_{i+1})}{t_{avg}(i+1, i+2) - t_{avg}(i, i_1)}. \tag{3}$$

In addition, we will say that Eq. (3) gives us the approximate acceleration at a time value midway in between $t_{avg}(i, i+1)$ and $t_{avg}(i+1, i+2)$. The plot of the approximate acceleration vs. time is shown in Fig. 4. We now have two fewer values of acceleration than the original number of position data points.

Discussion & Verification   In verifying our solution, we need to keep in mind that all experimental data is affected to one extent or another by measurement errors. Second, in addition to experimental errors, we need to note that there are consequences to treating *average* velocity values as *true* velocity values when computing the approximate acceleration. With these considerations in mind, we now must decide whether or not the plots of the x and y components of the acceleration are what they should *approximately* be, namely, 0 and $9.81\,\text{m/s}^2$, respectively. Clearly, the plots in Fig. 4 are not the plots of constant functions. On the other hand, can the plots in Fig. 4 be interpreted to provide the information we expect? Looking at the average values of the acceleration in the x and y directions, we see that $[(a_{approx})_x]_{avg} = 0.472\,\text{m/s}^2$ and $[(a_{approx})_y]_{avg} = -10.2\,\text{m/s}^2$, respectively. The x acceleration of $0.472\,\text{m/s}^2$ is close to 0, and the y acceleration of $-10.2\,\text{m/s}^2$ is close to the expected value of $-9.81\,\text{m/s}^2$. Unfortunately, the standard deviation of $(a_{approx})_x$ is $6.04\,\text{m/s}^2$, and the standard deviation of $(a_{approx})_y$ is $8.83\,\text{m/s}^2$. Therefore, the average values of the x and y components of the acceleration are not far from their expected values, but the large standard deviations tells us that we cannot attribute any significance to our results. The reason why the acceleration results are so bad is that we have progressed from position to velocity and then to acceleration by *numerical differentiation* instead of analyical differentiation. Numerical differentiation is inherently subject to considerable error. The reason for this is that while two curves may be close to one another, they may differ considerably in their slope, variation in slope, etc. We are seeing this as we progress from Fig. 2 to Fig. 4 in that the error continues to grow as we take higher derivatives. In the next example we will show that the opposite is true of *numerical integration*, i.e., the integration of discrete data. Numerical integration tends to smooth errors rather than magnify them.

EXAMPLE 2.7   *From Acceleration Data to Position*

Accelerometers have been mounted on the bicycle to study its motion along a velo-drome track. These accelerometers measure acceleration relative to a reference frame attached to the bike, but that data has been converted to a fixed xy reference frame, given in Table 1, and plotted in Fig. 2. Using the data in Table 1, and knowing that at $t = 0$, $x_0 = 208.9\,\text{ft}$, $y_0 = 0.0\,\text{ft}$, $\dot{x}_0 = 0.00\,\text{ft/s}$, and $\dot{y}_0 = 44.00\,\text{ft/s}$, determine

(a) The x and y components of velocity, as well as the speed, as a function of time.

(b) The trajectory of the bicycle in the given xy frame.

SOLUTION

Road Map   At the outset we know nothing about the shape of the track and the path of the bicycle. However, the ideas presented in this section will allow us to approximate both velocity and position at every data point so that we will be able to determine the bicycle's trajectory and "reconstruct" the shape of the track. Since we have acceleration data and we want velocity and position information, we need to integrate the acceleration data. In addition, since the data is discrete, we will have to resort to numerical integration. We will use an intuitive scheme called the *composite trapezoidal rule* (CTR).

The basic idea behind the CTR is that the area under a curve (i.e., its integral) can be approximated by the area under the trapezoids formed by connecting points on the curve. This can be seen in Fig. 3, where we have approximated the integral of $\sin x$ between 0 and π (the blue area) by summing the areas of the four trapezoids shown (hatched area), that is,

$$\int_0^\pi \sin x \, dx \approx \frac{\pi}{4}\left[\frac{\sin(0) + \sin\left(\frac{\pi}{4}\right)}{2}\right] + \frac{\pi}{4}\left[\frac{\sin\left(\frac{\pi}{4}\right) + \sin\left(\frac{\pi}{2}\right)}{2}\right]$$

$$+ \frac{\pi}{4}\left[\frac{\sin\left(\frac{\pi}{2}\right) + \sin\left(\frac{3\pi}{4}\right)}{2}\right] + \frac{\pi}{4}\left[\frac{\sin\left(\frac{3\pi}{4}\right) + \sin(\pi)}{2}\right]$$

$$= \frac{\pi/4}{2}\left[\sin(0) + 2\sin\left(\frac{\pi}{4}\right) + 2\sin\left(\frac{\pi}{2}\right) + 2\sin\left(\frac{3\pi}{4}\right) + \sin(\pi)\right]$$

$$= 1.896. \tag{1}$$

The exact value of the integral is 2. The approximation is not bad, and it improves as we add more trapezoids (by reducing the step size). Generalizing the above result, we can approximate the integral $\int_a^b f(x)\,dx$ via the CTR by dividing the interval $a \le x \le b$ into n equal segments, each $h = (b - a)/n$ in length, and applying the following formula:

$$\int_a^b f(x)\,dx = \frac{h}{2}\left[f(x_0) + 2f(x_1) + \cdots + 2f(x_i) + \cdots + 2f(x_{n-1}) + f(x_n)\right], \tag{2}$$

where $x_0 = a$, $x_n = b$, and $x_i = x_0 + ih$. Finally, before applying the CTR to the given acceleration data to find velocities and positions, we recall from calculus that if $f(t) = dF(t)/dt$, then $F(t) = F(t_0) + \int_{t_0}^t f(t)\,dt$. Therefore to compute $F(t)$ by integrating $f(t)$ over the interval $[t_0, t]$, we must add the value of $F(t)$ at $t = t_0$ to $\int_{t_0}^t f(t)\,dt$ (Section 2.2 is devoted entirely to ideas such as this).

Computation   Following the Road Map, we first apply the CTR to the acceleration data in Table 1 to obtain the velocity, paying attention to add to the integrals the initial value of the velocity components, i.e.,

$$v_x(t) = \dot{x}_0 + \int_0^t a_x(t)\,dt \qquad \text{and} \qquad v_y(t) = \dot{y}_0 + \int_0^t a_y(t)\,dt. \tag{3}$$

Figure 1

A bicycle in a velodrome.

Table 1

Raw data from the sensors on a bicycle for one complete circuit of the track. The acceleration data has measurement error (approximately $\pm 2\%$) associated with signal noise, mounting inaccuracies, and errors in manufacturing. Time is measured within 1 ms while acceleration is measured to a $10^{-2}\,\text{ft/s}^2$.

Time (s)	a_x (ft/s^2)	a_y (ft/s^2)
0.000	-9.39	0.12
1.864	-8.61	-3.37
3.728	-6.40	-6.62
5.592	-3.37	-8.47
7.456	0.16	-9.14
9.320	3.43	-8.49
11.184	6.71	-6.48
13.048	8.70	-3.54
14.912	9.09	-0.14
16.776	8.69	3.50
18.640	6.38	6.68
20.504	3.61	8.57
22.368	0.14	9.09
24.232	-3.61	8.58
26.096	-6.38	6.63
27.960	-8.62	3.66

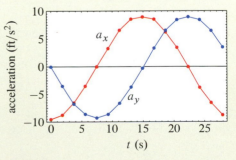

Figure 2

Plots of the acceleration data from Table 1.

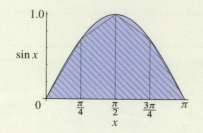

Figure 3
A plot of $\sin x$ and its approximation via the CTR for $0\,\text{rad} \leq x \leq \pi\,\text{rad}$.

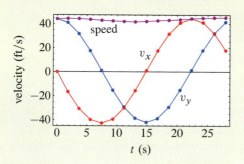

Figure 4
Plots of the x and y components of the velocities as computed by the CTR (Table 2). The speed is also shown.

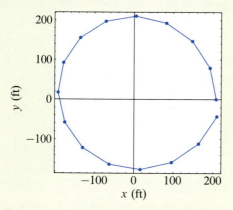

Figure 5
Plot of the position as computed by the CTR (Table 3).

In approximating the integrals in Eqs. (3) we used the smallest possible value of h allowed by the data, namely, $h = 1.864\,\text{s}$. The corresponding velocity results are shown in Table 2 and plotted in Fig. 4.

Table 2
The x and y components of the bicycle's velocity as computed by the CTR. We have used $v_{0x} = \dot{x}_0 = 0.00\,\text{ft/s}$, $v_{0y} = \dot{y}_0 = 44.00\,\text{ft/s}$, and $h = 1.864\,\text{s}$.

Time (s)	v_x (ft/s)	v_y (ft/s)
0.000	0.00	44.00
1.864	-16.78	40.97
3.728	-30.77	31.66
5.592	-39.87	17.60
7.456	-42.86	1.18
9.320	-39.52	-15.25
11.184	-30.07	-29.20
13.048	-15.70	-38.54
14.912	0.88	-41.97
16.776	17.45	-38.84
18.640	31.49	-29.35
20.504	40.80	-15.14
22.368	44.30	1.32
24.232	41.06	17.79
26.096	31.75	31.97
27.960	17.77	41.56

Table 3
The x and y components of the bicycle's position as computed by the CTR. We have used $x_0 = 208.9\,\text{ft}$, $y_0 = 0.0\,\text{ft}$, and $h = 1.864\,\text{s}$.

Time (s)	x (ft)	y (ft)
0.000	208.90	0.00
1.864	193.26	79.19
3.728	148.94	146.88
5.592	83.11	192.79
7.456	6.00	210.30
9.320	-70.77	197.18
11.184	-135.63	155.76
13.048	-178.29	92.62
14.912	-192.10	17.59
16.776	-175.02	-57.73
18.640	-129.41	-121.28
20.504	-62.03	-162.75
22.368	17.28	-175.63
24.232	96.84	-157.82
26.096	164.70	-111.44
27.960	210.85	-42.91

For example, the second and third entries for v_x are computed as follows (the first entry is simply the initial condition):

$$v_x(1.864\,\text{s}) = \frac{1.864}{2}(-9.39 - 8.61) = -16.78\,\text{ft/s},\tag{4}$$

$$v_x(3.728\,\text{s}) = \frac{1.864}{2}[-9.39 - 2(8.61) - 6.40] = -30.77\,\text{ft/s}.\tag{5}$$

We now apply the CTR to the velocity data in Table 2, again paying attention to account for the given initial position components as in

$$x(t) = x_0 + \int_0^t v_x(t)\,dt \qquad \text{and} \qquad y(t) = y_0 + \int_0^t v_y(t)\,dt.\tag{6}$$

Using the CTR to approximate the integrals in Eqs. (6), we obtain the position components shown in Table 3 and plotted in Fig. 5. For example, the second and third entries for y are computed as follows (the first entry is simply the initial condition):

$$y(1.864\,\text{s}) = \frac{1.864}{2}(44.00 + 40.97) = 79.19\,\text{ft},\tag{7}$$

$$y(3.728\,\text{s}) = \frac{1.864}{2}[44.00 + 2(40.97) + 31.66] = 146.9\,\text{ft}.\tag{8}$$

Discussion & Verification   The trajectory plot in Fig. 5 suggests that the bicycle's path is circular, which is consistent with the shape of a velodrome. The speed plot in Fig. 4 indicates that the bicycle is moving at a nearly constant speed. This result, combined with the path result, suggests that the bicycle is in a uniform circular motion, and this is consistent with the plots of the components of acceleration, which suggest that the acceleration components are periodic.

🔎 **A Closer Look**   It is immediately apparent that the integrated data is much "smoother" than the differentiated data in the previous example. Figures 4 and 5 don't have any of the "jaggedness" that can be found in Figs. 3 and 4 from Example 2.6.

PROBLEMS

Note: In all problems, all reference frames are stationary.

Problem 2.1

The position of a car traveling between two stop signs along a straight city block is given by $r = [9t - (45/2)\sin(2t/5)]$ m, where t denotes time and $0\,\text{s} \leq t \leq 17.7\,\text{s}$. Compute the displacement of the car between 2.1 and 3.7 s as well as between 11.1 and 12.7 s. For each of these time intervals compute the average velocity.

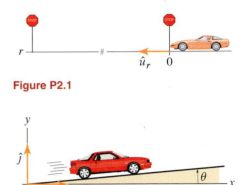

Figure P2.1

Problems 2.2 and 2.3

The position of the car relative to the coordinate system shown is

$$\vec{r}(t) = \left[\left(5.98t^2 + 0.139t^3 - 0.0149t^4\right)\hat{\imath} + \left(0.523t^2 + 0.0122t^3 - 0.00131t^4\right)\hat{\jmath}\right]\text{ft}.$$

Problem 2.2   Determine the velocity and acceleration of the car at $t = 15\,\text{s}$. In addition, again at $t = 15\,\text{s}$, determine the slope θ of the car's path relative to the coordinate system shown as well as the angle ϕ between velocity and acceleration.

Problem 2.3   Find the difference between the average velocity over the time interval $0\,\text{s} \leq t \leq 2\,\text{s}$ and the true velocity computed at the midpoint of the interval, i.e., at $t = 1\,\text{s}$. Repeat the calculation for the time interval $8\,\text{s} \leq t \leq 10\,\text{s}$. What do the results suggest about the approximation of the true velocity by the average velocity over different time intervals?

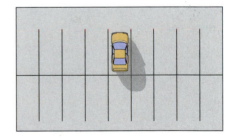

Figure P2.2 and P2.3

Problem 2.4

If $\vec{v}_{\text{avg}}$ is the average velocity of a point P over a given time interval, is $|\vec{v}_{\text{avg}}|$, the magnitude of the average velocity, equal to the average speed of P over the time interval in question?
Note: Concept problems are about *explanations*, not computations.

Problem 2.5

A car is seen parked in a given parking space at 8:00 A.M. on a Monday morning and is then seen parked in the same spot the next morning at the same time. What is the displacement of the car between the two observations? What is the distance traveled by the car during the two observations?
Note: Concept problems are about *explanations*, not computations.

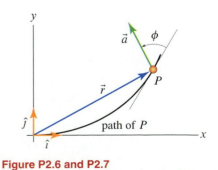

Figure P2.5

Problem 2.6

Let $\vec{r} = [t\,\hat{\imath} + (2 + 3t + 2t^2)\,\hat{\jmath}]$ m describe the motion of the point P relative to the Cartesian frame of reference shown. Determine an analytic expression of the type $y = y(x)$ for the trajectory of P.

Problem 2.7

Let $\vec{r} = [t\,\hat{\imath} + (2 + 3t + 2t^2)\,\hat{\jmath}]$ ft describe the motion of a point P relative to the Cartesian frame of reference shown. Recalling that for any two vectors $\vec{p}$ and $\vec{q}$ we have that $\vec{p} \cdot \vec{q} = |\vec{p}|\,|\vec{q}|\cos\beta$, where β is the angle formed by $\vec{p}$ and $\vec{q}$, and recalling that the velocity vector is *always* tangent to the trajectory, determine the function $\phi(x)$ describing the angle between the acceleration vector and the tangent to the path of P.

Figure P2.6 and P2.7

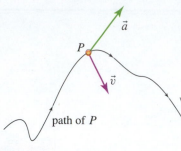

path of *P*

Figure P2.8 and P2.9

Problem 2.8

Is it possible for the vector $\vec{v}$ shown to represent the velocity of the point P?
Note: Concept problems are about *explanations*, not computations.

Problem 2.9

Is it possible for the vector $\vec{a}$ shown to be the acceleration of the point P?
Note: Concept problems are about *explanations*, not computations.

Problems 2.10 and 2.11

The motion of a point P with respect to a Cartesian coordinate system is described by
$\vec{r} = [2\sqrt{t}\,\hat{\imath} + (4\ln(t+1) + 2t^2)\,\hat{\jmath}]$ ft, where t is time expressed in s.

Problem 2.10   Determine P's displacement between times $t_1 = 4$ s and $t_2 = 6$ s.
In addition, determine the average velocity between t_1 and t_2.

Problem 2.11   Determine P's average acceleration between times $t_1 = 4$ s and
$t_2 = 6$ s.

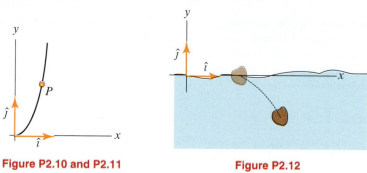

Figure P2.10 and P2.11   **Figure P2.12**

Problem 2.12

The motion of a stone thrown into a pond is described by

$$\vec{r}(t) = \left[\left(1.5 - 0.3e^{-13.6t}\right)\hat{\imath} + \left(0.094e^{-13.6t} - 0.094 - 0.72t\right)\hat{\jmath}\right] \text{m},$$

where t is time expressed in s, and $t = 0$ s is the time when the stone first hits the
water. Determine the stone's velocity and acceleration. In addition, find the initial angle
of impact θ of the stone with the water, i.e., the angle formed between the stone's
trajectory and the x axis at $t = 0$ s.

Problem 2.13

Two points P and Q happen to go by the same location in space (though at different
times).

(a) What must the paths of P and Q have in common if, at the location in question,
P and Q have identical speeds?

(b) What must the paths of P and Q have in common if, at the location in question,
P and Q have identical velocities?

Note: Concept problems are about *explanations*, not computations.

Problems 2.14 and 2.15

The position of point P is given by

$$\vec{r}(t) = 2.0\,[0.5 + \sin(\omega t)]\,\hat{\imath} + \left[9.5 + 10.5\sin(\omega t) + 4.0\sin^2(\omega t)\right]\hat{\jmath},$$

with $t \geq 0$, $\omega = 1.3\,\text{s}^{-1}$, and the position is measured in meters.

Problem 2.14   Find the trajectory of P in Cartesian components and then, using the x component of $\vec{r}(t)$, find the maximum and minimum values of x reached by P. The equation for the trajectory is valid for all values of x, yet the maximum and minimum values of x as given by the x component of $\vec{r}(t)$ are finite. What is the origin of this discrepancy?

Problem 2.15

(a) Plot the trajectory of P for $0 \leq t \leq 0.6\,\text{s}$, $0 \leq t \leq 1.4\,\text{s}$, $0 \leq t \leq 2.3\,\text{s}$, and $0 \leq t \leq 5\,\text{s}$.

(b) Plot the $y(x)$ trajectory for $-10 \leq x \leq 10\,\text{s}$.

(c) You will notice that the trajectory found in (b) does not agree with any of those found in (a). Explain this discrepancy by analytically determining the minimum and maximum values of x reached by P.

As you look at this sequence of plots, why does the trajectory change between some times and not others?

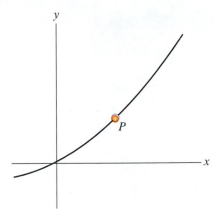

Figure P2.14 and P2.15

Problems 2.16 through 2.18

A bicycle is moving to the right at a speed $v_0 = 20\,\text{mph}$ on a horizontal and straight road. The radius of the bicycle's wheels is $R = 1.15\,\text{ft}$. Let P be a point on the periphery of the front wheel. One can show that the x and y coordinates of P are described by the following functions of time:

$$x(t) = v_0 t + R\sin(v_0 t / R) \quad \text{and} \quad y(t) = R\left[1 + \cos(v_0 t / R)\right].$$

Figure P2.16–P2.18

Problem 2.16   Determine the expressions for the velocity, speed, and acceleration of P as functions of time.

Problem 2.17   Determine the maximum and minimum speed achieved by P as well as the y coordinate of P when the maximum and minimum speeds are achieved. Finally, compute the acceleration of P when P achieves its maximum and minimum speeds.

Problem 2.18   Plot the trajectory of P for $0\,\text{s} < t < 1\,\text{s}$. For the same time interval, plot the speed as a function of time as well as the components of the velocity and acceleration of P.

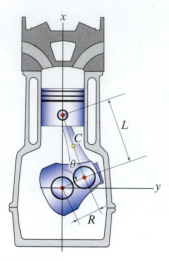

Figure P2.19–P2.21

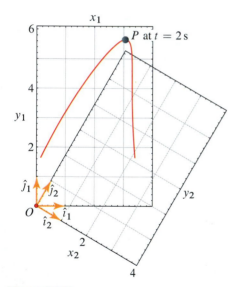

Figure P2.22

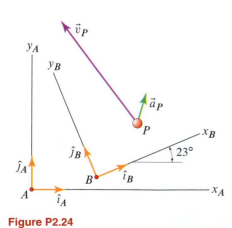

Figure P2.24

Problems 2.19 through 2.21

Point C is a point on the connecting rod of a mechanism called a *slider-crank*. The x and y coordinates of C can be expressed as follows: $x_C = R\cos\theta + \frac{1}{2}\sqrt{L^2 - R^2\sin^2\theta}$ and $y_C = (R/2)\sin\theta$, where θ describes the position of the crank. If the crank rotates at a constant rate, then we can express θ as $\theta = \omega t$, where t is time and ω is the crank's angular velocity. Let $R = 0.1\,\text{m}$, $L = 0.25\,\text{m}$, and $\omega = 250\,\text{rad/s}$.

Problem 2.19   Find expressions for the velocity, speed, and acceleration of C.

Problem 2.20   Determine the maximum and minimum speeds of C as well as C's coordinates when the maximum and minimum speeds are achieved. In addition, determine the acceleration of C when the speed is at a minimum.

Problem 2.21   Plot the trajectory of point C for $0\,\text{s} < t < 0.025\,\text{s}$. For the same interval of time, plot the speed as a function of time as well as the components of the velocity and acceleration of C.

Problem 2.22

The motion of a point P with respect to Cartesian frames 1 and 2 is described by

$$(\vec{r}_{P/O})_1 = \left[(t + \sin t)\,\hat{\imath}_1 + (2 + 4t - t^2)\,\hat{\jmath}_1\right]\text{m}$$

and

$$(\vec{r}_{P/O})_2 = \Big\{\left[(t + \sin t)\cos\theta + (2 + 4t - t^2)\sin\theta\right]\hat{\imath}_2$$
$$+ \left[-(t + \sin t)\sin\theta + (2 + 4t - t^2)\cos\theta\right]\hat{\jmath}_2\Big\}\,\text{m},$$

respectively, where t is time in seconds. Note that the two frames in this problem share the same origin, and therefore we are writing $(\vec{r}_{P/O})_1$ and $(\vec{r}_{P/O})_2$ to explicitly indicate that $(\vec{r}_{P/O})_1$ is expressed relative to frame 1 and $(\vec{r}_{P/O})_2$ is expressed relative to frame 2. Determine P's velocity and acceleration with respect to the two frames. In addition, determine the speed of P at time $t = 2\,\text{s}$, and verify that the speeds in the two frames are equal.

Problem 2.23

Let $\vec{r}_{P/A}$, $\vec{v}_{P/A}$, and $\vec{a}_{P/A}$ denote the position, velocity, and acceleration vectors of a point P with respect to the frame with origin at A. Let $\vec{r}_{P/B}$, $\vec{v}_{P/B}$, and $\vec{a}_{P/B}$ be the position, velocity, and acceleration vectors of the same point P with respect to the frame with origin at B. If frame B does not move relative to frame A, and if the frames are distinct, state whether or not each of the following relations is true and why.

(a) $\vec{r}_{P/A} - \vec{r}_{P/B} = \vec{0}$

(b) $\vec{v}_{P/A} - \vec{v}_{P/B} = \vec{0}$

(c) $\vec{v}_{P/A} \cdot \vec{a}_{P/B} = \vec{v}_{P/B} \cdot \vec{a}_{P/B}$

Note: Concept problems are about *explanations*, not computations.

Problem 2.24

The velocity of point P relative to frame A is $\vec{v}_{P/A} = (-14.9\,\hat{\imath}_A + 19.4\,\hat{\jmath}_A)\,\text{ft/s}$, and the acceleration of P relative to frame B is $\vec{a}_{P/B} = (3.97\,\hat{\imath}_B + 4.79\,\hat{\jmath}_B)\,\text{ft/s}^2$. Knowing that frames A and B do not move relative to one another, determine the expressions for the velocity of P in frame B and the acceleration of P in frame A. Verify that the speed of P and the magnitude of P's acceleration are the same in the two frames.

Problem 2.25

At the instant shown, when expressed via the $(\hat{u}_t, \hat{u}_n)$ component system, the airplane's velocity and acceleration are

$$\vec{v} = 135\,\hat{u}_t \text{ m/s} \quad \text{and} \quad \vec{a} = (-7.25\,\hat{u}_t + 182\,\hat{u}_n) \text{ m/s}^2.$$

Treating the $(\hat{u}_t, \hat{u}_n)$ and $(\hat{i}, \hat{j})$ component systems as stationary relative to one another, express the airplane's velocity and acceleration in the $(\hat{i}, \hat{j})$ component system. Determine the angle ϕ between the velocity and acceleration vectors, and verify that ϕ is the same in the $(\hat{u}_t, \hat{u}_n)$ and $(\hat{i}, \hat{j})$ component systems.

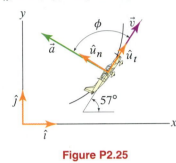

Figure P2.25

Problem 2.26

Two Coast Guard patrol boats P_1 and P_2 are stationary while monitoring the motion of a surface vessel A. The velocity of A with respect to P_1 is expressed by

$$\vec{v}_A = (-23\,\hat{i}_1 - 6\,\hat{j}_1) \text{ ft/s},$$

whereas the acceleration of A, expressed relative to P_2, is given by

$$\vec{a}_A = (-2\,\hat{i}_2 - 4\,\hat{j}_2) \text{ ft/s}^2.$$

Determine the velocity and the acceleration of A expressed with respect to the land-based component system $(\hat{i}, \hat{j})$.

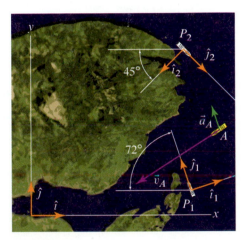

Figure P2.26

Problem 2.27

For a particle moving along a straight line, the table reports the particle's position x as a function of time. Determine the average velocity between every pair of consecutive time values for this motion, using Eq. (2.6). Provide a plot of the average velocity as a function of time.

t (s)	x (m)	t (s)	x (m)	t (s)	x (m)
0.00	0.000	1.00	1.344	2.00	1.193
0.20	0.331	1.20	1.458	2.20	0.963
0.40	0.645	1.40	1.500	2.40	0.686
0.60	0.928	1.60	1.468	2.60	0.375
0.80	1.165	1.80	1.364	2.80	0.046

Problem 2.28

Continue Prob. 2.27 by treating the average velocities as if they the were the true velocities, and compute the average accelerations corresponding to every pair of consecutive time values as was done in Example 2.6 on p. 43. Provide a plot of the average acceleration as a function of time.

Figure P2.29 and P2.30

Problem 2.29

The table gives the position vs. time data for a pendulum swinging in the xy plane. Compute the displacement between $t = 0.0$ s and $t = 0.539$ s and between $t = 0.0$ s and $t = 2.023$ s. Furthermore, compute the average velocity over the given time intervals. Knowing that the data in the table below concerns a swinging pendulum, interpret the result you obtain for the average velocity between $t = 0.0$ s and $t = 2.023$ s.

Time (s)	x (ft)	y (ft)	Time (s)	x (ft)	y (ft)
0.000	1.693	0.530	1.079	-1.679	0.491
0.135	1.641	0.439	1.214	-1.485	0.361
0.270	1.164	0.218	1.348	-0.898	0.129
0.405	0.627	0.052	1.483	-0.219	0.013
0.539	-0.222	0.005	1.618	0.586	0.038
0.674	-0.962	0.124	1.753	1.148	0.240
0.809	-1.489	0.343	1.888	1.542	0.435
0.944	-1.768	0.514	2.023	1.809	0.528

🖥 Problem 2.30 🖥

The table in Prob. 2.29 gives the position vs. time data for a pendulum swinging in the xy plane. Compute the components of the average velocity as well as the magnitude of the average velocity over each time step and plot these quantities vs. time. Furthermore, compute the components of the approximate acceleration as was done in Example 2.6 on p. 43 by using the average velocity data generated and plot the results vs. time. Finally compare the results to the plots of the components of the exact velocity and acceleration vs. time shown below. In these plots the vertical axes represent the quantity labeling each plot whereas the horizontal axes represent time expressed in seconds.

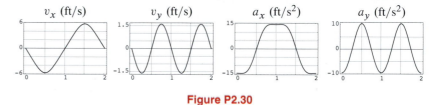

Figure P2.30

Problem 2.31

Let $f(t)$ be a function of time, and suppose that a table of values of $f(t)$ is provided for a sequence of equally spaced time instants. Then, for any three consecutive values of $f(t)$, i.e., $f(t_i)$, $f(t_{i+1})$, and $f(t_{i+2})$, you can approximate the value of the derivative of $f(t)$ with respect to time at $t = t_i$ by using the formula

$$\frac{df}{dt}(t_i) \approx \frac{-f(t_{i+2}) + 4f(t_{i+1}) - 3f(t_i)}{2\,\Delta t}, \tag{1}$$

where $\Delta t = t_{i+1} - t_i = t_{i+2} - t_{i+1}$. Use this formula to compute derivatives, and rework Example 2.6 to obtain new plots for the velocity and the acceleration. Does the formula given above allow you to obtain smoother plots for the velocity and acceleration with respect to those in Example 2.6?

Problem 2.32

Find the x and y components of the acceleration in Example 2.5 (except for the plots) by simply differentiating Eqs. (4) and (5) with respect to time. Verify that you get the results given in Example 2.5.

Problem 2.33 🌡

Find the x and y components of the acceleration in Example 2.5 (except for the plots) by differentiating the first of Eqs. (3) and the last of Eqs. (1) with respect to time and then solving the resulting two equations for $\ddot{x}$ and $\ddot{y}$. Verify that you get the results given in Example 2.5.

💻 Problem 2.34 💻

Pioneer 3 was a spin-stabilized spacecraft launched on 6 December 1958 by the U.S. Army Ballistic Missile agency in conjunction with NASA. It was a cone-shaped probe 58 cm high and 25 cm in diameter at its base. It was designed with a despin mechanism consisting of two 7 g masses (m in the figure) that could be spooled out to the end of two 150 cm wires when triggered by a hydraulic timer 10 h after launch. As they are deployed, the masses slow the spacecraft spin rate from its initial value to a desired one. The table below reports discrete acceleration vs. time data of one of the masses as it is deployed in a test run in which the spacecraft is kept with its z axis vertical and the masses deploy in the xy plane. Follow the steps described in Example 2.7, and reconstruct the velocity as well as the position of the deployed mass as a function of time. Finally plot the trajectory of the mass. Use the initial conditions $x(0) = 0.125\,\text{m}$, $y(0) = 0\,\text{m}$, $\dot{x}(0) = 0\,\text{m/s}$, and $\dot{y}(0) = 1.25\,\text{m/s}$.

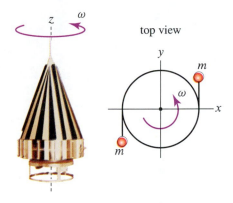

Figure P2.34

Time (s)	a_x (m/s^2)	a_y (m/s^2)	Time (s)	a_x (m/s^2)	a_y (m/s^2)
0.00	0.0	0.0	0.16	4.1	-78.7
0.02	-3.9	9.2	0.18	38.3	-79.6
0.04	-14.3	13.9	0.20	73.1	-64.9
0.06	-27.8	10.8	0.22	101.0	-34.7
0.08	-39.8	-1.1	0.24	116.0	7.5
0.10	-45.2	-20.6	0.26	112.0	55.5
0.12	-40.3	-43.7	0.28	88.5	101.0
0.14	-23.5	-65.0	0.30	46.1	135.0

Problem 2.35 🌡

The Center for Gravitational Biology Research at NASA's Ames Research Center runs a large centrifuge capable of $20g$ of acceleration ($12.5g$ is the maximum for human subjects). The distance from the axis of rotation to the cab at either A or B is $R = 25\,\text{ft}$. The trajectory of A is described by $y_A = \sqrt{R^2 - x_A^2}$ for $y_A \geq 0$ and by $y_A = -\sqrt{R^2 - x_A^2}$ for $y_A < 0$. If A moves at the constant speed $v_A = 120\,\text{ft/s}$, determine the velocity and acceleration of A when $x_A = -20\,\text{ft}$ and $y_A > 0$.

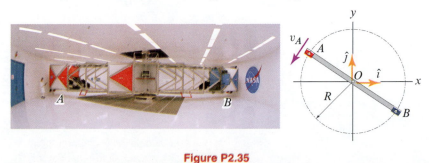

Figure P2.35

Problem 2.36

The orbit of a satellite A around planet B is the ellipse shown and is described by the equation $(x/a)^2 + (y/b)^2 = 1$, where a and b are the semimajor and semiminor axes of the ellipse, repectively. When $x = a/2$ and $y > 0$, the satellite is moving with a speed v_0 as shown. Determine the expression for the satellite's velocity $\vec{v}$ in terms of v_0, a, and b for $x = a/2$ and $y > 0$.

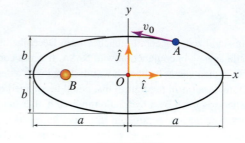

Figure P2.36

2.2 Elementary Motions

This section examines in detail how to relate acceleration to position and velocity in a variety of situations found in applications. To better focus on how these relations are built, here we avoid dealing with vector quantities and we examine only one-dimensional motions.

Figure 2.13
A car driving between two stop signs.

Driving down a city street

A car drives along a straight street between two stop signs (see Fig. 2.13). The car's velocity is given as

$$v = 9 - 9\cos\left(\tfrac{2}{5}t\right) \text{ m/s}, \quad 0\,\text{s} \le t \le 5\pi\,\text{s}, \tag{2.18}$$

which is plotted in Fig. 2.14. Given this information, we want to determine

1. The time it took to go from one stop sign to the other.

2. The distance between the two stop signs.

3. The acceleration at every instant along the way.

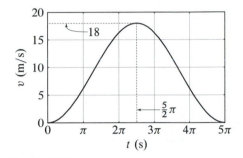

Figure 2.14
Velocity versus time curve for a car going between two stop signs.

We begin by observing that v in Eq. (2.18) is a scalar and therefore might not be used as a velocity, which is a vector. However, since the motion is one-dimensional, we can infer the direction of motion from the sign of v. Adopting this strategy and denoting the car's position by the coordinate s, we set

$$v = \dot{s}, \tag{2.19}$$

and by allowing v to take on both positive and negative values, we can then refer to v as the *velocity of the car*.

To answer question 1, since $v = 0$ at each of the stop signs, we can set to 0 the expression in Eq. (2.18) and solve for time, i.e.,

$$v = 9 - 9\cos\left(\tfrac{2}{5}t\right) = 0 \quad \Rightarrow \quad \cos\left(\tfrac{2}{5}t\right) = 1$$

$$\Rightarrow \quad \tfrac{2}{5}t = 0, 2\pi, 4\pi, \ldots \quad \Rightarrow \quad t = 0, 5\pi, 10\pi\,\text{s}, \ldots. \tag{2.20}$$

Since the motion starts at $t = 0$, the two times of interest are $t_0 = 0\,\text{s}$ and $t_1 = 5\pi\,\text{s}$. Thus, the answer to question 1 is that it takes $t_1 - t_0 = 5\pi\,\text{s} = 15.7\,\text{s}$ to go from the first stop sign to the second.

To answer question 2, recall that $v = \dot{s} = ds/dt$ so that we can write $ds = v\,dt$ and then use *indefinite* integration to obtain

$$\int ds = \int v(t)\,dt = \int \left[9 - 9\cos\left(\tfrac{2}{5}t\right)\right] dt. \tag{2.21}$$

Alternatively, we can use *definite* integration to obtain

$$\int_0^{s(t)} ds = \int_0^t v(t)\,dt = \int_0^t \left[9 - 9\cos\left(\tfrac{2}{5}t\right)\right] dt, \tag{2.22}$$

where the lower limits of integration indicate that we have set the origin of the s axis to be the car's position at $t = 0$, and the upper limits indicate that we wish to express s as a function of time.

> **Helpful Information**
>
> **Is there something wrong with Eq. (2.22)?** To be rigorous, Eq. (2.22) should be written as
>
> $$\int_0^{s(t)} d\sigma = \int_0^t v(\tau)\,d\tau$$
> $$= \int_0^t \left[9 - 9\cos\left(\tfrac{2}{5}\tau\right)\right] d\tau,$$
>
> where the variables of integration are distinct from the variables used in the limits of integration. The symbols chosen for the variables of integration do not change the integral, and this is why the variables of integration in a definite integral are called *dummy variables*. However, we feel that it is more meaningful to *keep* the variables of integration as s and t in Eq. (2.22) to remind us of their physical significance. We will adopt such a practice throughout this text since the use of the same symbol for both the variables and the limits of integration will be clear from the context.

From Eq. (2.22), we obtain

$$s\Big|_{s=0}^{s=s(t)} = \left[9t - \tfrac{45}{2}\sin\big(\tfrac{2}{5}t\big)\right]\Big|_{t=0}^{t=t} \quad \Rightarrow$$

$$s(t) = \left[9t - \tfrac{45}{2}\sin\big(\tfrac{2}{5}t\big)\right]\text{m}. \quad (2.23)$$

The result in Eq. (2.23) can also be obtained by indefinite integration. From Eq. (2.21) we have

$$s(t) = \left[9t - \tfrac{45}{2}\sin\big(\tfrac{2}{5}t\big) + C\right]\text{m}, \quad (2.24)$$

where C is the required constant of integration. To find C, recall that $s = 0$ for $t = 0$. Enforcing this condition, Eq. (2.24) yields

$$0 = 9(0) - \tfrac{45}{2}\sin\left[\tfrac{2}{5}(0)\right] + C \quad \Rightarrow \quad C = 0. \quad (2.25)$$

Substituting $C = 0$ in Eq. (2.24), we recover Eq. (2.23), as expected.

The formula in Eq. (2.23) gives the car's position for any time t. This allows us to compute the distance between the stop signs as $\Delta s = s(t_1) - s(t_0)$. Recalling that $t_0 = 0$, $t_1 = 5\pi$ s, and $s(0) = 0$, Δs is given by

$$\Delta s = 9(5\pi) - \tfrac{45}{2}\sin\left[\tfrac{2}{5}(5\pi)\right] = 45\pi\ \text{m} = 141\ \text{m}. \quad (2.26)$$

Finally, to answer question 3, we only need to differentiate v in Eq. (2.18) with respect to time:

$$a = \frac{dv}{dt} = \tfrac{18}{5}\sin\big(\tfrac{2}{5}t\big)\ \text{m/s}^2, \quad (2.27)$$

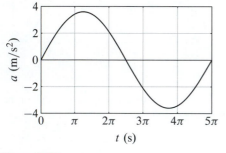

Figure 2.15
Acceleration versus time curve for the car driving between the two stop signs.

the plot of which can be found in Fig. 2.15. Notice that the acceleration is positive (i.e., the car gains speed) for the first $(5\pi/2)$ s, and then it is negative (i.e., the car slows down) until the car comes to a stop, as expected.

Rectilinear motion relations

The motion we have just studied is called *rectilinear* since it occurs along a straight line. The relations that govern it are useful in describing other types of one-dimensional motions even when the trajectory is not a straight line (see circular motion on p. 59). For this reason we are now going to investigate in detail how to relate acceleration, velocity, and position for rectilinear motions under various circumstances.

The signs of s, v, and a *are not* related.   We start with a simple observation about the problem we considered at the beginning of this section. Although the position s of the car was strictly increasing and the velocity v was always positive, the acceleration changed sign midway through the two stop signs. This tells us that, in general, the sign of s does not allow us to say *any-thing* about the signs of v and a. This can be seen more clearly in Fig. 2.16, where we consider four possible sign combinations for position and velocity. In Case 1, a particle with position $s > 0$ is moving to the right (i.e., s is increasing) so that $\dot{s}$ is positive. In Case 2, $\dot{s}$ is still positive, but now the particle is to the left of the origin so that $s < 0$. Cases 3 and 4 give us the other two possible sign combinations. Case 3 shows $s > 0$ with $\dot{s} < 0$, and Case 4 shows $s < 0$ with $\dot{s} < 0$. The same arguments apply to the acceleration so that in each of the four cases in Fig. 2.16, the acceleration could be either positive *or* negative.

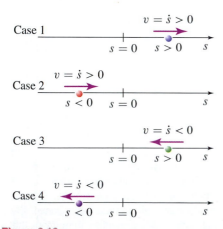

Figure 2.16
Rectilinear motion of a particle illustrating four possible relations between position and velocity.

Situations typically encountered in dynamics. In the example of the car traveling between stop signs, the known information was the velocity as a function of time. However, in most applications we are not given $v(t)$. Physical measurements* and Newton's second law usually provide us with $a(t)$, $a(v)$, or $a(s)$. The question then becomes, How do we calculate velocity and position starting from these forms of the acceleration?

If $a(t)$ is known

If we know the acceleration as a function of time $a(t)$, then we can determine $v(t)$ and $s(t)$ via time integration. Rewriting $a = dv/dt$ as $dv = a(t)\,dt$ and letting $v = v_0$ for $t = t_0$, we have

$$\int_{v_0}^{v} dv = \int_{t_0}^{t} a(t)\,dt, \tag{2.28}$$

or

$$v(t) = v_0 + \int_{t_0}^{t} a(t)\,dt. \tag{2.29}$$

Similarly, rewriting $v = ds/dt$ as $ds = v(t)\,dt$ and letting $s = s_0$ for $t = t_0$, we can determine $s(t)$ from Eq. (2.29) as follows:

$$\int_{s_0}^{s} ds = \int_{t_0}^{t} v(t)\,dt = \int_{t_0}^{t}\left[v_0 + \int_{t_0}^{t} a(t)\,dt\right]dt, \tag{2.30}$$

or

$$s(t) = s_0 + v_0(t - t_0) + \int_{t_0}^{t}\left[\int_{t_0}^{t} a(t)\,dt\right]dt. \tag{2.31}$$

If $a(v)$ is known

If we know the acceleration as a function of velocity $a(v)$, then finding v and s via integration is slightly more complicated. Observe that we can write

$$a(v) = \frac{dv}{dt} \quad \Rightarrow \quad dt = \frac{dv}{a(v)}. \tag{2.32}$$

If $a(v) \neq 0$ during the time interval considered, we can integrate the last expression in Eq. (2.32) to obtain time as a function of velocity $t(v)$, i.e.,

$$t(v) = t_0 + \int_{v_0}^{v} \frac{1}{a(v)}\,dv, \tag{2.33}$$

where v_0 is the value of v for $t = t_0$. The relation in Eq. (2.33) may seem strange since it yields $t(v)$. However, if $a(v) \neq 0$ during the time interval of interest, we can, in principle (though not always in practice), find $v(t)$.

---

* Accelerometers, which measure acceleration, are much more common than velocimeters.

Helpful Information

Integrals, integrands, and variables of integration. When we form an integral such as

$$\int_{t_0}^{t} \underbrace{a(t)}_{\substack{\text{both} \\ \text{have } t}} dt,$$

the integrand, which here is $a(t)$, must *always* involve only the variable with respect to which we are integrating (here t) and constants. For example, it would be acceptable if $a(t) = 3t^2 + 4$. However, if, say, $s(t)$ is not known and if the acceleration is given as $a(t) = 3t^2 + 4v(s)$, then the integral $\int[3t^2 + 4v(s)]\,dt$ *cannot* be evaluated because a depends on both t and s.

Helpful Information

Dummy variables can be useful. Looking at Eq. (2.30), we see that using dummy variables would be a real advantage here since there are so many t's involved. Using dummy variables, Eq. (2.30) would become

$$\int_{s_0}^{s} d\sigma = \int_{t_0}^{t} v(\tau)\,d\tau$$

$$= \int_{t_0}^{t}\left[v_0 + \int_{t_0}^{\tau} a(\xi)\,d\xi\right]d\tau.$$

Helpful Information

The chain rule. Since it is used very often, let's look more closely at the chain rule. Taking a bit of liberty, the chain rule can be presented as follows:

$$\frac{d(\text{Groucho})}{d(\text{Harpo})} = \frac{d(\text{Groucho})}{d(\text{Zeppo})} \frac{d(\text{Zeppo})}{d(\text{Harpo})}.$$

Although Zeppo did not appear on the left-hand side of the equation above, we were able to force him to appear on the right-hand side. This chain-rule-based "trick" will come in handy! The reason for using the Marx Brothers in this example is that the chain rule works only if all its terms are related to one another (mathematically, they must be functions of one another). Now, the connection with Eq. (2.34) is that we needed to make the variable s come into the picture even though, at first, it was not present in $a = dv/dt$. Therefore, we made v pose as our Groucho and t as our Harpo. Letting s take on the role of Zeppo, we were able to accomplish what we wanted!

To compute the position we can try to invert[*] $t(v)$ from Eq. (2.33) to find $v(t)$. Then we can try to integrate $v(t)$ with respect to t to obtain $s(t)$. Unfortunately, this is often difficult (or impossible) to do. Another approach is to obtain $s(v)$ instead of $s(t)$, using the chain rule of calculus, i.e.,

$$a = \frac{dv}{dt} = \frac{dv}{ds}\frac{ds}{dt} = v\frac{dv}{ds}, \tag{2.34}$$

which we can write as

$$ds = \frac{v}{a(v)}\,dv. \tag{2.35}$$

Setting $s = s_0$ when $v = v_0$, Eq. (2.35) can be integrated to obtain

$$s(v) = s_0 + \int_{v_0}^{v} \frac{v}{a(v)}\,dv. \tag{2.36}$$

If $a(s)$ is known

When the acceleration in known as a function of position, i.e., $a = a(s)$, we can again start from $a = v\,dv/ds$ given in Eq. (2.34) and, letting $v = v_0$ for $s = s_0$, obtain velocity as a function of position $v(s)$ as follows:

$$\int_{v_0}^{v} v\,dv = \int_{s_0}^{s} a(s)\,ds \quad \Rightarrow \quad \tfrac{1}{2}v^2 - \tfrac{1}{2}v_0^2 = \int_{s_0}^{s} a(s)\,ds, \tag{2.37}$$

or

$$v^2(s) = v_0^2 + 2\int_{s_0}^{s} a(s)\,ds. \tag{2.38}$$

Finally, once $v(s)$ is known through Eq. (2.38), we can obtain time as a function of position $t(s)$, starting from $v = ds/dt$ and then write

$$dt = \frac{ds}{v(s)} \quad \Rightarrow \quad \int_{t_0}^{t} dt = \int_{s_0}^{s} \frac{ds}{v(s)}, \tag{2.39}$$

where we have again let $s = s_0$ when $t = t_0$. Completing the integration of the left-hand side, we obtain

$$t(s) = t_0 + \int_{s_0}^{s} \frac{ds}{v(s)}. \tag{2.40}$$

What if a is constant?

If the acceleration is a constant, then the equations we have derived simplify *substantially*. The constant acceleration relations are important because there are many problems in dynamics in which the acceleration is constant. For example, in studying the motion of a projectile, we generally assume that the projectile's acceleration is constant.

---

[*] By *invert* we mean to solve $t(v)$ for v so that we have $v(t)$.

If the *acceleration is a constant a_c*, Eq. (2.29) becomes

$$v = v_0 + a_c(t - t_0) \quad \text{(constant acceleration)}, \qquad (2.41)$$

Eq. (2.31) becomes

$$s = s_0 + v_0(t - t_0) + \tfrac{1}{2}a_c(t - t_0)^2 \quad \text{(constant acceleration)}, \qquad (2.42)$$

and Eq. (2.38) becomes

$$v^2 = v_0^2 + 2a_c(s - s_0) \quad \text{(constant acceleration)}. \qquad (2.43)$$

Circular motion and angular velocity

The relationships for rectilinear motion are applicable to any one-dimensional motion. To demonstrate this idea, we will now apply them to a common one-dimensional curvilinear motion: *circular motion*.

In Fig. 2.17 we see a particle A moving in a circle of radius r and center O. Since r is constant, the position of A can be described via a single coordinate such as the oriented arc length s or the angle θ. Again referring to Fig. 2.17, if the line OA rotates through the angle $\Delta\theta$ in the time Δt, then we can define an average time rate of change of the angle θ as $\omega_{\text{avg}} = \Delta\theta/\Delta t$. Hence, following the development of Section 2.1 and letting $\Delta t \to 0$, we obtain the instantaneous time rate of change of θ, i.e., $\dot{\theta}$, called the *angular velocity*, as

$$\omega(t) = \lim_{\Delta t \to 0} \frac{\Delta\theta}{\Delta t} = \frac{d\theta(t)}{dt} = \dot{\theta}(t). \qquad (2.44)$$

We can then define *angular acceleration* α by differentiating Eq. (2.44) with respect to time, i.e.,

$$\alpha(t) = \frac{d\omega(t)}{dt} = \dot{\omega}(t) = \ddot{\theta}(t). \qquad (2.45)$$

When using the coordinate s, since $s = r\theta$ and r is constant, we can write

$$\dot{s} = r\dot{\theta} = \omega r \quad \text{and} \quad \ddot{s} = r\ddot{\theta} = \alpha r. \qquad (2.46)$$

Circular motion relations

All of the relationships we developed for rectilinear motion apply equally well to circular motion, except that we need to replace the rectilinear variables by their circular counterparts. For example, Eq. (2.29) becomes

$$\omega(t) = \omega_0 + \int_{t_0}^{t} \alpha(t)\, dt. \qquad (2.47)$$

Table 2.2 lists each kinematic variable in rectilinear motion and the corresponding kinematic variable for circular motion. Replacing each rectilinear motion variable with its circular motion counterpart in Eqs. (2.29)–(2.43), we obtain the corresponding circular motion equations. Finally, if the angular acceleration is constant, we can use the constant acceleration relations with a_c replaced by α_c.

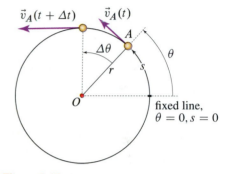

Figure 2.17
Particle A with speed $v_A = |\vec{v}_A|$ moving in a circle of radius r centered at O.

Table 2.2
Correspondence of kinematic variables between rectilinear and circular motion.

Kinematic variable	Rectilinear motion	Circular motion[a]
time	t	t
position	s	θ
velocity	v	ω
acceleration	a	α

[a] Except for time, each of these should have the word *angular* in front of its kinematic variable name.

End of Section Summary ——————

In this section we have developed relationships linking a single coordinate and its time derivatives. These relations have been categorized based on how the primary piece of information is provided:

1. If the acceleration is provided as a function of time, i.e., $a = a(t)$, for velocity and position, we have

> **Eqs. (2.29) and (2.31), p. 57**
>
> $$v(t) = v_0 + \int_{t_0}^{t} a(t)\, dt,$$
>
> $$s(t) = s_0 + v_0(t - t_0) + \int_{t_0}^{t} \left[\int_{t_0}^{t} a(t)\, dt \right] dt.$$

2. If the acceleration is provided as a function of velocity, i.e., $a = a(v)$, for time and position, we have

> **Eq. (2.33), p. 57, and Eq. (2.36), p. 58**
>
> $$t(v) = t_0 + \int_{v_0}^{v} \frac{1}{a(v)}\, dv,$$
>
> $$s(v) = s_0 + \int_{v_0}^{v} \frac{v}{a(v)}\, dv.$$

3. If the acceleration is provided as a function of position, i.e., $a = a(s)$, for velocity and time, we have

> **Eqs. (2.38) and (2.40), p. 58**
>
> $$v^2(s) = v_0^2 + 2\int_{s_0}^{s} a(s)\, ds,$$
>
> $$t(s) = t_0 + \int_{s_0}^{s} \frac{ds}{v(s)}.$$

4. If the acceleration is a constant a_c, for velocity and position, we have

> **Eqs. (2.41)–(2.43), p. 59**
>
> $$v = v_0 + a_c(t - t_0),$$
> $$s = s_0 + v_0(t - t_0) + \tfrac{1}{2}a_c(t - t_0)^2,$$
> $$v^2 = v_0^2 + 2a_c(s - s_0).$$

Circular motion.   For circular motion, the equations summarized in items 1–4 above hold as long as we use the replacement rules

> $$s \to \theta, \quad v \to \omega, \quad a \to \alpha,$$

where $\omega = \dot{\theta}$ and $\alpha = \ddot{\theta}$ are the *angular velocity* and *angular acceleration*, respectively.

EXAMPLE 2.8   *Measuring the Depth of a Well by Relating Time, Velocity, and Acceleration*

We can estimate the depth of a well by measuring the time it takes for a rock dropped from the top of the well to reach the water below. Assuming that gravity is the only force acting on the rock, estimate a well's depth under two different assumptions: the speed of sound is (a) finite and equal to $v_s = 340$ m/s and (b) infinite. Also, compare the two estimates to provide a "rule of thumb" as to when we can assume that the speed of sound is infinite.

SOLUTION

Road Map   In this problem our time measure is the sum of two parts: (1) the time taken by the rock to go from the top to the bottom of the well and (2) the time taken by sound to go from the bottom to the top of the well. The well's depth can be related to the first time by assuming that the rock travels at a constant acceleration, namely, $g = 9.81$ m/s^2. The well's depth can also be related to the second time by assuming that sound travels at a constant speed, which will be assumed to be finite in Part (a) and infinite in Part (b). By requiring that the two depth estimates be identical, we will be able to find a relation between the well's depth and the overall measured time.

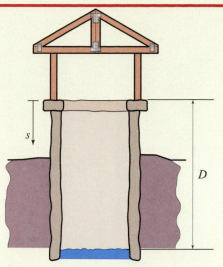

Figure 1
A well of depth D showing the positive direction of the coordinate s.

Part (a): Finite Sound Speed

Computation   Figure 1 shows a well of unknown depth D. Let t_m, t_i, and t_s be the (total) measured time, the time taken by the rock to fall the distance D and impact with the water, and the time it takes sound to go back up, respectively, so that

$$t_m = t_i + t_s. \qquad (1)$$

The motion of the rock falling the distance D is a rectilinear motion with constant acceleration $g = 9.81$ m/s^2. Hence, by choosing a coordinate axis pointing from the top to the bottom of the well, noting that the rock starts at $s_0 = 0$ m, and assuming that the rock is released with initial velocity v_0 equal to zero, Eq. (2.42) tells us that

$$D = \tfrac{1}{2}gt_i^2 \quad \Rightarrow \quad t_i = \sqrt{\frac{2D}{g}}. \qquad (2)$$

As soon as the rock hits the water, a sound wave traveling with a constant velocity $v_s = 340$ m/s, and therefore with constant acceleration $a_s = 0$ m/s^2, goes from the bottom of the well up to the observer's ear at $s = 0$. Equation (2.42) then tells us that

$$0 = D - v_s(t_m - t_i) \quad \Rightarrow \quad D = v_s t_s \quad \Rightarrow \quad t_s = \frac{D}{v_s}, \qquad (3)$$

where we have used Eq. (1) to write $t_s = t_m - t_i$. Next, using the expressions for t_i and t_s in Eqs. (2) and (3), respectively, Eq. (1) becomes

$$t_m = \sqrt{\frac{2D}{g}} + \frac{D}{v_s}. \qquad (4)$$

This equation can be solved for D to obtain (see the Helpful Information note in the margin for details)

$$\boxed{D = v_s t_m - \frac{v_s^2}{g}\left(\sqrt{1 + \frac{2t_m g}{v_s}} - 1\right).} \qquad (5)$$

——————————————— Part (b): Infinite Sound Speed ———————————————

Computation If the speed of sound were infinite, the last of Eqs. (3) would imply that $t_s = 0$. Hence, from Eq. (1), we see that $t_m = t_i$ so that the first of Eqs. (2) yields

$$D = \tfrac{1}{2}g t_m^2. \tag{6}$$

Discussion & Verification The result in Eq. (5) is dimensionally correct. Since the given data t_m, g, and v_s have dimensions of time (T), length over time squared (L/T^2), and length over time (L/T), respectively, then note that the argument of the square root term in Eq. (5) is nondimensional, i.e.,

$$\left[\frac{2t_m g}{v_s}\right] = [t_m][g]\frac{1}{[v_s]} = T\frac{L}{T^2}\frac{T}{L} = 1. \tag{7}$$

Consequently the dimensions of D in Eq. (5) are

$$[D] = \left[v_s t_m + \frac{v_s^2}{g}\right] = [v_s][t_m] + [v_s]^2\frac{1}{[g]} = \frac{L}{T}T + \frac{L^2}{T^2}\frac{T^2}{L} = L, \tag{8}$$

as expected. The result in Eq. (6) can be shown to be dimensionally correct in a similar manner.

🔍 **A Closer Look** We now compare the solutions with finite and infinite sound speed to understand under what conditions it is important to account for the finiteness of the speed of sound. Consider Fig. 2, which presents three curves derived under three different sets of assumptions (the curves have D on the horizontal axis to allow us to more easily make comments based on the well's depth). Specifically, Fig. 2 not only shows the functions in Eqs. (5) and (6), but also the solution we would obtain by taking into account both the finiteness of the speed of sound *as well as air resistance*. This latter curve was obtained by assuming that the rock used for the measurement (1) is spherical with a radius $r = 1$ cm and (2) is made out of granite, with a density of $2.75\,\text{g/cm}^3$; and it is subject to an aerodynamic drag force given by $F_D = C_D \rho A v^2/2$, where the dimensionless drag coefficient C_D was chosen to be equal to 0.3, ρ is the density of air at ground level, and A is the frontal area of the spherical rock.[*] Treating the curve obtained by this drag model as the "true" relation between D and t_m, observe that the red curve, representing Eq. (5), and the black curve, corresponding to Eq. (6), diverge from the true curve as the depth of the well increases. However, the three curves essentially coincide near the origin of the plot. Hence, one conclusion is that estimating the depth of a well while disregarding air resistance and the finiteness of the speed of sound is not that bad for shallow wells, where one could define *shallow* to mean, say, less than about 30 m. The second conclusion we can draw is that, by accounting for the finiteness of the speed of sound, our formula can now be applied for depths all the way up to 80 m without having to resort to complex theories of rock-air interaction. Finally, notice that the curves provided here allow us to get a quantitative appreciation for the error we would make in estimating the depth of the well depending on the curve used. For example, consider the case in which we measure $t_m = 4$ s. How deep is the well? Well, according to the red curve, the depth is roughly 70.5 m, whereas the black line indicates a depth of 78.5 m. Therefore, we could go with the estimate given by the black line (since it is easier to compute) with the knowledge that the error is of the order of 12%.

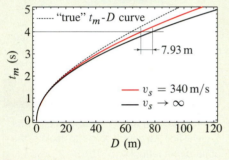

Figure 2
Measured time vs. well depth curves for various sets of assumptions.

—————————————————
[*] This formula for the aerodynamic drag is often discussed in fluid mechanics courses.

EXAMPLE 2.9   *Acceleration Function of Velocity: Descent of a Skydiver*

The skydiver shown in Figs. 1 and 2 has deployed his parachute after free-falling at $v_0 = 44.5\,\text{m/s}$. We will learn how to derive the governing equations for systems such as this in Chapter 3. For now, it suffices to say that the relevant forces on the skydiver are his total weight (i.e., his body weight and that of his equipment) and the drag force due to the parachute. If we model the drag force as being proportional to the square of the skydiver's velocity, or $F_d = C_d v^2$, where C_d denotes a drag coefficient,[*] Newton's second law tells us that the skydiver's acceleration is $a = g - C_d v^2/m$. Letting $C_d = 43.2\,\text{kg/m}$, $m = 110\,\text{kg}$, and $g = 9.81\,\text{m/s}^2$, determine

(a) The skydiver's velocity as a function of time.

(b) The terminal velocity reached by the skydiver.

(c) The skydiver's position as a function of time.

SOLUTION

Figure 1
A skydiver descending.

——————————————— **Part (a): From Acceleration to Velocity** ———————————————

Road Map   Since the acceleration is not given as a function of time, but as a function of velocity, i.e., $a = a(v)$, we cannot obtain $v(t)$ by integrating a with respect to time. However, recalling that $a = dv/dt$ can be rewritten as $dt = dv/a$, we can obtain time as a function of velocity, i.e., $t = t(v)$ and then we will try to invert this relationship to obtain $v = v(t)$. This is the strategy followed in developing Eq. (2.33) for the case when $a = a(v)$.

Computation   We begin by applying Eq. (2.33), or, equivalently, rewriting $a = dv/dt$ as $dt = dv/a(v)$ and integrating both sides to obtain

$$t(v) = \int_{v_0}^{v} \frac{dv}{g - C_d v^2/m}$$
$$= -\frac{1}{2}\sqrt{\frac{m}{gC_d}} \ln\left[\left(\frac{v\sqrt{C_d} - \sqrt{mg}}{v\sqrt{C_d} + \sqrt{mg}}\right)\left(\frac{v_0\sqrt{C_d} + \sqrt{mg}}{v_0\sqrt{C_d} - \sqrt{mg}}\right)\right], \qquad (1)$$

where we have set $v = v_0$ for $t = 0$, and where we note that this integral can be obtained using a comprehensive table of integrals[†] or a software package such as Mathematica. We now have $t(v)$, but we want $v(t)$. Hence, to invert Eq. (1), we first multiply both sides by $-2\sqrt{gC_d/m}$ and then exponentiate both sides to obtain

$$e^{-2t\sqrt{\frac{gC_d}{m}}} = \left(\frac{v\sqrt{C_d} - \sqrt{mg}}{v\sqrt{C_d} + \sqrt{mg}}\right)\left(\frac{v_0\sqrt{C_d} + \sqrt{mg}}{v_0\sqrt{C_d} - \sqrt{mg}}\right). \qquad (2)$$

Solving Eq. (2) for v and simplifying, we obtain

$$\boxed{v(t) = \sqrt{\frac{mg}{C_d}}\, \frac{v_0\sqrt{C_d} + \sqrt{mg} + \left(v_0\sqrt{C_d} - \sqrt{mg}\right)e^{-2t\sqrt{\frac{gC_d}{m}}}}{v_0\sqrt{C_d} + \sqrt{mg} - \left(v_0\sqrt{C_d} - \sqrt{mg}\right)e^{-2t\sqrt{\frac{gC_d}{m}}}},} \qquad (3)$$

the plot of which can be found in Fig. 3 for the parameters given. Notice that the skydiver starts out at $44.5\,\text{m/s}$ at $t = 0\,\text{s}$ and quickly (in about one second) slows down to approximately $5\,\text{m/s}$.

drag force F_d

weight force mg

Figure 2
Skydiver with drag and weight forces depicted.

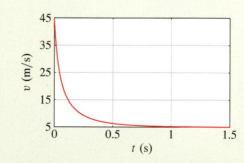

Figure 3
Velocity of the skydiver as he descends with his parachute deployed.

---

[*] The drag coefficient used in this problem is a condensed version of the drag coefficient found in Example 2.8, and they are related according to $C_d = \frac{1}{2}C_D \rho A$.

[†] See, for example, A. Jeffrey, *Handbook of Mathematical Formulas and Integrals*, 3rd ed., Academic Press, 2003; or R. J. Tallarida, *Pocket Book of Integrals and Mathematical Formulas*, 3rd ed., CRC Press, Boca Raton, FL, 1999. In addition, there are several Internet resources such as <http://en.wikipedia.org/wiki/Lists_of_integrals>.

Helpful Information

Another commonly accepted and consistent definition of *terminal velocity*. When a body is in free fall in a medium such as air (or water), the body experiences an aerodynamic (or fluid dynamic) resistance, called *drag*, that opposes gravity and increases with speed. If the body falls for enough time, the aerodynamic drag will end up equilibrating the force of gravity and the body will stop accelerating. The *terminal velocity* can be defined as the value of the velocity at which the acceleration becomes equal to zero.

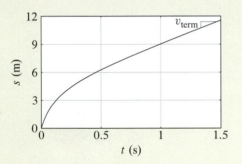

Figure 4
Position versus time of the skydiver as he descends with his parachute deployed.

──────────── **Part (b): Terminal Velocity** ────────────

Road Map The *terminal velocity* is defined as the velocity reached after an infinite amount of time. Therefore, we can answer the question in Part (b) by taking $v(t)$ from Eq. (3) and taking the limit as $t \to \infty$.

Computation Proceeding as described above, we obtain

$$
\begin{aligned}
v_{\text{term}} &= \lim_{t \to \infty} \sqrt{\frac{mg}{C_d}} \frac{v_0\sqrt{C_d} + \sqrt{mg} + (v_0\sqrt{C_d} - \sqrt{mg})e^{-2t\sqrt{\frac{gC_d}{m}}}}{v_0\sqrt{C_d} + \sqrt{mg} - (v_0\sqrt{C_d} - \sqrt{mg})e^{-2t\sqrt{\frac{gC_d}{m}}}} \\
&= \sqrt{\frac{mg}{C_d}} = \sqrt{\frac{(9.81)(110)}{43.2}} \text{ m/s} = 5.00 \text{ m/s}.
\end{aligned}
\tag{4}
$$

This agrees with the plot in Fig. 3.

────── **Part (c): Position vs. Time After Parachute Deployment** ──────

Road Map To obtain the skydiver's position vs. time, recall that $v(t) = ds/dt$, which can be rewritten as $ds = v(t)\,dt$ and can then be integrated to obtain the desired result.

Computation Given the form of $v(t)$ in Eq. (3), carrying out the required integration may appear challenging. However, this calculation can be easily tackled via the use of a table of integrals or a symbolic algebra package such as Mathematica. Regardless of the integration method, setting $s = 0$ for $t = 0$, we have

$$
\begin{aligned}
s(t) &= \int_0^t v(t)\,dt \\
&= \sqrt{\frac{mg}{C_d}}\left[t + \sqrt{\frac{m}{gC_d}}\ln\left(1 - \frac{v_0\sqrt{C_d} - \sqrt{mg}}{v_0\sqrt{C_d} + \sqrt{mg}}e^{-2t\sqrt{\frac{gC_d}{m}}}\right)\right] \\
&\quad - \frac{m}{C_d}\ln\left(1 - \frac{v_0\sqrt{C_d} - \sqrt{mg}}{v_0\sqrt{C_d} + \sqrt{mg}}\right),
\end{aligned}
\tag{5}
$$

the plot of which can be found in Fig. 4 for the parameters given. Notice that the plot of $s(t)$ vs. time quickly approaches a constant slope equal to the terminal velocity.

Discussion & Verification To check that our formulas are correct, we can start with verifying that they are dimensionally correct. Observe that C_d has dimensions of mass over length so that the terms $v\sqrt{C_d}$, $v_0\sqrt{C_d}$, and $\sqrt{mg}$ in Eqs. (3) and (5) are dimensionally homogeneous and the arguments of the exponential and logarithm functions in Eqs. (3) and (5) are nondimensional. This means that the dimensions of $v(t)$ in Eq. (3) are those of the term $\sqrt{mg/C_d}$, which has dimensions of length over time, as expected. As for Eq. (5), the dimensions of $s(t)$ are those of the terms $t\sqrt{mg/C_d}$ and m/C_d, both of which have dimensions of length, again as expected.

An additional verification is to differentiate Eq. (5) with respect to time to make sure that we recover Eq. (3). Finally, we could substitute Eq. (3) into the given expression for the acceleration and verify that we obtain the same expression resulting from differentiating Eq. (3) with respect to time. We leave these verifications to the reader.

A Closer Look This example shows that, even in seemingly simple problems, engineers are often confronted with considerable mathematical challenges. However, these challenges can often be tackled via a variety of computer tools. These tools relieve us from dealing with tedious and repetitive tasks so that we can focus on the more important aspects of the problem. In this example, the important aspects were the application of Eq. (2.33) to obtain $t(v)$ from $a(v)$, the inversion of $t(v)$ to obtain $v(t)$, the determination of v_{term}, and the time integration of $v(t)$ to obtain $s(t)$.

E X A M P L E   2.10   *Constant Angular Acceleration: Propeller and Supersonic Effects*

Referring to Fig. 2, the propeller shown has radius $r_p = 19$ ft and it rotates about its axis while keeping the propeller disk stationary.[*] Suppose that the propeller starts from rest with a constant angular acceleration $\alpha = 50$ rad/s^2.[†] Knowing that the speed of sound at sea level under standard conditions is $v_s = 1130$ ft/s, find

(a) ω_s, the angular speed of the propeller, expressed in rpm, at which 50% of the area of the propeller disk operates in the supersonic regime.

(b) t_s, the time it takes to achieve ω_s.

(c) θ_s, the number of revolutions experienced by the propeller in going from rest to ω_s.

SOLUTION

Road Map   As the propeller spins up, different points along a propeller's blade experience different speeds. This is so because the motion of each point on the propeller is circular, and while the angular velocity and acceleration for this motion are the same for all points, the distance from the axis of rotation is not. The overall motion is a constant *angular* acceleration motion, and we can use the formulas derived for this case.

Part (a): Calculation of ω_s

Computation   Referring to Fig. 3 and using Eq. (2.46) to describe the velocity of points in circular motion, we have that the speed $|\dot{s}|$ of a point at a distance r from the spin axis is

$$|\dot{s}| = r|\omega|, \tag{1}$$

which shows that $|\dot{s}|$ is proportional to the distance from the axis of rotation. Hence, if a point interior to the propeller disk moves at supersonic speeds, all of the points between it and the propeller's periphery will also move at supersonic speeds. Letting r_i be the radius of the inner disk of the propeller whose area is 50% of the total disk area (see Fig. 3) yields

$$\pi r_i^2 = \frac{\pi r_p^2}{2} \quad \Rightarrow \quad r_i = \frac{r_p}{\sqrt{2}} = 13.44 \text{ ft.} \tag{2}$$

If points on the circle of radius r_i have achieved the speed of sound, then combining Eq. (1) and the second of Eqs. (2), we have

$$v_s = r_i \omega_s \quad \Rightarrow \quad \boxed{\omega_s = \frac{v_s \sqrt{2}}{r_p} = 84.11 \text{ rad/s} = 803 \text{ rpm,}} \tag{3}$$

where, $\omega_s > 0$ since it is an angular *speed*.

Part (b): Calculation of t_s

Computation   To establish how long it takes to achieve ω_s, recall that the angular acceleration α is the time derivative of the angular velocity ω, i.e., $\alpha = \dot{\omega} = d\omega/dt$, so that we can write $d\omega = \alpha\, dt$. Therefore, observing that α is constant and letting $\omega_0 = 0$ be the initial angular velocity (the propeller starts from rest), we have

$$\int_{\omega_0}^{\omega} d\omega = \int_0^{t_s} \alpha\, dt \quad \Rightarrow \quad \omega_s = \omega_0 + \int_0^{t_s} \alpha\, dt \quad \Rightarrow \quad \omega_s = \alpha t_s, \tag{4}$$

---

[*] The propeller disk is the disk spanned by the propeller blades as they rotate.

[†] This is roughly what is takes to go from 0 to 1430 rpm in 3 s and is therefore very easy to achieve even with an average small-car engine.

Figure 1
V-22 Osprey aircraft.

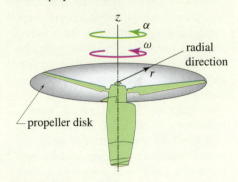

Figure 2
View of one of the engines and its companion propeller of the V-22 Osprey. The radius of the V-22 propellers is 19 ft.

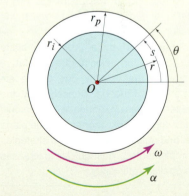

Figure 3
Definition of the coordinates s of a point a distance r from the center of rotation O. Using the angular coordinate θ, we have that $s = r\theta$. The shaded circle has an area equal to 50% of the total area of the propeller disk.

Combining the results in Eqs. (3) and (4), we have

$$t_s = \frac{v_s\sqrt{2}}{\alpha r_p} = 1.68\,\text{s.} \tag{5}$$

————————————— **Part (c): Calculation of** θ_s —————————————

Since a *revolution* is an angular displacement equal to 2π rad, we can calculate the number of revolutions experienced by the propeller by computing the difference in the propeller's angular position θ between $t = 0$ and $t = t_s$. Since the angular acceleration is constant, we can use Eq. (2.41) along with Table 2.2 to arrive at the following relationship between ω_s, α, and θ_s:

$$\omega_s^2 = \omega_0^2 + 2\alpha(\theta_s - \theta_0), \tag{6}$$

where θ_0 is the angular position of a point on the propeller disk at $t = 0$. Since all points on the propeller disk experience the same angular displacement, by choosing a point with $\theta_0 = 0$ and recalling $\omega_0 = 0$, Eq. (2.2) can be solved to obtain

$$\theta_s = \frac{\omega_s^2}{2\alpha} = 70.7\,\text{rad} = 11.3\,\text{rev.} \tag{7}$$

Discussion & Verification   Let L and T denote dimensions of length and time, respectively. Then, referring to Eq. (3), we see that the dimensions of ω_s are given by

$$[\omega_s] = \left[\frac{v_s}{r_p}\right] = [v_s]\frac{1}{[r_p]} = \frac{L}{T}\frac{1}{L} = \frac{1}{T}, \tag{8}$$

as expected. In addition, recalling that the unit of radian is nondimensional, we see that ω_s has the right dimensions and is expressed via appropriate units. Next, referring to Eq. (5), the dimensions of t_s are given by

$$[t_s] = \left[\frac{v_s}{\alpha r_p}\right] = [v_s]\frac{1}{[\alpha]}\frac{1}{[r_p]} = \frac{L}{T}\frac{1}{T^{-2}}\frac{1}{L} = T, \tag{9}$$

as expected. Considering Eq. (5) again, we see that t_s has been expressed via appropriate units. Finally, referring to Eq. (7), the dimensions of θ_s are given by

$$[\theta_s] = \left[\frac{\omega_s^2}{\alpha}\right] = [\omega_s^2]\frac{1}{[\alpha]} = \frac{1}{T^2}\frac{1}{T^{-2}} = 1, \tag{10}$$

as expected. Considering Eq. (7) again and recalling that the unit of radian is nondimensional, θ_s has been expressed via appropriate units.

As far as the numerical values of our results are concerned, because the propeller is accelerating from rest to the angular speed ω_s during the time interval $0 \le t \le t_s$, the angular displacement θ_s we computed must be smaller than the value $t_s\omega_s$, which represents the angular displacement the propeller would have experienced if it had been rotating at the *constant* angular speed ω_s for $0 \le t \le t_s$. Since $t_s\omega_s = 141$ rad, our expectation is met.

PROBLEMS

Problems 2.37 through 2.40

The following four problems refer to the car traveling between two stop signs presented at the beginning of this section on p. 55, in which the car's velocity is assumed to be given by $v = [9 - 9\cos(2t/5)]\,\text{m/s}$ for $0 \leq t \leq 5\pi$ s.

Problem 2.37 Determine v_{max}, the maximum velocity reached by the car. Furthermore, determine the position $s_{v_{max}}$ and the time $t_{v_{max}}$ at which v_{max} occurs.

Problem 2.38 Determine the time at which the brakes are applied and the car starts to slow down.

Problem 2.39 Determine the average velocity of the car between the two stop signs.

Problem 2.40 Determine $|a|_{max}$, the maximum of the magnitude of the acceleration reached by the car, and determine the position(s) at which $|a|_{max}$ occurs.

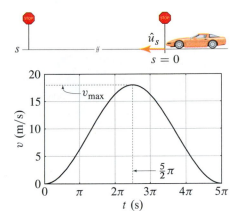

Figure P2.37–P2.40

Problem 2.41

A ring is thrown straight upward from a height $h = 2.5\,\text{m}$ off the ground and with an initial velocity $v_0 = 3.45\,\text{m/s}$. Gravity causes the ring to have a constant downward acceleration $g = 9.81\,\text{m/s}^2$. Determine h_{max}, the maximum height reached by the ring.

Problem 2.42

A ring is thrown straight upward from a height $h = 2.5\,\text{m}$ off the ground. Gravity causes the ring to have a constant downward acceleration $g = 9.81\,\text{m/s}^2$. Letting $d = 5.2\,\text{m}$, if the person at the window is to receive the ring in the gentlest possible manner, determine the initial velocity v_0 the ring must be given when first released.

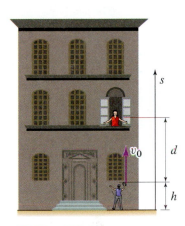

Figure P2.41 and P2.42

Problem 2.43

A car stops 4 s after the application of the brakes while covering a rectilinear stretch 337 ft long. If the motion occurred with a constant acceleration a_c, determine the initial speed v_0 of the car and the acceleration a_c. Express v_0 in mph and a_c in terms of g, the acceleration of gravity.

Figure P2.43

Problems 2.44 and 2.45

The motion of a peg sliding within a rectilinear guide is controlled by an actuator in such a way that the peg's acceleration takes on the form $\ddot{x} = a_0(2\cos 2\omega t - \beta \sin \omega t)$, where t is time, $a_0 = 3.5\,\text{m/s}^2$, $\omega = 0.5\,\text{rad/s}$, and $\beta = 1.5$.

Problem 2.44 Determine the expressions for the velocity and the position of the peg as functions of time if $\dot{x}(0) = 0\,\text{m/s}$ and $x(0) = 0\,\text{m}$.

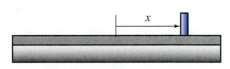

Figure P2.44 and P2.45

Problem 2.45 Determine the total distance traveled by the peg during the time interval $0\,\text{s} \leq t \leq 5\,\text{s}$ if $\dot{x}(0) = a_0\beta/\omega$.

Figure P2.46 and P2.47

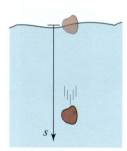

Figure P2.48–P2.52

Problem 2.46

Referring to Example 2.9 on p. 63, and defining *terminal velocity* as the velocity at which a falling object stops accelerating, determine the skydiver's terminal velocity without performing any integrations.

Problem 2.47

Referring to Example 2.9 on p. 63, determine the distance d traveled by the skydiver from the instant the parachute is deployed until the difference between the velocity and the terminal velocity is 10% of the terminal velocity.

Problems 2.48 and 2.49

The acceleration of an object in rectilinear free fall while immersed in a linear viscous fluid is $a = g - C_d v/m$, where g is the acceleration of gravity, C_d is a constant drag coefficient, v is the object's velocity, and m is the object's mass.

Problem 2.48   Letting $t_0 = 0$ and $v_0 = 0$, determine the velocity as a function of time and find the terminal velocity.

Problem 2.49   Letting $s_0 = 0$ and $v_0 = 0$, determine the position as a function of velocity.

Problem 2.50

A 1.5 kg rock is released from rest at the surface of a calm lake. If the resistance offered by the water as the rock falls is directly proportional to the rock's velocity, the rock's acceleration is $a = g - C_d v/m$, where g is the acceleration of gravity, C_d is a constant drag coefficient, v is the rock's velocity, and m is the rock's mass. Letting $C_d = 4.1$ kg/s, determine the rock's velocity after 1.8 s.

Problems 2.51 and 2.52

A 3.1 lb rock is released from rest at the surface of a calm lake, and its acceleration is $a = g - C_d v/m$, where g is the acceleration of gravity, $C_d = 0.27$ lb·s/ft is a constant drag coefficient, v is the rock's velocity, and m is the rock's mass.

Problem 2.51   Determine the depth to which the rock will have sunk when the rock achieves 99% of its terminal velocity.

Problem 2.52   Determine the rock's velocity after it drops 5 ft.

Problem 2.53

Suppose that the acceleration of an object of mass m along a straight line is $a = g - C_d v/m$, where the constants g and C_d are given and v is the object's velocity. If $v(t)$ is unknown and $v(0)$ is given, can you determine the object's velocity via the following integral?

$$v(t) = v(0) + \int_0^t \left(g - \frac{C_d}{m} v \right) dt$$

Note: Concept problems are about *explanations*, not computations.

Problem 2.54

A car travels on a rectilinear stretch of road at a constant speed $v_0 = 65$ mph. At $s = 0$ the driver applies the brakes hard enough to cause the car to skid. Assume that the car keeps sliding until it stops, and assume that throughout this process the car's acceleration is given by $\ddot{s} = -\mu_k g$, where $\mu_k = 0.76$ is the kinetic friction coefficient and g is the acceleration of gravity. Compute the car's stopping distance and time.

Figure P2.54

Problem 2.55

Heavy rains cause a particular stretch of road to have a coefficient of friction that changes as a function of location. Specifically, measurements indicate that the friction coefficient has a 3% decrease per meter. Under these conditions the acceleration of a car skidding while trying to stop can be approximated by $\ddot{s} = -(\mu_k - cs)g$ (the 3% decrease in friction was used in deriving this equation for acceleration), where μ_k is the friction coefficient under dry conditions, g is the acceleration of gravity, and c, with units of m^{-1}, describes the rate of friction decrement. Let $\mu_k = 0.5$, $c = 0.015\,\text{m}^{-1}$, and $v_0 = 45$ km/h, where v_0 is the initial velocity of the car. Determine the distance it will take the car to stop and the percentage of increase in stopping distance with respect to dry conditions, i.e., when $c = 0$.

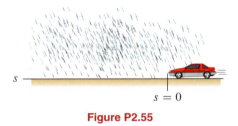

Figure P2.55

Problems 2.56 through 2.59

As you will learn in Chapter 3, the angular acceleration of a simple pendulum is given by $\ddot{\theta} = -(g/L)\sin\theta$, where g is the acceleration of gravity and L is the length of the pendulum cord.

Problem 2.56   Derive the expression of the angular velocity $\dot{\theta}$ as a function of the angular coordinate θ. The initial conditions are $\theta(0) = \theta_0$ and $\dot{\theta}(0) = \dot{\theta}_0$.

Problem 2.57   Let the length of the pendulum cord be $L = 1.5$ m. If $\dot{\theta} = 3.7$ rad/s when $\theta = 14°$, determine the maximum value of θ achieved by the pendulum.

Problem 2.58   The given angular acceleration remains valid even if the pendulum cord is replaced by a massless rigid bar. For this case, let $L = 5.3$ ft and assume that the pendulum is placed in motion at $\theta = 0°$. What is the minimum angular velocity at this position for the pendulum to swing through a full circle?

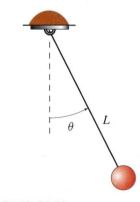

Figure P2.56–P2.59

Problem 2.59   Let $L = 3.5$ ft and suppose that at $t = 0$ s the pendulum's position is $\theta(0) = 32°$ with $\dot{\theta}(0) = 0$ rad/s. Determine the pendulum's period of oscillation, i.e., from its initial position back to this position.

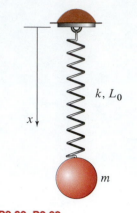

Figure P2.60–P2.62

Figure P2.63 and P2.64

Problems 2.60 through 2.62

As we will see in Chapter 3, the acceleration of a particle of mass m suspended by a linear spring with spring constant k and unstretched length L_0 (when the spring length is equal to L_0, the spring exerts no force on the particle) is given by $\ddot{x} = g - (k/m)(x - L_0)$.

Problem 2.60 Derive the expression for the particle's velocity $\dot{x}$ as a function of position x. Assume that at $t = 0$, the particle's velocity is v_0 and its position is x_0.

Problem 2.61 Let $k = 100\,\text{N/m}$, $m = 0.7\,\text{kg}$, and $L_0 = 0.75\,\text{m}$. If the particle is released from rest at $x = 0\,\text{m}$, determine the maximum length achieved by the spring.

Problem 2.62 Let $k = 8\,\text{lb/ft}$, $m = 0.048\,\text{slug}$, and $L_0 = 2.5\,\text{ft}$. If the particle is released from rest at $x = 0\,\text{ft}$, determine how long it takes for the spring to achieve its maximum length. *Hint:* A good table of integrals will come in handy.

Problems 2.63 and 2.64

Two masses m_A and m_B are placed at a distance r_0 from one another. Because of their mutual gravitational attraction, the acceleration of sphere B as seen from sphere A is given by

$$\ddot{r} = -G\left(\frac{m_A + m_B}{r^2}\right),$$

where G is the universal gravitational constant.

Problem 2.63 If the spheres are released from rest, determine

(a) The velocity of B (as seen by A) as a function of the distance r.

(b) The velocity of B (as seen by A) at impact if $r_0 = 7\,\text{ft}$, the weight of A is 2.1 lb, the weight of B is 0.7 lb, and

    (i) The diameters of A and B are $d_A = 1.5\,\text{ft}$ and $d_B = 1.2\,\text{ft}$, respectively.

    (ii) The diameters of A and B are infinitesimally small.

Problem 2.64 Assume that the particles are released from rest at $r = r_0$.

(a) Determine the expression relating their relative position r and time. *Hint:*

$$\int \sqrt{x/(1 - x)}\, dx = \sin^{-1}\left(\sqrt{x}\right) - \sqrt{x(1 - x)}.$$

(b) Determine the time it takes for the objects to come into contact if $r_0 = 3\,\text{m}$, A and B have masses of 1.1 and 2.3 kg, respectively, and

    (i) The diameters of A and B are $d_A = 22\,\text{cm}$ and $d_B = 15\,\text{cm}$, respectively.

    (ii) The diameters of A and B are infinitesimally small.

💡 Problem 2.65 💡

Suppose that the acceleration $\ddot{r}$ of an object moving along a straight line takes on the form

$$\ddot{r} = -G\left(\frac{m_A + m_B}{r^2}\right),$$

where the constants G, m_A, and m_B are known. If $\dot{r}(0)$ is given, under what conditions can you determine $\dot{r}(t)$ via the following integral?

$$\dot{r}(t) = \dot{r}(0) - \int_0^t G\,\frac{m_A + m_B}{r^2}\, dt$$

Note: Concept problems are about *explanations*, not computations.

Problems 2.66 and 2.67

If the truck brakes and the crate slides to the right relative to the truck, the horizontal acceleration of the crate is given by $\ddot{s} = -g\mu_k$, where g is the acceleration of gravity, $\mu_k = 0.87$ is the kinetic friction coefficient, and s is the position of the crate relative to a coordinate system attached to the ground (rather than the truck).

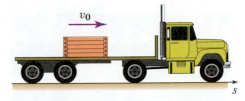

Figure P2.66 and P2.67

Problem 2.66 Assuming that the crate slides without hitting the right end of the truck bed, determine the time it takes to stop if its velocity at the start of the sliding motion is $v_0 = 55$ mph.

Problem 2.67 Assuming that the crate slides without hitting the right end of the truck bed, determine the distance it takes to stop if its velocity at the start of the sliding motion is $v_0 = 75$ km/h.

Problem 2.68

If the truck brakes hard enough that the crate slides to the right relative to the truck, the distance d between the crate and the front of the trailer changes according to the relation

$$\ddot{d} = \begin{cases} \mu_k g + a_T & \text{for } t < t_s, \\ \mu_k g & \text{for } t > t_s, \end{cases}$$

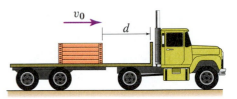

Figure P2.68

where t_s is the time it takes the truck to stop, a_T is the acceleration of the truck, g is the acceleration of gravity, and μ_k is the kinetic friction coefficient between the truck and the crate. Suppose that the truck and the crate are initially traveling to the right at $v_0 = 60$ mph and the brakes are applied so that $a_T = -10.0$ ft/s^2. Determine the minimum value of μ_k so that the crate does not hit the right end of the truck bed if the initial distance d is 12 ft. *Hint:* The truck stops *before* the crate stops.

Problem 2.69

Cars A and B are traveling at $v_A = 72$ mph and $v_B = 67$ mph, respectively, when the driver of car B applies the brakes abruptly, causing the car to slide to a stop. The driver of car A takes 1.5 s to react to the situation and applies the brakes in turn, causing car A to slide as well. If A and B slide with equal accelerations, i.e., $\ddot{s}_A = \ddot{s}_B = -\mu_k g$, where $\mu_k = 0.83$ is the kinetic friction coefficient and g is the acceleration of gravity, compute the minimum distance d between A and B at the time B starts sliding to avoid a collision.

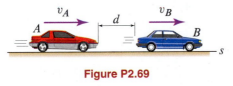

Figure P2.69

Problem 2.70

A hot air balloon is climbing with a velocity of 7 m/s when a sandbag (used as ballast) is released at an altitude of 305 m. Assuming that the sandbag is subject only to gravity and that therefore its acceleration is given by $\ddot{y} = -g$, g being the acceleration due to gravity, determine how long the sandbag takes to hit the ground and its impact velocity.

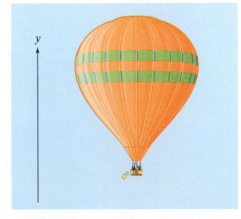

Figure P2.70

Figure P2.71

Problem 2.71

Approximately 1 h 15 min into the movie "King Kong" (the one directed by Peter Jackson), there is a scene in which Kong is holding Ann Darrow (played by the actress Naomi Watts) in his hand while swinging his arm in anger. A quick analysis of the movie indicates that at a particular moment Kong displaces Ann from rest by roughly 10 ft in a span of four frames. Knowing that the DVD plays at 24 frames per second and assuming that Kong subjects Ann to a constant acceleration, determine the acceleration Ann experiences in the scene in question. Express your answer in terms of the acceleration due to gravity g. Comment on what would happen to a person *really* subjected to this acceleration.

Problem 2.72

Derive the constant acceleration relation in Eq. (2.41), starting from Eq. (2.33). State what assumption you need to make about the acceleration a to complete the derivation. Finally, use Eq. (2.36), along with the result of your derivation, to derive Eq. (2.42). Be careful to do the integral in Eq. (2.36) before substituting your result for $v(t)$ (try it without doing so, to see what happens). After completing this problem, notice that Eqs. (2.41) and (2.42) are *not* subject to the same assumption you needed to make to solve both parts of this problem.

Problems 2.73 through 2.75

The spool of paper used in a printing process is unrolled with velocity v_p and acceleration a_p. The thickness of the paper is h, and the outer radius of the spool at any instant is r.

Figure P2.73–P2.75

Problem 2.73   If the velocity at which the paper is unrolled is *constant*, determine the angular acceleration α_s of the spool as a function of r, h, and v_p. Evaluate your answer for $h = 0.0048$ in., for $v_p = 1000$ ft/min, and two values of r, that is, $r_1 = 25$ in. and $r_2 = 10$ in.

Problem 2.74   If the velocity at which the paper is unrolled is *not constant*, determine the angular acceleration α_s of the spool as a function of r, h, v_p, and a_p. Evaluate your answer for $h = 0.0048$ in., $v_p = 1000$ ft/min, $a_p = 3$ ft/s^2, and two values of r, that is, $r_1 = 25$ in. and $r_2 = 10$ in.

Problem 2.75   If the velocity at which the paper is unrolled is *constant*, determine the angular acceleration α_s of the spool as a function of r, h, and v_p. Plot your answer for $h = 0.0048$ in. and $v_p = 1000$ ft/min as a function of r for 1 in. $\leq r \leq 25$ in. Over what range does α_s vary?

2.3   Projectile Motion

In this section we present a simple model to study the motion of projectiles: we model projectile motion as a constant acceleration motion.

Trajectory of a pumpkin used as a projectile

A pumpkin has been launched from a trebuchet (see Fig. 2.18) at the annual World Championship "Punkin Chunkin" competition in Millsboro, Delaware.[*] The pumpkin, shown in Fig. 2.19, is assumed to have been released at O with an initial speed v_0 and an elevation angle β. What is the trajectory of the pumpkin after it leaves the trebuchet? To answer this question, let's consider

Figure 2.18
A working trebuchet (pronounced *treb-yoo-shay*).

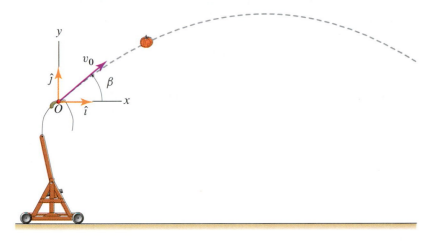

Figure 2.19. A pumpkin launched by a trebuchet showing the quantities described in the text. See Fig. 2.20 for some details on trebuchets.

the forces acting on the pumpkin while airborne. Referring to Fig. 2.21, these forces are the pumpkin's weight mg and the drag force $\vec{F}_d$ due to air resistance. As a first approximation, we will neglect air resistance so that the total force on the pumpkin is $\vec{F} = -mg\,\hat{j}$. Newton's second law is $\vec{F} = m\vec{a}$[†] so that, using the coordinate system in Fig. 2.19,

$$-mg\,\hat{j} = m(a_x\,\hat{i} + a_y\,\hat{j}) \quad \Rightarrow \quad -g\,\hat{j} = a_x\,\hat{i} + a_y\,\hat{j}. \qquad (2.48)$$

In Cartesian coordinates, $a_x = \ddot{x}$ and $a_y = \ddot{y}$ so that Eqs. (2.48) can be rewritten as

$$\ddot{x} = 0 \quad \text{and} \quad \ddot{y} = -g. \qquad (2.49)$$

Equations (2.48) express the fact that we have modeled the pumpkin's motion as a *constant acceleration* motion. Equations (2.49) say that *each* component of the acceleration is constant. Therefore, we can use the constant acceleration equations developed in Section 2.2 on a component-by-component

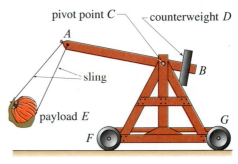

Figure 2.20
A trebuchet. The arm AB pivots clockwise about C due to the large counterweight D. As the arm reaches the vertical position, it launches the payload E to the right. The wheels at F and G are optional, but it turns out that when the wheels are present and the trebuchet is allowed to move, the launch distance is increased.

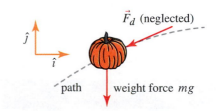

Figure 2.21
All the forces on the pumpkin as it flies through the air. In our model we neglect the drag force due to air resistance.

---

[*] Information about the World Championship "Punkin Chunkin" competition can be found on the web at <http://www.punkinchunkin.com/>.

[†] The application of Newton's second law is examined in detail in Chapter 3.

basis. Specifically, applying Eq. (2.42) (on p. 59) in the x direction yields

$$x(t) = x(0) + \dot{x}(0)(t - t_0) + \tfrac{1}{2}0\,(t - t_0)^2 \quad \Rightarrow \quad x(t) = v_0 \cos \beta\, t, \quad (2.50)$$

where $x(0) = 0$, $\dot{x}(0) = v_0 \cos \beta$, and $t_0 = 0$ (see Fig. 2.19). Similarly, for the y direction we have

$$y(t) = y(0) + \dot{y}(0)(t - t_0) - \tfrac{1}{2}g(t - t_0)^2$$
$$\Rightarrow \quad y(t) = v_0 \sin \beta\, t - \tfrac{1}{2}gt^2, \quad (2.51)$$

where $y(0) = 0$ and $\dot{y}(t) = v_0 \sin \beta$. We can now obtain the pumpkin's trajectory by eliminating time from Eqs. (2.50) and (2.51). Solving for t in Eq. (2.50) gives $t = x/(v_0 \cos \beta)$. Substituting this expression into Eq. (2.51) and simplifying, we have

$$y = (\tan \beta)x - \left(\frac{g \sec^2 \beta}{2v_0^2} \right)x^2, \quad (2.52)$$

which shows that the trajectory of the pumpkin is a parabola!

The simplicity of the pumpkin's motion results from studying the motion projectile without accounting for air resistance and the dependence of gravity on height. We now proceed to formally define such a motion and make a few remarks about its analysis.

Projectile motion

We define *projectile motion* as a motion in which the acceleration is *constant* and given by

$$a_{\text{horiz}} = 0 \quad \text{and} \quad a_{\text{vert}} = -g, \quad (2.53)$$

where, referring to Fig. 2.22, a_{horiz} and a_{vert} are the components of the acceleration in the horizontal and vertical directions, respectively, and where we have chosen the vertical direction to be positive *upward*. Our definition of projectile motion is a very simplified description of true projectile motion because the relations in Eq. (2.53) neglect air resistance and changes in gravitational attraction with changes in height. These effects will be considered in Chapters 3 and 5, respectively.

To describe the velocity, position, and trajectory of a projectile, we typically use a Cartesian coordinate system with axes parallel and perpendicular to the direction of gravity, as was done earlier in the analysis of the pumpkin problem. However, other choices are possible and, in some cases, more convenient (for a description of projectile motion using general Cartesian coordinate system see Example 2.13).

In deriving Eq. (2.52), we showed that the trajectory of a projectile is a parabola. This result is independent of coordinate system. However, the expression of the parabola does depend on the coordinate system used. The trajectory is of the form derived in Eq. (2.52), i.e., of the form $y = C_0 + C_1 x + C_2 x^2$ (C_0, C_1, and C_3 are constant coefficients),[*] if the y axis of our Cartesian coordinate system is parallel to the direction of gravity (see Example 2.13 for a case in which the y axis is not parallel to gravity).

---
[*] In Eq. (2.52) we found $C_0 = 0$, $C_1 = \tan \beta$, and $C_2 = -(g \sec^2 \beta)/(2v_0^2)$.

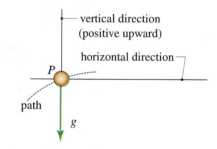

Figure 2.22
Acceleration of a point P in projectile motion.

Helpful Information

Trajectory of a projectile. The trajectory of a projectile is a parabola, and its mathematical expression is of the form $y = C_0 + C_1 x + C_2 x^2$ only when using a Cartesian coordinate system with the y axis parallel to the direction of gravity.

End of Section Summary

We defined *projectile motion* to be the motion of a particle in free flight, neglecting the forces due to air drag and neglecting changes in gravitational attraction with changes in height. In this case, referring to Fig. 2.23, the only force on the particle is the *constant* gravitational force, and the equations describing the motion are

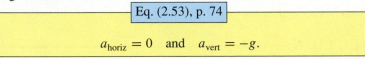

$$a_{\text{horiz}} = 0 \quad \text{and} \quad a_{\text{vert}} = -g.$$

Since both the number 0 and the acceleration due to gravity g are constants, that is, since the horizontal and vertical components of the acceleration are constant, the constant acceleration equations developed in Section 2.2 can be applied both in the vertical and in the horizontal directions.

In projectile motion the trajectory is a parabola. We showed that the trajectory of a projectile is a *parabola*. The mathematical form of the trajectory is of the type $y = C_0 + C_1 x + C_3 x^2$ if the motion is described using a Cartesian coordinate system with the y axis parallel to the direction of gravity.

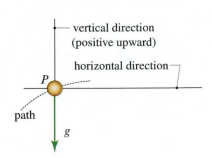

Figure 2.23
Figure 2.22 repeated. Acceleration of a point P in projectile motion.

EXAMPLE 2.11   *Range of Elevation Angles of a Projectile*

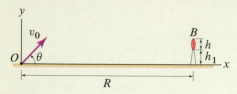

Figure 1
A projectile launched from O in an attempt to hit the target at B. Not drawn to scale.

A projectile is launched from O at speed $v_0 = 1100\,\text{ft/s}$ to hit a point B on a target that is $R = 1000\,\text{ft}$ away. The bottom of the target is $h_1 = 4\,\text{ft}$ above the ground, and the target is $h = 3\,\text{ft}$ high. Determine the range of angles at which the projectile can be fired in order to hit the target, and compare this with the angle subtended by the target as seen from O.

SOLUTION

Road Map   This is a projectile motion with given initial and final positions as well as initial speed v_0. Referring to Fig. 1, the components of the projectile's acceleration are $a_x = 0$ and $a_y = -g$. We will relate the projectile's time of flight t_f to y_B, the vertical position of B. We can then write the launch angle in terms of y_B and, in turn, infer the range of angles that allows the projectile to hit the target. Because R is so much larger than h and h_1, we should expect the range of angles we will find to be small (i.e., our aim will need to be very accurate). Therefore, in carrying out our calculations we will use more significant digits than we normally do. We will discuss the practical implications of this choice in the Discussion & Verification section of the example.

Computation   The initial and final x and y positions of B can be related to t_f by applying Eq. (2.42) (on p. 59) in the x and y directions

$$x_B = x_0 + v_{0x}t_f \quad \text{and} \quad y_B = y_0 + v_{0y}t_f - \tfrac{1}{2}gt_f^2. \tag{1}$$

Considering Fig. 1, we have $x_0 = y_0 = 0$, $v_{0x} = v_0\cos\theta$, $v_{0y} = v_0\sin\theta$, and $x_B = R$. We will set $y_B = h_1 = 4\,\text{ft}$ to find $\theta = \theta_{\min}$ and $y_B = h_1 + h = 7\,\text{ft}$ to find $\theta = \theta_{\max}$. Therefore, treating y_B as a known quantity, we have

$$R = v_0 t_f \cos\theta \quad \text{and} \quad y_B = v_0 t_f \sin\theta - \tfrac{1}{2}gt_f^2. \tag{2}$$

Equations (2) are two equations in the unknowns t_f and θ. Solving the first of Eqs. (2) for t_f and substituting the result into the second of Eqs. (2), we obtain

$$y_B = v_0\left(\frac{R}{v_0\cos\theta}\right)\sin\theta - \tfrac{1}{2}g\left(\frac{R}{v_0\cos\theta}\right)^2. \tag{3}$$

Since $\sin\theta/\cos\theta = \tan\theta$ and $1/\cos^2\theta = \sec^2\theta$, we have

$$y_B = R\tan\theta - \left(\frac{gR^2}{2v_0^2}\right)\sec^2\theta. \tag{4}$$

Finally, by noting that $\sec^2\theta = 1 + \tan^2\theta$ and then rearranging, Eq. (4) becomes

$$\left(\frac{gR^2}{2v_0^2}\right)\tan^2\theta - R\tan\theta + \left(y_B + \frac{gR^2}{2v_0^2}\right) = 0, \tag{5}$$

which is a quadratic equation in $\tan\theta$. Dividing through by the coefficient of the $\tan^2\theta$ term, we obtain

$$\tan^2\theta - \left(\frac{2v_0^2}{gR}\right)\tan\theta + \left(\frac{2y_B v_0^2}{gR^2} + 1\right) = 0. \tag{6}$$

Equation (6) can be solved for $\tan\theta$ to obtain the two solutions

$$\tan\theta = \frac{v_0^2 \pm \sqrt{v_0^4 - g(gR^2 + 2y_B v_0^2)}}{gR}, \tag{7}$$

which means that for each value of y_B, there are two possible values of the firing angle θ: θ_1 and θ_2. Substituting in values for all constants, including the two different values for y_B, we obtain

$$y_B = 4\,\text{ft} \quad \Rightarrow \quad \begin{cases} \theta_1 = 0.991678°, \\ \theta_2 = 89.2375041°, \end{cases} \tag{8}$$

$$y_B = 7\,\text{ft} \quad \Rightarrow \quad \begin{cases} \theta_1 = 1.163590°, \\ \theta_2 = 89.2374736°. \end{cases} \tag{9}$$

Equations (8) and (9) give us the values of θ needed to hit the bottom and top of the sign, respectively. Referring to Eq. (8), it is probably intuitive that if we chose $\theta_1 < \theta < \theta_2$, we would overshoot the bottom of the target whereas we would undershoot it for $\theta < \theta_1$ and $\theta > \theta_2$. For example, substituting $\theta = 45°$ (i.e., a value of θ between those in Eq. (8)) into Eq. (4) gives $y_B = 26.6\,\text{ft}$, which, as expected, is larger than 4 ft. Extending the discussion to Eq. (9), we can then say that if $\theta_1 < \theta < \theta_2$ in Eq. (9), we would overshoot the top of the sign whereas we would undershoot it for $\theta < \theta_1$ and $\theta > \theta_2$. We can therefore conclude that there are two ranges of firing angles such that our projectile will hit the target and that these ranges are given by

$$0.991678° \leq \theta \leq 1.163590° \tag{10}$$

and

$$89.23747360° \leq \theta \leq 89.23750406°. \tag{11}$$

Discussion & Verification   In obtaining the angle ranges in Eqs. (10) and (11) we went through a simple verification step by computing the answer for $\theta = 45°$, in which we saw that our results were as expected. To extend our discussion, let's now consider the size of the ranges in Eqs. (10) and (11):

$$\Delta\theta_1 = 1.163590° - 0.991678° = 0.171912°, \tag{12}$$

$$\Delta\theta_2 = 89.23747360° - 89.23750406° = -0.00003046°. \tag{13}$$

Equations (10) and (12) tell us that to hit our target, we need to elevate our launcher about 1° with an accuracy of 0.17°. Equations (11) and (13) tell us that we can also hit the target if we elevate our launcher to approximately 89.2°, but this time we have an *extremely small* margin of error. In fact, we need to be accurate to within three one-hundred thousandths of a degree! In addition, we also need to keep in mind that our model does not account for aerodynamic effects, which place an additional accuracy burden on our aim. These considerations tell us that, in practical applications, we need to rely on an *active guidance system* rather than the accuracy of the launch angles.

Let's complete this example by comparing the angle $\Delta\theta_1$ with the angle subtended by the target at a distance of 1000 ft. Referring to Fig. 2, we can see that the angle subtended is given by

$$\beta = \gamma - \phi, \tag{14}$$

where

$$\gamma = \tan^{-1}\left(\frac{7}{1000}\right) = 0.401064° \quad \text{and} \quad \phi = \tan^{-1}\left(\frac{4}{1000}\right) = 0.229182°, \tag{15}$$

so that $\beta = 0.171882°$. Since β is very close to the $\Delta\theta_1$ in Eq. (12) and since computing β is simpler than computing $\Delta\theta_1$, we might think that we could have computed β to approximate $\Delta\theta_1$. However, in general, the elevation angle ranges and angle subtended by the target can be substantially different.

Helpful Information

Number of digits in calculations. In the numerical calculations shown in this example, the differences in some of the numbers are *so* small that we are keeping many more digits than we normally would. If we did not, the differences would not be apparent.

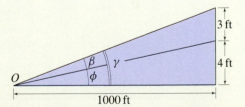

Figure 2
Not drawn to scale—the vertical dimension has been greatly exaggerated so that the angles can be easily seen.

EXAMPLE 2.12 *Initial Speed and Elevation Angle of a Projectile*

A baseball batter makes contact with a ball about 4 ft above the ground and hits it hard enough that it *just* clears the center field wall, which is 400 ft away and is 9 ft high. How fast must the ball be moving and at what angle must it be hit so that it just clears the center field wall as shown in Fig. 1?

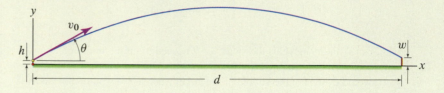

Figure 1. Side view, drawn to scale, of the given baseball field with all parameters defined. In the trajectory shown, the baseball was hit 4 ft off the ground, at 123.2 ft/s, and at a 30° angle so that it *just* clears a 9 ft fence that is 400 ft away.

SOLUTION

Road Map Referring to Fig. 2, we model the ball as a projectile with acceleration given by $a_x = 0$ and $a_y = -g$. We know the starting and ending locations of the projectile, and we wish to determine the v_0 and θ required to get it from start to finish. Therefore, we can proceed as in Example 2.11, i.e., by writing the projectile's x and y positions as a function of time and then, eliminating time, we will obtain an expression for v_0 in terms of θ.

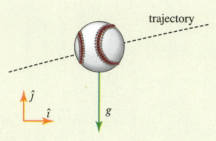

Figure 2
The only nonzero component of acceleration of the baseball.

Computation Since both components of acceleration are constant, we can apply the constant acceleration equation, Eq. (2.42) (on p. 59), in both the x and y directions to obtain

$$x = x_0 + v_{0x}t \qquad \Rightarrow \quad d = 0 + v_0 t \cos\theta, \tag{1}$$

$$y = y_0 + v_{0y}t + \tfrac{1}{2}a_y t^2 \quad \Rightarrow \quad w = h + v_0 t \sin\theta - \tfrac{1}{2}gt^2. \tag{2}$$

Equations (1) and (2) are two equations for the three unknowns v_0, θ, and t. Since we are interested in v_0 and θ, we can eliminate t from these two equations and then solve for v_0 as a function of θ to obtain

$$v_0 = d \sqrt{\frac{g}{2\cos\theta\,[(h-w)\cos\theta + d\sin\theta]}}. \tag{3}$$

This result tells us that there are infinitely many combinations of v_0 and θ that will *just* get the baseball over the center field fence.

Discussion & Verification The solution to the problem is an expression rather than a specific quantitative answer. To verify that Eq. (3) is correct, we first check that its dimensions are correct. Letting L and T denote dimensions of length and time, we have $[g] = L/T^2$ and $[h] = [w] = [d] = L$. Therefore, the dimensions of the argument of the square root in Eq. (3) are T^{-2}, so that the overall dimensions of the right-hand side of Eq. (3) are L/T, as expected. Further verification requires that we study the behavior of the expression in Eq. (3), as is done next.

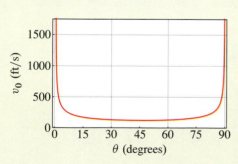

Figure 3
Plot of Eq. (3), that is, v_0 as a function of θ.

✎ **A Closer Look** For $d = 400$ ft, $h = 4$ ft, $w = 9$ ft, and $g = 32.2$ ft/s², the plot of the required v_0 for $0° \leq \theta \leq 90°$ is shown in Fig. 3. By careful inspection of the left side of the curve, we see that the curve approaches an asymptote value of θ other than

0°. This is so because the point at which the batter makes contact with the ball is lower than the height of the fence so that even if the batter hits the ball infinitely hard, there is an angle below which the ball will not clear the fence (we leave it to the reader to show that this angle is 0.716°). On the right side of the curve, we see that the required speed again approaches infinity, but now it is as the angle approaches 90°. Finally, it is also apparent that the optimal angle to hit the ball is near 45°, where by *optimal angle* we mean the angle corresponding to the smallest possible v_0. As it turns out, the optimal angle is not exactly at 45° since we are trying to get the maximum distance for the minimum speed between two points of *unequal* height. To find the optimal angle, we could differentiate Eq. (3) with respect to θ, set the result equal to zero, and then solve for the optimal angle. Equivalently, we can square both sides and differentiate that. Doing so and setting the result equal to zero, we have

$$\frac{d(v_0^2)}{d\theta}\bigg|_{\theta=\theta_0} = \frac{gd^2}{2}\left\{\frac{-d\cos^2\theta_0 + 2(h-w)\cos\theta_0\sin\theta_0 + d\sin^2\theta_0}{\cos^2\theta_0\,[(h-w)\cos\theta_0 + d\sin\theta_0]^2}\right\} = 0, \quad (4)$$

where we have replaced θ by θ_0, which is the optimal value of θ. Since the denominator and numerator do not go to zero at the same values of θ (if they did, we would have to apply l'Hopital's rule), we can solve Eq. (4) by setting the numerator of the fraction within curly braces to zero. Doing this and using some trigonometric identities,[*] we obtain

$$-d\cos(2\theta_0) + (h-w)\sin(2\theta_0) = 0, \quad (5)$$

or

$$\tan(2\theta_0) = \frac{d}{h-w} = -80. \quad (6)$$

Equation (6) has infinitely many solutions given by

$$2\theta_0 = 90.72° \pm n180°, \qquad n = 0, 1, \ldots, \infty, \quad (7)$$

but the only meaningful solution in this context is $2\theta_0 = 90.72°$, or $\theta_0 = 45.36°$. This means that the optimal angle to hit the ball when trying to hit it over an object that is higher than the initial position of the ball, is greater than 45°. From Eq. (5) we can also see that the optimal angle is less than 45° when trying to hit the ball over a lower object, and it is exactly equal to 45° only if $h = w$, i.e., if we are trying to hit the ball over an object whose height is the same as the ball's initial position.

The above discussion leads to the conclusion that the solution we derived does indeed have a behavior that matches our expectations and physical intuition.

---

[*] $\sin^2 x - \cos^2 x = -\cos(2x)$ and $2\sin x\cos x = \sin(2x)$.

EXAMPLE 2.13   *Projectile Motion in a General Cartesian Coordinate System*

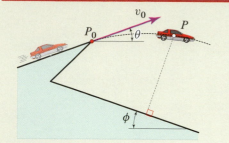

Figure 1
A movie stunt with a car jumping off a ridge.

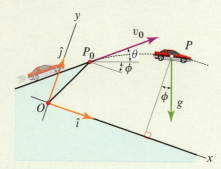

Figure 2
Acceleration in relation to the given coordinate system.

As part of a movie stunt, a car P runs off a ridge at the point P_0 with a speed v_0 as shown. Derive expressions for the velocity, position, and trajectory of the car such that it is easy to keep track of the perpendicular distance between the car and the lower incline.

SOLUTION

Road Map   We model the car's motion as a projectile motion. By choosing the Cartesian coordinate system in Fig. 2, the perpendicular distance between the car and the incline is directly provided by the y coordinate of the point P. Therefore, we will solve the problem by using the coordinate system shown. We begin by determining the components of the acceleration in the x and y directions. The x and y components of the acceleration are constant so that we can obtain velocity and position by direct application of constant acceleration equations. Finally, the trajectory is found by eliminating time from the description of the position.

Computation   Given the orientation of the y axis relative to gravity, we have

$$a_x = \ddot{x} = g \sin \phi \quad \text{and} \quad a_y = \ddot{y} = -g \cos \phi. \tag{1}$$

Since a_x and a_y are constant, the velocity components of P can be obtained as functions of time via a direct application of Eq. (2.41) on p. 59, i.e.,

$$v_x(t) = v_0 \cos(\theta + \phi) + (g \sin \phi)(t - t_0), \tag{2}$$
$$v_y(t) = v_0 \sin(\theta + \phi) - (g \cos \phi)(t - t_0), \tag{3}$$

where t_0 is the time at which P is at P_0 and where $v_0 \cos(\theta + \phi)$ and $v_0 \sin(\theta + \phi)$ are the x and y components of the initial velocity, respectively.

By direction application of Eq. (2.42) on p. 59, the position of P is given by

$$x(t) = x_0 + v_0 \cos(\theta + \phi)(t - t_0) + \tfrac{1}{2}(g \sin \phi)(t - t_0)^2, \tag{4}$$
$$y(t) = y_0 + v_0 \sin(\theta + \phi)(t - t_0) - \tfrac{1}{2}(g \cos \phi)(t - t_0)^2, \tag{5}$$

where x_0 and y_0 are the coordinates of the point P_0.

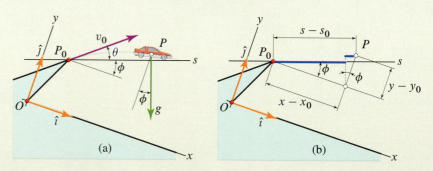

Figure 3. (a) Horizontal direction s and (b) displacements $x - x_0$ and $y - y_0$ in the x and y directions corresponding to the horizontal displacement $s - s_0$. The length of the blue segment to the left of the dotted line is $(x - x_0) \cos \phi$, and the length of the blue segment to the right of the dotted line is $(y - y_0) \sin \phi$.

To determine the trajectory, we need to eliminate time from Eqs. (4) and (5). To do so, we need to find an expression for time in terms of x and y. Referring to Fig. 3(a), we let s be the car's coordinate along the horizontal direction. Since the acceleration in the horizontal direction is zero, we have

$$\ddot{s} = 0 \quad \Rightarrow \quad \dot{s} = \text{constant} = v_{0s} \quad \Rightarrow \quad s - s_0 = v_{0s}(t - t_0), \tag{6}$$

where $v_{0s} = v_0 \cos\theta$ is the horizontal component of $\vec{v}_0$. Solving Eq. (6) for $t - t_0$, we have

$$t - t_0 = \frac{s - s_0}{v_{0s}} = \frac{s - s_0}{v_0 \cos\theta}. \tag{7}$$

Equation (7) is useful once rewritten in terms of x and y. Hence, referring to Fig. 3(b), using trigonometry, we can see that

$$s - s_0 = (x - x_0)\cos\phi + (y - y_0)\sin\phi. \tag{8}$$

Substituting Eq. (8) into Eq. (7), we have

$$t - t_0 = \frac{(x - x_0)\cos\phi + (y - y_0)\sin\phi}{v_0 \cos\theta}. \tag{9}$$

Finally, the trajectory is obtained by substituting Eq. (9) into Eq. (4) or (5). If we substitute Eq. (9) into Eq. (5), we obtain the trajectory of P in the following form:

$$y = y_0 + v_0 \sin(\theta + \phi)\left[\frac{(x - x_0)\cos\phi + (y - y_0)\sin\phi}{v_0 \cos\theta}\right]$$
$$- \tfrac{1}{2}g\cos\phi\left[\frac{(x - x_0)\cos\phi + (y - y_0)\sin\phi}{v_0 \cos\theta}\right]^2. \tag{10}$$

Discussion & Verification   Equations (2)–(5) were derived as direct applications of the constant acceleration equation and therefore are dimensionally correct. Because Eq. (9) was obtained by substituting Eq. (9) into Eq. (5) (without any additional simplifications or manipulations), to verify that Eq. (10) is dimensionally correct, we only need to verify that the expression for $t - t_0$ in Eq. (9) is dimensionally correct. Indeed Eq. (9) is correct since it is a fraction with a numerator with dimensions of length and a denominator with dimensions of length over time.

To further verify that Eq. (10) is correct, we can check that it is the equation of a parabola. One way to carry out this verification is to apply advanced notions from analytical geometry. We will use a different strategy based on observing what happens if we rotate our coordinate system so as to make the y axis parallel to the direction of gravity. In this case $\phi = 0$, $\cos\phi = 1$, and $\sin\phi = 0$, and Eq. (10) simplifies to

$$y = y_0 + \tan\theta(x - x_0) - \tfrac{1}{2}g\frac{(x - x_0)^2}{v_0^2 \cos^2\theta}, \tag{11}$$

i.e., it can be further reduced to the simple form $y = C_0 + C_1 x + C_3 x^2$ (with C_1, C_2, and C_3 constants), which we can recognize to be a parabola from elementary analytical geometry.

A Closer Look   This example is meant to illustrate that there are applications in which we may want to study projectile motion using a Cartesian coordinate system with axes not parallel to gravity. In this case, the expression for the projectile's trajectory can be complicated. A trajectory of the simple form $y = C_0 + C_1 x + C_3 x^2$ is only obtained when the projectile's motion is described using a Cartesian coordinate system with the y axis parallel to the direction of gravity.

PROBLEMS

Figure P2.77

Problem 2.76

The discussion in Example 2.12 revealed that the angle θ had to be greater than $\theta_{\min} = 0.716°$. Find an analytical expressions for $\theta_{\min}$ in terms of h, w, and d.

Problem 2.77

A stomp rocket is a toy consisting of a hose connected to a "blast pad" (i.e., an air bladder) at one end and to a short pipe mounted on a tripod at the other end. A rocket with a hollow body is mounted onto the pipe and is propelled into the air by "stomping" on the blast pad. Some manufactures claim that one can shoot a rocket over 200 ft in the air. Neglecting air resistance, determine the rocket's minimum initial speed such that it reaches a maximum flight height of 200 ft.

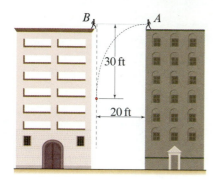

Figure P2.78

Problem 2.78

Stuntmen A and B are shooting a movie scene in which A needs to pass a gun to B. Stuntman B is supposed to start falling vertically precisely when A throws the gun to B. Treating the gun and the stuntman B as particles, find the velocity of the gun as it leaves A's hand so that B will catch it after falling 30 ft.

Problem 2.79

The jaguar A leaps from O at speed $v_0 = 6\,\text{m/s}$ and angle $\beta = 35°$ relative to the incline to try to intercept the panther B at C. Determine the distance R that the jaguar jumps from O to C (i.e., R is the distance between the two points of the trajectory that intersect the incline), given that the angle of the incline is $\theta = 25°$.

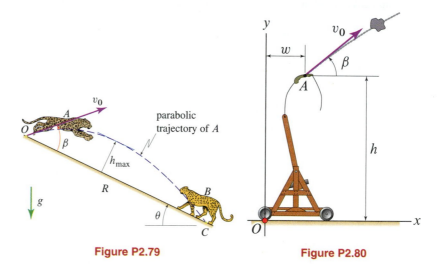

Figure P2.79   **Figure P2.80**

Problem 2.80

If the projectile is released at A with initial speed v_0 and angle β, derive the projectile's trajectory, using the coordinate system shown. Neglect air resistance.

Problem 2.81

A trebuchet releases a rock with mass $m = 50\,\text{kg}$ at point O. The initial velocity of the projectile is $\vec{v}_0 = (45\,\hat{\imath} + 30\,\hat{\jmath})\,\text{m/s}$. Neglecting aerodynamic effects, determine where the rock will land and its time of flight.

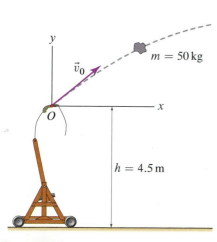

Figure P2.81

Problem 2.82

A golfer chips the ball into the hole on the fly from the rough at the edge of the green. Letting $\alpha = 4°$ and $d = 2.4$ m, verify that the golfer will place the ball within 10 mm of the hole if the ball leaves the rough with a speed $v_0 = 5.03$ m/s and an angle $\beta = 41°$.

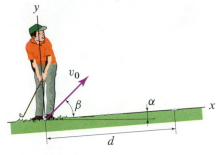

Figure P2.82

Problems 2.83 and 2.84

In a movie scene involving a car chase, a car goes over the top of a ramp at A and lands at B below.

Problem 2.83 If $\alpha = 20°$ and $\beta = 23°$, determine the distance d covered by the car if the car's speed at A is 45 km/h. Neglect aerodynamic effects.

Problem 2.84 Determine the speed of the car at A if the car is to cover distance $d = 150$ ft for $\alpha = 20°$ and $\beta = 27°$. Neglect aerodynamic effects.

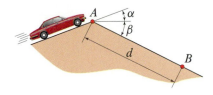

Figure P2.83 and P2.84

Problem 2.85

The M777 lightweight 155 mm howitzer is a piece of artillery whose rounds are ejected from the gun with a speed of 829 m/s. Assuming that the gun is fired over a flat battlefield and ignoring aerodynamic effects, determine (*a*) the elevation angle needed to achieve the maximum range, (*b*) the maximum possible range of the gun, and (*c*) the time it would take a projectile to cover the maximum range. Express the result for the range as a percentage of the actual maximum range of this weapon, which is 30 km for unassisted ammunition.

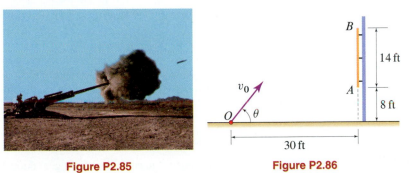

Figure P2.85   **Figure P2.86**

Problem 2.86

You want to throw a rock from point O to hit the vertical advertising sign AB, which is $R = 30$ ft away. You are able to throw a rock with the speed $v_0 = 45$ ft/s. The bottom of the sign is 8 ft off the ground and the sign is 14 ft tall. Determine the range of angles at which the projectile can be thrown in order to hit the target, and compare this with the angle subtended by the target as seen from an observer at point O. Compare your results with those found in Example 2.11.

Problem 2.87

Suppose that you can throw a projectile at a large enough v_0 so that it can hit a target a distance R downrange. Given that you know v_0 and R, determine the general expressions for the *two* distinct launch angles θ_1 and θ_2 that will allow the projectile to hit D. For $v_0 = 30$ m/s and $R = 70$ m, determine numerical values for θ_1 and θ_2.

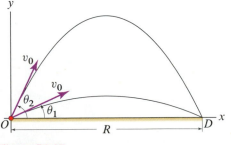

Figure P2.87

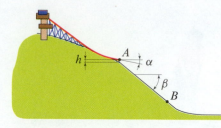

Figure P2.88

Problem 2.88

An alpine ski jumper can fly distances in excess of $100\,\text{m}^*$ by using his or her body and skis as a "wing" and therefore taking advantage of aerodynamic effects. With this in mind and assuming that a ski jumper could survive the jump, determine the distance the jumper could "fly" without aerodynamic effects, i.e., if the jumper were in free fall after clearing the ramp. For the purpose of your calculation, use the following typical data: $\alpha = 11°$ (slope of ramp at takeoff point A), $\beta = 36°$ (average slope of the hill),[†] $v_0 = 86\,\text{km/h}$ (speed at A), $h = 3\,\text{m}$ (height of takeoff point with respect to the hill). Finally, for simplicity, let the jump distance be the distance between the takeoff point A and the landing point B.

Problems 2.89 and 2.90

A soccer player practices kicking a ball from A directly into the goal (i.e., the ball does not bounce first) while clearing a 6 ft tall fixed barrier.

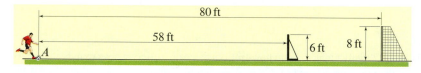

Figure P2.89 and P2.90

Problem 2.89
Determine the minimum speed that the player needs to give the ball to accomplish the task. *Hint:* To find $(v_0)_\text{min}$, consider the equation for the projectile's trajectory (see, e.g., Eq. (2.52)) for the case in which the ball reaches the goal at its base. Then solve this equation for the initial speed v_0 as a function of the initial angle θ, and finally find $(v_0)_\text{min}$ as you learned in calculus. Don't forget to check whether or not the ball clears the barrier.

Problem 2.90
Find the initial speed and angle that allow the ball to barely clear the barrier while barely reaching the goal at its base. *Hint:* As shown in Eq. (2.52), a projectile's trajectory can be given the form $y = C_1 x - C_2 x^2$ where the coefficients C_1 and C_2 can be found by forcing the parabola to go through two given points.

Problems 2.91 and 2.92

In a circus act a tiger is required to jump from point A to point C so that it goes through the ring of fire at B. *Hint:* As shown in Eq. (2.52), a projectile's trajectory can be given the form $y = C_1 x - C_2 x^2$ where the coefficients C_1 and C_2 can be found by forcing the parabola to go through two given points.

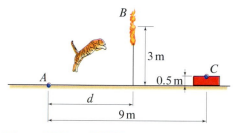

Figure P2.91 and P2.92

Problem 2.91
Determine the tiger's initial velocity if the ring of fire is placed at a distance $d = 5.5\,\text{m}$ from A. Furthermore, determine the slope of the tiger's trajectory as the tiger goes through the ring of fire.

Problem 2.92
Determine the tiger's initial velocity as well as the distance d so that the slope of the tiger's trajectory as the tiger goes through the ring of fire is completely horizontal.

---

[*] On March 20, 2005, using the very large ski ramp at Planica, Slovenia, Bjørn Einar Romøren of Norway set the world record by flying a distance of $239\,\text{m}$.

[†] While the given average slope of the landing hill is accurate, you should know that, according to regulations, the landing hill must have a curved profile. Here, we have chosen to use a landing hill with a *constant* slope of $36°$ to simplify the problem.

Problem 2.93

A jaguar A leaps from O at speed v_0 and angle β relative to the incline to attack a panther B at C. Determine an expression for the maximum *perpendicular* height h_{max} above the incline achieved by the leaping jaguar, given that the angle of the incline is θ.

Problems 2.94 and 2.95

The jaguar A leaps from O at speed v_0 and angle β relative to the incline to intercept the panther B at C. The distance along the incline from O to C is R, and the angle of the incline with respect to the horizontal is θ.

Problem 2.94 Determine an expression for v_0 as a function of β for A to be able to get from O to C.

Problem 2.95 Derive v_0 as a function of β to leap a given distance R along with the optimal value of launch angle β, i.e., the value of β necessary to leap a given distance R with the minimum v_0. Then plot v_0 as a function of β for $g = 9.81 \text{ m/s}^2$, $R = 7 \text{ m}$, and $\theta = 25°$, and find a numerical value of the optimal β and the corresponding value of v_0 for the given set of parameters.

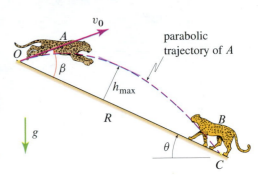

Figure P2.93–P2.95

Problems 2.96 and 2.97

A stomp rocket is a toy consisting of a hose connected to a blast pad (i.e., an air bladder) at one end and to a short pipe mounted on a tripod at the other end. A rocket with a hollow body is mounted onto the pipe and is propelled into the air by stomping on the blast pad.

Figure P2.96 and P2.97

Problem 2.96 If the rocket can be imparted an initial speed $v_0 = 120 \text{ ft/s}$, and if the rocket's landing spot at B is at the same elevation as the launch point, i.e., $h = 0 \text{ ft}$, neglect air resistance and determine the rocket's launch angle θ such that the rocket achieves the maximum possible range. In addition, compute R, the rocket's maximum range, and t_f, the corresponding flight time.

Problem 2.97 Assuming the rocket can be given an initial speed $v_0 = 120 \text{ ft/s}$, the rocket's landing spot at B is 10 ft higher than the launch point, i.e., $h = 10 \text{ ft}$, and neglecting air resistance, find the rocket's launch angle θ such that the rocket achieves the maximum possible range. In addition, as part of the solution, compute the corresponding maximum range and flight time. To do this:

(a) Determine the range R as a function of time.

(b) Take the expression for R found in (a), square it, and then differentiate it with respect to time to find the flight time corresponding to the maximum range and then that maximum range.

(c) Use the time found in (b) to then find the angle required to achieve the maximum range.

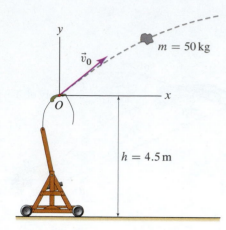

y

$\vec{v}_0$

$m = 50\,\mathrm{kg}$

O

x

$h = 4.5\,\mathrm{m}$

Figure P2.98–P2.101

Problem 2.98

A trebuchet releases a rock with mass $m = 50\,\mathrm{kg}$ at the point O. The initial velocity of the projectile is $\vec{v}_0 = (45\,\hat{\imath} + 30\,\hat{\jmath})\,\mathrm{m/s}$. If one were to model the effects of air resistance via a drag force directly proportional to the projectile's velocity, the resulting accelerations in the x and y directions would be $\ddot{x} = -(\eta/m)\dot{x}$ and $\ddot{y} = -g - (\eta/m)\dot{y}$, respectively, where g is the acceleration of gravity and $\eta = 0.64\,\mathrm{kg/s}$ is a viscous drag coefficient. Find an expression for the trajectory of the projectile.

Problem 2.99

Continue Prob. 2.98 and, for the case where $\eta = 0.64\,\mathrm{kg/s}$, determine the maximum height from the ground reached by the projectile and the time it takes to achieve it. Compare the result with what you would obtain in the absence of air resistance.

Problem 2.100

Continue Prob. 2.98 and, for the case where $\eta = 0.64\,\mathrm{kg/s}$, determine t_I and x_I, the value of t, and the x position corresponding to the projectile's impact with the ground.

Problem 2.101

With reference to Probs. 2.98 and 2.100, assume that an experiment is conducted so that the measured value of x_I is 10% smaller that what is predicted in the absence of viscous drag. Find the value of η that would be required for the theory in Prob. 2.98 to match the experiment.

Problem 2.102

Express the trajectory of the golf ball using the axes shown and in terms of initial speed v_0, initial angle β, slope α, and the acceleration of gravity g.

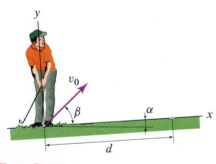

y

v_0

β

α

x

d

Figure P2.102

DESIGN PROBLEMS

Design Problem 2.1

In the manufacture of steel balls of the type used for ball bearings, it is important that their material properties be sufficiently uniform. One way to detect gross differences in their material properties is to observe how a ball rebounds when dropped on a hard strike plate. The rebound characteristics of a ball can be assessed via a quantity called the *coefficient of restitution* (COR).[*] Specifically, if $(v_n)_{strike}$ is the component of a ball's impact velocity normal to the strike plate, then the COR is given by

$$\text{COR} = \frac{|(v_n)_{\text{rebound}}|}{|(v_n)_{\text{strike}}|}, \qquad \text{(COR equation)}$$

where $(v_n)_{\text{rebound}}$ is the component of the rebound velocity normal to the strike plate.

Given that each ball has a radius $R = 0.2$ in., design a sorting device to select the balls with $0.800 < \text{COR} < 0.825$. The device consists of an incline defined by the angle θ and length L. The strike plate is placed at the bottom of a well with depth h and width w. Finally, at a distance ℓ from the incline, there is a thin vertical barrier with an gap of size d placed at a height b from the bottom of the well. Releasing a ball from rest at the top of the incline and assuming that the ball rolls without slip, we know the ball will reach the bottom of the incline with a speed[†]

$$v_0 = \sqrt{\tfrac{10}{7} g L \sin\theta},$$

where g is the acceleration due to gravity. After rolling off the incline, each ball will rebound off the strike plate in such as a way that the horizontal component of velocity is unaffected by the impact while the vertical component will behave as described in the COR equation. After rebounding, the balls to be isolated will pass through the vertical gap whereas the rest of the balls will not go through the gap. In your design, choose appropriate values of $L, \theta < 45°, h, w, d$, and ℓ to accomplish the desired task while ensuring that the overall dimensions of the device do not exceed 4 ft in both the horizontal and vertical directions.

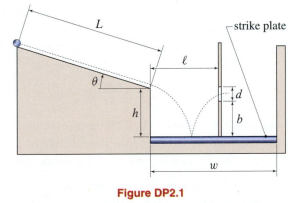

Figure DP2.1

---

[*] We will study the COR in detail in Section 5.2.
[†] We will see how to derive this formula in Chapters 7 and 8.

2.4   The Time Derivative of a Vector

Time derivatives of vector quantities are *everywhere* in dynamics. We have already seen the time derivative of position and velocity vectors. In coming chapters, we will also see the time derivatives of quantities such as angular velocity, momentum, and angular momentum, all of which are vectors. It is therefore worthwhile to devote a little time to the time derivative of a vector so as to strengthen our understanding of this operation.

Conceptually, the new idea we need to come away with is that the time derivative of a vector is intimately linked to the notion of *rotation*. We will see that vectors change with time in two ways:

1. Because they change in length.

2. Because they change in direction due to their *rotation*.

Consider an arbitrary vector $\vec{A}$ in the plane of the page at time t and time $t + \Delta t$, as shown in Fig. 2.24(a). The vector $\vec{A}$ is changing in length and

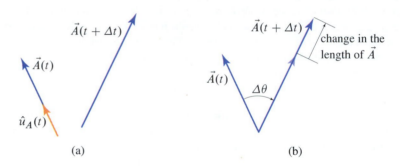

Figure 2.24. (a) Vector $\vec{A}$ changing from $\vec{A}(t)$ to $\vec{A}(t + \Delta t)$; (b) vectors $\vec{A}(t)$ and $\vec{A}(t + \Delta t)$ drawn with their tail points coinciding.

direction as well as in the position of its "tail." We will consider the change in $\vec{A}$ as depicted in Fig. 2.24(b), in which the tail points at t and $t + \Delta t$ are made to coincide for illustrative purposes. Since we can express $\vec{A}$ as a magnitude times a unit vector in the direction of $\vec{A}$, differentiating it with respect to time, we obtain

$$\dot{\vec{A}}(t) = \frac{d}{dt}\left[A(t)\,\hat{u}_A(t)\right] = \frac{dA}{dt}\,\hat{u}_A + A\frac{d\hat{u}_A}{dt} = \dot{A}\,\hat{u}_A + A\dot{\hat{u}}_A, \quad (2.54)$$

where A is the magnitude of $\vec{A}$, that is, $|\vec{A}|$; $\hat{u}_A$ is the unit vector in the direction of $\vec{A}$ (see Fig. 2.24(a)); and we have used the product rule of differentiation. Equation (2.54) shows that the time derivative of $\vec{A}$ consists of two parts:

1. The term $\dot{A}\,\hat{u}_A$, which is a vector in the direction of $\vec{A}$ measuring the time rate of change of the *magnitude* of $\vec{A}$, and

2. The term $A\dot{\hat{u}}_A$, which is the magnitude of $\vec{A}$ multiplied by the time derivative of the unit vector $\hat{u}_A$, measures the time rate of change of the *direction* of $\vec{A}$.

The vector $\dot{\hat{u}}_A$ can be written in a convenient way once we understand how to describe rotations and rotation rates in particular.

Rotation and angular velocity

Our objective now is to use vectors to describe rotations, and to do so we need to understand what distinguishes a rotation from other types of motion. We begin by considering Fig. 2.25, which shows the swinging motion of a door. Notice how the points on the hinge line *do not move*. This is what characterizes rotations: a body is rotating if it has a "hinge line," which is more properly called the *axis of rotation*. For general three-dimensional motions, the definition of axis of rotation allows for the axis to translate and change its orientation.

Now that we have the notion of axis of rotation, observe that we can use a unit vector to describe the orientation and direction of the axis of rotation via the right-hand rule, with the thumb pointing in the direction of the unit vector. It is this direction that we will call the *direction of rotation* (see Fig. 2.26). Using this idea and referring to Fig. 2.27, consider the vector $\vec{\gamma} = \gamma\,\hat{u}_\gamma$, where γ is an angle expressed in radians and $\hat{u}_\gamma$ is a unit vector parallel to the axis of rotation. Can we use the vector $\vec{\gamma}$ to denote the angular displacement corresponding to a rotation through the angle γ with the direction $\hat{u}_\gamma$? We can, but the problem with this representation is that it gives us a "vector" that does not act as a true vector. To understand this statement, consider Fig. 2.28, which

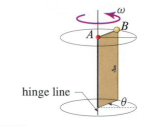

Figure 2.25
Depiction of the swinging motion of a door.

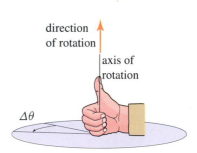

Figure 2.26
Right-hand rule used to depict the direction of rotation.

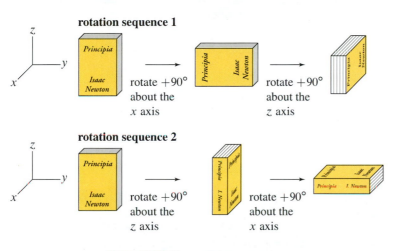

rotation sequence 1

rotate $+90°$ about the x axis

rotate $+90°$ about the z axis

rotation sequence 2

rotate $+90°$ about the z axis

rotate $+90°$ about the x axis

Figure 2.28. Two rotation sequences.

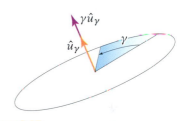

Figure 2.27
A possible, *but not recommended*, use of vectors to represent rotations.

shows two different rotation sequences of a book starting from the same initial position. In the first sequence, the book is rotated first by 90° about the positive x axis and then by 90° about the positive z axis. In the second sequence, the book is rotated first by 90° about the positive z axis and then by 90° about the positive x axis. Notice that all we have done is to reverse the two successive rotations between the two rotation sequences, and yet the book ends up in a different orientation. Now, say we use $\vec{\rho}_x$ to describe the rotation by 90° about the positive x axis and $\vec{\rho}_z$ to describe the rotation by 90° about the positive z axis. Then we would expect the vector sum $\vec{\rho}_x + \vec{\rho}_z$ to describe the first sequence and $\vec{\rho}_z + \vec{\rho}_x$ to describe the second. The problem is that vector algebra requires that $\vec{\rho}_x + \vec{\rho}_z = \vec{\rho}_z + \vec{\rho}_x$ even though we know that $\vec{\rho}_x + \vec{\rho}_z$ and $\vec{\rho}_z + \vec{\rho}_x$ are definitely different!

Another reason for not using vectors to represent angular displacements caused by rotations is that it would not work when discussing those general cases in which the axis of rotation changes with time.[*] This is so because it takes some finite amount of time to rotate through the angle γ, and for the vector $\vec{\gamma}$ to make sense, we would need to keep the axis of rotation fixed until the full angle γ were covered. The axis of rotation could be reoriented after γ is achieved, but this would prevent us from *continuously* changing the axis with time. This limitation actually points toward a solution to the problem. Using a calculus-inspired approach, we can view any rotation with a changing axis as an *infinite* sequence of *infinitesimal* rotations. According to this view, we can update the position and orientation of the axis of rotation between any two infinitesimal rotations while keeping the axis fixed during each rotation in the sequence. So, let's use the idea depicted in Fig. 2.27 to represent rotations, but let's make them infinitesimal. Referring to Fig. 2.29, we construct two sequences of rotations just as we did in Fig. 2.28. The difference now is that we will be making these rotations successively *smaller and smaller*. What Fig. 2.29 demonstrates is that as the rotations get smaller and smaller, the order in which these rotations are performed does not affect the final result. More

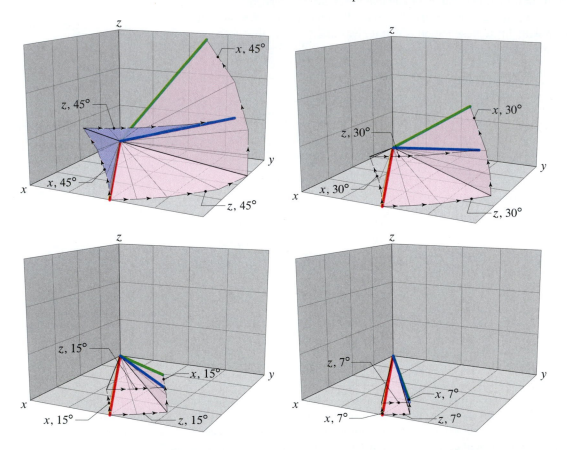

Figure 2.29. Each panel shows two rotation sequences. In the first sequence (purple surface), the red segment is moved into the blue segment by first rotating about the x axis and then about the z axis. In the second sequence (pink surface), the red segment is moved into the green segment by first rotating about the z axis and then about the x axis. Rotations are labeled so that, for example, the label $z, 30°$ indicates a rotation of $30°$ about the z axis. As rotations become smaller, i.e., in going from the top left to the lower right of the figure, it becomes harder to distinguish the blue and green segments; i.e., the final outcome of the rotation sequences tends to be the same.

---

[*] Mathematical objects called *quaternions* have been invented as a way to overcome the difficulties we are mentioning.

importantly, although this would be difficult to see directly from Fig 2.29, if $d\gamma\,\hat{u}_\gamma$ and $d\phi\,\hat{u}_\phi$ are two infinitesimal rotations, not only is the final outcome of $d\gamma\,\hat{u}_\gamma$ followed by $d\phi\,\hat{u}_\phi$ the same as that of $d\phi\,\hat{u}_\phi$ followed by $d\gamma\,\hat{u}_\gamma$, but that outcome is the vector $d\gamma\,\hat{u}_\gamma + d\phi\,\hat{u}_\phi$!

The final step in our discussion is to realize that by dividing an infinitesimal angular displacement by the infinitesimal amount of time it takes to perform it, we obtain the notion of *angular velocity*. This allows us to deal with the finite quantity $\dot{\vec{\gamma}}$ instead of $d\vec{\gamma}$ while we can always go back to the infinitesimal angular displacement by writing $d\vec{\gamma} = \dot{\vec{\gamma}}\,dt$.

In summary, given a *finite size* vector $\vec{\omega} = \omega\,\hat{u}_\omega$, we can view this vector as describing an *infinitesimal* rotation, i.e., an infinitesimal angular displacement

1. With axis parallel to $\hat{u}_\omega$,

2. Sweeping through an angle $|\vec{\omega}|\,dt$.

3. In a direction consistent with the right-hand rule.

The vector $\vec{\omega}$ represents an *angular velocity*, and as with other vectors, angular velocity vectors can be manipulated using standard vector operations. Finally, we wish to point out that since angles measured in radians are nondimensional, the dimensions of an angular velocity must be $1/T$, where T denotes the dimension of time.

Time derivative of a unit vector

We now go back to the time derivative of a unit vector and obtain a formula for it based on what we learned about rotations.

Referring to Fig. 2.30, given the unit vector $\hat{u}_A$, we express it as

$$\hat{u}_A = \cos\theta\,\hat{\imath} + \sin\theta\,\hat{\jmath} \tag{2.55}$$

so that

$$\dot{\hat{u}}_A = \dot{\theta}(-\sin\theta\,\hat{\imath} + \cos\theta\,\hat{\jmath}). \tag{2.56}$$

We can see that the term $-\sin\theta\,\hat{\imath} + \cos\theta\,\hat{\jmath}$ in Eq. (2.56) is a unit vector perpendicular to $\hat{u}_A$, and therefore $\dot{\hat{u}}_A$ is perpendicular to $\hat{u}_A$ (this is shown in Fig. 2.30). The term $\dot{\theta}$ in Eq. (2.56) is the rotation rate of $\hat{u}_A$. The infinitesimal angle $d\theta = \dot{\theta}\,dt$ is precisely the angle swept by the unit vector $\hat{u}_A$ during an infinitesimal time interval dt. But what is the corresponding axis of rotation? Using the right-hand rule and because both $\hat{u}_A$ and $\dot{\hat{u}}_A$ lie in the same plane, we conclude that the axis of rotation for $\hat{u}_A$ is perpendicular to the xy plane and oriented in the z direction. Hence, letting $\vec{\omega}_A$ denote the angular velocity *vector* of $\hat{u}_A$, we have

$$\vec{\omega}_A = \dot{\theta}\,\hat{k}. \tag{2.57}$$

Equation (2.57) implies that the vectors $\hat{u}_A$, $\dot{\hat{u}}_A$, and $\vec{\omega}_A$ are mutually perpendicular and that the vector $\vec{\omega}_A \times \hat{u}_A$ is parallel to $\dot{\hat{u}}_A$. We have

$$\vec{\omega}_A \times \hat{u}_A = \dot{\theta}\,\hat{k} \times (\cos\theta\,\hat{\imath} + \sin\theta\,\hat{\jmath})$$
$$= \dot{\theta}(\cos\theta\,\hat{k} \times \hat{\imath} + \sin\theta\,\hat{k} \times \hat{\jmath})$$
$$= \dot{\theta}(\cos\theta\,\hat{\jmath} - \sin\theta\,\hat{\imath}). \tag{2.58}$$

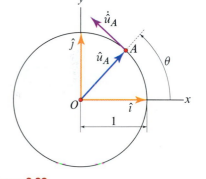

Figure 2.30
A rotating unit vector $\hat{u}_A$.

> **Helpful Information**
>
> **Vectors $\hat{u}_A$ and $\dot{\hat{u}}_A$ are mutually perpendicular.** One way to show that $\hat{u}_A$ and $\dot{\hat{u}}_A$ are mutually perpendicular starts with observing that $\hat{u}_A \cdot \hat{u}_A = 1 = $ constant, so that we have
>
> $$\frac{d}{dt}(\hat{u}_A \cdot \hat{u}_A) = 0 = \dot{\hat{u}}_A \cdot \hat{u}_A + \hat{u}_A \cdot \dot{\hat{u}}_A$$
> $$= 2\hat{u}_A \cdot \dot{\hat{u}}_A,$$
>
> which implies that $\hat{u}_A$ and $\dot{\hat{u}}_A$ are always perpendicular to one another.

Helpful Information

Have we really proved Eq. (2.59)? Our derivation of Eq. (2.59) does not constitute a mathematically rigorous proof. What we have really presented so far are simple and physically based arguments pointing to the idea that the time rate of change of a unit vector can always be expressed as the cross product of its angular velocity and the unit vector in question. However, Eq. (2.59) is valid under the most general of circumstances, and it can be proved in a rigorous way.

Equation (2.58) is a remarkable result because, by comparing it with Eq. (2.56), it tells us that $\vec{\omega} \times \hat{u}_A$ is not just parallel to $\dot{\hat{u}}_A$, it is $\dot{\hat{u}}_A$! That is,

$$\dot{\hat{u}}_A = \vec{\omega}_A \times \hat{u}_A. \tag{2.59}$$

It turns out that Eq. (2.59) is *universal*: whether in 2D or 3D, the time rate of change of a unit vector can always be represented as the cross product of the angular velocity of this vector with the unit vector in question.

So, we now have a general and physically motivated way to express the *time derivative of any unit vector* $\vec{u}$ as

$$\boxed{\dot{\hat{u}} = \vec{\omega}_u \times \hat{u},} \tag{2.60}$$

where $\vec{\omega}_u$ is the angular velocity of $\vec{u}$. Equation (2.60) is best remembered as

> *The time derivative of a unit vector is the angular velocity of the vector crossed with the vector itself.*

Time derivative of an arbitary vector

Using Eq. (2.60), the interpretation of the term $A\,\dot{\hat{u}}_A$ in Eq. (2.54) is now clear: it is the magnitude of $\vec{A}$ times $\vec{\omega}_A \times \hat{u}_A$ so that Eq. (2.54) becomes

$$\dot{\vec{A}}(t) = \dot{A}\hat{u}_A + A\vec{\omega}_A \times \hat{u}_A = \dot{A}\hat{u}_A + \vec{\omega}_A \times A\hat{u}_A, \tag{2.61}$$

or

$$\boxed{\dot{\vec{A}}(t) = \dot{A}\hat{u}_A + \vec{\omega}_A \times \vec{A}.} \tag{2.62}$$

Concept Alert

$|\dot{\vec{A}}|$ **and** $\dot{A}$ **are not the same thing.** In general, $\dot{A} \neq |\dot{\vec{A}}|$, where by $|\dot{\vec{A}}|$ we mean $|d\vec{A}/dt|$. The reason for this is that $\dot{\vec{A}}$ is made up of two vectors—$\dot{A}\hat{u}_A$ and $\vec{\omega}_A \times \vec{A}$—and the magnitude of $\dot{\vec{A}}$ is therefore a combination of the magnitudes of these two vectors. Since $\dot{A}\hat{u}_A$ is always perpendicular to $\vec{\omega}_A \times \vec{A}$, we can say that

$$|\dot{\vec{A}}| = \sqrt{\dot{A}^2 + |\vec{\omega}_A \times \vec{A}|^2}.$$

We now can interpret the time derivative of any vector as the time rate of change of the magnitude of the vector plus a time rate of change of direction of that vector

$$\underbrace{\dot{\vec{A}}(t)}_{\substack{\text{change} \\ \text{in } \vec{A}}} = \underbrace{\dot{A}\hat{u}_A}_{\substack{\text{change in} \\ \text{magnitude}}} + \underbrace{\vec{\omega}_A \times \vec{A}}_{\substack{\text{change in} \\ \text{direction}}}, \tag{2.63}$$

where the time rate of change of direction is given by the cross product of the vector's angular velocity and the vector itself. This relationship applies to *any* vector, and we will use it *extensively* for the remainder of the book. In particular, it will come in very handy when we need to find velocity vectors by differentiating position vectors and when we need to analyze the kinematics of finite size rigid bodies.

We end this section by mentioning that the direct application of the same ideas used to obtain Eq. (2.62) can be used to obtain the *second* derivative of an arbitrary vector with respect to time as

$$\boxed{\ddot{\vec{A}} = \ddot{A}\hat{u}_A + 2\vec{\omega}_A \times \dot{A}\hat{u}_A + \dot{\vec{\omega}}_A \times \vec{A} + \vec{\omega}_A \times (\vec{\omega}_A \times \vec{A}).} \tag{2.64}$$

Helpful Information

Cross products and 2D motion. In this section we have discovered that rotations and cross products give us a way to represent the time derivative of a vector. Rotations and cross products are intrinsically three-dimensional concepts (the cross product cannot be defined in a purely two-dimensional context). This means that when we apply Eqs. (2.60), (2.62), and (2.64) to study planar motions, these motions are viewed as special cases of three-dimensional motions.

Equation (2.64) can be interpreted by viewing $\vec{A}$ as the position vector of a point. According to this view, calling A the point in question, the second derivative of the vector $\vec{A}$ consists of the following quantities:

$\ddot{A}\hat{u}_A =$ the component of the acceleration of A in the direction of $\vec{A}$ if the direction of $\vec{A}$ were not to change

$2\vec{\omega}_A \times \dot{A}\,\hat{u}_A =$ the Coriolis acceleration of A, resulting from two equal, but different effects: (1) the change in direction of $\vec{A}$ due to $\vec{\omega}_A$ and (2) the effect of $\vec{\omega}_A$ on the component of the velocity of A in the direction of $\vec{A}$

$\dot{\vec{\omega}}_A \times \vec{A} =$ the component of the acceleration of A perpendicular to $\vec{A}$ if the magnitude of $\vec{A}$ were held fixed

$\vec{\omega}_A \times (\vec{\omega}_A \times \vec{A}) =$ the component of the acceleration of A perpendicular to $\vec{\omega}_A$ if the magnitude of $\vec{A}$ were held fixed

As demonstrated in Example 2.14, the term $\ddot{A}\,\hat{u}_A$ can also be interpreted as the acceleration of a point in rectilinear motion. In addition, as demonstrated in Example 2.15, the sum of the terms $\dot{\vec{\omega}}_A \times \vec{A}$ and $\vec{\omega}_A \times (\vec{\omega}_A \times \vec{A})$ can be interpreted as the acceleration of a point in circular motion.

Equation (2.64) will prove to be very useful in understanding the kinematics of rigid bodies, as we will see in Chapters 6 and 10.

End of Section Summary

This section considers the time derivative of a unit vector $\vec{u}$ and an arbitary vector $\vec{A}$. Vectors are characterized by their magnitude and direction, and we discovered that the time derivative of a vector consists of a part due to its change in length and a part due to its change in direction. Since a unit vector never changes its length, we discovered that its time derivative only consists of a contribution from a change in direction:

> **Eq. (2.60), p. 92**
>
> $$\underbrace{\dot{\hat{u}}(t)}_{\substack{\text{change} \\ \text{in } \vec{u}}} = \underbrace{\vec{\omega}_u \times \hat{u}}_{\substack{\text{change in} \\ \text{direction}}}.$$

For an arbitrary vector $\vec{A}$, the following relationship provides its time derivative:

> **Eq. (2.62), p. 92**
>
> $$\underbrace{\dot{\vec{A}}(t)}_{\substack{\text{change} \\ \text{in } \vec{A}}} = \underbrace{\dot{A}\,\hat{u}_A}_{\substack{\text{change in} \\ \text{magnitude}}} + \underbrace{\vec{\omega}_A \times \vec{A}}_{\substack{\text{change in} \\ \text{direction}}},$$

where, referring to Fig. 2.31, $\hat{u}_A$ is a unit vector in the direction of $\vec{A}$, $\dot{A} = d|\vec{A}|/dt$, and $\vec{\omega}_A$ is the angular velocity of the vector $\vec{A}$. For the same vector $\vec{A}$, the following relationship provides its second derivative with respect to time:

> **Eq. (2.64), p. 92**
>
> $$\ddot{\vec{A}} = \ddot{A}\,\hat{u}_A + 2\vec{\omega}_A \times \dot{A}\,\hat{u}_A + \dot{\vec{\omega}}_A \times \vec{A} + \vec{\omega}_A \times (\vec{\omega}_A \times \vec{A}).$$

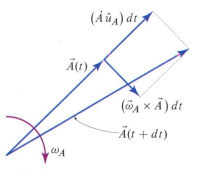

Figure 2.31
Depiction of the vector $\vec{A}$ at time t and at time $t + dt$ showing its change in magnitude and change in direction.

EXAMPLE 2.14   *Time Derivative of a Vector and Rectilinear Motion*

Consider rectilinear motion (see Eqs. (2.13), (2.16), and (2.17) as well as Section 2.2) using the notion of the time derivative of a vector as developed in the present section. In particular, determine expressions for the velocity and acceleration vectors of a particle whose motion is rectilinear.

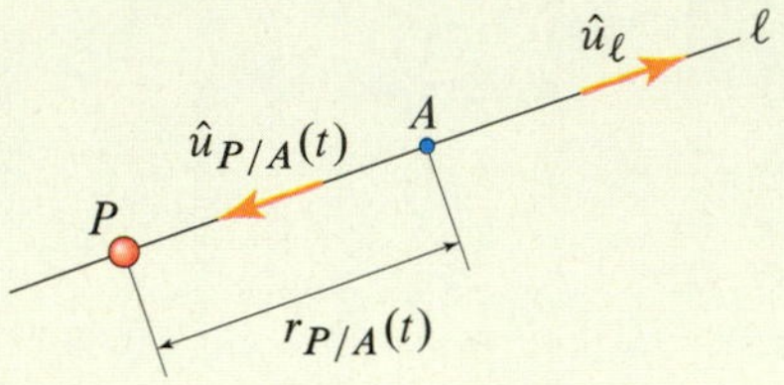

Figure 1
The point P moving in rectilinear motion along the line ℓ.

SOLUTION

Road Map   To describe rectilinear motion, we need a representation system compatible with the developments in this section. Referring to Fig. 1, let's

1. Choose a fixed point A along the trajectory to be the origin of our component system.

2. Call $\hat{u}_{P/A}$ the unit vector pointing *from A to P* (for an explanation of this notation see p. 10 of Section 1.2).

This representation will allow us to easily compute the time derivative of the position to find velocity and acceleration.

Computation   These definitions allow us to write the position vector of P as

$$\vec{r}_{P/A} = r_{P/A}\,\hat{u}_{P/A}, \tag{1}$$

where $r_{P/A}$ is the distance between P and A. Hence, we can calculate the velocity of P, using Eq. (2.62):

$$\vec{v}_P = \dot{r}_{P/A}\,\hat{u}_{P/A} + \vec{\omega}_r \times \vec{r}_{P/A}, \tag{2}$$

where $\vec{\omega}_r$ is the angular velocity vector of $\vec{r}_{P/A}$, which is the same as that of the unit vector $\hat{u}_{P/A}$, and by Eq. (2.60) we can write

$$\dot{\hat{u}}_{P/A} = \vec{\omega}_r \times \hat{u}_{P/A}. \tag{3}$$

Recall that the essence of the *rectilinearity* of P's motion makes $\hat{u}_{P/A}$ a constant (see the remark in the margin). Referring to Eq. (3) (and since $\hat{u}_{P/A}$ can never be zero), this means that, for rectilinear motions,

$$\vec{\omega}_r = \vec{0}. \tag{4}$$

Hence the expression for the velocity of P takes on the form

$$\boxed{\vec{v}_P = \dot{r}_{P/A}\,\hat{u}_{P/A}.} \tag{5}$$

To derive the acceleration vector of P, we can proceed in the same manner as for the velocity to obtain

$$\vec{a}_P = \dot{\vec{v}}_P = \ddot{r}_{P/A}\,\hat{u}_{P/A} + \vec{\omega}_v \times \vec{v}_P, \tag{6}$$

where $\vec{\omega}_v$ is the angular velocity of the vector $\vec{v}_P$. As with $\vec{r}_{P/A}$, the angular velocity of $\vec{v}_P$ is the same as that of $\hat{u}_{P/A}$, which, by Eq. (4), is zero. Therefore, the acceleration of P is

$$\boxed{\vec{a}_P = \ddot{r}_{P/A}\,\hat{u}_{P/A}.} \tag{7}$$

Discussion & Verification   We see that Eqs. (5) and (7), which describe one-dimensional motion, just give us the one-dimensional versions of Eqs. (2.16) and (2.17), on p. 34, which is what we expect should happen.

🔎 **A Closer Look**   Notice that Eq. (7) corresponds to the first term of Eq. (2.64). We can then interpret the first term of Eq. (2.64) simply as the acceleration of a point in rectilinear motion.

Helpful Information

In what way is $\hat{u}_{P/A}$ a constant? Referring to Fig. 1, and letting the *constant* unit vector $\hat{u}_\ell$ define the positive direction of the line along which P moves, we have

$$\hat{u}_{P/A}(t) = \begin{cases} \hat{u}_\ell & \text{when } P \text{ follows } A, \\ -\hat{u}_\ell & \text{when } P \text{ precedes } A. \end{cases}$$

When P coincides with A, then $\hat{u}_{P/A}$ is undefined. Hence we can say that $\hat{u}_{P/A}$ is constant except when P coincides with A.

EXAMPLE 2.15   *Time Derivative of a Vector and Circular Motion*

The point Q is moving on a circular path of radius r, which is centered at the fixed point O (see Fig. 1). Use the notion of the time derivative of a vector as developed in this section to determine expressions for the velocity and acceleration vectors of a particle whose motion is circular.

SOLUTION

Road Map   As with the previous example, we start by defining a vector to differentiate with respect to time. In this case, we choose the vector defining the position of Q relative to O. We will then determine the time derivatives of this vector, which will, in turn, give us the velocity and acceleration of the point Q as it moves in a circle.

Computation   Referring to Fig. 2, the position of Q relative to O is

$$\vec{r}_Q = r\,\hat{u}_Q, \tag{1}$$

where r is the radius of the circle and $\hat{u}_Q$ is the unit vector pointing from O to Q. Using Eq. (2.62) to compute the velocity of Q, we obtain

$$\vec{v}_Q = \dot{\vec{r}}_Q = \dot{r}\,\hat{u}_Q + \vec{\omega}_Q \times \vec{r}_Q = r\,\vec{\omega}_Q \times \hat{u}_Q, \tag{2}$$

where we have used the fact that r is constant and we have used Eq. (1) to obtain the last equality.

To determine $\vec{\omega}_Q$, which is the angular velocity of the unit vector $\hat{u}_Q$, observe that $\hat{u}_Q$ always lies in the xy plane, so that its axis of rotation must be parallel to the z axis. In addition, it is clear that the unit vector $\hat{u}_Q$ rotates at the rate $\dot{\beta}$, and so we must have

$$\vec{\omega}_Q = \omega_Q\,\hat{k} \quad \text{and} \quad \omega_Q = \dot{\beta}. \tag{3}$$

Substituting Eqs. (3) into Eq. (2), we obtain

$$\vec{v}_Q = r\dot{\beta}\,\hat{k} \times \hat{u}_Q, \tag{4}$$

which we will now evaluate in a couple of ways.

For the first way, referring to Fig. 2, we can write $\hat{u}_Q$ in terms of its Cartesian components as

$$\hat{u}_Q = \cos\beta\,\hat{i} + \sin\beta\,\hat{j}, \tag{5}$$

which, when substituted into Eq. (4), gives

$$\vec{v}_Q = r\dot{\beta}\,\hat{k} \times (\cos\beta\,\hat{i} + \sin\beta\,\hat{j}) = r\dot{\beta}(-\sin\beta\,\hat{i} + \cos\beta\,\hat{j}). \tag{6}$$

Again referring to Fig. 2, notice that the term $-\sin\beta\,\hat{i} + \cos\beta\,\hat{j}$ in Eq. (6) is equal to the unit vector $\hat{u}_\beta$, which is tangent to the circle at Q and pointing in the direction of increasing β. Therefore, $\vec{v}_Q$ can be given the following compact form

$$\boxed{\vec{v}_Q = r\dot{\beta}\,\hat{u}_\beta,} \tag{7}$$

which reminds us that the velocity vector is *always* tangent to the path and is directly proportional to both the angular velocity and the radius of the circular path.

We can also arrive at Eq. (7) by noting that the right-hand rule tells us that $\hat{k} \times \hat{u}_Q$ must be perpendicular to both $\hat{k}$ and $\hat{u}_Q$ and so points in the direction of $\hat{u}_\beta$. In

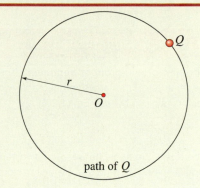

Figure 1
Point Q moving on a circle centered at O.

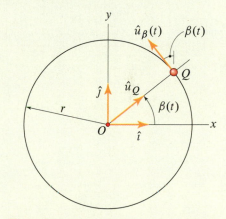

Figure 2
Particle Q moving along a circle centered at O.

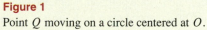

Helpful Information

Cross products. Since this example requires the computation of several cross products, here is a reminder of a visual aid from Chapter 1 that helps with cross products.

Arranging the three vectors $\hat{i}$, $\hat{j}$, and $\hat{k}$ as shown, then to calculate $\hat{j} \times \hat{k}$, just move around the circle, starting from $\hat{j}$ and going toward $\hat{k}$. Now notice that the next vector on the circle is $\hat{i}$, and in going from $\hat{j}$ to $\hat{k}$ we have moved in the direction of the arrow. Hence, $\hat{j} \times \hat{k} = +\hat{i}$. If instead we want to determine the outcome of $\hat{i} \times \hat{k}$, we need to notice that, moving along the circle starting from $\hat{i}$ and going toward $\hat{k}$, the subsequent vector is $\hat{j}$ and we move opposite to the arrow. Therefore, we have that $\hat{i} \times \hat{k} = -\hat{j}$.

addition, since $\hat{k}$ and $\hat{u}_Q$ are perpendicular to one another, the sine of the angle between them is equal to 1 and so

$$\hat{k} \times \hat{u}_Q = \hat{u}_\beta, \tag{8}$$

which, when substituted into Eq. (4), gives Eq. (7).

Next, we will obtain the acceleration by taking the time derivative of the velocity as given by Eq. (2):

$$\vec{a}_Q = r\left(\dot{\vec{\omega}}_Q \times \hat{u}_Q + \vec{\omega}_Q \times \dot{\hat{u}}_Q\right). \tag{9}$$

Equation (2.60) tells us that $\dot{\hat{u}}_Q = \vec{\omega}_Q \times \hat{u}_Q$. Hence, Eq. (9) becomes

$$\vec{a}_Q = r\left[\dot{\vec{\omega}}_Q \times \hat{u}_Q + \vec{\omega}_Q \times (\vec{\omega}_Q \times \hat{u}_Q)\right]. \tag{10}$$

Differentiating Eq. (3) with respect to time, we obtain

$$\dot{\vec{\omega}}_Q = \ddot{\beta}\,\hat{k} + \dot{\beta}\,\dot{\hat{k}} = \ddot{\beta}\,\hat{k}, \tag{11}$$

where we have used the fact that $\dot{\hat{k}} = \vec{0}$. Substituting Eqs. (3), (5), and (11) into Eq. (10) and carrying out all the cross products, we obtain the following expression for the acceleration of Q:

$$\vec{a}_Q = -r\left(\ddot{\beta}\sin\beta - \dot{\beta}^2\cos\beta\right)\hat{\imath} + r\left(\ddot{\beta}\cos\beta - \dot{\beta}^2\sin\beta\right)\hat{\jmath}. \tag{12}$$

Rearranging the acceleration terms in the following way

$$\vec{a}_Q = -\dot{\beta}^2\,r(\cos\beta\,\hat{\imath} + \sin\beta\,\hat{\jmath}) + \ddot{\beta}\,r(-\sin\beta\,\hat{\imath} + \cos\beta\,\hat{\jmath}) \tag{13}$$

allows us to write $\vec{a}_Q$ as

$$\boxed{\vec{a}_Q = -r\dot{\beta}^2\,\hat{u}_Q + r\ddot{\beta}\,\hat{u}_\beta.} \tag{14}$$

Note that we can also compute $\vec{a}_Q$ by differentiating Eq. (7) with respect to time to obtain

$$\vec{a}_Q = \dot{r}\dot{\beta}\,\hat{u}_\beta + r\ddot{\beta}\,\hat{u}_\beta + r\dot{\beta}\dot{\hat{u}}_\beta = r\ddot{\beta}\,\hat{u}_\beta + r\dot{\beta}\dot{\hat{u}}_\beta, \tag{15}$$

where we have used the fact that r is constant. Now, since $\dot{\hat{u}}_\beta = \vec{\omega}_\beta \times \hat{u}_\beta$, $\vec{\omega}_\beta = \vec{\omega}_Q = \dot{\beta}\,\hat{k}$, and $\dot{\beta}\,\hat{k} \times \hat{u}_\beta = -\dot{\beta}\,\hat{u}_Q$, Eq. (15) becomes

$$\vec{a}_Q = r\ddot{\beta}\,\hat{u}_\beta + (r\dot{\beta})(-\dot{\beta}\,\hat{u}_Q) = r\ddot{\beta}\,\hat{u}_\beta - r\dot{\beta}^2\,\hat{u}_Q, \tag{16}$$

which is identical to Eq. (14).

Discussion & Verification   Equation (14) tells us that, in a circular motion, the acceleration vector has two components:

1. A *tangential component*, which is *tangent* to the circular path and is proportional to the angular acceleration $\ddot{\beta}$.

2. A *radial component*, which is

   (a) Proportional to the *square* of the angular velocity.

   (b) *Always* directed *radially* inward toward the center of the circular trajectory.

Finally, observe that both the tangential and radial components of the acceleration are *directly proportional* to the radius of the circular trajectory.

 A Closer Look   This example helps us better understand the role played by the last two terms in Eq. (2.64) because it shows that the two terms in question describe the acceleration of a point in circular motion.

Helpful Information

Another cross product reminder. The magnitude of the cross product of two vectors is given by (see Eq. (1.20))

$$\left|\hat{k} \times \hat{u}_Q\right| = \left|\hat{k}\right|\left|\hat{u}_Q\right|\sin\theta,$$

where θ is the angle between the two vectors. Since both $\hat{k}$ and $\hat{u}_Q$ are unit vectors and $\hat{k}$ and $\hat{u}_Q$ are perpendicular to one another, this relation becomes

$$\left|\hat{k} \times \hat{u}_Q\right| = 1,$$

which is exactly what we have in Eq. (8).

EXAMPLE 2.16 *Time Derivative of a Vector Applied to a Tracking Problem*

It is known that an airplane B is flying at a constant speed v_0 and at a constant altitude h (see Fig. 1). The radar station at A tracks the plane by measuring the distance r between it and the plane, the rate at which r is changing, the antenna orientation θ, and the angular velocity of the antenna. Determine the relationships between those quantities that can be found by the tracking station and the speed and height of the airplane.

SOLUTION

Road Map   We will see that the height of the plane is found by using a simple geometric relationship. To obtain the relationship for speed that we seek, we will make use of an idea that we will call on many times: we can equate a *specific* expression for something (i.e., the velocity of the plane is horizontal) to a *general* expression for that something (i.e., the velocity of the plane using Eq. (2.62)).

Computation   Referring to Fig. 2, given that we know θ and r, and assuming that h_A is also known, finding the altitude of the plane h in terms of θ and r is an easy matter since

$$h = r \sin \theta + h_A. \tag{1}$$

To find the speed of the plane v_0, we need to first write its position and then differentiate it. Again referring to Fig. 2, the natural position vector to use is the position of the plane relative to the radar station, which can be represented as its magnitude times a unit vector parallel to $\vec{r}$ and pointing from A to B

$$\vec{r} = r \, \hat{u}_r, \tag{2}$$

where $r = |\vec{r}|$. Using this expression for $\vec{r}$, we can employ Eq. (2.62) to write the velocity of the plane as

$$\dot{\vec{r}} = \dot{r} \, \hat{u}_r + \vec{\omega}_r \times \vec{r} = \dot{r} \, \hat{u}_r + \vec{\omega}_r \times r\hat{u}_r. \tag{3}$$

To interpret $\vec{\omega}_r$ observe that the position vector $\vec{r}$ remains in the xy plane at all times, and its rotation rate is measured by $\dot{\theta}$. Using the right-hand rule tells us that $\vec{\omega}$ is given by

$$\vec{\omega} = \dot{\theta} \, \hat{k} = \omega \, \hat{k}, \tag{4}$$

where $\omega = \dot{\theta}$ will be negative if θ turns out to be decreasing. Substituting Eq. (4) into Eq. (3), we obtain

$$\dot{\vec{r}} = \dot{r} \, \hat{u}_r + \omega \hat{k} \times r\hat{u}_r = \dot{r} \, \hat{u}_r + r\omega \, \hat{u}_\theta, \tag{5}$$

where, referring to Fig. 2, we have used the fact that $\hat{k} \times \hat{u}_r = \hat{u}_\theta$ and we have simply defined $\hat{u}_\theta$ to be a unit vector perpendicular to $\hat{u}_r$ and pointing in the direction of increasing θ.[*]

Now, here is where we equate the general with the specific as mentioned in the Road Map. We *know* that the velocity of the plane is straight and horizontal, so it can be written as

$$\vec{v} = v_0 \, \hat{i}. \tag{6}$$

This *specific* relation for the velocity *must* equal the general relation for the velocity given in Eq. (5). Therefore, it must be true that

$$v_0 \, \hat{i} = \dot{r} \, \hat{u}_r + r\omega \, \hat{u}_\theta. \tag{7}$$

---

[*] We will see *a lot* more of these two unit vectors in Section 2.6.

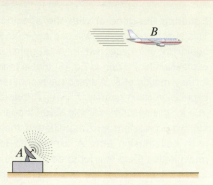

Figure 1
A radar station tracking a plane in flight.

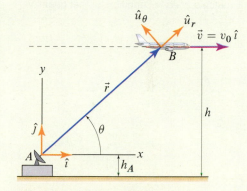

Figure 2
The given and defined dimensions and coordinate directions.

The problem with Eq. (7) is that it is written in two different component systems. For Eq. (7) to be practically useful, we need to write it using a single component system. Expressing $\hat{\imath}$ in terms of $\hat{u}_r$ and $\hat{u}_\theta$, we obtain

$$v_0(\cos\theta\,\hat{u}_r - \sin\theta\,\hat{u}_\theta) = \dot{r}\,\hat{u}_r + r\omega\,\hat{u}_\theta. \tag{8}$$

Now, we equate the coefficients of $\hat{u}_r$ and the coefficients of $\hat{u}_\theta$ to obtain the following two equations:

$$v_0\cos\theta = \dot{r}, \tag{9}$$

$$-v_0\sin\theta = r\omega. \tag{10}$$

Since we know r, θ, $\dot{r}$, and ω, Eqs. (9) and (10) are two equations for one unknown: v_0. We can solve both equations and find that

$$v_0 = \frac{\dot{r}}{\cos\theta} = -\frac{r\omega}{\sin\theta}, \tag{11}$$

where either expression would give v_0 from the radar measurements.

Discussion & Verification   Recalling that r has dimensions of length, θ is nondimensional, and ω has dimensions of 1 over time, we have that Eq. (11) is dimensionally correct since it has the dimensions of length over time, as expected. In addition, note that for the situation depicted in Fig. 2, we expect $\omega = \dot{\theta}$ to be negative, and this is consistent with Eq. (11). By solving Eq. (11) for ω, we obtain

$$\omega = -\frac{v_0\sin\theta}{r}, \tag{12}$$

which implies that $\omega < 0$ since $\sin\theta$ is positive, r is positive, and v_0 is positive because the plane is moving to the right. This result is also consistent with the right-hand rule since, for an airplane moving to the right, $\vec{r}$ rotates in the *negative z* direction.

A Closer Look   We stated that Eqs. (9) and (10) form a system of two equations in the one unknown v_0. Is there something wrong with this? The answer is no, and to understand this we need to go back and look more carefully at Eq. (5). This equation allows us to determine the velocity of the plane from radar measurements of r, θ, $\dot{r}$, and ω *no matter how the plane is moving* (see also marginal note entitled "Where is θ in Eq. (5)?"). However, in solving the problem we took advantage of the fact that the plane is flying in a specific *known* direction, and this resulted in having more information than is actually needed to solve the problem. So we have two equations to find one unknown. This situation is acceptable so long as the two equations we have do not contradict each other. Our solution is acceptable because Eqs. (9) and (10) are consistent with one another as they both imply that the plane is flying at a constant altitude.

PROBLEMS

Problem 2.103

Consider the vectors $\vec{a} = 2\,\hat{\imath} + 1\,\hat{\jmath} + 7\,\hat{k}$ and $\vec{b} = 1\,\hat{\jmath} + 2\,\hat{\jmath} + 3\,\hat{k}$. Compute the following quantities.

(a) $\vec{a} \times \vec{b}$

(b) $\vec{b} \times \vec{a}$

(c) $\vec{a} \times \vec{b} + \vec{b} \times \vec{a}$

(d) $\vec{a} \times \vec{a}$

(e) $(\vec{a} \times \vec{a}) \times \vec{b}$

(f) $\vec{a} \times (\vec{a} \times \vec{b})$

Parts (a)–(d) of this problem are meant to be a reminder that the cross product is an *anticommutative* operation while Parts (e) and (f) are meant to be a reminder that the cross product is an operation that is *not associative*.

Problem 2.104

Consider two vectors $\vec{a} = 1\,\hat{\imath} + 2\,\hat{\jmath} + 3\,\hat{k}$ and $\vec{b} = -6\,\hat{\imath} + 3\,\hat{\jmath}$.

(a) Verify that $\vec{a}$ and $\vec{b}$ are perpendicular to one another.

(b) Compute the vector triple product $\vec{a} \times (\vec{a} \times \vec{b})$.

(c) Compare the result from calculating $\vec{a} \times (\vec{a} \times \vec{b})$ with the vector $-|\vec{a}|^2\,\vec{b}$.

The purpose of this exercise is to show that as long as $\vec{a}$ and $\vec{b}$ are perpendicular to one another, you can always write $\vec{a} \times (\vec{a} \times \vec{b}) = -|\vec{a}|^2\,\vec{b}$. This identity turns out to be very useful in the study of the planar motion of rigid bodies.

Problem 2.105

Let $\vec{r}$ be the position vector of a point P with respect to a Cartesian coordinate system with axes x, y, and z. Let the motion of P be confined to the xy plane, so that $\vec{r} = r_x\,\hat{\imath} + r_y\,\hat{\jmath}$ (i.e., $\vec{r} \cdot \hat{k} = 0$). Also, let $\vec{\omega}_r = \omega_r\,\hat{k}$ be the angular velocity vector of the vector $\vec{r}$. Compute the outcome of the products $\vec{\omega}_r \times (\vec{\omega}_r \times \vec{r})$ and $\vec{\omega}_r \times (\vec{r} \times \vec{\omega}_r)$.

Problem 2.106

The three propellers shown are all rotating with the same *angular speed* of 1000 rpm about different coordinate axes.

(a) Provide the proper vector expressions for the *angular velocity* of each of the three propellers.

(b) Suppose that an identical propeller rotates at 1000 rpm about the axis ℓ oriented by the unit vector $\hat{u}_\ell$. Let any point P on ℓ have coordinates such that $x_P = y_P = z_P$. Find the vector representation of the angular velocity of this fourth propeller.

Express the answers using units of radians per second.

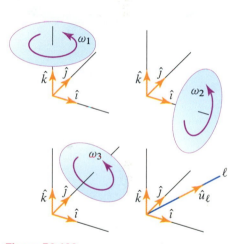

Figure P2.106

Problem 2.107

The propeller shown has a diameter of 38 ft and is rotating with a constant angular speed of 400 rpm. At a given instant, a point P on the propeller is at $\vec{r}_P = (12.5\,\hat{\imath} + 14.3\,\hat{\jmath})$ ft. Use Eqs. (2.62) and (2.64) to compute the velocity and acceleration of P, respectively.

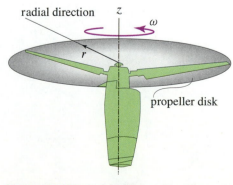

radial direction

propeller disk

Figure P2.107

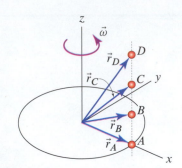

Figure P2.108

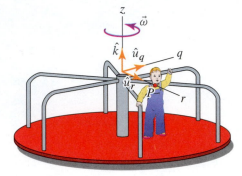

Figure P2.109

Problem 2.108

Consider the four points whose positions are given by the vectors $\vec{r}_A = \left(2\,\hat{\imath} + 0\,\hat{k}\right)$ m, $\vec{r}_B = \left(2\,\hat{\imath} + 1\,\hat{k}\right)$ m, $\vec{r}_C = \left(2\,\hat{\imath} + 2\,\hat{k}\right)$ m, and $\vec{r}_D = \left(2\,\hat{\imath} + 3\,\hat{k}\right)$ m. Knowing that the magnitude of these vectors is constant and that the angular velocity of these vectors at a given instant is $\vec{\omega} = 5\,\hat{k}$ rad/s, apply Eq. (2.62) to find the velocities $\vec{v}_A$, $\vec{v}_B$, $\vec{v}_C$, and $\vec{v}_D$. Explain why all the velocity vectors are the same even though the position vectors are not.

Problem 2.109

A child on a merry-go-round is moving radially outward at a constant rate of 4 ft/s. If the merry-go-round is spinning at 30 rpm, determine the velocity and acceleration of point P on the child when the child is 0.5 and 2.3 ft from the spin axis. Express the answers using the component system shown.

Problem 2.110

When a wheel rolls without slipping on a stationary surface, the point on the wheel that is in contact with the rolling surface has zero velocity. With this in mind, consider a nondeformable wheel rolling without slip on a flat stationary surface. The center of the wheel P is traveling to the right with a constant speed of 23 m/s. Letting $R = 0.35$ m, determine the angular velocity of the wheel, using the stationary component system shown.

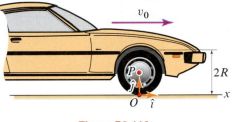

Figure P2.110

Problem 2.111

Starting with Eq. (2.62), show that the second derivative with respect to time of an arbitrary vector $\vec{A}$ is given by Eq. (2.64). Keep the answer in pure vector form, and do not resort to using components in any component system.

Problem 2.112

The radar station at O is tracking the meteor P as it moves through the atmosphere. At the instant shown, the station measures the following data for the motion of the meteor: $r = 21{,}000$ ft, $\theta = 40°$, $\dot{r} = -22{,}440$ ft/s, and $\dot{\theta} = -2.935$ rad/s. Use Eq. (2.62) to determine the magnitude and direction (relative to the xy coordinate system shown) of the velocity vector at this instant.

Problem 2.113

The radar station at O is tracking the meteor P as it moves through the atmosphere. At the instant shown, the station measures the following data for the motion of the meteor: $r = 21{,}000$ ft, $\theta = 40°$, $\dot{r} = -22{,}440$ ft/s, $\dot{\theta} = -2.935$ rad/s, $\ddot{r} = 187{,}500$ ft/s^2, and $\ddot{\theta} = -5.409$ rad/s^2. Use Eq. (2.64) to determine the magnitude and direction (relative to the xy coordinate system shown) of the acceleration vector at this instant.

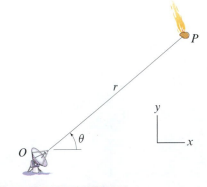

Figure P2.112 and P2.113

Problem 2.114

A plane B is approaching a runway along the trajectory shown while the radar antenna A is monitoring the distance r between A and B as well as the angle θ. If the plane has a constant approach speed v_0 as shown, use Eq. (2.62) to determine the expressions for $\dot{r}$ and $\dot{\theta}$ in terms of r, θ, v_0, and ϕ.

Problem 2.115

A plane B is approaching a runway along the trajectory shown with $\phi = 15°$, while the radar antenna A is monitoring the distance r between A and B as well as the angle θ. The plane has a constant approach speed v_0. In addition, when $\theta = 20°$, it is known that $\dot{r} = 216$ ft/s and $\dot{\theta} = -0.022$ rad/s. Use Eq. (2.62) to determine the corresponding values of v_0 and of the distance between the plane and the radar antenna.

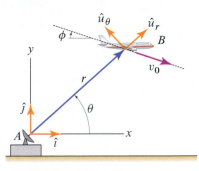

Figure P2.114 and P2.115

Problem 2.116

The end B of a robot arm is being extended with the constant rate $\dot{r} = 4$ ft/s. Knowing that $\dot{\theta} = 0.4$ rad/s and is constant, use Eqs. (2.62) and (2.64) to determine the velocity and acceleration of B when $r = 2$ ft. Express your answer using the component system shown.

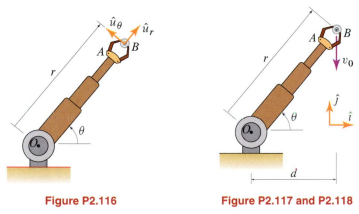

Figure P2.116   **Figure P2.117 and P2.118**

Problem 2.117

The end B of a robot arm is moving vertically down with a constant speed $v_0 = 2$ m/s. Letting $d = 1.5$ m, apply Eq. (2.62) to determine the rate at which r and θ are changing when $\theta = 37°$.

Problem 2.118

The end B of a robot arm is moving vertically down with a constant speed $v_0 = 6$ ft/s. Letting $d = 4$ ft, use Eqs. (2.62) and (2.64) to determine $\dot{r}$, $\dot{\theta}$, $\ddot{r}$, and $\ddot{\theta}$ when $\theta = 0°$.

Problem 2.119

A micro spiral pump consists of a spiral channel attached to a stationary plate. This plate has two ports, one for fluid inlet and the other for outlet, the outlet being farther from the center of the plate than the inlet. The system is capped by a rotating disk. The fluid trapped between the rotating disk and stationary plate is put in motion by the rotation of the top disk, which pulls the fluid through the spiral channel. With this in mind, consider a channel with geometry given by the equation $r = \eta\theta + r_0$, where $\eta = 12\,\mu$m

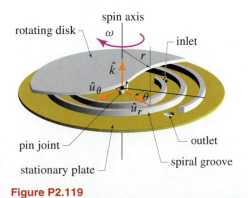

Figure P2.119

is called the polar slope, $r_0 = 146\,\mu m$ is the radius at the inlet, r is the distance from the spin axis, and θ is the angular position of a point in the spiral channel. If the top disk rotates with a constant angular speed $\omega = 30{,}000$ rpm, and assuming that the fluid particles in contact with the rotating disk are essentially stuck to it, determine the velocity and acceleration of one of such fluid particles when it is at $r = 170\,\mu m.$[*] Express the answer using the component system shown (which rotates with the top disk).

Problem 2.120

A disk rotates about its center, which is the fixed point O. The disk has a straight channel whose centerline passes by O and within which a collar A is allowed to slide. If, when A passes by O, the speed of A relative to the channel is $v = 14$ m/s and is increasing in the direction shown with a rate of 5 m/s^2, determine the acceleration of A given that $\omega = 4$ rad/s and is constant. Express the answer using the component system shown, which rotates with the disk. *Hint:* Apply Eq. (2.64) to the vector describing the position of A relative to O and then let $r = 0$.

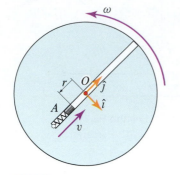

Figure P2.120

Problem 2.121

At the instant shown, the angular velocity and acceleration of the merry-go-round are as indicated in the figure. Assuming that the child is walking along a radial line, should the child walk outward or inward to make sure that he does not experience any sideways acceleration (i.e., in the direction of $\hat{u}_q$)?
Note: Concept problems are about *explanations*, not computations.

Problem 2.122

Assuming that the child shown is moving on the merry-go-round along a radial line, use Eq. (2.64) to determine the relation that ω, $\dot{\omega}$, r, and $\dot{r}$ must satisfy so that the child will not experience any sideways acceleration.

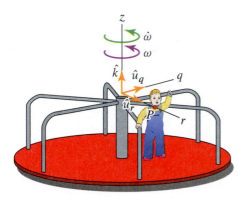

Figure P2.121 and P2.122

Problem 2.123

The mechanism shown is called a *swinging block* slider crank. First used in various steam locomotive engines in the 1800s, this mechanism is often found in door-closing systems. If the disk is rotating with a constant angular velocity $\dot{\theta} = 60$ rpm, $H = 4$ ft, $R = 1.5$ ft, and r is the distance between B and O, compute $\dot{r}$ and $\dot{\phi}$ when $\theta = 90°$. *Hint:* Apply Eq. (2.62) to the vector describing the position of B relative to O.

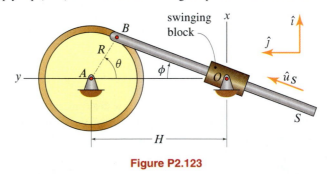

Figure P2.123

---

[*] The spiral pump was originally invented in 1746 by H. A. Wirtz, a Swiss pewterer from Zurich. Recently, the spiral pump concept has seen a comeback in microdevice design. The data used in this example is taken from M. I. Kilani, P. C. Galambos, Y. S. Haik, and C.-J. Chen, "Design and Analysis of a Surface Micromachined Spiral-Channel Viscous Pump," *Journal of Fluids Engineering*, **125**, pp. 339–344, 2003.

Problems 2.124 and 2.125

A sprinkler essentially consists of a pipe AB mounted on a hollow shaft. The water comes in the pipe at O and goes out the nozzles at A and B, causing the pipe to rotate. Assume that the particles of water move through the pipe at a constant rate *relative to the pipe* of 5 ft/s and that the pipe AB is rotating at a constant angular velocity of 250 rpm. In all cases, express the answers using the right-handed and orthonormal component system shown.

Problem 2.124 Determine the acceleration of the water particles when they are at $d/2$ from O (still within the horizontal portion of the pipe). Let $d = 7$ in.

Problem 2.125 Determine the acceleration of the water particles right before they are expelled at B. Let $d = 7$ in., $\beta = 15°$, and $L = 2$ in. *Hint:* In this case the vector describing the position of a water particle at B goes from O to B and is best written as $\vec{r} = r_B \hat{u}_B + r_z \hat{k}$.

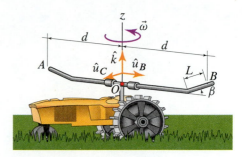

Figure P2.124 and P2.125

2.5 Planar Motion: Normal-Tangential Components

It is not always convenient to study motion using a Cartesian coordinate system. In this section and the next, we will learn about two additional ways to describe motion. Here we introduce a way to describe the velocity and acceleration vectors of a particle that is based *entirely* on the path of the particle.

Car racing and the *shape* of the track

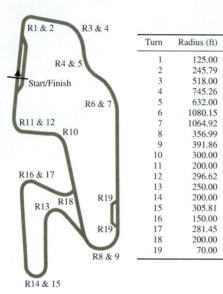

Turn	Radius (ft)
1	125.00
2	245.79
3	518.00
4	745.26
5	632.00
6	1080.15
7	1064.92
8	356.99
9	391.86
10	300.00
11	200.00
12	296.62
13	250.00
14	200.00
15	305.81
16	150.00
17	281.45
18	200.00
19	70.00

Figure 2.32
Turn radii of curvature for the Watkins Glen International racing circuit.

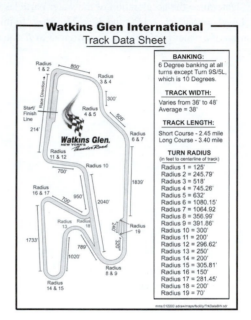

Figure 2.33. Elevation map for the Watkins Glen International racing circuit.

Figures 2.32 and 2.33 depict the geometry of the Watkins Glen International racing track, which is located in southern New York State and hosts road races for many different classes, including NASCAR and SCCA events. Figure 2.34 shows the circuit for the Monaco Grand Prix, a Formula 1 car race hosted in Monte Carlo (Principality of Monaco). These two tracks have different *shapes*. In fact, essentially all race car circuits have different shapes. Hence, we may ask the following question:

> What is it about a particular track *shape* that tests the ability of a driver and the engineering of a car?

To answer this question, we need to find a way to isolate the "dynamic character" that the track *shape* gives to a race. By *dynamic character*, we mean performance parameters such as the maximum speed or the maximum acceleration that a car can successfully achieve on a given track. We need to learn how to *describe motion* in such a way that pieces of information such as "the shape of the motion" are easily related to other pieces of information such as velocity or acceleration. To fulfill this task, we need to acquire some basic notions from the geometry of curves, and we will relate them to velocity and acceleration.

Figure 2.34
Monaco Grand Prix Formula 1 circuit.

Geometry of curves

To relate a point's velocity and acceleration to the shape of its trajectory, we need a few ideas from geometry.

Arc length. In Fig. 2.35 we see a point P traveling along a path. Instead of viewing the position $\vec{r}$ of P directly as a function of time, in geometry we prefer to view $\vec{r}$ as

$$\vec{r} = \vec{r}(s(t)), \qquad (2.65)$$

where $s(t)$ is the *distance traveled by P along its path from $t = 0$ to the current time t*. The variable s is called *arc length*. Since s is the *distance traveled*, $s \geq 0$ and, no matter which direction P moves along the path, s continues to increase.

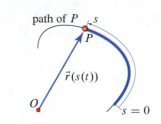

Figure 2.35
A point P moving along a path. Point O is a fixed reference point.

Unit tangent vector to the path. Referring to Fig. 2.36, since s measures distance traveled, for increasing s, as $\Delta s \to 0$, the vector $\Delta \vec{r} = \vec{r}(s + \Delta s) - \vec{r}(s)$ becomes tangent to the path while pointing in the direction of motion. In addition, as $\Delta s \to 0$, $\Delta s \approx |\Delta \vec{r}|$ and $|\Delta \vec{r}/\Delta s| \to 1$. Geometry then tells us that the vector $d\vec{r}(s)/ds$ has the following three properties: (1) it is *tangent* to the path, (2) it points in the direction of motion, and (3) $|d\vec{r}/ds| = 1$. Because of these properties, we write

$$\hat{u}_t(s) = \frac{d\vec{r}(s)}{ds} \quad \text{with} \quad |\hat{u}_t(s)| = 1, \qquad (2.66)$$

where the subscript t stands for *tangent*. The unit vector $\hat{u}_t(s)$ is called the *unit tangent vector* to the path and *always points in the direction of motion*.

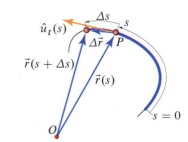

Figure 2.36
Definition of unit tangent vector.

Curvature. Unless the path is straight, the unit vector $\hat{u}_t(s)$ changes orientation along the path. In geometry, the rate of change of $\hat{u}_t(s)$ with respect to s, i.e., the vector $d\hat{u}_t(s)/ds$, is used to describe the curvature of the path. The *curvature* of the path, traditionally denoted by the Greek letter κ (kappa), is defined as

$$\kappa(s) = \left| \frac{d\hat{u}_t(s)}{ds} \right|. \qquad (2.67)$$

Notice that $\kappa(s)$ has dimensions of 1 over length. Also, if a line is straight, $\hat{u}_t(s) = $ constant and $\kappa(s) = 0$.

Principal unit normal to the path. Since $\hat{u}_t(s)$ is a *unit* vector for any value of s, then $\hat{u}_t(s) \cdot \hat{u}_t(s) = 1 = $ constant, and

$$\frac{d}{ds}(\hat{u}_t \cdot \hat{u}_t) = \frac{d\hat{u}_t}{ds} \cdot \hat{u}_t + \hat{u}_t \cdot \frac{d\hat{u}_t}{ds} = 2\hat{u}_t \cdot \frac{d\hat{u}_t}{ds} = 0, \qquad (2.68)$$

i.e., $\hat{u}_t$ and $d\hat{u}_t/ds$ are orthogonal. Therefore, when $\kappa \neq 0$, we can define a curvature-related unit normal to the path as follows:

$$\hat{u}_n(s) = \frac{d\hat{u}_t(s)/ds}{|d\hat{u}_t(s)/ds|} = \frac{1}{\kappa(s)} \frac{d\hat{u}_t(s)}{ds}, \qquad (2.69)$$

where the subscript n stands for *normal*. Referring to Fig. 2.37, the unit vector $\hat{u}_n(s)$ is called the *principal unit normal* to the path, and it *always points toward the concave side of the curve*. Because $\hat{u}_n(s)$ is only defined when $\kappa \neq 0$, we cannot use $\hat{u}_n(s)$ when dealing with straight lines or at inflection points of curved lines (i.e., where a line is locally straight).

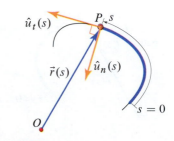

Figure 2.37
Principal unit normal to the path.

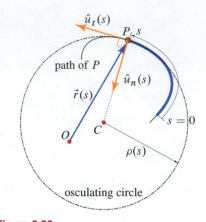

Figure 2.38
Osculating circle and radius of curvature.

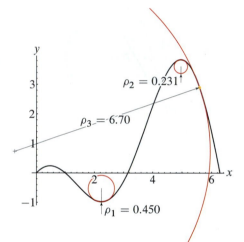

Figure 2.39
A 2D path showing its radius of curvature at three different points.

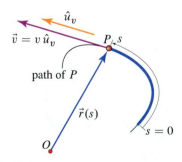

Figure 2.40
A point P moving along its path. Point O is a fixed reference point.

Osculating circle and radius of curvature.   The unit vectors $\hat{u}_t(s)$ and $\hat{u}_n(s)$ define a special tangent plane to the path called the *osculating plane*.[*] If the path is planar, then $\hat{u}_t(s)$ and $\hat{u}_n(s)$ lie in the plane of motion, which therefore coincides with the osculating plane. The direction perpendicular to the osculating plane is identified via the unit vector

$$\hat{u}_b(s) = \hat{u}_t(s) \times \hat{u}_n(s), \qquad (2.70)$$

which is called the *binormal* unit vector because it is normal to both $\hat{u}_t(s)$ and $\hat{u}_n(s)$. Referring to Fig. 2.38, if $\hat{u}_n(s)$ is defined at a point P, a very important result from geometry is that the osculating plane contains a circle called the *osculating circle* with the following three properties: (1) the circle is tangent to the path at P, (2) $\hat{u}_n(s)$ points toward the center C of the circle, and, most importantly, (3) the radius ρ of the circle is given by

$$\rho(s) = \frac{1}{\kappa(s)}. \qquad (2.71)$$

The radius $\rho(s)$ is called the *radius of curvature* of the path. In general, $\rho(s)$ changes along the path. If $\rho(s)$ is constant, then the path is a circle.

Radius of curvature in Cartesian coordinates.   In Cartesian coordinates, the planar path of a particle is usually given the form $y = y(x)$. In this case, geometry tells us that the radius of curvature at any position x is given by

$$\rho(x) = \frac{\left[1 + (dy/dx)^2\right]^{3/2}}{|d^2y/dx^2|}. \qquad (2.72)$$

As an example, Fig. 2.39 shows a plot of the path $y(x) = (1 - x)\sin x$ for $0 \le x \le 2\pi$. We have used Eq. (2.72) to compute the radii of curvature of $y(x)$ at three different points. The first point, the local minimum at $x = 2.24$, has $\rho_1 = 0.450$. The second, the local maximum at $x = 4.96$, has $\rho_2 = 0.231$. Finally, at $x = 5.60$, $\rho_3 = 6.70$. Figure 2.39 demonstrates the intuitive notion that the "tighter the bend" in the path, the smaller its radius of curvature and the "straighter the path," the larger its radius of curvature.

Normal-tangential components

We are now ready to tackle the question posed in the introduction about the relation between the shape of the path of a particle and the particle's velocity and acceleration.

Referring to Fig. 2.40, consider a point P moving along a path. Using the ideas discussed in Section 2.4, we can express the velocity of P as

$$\vec{v} = v\,\hat{u}_v, \qquad (2.73)$$

that is, as the product of its magnitude, namely, the speed v, and the unit vector $\hat{u}_v$ in the direction of $\vec{v}$. Using the geometrical concepts introduced earlier in this section, we can also write

$$\vec{v} = \frac{d\vec{r}(s(t))}{dt} = \frac{d\vec{r}}{ds}\frac{ds}{dt} = \dot{s}(t)\,\hat{u}_t(s(t)), \qquad (2.74)$$

---

[*]The adjective *osculating* originates from one of the Latin words for kiss. The osculating plane was so named to indicate that, of all the planes tangent to the path, it is the plane from which the path separates most gradually.

where in the last equality we have used Eq. (2.66). Since the velocity is always tangent to the path and, by definition, points in the direction of motion, then the unit vectors $\hat{u}_v$ and $\hat{u}_t$ must be equal to each other and (see Fig. 2.41)

$$\boxed{\vec{v} = v\,\hat{u}_t \quad \text{and} \quad v = \dot{s}.} \tag{2.75}$$

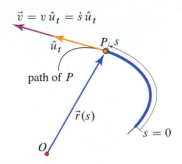

Figure 2.41
Velocity in normal-tangential components.

To obtain the acceleration of P, we differentiate the first of Eqs. (2.75) with respect to time, which yields

$$\vec{a} = \dot{\vec{v}} = \dot{v}\,\hat{u}_t + v\,\dot{\hat{u}}_t = \dot{v}\,\hat{u}_t + v\,\frac{d\hat{u}_t}{ds}\frac{ds}{dt}, \tag{2.76}$$

where in the last equality we have used the chain rule to account for the dependence of $\hat{u}_t$ on the arc length s. If at P the curvature $\kappa \neq 0$, then the vector $d\hat{u}_t/ds$ in Eq. (2.76) can be rewritten using Eqs. (2.69) and (2.71) as

$$\frac{d\hat{u}_t}{ds} = \frac{1}{\rho}\,\hat{u}_n. \tag{2.77}$$

Finally, since by the second of Eqs. (2.75) $\dot{s} = v$, and substituting Eq. (2.77) into Eq. (2.76), we have (see Fig. 2.42)

$$\boxed{\vec{a} = \dot{v}\,\hat{u}_t + \frac{v^2}{\rho}\,\hat{u}_n = a_t\,\hat{u}_t + a_n\,\hat{u}_n,} \tag{2.78}$$

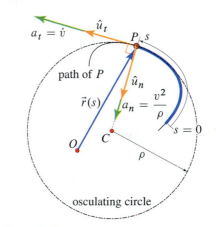

Figure 2.42
Acceleration in normal-tangential components.

where $a_t = \dot{v}$ and $a_n = v^2/\rho$ are the tangential and normal components of the acceleration, respectively.

Equations (2.75) and (2.78) tell us that if we represent the velocity and acceleration of a particle using the normal-tangential component system (sometimes called a path component system), i.e., a component system consisting of the unit vectors $\hat{u}_n$ and $\hat{u}_t$, then (1) the velocity has only one component, namely, v along the tangent direction and (2) the acceleration has only two components, one in the tangent direction and the other in the normal direction. The normal component of the acceleration is directly proportional to the square of the speed and is inversely proportional to the radius of curvature of the path. This result is precisely what we were looking for because it answers the question posed at the beginning of the section. Finally, since the derivation of Eqs. (2.75) and (2.78) did not require us to assume the motion to be planar, these relations are valid even in a full three-dimensional context.

Connection with the time derivative of a vector. We conclude our discussion by going back to Eq. (2.76) to obtain $\vec{a}$, using the ideas from Section 2.4. Doing so, i.e., applying Eq. (2.62) on p. 92, gives

$$\vec{a} = \dot{\vec{v}} = \dot{v}\,\hat{u}_t + \vec{\omega}_v \times \vec{v} = \dot{v}\,\hat{u}_t + v\vec{\omega}_v \times \hat{u}_t. \tag{2.79}$$

Comparing Eq. (2.79) with Eq. (2.78), we have

$$v\vec{\omega}_v \times \hat{u}_t = \frac{v^2}{\rho}\,\hat{u}_n. \tag{2.80}$$

Canceling v and noting from Eq. (2.70) that $\hat{u}_n = \hat{u}_b \times \hat{u}_t$, we obtain (see the Interesting Fact in the margin)

$$\vec{\omega}_v \times \hat{u}_t = \frac{v}{\rho}\,\hat{u}_b \times \hat{u}_t \quad \Rightarrow \quad \vec{\omega}_v = \frac{v}{\rho}\,\hat{u}_b, \tag{2.81}$$

that is, the rate of rotation of the velocity vector is directly proportional to the speed of the particle as well as the curvature (i.e., $1/\rho$) of the path.

Interesting Fact

The angular velocity of $\vec{v}$. In Eqs. (2.81), our conclusion about $\vec{\omega}_v$ is not entirely correct. If $\vec{\omega}_v = \frac{v}{\rho}\hat{u}_b + \beta\,\hat{u}_t$, where β is an arbitrary scalar, the first of Eqs. (2.81) is still satisfied since $\hat{u}_t \times \hat{u}_t = \vec{0}$. However, the term $\beta\,\hat{u}_t$ simply causes the vector $\vec{v}$ to rotate about itself. If the velocity vector is spinning about its own axis, then it is not changing as a vector, that is, it is not changing in direction or in magnitude. Therefore, whether we write $\vec{\omega}_v$ as $\frac{v}{\rho}\hat{u}_b$ or as $\frac{v}{\rho}\hat{u}_b + \beta\,\hat{u}_t$, we will always get the same result. What we have written in Eqs. (2.81) is the expression for $\vec{\omega}_v$ that actually contributes to $\vec{a}$.

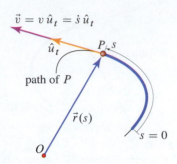

Figure 2.43
Figure 2.41 repeated. Representation of the velocity in normal-tangential component system.

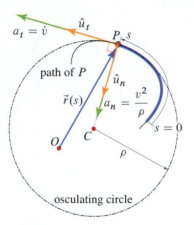

Figure 2.44
Figure 2.42 repeated. Acceleration in normal-tangential components.

End of Section Summary

In this section we have derived expressions for the velocity and acceleration of a point using the normal-tangential component system. Referring to Fig. 2.43, the velocity vector has the form

Eq. (2.75), p. 107

$$\vec{v} = v\,\hat{u}_t = \dot{s}\,\hat{u}_t,$$

where v is the speed, s is the arc length along the path, and $\hat{u}_t$ is the tangent unit vector at the point P.

Referring to Fig. 2.44, the acceleration vector in normal-tangential components has the form

Eq. (2.78), p. 107

$$\vec{a} = \dot{v}\,\hat{u}_t + \frac{v^2}{\rho}\,\hat{u}_n = a_t\,\hat{u}_t + a_n\,\hat{u}_n,$$

where $\dot{v} = a_t$ is the tangential component of acceleration and $v^2/\rho = a_n$ is the normal component of acceleration.

When we are using a Cartesian coordinate system, for a path expressed via a relation such as $y = y(x)$, the path's radius of curvature is given by ρ:

Eq. (2.72), p. 106

$$\rho(x) = \frac{\left[1 + (dy/dx)^2\right]^{3/2}}{\left|d^2y/dx^2\right|}.$$

EXAMPLE 2.17   *Relating the Shape of the Path to Acceleration*

Let's assume that by *lateral G-force* the Federation Internationale de l'Automobile (FiA) that compiled the map in Fig. 1 really meant to provide a measurement of the acceleration normal to the path of the racing cars expressed in "units of g," where g is the acceleration due to gravity. Use this information along with the reported speed to estimate the radius of curvature of the Monaco Formula 1 track (1) at the stretch preceding the Rascasse and (2) at the end of the Tunnel.

SOLUTION

Road Map   If we assume that a Formula 1 car can be modeled as a particle, then the solution to this problem is obtained by using the relation linking the speed to the component of the acceleration normal to the path, that is, $a_n = v^2/\rho$.

Computation   As indicated in the problem statement, if one then assumes that the lateral G-force is the component of the acceleration normal to the path, then at the end of the stretch preceding the Rascasse (see Fig. 2) we have

$$v = 141\,\text{km/h} = 39.17\,\text{m/s} \quad \text{and} \quad a_n = 1.5g = 14.72\,\text{m/s}^2, \tag{1}$$

so that

$$\boxed{\rho = \frac{v^2}{a_n} = 104\,\text{m.}} \tag{2}$$

Proceding in a similar way for the Tunnel (see Fig. 3), we have

$$v = 264\,\text{km/h} = 73.33\,\text{m/s} \quad \text{and} \quad a_n = 2.6g = 25.50\,\text{m/s}^2, \tag{3}$$

so that

$$\boxed{\rho = \frac{v^2}{a_n} = 211\,\text{m.}} \tag{4}$$

Discussion & Verification   The solution was obtained as a direct application of the formula for the normal component of the acceleration and is therefore elementary to verify that the dimensions are correct and proper units have been used.

🔎 **A Closer Look**   A basic question to ask is, How good are these estimates? After some research, the authors were able to obtain a map of the circuit from which the radii of curvature of most turns could be directly measured, although without taking into account changes in elevation along the curves. From this map, the radius of curvature in the first calculation was measured to be roughly 90 m, whereas the radius of curvature midway through the tunnel was found to be roughly 200 m. Hence, we can conclude that the gross estimates we have derived from the map in Fig. 1 and those that were obtained by direct (although still approximate) measurement are in reasonable agreement.

Figure 1
Formula 1 track at Monaco. The locations of the Tunnel and the Rascasse are indicated in red. Additional information is provided including typical speed, acceleration, and gear information at various important points along the track.

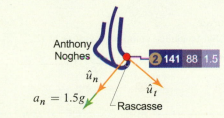

Figure 2
Normal tangential component system right before the Rascasse. The normal component of the acceleration is also shown. Refer to Fig. 1 for the key to the numbers shown in the blue boxes.

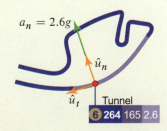

Figure 3
Normal tangential component system at the end of the Tunnel. The normal component of the acceleration is also shown. Refer to Fig. 1 for the key to the numbers shown in the blue boxes.

EXAMPLE 2.18 *Curvature and Projectile Motion*

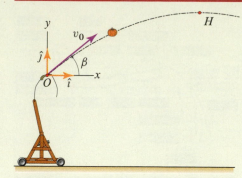

Figure 1
A pumpkin launched by a trebuchet.

A pumpkin has been launched from a trebuchet at the annual World Championship "Punkin Chunkin" competition in Millsboro, Delaware. The pumpkin, shown in Fig. 1, is assumed to have been released at O with an initial speed v_0 and an elevation angle β. Determine the time rate of change of the speed at the time of release and the radius of curvature of the pumpkin's trajectory at its highest point.

SOLUTION

Road Map By modeling the pumpkin's motion as a projectile motion, the pumpkin's acceleration is the acceleration of gravity g in the negative y direction. The key to the solution is then to compute the tangential and normal components of the acceleration since the rate of change of the speed is $\dot{v} = a_t$ and the radius of curvature ρ is such that $a_n = v^2/\rho$.

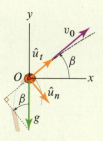

Figure 2
Normal-tangential component system at O.

Computation Referring to Fig. 2, recall that the velocity is always tangent to the path. Therefore, the tangent to the path at O is oriented at an angle β with respect to the x axis. Therefore we have

$$\dot{v}_O = a_{Ot} = -g \sin \beta, \tag{1}$$

where v_O and a_{Ot} denote the speed at O and the tangential component of the acceleration at O, respectively.

Considering the situation at the highest point on the trajectory, referring to Fig. 3, observe that the tangent to the trajectory at this point is completely horizontal. Therefore, calling H the highest point on the trajectory, since the pumpkin's acceleration is completely in the y direction, we have

$$a_{Ht} = 0 \quad \text{and} \quad a_{Hn} = g. \tag{2}$$

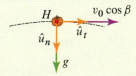

Figure 3
Normal-tangential component system at H.

Furthermore, the velocity at O is completely in the x direction. From projectile motion, we know that the horizontal component of the acceleration remains constant throughout the motion. Therefore we have

$$\vec{v}_O = v_{Ox}\,\hat{i} = v_0 \cos \beta\,\hat{i} \quad \Rightarrow \quad v_O = v_0 \cos \beta. \tag{3}$$

Using the normal component of Eq. (2.78), we have $a_n = v^2/\rho$. Therefore, using Eqs. (2) and (3), we have

$$\rho_H = \frac{v_H^2}{a_{Hn}} = \frac{v_0^2 \cos^2 \beta}{g}. \tag{4}$$

Discussion & Verification Both the results in Eqs. (1) and (4) are dimensionally correct. Referring to Eq. (1), observe that $\sin \beta$ is nondimensional and therefore $\dot{v}_O$ has the same dimensions of g, i.e., the dimensions of acceleration, as expected. As far as Eq. (4) is concerned, recall that the dimensions of v_0 are length over time and those of g are length over time squared. Therefore, ρ has dimensions of length, as expected. From Eq. (3) observe that $\dot{v}_O$ is negative. This is to be expected since, up until the projectile reaches point H, the speed of the projectile is expected to decrease. Finally, from Eq. (4), observe that ρ increases with the horizontal component of the initial velocity. This is to be expected since the "flatness" of the overall trajectory is governed by the horizontal component of the velocity. Hence, the solution we have obtained appears to be correct.

EXAMPLE 2.19   *Accelerations in Circular Motion*

The Center for Gravitational Biology Research at NASA's Ames Research Center runs a large centrifuge capable of simulating $20g$ of acceleration ($12.5g$ is the maximum for human subjects). The radius of the centrifuge is 29 ft, and the distance from the axis of rotation to the cab at either A or B is about 25 ft. Assuming that the centrifuge accelerates uniformly to reach its final speed and that it takes 12.5 s to do so, determine

(a) The angular velocity of the centrifuge ω_f required to maintain a final acceleration of $20g$ in cab A.

(b) The magnitude of the acceleration of cab A as a function of time from the moment the centrifuge starts spinning until its final speed is achieved.

Figure 1
The $20g$ centrifuge at the Center for Gravitational Biology Research, which is part of NASA's Ames Research Center in Moffett Field, California.

SOLUTION

Road Map   Since we know the magnitude of the final acceleration of A ($20g$), we can use the acceleration relationships developed in this section to determine the final value of the angular velocity of the centrifuge. Because the centrifuge accelerates uniformly during spin-up, we can use the constant acceleration relations from Section 2.2 to relate the centrifuge's final angular velocity to the needed angular acceleration during spin-up. Once the centrifuge's acceleration is known, we will be able to derive an expression for the acceleration of A as a function of time during spin-up.

Computation   The key element of the solution of this problem is the relation between the motion of A and the motion of the centrifuge as a whole. To establish this relation, consider Fig. 2 and observe that the path of A is a circle with center C on the spin axis of the centrifuge. Since $\hat{u}_n$ remains pointing toward C, its angular velocity is the angular velocity of the centrifuge, namely, $\omega\,\hat{u}_b$. Observing that $\hat{u}_t$ must remain perpendicular to $\hat{u}_n$, we see $\hat{u}_t$ must also rotate with angular velocity $\omega\,\hat{u}_b$. This angular velocity is the angular velocity $\vec{\omega}_v$ that appears in Eqs. (2.79)–(2.81). Consequently, using Eq. (2.81), whether during spin-up on not, the angular velocity of the centrifuge and the motion of A are related via the relation

$$\omega = \frac{v}{\rho}, \tag{1}$$

where v is the speed of A and ρ is the radius of the path of A.

When the magnitude of the acceleration of A reaches its final value $a_f = 20g$, the speed of A will become constant, i.e., $\dot{v}_f = 0$, and a_f will coincide with *just* the normal component of Eq. (2.78) so that

$$a_f = \frac{v_f^2}{\rho} = \rho\omega_f^2 \quad \Rightarrow \quad 20g = 20(32.2\,\text{ft/s}^2) = \rho\omega_f^2, \tag{2}$$

where we have used Eq. (1), ω_f is the final value of ω, and we have used $g = 32.2\,\text{ft/s}^2$. Since $\rho = 25$ ft, solving for ω_f, we obtain

$$\boxed{\omega_f = 5.075\,\text{rad/s} = 48.5\,\text{rpm}.} \tag{3}$$

We can now use Eq. (2.41), along with Table 2.2, to obtain the angular acceleration of the centrifuge during spin-up as

$$\omega_f = \omega_0 + \alpha t_f \quad \Rightarrow \quad 5.075\,\text{rad/s} = \alpha(12.5\,\text{s}), \quad \Rightarrow \quad \alpha = 0.406\,\text{rad/s}^2, \tag{4}$$

where we have used $\omega_0 = 0$.

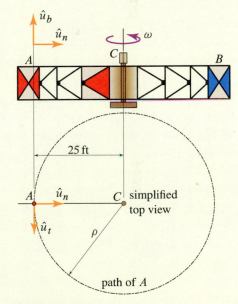

Figure 2
The $20g$ centrifuge showing the axis of rotation and the path of A as well as the normal-tangential component system at A.

To relate α to the acceleration of A, we rewrite Eq. (1) as $v = \rho\omega$, and differentiating with respect to time, we obtain

$$\dot{v} = \rho\dot{\omega} = \rho\alpha, \tag{5}$$

where we have used the fact that ρ is constant. The quantity $\dot{v}$ is the tangential component of the acceleration of A. The normal component of the acceleration of A is given by

$$a_n = \frac{v^2}{\rho} = \rho\omega^2 = \rho\alpha^2 t^2, \tag{6}$$

where we have used $\omega = \alpha t$ from Eq. (4), which is the value of ω at an arbitrary time t during spin-up. Therefore, the magnitude of the acceleration during spin-up is given by

$$a = \sqrt{a_t^2 + a_n^2} = \sqrt{\rho^2\alpha^2 + \rho^2\alpha^4 t^4} = \rho\alpha\sqrt{1 + \alpha^2 t^4}. \tag{7}$$

Recalling that $\rho = 25$ ft and using the result in Eq. (4), we have

$$\boxed{a = (10.15\,\text{ft/s}^2)\sqrt{1 + (0.1648\,\text{s}^{-4})t^4}.} \tag{8}$$

Discussion & Verification   Equations (2) and (7) present our results in symbolic form and allow us to easily verify that our results are dimensionally correct. In addition, the numerical form of the results presented in Eqs. (3) and (8) is expressed using appropriate and consistent units. Overall our solution indicates that the speed of A increases uniformly, and this is consistent with the given piece of information that the spin-up of the centrifuge occurs at a constant rate. Hence, our solution appears to be correct.

A Closer Look   The magnitude of the acceleration of A as given in Eq. (8) is an increasing function of time. Hence, a is largest at $t = 12.5$ s (the end of the spin-up) and its value is given by

$$a_{\max} = 644\,\text{ft/s}^2 = 20.0g, \tag{9}$$

which appears to be the same as the value of a_f. The result in Eq. (9) is somewhat unexpected because the value of a_f is based only on the normal acceleration of A at the end of spin-up whereas $a_{\max}$ includes the contributions of both the normal and tangential components of the acceleration of A. This indicates that the tangential component of acceleration contributes an insignificant amount to the total acceleration. When Eq. (7) is evaluated at the end of spin-up ($t = 12.5$ s), $a_t = 10.15\,\text{ft/s}^2$ (see the leading term on the right-hand side of Eq. (8)) and $a_n = 643.9\,\text{ft/s}^2$ so we can see that a_n is over 63 times larger than a_t.

We conclude by observing that our solution assumes that the angular acceleration is constant until $t = 12.5$ s, at which time it becomes zero so that the angular velocity becomes constant. This assumption is not entirely realistic since such abrupt changes in acceleration are not typically found in applications.

PROBLEMS

Problem 2.126

A particle P is moving along a path with the velocity shown. Is the sketch of the normal-tangential component system at P correct?
Note: Concept problems are about *explanations*, not computations.

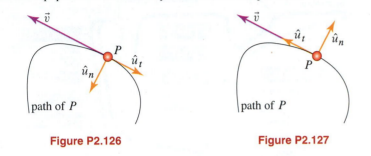

Figure P2.126   **Figure P2.127**

Problem 2.127

A particle P is moving along a path with the velocity shown. Is the sketch of the normal-tangential component system at P correct?
Note: Concept problems are about *explanations*, not computations.

Problem 2.128

A particle P is moving along a straight line with the velocity and acceleration shown. What is wrong with the unit vectors shown in the figure?
Note: Concept problems are about *explanations*, not computations.

Figure P2.128   **Figure P2.129**

Problem 2.129

A particle P is moving along some path with the velocity and acceleration shown. Can the path of P be the straight line shown?
Note: Concept problems are about *explanations*, not computations.

Problem 2.130

A particle P is moving along the curve C, whose equation is given by

$$(y^2 - x^2)(x - 1)(2x - 3) = 4(x^2 + y^2 - 2x)^2,$$

at a *constant* speed v_c. For any position on the curve C for which the radius of curvature is defined (i.e., *not* equal to infinity), what *must* be the angle ϕ between the velocity vector $\vec{v}$ and the acceleration vector $\vec{a}$?
Note: Concept problems are about *explanations*, not computations.

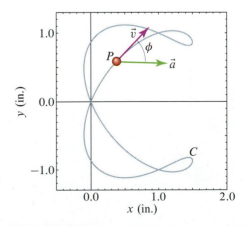

Figure P2.130

Problem 2.131

Making the same assumptions stated in Example 2.17, consider the map of the Formula 1 circuit at Hockenheim in Germany and estimate the radius of curvature of the curves Südkurve and Nordkurve (at the locations indicated in red).

Figure P2.131

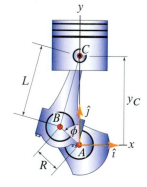

Figure P2.132

Figure P2.133

Problem 2.132

The motion of the piston C as a function of the crank angle ϕ and the lengths of the crank AB and connecting rod BC is given by $y_C = R \cos \phi + L \sqrt{1 - (R \sin \phi / L)^2}$ and $x_C = 0$. Using the component system shown, express $\hat{u}_t$, the unit vector tangent to the trajectory of C, as a function of the crank angle ϕ for $0 \le \phi \le 2\pi$ rad.
Note: Concept problems are about *explanations*, not computations.

Problem 2.133

An aerobatics plane initiates the basic loop maneuver such that, at the bottom of the loop, the plane is going 140 mph, while subjecting the plane to approximately $4g$ of acceleration. Estimate the corresponding radius of the loop.

Problem 2.134

Suppose that a highway exit ramp is designed to be a circular segment of radius $\rho = 130$ ft. A car begins to exit the highway at A while traveling at a speed of 65 mph and goes by point B with a speed of 25 mph. Compute the acceleration vector of the car as a function of the arc length s, assuming that the tangential component of the acceleration is constant between points A and B.

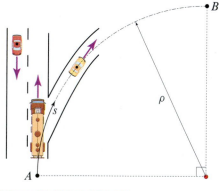

Figure P2.134 and P2.135

Problem 2.135

Suppose that a highway exit ramp is designed to be a circular segment of radius $\rho = 130$ ft. A car begins to exit the highway at A while traveling at a speed of 65 mph and goes by point B with a speed of 25 mph. Compute the acceleration vector of the car along the car's path as a function of the arc length s, assuming that between A and B the speed was controlled so as to maintain constant the rate dv/ds.

Problem 2.136

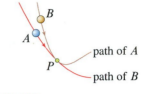

Particles A and B are moving in the plane with the same constant speed v, and their paths are tangent at P. Do these particles have zero acceleration at P? If not, do these particles have the same acceleration at P?

Note: Concept problems are about *explanations*, not computations.

Figure P2.136

Problem 2.137

Uranium is used in light water reactors to produce a controlled nuclear reaction for the generation of power. When first mined, uranium comes out as the oxide U_3O_8, which is 0.7% of the isotope U-235 and 99.3% of the isotope U-238.[*] To be used in a nuclear reactor, the concentration of U-235 must be in the 3–5% range.[†] The process of increasing the percentage of U-235 is called *enrichment*, and it is done in a number of ways. One method uses centrifuges, which spin at very high rates to create artificial gravity. In these centrifuges, the heavy U-238 atoms concentrate on the outside of the cylinder (where the acceleration is largest), and the lighter U-235 atoms concentrate near the spin axis. Before centrifuging, the uranium is processed into gaseous uranium hexafluoride or UF_6, which is then injected into the centrifuge. Assuming that the radius of the centrifuge is 20 cm and that it spins at 70,000 rpm, determine

(a) The velocity of the outer surface of the centrifuge.

(b) The acceleration in g experienced by an atom of uranium that is on the inside of the outer wall of the centrifuge.

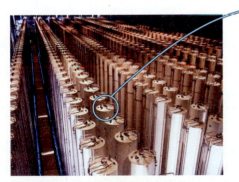
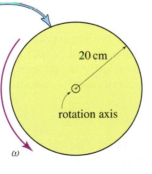

array of enrichment centrifuges   centrifuge cross section

Figure P2.137

Problem 2.138

Treating the center of the Earth as a fixed point, determine the magnitude of the acceleration of points on the surface of the earth as a function of the angle ϕ shown. Use $R = 6371$ km as the radius of the Earth.

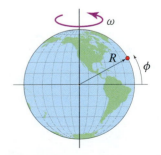

---

[*] The U-235 atom has 92 protons and 143 neutrons, giving an atomic mass of 235. The nucleus of U-238 also has 92 protons, but it has 146 neutrons, giving it an atomic mass of 238.

[†] For nuclear weapons, the concentration of U-235 must be about 90%.

Figure P2.138

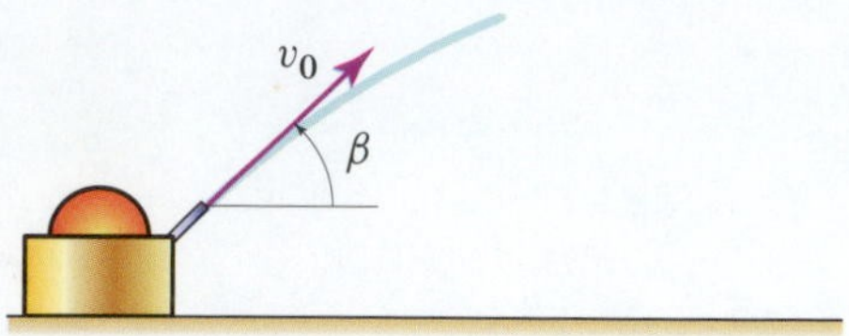

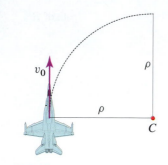

Figure P2.139 and P2.140

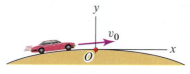

Figure P2.141

Problem 2.139

A water jet is ejected from the nozzle of a fountain with a speed $v_0 = 12$ m/s. Letting $\beta = 33°$, determine the rate of change of the speed of the water particles as soon as these are ejected as well as the corresponding radius of curvature of the water path.

Problem 2.140

A water jet is ejected from the nozzle of a fountain with a speed v_0. Letting $\beta = 21°$, determine v_0 so that the radius of curvature at the highest point on the water arch is 10 ft.

Problem 2.141

A jet is flying at a constant speed $v_0 = 750$ mph while performing a constant speed circular turn. If the magnitude of the acceleration needs to remain constant and equal to $9g$, where g is the acceleration due to gravity, determine the radius of curvature of the turn.

Problem 2.142

A car traveling with a speed $v_0 = 65$ mph almost loses contact with the ground when it reaches the top of the hill. Determine the radius of curvature of the hill at its top.

Figure P2.142 and P2.143

Problem 2.143

A car is traveling over a hill. If, using a Cartesian coordinate system with origin O at the top of the hill, the hill's profile is described by the function $y = -(0.003\,\text{m}^{-1})x^2$, where x and y are in meters, determine the minimum speed at which the car would lose contact with the ground at the top of the hill. Express the answer in km/h.

Problem 2.144

A race boat is traveling at a constant speed $v_0 = 130$ mph when it performs a turn with constant radius ρ to change its course by 90° as shown. The turn is performed while losing speed uniformly in time so that the boat's speed at the end of the turn is $v_f = 125$ mph. If the maximum allowed normal acceleration is equal to $2g$, where g is the acceleration due to gravity, determine the tightest radius of curvature possible and the time needed to complete the turn.

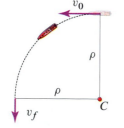

Figure P2.144 and P2.145

Problem 2.145

A race boat is traveling at a constant speed $v_0 = 130$ mph when it performs a turn with constant radius ρ to change its course by 90° as shown. The turn is performed while losing speed uniformly in time so that the boat's speed at the end of the turn is $v_f = 116$ mph. If the magnitude of the acceleration is not allowed to exceed $2g$, where g is the acceleration due to gravity, determine the tightest radius of curvature possible and the time needed to complete the turn.

Problem 2.146

A truck takes an exit ramp with a speed $v_0 = 55$ mph. The ramp is a circular arc with radius $\rho = 150$ ft. Determine the constant rate of change of the truck speed that will allow the truck to stop at B.

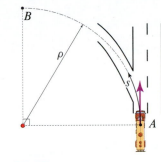

Figure P2.146

Problem 2.147

A jet is flying straight and level at a speed $v_0 = 1100$ km/h when it turns to change its course by $90°$ as shown. In an attempt to progressively tighten the turn, the speed of the plane is uniformly decreased in time while keeping the normal acceleration constant and equal to $8g$, where g is the acceleration due to gravity. At the end of the turn, the speed of the plane is $v_f = 800$ km/h. Determine the radius of curvature ρ_f at the end of the turn and the time t_f that the plane takes to complete its change in course.

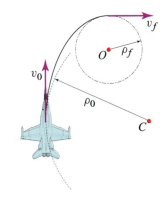

Problem 2.148

A car is traveling over a hill with a constant speed $v_0 = 70$ mph. Using the Cartesian coordinate system shown, the hill's profile is described by the function $y = -(0.0005 \text{ ft}^{-1})x^2$, where x and y are measured in feet. At $x = -300$ ft, the driver applies the brakes, causing a constant time rate of change of speed $\dot{v} = -3$ ft/s^2 until the car arrives at O. Determine the distance traveled while applying the brakes along with the time to cover this distance. *Hint:* To compute the distance traveled by the car along the car's path, observe that $ds = \sqrt{dx^2 + dy^2} = \sqrt{1 + (dy/dx)^2}\, dx$, and that

$$\int \sqrt{1 + C^2 x^2}\, dx = \frac{x}{2}\sqrt{1 + C^2 x^2} + \frac{1}{2C}\ln\left(Cx + \sqrt{1 + C^2 x^2}\right).$$

Figure P2.147

Problem 2.149

Recalling that a circle of radius R and center at the origin O of a Cartesian coordinate system with axes x and y can be expressed via the formula $x^2 + y^2 = R^2$, use Eq. (2.72) to verify that the radius of curvature of this circle is equal to R.

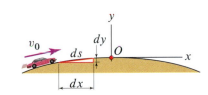

Figure P2.148

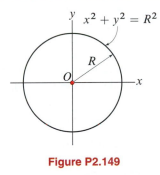

Figure P2.149

2.6 Planar Motion: Polar Coordinates

In this section we describe the position, velocity, and acceleration of a point moving on a plane when the point's coordinates are given relative to a polar coordinate system.

Tracking an airplane

An airplane flying along a straight path is tracked by a ground-based radar as shown in Fig. 2.45. The tracking data recorded consists of the distance r between the radar and the airplane as well as the antenna's orientation θ. How do we turn the data coming from the tracking station, namely, r and θ, and translate it into the object's position, velocity, and acceleration? In tackling this tracking problem, recall that we need a component system to describe vectors. Therefore, an element of the tracking problem is the following question: What component system should be used in relation to the data measured?

Referring to Fig. 2.45, it would seem that the obvious choice of coordinate system for describing the airplane's motion would be a Cartesian coordinate system since the airplane's path is a straight line and we could align our component system with that line. On the other hand, we have to remember that it is the radar station at A that is tracking the airplane, and the raw information collected consists of the distance r to the plane and the direction to the plane as defined by the angle θ. Let's see how we can translate $r(t)$ and $\theta(t)$ into $\vec{v}(t)$ and $\vec{a}(t)$ for the airplane.

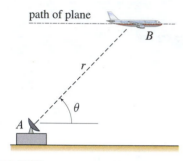

Figure 2.45
An airplane B is being tracked by a ground-based radar at A.

Polar coordinates and position, velocity, and acceleration

Referring to Fig. 2.46, the distance r between A and B and the angle θ, measured with respect to the x axis, unambiguously identify the position of B in the plane of motion. The quantities r and θ are the *polar coordinates** of B relative to the origin A and a reference line coinciding with the x axis. The airplane's position *vector* is then

$$\vec{r} = r\,\hat{u}_r,\qquad(2.82)$$

where $\hat{u}_r$ is the unit vector pointing from A to B. Although θ does not *explicitly* appear in Eq. (2.82), $\vec{r}$ does depend on θ because θ defines the direction of $\hat{u}_r$.

The time derivative of Eq. (2.82), along with the product rule, yields

$$\vec{v} = \dot{r}\,\hat{u}_r + r\,\dot{\hat{u}}_r.\qquad(2.83)$$

The time derivative of the unit vector $\hat{u}_r$ can be evaluated using Eq. (2.60) on p. 92, which allows us to rewrite Eq. (2.83) as

$$\vec{v} = \dot{r}\,\hat{u}_r + r\,\vec{\omega}_r \times \hat{u}_r,\qquad(2.84)$$

where $\vec{\omega}_r$ is the angular velocity of $\hat{u}_r$. Since θ increases in the counterclockwise direction, using the right-hand rule, we have $\vec{\omega}_r = \dot{\theta}\,\hat{k}$, where the

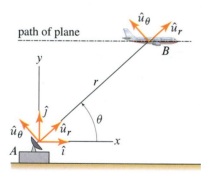

Figure 2.46
The polar coordinate system defining the position of the plane at B.

Common Pitfall

***r* does not stand for *radius*!** Sometimes the radial coordinate r is mistaken for the radius of curvature ρ of a point's path. The radius of curvature ρ is used in relation to the normal-tangential component system and, in general, does not have any direct relation with the coordinate r of a polar coordinate system.

---

* Polar coordinates are also called *radial-transverse coordinates*, in which *radial* refers to the r direction and *transverse* refers to the θ direction. In Section 2.8 we will see the cylindrical coordinate system, which is the three-dimensional generalization of the polar coordinate system.

unit vector $\hat{k}$ identifies the counterclockwise direction. Using the expression $\vec{\omega}_r = \dot{\theta}\,\hat{k}$, which is valid whether or not θ increases, the last term in Eq. (2.84) becomes $r\dot{\theta}\,\hat{k}\times\hat{u}_r = r\dot{\theta}\,\hat{u}_\theta$, where the unit vector $\hat{u}_\theta$ is given by $\hat{k}\times\hat{u}_r = \hat{u}_\theta$. Equation (2.84) then becomes

$$\vec{v} = \dot{r}\,\hat{u}_r + r\dot{\theta}\,\hat{u}_\theta = v_r\,\hat{u}_r + v_\theta\,\hat{u}_\theta, \qquad (2.85)$$

where

$$v_r = \dot{r} \quad \text{and} \quad v_\theta = r\dot{\theta} \qquad (2.86)$$

are the *radial* and *transverse components of the velocity*, respectively. The time derivative of Eq. (2.85) yields

$$\vec{a} = \ddot{r}\,\hat{u}_r + \dot{r}\,\dot{\hat{u}}_r + \dot{r}\dot{\theta}\,\hat{u}_\theta + r\ddot{\theta}\,\hat{u}_\theta + r\dot{\theta}\,\dot{\hat{u}}_\theta. \qquad (2.87)$$

Since $\dot{\hat{u}}_r = \vec{\omega}_r\times\hat{u}_r$, $\dot{\hat{u}}_\theta = \vec{\omega}_\theta\times\hat{u}_\theta$, and $\vec{\omega}_r = \vec{\omega}_\theta = \dot{\theta}\,\hat{k}$, Eq. (2.87) becomes

$$\vec{a} = \ddot{r}\,\hat{u}_r + \dot{r}\dot{\theta}\,\hat{k}\times\hat{u}_r + \dot{r}\dot{\theta}\,\hat{u}_\theta + r\ddot{\theta}\,\hat{u}_\theta + r\dot{\theta}^2\,\hat{k}\times\hat{u}_\theta. \qquad (2.88)$$

Noting that $\hat{k}\times\hat{u}_\theta = -\hat{u}_r$ and combining the coefficients of $\hat{u}_r$ and $\hat{u}_\theta$, we can rewrite Eq. (2.88) as

$$\vec{a} = (\ddot{r} - r\dot{\theta}^2)\hat{u}_r + (r\ddot{\theta} + 2\dot{r}\dot{\theta})\hat{u}_\theta = a_r\,\hat{u}_r + a_\theta\,\hat{u}_\theta, \qquad (2.89)$$

where

$$a_r = \ddot{r} - r\dot{\theta}^2 \quad \text{and} \quad a_\theta = r\ddot{\theta} + 2\dot{r}\dot{\theta} \qquad (2.90)$$

are the *radial* and *transverse components of the acceleration*, respectively.

Equations (2.82), (2.85), and (2.89) answer the question posed at the beginning of the section because they tell us how to use the $r(t)$ and $\theta(t)$ data obtained from the radar station to compute the position, velocity, and acceleration of the airplane.

🎓 Advanced Topic 🎓
A geometric view of the base vectors in polar coordinates

In discussing Eq. (2.15) on p. 34, we determined the coordinate lines of a Cartesian coordinate system. We now determine the coordinate lines of the polar coordinate system, using a similar procedure. Referring to Fig. 2.47, if we hold θ fixed, say, at $\theta = 30°$, and we let the coordinate r range from 0 to ∞ (r is a distance and cannot be negative), we trace a *straight line* moving away from the origin. If we hold r fixed, say, at $r = 2$ and vary θ, we trace a circle centered at the origin with radius equal to 2. This circle does have a positive direction corresponding to increasing values of θ. As we indicated while discussing Eq. (2.15), the unit vectors of a coordinate system are the tangents to the coordinate lines. Referring again to Fig. 2.47, at point Q with coordinates $(3, 60°)$ the unit vector tangent to the r coordinate line and pointing in the direction of increasing r is $\hat{u}_{r_Q}$. Similarly, the unit vector tangent to the θ coordinate line at Q and pointing in the direction of increasing θ is $\hat{u}_{\theta_Q}$. Contrary to what happens for the Cartesian coordinate systems, when we move to other points, say, P or S, we obtain unit vectors that do not have the same orientation as those at Q. We can say that in going from one point to another the base vectors rotate. In other words, the unit vector $\hat{u}_{r_Q}$ at Q can be obtained by rotating the unit vector $\hat{u}_{r_P}$ at P by 60°. To obtain the unit vector $\hat{u}_{\theta_Q}$ at Q, we need to rotate $\hat{u}_{\theta_P}$ by the same amount, namely, 60°. This is why, in deriving expressions for the velocity and acceleration in polar coordinates, we stated that the angular velocities of $\hat{u}_r$ and $\hat{u}_\theta$ are equal, that is, $\vec{\omega}_r = \vec{\omega}_\theta$.

> **🚨 Concept Alert**
>
> **Direction of $\hat{u}_r$ and $\hat{u}_\theta$.** The unit vector $\hat{u}_r$ always points away from the origin. The unit vector $\hat{u}_\theta$ always points in the direction of increasing θ.

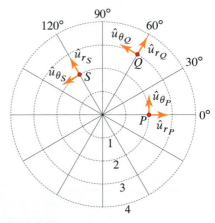

Figure 2.47
Coordinate lines of a polar coordinate system. The solid lines are *radial lines* whereas the dashed lines are *circumferential lines*.

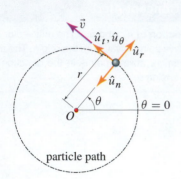

Figure 2.48
Particle moving on a circular path with both path and polar component systems shown.

Relation between path and polar components for circular motion

For circular motion problems there is a simple correspondence between path and polar components of vectors when the polar coordinate system used has its origin at the center of the path.

Referring to Fig. 2.48, consider the circular motion of a particle. Consider also a polar coordinate system with origin at the center of the path O. Finally consider the path component system defined by the particle's motion. Then

1. The two systems of unit vectors are related according to

$$\hat{u}_r = -\hat{u}_n \quad \text{and} \quad \hat{u}_\theta = \pm\hat{u}_t. \tag{2.91}$$

2. The components of velocity are related according to

$$\hat{u}_r = -\hat{u}_n \quad \Rightarrow \quad v_r = 0 \quad \Rightarrow \quad \dot{r} = 0, \tag{2.92}$$

$$\hat{u}_\theta = \pm\hat{u}_t \quad \Rightarrow \quad v_\theta = \pm v \quad \Rightarrow \quad |r\dot{\theta}| = v. \tag{2.93}$$

3. And the components of acceleration are related according to

$$\hat{u}_r = -\hat{u}_n \quad \Rightarrow \quad a_r = -a_n \quad \Rightarrow \quad \ddot{r} - r\dot{\theta}^2 = -\frac{v^2}{\rho} \tag{2.94}$$

$$\Rightarrow \quad r\dot{\theta}^2 = \frac{v^2}{\rho} \tag{2.95}$$

$$\Rightarrow \quad r^2\dot{\theta}^2 = v^2, \tag{2.96}$$

$$\hat{u}_\theta = \pm\hat{u}_t \quad \Rightarrow \quad a_\theta = \pm a_t \quad \Rightarrow \quad r\ddot{\theta} + 2\dot{r}\dot{\theta} = \pm\dot{v} \tag{2.97}$$

$$\Rightarrow \quad r\ddot{\theta} = \pm\dot{v}. \tag{2.98}$$

The $\pm$ sign in the above equations reminds us that the relation between $\hat{u}_\theta$ and $\hat{u}_t$ (and associated vector components) depends on the direction of motion as well as the convention chosen for the coordinate θ. In addition, since the origin of the polar coordinate system is the center of the path, r is constant so that $\dot{r} = \ddot{r} = 0$. Finally, notice that, in the case of circular motion, the radius r is also the path's radius of curvature ρ.

In summary, for circular motion, it is generally equally easy to use either polar or path components to model the motion (provided, of course, that the chosen polar coordinate system has its origin at the center of the circular path).

End of Section Summary

Referring to Fig. 2.49, r and θ are the polar coordinates of point P. The coordinate θ was chosen positive in the counterclockwise direction as viewed down the positive z axis. Using polar coordinates, the position vector of P is

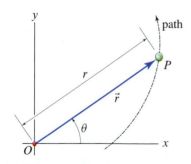

Figure 2.49
The position $\vec{r}$ of a particle defined using the polar coordinates r and θ.

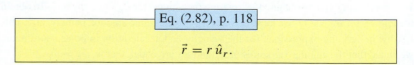

Eq. (2.82), p. 118

$$\vec{r} = r\,\hat{u}_r.$$

Differentiating Eq. (2.82) with respect to time, we obtain the following result for the velocity vector in polar coordinates

Eq. (2.85), p. 119

$$\vec{v} = \dot{r}\,\hat{u}_r + r\dot{\theta}\,\hat{u}_\theta = v_r\,\hat{u}_r + v_\theta\,\hat{u}_\theta,$$

where

Eqs. (2.86), p. 119

$$v_r = \dot{r} \quad \text{and} \quad v_\theta = r\dot{\theta}$$

are the radial and transverse components of the velocity, respectively.

Differentiating Eq. (2.85) with respect to time, we see that the acceleration vector in polar coordinates takes on the form

Eq. (2.89), p. 119

$$\vec{a} = (\ddot{r} - r\dot{\theta}^2)\hat{u}_r + (r\ddot{\theta} + 2\dot{r}\dot{\theta})\hat{u}_\theta = a_r\,\hat{u}_r + a_\theta\,\hat{u}_\theta,$$

where

Eqs. (2.90), p. 119

$$a_r = \ddot{r} - r\dot{\theta}^2 \quad \text{and} \quad a_\theta = r\ddot{\theta} + 2\dot{r}\dot{\theta}$$

are the radial and transverse components of the acceleration, respectively.

EXAMPLE 2.20   *Constant Velocity Motion in Polar Coordinates*

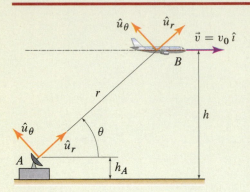

Figure 1
Radar station at A tracking a plane at B.

What relationships do the radar readings obtained by the station at A need to satisfy to conclude that the plane at B shown in Fig. 1 is flying straight and level at altitude h and at a constant speed v_0?

SOLUTION

Road Map   It is not hard to imagine that a radar station, such as the one at A, would record *discrete* readings of $r(t)$ and $\theta(t)$ for any object it is tracking. Given a tabulated list of r and θ as a function of time, the question is, How can we use these readings to determine whether or not a plane is flying (1) straight and level and (2) at a constant speed?

The solution strategy used in this example will be used frequently. We will use expressions for positions, velocities, and accelerations in a chosen coordinate system to give form to specific requirements (e.g., that the altitude be constant or that the speed be maintained). Once a requirement is obtained, it can be manipulated further (e.g., differentiated with respect to time) to obtain new relationships that will be consistent with the requirements' initial statements.

Computation   As far as assessing the plane's altitude is concerned, first we need to express the altitude using the given data, and then we need to *enforce* the requirement that the altitude be maintained constant. Hence, referring to Fig. 1, we have

$$h = r \sin\theta + h_A. \tag{1}$$

So, h will remain constant as long as

$$\boxed{r(t) \sin\theta(t) = \text{constant.}} \tag{2}$$

Since the speed is obtained as the magnitude of the velocity vector, to find what relationship must be satisfied in order for the plane to maintain the given constant speed, let's look at the expression for the plane's velocity. The velocity vector in polar coordinates is given by Eq. (2.85), that is, $\vec{v} = \dot{r}\,\hat{u}_r + r\dot{\theta}\,\hat{u}_\theta$, and so its magnitude is given by the square root of the sum of the squares of its components, i.e.,

$$|\vec{v}| = v = \sqrt{\dot{r}^2 + r^2\dot{\theta}^2}. \tag{3}$$

Therefore, for the speed to be a constant equal to v_0 we must have

$$\boxed{v_0 = \sqrt{\dot{r}^2 + r^2\dot{\theta}^2} = \text{constant.}} \tag{4}$$

Discussion & Verification   Since r has dimensions of length and θ is nondimensional, then $\dot{r}$ and $r\dot{\theta}$ both have dimensions of length over time, i.e., of velocity. Hence, we can conclude that Eq. (4) is dimensionally correct.

🔎 **A Closer Look**   While Eq. (4) gives the relationship we wanted, it is often desirable to have relations that require as few mathematical operations as possible. For example, it turns out that there is no need to compute a square root to verify whether or not the speed is a constant—if it is true that v_0 is a constant, then we necessarily must have that v_0^2 is also a constant. Therefore, as long as

$$\boxed{\dot{r}^2(t) + r^2(t)\dot{\theta}^2(t) = \text{constant,}} \tag{5}$$

we can conclude that the plane is flying at a constant speed.

EXAMPLE 2.21   *Rectilinear Motion & Polar Coordinates*

As a part of an assembly process, the end effector A on the robotic arm in Fig. 2 needs to move the gear B along the vertical line shown in a specified fashion. Arm OA can vary its length by telescoping via internal actuators. A motor at O allows the arm to pivot in the vertical plane. When $\theta = 50°$, B is moving downward with a speed $v_0 = 8$ ft/s and a downward acceleration with magnitude $a_0 = 0.5$ ft/s. At this instant, determine the required length of the arm, the rate at which the arm is extending, and its rotation rate $\dot{\theta}$. In addition, determine the second time derivatives of both the arm's length and the angle θ.

Figure 1
A robotic arm.

SOLUTION

Road Map We know the gear's path, velocity, and acceleration as well as the angle θ at the instant shown. Therefore, we can determine the length r at this instant via simple trigonometry. As far as determining $\dot{r}$ and $\dot{\theta}$ is concerned, observe that the known velocity of B is easily written in terms of the unit vector $\hat{\jmath}$ shown in Fig. 3. Once this is done, we can use trigonometry to rewrite the velocity of B via $\hat{u}_r$ and $\hat{u}_\theta$. Then $\dot{r}$ and $\dot{\theta}$ are found by equating this *specific* expression for the velocity of B to the general expression of the velocity vector in polar coordinates. Finally, $\ddot{r}$ and $\ddot{\theta}$ are found by applying to the acceleration the same strategy just described for the velocity.

Computation We know that $\theta = 50°$ at this instant, so the geometry in Fig. 3 tells us that

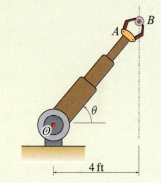

$$\boxed{r = \frac{4\,\text{ft}}{\cos\theta} = 6.22\,\text{ft}.} \tag{1}$$

Since B moves vertically downward, the velocity of the end effector is

$$\vec{v} = -v_0\,\hat{\jmath} = -(8\,\text{ft/s})\,\hat{\jmath}. \tag{2}$$

Equating Eq. (2) with the expression of the velocity in polar coordinates, we have

$$-(8\,\text{ft/s})\,\hat{\jmath} = \dot{r}\,\hat{u}_r + r\dot{\theta}\,\hat{u}_\theta. \tag{3}$$

Now, we can write $\hat{\jmath}$ in terms of the polar unit vectors as

$$\hat{\jmath} = \sin\theta\,\hat{u}_r + \cos\theta\,\hat{u}_\theta = \sin 50°\,\hat{u}_r + \cos 50°\,\hat{u}_\theta. \tag{4}$$

Substituting Eq. (4) into Eq. (3) and equating components, we have

$$\dot{r} = -8\sin 50°\,\text{ft/s} \quad\text{and}\quad r\dot{\theta} = -8\cos 50°\,\text{ft/s}, \tag{5}$$

which, by using the result in Eq. (1), can then be solved for $\dot{r}$ and $\dot{\theta}$ to obtain

$$\boxed{\dot{r} = -6.13\,\text{ft/s} \quad\text{and}\quad \dot{\theta} = -0.826\,\text{rad/s}.} \tag{6}$$

Dealing with the acceleration of B as we have done for the velocity, we have

$$\vec{a} = -a_0\,\hat{\jmath} = -\left(0.5\,\text{ft/s}^2\right)\hat{\jmath}. \tag{7}$$

Equating Eq. (7) with the expression for acceleration in polar coordinates, we have

$$-(0.5\,\text{ft/s}^2)\,\hat{\jmath} = (\ddot{r} - r\dot{\theta}^2)\,\hat{u}_r + (r\ddot{\theta} + 2\dot{r}\dot{\theta})\,\hat{u}_\theta. \tag{8}$$

Using Eq. (4) again and equating components, we have

$$\ddot{r} - r\dot{\theta}^2 = -0.5\sin 50°\,\text{ft/s}^2 \quad\text{and}\quad r\ddot{\theta} + 2\dot{r}\dot{\theta} = -0.5\cos 50°\,\text{ft/s}^2. \tag{9}$$

Figure 2
Schematic of robot arm shown in Fig. 1.

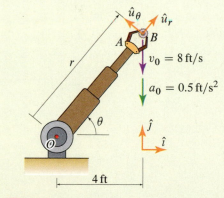

Figure 3
Robotic arm showing the polar and Cartesian coordinate systems we will use as well as the velocity and acceleration of the end effector.

Solving Eqs. (9) for $\ddot{r}$ and $\ddot{\theta}$, we have

$$\ddot{r} = -0.5\sin 50°\,\text{ft/s}^2 + r\dot{\theta}^2 \quad\text{and}\quad \ddot{\theta} = -\frac{1}{r}(0.5\cos 50°\,\text{ft/s}^2 + 2\dot{r}\dot{\theta}). \tag{10}$$

Substituting in Eqs. (10) the results we have already obtained for r, $\dot{r}$, and $\dot{\theta}$, we have

$$\boxed{\ddot{r} = 3.87\,\text{ft/s}^2 \quad\text{and}\quad \ddot{\theta} = -1.68\,\text{rad/s}^2.} \tag{11}$$

Discussion & Verification   Our results are dimensionally correct, and appropriate units have been used. To understand whether or not the signs of our results are correct, we begin by noticing that in Eqs. (6) $\dot{r} < 0$ and $\dot{\theta} < 0$, which means that the arm is actually getting shorter while rotating clockwise. Intuition tells us that if the end effector is moving straight down, then the arm does need to be get shorter and rotate clockwise. Fortunately, both of these observations are consistent with the results in Eq. (6).

As for the sign of $\ddot{r}$ in Eqs. (11), the fact that $\ddot{r} > 0$ indicates that while the arm is getting shorter (i.e., $\dot{r} < 0$), the rate at which this happens is decreasing. That is, since $\ddot{r}$ has a sign opposite to $\dot{r}$, we should expect that as B keeps moving down, the length of the arm will stop shortening. This result is correct because the length of the arm will stop shortening when the arm becomes horizontal, and then it will start increasing for negative values of θ. As far as the sign of $\ddot{\theta}$ is concerned, our result indicates that the rate of clockwise rotation is increasing. This result is to be expected even if the acceleration of B were equal to zero. In this case, i.e., if $a_\theta = 0 = r\ddot{\theta} + 2\dot{r}\dot{\theta}$, the sign of $\ddot{\theta}$ is opposite to the sign of the product $\dot{r}\dot{\theta}$. For us, such a product is positive because we found that both $\dot{r}$ and $\dot{\theta}$ were negative. In our case $a_\theta \neq 0$, but B is accelerating downward and therefore provides no contribution to $\ddot{\theta}$ in the counterclockwise direction.

EXAMPLE 2.22   *Acceleration During Orbital Motion*

For a satellite orbiting a planet, empirical observations (see Kepler's laws in Chapter 1) tell us that in the polar coordinate system shown in Fig. 1, the quantity $r^2\dot\theta$ remains constant throughout the satellite's motion. Show that for such a motion, the satellite's acceleration is purely in the radial direction; i.e., the transverse component of the satellite's acceleration is equal to zero.

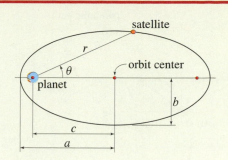

Figure 1
A satellite orbiting a planet. The orbit is an ellipse with major and minor semiaxes equal to a and b, respectively. The length $c = \sqrt{a^2 - b^2}$ denotes the distance between either focus (one of which is occupied by the planet) and the orbit center.

SOLUTION

Road Map We need to show that $a_\theta = 0$. To do this, we can determine how the expression for a_θ given by Eq. (2.89) relates to the law of Kepler stating that $r^2\dot\theta$ is constant.

Computation Equation (2.89) tells us that in polar coordinates the r and θ components of the satellite's acceleration are given by

$$a_r = \ddot r - r\dot\theta^2 \quad \text{and} \quad a_\theta = r\ddot\theta + 2\dot r\dot\theta. \tag{1}$$

Kepler's observations indicate that

$$r^2\dot\theta = K, \tag{2}$$

where K is a constant whose value depends on the particular orbit followed by the satellite. Now, if we differentiate Eq. (2) with respect to time, we obtain

$$2r\dot r\dot\theta + r^2\ddot\theta = 0, \tag{3}$$

where we have used the fact that K is constant so that $\dot K = 0$. Equation (3) tells us that

$$2r\dot r\dot\theta + r^2\ddot\theta = r(r\ddot\theta + 2\dot r\dot\theta) = 0 \quad \Rightarrow \quad r\ddot\theta + 2\dot r\dot\theta = 0, \tag{4}$$

where we have used the fact that r is never zero to obtain the final expression. Comparing Eq. (4) with Eq. (1), we see that

$$a_\theta = 0, \tag{5}$$

which is what we set out to show.

Discussion & Verification The derivation of our result is elementary, and it is correct because we have correctly applied the chain and product rules of calculus in taking the derivative of Eq. (2).

🔍 **A Closer Look** The result we have obtained is based on astronomical observations that predate the work of Newton. From a historical viewpoint this is important because, in formulating his second law of motion and his law of universal gravitation, Newton needed to formulate a theory consistent with Kepler's observations. Therefore it is not by chance that Newton's law of gravitation demands that the force of gravity between two particles be directed along the line connecting the particles. This requirement, along with Newton's second law, $\vec F = m\vec a$, causes the acceleration of the planet in Fig. 1 to be *completely* along the radial line connecting the satellite and the planet (i.e., $a_\theta = 0$) so that, overall, both the universal law of gravitation as well as Newton's second law are consistent with Kepler's observations.

We will come across Eq. (2) again in Section 5.3 because Eq. (2) is also the mathematical expression of the fact that the angular momentum of the satellite, computed relative to the planet, is conserved. More generally, Eq. (2) expresses the conservation of angular momentum for a particle moving under the action of a central force (provided that the origin of polar coordinate system used is the center of the force).

EXAMPLE 2.23   *Projectile Motion in Polar Coordinates*

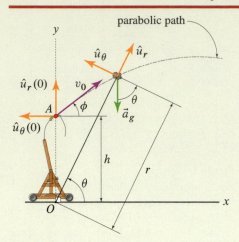

Figure 1
Projectile motion in polar coordinates. The point O is the origin of the coordinate system. The point A is the point at which the object was released, and it is taken to be the initial position of the projectile.

In Section 2.3, we saw that the equations describing the motion of a projectile in Cartesian coordinates took the *simple* form $\ddot{x} = 0$ and $\ddot{y} = -g$. Referring to Fig. 1, revisit the projectile problem and derive the equations describing the motion of a projectile released at A, using polar coordinates. In addition, derive the expressions for the initial conditions of the motion in the same polar coordinate system, given that the projectile is launched from point A at speed v_0 in the direction shown in Fig. 1.

SOLUTION

Road Map   What we want to show is that any problem can be formulated using any coordinate system we choose. However, this freedom of choice comes with the consequence that if we do not choose wisely, the mathematical complexity of the problem can be substantial.

Since we know the acceleration of the projectile at every point along its path, the idea is to relate that known acceleration to the direction of a_r and a_θ at every point. A similar idea will apply to finding the initial conditions; that is, we know the initial conditions relative to point A, and we just need to translate those to the polar coordinate system whose origin is at point O.

Computation   We begin by finding the polar components of the projectile's acceleration due to gravity, which we denote by $\vec{a}_g$. The components of this vector along the radial and transverse directions are

$$a_{gr} = \vec{a}_g \cdot \hat{u}_r = -g \sin\theta, \tag{1}$$

$$a_{g\theta} = \vec{a}_g \cdot \hat{u}_\theta = -g \cos\theta, \tag{2}$$

where g is the acceleration due to gravity. Equating the *general* expressions for the components of acceleration in polar coordinates given by Eq. (2.90) to the corresponding components in Eqs.(1) and (2) gives

$$\ddot{r} - r\dot{\theta}^2 = -g \sin\theta, \tag{3}$$

$$r\ddot{\theta} + 2\dot{r}\dot{\theta} = -g \cos\theta. \tag{4}$$

Equations (3) and (4) form a system of *coupled*, *nonlinear* ordinary differential equations which, mathematically, are *far* more complex than the equations we obtained in Cartesian coordinates.

To integrate Eqs. (3) and (4) so as to determine the motion of the projectile, we would need to complement these differential equations with corresponding *initial conditions*. These conditions consist of the position and velocity of A at time $t = 0$ expressed in polar coordinates. Referring to Fig. 1, we see that at time $t = 0$ the projectile is at A and therefore we have

$$r(0) = h \quad \text{and} \quad \theta(0) = \frac{\pi}{2}. \tag{5}$$

For the velocity at $t = 0$ we have

$$\vec{v}(0) = v_0 \sin\phi\, \hat{u}_r(0) - v_0 \cos\phi\, \hat{u}_\theta(0), \tag{6}$$

where the quantities v_0 and ϕ are known. Equating Eq. (6) to the general expression for the velocity vector in polar coordinates given by Eq. (2.85), we have

$$\vec{v}(0) = \dot{r}(0)\, \hat{u}_r(0) + r(0)\dot{\theta}(0)\, \hat{u}_\theta(0) = v_0 \sin\phi\, \hat{u}_r(0) - v_0 \cos\phi\, \hat{u}_\theta(0), \tag{7}$$

Concept Alert

Unit vectors in path and polar component systems are functions of time. When we express vectors using normal-tangential components or polar components, it is important to keep in mind that the unit vectors of these component systems are functions of time. This is why, in expressing the velocity at time $t = 0$, we had to use the unit vectors $\hat{u}_r$ and $\hat{u}_\theta$ at $t = 0$.

that is,

$$\dot{r}(0) = v_0 \sin\phi \quad \text{and} \quad r(0)\dot{\theta}(0) = -v_0 \cos\phi. \tag{8}$$

In summary, the initial conditions for this problem take on the form

$$r(0) = h, \qquad \theta(0) = \frac{\pi}{2}, \tag{9}$$

$$\dot{r}(0) = v_0 \sin\phi, \qquad \dot{\theta}(0) = -\frac{v_0}{h}\cos\phi, \tag{10}$$

where we have used the fact that $r(0) = h$ from Eq. (5).

Discussion & Verification Recalling that r has dimensions of length and that θ is nondimensional, we can verify that the left-hand sides of Eqs. (3) and (4) have dimensions of length over time squared, as expected since the dimensions of the right-hand sides of these equations are those of the acceleration of gravity g. We can verify the dimensional correctness of Eqs. (9) and (10) in a similar manner.

A Closer Look Equations (3), (4), (9), and (10) are much more complicated-looking than the corresponding equations in Cartesian coordinates (i.e., $\ddot{x} = 0$ and $\ddot{y} = -g$, along with the initial conditions $x(0) = 0$ and $y(0) = h$). However, the trajectory obtained by solving Eqs. (3) and (4) is identical to that obtained by solving the corresponding equations in Cartesian coordinates (i.e., $\ddot{x} = 0$ and $\ddot{y} = -g$); that is, we would obtain exactly the same parabola in either case (provided, of course, that the same initial position and velocity are used).

A question frequently asked by students is, Can I solve this problem using "this or that" coordinate system? The answer to this question is, in general, yes. However, whether you are solving the problem analytically or numerically, the coordinate system chosen *does* make a difference in how involved the problem's solution becomes. In this example, we have derived (although not solved) the equations governing the motion of a projectile. Clearly, the choice of coordinate system did not change the underlying physics of the problem. However, the resulting system of equations cannot be easily solved by hand because they are nonlinear and, above all, coupled; that is, the equation containing $\ddot{r}$ also contains θ and $\dot{\theta}$, and the equation containing $\ddot{\theta}$ also contains r and $\dot{r}$. Furthermore, even the derivation of the initial conditions required several steps to complete.

Another lesson to be learned concerns the polar coordinate system in particular. We should be aware of the fact that this problem becomes *ill-posed* if point A, the initial position of the projectile, is chosen to coincide with point O, i.e., the origin of the chosen coordinate system. In this case, although we could still write Eqs. (3) and (4), we can no longer write meaningful initial conditions because the value of θ for a point at the origin is arbitrary and (as h would be equal to zero) $\dot{\theta}$ becomes undefined. This implies that when using the polar coordinate system, we should choose the origin of the coordinate system so as *not* to coincide with points where initial conditions are specified.

PROBLEMS

💡 Problem 2.150 💡

A particle P is moving along a path with the velocity shown. Discuss in detail whether or not there are incorrect elements in the sketch of the polar component system at P.
Note: Concept problems are about *explanations*, not computations.

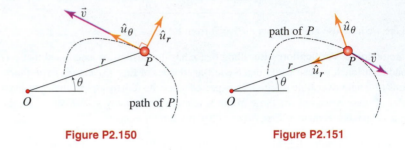

Figure P2.150   Figure P2.151

💡 Problem 2.151 💡

A particle P is moving along a path with the velocity shown. Discuss in detail whether or not there are incorrect elements in the sketch of the polar component system at P.
Note: Concept problems are about *explanations*, not computations.

💡 Problem 2.152 💡

A particle P is moving along a path with the velocity shown. Discuss in detail whether or not there are incorrect elements in the sketch of the polar component system at P.
Note: Concept problems are about *explanations*, not computations.

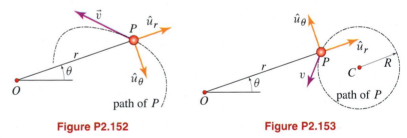

Figure P2.152   Figure P2.153

💡 Problem 2.153 💡

A particle P is moving along a circle with center C and radius R in the direction shown. Letting O be the origin of a polar coordinate system with the coordinates r and θ shown, discuss in detail whether or not there are incorrect elements in the sketch of the polar component system at P.
Note: Concept problems are about *explanations*, not computations.

Problem 2.154 🌡️

A radar station is tracking a plane flying at a constant altitude with a speed $v_0 = 550$ mph. If at a given instant $r = 7$ mi and $\theta = 32°$, determine the corresponding values of $\dot{r}$, $\dot{\theta}$, $\ddot{r}$, and $\ddot{\theta}$.

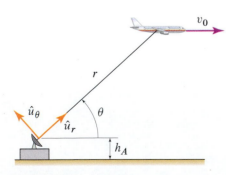

Figure P2.154

Problem 2.155

During a given time interval, a radar station tracking an airplane records the readings

$$\dot{r}(t) = [449.8 \cos \theta(t) + 11.78 \sin \theta(t)] \text{ mph},$$

$$r(t)\dot{\theta}(t) = [11.78 \cos \theta(t) - 449.8 \sin \theta(t)] \text{ mph},$$

where t denotes time. Determine the speed of the plane. Furthermore, determine whether the plane being tracked is ascending or descending and the corresponding climbing rate (i.e., the rate of change of the plane's altitude) expressed in ft/s.

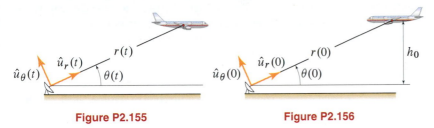

Figure P2.155   **Figure P2.156**

Problem 2.156

At a given instant, an airplane flying at an altitude $h_0 = 10{,}000$ ft begins its descent in preparation for landing when it is $r(0) = 20$ mi from the radar station at the destination's airport. At that instant, the aiplane's speed is $v_0 = 300$ mph, the climb rate is -5 ft/s, and the horizontal component of velocity is decreasing steadily at a rate of 15 ft/s^2. Determine the $\dot{r}$, $\dot{\theta}$, $\ddot{r}$, and $\ddot{\theta}$ that would be observed by the radar station.

Problem 2.157

At a given instant, the merry-go-round is rotating with an angular velocity $\omega = 20$ rpm while the child is moving radially outward at a constant rate of 0.7 m/s. Assuming that the angular velocity of the merry-go-round remains constant, i.e., $\alpha = 0$, determine the magnitudes of the speed and of the acceleration of the child when he is 0.8 m away from the spin axis.

Problem 2.158

At a given instant, the merry-go-round is rotating with an angular velocity $\omega = 18$ rpm, and it is slowing down at a rate of 0.4 rad/s^2. When the child is 2.5 ft away from the spin axis, determine the time rate of change of the child's distance from the spin axis so that the child experiences no transverse acceleration while moving along a radial line.

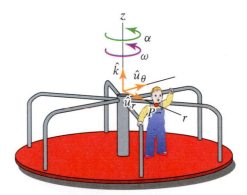

Figure P2.157–P2.159

Problem 2.159

At a given instant, the merry-go-round is rotating with an angular velocity $\omega = 18$ rpm. When the child is 0.45 m away from the spin axis, determine the second derivative with respect to time of the child's distance from the spin axis so that the child experiences no radial acceleration.

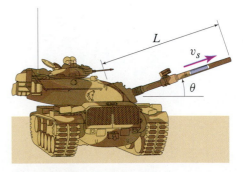

Problem 2.160

The cutaway of the gun barrel shows a projectile that, upon exit, moves with a speed $v_s = 5490$ ft/s relative to the gun barrel. The length of the gun barrel is $L = 15$ ft. Assuming that the angle θ is increasing at a constant rate of 0.15 rad/s, determine the

Figure P2.160

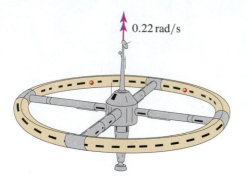

0.22 rad/s

Figure P2.161 and P2.162

speed of the projectile right when it leaves the barrel. In addition, assuming that the projectile acceleration along the barrel is constant and that the projectile starts from rest, determine the magnitude of the acceleration upon exit.

Problem 2.161

A space station is rotating in the direction shown at a constant rate of 0.22 rad/s. A crew member travels from the periphery to the center of the station through one of the radial shafts at a constant rate of 1.3 m/s (relative to the shaft) while holding onto a handrail in the shaft. Taking $t = 0$ to be the instant at which travel through the shaft begins and knowing that the radius of the station is 200 m, determine the velocity and acceleration of the crew member as a function of *time*. Express your answer using a polar coordinate system with origin at the center of the station.

Problem 2.162

Solve Prob. 2.161 and express your answers as a function of *position* along the shaft traveled by the astronaut.

Problem 2.163

A person driving along a rectilinear stretch of road is fined for speeding, having been clocked at 75 mph when the radar gun was pointing as shown. The driver claims that, because the radar gun is off to the side of the road instead of directly in front of his car, the radar gun overestimates his speed. Is he right or wrong and why?
Note: Concept problems are about *explanations*, not computations.

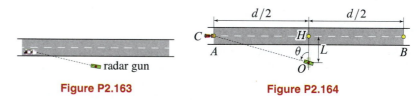

Figure P2.163   **Figure P2.164**

Problem 2.164

A motion tracking camera is placed along a rectilinear stretch of a racetrack (the figure is not to scale). A car C enters the stretch at A with a speed $v_A = 110$ mph and accelerates uniformly in time so that at B it has a speed $v_B = 175$ mph, where $d = 1$ mi. Letting the distance $L = 50$ ft, if the camera is to track the motion of C, determine the camera's angular velocity as well as the time rate of change of the angular velocity when the car is at A and at H.

Problem 2.165

The radar station at O is tracking a meteor P as it moves through the atmosphere. At the instant shown, the station measures the following data for the motion of the meteor: $r = 21{,}000$ ft, $\theta = 40°$, $\dot{r} = -22{,}440$ ft/s, $\dot{\theta} = -2.935$ rad/s, $\ddot{r} = 187{,}500$ ft/s^2, and $\ddot{\theta} = -5.409$ rad/s^2.

(a) Determine the magnitude and direction (relative to the xy coordinate system shown) of the velocity vector at this instant.

(b) Determine the magnitude and direction (relative to the xy coordinate system shown) of the acceleration vector at this instant.

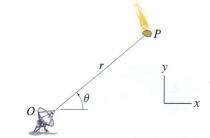

Figure P2.165

Problems 2.166 and 2.167

As a part of an assembly process, the end effector at A on the robotic arm needs to move the gear at B along the vertical line shown with some known velocity v_0 and acceleration a_0. Arm OA can vary its length by telescoping via internal actuators, and a motor at O allows it to pivot in the vertical plane.

Problem 2.166 When $\theta = 50°$, it is required that $v_0 = 8$ ft/s (down) and that it be slowing down at $a_0 = 2$ ft/s². Using $h = 4$ ft, determine, at this instant, the values for $\ddot{r}$ (the extensional acceleration) and $\ddot{\theta}$ (the angular acceleration).

Problem 2.167 Letting v_0 and a_0 be positive if the gear moves and accelerates upward, determine expressions for r, $\dot{r}$, $\ddot{r}$, θ, and $\ddot{\theta}$ that are valid for any value of θ.

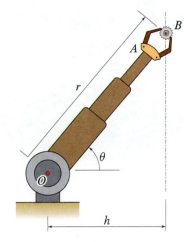

Problem 2.168

The time derivative of the acceleration, i.e., $\dot{\vec{a}}$, is usually referred to as the *jerk* because "jerky" motion is generally associated with quickly changing acceleration.[*] Starting from Eq. (2.89), compute the jerk in polar coordinates.

Figure P2.166 and P2.167

Problems 2.169 and 2.170

In the cutting of sheet metal, the robotic arm OA needs to move the cutting tool at C counterclockwise at a constant speed v_0 along a circular path of radius ρ. The center of the circle is located in the position shown relative to the base of the robotic arm at O.

Problem 2.169 When the cutting tool is at D ($\phi = 0$), determine r, $\dot{r}$, $\dot{\theta}$, $\ddot{r}$, and $\ddot{\theta}$ as functions of the given quantities (i.e., d, h, ρ, v_0).

Problem 2.170 For all positions along the circular cut (i.e., for any value of ϕ), determine r, $\dot{r}$, $\dot{\theta}$, $\ddot{r}$, and $\ddot{\theta}$ as functions of the given quantities (i.e., d, h, ρ, v_0). These quantities can be found "by hand," but it is tedious, so you might consider using symbolic algebra software such as Mathematica or Maple.

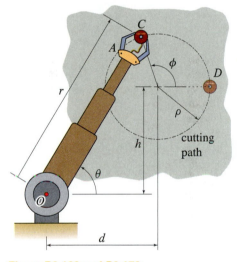

Problem 2.171

Considering the system analyzed in Example 2.23, let $h = 15$ ft, $v_0 = 55$ mph, and $\phi = 25°$. Plot the trajectory of the projectile in two different ways: (1) by solving the projectile motion problem using Cartesian coordinates and plotting y versus x and (2) by using a computer to solve Eqs. (3), (4), (9), and (10) in Example 2.23. You should, of course, get the same trajectory regardless of the coordinate system used.

Figure P2.169 and P2.170

Problem 2.172

The *reciprocating rectilinear motion* mechanism shown consists of a disk pinned at its center at A that rotates with a constant angular velocity ω_{AB}, a slotted arm CD that is pinned at C, and a bar that can oscillate within the guides at E and F. As the disk rotates, the peg at B moves within the slotted arm, causing it to rock back and forth. As the arm rocks, it provides a slow advance and a quick return to the reciprocating bar due to the change in distance between C and B. Letting $\theta = 30°$, $\omega_{AB} = 50$ rpm = constant, $R = 0.3$ ft, and $h = 0.6$ ft, determine $\dot{\phi}$ and $\ddot{\phi}$, i.e., the angular velocity and angular acceleration of the slotted arm CD, respectively.

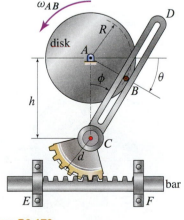

[*] For passengers riding in vehicles, high values of jerk usually make for an uncomfortable ride. Elevator manufacturers are very interested in jerk since they want to move passengers quickly from one floor to another without large changes in acceleration.

Figure P2.172

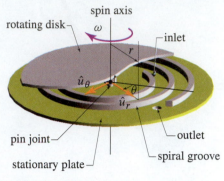

Figure P2.173 and P2.174

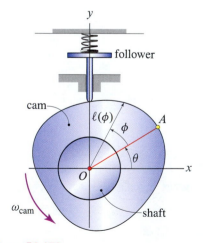

Figure P2.176

Problems 2.173 and 2.174

A micro spiral pump[*] consists of a spiral channel attached to a stationary plate. This plate has two ports, one for fluid inlet and another for outlet, the outlet being farther from the center of the plate than the inlet. The system is capped by a rotating disk. The fluid trapped between the rotating disk and the stationary plate is put in motion by the rotation of the top disk, which pulls the fluid through the spiral channel.

Problem 2.173 Consider a spiral channel with the geometry given by the equation $r = \eta\theta + r_0$, where $r_0 = 146\,\mu\text{m}$ is the starting radius, r is the distance from the spin axis, and θ is the angular position of a point in the spiral channel. Assume that the radius at the outlet is $r_{\text{out}} = 190\,\mu\text{m}$, that the top disk rotates with a constant angular speed ω, and that the fluid particles in contact with the rotating disk are essentially stuck to it. Determine the constant η and the value of ω (in rpm) such that after 1.25 rev of the top disk, the speed of the particles in contact with this disk is $v = 0.5\,\text{m/s}$ at the outlet.

Problem 2.174 Consider a spiral channel with the geometry given by the equation $r = \eta\theta + r_0$, where $\eta = 12\,\mu\text{m}$ is called the polar slope, $r_0 = 146\,\mu\text{m}$ is the starting radius, r is the distance from the spin axis, and θ is the angular position of a point in the spiral channel. If the top disk rotates with a constant angular speed $\omega = 30{,}000\,\text{rpm}$, and assuming that the fluid particles in contact with the rotating disk are essentially stuck to it, use the polar coordinate system shown and determine the velocity and acceleration of one fluid particle when it is at $r = 170\,\mu\text{m}$.

Problem 2.175

The mechanism shown is called a *swinging block* slider crank. First used in various steam locomotive engines in the 1800s, this mechanism is often found in door-closing systems. If the disk is rotating with a constant angular velocity $\dot{\theta} = 60\,\text{rpm}$, $H = 4\,\text{ft}$, $R = 1.5\,\text{ft}$, and r denotes the distance between B and O, compute $\dot{r}$, $\dot{\phi}$, $\ddot{r}$, and $\ddot{\phi}$ when $\theta = 90°$.

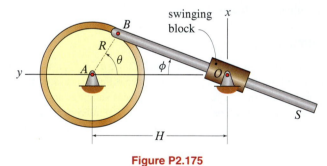

Figure P2.175

Problem 2.176

The cam is mounted on a shaft that rotates about O with constant angular velocity ω_{cam}. The profile of the cam is described by the function $\ell(\phi) = R_0(1 + 0.25\cos^3\phi)$, where the angle ϕ is measured relative to the segment OA, which rotates with the cam. Letting $\omega_{\text{cam}} = 3000\,\text{rpm}$ and $R_0 = 3\,\text{cm}$, determine the velocity and acceleration of

---

[*] The spiral pump was originally invented in 1746 by H. A. Wirtz, a Swiss pewterer from Zurich. Recently, the spiral pump concept has seen a comeback in microdevice design. Some of the data used in this problem is taken from M. I. Kilani, P. C. Galambos, Y. S. Haik, and C.-J. Chen, "Design and Analysis of a Surface Micromachined Spiral-Channel Viscous Pump," *Journal of Fluids Engineering*, **125**, pp. 339–344, 2003.

the follower when $\theta = 33°$. Express the acceleration of the follower in terms of g, the acceleration due to gravity.

Problem 2.177

The collar is mounted on the horizontal arm shown, which is originally rotating with the angular velocity ω_0. Assume that after the cord is cut, the collar slides along the arm in such a way that the collar's total acceleration is equal to zero. Determine an expression of the radial component of the collar's velocity as a function of r, the distance from the spin axis. *Hint:* Using polar coordinates, observe that $d(r^2\dot{\theta})/dt = ra_\theta$.

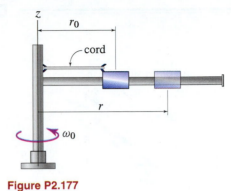

Figure P2.177

Problem 2.178

Particle A slides over the semicylinder while pushed by the arm pinned at C. The motion of the arm is controlled such that it starts from rest at $\theta = 0$, ω increases uniformly as a function of θ, and $\omega = 0.5\,\text{rad/s}$ for $\theta = 45°$. Letting $R = 4\,\text{in.}$, determine the speed and the magnitude of the acceleration of A when $\phi = 32°$.

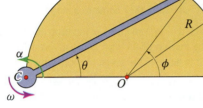

Figure P2.178

Problem 2.179

A satellite is moving along the elliptical orbit shown. Using the polar coordinate system in the figure, the satellite's orbit is described by the equation

$$r(\theta) = 2b^2 \frac{a + \sqrt{a^2 - b^2}\cos\theta}{a^2 + b^2 - (a^2 - b^2)\cos(2\theta)},$$

which implies the following identity

$$\frac{rr'' - 2(r')^2 - r^2}{r^3} = -\frac{a}{b^2},$$

where the prime indicates differentiation with respect to θ. Using this identity and knowing that the satellite moves so that $K = r^2\dot{\theta}$ with K constant (i.e., according to Kepler's laws), show that the radial component of acceleration is proportional to $-1/r^2$, which is in agreement with Newton's universal law of gravitation.

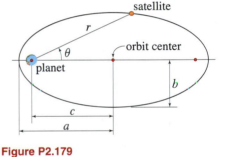

Figure P2.179

DESIGN PROBLEMS

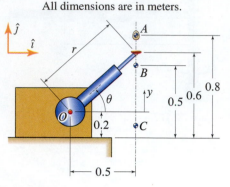

All dimensions are in meters.

Figure DP2.2 and DP2.3

Design Problem 2.2

As a part of a robotics competition, a robotic arm is to be designed so as to catch an egg without breaking it. The egg is released at point A from rest while the arm is initially also at rest in the position shown. The arm starts moving when the egg is released, and it catches the egg at B with the same speed and acceleration that the egg has at that instant (matching the speed avoids impact and matching the acceleration keeps them together after impact).

Assume that the arm will start slowing the egg down at B and will bring the egg to a complete stop at C. Also assume that the vertical rate of change of deceleration (the time derivative of the acceleration) felt by the egg remains constant and that the egg arrives at C with zero acceleration. For the motion between B and C, determine

(a) The rate of change of the deceleration.

(b) The function $y(t)$ of the vertical motion.

(c) The time it takes for the arm to bring the egg to a stop.

(d) The vertical position C at which they come to a stop.

(e) Then, using the fact that $\dot{\vec{a}} \cdot \hat{\imath}$ is 0 and $\dot{\vec{a}} \cdot \hat{\jmath}$ is the time rate of change of the vertical acceleration due to the motion of the arm along a vertical line, plot the functions $r(t)$ and $\theta(t)$ required to achieve the given motion from B to C.

(f) Finally, use the geometrical constraints $0.5 \tan \theta = y$ and $r^2 = y^2 + (0.5)^2$ to determine analytical expressions for $r(t)$ and $\theta(t)$, and compare the plots of these analytical expressions with the plots found in Part (e). They should, of course, be the same.

Design Problem 2.3

As a part of a robotics competition, a robotic arm is to be designed so as to catch an egg without breaking it. The egg is released at point A from rest while the arm is initially also at rest in the position shown. The arm starts moving when the egg is released, and it is to catch the egg at point B in such a way as to avoid any impact between the egg and the robot hand. See Design Problem 2.2 for how the robot arm needs to be moving for it to *catch* the egg. With this in mind, find an acceleration profile (i.e., $\ddot{r}(t)$ and $\ddot{\theta}(t)$) of the arm as it moves from its initial position to B that satisfies this condition.

2.7 Relative Motion Analysis and Differentiation of Geometrical Constraints

In this section we discuss the concepts of relative motion and differentiation of constraints. These concepts are used in the solution of problems with multiple moving objects and are important in the development of rigid body kinematics. We will study relative motion using frames of reference that only translate relative to one another. The general case, which includes frames that also rotate relative to one another, is presented in Section 6.4. As for the moving target problem stated below, we present its solution in Example 2.26.

Hitting a moving target

Referring to Fig. 2.50, consider an action movie scene in which the railcar A and the vehicle B are about to collide at the railroad crossing C. The railcar A is traveling with a constant speed of $18\,\text{m/s}$ while B, which is loaded with explosives, is approaching the crossing at a constant speed of $40\,\text{m/s}$. The movie's hero is on A and needs to aim a gun at B and fire it to destroy B before the collision. The gun can fire a projectile P at $300\,\text{m/s}$, and our hero is supposed to fire the gun exactly 4 s from the crossing. If you were the technical advisor on this film, *in what direction, relative to the railcar, would you tell our hero to point the gun on A in order to hit B?*

The questions posed by this problem require that we characterize the motion of B *from the viewpoint of the hero*; that is, the speed of the projectile must be interpreted as the *speed of the projectile as seen by the hero*. On the other hand, the velocities of A and B are given by an observer who is not moving relative to the tracks and the road and thus sees *both A and B moving*. The observer who is not moving relative to the tracks and road will be referred to as *stationary*, and the hero at A is then a *moving observer*, i.e., an observer whose frame of reference, here the railcar A, is moving. Based on these observations, the problem's solution is found by translating position, velocity, and acceleration information from the stationary frame to the moving frame. Translating information between reference frames in motion relative to one another is what the analysis of *relative motion* is about.

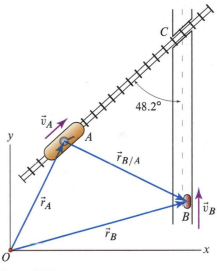

Figure 2.50
Railcar A and vehicle B approaching a rail crossing. *Not drawn to scale.*

Relative motion

Referring to Fig. 2.51, consider two points A and B moving on separate paths in a plane. We attach to the plane of motion a Cartesian frame of reference with axes X and Y and base vectors $\hat{I}$ and $\hat{J}$. We refer to this frame as the *stationary frame*. To particle A we attach a Cartesian frame of reference with axes x and y and base vectors $\hat{i}$ and $\hat{j}$. We refer to this second frame as the *moving frame*. Using Eq. (1.16) on p. 11, in the XY frame the position of B relative to A is

$$\vec{r}_{B/A} = (X_B - X_A)\,\hat{I} + (Y_B - Y_A)\,\hat{J}, \qquad (2.99)$$

where (X_A, X_B) and (X_B, Y_B) are the coordinates of A and B relative to the XY frame, respectively, and where we recall that the subscript B/A is read "B relative to A" (see discussion of Eq. (1.16) on p. 11).

The position of B relative to A as viewed by the xy frame is

$$\vec{r}_{B/A} = x_B\,\hat{i} + y_B\,\hat{j}, \qquad (2.100)$$

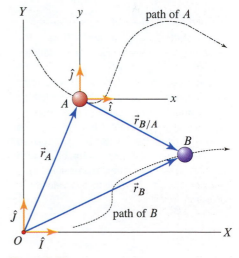

Figure 2.51
Two particles A and B and the definition of their relative position vector $\vec{r}_{B/A}$.

where x_B and y_B are the coordinates of B relative to the xy frame. Since the xy frame is Cartesian, when the moving observer differentiates the vector $\vec{r}_{B/A}$ with respect to time, this observer obtains

$$\left(\dot{\vec{r}}_{B/A}\right)_{xy \text{ frame}} = \dot{x}_B\,\hat{\imath} + \dot{y}_B\,\hat{\jmath}. \tag{2.101}$$

By contrast, when the stationary observer computes the same time derivative, the result is

$$\left(\dot{\vec{r}}_{B/A}\right)_{XY \text{ frame}} = \left(\dot{X}_B - \dot{X}_A\right)\hat{I} + \left(\dot{Y}_B - \dot{Y}_A\right)\hat{J}$$
$$= \dot{x}_B\,\hat{\imath} + x_B\,\dot{\hat{\imath}} + \dot{y}_B\,\hat{\jmath} + y_B\,\dot{\hat{\jmath}}, \tag{2.102}$$

where the last equality in Eq. (2.102) is obtained by letting the stationary observer compute the time derivative of Eq. (2.100). Applying Eq. (2.60) on p. 92 gives $\dot{\hat{\imath}} = \vec{\omega}_{\hat{\imath}} \times \hat{\imath}$, and $\dot{\hat{\jmath}} = \vec{\omega}_{\hat{\jmath}} \times \hat{\jmath}$, where $\vec{\omega}_{\hat{\imath}} = \vec{\omega}_{\hat{\jmath}}$ is the angular velocity of the xy frame as measured by the stationary observer. In this section, we consider only the case in which the xy and XY frames do not rotate relative to one another, so that $\vec{\omega}_{\hat{\imath}} = \vec{\omega}_{\hat{\jmath}} = \vec{0}$ and the stationary observer obtains

$$\dot{\hat{\imath}} = \vec{0} \quad \text{and} \quad \dot{\hat{\jmath}} = \vec{0}. \tag{2.103}$$

Consequently, substituting Eqs. (2.103) into Eq. (2.102) and comparing the result with Eq. (2.101), we have

$$\left(\dot{\vec{r}}_{B/A}\right)_{xy \text{ frame}} = \left(\dot{\vec{r}}_{B/A}\right)_{XY \text{ frame}}. \tag{2.104}$$

Equation (2.104) states that the time rate of change of the vector $\vec{r}_{B/A}$ is the same for the stationary and moving observers when these observers only translate relative to one another.

To relate position, velocity, and acceleration between the xy and XY reference frames, we now consider the vector triangle OAB, for which we have (see Fig. 2.51)

$$\boxed{\vec{r}_B = \vec{r}_A + \vec{r}_{B/A},} \tag{2.105}$$

where $\vec{r}_A$ and $\vec{r}_B$ are the position vectors of A and B relative to the XY reference frame, respectively. The time derivative of Eq. (2.105) gives

$$\boxed{\vec{v}_B = \vec{v}_A + \vec{v}_{B/A},} \tag{2.106}$$

where $\vec{v}_{B/A} = d\vec{r}_{B/A}/dt$ is the relative velocity of B with respect to A. Differentiating Eq. (2.106) with respect to time, we have

$$\boxed{\vec{a}_B = \vec{a}_A + \vec{a}_{B/A},} \tag{2.107}$$

where $\vec{a}_{B/A} = d^2\vec{r}_{B/A}/dt^2$ is the relative acceleration of B with respect to A. Because of Eq. (2.104), $\vec{v}_{B/A}$ and $\vec{a}_{B/A}$ are identical in the xy and XY frames. Equation (2.106) says that *the velocity of B, as seen by the stationary observer, is equal to the velocity of A, as seen by the stationary observer, plus the relative velocity of B with respect to A, which is the velocity of B as seen by the moving observer.* By replacing *velocity* with *acceleration*, Eq. (2.107) can be read in a similar way.

Differentiation of geometrical constraints

The relative motion equations have an important application in the analysis of constrained systems. As an example of a constrained system, consider the pulley systems in Fig. 2.52, for which we want to determine how the velocity and acceleration of block P are related to the velocity and acceleration of block Q under the constraint that the cords in the system are *inextensible*. Although for simple pulley systems we can sometimes intuit these relationships, here we want to develop a systematic approach applicable to systems of any complexity.

The key to the analysis of *any* pulley system is the notion of cord (or cable or rope) length and its first and second time derivatives. In Fig. 2.52 there are three cords. However, cords GI and JH simply keep block P attached to pulley G and pulley H attached to the fixed ceiling at J, respectively. Therefore

$$y_{I/G} = \text{constant} \quad \Rightarrow \quad v_{I/G} = 0 \quad \Rightarrow \quad a_{I/G} = 0 \quad (2.108)$$

$$\Rightarrow \quad v_P = v_G \quad \Rightarrow \quad a_P = a_G, \quad (2.109)$$

$$y_{H/J} = \text{constant} \quad \Rightarrow \quad v_{H/J} = 0 \quad \Rightarrow \quad a_{H/J} = 0 \quad (2.110)$$

$$\Rightarrow \quad v_H = v_J = 0 \quad \Rightarrow \quad a_H = a_J = 0, \quad (2.111)$$

where Eqs. (2.111) hold because J is a fixed point. Recognizing these relations, we now only need to worry about the cord $ABCDEF$.

Let the length of cord $ABCDEF$ be L, which is a constant since all cords are assumed inextensible. We now express L in terms of the quantities shown in Fig. 2.52, i.e.,

$$L = \overline{AB} + \widehat{BC} + \overline{CD} + \widehat{DE} + \overline{EF}, \quad (2.112)$$

where $\widehat{BC}$ and $\widehat{DE}$ are the lengths of the *curved* segments of cord that wrap around pulleys G and H, respectively, and the letters with overbars represent the lengths of the corresponding *straight* line segments. Observe that the length of each cord segment either is *constant* or can be *written in terms of the coordinates of blocks P and Q*. Let's write them each out as follows:

$$\overline{AB} = y_P - \overline{GI}, \quad \overline{CD} = y_P - \overline{GI} - \overline{JH}, \quad \overline{EF} = y_Q - \overline{JH}, \quad (2.113)$$

and then Eq. (2.112) becomes

$$L = 2y_P + y_Q - 2\overline{GI} - 2\overline{JH} + \widehat{BC} + \widehat{DE}. \quad (2.114)$$

We now differentiate Eq. (2.114) with respect to time and recognize that

- $d(\overline{GI})/dt = dy_{I/G}/dt = v_{I/G} = 0$, due to Eq. (2.108).

- $d(\overline{JH})/dt = dy_{H/J}/dt = v_{H/J} = 0$, due to Eq. (2.110).

- $d(\widehat{BC})/dt$ and $d(\widehat{DE})/dt$ are both zero because the *quantity* of cord wrapped around any of the pulleys is always the same.

Therefore we obtain

$$\dot{L} = 2\dot{y}_P + \dot{y}_Q. \quad (2.115)$$

Recalling that L is constant because the cord is inextensible, $\dot{L} = 0$ and Eq. (2.115) yields the relation between the velocities of blocks P and Q, i.e.,

$$2\dot{y}_P + \dot{y}_Q = 0 \quad \text{or} \quad 2v_P + v_Q = 0. \quad (2.116)$$

A few observations about Eq. (2.116) are now in order.

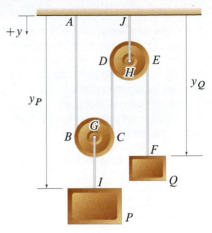

Figure 2.52
Simple pulley system used to demonstrate the principles behind the differentiation of geometric constraints.

Helpful Information

What about the cord wrapped around the pulleys? The fact that the time derivative of the segment of cord wrapped around each pulley is zero was crucial to our simplification of the kinematic equations. Observe that it is not the *part* of the cord that is constant, but the *quantity* of cord that is important. Thus, even though the part of the cord that is touching each pulley is changing, the *length* of cord that is touching each pulley is not. This length is equal to $\pi \times$ the radius of the pulley and is therefore constant.

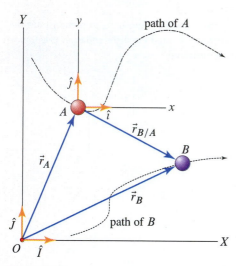

Helpful Information

What if the cord length is not constant? In some problems the cord length is not constant, such as when a motor or winch is pulling in or letting out cord. In these cases, we modify the kinematics by setting the time derivative of the appropriate cord length equal to the rate at which the cord is becoming longer or shorter, i.e.,

$$\dot{L} = \begin{cases} + \text{ rate of increase,} \\ \quad \text{or} \\ - \text{ rate of decrease.} \end{cases}$$

Figure 2.53

Figure 2.51 repeated. Two particles A and B and the definition of their relative position vector $\vec{r}_{B/A}$.

- Equation (2.116) says that if P is moving, say, at $4\,\text{m/s}$ *down*, then Q must be moving at $8\,\text{m/s}$ *up*. This is so because $v_Q = -2v_P$ and we have defined *both* y_P and y_Q to be positive downward.

- We can differentiate Eq. (2.116) once with respect to time to obtain a relationship between the accelerations of blocks P and Q

$$2a_P + a_Q = 0. \tag{2.117}$$

- We did not need to know the length of each cord—we only needed to know that each length was constant. This will frequently be the case.

End of Section Summary

Relative motion.   Referring to Fig. 2.53, consider the planar motion of points A and B. The positions of A and B relative to the XY frame are $\vec{r}_A$ and $\vec{r}_B$, respectively. Attached to A there is a frame xy that translates but does not rotate relative to frame XY. In either frame, the position of B relative to A is given by $\vec{r}_{B/A}$. Using vector addition, the vectors $\vec{r}_A$, $\vec{r}_B$, and $\vec{r}_{B/A}$ are related as follows:

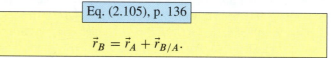

Eq. (2.105), p. 136

$$\vec{r}_B = \vec{r}_A + \vec{r}_{B/A}.$$

The first and second time derivatives of the above equation are, respectively,

Eqs. (2.106) and (2.107), p. 136

$$\vec{v}_B = \vec{v}_A + \vec{v}_{B/A},$$
$$\vec{a}_B = \vec{a}_A + \vec{a}_{B/A},$$

where $\vec{v}_{B/A} = \dot{\vec{r}}_{B/A}$ and $\vec{a}_{B/A} = \ddot{\vec{r}}_{B/A}$ are the relative velocity and acceleration of B with respect to A, respectively. In general, the vectors $\vec{v}_{B/A}$ and $\vec{a}_{B/A}$ computed by the xy observer are different from the vectors $\vec{v}_{B/A}$ and $\vec{a}_{B/A}$ computed by the XY observer. However, the xy and XY observers compute the same $\vec{v}_{B/A}$ and the same $\vec{a}_{B/A}$ if these observers do not rotate relative to one another.

Constrained motion.   There are very few dynamics problems in which the motion is *not* constrained in some way. In certain classes of systems, constraints are described by geometrical relations between points in the system. We analyzed a pulley system whose motion was constrained by the inextensibility of the cords in the system. The key to constrained motion analysis is the awareness that we can differentiate the equations describing the geometrical constraints to obtain velocities and accelerations of points of interest. In the case of the pulley system studied in this section, it was the time rate of change of the cord length that we used. For an inextensible cord, the cord length is constant so that the time rate of change of the cord length is zero.

EXAMPLE 2.24   *Relative Speed and Acceleration*

The driver of car B sees a police car P and applies his brakes, causing the car to decelerate at a constant rate of 25 ft/s^2. At the same time, the police car is traveling at a constant speed $v_P = 35$ mph, and using a radar gun, the police officer sees B coming toward her at 65 mph when $\theta = 22°$. At the instant that the radar gun measurement was taken, determine the corresponding true speed of B and the magnitude of the relative acceleration of B with respect to P.

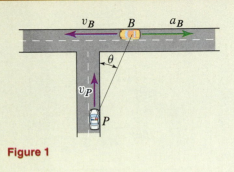

Figure 1

SOLUTION

Road Map   We need to find the speed of B with respect to the road, which we choose as our stationary frame. The speed measured by the radar gun is relative to the moving observer P and is the component of the velocity of B relative to P along the line connecting B and P. Hence, we will find the velocity of B relative to P and then consider the component of this velocity along the line PB. As for the relative acceleration of B with respect to P, we can calculate it as a direct application of Eq. (2.107).

Computation   Letting v_B be the speed of B, referring to Fig. 2, the velocity of B relative to P is

$$\vec{v}_{B/P} = \vec{v}_B - \vec{v}_P = -v_B \, \hat{\imath} - v_P \, \hat{\jmath}. \tag{1}$$

Denoting the radar gun speed reading by v_r, noting that v_r is negative because the police officer sees B coming *toward* her, and observing that $\hat{u}_{B/P} = \sin\theta \, \hat{\imath} + \cos\theta \, \hat{\jmath}$ (the unit vector pointing from P to B), we have

$$-v_r = \vec{v}_{B/P} \cdot \hat{u}_{B/P} = -v_B \sin\theta - v_P \cos\theta. \tag{2}$$

Solving Eq. (2) for v_B, we have

$$\boxed{v_B = \frac{v_r - v_P \cos\theta}{\sin\theta} = 86.9 \, \text{mph.}} \tag{3}$$

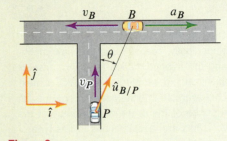

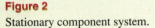

Figure 2
Stationary component system.

Applying Eq. (2.107), the acceleration of B relative to B is

$$\vec{a}_{B/P} = \vec{a}_B - \vec{a}_P = (25 \, \text{ft/s}^2) \, \hat{\imath}, \tag{4}$$

since the police car is traveling at a constant velocity. Therefore we have

$$\boxed{|\vec{a}_{B/P}| = 25 \, \text{ft/s}^2.} \tag{5}$$

Discussion & Verification   Since the terms $\cos\theta$ and $\sin\theta$ are nondimensional, the result in Eq. (3) is dimensionally correct. The result in Eq. (5) is also dimensionally correct since it was derived as a direct application of a dimensionally correct formula for the relative acceleration.

EXAMPLE 2.25   *Relative Motion in a Car Shock Absorbing System*

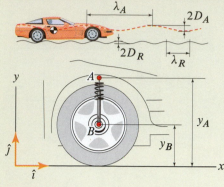

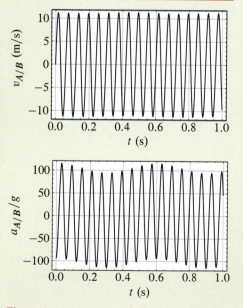

Figure 1

Very simple schematic of a car's shock absorbing system. Point A is attached to the frame of the car.

Helpful Information

Our model. Real roads do not undulate sinusoidally, real wheels are deformable, and the relative motion of A with respect to B should be studied in a 3D context. This example is only meant to help us get a sense of the order of magnitude of the velocities and accelerations in this type of system.

Consider a car traveling over an undulating road (Fig. 1) such that points A and B have the same constant horizontal velocity v_0. Determine, as functions of time, the relative position, velocity, and acceleration of A with respect to B. Assume that A and B move such that $y_A = h_A + D_A \sin(2\pi x_A/\lambda_A)$ and $y_B = R_W + D_R \sin(2\pi x_B/\lambda_R)$, respectively. Let h_A and R_w be the height of the top of the suspension and the center of the wheel, respectively, when the car is not moving.

SOLUTION

Road Map   Once we describe the position of points A and B, we can find their relative position by subtracting them. The relative velocity and acceleration can then be found by differentiating the relative position with respect to time.

Computation   Using the coordinate system shown in Fig. 1, we have

$$\vec{r}_A = x_A\,\hat{\imath} + y_A\,\hat{\jmath} \quad \text{and} \quad \vec{r}_B = x_B\,\hat{\imath} + y_B\,\hat{\jmath}. \tag{1}$$

Using Eq. (2.105), the relative position becomes

$$\vec{r}_{A/B} = \vec{r}_A - \vec{r}_B = (x_A - x_B)\,\hat{\imath} + (y_A - y_B)\,\hat{\jmath} = y_{A/B}\,\hat{\jmath}, \tag{2}$$

where $x_A = x_B$ since A and B move with the same constant horizontal velocity. Using the given trajectories of A and B, the relative position vector $\vec{r}_{A/B}$ takes on the form

$$\vec{r}_{A/B} = y_{A/B}\,\hat{\jmath} = \left[h_A - R_W + D_A \sin\!\left(\frac{2\pi v_0 t}{\lambda_A}\right) - D_R \sin\!\left(\frac{2\pi v_0 t}{\lambda_R}\right) \right] \hat{\jmath}, \tag{3}$$

where we have used $x_A = x_B = v_0 t$ since the vehicle is moving at the constant speed v_0. We have set, without loss of generality, $x_A(0) = x_B(0) = 0$.

The relative velocity and acceleration are obtained by differentiating Eq. (3) with respect to time, using the chain rule, and noting that $\dot{x}_A = v_0$ and $\ddot{x}_A = 0$, i.e.,

$$\dot{y}_{A/B} = \frac{dy_{A/B}}{dx_A}\frac{dx_A}{dt} = \frac{dy_{A/B}}{dx_A}\dot{x}_A = \frac{dy_{A/B}}{dx_A}v_0, \tag{4}$$

$$\ddot{y}_{A/B} = v_0\frac{d}{dt}\left(\frac{dy_{A/B}}{dx_A}\right) = v_0\frac{d}{dx_A}\left(\frac{dy_{A/B}}{dx_A}\right)\frac{dx_A}{dt} = \frac{d^2 y_{A/B}}{dx_A^2}v_0^2, \tag{5}$$

so that

$$\vec{v}_{A/B} = \dot{y}_{A/B}\,\hat{\jmath} = \frac{dy_{A/B}}{dx_A}v_0\,\hat{\jmath} \quad \text{and} \quad \vec{a}_{A/B} = \ddot{y}_{A/B}\,\hat{\jmath} = \frac{d^2 y_{A/B}}{dx_A^2}v_0^2\,\hat{\jmath}. \tag{6}$$

Substituting Eq. (3) into Eqs. (6), we obtain

$$\vec{v}_{A/B} = 2\pi v_0\left[\frac{D_A}{\lambda_A}\cos\!\left(\frac{2\pi x_A}{\lambda_A}\right) - \frac{D_R}{\lambda_R}\cos\!\left(\frac{2\pi x_B}{\lambda_R}\right)\right]\hat{\jmath}, \tag{7}$$

$$\vec{a}_{A/B} = -4\pi^2 v_0^2\left[\frac{D_A}{\lambda_A^2}\sin\!\left(\frac{2\pi x_A}{\lambda_A}\right) - \frac{D_R}{\lambda_R^2}\sin\!\left(\frac{2\pi x_B}{\lambda_R}\right)\right]\hat{\jmath}. \tag{8}$$

Discussion & Verification   The verification of the dimensional correctness of the result is left to the reader as an exercise.

Figure 2

Plots of relative velocity and acceleration. The top plot shows acceleration values in g. The following values were used: $D_A = D_R = 0.1$ m, $\lambda_A = 10$ m, $\lambda_R = 1$ m, $v_0 = 60$ km/h.

🔎 **A Closer Look**   Figure 2 shows a representation of Eqs. (7) and (8) for a specific choice of the relevant geometrical parameters. Notice that the magnitudes of the vertical velocity are of the same order as the horizontal speed of the car. Furthermore, the maximum acceleration exceeds $100g$!

EXAMPLE 2.26 *Analysis of a Moving Target Problem*

In an action movie scene, railcar A and vehicle B are 4 s away from colliding at the crossing C (Fig. 1). At this instant the movie's hero is traveling on the railcar and is set to fire a gun at B to destroy B before the collision. A and B move with constant speeds $v_A = 18$ m/s and $v_B = 40$ m/s, respectively, and the gun's projectile P travels at 300 m/s. In what direction θ would you tell the hero on A to point the gun in order to hit B?

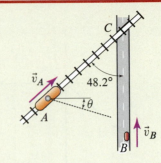

Figure 1
Railcar and robot approaching a rail crossing. The figure is *not drawn to scale*.

SOLUTION

Road Map Let $\vec{v}_{P/A}$ be the velocity vector of the projectile P relative to railcar A. The magnitude of $\vec{v}_{P/A}$ is 300 m/s, and we want to solve for $\vec{v}_{P/A}$, which provides the direction the hero points the gun and can be described by the angle θ in Fig. 1. To find θ, we will use relative kinematics to relate $\vec{v}_{P/A}$, which is measured relative to A, with $\vec{v}_A$ and $\vec{v}_B$, which are relative to the ground. If $\vec{r}_P(t)$ and $\vec{r}_B(t)$ are the positions of the projectile P and of the target B as a function of time, respectively, then to "hit the target" means that there must be a time t_h such that P and B occupy the same position, i.e.,

$$\vec{r}_P(t_h) = \vec{r}_B(t_h). \tag{1}$$

Our strategy will be to describe the velocities of P and B, which we will integrate with respect to time to obtain $\vec{r}_P(t)$ and $\vec{r}_B(t)$ so that we can satisfy Eq. (1).

Computation Referring to Fig. 2, we define a reference frame with origin at C and a y axis coinciding with the path of B. We will refer to this frame as *stationary*. We also observe that, once fired, P travels with a constant velocity $\vec{v}_P$ relative to the stationary frame. Thus, if we let t_f be the time of firing, P moves along the segment joining the position of A at t_f to the position of B at t_h. If we Let $\vec{v}_P$ and $\vec{v}_B$ be the velocities of P and B, respectively, in the stationary frame, they can be written as

$$\vec{v}_P = v_{Px}\,\hat{i} + v_{Py}\,\hat{j} \quad \text{and} \quad \vec{v}_B = v_B\,\hat{j}, \tag{2}$$

where v_{Px} and v_{Py} are unknown and $v_B = 40$ m/s. Since $\vec{v}_P$ and $\vec{v}_B$ are constant, applying constant acceleration equations component by component, Eq. (2) becomes

$$\vec{r}_P(t) = [r_{Px}(t_f) + v_{Px}(t - t_f)]\,\hat{i} + [r_{Py}(t_f) + v_{Py}(t - t_f)]\,\hat{j}, \tag{3}$$

$$\vec{r}_B(t) = [r_{By}(t_f) + v_B(t - t_f)]\,\hat{j}, \tag{4}$$

where $\vec{r}_P(t_f) = r_{Px}(t_f)\,\hat{i} + r_{Py}(t_f)\,\hat{j}$ and $\vec{r}_B(t_f) = r_{By}(t_f)\,\hat{j}$ are the positions of P and B at the time of firing, respectively.

Since P is fired from A, for $t = t_f$ we must have $\vec{r}_P(t_f) = \vec{r}_A(t_f)$. In addition, at $t = t_f$, A and B are 4 s away from C, and so we have (see Fig. 3)

$$\vec{r}_P(t_f) = \vec{r}_A(t_f) = -d\sin\beta\,\hat{i} - d\cos\beta\,\hat{j} \quad \text{with} \quad d = v_A(4\,\text{s}) = 72\,\text{m}, \tag{5}$$

and

$$\vec{r}_B(t_f) = -\ell\,\hat{j} \quad \text{with} \quad \ell = v_B(4\,\text{s}) = 160\,\text{m}. \tag{6}$$

Since $\vec{v}_P$ is constant, its value is determined by the conditions at A (i.e., at $t = t_f$). Referring to Fig. 4, relative kinematics says that at the time of firing we have

$$\vec{v}_P = \vec{v}_A + \vec{v}_{P/A} = v_A(\sin\beta\,\hat{i} + \cos\beta\,\hat{j}) + v_{P/A}(\cos\theta\,\hat{i} - \sin\theta\,\hat{j}),$$

$$= (v_A\sin\beta + v_{P/A}\cos\theta)\,\hat{i} + (v_A\cos\beta - v_{P/A}\sin\theta)\,\hat{j}, \tag{7}$$

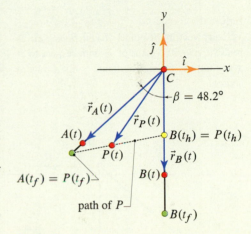

Figure 2
Position vectors of points A, B, and P at a generic time t. The figure is *not drawn to scale*.

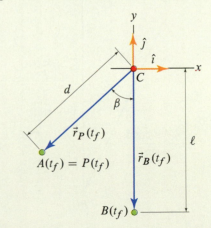

Figure 3
Position vectors of points P and B at the time of firing. The figure is *not drawn to scale*.

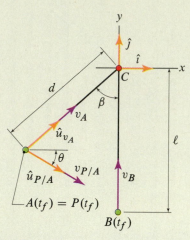

Figure 4
Velocity vectors of points A and B as well as relative velocity vector of P with respect to A at the time of firing. The figure is *not drawn to scale*.

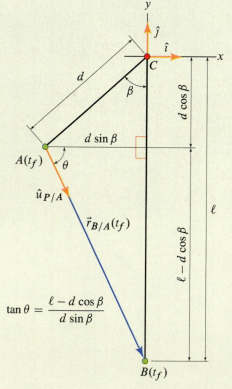

Figure 5
Position of B relative to A at the time of firing. This figure *is* drawn to scale.

Using Eqs. (5)–(7), we can then rewrite Eqs. (3) and (4) as

$$\vec{r}_P(t) = [-d\sin\beta + (v_A\sin\beta + v_{P/A}\cos\theta)(t - t_f)]\,\hat{\imath}$$
$$+ [-d\cos\beta + (v_A\cos\beta - v_{P/A}\sin\theta)(t - t_f)]\,\hat{\jmath}, \tag{8}$$
$$\vec{r}_B(t) = [-\ell + v_B(t - t_f)]\,\hat{\jmath}. \tag{9}$$

Now we are ready to enforce the condition in Eq. (1), i.e., to set $\vec{r}_P(t_h) = \vec{r}_B(t_h)$. Letting $t = t_h$ in Eqs. (8) and (9) and setting $\vec{r}_P(t_h) = \vec{r}_B(t_h)$ component by component, we have

$$-d\sin\beta + (v_A\sin\beta + v_{P/A}\cos\theta)(t_h - t_f) = 0, \tag{10}$$
$$-d\cos\beta + (v_A\cos\beta - v_{P/A}\sin\theta)(t_h - t_f) = -\ell + v_B(t_h - t_f). \tag{11}$$

Solving Eq. (10) for $t_h - t_f$ and substituting the result in Eq. (11), we get

$$-d\cos\beta + (v_A\cos\beta - v_{P/A}\sin\theta)\frac{d\sin\beta}{v_A\sin\beta + v_{P/A}\cos\theta}$$
$$= -\ell + v_B\frac{d\sin\beta}{v_A\sin\beta + v_{P/A}\cos\theta}. \tag{12}$$

Multiplying through by the denominator of the two fractions and simplifying give

$$-dv_{P/A}(\cos\theta\cos\beta + \sin\theta\sin\beta) = -\ell(v_A\sin\beta + v_{P/A}\cos\theta) + v_Bd\sin\beta. \tag{13}$$

Now, recall that we want to solve for θ, and so isolating $\sin\theta$ and $\cos\theta$ on one side the equation, we obtain

$$(\ell - d\cos\beta)\cos\theta - d\sin\beta\sin\theta = \frac{\sin\beta}{v_{P/A}}(v_Bd - v_A\ell). \tag{14}$$

Now recall that, at $t = t_f$, A and B are 4 s away from colliding at C, and therefore we must have

$$4\,\text{s} = d/v_A = \ell/v_B \quad\Rightarrow\quad v_A\ell = v_Bd \quad\Rightarrow\quad v_Bd - v_A\ell = 0. \tag{15}$$

Therefore, Eq. (14) becomes

$$(\ell - d\cos\beta)\cos\theta - d\sin\beta\sin\theta = 0, \tag{16}$$

which can be solved for θ to obtain

$$\boxed{\theta = \tan^{-1}\left(\frac{\ell - d\cos\beta}{d\sin\beta}\right) = 64.4°.} \tag{17}$$

Discussion & Verification The argument of the inverse tangent in Eq. (17) is nondimensional, as it should be.

Observe that our result is independent of $v_{P/A}$. This result is unusual in moving target problems, and in general, it suggests that an observer on B sees P travel along a straight line that does not change its orientation, almost as if B were stationary and P traveled along the line connecting A and B at the time of firing. Although unusual, in this particular problem this result is correct, but only because *A and B are on a collision course while traveling at constant velocity*. This means that A and B see each other moving directly toward each other along a straight line whose orientation relative to A or B remains fixed. In turn, this means that when the hero aims the gun at B, the hero must be aiming directly at B. This can be seen by referring to Fig. 5, which also allows us to obtain the result found in Eq. (17) by using simple trigonometry.

EXAMPLE 2.27   *Pulley System Analysis*

Figure 1 shows a pulley system for which we want to determine how the velocity and acceleration of block P are related to the velocity and acceleration of block Q when they are connected by the pulleys and *inextensible* cords shown. Find the equations relating v_P to v_Q and a_P to a_Q.

SOLUTION

Road Map   The key to the analysis of *any* pulley system is the notion of cord (or cable or rope) length and its first and second time derivatives. We note that in Fig. 1 there are a total of four cords, but two of the cords, that is, cords EF and MN, simply keep block P attached to pulley E and pulley N attached to the fixed ceiling, respectively. Kinematically, we are saying that

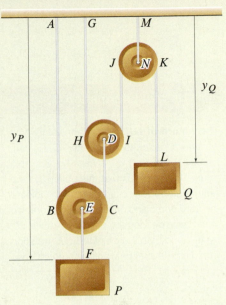

Figure 1
A pulley system for which we want to relate the motion of block P to that of block Q.

$$y_{F/E} = \text{constant} \quad \Rightarrow \quad v_{F/E} = 0 \quad \Rightarrow \quad a_{F/E} = 0 \tag{1}$$
$$\Rightarrow \quad v_P = v_F = v_E \quad \Rightarrow \quad a_P = a_F = a_E, \tag{2}$$
$$y_{N/M} = \text{constant} \quad \Rightarrow \quad v_{N/M} = 0 \quad \Rightarrow \quad a_{N/M} = 0 \tag{3}$$
$$\Rightarrow \quad v_M = v_N = 0 \quad \Rightarrow \quad a_M = a_N = 0. \tag{4}$$

Recognizing these relations, we now only need to address the two cords: $ABCD$ and $GHIJKL$.

Computation   Let the length of cord $ABCD$ be L_1 and the length of cord $GHIJKL$ be L_2. Notice that in the situation shown in Fig. 1, the cords have constant length since we are told they are inextensible. Let's express these lengths in terms of the quantities shown in Fig. 1. The lengths L_1 and L_2 can be written as

$$L_1 = \overline{AB} + \overparen{BC} + \overline{CD}, \tag{5}$$
$$L_2 = \overline{GH} + \overparen{HI} + \overline{IJ} + \overparen{JK} + \overline{KL}, \tag{6}$$

where $\overparen{BC}$, $\overparen{HI}$, and $\overparen{JK}$ represent the lengths of the *curved* segments of cord that wrap around pulleys E, D, and N, respectively, and the letters with overbars represent the lengths of the corresponding *straight* line segments. Observe that the lengths of each of the cord segments either are *constant* or can be *written in terms of the coordinates describing the positions of the blocks* P and Q. Hence, we can write

$$\overline{AB} = y_P - \overline{EF}, \qquad\qquad \overline{CD} = y_P - \overline{EF} - \overline{GH}, \tag{7}$$
$$\overline{IJ} = \overline{GH} - \overline{MN}, \qquad\qquad \overline{KL} = y_Q - \overline{MN}, \tag{8}$$

and then Eqs. (5) and (6) become, respectively,

$$L_1 = 2y_P - \overline{GH} - 2\overline{EF} + \overparen{BC}, \tag{9}$$
$$L_2 = y_Q + 2\overline{GH} + \overparen{HI} + \overparen{JK} - 2\overline{MN}. \tag{10}$$

If we now differentiate Eqs. (9) and (10) with respect to time and recognize that

- $d(\overline{EF})/dt = dy_{F/E}/dt = v_{F/E} = 0$ due to Eq. (1).

- $d(\overline{MN})/dt = dy_{N/M}/dt = v_{N/M} = 0$ due to Eq. (3).

- $d(\overparen{BC})/dt, d(\overparen{HI})/dt$, and $d(\overparen{JK})/dt$ are all zero because the *quantity* of cord wrapped around any of the pulleys is always the same.

> ## Concept Alert
>
> **Does the length of the cord matter?**
> Again, it is not the length of the cord that is important; only its rate of change of length matters.

then we obtain

$$\dot{L}_1 = 2\dot{y}_P - \frac{d(\overline{GH})}{dt} = 2\dot{y}_P - \dot{\overline{GH}}, \tag{11}$$

$$\dot{L}_2 = \dot{y}_Q + 2\frac{d(\overline{GH})}{dt} = \dot{y}_Q + 2\dot{\overline{GH}}. \tag{12}$$

We now make the following observations:

1. Since we have stated that the cord is inextensible, its length must be constant and therefore the time derivative of its length must be zero, i.e., $\dot{L}_1 = \dot{L}_2 = 0$.

2. We still have $\dot{\overline{GH}}$ hanging around in Eqs. (11) and (12). However, both equations involve $\dot{\overline{GH}}$ so it can be eliminated from them.

Using these two items, Eqs. (11) and (12) become the one relationship between the velocity of block P and block Q

$$\boxed{4\dot{y}_P + \dot{y}_Q = 0 \quad \text{or} \quad 4v_P + v_Q = 0,} \tag{13}$$

which can then be differentiated to the acceleration relationship as

$$\boxed{4a_P + a_Q = 0.} \tag{14}$$

Discussion & Verification   Equations (13) and (14) are dimensionally homogeneous, as they should be. In addition, notice that Eq. (13) says that if P is moving at $4\,\text{m/s}$ *down*, then Q must be moving at $16\,\text{m/s}$ *up*. This is so because we have defined *both* y_P and y_Q to be positive downward and $v_Q = -4v_P$. This result is consistent with the geometry of the pulley system in Fig. 1.

⊕ Helpful Information

The nature of $\dot{\overline{GH}}$. The manner in which we were able to eliminate $\dot{\overline{GH}}$ from Eqs. (11) and (12) will always be available to us. The reason is that this pulley system really has *three* important moving objects, the third one being pulley D. Therefore, we could have defined the position of pulley D by y_D, and then the awkward $\overline{GH}$ notation would not have appeared in the equations.

EXAMPLE 2.28 *Constrained Motion of a Ladder*

A worker on a maintenance crew has decided that it would be easier to let his truck "do the lifting" than to do it himself. With this in mind, he decides to prop his ladder up against the building and stand on the ladder's end at A while his buddy starts the truck from rest and backs it up with constant acceleration a_T in the direction shown in Fig. 1. Given a sufficiently strong ladder and that there are no windows in the way, this will cause the worker at A to rise up the wall of the building. Observing that end B of the ladder is a distance h off the ground, the length of the ladder is L, and θ starts at θ_0, determine the velocity and acceleration of the worker at A as a function of the angle θ and, as a part of the solution, determine how $\dot\theta$ and $\ddot\theta$ relate to a_T.

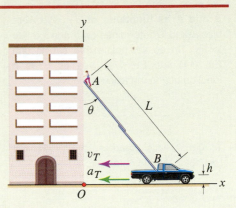

Figure 1
The quantities v_T and a_T denote the components of the velocity and acceleration, respectively, of the truck in the direction of the truck's motion, which is to the left.

SOLUTION

Road Map We can do the required kinematic analysis of this system by noticing the following geometric constraints:

1. The length of the ladder is a constant L.

2. End A of the ladder is constrained to move only in the vertical direction by the wall.

3. End B of the ladder is constrained to move only in the horizontal direction by the motion of the truck.

All of these can be incorporated into one relationship by noting that point A, point B, and the wall of the building will always form a right triangle so that

$$(y_A - h)^2 + x_B^2 = L^2. \qquad (1)$$

Equation (1) is the key to our analysis.

Computation Since Eq. (1) is true for any value of y_A and x_B, we can differentiate it with respect to time to obtain

$$(y_A - h)\dot y_A + x_B \dot x_B = 0, \qquad (2)$$

where we have used the fact that the length of the ladder is constant so that its time derivative is zero. Rearranging Eq. (2), we obtain

$$\dot y_A = -\left(\frac{x_B}{y_A - h}\right)\dot x_B. \qquad (3)$$

This does not give us the velocity of A in the form we seek since we can still write x_B and y_A in terms of the angle θ. Noting that

$$x_B = L\sin\theta, \qquad y_A - h = L\cos\theta, \qquad (4)$$

and $\dot x_B = -v_T$, we see that Eq. (3) becomes

$$\dot y_A = v_A = v_T \tan\theta. \qquad (5)$$

We need to take this a step further since we know only a_T, the constant acceleration of the truck, and not v_T. Let's see how.

We begin by applying the chain rule to v_T to get a_T, which gives

$$a_T = \frac{dv_T}{dt} = \frac{dv_T}{d\theta}\frac{d\theta}{dt} = \dot\theta\,\frac{dv_T}{d\theta}, \qquad (6)$$

Helpful Information

Signs of $\dot x_B$ and $\dot y_A$. We must have the minus sign in the relation $\dot x_B = -v_T$ since we have defined positive x_B to be to the right and since we have designated positive v_T to be to the left. In addition, since positive y is upward and $0° < \theta < 90°$, Eq. (3) tells us that when the truck is backing up, i.e., when $v_T > 0$, then the worker at A must be moving upward.

🧭 Helpful Information

$\dot{\theta}$ **and** $\ddot{\theta}$ **as functions of** a_T. We were also asked to determine how $\dot{\theta}$ and $\ddot{\theta}$ relate to a_T. To do so, notice that substituting Eq. (11) into Eq. (7) gives us $\dot{\theta}$ as a function of θ in the following form:

$$\dot{\theta} = \frac{-\sqrt{2a_T L(\sin\theta_0 - \sin\theta)}}{L\cos\theta}.$$

To get $\ddot{\theta}$, let's differentiate Eq. (7) with respect to time to obtain

$$\ddot{\theta} = \frac{-a_T L\cos\theta - v_T L\dot{\theta}\sin\theta}{L^2\cos^2\theta},$$

which, upon substituting in Eq. (11), becomes

$$\ddot{\theta} = -\frac{a_T}{L}\sec\theta$$
$$+ \frac{2a_T\sin\theta}{L\cos^3\theta}(\sin\theta_0 - \sin\theta).$$

in which $\dot{\theta}$ can be found by taking the time derivative of x_B in Eqs. (4), that is,

$$\dot{x}_B = L\dot{\theta}\cos\theta \quad\Rightarrow\quad \dot{\theta} = \frac{\dot{x}_B}{L\cos\theta} = \frac{-v_T}{L\cos\theta}. \tag{7}$$

Substituting $\dot{\theta}$ from Eq. (7) into Eq. (6), we obtain

$$a_T = \frac{-v_T}{L\cos\theta}\frac{dv_T}{d\theta} \quad\Rightarrow\quad a_T L\cos\theta\, d\theta = -v_T\, dv_T. \tag{8}$$

Now, since a_T is constant, we can integrate this as follows

$$a_T L\int_{\theta_0}^{\theta}\cos\theta\, d\theta = -\int_0^{v_T} v_T\, dv_T, \tag{9}$$

$$a_T L(\sin\theta - \sin\theta_0) = -\tfrac{1}{2}v_T^2, \tag{10}$$

which gives the following for v_T as a function of θ

$$v_T^2 = 2a_T L(\sin\theta_0 - \sin\theta). \tag{11}$$

Finally, substituting Eq. (11) into Eq. (5), we obtain the final result for v_A

$$\boxed{v_A = \sqrt{2a_T L(\sin\theta_0 - \sin\theta)}\,\tan\theta.} \tag{12}$$

To obtain the acceleration of the worker at A, we could differentiate Eq. (2), (3), or (5), but we will work with Eq. (5) since it is the simplest of the three. Doing so, we obtain

$$\dot{v}_A = a_A = \dot{v}_T\tan\theta + v_T\dot{\theta}\sec^2\theta \tag{13}$$

$$= a_T\tan\theta + v_T\dot{\theta}\sec^2\theta, \tag{14}$$

so that, after we substitute in $\dot{\theta}$ from Eq. (7), Eq. (14) becomes

$$a_A = a_T\tan\theta - \frac{v_T^2\sec^2\theta}{L\cos\theta} = a_T\tan\theta - \frac{v_T^2}{L}\sec^3\theta. \tag{15}$$

Substituting Eq. (11) into Eq. (15), we obtain

$$\boxed{a_A = a_T\left[\tan\theta - 2\sec^3\theta(\sin\theta_0 - \sin\theta)\right].} \tag{16}$$

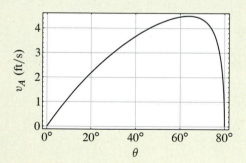

Figure 2
Plot of v_A as given by Eq. (12) for $L = 28$ ft, $a_T = 1$ ft/s^2, and $\theta_0 = 80°$.

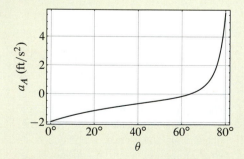

Figure 3
Plot of a_A as given by Eq. (16) for $a_T = 1$ ft/s^2 and $\theta_0 = 80°$.

Discussion & Verification   Our result was obtained in analytical form, and it is dimensionally correct. To facilitate the discussion of our result, plots of the functions in Eqs. (12) and (16) can be found in Figs. 2 and 3, respectively, for specified values of the parameters. Keep in mind that both Figs. 2 and 3 should be read from right to left since θ decreases as the truck backs up. It should be observed that the velocity of the person at A goes to zero as the ladder reaches the vertical position. This is to be expected, since the ladder has a finite length and as the ladder becomes more and more vertical, the amount of vertical displacement possible decreases. It is interesting that the acceleration of the person at A is quite large when the truck starts moving, i.e., when $\theta = \theta_0 = 80°$. We should also mention that Eq. (12) can also be found by using constant acceleration equations from Section 2.2.

PROBLEMS

💡 Problem 2.180 💡

Reference frame A is translating relative to reference frame B. Both frames track the motion of a particle C. If at one instant the velocity of particle C is the same in the two frames, what can you infer about the motion of frames A and B at that instant?
Note: Concept problems are about *explanations*, not computations.

💡 Problem 2.181 💡

Reference frame A is translating relative to reference frame B with velocity $\vec{v}_{A/B}$ and acceleration $\vec{a}_{A/B}$. A particle C appears to be stationary relative to frame A. What can you say about the velocity and acceleration of particle C relative to frame B?
Note: Concept problems are about *explanations*, not computations.

💡 Problem 2.182 💡

Reference frame A is translating relative to reference frame B with constant velocity $\vec{v}_{A/B}$. A particle C appears to be in uniform rectilinear motion relative to frame A. What can you say about the motion of particle C relative to frame B?
Note: Concept problems are about *explanations*, not computations.

Problem 2.183

A skier is going down an undulating *slope* with moguls. Let the skis be short enough for us to assume that the skier's feet are tracking the moguls' profile. Then if the skier is skilled enough to maintain her hips on a straight line trajectory and vertically aligned over her feet, determine the velocity and acceleration of her hips relative to her feet when her speed is equal to $15\,\text{km/h}$. For the profile of the moguls, use the formula $y(x) = h_I - 0.15x + 0.125\sin(\pi x/2)\,\text{m}$, where h_I is the elevation at which the skier starts the descent.

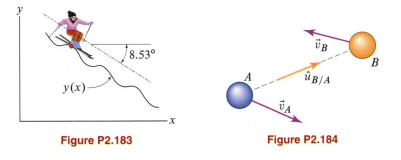

Figure P2.183   **Figure P2.184**

Problem 2.184

Two particles A and B are moving in a plane with arbitrary velocity vectors $\vec{v}_A$ and $\vec{v}_B$, respectively. Letting the *rate of separation* (ROS) be defined as the component of the relative velocity vector along the line connecting particles A and B, determine a general expression for ROS. Express your result in terms of $\vec{r}_{B/A} = \vec{r}_B - \vec{r}_A$, where $\vec{r}_A$ and $\vec{r}_B$ are the position vectors of A and B, respectively, relative to some chosen fixed point in the plane of motion.

Problem 2.185

Three vehicles A, B, and C are in the positions shown and are moving with the indicated directions. We define the *rate of separation* (ROS) of two particles P_1 and P_2 as

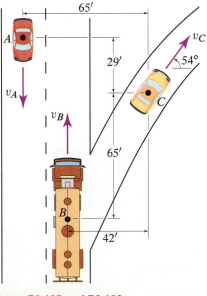

Figure P2.185 and P2.186

the component of the relative velocity of, say, P_2 with respect to P_1 in the direction of the relative position vector of P_2 with respect to P_1, which is along the line that connects the two particles. At the given instant, determine the rates of separation ROS$_{AB}$ and ROS$_{CB}$, that is, the rate of separation between A and B and between C and B. Let $v_A = 60$ mph, $v_B = 55$ mph, and $v_C = 35$ mph. Furthermore, treat the vehicles as particles and use the dimensions shown in the figure.

Problem 2.186

Car A is moving at a constant speed $v_A = 75$ km/h, while car C is moving at a constant speed $v_C = 42$ km/h on a circular exit ramp with radius $\rho = 80$ m. Determine the velocity and acceleration of C relative to A.

Problem 2.187

A remote control boat, capable of a maximum speed of 10 ft/s in still water, is made to cross a stream with a width $w = 35$ ft that is flowing with a speed $v_W = 7$ ft/s. If the boat starts from point O and keeps its orientation parallel to the cross-stream direction, find the location of point A at which the boat reaches the other bank while moving at its maximum speed. Furthermore, determine how long the crossing requires.

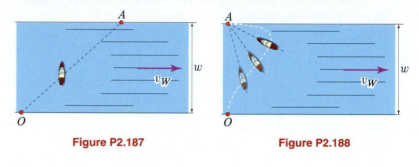

Figure P2.187   **Figure P2.188**

Problem 2.188

A remote control boat, capable of a maximum speed of 10 ft/s in still water, is made to cross a stream of width $w = 35$ ft that is flowing with a speed $v_W = 7$ ft/s. The boat is placed in the water at O, and it is *intended* to arrive at A by using a homing device that makes the boat always point toward A. Determine the time the boat takes to get to A and the path it follows. Also, consider a case in which the maximum speed of the boat is equal to the speed of the current. In such a case, does the boat ever make it to point A? *Hint:* To solve the problem, write $\vec{v}_{B/W} = v_{B/W}\,\hat{u}_{A/B}$, where the unit vector $\hat{u}_{A/B}$ always points from the boat to point A and is therefore a function of time.

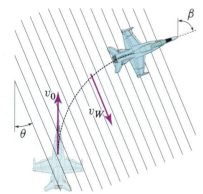

Figure P2.189

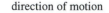
direction of motion

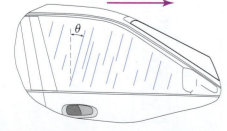

Figure P2.190

Problem 2.189

A plane is initially flying north with a speed $v_0 = 430$ mph relative to the ground while the wind has a constant speed $v_W = 12$ mph forming an angle $\theta = 23°$ with the north-south direction. The plane performs a course change of $\beta = 75°$ eastward while maintaining a constant reading of the airspeed indicator. Letting $\vec{v}_{P/A}$ be the velocity of the airplane relative to the air and assuming that the airspeed indicator measures the magnitude of the component of $\vec{v}_{P/A}$ in the direction of motion of the airplane, determine the speed of the plane relative to the ground after the course correction.

Problem 2.190

An interesting application of the relative motion equations is the experimental determination of the speed at which rain falls. Say you perform an experiment in your car in which you park your car in the rain and measure the angle the falling rain makes on your side window. Let this angle be $\theta_{\text{rest}} = 20°$. Next you drive forward at 25 mph and measure the new angle $\theta_{\text{motion}} = 70°$ that the rain makes with the vertical. Determine the speed of the falling rain.[*]

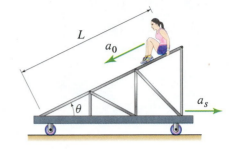

Problem 2.191

A woman is sliding down an incline with a constant acceleration of $a_0 = 2.3\,\text{m/s}^2$ relative to the incline. At the same time the incline is accelerating to the right at $1.2\,\text{m/s}^2$ relative to the ground. Letting $\theta = 34°$ and $L = 4\,\text{m}$ and assuming that both the woman and the incline start from rest, determine the horizontal distance traveled by the woman with respect to the ground when she reaches the bottom of the slide.

Figure P2.191

Problem 2.192

The pendulum bob A swings about O, which is a fixed point, while bob B swings about A. Express the components of the acceleration of B relative to the component system shown with origin at the fixed point O in terms of L_1, L_2, θ, ϕ, and the necessary time derivatives of ϕ and θ.

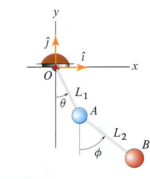

Problem 2.193

Revisit Example 2.26 in which the movie's hero is traveling on train car A with constant speed $v_A = 18\,\text{m/s}$ while the target B is moving at a constant speed $v_B = 40\,\text{m/s}$ (so that $a_B = 0$). Recall that 4 s before an otherwise inevitable collision between A and B, a projectile P traveling at a speed of $300\,\text{m/s}$ relative to A is shot toward B. Take advantage of the solution in Example 2.26, and determine the time it takes the projectile P to reach B and the projectile's distance traveled.

Figure P2.192

Problem 2.194

Consider the following variation of the problem in Example 2.26 in which a movie hero needs to destroy a mobile robot B, except this time they are not going to collide at C. Assume that the hero is traveling on the train car A with constant speed $v_A = 18\,\text{m/s}$ while the robot B travels at a constant speed $v_B = 50\,\text{m/s}$. In addition, assume that at time $t = 0\,\text{s}$ the train car A and the robot B are 72 and 160 m away from C, respectively. To prevent B from reaching its intended target, at $t = 0\,\text{s}$ the hero fires a projectile P at B. If P can travel at a constant speed of $300\,\text{m/s}$ relative to the gun, determine the orientation θ that must be given to the gun to hit B. *Hint:* An equation of the type $\sin \beta \pm A \cos \beta = C$ has the solution $\beta = \mp \gamma + \sin^{-1}(C \cos \gamma)$, if $|C \cos \gamma| \le 1$, where $\gamma = \tan^{-1} A$.

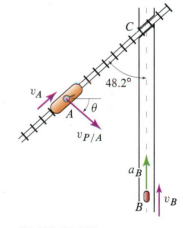

Problem 2.195

Consider the following variation of the problem in Example 2.26 in which a movie hero needs to destroy a mobile robot B. As was done in that problem, assume that the movie hero is traveling on the train car A with constant speed $v_A = 18\,\text{m/s}$ and that, 4 s before an otherwise inevitable collision at C, the hero fires a projectile P traveling at $300\,\text{m/s}$ relative to A. Differently from the problem in Example 2.26, assume that the robot B travels with a constant acceleration $a_B = 10\,\text{m/s}^2$ and that $v_B(0) = 20\,\text{m/s}$,

Figure P2.193–P2.195

[*] It turns out that a small raindrop in a light rain (0.04 in./h of rain) is about 0.5 mm across and falls at about 2 m/s. A 2.6 mm raindrop in a moderate rain (0.25 in./h of rain) falls at about 7.6 m/s, and a large 4 mm raindrop in a thunderstorm (1.0 in./h of rain) falls at 8.8 m/s.

where $t = 0$ is the time of firing. Determine the orientation θ of the gun fired by the hero so that B can be destroyed before the collision at C.

Problem 2.196

A park ranger R is aiming a rifle armed with a tranquilizer dart at a bear (the figure is not to scale). The bear is moving in the direction shown at a constant speed $v_B = 25$ mph. The ranger fires the rifle when the bear is at C at a distance of 150 ft. Knowing that $\alpha = 10°$, $\beta = 108°$, the dart travels with a constant speed of 425 ft/s, and the dart and the bear are moving in a horizontal plane, determine the orientation θ of the rifle so that the ranger can hit the bear. *Hint:* An equation of the type $\sin \beta \pm A \cos \beta = C$ has the solution $\beta = \mp \gamma + \sin^{-1}(C \cos \gamma)$, if $|C \cos \gamma| \leq 1$, where $\gamma = \tan^{-1} A$.

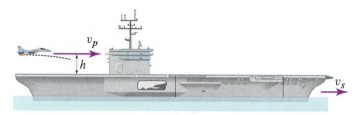

Figure P2.196

Problem 2.197

An airplane flying horizontally with a speed $v_p = 110$ km/h relative to the water, drops a crate onto a carrier when vertically over the back end of the ship, which is traveling at a speed $v_s = 26$ km/h relative to the water. If the plane drops the crate from a height $h = 20$ m, at what distance from the back of the ship will the crate first land on the deck of the ship?

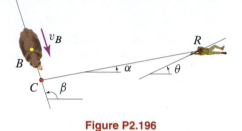

Figure P2.197 and P2.198

Problem 2.198

An airplane flying horizontally with a speed v_p relative to the water drops a crate onto a carrier when vertically over the back end of the ship, which is traveling at a speed $v_s = 32$ mph relative to the water. The length of the carrier's deck is $\ell = 1000$ ft, and the drop height is $h = 50$ ft. Determine the maximum value of v_p so that the crate will first impact within the rear half of the deck.

Problem 2.199

The object in the figure is called a *gun tackle*, and it used to be very common on sailboats to help in the operation of front-loaded guns. If the end at A is pulled down at a speed of 1.5 m/s, determine the velocity of B. Neglect the fact that some portions of the rope are not vertically aligned.

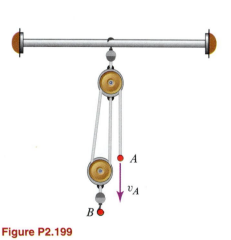

Figure P2.199

Problem 2.200

The gun tackle shown in the figure is operated with the help of a horse. If the horse moves to the right at a speed of $7\,\text{ft/s}$, determine the velocity and acceleration of B when the horizontal distance from B to A is $15\,\text{ft}$. Except for the part of the rope attached to the horse, neglect the fact that some portions are not vertically aligned.

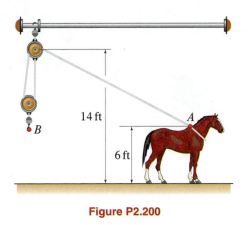

14 ft

A

B

6 ft

Figure P2.200

Problem 2.201

The figure shows an inverted gun tackle with snatch block, which used to be common on sailboats. If the end at A is pulled at a speed of $1.5\,\text{m/s}$, determine the velocity of B. Neglect the fact that some portions of the rope are not vertically aligned.

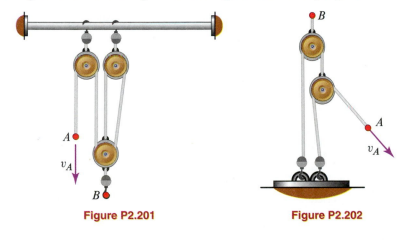

B

A

v_A

A

v_A

B

Figure P2.201   **Figure P2.202**

Problem 2.202

In maritime speak, the system in the figure is often called a *whip-upon-whip purchase* and is used for controlling certain types of sails on small cutters (by attaching point B to the sail to be unfurled). If the end of the rope at A is pulled with a speed of $4\,\text{m/s}$, determine the velocity of B. Neglect the fact that some portions of the rope are not vertically aligned.

Problem 2.203

The pulley system shown is used to store a bicycle in a garage. If the bicycle is hoisted via a winch that winds the rope at a rate $v_0 = 5\,\text{in./s}$, determine the vertical speed of the bicycle.

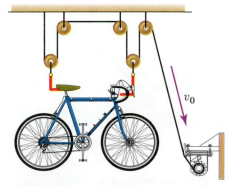

v_0

Figure P2.203

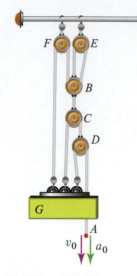

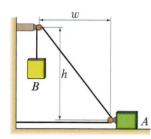

Figure P2.205 and P2.206

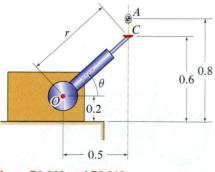

Figure P2.207 and P2.208

All dimensions are in meters.

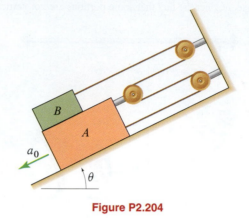

Figure P2.209 and P2.210

Problem 2.204

Block A is released from rest and starts sliding down the incline with an acceleration $a_0 = 3.7 \, \text{m/s}^2$. Determine the acceleration of block B relative to the incline. Also, determine the time needed for B to move a distance $d = 0.2 \, \text{m}$ relative to A.

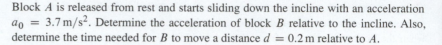

Figure P2.204

Problem 2.205

Assuming that all ropes are vertically aligned, determine the velocity and acceleration of the load G if $v_0 = 3 \, \text{ft/s}$ and $a_0 = 1 \, \text{ft/s}^2$.

Problem 2.206

The load G is initially at rest when the end A of the rope is pulled with the constant acceleration a_0. Determine a_0 so that G is lifted 2 ft in 4.3 s.

Problem 2.207

At the instant shown, block A is moving at a constant speed $v_0 = 3 \, \text{m/s}$ to the left and $w = 2.3 \, \text{m}$. Using $h = 2.7 \, \text{m}$, determine how much time is needed to lower B 0.75 m from this position.

Problem 2.208

At the instant shown, $h = 10 \, \text{ft}$, $w = 8 \, \text{ft}$, and block B is moving with a speed $v_0 = 5 \, \text{ft/s}$ and an acceleration $a_0 = 1 \, \text{ft/s}^2$, both downward. Determine the velocity and acceleration of block A.

Problem 2.209

As a part of a robotics competition, a robotic arm is to be designed so as to catch an egg without breaking it. The egg is released from rest at $t = 0$ from point A while the arm C is initially also at rest in the position shown. The arm starts moving when the egg is released, and it is to catch the egg at some point in such a way as to avoid any impact between the egg and the robot hand. The arm catches the egg without any impact by specifying that the arm and the egg have to be at the same position at the same time with identical velocities. A student proposes that this can be done by specifying a constant value of $\ddot{\theta}$ for which (after a fair bit of work) it is found that the arm catches the egg at $t = 0.4391 \, \text{s}$ for $\ddot{\theta} = -13.27 \, \text{rad/s}^2$. Using these values of t and $\ddot{\theta}$, determine the acceleration of both the arm and the egg at the time of catch. Once you have done this, explain whether or not using a constant value of $\ddot{\theta}$, as has been proposed, is an acceptable strategy.

💡 Problem 2.210 💡

Referring to the problem of a robot arm catching an egg (Prob. 2.209), the strategy is that the arm and the egg must have the same velocity and the same position at the same time for the arm to gently catch the egg. In addition, what should be true about the accelerations of the arm and the egg for the catch to be successful *after* they rendezvous with the same velocity at the same position and time? Describe what happens if the accelerations of the arm and egg do not match.

Note: Concept problems are about *explanations*, not computations.

Problem 2.211

The piston head at C is constrained to move along the y axis. Let the crank AB be rotating counterclockwise at a constant angular speed $\dot{\theta} = 2000$ rpm, $R = 3.5$ in., and $L = 5.3$ in. Determine the velocity of C when $\theta = 35°$.

Problem 2.212

The piston head at C is constrained to move along the y axis. Let the crank AB be rotating counterclockwise at a constant angular speed $\dot{\theta} = 2000$ rpm, $R = 3.5$ in., and $L = 5.3$ in. Determine expressions for the velocity and acceleration of C as a function of θ and the given parameters.

Problem 2.213

Let $\vec{\omega}_{BC}$ denote the angular velocity of the relative position vector $\vec{r}_{C/B}$. As such, $\vec{\omega}_{BC}$ is also the angular velocity of the connecting rod BC. Using the concept of time derivative of a vector given in Section 2.4 on p. 88, determine the component of the relative velocity of C with respect to B along the direction of the connecting rod BC.

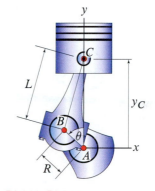

Figure P2.211–P2.213

Problems 2.214 and 2.215

In the cutting of sheet metal, the robotic arm OA needs to move the cutting tool at C counterclockwise at a constant speed v_0 along a circular path of radius ρ. The center of the circle is located in the position shown relative to the base of the robotic arm at O.

Problem 2.214 For all positions along the circular cut (i.e., for any value of ϕ), determine r, $\dot{r}$, and $\dot{\theta}$ as functions of the given quantities (i.e., d, h, ρ, v_0). Use one or more geometric constraints and their derivatives to do this. These quantities can be found "by hand," but it is tedious, so you might consider using symbolic algebra software such as Mathematica or Maple.

Problem 2.215 For all positions along the circular cut (i.e., for any value of ϕ), determine $\ddot{r}$ and $\ddot{\theta}$ as functions of the given quantities (i.e., d, h, ρ, v_0). These quantities can be found by hand, but it is very tedious, so you might consider using symbolic algebra software such as Mathematica or Maple.

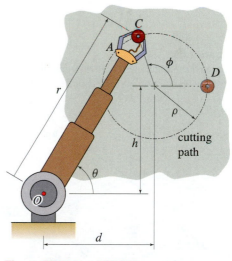

Figure P2.214 and P2.215

Figure 2.54
A skysurfer inverted and spinning.

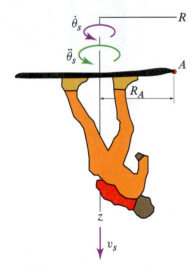

Figure 2.55
Skysurfer and the coordinate system chosen to describe the motion of point A on the sky board.

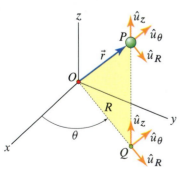

Figure 2.56
Coordinate directions in the cylindrical coordinate system. As with polar coordinates, defined in Section 2.6, the θ coordinate is implicitly contained in $\hat{u}_R$.

2.8   Motion in Three Dimensions

So far in Chapter 2, we have primarily focused on planar motion, which, while occurring in a three-dimensional environment, is such that we can describe the position of moving points using only two coordinates. In this section we consider the motion of points that are not constrained to move in a plane and therefore require that we use three coordinates when describing their position.

Skydiving acrobatics

The skysurfer shown in Fig. 2.54 has jumped out of an airplane and is falling straight down while rotating about the vertical axis that goes through her or his mass center. Referring to Fig. 2.55, suppose that we place an accelerometer at the end of the sky board at point A. This device can tell us the acceleration $\vec{a}_A$ of A at any instant and we can integrate $\vec{a}_A$ with respect to time to provide the velocity $\vec{v}_A$ (assuming we know the initial velocity of A). Our goal is to relate the given measurements of $\vec{a}_A$ and $\vec{v}_A$ to the rates at which the surfer is rotating and falling. We will assume that the spin axis, i.e., the z axis in Fig. 2.55, goes through the mass center of the surfer and that the mass center has reached terminal velocity during descent, i.e., $v_s = $ constant. Describing the orientation of the sky board via an angle θ_s, we can characterize the skysurfer's spin rates via the time derivatives of θ_s, i.e., $\dot{\theta}_s$ and $\ddot{\theta}_s$. Hence, how are $\vec{v}_A$ and $\vec{a}_A$ related to v_s, $\dot{\theta}_s$, and $\ddot{\theta}_s$?

In confronting this problem, we immediately observe that the curve representing the trajectory of A is not a planar curve. Therefore, the first step in solving our problem is to describe the three-dimensional motion of A via an appropriate choice of coordinate system. Observe that if the surfer were not falling, his or her spinning motion would easily be described using a polar coordinate system with origin on the z axis. Intuition then suggests that the motion of the skysurfer can be conveniently described by simply "adding a z axis" to the familiar polar coordinate system. Hence, we will consider the suggested three-dimensional extension of the polar coordinate system. After doing so, we will come back to the study of the surfer's motion and provide an answer to our question. Finally, we will consider other common approaches to describing three-dimensional motion.

Position, velocity, & acceleration in cylindrical coordinates

The three-dimensional extension of the polar coordinate system is called the *cylindrical coordinate system*. Figure 2.56 shows the coordinate directions for a cylindrical coordinate system with origin at O and with the coordinate θ defined to be positive counterclockwise as viewed down the positive z axis. If θ is defined to be positive in the clockwise direction, the results that follow will be different. Referring to Fig. 2.56, the *cylindrical coordinates* of a point P are

$$R, \theta, \text{ and } z, \tag{2.118}$$

where R and θ are the polar coordinates of the projection of point P onto the $R\theta$ plane, namely, point Q, and the coordinate z is equal to the distance between P and Q taken positive or negative depending on whether or not P is reached from Q by moving in the positive z direction.

To describe vectors we introduce the following set of three mutually orthogonal base vectors

$$\hat{u}_R, \hat{u}_\theta, \text{ and } \hat{u}_z \text{ with } \hat{u}_z = \hat{u}_R \times \hat{u}_\theta, \tag{2.119}$$

where $\hat{u}_R$ and $\hat{u}_\theta$ are determined as in the case of the polar coordinate system and are therefore parallel to the $R\theta$ plane, whereas $\hat{u}_z$ is a unit vector parallel to the z axis and pointing in the positive z direction. It follows that the position *vector* of P relative to the origin O is given by

$$\boxed{\vec{r} = R\,\hat{u}_R + z\,\hat{u}_z.} \tag{2.120}$$

Equation (2.120) expresses the fact that we can locate a point by tracking the point's projection on the $R\theta$ plane, given by the term $R\,\hat{u}_R$, and then adding to this the point's elevation z with respect to the $R\theta$ plane. We observe that the coordinates R and z appear explicitly in the expression for the position vector in Eq. (2.120), while the coordinate θ appears *implicitly* in the unit vector $\hat{u}_R$.

Differentiating Eq. (2.120) with respect to time, we obtain the following result for the velocity in cylindrical coordinates

$$\boxed{\vec{v} = \dot{R}\,\hat{u}_R + R\dot{\theta}\,\hat{u}_\theta + \dot{z}\,\hat{u}_z = v_R\,\hat{u}_R + v_\theta\,\hat{u}_\theta + v_z\,\hat{u}_z,} \tag{2.121}$$

where

$$\boxed{v_R = \dot{R}, \quad v_\theta = R\dot{\theta}, \quad \text{and} \quad v_z = \dot{z}} \tag{2.122}$$

are the cylindrical components of the velocity vector. In writing Eq. (2.121), we have used Eq. (2.85) on p. 119 and the fact that $\hat{u}_z$ is constant because the z axis is fixed. We obtain the acceleration in cylindrical coordinates by differentiating Eq. (2.121) with respect to time, i.e.,

$$\boxed{\begin{aligned} \vec{a} &= \left(\ddot{R} - R\dot{\theta}^2\right)\hat{u}_R + \left(R\ddot{\theta} + 2\dot{R}\dot{\theta}\right)\hat{u}_\theta + \ddot{z}\,\hat{u}_z \\ &= a_R\,\hat{u}_R + a_\theta\,\hat{u}_\theta + a_z\,\hat{u}_z, \end{aligned}} \tag{2.123}$$

where

$$\boxed{a_R = \ddot{R} - R\dot{\theta}^2, \quad a_\theta = R\ddot{\theta} + 2\dot{R}\dot{\theta}, \quad \text{and} \quad a_z = \ddot{z}} \tag{2.124}$$

are the cylindrical components of the acceleration vector. In writing Eq. (2.123), we have used Eq. (2.89) on p. 119 and the constancy of $\hat{u}_z$.

Skydiving acrobatics revisited

Now that the cylindrical coordinate system has been formally introduced, we go back to the skysurfer problem. Recall that we expected to describe the skysurfer's spinning motion via some angle θ_s such that the rates $\dot{\theta}_s$ and $\ddot{\theta}_s$ would describe the angular velocity and acceleration of the skysurfer, respectively. Using a cylindrical coordinate system, we can do so by identifying θ_s with the θ coordinate of point A. Then, using Eqs. (2.121) and (2.123) to express $\vec{v}_A$ and $\vec{a}_A$, we have

$$\vec{v}_A = R_A\dot{\theta}_s\,\hat{u}_\theta + v_s\,\hat{u}_z, \tag{2.125}$$

where $\dot{R}_A = 0$ because R_A is constant. For the acceleration, we obtain

$$\vec{a}_A = -R_A \dot{\theta}_s^2 \, \hat{u}_R + R_A \ddot{\theta}_s \, \hat{u}_\theta, \tag{2.126}$$

where we have used the fact that the skysurfer has reached terminal velocity so that $\ddot{z} = d\dot{z}/dt = dv_s/dt = 0$. Our next step is to carefully consider the information that the accelerometer at A provides. In general, accelerometers give acceleration information along three mutually orthogonal axes. These accelerations can be integrated with respect to time within the accelerometer to also provide velocities along the same three mutually perpendicular axes. If we are careful to mount the accelerometer so that the axes are aligned as shown in Fig. 2.57, then we can say that the velocity and acceleration data from the

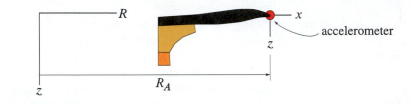

Figure 2.57. Blowup of the accelerometer and sky board from the skysurfer in Fig. 2.55.

accelerometer along its x, y, and z axes, which we write, respectively, as v_{Ax}, v_{Ay}, v_{Az}, a_{Ax}, a_{Ay}, and a_{Az}, can be related to Eqs. (2.125) and (2.126) via the relations

$$v_{Ax} = v_R = 0, \qquad\qquad a_{Ax} = a_R = -R_A \dot{\theta}_s^2, \tag{2.127}$$

$$v_{Ay} = v_\theta = R_A \dot{\theta}_s, \qquad\qquad a_{Ay} = a_\theta = R_A \ddot{\theta}_s, \tag{2.128}$$

$$v_{Az} = v_z = v_s, \qquad\qquad a_{Az} = a_z = 0. \tag{2.129}$$

These equations allow us to determine R_A, v_s, $\dot{\theta}_s$, and $\ddot{\theta}_s$ by solving the four equations consisting of the second of Eqs. (2.127), and Eqs. (2.128) and the first of Eqs. (2.129). These solutions are

$$\dot{\theta}_s = -\frac{a_{Ax}}{v_{Ay}}, \quad \ddot{\theta}_s = -\frac{a_{Ax} a_{Ay}}{v_{Ay}^2}, \quad R_A = -\frac{v_{Ay}^2}{a_{Ax}}, \quad v_s = v_{Az}. \tag{2.130}$$

Equations (2.130) show that, with the accelerometer mounted on the end of the skyboard, we can determine how fast the skysurfer is falling, his or her rotation rates, and even the distance from the accelerometer to the axis of rotation.

Position, velocity, & acceleration in spherical coordinates

Another coordinate system that is often used in the study of three-dimensional motion, such as, for example, in solving problems involving navigation and motion relative to the Earth, is the *spherical coordinate system*.

Referring to Fig. 2.58, the position of a point P in a three-dimensional space can be described via the three quantities r, ϕ, and θ. These quantities are called the *spherical coordinates* of P: the angles ϕ and θ give orientation to the line going from the origin O to P, while r is the distance between P and O. In Fig. 2.58 we have also indicated three mutually orthogonal unit vectors

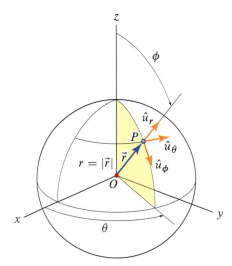

Figure 2.58
Definition of the spherical coordinates and corresponding orthonormal base vectors.

$\hat{u}_r$, $\hat{u}_\phi$, and $\hat{u}_\theta$. These unit vectors are oriented such that $\hat{u}_r$ points from O to P, while $\hat{u}_\phi$ and $\hat{u}_\theta$ point in the direction of growing ϕ and θ, respectively, with $\hat{u}_\phi \times \hat{u}_\theta = \hat{u}_r$. If ϕ and θ as well as the unit vectors $\hat{u}_r$, $\hat{u}_\phi$, and $\hat{u}_\theta$ are not defined as given here, the results that follow may not be applicable.

In spherical coordinates, the position vector of P relative to the origin O is given by

$$\vec{r} = r\,\hat{u}_r. \tag{2.131}$$

Differentiating $\vec{r}$ with respect to time, we obtain

$$\vec{v} = \dot{\vec{r}} = \dot{r}\,\hat{u}_r + r\,\dot{\hat{u}}_r, \tag{2.132}$$

which implies that to describe the velocity vector we need to express $\dot{\hat{u}}_r$ in terms of the base vectors $\hat{u}_r$, $\hat{u}_\theta$, and $\hat{u}_\phi$. Recall that the time derivative of a unit vector is equal to the angular velocity of the unit vector crossed with the vector itself, as shown in Eq. (2.60) on p. 92. Therefore, we need to find the angular velocity of the unit base vectors to obtain $\dot{\hat{u}}_r$. To do this, we note that if the coordinates θ and ϕ undergo infinitesimal changes, the angular velocity corresponding to these infinitesimal changes is (see the discussion starting on p. 89)

$$\begin{aligned}\vec{\omega} &= \dot{\phi}\,\hat{u}_\theta + \dot{\theta}\,\hat{u}_z \\ &= \dot{\theta}\cos\phi\,\hat{u}_r - \dot{\theta}\sin\phi\,\hat{u}_\phi + \dot{\phi}\,\hat{u}_\theta,\end{aligned} \tag{2.133}$$

where Eq. (2.133) is obtained by observing that $\hat{u}_z = \cos\phi\,\hat{u}_r - \sin\phi\,\hat{u}_\phi$.

Using the result in Eq. (2.133) along with Eq. (2.60) on p. 92, we can find the time derivative of each of the unit vectors $\hat{u}_r$, $\hat{u}_\theta$, and $\hat{u}_\phi$ as

$$\dot{\hat{u}}_r = \vec{\omega} \times \hat{u}_r = \dot{\phi}\,\hat{u}_\phi + \dot{\theta}\sin\phi\,\hat{u}_\theta, \tag{2.134}$$

$$\dot{\hat{u}}_\phi = \vec{\omega} \times \hat{u}_\phi = -\dot{\phi}\,\hat{u}_r + \dot{\theta}\cos\phi\,\hat{u}_\theta, \tag{2.135}$$

$$\dot{\hat{u}}_\theta = \vec{\omega} \times \hat{u}_\theta = -\dot{\theta}\sin\phi\,\hat{u}_r - \dot{\theta}\cos\phi\,\hat{u}_\phi. \tag{2.136}$$

Substituting Eq. (2.134) in Eq. (2.132), we obtain the velocity in spherical coordinates as

$$\vec{v} = \dot{r}\,\hat{u}_r + r\dot{\phi}\,\hat{u}_\phi + r\dot{\theta}\sin\phi\,\hat{u}_\theta = v_r\,\hat{u}_r + v_\phi\,\hat{u}_\phi + v_\theta\,\hat{u}_\theta, \tag{2.137}$$

where

$$v_r = \dot{r}, \quad v_\phi = r\dot{\phi}, \quad \text{and} \quad v_\theta = r\dot{\theta}\sin\phi \tag{2.138}$$

are the spherical components of the velocity vector. Differentiating Eq. (2.137) with respect to time and using Eqs. (2.134)–(2.136) give the acceleration as

$$\begin{aligned}\vec{a} &= \left(\ddot{r} - r\dot{\phi}^2 - r\dot{\theta}^2\sin^2\phi\right)\hat{u}_r \\ &\quad + \left(r\ddot{\phi} + 2\dot{r}\dot{\phi} - r\dot{\theta}^2\sin\phi\cos\phi\right)\hat{u}_\phi \\ &\quad + \left(r\ddot{\theta}\sin\phi + 2\dot{r}\dot{\theta}\sin\phi + 2r\dot{\phi}\dot{\theta}\cos\phi\right)\hat{u}_\theta \\ &= a_r\,\hat{u}_r + a_\phi\,\hat{u}_\phi + a_\theta\,\hat{u}_\theta,\end{aligned} \tag{2.139}$$

where

$$\begin{aligned}a_r &= \ddot{r} - r\dot{\phi}^2 - r\dot{\theta}^2\sin^2\phi, \\ a_\phi &= r\ddot{\phi} + 2\dot{r}\dot{\phi} - r\dot{\theta}^2\sin\phi\cos\phi, \\ a_\theta &= r\ddot{\theta}\sin\phi + 2\dot{r}\dot{\theta}\sin\phi + 2r\dot{\phi}\dot{\theta}\cos\phi\end{aligned} \tag{2.140}$$

are the spherical components of the acceleration vector.

Helpful Information

Where are ϕ and θ? We have seen this before, that is, the position vector *seems* not to depend on all of the "coordinates". In this case, the position vector $\vec{r} = r\,\hat{u}_r$ seems not to be a function of either θ or ϕ. However, the angles θ and ϕ are *implicitly* contained in the definition of the direction of $\hat{u}_r$.

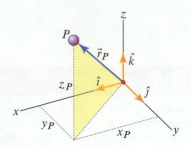

Figure 2.59
The Cartesian coordinate system and its unit vectors in three dimensions.

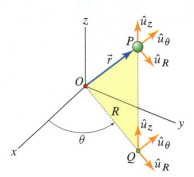

Figure 2.60
Figure 2.56 repeated. Coordinate directions in the cylindrical coordinate system. As with polar coordinates defined in Section 2.6, the θ coordinate is implicitly contained in $\hat{u}_r$.

Position, velocity, & acceleration in Cartesian coordinates

The Cartesian coordinate system (discussed in detail in Section 1.2) and its unit vectors in three dimensions are defined as shown in Fig. 2.59. Because the directions of its unit vectors do not change, we obtain the following relations for position, velocity, and acceleration, respectively, in Cartesian coordinates:

$$\vec{r}(t) = x(t)\,\hat{\imath} + y(t)\,\hat{\jmath} + z(t)\,\hat{k}, \tag{2.141}$$

$$\vec{v}(t) = \dot{x}\,\hat{\imath} + \dot{y}\,\hat{\jmath} + \dot{z}\,\hat{k} = v_x\,\hat{\imath} + v_y\,\hat{\jmath} + v_z\,\hat{k}, \tag{2.142}$$

$$\vec{a}(t) = \ddot{x}\,\hat{\imath} + \ddot{y}\,\hat{\jmath} + \ddot{z}\,\hat{k} = a_x\,\hat{\imath} + a_y\,\hat{\jmath} + a_z\,\hat{k}. \tag{2.143}$$

End of Section Summary

Cylindrical coordinates.   Referring to Fig. 2.60, the position of a point P in three dimensions can be described via the three quantities R, θ, and z that are the point's cylindrical coordinates. In addition, we see that cylindrical coordinates involve the orthogonal triad of unit vectors $\hat{u}_R$, $\hat{u}_\theta$, and $\hat{u}_z$. Using this triad, the position in cylindrical coordinates is given by

> Eq. (2.120), p. 155
>
> $$\vec{r} = R\,\hat{u}_R + z\,\hat{u}_z.$$

The velocity vector in cylindrical coordinates is given by

> Eq. (2.121), p. 155
>
> $$\vec{v} = \dot{R}\,\hat{u}_R + R\dot{\theta}\,\hat{u}_\theta + \dot{z}\,\hat{u}_z = v_R\,\hat{u}_R + v_\theta\,\hat{u}_\theta + v_z\,\hat{u}_z,$$

where

> Eqs. (2.122), p. 155
>
> $$v_R = \dot{R}, \quad v_\theta = R\dot{\theta}, \quad \text{and} \quad v_z = \dot{z}$$

are the components of the velocity vector in the $\hat{u}_R$, $\hat{u}_\theta$, and $\hat{u}_z$ directions, respectively. Finally, the acceleration vector in cylindrical coordinates is given by

> Eq. (2.123), p. 155
>
> $$\vec{a} = \left(\ddot{R} - R\dot{\theta}^2\right)\hat{u}_R + \left(R\ddot{\theta} + 2\dot{R}\dot{\theta}\right)\hat{u}_\theta + \ddot{z}\,\hat{u}_z$$
> $$= a_R\,\hat{u}_R + a_\theta\,\hat{u}_\theta + a_z\,\hat{u}_z,$$

where

> Eqs. (2.124), p. 155
>
> $$a_R = \ddot{R} - R\dot{\theta}^2, \quad a_\theta = R\ddot{\theta} + 2\dot{R}\dot{\theta}, \quad \text{and} \quad a_z = \ddot{z}$$

are the components of the acceleration vector in the $\hat{u}_R$, $\hat{u}_\theta$, and $\hat{u}_z$ directions, respectively.

Spherical coordinates.   Referring to Fig. 2.61, the position of a point P in three dimensions can be described via the three quantities r, θ, and ϕ that are the point's spherical coordinates. In addition, spherical coordinates involve the orthogonal triad of unit vectors $\hat{u}_r$, $\hat{u}_\phi$, and $\hat{u}_\theta$, with $\hat{u}_\phi \times \hat{u}_\theta = \hat{u}_r$. Using this triad, the position vector in spherical coordinates is given by

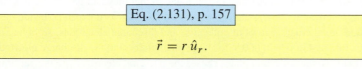

Eq. (2.131), p. 157

$$\vec{r} = r\,\hat{u}_r.$$

The velocity vector in spherical coordinates is given by

Eq. (2.137), p. 157

$$\vec{v} = \dot{r}\,\hat{u}_r + r\dot{\phi}\,\hat{u}_\phi + r\dot{\theta}\sin\phi\,\hat{u}_\theta = v_r\,\hat{u}_r + v_\phi\,\hat{u}_\phi + v_\theta\,\hat{u}_\theta,$$

where

Eqs. (2.138), p. 157

$$v_r = \dot{r}, \quad v_\phi = r\dot{\phi}, \quad \text{and} \quad v_\theta = r\dot{\theta}\sin\phi$$

are the components of the velocity vector in the $\hat{u}_r$, $\hat{u}_\phi$, and $\hat{u}_\theta$ directions, respectively. Finally, the acceleration vector in spherical coordinates is given by

Eq. (2.139), p. 157

$$\begin{aligned}
\vec{a} &= \left(\ddot{r} - r\dot{\phi}^2 - r\dot{\theta}^2\sin^2\phi\right)\hat{u}_r \\
&\quad + \left(r\ddot{\phi} + 2\dot{r}\dot{\phi} - r\dot{\theta}^2\sin\phi\cos\phi\right)\hat{u}_\phi \\
&\quad + \left(r\ddot{\theta}\sin\phi + 2\dot{r}\dot{\theta}\sin\phi + 2r\dot{\phi}\dot{\theta}\cos\phi\right)\hat{u}_\theta \\
&= a_r\,\hat{u}_r + a_\phi\,\hat{u}_\phi + a_\theta\,\hat{u}_\theta,
\end{aligned}$$

where

Eqs. (2.140), p. 157

$$\begin{aligned}
a_r &= \ddot{r} - r\dot{\phi}^2 - r\dot{\theta}^2\sin^2\phi, \\
a_\phi &= r\ddot{\phi} + 2\dot{r}\dot{\phi} - r\dot{\theta}^2\sin\phi\cos\phi, \\
a_\theta &= r\ddot{\theta}\sin\phi + 2\dot{r}\dot{\theta}\sin\phi + 2r\dot{\phi}\dot{\theta}\cos\phi
\end{aligned}$$

are the components of the acceleration vector in the $\hat{u}_r$, $\hat{u}_\phi$, and $\hat{u}_\theta$ directions, respectively.

Cartesian coordinates.   Referring to Fig. 2.62, in the three-dimensional Cartesian coordinate system, the position, velocity, and acceleration, respectively, are given in component form as

Eqs. (2.141)–(2.143), p. 158

$$\begin{aligned}
\vec{r}(t) &= x(t)\,\hat{\imath} + y(t)\,\hat{\jmath} + z(t)\,\hat{k}, \\
\vec{v}(t) &= \dot{x}\,\hat{\imath} + \dot{y}\,\hat{\jmath} + \dot{z}\,\hat{k} = v_x\,\hat{\imath} + v_y\,\hat{\jmath} + v_z\,\hat{k}, \\
\vec{a}(t) &= \ddot{x}\,\hat{\imath} + \ddot{y}\,\hat{\jmath} + \ddot{z}\,\hat{k} = a_x\,\hat{\imath} + a_y\,\hat{\jmath} + a_z\,\hat{k}.
\end{aligned}$$

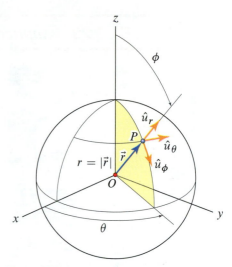

Figure 2.61
Figure 2.58 repeated. Definition of the unit vectors in spherical coordinates.

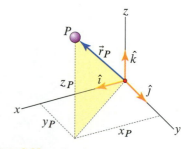

Figure 2.62
Figure 2.59 repeated. The Cartesian coordinate system and its unit vectors in three dimensions.

EXAMPLE 2.29 *Application of Cartesian Coordinates*

Figure 1
Photos of an electrostatic precipitator from a manufacturing plant. The top photo shows the outside of a precipitator with the air inlet in the middle. The bottom photo shows the collecting plates on the inside of the precipitator.

Electrostatic precipitators are used to "scrub" or clean the emissions from coal-fired power plants (see Fig. 1). They work by sending the particulate-laden flue gas through a large structure that slows down the particles so that they can be efficiently collected. The gas enters at 50–60 ft/s and slows down to 3–6 ft/s as it expands. Inside the main structure are alternating rows of collection and discharge electrodes. Applying a high voltage (about 45,000–70,000 V for large precipitators) to the discharge electrodes located between the collection plates causes them to emit electrons into the gas, thus ionizing the immediate area. When the flue gas passes through this "corona," the particles become negatively charged and are then attracted to the positively charged collecting plates from which they can be collected and removed.

Using the geometry defined in Fig. 2, determine where the particle P will land on the collecting plate given that $\theta_1 = 30°$, $\theta_2 = 55°$, $d = 1.5$ ft, $|\vec{v}_P| = v_P = 5$ ft/s and that gravity acts in the $-z$ direction. Assume that the collecting plate imposes a constant y component of acceleration on the particle of $a_{ep} = 250$ ft/s^2.

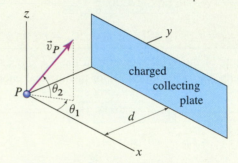

Figure 2. Schematic of a particle moving near a positively charged collecting plate in an electrostatic precipitator. Collection efficiencies of modern precipitators range from 99.5–99.9%.

SOLUTION

Road Map The key to the solution is to first determine the time it takes the particle to get to the plate. We can find that time since we know the acceleration in the y direction and we can find the initial velocity in the y direction. Once that time is known, we can use constant acceleration kinematics to find the position at which the particle hits the plate.

Computation We begin by finding the components of the initial velocity of P in all three Cartesian coordinate directions. Referring to Fig. 3, we can see that the initial component of $\vec{v}_P$ in the xy plane is $v_P \cos\theta_2$ and that the initial z component of $\vec{v}_P$ is given by

$$v_{Pz} = v_P \sin\theta_2. \tag{1}$$

Now that we have the component of v_P in the xy plane, it is easier to see that the initial x and y components are given by

$$v_{Px} = (v_P \cos\theta_2) \cos\theta_1, \tag{2}$$

$$v_{Py} = (v_P \cos\theta_2) \sin\theta_1. \tag{3}$$

Next, we are told that the acceleration of the particle is

$$\vec{a}_P = a_{ep}\,\hat{j} - g\,\hat{k} = \left(250\,\hat{j} - 32.2\,\hat{k}\right) \text{ ft/s}^2, \tag{4}$$

which is constant in all component directions.

Now that we have all accelerations and the initial velocities, we can use the y component to determine the time it takes for the particle to hit the plate. Using Eq. (2.42), we have

$$d = v_{Py}t_c + \tfrac{1}{2}a_{ep}t_c^2, \tag{5}$$

where t_c is the time for P to reach the collector plate. Equation (5) can be solved for t_c to obtain

$$t_c = -\frac{v_P}{a_{ep}}\cos\theta_2\sin\theta_1 \pm \sqrt{2da_{ep} + v_P^2\cos^2\theta_2\sin^2\theta_1}, \tag{6}$$

where Eq. (3) has been used. Substituting in the given parameters, we find that t_c equals either 0.1040 s or −0.1154 s, with the physically appropriate answer being the positive one

$$t_c = 0.1040\,\text{s}. \tag{7}$$

Knowing the time to reach the collector, we can now determine the distance traveled in the x and z directions. Again applying Eq. (2.42), but now in the x and z directions, we obtain

$$x_c = v_{Px}t_c = 0.258\,\text{ft}, \tag{8}$$

$$z_c = v_{Pz}t_c - \tfrac{1}{2}gt_c^2 = 0.252\,\text{ft}. \tag{9}$$

Figure 4 shows the trajectory of the particle that has been attracted to the collecting plate.

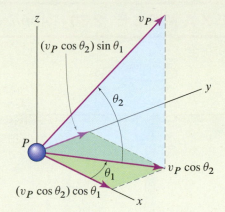

Figure 3

Graphical depiction of the initial velocity of the particle in the precipitator.

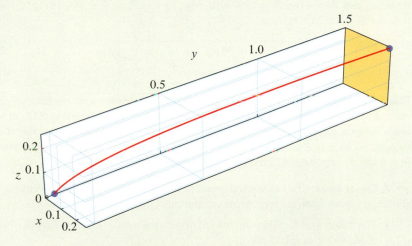

Figure 4. Trajectory of the particle P. All dimensions are in feet.

Discussion & Verification   Figure 4 is very helpful in assessing whether or not our solution is reasonable. Given that the initial speed of the particle is 5 ft/s and that it is accelerating toward the collecting plate at 250 ft/s², it seems reasonable that it would not move very far down the precipitator (i.e., in the x direction) before hitting the collecting plate.

> **Interesting Fact**
>
> **Is 250 ft/s² a reasonable accleration?**
> We stated that the particle is accelerating at 250 ft/s² toward the collector plate. This is reasonable when we consider that the particles collected on these plates are *very* small. For example, in some electrostatic precipitators, the particles range from smaller than 1 μm to up to 3 μm in diameter. A 1 μm particle of sodium would have a mass of about 10^{-15} kg, and so to generate an acceleration of 250 ft/s², which is equal to 76.2 m/s², requires a force of only about 8×10^{-14} N.

E X A M P L E 2.30   *Application of Cylindrical Coordinates*

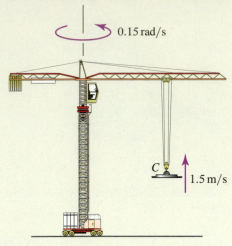

Figure 1
A top-slewing crane. This kind of crane is very common on big construction sites.

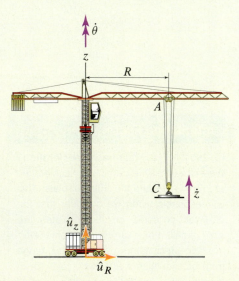

Figure 2
The top-slewing crane with a cylindrical coordinate system defined.

A top-slewing crane (also called a tower crane) is lifting an object C at a constant rate of 1.5 m/s while rotating at a constant rate of 0.15 rad/s in the direction shown in Fig. 1. If the distance between the object and the axis of rotation of the crane's boom is 45 m, find the velocity and acceleration of C, assuming that the swinging motion of C can be neglected.

SOLUTION

Road Map   This problem is readily solved using cylindrical coordinates since we are provided with quantities that are naturally defined in a cylindrical coordinate system with its origin placed at the intersection of the $\hat{u}_R$ and $\hat{u}_z$ vectors at the base of the crane (see Fig. 2).

Computation   According to our choice of cylindrical coordinates, the velocity and acceleration vectors of C are given by Eqs. (2.121) and (2.123), respectively, which are

$$\vec{v}_C = \dot{R}\,\hat{u}_R + R\dot{\theta}\,\hat{u}_\theta + \dot{z}\,\hat{u}_z, \tag{1}$$

$$\vec{a}_C = \left(\ddot{R} - R\dot{\theta}^2\right)\hat{u}_R + \left(R\ddot{\theta} + 2\dot{R}\dot{\theta}\right)\hat{u}_\theta + \ddot{z}\,\hat{u}_z. \tag{2}$$

Recalling that $R = 45$ m, $\dot{R} = 0$, $\ddot{R} = 0$, $\dot{z} = 1.5$ m/s, $\ddot{z} = 0$, $\dot{\theta} = 0.15$ rad/s, and $\ddot{\theta} = 0$, Eqs. (1) and (2) become

$$\vec{v}_C = (6.75\,\hat{u}_\theta + 1.5\,\hat{u}_z)\,\text{m/s}, \tag{3}$$

$$\vec{a}_C = -1.01\,\hat{u}_R\,\text{m/s}^2. \tag{4}$$

Discussion & Verification   The solution seems reasonable since the motion of C is the composition of a uniform circular motion with angular velocity equal to 0.15 rad/s and a uniform rectilinear motion in the positive z direction. We were expecting the velocity not to have a radial component, while the acceleration was expected to have only a radial component, which is what Eqs. (3) and (4) reflect.

🔍 **A Closer Look**   This example demonstrates how easy it can be to find velocities and accelerations when the appropriate component system is used. Even if:

- The trolley at A were moving inward or outward with *any* known $R(t)$ (so that R, $\dot{R}$, and $\ddot{R}$ were known).

- The payload C were moving up or down at *any* known $z(t)$ (so that $\dot{z}$ and $\ddot{z}$ were known).

- The tower crane were rotating with *any* known $\theta(t)$ (so that $\dot{\theta}$ and $\ddot{\theta}$ were known),

then we would still be able to easily determine the values of all the quantities in Eqs. (1) and (2) to find the velocity and acceleration of the payload at C.

EXAMPLE 2.31   *Application of Spherical Coordinates*

Revisit Example 2.20 and determine, using spherical coordinates, what relationships the radar readings obtained by the station at A need to satisfy for you to conclude that the jet at B shown in Fig. 1 is flying

(a) At a level altitude.

(b) In a straight line at constant speed v_0.

SOLUTION

Road Map   As with Example 2.20, the idea is to write the imposed constraints in terms of the chosen coordinate system. For this system, that means that the altitude is constant and that the velocity vector is constant.

Computation   Given the coordinate system indicated in the figure, the plane's altitude h is given by

$$h = r \cos \phi + h_A, \tag{1}$$

where h_A is the elevation of the radar antenna above ground. Since h_A is constant, the plane can be said to be flying at a constant altitude as long as

$$r(t) \cos \phi(t) = \text{constant}. \tag{2}$$

However, contrary to what we saw in Example 2.20, this relationship is not sufficient to guarantee that the trajectory of the plane is a straight line since Eq. (2) is satisfied by *any* trajectory lying on a plane h above the ground.

   If the plane being tracked is flying along a straight line *and* at a constant speed, then its velocity vector must be constant. While it is tempting to say that constant velocity means that

$$\dot{r} = \text{constant}, \tag{3}$$

$$r\dot{\phi} = \text{constant}, \tag{4}$$

$$r\dot{\theta} \sin \phi = \text{constant}, \tag{5}$$

that is, that each component of the velocity must be constant, this is not the case since the base vectors of a spherical coordinate system change direction as the plane moves (see marginal note). Therefore, in order for $\vec{v}$ to be constant, the airplane's acceleration must be equal to zero, i.e.,

$$
\begin{aligned}
\ddot{r} - r\dot{\phi}^2 - r\dot{\theta}^2 \sin^2 \phi &= 0, \\
r\ddot{\phi} + 2\dot{r}\dot{\phi} - r\dot{\theta}^2 \sin \phi \cos \phi &= 0, \\
r\ddot{\theta} \sin \phi + 2\dot{r}\dot{\theta} \sin \phi + 2r\dot{\phi}\dot{\theta} \cos \phi &= 0.
\end{aligned}
\tag{6}
$$

   Finally, to measure the airplane's speed v_0, we need to compute the magnitude of $\vec{v}$, which is given by the square root of the sum of the squares of the components of the velocity vector, that is,

$$v_0 = \sqrt{\dot{r}^2 + (r\dot{\phi})^2 + (r\dot{\theta} \sin \phi)^2}. \tag{7}$$

Discussion & Verification   Our results are correct since they were obtained in symbolic form by a direct application of Eqs. (2.140) and (2.138), respectively, i.e., without additional manipulations that could have introduced errors.

🔎 **A Closer Look**   Verifying that the speed is constant does not allow us to say that the plane's trajectory is straight or level.

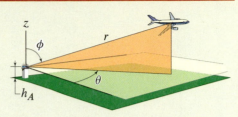

Figure 1
Airliner flying straight and level at altitude h.

🧭 Helpful Information

What does *constant velocity* mean? We know that mathematically, constant velocity means that $\dot{\vec{v}} = \vec{0}$. We know that this means that the velocity vector doesn't change magnitude or direction. In spherical coordinates, this relationship takes the form

$$\dot{\vec{v}} = \frac{d}{dt} \left(\dot{r}\,\hat{u}_r + r\dot{\phi}\,\hat{u}_\phi + r\dot{\theta} \sin \phi\,\hat{u}_\theta \right),$$

When this derivative is expanded, we must differentiate the scalar coefficients of the unit vectors *as well as the unit vectors themselves*. So, as we have already seen, constant scalar components of a vector do not imply that the vector is constant since the unit vectors can change direction. This is why Eqs. (3)–(5) are not sufficient to say that $\vec{v}$ is constant. In addition, this is why it *is* sufficient in Cartesian coordinates to say that the constancy of the scalar coefficients is sufficient to say that a vector is constant.

PROBLEMS

Problem 2.216

Although point P is moving on a sphere, its motion is being studied with the *cylindrical* coordinate system shown. Discuss in detail whether or not there are incorrect elements in the sketch of the cylindrical component system at P.
Note: Concept problems are about *explanations*, not computations.

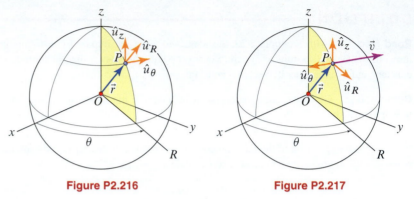

Figure P2.216 **Figure P2.217**

Problem 2.217

Although point P is moving on a sphere, its motion is being studied with the *cylindrical* coordinate system shown. Discuss in detail whether or not there are incorrect elements in the sketch of the cylindrical component system at P.
Note: Concept problems are about *explanations*, not computations.

Problem 2.218

Discuss in detail whether or not (a) there are incorrect elements in the sketch of the cylindrical component system at P and (b) the formulas for the velocity and acceleration components derived in the section can be used with the coordinate system shown.
Note: Concept problems are about *explanations*, not computations.

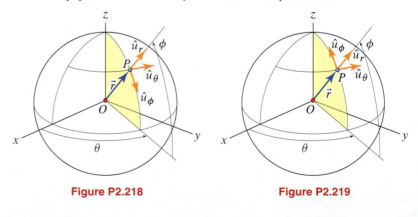

Figure P2.218 **Figure P2.219**

Problem 2.219

Discuss in detail whether or not (a) there are incorrect elements in the sketch of the cylindrical component system at P and (b) the formulas for the velocity and acceleration components derived in the section can be used with the coordinate system shown.
Note: Concept problems are about *explanations*, not computations.

Problem 2.220

A top-slewing crane is lifting an object C at a constant rate of $5.3\,\text{ft/s}$ while rotating at a constant rate of $0.12\,\text{rad/s}$ about the vertical axis. If the distance between the object and the axis of rotation of the crane's boom is currently $46\,\text{ft}$ and it is being reduced at a constant rate of $6.5\,\text{ft/s}$, find the velocity and acceleration of C, assuming that the swinging motion of C can be neglected.

Problem 2.221

An airplane is flying horizontally at a speed $v_0 = 320\,\text{mph}$ while its propellers rotate at an angular speed $\omega = 1500\,\text{rpm}$. If the propellers have a diameter $d = 14\,\text{ft}$, determine the magnitude of the acceleration of a point on the periphery of the propeller blades.

Problem 2.222

A particle is moving over the surface of a right cone with angle β and under the constraint that $R^2\dot{\theta} = K$, where K is a constant. The equation describing the cone is $R = z\tan\beta$. Determine the expressions for the velocity and the acceleration of the particle in terms of K, β, z, and the time derivatives of z.

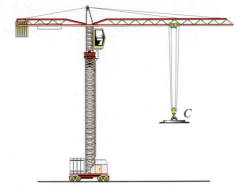

Figure P2.220

Figure P2.221

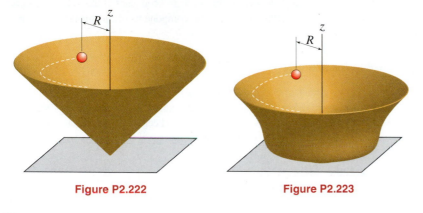

Figure P2.222　　　　　　　　**Figure P2.223**

Problem 2.223

Solve Prob. 2.222 for general surfaces of revolution; that is, R is no longer equal to $z\tan\beta$, but is now an arbitrary function of z, that is, $R = f(z)$. The expressions you are required to find will contain K, $f(z)$, derivatives of $f(z)$ with respect to z, as well as derivatives of z with respect to time.

Problem 2.224

Revisit Example 2.31, and assuming that the plane is accelerating, determine the relation that the radar readings obtained by the station at A need to satisfy for you to conclude that the jet is flying along a straight line whether at constant altitude or not.

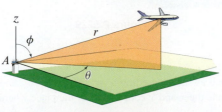

Figure P2.224

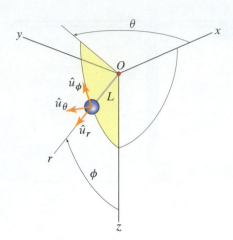

Figure P2.225

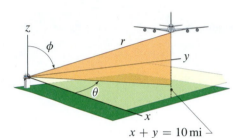

Figure P2.226

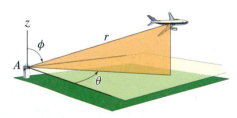

Figure P2.227 and P2.228

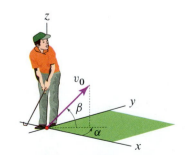

Figure P2.230

Problem 2.225

The system depicted in the figure is called a *spherical pendulum*. The fixed end of the pendulum is at O. Point O behaves as a spherical joint; i.e., the location of O is fixed while the pendulum's cord can swing in any direction in the three-dimensional space. Assume that the pendulum's cord has a constant length L, and use the coordinate system depicted in the figure to derive the expression for the acceleration of the pendulum.

Problem 2.226

An airplane is traveling at a constant altitude of 10,000 ft, with a constant speed of 450 mph, within the plane whose equation is given by $x + y = 10$ mi and in the direction of increasing x. Find the expressions for $\dot{r}$, $\dot{\theta}$, $\dot{\phi}$, $\ddot{r}$, $\ddot{\theta}$, and $\ddot{\phi}$ that would be measured when the airplane is closest to the radar station.

Problem 2.227

An airplane is being tracked by a radar station at A. At the instant $t = 0$, the following data is recorded: $r = 15$ km, $\phi = 80°$, $\theta = 15°$, $\dot{r} = 350$ km/h, $\dot{\phi} = -0.002$ rad/s, $\dot{\theta} = 0.003$ rad/s. If the airplane were flying so as to keep each of the velocity components constant for a few minutes, determine the components of the airplane's acceleration when $t = 30$ s.

Problem 2.228

An airplane is being tracked by a radar station at A. At the instant $t = 0$, the following data is recorded: $r = 15$ km, $\phi = 80°$, $\theta = 15°$, $\dot{r} = 350$ km/h, $\dot{\phi} = -0.002$ rad/s, $\dot{\theta} = 0.003$ rad/s. If the airplane were flying so as to keep each of the velocity components constant, plot the trajectory of the airplane for $0 < t < 150$ s.

Problem 2.229

A carnival ride called the *octopus* consists of eight arms that rotate about the z axis at the constant angular velocity $\dot{\theta} = 6$ rpm. The arms have a length $L = 22$ ft and form an angle ϕ with the z axis. Assuming that ϕ varies with time as $\phi(t) = \phi_0 + \phi_1 \sin \omega t$ with $\phi_0 = 70.5°$, $\phi_1 = 25.5°$, and $\omega = 1$ rad/s, determine the magnitude of the acceleration of the outer end of an arm when ϕ achieves its maximum value.

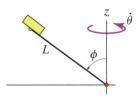

Figure P2.229

Problem 2.230

A golfer chips the ball as shown. Letting $\alpha = 23°$, $\beta = 41°$, and the initial speed be $v_0 = 6$ m/s, determine the x and y coordinates of the place where the ball will land.

Problem 2.231 🌡

In a racquetball court, at point P with coordinates $x_P = 35$ ft, $y_P = 16$ ft, and $z_P = 1$ ft, a ball is imparted a speed $v_0 = 90$ mph and a direction defined by the angles $\theta = 63°$ and $\beta = 8°$ (β is the angle formed by the initial velocity vector and the xy plane). The ball bounces off the left vertical wall to then then hit the front wall of the court. Assume that the rebound off the left vertical wall occurs such that (1) the component of the ball's velocity tangent to the wall before and after rebound is the same and (2) the component of velocity normal to the wall right after impact is equal in magnitude and opposite in direction to the same component of velocity right before impact. Accounting for the effect of gravity, determine the coordinates of the point on the front wall that will be hit by the ball after rebounding off the left wall.

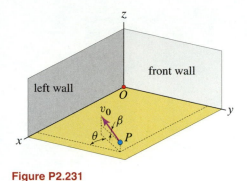

Figure P2.231

2.9 Chapter Review

In this chapter we have presented some basic definitions needed to study the motion of objects. We have also developed some basic tools for the analysis of motion both in two and three dimensions. We now present a concise summary of the material covered in this chapter.

Position, velocity, acceleration

Position. The position of a point is a *vector* going from the origin of the chosen frame of reference to the point in question. The position vectors of a point measured by different reference frames are different from one another.

Trajectory. The trajectory of a moving point is the line traced by the point during its motion. Another name for trajectory is *path*.

Displacement. The displacement between positions A and B is the vector going from A to B. In general, the magnitude of the displacement between two positions is not the distance traveled along the path between these positions.

Velocity. The velocity vector is the time rate of change of the position vector. The velocity vector is the same with respect to any two reference frames that do not move relative to one another. The velocity is *always* tangent to the path.

Speed. The speed is the magnitude of the velocity and is a nonnegative scalar quantity.

Acceleration. The acceleration vector is the time rate of change of the velocity vector. As with the velocity, the acceleration is the same with respect to any two reference frames that are not moving relative to one another. Contrary to what happens for the velocity, the acceleration vector is, in general, not tangent to the trajectory.

Cartesian coordinates. The Cartesian coordinates of a particle P moving along some path are shown in Fig. 2.63. The position vector is given by

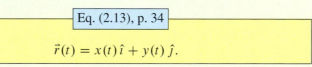

$$\vec{r}(t) = x(t)\,\hat{\imath} + y(t)\,\hat{\jmath}.$$

In Cartesian components, the velocity and acceleration vectors are given by

Eqs. (2.16) and (2.17), p. 34

$$\vec{v}(t) = \dot{x}(t)\,\hat{\imath} + \dot{y}(t)\,\hat{\jmath} = v_x(t)\,\hat{\imath} + v_y(t)\,\hat{\jmath},$$
$$\vec{a}(t) = \ddot{x}(t)\,\hat{\imath} + \ddot{y}(t)\,\hat{\jmath} = \dot{v}_x(t)\,\hat{\imath} + \dot{v}_y(t)\,\hat{\jmath} = a_x(t)\,\hat{\imath} + a_y(t)\,\hat{\jmath}.$$

Elementary motions and projectile motion

In applications, the acceleration of a point can be found as a function of time, position, and sometimes velocity.

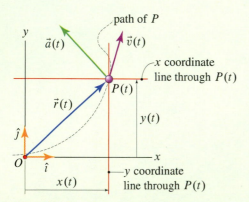
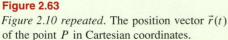

Figure 2.63
Figure 2.10 repeated. The position vector $\vec{r}(t)$ of the point P in Cartesian coordinates.

1. If the acceleration is provided as a function of time, i.e., $a = a(t)$, for velocity and position, we have

Eqs. (2.29) and (2.31), p. 57

$$v(t) = v_0 + \int_{t_0}^{t} a(t)\,dt,$$

$$s(t) = s_0 + v_0(t - t_0) + \int_{t_0}^{t}\left[\int_{t_0}^{t} a(t)\,dt\right]dt.$$

2. If the acceleration is provided as a function of velocity, i.e., $a = a(v)$, for time and position, we have

Eq. (2.33), p. 57, and Eq. (2.36), p. 58

$$t(v) = t_0 + \int_{v_0}^{v} \frac{1}{a(v)}\,dv,$$

$$s(v) = s_0 + \int_{v_0}^{v} \frac{v}{a(v)}\,dv.$$

3. If the acceleration is provided as a function of position, i.e., $a = a(s)$, for velocity and time, we have

Eqs. (2.38) and (2.40), p. 58

$$v^2(s) = v_0^2 + 2\int_{s_0}^{s} a(s)\,ds,$$

$$t(s) = t_0 + \int_{s_0}^{s} \frac{ds}{v(s)}.$$

4. If the acceleration is a constant a_c, for velocity and position, we have

Eqs. (2.41)–(2.43), p. 59

$$v = v_0 + a_c(t - t_0),$$

$$s = s_0 + v_0(t - t_0) + \tfrac{1}{2}a_c(t - t_0)^2,$$

$$v^2 = v_0^2 + 2a_c(s - s_0).$$

Circular motion. For circular motion, the equations summarized in items (1)–(4) above hold as long as we use the replacement rules

$$s \to \theta, \quad v \to \omega, \quad a \to \alpha,$$

where $\omega = \dot{\theta}$ and $\alpha = \ddot{\theta}$ are the *angular velocity* and *angular acceleration*, respectively.

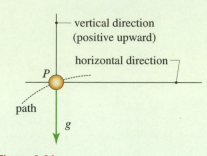

Projectile motion.   A common application of the constant acceleration equations summarized above is the study of projectile motion. We defined *projectile motion* as the motion of a particle in free flight, neglecting the forces due to air drag and neglecting changes in gravitational attraction with changes in height. In this case, referring to Fig. 2.64, the only force on the particle is the *constant* gravitational force, and the equations describing the motion are

$$\boxed{\text{Eqs. (2.53), p. 74}}$$
$$a_{\text{horiz}} = 0 \quad \text{and} \quad a_{\text{vert}} = -g.$$

Figure 2.64
Figure 2.22 repeated. Acceleration of a point P in projectile motion.

We showed that the trajectory of a projectile is a *parabola*. The mathematical form of the trajectory is of the type $y = C_0 + C_1 x + C_3 x^2$ if the motion is described using a Cartesian coordinate system with the y axis parallel to the direction of gravity.

Time derivative of a vector

A very important skill in the study of motion is the ability to compute the time derivative of vector quantities. Vectors are characterized by their magnitude and direction, and therefore the time derivative of a vector consists of a change in length and a change in direction. In the case of a unit vector, its time derivative consists of only a change in direction:

$$\boxed{\text{Eq. (2.60), p. 92}}$$
$$\underbrace{\dot{\hat{u}}(t)}_{\substack{\text{change} \\ \text{in } \vec{u}}} = \underbrace{\vec{\omega}_u \times \hat{u}}_{\substack{\text{change in} \\ \text{direction}}}.$$

In the case of an arbitrary vector $\vec{A}$ we have

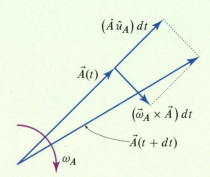

$$\boxed{\text{Eq. (2.62), p. 92}}$$
$$\underbrace{\dot{\vec{A}}(t)}_{\substack{\text{change} \\ \text{in } \vec{A}}} = \underbrace{\dot{A}\,\hat{u}_A}_{\substack{\text{change in} \\ \text{magnitude}}} + \underbrace{\vec{\omega}_A \times \vec{A}}_{\substack{\text{change in} \\ \text{direction}}},$$

where, referring to Fig. 2.65, $\hat{u}_A$ is a unit vector in the direction of $\vec{A}$, $\dot{A} = d|\vec{A}|/dt$, and $\vec{\omega}_A$ is the angular velocity of the vector $\vec{A}$. For the same vector $\vec{A}$, the following relationship provides its second derivative with respect to time:

Figure 2.65
Figure 2.31 repeated. Depiction of the vector $\vec{A}$ at time t and at time $t + dt$ showing its change in magnitude and change in direction.

$$\boxed{\text{Eq. (2.64), p. 92}}$$
$$\ddot{\vec{A}} = \ddot{A}\,\hat{u}_A + 2\vec{\omega}_A \times \dot{A}\,\hat{u}_A + \dot{\vec{\omega}}_A \times \vec{A} + \vec{\omega}_A \times (\vec{\omega}_A \times \vec{A}).$$

Normal-tangential components and polar coordinates

While we can always represent vectors using Cartesian components, it is sometimes useful to represent vector quantities in other ways.

Normal-tangential component system.   Referring to Fig. 2.66, the velocity vector has the form

Eq. (2.75), p. 107

$$\vec{v} = v\,\hat{u}_t = \dot{s}\,\hat{u}_t,$$

where v is the speed, s is the arc length along the path, and $\hat{u}_t$ is the tangent unit vector at the point P.

Referring to Fig. 2.67, the acceleration vector in normal-tangential components has the form

Eq. (2.78), p. 107

$$\vec{a} = \dot{v}\,\hat{u}_t + \frac{v^2}{\rho}\,\hat{u}_n = a_t\,\hat{u}_t + a_n\,\hat{u}_n,$$

where $\dot{v} = a_t$ is the tangential component of acceleration and $v^2/\rho = a_n$ is the normal component of acceleration.

When we are using a Cartesian coordinate system, for a path expressed via a relation such as $y = y(x)$, the path's radius of curvature is given by ρ:

Eq. (2.72), p. 106

$$\rho(x) = \frac{\left[1 + (dy/dx)^2\right]^{3/2}}{\left|d^2y/dx^2\right|}.$$

Polar coordinates.   Referring to Fig. 2.68, r and θ are the polar coordinates of point P. The coordinate θ was chosen positive in the counterclockwise direction as viewed down the positive z axis. Using polar coordinates, the position vector of P is

Eq. (2.82), p. 118

$$\vec{r} = r\,\hat{u}_r.$$

Differentiating the position with respect to time, we obtain the following result for the velocity vector in polar coordinates

Eq. (2.85), p. 119

$$\vec{v} = \dot{r}\,\hat{u}_r + r\dot{\theta}\,\hat{u}_\theta = v_r\,\hat{u}_r + v_\theta\,\hat{u}_\theta,$$

where

Eqs. (2.86), p. 119

$$v_r = \dot{r} \quad \text{and} \quad v_\theta = r\dot{\theta}$$

are the radial and transverse components of the velocity, respectively.

Differentiating the velocity with respect to time, the acceleration vector in polar coordinates takes on the form

Eq. (2.89), p. 119

$$\vec{a} = (\ddot{r} - r\dot{\theta}^2)\hat{u}_r + (r\ddot{\theta} + 2\dot{r}\dot{\theta})\hat{u}_\theta = a_r\,\hat{u}_r + a_\theta\,\hat{u}_\theta,$$

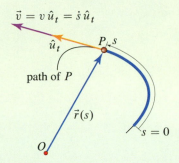

Figure 2.66
Figure 2.41 repeated. Representation of the velocity in normal-tangential component system.

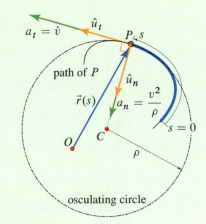

Figure 2.67
Figure 2.42 repeated. Acceleration in normal-tangential components.

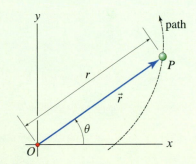

Figure 2.68
Figure 2.49 repeated. The position $\vec{r}$ of a particle defined using the polar coordinates r and θ.

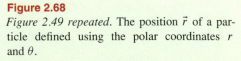

where

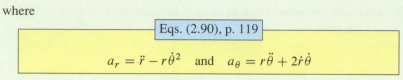

$$a_r = \ddot{r} - r\dot{\theta}^2 \quad \text{and} \quad a_\theta = r\ddot{\theta} + 2\dot{r}\dot{\theta}$$

are the radial and transverse components of the acceleration, respectively.

Relative motion

In general, physical systems consist of several moving parts. To study the motion of these systems, it is important to be able to describe the motion of one object relative to another. Referring to Fig. 2.69, consider the planar motion of points A and B. The positions of A and B relative to the XY frame are $\vec{r}_A$ and $\vec{r}_B$, respectively. Attached to A there is a frame xy that translates but does not rotate relative to frame XY. In either frame, the position of B relative to A is given by $\vec{r}_{B/A}$. Using vector addition, the vectors $\vec{r}_A$, $\vec{r}_B$, and $\vec{r}_{B/A}$ are related as follows:

Eq. (2.105), p. 136

$$\vec{r}_B = \vec{r}_A + \vec{r}_{B/A}.$$

The first and second time derivatives of the above equation are, respectively,

Eqs. (2.106) and (2.107), p. 136

$$\vec{v}_B = \vec{v}_A + \vec{v}_{B/A},$$
$$\vec{a}_B = \vec{a}_A + \vec{a}_{B/A},$$

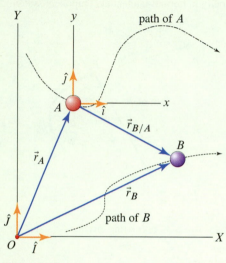

Figure 2.69
Figure 2.51 repeated. Two particles A and B and the definition of their relative position vector $\vec{r}_{B/A}$.

where $\vec{v}_{B/A} = \dot{\vec{r}}_{B/A}$ and $\vec{a}_{B/A} = \ddot{\vec{r}}_{B/A}$ are the relative velocity and acceleration of B with respect to A, respectively. In general, the vectors $\vec{v}_{B/A}$ and $\vec{a}_{B/A}$ computed by the xy observer are different from the vectors $\vec{v}_{B/A}$ and $\vec{a}_{B/A}$ computed by the XY observer. However, the xy and XY observers compute the same $\vec{v}_{B/A}$ and the same $\vec{a}_{B/A}$ if these observers do not rotate relative to one another.

Constrained motion. There are very few dynamics problems in which the motion is *not* constrained in some way. Constraints are often described by geometrical relations between points in the system. We analyzed a pulley system whose motion was constrained by the inextensibility of the cords in the system. The key to constrained motion analysis is the awareness that we can differentiate the equations describing the geometrical constraints to obtain velocities and accelerations of points of interest. In the case of the pulley system studied in this section, it was the time rate of change of the cord length that we used. For an inextensible cord, the cord length is constant so that the time rate of change of the cord length is zero.

Motion in three dimensions

When the motion of points in a system is not constrained to be planar, we need to use coordinate systems that are fully three-dimensional. We studied three such coordinate systems.

Cylindrical coordinates.   Referring to Fig. 2.70, the position of a point P in three dimensions can be described via the three quantities R, θ, and z that are the point's cylindrical coordinates. In addition, we see that cylindrical coordinates involve the orthogonal triad of unit vectors $\hat{u}_R$, $\hat{u}_\theta$, and $\hat{u}_z$. Using this triad, the position in cylindrical coordinates is given by

Eq. (2.120), p. 155

$$\vec{r} = R\,\hat{u}_R + z\,\hat{u}_z.$$

The velocity vector in cylindrical coordinates is given by

Eq. (2.121), p. 155

$$\vec{v} = \dot{R}\,\hat{u}_R + R\dot{\theta}\,\hat{u}_\theta + \dot{z}\,\hat{u}_z = v_R\,\hat{u}_R + v_\theta\,\hat{u}_\theta + v_z\,\hat{u}_z,$$

where

Eqs. (2.122), p. 155

$$v_R = \dot{R}, \quad v_\theta = R\dot{\theta}, \quad \text{and} \quad v_z = \dot{z}$$

are the components of the velocity vector in the $\hat{u}_R$, $\hat{u}_\theta$, and $\hat{u}_z$ directions, respectively. Finally, the acceleration vector in cylindrical coordinates is given by

Eq. (2.123), p. 155

$$\vec{a} = \left(\ddot{R} - R\dot{\theta}^2\right)\hat{u}_R + \left(R\ddot{\theta} + 2\dot{R}\dot{\theta}\right)\hat{u}_\theta + \ddot{z}\,\hat{u}_z$$
$$= a_R\,\hat{u}_R + a_\theta\,\hat{u}_\theta + a_z\,\hat{u}_z,$$

where

Eqs. (2.124), p. 155

$$a_R = \ddot{R} - R\dot{\theta}^2, \quad a_\theta = R\ddot{\theta} + 2\dot{R}\dot{\theta}, \quad \text{and} \quad a_z = \ddot{z}$$

are the components of the acceleration vector in the $\hat{u}_R$, $\hat{u}_\theta$, and $\hat{u}_z$ directions, respectively.

Spherical coordinates.   Referring to Fig. 2.71, the position of a point P in three dimensions can be described via the three quantities r, θ, and ϕ that are the point's spherical coordinates. In addition, spherical coordinates involve the orthogonal triad of unit vectors $\hat{u}_r$, $\hat{u}_\phi$, and $\hat{u}_\theta$, with $\hat{u}_\phi \times \hat{u}_\theta = \hat{u}_r$. Using this triad, the position vector in spherical coordinates is given by

Eq. (2.131), p. 157

$$\vec{r} = r\,\hat{u}_r.$$

The velocity vector in spherical coordinates is given by

Eq. (2.137), p. 157

$$\vec{v} = \dot{r}\,\hat{u}_r + r\dot{\phi}\,\hat{u}_\phi + r\dot{\theta}\sin\phi\,\hat{u}_\theta = v_r\,\hat{u}_r + v_\phi\,\hat{u}_\phi + v_\theta\,\hat{u}_\theta,$$

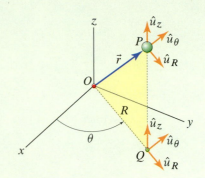

Figure 2.70
Figure 2.56 repeated. Coordinate directions in the cylindrical coordinate system. As with polar coordinates defined in Section 2.6, the θ coordinate is implicitly contained in $\hat{u}_r$.

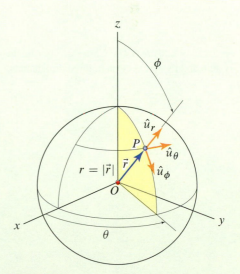

Figure 2.71
Figure 2.58 repeated. Definition of the unit vectors in spherical coordinates.

where

$$
\boxed{\text{Eqs. (2.138), p. 157}}
$$

$$
v_r = \dot{r}, \quad v_\phi = r\dot{\phi}, \quad \text{and} \quad v_\theta = r\dot{\theta}\sin\phi
$$

are the components of the velocity vector in the $\hat{u}_r$, $\hat{u}_\phi$, and $\hat{u}_\theta$ directions, respectively. Finally, the acceleration vector in spherical coordinates is given by

$$
\boxed{\text{Eq. (2.139), p. 157}}
$$

$$
\begin{aligned}
\vec{a} &= \left(\ddot{r} - r\dot{\phi}^2 - r\dot{\theta}^2\sin^2\phi\right)\hat{u}_r \\
&\quad + \left(r\ddot{\phi} + 2\dot{r}\dot{\phi} - r\dot{\theta}^2\sin\phi\cos\phi\right)\hat{u}_\phi \\
&\quad + \left(r\ddot{\theta}\sin\phi + 2\dot{r}\dot{\theta}\sin\phi + 2r\dot{\phi}\dot{\theta}\cos\phi\right)\hat{u}_\theta \\
&= a_r\,\hat{u}_r + a_\phi\,\hat{u}_\phi + a_\theta\,\hat{u}_\theta,
\end{aligned}
$$

where

$$
\boxed{\text{Eqs. (2.140), p. 157}}
$$

$$
\begin{aligned}
a_r &= \ddot{r} - r\dot{\phi}^2 - r\dot{\theta}^2\sin^2\phi, \\
a_\phi &= r\ddot{\phi} + 2\dot{r}\dot{\phi} - r\dot{\theta}^2\sin\phi\cos\phi, \\
a_\theta &= r\ddot{\theta}\sin\phi + 2\dot{r}\dot{\theta}\sin\phi + 2r\dot{\phi}\dot{\theta}\cos\phi
\end{aligned}
$$

are the components of the acceleration vector in the $\hat{u}_r$, $\hat{u}_\phi$, and $\hat{u}_\theta$ directions, respectively.

Cartesian coordinates.   Referring to Fig. 2.72, in the three-dimensional Cartesian coordinate system, the position, velocity, and acceleration, respectively, are given in component form as

$$
\boxed{\text{Eqs. (2.141)–(2.143), p. 158}}
$$

$$
\begin{aligned}
\vec{r}(t) &= x(t)\,\hat{i} + y(t)\,\hat{j} + z(t)\,\hat{k}, \\
\vec{v}(t) &= \dot{x}\,\hat{i} + \dot{y}\,\hat{j} + \dot{z}\,\hat{k} = v_x\,\hat{i} + v_y\,\hat{j} + v_z\,\hat{k}, \\
\vec{a}(t) &= \ddot{x}\,\hat{i} + \ddot{y}\,\hat{j} + \ddot{z}\,\hat{k} = a_x\,\hat{i} + a_y\,\hat{j} + a_z\,\hat{k}.
\end{aligned}
$$

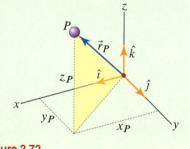

Figure 2.72

Figure 2.59 repeated. The Cartesian coordinate system and its unit vectors in three dimensions.

REVIEW PROBLEMS

Problem 2.232

The figure shows the displacement vector of a point P between two time instants t_1 and t_2. Is it possible for the vector $\vec{v}_{\text{avg}}$ shown to be the average velocity of P over the time interval $[t_1, t_2]$?

Note: Concept problems are about *explanations*, not computations.

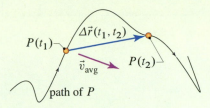

Figure P2.232

Problem 2.233

The motion of a point P with respect to a Cartesian coordinate system is described by $\vec{r} = \{2\sqrt{t}\,\hat{\imath} + [4\ln(t+1) + 2t^2]\,\hat{\jmath}\}$ ft, where t is time expressed in seconds. Determine the average velocity between $t_1 = 4$ s and $t_2 = 6$ s. Then find the time $\bar{t}$ for which the x component of P's velocity is *exactly* equal to the x component of P's average velocity between times t_1 and t_2. Is it possible to find a time at which P's velocity and P's average velocity are exactly equal? Explain why. *Hint:* Velocity is a vector.

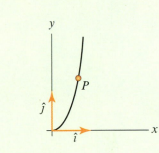

Figure P2.233

Problem 2.234

The velocity and acceleration of point P expressed relative to frame A at some time t are

$$\vec{v}_{P/A} = (12.5\,\hat{\imath}_A + 7.34\,\hat{\jmath}_A)\ \text{m/s} \quad \text{and} \quad \vec{a}_{P/A} = (7.23\,\hat{\imath}_A - 3.24\,\hat{\jmath}_A)\ \text{m/s}^2.$$

Knowing that frame B does not move relative to frame A, determine the expressions for the velocity and acceleration of P with respect to frame B. Verify that the speed of P and the magnitude of P's acceleration are the same in the two frames.

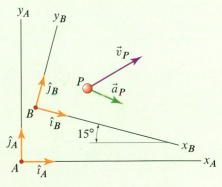

Figure P2.234

Problems 2.235 and 2.236

A dynamic fracture model proposed to explain the behavior of cracks propagating at high velocity views the crack path as a *wavy path.*[*] In this model, a crack tip appearing to travel along a straight path actually travels at roughly the speed of sound along a wavy path. Let the wavy path of the crack tip be described by the function $y = h\sin(2\pi x/\lambda)$, where h is the amplitude of the crack tip fluctuations in the direction perpendicular to the crack plane and λ is the corresponding period. Assume that the crack tip travels along the wavy path at a constant speed v_s (e.g., the speed of sound).

Problem 2.235 Find the expression for the x component of the crack tip velocity as a function of v_s, λ, h, and x.

Problem 2.236 Denote the *apparent* crack tip velocity by v_a, and define it as the average value of the x component of the crack velocity, that is,

$$v_a = \frac{1}{\lambda}\int_0^\lambda v_x\,dx.$$

In dynamic fracture experiments on polymeric materials, $v_a = 2v_s/3$, v_s is found to be close to 800 m/s, and λ is of the order of 100 μm. What value of h would you expect to find in the experiments if the wavy crack theory were confirmed to be accurate?

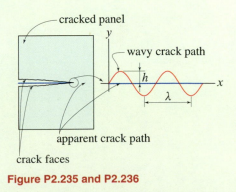

Figure P2.235 and P2.236

---

[*]In dynamic fracture the structural failure of a material occurs at speeds close to the speed of sound (in that material). This field of study is very important in the design of impact- and/or blast-resistant structures. The model mentioned in these problems is due to H. Gao, "Surface Roughening and Branching Instabilities in Dynamic Fracture," *Journal of the Mechanics and Physics of Solids,* **41**(3), pp. 457–486, 1993.

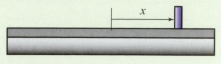

Figure P2.237

🖥 Problem 2.237 🖥

The motion of a peg sliding within a rectilinear guide is controlled by an actuator in such a way that the peg's acceleration takes on the form $\ddot{x} = a_0(2\cos 2\omega t - \beta \sin \omega t)$, where t is time, $a_0 = 3.5\,\text{m/s}^2$, $\omega = 0.5\,\text{rad/s}$, and $\beta = 1.5$. Determine the total distance traveled by the peg during the time interval $0\,\text{s} \le t \le 5\,\text{s}$ if $\dot{x}(0) = a_0\beta/\omega + 0.3\,\text{m/s}$. When compared with Prob. 2.45, why does the addition of $0.3\,\text{m/s}$ in the initial velocity turn this into a problem that requires a computer for the solution?

Problem 2.238

The acceleration of an object in rectilinear free fall while immersed in a linear viscous fluid is $a = g - C_d v/m$, where g is the acceleration of gravity, C_d is a constant drag coefficient, v is the object's velocity, and m is the object's mass. Letting $v = 0$ and $s = 0$ for $t = 0$, where s is position and t is time, determine the position as a function of time.

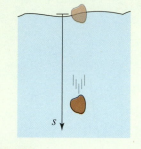

Figure P2.238

Problem 2.239

Heavy rains cause a particular stretch of road to have a coefficient of friction that changes as a function of location. Under these conditions the acceleration of a car skidding while trying to stop can be approximated by $\ddot{s} = -(\mu_k - cs)g$, where μ_k is the friction coefficient under dry conditions, g is the acceleration of gravity, and c, with units of m^{-1}, describes the rate of friction decrement. Let $\mu_k = 0.5$, $c = 0.015\,\text{m}^{-1}$, and $v_0 = 45\,\text{km/h}$, where v_0 is the initial velocity of the car. Determine the time it will take the car to stop and the percentage of increase in stopping time with respect to dry conditions, i.e., when $c = 0$.

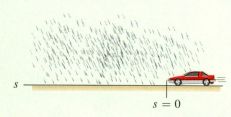

Figure P2.239

Problem 2.240

The acceleration of a particle of mass m suspended by a linear spring with spring constant k and unstretched length L_0, (when the spring length is equal to L_0, the spring exerts no force on the particle) is given by $\ddot{x} = g - (k/m)(x - L_0)$. Assuming that at $t = 0$ the particle is at rest and its position is $x = 0\,\text{m}$, derive the expression of the particle's position x as a function of time. *Hint:* A good table of integrals will come in handy.

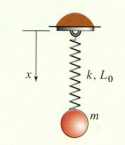

Figure P2.240

Problem 2.241

In a movie scene involving a car chase, a car goes over the top of a ramp at A and lands at B below. Letting $\alpha = 18°$ and $\beta = 25°$, determine the speed of the car at A if the car is to be airborne for a full 3 s. Furthermore, determine the distance d covered by the car during the stunt as well as the impact speed and angle at B. Neglect aerodynamic effects. Express your answer using the U.S. Customary system of units.

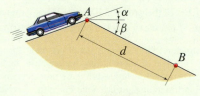

Figure P2.241

Problem 2.242

Consider the problem of launching a projectile a distance R from O to D with a known launch speed v_0. It is probably clear to you that you also need to know the launch angle θ if you want the projectile to land exactly at R. But it turns out that the condition determining whether or not v_0 is large enough to get to R does not depend on θ. Determine this condition on v_0. *Hint:* Find v_0 as a function of R and θ, and then remember that the sine function is bounded by 1.

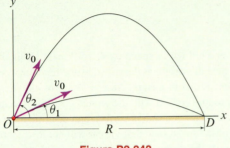

Figure P2.242

Problem 2.243

A skater is spinning with her arms completely stretched out and with an angular velocity $\omega = 60$ rpm. Letting $r_b = 0.55$ ft, and $\ell = 2.2$ ft and neglecting the change in ω as the skater lowers her arms, determine the velocity and acceleration of the hand A right when $\beta = 0°$ if the skater lowers her arms at the constant rate $\dot{\beta} = 0.2$ rad/s. Express the answers using the component system shown, which rotates with the skater and for which the unit vector $\hat{\jmath}$ (not shown) is such that $\hat{\jmath} = \hat{k} \times \hat{\imath}$.

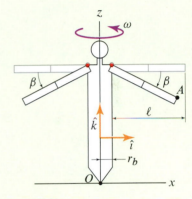

Figure P2.243

Problem 2.244

A roller coaster travels over the top A of the track section shown with a speed $v = 60$ mph. Compute the largest radius of curvature ρ at A such that the passengers on the roller coaster will experience weightlessness at A.

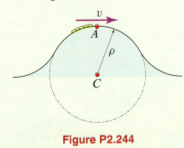

Figure P2.244

Problem 2.245

Determine, as a function of the latitude λ, the normal acceleration of the point P on the surface of the Earth due to the spin ω_E of Earth about its axis. In addition, determine the normal acceleration of the Earth due to its rotation about the Sun. Using these results, determine the latitude above which the acceleration due to the orbital motion of the Earth is more significant than the acceleration due to the spin of the Earth about its axis. Use $R_E = 6371$ km for the mean radius of the Earth, and assume the Earth's orbit about the Sun is circular with radius $R_O = 1.497 \times 10^8$ km.

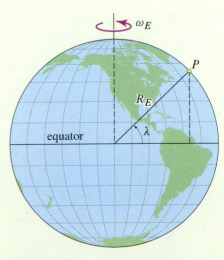

Figure P2.245

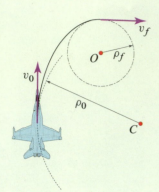

Figure P2.246

Problem 2.246

A jet is flying straight and level at a speed $v_0 = 1100$ km/h when it turns to change its course by 90° as shown. The turn is performed by decreasing the path's radius of curvature uniformly as a function of the position s along the path while keeping the normal acceleration constant and equal to $8g$, where g is the acceleration due to gravity. At the end of the turn, the speed of the plane is $v_f = 800$ km/h. Determine the radius of curvature ρ_f at the end of the turn and the time t_f that the plane takes to complete its change in course.

Problem 2.247

A car is traveling over a hill with a speed $v_0 = 160$ km/h. Using the Cartesian coordinate system shown, the hill's profile is described by the function $y = -(0.003 \text{ m}^{-1})x^2$, where x and y are measured in meters. At $x = -100$ m, the driver realizes that her speed will cause her to lose contact with the ground once she reaches the top of the hill at O. Verify that the driver's intuition is correct, and determine the minimum constant time rate of change of the speed such that the car will not lose contact with the ground at O. *Hint:* To compute the distance traveled by the car along the car's path, observe that $ds = \sqrt{dx^2 + dy^2} = \sqrt{1 + (dy/dx)^2}\, dx$ and that

$$\int \sqrt{1 + C^2 x^2}\, dx = \frac{x}{2}\sqrt{1 + C^2 x^2} + \frac{1}{2C}\ln\left(Cx + \sqrt{1 + C^2 x^2}\right).$$

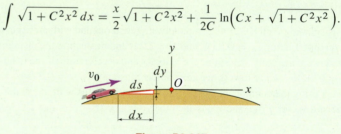

Figure P2.247

Problem 2.248

The mechanism shown is called a *swinging block* slider crank. First used in various steam locomotive engines in the 1800s, this mechanism is often found in door-closing systems. Let $H = 1.25$ m, $R = 0.45$ m, and r denote the distance between B and O. Assuming that the speed of B is constant and equal to 5 m/s, determine $\dot{r}$, $\dot{\phi}$, $\ddot{r}$, and $\ddot{\phi}$ when $\theta = 180°$.

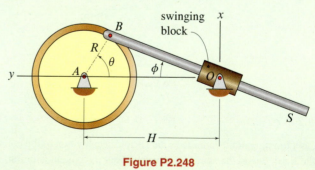

Figure P2.248

Problem 2.249

The cam is mounted on a shaft that rotates about O with constant angular velocity ω_{cam}. The profile of the cam is described by the function $\ell(\phi) = R_0(1 + 0.25 \cos^3 \phi)$,

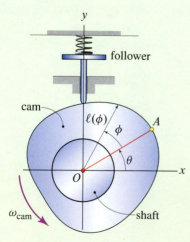

Figure P2.249

where the angle ϕ is measured relative to the segment OA, which rotates with the cam. Letting $R_0 = 3$ cm, determine the maximum value of angular velocity ω_{max} such that the maximum speed of the follower is limited to 2 m/s. In addition, compute the smallest angle θ_{min} for which the follower achieves it maximum speed.

Problem 2.250

A car is traveling at a constant speed $v_0 = 210$ km/h along a circular turn with radius $R = 137$ m (the figure is not to scale). The camera at O is tracking the motion of the car. Letting $L = 15$ m, determine the camera's rotation rate as well as the corresponding time rate of change of the rotation rate when $\phi = 30°$.

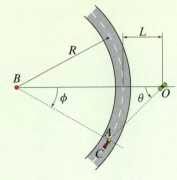

Figure P2.250

Problems 2.251 through 2.253

A fountain has a spout that can rotate about O and whose angle β is controlled so as to vary with time according to $\beta = \beta_0[1 + \sin^2(\omega t)]$, with $\beta_0 = 15°$ and $\omega = 0.4\pi$ rad/s. The length of the spout is $L = 1.5$ ft, the water flow through the spout is constant, and the water is ejected at a speed $v_0 = 6$ ft/s, measured relative to the spout.

Problem 2.251   Determine the largest speed with which the water particles are released from the spout.

Problem 2.252   Determine the magnitude of the acceleration immediately before release when $\beta = 15°$.

Problem 2.253   Determine the highest position reached by the resulting water arc.

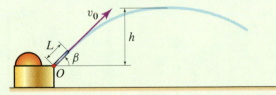

Figure P2.251–P2.253

Problem 2.254

A plane is initially flying north with a speed $v_0 = 430$ mph relative to the ground while the wind has a constant speed $v_W = 12$ mph in the north-south direction. The plane performs a circular turn with radius of $\rho = 0.45$ mi. Assume that the airspeed indicator on the plane measures the absolute value of the component of the relative velocity of the plane with respect to the air in the direction of motion. Then determine the value of the tangential component of the airplane's acceleration when the airplane is halfway through the turn, assuming that the airplane maintains constant the reading of the airspeed indicator.

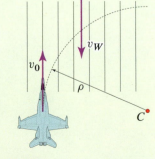

Figure P2.254

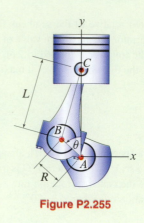

Figure P2.255

Problem 2.255

The piston head at C is constrained to move along the y axis. Let the crank AB be rotating counterclockwise at a constant angular speed $\dot\theta = 2000$ rpm, $R = 3.5$ in., and $L = 5.3$ in. Obtain the angular velocity of the connecting rod BC by differentiating the relative position vector of C with respect to B when $\theta = 35°$. *Hint:* You will also need to determine the velocity of B and enforce the constraint that demands that C move only along the y axis.

Problem 2.256

A child A is swinging from a swing that is attached to a trolley that is free to move along a fixed rail. Letting $L = 3.2$ m, if at a given instant $a_B = 3.4$ m/s^2, $\theta = 23°$, $\dot\theta = 0.45$ rad/s, and $\ddot\theta = -0.2$ rad/s^2, determine the magnitude of the acceleration of the child relative to the rail at that instant.

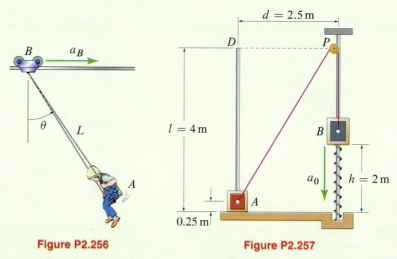

Figure P2.256 **Figure P2.257**

Problem 2.257

Block B is released from rest at the position shown, and it has a constant acceleration downward $a_0 = 5.7$ ft/s^2. Determine the velocity and acceleration of block A at the instant that B touches the floor.

Problem 2.258

At a given instant, an airplane is flying horizontally with speed $v_0 = 290$ mph and acceleration $a_0 = 12$ ft/s^2. At the same time, the airplane's propellers rotate at an angular speed $\omega = 1500$ rpm while accelerating at a rate $\alpha = 0.3$ rad/s^2. Knowing that propeller diameter is $d = 14$ ft, determine the magnitude of the acceleration of a point on the periphery of the propellers at the given instant.

Figure P2.258

Problem 2.259

A golfer chips the ball as shown. Treating α, β, and the initial speed v_0 as given, find an expression for the radius of curvature of the ball's trajectory as a function of time and the given parameters. *Hint:* Use the Cartesian coordinate system shown to determine the acceleration and the velocity of the ball. Then reexpress these quantities, using normal-tangential components.

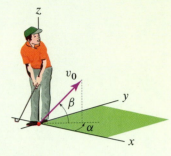

Figure P2.259

Problem 2.260

A carnival ride called the octopus consists of eight arms that rotate about the z axis with a constant angular velocity $\dot{\theta} = 6$ rpm. The arms have a length $L = 8$ m and form an angle ϕ with the z axis. Assuming that ϕ varies with time as $\phi(t) = \phi_0 + \phi_1 \sin \omega t$ with $\phi_0 = 70.5°$, $\phi_1 = 25.5°$, and $\omega = 1$ rad/s, determine the magnitude of the acceleration of the outer end of an arm when ϕ achieves its minimum value.

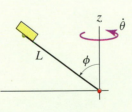

Figure P2.260

Force and Acceleration Methods for Particles

In this chapter we show how Newton's second law, $\vec{F} = m\vec{a}$, is applied to study the motion of bodies that are modeled as particles. Throughout this chapter, Newton's second law is viewed as an *axiom*, i.e., a statement we treat as true and not derivable from other principles. As such, there is little to explain about the law itself. The chapter will therefore emphasize *how* $\vec{F} = m\vec{a}$ is applied, and it begins the study of *kinetics*, namely, the study of the forces that cause and are caused by motion. By the time we complete this chapter, we will be able to use the kinematics (discussed in Chapter 2) along with Newton's second law to either (1) predict the motion of a particle system caused by given forces or (2) determine the forces needed for a particle system to move in a prescribed way.

3.1 Rectilinear Motion

A chameleon capturing an insect

Suppose that a chameleon is propelling its long tongue out to snatch an insect for a meal (see Fig. 3.1). This process occurs so fast that a high-speed video camera needs to be used to capture the event. Since it is the "stickiness" of the chameleon's tongue that allows it to latch onto the insect, the question is, Can we use the video to estimate how much stickiness, i.e., force, is required to get the insect where the chameleon would like it to be?

Figure 3.1
A chameleon capturing an insect.

We begin by describing the insect's motion based on the available video data, which tells us that it takes 0.15 s for the chameleon to completely retrieve the insect. If the insect is not initially moving, then the initial speed of the insect is zero. The final speed of the insect must be zero since it ends up in the chameleon's mouth for ingestion. You may recall the problem we studied in Section 2.2 in which a car travels between two stop signs—the motion of the insect is not unlike that of the car. As for the car, there are three considerations we need to take into account in generating the velocity versus time profile:

1. The time it takes for the chameleon to retrieve the insect.

2. The velocity, which must be zero at the start and end of the time interval.

3. The distance traveled by the insect from start to finish.

Modeling the velocity profile. The velocity profile in Fig. 3.2 is of the form

$$v(t) = A[1 - \cos(Bt)].$$

Since the above expression is such that $v = 0$ for $t = 0$, the coefficients A and B can be found by satisfying the remaining two criteria listed. To make v zero at $t = 0.15\,\text{s}$, we must have $B(0.15\,\text{s}) = 2\pi$, and so $B = 41.89\,\text{s}^{-1}$. For the distance traveled to be $0.3\,\text{m}$, the integral of $v(t)$ over the interval $[0, 0.15\,\text{s}]$ must be equal to $0.3\,\text{m}$, i.e.,

$$\int_0^{0.15\,\text{s}} A[1 - \cos(41.89\,t)]\,dt = 0.3\,\text{m}$$

$$\Rightarrow \quad (0.15\,\text{s})A = 0.3\,\text{m},$$

so that $A = 2\,\text{m/s}$.

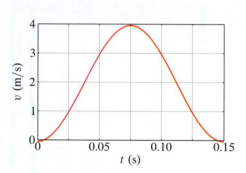

Figure 3.2
Velocity versus time profile for the tip of the chameleon's tongue.

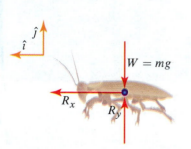

Figure 3.3
FBD of the insect retrieved by the chameleon. Note that the insect is shown "in transparency" and that the forces are applied to a "blue dot." This is done to emphasize that our original object (the insect) is modeled as a particle (the blue dot).

The distance traveled needs to be determined experimentally and is found to be approximately 0.3 m. Since we want a smooth curve for the velocity profile, we will approximate the velocity via a trigonometric function. Using the above three criteria, we will approximate the velocity profile as (see the Helpful Information marginal note)

$$v(t) = 2[1 - \cos(41.89\,t)]\,\text{m/s}, \tag{3.1}$$

a plot of which is shown in Fig. 3.2.

Now that we have a model of v vs. t, we can proceed with the determination of the force. We first identify the system to analyze and then draw a free body diagram (FBD) of this system. In this case, it is the insect we consider since it is the force *on the insect* that we want to determine. Again, the video shows that the insect moves in an almost straight line and that the insect does not touch the ground as it is retrieved. Modeling the insect as a particle, its FBD while being retrieved is shown in Fig. 3.3. Note that, in the FBD, the insect is shown "in transparency" and that a "blue dot" is shown to be the point of application of the forces. This is done to emphasize that our original object (the insect) is modeled as a particle (the blue dot). We will use this graphical device throughout the book whenever an object is modeled as a particle.

Since we know that the insect moves in a horizontal straight line, we choose the x axis to be along that line. Consequently, the y component of the insect's acceleration is zero. In addition, Newton's second law, $\vec{F} = m\vec{a}$, tells us that the sum of all the forces acting on the insect is balanced by, or is equal to, the insect's mass times the acceleration of the insect, i.e.,

$$\vec{R} + \vec{W} = m\vec{a}, \tag{3.2}$$

where $\vec{R}$ is the force exerted by the chameleon on the insect and $\vec{W}$ is the insect's weight, which can be expressed as $\vec{W} = -mg\,\hat{j}$. Since Newton's second law is a vector equation, we consider its components separately to obtain

$$\sum F_x = ma_x \quad \Rightarrow \quad R_x = ma_x, \tag{3.3}$$

$$\sum F_y = ma_y \quad \Rightarrow \quad R_y - mg = 0. \tag{3.4}$$

Equation (3.4) tells us that $R_y = mg$ or that the y component of the sticky force simply needs to support the weight of the insect. Since chameleons like to eat cockroaches, we will let the insect be a cockroach, which typically has a mass $m = 6\,\text{g}$. When Eq. (3.1) is differentiated with respect to time and substituted into Eq. (3.3), we obtain

$$R_x = m[83.78 \sin(41.89\,t)]\,\text{N}, \tag{3.5}$$

which, since $R_y = mg$, gives a total required force of

$$|\vec{R}| = \sqrt{R_x^2 + R_y^2} = \sqrt{[83.78m \sin(41.89\,t)]^2 + (mg)^2}. \tag{3.6}$$

The value of $|\vec{R}|$ given by Eq. (3.6), divided by the cockroach's weight, has been plotted in Fig. 3.4. The plot shows that the maximum force required is almost $9mg$, and this situation occurs at two different points in time. Referring to Eq. (3.5), the first time the maximum force is achieved corresponds to a positive value of R_x so that the force is directed toward the chameleon. The second maximum corresponds to a negative value of R_x, which is needed to slow down the cockroach before the cockroach comes to a stop in the chameleon's mouth. The maximum value of $|\vec{R}|$ can be calculated by differentiating Eq. (3.6) with respect to time and setting the result equal to zero, which tells us that $R_{max} = 0.5061$ N. This value occurs at $t_1 = 0.03750$ s and $t_2 = 0.1125$ s.

We complete our discussion by determining the acceleration experienced by the cockroach. We saw that $a_y = 0$ and $a_x = 83.78 \sin(41.89t)$ m/s^2 so that a_{max} occurs when a_x is largest.[*] Differentiating a_x and setting it equal to zero to find the time t_m at which the acceleration is a maximum, we obtain

$$\frac{da_x}{dt} = 3509.5 \cos(41.89\,t_m)\,\text{m/s}^3 = 0, \qquad (3.7)$$

or $41.89 t_m = (2n + 1)\pi/2\,(n = 0, 1, 2, \dots)$ so that $(t_m)_1 = 0.03750$ s and $(t_m)_2 = 0.1125$ s (these are the only two times that fall within the 0.15 s interval). Substituting t_m into a_x, we obtain 83.78 m/s^2, and thus we see that the insect experiences $83.78/9.81 = 8.54g$ of acceleration. Notice that the maximum force occurs at the same times as the maximum acceleration. In one-dimensional problems, the maximum force and maximum acceleration will always occur at the same time.

Remarks about the chameleon problem

The chameleon problem illustrates how it is possible to use Newton's second law to compute a force on an object whose motion is given. Newton's second law can also be applied to solve the reverse problem; i.e., given the forces acting on an object, we can find the resulting motion. For this reason, it is important to know how to mathematically represent the various types of forces we often encounter in applications. We begin our formal discussion of the application of Newton's second law with a review of two forces commonly found in engineering applications: friction and spring forces.

Another remark about the chameleon problem concerns the fact that its solution included the sketch of an FBD along with a choice of component system to properly express the kinematics as well as the forces in the problem. These elements, i.e., the FBD and choice of a component system, the expression of the forces, and the kinematics are common elements to any problem involving the use of Newton's second law. We will discuss these steps in a systematic fashion after reviewing friction and spring forces.

Friction

Friction plays an important role in many engineering problems and is still the object of active research as it is an extremely complex phenomenon. Here we only review an elementary model of friction[†] called the *Coulomb friction model*, which was first published by Charles Augustin Coulomb (1736–1806) around 1773. While this model is very simple, it can still be considered accurate in many situations.

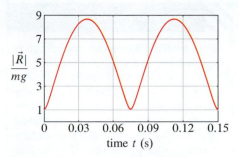

Figure 3.4
Required force on the insect as it is pulled into the mouth of the chameleon.

---

[*] Keep in mind that $a_{max} \neq R_{max}/m$ since R_{max} includes the weight force from R_y, even though the insect is not accelerating in the y direction.

[†] A treatment of friction can be found in Chapter 9 of M. E. Plesha, G. L. Gray, and F. Costanzo, *Engineering Mechanics: Statics*, McGraw-Hill, Dubuque, 2010.

///
Common Pitfall

About friction. A common misconception regarding the Coulomb friction model is that $F = \mu_s N$ when the body is not moving and $F = \mu_k N$ when the body is moving. When a wheel rolls without slip, it is moving, but the friction force F and the normal force N between the wheel and the surface on which it rolls are such that $|F| \leq \mu_s |N|$. In addition, the only time we can say that $F = \mu_k N$ is when an object is *sliding* over the surface on which it rests, that is, the object and the surface have different velocities. When a body is not moving relative to the surface on which it rests, then all we can say is that $|F| \leq \mu_s |N|$ and nothing more, unless we know that slip is impending, in which case we can say that $|F| = \mu_s |N|$.

The friction force is generally defined to be the component of the contact force between two objects that is tangent to the contact surface. The Coulomb friction model states that when two objects slide relative to one another, the friction force between the two objects is equal to the normal force between the objects times a parameter called the *coefficient of kinetic friction*. When the objects do not slide relative to one another, the friction force is bounded by the value of the normal force times a parameter called the *coefficient of static friction*. The friction force acts to oppose relative motion or tendency for relative motion between the objects that are in contact. The Coulomb friction model specifies that all friction falls within one of the following three conditions:

No slip   When two objects do not move (or do not *slip* or do not *slide*) relative to one another, then the friction force satisfies the inequality

$$|F| \leq \mu_s |N|, \tag{3.8}$$

where μ_s is the static friction coefficient and N is the normal force between the two bodies. The direction of F on one body is equal and opposite to the direction of F on the other body (due to Newton's third law). In addition, the direction of F on a given body is *opposite* to the direction that the velocity vector of that body would have, relative to the other body, in the absence of friction. Note that the absolute value signs around F and N are important. Often the direction of the friction force is one of the unknowns of the problem, and writing $|F|$ allows us not to worry about having to guess the "right" direction for F. As far as N is concerned, usually it is assumed to be positive when "in compression," and at least for equilibrium problems, the sign of N is easy to predict. However, in dynamics there are important examples, such as mechanisms with pegs moving along slotted guides, in which the force N can change sign with time. Therefore, we write $|N|$ instead of N to make sure that Eq. (3.8) is always evaluated correctly.

Impending slip   When slip between two bodies is impending, that is, when one body is *just about* to slip relative to the other, then the friction and normal forces are related according to

$$F = \mu_s N, \tag{3.9}$$

where the direction of F must be consistent with the fact that the friction force opposes the tendency for relative motion of A and B.

Slip   When two bodies A and B slide relative to one another (see Fig. 3.5), then the friction force on body A due to body B is

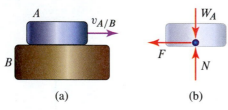

(a)   (b)

Figure 3.5
(a) Body A sliding on body B; the relative velocity is given by $v_{A/B}$. (b) FBD of body A showing the direction of kinetic friction. Body A has been modeled as a particle.

$$\vec{F} = -\mu_k |N| \frac{\vec{v}_{A/B}}{|\vec{v}_{A/B}|}, \tag{3.10}$$

where μ_k is the kinetic friction coefficient and $\mu_k \leq \mu_s$. Notice that the direction of $\vec{F}$ on body A is *opposite* to the direction of the velocity of A relative to B. Equation (3.10) is usually written as

$$F = \mu_k N, \tag{3.11}$$

where the direction of F must be consistent with the fact that the friction force opposes relative motion of A and B.

Springs

Springs, which are often thought of as having the coiled shape shown in Fig. 3.6, come in myriad shapes, sizes, and materials. These include bungee cords, beams made of metal or other materials, plates and shells, and torsional rods.[*]

Referring to Fig. 3.7, from a geometric viewpoint, we describe a spring in terms of its length. The *unstretched length of the spring* is the length of the spring when no force is applied to the spring. Letting L and L_0 denote the current and unstretched lengths of a spring, respectively, we define the *stretch*, denoted by the Greek letter δ, to be the quantity

$$\delta = L - L_0. \tag{3.12}$$

We say that a spring is stretched or compressed depending on whether $\delta > 0$ or $\delta < 0$, respectively.

In the vast majority of the applications we consider, it is assumed that springs have negligible mass. This assumption causes the force internal to a spring to be equal to the external force applied to the spring. Unless stated otherwise, we will always assume that springs are massless.

Linear elastic springs

A spring is said to be *linear elastic* if the internal force in the spring is linearly related to the amount the spring is stretched or compressed. Referring to Fig. 3.8, the force F_s required to stretch a linear elastic spring by an amount δ is given by

$$F_s = k\delta = k(L - L_0), \tag{3.13}$$

where k is the *spring constant* and has dimensions of force over length. Unless we indicate otherwise, we will always assume that springs are linear elastic.

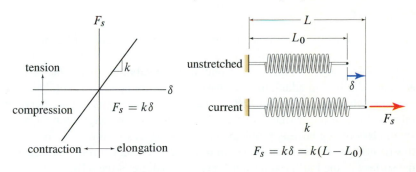

Figure 3.8. Spring law for a linear elastic spring.

Spring nonlinearities

While we will be dealing almost exclusively with linear elastic springs, nonlinearities in elastic structures (e.g., springs) are important in engineering dynamics. As a simple example, an elastic band is highly nonlinear over a wide range of its possible stretch. Figure 3.9 shows a qualitative plot of the force in an

[*]See, for example, A. M. Wahl, *Mechanical Springs*, 2nd ed. McGraw-Hill, New York, 1963. Reprinted in 1991 by the Spring Manufacturers Institute, Inc., Rolling Meadows, IL.

Figure 3.6
Simple coil springs of the type that can be found in a pen.

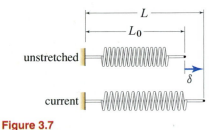

Figure 3.7
Stretch of a spring.

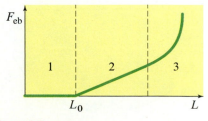

Figure 3.9
Qualitative depiction of the *nonlinear* force vs. displacement curve for an elastic band.

elastic band as a function of its length. When the length of the elastic band L is less than its unstretched length L_0, then the force in the elastic band is zero (region 1). When its length reaches its unstretched length, then the force in the elastic band starts increasing from zero and the force-displacement curve is close to linear (region 2). If we keep stretching the elastic band, then the force starts increasing very quickly (region 3) until we can't stretch it any farther (the vertical part of the curve at the right end) before it breaks. Throughout this book, in examples and homework problems, we will occasionally encounter nonlinear springs, and we will discover the amazing ways they can affect a system's behavior.

Springs and mechanical system modeling

In addition to being an actual structural element, a spring can be thought of in more abstract terms as any element providing a force between two points that depends on only the distance between the points. For example, in many substances, we know that the force between two atoms depends on only the distance between the atoms. In these cases we say that the atoms behave as if they were connected via (nonlinear) springs, and indeed we model atomic bonds as springs in molecular and lattice dynamics. Another example is provided by the universal law of gravitation that tells us that the gravitational attraction between two bodies is inversely proportional to the square of the distance between the bodies. Again, we can view the bodies in question as connected by a (nonlinear) spring. More generally, we use springs when modeling elastic systems, where an elastic system is any physical system in which the forces internal to the system depend on only the relative position of the points in the system. Since almost any mechanical system displays some elasticity, we can say that "springs are everywhere," whether or not they appear in the traditional coil shape shown in Fig. 3.6.

Inertial reference frames

The use of Newton's second law *requires* that the acceleration $\vec{a}$ be measured with respect to an *inertial frame of reference*. An *inertial reference frame* is one in which Newton's first and second laws are valid, at least to the level of accuracy desired. In addition, if we have found a frame that *acceptably* satisfies this definition (i.e., it is inertial), then any frame that is not accelerating relative to such an inertial frame is also inertial. Interestingly, for all but a small class of engineering problems (e.g., as long as we stay away from relativistic effects or are not interested in orbital mechanics), a frame attached to the surface of the Earth can be considered inertial. Let's see why.

One *very good* inertial frame is one that has its origin at the center of mass of our solar system (which is very near the center of the sun). Relative to this inertial reference frame, a reference frame attached to the surface of the Earth is accelerating due to a number of motions, including the rotation of the Earth on its axis and the orbit of the Earth about the sun. The rotation of the Earth on its axis is, by far, the largest contributor, as long as the reference frame is placed sufficiently far from the poles. Referring to Fig. 3.10, to compute the magnitude of the acceleration of the point P, we will take the mean radius of the Earth to be 6371 km. Since the Earth spins at 1 rev/day, its angular velocity is 0.00007272 rad/s, and so, referring to Fig. 3.10, we see that the

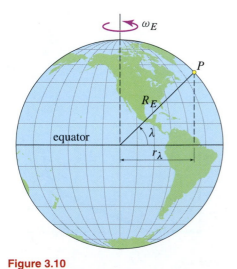

Figure 3.10
View of the spinning Earth showing how the latitude λ is measured.

normal acceleration due to the rotation of the Earth is equal to

$$a_n = r_\lambda \omega_E^2 = (R_E \cos \lambda)\, \omega_E^2 = (6{,}371{,}000\,\text{m})\left(0.00007272\,\tfrac{\text{rad}}{\text{s}}\right)^2 \cos \lambda$$
$$= 0.03369 \cos \lambda\,\text{m/s}^2. \tag{3.14}$$

For a midlatitude location like State College, Pennsylvania, United States, which is at latitude 40.8° north, the acceleration given by Eq. (3.14) is approximately $0.0255\,\text{m/s}^2$ — this is negligible in most engineering applications.

Governing equations, equations of motion, and degrees of freedom

The application of Newton's second law to the solution of dynamics problems has been accompanied by the creation of a rich terminology used to identify the many aspects of these problems. Here we present some of the terms that are most frequently used in practice.

Newton's second law is a balance principle.   Newton's second law, which we have written as

$$\boxed{\vec{F} = m\vec{a},} \tag{3.15}$$

is the first *balance principle* we encounter in dynamics. A *balance principle*, or a *balance law*, is a relation that equates the change in a fundamental property of a physical system, such as its mass or charge, with the cause of that change. In the case of Newton's second law, the balance in question is that between the total force $\vec{F}$ acting on a particle of mass m and the change in the state of motion of the particle as measured by the quantity $m\vec{a}$. As long as we model a physical body as particle, then the motion of our object must conform to Newton's second law no matter the specific makeup of the body. As such, Newton's second law is central to the modeling process of physical bodies. However, when we write $\vec{F} = m\vec{a}$, it is important to understand that Newton's second law provides neither $\vec{F}$ nor $m\vec{a}$: it simply requires these two quantities balance one another. Newton's second law is only one of the contributing elements to the solution of a problem — the other two essential elements are force laws, which describe $\vec{F}$ in detail, and the kinematic equations, which described $\vec{a}$ in detail, as discussed next.

Force laws and kinematic equations.   In solving a problem, expressions for all the forces (and moments) and for all the acceleration terms must be provided. The forces appearing in Newton's second law often contain kinematic variables mixed with quantities we call *material* (or *constitutive*) *parameters*. For example, when the force of a spring F_s is expressed by the familiar form $F_s = kx$, F_s is a function of the kinematic variable "position" via x and of a parameter k that depends on the material makeup of the spring. Another example is the force describing air resistance. This force depends on the velocity vector and a drag coefficient. In this case, it is the velocity vector that is the kinematic variable, and the drag coefficient is the material parameter. From now on, we will call *force laws* those equations that provide the specific forms of forces in terms of material parameters and kinematic variables. The equations describing the acceleration terms appearing in $\vec{F} = m\vec{a}$, via the use of a specific component system, will be called *kinematic equations*.

Interesting Fact

Acceleration due to the Earth's rotation about its axis vs. acceleration due to the Earth's orbital motion about the Sun. From Fig. 3.10 and Eq. (3.14) we can see that, assuming that the center of the Earth does not move, the normal component of acceleration for point P at latitude λ must be

$$a_n = r_\lambda \omega_E^2 = (R_E \cos \lambda)\, \omega_E^2.$$

Thus we see that a_n is largest when we are at the equator ($\lambda = 0°$) and is smallest when we are at one of the poles ($\lambda = \pm 90°$). Assuming that the Sun is the origin of an inertial frame, orbital mechanics tells us that the normal acceleration due to the Earth's orbit about the Sun is approximately equal to $0.005941\,\text{m/s}^2$. Therefore, the acceleration due to the orbital motion of the Earth is more significant than the acceleration due to the rotation of the Earth about its axis only for latitudes above $\lambda = 80°$.

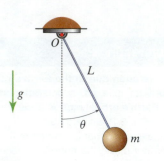

Figure 3.11
As the pendulum swings, the tension in the pendulum cord changes. The equation describing the tension (as a function of position or time) is an example of a constraint force equation.

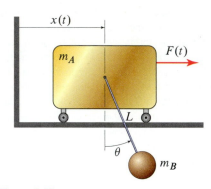

Figure 3.12
An illustration of *degrees of freedom* using a rolling cart with an attached pendulum.

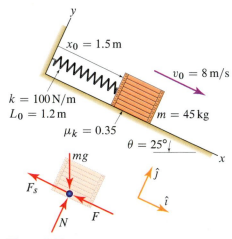

Figure 3.13
A crate sliding down an incline; x_0 and v_0 are the crate's initial position and speed, respectively. The crate has been modeled as a particle.

Governing equations. We call *governing equations* all of the equations needed to solve a specific problem. In a particle problem, the governing equations are Newton's second law, the force laws, and the kinematic relations taken together.

Equations of motion. The motion of a particle is the position vector of the particle as a function of time. For more complex models, such as a rigid body, the motion is the set of functions capable of uniquely specifying the position of the body as a function of time. The *equations of motion* of an object are those equations that allow us to determine the object's motion. In this book, equations of motion take on the form of ordinary differential equations and, in the case of a particle, are typically derived by substituting the force laws and the kinematic equations into Newton's second law.

Constraint force equations. The governing equations can be manipulated to derive expressions for forces (and moments) acting on the system as a function of the motion. An example of this type of equation is the expression for the tension in a pendulum cord as a function of the position of the pendulum's bob (see Fig. 3.11). We will call these equations *constraint force equations*. Referring to the chameleon problem at the beginning of the section, another example of a constraint force equation is Eq. (3.5), which describes the force exerted on the insect by the chameleon's tongue necessary for the motion of the insect to be as observed.

Degrees of freedom. Mathematically, a system's *degrees of freedom* are the independent coordinates needed to uniquely specify the position of a system. A more physically intuitive way of thinking about degrees of freedom is to think of the *number of degrees of freedom* as the number of different coordinates in a system that must be fixed in order to keep the system from moving. For example, the simple pendulum in Fig. 3.11 has one degree of freedom since we only need to fix the coordinate θ to keep the pendulum from moving. The cart and pendulum shown in Fig. 3.12 has two degrees of freedom since we would need to fix the two coordinates x and θ to completely prevent the system from moving. It is useful to identify the degrees of freedom of a system because *the number of equations of motion of a system is equal to its number of degrees of freedom.*

■ **Mini-Example.** To illustrate the terminology introduced above, let's determine the motion of the crate in Fig. 3.13 and the force in the spring as a function of time from the moment of release until the crate first stops.
Solution. Since the crate's motion is rectilinear, we choose the Cartesian coordinate system shown in Fig. 3.13, and using the crate's FBD, the balance principle we apply to the crate is Newton's second law, which yields

$$\sum F_x: \quad mg\sin\theta - F - F_s = ma_x, \tag{3.16}$$

$$\sum F_y: \quad N - mg\cos\theta = ma_y, \tag{3.17}$$

where F_s and F are the spring and friction forces, respectively. We cannot obtain the information we seek directly from these equations because we still need to describe the forces acting on the system as well as the motion's geometry. We begin with describing F_s and F via what we called *force laws*. These laws are

$$F_s = k(x - L_0) \quad \text{and} \quad F = \mu_k N, \qquad (3.18)$$

where k and L_0 are the spring constant and the spring's unstretched length, respectively, and μ_k is the kinetic friction coefficient between the crate and the incline. Next, we describe the acceleration via the *kinematic equations*:

$$a_x = \ddot{x} \quad \text{and} \quad a_y = \ddot{y} \quad \text{with} \quad y = 0 = \text{const.} \quad \Rightarrow \quad a_y = 0. \quad (3.19)$$

We now have as many equations as we need to solve the problem. Equations (3.16)–(3.19) are the system's *governing equations* because they completely describe the system's behavior. Next we derive the crate's equations of motion. Since the crate does not move in the y direction and it is sufficient to fix the coordinate x to keep the crate from moving, the crate has only one *degree of freedom*. This means that we need to derive only one equation of motion for $x(t)$. To obtain this equation, we substitute Eqs. (3.19) into Eq. (3.17) and solve for N to obtain $N = mg \cos \theta$, which is the constraint force equation for this system. Finally, substituting this result as well as Eqs. (3.18) and a_x from Eqs. (3.19) into Eq. (3.16), we obtain

$$m\ddot{x}(t) + kx(t) = mg(\sin \theta - \mu_k \cos \theta) + kL_0. \qquad (3.20)$$

In the above equation, since $\ddot{x}(t)$ is the second time derivative of $x(t)$, it is *not* independent of $x(t)$. Because Eq. (3.20) contains both the function $x(t)$ and its second derivative, this equation is a *differential equation* with $x(t)$ as the only unknown. This makes Eq. (3.20) the *equation of motion* for this problem. Since the highest derivative in Eq. (3.20) is the *second* time derivative of $x(t)$, the theory of differential equations tells us that we need *two* extra conditions to fully solve the problem. In our problem, these two conditions are obtained by ensuring that the position and velocity of the crate at time $t = 0$ are those indicated in the problem's statement (see Fig. 3.13), namely, $x(0) = x_0 = 1.5\,\text{m}$ and $\dot{x}(0) = v_0 = 8\,\text{m/s}$. The theory of differential equations then tells us that our solution is given by

$$x(t) = A \sin \sqrt{\frac{k}{m}}\, t + B \cos \sqrt{\frac{k}{m}}\, t + \frac{mg}{k}(\sin \theta - \mu_k \cos \theta) + L_0, \quad (3.21)$$

where A and B are determined using the above-mentioned conditions at $t = 0$. This solution is valid only as long as it remains consistent with the FBD used to derive it. Specifically, it is valid only from the time of release until the crate reaches its lowest position. After the crate stops, the crate's FBD will change in that the friction force changes (1) direction and (2) type, i.e., becomes static, at least for the time instant when $\dot{x} = 0$. To predict the crate's motion after it comes to a stop for the first time, we would have to sketch a new FBD and start the solution process anew. Finally, the spring force F_s is obtained by substituting Eq. (3.21) into the expression for the spring force in Eq. (3.18). ■

A recipe for applying Newton's second law

We now present a "recipe" for the application of Newton's second law. This recipe provides a framework that can be applied to any kinetics problem in dynamics and is used in all areas and at all levels of mechanics to solve problems.

Ingredients

Every time we solve a problem in dynamics, there will *always* be three groups of equations from which the problem's governing equations are obtained:

1. **Balance principles**
   For the chameleon example, the balance principle we applied was Newton's second law, expressed via Eqs. (3.3) and (3.4). In Chapters 4 and 5, we will use balance principles such as the balance of energy, linear momentum, and/or angular momentum.

2. **Force laws**
   These are equations describing the *forces* acting on the system. The force laws we most frequently encounter in this textbook are (1) gravity, (2) friction laws, (3) spring force laws, and (4) resistance laws due to fluids. Sometimes the description of a force is so elementary that we do not write an explicit equation for it. For example, referring to Fig. 3.13, the force law describing the gravitational attraction between the crate and the Earth was indicated as mg directly on the FBD. In some cases, a force might be one of the unknowns of the problem. This was the case in the chameleon example. In such cases, we indicate the force in question on the FBD and then use the rest of the governing equations to determine it.

3. **Kinematic equations**
   These are the equations that provide a geometric description of the motion. In the chameleon example, the kinematic relation is the velocity equation (Eq. (3.1)), whose derivative is used to obtain the required acceleration.

Preparation

Now that we have the ingredients, we can combine them. The steps outlined below are not meant to be rigidly followed: they are steps that the authors find themselves using in the solution of problems. In the end, the way the ingredients are combined depends on the person solving the problem.

> **Helpful Information**
>
> **What is a good model?** A good model captures all of the relevant physics of a problem with the least mathematical complexity. We must decide what is a "good enough answer," and, in turn, this depends on what we want to do with a model. Knowing what forces to include in the model is a function of experience as well as desired accuracy. In general, an "all-inclusive" model is almost never a good model. In addition, the more sophisticated a model is, the more computationally involved the solution process will be, so that a cost/benefit assessment is often a necessary part of the modeling process.

Step 1. Road map and modeling: *Review the given information, identify the system, state assumptions, sketch the FBD, and identify a problem-solving strategy.* In the chameleon example, reviewing the given information helped us realize that the problem was of the type "*given* the insect's motion, *find* the forces acting on it" and told us that the system to analyze was the insect. In drawing an FBD of the insect, we *modeled* it as a particle and also decided what forces to include and what forces to neglect; that is, we included the insect's weight and the force between the insect and the chameleon's tongue. Choosing the relevant forces is the result of assumptions, which should be stated, about what physical phenomena really matter in the problem. Since forces are vectors, a necessary element of the FBD and the modeling process is the choice of a component system, which must be the same as that used to describe the kinematics. *The FBD (and the associated component system) is the most important element of the solution because it visualizes the overall model.* You should always make sure that the FBD is easy to read, with all forces (and, in later chapters, moments) easily distinguishable from one another, unambiguously drawn, and with a clear indication of their positive direction. Once a clear physical picture of the problem is formed, we need to decide on a solution strategy. In the chameleon example, the solution strategy consisted of a direct application of Newton's second law.

Step 2. Governing equations: *Write the balance principles, force laws, and kinematic equations.* Once an FBD is sketched, we assemble the equations that translate our model into mathematics. These are the problem's *governing equations*, which consist of balance principles, force laws, and kinematic equations. The *balance principles*, which in this chapter will always be given by Newton's second law, are the equations that most immediately follow from the FBD. As in the chameleon problem, this means summing forces in the component directions and setting these sums equal to the corresponding component of $m\vec{a}$. In general, these equations look like

$$\sum F_a = ma_a, \quad \sum F_b = ma_b, \quad \text{and} \quad \sum F_c = ma_c, \tag{3.22}$$

where a, b, and c are the orthogonal directions of the chosen component system, and we usually need only two directions for planar problems. Next, we must describe the left-hand side of Newton's second law. For example, F_a, F_b, and F_c may consist of forces due to gravity, friction, springs, etc. The equations that describe these phenomena are the *force laws*. We must then properly describe the right-hand side of Newton's second law, namely, the components of the acceleration. This is done by writing the *kinematic equations*.

Step 3. Computation: *Solve the assembled system of equations.* The solution of the governing equations requires us to assess whether or not the equation system we have is *solvable*. The balance principles, force laws, and kinematics equations always yield *independent equations*. If the system of equations we assemble is not solvable, we need to go back and check (1) whether or not the model is complete, (2) whether or not all of the forces have been properly described, and (3) whether or not all of the kinematics have been properly described.

Step 4. Discussion and verification: *Study the solution and do a "sanity check."* Once a solution is obtained, we should make sure that it *makes physical sense*. This means checking the correctness of the units used to express the solution as well as the order of magnitude of the results. Finally, the solution should be reconciled with practical experience of how the physical world behaves. Sometimes this is as easy as confirming that the solution matches our physical intuition. At other times, our physical intuition will not be met, and this means either that we have made a mistake or that our physical intuition is incorrect. A problem should not be declared solved until everything fits.

End of Section Summary

Applying Newton's second law. In this section we developed a four-step problem-solving procedure for applying Newton's second law to mechanical systems. The central element of this procedure is the derivation of the governing equations, which originate from, at most, the

1. Balance principles.

2. Force laws.

3. Kinematic equations.

In this chapter, the balance principle is Newton's second law, which we write as

Eq. (3.15), p. 189

$$\vec{F} = m\vec{a}.$$

We apply Newton's second law in component form as

Eqs. (3.22), p. 193

$$\sum F_a = ma_a, \quad \sum F_b = ma_b, \quad \text{and} \quad \sum F_c = ma_c,$$

where a, b, and c are the orthogonal directions of the chosen component system and we usually need only two directions for planar problems. The four-step "recipe" we presented for applying Newton's second law, outlined in the margin note, is given as a guide to the *order* in which things should be done, and we will use it consistently in each example we present.

Solvability of a system of equations. We need as many equations as we have unknowns for both algebraic and differential equations. It is important to keep in mind that for differential equations, a function and its time derivatives, say, $x(t)$ and $\dot{x}(t)$, are *not* two different unknowns since $\dot{x}(t)$ is not independent of $x(t)$.

Inertial reference frames. An *inertial reference frame* is a frame in which Newton's laws of motion provide predictions that agree with experimental verification. For most engineering problems, a frame attached to the surface of the Earth can be considered to be inertial. Frames that are not accelerating with respect to a given inertial frame are also inertial.

Governing equations and equations of motion. The *governing equations* for a system consist of the (1) balance principles, (2) force laws, and (3) kinematic equations. The *equations of motion* are differential equations derived from the governing equations that allow for the determination of the motion.

Degrees of freedom. A system's *degrees of freedom* are the independent coordinates needed to completely describe a system's position. The number of degrees of freedom is equal to the number of different coordinates in a system that must be fixed in order to keep the system from moving. The required number of equations of motion is equal to the number of degrees of freedom for a system.

Friction. We will include friction via the Coulomb friction model. According to this model, in the absence of slip, the magnitude of the friction force F satisfies the following inequality:

Eq. (3.8), p. 186

$$|F| \le \mu_s |N|,$$

where μ_s is the *coefficient of static friction* and N is the force normal to the contact surface. The relation

Eq. (3.9), p. 186

$$F = \mu_s N$$

defines the case of impending slip.

  If A and B are two objects sliding with respect to one another, the magnitude of the friction force exerted by B onto A is given by

Eq. (3.11), p. 186

$$F = \mu_k N,$$

where μ_k is the *coefficient of kinetic friction* and where the direction of the friction force must be consistent with the fact that friction opposes the relative motion of A and B.

Springs.   A spring is said to be *linear elastic* if the internal force in the spring is linearly related to the amount the spring is stretched or compressed. These are the springs that will be commonly used in this textbook. The force F_s required to stretch a linear elastic spring by an amount δ is given by

Eq. (3.13), p. 187

$$F_s = k\delta = k(L - L_0),$$

where k is the *spring constant* and where L and L_0 are the current and unstretched lengths of the spring, respectively. When $\delta > 0$, the spring is said to be stretched; and when $\delta < 0$, the spring is said to be compressed.

EXAMPLE 3.1   *Friction and Impending Slip*

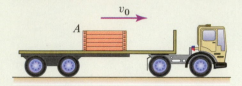

Figure 1
A truck hauling a large crate.

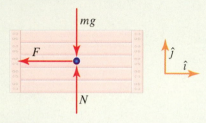

Figure 2
FBD of the crate.

Common Pitfall

Newton's Second Law and Inertial Frames. The application of Newton's second law requires the use of an inertial frame of reference. Therefore, the component system shown in the FBD in Fig. 2 is to be understood as originating from an xy coordinate system fixed with the ground, which we assume to be an inertial reference frame. It would be a mistake to choose a coordinate system moving with the truck because the truck is decelerating with respect to the ground and therefore is not an inertial reference frame.

The truck shown in Fig. 1 is traveling at $v_0 = 100\,\text{km/h}$ when the driver slams on the brakes and comes to a stop as quickly as possible. If the coefficient of static friction between the crate A and the bed of the truck is 0.35, determine the minimum stopping distance $d_{\min}$ and the minimum stopping time $t_{\min}$ of the truck such that the crate does not slide forward on the truck.

SOLUTION

Road Map & Modeling   The truck must stop as fast as possible without causing the *crate* to slide. Therefore, the system to analyze is the crate, which we model as a particle. In the FBD shown in Fig. 2, we have assumed that the only relevant forces are the crate's weight mg, the normal force N between the crate and the truck, and the friction force F between the crate and the truck. The force F points left because if the crate were to slip, it would slip to the right relative to the truck and F must oppose this motion. Since the motion is rectilinear, we have chosen a Cartesian coordinate system with the x axis parallel to the direction of motion. The maximum deceleration without sliding corresponds to the maximum possible friction force acting on the crate, which is the force corresponding to *impending slip*. Our solution strategy will be to relate the maximum friction force to the acceleration via Newton's second law. Once we have an expression for the maximum acceleration, we will apply our knowledge of kinematics to compute $d_{\min}$ and $t_{\min}$.

Governing Equations

Balance Principles   Referring to Fig. 2, Newton's second law yields

$$\sum F_x: \qquad -F = ma_x, \tag{1}$$

$$\sum F_y: \qquad N - mg = ma_y. \tag{2}$$

Force Laws   The friction law for *impending slip* is

$$F = \mu_s N. \tag{3}$$

Kinematic Equations   The kinematic equations are

$$a_x = a_{\max} \quad \text{and} \quad a_y = 0, \tag{4}$$

which express the fact that while a_x is still unknown, we are seeking the maximum value of acceleration and the fact that the crate is not moving vertically.

Computation   Equations (1)–(4) are four equations in the unknowns N, a_y, $a_{\max}$, and F. Equation (2) and the second of Eqs. (4) tell us that $N = mg$. Substituting this result into Eq. (3) and, in turn, substituting that into Eq. (1), we obtain

$$-\mu_s mg = ma_{\max} \quad \Rightarrow \quad a_{\max} = -\mu_s g, \tag{5}$$

where the negative sign indicates that the crate (and truck) is decelerating when the driver applies the brakes.

Since the maximum deceleration is constant (both μ_s and g are constants), we can determine the stopping distance by using Eq. (2.43) on p. 59:

$$v^2 = v_0^2 + 2a_c(x - x_0) \quad \Rightarrow \quad 0 = v_0^2 - 2\mu_s g d_{\min}. \tag{6}$$

so that

$$d_{min} = \frac{v_0^2}{2\mu_s g} = 112\,\text{m}, \tag{7}$$

where we have used the conversion $100\,\text{km/h} = 27.78\,\text{m/s}$. To determine the stopping time, we apply Eq. (2.41) on p. 59:

$$v = v_0 + a_c t \quad \Rightarrow \quad 0 = v_0 - \mu_s g t_{min}, \tag{8}$$

so that

$$t_{min} = \frac{v_0}{\mu_s g} = 8.09\,\text{s}. \tag{9}$$

Discussion & Verification   The symbolic forms of the results in Eqs. (7) and (9) have the correct dimensions for d_{min} and t_{min}. Also the final numerical results have been expressed using appropriate units. As far as the values we have obtained for d_{min} and t_{min} are concerned, these are certainly reasonable since, in absolute value, the maximum deceleration possible under the stated assumptions is roughly one-third of the acceleration of gravity.

A Closer Look   The model used here, which is based on the Coulomb friction model, is such that the deceleration we calculated is completely independent of the crate's mass. Therefore, the crate could have been twice as heavy, and we would have obtained the same answer. This is a basic property of this friction model. Another important observation to make is that, again because of the friction model used, the minimum time required to stop is linearly proportional to the initial speed v_0, and the minimum distance required to stop is proportional to the initial speed squared v_0^2. This is shown in Figs. 3 and 4, respectively.

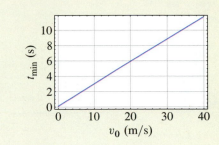

Figure 3
The minimum stopping time of the crate as a function of the initial speed of the truck.

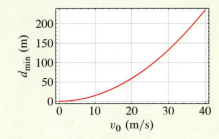

Figure 4
The minimum stopping distance of the crate as a function of the initial speed of the truck.

EXAMPLE 3.2   *Transition from Static to Kinetic Friction*

Figure 1

A person pushing a crate of mass m over a rough surface with a time-dependent force $P(t)$.

A simple problem of practical importance that illustrates the nature of friction is the study of the initiation of motion of an object on a rough surface. For the purpose of this example, let's look at what happens when a person pushes a crate of mass m on a rough floor as shown in Fig. 1. Letting μ_s and μ_k, with $\mu_k < \mu_s$, be the static and kinetic friction coefficients between the crate and the ground, respectively, consider a case in which the force $P(t)$ exerted by the person increases linearly with time; i.e., $P(t) = P_0 t$. For this case, determine the time at which the motion begins and the friction force as a function of time.

SOLUTION

Road Map & Modeling   We are given one of the forces applied to the crate as a function of time. To determine the time at which the motion starts, we need to determine when the horizontal component of the applied force P overcomes the maximum friction force allowed by the *static* friction coefficient, which is done by applying Newton's second law. Referring to Fig. 2, we model the crate as a particle under the action of gravity mg, the friction force F, the normal force N between the crate and the ground, and the force $P(t)$ applied by the person. The force $P(t)$ has been plotted as a function of time in Fig. 3. For the sake of generality, the direction of the applied force $P(t)$ is assumed to form a generic angle θ with the horizontal direction.

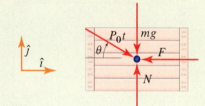

Figure 2

FBD of the crate.

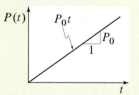

Figure 3

Plot of $P(t)$ versus time.

Governing Equations

Balance Principles   Based on the FBD in Fig. 2, Newton's second law gives

$$\sum F_x: \qquad P_0 t \cos\theta - F = ma_x, \tag{1}$$

$$\sum F_y: \quad N - P_0 t \sin\theta - mg = ma_y. \tag{2}$$

Force Laws   The force law for friction will depend on the value of F. If there were no friction, the crate would move to the right, and so the force F must point to the left, as indicated on the FBD in Fig. 2. In addition, $F \leq \mu_s N$ when the crate is not moving, and $F = \mu_k N$ after the crate starts moving. As long as the component of $P(t)$ propelling the crate forward is less than the maximum possible static friction, the crate will not move, that is, as long as $P(t)\cos\theta \leq \mu_s N$. Once $P(t)\cos\theta > \mu_s N$, then the crate starts to move, kinetic friction comes into play, and the friction force *immediately* drops to $\mu_k N$ (since $\mu_k < \mu_s$). Therefore, to compute the time at which the motion starts, the force law to use is

$$F = \mu_s N. \tag{3}$$

Once the crate starts moving, this equation will need to be replaced by

$$F = \mu_k N. \tag{4}$$

The force law describing the weight of the crate is elementary and has been indicated directly on the FBD as mg. We do not have an explicit force law for N since its value is determined by the fact that the crate cannot move in the y direction.

Kinematic Equations   Since the crate does not move in the y direction, we have

$$a_y = 0. \tag{5}$$

In the x direction, before the motion starts, we have

$$a_x = 0. \tag{6}$$

Once the motion starts, a_x becomes an unknown.

Computation   Substituting Eq. (5) into Eq. (2) and solving for N, we obtain

$$N = P_0 t \sin \theta + mg. \tag{7}$$

Recalling that $a_x = 0$ until the motion begins, we solve Eq. (1) for F and set the result equal to $\mu_s N$, where N comes from Eq. (7). This yields

$$\mu_s (P_0 t_s \sin \theta + mg) = P_0 t_s \cos \theta, \tag{8}$$

where t_s is the time at which motion begins. Solving for t_s in Eq. (8), we obtain

$$t_s = \frac{mg}{P_0} \left(\frac{\mu_s}{\cos \theta - \mu_s \sin \theta} \right). \tag{9}$$

Once $t = t_s$, the crate starts to move, and as indicated in Eq. (4), F immediately drops from $\mu_s N$ to $\mu_k N$. Also, for $t > t_s$, a_x is no longer known. Therefore, for $t > t_s$, we have four equations—Eqs. (1), (2), (4), and (5)—to solve for the four unknowns a_x, a_y, F, and N. Solving, we obtain

$$\left. \begin{aligned} a_x &= \frac{P_0 t}{m} (\cos \theta - \mu_k \sin \theta) - \mu_k g, \\ a_y &= 0, \\ F &= \mu_k (mg + P_0 t \sin \theta), \\ N &= mg + P_0 t \sin \theta. \end{aligned} \right\} \text{ for } t > t_s. \tag{10}$$

Discussion & Verification   Observing that the quantity P_0 has dimensions of force per unit time, we can easily verify that the dimensions of the results in Eqs. (9) and (10) are correct.

🔍 **A Closer Look**   This example demonstrates several features of the Coulomb friction model. First, Eq. (9) tells us that the crate will never move if $\cos \theta \le \mu_s \sin \theta$ (since the denominator becomes negative). Also, the closer $\cos \theta$ is to $\mu_s \sin \theta$, the closer the denominator in Eq. (9) is to zero and the longer it takes to get the crate moving. 💻➡ An interesting representation of Eq. (9) can be found in Fig. 4, which shows t_s as a function of θ and μ_s. The black curve at the edge of the red region represents the θ and μ_s values for which the denominator of Eq. (9) is zero. Since the crate only moves for positive values of t_s, no matter how long we wait or how hard we push, the crate will never move for any θ and μ_s values above and to the right of the black curve. As we approach the black curve from below, it takes longer and longer to get the crate moving. For example, as θ approaches 90°, it becomes impossible to move the crate no matter how small μ_s is, unless $\mu_s = 0$. ⬅💻 Since friction is the key ingredient in this problem, it is illustrative to plot the friction force F as a function of time. Until the crate starts to move, Eq. (1) gives F as

$$F = P_0 t \cos \theta, \quad \text{for } t < t_s. \tag{11}$$

As we have just seen, Eq. (10) gives us F for $t > t_s$. A plot of F vs. t for some particular values of the system parameters is shown in Fig. 5. Notice the discontinuity in the friction force when the crate starts to slip at $t = t_s$. This is a general feature of Coulomb friction and will always happen as long as $\mu_s \ne \mu_k$. Finally, note that F keeps increasing with time after the crate starts moving because the value of N increases to balance the corresponding increasing value of the vertical component of $P(t)$.

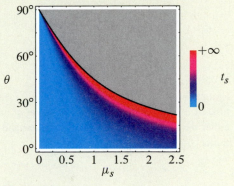

Figure 4

A contour plot of the term in parentheses in Eq. (9). The factor mg/P_0 has been ignored since it simply multiplies the time by a constant factor and we do not have specific numbers for m and P_0. Note that combinations of θ and μ_s lying in the gray area above the black curve correspond to values for which the crate will never move.

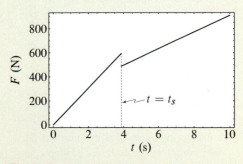

Figure 5

Friction F vs. time t. The parameters used were $m = 50\,\text{kg}$, $g = 9.81\,\text{m/s}^2$, $P_0 = 200\,\text{N/s}$, $\theta = 40°$, $\mu_s = 0.6$, and $\mu_k = 0.45$.

E X A M P L E   3.3   *Motion under the Action of Spring Forces*

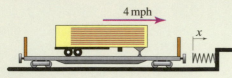

Figure 1
Railcar running into a large spring.

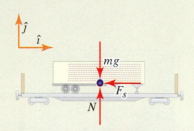

Figure 2
FBD of the railcar shown in Fig. 1, where m is either the mass of the railcar or the railcar and trailer.

A 60 ton railcar and its cargo, a 27 ton trailer, are moving to the right at 4 mph, as shown in Fig. 1, when they encounter a large linear spring that has been designed to stop a 60 ton railcar moving at 5 mph in a distance of 3 ft when it is initially uncompressed. If the trailer does not slip relative to the railcar and if the spring is initially uncompressed, determine (a) how much the spring compresses in stopping the 87 ton loaded railcar and (b) how long it takes for the spring to stop the railcar.

SOLUTION

Road Map & Modeling   The given information consists of (1) the conditions under which the railcar and its cargo impact the spring and (2) the spring's design criterion. We need to compute the stopping distance and time of the railcar and its cargo, and so "railcar and cargo" is the system to analyze. We will model this system as a particle, and its FBD is shown in Fig. 2. We have assumed that the only relevant forces are gravity, the reaction with the rails, and the spring force. Note that we are not given direct information about the spring constant or its unstretched length. However, using the stated design criteria in conjunction with Newton's second law, we will be able to first determine the spring's force law and then use the result to compute the system's acceleration. Once we have the system's acceleration, we will need to apply kinematics to obtain the corresponding stopping time and distance information.

Governing Equations

Balance Principles   Based on the FBD in Fig. 2, Newton's second law gives

$$\sum F_x: \qquad -F_s = ma_x, \tag{1}$$

$$\sum F_y: \quad N - mg = ma_y. \tag{2}$$

Force Laws   While the spring constant k is unknown, the spring's force law can still be given the form

$$F_s = kx, \tag{3}$$

where x is measured from the uncompressed end position of the spring.

Kinematic Equations   The kinematic relations are

$$a_x = \ddot{x} \quad \text{and} \quad a_y = 0, \tag{4}$$

where we have chosen to represent a_x as $\ddot{x}$ because the problem is asking us to relate acceleration to position, and where the second of Eqs. (4) states that there is no motion in the vertical direction.

Computation   Since there is no motion in the y direction, we will disregard the equations in the vertical direction. Next, substituting the first of Eqs. (4) and Eq. (3) into Eq. (1) and rearranging, we obtain

$$\ddot{x} + \frac{k}{m}x = 0. \tag{5}$$

Rearranging Eq. (5) and making use of the chain rule, we obtain

$$\ddot{x} = \dot{x}\frac{d\dot{x}}{dx} = -\frac{k}{m}x \quad \Rightarrow \quad \dot{x}\,d\dot{x} = -\frac{k}{m}x\,dx, \tag{6}$$

which can be integrated as

$$\int_{v_i}^{v} \dot{x}\,d\dot{x} = -\int_{x_i}^{x} \frac{k}{m}x\,dx \quad \Rightarrow \quad v^2 - v_i^2 = -\frac{k}{m}(x^2 - x_i^2), \tag{7}$$

where v is the speed of the railcar at the location x and where the subscript i stands for *initial*. Letting v_f be the speed corresponding to the final position x_f, solving Eq. (7) for k, and substituting in numbers corresponding to the spring's design criteria, we obtain

$$k = \frac{m_r \left(v_f^2 - v_i^2\right)}{x_i^2 - x_f^2} = 22{,}270\,\text{lb/ft},   \tag{8}$$

where we have used the unit conversions $5\,\text{mph} = 7.333\,\text{ft/s}$ and $1\,\text{ton} = 2000\,\text{lb}$, m_r is the mass of the railcar, $v_f = 0$, $x_i = 0$, and $x_f = 3\,\text{ft}$.

──────────────── **Part (a)** ────────────────

Now that we have k, we can determine how far the spring compresses with the loaded railcar by applying Eq. (7) again. Letting $v = v_f$ for $x = x_f$, solving Eq. (7) for x_f, and substituting in the numbers for the loaded railcar, we obtain

$$\boxed{x_f = \sqrt{x_i^2 - \frac{m_t}{k}\left(v_f^2 - v_i^2\right)} = 2.890\,\text{ft},}   \tag{9}$$

where we have used the unit conversion $4\,\text{mph} = 5.867\,\text{ft/s}$ and m_t is the total mass of the railcar and the trailer.

──────────────── **Part (b)** ────────────────

To determine the stopping time t_f, we go back to Eq. (7), solve it for v, and rearrange the result to integrate it with respect to time, i.e.,

$$v = \frac{dx}{dt} = \sqrt{v_i^2 - \frac{k}{m_t}\left(x^2 - x_i^2\right)} \quad\Rightarrow\quad \int_{x_i}^{x_f} \frac{dx}{\sqrt{v_i^2 - \frac{k}{m_t}\left(x^2 - x_i^2\right)}} = \int_{t_i}^{t_f} dt.   \tag{10}$$

Using symbolic algebra software or a table of integrals, we obtain

$$t_f - t_i = \sqrt{\frac{m_t}{k}} \left[\sin^{-1}\left(\frac{\sqrt{\frac{k}{m_t}}\,x_f}{\sqrt{v_i^2 + \frac{k}{m_t}x_i^2}} \right) - \sin^{-1}\left(\frac{\sqrt{\frac{k}{m_t}}\,x_i}{\sqrt{v_i^2 + \frac{k}{m_t}x_i^2}} \right) \right] \quad\Rightarrow\quad t_f = 0.774\,\text{s},   \tag{11}$$

where we substituted in $x_f = 2.890\,\text{ft}$, $v_i = 5.867\,\text{ft/s}$, $k = 22{,}270\,\text{lb/ft}$, $m_t = (174{,}000/32.2)\,\text{slug}$, $t_i = 0\,\text{s}$, and $x_i = 0\,\text{ft}$.

Discussion & Verification   The result in Eq. (9) tells us that a 174,000 lb loaded railcar traveling at 4 mph compresses the spring a little less than 3 ft. This result seems reasonable. Although the loaded car is 45% heavier than when it is empty, the spring compression should have not been expected to necessarily increase because the loaded car is moving with a speed that is only 80% of the design speed for an impact between an unloaded railcar and the bumper. Clearly the increase in weight is larger than the decrease in speed; however, the spring compression depends on the *square* of the speed, i.e., in this problem, speed is a more significant factor than weight.

Helpful Information

Integral in Eq. (10). In computing the integral in Eq. (10), it might help to notice that if we let

$$a = \sqrt{v_i^2 + \frac{k}{m_t}x_i^2} \quad \text{and} \quad u = \sqrt{\frac{k}{m_t}}\,x,$$

then Eq. (10), with lower limits set equal to zero, becomes

$$\int_0^{t_f} dt = \sqrt{\frac{m_t}{k}} \int_0^{\sqrt{\frac{k}{m_t}}\,x_f} \frac{du}{\sqrt{a^2 - u^2}},$$

or

$$t_f = \sqrt{\frac{m_t}{k}}\, \sin^{-1}\left(\frac{u}{a}\right)\Bigg|_0^{\sqrt{\frac{k}{m_t}}\,x_f}$$

$$= \sqrt{\frac{m_t}{k}}\, \sin^{-1}\left(\frac{\sqrt{\frac{k}{m_t}}\,x_f}{\sqrt{v_i^2 + \frac{k}{m_t}x_i^2}} \right).$$

PROBLEMS

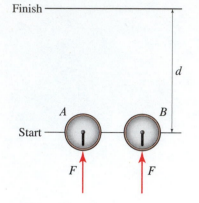

Figure P3.1

Figure P3.3

Problem 3.1

Two curling stones A and B, with masses m and $4m$, respectively, are pushed by two identical forces F over the distance d. Which stone arrives first to the finish line?
Note: Concept problems are about *explanations*, not computations.

Problem 3.2

An object is lowered very slowly onto a conveyer belt. What is the direction of the friction force acting on the object at the instant the object touches the belt?

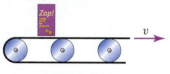

Figure P3.2

Note: Concept problems are about *explanations*, not computations.

Problem 3.3

A person is trying to move a heavy crate by pushing on it. While the person is pushing, what is the net or total force acting on the crate if the crate does not move?
Note: Concept problems are about *explanations*, not computations.

Problem 3.4

A person is lifting a 75 lb crate A by applying a constant force $P = 40$ lb to the pulley system shown. Neglecting friction and the inertia of the pulleys, determine the acceleration of the crate.

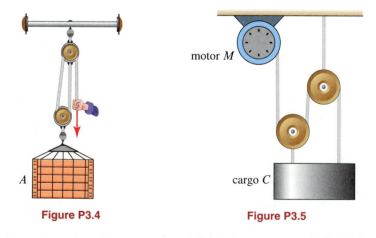

Figure P3.4   **Figure P3.5**

Problem 3.5

The motor M is at rest when someone flips a switch and it starts pulling in the rope. The acceleration of the rope is uniform and such that it takes 1 s to achieve a retraction rate of 4 ft/s. After 1 s the retraction rate becomes constant. Determine the tension in the rope during and after the initial 1 s interval. The cargo C weighs 130 lb, the weight of the ropes and pulleys is negligible, and friction in the pulleys is negligible.

Problem 3.6

A hammer hits a mass m on the end of a metal bar. In Chapter 5 we will see that this imparts an instantaneous initial velocity v_0 at $x = 0$ to the mass. Treating the bar as a massless spring, determine the equation of motion of the mass m. The equivalent spring constant of a bar in compression is given by $k_{eq} = EA/L$, where E is Young's modulus of the bar, A is the cross-sectional area of the bar, and L the length of the bar.[*]

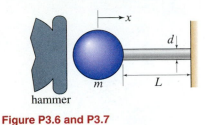

Figure P3.6 and P3.7

Problem 3.7

For the mass described in Prob. 3.6:

(a) Integrate the equation of motion to determine the speed of the mass $v(x)$ as a function of x.

(b) Integrate the result found in Part (a) to obtain the position of the mass as a function of time $x(t)$ from the initial time up until the mass stops for the first time.

Problem 3.8

As the skydiver moves downward with a speed v, the air drag exerted by the parachute on the skydiver has a magnitude $F_d = C_d v^2$ (C_d is a drag coefficient) and a direction opposite to the direction of motion. Determine the expression of the skydiver's acceleration in terms of C_d, v, the mass of the skydiver m, and the acceleration due to gravity.

Problems 3.9 through 3.11

The truck shown is traveling at $v_0 = 60$ mph when the driver applies the brakes to come to a stop. The deceleration of the truck is constant and the truck comes to a complete stop after braking for a distance of 350 ft. Treat the crate as a particle so that tipping can be neglected.

Figure P3.8

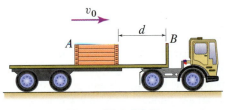

Figure P3.9–P3.11

Problem 3.9   Determine the minimum coefficient of static friction between the crate A and the truck so that the crate does *not* slide relative to the truck.

Problem 3.10   If the coefficient of kinetic friction between the crate A and the bed of the truck is 0.3 and static friction is not sufficient to prevent slip, determine the minimum distance d between the crate and the truck B so that the crate never hits the truck at B.

Problem 3.11   If the coefficient of kinetic friction between the crate A and the bed of the truck is 0.3, static friction is not sufficient to prevent slip, and the distance d from the front of the crate to the truck at B is 10 ft, determine the speed *relative to the truck* with which the crate strikes the truck at B.

---

[*] Those of you who have had a course in strength of materials may recognize that $\sigma = F/A = E\epsilon = E\Delta L/L$. Therefore, $F = (EA/L)\Delta L = (EA/L)x$. Since $F = k_{eq}x$ for this system, we see that $k_{eq} = EA/L$.

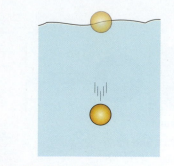

Figure P3.12 and P3.13

Problem 3.12

A metal ball with mass $m = 0.15\,\text{kg}$ is dropped from rest in a fluid. The magnitude of the resistance due to the fluid is given by $C_d v$, where C_d is a drag coefficient and v is the ball's speed. If $C_d = 2.1\,\text{kg/s}$, determine the ball's speed 4 s after release.

Problem 3.13

A metal ball weighing 0.35 lb is dropped from rest in a fluid. The magnitude of the resistance due to the fluid is given by $C_d v$, where C_d is a drag coefficient and v is the ball's speed. It is observed that 2 s after release the speed of the ball is 25 ft/s. Determine the value of C_d.

Problem 3.14

A horse is lifting a 500 lb crate by moving to the right at a constant speed $v_0 = 3\,\text{ft/s}$. Observing that B is fixed and letting $h = 6\,\text{ft}$ and $\ell = 14\,\text{ft}$, determine the tension in the rope when the horizontal distance d between B and point A on the horse is 10 ft.

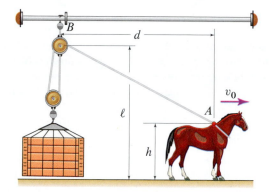

Figure P3.14

Problem 3.15

Figure P3.15

The centers of two spheres A and B with masses $m_A = 1\,\text{kg}$ and $m_B = 2\,\text{kg}$ are a distance $r_0 = 1\,\text{m}$ apart. B is fixed in space, and A is initially at rest. Using Eq. (1.6) on p. 5, which is Newton's universal law of gravitation, determine the speed with which A impacts B if the radii of the two spheres are $r_A = 0.05\,\text{m}$ and $r_B = 0.15\,\text{m}$. Assume that the two masses are infinitely far from any other mass so that they are only influenced by their mutual attraction.

Problems 3.16 and 3.17

Spring scales work by measuring the displacement of a spring that supports both the platform and the object, of mass m, whose weight is being measured. Neglect the mass of the platform on which the mass sits and assume that the spring is uncompressed before the mass is placed on the platform. In addition, assume that the spring is linear elastic with spring constant k.

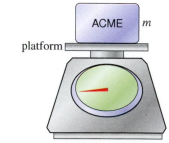

Figure P3.16 and P3.17

Problem 3.16   If the mass m is gently placed on the spring scale (i.e., it is dropped from zero height above the scale), determine the *maximum reading* on the scale after the mass is released.

Problem 3.17   If the mass m is gently placed on the spring scale (i.e., it is dropped from zero height above the scale), determine the *maximum speed* attained by the mass m as the spring compresses.

Problem 3.18

A scale is to be used on a wood countertop made from some very nice Brazilian cherry. To protect the countertop, the owner attaches self-sticking felt pads to the feet of the scale. When the weight was placed on the scale before the felt pads were applied, the scale read a certain value. Will the value be higher, lower, or the same when the same weight is placed on the scale but with the felt pads between the scale and the countertop? Ignore the transient dynamic effects that occur immediately after the weight is placed on the scale.

Note: Concept problems are about *explanations*, not computations.

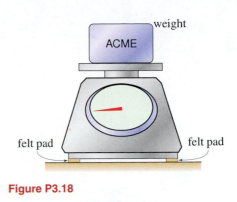

Figure P3.18

Problems 3.19 and 3.20

A vehicle is stuck on the railroad tracks as a 430,000 lb locomotive is approaching with a speed of 75 mph. As soon as the problem is detected, the locomotive's emergency brakes are activated, locking the wheels and causing the locomotive to slide.

Problem 3.19   If the coefficient of kinetic friction between the locomotive and the track is 0.45, what is the minimum distance d at which the brakes must be applied to avoid a collision? What would that distance be if instead of a locomotive, there was a 30×10^6 lb train? Treat the locomotive and the train as particles, and assume that the railroad tracks are rectilinear and horizontal.

Figure P3.19 and P3.20

Problem 3.20   Continue Prob. 3.19 and determine the time required to stop the locomotive.

Problem 3.21

A car is driving down a 23° rough incline at 55 km/h when its brakes are applied. Treating the car as a particle and neglecting all forces except gravity and friction, determine the stopping distance if

(a) The tires slide and the coefficient of kinetic friction between the tires and the road is 0.7.

(b) The car is equipped with antilock brakes and the tires do not slide. Use 0.9 for the coefficient of static friction between the tires and the road.

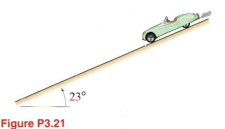

Figure P3.21

Problems 3.22 through 3.24

Car bumpers are designed to limit the extent of damage to the car in the case of low-velocity collisions. Consider a 3300 lb passenger car impacting a concrete barrier while traveling at a speed of 4.0 mph. Model the car as a particle, and consider two types of bumper: (1) a simple linear spring with constant k and (2) a linear spring of constant k in parallel with a shock absorbing unit generating a nearly constant force of 700 lb over 0.25 ft.

Problem 3.22   If the bumper is of type 1 and if $k = 6500$ lb/ft, find the spring compression necessary to stop the car.

Problem 3.23   If the bumper is of type 1, find the value of k necessary to stop the car when the bumper is precompressed 0.25 ft.

Problem 3.24   If the bumper is of type 2, find the value of k necessary to stop the car when the bumper is precompressed 0.25 ft.

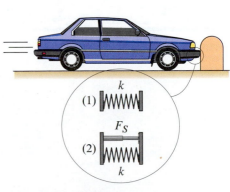

Figure P3.22–P3.24

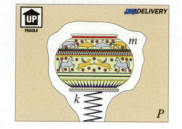

Figure P3.26

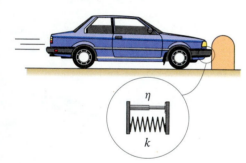

Figure P3.27 and P3.28

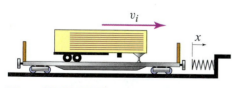

Figure P3.29–P3.31

Problem 3.25

What would it mean for the static or kinetic friction coefficients to be negative? Is this possible? Can either the static or kinetic friction coefficients be greater than 1? If yes, explain and give an example.

Note: Concept problems are about *explanations*, not computations.

Problem 3.26

Packages for transporting delicate items (e.g., a laptop or glass) are designed to "absorb" some of the energy of the impact in order to protect the contents. These energy absorbers can get pretty complicated to model (e.g., the mechanics of styrofoam peanuts is not easy), but we can begin to understand how they work by modeling them as a linear elastic spring of constant k that is placed between the contents (an expensive vase) of mass m and the package P. Assuming that $m = 3$ kg and that the box is dropped from a height of 1.5 m, determine the maximum displacement of the vase relative to the box and the maximum force on the vase if $k = 3500$ N/m. Treat the vase as a particle and neglect all forces except for gravity and the spring force. Assume that the spring relaxes after the box is dropped and that it does not oscillate.

Problems 3.27 and 3.28

Car bumpers are designed to limit the extent of damage to the car in the case of low-velocity collisions. Consider a passenger car impacting a concrete barrier while traveling at a speed of 4.0 mph. Model the car as a particle of mass m, and assume that the bumper has a spring element in parallel with a shock absorber so that the overall force exerted by the bumper is $F_B = k\delta + \eta\dot{\delta}$, where k, δ, and η denote the spring constant, the spring compression, and the bumper damping coefficient, respectively.

Problem 3.27   Derive the equations of motion for the car during the collision.

Problem 3.28   Let the weight of the car be 3300 lb, $k = 6500$ lb/ft, and $\eta = 300$ lb·s/ft, and let the car be traveling at 4.0 mph at impact. Determine the maximum compression of the bumper necessary to bring the car to a stop. Also determine the time required to stop the car.

Problems 3.29 through 3.31

A railcar with an overall mass of 75,000 kg traveling with a velocity v_i is approaching a barrier equipped with a bumper consisting of a nonlinear spring whose force vs. compression law is given by $F_s = \beta x^3$, where $\beta = 640 \times 10^6$ N/m³ and x is the compression of the bumper.

Problem 3.29   Treating the system as a particle and assuming that the contact between the railcar and rails is frictionless, determine the maximum value of v_i so that the compression of the bumper is limited to 20 cm.

Problem 3.30   Treating the system as a particle, assuming that the contact between railcar and rails is frictionless, and letting $v_i = 6$ km/h, determine the bumper compression necessary to bring the railcar to a stop.

Problem 3.31   Treating the system as a particle, assuming that the contact between the railcar and rails is frictionless, and letting $v_i = 6$ km/h, determine how long it takes for the bumper to bring the railcar to a stop.

Problem 3.32

A 6 lb collar is constrained to travel along a rectilinear and frictionless bar of length $L = 5$ ft. The springs attached to the collar are identical, and are unstretched when the collar is at B. Treating the collar as a particle, neglecting air resistance, and knowing that at A the collar is moving to the right with a speed of 11 ft/s, determine the linear spring constant k so that the collar reaches D with zero speed. Points E and F are fixed.

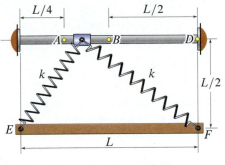

Figure P3.32 and P3.33

Problem 3.33

A 10 lb collar is constrained to travel along a rectilinear and frictionless bar of length $L = 5$ ft. The springs attached to the collar are identical, they have a spring constant $k = 4$ lb/ft, and they are unstretched when the collar is at B. Treating the collar as a particle, neglecting air resistance, and knowing that at A the collar is moving to the right with a speed of 14 ft/s, determine the speed with which the collar arrives at D. Points E and F are fixed.

Problem 3.34

An 11 kg collar is constrained to travel along a rectilinear and frictionless bar of length $L = 2$ m. The springs attached to the collar are identical, and they are unstretched when the collar is at B. Treating the collar as a particle, neglecting air resistance, and knowing that at A the collar is moving upward with a speed of 23 m/s, determine the linear spring constant k so that the collar reaches D with zero speed. Points E and F are fixed.

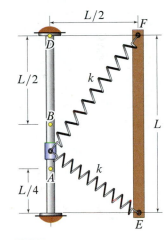

Figure P3.34

Problem 3.35

Derive the equation of motion of the mass m released from rest at $x = x_0$ from the slingshot. Assume that the cords connecting the mass to the slingshot are linear springs with spring constant k and unstretched length L_0. In addition, assume that the mass is equidistant from the two supports and that the mass and both springs lie in the xy plane. Ignore gravity and assume $L > L_0$.

Problem 3.36

Determine the speed of the mass m when it reaches $x = 0$ if it is released from rest at $x = x_0$ from the slingshot. Assume that the cords connecting the mass to the slingshot are linear springs with spring constant k and unstretched length L_0. Ignore gravity and assume $L > L_0$.

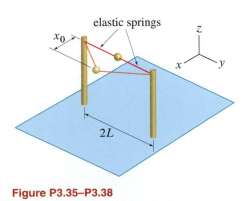

Figure P3.35–P3.38

Problem 3.37

Given the approximation

$$\frac{1}{\sqrt{(x/L)^2 + 1}} \approx 1 - \frac{(x/L)^2}{2}, \tag{1}$$

show that the equation of motion for the mass m when it is released from rest at $x = x_0$ from the slingshot can be written as

$$\ddot{x} + \omega_0^2 x \left(1 + \Lambda x^2\right) = 0, \tag{2}$$

where

$$\omega_0^2 = \frac{2k}{m}\left(\frac{L - L_0}{L}\right) \quad \text{and} \quad \Lambda = \frac{L_0}{2L^2(L - L_0)}. \tag{3}$$

Assume that the cords connecting the mass to the slingshot are linear springs with spring constant k, length L, and unstretched length L_0. Ignore gravity and assume $L > L_0$. Equation (2) is a famous differential equation in mechanics called *Duffing's equation*.

Problem 3.38

The force on the mass for the slingshot in Prob. 3.35 and the force on the mass in the Duffing equation obtained from the slingshot equation of motion (and defined by Eqs. (2) and (3)) can be plotted as a function of x. The nature of that force depends on whether or not the springs are initially stretched ($L > L_0$) or initially compressed ($L < L_0$). The figure shows the elastic restoring force on the mass m as a function of the displacement x for four different cases:

- —— Force for slingshot with $L = 2$ and $L_0 = 1$.
- —— Force for slingshot with $L = 1$ and $L_0 = 2$.
- —— Force for the Duffing equation with $L = 2$ and $L_0 = 1$.
- —— Force for the Duffing equation with $L = 1$ and $L_0 = 2$.

For small x, the force given by the Duffing equation is a good approximation to the force in the slingshot. Explain which of the curves corresponds to a "hardening" spring (a spring that gets stiffer as you pull it) and which corresponds to a "softening" spring (a spring that gets less stiff as you pull it), and explain physically why we see this behavior.

Note: Concept problems are about *explanations*, not computations.

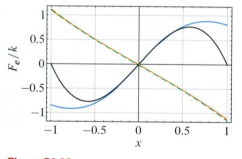

Figure P3.38

3.2   Curvilinear Motion

This section builds on what we learned in Section 3.1 by applying Newton's second law to curvilinear motion problems. No new material is introduced, thus allowing us to strengthen our modeling and problem-solving skills. Here we want to reinforce the idea that the application of Newton's second law to curvilinear problems is formally identical to what we saw in Section 3.1. The central element of a solution remains the FBD of the system analyzed. Then, after we choose a convenient component system, the solution of a problem still consists of applying Newton's second law, deriving the force laws, and making sure that we use the kinematic equations to describe position, velocity, and acceleration in Newton's second law and the force laws.

Newton's second law in 2D and 3D component systems

When we apply the vector equation $\vec{F} = m\vec{a}$ in the solution of engineering problems, we do it in component form, using Eq. (3.22) on p. 193, after choosing a convenient component system. For curvilinear problems we can choose our component system from those described in Chapter 2.* In this book we consider planar problems treatable via purely two-dimensional component systems as well as planar problems requiring the analysis of forces normal to the plane of motion and hence needing three-dimensional component systems. We also consider three-dimensional problems. Therefore, we now present expressions of the component form of Newton's second law, using the two- and three-dimensional component systems we have introduced in Chapter 2. The use of these component systems will be illustrated in the examples and homework problems.

Two-dimensional component systems

Cartesian components.   Referring to Fig. 3.14, Newton's second law in planar Cartesian components is

$$\sum F_x = ma_x \quad \text{and} \quad \sum F_y = ma_y. \tag{3.23}$$

The associated kinematic equations are given by Eqs. (2.17), which imply

$$a_x = \ddot{x} \quad \text{and} \quad a_y = \ddot{y}. \tag{3.24}$$

Path components.   Referring to Fig. 3.15, Newton's second law in planar path components is

$$\sum F_n = ma_n \quad \text{and} \quad \sum F_t = ma_t. \tag{3.25}$$

The associated kinematic equations are given by Eqs. (2.78), which imply

$$a_n = \frac{v^2}{\rho} \quad \text{and} \quad a_t = \dot{v}. \tag{3.26}$$

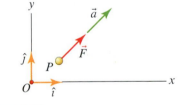

Figure 3.14
Force, acceleration, and a Cartesian component system.

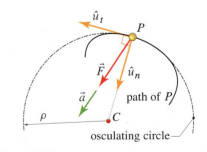

Figure 3.15
Force, acceleration, and a path component system.

---

* Chapter 2 covers some of the coordinate and component systems most frequently used in engineering. However, other coordinate systems are possible and are indeed used in applications.

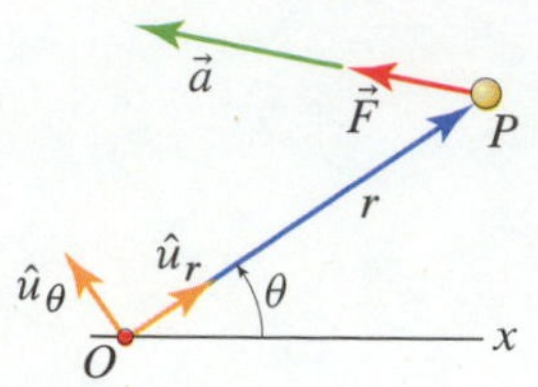

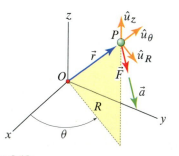

Figure 3.16
Force, acceleration, and a polar component system.

Polar components.   Referring to Fig. 3.16, Newton's second law in planar polar components is

$$\sum F_r = ma_r \quad \text{and} \quad \sum F_\theta = ma_\theta. \tag{3.27}$$

The associated kinematic equations are given by Eqs. (2.90), which are

$$a_r = \ddot{r} - r\dot{\theta}^2 \quad \text{and} \quad a_\theta = r\ddot{\theta} + 2\dot{r}\dot{\theta}. \tag{3.28}$$

Three-dimensional component systems

Cartesian components.   Referring to Fig. 3.17, Newton's second law in

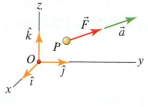

Figure 3.17. Force, acceleration, and a 3D Cartesian component system.

three-dimensional Cartesian components is

$$\sum F_x = ma_x, \quad \sum F_y = ma_y, \quad \text{and} \quad \sum F_z = ma_z. \tag{3.29}$$

The associated kinematic equations are given by Eqs. (2.143), which imply

$$a_x = \ddot{x}, \quad a_y = \ddot{y}, \quad \text{and} \quad a_z = \ddot{z}. \tag{3.30}$$

Path components.   Referring to Fig. 3.18, Newton's second law in three-dimensional path components is

$$\sum F_n = ma_n, \quad \sum F_t = ma_t, \quad \text{and} \quad \sum F_b = ma_b. \tag{3.31}$$

The associated kinematic equations are given by Eqs. (2.78), which imply

$$a_n = \frac{v^2}{\rho}, \quad a_t = \dot{v}, \quad \text{and} \quad a_b = 0, \tag{3.32}$$

where $a_b = 0$ because, by definition, $\hat{u}_b$ is perpendicular to the plane containing the velocity *and* the acceleration vectors (see Section 2.5).

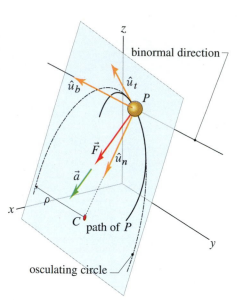

Figure 3.18
Force, acceleration, and a 3D path component system.

Cylindrical components.   Referring to Fig. 3.19, Newton's second law in three-dimensional cylindrical components is

$$\sum F_R = ma_R, \quad \sum F_\theta = ma_\theta, \quad \text{and} \quad \sum F_z = ma_z. \tag{3.33}$$

The associated kinematic equations are given by Eqs. (2.124), which are

$$a_R = \ddot{R} - R\dot{\theta}^2, \quad a_\theta = R\ddot{\theta} + 2\dot{R}\dot{\theta}, \quad \text{and} \quad a_z = \ddot{z}. \tag{3.34}$$

Figure 3.19
Force, acceleration, and a cylindrical component system.

Spherical components.   Referring to Fig. 3.20, Newton's second law in three-dimensional spherical components is

$$\sum F_r = ma_r, \quad \sum F_\phi = ma_\phi, \quad \text{and} \quad \sum F_\theta = ma_\theta. \tag{3.35}$$

The associated kinematic equations are given by Eqs. (2.140), which are

$$
\begin{aligned}
a_r &= \ddot{r} - r\dot{\phi}^2 - r\dot{\theta}^2 \sin^2\phi, \\
a_\phi &= r\ddot{\phi} + 2\dot{r}\dot{\phi} - r\dot{\theta}^2 \sin\phi \cos\phi, \\
a_\theta &= r\ddot{\theta}\sin\phi + 2\dot{r}\dot{\theta}\sin\phi + 2r\dot{\phi}\dot{\theta}\cos\phi.
\end{aligned}
\tag{3.36}
$$

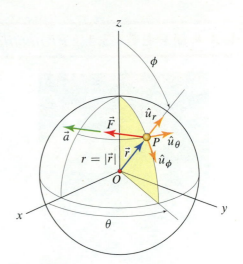

Figure 3.20
Force, acceleration, and a spherical component system.

EXAMPLE 3.4 *Projectile Motion with Drag*

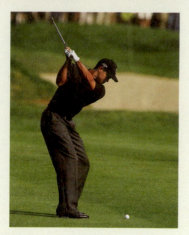

Figure 1
Tiger Woods swinging an iron.

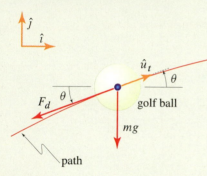

Figure 2
FBD of a golf ball in flight, modeled as a particle and subject to gravity and the aerodynamic drag force F_d. The unit vector $\hat{u}_t$ is tangent to the ball's path and indicates the direction of the ball's velocity.

The projectile motion model presented in Section 2.3, in which a projectile is subject only to constant gravity and the trajectory is a parabola, is not adequate for studying the trajectory of objects like golf balls. In general, the trajectory of a golf ball is not a parabola and is significantly affected by many factors such as the dimple pattern on the ball, the spin imparted to the ball, the relative humidity of the atmosphere, the local density of the air, etc. Accounting for all of these factors is beyond the scope of this book. Here we can consider a simple improvement on the model in Section 2.3 by lumping the effects just mentioned into an aerodynamic drag, which we assume to be proportional to the square of the ball's speed and directed opposite to the ball's velocity. Use this model to derive the equations of motion of the golf ball.

SOLUTION

Road Map & Modeling   Referring to Fig. 2, we have modeled the golf ball as a particle subject to its own weight mg and a drag force F_d. We chose a Cartesian component system with the y axis parallel to gravity because it simplifies the representation of the weight force and it allows for an easy comparison between the current model and that in Section 2.3. The direction of the drag force is opposite to the ball's velocity, which is in the $\hat{u}_t$ direction. Although the orientation of $\hat{u}_t$ is currently unknown, for convenience we have oriented it via the time-dependent angle θ. As we saw in Section 3.1, the ball's equations of motion are derived by combining Newton's second law with the force laws and the kinematic equations.

Governing Equations

Balance Principles   Using the FBD in Fig. 2, Newton's second law, in component form, gives

$$\sum F_x: \qquad -F_d \cos\theta = ma_x, \tag{1}$$

$$\sum F_y: \quad -F_d \sin\theta - mg = ma_y. \tag{2}$$

Force Laws   Because F_d is proportional to the square of the speed, we have

$$F_d = C_d v^2, \tag{3}$$

where C_d is a *drag coefficient*[*] and v is the ball's speed.

Kinematic Equations   In the chosen component system, we have

$$a_x = \ddot{x} \quad \text{and} \quad a_y = \ddot{y}. \tag{4}$$

In addition, because the ball's velocity can be written as $\vec{v} = \dot{x}\,\hat{\imath} + \dot{y}\,\hat{\jmath}$ and the speed is the magnitude of $\vec{v}$, we have

$$v = \sqrt{\dot{x}^2 + \dot{y}^2}. \tag{5}$$

Finally, given that θ is the orientation of $\vec{v}$ relative to the x direction, we have

$$\cos\theta = \frac{\dot{x}}{v} = \frac{\dot{x}}{\sqrt{\dot{x}^2 + \dot{y}^2}} \quad \text{and} \quad \sin\theta = \frac{\dot{y}}{v} = \frac{\dot{y}}{\sqrt{\dot{x}^2 + \dot{y}^2}}, \tag{6}$$

where the components of $\vec{v}$ are depicted in Fig. 3.

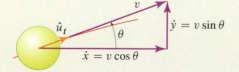

Figure 3
Components of the velocity vector of the golf ball.

---

[*] The coefficient C_d in Eq. (3) has dimensions of mass over length and should not be confused with the nondimensional drag coefficient normally used in aerodynamics.

Computation   Substituting Eqs. (3)–(6) into Eqs. (1) and (2), we obtain

$$-C_d \dot{x} \sqrt{\dot{x}^2 + \dot{y}^2} = m\ddot{x}, \qquad (7)$$

$$-C_d \dot{y} \sqrt{\dot{x}^2 + \dot{y}^2} - mg = m\ddot{y}. \qquad (8)$$

Equations (7) and (8) are the equations of motion for this problem.

Discussion & Verification   Since the dimensions of C_d are mass over length, Eqs. (7) and (8) are dimensionally correct. In addition, referring to Eq. (7), note that for a positive $\dot{x}$, the left-hand side is negative, thus indicating that the effect of the drag is that of slowing the particle down, as expected. A similar argument can be made for Eq. (8). Finally, note that if we set $C_d = 0$ in Eqs. (7) and (8), we recover the equations of motion of a projectile according to the model in Section 2.3, namely, $\ddot{x} = 0$ and $\ddot{y} = -g$. Therefore, we can conclude that, overall, the equations of motion we have derived appear to be correct.

A Closer Look   When $C_d \neq 0$, Eqs. (7) and (8) cannot be solved analytically. However, they can easily be solved numerically using mathematical software as long as we provide values for m, g, and C_d as well as the initial position and velocity of the ball. A typical golf ball weighs 1.61 oz, corresponding to a mass of 3.125×10^{-3} slug. To estimate C_d, the authors have calibrated the current model, using experimental data pertaining to the swing of Tiger Woods who, using a driver, imparted to a ball an initial speed of 186 mph and an upward direction of 11.2° with respect to the horizontal, and caused the ball to travel on the fly a distance of roughly 270 yd, or 810 ft. Using this information and with the help of a computer, we estimated that $C_d = 4.71 \times 10^{-7}$ lb·s²/ft². Figure 4 shows a comparison between the trajectory corresponding to the stated values of C_d and the initial conditions (the origin of the chosen xy coordinate system was used as the ball's initial position) and a trajectory with the same initial conditions but with $C_d = 0$. The reduction of the range of the ball due to drag is about 8%.

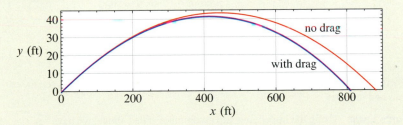

Figure 4. Trajectories of a golf ball with and without drag computed using the parameters and initial conditions stated in the text. To allow the two trajectories to be easily distinguished, the x and y axes have been given different scales.

EXAMPLE 3.5   *Maximum Speed Allowed by Friction*

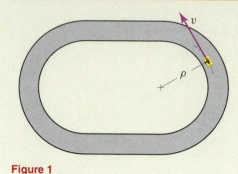

Figure 1
Top view of the racetrack with banked turns.

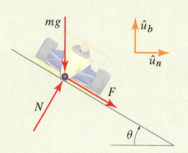

Figure 2
FBD of the car on the banked turn. The car's velocity vector is pointing out of the page.

A race car moves at a constant speed v along a banked turn on the track shown. Let the bank angle and turn radius of curvature be those of the Talladega Superspeedway in East Aboga, Alabama, which means that ρ is 1100 ft and the turn bank angle is 33°. For this turn, determine the *maximum* value of v such that the car does not slide. Assume that the static friction coefficient between the car and track is $\mu_s = 0.9$.

SOLUTION

Road Map & Modeling   If the maximum value of v for the car not to slip were exceeded, the car would slip toward the *outside* of the track. Thus, the component of the friction force keeping the car from sliding sideways points toward the inside of the track. Hence, by modeling the car as a particle and neglecting all forces except friction and gravity, the car's FBD is that shown in Fig. 2. To find $v_{\max}$, we will enforce Newton's second law along with the condition that the car is not to slip, which implies that the car moves along a circle with radius ρ with no motion in the vertical direction. Given that we need to find the *maximum* possible speed for the car not to slip, we will work under the assumption that slip is impending. This is a "working assumption," that is, it will allow us to obtain a solution, but that solution may not necessarily be physically achievable. After obtaining our solution, we will need to verify that the solution is compatible with the working assumption in question. Finally, since the path of the car is circular, the use of path components or cylindrical components is equally convenient. In Fig. 2 we have used path components (you will have the opportunity to use cylindrical components in Prob. 3.55).

Governing Equations

Balance Principles   Applying Newton's second law to the FBD in Fig. 2 gives

$$\sum F_n: \qquad F\cos\theta + N\sin\theta = ma_n, \tag{1}$$

$$\sum F_t: \qquad F_t = ma_t, \tag{2}$$

$$\sum F_b: \quad -mg + N\cos\theta - F\sin\theta = ma_b, \tag{3}$$

where F_t is the "driving force" in the tangent direction due to friction between the car's tires and the track.

Force Laws   Using the impending slip assumption stated above, we have

$$F = \mu_s N. \tag{4}$$

Kinematic Equations   In path components, the acceleration has only components in the nt plane. Therefore we have

$$a_n = v_{\max}^2/\rho, \quad a_t = \dot{v} = 0, \quad \text{and} \quad a_b = 0, \tag{5}$$

where we have enforced the fact that the car moves with constant speed when writing $a_t = 0$.

Computation   Equations (1)–(5) are the governing equations for this problem. The quantities g, θ, and ρ are known, and so we have seven equations in the seven unknowns F_t, N, F, a_t, a_b, a_n, and $v_{\max}$. These equations can be solved by hand or with mathematical software such as Mathematica or Maple to obtain

$$a_t = 0 = a_b, \tag{6}$$

$$a_n = \frac{g(\mu_s + \tan\theta)}{1 - \mu_s \tan\theta} = 120\,\text{ft/s}^2, \tag{7}$$

$$F_t = 0, \tag{8}$$

$$F = \frac{\mu_s mg}{\cos\theta - \mu_s \sin\theta} = (83.2\,\text{ft/s}^2)\,m, \tag{9}$$

$$N = \frac{mg}{\cos\theta - \mu_s \sin\theta} = (92.4\,\text{ft/s}^2)\,m, \tag{10}$$

$$v_{\max} = \sqrt{\rho g}\,\sqrt{\frac{\mu_s + \tan\theta}{1 - \mu_s \tan\theta}} = 363\,\text{ft/s} = 248\,\text{mph}. \tag{11}$$

Discussion & Verification   The results in Eqs. (6)–(11) are dimensionally correct and expressed using appropriate units. Hence, we need to answer whether or not the solution is consistent with the impending slip assumption used to derive it. The answer is yes. Specifically, notice that the obtained values of F and N are positive; i.e., F and N point exactly as shown on the FBD. In addition, the value of a_n is positive, as it should be, since the normal component of acceleration can only point toward the center of the circular path. Hence, we can conclude that the solution is physically acceptable, that the working assumption was correct, and that the problem is solved.

A Closer Look   To better understand the verification of the impending slip condition just presented, suppose that the given value of μ_s were 1.6 instead of 0.9. Then $\mu_s > 1/\tan\theta = 1.54$ and the term $1 - \mu_s \tan\theta$ in the expression for $v_{\max}$ would be negative so that $v_{\max}$ would involve the square root of a negative number. This result tells us that if $\mu_s > 1/\tan\theta$, then the original impending slip assumption becomes incorrect! The question now is, If $\mu_s > 1/\tan\theta$, how do we solve the problem? To answer this question, we need to analyze the impending slip solution more closely. Consider the plot of $v_{\max}$ vs. μ_s given by Eq. (11) and shown in Fig. 3. As μ_s approaches $1/\tan\theta$, $v_{\max}$ becomes infinite. This implies that we would never need a μ_s larger than 1.54 no matter how fast we wanted to go, and it indicates that for $\mu_s > 1/\tan\theta$, the working assumption must change from impending slip to no slip. Under this new assumption, Eq. (4) is replaced by the inequality $|F| \le \mu_s |N|$, and we solve Eqs. (1)–(3) as well as Eqs. (5) for $F_t, N, F, a_t, a_b,$ and a_n as functions of v:

$$a_t = 0 = a_b, \tag{12}$$

$$a_n = v^2/\rho, \tag{13}$$

$$F_t = 0, \tag{14}$$

$$F = \frac{mv^2}{\rho}\cos\theta - mg\sin\theta, \tag{15}$$

$$N = \frac{mv^2}{\rho}\sin\theta + mg\cos\theta. \tag{16}$$

By using Eqs. (15) and (16), and recalling that this solution assumes $\mu_s > 1/\tan\theta$, it is possible to verify that the no slip condition is indeed satisfied for any value of v, i.e., $v_{\max}$ is not restricted by insufficient friction. Going back to Fig. 3, observe that since μ_s is a positive quantity, the portion of the figure shaded in yellow concerns situations that are not physically achievable. However, the plot does indicate that if there were no friction ($\mu_s = 0$), the car would not slip as long as it traveled at roughly 150 ft/s.

Helpful Information

Why is the mass still in the governing equations? The mass of the car still appears in the governing equations, but we've made no mention of the mass — we did not state whether the mass is known or unknown. The solution of Eqs. (3)–(4) shows that only F and N depend on m. However, we were able to find the maximum speed $v_{\max}$ without needing to know m. This is a characteristic of some dynamics problems — the mass sometimes does not play a role in part or all of the solution. As with this problem, it is a simple matter to call the mass m and carry through the solution to see what happens in the end.

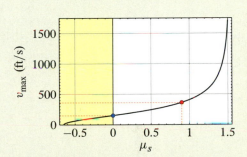

Figure 3
Plot of $v_{\max}$ vs. μ_s as given by Eq. (11), i.e., under the impending slip condition assumption. The red dot corresponds to $\mu_s = 0.9$ used in the example. The blue dot corresponds to $\mu_s = 0$.

E X A M P L E 3.6   *Analysis of Circular Motion*

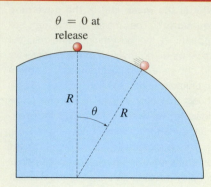

$\theta = 0$ at release

Figure 1
A small sphere sliding down a smooth semi-cylindrical surface.

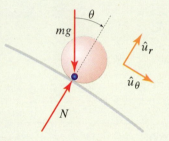

Figure 2
FBD of the particle when it is at an arbitrary position on the semicylinder.

A small sphere is at rest at the top of a frictionless semicylindrical surface. The sphere is given a *slight* nudge to the right so that it slides along the surface. Determine the angle θ at which the sphere separates from the surface.

SOLUTION

Road Map & Modeling   It may not be immediately obvious that the sphere will separate from the surface before θ reaches 90°. However, using a small ball, it is easy to set up a simple experiment to verify the claim. Let θ_s denote the value of θ at which separation occurs. Referring to Fig. 2, to analytically determine θ_s, we can model the sphere as a particle and sketch its FBD, which we have drawn at an arbitrary value of θ given that θ_s is not known yet. For simplicity, we have accounted for only the sphere's weight mg and the normal force N between the sphere and the sliding surface. Friction was omitted, given that the sliding surface is frictionless.[*] Until the sphere separates, the sphere's motion is circular and we will use polar coordinates to study it. To find θ_s, recall that *separation* implies "lack of contact," a condition for which $N = 0$. Therefore, we will solve the problem by applying Newton's second law along with kinematics to determine N as a function of θ. Then we will find θ_s by finding the value of θ for which $N = 0$.

Governing Equations

Balance Principles   Using the FBD in Fig. 2, Newton's second law yields

$$\sum F_r: \quad N - mg\cos\theta = ma_r, \tag{1}$$

$$\sum F_\theta: \quad mg\sin\theta = ma_\theta. \tag{2}$$

Force Laws   Since the only force law needed in this problem is that describing gravity, we can say that all force laws are accounted for on the FBD.

Kinematic Equations   The kinematics relations are

$$a_r = \ddot{r} - r\dot{\theta}^2 = -R\dot{\theta}^2, \tag{3}$$

$$a_\theta = r\ddot{\theta} + 2\dot{r}\dot{\theta} = R\ddot{\theta}, \tag{4}$$

where we have let $r = R = $ constant since the path is circular.

Computation   Substituting Eqs. (3) and (4) into Eqs. (1) and (2) and simplifying, we obtain

$$N - mg\cos\theta = -mR\dot{\theta}^2, \tag{5}$$

$$g\sin\theta = R\ddot{\theta}. \tag{6}$$

Equation (5) gives us N as a function of both θ and $\dot{\theta}$. Because we need N as a function of only θ, we need to find an expression for $\dot{\theta}$ as a function of θ. We can find this expression by considering Eq. (6), which gives us $\ddot{\theta}$ as a function of θ. To turn this relation into an expression of $\dot{\theta}$ as a function of θ, we begin by rewriting $\ddot{\theta}$, using the chain rule, and then using that in Eq. (6), i.e.,

$$\ddot{\theta} = \frac{d\dot{\theta}}{d\theta}\frac{d\theta}{dt} = \dot{\theta}\frac{d\dot{\theta}}{d\theta} \quad \Rightarrow \quad \dot{\theta}\frac{d\dot{\theta}}{d\theta} = \frac{g}{R}\sin\theta. \tag{7}$$

_____

[*] We will revisit this problem in Example 7.5 in Chapter 7, where we will model the sphere as a rigid body and account for friction, which causes the sphere to rotate over the surface of the cylinder.

Multiplying both sides of Eq. (7) by $d\theta$ and then integrating the result from the top of the cylinder to an arbitrary position, we obtain

$$\int_0^{\dot\theta} \dot\theta \, d\dot\theta = \int_0^{\theta} \frac{g}{R} \sin\theta \, d\theta, \tag{8}$$

$$\frac{\dot\theta^2}{2} = -\frac{g}{R}\cos\theta \Big|_0^{\theta} = \frac{g}{R}(1 - \cos\theta), \tag{9}$$

$$\dot\theta^2 = \frac{2g}{R}(1 - \cos\theta), \tag{10}$$

where we used the fact that $\dot\theta = 0$ when $\theta = 0$ to obtain the lower limits of integration. Substituting Eq. (10) into Eq. (5), we obtain N as a function of θ

$$N = mg(3\cos\theta - 2), \tag{11}$$

which has been plotted in Fig. 3. Force N goes to zero when

$$3\cos\theta_s - 2 = 0 \quad \Rightarrow \quad \cos\theta_s = \frac{2}{3} \tag{12}$$

$$\Rightarrow \quad \theta_s = \pm 48.2° \pm n360°, \; n = 0, 1, \dots, \infty. \tag{13}$$

Since we are only interested in solutions in the range $0° \leq \theta \leq 90°$, the only physically meaningful solution is

$$\boxed{\theta_s = 48.2°.} \tag{14}$$

Discussion & Verification   The result in Eq. (11) is dimensionally correct since the term mg has dimensions of force and the term in parentheses is nondimensional. In addition, the result appears to be reasonable because, as intuition suggests, N is equal to mg when $\theta = 0$ and it decreases as θ increases.

🔍 **A Closer Look**   Figure 3 shows that for $\theta > 48.2°$, the normal force becomes negative. What does this mean physically? Since the governing equations are written under the assumption that the sphere moves along the circular path with radius equal to R, the solution yields the value of N that is required for the sphere to move as prescribed. This means that for $48.2° < \theta \leq 90°$, for the sphere to remain in contact with the sliding surface, the surface must actually *pull the sphere in*. At $\theta = 90°$, the mass must be pulled in at *twice its weight*. Finally, notice that even though the normal force is a function of the weight of the sphere, the angle at which the sphere leaves the surface is independent of m and g.

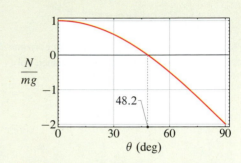

Figure 3
The normal force N (normalized by the weight of the particle mg) as a function of θ as the particle slides down the semicylinder.

EXAMPLE 3.7   *Central Force Motion*

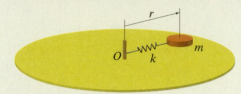

Figure 1
A disk of mass m moving on a smooth horizontal plane. The disk's motion is constrained only by the linear elastic spring.

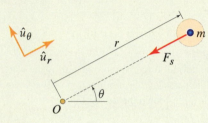

Figure 2
FBD of the system as viewed from above.

Consider a disk of mass m at one end of a linear elastic spring, the other end of which is pinned. The disk is free to move in the smooth horizontal plane. Determine the equations of motion of the disk for any values of the initial position and velocity as well as the system parameters, namely, the spring constant k, the unstretched length of the spring r_u, and the mass of the disk m. After doing this, plot the disk's trajectory using $r(0) = 0.35$ m, $\theta(0) = 0$ rad, $\dot{r}(0) = 0$ m/s, $\dot{\theta}(0) = 0.5$ rad/s, $r_u = 0.25$ m, and four different values of k/m: 5, 20, 100, and 500 s^{-2}. Do each of the four plots for $0 \le t \le 10$ s.

SOLUTION

Road Map & Modeling   To find the equations of motion, we will apply Newton's second law to the disk and combine it with the force laws and kinematic equations. How many equations of motion should we be looking for? The answer is that we need to find as many equations of motion as there are degrees of freedom, which, in this case, is equal to two because the disk's motion is planar. Referring to Fig. 2 given that the motion of the disk occurs over a *smooth* horizontal plane, we will model the disk as a particle subject only to the spring force F_s. The motion is planar, and while we could use Cartesian coordinates to describe the motion, the fact that F_s always points toward O makes it convenient to use a polar coordinate system with origin at O.

Governing Equations

Balance Principles   Referring to the FBD in Fig. 2, Newton's second law yields

$$\sum F_r: \quad -F_s = ma_r, \tag{1}$$

$$\sum F_\theta: \quad 0 = ma_\theta. \tag{2}$$

Force Laws   The only force law needed is that of the linear spring, i.e.,

$$F_s = k(r - r_u). \tag{3}$$

Kinematic Equations   The kinematic equations in polar coordinates are

$$a_r = \ddot{r} - r\dot{\theta}^2, \tag{4}$$

$$a_\theta = r\ddot{\theta} + 2\dot{r}\dot{\theta}. \tag{5}$$

Computation   Substituting Eqs. (3)–(5) into Eqs. (1) and (2) and rearranging, we obtain the equations of motion for this system as[*]

$$\boxed{\begin{aligned} \ddot{r} - r\dot{\theta}^2 + \frac{k}{m}(r - r_u) = 0, \\ r\ddot{\theta} + 2\dot{r}\dot{\theta} = 0, \end{aligned}}\begin{aligned} \tag{6} \\ \tag{7} \end{aligned}$$

where we have divided m out of Eq. (7). Solving Eqs. (6) and (7) for $r(t)$ and $\theta(t)$ would tell us the motion of the mass at the end of the spring for all time. Unfortunately, these equations cannot be solved analytically. Therefore, we must resort to computer solutions.

🖥️ ➡ To solve Eqs. (6) and (7) on a computer, we will need to use the values for all the constants in the system, as well as the initial conditions, that are given in the

---

[*] Notice that the spring constant k and the mass m only appear as a ratio in Eq. (6). Therefore, we don't need to specify k and m individually, we need only specify their ratio.

Helpful Information

Integral of Eq. (7). Although it takes some experience to see this, Eq. (7) can be written as

$$r\ddot{\theta} + 2\dot{r}\dot{\theta} = \frac{1}{r}\frac{d}{dt}(r^2\dot{\theta}) = 0.$$

This equation says that the time derivative of $r^2\dot{\theta}$ is equal to zero. Therefore, $r^2\dot{\theta}$ is a constant. The fact that there is this quantity associated with the motion that does not change is referred to as a *constant of the motion* or as an *integral of the equations of the motion*. In this case, this constant arises from the fact that the force acting on the disk is a *central force*; that is, it always points toward one central location, namely, the pin at O (see Fig. 1). In Chapter 5 we will see that a central force always leads to a constant *angular momentum* and that the quantity $r^2\dot{\theta}$ is simply the system's angular momentum (per unit mass).

problem statement. We will compute four different solutions, all of which will have the same initial conditions, which are given as $r(0) = 0.35\,\text{m}$, $\theta(0) = 0\,\text{rad}$, $\dot{r}(0) = 0\,\text{m/s}$, and $\dot{\theta}(0) = 0.5\,\text{rad/s}$. In addition, all of the four simulations will have the same value of the spring's unstretched length, namely, $r_u = 0.25\,\text{m}$. The simulations in question will differ only in the values chosen for the ratio k/m, which we set to 5, 20, 100, and $500\,\text{s}^{-2}$. Using mathematical software and integrating our equations for $0 \leq t \leq 10\,\text{s}$, we obtain the trajectories shown in Fig. 3.

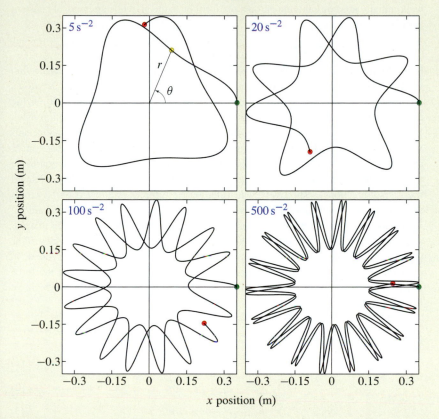

Figure 3. Trajectories of the disk for four different values of k/m, which are shown in the upper left corner of each figure. The values of x and y are computed as $r\cos\theta$ and $r\sin\theta$, respectively. The green dots indicate the start of each trajectory, and the red dots indicate the end after $10\,\text{s}$.

Discussion & Verification   The "answer" to this example is not a number, but is a pair of ordinary differential equations given by Eqs. (6) and (7). The number of equations of motion matches the number of degrees of freedom, which is equal to two because the disk is free to move in the horizontal plane, which is two-dimensional. Since the equation of motion we obtained cannot be solved analytically, we had to resort to a computer solution for specific values of the system parameters and over a specific interval of time. We encourage you to try some solutions on your own.

EXAMPLE 3.8   *Coriolis Effect*

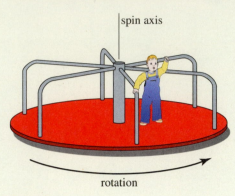

spin axis

rotation

Figure 1
A small child walking radially on a spinning merry-go-round.

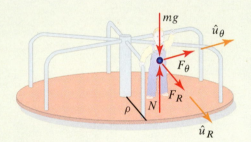

mg

$\hat{u}_\theta$

F_θ

F_R

ρ N

$\hat{u}_R$

Figure 2
FBD of the child walking radially outward on the merry-go-round.

When one of the authors was a child, he and some of his friends would go to the playground and play on a merry-go-round similar to that shown in Fig. 1. In walking over the platform while spinning, not only did we feel thrown radially outward but also we felt thrown sideways. Investigate this motion, and determine the forces required to walk radially at a constant rate v_0 on a platform of radius ρ, while spinning at constant angular rate ω_0.

SOLUTION

Road Map & Modeling   We are told how a child moves, and we are asked to find what forces are necessary for the motion to occur as described. Recalling Newton's second law, $\vec{F} = m\vec{a}$, in this problem we are given $\vec{a}$ and asked to find $\vec{F}$. Referring to Fig. 2, we will model the child as a particle and describe his or her motion using a cylindrical coordinate system with origin on the spin axis of the merry-go-round. We assume that the child is subject to gravity, namely, mg, the normal reaction N at the floor, as well as any other forces that might be required for the motion to be as described. This is why the FBD includes the forces F_R and F_θ, in the radial and transverse directions, respectively. The value of these forces is dictated by Newton's second law and by kinematics. The specifics of how these forces are generated are not important from the viewpoint of the problem's solution. However, we can say that F_R and F_θ will be due to friction between the child and the spinning platform as well as the fact that the child can hold onto the railings mounted on the merry-go-round. In preparation for the Discussion & Verification of the solution, we note that the force required to maintain a given motion has a direction *opposite* to that along which we feel "thrown."[*]

Governing Equations

Balance Principles   Newton's second law tells us that

$$\sum F_R: \qquad F_R = ma_R, \tag{1}$$

$$\sum F_\theta: \qquad F_\theta = ma_\theta, \tag{2}$$

$$\sum F_z: \quad N - mg = ma_z. \tag{3}$$

Force Laws   All known forces are accounted for on the FBD. The forces N, F_R, and F_θ are unknowns of the problem.

Kinematic Equations   Using cylindrical components and accounting for the given motion, we have

$$a_R = \ddot{R} - R\dot{\theta}^2 = -R\omega_0^2, \tag{4}$$

$$a_\theta = R\ddot{\theta} + 2\dot{R}\dot{\theta} = 2v_0\omega_0, \tag{5}$$

$$a_z = \ddot{z} = 0. \tag{6}$$

Computation   Equations (3) and (6) simply tell us that $N = mg$. Equations (1), (2), (4), and (5) tell us that

$$F_R = -mR\omega_0^2, \tag{7}$$

$$F_\theta = 2mv_0\omega_0, \tag{8}$$

which are the forces required to keep the child moving in the radial direction at a constant speed as the merry-go-round spins at a constant angular velocity.

---

[*] For example, the force *on* you as you round a curve in a car is in the same direction as your acceleration, i.e., toward the center of the curve. However, the physical sensation you experience is that of being thrown *away* from the center of the curve.

Discussion & Verification   Equation (7) says that F_R is independent of how fast the child walks and that it always acts inward given that m, r, and ω_0^2 are positive. As intuition might suggest, this corresponds to a physical sensation of being "thrown" outward. By contrast, F_θ in Eq. (8) can be positive or negative because v_0 and ω_0 can each be positive or negative,[*] thus leading to four possible cases. In each of the cases, when we use right and left to indicate direction, they will be relative to the child, who we assume is facing in the direction of motion.

Case 1   In this case, let $v_0 > 0$ and $\omega_0 > 0$, so that the child is walking *outward* and the merry-go-round is spinning as shown in Fig. 1, i.e., counterclockwise (ccw) as viewed from above. This situation is shown in Fig. 3(a). For this case, Eq. (8) tells us that $F_\theta > 0$ so that the force *applied to the child* while walking *outward* will be to his or her *left*. The child will then feel as though thrown to the *right*.

Case 2   In this case, $v_0 < 0$ and $\omega_0 > 0$, so that the child is walking *inward* and the merry-go-round is spinning counterclockwise (ccw). This situation is shown in Fig. 3(b). For this case, Eq. (8) tells us that $F_\theta < 0$ so that the force *applied to the child* while walking *inward* will again be to her or his *left*. Once again, the child will then feel as though thrown to the *right*.

Case 3   In this case, $v_0 > 0$ and $\omega_0 < 0$, so that the child is walking *outward* and the merry-go-round is now spinning clockwise (cw). This situation is shown in Fig. 3(c). For this case, Eq. (8) tells us that $F_\theta < 0$ so that the force *applied to the child* while walking *outward* will now be to his or her *right*. This time, the child will then feel as though thrown to the *left*.

Case 4   In this case, $v_0 < 0$ and $\omega_0 < 0$, so that the child is walking *inward* and the merry-go-round is again spinning clockwise (cw). This situation is shown in Fig. 3(d). For this case, Eq. (8) tells us that $F_\theta > 0$ so that the force *applied to the child* while walking *inward* will now be to her or his *right*, and the child will again feel as though thrown to the *left*.

The analysis of the above four cases tells us that whether the child walks inward or outward, as long as the merry-go-round spins counterclockwise (ccw), the child is always thrown to the right. Similarly, whether or not the child walks inward or outward, as long as the merry-go-round spins clockwise (cw), the child is always thrown to the left. It turns out that we can generalize these results for a person walking in *any* direction to state the following:

> *No matter which direction a person walks on a rotating reference frame, if the frame is rotating counterclockwise (ccw), then the person is thrown to his or her right, whereas the person is thrown to her or his left if the reference frame is rotating clockwise (cw).*

This result is a little difficult to prove in general with what we know now, but it will not be difficult to show once we learn about the kinematics of motion relative to rotating reference frames, which we will cover in Section 6.4.

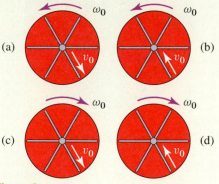

Figure 3
(a) Case 1: $v_0 > 0$, $\omega_0 > 0$; (b) Case 2: $v_0 < 0$, $\omega_0 > 0$; (c) Case 3: $v_0 > 0$, $\omega_0 < 0$; (d) Case 4: $v_0 < 0$, $\omega_0 < 0$.

Helpful Information

Coriolis acceleration comes from *moving* on a *rotating* frame. The F_θ in this problem comes from the $2\dot{R}\dot{\theta}$ component of a_θ. For this component of a_θ to be nonzero, *both* ω_0 and v_0 must be nonzero. Physically, this means that for $2\dot{R}\dot{\theta}$ to be nonzero, we must be *moving* ($\dot{R} \neq 0$) on a *rotating* reference frame ($\dot{\theta} \neq 0$).

---

[*] The expression for F_θ originates from the $2\dot{R}\dot{\theta}$ component of a_θ. As we discussed in Section 2.4 (see discussion of Eq. (2.64) on p. 92), this component of acceleration is often referred to as the *Coriolis component of acceleration*.

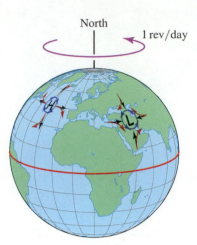

Figure 3.21
Direction of the rotation of the Earth.

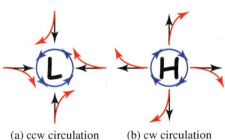

(a) ccw circulation   (b) cw circulation

Figure 3.22
Circulation around (a) low- and (b) high-pressure systems in the northern hemisphere.

Coriolis component of acceleration

The $2\dot{r}\dot{\theta}$ component of acceleration is often referred to as the *Coriolis acceleration*, though we prefer to call it the Coriolis *component* of acceleration. It is named after Gaspard-Gustave Coriolis (1792–1843) who studied mechanics and mathematics in France. He gave the terms *work* and *kinetic energy* their present scientific meaning, and in an 1835 publication, Coriolis showed that Newton's laws may be applied in a rotating reference frame as long as an "extra" acceleration is added to the equations of motion. The phenomenon in which a person feels as though she or he is being thrown sideways when moving on a rotating platform is a consequence of this acceleration. In fact, this acceleration gives rise to the ccw circulation around low-pressure systems and cw circulation around high-pressure systems in the northern hemisphere. We can understand this circulation around high and low pressure systems if we view the spin of the Earth as we did the spin of the merry-go-round in Fig. 3 of Example 3.8. This "merry-go-round view" of the Earth will require us to look *down* the North Pole of the Earth as it spins on its axis (see Fig. 3.21). Looking down the North Pole, we see that the Earth is spinning in a counterclockwise direction. From what we learned in Example 3.8, this would mean that if something were to move on the surface of the Earth in the northern hemisphere, it would be deflected to the right (i.e., it would take a force directed to the left to keep it moving in a straight line). Now, with reference to Fig. 3.22, which depicts low- and high-pressure systems in the northern hemisphere, we know that low- and high-pressure systems are so named because the air pressure in them is lower and higher, respectively, than the surrounding air pressure. Since air flows from areas of higher pressure to areas of lower pressure, we see that air will try to flow *into* the low from its surroundings and *away* from the high to its surroundings (the black arrows in Fig. 3.22). As air flows into the low, the Coriolis component of acceleration causes it to deflect to the right (the red arrows in Fig. 3.22). Since every "particle" of air is deflected to the right, the overall motion becomes that of a circulation *around* the low—a circulation in the ccw direction (Fig. 3.22(a)). A similar argument tells us that the circulation around a high must be in the cw direction (Fig. 3.22(b)). Of course, if one views the Earth from the southern hemisphere, the Earth appears to be rotating cw, so in the southern hemisphere objects are deflected to the left and the circulation around lows and highs is cw and ccw, respectively.

You might be wondering why you don't have to worry about the Coriolis component of acceleration when you play baseball or basketball—it is there, it is just that the $\dot{\theta}$ for the Earth is *really* small. It is easy to show that 1 rev/day is 0.00007272 rad/s. Although it is present when you shoot a basketball, this acceleration would need to act for a *long* time to have a visible effect—the sort of times (days) experienced by a particle of air moving through the atmosphere. That is not to say that engineers don't need to worry about the Coriolis component of acceleration—they do. However, if the kinematic analysis of a system is carried out using the ideas presented in Chapter 2, the Coriolis component of acceleration will *always* be taken into account "automatically."

PROBLEMS

💡 Problem 3.39 💡

A train is traveling with constant speed v_0 along the level track OAB, which lies in the horizontal plane. The section between points O and A is straight, and the section between points A and B is circular with radius of curvature ρ. The train starts at point O at time $t = 0$, reaches point A at time t_A, and reaches point B at time t_B. For this motion, sketch the magnitude of the acceleration vector as a function of time. Would you want to design a train track with this shape?

Note: Concept problems are about *explanations*, not computations.

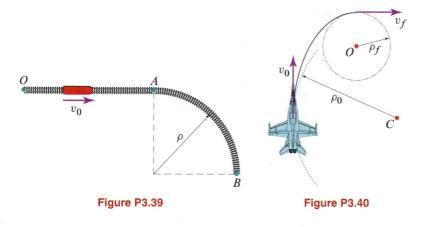

Figure P3.39   Figure P3.40

💡 Problem 3.40 💡

A plane is turning along a horizontal path at constant speed. What is the total force acting on the plane in the direction of the path?

Note: Concept problems are about *explanations*, not computations.

💡 Problem 3.41 💡

Body D is in equilibrium when cable AC suddenly breaks. If B is fixed, does the tension in the inextensible cable BC increase or decrease at the instant of release?

Note: Concept problems are about *explanations*, not computations.

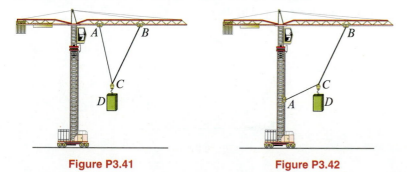

Figure P3.41   Figure P3.42

💡 Problem 3.42 💡

Body D is in equilibrium when cable AC suddenly breaks. If B is fixed, does the tension in the inextensible cable BC increase or decrease at the instant of release?

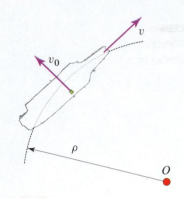

Figure P3.43

Note: Concept problems are about *explanations*, not computations.

💡 Problem 3.43 💡

An aircraft carrier is turning with a constant speed v along a circular path with radius ρ and center O. During the maneuver, a forklift is being driven across the deck of the ship at a constant speed v_0, relative to the deck. Does the friction force between the forklift and the deck have a component perpendicular to the relative velocity of the forklift with respect to the deck?

Note: Concept problems are about *explanations*, not computations.

Problem 3.44 🌡

A jet is coming out of a dive, and a sensor in the pilot's seat measures a force of 800 lb for a pilot whose weight is 180 lb. If the jet's instruments indicate that the plane is traveling at 850 mph, determine the radius of curvature of the plane's path at this instant.

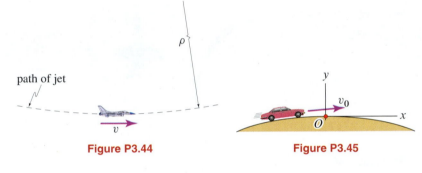

Figure P3.44   **Figure P3.45**

Problem 3.45 🌡

A car traveling over a hill at constant speed starts to lose contact with the ground at O. If the radius of curvature of the hill is 282 ft, determine the speed of the car at O.

Problem 3.46 🌡

A 950 kg aerobatics plane initiates the basic loop maneuver at the bottom of a loop with radius $\rho = 110$ m and a speed of 225 km/h. At this instant, determine the magnitude of the plane's acceleration, expressed in terms of g, the acceleration due to gravity, and the magnitude of the lift provided by the wings.

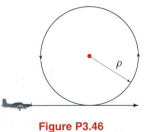

Figure P3.46

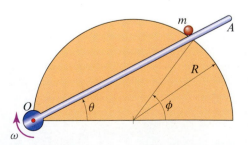

Figure P3.47

Problem 3.47 🌡

The ball of mass m is guided along the vertical circular path of radius $R = 1$ m using the arm OA. If the arm starts from $\phi = 90°$ and rotates clockwise with a constant angular velocity $\omega = 0.87$ rad/s, determine the angle ϕ at which the particle starts to leave the surface of the semicylinder. Neglect all friction forces acting on the ball, neglect the thickness of the arm OA, and treat the ball as a particle.

Problem 3.48

Referring to Example 3.7, instead of using polar coordinates as was done in that example, work the problem using a Cartesian coordinate system with origin at O (cf. Fig. 1 of Example 3.7) and derive the problem's equations of motion.

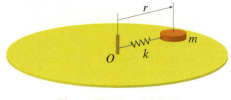

Figure P3.48 and P3.49

Problem 3.49

Continue Prob. 3.48 and using mathematical software, numerically solve the equations of motion. Use the same parameters and initial conditions that were used in Example 3.7, and compare your results with those presented in that example.

Problem 3.50

Derive the equations of motion for the pendulum supported by a linear spring of constant k and unstretched length r_u. Neglect friction at the pivot O, the mass of the spring, and air resistance. Treat the pendulum bob as a particle of mass m and use polar coordinates.

Problem 3.51

Using mathematical software, solve the equations of motion for the pendulum supported by a linear spring of constant k (derived in Prob. 3.50). Plot the trajectory of the mass m in the vertical plane for a number of different values of k/m. The unstretched length of the spring is 0.25 m, and the mass is released when the pendulum is vertical, the spring is stretched 0.75 m, and the mass is moving to the right at 1 m/s.

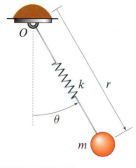

Figure P3.50 and P3.51

Problem 3.52

Revisiting Example 3.4, assume that the drag force in the x direction is proportional to the square of the x component of velocity, but that the ball's trajectory is shallow enough to neglect the drag force in the y direction. Using this assumption, letting O be the initial position of the ball, and letting the ball's initial velocity have a magnitude v_0 and form an angle θ_0 with the x axis, determine an expression for the trajectory of the ball of the form $y = y(x)$.

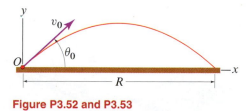

Figure P3.52 and P3.53

Problem 3.53

Revisit Example 3.4 and assume that the trajectory of the ball is shallow enough that the effects of the drag force in the y direction can be neglected and that the component of the drag force in the x direction is proportional to the square of the x component of velocity. Using this assumption and the same drag coefficient discussed in the example ($C_d = 4.71 \times 10^{-7}$ lb·s^2/ft^2), compute the horizontal distance R traveled by a 1.61 oz

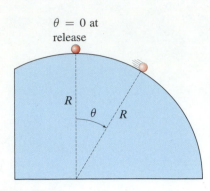

$\theta = 0$ at
release

Figure P3.54

golf ball subject to the same initial conditions given in the example, i.e., the initial velocity has a magnitude $v_0 = 187$ mph and an initial orientation $\theta_0 = 11.2°$.

Problem 3.54

Revisit Example 3.6 on p. 216 by letting the sphere be released at $\theta = 0$ with a speed $v_0 = 0.5$ m/s. Neglecting friction, compute the angle at which the sphere separates from the cylinder if $R = 1.35$ m.

Problem 3.55

Using a cylindrical component system whose origin is at the center of curvature of the path of the car, derive the governing equations for Example 3.5 on p. 214. Solve the equations and verify that you get the same solution.

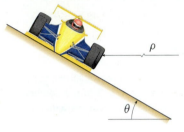

Figure P3.55–P3.57

Problem 3.56

A race car is traveling at a constant speed over a circular banked turn. Oil on the track has caused the static friction coefficient between the tires and the track to be $\mu_s = 0.2$. If the radius of the car's trajectory is $\rho = 320$ m and the bank angle is $\theta = 33°$, determine the range of speeds within which the car must travel not to slide sideways.

Problem 3.57

A race car is traveling at a constant speed $v = 200$ mph over a circular banked turn. Let the weight of the car be $W = 4000$ lb, the radius of the car's trajectory be $\rho = 1100$ ft, the bank angle be $\theta = 33°$, and the coefficient of static friction between the car and the track be $\mu_s = 1.7$. Determine the component of the friction force perpendicular to the direction of motion.

Problem 3.58

The cutaway of the gun barrel shows a projectile moving through the barrel. If the projectile's exit speed is $v_s = 1675$ m/s (relative to the barrel), the projectile's mass is 18.5 kg, the length of the barrel is $L = 4.4$ m, the acceleration of the projectile down the gun barrel is constant, and θ is increasing at a constant rate of 0.18 rad/s, determine

(a) The acceleration of the projectile.

(b) The pressure force acting on the back of the projectile.

(c) The normal force on the gun barrel due to the projectile.

as the projectile leaves the gun, but while it is still in the barrel. Assume that the projectile exits the barrel when $\theta = 20°$, and ignore friction between the projectile and the barrel.

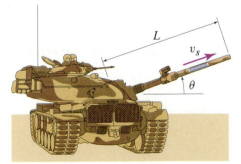

Figure P3.58

Problems 3.59 and 3.60

A simple sling can be built by placing a projectile in a tube and then spinning it. Consider a simple model in which the tube is pinned as shown and is rotated about the pin in the *horizontal* plane at constant angular velocity ω. Assume that there is no friction between the projectile and the inside of the tube and that the projectile is initially kept fixed at a distance d from the open end of the tube.

Problem 3.59 After the projectile is released, compute the normal force exerted by the inside of the tube on the projectile as a function of position of the projectile along the tube.

Problem 3.60 Leting $d = 3$ ft and $L = 7$ ft, determine the value of the tube's (constant) angular velocity ω if, after release, the projectile exits the tube with a (total) speed of 90 ft/s.

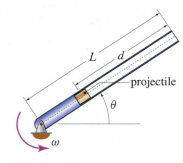

Figure P3.59 and P3.60

Problem 3.61

The trolley T moves along rails on the horizontal truss of the tower crane. The trolley and the load of mass m are both initially at rest (with $\theta = 0$) when the trolley starts moving to the right with constant acceleration $a_T = g$. Determine

(a) The maximum angle $\theta_{\max}$ achieved by the load m.

(b) The tension in the supporting cable as a function of θ.

Treat the load m as a particle, and ignore the mass of the supporting cable.

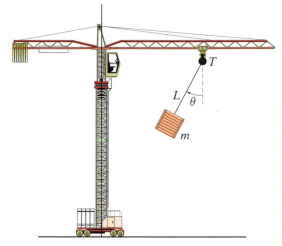

Figure P3.61

💡 Problem 3.62 💡

If the particle is constrained to only move back and forth in the plane of the center of the bowl, how many degrees of freedom must it have? Recall that this will also be the number of equations of motion of the particle.

Note: Concept problems are about *explanations*, not computations.

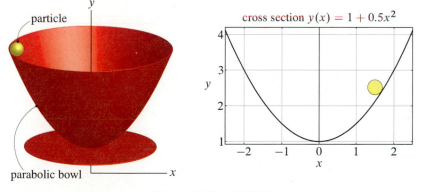

Figure P3.62 and P3.63

Problem 3.63

Derive the equation(s) of motion for a particle moving on the inner surface of the parabolic bowl shown. Assume that the particle only moves in the vertical xy plane that goes through the center of the bowl, that the equation of the parabola is $y(x) = 1 + 0.5x^2$, and that gravity is acting in the $-y$ direction. The bowl's cross section is shown on the right side of the figure.

Problem 3.64 🌡

A particle moves over the inner surface of an inverted cone. Assuming that the cone's surface is frictionless and using the cylindrical coordinate system shown, show that the particle's equations of motion are

$$\ddot{R}(1 + \cot^2 \phi) - R\dot{\theta}^2 = -g \cot \phi,$$
$$R\ddot{\theta} + 2\dot{R}\dot{\theta} = 0.$$

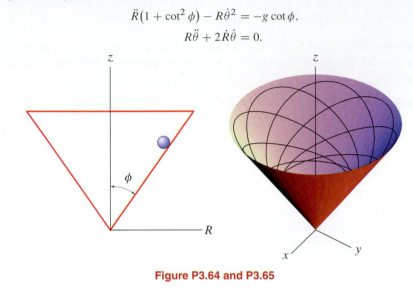

Figure P3.64 and P3.65

🖥 Problem 3.65 🖥

Continue Prob. 3.64 by integrating the particle's equations of motion and plotting its trajectory for $0 \leq t \leq 25$ s. Use the following parameter values and initial conditions: $g = 9.81 \, \text{m/s}^2$, $\phi = 30°$, $\theta(0) = 0°$, $\dot{\theta}(0) = 1.00 \, \text{rad/s}$, $R(0) = 5 \, \text{m}$, $\dot{R}(0) = 0 \, \text{m/s}$.

Problem 3.66 🌡

The pendulum is released from rest when $\theta = 0°$. If the string holding the pendulum bob breaks when the tension is twice the weight of the bob, at what angle does the string break? Treat the pendulum as a particle, ignore air resistance, and let the string be inextensible and massless.

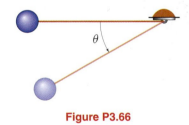

Figure P3.66

Problem 3.67 🌡

The package handling system is designed to launch the small package of mass m from A, using a compressed linear spring of constant k. After launch, the package slides along the track until it lands on the conveyor belt at B. The track has small, well-oiled rollers, making any friction between the packages and the track negligible. Modeling the package as a particle, determine the minimum initial compression of the spring so that the package gets to B without separating from the track, and determine the corresponding speed with which the package reaches the conveyor at B.

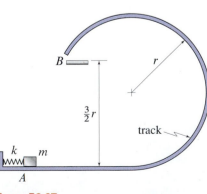

Figure P3.67

Problems 3.68 and 3.69

A satellite orbits the Earth as shown. Model the satellite as a particle, and assume that the center of the Earth can be chosen as the origin of an inertial frame of reference.

Problem 3.68 Using a polar coordinate system and letting m_e be the mass of the Earth, determine the equations of motion of the satellite.

Problem 3.69 The minimum and maximum distances from the center of the Earth are $R_P = 4.5 \times 10^7$ m and $R_A = 6.163 \times 10^7$ m, respectively, where the subscripts P and A stand for *perigee* (the point on the orbit closest to Earth) and *apogee* (the point on the orbit farthest from Earth), respectively. If the satellite's speed at P is $v_P = 3.2 \times 10^3$ m/s, determine the satellite's speed at A.

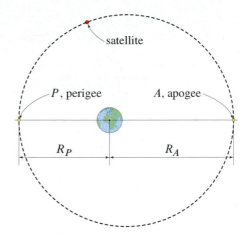

Figure P3.68 and P3.69

Problem 3.70

The disk shown, weighing 3 lb, rotates about O by sliding without friction over the horizontal surface shown. The spring is linear with constant k and unstretched length $L_0 = 0.75$ ft. The maximum distance achieved by the disk from point O is $d_{max} = 1.85$ ft while traveling at a speed $v_0 = 20$ ft/s. Determine the value of k such that the minimum distance between the disk and O is $d_{min} = d_{max}/2$.

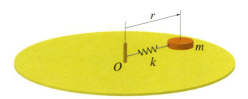

Figure P3.70

Problems 3.71 and 3.72

A pendulum with cord length $L = 6$ ft and a bob weighing 3 lb is released from rest at an angle θ_i. Once the pendulum has swung to the vertical position (i.e., $\theta = 0$), its cord runs into a small fixed obstacle. In solving this problem, neglect the size of the obstacle; model the pendulum's bob as a particle and the pendulum's cord as massless and inextensible; and let gravity and the tension in the cord be the only relevant forces.

Problem 3.71 What is the maximum height reached by the pendulum, measured from its lowest point, if $\theta_i = 20°$?

Problem 3.72 If the bob is released from rest at $\theta_i = 90°$, at what angle ϕ does the cord go slack?

Problems 3.73 through 3.76

A spherical pendulum is suspended from point O and put in motion.

Problem 3.73 Derive the pendulum's equations of motion using Cartesian coordinates.

Problem 3.74 Derive the pendulum's equations of motion using cylindrical coordinates.

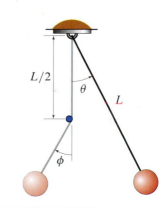

Figure P3.71 and P3.72

Problem 3.75 Derive the pendulum's equations of motion using spherical coordinates.

Problem 3.76 At the lowest point on its trajectory, the pendulum in the figure has a speed $v_0 = 2.5$ ft/s while $\phi_0 = 15°$. Letting $L = 2$ ft, plot the trajectory of the pendulum bob for 5 s.

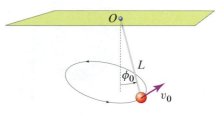

Figure P3.73–P3.76

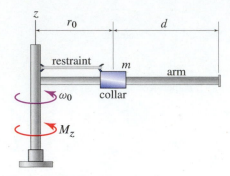

Figure P3.77–P3.80

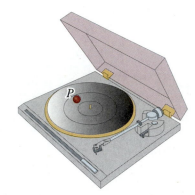

Figure P3.81

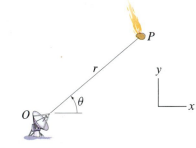

Figure P3.82

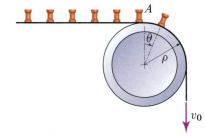

Figure P3.83

Problems 3.77 through 3.80

Consider a collar with mass m that is free to slide with no friction along a rotating arm, which has negligible mass. The system is initially rotating with an angular velocity ω_0 while the collar is kept a distance r_0 away from the z axis. At some point, the restraint keeping the collar in place is removed so that the collar is allowed to slide.

Problem 3.77 Derive the collar's equations of motion when $M_z = 0$.

Problem 3.78 If no external forces and moments are applied to the system, what are the radial speed and the total speed of the collar when it reaches the end of the arm? Use $m = 2\,\text{kg}$, $\omega_0 = 1\,\text{rad/s}$, $r_0 = 0.5\,\text{m}$, and $d = 1\,\text{m}$.

Problem 3.79 Compute the moment M_z that you would need to apply to the arm, as a function of r, to keep the arm rotating at a *constant* angular velocity ω_0. In addition, determine the radial speed as well as the total speed with which the collar would reach the end of the arm with this moment applied. *Hint:* The moment M_z applied to the arm of negligible mass is equivalent to a force M_z/r applied to the collar in the plane of motion and perpendicular to the arm.

Problem 3.80 Compute and plot the collar's trajectory from the moment of release until the collar reaches the end of the arm. Use the parameters and initial conditions given in Prob. 3.78.

Problem 3.81

The particle P is placed on the turntable, and both are initially at rest. The turntable is then turned on so that the disk starts spinning. Assuming that there is no friction between the turntable disk and the particle, what will be the motion of the particle after the disk starts spinning? Explain.
Note: Concept problems are about *explanations*, not computations.

Problem 3.82

The radar station at O is tracking the meteor P as it moves through the atmosphere. The data measured by the radar station indicates that the acceleration vector of P is almost exactly in the opposite direction of the velocity vector. Explain why this is what we should expect.
Note: Concept problems are about *explanations*, not computations.

Problem 3.83

The conveyer belt moves parts, each with mass m, at a constant speed v_0. When the parts get to A, they begin moving over a circular path of radius ρ. If the coefficient of static friction between the belt and the parts is μ_s, determine the angle θ at which the parts will start to slide on the belt. Neglect the size of the parts. After determining θ in terms of μ_s, v_0, ρ, and g, evaluate θ for $\mu_s = 0.6$, $v_0 = 3\,\text{mph}$, and $\rho = 14\,\text{in}$.

DESIGN PROBLEMS

Design Problem 3.1

The mechanism in the figure needs to be designed so as to capture parcels at A and to deliver them at B. The impact speed at B must not exceed $1.5\,\text{m/s}$. The parcels have masses ranging between 5 and $10\,\text{kg}$ incoming at speeds between 3 and $6\,\text{m/s}$. Determine the acceptable ranges of the spring constant, free length of the spring, as well as the kinetic friction coefficient between the sliding surface and the cradle to achieve the desired result.

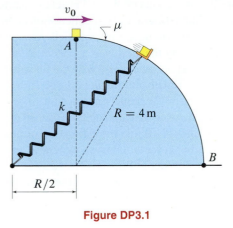

Figure DP3.1

Design Problem 3.2

Consider the cam-follower system in the figure. The cam is on a shaft that rotates with speeds up to $3000\,\text{rpm}$. The profile of the cam is described by the function $\ell(\phi) = R_0(1 + 0.25\cos^3\phi)$, where the angle ϕ is measured relative to the segment OA, which rotates with the cam, and where $R_0 = 3\,\text{cm}$. The follower has a mass of $90\,\text{g}$. Assuming that the contact between cam and follower is frictionless, choose appropriate values of the spring constant k and the spring's unstretched length L_0 such that the follower always remains in contact with the cam while minimizing the value of the contact force between the follower and the cam.

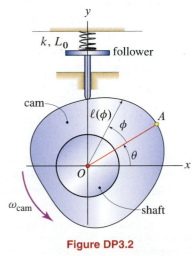

Figure DP3.2

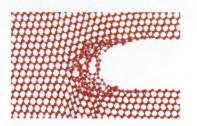

Figure 3.23

Molecular dynamics simulation of fracture in silicon.

Interesting Fact

Molecular dynamics simulations. MD simulations allow us to predict the mechanical, electrical, and chemical properties of nanomaterials, including friction properties of nanoscale materials that are in contact with one another. (This field is called *nanotribology*, and it is becoming an increasingly important research area. See J. Krim, "Friction at the Atomic Scale," *Scientific American*, **275**(4), pp. 74–80, October 1996.) Keeping in mind that a mere 1 g of Ni has approximately 10^{22} atoms (this is probably greater than the number of stars in the universe), performing MD simulations for a meaningfully large amount of material is a formidable computational task. However, due to recent advances in numerical computing as well as in computer hardware, it is now possible to carry out MD simulations of systems consisting of millions of atoms.

3.3   Systems of Particles

Engineering materials one atom at a time

Nanotechnology[*] is the development of processes and devices in which a material's behavior is engineered at length scales smaller than 100 nm (1 nm = 10^{-9} m), often by manipulating individual atoms. As with any field of science and engineering, it is desirable to model nanoelectromechanical systems to predict and understand their behavior. It turns out that the particle dynamics we study in this book plays an important role in providing some of the modeling capability needed in nanotechnology. Particle dynamics is a core component of a nanoscale modeling approach called *molecular dynamics* (MD), in which the force laws describing atomic interactions are used together with Newton's second law to study the motion of a system of atoms. Then, by combining concepts from statistical mechanics with the knowledge of the system's motion, it is possible to estimate and predict a large variety of physical properties. As an example, Fig. 3.23 depicts a snapshot of an MD simulation of fracture in silicon that allows us to estimate the fracture toughness of the material under various conditions. While the estimation of material properties is outside the scope of this book, we will now show how to carry out a simulation predicting the motion of a system of three Ni (nickel) atoms. The process followed for such a small system, when automated on a computer and applied to systems ranging from thousands to millions of atoms, is very similar to what is actually done in MD. To set up the system's equations of motion, we will use some ideas that are not needed in the study of single-particle motion but become important in the study of *systems of particles*.

Setting up a molecular dynamics simulation of three nickel atoms

To derive the equations needed to carry out an MD simulation of a system consisting of three nickel atoms, we start with the modeling of interatomic forces. As atoms get closer to one another, they repel one another *very* strongly. When they are far apart, their attraction is *very* weak. Deriving accurate relationships able to account for these characteristics typically requires the application of quantum mechanics and is beyond the scope of this book. Therefore, we will borrow the description of the interatomic force from physics. For elements such as Ni, there is a famous relation that can be found in most solid state physics books[†] stating that the force $\vec{f}_{ij}$ on atom i due to atom j can be approximated by

$$\vec{f}_{ij} = 24\epsilon\left(\frac{\sigma^6}{r_{ij}^7} - \frac{2\sigma^{12}}{r_{ij}^{13}}\right)\hat{u}_{j/i}, \qquad (3.37)$$

where r_{ij} is the distance between atoms i and j, $\hat{u}_{j/i}$ is a unit vector directed from atom i to atom j, and ϵ and σ are material-specific parameters with dimensions of energy and length, respectively. For atomic Ni, the parameter

---

[*] See, for example, G. Stix, "Little Big Science," *Scientific American*, **285**(3), September 2001, pp. 32–37. The September 2001 issue of *Scientific American* is devoted to nanotechnology.
[†] See, for example, C. Kittel, *Introduction to Solid State Physics*, 7th ed., Wiley, New York, 1995.

values are $\epsilon = 0.74\,\text{eV}$ and $\sigma = 2.22\,\text{Å}$.[†] The expression in Eq. (3.37) is the famous Lennard-Jones force law, and its component in the $\hat{u}_{j/i}$ direction has been plotted in Fig. 3.24.

Balance Principles.   We will assume that the three Ni atoms are released from rest in the positions shown in Fig. 3.25. We will ignore gravity, given that the weight forces are much smaller in magnitude than the interatomic forces. To see this, referring to Fig. 3.25, observe that the interatomic forces are on the order of 10^{-9} N. In addition, since the mass of one Ni atom is 9.749×10^{-26} kg, the weight of one Ni atom is 9.564×10^{-25} N. Therefore, the interatomic forces are 10^{15} times larger than the weight forces (the gravitational attraction among the atoms is even smaller than that between the atoms and the Earth).

To derive equations of motion that are valid at any instant, we need to consider an arbitrary position of the system such as that in Fig. 3.26. Using the FBDs in Fig. 3.26 and applying Newton's second law to each particle, we have

$$\vec{f}_{12} + \vec{f}_{13} = m_1\vec{a}_1, \tag{3.38}$$

$$\vec{f}_{21} + \vec{f}_{23} = m_2\vec{a}_2, \tag{3.39}$$

$$\vec{f}_{31} + \vec{f}_{32} = m_3\vec{a}_3. \tag{3.40}$$

Referring to Fig. 3.26, we can write Eq. (3.38) in components as

$$f_{12}\cos\theta_{12} + f_{13}\cos\theta_{13} = m_1 a_{1x}, \tag{3.41}$$

$$f_{12}\sin\theta_{12} + f_{13}\sin\theta_{13} = m_1 a_{1y}, \tag{3.42}$$

where $f_{ij} = |\vec{f}_{ij}|$. Similarly, we rewrite Eq. (3.39) as

$$-f_{12}\cos\theta_{12} + f_{23}\cos\theta_{23} = m_2 a_{2x}, \tag{3.43}$$

$$-f_{12}\sin\theta_{12} + f_{23}\sin\theta_{23} = m_2 a_{2y}, \tag{3.44}$$

and Eq. (3.40) as

$$-f_{13}\cos\theta_{13} - f_{23}\cos\theta_{23} = m_3 a_{3x}, \tag{3.45}$$

$$-f_{13}\sin\theta_{13} - f_{23}\sin\theta_{23} = m_3 a_{3y}. \tag{3.46}$$

Notice that, in writing Eqs. (3.41)–(3.46), we have also enforced Newton's third law, which says that $\vec{f}_{ij} = -\vec{f}_{ji}$.

Force Laws.   Using Eq. (3.37) to express f_{12}, f_{13}, and f_{23}, we have

$$f_{12} = 24\epsilon\left(\frac{\sigma^6}{r_{12}^7} - \frac{2\sigma^{12}}{r_{12}^{13}}\right), \tag{3.47}$$

$$f_{13} = 24\epsilon\left(\frac{\sigma^6}{r_{13}^7} - \frac{2\sigma^{12}}{r_{13}^{13}}\right), \tag{3.48}$$

$$f_{23} = 24\epsilon\left(\frac{\sigma^6}{r_{23}^7} - \frac{2\sigma^{12}}{r_{23}^{13}}\right). \tag{3.49}$$

[†] The symbol eV stands for *electronvolt*, a unit of energy, where $1\,\text{eV} = 1.6021892\times10^{-19}$ J. The symbol Å stands for *angstrom*, a unit of length, where $1\,\text{Å} = 10^{-10}$ m.

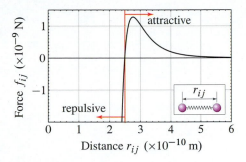

Figure 3.24
The interaction force between two nickel particles as described by the Lennard-Jones potential, where the values used for ϵ and σ are those for Ni.

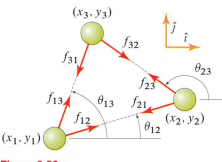

Figure 3.25
Initial positions of the three Ni atoms under consideration. Their initial velocities are zero.

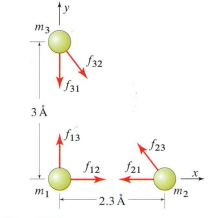

Figure 3.26
The Ni atoms in arbitrary positions. The position of m_1 is (x_1, y_1), of m_2 is (x_2, y_2), and of m_3 is (x_3, y_3).

Kinematic Equations.   Since we chose a Cartesian component system, the components of the acceleration of the three particles are given by

$$a_{1x} = \ddot{x}_1, \qquad a_{1y} = \ddot{y}_1, \tag{3.50}$$

$$a_{2x} = \ddot{x}_2, \qquad a_{2y} = \ddot{y}_2, \tag{3.51}$$

$$a_{3x} = \ddot{x}_3, \qquad a_{3y} = \ddot{y}_3. \tag{3.52}$$

If we substitute Eqs. (3.47)–(3.52) into Eqs. (3.41)–(3.46), we obtain a system of six equations in the following unknowns: $\ddot{x}_1$, $\ddot{y}_1$, $\ddot{x}_2$, $\ddot{y}_2$, $\ddot{x}_3$, $\ddot{y}_3$, θ_{12}, θ_{13}, θ_{23}, r_{12}, r_{13}, and r_{23}. We need to reduce this list to six, and we do it by recognizing that θ_{12}, θ_{13}, θ_{23}, r_{12}, r_{13}, and r_{23} can all be written in terms of the particles' positions. Referring to Fig. 3.26, the needed additional kinematic relations are

$$r_{12} = \sqrt{(x_2 - x_1)^2 + (y_2 - y_1)^2}, \tag{3.53}$$

$$r_{13} = \sqrt{(x_3 - x_1)^2 + (y_3 - y_1)^2}, \tag{3.54}$$

$$r_{23} = \sqrt{(x_3 - x_2)^2 + (y_3 - y_2)^2}, \tag{3.55}$$

and

$$\cos\theta_{12} = \frac{x_2 - x_1}{r_{12}}, \qquad \sin\theta_{12} = \frac{y_2 - y_1}{r_{12}}, \tag{3.56}$$

$$\cos\theta_{13} = \frac{x_3 - x_1}{r_{13}}, \qquad \sin\theta_{13} = \frac{y_3 - y_1}{r_{13}}, \tag{3.57}$$

$$\cos\theta_{23} = \frac{x_3 - x_2}{r_{23}}, \qquad \sin\theta_{23} = \frac{y_3 - y_2}{r_{23}}. \tag{3.58}$$

Equations (3.41)–(3.58) are the system's governing equations. The equations of motion are obtained by first substituting Eqs. (3.53)–(3.55) into Eqs. (3.47)–(3.49) and (3.56)–(3.58) and then substituting the result into Eqs. (3.41)–(3.46). For particle 1, we obtain the following two equations:

$$\frac{m_1}{24\epsilon\sigma^6} \ddot{x}_1 = \frac{(x_2 - x_1)\left\{\left[(x_2 - x_1)^2 + (y_2 - y_1)^2\right]^3 - 2\sigma^6\right\}}{\left[(x_2 - x_1)^2 + (y_2 - y_1)^2\right]^7}$$
$$+ \frac{(x_3 - x_1)\left\{\left[(x_3 - x_1)^2 + (y_3 - y_1)^2\right]^3 - 2\sigma^6\right\}}{\left[(x_3 - x_1)^2 + (y_3 - y_1)^2\right]^7}, \tag{3.59}$$

$$\frac{m_1}{24\epsilon\sigma^6} \ddot{y}_1 = \frac{(y_2 - y_1)\left\{\left[(x_2 - x_1)^2 + (y_2 - y_1)^2\right]^3 - 2\sigma^6\right\}}{\left[(x_2 - x_1)^2 + (y_2 - y_1)^2\right]^7}$$
$$+ \frac{(y_3 - y_1)\left\{\left[(x_3 - x_1)^2 + (y_3 - y_1)^2\right]^3 - 2\sigma^6\right\}}{\left[(x_3 - x_1)^2 + (y_3 - y_1)^2\right]^7}. \tag{3.60}$$

The equations of motion for particles 2 and 3 look similar.

The equations we have derived are nonlinear differential equations that can be solved numerically. Letting $m_1 = m_2 = m_3 = 9.749 \times 10^{-26}$ kg, $\epsilon = 1.1856 \times 10^{-19}$ J, and $\sigma = 2.22 \times 10^{-10}$ m, enforcing the initial conditions specified in Fig. 3.25, and solving the equations numerically, we obtain the trajectories shown in Fig. 3.27. These trajectories were obtained by integrating the equations of motion for 2×10^{-12} s (2 ps).

We conclude the analysis of motion of our three Ni atom system with two important remarks:

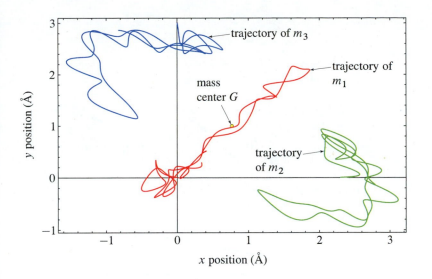

Figure 3.27. Trajectories of the three Ni atoms for 2×10^{-12} s.

1. The solution just completed showed us that the problem-solving approach developed in Section 3.1, in which we apply the balance principles and write the force laws and the kinematics equations, can be applied in the solution of problems involving more than one particle. However, we did have to introduce one new ingredient in our solution process — in addition to Newton's second law we had to apply Newton's third law. As we will demonstrate in the examples, we do not normally write special equations to enforce Newton's third law. Rather, we typically account for Newton's third law directly on the FBDs of the various bodies in the system.

2. In Fig. 3.27, besides the trajectories of the three atoms, we have also plotted the trajectory of the system's center of mass, which we have calculated via the relations

$$x_G = \frac{m_1 x_1 + m_2 x_2 + m_3 x_3}{m_1 + m_2 + m_3}, \tag{3.61}$$

$$y_G = \frac{m_1 y_1 + m_2 y_2 + m_3 y_3}{m_1 + m_2 + m_3}. \tag{3.62}$$

Our solution reveals that the trajectory of the center of mass G consists of the single yellow dot near the center of Fig. 3.27. This tells us that the mass center does not move during the simulation even though atoms 1, 2, and 3 keep changing their positions. To explain this behavior, we will now consider more carefully the applications of Newton's laws of motion to particle systems and will develop some new concepts.

Newton's second law for systems of particles

Referring to Fig. 3.28, consider a collection of n particles. The quantities m_i and $\vec{r}_i$ denote the mass and the position of particle i, respectively. A force acting on a particle of the system is called *internal* if it arises from the interaction between that particle and other particles within the system; otherwise

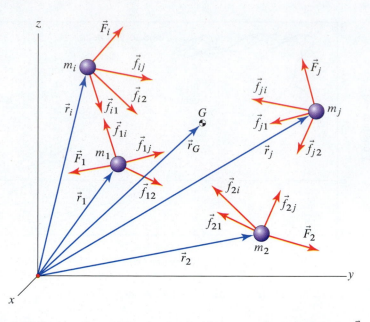

Figure 3.28. System of n particles showing position vectors $\vec{r}_i$, external forces $\vec{F}_i$, and internal forces $\vec{f}_{ij}$.

the force is called *external*. The force denoted by $\vec{F}_i$ is the *sum* of all external forces acting on particle i (e.g., gravity, air drag, etc.), and it would exist even if the rest of the particles were removed from the system. With $\vec{f}_{ij}$ we denote the force on particle i due to particle j. Therefore, the forces $\vec{f}_{i1}, \vec{f}_{i2}, \ldots, \vec{f}_{in}$ are the *internal* forces acting on particle i.

Applying Newton's second law to each particle within the system, we have

$$\vec{F}_1 + \sum_{\substack{j=1 \\ j \neq 1}}^{n} \vec{f}_{1j} = m_1 \vec{a}_1, \tag{3.63}$$

$$\vec{F}_2 + \sum_{\substack{j=1 \\ j \neq 2}}^{n} \vec{f}_{2j} = m_2 \vec{a}_2, \tag{3.64}$$

$$\vdots$$

$$\vec{F}_n + \sum_{\substack{j=1 \\ j \neq n}}^{n} \vec{f}_{nj} = m_n \vec{a}_n, \tag{3.65}$$

where we have noted that in equation i, we cannot have $j = i$ because we do not allow a particle to interact with itself.

To understand the effect of Newton's third law on the system as a whole, we sum Eqs. (3.63)–(3.65) and obtain

$$\sum_{i=1}^{n} \vec{F}_i + \sum_{i=1}^{n} \sum_{\substack{j=1 \\ j \neq i}}^{n} \vec{f}_{ij} = \sum_{i=1}^{n} m_i \vec{a}_i. \tag{3.66}$$

The first term in Eq. (3.66), $\sum_{i=1}^{n} \vec{F}_i$, is the sum of all external forces acting on the system of particles, which we will denote by $\vec{F}$, i.e.,

$$\vec{F} = \sum_{i=1}^{n} \vec{F}_i. \tag{3.67}$$

The second term is the sum of all internal forces. Referring to Fig. 3.28, we observe that for every $\vec{f}_{ij}$ in the system there must be an $\vec{f}_{ji}$. By Newton's third law, these forces are equal and opposite and they will *always* cancel one another. For example, if $n = 3$, the second term in Eq. (3.66) yields

$$\sum_{i=1}^{3} \sum_{\substack{j=1 \\ j \neq i}}^{3} \vec{f}_{ij} = \vec{f}_{12} + \vec{f}_{13} + \vec{f}_{21} + \vec{f}_{23} + \vec{f}_{31} + \vec{f}_{32} = \vec{0}, \tag{3.68}$$

given that $\vec{f}_{12} = -\vec{f}_{21}$, $\vec{f}_{13} = -\vec{f}_{31}$, and $\vec{f}_{23} = -\vec{f}_{32}$. Therefore, by Newton's third law, the second sum on the left-hand side of Eq. (3.66) *always* vanishes, i.e.,

$$\sum_{i=1}^{n} \sum_{\substack{j=1 \\ j \neq i}}^{n} \vec{f}_{ij} = \vec{0}. \tag{3.69}$$

Finally, the last term in Eq. (3.66), $\sum_{i=1}^{n} m_i \vec{a}_i$, can be written as

$$\sum_{i=1}^{n} m_i \vec{a}_i = \sum_{i=1}^{n} m_i \frac{d^2 \vec{r}_i}{dt^2} = \frac{d^2}{dt^2} \sum_{i=1}^{n} m_i \vec{r}_i, \tag{3.70}$$

where, in moving d^2/dt^2 outside the summation, we used the fact that the mass of each particle is constant. We now recall that the term $\sum_{i=1}^{n} m_i \vec{r}_i$ defines the position of the system's center of mass through the relation

$$m \vec{r}_G = \sum_{i=1}^{n} m_i \vec{r}_i, \tag{3.71}$$

where $m = \sum_{i=1}^{n} m_i$ is the total mass of the system and $\vec{r}_G$ is the position of the mass center G (see Fig. 3.28). Using Eq. (3.71), Eq. (3.70) becomes

$$\sum_{i=1}^{n} m_i \vec{a}_i = \frac{d^2}{dt^2} (m \vec{r}_G) = m \frac{d^2 \vec{r}_G}{dt^2} = m \vec{a}_G, \tag{3.72}$$

where we have used the fact that, because each mass m_i is constant, the total mass m must also be constant. Substituting Eqs. (3.67), (3.69), and (3.72) into Eq. (3.66), we obtain

$$\vec{F} = \sum_{i=1}^{n} m_i \vec{a}_i = m \vec{a}_G. \tag{3.73}$$

Equation (3.73) says that the mass center of a system of particles moves as if it were a particle of mass m subjected to *only* the total external force $\vec{F}$. Consequently,

- If $\vec{F} = \vec{0}$, then the mass center moves with constant velocity. Therefore, if $\vec{F} = \vec{0}$ *and* the mass center is initially at rest, the mass center will remain at rest.

- If $\vec{F} = \vec{0}$, then $m\vec{a}_G = \vec{0}$ and its time integral must be constant, i.e.,

$$\int m\vec{a}_G \, dt = m\vec{v}_G = \text{constant}, \tag{3.74}$$

which is a statement of the *conservation of linear momentum* for a system of particles. We will discuss this conservation law in Section 5.1.

These considerations explain the behavior of the mass center of the three Ni atom system studied earlier in this section (see Fig. 3.27). Since the system was released from rest, its mass center was also initially at rest. This, along with the fact that the particle system in question was not subjected to any external force tell us that the system's center of mass had to remain at rest.

End of Section Summary

In practice, applying Newton's second law to a system of particles is essentially identical to applying it to a single particle — we write $(\vec{F}_i)_{\text{tot}} = m_i\vec{a}_i$, $i = 1, \ldots, n$, for each of the n particles in the system, where $(\vec{F}_i)_{\text{tot}}$ is the total force acting on particle i, consisting of both the external and internal forces acting on the particle. In addition to Newton's second law, we need to enforce Newton's third law, which is crucial to correctly account for the effect of the internal forces.

Motion of the mass center.   The motion of the mass center of a system of particles is governed by the relation

Eq. (3.73), p. 237

$$\vec{F} = \sum_{i=1}^{n} m_i\vec{a}_i = m\vec{a}_G,$$

where $\vec{F}$ is the total external force on the system, m is the total mass of the system, and $\vec{a}_G$ is the acceleration of the center of mass of the system. The above equation has some important consequences:

- If $\vec{F} = \vec{0}$, then the mass center moves with constant velocity; and if the mass center is initially at rest, it will remain at rest.

- If $\vec{F} = \vec{0}$, then we can conclude that $m\vec{a}_G$ must be zero and its time integral must be a constant.

EXAMPLE 3.9   *Motion with Conservation of Linear Momentum*

The canoeist P has paddled her canoe D to the dock, as shown in Fig. 1. After the front end of her canoe reaches the end of the dock, she decides to walk the distance L from the back of the canoe to the front and exit onto the dock. Assuming that the canoe can slide in the water with negligible resistance, determine the distance between the person and the dock when she reaches the front end of the canoe to determine whether or not she will be able to exit onto the dock without getting wet. Assume that the person weighs 175 lb, the canoe weighs 33 lb, and $L = 10$ ft.

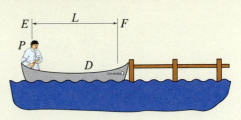

Figure 1
A canoeist who has just reached the dock and is about to walk the length of the canoe in order to exit onto the dock.

SOLUTION

Road Map & Modeling   We will view the person and canoe as a "particle system." In Fig. 2, we have drawn the FBD of the system *as a whole*, and therefore we have indicated only the forces external to the system. Our model includes the weight of both the person and the canoe as well as the buoyancy force F_B. We have neglected any horizontal resistance offered by the water and/or air. Because the total external horizontal force is equal to zero and the system is initially at rest, the position of the person and the canoe will change so as to preserve the initial position of the mass center. To solve the problem, we will write Newton's second law for the system *as a whole* and integrate it with respect to time to determine the position of each part of the system.

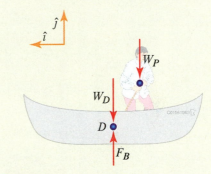

Figure 2
FBD of the canoe and canoeist. We have modeled the canoe and the canoeist as particles.

Governing Equations

Balance Principles   Referring to the FBD in Fig. 2, applying the first of Eqs. (3.73) in the x direction, we have

$$\sum F_x: \quad 0 = m_P a_{Px} + m_D a_{Dx}, \tag{1}$$

where a_{Px} and a_{Dx} are the x components of the acceleration of P and D, respectively.

Force Laws   All forces are accounted for on the FBD.

Kinematic Equations   Because our analysis is limited to the x direction, the kinematic relations we need are

$$a_{Px} = \dot{v}_{Px}, \quad v_{Px} = \dot{x}_P, \quad a_{Dx} = \dot{v}_{Dx}, \quad \text{and} \quad v_{Dx} = \dot{x}_D, \tag{2}$$

where, referring to Fig. 3, the coordinate x is measured from the end of the dock.

Computation   Integrating Eq. (1) with respect to time yields

$$m_P v_{Px} + m_D v_{Dx} = C_1, \tag{3}$$

where C_1 is a constant of integration. This equation says that *for any arbitrary position of P and D*, the sum of the product of their masses and the x component of their velocities is constant. Hence, we can evaluate C_1 by recalling that both the canoe D and the canoeist P have zero velocity after they reach the dock and before the canoeist starts to walk the length of the canoe, that is,

$$m_P \cdot 0 + m_D \cdot 0 = C_1 \quad \Rightarrow \quad C_1 = 0. \tag{4}$$

Next, substituting $C_1 = 0$ into Eq. (3) and integrating with respect to time, we have

$$m_P x_P + m_D x_D = C_2, \tag{5}$$

> ### Helpful Information
>
> **FBDs and external forces.** The contact force between the canoeist and the canoe is not shown in Fig. 2 because this force is *internal* to the system. It is very important to remember that the *only* forces appearing on a system's FBD are the *external* forces.

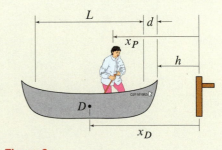

Figure 3
Definitions of distances and coordinate directions for the problem's kinematics.

where C_2 is a second constant of integration. Referring to Fig. 3, to find C_2 we evaluate Eq. (5) when $h = 0$, i.e., when the canoe is touching the dock and the canoeist is at the left end of the canoe. This yields

$$m_P(d + L) + m_D(d + L/2) = C_2, \tag{6}$$

where d is the constant distance from the end of the canoe to the point in the canoe at which the canoeist would step out. Substituting the expression for C_2 given by Eq. (6) into Eq. (5), we have that the positions of the canoeist and the canoe are related as follows:

$$m_P x_P + m_D x_D = m_P(d + L) + m_D(d + L/2). \tag{7}$$

We can now determine where the canoe ends up when the canoeist walks to the front end. When the canoeist is at the front end of the canoe, we must have $x_{D/P} = x_D - x_P = L/2$. Therefore, $x_D = x_P + L/2$, and Eq. (7) becomes

$$m_P x_P + m_D(x_P + L/2) = m_P(d + L) + m_D(d + L/2). \tag{8}$$

Solving for x_P, we have

$$x_P = \frac{m_P(d + L) + m_D d}{m_P + m_D}. \tag{9}$$

Finally, since $h = x_P - d$ when the canoeist is at the front end of the canoe, for this situation we obtain

$$\boxed{h = \frac{m_P}{m_P + m_D} L = 8.41 \text{ ft},} \tag{10}$$

where we have used the given data to obtain the final numerical answer. The result in Eq. (10) indicates that the poor canoeist will not be able to step from the canoe onto the dock without getting wet!

Discussion & Verification   The result in Eq. (10) has the dimension of length, as it should. The distance h is smaller than the length of the canoe and is therefore within reason.

🔍 **A Closer Look**   One solution to the problem of the canoe moving away from the dock would be for someone on the dock to tie the canoe to the dock after the canoe arrives. We will explore this idea in Example 5.5.

An interesting feature of Eq. (10) is that the speed with which the person walks along the canoe from one end to the other does not appear in the solution. In addition, it would be reasonable to expect that if the canoe were tied to the dock with a rope and then the canoeist walked the length of the canoe, there would be a tension in the rope. It turns out that it is not the speed that determines the tension in the rope — it is the acceleration of the person that determines the rope's tension. We will see an interesting variation of this problem in Chapter 5 when we study momentum methods for systems of particles (see Examples 5.4 and 5.5).

EXAMPLE 3.10 *Pulley System Analysis*

The pulley system in Fig. 1 is designed to lift a heavy load at A by attaching a mass at P (or, alternatively, though not equivalently, by pulling on the cable at P with some force). Assuming that the payload A has mass m_A and the pulley housing B (which includes the pulleys) has mass m_B, determine the accelerations of A and P if a mass m_P is attached at P.

SOLUTION

Road Map & Modeling Because we must determine the acceleration of specific elements of the system, we need to isolate these elements and sketch an FBD for each (as opposed to sketching an FBD for the whole system, as was done in Example 3.9). We will model each moving element as a particle and sketch its corresponding FBD. These elements are the weights A and P as well as the pulley housing B. We assume that the pulleys are frictionless and massless. We also assume that the cables are massless and inextensible so that the tension in each cable is the same throughout the cable. Then, assuming that gravity and the tension in the cables are the only relevant forces, the FBDs for the selected elements are those in Fig. 2. Notice that in Fig. 2 only the vertical direction is explicitly given a positive direction (via the unit vector $\hat{\jmath}$) since the problem is one-dimensional.

Governing Equations

Balance Principles Using the FBDs in Fig. 2, application of Newton's second law to A, B, and P yields

$$\left(\sum F_y\right)_A: \qquad m_A g - T_H = m_A a_A, \tag{1}$$

$$\left(\sum F_y\right)_B: \quad m_B g + T_H - 4T_G = m_B a_B, \tag{2}$$

$$\left(\sum F_y\right)_P: \qquad m_P g - T_G = m_P a_P. \tag{3}$$

Force Laws The weights of A, B, and P are already accounted for in Fig. 2. As far as the cables are concerned, their inextensibility is enforced via kinematic constraints instead of force laws.

Kinematic Equations Since the cables are inextensible, from Fig. 3 we have

$$L_G = 4y_B + y_P \quad \Rightarrow \quad 4a_B + a_P = 0, \tag{4}$$

where L_G denotes the length of cable G. In addition, the inextensibility of the cable connecting A and B demands that

$$a_A = a_B. \tag{5}$$

Computation Equations (1)–(5) provide a system of five equations in the five unknowns a_A, a_B, a_P, T_G, and T_H that can be solved to obtain

$$a_A = -\frac{4m_P - (m_A + m_B)}{16m_P + (m_A + m_B)} g = a_B, \tag{6}$$

$$a_P = \frac{4[4m_P - (m_A + m_B)]}{16m_P + (m_A + m_B)} g, \tag{7}$$

$$T_G = \frac{5m_P (m_A + m_B)}{16m_P + (m_A + m_B)} g, \tag{8}$$

$$T_H = \frac{20m_A m_P}{16m_P + (m_A + m_B)} g. \tag{9}$$

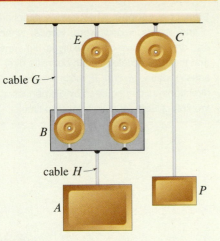

cable G

cable H

Figure 1
A pulley system designed for lifting large loads.

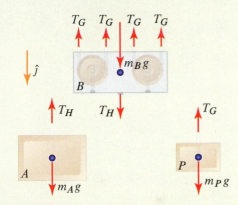

$T_G \quad T_G \quad T_G \quad T_G$

$m_B g$

$\hat{\jmath}$

$T_H \qquad T_H \qquad\qquad T_G$

$m_A g$

$m_P g$

Figure 2
FBD of mass A, mass P, and pulley housing B.

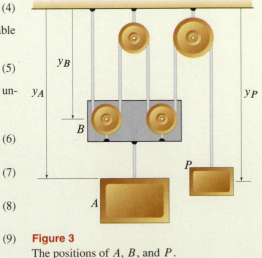

y_B

y_A

y_P

Figure 3
The positions of A, B, and P.

Discussion & Verification   The fractions in Eqs. (6) and (7) are dimensionless, so the results in these equations have the dimensions of g, i.e., acceleration, as they should. The fractions in Eqs. (8) and (9) have dimensions of mass so the results in these equations have dimensions of force, as they should. Equations (6) and (7) tell us that the accelerations of A and P are in the opposite direction, again as expected. Finally, the cable tension T_G is *always* positive so the cable never goes slack, as it should. Hence, our solution seems to be correct.

A Closer Look   Focusing on either Eq. (6) or (7), we see that the "bang for the buck" that we get with the mass at P is 4 times its mass since the numerator of either of these equations contains the term $4m_P - (m_A + m_B)$. This term is positive if $m_P > \frac{1}{4}(m_A + m_B)$; and furthermore, if $m_P > \frac{1}{4}(m_A + m_B)$, then $a_P > 0$ (otherwise, $a_P < 0$). Thus, we see that it is not m_P that must be greater than the combined mass of A and B for P to pull A up: m_P need only be greater than $\frac{1}{4}(m_A + m_B)$!

EXAMPLE 3.11   *Sliding with Friction*

A pair of stacked books with masses $m_1 = 1.5\,\text{kg}$ and $m_2 = 1\,\text{kg}$ is thrown on a table (Fig. 1). The books strike the table with essentially zero vertical speed and their common horizontal speed is $v_0 = 0.75\,\text{m/s}$. Letting $\mu_{k1} = 0.45$ be the kinetic friction between the bottom book and the table, and letting $\mu_{s2} = 0.4$ and $\mu_{k2} = 0.3$ be the coefficients of static and kinetic friction between the two books, respectively, determine the books' final positions.

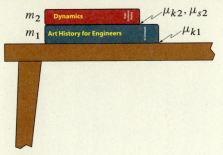

Figure 1
A pair of stacked books thrown horizontally on a rough table.

SOLUTION

Road Map & Modeling   When the books strike the table, the bottom book *must* slide on the table because it would otherwise experience an infinite deceleration in going from v_0 to zero. However, there is no similar argument telling us that the top book must slide relative to the bottom book. Hence, we begin by *assuming* that m_2 does *not* slip on m_1, and then we compute the corresponding solution. After this solution is obtained, we will check whether or not it is consistent with the assumption used to compute it. If not, we will conclude that the books slide relative to one another, and we will have to compute a new solution. We model the books as particles, and since we need to determine their individual positions, we draw an FBD of each, shown in Fig. 2, which also shows our choice of component system. We have chosen the x axis to be parallel to the trajectory of the books. In addition, the origin of the chosen system is taken to be the point at which the books first impact the table. In drawing the FBDs, we have assumed that the only relevant forces are the weights, the normal reactions, and friction.

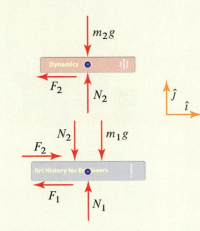

Figure 2
FBDs for the two books modeled as particles.

Governing Equations

Balance Principles   Referring to Fig. 2, Newton's second law yields

$$\left(\sum F_x\right)_2: \qquad\qquad -F_2 = m_2 a_{2x}, \tag{1}$$

$$\left(\sum F_y\right)_2: \qquad\qquad N_2 - m_2 g = m_2 a_{2y}, \tag{2}$$

$$\left(\sum F_x\right)_1: \qquad\qquad F_2 - F_1 = m_1 a_{1x}, \tag{3}$$

$$\left(\sum F_y\right)_1: \qquad N_1 - N_2 - m_1 g = m_1 a_{1y}. \tag{4}$$

Force Laws   Using the no slip assumption between books 1 and 2 and the sliding assumption between book 1 and the table, the friction laws are

$$F_1 = \mu_{k1} N_1 \quad \text{and} \quad |F_2/N_2| < \mu_{s2}. \tag{5}$$

Kinematic Equations   Letting a be the common acceleration of the books, we have

$$a_{2x} = a, \quad a_{2y} = 0, \quad a_{1x} = a, \quad \text{and} \quad a_{1y} = 0. \tag{6}$$

Computation   The first of Eqs. (5), along with the equations resulting from substituting Eqs. (6) into Eqs. (1)–(4), forms a system of five equations in the five unknowns F_1, F_2, a, N_1, and N_2. Since our first objective is to verify whether or not the no slip assumption is correct, we begin by solving for just F_2 and N_2:

$$F_2 = \mu_{k1} m_2 g = 4.414\,\text{N} \quad \text{and} \quad N_2 = m_2 g = 9.81\,\text{N}. \tag{7}$$

Discussion & Verification   Substituting the results of Eqs. (7) into the inequality in Eq. (5), we have

$$|F_2/N_2| = 0.45 \not< \mu_{s2} = 0.4, \tag{8}$$

which means that the no slip assumption is incorrect and that m_1 *does* slip over m_2.

—————————————————— m_2 **slides over** m_1 ——————————————————

We now rework the problem, assuming that the books slide relative to one another. The FBDs in Fig. 2 still apply, so that Eqs. (1)–(4) are still valid. However, the force laws and corresponding kinematic equations need to reflect the new working assumption.

Force Laws   The force laws are now

$$F_1 = \mu_{k1} N_1 \quad \text{and} \quad F_2 = \mu_{k2} N_2. \tag{9}$$

Kinematic Equations   The x components of acceleration of m_1 and m_2 are now different, so we have

$$a_{2x} = a_2, \quad a_{2y} = 0, \quad a_{1x} = a_1, \quad \text{and} \quad a_{1y} = 0. \tag{10}$$

Computation   The relations obtained by substituting Eqs. (10) into Eqs. (1)–(4), along with Eqs. (9), form a system of six equations in the six unknowns a_1, a_2, F_1, F_2, N_1, and N_2. Solving this system, we have

$$a_1 = -(g/m_1)[\mu_{k1}(m_1 + m_2) - \mu_{k2}m_2] = -5.396\,\text{m/s}^2, \tag{11}$$

$$a_2 = -\mu_{k2}g = -2.943\,\text{m/s}^2, \tag{12}$$

$$F_1 = \mu_{k1}g(m_1 + m_2) = 11.04\,\text{N}, \tag{13}$$

$$F_2 = \mu_{k2}m_2 g = 2.943\,\text{N}, \tag{14}$$

$$N_1 = g(m_1 + m_2) = 24.52\,\text{N}, \tag{15}$$

$$N_2 = gm_2 = 9.81\,\text{N}. \tag{16}$$

Now that we know a_1 and a_2, and recalling that v_0 is the initial speed of both books, we can compute how far each of them slides, using constant acceleration kinematics from Section 2.2. This yields

$$0 = v_0^2 + 2a_1 x_1 = v_0^2 - \frac{2g}{m_1}[\mu_{k1}(m_1 + m_2) - \mu_{k2}m_2]x_1, \tag{17}$$

$$0 = v_0^2 + 2a_2 x_2 = v_0^2 - 2\mu_{k2}gx_2. \tag{18}$$

Solving for x_1 and x_2 and using the given parameters, we have

$$\boxed{x_1 = \frac{m_1 v_0^2}{2g[\mu_{k1}(m_1 + m_2) - \mu_{k2}m_2]} = 0.0521\,\text{m},} \tag{19}$$

$$\boxed{x_2 = \frac{v_0^2}{2g\mu_{k2}} = 0.0956\,\text{m}.} \tag{20}$$

Referring to Fig. 3, our solution implies that book 2, as viewed by an inertial observer, i.e., someone sitting on the table, slides 9.6 cm, but *relative* to the bottom book, it slides

$$\boxed{x_{\text{top/bottom}} = x_{2/1} = x_2 - x_1 = 0.0434\,\text{m}.} \tag{21}$$

Discussion & Verification   The signs of the results in Eqs. (19)–(21) are what we expect given that both books move to the right, and book 2 stops to the right of book 1. Also, the distances slid by the books are on the order of centimeters, which seems reasonable. Because their accelerations are constant and the two books start their motion from the same position with identical speeds, the fact that $x_1 < x_2$ implies that book 1 stops first, i.e., prior to book 2 time-wise. This tells us that the friction laws used in the present solution remained valid throughout the motion of both books. Had the situation been reversed, we would have had to compute a new solution, although such a scenario is not achievable using the Coulomb friction model (see the Interesting Fact note in the margin). In conclusion, the current solution appears to be correct.

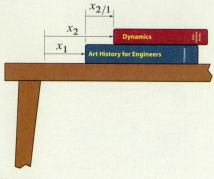

Figure 3
Relative position kinematics of the two books.

Interesting Fact

Some solutions are not possible? Interestingly, the model we use for friction (see p. 185) does not allow for some motions that, according to common experience, seem to be physically possible. For example, since the normal force is constant, the Coulomb model implies that the friction is constant while the books are sliding. Therefore, it does not allow for the top book to stop sliding relative to the bottom book before the bottom book stops sliding relative to the ground, since this would require a nonconstant acceleration of the top book. This illustrates how engineers must be aware of the limitations of the models they are using.

PROBLEMS

Problem 3.84

The driver of the truck suddenly applies the brakes, and the truck comes to a stop. During braking, either the crate slides or it does not. Considering the forces acting on the truck during braking, will the truck stop in a shorter distance (or time) if the crate slides, or will the distance (or time) be shorter if it does not? Justify your answer.
Note: Concept problems are about *explanations*, not computations.

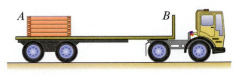

Figure P3.84

Problem 3.85

A car is being pulled to the right in the two ways shown. Neglecting the inertia of the pulleys and rope as well as any friction in the pulleys, if the car is allowed to roll freely, will the acceleration of the car in (a) be smaller, equal to, or larger than the acceleration of the car in (b)?
Note: Concept problems are about *explanations*, not computations.

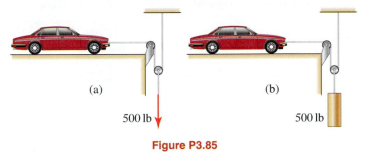

(a)   (b)

500 lb   500 lb

Figure P3.85

Problem 3.86

Particles A and B, which are connected via a massless linear spring, have been thrown up in the air and are moving under the action of the spring force and their own weight. Assuming that no other forces are affecting the motion of the particles, what will be the acceleration of their center of mass?
Note: Concept problems are about *explanations*, not computations.

Figure P3.86

Problem 3.87

A person lifts the 80 kg load A by pulling down on the rope with a constant force F as shown. Neglecting any source of friction as well as the inertia of the ropes and the pulleys, determine F if A accelerates upward at $0.5 \, \text{m/s}^2$.

Problem 3.88

The load A weighs 185 lb. Neglecting any source of friction as well as the inertia of the ropes and the pulleys, determine the acceleration of A if a person pulls down on the rope with a constant force $F = 185$ lb as shown.

Problem 3.89

A person lifts the 80 kg load A by pulling down on the rope with a constant force F as shown. Neglecting friction, the inertia of the ropes, and the rotational inertia of the

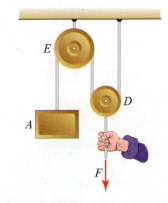

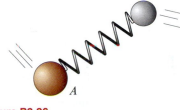

Figure P3.87–P3.89

pulleys, but accounting for the fact that pulley D has a mass $m_D = 8\,\text{kg}$, determine F if A accelerates upward at $2.5\,\text{m/s}^2$.

Problem 3.90

Two particles A and B with masses m_A and m_B, respectively, are placed at a distance r_0 from one another. Assuming that the only force acting on the masses is their mutual gravitational attraction, determine the acceleration of particle B as seen from particle A.

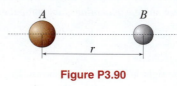

Figure P3.90

Problem 3.91

Revisit Example 3.10 and determine the expression for the acceleration of A if the load at P is replaced by a force with magnitude equal to the load's weight, i.e., $F = m_P g$.

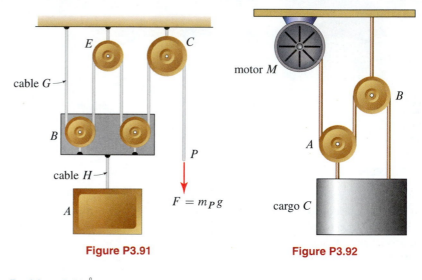

Figure P3.91   **Figure P3.92**

Problem 3.92

The motor M is at rest when someone flips a switch and it starts pulling in the rope. The acceleration of the rope is uniform and it takes 1 s to achieve a retraction rate of 4 ft/s. After 1 s the retraction rate becomes constant. Determine the tension in the cable during and after the initial 1 s interval. The cargo C weighs 130 lb, pulleys A and B each weigh 12 lb, and the weight of the ropes is negligible. Neglect friction in the pulleys and the rotational inertia of the pulleys.

Problem 3.93

As seen in Fig. P3.93(a), a window washing platform is controlled via the two pulley systems at AB and CD. The workers E and F can raise and lower the platform P by pulling on the cables H and I, respectively. The weight of each of the workers is 185 lb, and the platform P weighs 200 lb. A schematic representation of the pulley system is shown in Fig. P3.93(b). If the workers start from rest and, in 1.5 s, uniformly start pulling the cable in at 2.5 ft/s, determine the force each worker must exert on

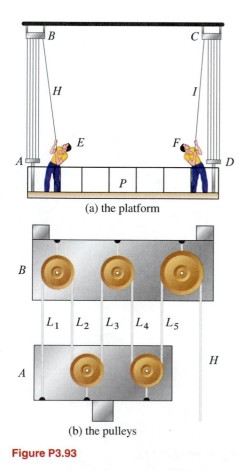

(a) the platform

(b) the pulleys

Figure P3.93

the cables H and I during that 1.5 s. Neglect the mass of the pulleys, friction in the pulleys, and the mass of the cable.

Problem 3.94

Revisit Example 3.11 and assume that the static coefficient of friction between the two books is $\mu_{s2} = 0.55$, while all other parameters stay as specified in the example. Determine the acceleration of each of the books.

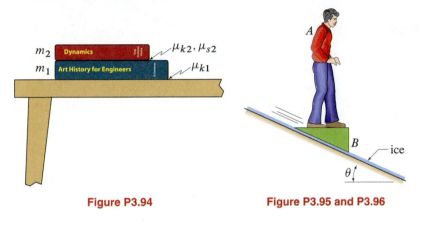

Figure P3.94   **Figure P3.95 and P3.96**

Problem 3.95

A person A is trying to keep his balance while on a sled B that is sliding down an icy incline. Letting $m_A = 78$ kg and $m_B = 25$ kg be the masses of A and B, respectively, and assuming that there is enough friction between A and B for A not to slide with respect to B, determine the value of the normal reaction force between A and B as well as the magnitude of their acceleration if $\theta = 20°$. Friction between the sled and the incline is negligible.

Problem 3.96

A person A is trying to keep his balance while on a sled B that is sliding down an icy incline. The weights of A and B are $W_A = 181$ lb and $W_B = 50$ lb, respectively. Determine the minimum coefficient of static friction μ_s between A and B required for A not to slide with respect to B if $\theta = 23°$. Friction between the sled and the incline is negligible.

Problem 3.97

A force F_0 of 400 lb is applied to block A. Letting the weights of A and B be 55 and 73 lb, respectively, and letting the static *and* kinetic friction coefficients between blocks A and B be $\mu_1 = 0.25$, and the static *and* kinetic friction coefficients between block B and the ground be $\mu_2 = 0.45$, determine the accelerations of both blocks.

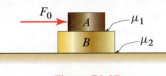

Figure P3.97

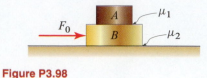

Figure P3.98

Problem 3.98

A force F_0 of 400 lb is applied to block B. Letting the weights of A and B be 55 and 73 lb, respectively, and letting the static *and* kinetic friction coefficients between blocks A and B be $\mu_1 = 0.25$, and the static *and* kinetic friction coefficients between block B and the ground be $\mu_2 = 0.45$, determine the accelerations of both blocks.

Problems 3.99 and 3.100

Spring scales work by measuring the displacement of a spring that supports both the platform of mass m_p and the object, of mass m, whose weight is being measured. Most scales read zero when no mass m has been placed on them; that is, they are calibrated so that the weight reading neglects the mass of the platform m_p. Assume that the spring is linear elastic with spring constant k.

Figure P3.99 and P3.100

Problem 3.99   If the mass m is gently placed on the spring scale (i.e., it is dropped from zero height above the scale), determine the *maximum reading* on the scale after the mass is released.

Problem 3.100   If the mass m is gently placed on the spring scale (i.e., it is dropped from zero height above the scale), determine the expression for *maximum speed* attained by the mass m as the spring compresses.

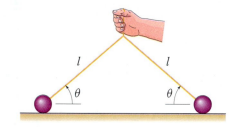

Figure P3.101

Problem 3.101

Two identical balls, each of mass m, are connected by a string of negligible mass and length $2l$. A short string is attached to the middle of the string connecting the two balls and is pulled vertically with a constant force P. If the system starts from rest at $\theta = \theta_0$ and assuming that the balls only move in the horizontal direction, determine the expression for the speed of the two balls as θ approaches $90°$. Neglect the size of the balls as well as friction between the balls and the surface on which they slide.

Problem 3.102

A simple elevator consists of a 15,000 kg car A connected to a 12,000 kg counterweight B. Suppose that a failure occurs when the car is at rest and 50 m above its buffer, causing the elevator car to fall. Model the car and the counterweight as particles and the cord as massless and inextensible; and model the action of the emergency brakes using a Coulomb friction model with kinetic friction coefficient $\mu_k = 0.5$ and a normal force equal to 35% of the car's weight. Determine the speed with which the car impacts the buffer.

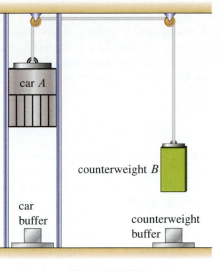

Figure P3.102

Problem 3.103

The double pendulum shown consists of two particles with masses $m_1 = 7.5\,\text{kg}$ and $m_2 = 12\,\text{kg}$ connected by two inextensible cords of length $L_1 = 1.4\,\text{m}$ and $L_2 = 2\,\text{m}$ and negligible mass. If the system is released from rest when $\theta = 10°$ and $\phi = 20°$, determine the tension in the two cords at the instant of release.

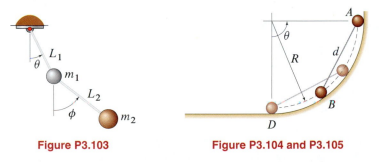

Figure P3.103   **Figure P3.104 and P3.105**

Problems 3.104 and 3.105

Two small spheres A and B, each of mass m, are attached at either end of a rod of length d. The system is released from rest in the position shown. Neglect friction, treat the spheres as particles (assume that their diameter is negligible), neglect the mass of the rod, assume the rod is rigid, and assume that $d < R$. *Hint:* The force that the rod exerts on either ball has the same direction as the rod itself.

Problem 3.104   Using the angle θ as the dependent variable, derive the equation of motion for the particle system from the moment of release until particle B reaches point D.

Problem 3.105   Determine the expression for the speed of the spheres immediately before B reaches point D.

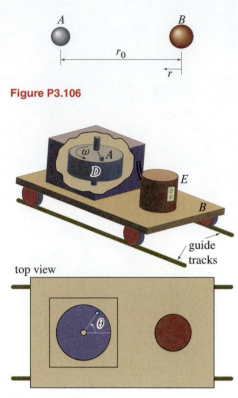

Figure P3.106

top view

Figure P3.107

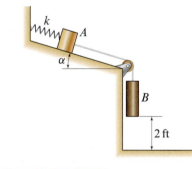

Figure P3.108–P3.111

Problem 3.106

Two particles A and B with masses m_A and m_B, respectively, are a distance r_0 apart, and both masses are initially at rest. Using Eq. (1.6) on p. 5, determine the amount of time it takes for the two masses to come into contact if $m_A = 1$ kg, $m_B = 2$ kg, and $r_0 = 1$ m. Assume that the two masses are out in space and infinitely far from any other mass.

Problem 3.107

Energy storage devices that use spinning flywheels to store energy are becoming available.[†] To store as much energy as possible, it is important that the flywheel spin as fast as possible. Unfortunately, if it spins too fast, internal stresses in the flywheel cause it to come apart catastrophically. Therefore, it is important to keep the speed at the periphery of the flywheel below about 1000 m/s. In addition, it is critical that the flywheel be as balanced as possible to avoid the tremendous vibrations that would otherwise result. With this in mind, let the flywheel D, whose diameter is 0.3 m, rotate at $\omega = 60,000$ rpm. In addition, assume that the cart B is constrained to move rectilinearly along the guide tracks. Given that the flywheel is not perfectly balanced, that the unbalanced weight A has mass m_A, and that the total mass of the flywheel D, cart B, and electronics package E is m_B, determine the constraint force between the wheels of the cart and the guide tracks as a function of θ, the masses, the diameter, and the angular speed of the flywheel. What is the *maximum* constraint force between the wheels of the cart and the guide tracks? Finally, evaluate your answers for $m_A = 1$ g (about the mass of a paper clip) and $m_B = 70$ kg (the mass of the flywheel might be about 40 kg). Assume that the unbalanced mass is at the periphery of the flywheel.

Problems 3.108 through 3.111

Two blocks A and B weighing 123 and 234 lb, respectively, are released from rest as shown. At the moment of release the spring is unstretched. In solving these problems, model A and B as particles, neglect air resistance, and assume that the cord is inextensible. *Hint:* If B hits the ground, then its maximum displacement is equal to the distance between the initial position of B and the ground.

Problem 3.108 Determine the maximum displacement and the maximum speed of block B if $\alpha = 0°$, the contact between A and the surface is frictionless, and the spring constant is $k = 30$ lb/ft.

Problem 3.109 Determine the maximum displacement and the maximum speed of block B if $\alpha = 20°$, the contact between A and the incline is frictionless, and the spring constant is $k = 30$ lb/ft.

Problem 3.110 Determine the maximum displacement and the maximum speed of block B if $\alpha = 20°$, the contact between A and the incline is frictionless, and the spring constant is $k = 300$ lb/ft.

Problem 3.111 Determine the maximum displacement and the maximum speed of block B if $\alpha = 35°$, the static and kinetic friction coefficients are $\mu_s = 0.25$ and $\mu_k = 0.2$, respectively, and the spring constant is $k = 25$ lb/ft.

[†] You will see the details in Chapter 8, but for now you can think of this as a bunch of particles moving in circles, each of which stores $\frac{1}{2}mv^2$ in kinetic energy.

Problem 3.112

A 62 kg woman A sits atop the 60 kg cart B, both of which are initially at rest. If the woman slides down the frictionless incline of length $L = 3.5$ m, determine the velocity of both the woman and the cart when she reaches the bottom of the incline. Ignore the mass of the wheels on which the cart rolls and any friction in their bearings. The angle $\theta = 26°$.

Problems 3.113 and 3.114

In the ride shown, a person A sits in a seat that is attached via a cable of length L to a freely moving trolley B of mass m_B. The total mass of the person and the seat is m_A. The trolley is constrained by the beam to move in only the horizontal direction. The system is released from rest at the angle $\theta = \theta_0$ and is allowed to swing in the vertical plane. Neglect the mass of the cable, and treat the person and the seat as a single particle.

Problem 3.113 Derive the system's equations of motion, using the position of the trolley and the angle θ as dependent variables.

Problem 3.114 Derive the system's equations of motion, using the position of the trolley and the angle θ as dependent variables, and then use a computer to solve these equations for one full period/cycle of the motion. Plot the speed of the trolley and the speed of the person vs. the angle θ for $m_A = 45$ kg, $m_B = 10$ kg, $L = 3$ m, and $\theta_0 = 70°$.

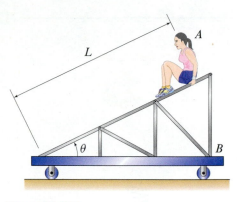

Figure P3.112

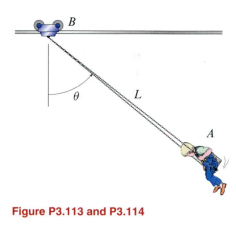

Figure P3.113 and P3.114

DESIGN PROBLEMS

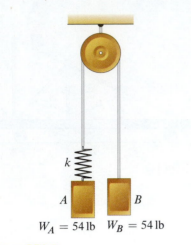

Figure DP3.3

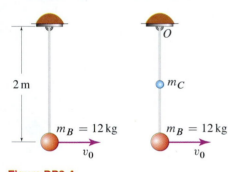

Figure DP3.4

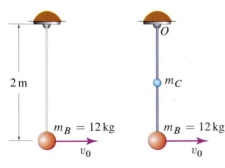

Figure DP3.5

Design Problem 3.3

In pulley problems, we have assumed that the ropes are inextensible and massless. In some engineering problem, such as in the design of fast elevators, this may not be a reasonable assumption. Here we will confront one aspect of any design process concerning the *quality* of the information provided by models of different complexity. In this problem, we will consider the effects of the rope's deformation. However, to avoid excessive complexity, we will model the rope's elasticity as being "lumped" at one end.

Let the system be released from rest and the velocity of A be controlled so that it accelerates uniformly downward to a speed of $3\,\text{m/s}$ in $1.2\,\text{s}$. Over this time interval, determine and plot, for different values of k, the position of B as a function of time as well as the tension in the rope as a function of time. Keep in mind that the inextensible rope model can be viewed as a special case of the deformable rope model with an infinite stiffness k. Consider various values of k, starting with extremely large values (to simulate infinity) and gradually decreasing the value of k to understand at what point the two models provide significantly different answers (you will need to decide what *significant* means in this context).

Design Problem 3.4

In this problem we will explore an assumption that we have made, namely, that we can neglect the mass of cables or ropes in pulley systems.

Consider the simple pendulum in the left part of the figure. Let the pendulum bob be set in motion in the position shown with a speed $v_0 = 3\,\text{m/s}$. Determine the motion of the pendulum bob and, in particular, its maximum displacement from its starting position. Next consider a double pendulum with an equally long cord but with a cord of mass m_C. In order to gain an appreciation for the effect of the mass of the cord on the pendulum's motion, let the entire mass of the cord be lumped at its midpoint (we then neglect the mass of the two cords connecting O to m_C and m_C to m_B). Next let this double pendulum be set in motion in the same way as in the previous case. Again, determine the motion of the pendulum bob m_B, and determine the value of m_C necessary to cause a 10% difference in the maximum displacement from its starting position.

Design Problem 3.5

In this problem we will explore an assumption that we have made, namely, that we can neglect the mass of cables or ropes in pulley systems.

Consider the simple pendulum in the left part of the figure. Let the pendulum bob be set in motion in the position shown with a speed $v_0 = 3\,\text{m/s}$. Determine the motion of the pendulum bob and, in particular, its maximum displacement from its starting position. Next consider a pendulum with an equally long cord but with a cord of mass m_C. In order to gain an appreciation for the effect of the mass of the cord on the pendulum's motion, let the entire mass of the cord be lumped at its midpoint (we then neglect the mass of the single rigid rod connecting O to m_C to m_B). Next let this rigid pendulum be set in motion in the same way as in the previous case. Again, determine the motion of the pendulum bob m_B, and determine the value of m_C necessary to cause a 10% difference in the maximum displacement from its starting position. *Hint:* Assume that the rigid bar can provide only forces that are parallel to the bar itself.

3.4   *Chapter Review*

In this chapter we have presented a general approach to the solution of kinetics problems involving the motion of a single particle or a system of particles. We now present a concise summary of the material covered in this chapter.

Application of Newton's second law

Applying Newton's second law.   In this chapter we developed a four-step problem-solving procedure for applying Newton's second law to mechanical systems. The central element of this procedure is the derivation of the governing equations, which originate from, at most, the

1. Balance principles.

2. Force laws.

3. Kinematic equations.

In this chapter, the balance principle was Newton's second law, which we write as

> Eq. (3.15), p. 189
>
> $$\vec{F} = m\vec{a}.$$

We applied Newton's second law in component form as

> Eq. (3.22), p. 193
>
> $$\sum F_a = ma_a, \quad \sum F_b = ma_b, \quad \text{and} \quad \sum F_c = ma_c,$$

where a, b, and c are the orthogonal directions of the chosen component system, and we usually need only two directions for planar problems. The four-step "recipe" we presented for applying Newton's second law, outlined in the margin note, is given as a guide to the *order* in which things should be done, and we will use it consistently in each example we present.

Solvability of a system of equations.   We need as many equations as we have unknowns for both algebraic and differential equations. It is important to keep in mind that for differential equations, a function and its time derivatives, say, $x(t)$ and $\dot{x}(t)$, are *not* two different unknowns since $\dot{x}(t)$ is not independent of $x(t)$.

Inertial reference frames.   An *inertial reference frame* is a frame in which Newton's laws of motion provide predictions that agree with experimental verification. For most engineering problems, a frame attached to the surface of the Earth can be considered to be inertial. Frames that are not accelerating with respect to a given inertial frame are also inertial.

Governing equations and equations of motion.   The *governing equations* for a system consist of the (1) balance principles, (2) force laws, and (3) kinematic equations. The *equations of motion* are differential equations derived from the governing equations that allow for the determination of the motion.

Helpful Information

We will structure the solution of kinetics problems using the following four steps:

1. *Road Map & Modeling:* Identify data and unknowns, identify the system, state assumptions, sketch the FBD, and identify a problem-solving strategy.

2. *Governing Equations:* Write the balance principles, force laws, and kinematic equations.

3. *Computation:* Solve the assembled system of equations.

4. *Discussion & Verification:* Study the solution and perform a "sanity check."

Degrees of freedom.   A system's *degrees of freedom* are the independent coordinates needed to completely describe a system's position. The number of degrees of freedom is equal to the number of different coordinates in a system that must be fixed in order to keep the system from moving. The required number of equations of motion is equal to the number of degrees of freedom for a system.

Friction.   We will account for friction via the Coulomb friction model. According to this model, in the absence of slip, the magnitude of the friction force F satisfies the inequality

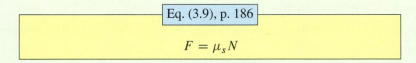

Eq. (3.8), p. 186

$$|F| \leq \mu_s |N|,$$

where μ_s is the *coefficient of static friction* and N is the force normal to the contact surface. The relation

Eq. (3.9), p. 186

$$F = \mu_s N$$

defines the case of impending slip.

  If A and B are two objects sliding with respect to one another, the magnitude of the friction force exerted by B onto A is given by

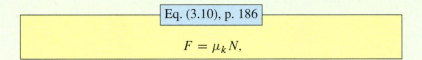

Eq. (3.10), p. 186

$$F = \mu_k N,$$

where μ_k is the *coefficient of kinetic friction* and where the direction of the friction force must be consistent with the fact that friction opposes the relative motion of A and B.

Springs.   A spring is said to be *linear elastic* if the internal force in the spring is linearly related to the amount the spring is stretched or compressed. The force F_s required to stretch or compress a linear elastic spring by an amount δ is given by

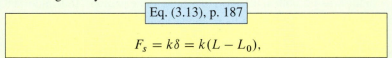

Eq. (3.13), p. 187

$$F_s = k\delta = k(L - L_0),$$

where k is the *spring constant* and where L and L_0 are the current and unstretched lengths of the spring, respectively. When $\delta > 0$, the spring is said to be stretched; and when $\delta < 0$, the spring is said to be compressed.

Systems of particles

In practice, applying Newton's second law to a system of particles is essentially identical to applying it to a single particle — we write $(\vec{F}_i)_{\text{tot}} = m_i \vec{a}_i, i = 1, \dots, n$, for each of the n particles in the system, where $(\vec{F}_i)_{\text{tot}}$ is the total force acting on particle i, thus including both the external and internal forces acting on the particle. In addition to Newton's second law, we need to enforce Newton's third law, which is crucial to account for the effect of the internal forces.

Motion of the mass center.   The motion of the mass center of a system of particles is governed by the relation

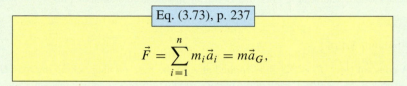

Eq. (3.73), p. 237

$$\vec{F} = \sum_{i=1}^{n} m_i \vec{a}_i = m\vec{a}_G,$$

where $\vec{F}$ is the total external force on the system, m is the total mass of the system, and $\vec{a}_G$ is the acceleration of the center of mass of the system. The above equation has some important consequences:

- If $\vec{F} = \vec{0}$, then the mass center moves with constant velocity; and if the mass center is initially at rest, it will remain at rest.

- If $\vec{F} = \vec{0}$, then we can conclude that $m\vec{a}_G$ must be zero and its time integral must be a constant.

REVIEW PROBLEMS

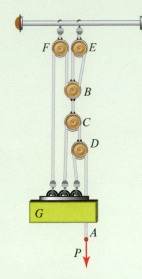

Figure P3.115 and P3.116

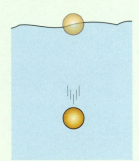

Figure P3.117 and P3.118

Problem 3.115

A constant force P is applied at A to the rope running behind the load G, which has a mass of 300 kg. Assuming that any source of friction and the inertia of the pulleys can be neglected, determine P such that G has an upward acceleration of $1\,\text{m/s}^2$.

Problem 3.116

A constant force $P = 300\,\text{lb}$ is applied at A to the rope running behind the load G, which weighs 1000 lb. If each of the pulleys weighs 7 lb, and assuming that any source of friction and the rotational inertia of the pulleys can be neglected, determine the acceleration of G and the tension in the rope connecting pulleys B and C.

Problem 3.117

A metal ball weighing 0.2 lb is dropped from rest in a fluid. If the magnitude of the resistance due to the fluid is given by $C_d v$, where $C_d = 0.5\,\text{lb·s/ft}$ is a drag coefficient and v is the ball's speed, determine the depth at which the ball will have sunk when the ball achieves a speed of 0.3 ft/s.

🖥 Problem 3.118 🖥

A metal ball weighing 0.2 lb is dropped from rest in a fluid. It is observed that after falling 1 ft, the ball has a speed of 2.25 ft/s. If the magnitude of the resistance due to the fluid is given by $C_d v$, where C_d is a drag coefficient and v is the ball's speed, determine the value of C_d.

Problem 3.119

Two particles A and B, with masses m_A and m_B, respectively, are a distance r_0 apart. Particle B is fixed in space, and A is initially at rest. Using Eq. (1.6) on p. 5 and assuming that the diameters of the masses are negligible, determine the time it takes for the two particles to come into contact if $m_A = 1\,\text{kg}$, $m_B = 2\,\text{kg}$, and $r_0 = 1\,\text{m}$. Assume that the two masses are infinitely far from any other mass.

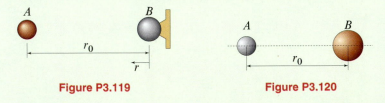

Figure P3.119   Figure P3.120

Problem 3.120

The centers of two spheres A and B, with weights $W_A = 3\,\text{lb}$ and $W_B = 7\,\text{lb}$, respectively, are a distance $r_0 = 5\,\text{ft}$ apart when they are released from rest. Using Eq. (1.6) on p. 5, determine the speed with which they collide if the diameters of spheres A and B are $d_A = 2.5\,\text{in.}$ and $d_B = 4\,\text{in.}$, respectively. Assume that the two masses are infinitely far from any other mass.

Problem 3.121

A roller coaster goes over the top A of the track shown with a speed $v = 135\,\text{km/h}$. If the radius of curvature at A is $\rho = 60\,\text{m}$, determine the minimum force that a restraint must apply to a person with a mass of 85 kg to keep the person on his or her seat.

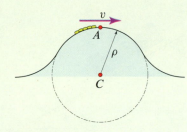

Figure P3.121

Problem 3.122

A 50,000 lb aircraft is flying along a rectilinear path at a constant altitude with a speed $v = 720\,\text{mph}$ when the pilot initiates a turn by banking the plane $20°$ to the right. Assuming that the initial rate of change of speed is negligible, determine the components of the acceleration of the aircraft right at the beginning of the turn if the pilot does not adjust the attitude of the aircraft so that the magnitude of the lift remains the same as when the plane is flying straight and the aerodynamic drag remains in the horizontal plane. In addition, determine the radius of curvature at the beginning of the turn.

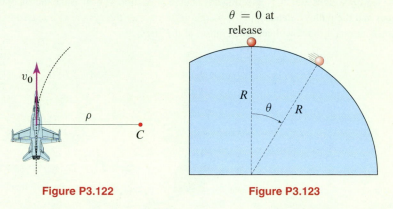

Figure P3.122   **Figure P3.123**

Problem 3.123

Referring to Example 3.6 on p. 216, let $R = 1.25\,\text{ft}$ and let the angle at which the sphere separates from the cylinder be $\theta_s = 34°$. If the sphere were placed in motion at the very top of the cylinder, determine the sphere's initial speed.

Problem 3.124

Revisit Example 3.4 and assume that the drag force acting on the ball has the form $\vec{F}_d = -\eta\vec{v}$, where $\vec{v}$ is the velocity of the ball and η is a drag coefficient. Determine the trajectory of the ball, expressing it in the form $y = y(x)$.

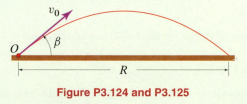

Figure P3.124 and P3.125

Problem 3.125

Revisit Example 3.4 and assume that the drag force acting on the ball has the form $\vec{F}_d = -\eta\vec{v}$, where $\vec{v}$ is the velocity of the ball and η is a drag coefficient. Determine the value of η such that a 1.61 oz ball has a range $R = 270\,\text{yd}$ when put in motion with an initial velocity of magnitude $v_0 = 186\,\text{mph}$ and initial direction $\beta = 11.2°$.

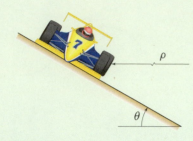

Figure P3.126

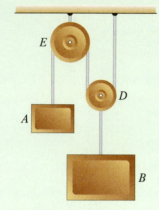

Figure P3.127 and P3.128

Problem 3.126

Referring to Example 3.5 on p. 214, show that, for $\theta = 33°$ and under the assumption that $\mu_s > 1/\tan\theta$, the no slip solution in Eqs. (15) and (16) satisfies the no slip condition $|F| \le \mu_s|N|$ for any value of the car's speed.

Problem 3.127

The load B has a mass $m_B = 250\,\text{kg}$, and the load A has a mass $m_A = 120\,\text{kg}$. Let the system be released from rest, and neglecting any source of friction as well as the inertia of the ropes and the pulleys, determine the acceleration of A and the tension in the cord to which A is attached.

Problem 3.128

The load B weighs 300 lb. Neglecting any source of friction as well as the inertia of the ropes and the pulleys, determine the weight of A if, after the system is released from rest, B moves upward with an acceleration of $0.75\,\text{ft/s}^2$.

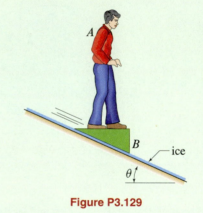

Figure P3.129

Problem 3.129

A person A is trying to keep his balance while on a sled B that is sliding down an icy incline. Let $m_A = 78\,\text{kg}$ and $m_B = 25\,\text{kg}$ be the masses of A and B, respectively. In addition, let the static and kinetic friction coefficients between A and B be $\mu_s = 0.4$ and $\mu_k = 0.35$, respectively. Determine the acceleration of A if $\theta = 23°$. Friction between the sled and the incline is negligible.

Problem 3.130

Derive the equations of motion for the double pendulum shown.

🖥 Problem 3.131 🖥

Derive the equations for the double pendulum shown. After doing so, let $L_1 = 1.4\,\text{m}$, $L_2 = 2\,\text{m}$, $m_1 = 7.5\,\text{kg}$, and $m_2 = 12\,\text{kg}$, and release the pendulum from rest with $\theta(0) = 25°$ and $\phi(0) = -37°$. Integrate the equations of motion and plot the trajectory of each of the particles for at least 5 s.

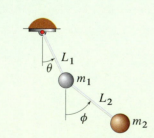

Figure P3.130 and P3.131

4

Energy Methods for Particles

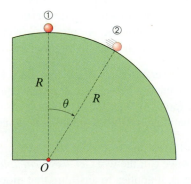

This chapter presents the concepts of *work of a force* and *kinetic energy* of a particle. These two quantities play a crucial role in a balance law called the work-energy principle, which is intimately related to Newton's second law. We will see that the work-energy principle can be derived from $\vec{F} = m\vec{a}$ by integrating $\vec{F} = m\vec{a}$ with respect to position. We sometimes refer to the work-energy principle as the "pre-integrated" form of Newton's second law to remind us of the connection between these two fundamental laws of physics. The chapter will conclude with a presentation of the concepts of *power* and *efficiency*, which are important in measuring the performance of engines and machines.

4.1 Work-Energy Principle for a Particle

Relating changes in speed to changes in position

To find the point at which a small sphere sliding down a semicylinder separates from the surface (see Fig. 4.1), in Example 3.6 on p. 216, we (1) wrote $\vec{F} = m\vec{a}$ for the sphere in polar coordinates, (2) applied the chain rule and integrated the θ component of $\vec{F} = m\vec{a}$ with respect to θ, (3) obtained a relation

Figure 4.1. A particle sliding on a frictionless semicylindrical surface. At ①, the particle has a known position and speed, and we wish to find its speed at ②.

259

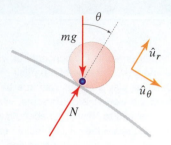

Figure 4.2
FBD of the particle at ② of Fig. 4.1.

for the angular speed as a function of θ, and (4) used this relation to solve for the value of θ at which the normal force between the sphere and the sliding surface became zero. Fortunately, there is a way to "condense" Steps 1–3 into a single step. To understand how this is possible, we need to reexamine the steps we followed in solving Example 3.6.

We begin by denoting a specific position of the particle by a number inside a circle, so that position 1 is denoted by ①, position 2 by ②, etc. Referring to Fig. 4.1, we will release the particle at ① and we will find its speed at ②. Let $\theta = \theta_1$ and $\dot{\theta} = \dot{\theta}_1$ at ①, and $\theta = \theta_2$ and $\dot{\theta} = \dot{\theta}_2$ at ②. Referring to Fig. 4.2, Newton's second law gives

$$\sum F_r: \quad N - mg\cos\theta = ma_r = -mR\dot{\theta}^2, \tag{4.1}$$

$$\sum F_\theta: \quad mg\sin\theta = ma_\theta = mR\ddot{\theta}, \tag{4.2}$$

where $a_r = \ddot{r} - r\dot{\theta}^2 = -R\dot{\theta}^2$ and $a_\theta = r\ddot{\theta} + 2\dot{r}\dot{\theta} = R\ddot{\theta}$ because $r = R = $ constant. Equation (4.1) gives us N as a function of θ once we know $\dot{\theta}$ as a function of θ. Equation (4.2) gives us $\dot{\theta}$ as a function of θ. Recalling that $\ddot{\theta} = \dot{\theta}d\dot{\theta}/d\theta$, and integrating Eq. (4.2) between ① and ②, we have

$$\int_{\dot{\theta}_1}^{\dot{\theta}_2} mR\dot{\theta}\, d\dot{\theta} = \int_{\theta_1}^{\theta_2} mg\sin\theta\, d\theta$$

$$\Rightarrow \quad \tfrac{1}{2}mR\big(\dot{\theta}_2^2 - \dot{\theta}_1^2\big) = -mg(\cos\theta_2 - \cos\theta_1). \tag{4.3}$$

Multiplying Eq. (4.3) through by R and rearranging, we obtain

$$\tfrac{1}{2}mR^2\dot{\theta}_2^2 - \tfrac{1}{2}mR^2\dot{\theta}_1^2 = mg[R(\cos\theta_1 - \cos\theta_2)]. \tag{4.4}$$

Finally, since $\dot{r} = 0$, the velocity of the sphere is $\vec{v} = R\dot{\theta}\,\hat{u}_\theta$ and its speed is $v = |R\dot{\theta}|$. Using this result, Eq. (4.4) can be given the following form:

$$\tfrac{1}{2}mv_2^2 - \tfrac{1}{2}mv_1^2 = mg[R(\cos\theta_1 - \cos\theta_2)]. \tag{4.5}$$

Equation (4.5) is a remarkable result. Let's see why.

- From introductory physics, we recognize the terms on the left-hand side of Eq. (4.5) as *kinetic energy* terms.

- The term $mg[R(\cos\theta_1 - \cos\theta_2)]$ consists of the product of the force mg with the vertical distance traveled by the particle in going from ① to ②. Hence, again recalling introductory physics, we can view these terms as representing the *work* of the weight in going from ① to ②.

- Based on these observations, we see that Eq. (4.5) indicates that the change in kinetic energy between ① and ② is equal to, or *balanced by*, the work of the force mg over the distance $R(\cos\theta_1 - \cos\theta_2)$, which is the distance separating ① from ② in the direction of gravity.

As it turns out, Eq. (4.5) is due to one of the most profound results in mechanics: a balance law we call the *work-energy principle*. This balance law is *universally valid*; that is, it applies to anything that moves according to Newton's laws of motion and not just to the sphere in this example. One of the practical implications of this realization is that we could have approached the solution of our example by simply writing Eq. (4.5) from the outset instead of starting from $\vec{F} = m\vec{a}$ and integrating; it is for this reason that we said that Steps 1–3 can be "condensed" into a single step, namely, the statement of the work-energy principle.

We will now proceed to show that the result obtained here holds for any moving particle. In doing so, we will present formal definitions of work and kinetic energy.

Work-energy principle and its relation with $\vec{F} = m\vec{a}$

Referring to Fig. 4.3, consider a particle of mass m moving along a path $\mathcal{L}_{1\text{-}2}$ between points P_1 and P_2 under the action of a force $\vec{F}$. By Newton's second law we have $\vec{F} = m\vec{a}$, where $\vec{a}$ is the acceleration of the particle. Dotting both sides of $\vec{F} = m\vec{a}$ with the particle's infinitesimal displacement $d\vec{r}$, we have

$$\vec{F} \cdot d\vec{r} = m\vec{a} \cdot d\vec{r}. \tag{4.6}$$

Recalling that $d\vec{r} = \vec{v}\,dt$ and $\vec{a} = d\vec{v}/dt$, we can rewrite Eq. (4.6) as

$$\vec{F} \cdot d\vec{r} = m\frac{d\vec{v}}{dt} \cdot \vec{v}\,dt = m\vec{v} \cdot d\vec{v}, \tag{4.7}$$

where the last expression was obtained by canceling dt. We now observe that the differential of $\frac{1}{2}m\vec{v} \cdot \vec{v}$ is equal to the last term in Eq. (4.7), that is,

$$d\left(\tfrac{1}{2}m\vec{v} \cdot \vec{v}\right) = \tfrac{1}{2}m\left(d\vec{v} \cdot \vec{v} + \vec{v} \cdot d\vec{v}\right) = \tfrac{1}{2}m\left(2\vec{v} \cdot d\vec{v}\right) = m\vec{v} \cdot d\vec{v}, \tag{4.8}$$

where $d(\ )$ indicates the differential of the quantity in parentheses. Substituting Eq. (4.8) into Eq. (4.7) and integrating along the path of the particle from the initial point P_1 to the final point P_2, we obtain

$$\int_{\mathcal{L}_{1\text{-}2}} \vec{F} \cdot d\vec{r} = \int_{\mathcal{L}_{1\text{-}2}} d\left(\tfrac{1}{2}m\vec{v} \cdot \vec{v}\right). \tag{4.9}$$

Both integrals in Eq. (4.9) are *path* or *line* integrals, which you have probably studied in your calculus courses. The integral on the right-hand side of Eq. (4.9) is the line integral of an *exact differential* (see the Helpful Information marginal note) and can be written as

$$\int_{\mathcal{L}_{1\text{-}2}} d\left(\tfrac{1}{2}m\vec{v} \cdot \vec{v}\right) = \tfrac{1}{2}m\vec{v}_2 \cdot \vec{v}_2 - \tfrac{1}{2}m\vec{v}_1 \cdot \vec{v}_1, \tag{4.10}$$

where $\vec{v}_1$ and $\vec{v}_2$ are the velocities of the particle at points P_1 and P_2, respectively. Since $\vec{v} \cdot \vec{v} = v^2$, substituting Eq. (4.10) into Eq. (4.9) yields

$$\boxed{\int_{\mathcal{L}_{1\text{-}2}} \vec{F} \cdot d\vec{r} = \tfrac{1}{2}mv_2^2 - \tfrac{1}{2}mv_1^2.} \tag{4.11}$$

We now introduce the following definitions:

$$U_{1\text{-}2} = \int_{\mathcal{L}_{1\text{-}2}} \vec{F} \cdot d\vec{r} = \text{the } \textit{work} \text{ done by } \vec{F} \text{ on the particle} \tag{4.12}$$
in moving from P_1 to P_2 along the path $\mathcal{L}_{1\text{-}2}$,

$$T = \tfrac{1}{2}mv^2 = \text{the } \textit{kinetic energy} \text{ of the particle.} \tag{4.13}$$

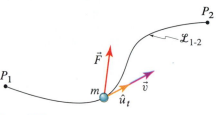

Figure 4.3
Particle of mass m moving along the path $\mathcal{L}_{1\text{-}2}$ while being subjected to the force $\vec{F}$.

Ⓝ Helpful Information

Line integral of an exact differential. All introductory calculus textbooks discuss line integrals (see, e.g., G. B. Thomas, Jr., and R. L. Finney, *Calculus and Analytic Geometry*, 9th ed., Addison-Wesley, Boston, 1996). A theorem pertaining to line integrals states that if $\varphi = \varphi(\vec{r})$, then

$$\int_{\mathcal{L}_{1\text{-}2}} d\varphi = \varphi(\vec{r}_2) - \varphi(\vec{r}_1),$$

where $\vec{r}_1$ is the initial point of $\mathcal{L}_{1\text{-}2}$ and $\vec{r}_2$ is the endpoint of $\mathcal{L}_{1\text{-}2}$. The only restrictions on this result are that $\varphi(\vec{r})$ must be continuous and single-valued in the region containing $\mathcal{L}_{1\text{-}2}$ and that $\mathcal{L}_{1\text{-}2}$ must be smooth. It is this theorem that we apply in Eq. (4.10).

By using these definitions, it can be seen why we refer to Eq. (4.12) as the *work-energy principle*. In words, we interpret the work-energy principle as saying that *the change in kinetic energy of a particle is equal to the work done on that particle*. Observe that both work and kinetic energy are *scalar* quantities. In addition, observe that, by definition, the kinetic energy is *never* negative. Using the definitions in Eqs. (4.12) and (4.13), the work-energy principle can be given the form

$$T_1 + U_{1\text{-}2} = T_2. \tag{4.14}$$

Units for work and kinetic energy

The dimension of both work and kinetic energy are (force) × (length) or, equivalently, (mass) × (length)^2/(time)^2. Table 4.1 gives the units of energy we will use in both the U.S. Customary system as well as the SI system. The choice

Table 4.1. Units of work, energy, and moment.

Quantity	U.S. Customary	SI
energy or work	ft·lb	J (joule)
moment	ft·lb	N·m

for the U.S. Customary system is somewhat arbitrary since ft·lb is the same as lb·ft. We have also included the units of a moment in Table 4.1. From a dimensional viewpoint, energy, work, and moment are equivalent. This does not pose a problem in the SI system since, when referring to energy, the unit N·m is called a *joule* and is written with the symbol J. By contrast, in the U.S. Customary system the unit ft·lb is used for moment, work, and energy.

Work of a force

Having derived the work-energy principle, it is now important that we develop a physical intuition for what it means for a force to do work. Referring to Fig. 4.4, observe that the work of a force $\vec{F}$ can be written in the following ways

$$U_{1\text{-}2} = \int_{\mathscr{L}_{1\text{-}2}} \vec{F} \cdot d\vec{r} = \int_{t_1}^{t_2} \vec{F} \cdot \vec{v}\, dt = \int_{t_1}^{t_2} \vec{F} \cdot v\, \hat{u}_t\, dt \tag{4.15}$$

$$= \int_{t_1}^{t_2} \vec{F} \cdot \frac{ds}{dt}\, \hat{u}_t\, dt = \int_{s_1}^{s_2} \vec{F} \cdot \hat{u}_t\, ds, \tag{4.16}$$

where t_1 and t_2 are the times at which the particle is at P_1 and P_2, respectively, and we have used $d\vec{r} = \vec{v}\, dt$. In Eqs. (4.15) and (4.16) we have also used the arc length s along with its companion normal-tangential component system. Specifically, s_1 and s_2 are the values of the arc length at P_1 and P_2, respectively; $\hat{u}_t$ is the tangent unit vector (pointing in the direction of motion); and $\vec{v} = v\, \hat{u}_t = (ds/dt)\, \hat{u}_t$.

The last expression in Eq. (4.16) tells us that the work done by the force $\vec{F}$ in moving along the path $\mathscr{L}_{1\text{-}2}$ *depends only on the component of $\vec{F}$ in the direction of motion*, i.e., $\vec{F} \cdot \hat{u}_t$. Equations (4.15) and (4.16) also tell us that

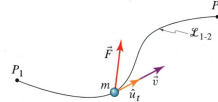

Figure 4.4
Figure 4.3 repeated. Particle of mass m moving along the path $\mathscr{L}_{1\text{-}2}$ while being subjected to the force $\vec{F}$.

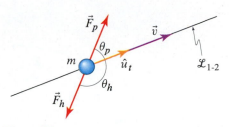

- The sign of the work is determined by the sign of $\vec{F} \cdot \hat{u}_t$: it is positive if $\vec{F}$ promotes the motion, whereas it is negative if $\vec{F}$ hinders the motion (see Fig. 4.5).

- Since the second integral in Eq. (4.15) is with respect to time, $\vec{F} \cdot \vec{v}$ can be interpreted as the *rate of work* done by the force $\vec{F}$. We will explore this idea in Section 4.4 where we study power and efficiency.

- Forces of constraint, such as normal forces, *never* contribute to $U_{1\text{-}2}$ because they are always perpendicular to the path.*

Figure 4.5
The angle θ_p between the force $\vec{F}_p$ and the velocity vector $\vec{v}$ is acute. Hence $\cos \theta_p$ is positive and the work of $\vec{F}_p$ is positive. By contrast, the angle θ_h between the force $\vec{F}_h$ and the velocity vector $\vec{v}$ is obtuse. Hence, $\cos \theta_h$ is negative and the work of $\vec{F}_h$ is negative.

Work of a constant force

Here we consider a simple two-dimensional example in which we compute the work done by a constant force. The generalization to three dimensions is not difficult. We will consider more complex examples in Section 4.2 where we will compute the work of forces such as spring forces and gravitation as given by Newton's universal law of gravitation.

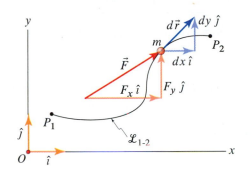

Figure 4.6. A constant force $\vec{F}$ acting on particle.

Referring to Fig. 4.6, we can express a force $\vec{F}$ in two dimensions in Cartesian components as $\vec{F} = F_x \hat{\imath} + F_y \hat{\jmath}$. We will assume that $\vec{F}$ is *constant*. Letting $d\vec{r}$ denote the infinitesimal displacement of the point of application of $\vec{F}$, in Cartesian components we can write $d\vec{r}$ as $d\vec{r} = dx\,\hat{\imath} + dy\,\hat{\jmath}$. We then have the following expression for the work done by $\vec{F}$:

$$
\begin{aligned}
U_{1\text{-}2} &= \int_{\mathscr{L}_{1\text{-}2}} \vec{F} \cdot d\vec{r} = \int_{\mathscr{L}_{1\text{-}2}} (F_x \hat{\imath} + F_y \hat{\jmath}) \cdot (dx\,\hat{\imath} + dy\,\hat{\jmath}) \\
&= \int_{x_1}^{x_2} F_x\,dx + \int_{y_1}^{y_2} F_y\,dy = F_x \int_{x_1}^{x_2} dx + F_y \int_{y_1}^{y_2} dy \\
&= F_x(x_2 - x_1) + F_y(y_2 - y_1) = \vec{F} \cdot (\vec{r}_2 - \vec{r}_1). \tag{4.17}
\end{aligned}
$$

Equation (4.17) tells us that the work of a constant force only depends on the endpoints of the path over which the force acts and not the path itself.

> **Common Pitfall**
>
> **Limits and variables of integration.** In Eq. (4.17) we start with a line integral over $\mathscr{L}_{1\text{-}2}$ and end up with simple integrals in dx and dy. Changes of variables of integration as shown in Eq. (4.17) are very common in this section, and it is crucial that the variable of integration in an integral always be matched by corresponding limits of integration. For example, in Eqs. (4.15) and (4.16) the limits of integration are t_1 and t_2 when the integral is in dt, whereas the limits of integration are s_1 and s_2 when the integral is in ds.

---

* If the constraint is *moving* and it has a component of velocity normal to the constraint surface, then Eq. (4.15) tells us that the constraint force *will* do work. We will not consider such cases in this book, as this is a topic for advanced dynamics courses.

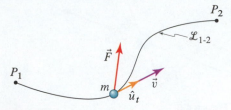

Figure 4.7

Figure 4.3 repeated. Particle of mass m moving along the path $\mathscr{L}_{1\text{-}2}$ while being subjected to the force $\vec{F}$.

Helpful Information

Other forms of the work of a force. By recalling that position and velocity are related by the expression $d\vec{r} = \vec{v}\,dt$ and that, in normal-tangential components, the velocity can be expressed as $\vec{v} = v\,\hat{u}_t$, we can obtain the following two useful forms of the work of a force:

$$U_{1\text{-}2} = \int_{t_1}^{t_2} \vec{F} \cdot \vec{v}\,dt = \int_{s_1}^{s_2} \vec{F} \cdot \hat{u}_t\,ds,$$

where t is time, s is the arc length along the path, and $\hat{u}_t$ is the unit vector tangent to the path and pointing in the direction of motion.

End of Section Summary

The work-energy principle is expressed by the following equation:

Eq. (4.14), p. 262

$$T_1 + U_{1\text{-}2} = T_2,$$

where $U_{1\text{-}2}$ is the *work* and is defined as (see Fig. 4.7)

Eq. (4.12), p. 261

$$U_{1\text{-}2} = \int_{\mathscr{L}_{1\text{-}2}} \vec{F} \cdot d\vec{r}$$

and T is the *kinetic energy* and is defined as

Eq. (4.13), p. 261

$$T = \tfrac{1}{2}mv^2.$$

Work and kinetic energy have the following basic properties:

- Work and kinetic energy are *scalar* quantities.

- The work depends on only the component of $\vec{F}$ in the direction of motion, that is, on $\vec{F} \cdot \hat{u}_t$, and its sign is determined by the sign of $\vec{F} \cdot \hat{u}_t$.

- The kinetic energy is *never* negative.

EXAMPLE 4.1   *Relating Kinetic Energy to Average Force*

An F/A-18 Hornet (see Fig. 1) takes off from an aircraft carrier using two separate propulsion systems: its two jet engines and a steam-powered catapult. During launch, a fully loaded Hornet weighing 50,000 lb goes from 0 (relative to the carrier) to 165 mph (relative to surface of the Earth) in a distance of 300 ft (relative to the carrier) while each of its two engines is at full power, generating about 22,000 lb of thrust, and the catapult is engaged. For a *stationary* aircraft carrier, neglecting aerodynamic forces, determine:

(a) The total work done on the aircraft during launch.

(b) The work done by the catapult on the aircraft during launch.

(c) The average force exerted by the catapult on the aircraft.

Figure 1
An F/A-18 Hornet taking off from an aircraft carrier.

SOLUTION

Road Map & Modeling   Referring to the FBD in Fig. 2, we will model the airplane as a particle under the action of gravity, the engines' thrust F_T, the force of the catapult F_C, and the normal reaction between the aircraft and the deck. Based on this model, the work-energy principle tells us that the work done on the airplane is equal to the airplane's change in kinetic energy. We can calculate the kinetic energy because we are given the airplane's weight and change in speed. Since the engines' thrust is constant and we know the takeoff distance, we can compute the work done by the engines via a direct application of the definition of work. Subtracting the work of the engines from the total work will allow us to find the work of the catapult.

Governing Equations

Balance Principles   The work-energy principle gives

$$T_1 + U_{1\text{-}2} = T_2, \qquad (1)$$

where ① is at the start of the launch, ② is right before takeoff, $U_{1\text{-}2}$ is the total work done on the aircraft between ① and ②, and T_1 and T_2 are the aircraft's kinetic energy at ① and ②, respectively, which are given by

$$T_1 = \tfrac{1}{2}mv_1^2 \quad \text{and} \quad T_2 = \tfrac{1}{2}mv_2^2, \qquad (2)$$

where v_1 and v_2 are the speed of the aircraft at ① and ②, respectively.

Force Laws   Both gravity and the engines' thrust are modeled as constant forces. Because we are only interested in the average value of the catapult force, we will model F_C as a constant. Since the work is computed using forces and force laws, in work-energy problems, we will always determine expressions for the work in the Force Laws portion of our solution procedure. Hence, the expressions for the work done by the engines and catapult are

$$(U_{1\text{-}2})_{\text{engines}} = \int_{\mathcal{L}_{1\text{-}2}} \vec{F}_T \cdot d\vec{r} = \int_{0\,\text{ft}}^{300\,\text{ft}} (44{,}000\,\text{lb})\,dx = 1.32 \times 10^7\,\text{ft·lb}, \qquad (3)$$

$$(U_{1\text{-}2})_{\text{catapult}} = \int_{\mathcal{L}_{1\text{-}2}} \vec{F}_C \cdot d\vec{r} = \int_{0\,\text{ft}}^{300\,\text{ft}} F_C\,dx = (300\,\text{ft})\,F_C, \qquad (4)$$

where $\mathcal{L}_{1\text{-}2}$ is the 300 ft rectilinear stretch traveled by the plane between ① and ②. Note that the aircraft's weight and the normal reaction N do no work since the motion of the airplane is assumed to be horizontal and therefore perpendicular to these forces.

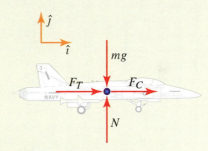

Figure 2
FBD of the aircraft as it is propelled by the jet engines and by the catapult.

Kinematic Equations   Recalling the definition of ① and ②, we have

$$v_1 = 0 \quad \text{and} \quad v_2 = 165\,\text{mph} = 242.0\,\text{ft/s}. \tag{5}$$

Computation   Combining Eqs. (2) and (5) with Eq. (1) and solving for $U_{1\text{-}2}$, we find the total work done on the aircraft during launch to be

$$\boxed{U_{1\text{-}2} = T_2 - T_1 = 4.547 \times 10^7 \,\text{ft·lb.}} \tag{6}$$

Therefore, the work done by the catapult is the total work done minus the work done by the engines, that is,

$$(U_{1\text{-}2})_{\text{catapult}} = U_{1\text{-}2} - (U_{1\text{-}2})_{\text{engines}}. \tag{7}$$

Substituting Eqs. (3) and (6) in Eq. (7), we have

$$\boxed{(U_{1\text{-}2})_{\text{catapult}} = 3.227 \times 10^7 \,\text{ft·lb.}} \tag{8}$$

Finally, solving Eq. (4) for F_C and using the result in Eq. (8), we have

$$\boxed{F_C = \frac{(U_{1\text{-}2})_{\text{catapult}}}{300\,\text{ft}} = 108,000\,\text{lb.}} \tag{9}$$

Discussion & Verification   The calculation of the total work done on the aircraft and the work done by the catapult was done via a direct application of the work-energy principle, and therefore we need only verify that proper units were used to express our result, which is indeed the case. As far as the result in Eq. (9) is concerned, since the dimensions of work are force times length, we know that the dimensions of our result are correct and proper units were used to express it.

🔍 **A Closer Look**   This problem emphasizes an important point: *no matter what forces are acting on a particle, the work done by all forces is equal to the change in kinetic energy of the body.*

The amount of work done on the F/A-18 is equivalent to pushing a 70 lb wooden crate over level concrete ($\mu_k \approx 0.6$) at a constant speed for more than 205 miles! Also, notice that the force due to the catapult is more than twice that provided by the aircraft's engines. This tells us that an airplane such as an F/A-18 could not take off unassisted from an aircraft carrier.

EXAMPLE 4.2   *Relating Speed to Position*

The block of mass $m = 20\,\text{kg}$ is connected via a pulley at D to the winch at E by an inextensible cord. The winch is able to exert a constant force $P = 130\,\text{N}$ on the cord. The friction between the block and the horizontal bar on which the block slides is negligible. Letting $h = 0.8\,\text{m}$, if the block starts from rest in the position shown, determine the speed of the block when it has moved the distance $d = 1.15\,\text{m}$ and is directly under the pulley.

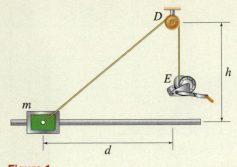

Figure 1
Winch and block system geometry.

SOLUTION

Road Map & Modeling   Since we are interested in relating speed to position, we will solve this problem by using the work-energy principle. Referring to the FBD in Fig. 2, we will model the block as a particle subject to its own weight, the normal reaction between the block and the horizontal guide, as well as the force in the cord. Notice that neither the weight nor the force N does any work, given that the block's motion is in the horizontal direction.

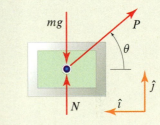

Figure 2
FBD of the block in a generic position between its initial and final positions.

Governing Equations

Balance Principles   Referring to Fig. 3 and applying the work-energy principle between ① and ②, we obtain

$$T_1 + U_{1\text{-}2} = T_2. \tag{1}$$

The kinetic energies are given by

$$T_1 = \tfrac{1}{2}mv_1^2 \quad \text{and} \quad T_2 = \tfrac{1}{2}mv_2^2, \tag{2}$$

where v_1 and v_2 are the block's speed at ① and ②, respectively.

Force Laws   The only force doing work is the force P, whose magnitude is constant and given. As mentioned in Example 4.1, when using the work-energy principle, we devote the Force Laws step of our solution procedure to the computation of the work done by the forces appearing in the FBD. Hence, we have

$$U_{1\text{-}2} = \int_{\mathscr{L}_{1\text{-}2}} \vec{F} \cdot d\vec{r} = \int_d^0 -P\cos\theta\, dx, \tag{3}$$

where we have used $\vec{F} = -P\cos\theta\,\hat{\imath} + P\sin\theta\,\hat{\jmath}$ and $d\vec{r} = dx\,\hat{\imath}$.

Kinematic Equations   Recalling that the block is released from rest and that the speed at ② is the unknown of this problem, we have

$$v_1 = 0. \tag{4}$$

In addition, observe that the integral in Eq. (3) contains the variable θ in the integrand and uses the variable x as the variable of integration. To compute this integral, we need to express either the variable θ as a function of x or the variable x as a function of θ. We will choose the latter strategy (the final result must be the same no matter which strategy we use) and will use differentiation of constraints (see Section 2.7 on p. 135) to obtain the relation we need. Noticing that $x\tan\theta = h$ and then taking the differential of this relation, we have

$$\tan\theta\, dx + x\sec^2\theta\, d\theta = \tan\theta\, dx + \frac{h}{\tan\theta}\sec^2\theta\, d\theta = 0, \tag{5}$$

where we have used the fact that $x = h/\tan\theta$ and that $dh = 0$ since h is a constant. Solving for dx, we find that

$$dx = \frac{-h}{\sin^2\theta}\, d\theta. \tag{6}$$

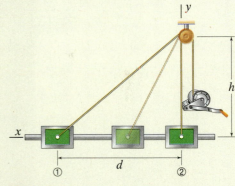

Figure 3
Winch and block system showing ① and ② and the Cartesian coordinate system used.

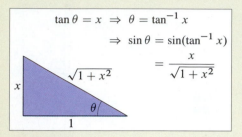

Figure 4
Demonstration of the trigonometric identity used in Eq. (8).

Computation   Substituting Eq. (6) into Eq. (3), we have

$$U_{1\text{-}2} = \int_{\tan^{-1}(h/d)}^{\pi/2} P\left(\frac{h\cos\theta}{\sin^2\theta}\right) d\theta, \tag{7}$$

where we have used the fact that $\theta = \tan^{-1}(h/d)$ when $x = d$ and $\theta = \pi/2\,\text{rad}$ when $x = 0$. Evaluating the integral in Eq. (7), we obtain

$$U_{1\text{-}2} = -P\left.\frac{h}{\sin\theta}\right|_{\tan^{-1}(h/d)}^{\pi/2} = -P\left(h - \sqrt{d^2 + h^2}\right), \tag{8}$$

where we have used the identity $\sin(\tan^{-1} x) = x/\sqrt{1 + x^2}$ (see Fig. 4).

Combining Eqs. (2), (4), and (8) with Eq. (1), we obtain

$$-P\left(h - \sqrt{d^2 + h^2}\right) = \tfrac{1}{2}mv_2^2, \tag{9}$$

which, upon solving for v_2, gives

$$v_2 = \sqrt{\frac{2P}{m}\left(\sqrt{d^2 + h^2} - h\right)} = 2.79\,\text{m/s}. \tag{10}$$

Discussion & Verification   Observing that the term P/m has dimensions of acceleration and that the term $\sqrt{d^2 + h^2} - h$ has dimensions of length, we see that v_2 has dimensions of length over time, as it should. Furthermore, the final numerical result was expressed with appropriate units. Also, observe that the argument of the square root in Eq. (10) contains the term $P(\sqrt{d^2 + h^2} - h)$, which can be interpreted as the work of a *constant* force (i.e., constant in both magnitude and direction) with magnitude P along the distance $\Delta = \sqrt{d^2 + h^2} - h$. What makes Δ interesting is that it corresponds to the amount of cord that is wound onto the winch as the block goes from ① to ②. In fact, this is the work done by the tension in the vertical branch of the cord, i.e., the portion of the cord that goes from the pulley to the winch. Hence, overall our result appears to be correct.

A Closer Look   One question we should consider is whether or not it is possible to obtain the result in Eq. (10) without resorting to the somewhat involved integration in Eqs. (7) and (8), but rather by computing $U_{1\text{-}2}$ in some more physically based way. We will explore this question in Example 4.3.

EXAMPLE 4.3   *Choosing a Convenient FBD to Relate Speed to Position*

Let's revisit Example 4.2 and again determine the speed of the block when it has moved the distance d and is directly under the pulley. However, instead of focusing on the FBD of only the block, solve the problem by using an FBD that includes the pulley at D.

SOLUTION

Road Map & Modeling   As in Example 4.2, we will apply the work-energy principle, but we will use the FBD suggested in the problem statement, which is shown in Fig. 2. Again, the only force doing work on the block is P because the point of application of the reactions R_x and R_y is fixed (hence, R_x and R_y do no work) and the forces mg and N are perpendicular to the direction of motion of the block.

Governing Equations

Balance Principles   Letting ① be the position of the block at the instant of release and ② be the position of the block when the block is directly under the pulley, applying the work-energy principle between ① and ②, we obtain

$$T_1 + U_{1\text{-}2} = T_2. \tag{1}$$

The block's kinetic energies are given by

$$T_1 = \tfrac{1}{2}mv_1^2 \quad \text{and} \quad T_2 = \tfrac{1}{2}mv_2^2, \tag{2}$$

where v_1 and v_2 are the block's speed at ① and ②, respectively.

Force Laws   This time $U_{1\text{-}2}$ takes on the following simple form:

$$U_{1\text{-}2} = P\Delta, \tag{3}$$

where Δ is the amount of cord wound up on the winch as the block moves from ① to ② and is therefore given by (see Fig. 3)

$$\Delta = L_① - L_② = \left(\sqrt{d^2 + h^2} + \ell\right) - (h + \ell) = \sqrt{d^2 + h^2} - h, \tag{4}$$

where $L_①$ and $L_②$ are the length of the cord at ① and ②, respectively.

Kinematic Equations   Recalling that m starts from rest and that v_2 is the unknown of this problem, we have

$$v_1 = 0. \tag{5}$$

Computation   Substituting Eqs. (2)–(5) into Eq. (1), we obtain

$$P\left(\sqrt{d^2 + h^2} - h\right) = \tfrac{1}{2}mv_2^2, \tag{6}$$

which, upon solving for v_2, gives

$$\boxed{v_2 = \sqrt{\frac{2P}{m}\left(\sqrt{d^2 + h^2} - h\right)}.} \tag{7}$$

Discussion & Verification   As expected, we obtained the same result found in Example 4.2.

🔎 **A Closer Look**   The approach followed in this example is *much* more straightforward than that in Example 4.2 even though the *only* difference between the two examples is how we computed $U_{1\text{-}2}$. The lesson here is that judiciously selecting the system to analyze can save significant time and effort.

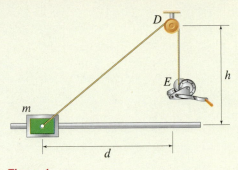
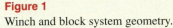

Figure 1
Winch and block system geometry.

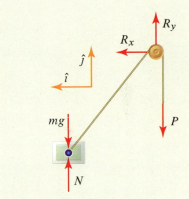

Figure 2
FBD of the block, cord, and pulley as the block moves from ① to ②.

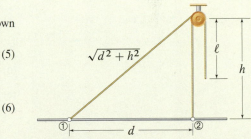
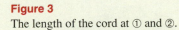

Figure 3
The length of the cord at ① and ②.

PROBLEMS

💡 Problem 4.1 💡

A rocket lifts off with an acceleration a. During liftoff, in terms of absolute values, is the work done on an astronaut by gravity larger than, equal to, or smaller than the work done by the normal reaction between the astronaut and her or his seat?

Note: Concept problems are about *explanations*, not computations.

Figure P4.1

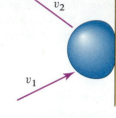

Figure P4.2

💡 Problem 4.2 💡

A soft rubber ball bounces against a wall. Assuming that the wall's deformation due to the ball's impact is negligible, does the contact force due to the wall do positive work, no work, or negative work on the ball?

Note: Concept problems are about *explanations*, not computations.

Problem 4.3 🌡

Determine the kinetic energy of the bodies listed below, when modeled as particles. Express all answers using both U.S. Customary units and SI units.

(a) A .30-06 bullet weighing 150 gr (1 lb = 7000 gr) and traveling at 3000 ft/s.

(b) A 25 kg child traveling in a car at 45 km/h.

(c) A 415,000 lb locomotive traveling at 75 mph.

(d) A 20 g metal fragment from a space vehicle traveling at 8000 km/s.

(e) A 3000 lb car traveling at 60 mph.

Figure P4.4 and P4.5

Problems 4.4 and 4.5

Consider a 3000 lb car whose speed is increased by 30 mph.

Problem 4.4 🌡 Modeling the car as a particle and assuming that the car is traveling on a rectilinear and horizontal stretch of road, determine the amount of work done on the car throughout the acceleration process if the car starts from rest.

Problem 4.5 🌡 Modeling the car as a particle and assuming that the car is traveling on a rectilinear and horizontal stretch of road, determine the amount of work done on the car throughout the acceleration process if the car has an initial speed of 45 mph.

Problem 4.6

A 75 kg skydiver is falling at a speed of 250 km/h when the parachute is deployed, allowing the skydiver to land at a speed of 4 m/s. Modeling the skydiver as a particle, determine the total work done on the skydiver from the moment of parachute deployment until landing.

Figure P4.6

Problems 4.7 and 4.8

Consider a 1500 kg car whose speed is increased by 45 km/h over a distance of 50 m while traveling up an incline with a 15% grade.

100   15

Figure P4.7 and P4.8

Problem 4.7   Modeling the car as a particle, determine the work done on the car if the car starts from rest.

Problem 4.8   Modeling the car as a particle, determine the work done on the car if the car has an initial speed of 60 km/h.

Problem 4.9

A 350 kg crate is sliding down a rough incline with a constant speed $v = 7$ m/s. Assuming that the angle of the incline is $\theta = 33°$ and that the only forces acting on the crate are gravity, friction, and the normal force between the crate and the incline, determine the work done by friction over every meter slid by the crate.

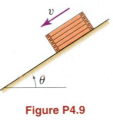

Figure P4.9

Problem 4.10

A vehicle A is stuck on the railroad tracks as a train B approaches with a speed of 120 km/h. As soon as the problem is detected, the train's emergency brakes are activated, locking the wheels and causing the train's wheels to slide relative to the tracks. If the coefficient of kinetic friction between the wheels and the track is 0.2, determine the minimum distance $d_{\min}$ at which the brakes must be applied to avoid a collision under the following circumstances:

Figure P4.10

(a) The train consists of just a 195,000 kg locomotive.

(b) The train consists of a 195,000 kg locomotive and a string of cars whose mass is 10×10^6 kg, all of which are able to apply brakes and lock their wheels.

(c) The train consists of a 195,000 kg locomotive and a string of cars whose mass is 10×10^6 kg, but only the locomotive is able to apply its brakes and lock its wheels.

Treat the train as a particle, and assume that the railroad tracks are rectilinear and horizontal.

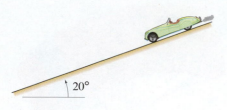

Figure P4.11 and P4.12

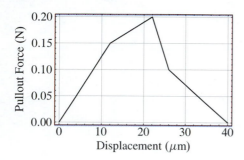

Figure P4.13

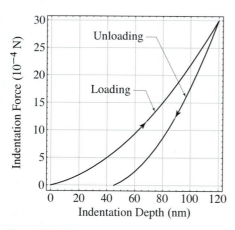

Figure P4.14

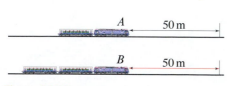

Figure P4.15

Problems 4.11 and 4.12

A classic car is driving down a 20° incline at 45 km/h when the brakes are applied. Treat the car as a particle, and neglect all forces except gravity and friction.

Problem 4.11 Determine the stopping distance if the tires slide and the coefficient of kinetic friction between the tires and the road is 0.7.

Problem 4.12 Determine the minimum stopping distance if the car is retrofitted with antilock brakes and the tires do not slide. Use 0.9 for the coefficient of static friction between the tires and the road.

Problem 4.13

Many advanced materials consist of fibers (made of glass, Kevlar, carbon, etc.) placed within a matrix (such as epoxy, a titanium alloy, etc.). For these materials it is important to assess the bond strength between fibers and matrix, and this is often done via a *pullout test*, in which the tip of a fiber is exposed, the material sampled is properly clamped, and the fiber is pulled out of the matrix. The data collected often consists of a graph like the one shown, in which the force exerted on the fiber is recorded as a function of pullout displacement. With this in mind, the interface toughness assessment process may require a measure of the *energy* expended to pull out the fiber. Use the force-displacement graph shown, which is typical for a glass-fiber reinforced epoxy, to measure the total pullout energy. *Hint:* The work of the pullout force is given by the area under the curve.

Problem 4.14

Structural components subjected to large contact forces (e.g., brake disks) are often made of high-grade steel coated with a thin film of a very hard material such as diamond. A common test to assess the mechanical properties (hardness, elastic moduli, etc.) of the coating is the *nanoindentation test*, which, roughly speaking, consists of making a controlled dent in the film using a nail-like object, called an *indentor*. During the test one measures the force applied to the indentor and the indentation depth. In the graph shown, you can see the curves interpolating the loading and unloading data for a diamond film on steel. Since the unloading curve does not go back to the origin of the plot, the film is permanently deformed during the indentation process (i.e., an actual permanent dent remains). Determine the energy lost to permanent deformation if the indentation force is given by

$$F_I = \begin{cases} \frac{25}{4}x + \frac{5}{32}x^2 & \text{during loading,} \\ 300 - \frac{145}{6}x + \frac{7}{18}x^2 & \text{during unloading,} \end{cases}$$

where F_I is expressed in μN and x (indentation depth) in nm.

Problem 4.15

Two identical locomotives A and B are coupled with 1 and 2 passenger cars, respectively. Suppose that each passenger car is identical to the others in all respects. If each train (locomotive plus cars) starts from rest, each locomotive exerts the maximum tractive effort (traction force), and assuming that gravity and the tractive effort are the only relevant forces acting on the trains, which of the two trains will have greater kinetic energy after the locomotives have moved 50 m along a horizontal and rectilinear stretch? **Note:** Concept problems are about *explanations*, not computations.

Problem 4.16

An F/A-18 Hornet takes off from an aircraft carrier, using two separate propulsion systems: its two jet engines and a steam-powered catapult. During launch, a fully loaded Hornet weighing about 50,000 lb goes from 0 mph (relative to the aircraft carrier) to 165 mph (measured relative to surface of the Earth) in a distance of 300 ft (measured relative to the aircraft carrier) while each of its two engines is at full power generating about 22,000 lb of thrust. Assuming that the aircraft carrier is traveling in the same direction as the takeoff direction and at a speed of 30 knot, determine:

(a) The total work done on the aircraft during launch.

(b) The work done by the catapult on the aircraft during launch.

(c) The force exerted by the catapult on the aircraft.

In solving this problem, model the aircraft as a particle; assume that its trajectory is horizontal and that the catapult assists the aircraft the full 300 ft needed for takeoff; and finally, let all forces be constant and neglect air resistance and friction.

Figure P4.16

Problem 4.17

Rubber bumpers are commonly used in marine applications to keep boats and ships from getting damaged by docks. Treating the boat C as a particle, neglecting its vertical motion, and neglecting the drag force between the water and the boat C, what is the maximum speed of the boat at impact with the bumper B so that the deflection of the bumper is limited to 6 in.? The weight of the boat is 70,000 lb, and the force compression profile for the rubber bumper is given by $F_B = \beta x^3$, where $\beta = 3.5 \times 10^6$ lb/ft^3 and x is the compression of the bumper.

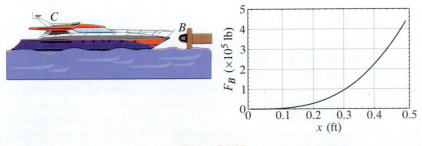

Figure P4.17

💡 Problem 4.18 💡

Two identical cars travel at a speed of 60 mph, one along a newly asphalt-paved straight and horizontal road and the other on a straight and horizontal dirt road. If brakes are applied and if the second car slips during the braking process, what difference will there be in the amount of work done to stop each car?

Note: Concept problems are about *explanations*, not computations.

Problems 4.19 and 4.20

Packages for transporting delicate items (e.g., a laptop or glass) are designed to "absorb" some of the energy of the impact in order to protect their contents. These energy absorbers can get pretty complicated (e.g., the mechanics of Styrofoam peanuts is not easy), but we can begin to understand how they work by modeling them as a linear elastic spring of constant k that is placed between the contents (an expensive vase) of mass m and the package P. Assume that the vase weighs 6 lb and that the box is dropped from a height of 5 ft. Treat the vase as a particle, and neglect all forces except for gravity and the spring force.

Figure P4.19 and P4.20

Problem 4.19 🌡   Determine the maximum displacement of the vase relative to the box and the maximum force on the vase if $k = 264$ lb/ft.

💻 Problem 4.20 💻   Plot the maximum displacement of the vase relative to the box and the maximum force on the vase as a function of the linear elastic spring constant k. What do these plots tell you about the problem you would encounter in trying to minimize the force on the vase?

💡 Problem 4.21 💡

A block A moves horizontally under the action of a force F whose line of action is parallel to the motion. If the kinetic energy of A as a function of x is that shown (x_1 and x_2 are extrema for T_A), what can you say about the *sign* of F for $x_0 < x < x_1$ and $x_1 < x < x_2$?

Note: Concept problems are about *explanations*, not computations.

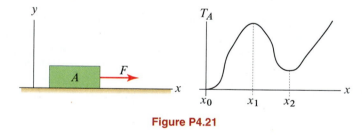

Figure P4.21

Problem 4.22 🌡

While the stiffness of an elastic cord can be quite constant (i.e., the force versus displacement curve is a straight line) over a large range of stretch, as a bungee cord is stretched, it softens; that is, the cord tends to get less stiff as it gets longer. Assuming a softening force-displacement relation of the form $k\delta - \beta\delta^3$, where δ (measured in ft) is the displacement of the cord from its unstretched length, considering a bungee cord whose unstretched length is 150 ft, and letting $k = 2.58$ lb/ft, determine the value of the constant β such that a bungee jumper weighing 170 lb and starting from rest gets to the bottom of a 400 ft tower with zero speed.

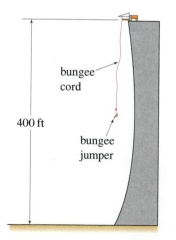

Figure P4.22

Problems 4.23 through 4.25

Car bumpers are designed to limit the extent of damage to the car in the case of low-velocity collisions. Consider a 1420 kg passenger car impacting a concrete barrier while traveling at a speed of 5.0 km/h. Model the car as a particle and consider two bumper models: (1) a simple linear spring with constant k and (2) a linear spring of constant k in parallel with a shock absorbing unit generating a nearly constant force $F_S = 2000$ N over 10 cm.

Problem 4.23 If the bumper is of type (1) and if $k = 9 \times 10^4$ N/m, find the spring compression (distance) necessary to stop the car.

Problem 4.24 If the bumper is of type (1), find the value of k necessary to stop the car when the bumper is compressed 10 cm.

Problem 4.25 If the bumper is of type (2), find the value of k necessary to stop the car when the bumper is compressed 10 cm.

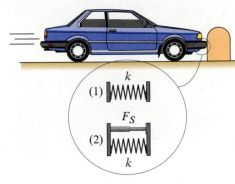

Figure P4.23–P4.25

Figure 4.8
Example of a gondola ropeway.

Figure 4.9
Example of a chairlift.

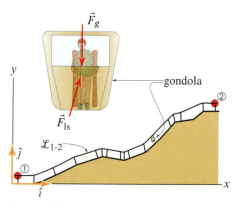

Figure 4.10
Schematic representation of a gondola ropeway along with the FBD of a passenger. The force on the passenger due to the lift system is $\vec{F}_{ls}$.

4.2 Conservative Forces and Potential Energy

Lifting people up a mountain

Figures 4.8 and 4.9 show two ropeways (also called ski lifts or simply lifts), which are transportation systems carrying people between two different elevations. There are several kinds of ropeways. For example, gondola systems (see Fig. 4.8) can have a cabin capacity of up to 15 people, a transport capacity of up to 5000 persons/h with a line speed of 6 m/s over distances of a few kilometers, and with changes of elevation in excess of 1000 m. Table 4.2 contains a summary of publicly available technical data for three gondola lifts. Based on this data, one of the questions we can ask is:

> What fraction of the energy available to operate a lift is expended in lifting people?

Answering this question will give us an idea of the "energy cost" of the rest of the lift operation (e.g., for overcoming friction in the entire drive system).

Table 4.2. Technical data for three gondola ropeways. The data was obtained from Leitner AG Ropeways at `http://www.leitner-lifts.com`. The rightmost column reports the amount of energy that can be provided to the lift system every second.

Location	Length (m)	Elevation (m)	Capacity (person/h)	Energy per unit time (kJ/s)
Hochgurgl (AT)	2216	554	2489	980
Verbier (CH)	1826	264	2400	800
Bruneck (IT)	4060	1314	1700	1560

We begin by considering a person in a gondola going from ① to ② along a path $\mathcal{L}_{1\text{-}2}$ that depends on the landscape (Fig. 4.10). Modeling the person as a particle, the work-energy principle tells us that $T_1 + U_{1\text{-}2} = T_2$, where T_1 and T_2 are the person's kinetic energy at ① and ②, respectively, and $U_{1\text{-}2}$ is the work done on the person by *all the forces* acting on him or her. Since the ropeway moves at an essentially constant speed, $T_1 = T_2$ and $U_{1\text{-}2} = 0$. Referring to the person's FBD in Fig. 4.10, we observe that

$$U_{1\text{-}2} = (U_{1\text{-}2})_{ls} + (U_{1\text{-}2})_g, \qquad (4.18)$$

where $(U_{1\text{-}2})_{ls}$ and $(U_{1\text{-}2})_g$ are the work done by the lift system and by gravity, respectively. Hence, given that $U_{1\text{-}2} = 0$, we can compute $(U_{1\text{-}2})_{ls}$ by computing the work of gravity, i.e.,

$$(U_{1\text{-}2})_{ls} = -(U_{1\text{-}2})_g \quad \text{with} \quad (U_{1\text{-}2})_g = \int_{\mathcal{L}_{1\text{-}2}} \vec{F}_g \cdot d\vec{r}. \qquad (4.19)$$

Notice that *our data does not tell us what $\mathcal{L}_{1\text{-}2}$ is!* As it turns out, our calculation does not require an exact knowledge of $\mathcal{L}_{1\text{-}2}$ because the force of gravity is not "sensitive" to the shape of $\mathcal{L}_{1\text{-}2}$. To see this, let's describe gravity as the constant force $\vec{F}_g = -mg\,\hat{j}$. Hence, since $d\vec{r} = dx\,\hat{i} + dy\,\hat{j}$, applying the second of Eqs. (4.19), we have

$$(U_{1\text{-}2})_g = \int_{\mathcal{L}_{1\text{-}2}} -mg\,\hat{j} \cdot (dx\,\hat{i} + dy\,\hat{j}) = -mg \int_{y_1}^{y_2} dy, \qquad (4.20)$$

or

$$\boxed{(U_{1\text{-}2})_g = -mg(y_2 - y_1).} \qquad (4.21)$$

Equation (4.21) says that the work done by gravity on a particle moving along an *arbitrary* path is equal to the particle's weight times the particle's change in height (i.e., $y_1 - y_2$). The fact that this work is independent of *how* ② is reached from ① is an important property, which we will discuss in greater detail soon.

Going back to calculating the energy required to operate a lift, referring to Table 4.2, we consider, for example, the ropeway at Hochgurgl, which, at full capacity, can lift 2489 people every hour over a change in elevation $y_2 - y_1 = 554\,\text{m}$ (1820 ft). To use this data along with Eq. (4.21), we need a value for the mass of a person. A reasonable choice for the mass of the average passenger, along with his or her equipment, is 95 kg (i.e., approximately 210 lb; see the Interesting Fact note in the margin for a justification). Thus, by using Eq. (4.21) and the first of Eqs. (4.19), the work done in an hour at full capacity is

$$(U_{1\text{-}2})_{ls} = 2489(95\,\text{kg})(9.81\,\text{m/s}^2)(554\,\text{m}) = 1.285{\times}10^9\,\text{J}. \qquad (4.22)$$

To answer our original question, the energy in Eq. (4.22) needs to be divided by the energy that the lift's power supply can provide in the course of an hour. Using Table 4.2 again, at Hochgurgl the maximum hourly available energy is $(980\,\text{kJ/s})(3600\,\text{s}) = 3.528{\times}10^9\,\text{J}$ so that

$$\frac{\text{energy required to lift people}}{\text{energy available}} = \frac{1.285{\times}10^9\,\text{J}}{3.528{\times}10^9\,\text{J}} = 0.364. \qquad (4.23)$$

For the ropeways at Verbier and Bruneck, the ratio in Eq. (4.23) is 0.205 and 0.371, respectively, indicating that there is much more that needs to be accounted for in operating a lift than simply lifting people!

In computing the work done by a lift system, we learned that the work done by gravity in going from ① to ② depends only on the initial and final positions but is otherwise independent of the path $\mathcal{L}_{1\text{-}2}$. There are other forces that have the distinctive property that their work is independent of the path followed in going from an initial to a final position. We now consider another important example of this type of force, namely, a central force, which includes as special cases both the force of a spring and gravity as described by the universal law of gravitation.

Work of a central force

A *central force* $\vec{F}_c(r)$ acting on a particle P is a force whose line of action always passes through the same point O (see Fig. 4.11) and whose magnitude is a function of the distance r from O to P. Spring forces and Newton's universal law of gravitation can be treated as central forces. We now derive the general expression of the work of a central force and then apply this result to the case of spring forces and gravity.

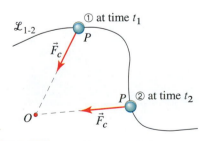

Figure 4.11
Particle P moving under the action of a central force $\vec{F}_c$.

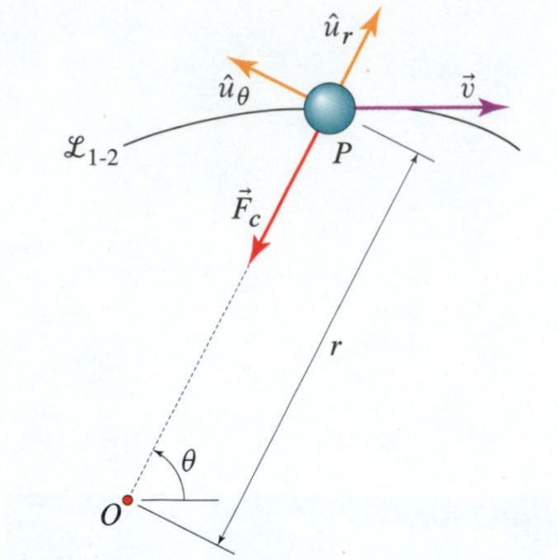

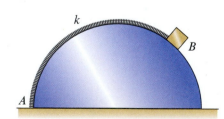

Figure 4.12

A particle P under the action of a central force $\vec{F}_c$. The figure also shows the velocity vector of P as well as the polar coordinate system used to describe $\vec{F}_c$ and the motion of P.

Referring again to Fig. 4.11, we wish to determine the work done by $\vec{F}_c(r)$ on the particle P as it moves along the path $\mathscr{L}_{1\text{-}2}$ from ①, at time t_1, to ②, at time t_2. Figure 4.12 shows the polar coordinate system we will use to describe the motion of P and the central force $\vec{F}_c$. Using the component system shown and applying Eq. (4.15) on p. 262, the work done by $\vec{F}_c$ is given by

$$(U_{1\text{-}2})_c = \int_{t_1}^{t_2} \vec{F} \cdot \vec{v}\, dt = \int_{t_1}^{t_2} F_c(r)\,\hat{u}_r \cdot \left(\dot{r}\,\hat{u}_r + r\dot{\theta}\,\hat{u}_\theta\right) dt, \quad (4.24)$$

where we have used $\vec{F}_c = F_c(r)\,\hat{u}_r$ and Eq. (2.85) on p. 119, which is the expression of the velocity in polar coordinates. Expanding the last dot product in Eq. (4.24) and noting that $\dot{r}\, dt = (dr/dt)\, dt = dr$, we obtain

$$(U_{1\text{-}2})_c = \int_{r_1}^{r_2} F_c(r)\, dr. \quad (4.25)$$

Work done by a spring force. Figure 4.13(a) shows a linear spring connecting a particle P to the fixed point O while P moves in a plane from point ① to ② along the arbitrary path $\mathscr{L}_{1\text{-}2}$. As can be seen in Fig. 4.13(b), the spring force always acts toward or away from O. For a spring whose current length

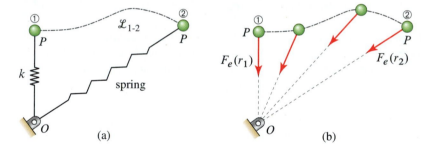

Figure 4.13. (a) A particle P moving along the path $\mathscr{L}_{1\text{-}2}$ from ① to ②. One end of the linear spring is attached to P and the other to the fixed point O. (b) Direction of the spring force as P moves along $\mathscr{L}_{1\text{-}2}$.

is r and whose unstretched length is L_0, the spring's stretch is $\delta = r - L_0$ and the corresponding force is $\vec{F}_e = -k(r - L_0)\,\hat{u}_r$ (see Eq. (3.13) on p. 187; the subscript e stands for elastic). Thus, using Eq. (4.25), we have

$$(U_{1\text{-}2})_e = \int_{r_1}^{r_2} -k(r - L_0)\, dr = -\tfrac{1}{2}k\left[(r_2 - L_0)^2 - (r_1 - L_0)^2\right], \quad (4.26)$$

or

$$(U_{1\text{-}2})_e = -\tfrac{1}{2}k\left(\delta_2^2 - \delta_1^2\right), \quad (4.27)$$

where $\delta_1 = r_1 - L_0$ and $\delta_2 = r_2 - L_0$. As with the work done by a constant force, the work done by a spring force acting on a particle only depends on the endpoints of the path followed by the particle.

Referring to Fig. 4.14, we see that in some cases spring forces are not really central forces. However, even in these cases it can be proved that the work done by a spring always takes on the form given in Eq. (4.27).

Figure 4.14

A block B on a semicylinder connected to point A via a linear spring with stiffness k that is lying over the surface of the semicylinder

Work done by the force of gravity. Referring to Fig. 4.15, consider the gravitational force on a particle B of mass m_B exerted by a particle A of mass m_A whose position is fixed. This force is given by Newton's universal law of gravitation (see Eq. (1.6) on p. 5), that is,*

$$\vec{F}_{BA} = -\frac{Gm_A m_B}{r^2}\,\hat{u}_r. \tag{4.28}$$

Equation (4.25) then gives

$$(U_{1\text{-}2})_G = \int_{r_1}^{r_2} -\frac{Gm_A m_B}{r^2}\,dr = -Gm_A m_B \int_{r_1}^{r_2} \frac{dr}{r^2}, \tag{4.29}$$

which yields

$$(U_{1\text{-}2})_G = -Gm_A m_B\left(-\frac{1}{r_2}+\frac{1}{r_1}\right). \tag{4.30}$$

Again we see that the work done depends only on the initial and final positions of the particle.

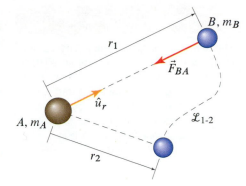

Conservative forces and potential energy

In computing the work done on a particle by spring and gravitational forces, we found that this work depended *only* on the initial and final positions of the particle, as opposed to depending on the path in some intrinsic way. As we saw in the marginal note "Line integral of an exact differential" on p. 261, if a path or line integral depends on only the limits of integration, then the integrand must be the exact differential of some function. We can therefore say that, for certain types of forces, we can write the work as

$$U_{1\text{-}2} = \int_{\mathscr{L}_{1\text{-}2}} \vec{F}\cdot d\vec{r} = \int_{\mathscr{L}_{1\text{-}2}} -(dV) = -\left[V(\vec{r}_2)-V(\vec{r}_1)\right]$$

$$= -(V_2 - V_1), \tag{4.31}$$

where $\vec{r}_1$ and $\vec{r}_2$ are the position vectors corresponding to ① and ②, respectively, and where V is a scalar function of position called the *potential energy* of the force $\vec{F}$. Forces for which Eq. (4.31) is true, i.e., forces for which there exists a scalar potential energy, are called *conservative forces*.

If all the forces doing work on a physical system are conservative, then the statement of the work-energy principle can be given a form that reflects the conservative nature of the forces. Specifically, substituting Eq. (4.31) into Eq. (4.14) on p. 262, we obtain the form of the work-energy principle called *conservation of mechanical energy*:

$$T_1 + V_1 = T_2 + V_2. \tag{4.32}$$

Systems for which Eq. (4.32) holds are called *conservative systems*. Equation (4.32) states that the total mechanical energy, that is, the potential plus kinetic energy, is conserved between any two points on the path $\mathscr{L}_{1\text{-}2}$, as long as all forces doing work are conservative.

With the help of Eq. (4.31), we will now determine V for some of the conservative forces we have encountered.

---

* Notice that Eq. (4.28) has a different sign than does Eq. (1.6). This is so because the unit vector $\hat{u}_r$ in Eq. (4.28) points from A to B.

Interesting Fact

Why the minus sign in Eq. (4.31)? The minus sign is there so that *the potential energy V can be viewed as a measure of the potential or capacity to do positive work* in returning to the zero potential energy state. This can be seen by noticing that Eq. (4.31) states that $U_{1\text{-}2} = V_1 - V_2$ so that the work is positive if V decreases and negative if V increases.

Common Pitfall

Importance of the path in computing work. Equation (4.12) on p. 261 implies that, *in general*, the work of a force *depends on the path*! In computing the work of constant and central forces, we discovered that the work done by these forces depends on only the endpoints of the path. What this means is that there are some common and important forces for which there is no path dependence. However, path independence should not be assumed to be a common event. In many applications, such as in the presence of friction, the opposite is true.

Potential energy of a constant gravitational force. In Eq. (4.21) we saw that the work done by a constant gravitational force on a particle of mass m is $U_{1\text{-}2} = -mg(y_2 - y_1)$. Comparing this expression with Eq. (4.31) yields

$$V_g = mgy, \tag{4.33}$$

where V_g is the potential energy corresponding to the constant gravitational force mg and y is the vertical position measured from the $y = 0$ or *datum line*[*] with y increasing in the direction *opposite* to gravity. Since it is only the change in height that determines the work done by a constant gravitational force, the vertical position of the datum line is entirely arbitrary.

Potential energy of a spring force. The work done by a spring force was found to be $U_{1\text{-}2} = -\frac{1}{2}k\left(\delta_2^2 - \delta_1^2\right)$. Comparing this with Eq. (4.31), we see that

$$V_e = \tfrac{1}{2}k\delta^2, \tag{4.34}$$

where V_e is the elastic potential energy of the linear spring force and δ is *the distance the spring is stretched or compressed from its unstretched length.*

Potential energy of the force of gravity. In Eq. (4.30), we determined the work done by the force of gravity on a particle B due to its interaction with a particle A to be $U_{1\text{-}2} = -Gm_Am_B(-1/r_2 + 1/r_1)$. Comparing this expression with Eq. (4.31), we see that the potential energy associated with a gravitational force between A and B is given by

$$V_G = -\frac{Gm_Am_B}{r}, \tag{4.35}$$

where r is the distance between A and B.

■ **Mini-Example.** Revisit the problem at the beginning of Section 4.1 in which a particle slides down a frictionless semicylinder. Show that Eq. (4.4) can be obtained *directly* by applying the work-energy principle.
Solution. Referring to Fig. 4.16, we will use a polar coordinate system with origin at O. Applying Eq. (4.32), we have

$$T_1 + V_1 = T_2 + V_2. \tag{4.36}$$

Now we need to provide expressions for the kinetic energies T_1 and T_2 and potential energies V_1 and V_2. As long as the particle remains in contact with the surface, we have that its velocity is described by

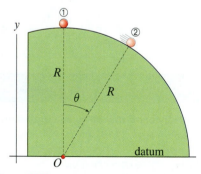

Figure 4.16
A particle sliding on a frictionless semicylindrical surface.

$$\vec{v} = R\dot{\theta}\,\hat{u}_\theta \quad \Rightarrow \quad T_1 = \tfrac{1}{2}m(R\dot{\theta}_1)^2 \quad \text{and} \quad T_2 = \tfrac{1}{2}m(R\dot{\theta}_2)^2. \tag{4.37}$$

To evaluate V_1 and V_2, we need to rely on the particle's FBD in Fig. 4.17. As we have just seen, the work of the weight force can be accounted for via the potential energy functions $V_1 = mgy_1$ and $V_2 = mgy_2$ where, as shown in Fig. 4.16, we have selected the datum line to coincide with the bottom of the semicylinder. In addition, we need not worry about N as we have already seen

---

[*] The word *datum* comes from the Latin verb *dare*, which means to give, and it is used in surveying, mapping, and geology to designate a given point, line, or surface used as a reference.

that normal forces do no work. Using Eqs. (4.37) and the expressions above for V_1 and V_2, Eq. (4.36) becomes

$$\tfrac{1}{2}m(R\dot{\theta}_1)^2 + mgy_1 = \tfrac{1}{2}m(R\dot{\theta}_2)^2 + mgy_2. \tag{4.38}$$

Noticing that $y_1 = R\cos\theta_1$ and $y_2 = R\cos\theta_2$, Eq. (4.38) becomes

$$\tfrac{1}{2}m(R\dot{\theta}_1)^2 + mgR\cos\theta_1 = \tfrac{1}{2}m(R\dot{\theta}_2)^2 + mgR\cos\theta_2, \tag{4.39}$$

which, with a little rearranging, is identical to Eq. (4.4). What should be noticed is that, by applying the work-energy principle, we were able to obtain Eq. (4.4) without explicitly integrating Newton's second law! ⎯⎯⎯⎯⎯ ▪

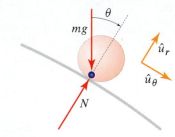

Figure 4.17
Figure 4.2 repeated. FBD of the particle at ② of Fig. 4.16.

🎓 Advanced Topic 🎓
When is a force conservative?

Is there a way to tell whether or not a force is conservative? That is, can we determine whether or not an associated potential energy exists? If we establish that a potential energy exists, can we find it? Also, given a potential energy, can we find the associated force? Let's try to answer these questions.

Equation (4.31) implies that the infinitesimal work of a conservative force $\vec{F}$ can be written as

$$\vec{F} \cdot d\vec{r} = -dV, \tag{4.40}$$

where we recall that a conservative force is only a function of position. In terms of Cartesian components, we can write the differential of V as

$$dV = \frac{\partial V}{\partial x}\,dx + \frac{\partial V}{\partial y}\,dy + \frac{\partial V}{\partial z}\,dz \tag{4.41}$$

and $\vec{F} \cdot d\vec{r}$ as

$$\vec{F} \cdot d\vec{r} = \left(F_x\,\hat{\imath} + F_y\,\hat{\jmath} + F_z\,\hat{k}\right) \cdot \left(dx\,\hat{\imath} + dy\,\hat{\jmath} + dz\,\hat{k}\right) = F_x\,dx + F_y\,dy + F_z\,dz. \tag{4.42}$$

Inserting Eqs. (4.41) and (4.42) into Eq. (4.40), we see that

$$F_x\,dx + F_y\,dy + F_z\,dz = -\frac{\partial V}{\partial x}\,dx - \frac{\partial V}{\partial y}\,dy - \frac{\partial V}{\partial z}\,dz. \tag{4.43}$$

Since dx, dy, and dz are independent of each other, for Eq. (4.43) to be satisfied, it must be true that

$$F_x = -\frac{\partial V}{\partial x}, \quad F_y = -\frac{\partial V}{\partial y}, \quad \text{and} \quad F_z = -\frac{\partial V}{\partial z}. \tag{4.44}$$

Some of you may recognize that, except for the sign, the right-hand sides of Eqs. (4.44) are the components of the gradient of V, so we can say

$$\boxed{\vec{F} = -\left(\frac{\partial V}{\partial x}\,\hat{\imath} + \frac{\partial V}{\partial y}\,\hat{\jmath} + \frac{\partial V}{\partial z}\,\hat{k}\right) = -\nabla V,} \tag{4.45}$$

where ∇V is the gradient of V. Equation (4.45) answers one of our questions; that is, given a potential V, we can find $\vec{F}$ by computing the negative of the gradient of V.

If we take the curl of $\vec{F}$, which is defined to be $\nabla \times \vec{F}$ (also written as curl $\vec{F}$), in Cartesian components, we obtain

$$\nabla \times \vec{F} = \left(\frac{\partial}{\partial x}\,\hat{\imath} + \frac{\partial}{\partial y}\,\hat{\jmath} + \frac{\partial}{\partial z}\,\hat{k}\right) \times \left(F_x\,\hat{\imath} + F_y\,\hat{\jmath} + F_z\,\hat{k}\right)$$

$$= \left(\frac{\partial F_z}{\partial y} - \frac{\partial F_y}{\partial z}\right)\hat{\imath} + \left(\frac{\partial F_x}{\partial z} - \frac{\partial F_z}{\partial x}\right)\hat{\jmath} + \left(\frac{\partial F_y}{\partial x} - \frac{\partial F_x}{\partial y}\right)\hat{k}. \tag{4.46}$$

Interesting Fact

Gradient operator tidbits. The vector differential operator ∇ should be read as *nabla* or *del*. The gradient of the scalar function V is often written grad V. Finally, the gradient operator in cylindrical coordinates can be written as

$$\nabla = \frac{\partial}{\partial r}\,\hat{u}_r + \frac{1}{r}\frac{\partial}{\partial\theta}\,\hat{u}_\theta + \frac{\partial}{\partial z}\,\hat{u}_z.$$

Substituting Eqs. (4.44) into Eq. (4.46), we obtain

$$\nabla \times \vec{F} = \left(-\frac{\partial^2 V}{\partial z\,\partial y} + \frac{\partial^2 V}{\partial y\,\partial z}\right)\hat{\imath} + \left(-\frac{\partial^2 V}{\partial x\,\partial z} + \frac{\partial^2 V}{\partial z\,\partial x}\right)\hat{\jmath}$$
$$+ \left(-\frac{\partial^2 V}{\partial y\,\partial x} + \frac{\partial^2 V}{\partial x\,\partial y}\right)\hat{k} = \vec{0}, \quad (4.47)$$

where the equality to zero holds as long as the partial derivatives in Eq. (4.47) are continuous so that we can interchange the order of differentiation of the second derivatives of V, that is,

$$\frac{\partial^2 V}{\partial z\,\partial y} = \frac{\partial^2 V}{\partial y\,\partial z}, \quad \frac{\partial^2 V}{\partial x\,\partial z} = \frac{\partial^2 V}{\partial z\,\partial x}, \quad \text{and} \quad \frac{\partial^2 V}{\partial y\,\partial x} = \frac{\partial^2 V}{\partial x\,\partial y}. \quad (4.48)$$

Equation (4.47) helps us answer the other question we asked because it shows that *the curl of a conservative force is necessarily equal to zero*. It is possible to prove that the converse of this statement is also true, i.e., *if the curl of a force is equal to zero, then the force is conservative.*

Work-energy principle for any type of force

Equation (4.14) is the *most* general form of the work-energy principle since it applies to any system. However, when all the forces doing work on a system are conservative, then we can use changes in potential energy to find the work done by a force (see Eq. (4.31)) instead of computing the work from its integral definition (see Eq. (4.12)). When a problem involves both conservative *and* nonconservative forces, it is useful to have a form of the work-energy principle that still takes advantage of Eq. (4.31) for computing the work of the conservative forces. To derive such a form of the work-energy principle, we begin by letting $U_{1\text{-}2} = (U_{1\text{-}2})_{\text{c}} + (U_{1\text{-}2})_{\text{nc}}$ and then writing the work-energy principle as

$$T_1 + (U_{1\text{-}2})_{\text{c}} + (U_{1\text{-}2})_{\text{nc}} = T_2, \quad (4.49)$$

where $(U_{1\text{-}2})_{\text{c}}$ and $(U_{1\text{-}2})_{\text{nc}}$ are the work contributions due to the conservative and nonconservative forces, respectively, and where the subscripts c and nc stand for conservative and nonconservative, respectively. Using Eq. (4.31), we can write $(U_{1\text{-}2})_{\text{c}} = -(V_2 - V_1)$ so that Eq. (4.49) becomes

$$\boxed{T_1 + V_1 + (U_{1\text{-}2})_{\text{nc}} = T_2 + V_2.} \quad (4.50)$$

Equation (4.50) is just as general as Eq. (4.14), but it is easier to apply since the work done by conservative forces, such as spring and gravitational forces, can be included in the V_1 and V_2 terms.

Important note. The *only* conservative forces we will include in the potential energy V are spring forces and gravitational forces. While there are some conservative forces, such as constant forces, that could be included in V, we will not do so here to avoid confusion with forces such as constant frictional forces.

End of Section Summary

In this section we discovered that there are forces whose work, in taking a particle from an initial to a final position, does not depend on the path connecting these positions. Specifically, we derived the following expressions.

Work of a constant gravitational force. For a particle of mass m moving in a constant gravitational field from ① to ②, the work done on the particle is

Eq. (4.21), p. 277

$$(U_{1\text{-}2})_g = -mg(y_2 - y_1),$$

for which gravity must act in the $-y$ direction.

Work done by a spring force. For a particle subject to the force of a linear elastic spring with constant k, the work done on the particle is

Eq. (4.27), p. 278

$$(U_{1\text{-}2})_e = -\tfrac{1}{2}k\left(\delta_2^2 - \delta_1^2\right),$$

where δ_1 and δ_2 are the stretch of the spring at ① and ②, respectively.

Conservative Systems. A force is said to be *conservative* if its work depends on only the initial and final position of its point of application but is otherwise independent of the path connecting these positions. The work of a conservative force can be characterized via a scalar potential energy function V. A *conservative system* is one for which all the forces doing work are conservative. For conservative systems, the work-energy principle becomes the *conservation of mechanical energy*, which is given by

Eq. (4.32), p. 279

$$T_1 + V_1 = T_2 + V_2.$$

Important conservative forces include elastic spring forces and gravitational forces. The *potential energy of a linear elastic spring force* is given by

Eq. (4.34), p. 280

$$V_e = \tfrac{1}{2}k\delta^2,$$

the *potential energy of a constant gravitational force* is given by

Eq. (4.33), p. 280

$$V_g = mgy,$$

and the *potential energy of the force of gravity* is given by

Eq. (4.35), p. 280

$$V_G = -\frac{Gm_A m_B}{r}.$$

> **Concept Alert**
>
> **Elastic potential energy is never negative.** Whether a spring is elongated or compressed, i.e., whether δ is positive or negative, the potential energy of a spring is never negative!

> **Interesting Fact**
>
> **Gravitational potential energy of spheres.** Equation (4.35) is valid for finite size bodies (that is, other than particles) as long as r is measured from their center and they both have spherically symmetric mass distributions. Therefore, it is valid for almost all large celestial bodies since their mass distribution is generally very close to spherically symmetric and because they are so far from one another.

Finally, if we have a system in which there are both conservative and non-conservative forces doing work, we *can* use Eq. (4.14) on p. 262, that is, $T_1 + U_{1\text{-}2} = T_2$, *or* we can take advantage of the potential energies of the conservative forces by writing the work-energy principle as

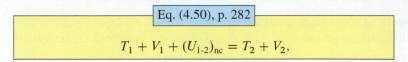

Eq. (4.50), p. 282

$$T_1 + V_1 + (U_{1\text{-}2})_{\text{nc}} = T_2 + V_2,$$

where $(U_{1\text{-}2})_{\text{nc}}$ is the work done by nonconservative forces. In this book, the only conservative forces we will include in V are linear elastic spring forces and gravitational forces.

EXAMPLE 4.4   *Demolishing a Building: Speed of a Wrecking Ball*

A 2500 lb wrecking ball A is released from rest with $\beta = 27°$ as shown in Fig. 1. When the ball hits the structure to be demolished, the cable holding the wrecking ball forms an angle $\gamma = 11°$ with respect to the vertical. If the length L of the cable is 30 ft, determine the speed with which the wrecking ball impacts the structure.

SOLUTION

Road Map & Modeling   This problem requires that we relate a change in the position of the wrecking ball with a corresponding change in speed and therefore is an ideal candidate for the application of the work-energy principle. Referring to Fig. 2, we will model the wrecking ball as a particle subject to its own weight and the tension in the cable to which the ball is attached. We will assume that the cable is inextensible and that the end at O is fixed. Referring to Fig. 3, we will denote the release position as ① and the position right before impact with the structure as ②. Notice that, since the ball's trajectory is a circle centered at O, the cable tension P is always perpendicular to the trajectory of the ball and therefore does not do any work. The only force doing work is gravity, which is a conservative force.

Governing Equations

Balance Principles   We now apply the principle of conservation of mechanical energy between ① and ②, which reads

$$T_1 + V_1 = T_2 + V_2, \tag{1}$$

where

$$T_1 = \tfrac{1}{2}mv_1^2 \quad \text{and} \quad T_2 = \tfrac{1}{2}mv_2^2, \tag{2}$$

and where m is the mass of the wrecking ball and v_1 and v_2 are its speed at ① and ②, respectively.

Force Laws   Recalling that only gravity does work, and choosing the datum line as shown in Fig. 3, we have

$$V_1 = -mgL\cos\beta \quad \text{and} \quad V_2 = -mgL\cos\gamma. \tag{3}$$

Kinematic Equations   Since the system is released from rest, we have

$$v_1 = 0. \tag{4}$$

Computation   Substituting Eqs. (2)–(4) into Eq. (1), we have

$$-mgL\cos\beta = \tfrac{1}{2}mv_2^2 - mgL\cos\gamma, \tag{5}$$

which, after solving for v_2, yields

$$\boxed{v_2 = \sqrt{2gL(\cos\gamma - \cos\beta)} = 13.2\,\text{ft/s}.} \tag{6}$$

Discussion & Verification   The dimensions of the term under the square root sign are length squared over time squared. Therefore, our final result is dimensionally correct, and it has been expressed with correct units.

A Closer Look   Notice that the quantity $L(\cos\gamma - \cos\beta)$ corresponds to the vertical drop of the ball. Hence, the speed achieved by the ball is precisely what the ball would achieve if it were dropped from a height equal to $L(\cos\gamma - \cos\beta)$.

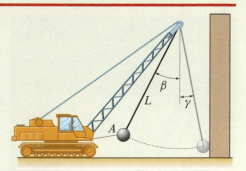

Figure 1

Figure 2
FBD of the wrecking ball.

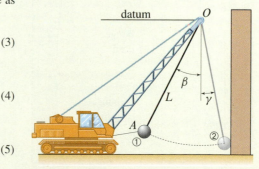

Figure 3
Definition of ① and ② as well as of the datum line for gravitational potential energy.

EXAMPLE 4.5   *Bungee Jumping: Conservation of Mechanical Energy*

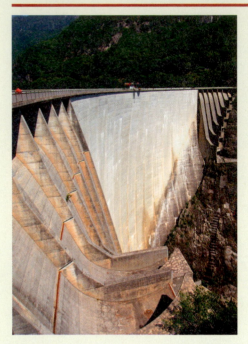

Figure 1
The Verzasca dam in southern Switzerland, which was the location for the bungee jumping scene in the 1995 James Bond film *GoldenEye*.

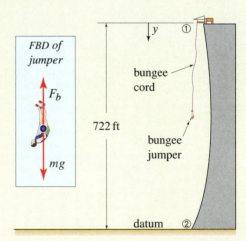

Figure 2
Profile of the Verzasca dam showing the jump platform as well as the bungee jumper. ① and ② are also defined. The blue inset shows the FBD after the jumper has fallen a distance greater than the unstretched length of the cord.

The mechanics of bungee jumping are rather straightforward, but a miscalculation can have dire consequences. Let's consider the highest bungee jumping location in the world—the Verzasca dam in southern Switzerland. The jump height off of the Verzasca dam is 722 ft. Given a jumper weighing 170 lb, determine the relationship between the stiffness k of the bungee cord and its unstretched length L_0; i.e., find k as a function of L_0, so that the jumper has zero speed at the bottom of the dam. In addition, determine the unstretched length so that acceleration of the jumper does not exceed $4g$ during the jump.

SOLUTION

Road Map & Modeling   Referring to the FBD in Fig. 2, we model the jumper as a particle subject to gravity and the force F_b of the bungee cord, which we model as a linear elastic spring. The jumper begins the jump at ① with zero speed and ends the jump at ② with zero speed (see Fig. 2). Since we know the jumper's speed at ① and ② and we know all the forces doing work on the jumper, we can apply the work-energy principle to the jumper to determine the relationship between the bungee stiffness and its unstretched length. We will then apply Newton's second law to the jumper to determine the maximum acceleration so that we can find k and L_0 for the bungee cord. Note that all the forces acting on the jumper are conservative and that F_b is zero until the jumper falls a distance equal to the unstretched length of the cord.

Governing Equations

Balance Principles   Applying the work-energy principle between ① and ②, we have

$$T_1 + V_1 = T_2 + V_2, \tag{1}$$

where we have used the fact that all forces doing work are conservative. The kinetic energies can be written as

$$T_1 = \tfrac{1}{2}mv_1^2 \quad \text{and} \quad T_2 = \tfrac{1}{2}mv_2^2, \tag{2}$$

where m is the jumper's mass and v_1 and v_2 are the jumper's speed at ① and ②, respectively. Since we also need to determine the jumper's acceleration, we will write Newton's second law for the jumper in the y direction as

$$\sum F_y: \quad mg - F_b = ma_y, \tag{3}$$

where we note that $F_b = 0$ until the bungee cord engages.

Force Laws   If we place the datum line for gravitational potential energy at ②, then

$$V_1 = mgh \quad \text{and} \quad V_2 = \tfrac{1}{2}k(h - L_0)^2, \tag{4}$$

where $h = 722\,\text{ft}$, $mg = 170\,\text{lb}$, and we have accounted for the potential energy of the bungee cord in V_2. We will also need the force law for the bungee cord, which is given by

$$F_b = k\delta = k(y - L_0), \tag{5}$$

where we note that $F_b = 0$ when $y \le L_0$.

Kinematic Equations   Since the jumper starts and ends the jump with zero speed, we have

$$v_1 = v_2 = 0. \tag{6}$$

Computation   Substituting Eqs. (2), (4), and (6) into Eq. (1) and solving for k, we obtain

$$k = \frac{2mgh}{(h - L_0)^2}. \tag{7}$$

Equation (7) gives the desired k as a function of L_0, a plot of which is shown in Fig. 3.

We now want to design the bungee system so that the maximum acceleration of a jumper weighing 170 lb does not exceed $a_{max} = 4g$. As with any design problem, there are infinitely many solutions that will satisfy the criteria that the jumper have zero speed at ② and that the jumper's acceleration not exceed $4g$. Referring to the FBD inset in Fig. 2, we know that until the spring engages, the jumper will be in free fall and his acceleration will be g downward. Once the spring engages, the acceleration is determined by solving Eqs. (3) and (5) for a_y, which gives

$$a_y = g - \frac{k}{m}(y - L_0). \tag{8}$$

Recall that Eq. (7) provides a relation between k and L_0 that ensures that the jumper has zero speed at ②. Therefore, using Eq. (7), Eq. (8) becomes

$$a_y = g - \frac{2gh(y - L_0)}{(h - L_0)^2} \quad \text{or} \quad \frac{a_y}{g} = 1 - \frac{2h(y - L_0)}{(h - L_0)^2}, \quad y > L_0. \tag{9}$$

We will choose the following value of the unstretched length of the bungee cord:

$$L_0 = h/2 = 361 \text{ ft.} \tag{10}$$

We now need to verify that the design criterion requiring $a_y < 4g$ is always met. To do so, using the chosen L_0 and referring to Fig. 4, we plot a_y/g versus y. The plot shows that the chosen value of L_0 is such that our design goal is met.

Discussion & Verification   The right-hand side of Eq. (7) has dimensions of force over length and therefore has the proper dimensions for k. Equation (7) also implies that the stiffness of the cord must be proportional to the jumper's weight, which is to be expected. In addition, Eq. (7) and the corresponding plot in Fig. 3 demonstrate an interesting aspect of the required stiffness k. If the bungee cord has a very short unstretched length (i.e., the denominator in Eq. (7) will be large), then the stiffness required to stop the jumper after falling 722 ft is very small and it increases very slowly until L_0 reaches about 500 ft. As L_0 gets closer to the jump height of 722 ft (i.e., the denominator in Eq. (7) will be small), the spring has to get very stiff to be able to stop the jumper in time and, at $L_0 = 722$ ft, the stiffness becomes infinite.

Hence, overall our solution seems to be correct. As far as the value of L_0 is concerned, we have already verified that it satisfies the required criteria.

🔍 **A Closer Look**   Figure 4 shows the acceleration of the jumper as a function of distance from the top of the dam for $L_0 = 361$ ft. This figure demonstrates that the largest acceleration experienced by the jumper is $3g$, and this occurs at the very end of the jump (that is, at $y = 722$ ft, the acceleration is $3g$ upward). As L_0 increases, the maximum acceleration experienced by the jumper increases so that, for example, when $L_0 = 650$ ft, we find that the maximum acceleration of the jumper is over $19g$!

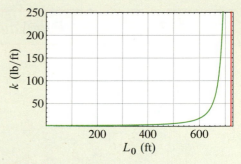

Figure 3
Bungee stiffness k as a function of its unstretched length L_0 required to achieve zero speed at the bottom of the jump (dark green curve). The red vertical line is at $L_0 = 722$ ft.

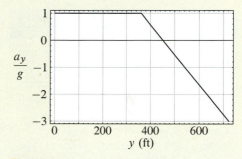

Figure 4
The acceleration of the bungee jumper as a function of y for $L_0 = h/2 = 361$ ft.

| Interesting Fact

Bungee cord stiffness and unstretched length. We have not taken into account the fact that a real bungee cord has limits to the amount that it can stretch. For example, for $L_0 = 1$ ft, Eq. (7) tells us that $k = 0.472$ lb/ft for $h = 722$ ft and $mg = 170$ lb. In fact, increasing L_0 to 400 ft only increases k to 2.37 lb/ft. Of course, the 1 ft bungee cord would have to stretch 721 ft or 72,100%, and the 400 ft bungee cord would only have to stretch 300 ft or 80.5%. While a rubber band can be easily stretched to a little less than twice its original length, we are not aware of a rubber band that can stretch to over 700 times its original length!

E X A M P L E   4.6   *Pole Vaulting: Turning Speed into Height*

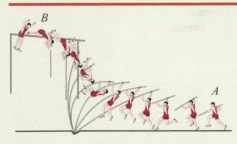

Figure 1
Sequence of images showing the positions of a vaulter during a vault. The blue pole has been shortened in some of the frames for clarity.

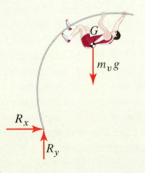

Figure 2
FBD of the vaulter and pole during the vault. The weight force acting on the vaulter acts at his mass center G.

A pole vaulter relies on several skills to achieve maximum height during a vault, but one of the most important is running speed. Speed is important because it is vital that the vaulter turn kinetic energy (running speed) into potential energy (vault height). Figure 1 shows the final moments of a vault during which the forward speed of the vaulter at A is turned into height at B. Modeling the vaulter as a particle, determine the maximum possible height that can be achieved by the "world's fastest pole vaulter," using the following data and assumptions:

1. At the time this book was written, the men's world record in the 100 m dash was 9.72 s, which was established by Usain Bolt of Jamaica on May 31, 2008; the record in the 200 m was 19.32 s, which was set on August 1, 1996, by Michael Johnson of the United States.

2. The pole vaulter is 6 ft tall with mass center at 55% of body height as measured from the ground.

3. *All* of the vaulter's kinetic energy is converted to potential energy during the vault, and no energy is lost from the system consisting of the pole and the vaulter.

4. The vaulter does no work during the vault, and his velocity is zero when he reaches the peak height of the vault.

SOLUTION

Road Map & Modeling   We will assume that the vault begins when the vaulter has reached maximum speed. To determine that top speed, we will use the 100 and 200 m dash data as follows. We assume that the vaulter starts from rest and accelerates to top speed with an acceleration that is approximately the same for both races. Since the runner has reached top speed by the time he reaches 100 m, we will assume that he runs at a uniform speed between 100 and 200 m. Therefore, we can approximate his maximum speed as

$$v_{\text{max}} = \frac{200 - 100}{19.32 - 9.72} \text{ m/s} = 10.42 \text{ m/s.} \tag{1}$$

To complete our model, we need to determine the forces acting on the vaulter. Referring to Fig. 2, we have treated the vaulter as a particle, and we have neglected air resistance as well as any moment that might be applied to the bottom of the pole by the box during the vault.* Note that in the FBD we have not included the weight of the pole. We will do so in Prob. 4.63 of Section 4.3 (after covering the work-energy principle for particle systems) to see whether or not it is important to account for the weight of the pole in our analysis. Observe that the point of application of the forces R_x and R_y is fixed so that these forces do no work. Hence, the only force doing work is gravity, which is a conservative force for which we know the potential energy function. Consequently, given that we want to relate changes in speed to changes in position, we will be able to solve this problem by applying the work-energy principle.

Governing Equations

Balance Principles   All the forces doing work are conservative, and the work-energy principle is

$$T_1 + V_1 = T_2 + V_2, \tag{2}$$

_____

* The box is the 20 cm depression in which the vaulter plants the pole to begin the vault.

where, referring to Fig. 3, ① corresponds to the instant the vaulter plants the pole and ② corresponds to the peak height of the vault. The vaulter's kinetic energies at ① and ② are given by

$$T_1 = \tfrac{1}{2}m_v v_1^2 \quad \text{and} \quad T_2 = \tfrac{1}{2}m_v v_2^2, \tag{3}$$

where m_v is the vaulters's mass and v_1 and v_2 are the vaulter's speed at ① and ②, respectively.

Force Laws   We will include all potential energy and work terms here. Recalling that the vaulter's weight is the only force doing work, we have (see Fig. 3)

$$V_1 = m_v gd \quad \text{and} \quad V_2 = m_v g(d + h), \tag{4}$$

where we have used the ground as our datum line for gravitational potential energy.

Kinematic Equations   Based on the Road Map & Modeling discussion, and ignoring any horizontal component of velocity the vaulter might have at ②, we have

$$v_1 = v_{\text{max}} \quad \text{and} \quad v_2 = 0. \tag{5}$$

Computation   Substituting Eqs. (3)–(5) into Eq. (2) and simplifying, we have

$$\tfrac{1}{2}v_{\text{max}}^2 = gh \quad \Rightarrow \quad h = \frac{v_{\text{max}}^2}{2g} = 5.534\,\text{m}, \tag{6}$$

where we have used the value of v_{max} in Eq. (1). Since at ① the vaulter's mass center is a distance $d = 0.55(6)\,\text{ft} = 3.3\,\text{ft} = 1.006\,\text{m}$ above the ground, we can add the distance d to h to find that the maximum height achievable by a pole vaulter is

$$\boxed{h_{\text{max}} = 6.54\,\text{m}.} \tag{7}$$

Discussion & Verification   Given that the right-hand side of Eq. (6) has dimensions of length, we can say that our result has the correct dimensions. As far as the correctness of the value we have obtained, it is larger than the current world record but not far from it (see discussion below). Hence, overall, our solution appears to be correct.

🖉 **A Closer Look**   At the time this book was written, the world record in the pole vault was 6.14 m (see marginal note entitled "World records"). The height we obtained is not far from this (it is 8% higher). There are many factors that we have not included in our analysis. Here are a few of these factors, which may contribute to our result being *too high*.

- Vaulters are not able to take advantage of all of their speed at the pole plant since they must leap at a takeoff angle of 15–20°.

- While elite male vaulters run faster than 9.0 m/s during the last 5 m of their approach sprint, our estimate of 10.42 m/s was too high.

- Vaulters cannot have zero velocity at the maximum height because they need some horizontal speed to clear the bar.

On the other hand, we have not taken into account the fact that the vaulter can do work during the vault between ① and ②, for example, by pushing with the arms against the pole. This allows the vaulter to go higher than our prediction would suggest, and this additional work can generally add about 0.8 m to the vault.

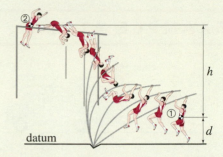

Figure 3
Sequence showing ① and ② within the vault.

> **Interesting Fact**
>
> **Fiberglass poles.** Fiberglass poles were introduced in the early 1960s. Before that, aluminum, steel, or bamboo poles were used. With fiberglass poles, vault heights and records increased dramatically. From 1940 to 1960, the world record increased from 4.7 to 4.8 m whereas from 1960 to 1963 it increased from 4.8 to 5.1 m. Fiberglass poles were thought to "catapult" the vaulter over the bar, but studies showed otherwise. There are two reasons why fiberglass poles facilitate performances. First, a flexible pole allows for a higher grip height on the pole. A vaulter's grip height is determined by the kinetic energy at takeoff—the higher the kinetic energy, the longer the pole vaulter can rotate to vertical. Second, flexible poles allow vaulters to have a lower takeoff angle since flexible poles can accept and return much more energy than a stiff pole.

> **Interesting Fact**
>
> **World records.** At the time this book was written, the world record in the pole vault was 6.14 m = 20.14 ft, set by Sergey Bubka of the Ukraine in 1994. For comparison, the high jump world record was 2.45 m = 8.04 ft, set by Javier Sotomayor of Cuba in 1993. If all that mattered were kinetic energy, these two records would be the same. However, without a pole this energy cannot be used to gain elevation, and that's why horizontal speed does not play a strong role in the high jump.

EXAMPLE 4.7   *Sliding With and Without Friction*

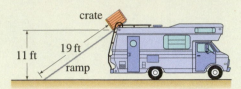

Figure 1
The RV with the ramp and its dimensions.

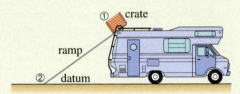

Figure 2
FBD of the crate as it slides down the ramp.

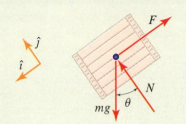

Figure 3
Definition of ① and ② for the work-energy principle.

A crate is unloaded from the top of a recreational vehicle (RV) using the ramp shown in Fig. 1. Determine the speed of the crate as it hits the ground for each of the following two cases if the crate is released from rest in the position shown and if

(a) there is negligible friction between the crate and the ramp; and

(b) the kinetic friction coefficient between the crate and the ramp is $\mu_k = 0.32$.

Assume that static friction is insufficient to prevent slipping.

SOLUTION

Road Map & Modeling   Referring to Fig. 2, we will model the crate as a particle subject to its own weight, the normal reaction between the crate and the slide, and the friction force F. The force F will be set to zero when we consider the case of frictionless contact between crate and slide. Since we need to relate a change in the crate's position to a corresponding change in speed, we will apply the work-energy principle. However, since one of the forces doing work is the friction force F, we will also apply Newton's second law in the direction perpendicular to the slide to determine the relation between the normal reaction N and F. We define ① to be the point of release of the crate and ② to be just before the crate hits the ground (see Fig. 3). The length of the ramp l is 19 ft and the vertical drop h is 11 ft. We will solve Parts (a) and (b) at the same time since the solution for Part (a) is a special case of the solution to case (b). Finally, since it will be used later in the calculations, we note that the angle θ appearing in the FBD of Fig. 2 is such that

$$\sin\theta = 11/19 \quad \Rightarrow \quad \theta = 35.38°. \tag{1}$$

Governing Equations

Balance Principles   The work-energy principle, applied between ① and ②, reads

$$T_1 + V_1 + (U_{1\text{-}2})_{nc} = T_2 + V_2, \tag{2}$$

The kinetic energies can be written as

$$T_1 = \tfrac{1}{2}mv_1^2 \quad \text{and} \quad T_2 = \tfrac{1}{2}mv_2^2, \tag{3}$$

where m is the crate's mass and v_1 and v_2 are the crate's speed at ① and ②, respectively. Referring to the FBD in Fig. 2, Newton's second law in the y direction gives

$$\sum F_y: \quad N - mg\cos\theta = ma_y. \tag{4}$$

Force Laws   By setting the datum line at ②, the potential energies in Eq. (2) are

$$V_1 = mgh \quad \text{and} \quad V_2 = 0. \tag{5}$$

The work done by the friction force F is

$$(U_{1\text{-}2})_{nc} = \int_{\mathcal{L}_{1\text{-}2}} -F\,\hat{\imath} \cdot d\vec{r} = \int_0^l -F\,\hat{\imath} \cdot dx\,\hat{\imath} = \int_0^l -F\,dx, \tag{6}$$

where F is related to the normal force N by the Coulomb friction law for sliding, i.e.,

$$F = \mu_k N. \tag{7}$$

Kinematic Equations   Since the crate slides along the ramp, we have

$$a_y = 0. \tag{8}$$

To compute the kinetic energies at ① and ②, we note that

$$v_1 = 0 \tag{9}$$

and that v_2 is the main unknown of the problem.

Computation   Using Eqs. (4), (7), and (8), we have

$$F = \mu_k mg \cos \theta. \tag{10}$$

Substituting Eq. (10) into Eq. (6), we then have

$$(U_{1\text{-}2})_{\text{nc}} = -\mu_k mgl \cos \theta. \tag{11}$$

Finally, substituting Eqs. (3), (5), (9), and (11) into Eq. (2), we obtain

$$mgh - \mu_k mgl \cos \theta = \tfrac{1}{2} mv_2^2. \tag{12}$$

For case (a), we have $\mu_k = 0$ so that Eq. (12) yields

$$\boxed{v_2 = \sqrt{2gh} = 26.6\,\text{ft/s}.} \tag{13}$$

For case (b), μ_k is different from zero and equal to the given value of 0.32, so that, from Eq. (12), we have

$$\boxed{v_2 = \sqrt{2g(h - \mu_k l \cos \theta)} = 19.7\,\text{ft/s}.} \tag{14}$$

Discussion & Verification   Given that the terms under the square root sign in Eqs. (13) and (14) have dimensions of acceleration times length, our results are dimensionally correct. In addition, the result in Eq. (13) is larger than that in Eq. (14), as expected given that the effect of friction is to oppose the motion of the crate. Therefore, our solution is reasonable.

🔍 **A Closer Look**   The result in Eq. (13) corresponds to the speed that the crate would have if the crate had been *dropped* from the top of the RV. While perhaps surprising, this result is indeed correct. In fact, in Part (a), the only two forces acting on the crate are the weight force mg and the normal force N. The normal force does no work, and the work done by the weight force can be easily computed as the force times the component of displacement in the direction of the force (since the weight, in this problem, is a constant force); this component is the vertical drop h. Therefore, the work done by mg is mgh, and that is exactly what it would be if we had dropped the crate from the top of the RV.

EXAMPLE 4.8   *Cyclic Work to Promote Swinging Motion*

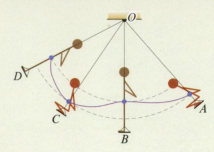

Figure 1
One possible stand-squat sequence during a
half swing. You are squatting at A, standing at
B, squatting at C, and then standing again at
the top of the swing at D. The blue dots indicate the position of your mass center, and the
purple line is the trajectory of your mass center.
The two dashed lines indicate arcs of nearest
and farthest distance of your mass center from
point O.

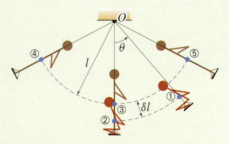

Figure 2
The five key positions of the pumping motion,
along with the dimensions of the movement of
the mass center.

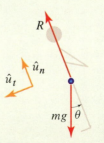

Figure 3
FBD of you on the swing at an arbitrary angle.

Figure 1 shows you swinging in the standing and squatting positions. As you probably
know, if you raise (stand) and lower (squat) yourself as you swing, you can increase
the amplitude of swinging with each cycle. While the analysis of the motion shown in
Fig. 1 can be very involved,[*] we can create a simple model to investigate how pumping
can increase the amplitude of each swing cycle.

We will analyze the pumping motion depicted in Fig. 2. This motion consists of
the following sequence of events:

1. At ①, that is, at the top of a backswing, the speed is zero and you are squatting.

2. At ②, which occurs at the lowest point in the swing, you stand upright to move
   to ③. This motion is assumed to occur instantaneously.

3. You next swing from ③ to ④ and then back to ⑤ while standing upright.

4. At ⑤, you then instantaneously squat and another cycle begins.

The goal is to find θ_5 as a function of θ_1 due to the pumping motion described above.
When you are standing, your mass center is a distance l from the fixed point O; and
when you are squatting, your mass center is a distance $l + \delta l$ from O.

SOLUTION

Road Map & Modeling   As you alternately stand and squat, you will be doing work,
and it is this work that will increase the swing angle with each cycle. We will write
the work-energy principle to relate the potential energy at the beginning and end of the
swing to the work done during the swing. You will be modeled as a particle located
at your mass center, which moves as described in Fig. 2. In applying the work-energy
principle between ① and ⑤, we will need to calculate the work done as you go from
squatting to standing from ② to ③. Your speed at both ① and ⑤ is zero. We will ignore
any drag and dissipative forces. Your FBD in an arbitrary position is shown in Fig. 3.
This figure shows us that the weight force mg must do work (which we will compute
via potential energy terms). Although possibly not obvious, the force R also does work
because it is responsible for lifting your mass center from $l + \delta l$ up to l in going from
② to ③.

Governing Equations

Balance Principles   The work-energy principle, applied between ① and ⑤, reads

$$T_1 + V_1 + (U_{1\text{-}5})_{nc} = T_5 + V_5. \tag{1}$$

The kinetic energies are given by

$$T_1 = \tfrac{1}{2}mv_1^2 \quad \text{and} \quad T_5 = \tfrac{1}{2}mv_5^2, \tag{2}$$

where m is the mass of the person and v_1 and v_5 are the speed of the person at ① and
⑤, respectively. In addition, referring to Fig. 3, Newton's second law applied at ② (i.e.,
at $\theta = 0$) gives

$$\sum F_n: \quad R - mg = ma_n. \tag{3}$$

---

[*] See, for example, W. B. Case and M. A. Swanson, "The Pumping of a Swing from the Seated
Position," *American Journal of Physics*, **58**(5), 1990, pp. 463–467, or W. B. Case, "The Pumping
of a Swing from the Standing Position," *American Journal of Physics*, **64**(3), 1996, pp. 215–220.

Force Laws   Choosing the datum for the gravitational potential energy at ②, we have

$$V_1 = mg(l + \delta l)(1 - \cos\theta_1) \quad \text{and} \quad V_5 = mgl(1 - \cos\theta_5). \tag{4}$$

In addition, recalling that $(U_{1\text{-}5})_{nc}$ is the work done by R in lifting the mass center from ② to ③, we have

$$(U_{1\text{-}5})_{nc} = R\,\delta l, \tag{5}$$

where we *assume* that R is constant throughout the lift.

Kinematic Equations   At ① and ⑤, we have

$$v_1 = 0 \quad \text{and} \quad v_5 = 0. \tag{6}$$

In addition, the acceleration a_n at ② is given by

$$a_n = \frac{v_2^2}{l + \delta l}. \tag{7}$$

Computation   Combining Eqs. (3)–(7) with Eq. (1), we obtain

$$mg(l + \delta l)(1 - \cos\theta_1) + m\left(g + \frac{v_2^2}{l + \delta l}\right)\delta l = mgl(1 - \cos\theta_5). \tag{8}$$

Applying the work-energy principle between ① and ②, during which only the weight force does work, we can relate v_2 to θ_1 as follows

$$T_1 + V_1 = T_2 + V_2 \quad \Rightarrow \quad mg(l + \delta l)(1 - \cos\theta_1) = \tfrac{1}{2}mv_2^2, \tag{9}$$

which, by solving for $v_2^2/(l + \delta l)$, yields

$$\frac{v_2^2}{l + \delta l} = 2g(1 - \cos\theta_1). \tag{10}$$

Substituting Eq. (10) into Eq. (8) and solving for θ_5, we obtain the final result

$$\boxed{\theta_5 = \cos^{-1}\left[\cos\theta_1 + \frac{\delta l}{l}(3\cos\theta_1 - 4)\right].} \tag{11}$$

Discussion & Verification   The argument of the inverse cosine function in Eq. (11) is nondimensional as it should be. In addition, recalling that the cosine function varies only between -1 and 1, we observe that the term $3\cos\theta_1 - 4$ is always negative. In turn this causes the value of θ_5 to be greater than θ_1 whenever δl is positive. This result confirms that the pumping action will result in an increased swing angle.

🔍 **A Closer Look**   Now that we have the angle at the end of a swing cycle θ_5 in terms of the angle at the beginning of a swing cycle θ_1, we can plug in some numbers to see if our pumping scheme works. For example, say that $l = 1.8\,\text{m}$, $\delta l = 0.2\,\text{m}$, and $\theta_1 = 40°$. With these numbers, Eq. (11) tells us that $\theta_5 = 54.8°$, thus confirming that we get an increase in swing angle for every cycle (see Figs. 4 and 5). However, notice that Figs. 4 and 5 tell us that if $\theta_1 = 0°$, then by simply standing up (while swinging at zero speed), you will somehow swing to $\theta_5 \approx 27°$! This result is clearly at odds with what happens in reality, and it indicates that our model has some strong limitations. In particular, as discussed in the Helpful Information marginal note, our model cannot be expected to be accurate unless the ratio $\delta l/l$ is sufficiently small. In addition, our model does implicitly assume that $\theta_1 \neq 0$; i.e., that our motion does not start from static equilibrium. The lesson to learn here is that when we construct a physical model, we must be aware of the limitations of the model when interpreting the model's predictions.

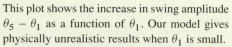

Helpful Information

Is it correct to assume that R is constant? Equations (3) and (7) imply that R cannot be constant between ② and ③ because R depends on a_n. In turn, a_n depends on the distance to point O and on v_2, both of which vary as the particle is lifted from ② to ③. The distance from the particle to O changes from $l + \delta l$ to l. In addition, applying the work-energy principle from ② to ③, we know that there must be a change in speed between these two positions because R does work between them. However, it can be argued that the assumption that R is constant between ② and ③ is acceptable as long as δl is small compared to l.

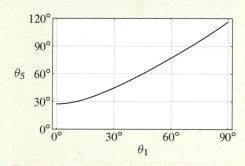

Figure 4
The angle θ_5 as a function of θ_1. This figure, along with Fig. 5, shows that our model gives physically unrealistic results when θ_1 is small.

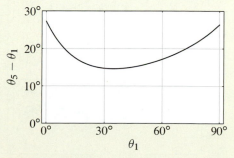

Figure 5
This plot shows the increase in swing amplitude $\theta_5 - \theta_1$ as a function of θ_1. Our model gives physically unrealistic results when θ_1 is small.

EXAMPLE 4.9   *A Conservative Force Field*

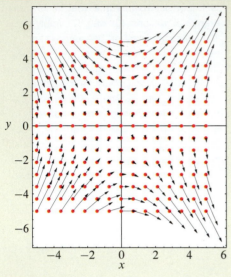

Figure 1
Graphical representation of the force field given in Eq. (1). If each red dot were a particle, then the arrow on each of those dots would represent the force vector on it. In addition, the length of each arrow is proportional to the magnitude of the force.

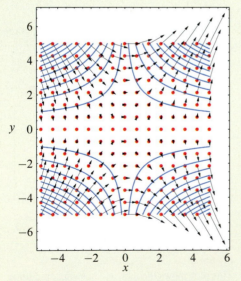

Figure 2
Figure 1 with constant values of V superimposed (i.e., the blue lines).

The force field shown in Fig. 1 is mathematically represented by

$$\vec{F} = y^2\,\hat{\imath} + 2xy\,\hat{\jmath}. \tag{1}$$

Determine the potential energy of $\vec{F}$.

SOLUTION

Road Map & Modeling   By computing the curl of the given force field, we can easily verify that curl $\vec{F} = \vec{0}$. Therefore we know that there is a potential V whose gradient is $-\vec{F}$. With this in mind, our strategy will be to use Eqs. (4.44) on p. 281 since they tell us how the components of $\vec{F}$ relate to V. We can view each of these equations as a simple first-order differential equation that we integrate.

Governing Equations   From Eq. (1), the components of $\vec{F}$ are

$$F_x = y^2 \quad \text{and} \quad F_y = 2xy. \tag{2}$$

Using Eqs. (4.44), the relation between the components of $\vec{F}$ and V is

$$F_x = -\frac{\partial V}{\partial x} \quad \text{and} \quad F_y = -\frac{\partial V}{\partial y}. \tag{3}$$

Computation   Equating the first of Eqs. (2) with the first of Eqs. (3), we have

$$y^2 = -\frac{\partial V}{\partial x}, \tag{4}$$

which we integrate to obtain

$$V = -xy^2 + f(y), \tag{5}$$

where $f(y)$ is an arbitrary function of y. Next, we equate the second of Eqs. (2) with the second of Eqs. (3) to obtain

$$2xy = -\frac{\partial V}{\partial y}, \tag{6}$$

which we integrate to obtain

$$V = -xy^2 + g(x), \tag{7}$$

where $g(x)$ is an arbitrary function of x. Comparing Eqs. (5) and (7), we see that

$$V = -xy^2 + f(y) \quad \text{and} \quad V = -xy^2 + g(x), \tag{8}$$

and the only way that both of these can be true is if $V = -xy^2 + \text{constant}$. Since only differences in potential energies matter, we can set this constant equal to zero to obtain

$$\boxed{V = -xy^2.} \tag{9}$$

Discussion & Verification   Equation (9) for different values of V has been plotted on top of Fig. 1, and the result is shown in Fig. 2. Notice that the blue lines, that is, lines of constant V, are perpendicular to the force vectors. This is what we should expect since the force field $\vec{F}$ is the negative of the gradient of V. Recall from calculus that the gradient of a scalar function gives, at each point, the direction of greatest change of that function. This agrees with our result since the direction of greatest change of V must be orthogonal to lines of constant V.

= PROBLEMS =

Problem 4.26

The pendulum shown is put in motion with a speed v_0 when $\theta = 0°$. Letting $L = 2$ ft, determine v_0 if the pendulum first comes to a stop at $\theta = 47°$.

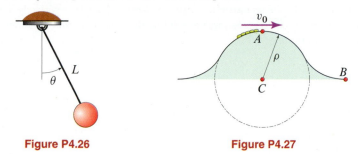

Figure P4.26   **Figure P4.27**

Problem 4.27

Point A is the highest point along the roller coaster ride section shown. The inscribed circle at A has radius $\rho = 25$ m and center at C. If point B is on a horizontal line going through C and if the roller coaster has a speed $v_0 = 45$ km/h at A, neglecting friction and air drag and treating the roller coaster as a particle, determine the speed of the roller coaster at B.

Problem 4.28

Assuming that the plunger of a pinball machine has negligible mass and that friction is negligible, determine the spring constant k such that a 2.85 oz ball is released with a speed $v = 15$ ft/s, after pulling back the plunger 2 in. from its rest position, i.e., from the position in which the spring is uncompressed.

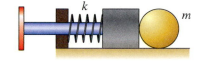

Figure P4.28

Problems 4.29 and 4.30

Consider a 3300 lb car whose speed is increased by 35 mph over a distance of 200 ft while traveling up a rectilinear incline with a 15% grade. Model the car as a particle, assume that the tires do not slip, and neglect *all* sources of frictional losses and drag.

Figure P4.29 and P4.30

Problem 4.29 Determine the work done on the car by the engine if the car starts from rest.

Problem 4.30 Determine the work done on the car by the engine if the car has an initial speed of 30 mph.

Problem 4.31

A classic car is driving down an incline at 60 km/h when its brakes are applied. Treating the car as a particle, neglecting all forces except gravity and friction, and assuming

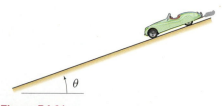

Figure P4.31

that the tires slip, determine the coefficient of kinetic friction if the car comes to a stop in 55 m and $\theta = 20°$.

Problem 4.32

A 75 kg skydiver is falling at a speed of 250 km/h when, at a height of 245 m, the parachute is deployed, allowing the skydiver to land at a speed of 4 m/s. Modeling the skydiver as a particle and assuming that the skydiver follows a perfectly vertical trajectory, determine the average force exerted by the parachute from the moment of deployment until landing.

Problem 4.33

Packages for transporting delicate items (e.g., a laptop or glass) are designed to "absorb" some of the energy of the impact in order to protect their contents. These energy absorbers can get very complicated (e.g., the mechanics of Styrofoam peanuts can be complex), but we can begin to understand how they work by modeling them as a linear elastic spring of constant k that is placed between the contents (an expensive vase) of mass m and the package P. Assume that the vase's mass is 3 kg and that the box is dropped from rest from a height of 1.5 m. Treating the vase as a particle and neglecting all forces except for gravity and the spring force, determine the value of the spring constant k so that the maximum displacement of the vase relative to the box is 0.15 m. Assume that the spring relaxes after the box is dropped and that it does not oscillate.

Figure P4.32

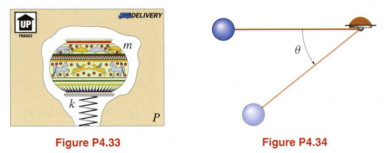

Figure P4.33   **Figure P4.34**

Problem 4.34

The pendulum is released from rest when $\theta = 0°$. If the string holding the pendulum bob breaks when the tension is twice the weight of the bob, at what angle does the string break? Treat the pendulum as a particle, ignore air resistance, and let the string be inextensible and massless.

Problem 4.35

The force acting on a stationary electric charge q_A interacting with a charge q_B is described by Coulomb's law and takes the form

$$\vec{F} = k \frac{q_A q_B}{r^2} \hat{u}_r,$$

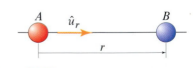

Figure P4.35

where $k = 8.9875 \times 10^9 \text{ N} \cdot \text{m}^2/\text{C}^2$ (C is the symbol for coulomb, the unit used to measure electric charge) is a constant and r is the distance between A and B. This force law is mathematically very similar to Newton's universal gravitation law. With this in mind, determine an expression of the electrostatic potential energy, choosing the datum at infinity, i.e., such that the potential energy is equal to zero when the two charges are separated by an infinite distance.

Problem 4.36

The force-compression profile of a rubber bumper B is given by $F_B = \beta x^3$, where $\beta = 3.5 \times 10^6$ lb/ft^3 and x is the bumper's compression measured in the horizontal direction. Determine the expression for the potential energy of the bumper B. In addition, if the cruiser C weighs 70,000 lb and impacts B with a speed of 5 ft/s, determine the compression required to bring C to a stop. Model C as a particle and neglect C's vertical motion as well as the drag force between the water and the cruiser C.

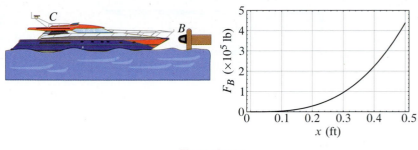

Figure P4.36

Problem 4.37

The force law between two Ni atoms was given in Eq. (3.37) on p. 232. Using this equation, determine the potential between two Ni atoms. Assume the potential between the two atoms is zero when the distance between those two atoms is infinite.

Problem 4.38

A satellite orbits the Earth along the orbit shown. The minimum and maximum distances from the center of the Earth are $R_P = 4.5 \times 10^7$ m and $R_A = 6.163 \times 10^7$ m, respectively, where the subscripts P and A stand for *perigee* (the point on the orbit closest to Earth) and *apogee* (the point on the orbit farthest from Earth), respectively. Modeling the satellite as a particle and assuming that the center of the Earth can be chosen as the origin of an inertial frame of reference, if the satellite's speed at P is $|\vec{v}|_P = 3.2 \times 10^3$ m/s, determine the satellite's speed at A.

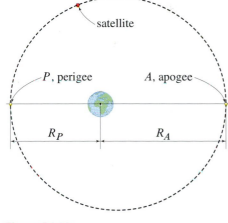

Figure P4.38

Problem 4.39

The package handling system is designed to launch the small package of mass m from A by using a compressed linear spring of constant k. After launch, the package slides along the track until it lands on the conveyor belt at B. The track has small, well-oiled rollers so that you can neglect any energy loss due to the movement of the package along the track. Modeling the package as a particle, determine the minimum initial compression of the spring so that the package gets to B without separating from the track at C. Finally, determine the speed with which the package reaches the conveyor at B.

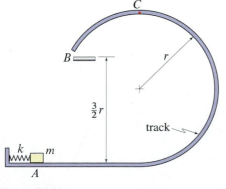

Figure P4.39

Problem 4.40

Compressed gas is used in many circumstances to propel objects within tubes. For example, you can still find pneumatic tubes in use in many banks to send and receive items to drive-through tellers,* and it is compressed gas that propels a bullet out of a gun barrel. Let the cross-sectional area of the tube be given by A and the position of the cylinder by s, and assume that the compressed gas is an ideal gas at constant temperature so that the pressure P times the volume Ω is a constant, i.e., $P\Omega = $ constant. Show that the potential energy of this compressed gas is given by $V = -P_0 s_0 A \ln(s/s_0)$, where P_0 is the initial pressure and s_0 is the initial value of s. Model the cylinder as a particle, and assume that the forces resisting the motion of the cylinder are negligible.

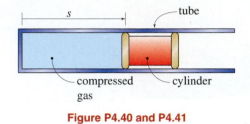

Figure P4.40 and P4.41

Problem 4.41

When a gun fires a bullet, the gun barrel acts as the tube, and the bullet acts as the cylinder in Prob. 4.40. Using the assumptions and result of that problem, determine the velocity of a bullet at the end of a 24 in. gun barrel, given that the bore diameter is 0.458 in., the bullet weight is 300 gr (7000 gr = 1 lb), the initial firing pressure is 27,000 psi, and the initial distance between the back of the bullet and the back wall of the firing chamber (i.e., s_0 in Prob. 4.40) is

(a) 1.855 in. (this distance is realistic and accurate),

(b) 1.5 in., and

(c) explain why the velocity of the bullet at the end of the gun barrel is lower in Part (b) when compared to Part (a).

Problem 4.42

The resistance of a material to fracture is assessed via a fracture test. One such test is the *Charpy impact test*, in which the fracture toughness is assessed by measuring the energy required to break a specimen of a specified geometry. This is done by releasing a heavy pendulum from rest at an angle θ_i and by measuring the maximum swing angle θ_f reached by the pendulum after the specimen is broken. Suppose that in an experiment $\theta_i = 45°$, $\theta_f = 23°$, the weight of the pendulum's bob is 3 lb, and the length of the pendulum is 3 ft. Neglecting the mass of any other component of the testing apparatus, assuming that the pendulum's pivot is frictionless, and treating the pendulum's bob as a particle, determine the fracture energy of the specimen tested. Assume that the fracture energy is the energy required to break the specimen.

Problems 4.43 and 4.44

A pendulum with mass $m = 1.4$ kg and length $L = 1.75$ m is released from rest at an angle θ_i. Once the pendulum has swung to the vertical position (i.e., $\theta = 0$), its cord runs into a small fixed obstacle. In solving this problem, neglect the size of the obstacle,

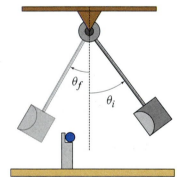

Figure P4.42

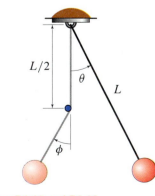

Figure P4.43 and P4.44

* Pneumatic tubes were used extensively by postal services in the late 19th and early 20th centuries. By the early 20th century, the cities of Philadelphia, New York, Boston, and Chicago had pneumatic networks, as did Paris, Berlin, and London. Their use for mail did not last past the end of World War I (1918), but almost every department store in the United States had tubes carrying cash and paperwork in the 1920s and 1930s. They still find use in hospitals for the delivery of medication and other items. See Robin Pogrebin, "Underground Mail Road," *New York Times*, May 7, 2001.

model the pendulum's bob as a particle, model the pendulum's cord as massless and inextensible, and let gravity and the tension in the cord be the only relevant forces.

Problem 4.43 What is the maximum height, measured from its lowest point, reached by the pendulum if $\theta_i = 20°$?

Problem 4.44 If the bob is released from rest at $\theta_i = 90°$, at what angle ϕ does the cord go slack?

Problem 4.45

While the stiffness of an elastic cord can be nearly constant (i.e., the force versus displacement curve is a straight line) over a large range of stretch, as a bungee cord is stretched, it softens; that is, the cord tends to get less stiff as it gets longer. Assuming a softening force-displacement relation of the form $k\delta - \beta\delta^3$, where $k = 2.58\,\text{lb/ft}$ and $\beta = 0.000013\,\text{lb/ft}^3$ and where δ (measured in ft) is the displacement of the cord from its unstretched length, and considering a bungee cord whose unstretched length is 150 ft, determine

(a) the expression of the cord's potential energy as a function of δ;

(b) the velocity at the bottom of a 400 ft tower of a bungee jumper weighing 170 lb and starting from rest;

(c) the maximum acceleration, expressed in g's, felt by the bungee jumper in question.

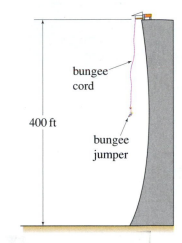

Figure P4.45

Problems 4.46 through 4.48

A crate, initially traveling horizontally with a speed of 18 ft/s, is made to slide down a 14 ft chute inclined at 35°. The surface of the chute has a coefficient of kinetic friction μ_k, and at its lower end, it smoothly lets the crate onto a horizontal trajectory. The horizontal surface at the end of the chute has a coefficient of kinetic friction μ_{k2}. Model the crate as a particle, and assume that gravity and the contact forces between the crate and the sliding surface are the only relevant forces.

Problem 4.46 If $\mu_k = 0.35$, what is the speed with which the crate reaches the bottom of the chute (immediately before the crate's trajectory becomes horizontal)?

Problem 4.47 Find μ_k such that the crate's speed at the bottom of the chute (immediately before the crate's trajectory becomes horizontal) is 15 ft/s.

Problem 4.48 Let $\mu_k = 0.5$ and suppose that once the crate reaches the bottom of the chute and after sliding horizontally for 5 ft, the crate runs into a bumper. If the weight of the crate is $W = 110\,\text{lb}$, $\mu_{k2} = 0.33$, and you model the bumper as a linear spring with constant k and neglect the mass of the bumper, determine the value of k so that the crate comes to a stop 2 ft after impacting with the bumper.

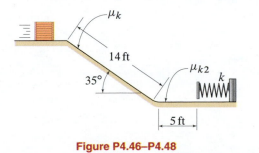

Figure P4.46–P4.48

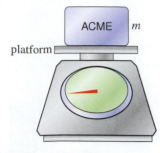

Figure P4.49 and P4.50

Problems 4.49 and 4.50

Spring scales work by measuring the displacement of a spring that supports both the platform and the object, of mass m, whose weight is being measured. Neglect the mass of the platform on which the mass sits, and assume that the spring is uncompressed before the mass is placed on the platform. In addition, assume that the spring is linear elastic with spring constant k. You may have solved these same problems using Newton's second law when doing Prob. 3.16 and 3.17 — here use the work-energy principle to solve them.

Problem 4.49 ǀ If the mass m is gently placed on the spring scale (i.e., it is dropped from zero height above the scale), determine the *maximum reading* on the scale after the mass is released.

Problem 4.50 ǀ If the mass m is gently placed on the spring scale (i.e., it is dropped from zero height above the scale), determine the expression for *maximum velocity* attained by the mass m as the spring compresses.

Problem 4.51 ǀ

A 6 lb collar is constrained to travel along a rectilinear and frictionless bar of length $L = 5$ ft. The springs attached to the collar are identical, and they are unstretched when the collar is at B. Treating the collar as a particle, neglecting air resistance, and knowing that at A the collar is moving to the right with a speed of 11 ft/s, determine the linear spring constant k so that the collar reaches D with zero speed. Points E and F are fixed.

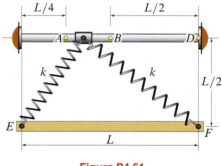

Figure P4.51

Problem 4.52 ǀ

A 3 kg collar is constrained to travel in the *horizontal plane* along a frictionless ring of radius $R = 0.75$ m. The spring attached to the collar has a spring constant $k = 21$ N/m. Treating the collar as a particle, neglecting air resistance, and knowing that at A the collar is at rest, determine the spring's unstretched length if the collar is to reach point B with a speed of 2 m/s.

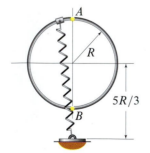

Figure P4.52

DESIGN PROBLEMS

Design Problem 4.1

In practice, during a jump, bungee cords stretch to 2–4 times their unstretched length, and a jumper feels no more than 2.5–$3.5g$ of acceleration. Assume the force in the bungee cord has the mathematical form $k\delta - \beta\delta^3$, where k and β are constants and δ is the amount of stretch in the cord past its unstretched length. Design a cord (that is, design the constants k and β) so that the bungee cord stretches 2.5 times its unstretched length, the acceleration of the jumper does not exceed $3g$, and the bungee cord has zero stiffness at the bottom of a 400 ft drop.

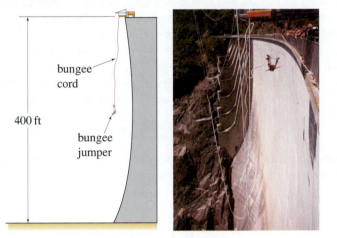

Figure DP4.1

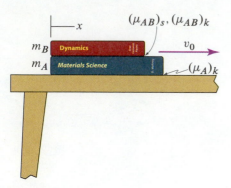

Figure 4.18
A modified version of Fig. 1 from Example 3.11, in which books A and B were labeled 1 and 2, respectively. The system parameters are $m_A = 1.5\,\text{kg}$, $m_B = 1.0\,\text{kg}$, $(\mu_A)_k = 0.45$, $(\mu_{AB})_k = 0.3$, and $(\mu_{AB})_s = 0.4$.

Helpful Information

The meaning of ① and ②. In studying particle systems, when we say that "① (or ②, etc.) is the position at which condition X happens," we mean that "① (or ②, etc.) is the collection of the *positions* of each individual system element corresponding to condition X being true for each element of the system." However, we are not saying that the condition in question is achieved by each element at the same time. For example, books A and B do not achieve ②, i.e., do not stop, at the same instant.

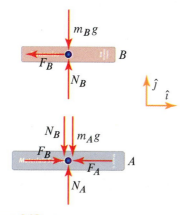

Figure 4.19
FBDs of the two books shown in Fig. 4.18.

4.3   Work-Energy Principle for Systems of Particles

This section presents the application of the work-energy principle to systems of particles. We will discover a fundamental new concept pertaining to systems of particles that is absent in the single-particle case — internal work.

Determining how far two stacked books slide on a table

Referring to Fig. 4.18, we revisit Example 3.11 on p. 243, in which a student has thrown a pair of stacked books on a table, both with an initial horizontal velocity of $v_0 = 0.75\,\text{m/s}$. As in Example 3.11, we want to determine the final positions of the books if they hit at $x = 0$. However, here we solve the problem by applying the work-energy principle. We denote as ① and ② the positions at which the books first hit the table and come to a stop, respectively (see the Helpful Information marginal note). By modeling the books as particles and referring to the FBD of B in Fig. 4.19, application of the work-energy principle to B yields

$$T_{B1} + V_{B1} + (U_{1\text{-}2})_{\text{nc}}^{B} = T_{B2} + V_{B2}. \tag{4.51}$$

Only the friction force F_B does work on B. Hence, observing that F_B is a nonconservative force, we see that the terms in Eq. (4.51) are

$$T_{B1} = \tfrac{1}{2}m_B v_0^2, \quad T_{B2} = 0, \quad V_{B1} = 0, \quad V_{B2} = 0, \tag{4.52}$$

$$(U_{1\text{-}2})_{\text{nc}}^{B} = \int_{0}^{x_{B2}} -F_B \, dx_B. \tag{4.53}$$

Similarly, applying the work-energy principle to A, we obtain (see Fig. 4.19)

$$T_{A1} + V_{A1} + (U_{1\text{-}2})_{\text{nc}}^{A} = T_{A2} + V_{A2}, \tag{4.54}$$

where

$$T_{A1} = \tfrac{1}{2}m_A v_0^2, \quad T_{A2} = 0, \quad V_{A1} = 0, \quad V_{A2} = 0, \tag{4.55}$$

$$(U_{1\text{-}2})_{\text{nc}}^{A} = \int_{0}^{x_{A2}} (F_B - F_A) \, dx_A. \tag{4.56}$$

As we saw in Example 3.11, A slides relative to the table, B slides relative to A, and A stops before B. Hence, the force laws for friction are as follows:

$$F_A = (\mu_A)_k N_A = (\mu_A)_k (m_A + m_B)g, \tag{4.57}$$

$$F_B = (\mu_{AB})_k N_B = (\mu_{AB})_k m_B g, \tag{4.58}$$

where we have summed forces in the vertical direction on A and B to obtain N_A and N_B, respectively. Substituting Eqs. (4.52), (4.53), and (4.58) into Eq. (4.51), we obtain

$$\tfrac{1}{2}m_B v_0^2 - (\mu_{AB})_k m_B g x_{B2} = 0, \tag{4.59}$$

and substituting Eqs. (4.55)–(4.58) into Eq. (4.54), we obtain

$$\tfrac{1}{2}m_A v_0^2 + [(\mu_{AB})_k m_B g - (\mu_A)_k (m_A + m_B)g]x_{A2} = 0. \tag{4.60}$$

Equations (4.59) and (4.60) can be solved for x_{B2} and x_{A2}, respectively, to obtain $x_{B2} = 0.09557$ m and $x_{A2} = 0.05213$ m, which, not surprisingly, are exactly what we obtained in Example 3.11.

Having applied the work-energy principle to A and B individually, we now focus on the question of what form the work-energy principle for A and B as a system should take. We begin answering this question by summing Eqs. (4.59) and (4.60), which, after rearranging, yields

$$\tfrac{1}{2}m_A v_0^2 + \tfrac{1}{2}m_B v_0^2 - (\mu_{AB})_k m_B g(x_{B2} - x_{A2})$$
$$- (\mu_A)_k(m_A + m_B)g x_{A2} = 0. \quad (4.61)$$

The first two terms in Eq. (4.61) represent the individual kinetic energies of A and B, respectively. It seems natural to view the sum of the individual kinetic energies as the *total kinetic energy of the system* since this would ensure that the system's kinetic energy (1) is never negative (as in the case of a single particle) and (2) is sensitive to the motion of each part of the system (see the Helpful Information note in the margin). Next, the term $-(\mu_A)_k(m_A + m_B)g x_{A2}$ is the work of the force $F_A = (\mu_A)_k(m_A + m_B)g$ over the distance x_{A2}. Referring to the system's FBD in Fig. 4.20, since F_A is an *external* force, the term $-F_A x_{A2}$ can be interpreted as the work of the forces external to the system. Finally, the term $-(\mu_{AB})_k m_B g(x_{B2} - x_{A2})$ is the work of the force $F_B = (\mu_{AB})_k m_B g$ over the distance $x_{B2} - x_{A2}$. What is interesting about the work term $-F_B(x_B - x_A)$ is that

1. F_B, being an internal force, *does not appear in the system's FBD*; and

2. the distance $x_B - x_A$ is not the distance traveled by B: it is the distance *traveled by B relative to A!*

These observations tell us that, contrary to what happens in writing Newton's laws, internal forces do play an important role in writing the work-energy principle for a system. In addition, by observing that if A and B had moved together, i.e., if $x_{B2} = x_{A2}$, then the work of the internal forces would have been zero, it seems that the presence of relative motion is needed for internal forces to do work. This is an important point because in many problems we will be able to write the statement of the work-energy principle for a system without worrying about the work of internal forces, either because there is no relative motion or because the relative motion is such that the internal work is equal to zero.

We now turn to a more formal development to put all of the observations made thus far on a firm foundation.

Internal work and work-energy principle for a system

Referring to Fig. 4.21, consider a system of n particles. The ith particle, whose mass is m_i and position is $\vec{r}_i$, is acted upon by an total *external force* $\vec{F}_i$ and interacts with the other $n - 1$ particles through *internal forces* labeled $\vec{f}_{ij}$ ($j = 1, \ldots, n$, $i \neq j$). Applying Newton's second law to the ith particle, we have

$$\vec{F}_i + \sum_{j=1}^{n} \vec{f}_{ij} = m_i \vec{a}_i, \quad i \neq j. \quad (4.62)$$

As we did for a single particle, we take the dot product of this equation with $d\vec{r}_i$ and integrate along $(\mathcal{L}_{1\text{-}2})_i$, the path traveled by particle i in going from

Helpful Information

About the kinetic energy of a system. Let m and v_G be the system's total mass and the speed of the center of mass, respectively. Would it be meaningful to compute the system's kinetic energy as $\tfrac{1}{2}mv_G^2$? The answer to this question is *no!* To understand why this is not a good idea, consider an object spinning about its center of mass, which is fixed. In this case, the formula $\tfrac{1}{2}mv_G^2$ would tell us that the body's kinetic energy is equal to zero even if the system as a whole is moving. Since changes in kinetic energy tell us how much work is done on the system, it is crucial that the kinetic energy account for the motion of each part of the system and not just the mass center.

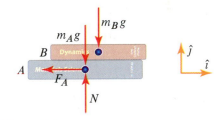

Figure 4.20
FBD of both books as the slide on the surface of the table.

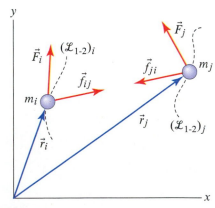

Figure 4.21
Particles i and j of a system of n particles.

① to ②. There will be n of these equations (one for each particle). Summing all n of them gives

$$\sum_{i=1}^{n} \int_{(\mathcal{L}_{1\text{-}2})_i} \left(\vec{F}_i + \sum_{j=1}^{n} \vec{f}_{ij} \right) \cdot d\vec{r}_i = \sum_{i=1}^{n} \int_{(\mathcal{L}_{1\text{-}2})_i} m_i \frac{d\dot{\vec{r}}_i}{dt} \cdot d\vec{r}_i, \qquad (4.63)$$

where we have written $\vec{a}_i = d\dot{\vec{r}}_i / dt$. Proceeding as we did for a single particle, we can write the right-hand side of Eq. (4.63) as

$$\sum_{i=1}^{n} \int_{(\mathcal{L}_{1\text{-}2})_i} m_i \frac{d\dot{\vec{r}}_i}{dt} \cdot d\vec{r}_i = \sum_{i=1}^{n} \int_{(\mathcal{L}_{1\text{-}2})_i} m_i \vec{v}_i \cdot d\vec{v}_i$$

$$= \sum_{i=1}^{n} \tfrac{1}{2} m_i \vec{v}_i \cdot \vec{v}_i \Big|_1^2$$

$$= \left(\sum_{i=1}^{n} \tfrac{1}{2} m_i v_i^2 \right)_2 - \left(\sum_{i=1}^{n} \tfrac{1}{2} m_i v_i^2 \right)_1. \qquad (4.64)$$

If we define the *kinetic energy of the particle system* to be the sum of the kinetic energies of each part of the system, then the quantity

$$T = \sum_{i=1}^{n} \tfrac{1}{2} m_i v_i^2 \qquad (4.65)$$

in Eq. (4.64) can be rewritten as

$$\sum_{i=1}^{n} \int_{(\mathcal{L}_{1\text{-}2})_i} m_i \frac{d\dot{\vec{r}}_i}{dt} \cdot d\vec{r}_i = T_2 - T_1. \qquad (4.66)$$

Going back to Eq. (4.63), the left-hand side of this equation can be written as

$$\sum_{i=1}^{n} \int_{(\mathcal{L}_{1\text{-}2})_i} \left(\vec{F}_i + \sum_{j=1}^{n} \vec{f}_{ij} \right) \cdot d\vec{r}_i = \overbrace{\sum_{i=1}^{n} \int_{(\mathcal{L}_{1\text{-}2})_i} \vec{F}_i \cdot d\vec{r}_i}^{(U_{1\text{-}2})_{\text{ext}}}$$

$$+ \overbrace{\sum_{i=1}^{n} \sum_{\substack{j=1 \\ j \neq i}}^{n} \int_{(\mathcal{L}_{1\text{-}2})_i} \vec{f}_{ij} \cdot d\vec{r}_i}^{(U_{1\text{-}2})_{\text{int}}}, \qquad (4.67)$$

where $(U_{1\text{-}2})_{\text{ext}}$ and $(U_{1\text{-}2})_{\text{int}}$ represent the work done on the system by *external* and *internal* forces, respectively. In discussing Newton's second law for systems, we saw that terms similar to the double sum in Eq. (4.67) vanish because of Newton's third law. However, in this case the double sum does not vanish and remains an important contribution to the work-energy principle. In fact, for each pair of particles i and j, the double sum that defines the term $(U_{1\text{-}2})_{\text{int}}$ contains the terms

$$\vec{f}_{ij} \cdot d\vec{r}_i + \vec{f}_{ji} \cdot d\vec{r}_j = \vec{f}_{ij} \cdot (d\vec{r}_i - d\vec{r}_j) = \vec{f}_{ij} \cdot d\vec{r}_{i/j}, \qquad (4.68)$$

where we have used the fact that $\vec{f}_{ij} = -\vec{f}_{ji}$ by Newton's third law and where, using relative kinematics, $d\vec{r}_{i/j} = d\vec{r}_i - d\vec{r}_j$ is the infinitesimal *relative* displacement of particle i with respect to particle j. Equation (4.68) tells us that for internal forces to do no work, either there must be no relative motion or the component of $d\vec{r}_{i/j}$ along $\vec{f}_{ij}$ must be zero for all particle pairs.

Summarizing, we use Eqs. (4.66) and (4.67) to write the work-energy principle for a system of particles as

$$T_1 + (U_{1\text{-}2})_{\text{ext}} + (U_{1\text{-}2})_{\text{int}} = T_2 \qquad (4.69)$$

or as

$$T_1 + V_1 + (U_{1\text{-}2})_{\text{nc}}^{\text{ext}} + (U_{1\text{-}2})_{\text{nc}}^{\text{int}} = T_2 + V_2 \qquad (4.70)$$

when there are also conservative forces doing work on the system.

Equation (4.69) (or Eq. (4.70)) formalizes what we had discovered in the analysis of the sliding books problem earlier in this section. That is, in applying the work-energy principle to particle systems, it is important to account for the work of internal forces. In addition, in view of the discussion following Eq. (4.68), there are situations in which the work of the internal forces vanishes even when there is relative motion. We will demonstrate one important application of this idea in some of the examples at the end of the section.

Kinetic energy for a system of particles

Here we consider a way of expressing the kinetic energy of a particle system that will facilitate understanding of the expression for the kinetic energy of a rigid body in Chapter 8.

We have defined the kinetic energy for a system of particles as

$$T = \sum_{i=1}^{n} \tfrac{1}{2} m_i \vec{v}_i \cdot \vec{v}_i. \qquad (4.71)$$

Now, referring to Fig. 4.22, notice that $\vec{r}_i = \vec{r}_G + \vec{r}_{i/G}$, where $\vec{r}_G$ and $\vec{r}_{i/G}$ are the position of the center of mass G and the position of particle i relative to G, respectively. Differentiating this expression for $\vec{r}_i$ with respect to time gives $\vec{v}_i = \vec{v}_G + \vec{v}_{i/G}$, which allows us to write Eq. (4.71) as

$$T = \sum_{i=1}^{n} \tfrac{1}{2} m_i \left(\vec{v}_G + \vec{v}_{i/G}\right) \cdot \left(\vec{v}_G + \vec{v}_{i/G}\right)$$

$$= \sum_{i=1}^{n} \tfrac{1}{2} m_i \left(\vec{v}_G \cdot \vec{v}_G + 2\vec{v}_G \cdot \vec{v}_{i/G} + \vec{v}_{i/G} \cdot \vec{v}_{i/G}\right)$$

$$= \tfrac{1}{2} m \underbrace{\vec{v}_G \cdot \vec{v}_G}_{v_G^2} + \underbrace{\left(\sum_{i=1}^{n} m_i \vec{v}_{i/G}\right)}_{\frac{d}{dt}\sum_{i=1}^{n} m_i \vec{r}_{i/G}} \cdot \vec{v}_G + \sum_{i=1}^{n} \tfrac{1}{2} m_i \underbrace{\vec{v}_{i/G} \cdot \vec{v}_{i/G}}_{v_{i/G}^2}, \qquad (4.72)$$

where m is the system's total mass. Referring to the second term on the last line of Eq. (4.72), using the definition of center of mass in Eq. (3.71) on p. 237, and

Work done by internal forces. As we will see in Example 4.12, not all internal forces do work. In fact, the only forces that do internal work are forces that cause deformation, such as spring forces (though we normally compute their work via their potential energy) and friction forces. Though not obvious, even *very* smooth surfaces exhibit roughness at small scales (see, e.g., S. Chandrasekaran, J. Check, S. Sundararajan, and P. Shrotriya, "The Effect of Anisotropic Wet Etching on the Surface Roughness Parameters and Micro/nanoscale Friction Behavior of Si(100) Surfaces," *Sensors and Actuators A*, **121**, 2005, pp. 121–130). The image below shows two objects sliding past one another.

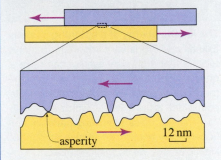

While they appear smooth, magnification of the interface shows that the surfaces are rough and that they only touch over a small percentage of their apparent area. The protrusions are called *asperities*, and it is these asperities that break and/or deform as the surfaces slide past one another.

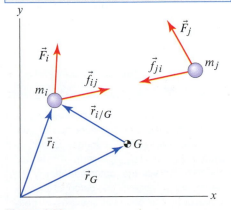

Figure 4.22
Particles i and j of a system of n particles with center of mass at G.

recalling that $\vec{r}_{i/G}$ describes the position of each particle *relative to the mass center* (see also the Helpful Information marginal note), we have

$$\frac{d}{dt}\sum_{i=1}^{n} m_i \vec{r}_{i/G} = \frac{d}{dt}\left(m\,\vec{r}_{G/G}\right). \tag{4.73}$$

Since the term $\vec{r}_{G/G}$ is the position of the mass center relative to the mass center itself, the left-hand side of Eq. (4.73) must always be equal to zero. In conclusion, Eq. (4.72) takes on the form

$$T = \tfrac{1}{2}mv_G^2 + \tfrac{1}{2}\sum_{i=1}^{n} m_i v_{i/G}^2, \tag{4.74}$$

which states that *the kinetic energy of a system of particles depends on the motion of the mass center of the system of particles as well as the motion of all the particles relative to the mass center.*

End of Section Summary

In this section, we developed the work-energy principle for a system of particles, and we discovered that internal forces play an important role in writing this principle. We defined the kinetic energy of a particle system as

$$\boxed{\text{Eq. (4.65), p. 304}}$$

$$T = \sum_{i=1}^{n} \tfrac{1}{2}m_i v_i^2,$$

where m_i and v_i are the mass and the speed of the ith particle in the system, respectively. Then the work-energy principle for a system of particles was found to be

$$\boxed{\text{Eq. (4.70), p. 305}}$$

$$T_1 + V_1 + (U_{1\text{-}2})_{\text{nc}}^{\text{ext}} + (U_{1\text{-}2})_{\text{nc}}^{\text{int}} = T_2 + V_2,$$

where V is the potential energy of the system, which may have contributions from both *external* and *internal* forces. The kinetic energy of a system can also be expressed in terms of the speed of the mass center and the relative velocity of the particles in the system with respect to the mass center, that is,

$$\boxed{\text{Eq. (4.74), p. 306}}$$

$$T = \tfrac{1}{2}mv_G^2 + \tfrac{1}{2}\sum_{i=1}^{n} m_i v_{i/G}^2.$$

EXAMPLE 4.10   *Conservation of Energy in a Particle System*

Analyze the simple catapult in Fig. 1, which consists of the two blocks A and B connected via the pulley at P. The stretched spring attached to B, as well as the weight of B falling through the distance h, will launch A. Let B, with $m_B = 8$ kg, be released from rest at a distance $h = 2$ m above the ground. At the time of release, A, with $m_A = 1$ kg, is in the position shown in Fig. 1. The spring attached to B is linear with constant $k = 40$ N/m, and it is unstretched when B reaches the ground. Using the dimensions shown in Fig. 1, determine the speeds of A and B immediately before B hits the ground.

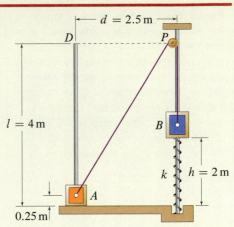

Figure 1
Proposed catapult system with masses A and B connected via the pulley at P. The dimensions shown are those at release.

SOLUTION

Road Map & Modeling   We need to relate changes of position to changes in speed, and so we will apply the work-energy principle. We will assume that the inertia and size of the pulley are negligible, the system is frictionless, and the cord is massless as well as inextensible. Referring to Fig. 2, and modeling A and B as particles, we assume that the only external forces acting on the system are gravity, the spring force, the reactions at P, and the normal forces at A and B. The reactions at P do no work since their point of application is fixed. The gravitational and spring forces are conservative. While the rope provides an internal force, for the rope to do work the two ends of the rope need to move relative to one another in the direction of the tension (see discussion of Eq. (4.68) on p. 304), i.e., the rope would have to stretch. Since the rope is inextensible, its tension does no work. We define ① to be at release (from rest) and ② to be the position at which B hits the ground and the spring becomes unstretched.

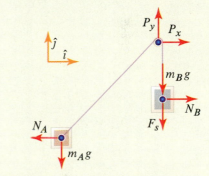

Figure 2
A FBD of the system shown in Fig. 1.

Governing Equations

Balance Principles   From the discussion above, we see that $(U_{1\text{-}2})_{nc}^{ext}$ and $(U_{1\text{-}2})_{nc}^{int}$ in Eq. (4.70) are both zero. Therefore, the work-energy principle between ① and ② becomes

$$T_1 + V_1 = T_2 + V_2, \tag{1}$$

The system's kinetic energies can be written as

$$T_1 = \tfrac{1}{2}m_A v_{A1}^2 + \tfrac{1}{2}m_B v_{B1}^2 \quad \text{and} \quad T_2 = \tfrac{1}{2}m_A v_{A2}^2 + \tfrac{1}{2}m_B v_{B2}^2, \tag{2}$$

where v_{A1} and v_{B1} are the speeds of A and B, respectively, at ① and v_{A2} and v_{B2} are the speeds of A and B, respectively, at ②.

Force Laws   The potential energies in Eq. (1) are made up of the gravitational potential energies of m_A and m_B as well as the elastic potential energy of the spring. Therefore, referring to Figs. 1 and 3, we have

$$V_1 = \tfrac{1}{2}kh^2 + m_A g y_{A1} + m_B g y_{B1} \quad \text{and} \quad V_2 = m_A g y_{A2} + m_B g y_{B2}, \tag{3}$$

where we have used the fact that the spring is unstretched at ②.

Kinematic Equations   At ① we have

$$v_{A1} = 0 \quad \text{and} \quad v_{B1} = 0. \tag{4}$$

As far as ② is concerned, we note that v_{A2} and v_{B2} are the unknowns of the problem. In addition we note that both the position and speed of A are related to the position and speed of B. To determine these reations, referring to Fig. 3, we observe that the cord's length is

$$L = \sqrt{(l - y_A)^2 + d^2} + (l - y_B). \tag{5}$$

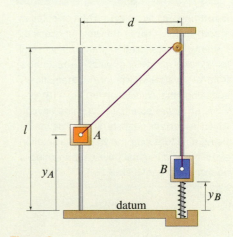

Figure 3
Definition of y_A, y_B, and the datum line for the gravitational potential energy.

Since the cord is inextensible, differentiating Eq. (5) with respect to time, we have

$$0 = \frac{-(l - y_A)v_{Ay}}{\sqrt{(l - y_A)^2 + d^2}} - v_{By} \quad \Rightarrow \quad \left| -\frac{l - y_A}{\sqrt{(l - y_A)^2 + d^2}} \right| v_A = v_B, \quad (6)$$

where the introduction of the absolute value in the second equation allows us to replace v_{Ay} and v_{By} with the *speeds* v_A and v_B, respectively. To determine the relationship between the displacement of A and the displacement of B, we can evaluate Eq. (5) at ① and ② as follows:

$$L = \sqrt{(l - y_{A1})^2 + d^2} + (l - y_{B1}) = \sqrt{(l - y_{A2})^2 + d^2} + (l - y_{B2}). \quad (7)$$

The second equality in Eq. (7) implies that

$$y_{B1} - y_{B2} = \sqrt{(l - y_{A1})^2 + d^2} - \sqrt{(l - y_{A2})^2 + d^2}. \quad (8)$$

Recalling that $y_{B1} - y_{B2} = h$ and y_{A1} is known and solving Eq. (8) for y_{A2}, we have

$$y_{A2} = l - \sqrt{h^2 + (l - y_{A1})^2 - 2h\sqrt{(l - y_{A1})^2 + d^2}}. \quad (9)$$

Computation   Substituting Eqs. (2)–(4) into Eq. (1), we obtain

$$\tfrac{1}{2}kh^2 + m_A g y_{A1} + m_B g y_{B1} = \tfrac{1}{2}m_A v_{A2}^2 + \tfrac{1}{2}m_B v_{B2}^2 + m_A g y_{A2} + m_B g y_{B2}. \quad (10)$$

Recalling again that $y_{B1} - y_{B2} = h$, substituting Eqs. (6) into Eq. (10), and solving for v_{A2} yield

$$v_{A2} = \sqrt{\frac{kh^2 + 2m_A g(y_{A1} - y_{A2}) + 2m_B gh}{m_A + m_B(l - y_{A2})^2 / [(l - y_{A2})^2 + d^2]}}. \quad (11)$$

Recalling that $l = 4\,\text{m}$, $y_{A1} = 0.25\,\text{m}$, $h = 2\,\text{m}$, and $d = 2.5\,\text{m}$, from Eq. (9) we see that $y_{A2} = 3.814\,\text{m}$. Using this result, along with the given data, from Eqs. (11) and (6) we see that

$$\boxed{v_{A2} = 19.7\,\text{m/s} \quad \text{and} \quad v_{B2} = 1.46\,\text{m/s}.} \quad (12)$$

Discussion & Verification   The term multiplying v_A in Eq. (6) is nondimensional, which implies that Eq. (6) is dimensionally correct. The argument of the square root in Eq. (11) has dimensions of energy divided by mass, i.e., speed squared. Therefore, Eq. (11) is also dimensionally correct. To assess whether or not the value for v_{A2} (and therefore the corresponding value of v_{B2}) in Eq. (12) is reasonable, we can say that v_{A2} should not exceed the speed that we would obtain if *all* of the potential energy of B were turned into the kinetic energy of A. Such a value is obtained as follows:

$$\tfrac{1}{2}kh^2 + m_B gh = \tfrac{1}{2}m_A v_{A2}^2 \quad \Rightarrow \quad v_{A2} = 21.8\,\text{m/s}. \quad (13)$$

Because the result in Eq. (12) is less than that of Eq. (13), we can say that our answer behaves as expected. Interestingly, even without the spring (i.e., for $k = 0$), we achieve $v_{A2} = 15.3\,\text{m/s}$ with $v_{B2} = 1.14\,\text{m/s}$. If we were to orient the shaft holding A horizontally, then we wouldn't have to worry about the increase in potential energy of A and we would achieve an even larger v_{A2}.

EXAMPLE 4.11　*Why Does a Snapped Towel Hurt?*

If you have ever been struck by the end of a snapped towel, you know that it stings and that there is often a loud crack that can be heard at the end of the snap. We would like to construct a model capable of explaining these observations.

　　To explain the discomfort and sound associated with a snapped towel, we are going to look at the simple model of the towel shown in Fig. 2. Using this string-based model, we attach one end to the ceiling, fold the string back on itself, and then release the unattached end and let it fall freely. For a string of length L and mass m, determine the speed of the free end of the string as a function of its distance below the ceiling.

Figure 1
A sequence of photos showing one of your authors snapping a towel. The exposure time of each photo is 0.0008 s. *Top frame:* preparing for the snap. *Middle frame:* the towel is halfway through the forward part of the snap. The towel is bent like the initial state of our string. *Bottom frame:* the towel has reached its farthest extent. Notice that the last few inches of the towel are blurry due to the high speed at which this part of the towel is moving.

SOLUTION

Road Map & Modeling　We are interested in the speed of the free end after it has fallen a given distance. Therefore, the work-energy principle would seem to be the ideal approach to obtain the solution. While the string is a not a system of particles (it is a system in which the mass is distributed in space), we can still apply the concepts for systems of particles. We will do so by applying the work-energy principle to the equivalent mass center of the system. We start by assuming that the string is inextensible. Then, referring to the FBD in Fig. 3, we *view* the string as a system of two particles (whose mass sums to that of the string), one particle being the left branch and the other the right. These particles will be placed at the midpoint or mass center of their corresponding branches. Figure 3 shows the string in an arbitrary position, which we will call ②, and we will let ① be at release. Notice that the only forces acting on the string are the weight of each segment and the reaction at the support. We assume that no other forces are acting on the system, which means that the weight forces $m_L g$ and $m_R g$ are the only forces doing work on the string — we have a conservative system.

Governing Equations

Balance Principles　The system is conservative, so the work-energy principle is given by

$$T_1 + V_1 = T_2 + V_2. \tag{1}$$

The kinetic energies are given by

$$T_1 = \tfrac{1}{2}m_{L1}v_{L1}^2 + \tfrac{1}{2}m_{R1}v_{R1}^2 \quad \text{and} \quad T_2 = \tfrac{1}{2}m_{L2}v_{L2}^2 + \tfrac{1}{2}m_{R2}v_{R2}^2, \tag{2}$$

where m_L and m_R are the masses of the left and right branches of the string, respectively, and where v_L and v_R are the speeds of the mass centers of the left and right branches of the string, respectively. In all cases, the second subscript indicates the position.

Force Laws　We assume that the string is inextensible and that it does not dissipate any energy. To determine the system's potential energy, it is useful to make use of the mass per unit length of string $\rho = m/L$. Letting ℓ_L and ℓ_R be the lengths of the left and right branches of the string, respectively, and defining the datum for gravitational potential energy to be at the top of the fixed portion of string, we have

$$V_1 = -m_{L1}gL/4 - m_{R1}gL/4 = -(m_{L1} + m_{R1})gL/4, \tag{3}$$
$$V_2 = -m_{L2}g\ell_L/2 - m_{R2}g(y + \ell_R/2), \tag{4}$$

where the masses of the left and right branches of the string at ① and ② are given by

$$m_{L1} = \rho L/2, \qquad m_{R1} = \rho L/2, \qquad m_{L2} = \rho \ell_L, \qquad m_{R2} = \rho \ell_R, \tag{5}$$

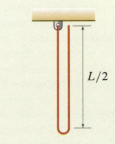

Figure 2
The initial configuration of the string.

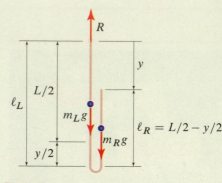

Figure 3
FBD and kinematics of the string after the free end has fallen a distance y, where $m_L g$ is the weight of the left segment and $m_R g$ is the weight of the right.

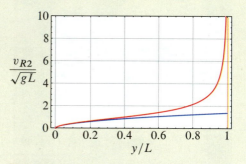

Figure 4
Nondimensional speed of the falling string (red line) and a particle in free fall (blue line) as a function of nondimensional distance. The orange vertical line represents $y = L$.

Interesting Fact

Cracking a whip. When a whip is cracked, the tip reaches a supersonic velocity for a duration of 1.2 ms. In a distance of 45 cm, the tip is accelerated from $M = 1$ to $M = 2.2$ (M is the Mach number, which is the ratio of the object's speed to the speed of sound in the surrounding medium). Assuming a uniform acceleration during this period and the speed of sound in air of 345 m/s, the acceleration of the tip exceeds 51,000 g! See P. Krehl, S. Engemann, and D. Schwenkel, "The Puzzle of Whip Cracking — Uncovered by a Correlation of Whip-Tip Kinematics with Shock Wave Emission," *Shock Waves*, **8**(1), 1998, pp. 1–9.

so that V_1 and V_2 become

$$V_1 = -\tfrac{1}{4}\rho g L^2 \quad \text{and} \quad V_2 = -\rho g \ell_L^2/2 - \rho g \ell_R (y + \ell_R/2). \tag{6}$$

Kinematic Equations   Referring to Fig. 3, the total length of the string L and the fall distance y are related to the length of the left and right segments as follows:

$$L = \ell_L + \ell_R \quad \text{and} \quad \ell_L = y + \ell_R. \tag{7}$$

Solving Eqs. (7) for ℓ_L and ℓ_R, we obtain

$$\ell_L = L/2 + y/2 \quad \text{and} \quad \ell_R = L/2 - y/2, \tag{8}$$

which are shown in Fig. 3.

Since the string is released from rest, at ① we have

$$v_{L1} = 0 \quad \text{and} \quad v_{R1} = 0. \tag{9}$$

Due to the inextensibility of the string, all points on the left segment of string must have a common velocity (as must all points of the right segment). Therefore, since the left branch is attached to the fixed ceiling, we must have

$$v_{L2} = 0. \tag{10}$$

Computation   By substituting Eqs. (5), (8), (9), and (10) into Eq. (2) and then simplifying, the kinetic energies become

$$T_1 = 0 \quad \text{and} \quad T_2 = \tfrac{1}{2}\rho \ell_R v_{R2}^2 = \tfrac{1}{4}\rho(L - y)v_{R2}^2. \tag{11}$$

Substituting Eqs. (8) into the second of Eqs. (6) and simplifying, we obtain the final form for V_2 to be

$$V_2 = -\frac{\rho g}{4}\left(L^2 + 2Ly - y^2\right). \tag{12}$$

Substituting the first of Eqs. (6), Eqs. (11), and Eq. (12) into Eq. (1) and solving for v_{R2}, we obtain

$$\boxed{v_{R2} = \sqrt{gy\left(\frac{2L - y}{L - y}\right)}.} \tag{13}$$

Discussion & Verification   The argument of the square root in Eq. (13) has dimensions of acceleration times length, which is the same as speed squared. Therefore, the answer in Eq. (13) is dimensionally correct. In addition, notice that the value of the speed increases as y increases, as should be expected. Hence, overall our solution appears to be correct.

🔎 **A Closer Look**   When we let $y \to L$ in Eq. (13), that is, as the string approaches the end of its fall, we see that $v_{R2} \to \infty$ (see Fig. 4). This is an amazing result! Simply take a string, bend it, and drop it as we have done here, and the speed of the tip of the string becomes infinite at the end of the fall. While we know that an infinite speed is not achieved in reality, what the model does imply is that the tip of the string is going *very* fast. In fact, when a real snapped towel is aided by the person snapping it, the loud crack produced is a consequence of the shock wave generated by the tip of the towel when the speed of the tip becomes larger than the speed of sound.[*] This also applies to the cracking of a whip.

---
[*] N. Lee, S. Allen, E. Smith, and L. M. Winters, "Does the Tip of a Snapped Towel Travel Faster Than Sound?" *The Physics Teacher*, **31**(6), 1993, pp. 376–377.

EXAMPLE 4.12   *Internal Work due to Friction*

The two blocks A and B in Fig. 1 are connected via an inextensible cord and the pulley system shown. There is negligible friction between A and the incline, and the static and kinetic coefficients of friction between A and B are μ_s and μ_k, respectively. Assuming that μ_s is insufficient to prevent slip and that the system is released from rest, determine the velocity of A and B after B has moved *up the incline* a distance d *relative to A*. Solve the problem by using the following parameters: $m_A = 4\,\text{kg}$, $m_B = 1\,\text{kg}$, $\theta = 30°$, $d = 0.35\,\text{m}$, and $\mu_k = 0.1$.

SOLUTION

Road Map & Modeling   We will apply the work-energy principle since we need to relate changes in position to changes in velocity. As we did in Example 4.3, we will model enough of the system so that we don't have to worry about the tension in the cord. Referring to Fig. 2, we have sketched the system's FBD, which includes A and B modeled as particles, the cord, and the pulleys. The contact forces between A and B are not shown since they are internal to the system. We will treat the pulleys as frictionless and the cord as inextensible as well as massless. Since the pulleys are fixed, the reactions at the pulleys do no work. Also, the cord tension does no work because the cord is inextensible (Prob. 4.59 verifies this). However, since A and B slide relative to one another, we need to include the work done by the internal friction force between A and B, so we have also drawn the FBD of B in Fig. 2 so that we can find this friction force. Finally, we will let ① be when the system is released and ② be when B has moved the distance d up the incline relative to A.

Governing Equations

Balance Principles   The work-energy principle for the system in Fig. 2 is

$$T_1 + V_1 + (U_{1\text{-}2})_{\text{nc}}^{\text{int}} = T_2 + V_2, \qquad (1)$$

where $(U_{1\text{-}2})_{\text{nc}}^{\text{int}}$ is the work done by friction. The kinetic energies are given by

$$T_1 = \tfrac{1}{2}m_A v_{A1}^2 + \tfrac{1}{2}m_B v_{B1}^2 \quad \text{and} \quad T_2 = \tfrac{1}{2}m_A v_{A2}^2 + \tfrac{1}{2}m_B v_{B2}^2, \qquad (2)$$

where v_A and v_B are the speed of A and B, respectively, and the second subscript on the speeds indicates position. Again referring to Fig. 2, we sum forces in the y direction on B and obtain

$$\sum F_y: \quad N_B - m_B g \cos\theta = 0, \qquad (3)$$

where we have set $a_{By} = 0$ because there is no motion in the y direction.

Force Laws   For the system's potential energy we can write

$$V_1 = V_{A1} + V_{B1} = 0, \qquad (4)$$
$$V_2 = V_{A2} + V_{B2} = -m_A g\,\Delta x_A \sin\theta - m_B g\,\Delta x_B \sin\theta, \qquad (5)$$

where we have placed the potential energy datum at ① and where Δx_A and Δx_B are the displacements parallel to the incline of A and B, respectively, between ① and ②. Both Δx_A and Δx_B are unknown, and we will use kinematics to relate them to the given distance d.

     The work of the internal friction force F_B is given by

$$(U_{1\text{-}2})_{\text{nc}}^{\text{int}} = -\int_0^d F_B\,dx. \qquad (6)$$

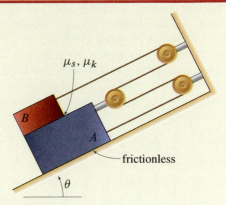

Figure 1
The two-block system connected via pulleys.

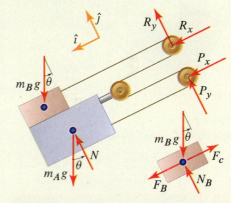

Figure 2
FBD of the system and of just B.

🧭 Helpful Information

Why choose the FBD of B when determining F_B? Could we have chosen the FBD of A to find F_B? The easiest way to find out is to try it, but we would quickly discover that the FBD of A involves the unknown N_B and that N_B cannot be found until we draw the FBD of B. By contrast, starting with Eq. (3) immediately leads to F_B via Eq. (7).

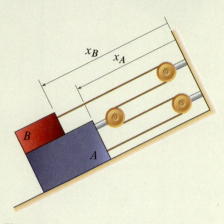

Figure 3
Definition of x_A and x_B for the two blocks.

Finally, since we know that μ_s is insufficient to prevent slipping, we know that

$$F_B = \mu_k N_B. \tag{7}$$

Kinematic Equations   Since the system starts from rest, we have

$$v_{A1} = 0 \quad \text{and} \quad v_{B1} = 0. \tag{8}$$

In addition, we can relate the motions of A and B using pulley kinematics (see Section 2.7 on p. 135). Referring to Fig. 3, we see that for arbitrary x_A and x_B

$$3x_A + x_B = L \quad \Rightarrow \quad 3v_A = -v_B, \tag{9}$$

where L is constant and where v_A and v_B represent the velocity of A and B, respectively, since the motion is one-dimensional. From Eq. (9), at ② we have

$$3v_{A2} = -v_{B2} \quad \text{and} \quad 3\Delta x_A = -\Delta x_B. \tag{10}$$

Finally, we know that A displaces a distance d relative to B. Therefore, we can write

$$\Delta x_{A/B} = d = \Delta x_A - \Delta x_B. \tag{11}$$

Solving Eqs. (10) and (11) for Δx_A and Δx_B, we obtain

$$\Delta x_A = \tfrac{1}{4}d \quad \text{and} \quad \Delta x_B = -\tfrac{3}{4}d. \tag{12}$$

Computation   Substituting Eqs. (3) and (7) into Eq. (6), we have

$$(U_{1\text{-}2})_{\text{nc}}^{\text{int}} = -\mu_k m_B g d \cos\theta. \tag{13}$$

Next, we substitute the first of Eqs. (10) into Eq. (2) to obtain

$$T_2 = \tfrac{1}{2}m_A v_{A2}^2 + \tfrac{1}{2}m_B(-3v_{A2})^2 = \tfrac{1}{2}(m_A + 9m_B)v_{A2}^2, \tag{14}$$

and then substitute Eqs. (12) into Eq. (5) to obtain

$$V_2 = \tfrac{1}{4}(3m_B - m_A)g d \sin\theta. \tag{15}$$

We then take Eqs. (4), (8), and (13)–(15) and substitute them into Eq. (1) to obtain one equation for v_{A2}

$$-\mu_k m_B g d \cos\theta = \tfrac{1}{2}(m_A + 9m_B)v_{A2}^2 + \tfrac{1}{4}(3m_B - m_A)g d \sin\theta, \tag{16}$$

which gives

$$v_{A2} = \pm\sqrt{\frac{2gd}{m_A + 9m_B}\left[\tfrac{1}{4}(m_A - 3m_B)\sin\theta - \mu_k m_B \cos\theta\right]}. \tag{17}$$

Observing that A moves up the incline and B moves down the incline, using Eqs. (17) and (9), as well as the given data, we have

$$\boxed{v_{A2} = 0.142\,\text{m/s} \quad \text{and} \quad v_{B2} = -0.427\,\text{m/s.}} \tag{18}$$

Discussion & Verification   The dimensions of the argument of the square root in Eq. (17) are energy divided by mass, i.e., speed squared. Hence, the expression for v_{A2} in Eq. (17) is dimensionally correct. The expression of v_B in Eqs. (9) is also dimensionally correct. As far as the values in Eqs. (18) are concerned, since B acts as a counterweight for A, a way to check the reasonableness of our result is to compute the speed that A would have if it were disconnected from B and moved a distance $d/4$ down the incline without friction. This speed, given by $(v_{A2})_{\text{no friction}} = \sqrt{2(d/4)g\sin\theta} = 0.926\,\text{m/s}$, is an upper bound for v_{A2}. That is, v_{A2} must be less than $0.926\,\text{m/s}$, as indeed it is.

🔎 **A Closer Look**   Although the solution of the problem is complete, there is an additional issue worth considering: what if we had not been told that the block B was sliding up the incline? In this case we would have had to solve a companion statics problem to determine whether or not slip was occurring at all and, in case of slip, in which direction. We will briefly consider this problem, and we will determine the critical value of μ_s for slipping to occur at all.

Referring to Fig. 4, we can write Newton's second law for A as

$$\sum F_x: \quad m_A g \sin\theta - F_B - 3F_c = m_A a_{Ax} = 0, \tag{19}$$

$$\sum F_y: \quad N - N_B - m_A g \cos\theta = m_A a_{Ay} = 0, \tag{20}$$

and for B as

$$\sum F_x: \quad F_B - F_c + m_B g \sin\theta = m_B a_{Bx} = 0, \tag{21}$$

$$\sum F_y: \quad N_B - m_B g \cos\theta = m_B a_{By} = 0, \tag{22}$$

where we have set all the accelerations to zero given that we are analyzing the static equilibrium of the system. Solving Eqs. (19)–(22) for N, N_B, F_B, and F_c, we obtain

$$N = (m_A + m_B)g\cos\theta, \qquad N_B = m_B g\cos\theta, \tag{23}$$

$$F_B = \tfrac{1}{4}(m_A - 3m_B)g\sin\theta, \qquad F_c = \tfrac{1}{4}(m_A + m_B)\sin\theta. \tag{24}$$

The critical value of μ_s for slip is equal to F_B/N_B, which can easily be found from Eqs. (23) and (24) to be

$$\frac{F_B}{N_B} = \frac{m_A - 3m_B}{4m_B}\tan\theta = \tfrac{1}{4}(\nu - 3)\tan\theta, \tag{25}$$

where we have defined ν to be the ratio of m_A to m_B. 🖥️ ⇥ A contour plot of F_B/N_B as given by Eq. (25) can be found in Fig. 5. This single plot provides a wealth of information.

- For $\nu > 3$, we find that $F_B/N_B > 0$ and so the blocks always will slip as indicated in the problem statement, that is, B will slip up the incline relative to A. Conversely, a negative value of F_B/N_B tells us that slipping occurs in the direction opposite to that assumed. This is true for the entire left side of the plot for which $\nu < 3$ and for any value of θ. In either case, slipping depends on the value of μ_s. Specifically, if $\mu_s > (\mu_s)_{\text{crit}}$, then it *will not* slip; and if $\mu_s < (\mu_s)_{\text{crit}}$, then it *will* slip. Note that the value of ν separating the two behaviors discussed above, i.e., $\nu = 3$, comes from the pulley arrangement; i.e., the force applied to A is 3 times that applied to B.

- As θ, the incline angle, increases, the value of μ_s to keep the blocks from sliding relative to one another increases rapidly and is infinite for $\theta = 90°$. Notice that all curves in Fig. 5 converge to the point at $\nu = 3$ and $\theta = 90°$. This is so because, with the given pulley arrangement, the system is perfectly balanced when m_A is 3 times m_B, and so it will not slip for *any* value of μ_s. ⇤ 🖥️

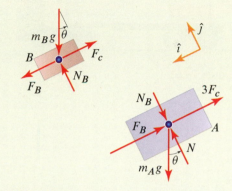

Figure 4

FBDs of A and B to determine the slip condition.

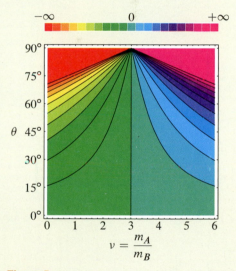

Figure 5

A contour plot of $F_B/N_B = (\mu_s)_{\text{crit}}$ as a function of the incline angle θ and $\nu = m_A/m_B$ from Eq. (25).

PROBLEMS

Figure P4.53

💡 Problem 4.53 💡

The driver of the truck suddenly applies its brakes, and the truck comes to a stop under the action of a constant braking force. During braking, either the crate slides or it does not. Considering the work-energy principle applied to the truck during braking, will the truck stop in a shorter distance (or time) if the crate slides, or will the distance (or time) be shorter if it does not? Assume that the truck bed is long enough that you don't have to worry about whether or not the crate hits the truck. Justify your answer.
Note: Concept problems are about *explanations*, not computations.

Problem 4.54

Consider a pulley system in which bodies A and B have masses $m_A = 2\,\text{kg}$ and $m_B = 10\,\text{kg}$. If the system is released from rest, neglecting all sources of friction as well as the inertia of the pulleys, determine the speeds of A and B after B has displaced a distance of 0.6 m downward.

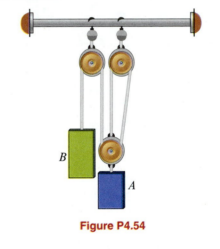

Figure P4.54

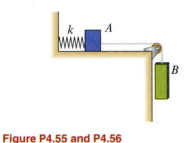

Figure P4.55 and P4.56

Problem 4.55

Blocks A and B, weighing 7 and 15 lb, respectively, are released from rest when the spring is unstretched. If all sources of friction are negligible and $k = 12\,\text{lb/ft}$, determine the maximum vertical displacement of B from the release position, assuming that A never leaves the horizontal surface shown and the cord connecting A and B is inextensible.

Problem 4.56

Blocks A and B are released from rest when the spring is unstretched. Block A has a mass $m_A = 2\,\text{kg}$, and the linear spring has stiffness $k = 7\,\text{N/m}$. If all sources of friction are negligible, determine the mass of block B such that B has a speed $v_B = 1.5\,\text{m}$ after moving 1.2 m downward, assuming that A never leaves the horizontal surface shown and the cord connecting A and B is inextensible.

Problem 4.57

A 700 lb floating platform is at rest when a 200 lb crate is thrown onto it with a horizontal speed $v_0 = 12$ ft/s. Once the crate stops sliding relative to the platform, the platform and the crate move with a speed $v = 2.667$ ft/s. Neglecting the vertical motion of the system as well as any resistance due to the relative motion of the platform with respect to the water, determine the distance that the crate slides relative to the platform if the coefficient of kinetic friction between the platform and the crate is $\mu_k = 0.25$.

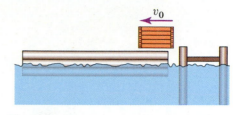

Figure P4.57

Problem 4.58

Two small spheres A and B, each of mass m, are attached at either end of a stiff light rod of length d. The system is released from rest in the position shown. Determine the speed of the spheres when A has reached point D, and determine the normal force between sphere A and the surface on which it is sliding immediately before it reaches point D. Neglect friction, treat the spheres as particles (assume that their diameter is negligible), and neglect the mass of the rod.

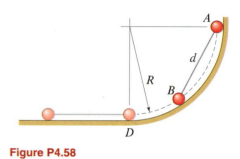

Figure P4.58

Problem 4.59

Solve Example 4.12 by applying the work-energy principle to each block individually, and show that the net work done by the cord on the two blocks is zero.

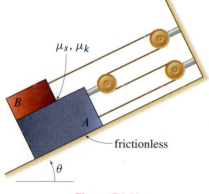

Figure P4.59

Problem 4.60

Consider a simple elevator design in which a 15,000 kg car A is connected to a 12,000 kg counterweight B. Suppose that a failure of the drive system occurs (the failure does not affect the rope connecting A and B) when the car is at rest and 50 m above its buffer, causing the elevator car to fall. Model the car and the counterweight as particles and the cord as massless and inextensible, and model the action of the emergency brakes using a Coulomb friction model with kinetic friction coefficient $\mu_k = 0.5$ and a normal force equal to 35% of the car's weight. Determine the speed with which the car impacts the buffer.

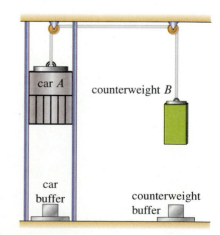

Figure P4.60

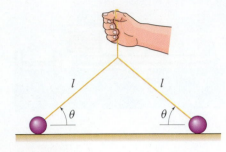

Figure P4.61

Problem 4.61

Two identical balls, each of mass m, are connected by a string of negligible mass and length $2l$. A short string is attached to the first at its middle and is pulled vertically with a constant force P (exerted by the hand). If the system starts at rest when $\theta = \theta_0$, determine the speed of the two balls as θ approaches $90°$. Neglect the size of the balls as well as friction between the balls and the surface on which they slide.

Problem 4.62

Consider the simple catapult shown in the figure with an 800 lb counterweight A and a 150 lb projectile B. If the system is released from rest as shown, determine the speed of the projectile after the arm rotates (counterclockwise) through an angle of $110°$. Model A and B as particles, neglect the mass of the catapult's arm, and assume that friction is negligible. The catapult's frame is fixed with respect to the ground, and the projectile does not separate from the arm during the motion considered.

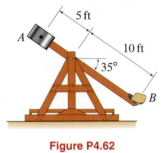

Figure P4.62

Problem 4.63

Revisit Example 4.6 on p. 288 and determine the maximum height reached by the pole vaulter, but this time include the mass of the pole in your analysis. To solve the problem, use the following data: the maximum speed achieved by the pole vaulter and pole at the time the vault begins is $v_{\max} = 34.19$ ft/s; the pole vaulter is 6 ft tall with mass center at 55% of body height (as measured from the ground) and weighs 190 lb; the pole is uniform, has a weight of 5.8 lb, and is 17.1 ft long. In addition, assume that as the pole vaulter sprints before the vault, the pole is carried horizontally at the same height as the vaulter's center of mass relative to the ground. Explain why the vault height you find when the pole is included is higher than the vault height determined in Example 4.6, when the pole was not included.

Figure P4.63

Problems 4.64 and 4.65

Two blocks *A* and *B* weighing 123 and 234 lb, respectively, are released from rest as shown. At the moment of release the spring is unstretched. In solving these problems, model *A* and *B* as particles, neglect air resistance, and assume that the cord is inextensible.

Problem 4.64 Determine the maximum speed attained by block *B* and the distance from the floor where the maximum speed is achieved if $\alpha = 20°$, the contact between *A* and the incline is frictionless, and the spring constant is $k = 30\,\text{lb/ft}$.

Problem 4.65 Determine the maximum displacement of block *B* if $\alpha = 20°$, the contact between *A* and the incline is frictionless, and the spring constant is $k = 300\,\text{lb/ft}$.

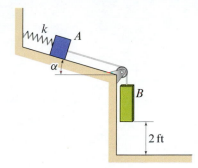

Figure P4.64 and P4.65

Problem 4.66

Spring scales work by measuring the displacement of a spring that supports both the platform of mass m_p and the object of mass m, whose weight is being measured. In your solution, note that most scales read zero when no mass m has been placed on them; that is, they are calibrated so that the weight reading neglects the mass of the platform m_p. Assume that the spring is linear elastic with spring constant k. If the mass m is gently placed on the spring scale (i.e., it is dropped from zero height above the scale), determine the *maximum reading* on the scale after the mass is released.

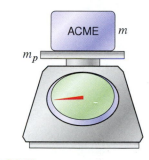

Figure P4.66

Problem 4.67

The rope of mass m and length l is released from rest with a *very* small amount of it hanging over the edge (i.e., $s > 0$ but it is very close to zero). Determine the speed of the rope as a function of s. Assume the surface is smooth.

Figure P4.67

Problem 4.68

Consider a system of four identical particles A, B, D, and E, each with a mass of 2.3 kg. These particles are mounted on a wheel of negligible mass whose hub is at O and whose radius is $R = 0.75\,\text{m}$. The wheel is mounted on a cart H of negligible mass. Suppose that the wheel is spinning counterclockwise with an angular speed $\omega = 3\,\text{rad/s}$ and that the cart is moving to the right with a speed of $v_H = 11\,\text{m/s}$. Compute the kinetic energy of the system, using

(a) Eq. (4.71) on p. 305,

(b) Eq. (4.74) on p. 306.

Show that the result is the same in both cases.

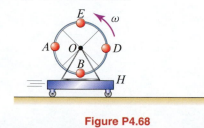

Figure P4.68

💻 **Problem 4.69** 💻

Reexamine Example 4.10 by choosing ① to be the same as that in the example and ② to be an arbitrary position after release. Determine the speed of A as a function of y_A, and plot the kinetic energy of A as a function of y_A. Use this plot to argue that the cord connecting A and B never goes slack between the instant of release and the impact of B with the ground. *Hint:* To argue that the cord remains taut between the positions of interest, you may want to look at concept Prob. 4.21 on p. 274.

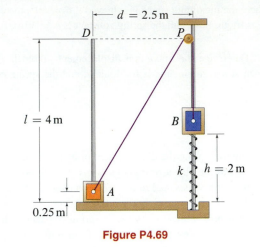

Figure P4.69

=== **DESIGN PROBLEMS** ===

Design Problem 4.2

The plunger (the rod and rectangular block attached to the end of it) of a pinball machine has weight $W_p = 5$ oz. The weight of the ball is $W_b = 2.85$ oz, and the machine is inclined at $\theta = 8°$ with the horizontal. Design the springs k_1 and k_2 (i.e., their stiffnesses and unstretched lengths) so that the ball separates from the plunger with a speed $v = 15$ ft/s after pulling back the plunger 2 in. from its rest position and so that the plunger comes no closer than 0.5 in. from the stop. Note that as part of your design, you will also need to specify reasonable values for the dimensions ℓ_1 and ℓ_2.

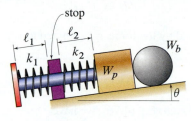

Figure DP4.2

4.4   Power and Efficiency

Power developed by a force

The concept of *mechanical power* arises from a need to quantify the ability of a motor or other machine (see Fig. 4.23) to do mechanical work or provide energy in a given amount of time. To make this concept precise, referring to Fig. 4.24, we consider a force $\vec{F}$ whose point of application undergoes an infinitesimal displacement $d\vec{r}$ in a corresponding infinitesimal time interval dt. The infinitesimal work done by $\vec{F}$ during the time interval dt is

$$dU = \vec{F} \cdot d\vec{r}. \tag{4.75}$$

The *power* P developed by the force $\vec{F}$ is defined as time rate at which the $\vec{F}$ does work, that is,

$$\text{power} = P = \frac{dU}{dt}. \tag{4.76}$$

Substituting Eq. (4.75) into Eq. (4.76), we have

$$P = \frac{\vec{F} \cdot d\vec{r}}{dt} = \vec{F} \cdot \frac{d\vec{r}}{dt}. \tag{4.77}$$

Recalling that the velocity of the point of application of $\vec{F}$ is $\vec{v} = d\vec{r}/dt$, we can then rewrite Eq. (4.77) as

$$P = \vec{F} \cdot \vec{v}. \tag{4.78}$$

Dimensions and units of power

Because of its definition, the power developed by a force is a scalar quantity with dimensions of (force) × (length)/(time) or, equivalently, (mass) × (length)^2/(time)^3.

In the SI system, the unit of measure for power is called the *watt*, which is abbreviated via the symbol W and is defined as follows:

$$1\,\text{W} = 1\,\text{N}\cdot(1\,\text{m/s}) = 1\,\text{J/s}. \tag{4.79}$$

In the U.S. Customary system, the unit of measure for power is called *horsepower*, which is abbreviated by the symbol hp and is defined as follows:

$$1\,\text{hp} = 550\,\text{ft}\cdot\text{lb/s}. \tag{4.80}$$

Expressed to four significant figures, the conversion from horsepower to watt is $1\,\text{hp} = 745.7\,\text{W}$.

Efficiency

When a motor or a machine does work, it does so by using some form of energy supply, such as organic fuel, chemical batteries, electricity, etc. Therefore an important measure of the performance of a motor or a machine is the ratio between its ability to *provide* power (i.e., power output) and the power *needed*

Figure 4.23
The 2009 Ferrari Formula One race car. Machines such as this car produce a tremendous amount of power for their size.

Figure 4.24
A force acting on a point moving along a path.

for its operation (i.e., power input). Such a measure is called *efficiency*, which is often denoted via the Greek letter ϵ (epsilon) and is defined as follows:

$$\epsilon = \frac{\text{power output}}{\text{power input}}. \qquad (4.81)$$

Because of its definition, efficiency is a nondimensional scalar quantity. Given that some of the power supplied to a motor or a machine is used to move the internal components of the motor or the machine, thus overcoming internal frictional resistance, the power output of a motor or machine is always smaller than the power input. Therefore the efficiency of a motor or a machine is a number that is always smaller than 1.

EXAMPLE 4.13   *Computing Power*

Figure 1
An F/A-18 taking off from an aircraft carrier.

In Example 4.1 on p. 265 we considered the takeoff of a 50,000 lb F/A-18 Hornet (see Fig. 1) from a stationary aircraft carrier. The F/A-18 goes from 0 to 165 mph in a distance of 300 ft, assisted by a catapult and with each of its two engines generating a constant thrust of 22,000 lb. Neglecting aerodynamic forces, in Example 4.1 we determined that the work done on the aircraft by the catapult was $(U_{1\text{-}2})_{\text{catapult}} = 3.227 \times 10^7$ ft·lb. Continuing Example 4.1, determine the power developed by the thrust generated by the engines at the beginning and at the end of the takeoff. In addition, determine the average power supplied to the aircraft by the catapult, knowing that the takeoff occurs in roughly 2 s.

SOLUTION

Road Map & Modeling   Denoting the beginning and the end of the takeoff by ① and ②, respectively, we know the thrust of the engines as well as the velocity of the aircraft at ① and ②. Therefore, we can compute the power developed by the engines' thrust via the relation expressing the power developed by a force in terms of the force itself and the velocity vector. To compute the *average* power of the catapult, we will compute the time average of the power developed by the force exerted by the catapult on the aircraft and relate that to the given work done by the catapult.

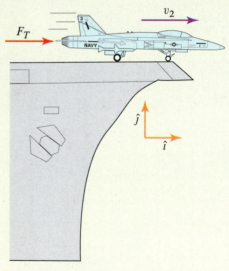

Figure 2
Diagram showing the aircraft's velocity vector and the engines' thrust vector at ②.

Computation   Recall that the quantity $\vec{F} \cdot \vec{v}$ is the power developed by a force $\vec{F}$ whose point of application moves with velocity $\vec{v}$. Since the plane starts from rest, i.e., $\vec{v}_1 = \vec{0}$, letting P_{T1} denote the power developed by the engines' thrust at ①, we must have

$$P_{T1} = 0. \tag{1}$$

Referring to Fig. 2, at ② we will assume that the plane is moving horizontally with a velocity $\vec{v}_2 = v_2\,\hat{\imath}$, where $v_2 = 165$ mph. In addition, assuming that the engines' thrust is also horizontal, the total thrust of the engines is $\vec{F}_T = F_T\,\hat{\imath}$ with $F_T = 44{,}000$ lb. Therefore, letting P_{T2} be the power developed by $\vec{F}_T$ at ②, we have

$$P_{T2} = F_T\,\hat{\imath} \cdot v_2\,\hat{\imath} = F_T v_2 = 19{,}400\,\text{hp}. \tag{2}$$

Finally, letting P_{cat} and $(P_{\text{avg}})_{\text{cat}}$ be the instantaneous and the average power supplied to the aircraft by the catapult between ① and ②, respectively, we have

$$(P_{\text{avg}})_{\text{cat}} = \frac{1}{t_2 - t_1}\int_{t_1}^{t_2} P_{\text{cat}}\,dt = \frac{1}{t_2 - t_1}\int_{t_1}^{t_2} \frac{dU_{\text{cat}}}{dt}\,dt = \frac{(U_{1\text{-}2})_{\text{cat}}}{t_2 - t_1}. \tag{3}$$

Recalling that $t_2 - t_1 = 2$ s and that $(U_{1\text{-}2})_{\text{cat}} = 3.227 \times 10^7$ ft·lb, we have

$$(P_{\text{avg}})_{\text{cat}} = 29{,}300\,\text{hp}. \tag{4}$$

Discussion & Verification   The results in Eqs. (1) and (2) are obtained as the direct application of a known formula. Hence, we need only verify that we used the correct units, which we have since we expressed our result in horsepower and the data was provided in U.S. Customary units. The result in Eq. (3) was obtained by applying the definition of time average. Keeping this in mind, intuition suggests that the notion of "average power" should correspond to the ratio between the work done by the catapult and the time taken to do such work, which is exactly what our calculation yielded.

EXAMPLE 4.14   *Assessing Lift Speed Given the Efficiency*

An electric motor with an efficiency of 85% drives the pulley system shown to lift a 400 kg crate. After the motor is switched on and the system has reached a constant lifting speed, the motor is drawing 15 kW. Assume that the motor is operating at its rated efficiency, that the frictional losses in the pulley system are negligible, and that the weights of the pulleys A and B are negligible. Determine the speed with which the crate is being lifted by considering

(a) an FBD of the system that includes the crate and both pulleys A and B,

(b) an FBD of the crate that has been separated from the system at the hook H.

Neglect the size of the pulleys in your solution.

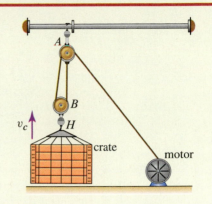

Figure 1
Pulley system lifting a crate with speed v_c.

SOLUTION

Road Map & Modeling   Using the knowledge of the input power and efficiency, we can calculate the output power of the motor. This power corresponds to the power used by the pulley system to lift the crate. Since there are no power losses within the pulley system, the output power can be viewed as either *the power of the cable between pulley A and the motor* or as *the power developed by the tension in the hook H attached to the crate*. Parts (a) and (b) of the problem statement reflect these two viewpoints. Hence, we need to use Newton's second law to determine the tension in the cable attached to the motor and then we will be able to compute the lifting speed by dividing the power developed by this force by the magnitude of the force. We will perform a similar calculation for the hook attached to the crate.

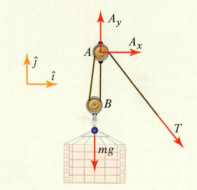

Figure 2
FBD of the pulleys and crate.

─────────── **Solution for Part (a)** ───────────

Governing Equations

Balance Principles   The FBD of the pulleys and crate is shown in Fig. 2, in which the crate has been modeled as a particle and where T is the tension in the cable being pulled in by the motor. While it is the power developed by the tension T that we want to find, this FBD does not allow us to find T itself. Instead, applying Newton's second law to the FBD in Fig. 3(a), we find

$$\sum F_y: \quad 2T - mg = ma_y. \tag{1}$$

Force Laws   All forces are accounted for on the FBD.

Kinematic Equations   Since the crate is moving with constant velocity, we have

$$a_y = 0. \tag{2}$$

To find the speed of the point of application of the tension in the cable between the motor and the pulley at A, we need to refer to the pulley kinematics in Fig. 3(b) and write

$$2\ell_c + \ell_m = \text{constant} \quad \Rightarrow \quad 2\dot{\ell}_c = -\dot{\ell}_m \quad \Rightarrow \quad 2v_c = \dot{\ell}_m, \tag{3}$$

where we have used the fact that $v_c = -\dot{\ell}_c$ and we note that when $\dot{\ell}_m > 0$, the motor is winding in cable.

Computation   Substituting Eq. (2) into Eq. (1), we find that

$$T = mg/2. \tag{4}$$

We need to compute the power provided by the motor to the pulley system. Letting P_i and P_o be the motor power input and output, respectively, we have

$$P_i = 15\,\text{kW} \quad \text{and} \quad P_o = \epsilon P_i = 12.75\,\text{kW}. \tag{5}$$

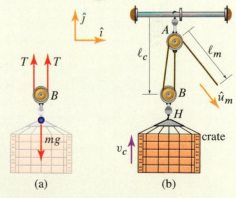

Figure 3
(a) FBD of the crate and the pulley at B.
(b) Kinematics of the pulley system connecting the motor to the crate.

Recalling that P_o is the power developed by the tension T, and that T is in the same direction as the velocity of the cable winding up on the motor, we then must have

$$P_o = T\,\hat{u}_m \cdot \dot{\ell}_m\,\hat{u}_m = T\dot{\ell}_m = \frac{mg}{2}(2v_c) = mgv_c, \qquad (6)$$

where we have used Eqs. (3) and (4). Solving Eq. (6) for v_c, we have

$$\boxed{v_c = \frac{P_o}{mg} = 3.25\,\text{m/s},} \qquad (7)$$

where we have used P_o from Eqs. (5).

────────────────────────── **Solution for Part (b)** ──────────────────────────

Governing Equations

Balance Principles   The FBD of the crate is shown in Fig. 4, in which the crate has been modeled as a particle subject to its own weight and the tension F_H in the hook. Using the FBD in Fig. 4, Newton's second law yields

$$\sum F_y: \quad F_H - mg = ma_y. \qquad (8)$$

Force Laws   All forces are accounted for on the FBD.

Kinematic Equations   Since the crate is moving with constant velocity, we have

$$a_y = 0. \qquad (9)$$

Computation   Substituting Eq. (9) into Eq. (8) and solving for F_H, we have

$$F_H = mg. \qquad (10)$$

Again, we need to compute the power provided by the motor to the pulley system. Recalling that P_o is the power developed by the tension F_H, and that F_H is in the same direction as the velocity of the crate, we then must have

$$P_o = F_H\,\hat{j} \cdot v_c\,\hat{j} = F_H v_c, \qquad (11)$$

where v_c is the speed of the crate. Solving Eq. (11) for v_c, we have

$$\boxed{v_c = \frac{P_o}{F_H} = \frac{P_o}{mg} = 3.25\,\text{m/s},} \qquad (12)$$

where we have used P_o from Eqs. (5).

──

Discussion & Verification   In Parts (a) and (b), P_o has dimensions of force times length divided by time, so the results in Eqs. (7) and (12) have the expected dimensions of length over time. In addition, given that the data was provided using SI units, the results in question are expressed using appropriate units. As far as the value of v_c is concerned, as expected, it is smaller than the maximum possible value of lifting speed given by $P_i/mg = 3.82\,\text{m/s}$. Hence, overall our solution appears to be correct.

🔍 A Closer Look   Notice that in Part (a) of this example, we had to analyze the pulley kinematics, but we did not have to do so in Part (b). That is so because in pulley systems, the power supplied determines the speed of the payload, regardless of the pulley arrangement. This stems from the fact that power is the product of force and velocity and the nature of pulley systems is to keep that product constant.

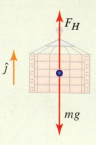

F_H

$\hat{j}$

mg

Figure 4
FBD of the crate.

PROBLEMS

Problem 4.70

Consider a battery that can power a machine whose operation requires 200 W of power. If the battery can keep the machine operating for 8 h and supply energy at a constant rate, determine the amount of energy initially stored in the battery pack.

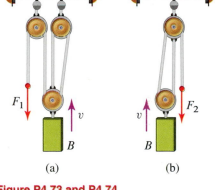

Figure P4.70

Problem 4.71

A person takes 16 s to lift a 200 lb crate to a height of 30 ft. Determine the average power supplied by the person.

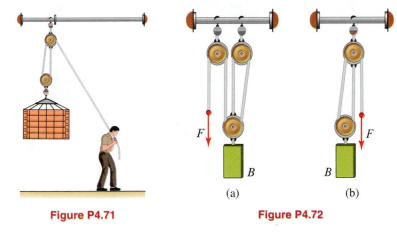

Figure P4.71   **Figure P4.72**

Problem 4.72

The weights B in the pulley systems shown are identical. Assume that the two systems are released from rest, there are no energy losses in the pulleys, and the same constant force F is applied. After 1 s, will the power developed by the force F for system (a) be smaller than, equal to, or greater than that of system (b)?
Note: Concept problems are about *explanations*, not computations.

Figure P4.73 and P4.74

Problem 4.73

The weights B in the pulley systems shown are identical and are being lifted at the same constant speed v. Assume there are no energy losses in the pulleys. Is the power developed by force F_1 in (a) smaller than, equal to, or greater than that developed by force F_2 in (b)?
Note: Concept problems are about *explanations*, not computations.

Problem 4.74

For the two pulley systems shown, the block B has a mass $m_B = 254$ kg and is being lifted with a constant speed $v = 3$ m/s. Determine the power developed by the forces F_1 and F_2.

Problem 4.75

A cyclist is riding with a speed $v = 18$ mph over an inclined road with $\theta = 15°$. Neglecting aerodynamic drag, if the cyclist were to keep his power output constant, what speed would he attain if θ were equal to $20°$?

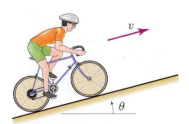

Figure P4.75

Figure P4.76

Figure P4.77

Photo © 2006 Mazda Motor of America, Inc.
Used by permission.

Problem 4.76

A 1500 kg car is traveling up the slope shown at a constant speed. Knowing that the maximum power output of the car is 160 hp, at what speed can the car travel if air resistance is negligible? Also, knowing that 1 L of regular gasoline provides 34.8 MJ of energy, how many liters of gasoline will be required in 1 h if the engine has an efficiency $\epsilon = 0.20$?

Problem 4.77

A 2600 lb car on straight horizontal road goes from 0 to 60 mph in 7 s. Neglecting aerodynamic drag, if the force propelling the car is constant during the acceleration phase, determine the power developed by this force 7 s after the beginning of the motion.

Problem 4.78

The height h of the aerobic stepper shown is 25 cm. A 65 kg person goes up and down the step once every 2 s. Accounting only for the work done in lifting her body and assuming that a muscle efficiency is 25%, determine the calories (1 C = 4.184 kJ) burned by the person in 1 h.

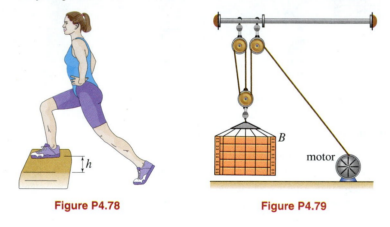

Figure P4.78   **Figure P4.79**

Problem 4.79

The motor powering the pulley system shown is drawing 6 kW of power. The crate B has a mass of 250 kg. Determine the speed of the crate if the crate is lifted at a constant speed and the motor's efficiency is $\epsilon = 0.87$.

Problem 4.80

Assuming that the motor shown has an efficiency $\epsilon = 0.80$, determine the power to be supplied to the motor if it is to pull a 250 lb crate up the incline with a constant speed $v = 7$ ft/s. Assume that the kinetic friction coefficient between the slide and the crate is $\mu_k = 0.25$ and that $\theta = 28°$.

Problem 4.81

Assume that the motor shown has an efficiency $\epsilon = 0.82$ and that it draws 6 hp. Determine the constant speed with which the motor can pull a 300 lb crate up the incline if the kinetic friction coefficient between the slide and the crate is $\mu_k = 0.45$ and $\theta = 32°$.

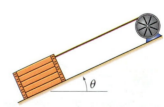

Figure P4.80 and P4.81

4.5   Chapter Review

In this chapter we have introduced the concepts of work of a force, kinetic energy of a particle, and kinetic energy of a particle system. We have shown that the work done on a system is equal to the change in kinetic energy of the system. This relationship is referred to as the work-energy principle, and it is a direct consequence of Newton's second law. Finally, we have introduced the concepts of power developed by a force and efficiency of a motor or a machine.

Work-energy principle for a particle

The *work-energy principle* is expressed by the following equation:

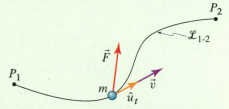

Figure 4.25
Figure 4.3 repeated. Particle of mass m moving along the path $\mathcal{L}_{1\text{-}2}$ while being subjected to the force $\vec{F}$.

> **Eq. (4.14), p. 262**
> $$T_1 + U_{1\text{-}2} = T_2,$$

where $U_{1\text{-}2}$ is the *work* done by the force $\vec{F}$ (see Fig. 4.25) and is defined as

> **Eq. (4.12), p. 261**
> $$U_{1\text{-}2} = \int_{\mathcal{L}_{1\text{-}2}} \vec{F} \cdot d\vec{r}$$

and T is the particle's *kinetic energy*, which is defined as

> **Eq. (4.13), p. 261**
> $$T = \tfrac{1}{2}mv^2,$$

where m is the mass of the particle and v is its speed. Work and kinetic energy have the following basic properties:

- Work and kinetic energy are *scalar* quantities.

- The work depends on only the component of $\vec{F}$ in the direction of motion, that is, on $\vec{F} \cdot \hat{u}_t$, and its sign is determined by the sign of $\vec{F} \cdot \hat{u}_t$.

- The kinetic energy is *never* negative.

Conservative forces and potential energy

In Section 4.2 we discovered that there are forces whose work, in taking a particle from an initial to a final position, does not depend on the path connecting these positions. Specifically, we derived the following expressions.

Work of a constant gravitational force.   For a particle of mass m moving in a constant gravitational field from ① to ②, the work done on the particle is

> **Eq. (4.21), p. 277**
> $$(U_{1\text{-}2})_g = -mg(y_2 - y_1),$$

for which gravity must act in the $-y$ direction.

Helpful Information

Other forms of the work of a force. By recalling that position and velocity are related by the expression $d\vec{r} = \vec{v}\, dt$ and that, in normal-tangential components, the velocity can be expressed as $\vec{v} = v\,\hat{u}_t$, we can obtain the following two useful forms of the work of a force:

$$U_{1\text{-}2} = \int_{t_1}^{t_2} \vec{F} \cdot \vec{v}\, dt = \int_{s_1}^{s_2} \vec{F} \cdot \hat{u}_t\, ds,$$

where t is time, s is the arc length along the path, and $\hat{u}_t$ is the unit vector tangent to the path and pointing in the direction of motion.

Work done by a spring force. For a particle subject to the force of a linear elastic spring with constant k, the work done on the particle is

Eq. (4.27), p. 278

$$(U_{1\text{-}2})_e = -\tfrac{1}{2}k\big(\delta_2^2 - \delta_1^2\big),$$

where δ_1 and δ_2 are the amount the spring is stretched (or compressed) at ① and ②, respectively.

Conservative systems. A force is said to be *conservative* if its work depends on only the initial and final positions of its point of application but is otherwise independent of the path connecting these positions. The work of a conservative force can be characterized via a scalar potential energy function. A *conservative system* is one for which all the forces doing work are conservative. For conservative systems, the work-energy principle becomes the *conservation of mechanical energy*, which is given by

Eq. (4.32), p. 279

$$T_1 + V_1 = T_2 + V_2.$$

Important conservative forces include elastic spring forces and gravitational forces (constant or not). The *potential energy of a linear elastic spring force* is given by

Eq. (4.34), p. 280

$$V_e = \tfrac{1}{2}k\delta^2.$$

The *potential energy of a constant gravitational force* is given by

Eq. (4.33), p. 280

$$V_g = mgy,$$

where the positive y direction is opposite to gravity. The *potential energy of the force of gravity* is given by

Eq. (4.35), p. 280

$$V_G = -\frac{Gm_A m_B}{r}.$$

In a system for which there are both conservative and nonconservative forces doing work, we can use Eq. (4.14) on p. 262 or we can take advantage of the potential energies of the conservative forces by writing the work-energy principle as

Eq. (4.50), p. 282

$$T_1 + V_1 + (U_{1\text{-}2})_{\text{nc}} = T_2 + V_2,$$

where $(U_{1\text{-}2})_{\text{nc}}$ is the work done by nonconservative forces.

⚓ Concept Alert

Elastic potential energy is never negative. Whether a spring is elongated or compressed, i.e., whether δ is positive or negative, the potential energy of a spring is never negative!

Interesting Fact

Gravitational potential energy of spheres. Equation (4.35) is valid for finite size bodies (that is, other than particles) as long as r is measured from their centers and they both have spherically symmetric mass distributions. Therefore, it is valid for almost all large celestial bodies since their mass distribution is generally very close to spherically symmetric and because they are so far from one another.

Work-energy principle for systems of particles

In Section 4.3, we developed the work-energy principle for a system of particles and we discovered that internal forces play an important role in writing
this principle. We defined the kinetic energy of a particle system as the sum of
the individual kinetic energies of each particle, that is,

Eq. (4.65), p. 304

$$T = \sum_{i=1}^{n} \tfrac{1}{2} m_i v_i^2,$$

where m_i and v_i are the mass and the speed of the ith particle in the system.
The work-energy principle for a system of particles was found to be

Eq. (4.70), p. 305

$$T_1 + V_1 + (U_{1\text{-}2})_{\text{nc}}^{\text{ext}} + (U_{1\text{-}2})_{\text{nc}}^{\text{int}} = T_2 + V_2,$$

where V is the potential energy of the system, which may have contributions
from both *external* and *internal* forces.

The kinetic energy of a system can also be expressed in terms of the speed
of the mass center and the speed of each of the particles relative to the mass
center. Specifically, we have

Eq. (4.74), p. 306

$$T = \tfrac{1}{2} m v_G^2 + \tfrac{1}{2} \sum_{i=1}^{n} m_i v_{i/G}^2.$$

Power and efficiency

In Section 4.4 we defined the *power P developed by a force* $\vec{F}$ whose point of
application moves with a velocity $\vec{v}$ as

Eq. (4.78), p. 320

$$P = \vec{F} \cdot \vec{v},$$

and we observed that power has units of work per unit time.

In addition to power, Section 4.4 introduced the notion of *efficiency*, which
is an important concept used to assess the performance of motors and/or machines. Given a machine or a motor whose operation requires a certain power
input, efficiency, usually denoted by the Greek letter ϵ (epsilon), is defined as

Eq. (4.81), p. 321

$$\epsilon = \frac{\text{power output}}{\text{power input}},$$

where power output is the work done by the machine or motor per unit time.

Common Pitfall

**Internal work due to nonconservative
forces.** Don't forget that *internal work* due
to nonconservative forces needs to be ac-
counted for in the $(U_{1\text{-}2})_{\text{nc}}^{\text{int}}$ term when we
write the work-energy principle. The most
important example of this type of work is
due to internal friction.

Common Pitfall

Internal work due to conservative forces.
Don't forget that *internal work* due to conservative forces (e.g., springs) *is not* in-
cluded in the $(U_{1\text{-}2})_{\text{nc}}^{\text{int}}$ term — it is accounted for via the potential energy V,
where, for a system, the potential energy
V includes the potential energies of both
the external and the internal conservative
forces.

REVIEW PROBLEMS

Problem 4.82

Rubber bumpers are commonly used in marine applications to keep boats and ships from getting damaged by docks. Consider a boat with a mass of 35,000 kg and a bumper with a force compression profile given by $F_B = \beta x^3$, where β is a constant and x is the compression of the bumper. Treating the boat C as a particle, neglecting its vertical motion, and neglecting the drag force between the water and the boat C, determine β so that if the boat were to impact the bumper with a speed of 4 m/s, the maximum compression of the bumper would be 18 cm.

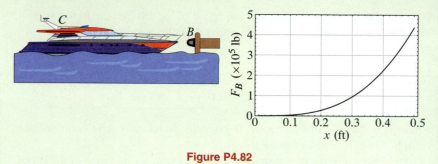

Figure P4.82

Problem 4.83

Starting from the position shown, assume that each horse moves to the right in such a way that the tension in the cord is the same in cases (a) and (b) and remains constant. Knowing that $\beta < \gamma$ and that in both cases (a) and (b) the horse advances by an equal amount L, determine which of the following statements is true: (1) the tension in the cord does more work in (a) than in (b); (2) the tension in the cord does exactly the same amount of work in (a) as in (b); (3) the tension in the cord does less work in (a) than in (b).

Note: Concept problems are about *explanations*, not computations.

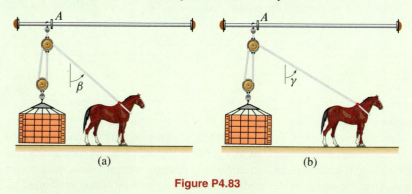

Figure P4.83

Problem 4.84

A metal ball with mass $m = 0.15$ kg is dropped from rest in a fluid. The magnitude of the resistance due to the fluid is given by $C_d v$, where C_d is a drag coefficient and v is the ball's speed. If $C_d = 2.1$ kg/s, determine the total work done on the ball from the moment of release until the ball achieves 99% of terminal velocity.

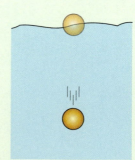

Figure P4.84 and P4.85

Problem 4.85

A metal ball weighing 0.2 lb is dropped from rest in a fluid. If the magnitude of the resistance due to the fluid is given by $C_d v$, where $C_d = 0.5\ \text{lb·s/ft}$ is a drag coefficient and v is the ball's speed, determine the work done by the drag force during the first 2 s of the ball's motion.

Problem 4.86

An 11 kg collar is constrained to travel along a rectilinear and frictionless bar of length $L = 2\,\text{m}$ that lies in the vertical plane. The springs attached to the collar are identical, and they are unstretched when the collar is at B. Treating the collar as a particle, neglecting air resistance, and knowing that at A the collar is moving upward with a speed of 23 m/s, determine the spring constant k so that the collar reaches D with zero speed. Points E and F are fixed.

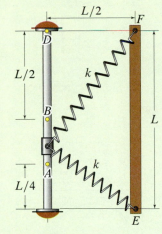

Figure P4.86

Problem 4.87

A 7 lb collar is constrained to travel along a frictionless vertical ring of radius $R = 1\,\text{ft}$. The spring attached to the collar has a spring constant $k = 20\,\text{lb/ft}$. Treating the collar as a particle, neglecting air resistance, and knowing that, while at rest at A, the collar is displaced gently to the left, determine the spring's unstretched length if the collar is to reach point B with a speed of 15 ft/s.

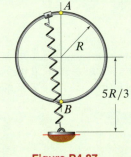

Figure P4.87

Problem 4.88

Strength of materials tells us that if a load P is applied at the free end of a cantilevered beam, then the tip displacement δ is given by $\delta = PL^3/(3EI_{cs})$, where L is the length of the beam and E and I_{cs} are constants that depend on the material makeup and the geometry of the cross section, respectively. Determine an expression for the potential energy of a cantilevered beam loaded as shown.

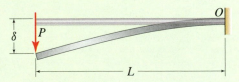

Figure P4.88

Problem 4.89

Consider the catapult shown in the figure with a 1200 kg counterweight A and a 330 kg projectile B. If the system is released from rest as shown, determine the speed of the projectile after the arm rotates (counterclockwise) through an angle of 110°. Model A and B as particles; assume that the catapult's arm has negligible mass and that friction is negligible. In addition, assume that the cord has negligible mass, is inextensible, and is always vertical. The catapult's frame is fixed with respect to the ground, and the projectile does not separate from the arm during the motion considered.

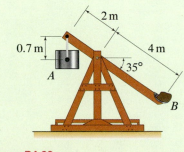

Figure P4.89

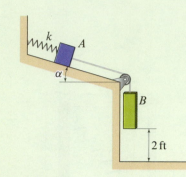

Figure P4.90

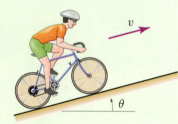

Figure P4.91

Figure P4.92

Problem 4.90

Two blocks A and B weighing 123 and 234 lb, respectively, are released from rest as shown. At the moment of release the spring is unstretched. Model A and B as particles, neglect air resistance, and assume that the cord is inextensible. Determine the maximum speed attained by block B and the distance from the floor where the maximum speed is achieved if $\alpha = 35°$, the static and kinetic friction coefficients are $\mu_s = 0.25$ and $\mu_k = 0.2$, respectively, and the spring constant is $k = 25$ lb/ft.

Problem 4.91

Spring scales work by measuring the displacement of a spring that supports both the platform of mass m_p and the object, of mass m, whose weight is being measured. In your solution, note that most scales read zero when no mass m has been placed on them; that is, they are calibrated so that the weight reading neglects the mass of the platform m_p. Assume that the spring is linear elastic with spring constant k. If the mass m is gently placed on the spring scale (i.e., it is dropped from zero height above the scale), determine the *maximum velocity* attained by the mass m as the spring compresses.

Problem 4.92

A cyclist is riding with a constant speed $v = 25$ km/h over an inclined road with $\theta = 15°$. The combined mass of the cyclist and bicycle is 85 kg. Neglecting aerodynamic drag, determine the number of calories (1 C = 4.184 kJ) burned by the cyclist in the course of 15 min if his muscle efficiency is 25%.

Problem 4.93

For the two pulley systems shown, the same block B, weighing 150 lb, is being lifted by applying the same constant force $F = 80$ lb. If the two systems are released from rest, determine the power developed by the force F in the two cases 1 s after release.

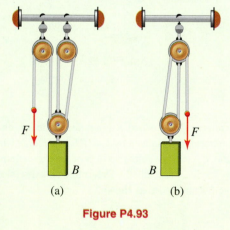

(a)   (b)

Figure P4.93

Momentum Methods for Particles

The chapter begins by discussing the concepts of momentum and impulse. We will derive a new balance law, the impulse-momentum principle, that *integrates* Newton's second law for a particle with respect to time. This balance law will be essential for the study of impacts.

Next, we will study a new concept that turns out to be fundamental to mechanics — the concept of angular momentum. We will see not only that the concept of angular momentum is useful for solving problems like those found in orbital mechanics, but also that it is fundamental in extending the applicability of Newton's laws to rigid bodies. Finally, we close with the study of mass flows, which is again a topic for which the idea of momentum will be essential.

5.1   Momentum and Impulse

Automobile collision reconstruction

Figure 5.1 displays the scene of an automobile collision. A preliminary investigation has failed to reveal the cause of the crash, and a collision expert is therefore hired to *reconstruct* the crash. To do this, the collision expert collects data on the length and orientation of the skid marks; the type, the weights, and the state of damage of the vehicles; as well as the roughness of the pavement. In this section, we will learn some of the physical principles used by forensic engineers to explain how the collision occurred.

Using the work-energy principle, the forensic engineer will be able to estimate the cars' velocities right after impact, knowing how far the cars moved after the impact as well as the properties of the tires and pavement. However, it is the determination of the cars' velocities right before the impact that is of interest in understanding the liability for a collision. As we saw in Chapter 3, to determine the change in velocity of an object between two different times, we must have a *detailed* knowledge of the forces acting on the object as well as knowledge of the time over which those forces act. However, in a crash a vehicle's speed can be reduced by 45 mph in less than 0.2 s. Actually, as collisions go, 0.2 s is a long time, and many impacts (e.g., when a bat hits a baseball or the head of a golf club hits a golf ball) typically last less than 100 ms. For example, as it can be seen in Fig. 5.2, the impact of a racquetball

Figure 5.1
Scene after a car crash.

Interesting Fact

Do collision reconstructors really exist? Collision reconstructors are more properly called forensic engineers, and they play an important role as expert witnesses in courts of law.

333

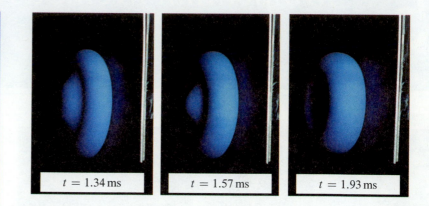

| $t = 1.34$ ms | $t = 1.57$ ms | $t = 1.93$ ms |

Figure 5.2. High speed photography sequence capturing the rebound of a racquetball against a wall.

against a wall takes about 0.004 s (4 ms). The message here is that forces, be they large (as in the examples given above) or small, change both the speed and direction that objects move. That is, forces change the *velocity* of objects. It is this concept that we will explore in this section.

Impulse-momentum principle

We will now consider how a change in velocity is related to the time interval over which that change takes place. To do this, considering a particle of mass m, we now integrate Newton's second law with respect to time over a time interval $t_1 \leq t \leq t_2$:

$$\int_{t_1}^{t_2} \vec{F} \, dt = \int_{t_1}^{t_2} m\vec{a} \, dt. \tag{5.1}$$

Since $\vec{a} \, dt = d\vec{v}$ (and m, for a particle, is constant), we obtain

$$\int_{t_1}^{t_2} \vec{F} \, dt = m\vec{v}(t_2) - m\vec{v}(t_1). \tag{5.2}$$

Equation (5.2) states that a change in the quantity $m\vec{v}$ over a certain time interval is related to the *time integral* of the force over that time interval. To better understand the meaning of Eq. (5.2) and to see how it can help us deal with phenomena such as collisions, we will introduce some basic terminology.

Linear momentum of a particle

Since the quantity $m\vec{v}$ plays a prominent role in Eq. (5.2), we give this quantity its own name and symbol. Let the vector quantity $\vec{p}(t)$ be defined as

$$\boxed{\vec{p}(t) = m\vec{v}(t).} \tag{5.3}$$

The quantity $\vec{p}(t)$ is called *linear momentum* or simply *momentum* of the particle of mass m. By definition, the momentum has dimensions of mass times length *divided by* time:

$$[\vec{p}] = [M][L][T]^{-1}. \tag{5.4}$$

Consequently, the units in the SI system are kg·m/s and in the U.S. Customary system are lb·s or slug·ft/s.

Linear impulse of a force

The quantity on the left-hand side of Eq. (5.2), that is,

$$\int_{t_1}^{t_2} \vec{F}(t)\, dt, \tag{5.5}$$

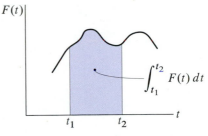

$$F(t)$$
$$\int_{t_1}^{t_2} F(t)\, dt$$

Figure 5.3
Geometric interpretation of the concept of impulse of a force.

is called the *linear impulse* (or simply *impulse*) *of the force* $\vec{F}(t)$ *between times* t_1 *and* t_2. Because it is an integral over time, any component of the impulse can be given a graphical interpretation as the *area under the curve* of a force versus time plot, as shown in Fig. 5.3.

The impulse has dimensions of force times time or mass times length divided by time. Therefore, the units of impulse are the same as for momentum. However, it is common to express the impulse using units of N·s in the SI system and lb·s in the U.S. Customary system.

Impulse-momentum principle for a particle

Using the definition of momentum, Eq. (5.2) can be written as

$$\int_{t_1}^{t_2} \vec{F}(t)\, dt = \vec{p}(t_2) - \vec{p}(t_1) \quad \text{or} \quad \vec{p}(t_1) + \int_{t_1}^{t_2} \vec{F}(t)\, dt = \vec{p}(t_2). \tag{5.6}$$

Equation (5.6) states that *a particle's change in momentum over an interval of time is equal to the impulse imparted to that particle during that same time interval*. The expression in Eq. (5.6) is called the *impulse-momentum principle* and, as we will see later in this chapter, will allow us to solve collision problems.

Equation (5.6) can be written in differential form as

$$\vec{F} = \dot{\vec{p}}, \tag{5.7}$$

which can be viewed as a restatement of Newton's second law.

Average force

Suppose we want to know the force causing a known change in momentum. This problem may be impossible to solve since the change of momentum between two instants t_1 and t_2 is not enough to allow the reconstruction of the force $\vec{F}(t)$ for every instant in time between t_1 and t_2. This can be seen in Fig. 5.4 where there is a time-varying force and its average over some time interval — we need to know the value of the force at every instant in time to know the area under the force versus time curve. However, we can calculate the *average force* that caused this change of momentum. By definition, the average force over a certain time interval is the value of the force if it were constant, i.e.,

$$\vec{F}_{\text{avg}} = \frac{\displaystyle\int_{t_1}^{t_2} \vec{F}(t)\, dt}{t_2 - t_1} \tag{5.8}$$

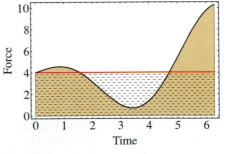

Figure 5.4
A force (the black line) and its average (the red line) over a certain time interval. The area under the force curve (light brown) equals the area under the average force curve (dashed area).

Therefore, using Eqs. (5.6), the average force can be calculated by applying the following formula:

$$\vec{F}_{\text{avg}} = \frac{\vec{p}(t_2) - \vec{p}(t_1)}{t_2 - t_1}. \tag{5.9}$$

The ability to calculate the average force needed to accomplish a given change in momentum is very useful since it tells us the magnitude of the actual force must be at least as large as or larger than that of the average during some part of the time interval.

Impulse-momentum principle for systems of particles

To extend the impulse-momentum principle to a system of particles, we need to revisit some of the ideas discussed in Chapter 3 and to distinguish between closed and open systems.

Closed systems are systems that do not exchange mass with their surroundings, whereas *open systems* do exchange mass with their surroundings. The mass of a closed system is necessarily constant, whereas an open system can have constant or variable mass. The distinction between these two types of systems becomes particularly important when we are talking about systems of particles and, in particular, about *mass flows* and *variable mass systems*, which are open by definition (see Section 5.5).

We consider a closed system of N particles (N is constant), and we view the total force on each particle as the sum of two parts (see Fig. 5.5):

1. an *external* force $\vec{F}_i$ due to the interaction of particle i with the physical bodies that do not belong to the system,

2. an *internal* force $\sum_{j=1}^{N} \vec{f}_{ij}$ due to the interaction between particle i and all of the other particles in the system ($\vec{f}_{ii} = \vec{0}$ since a particle does not exert a force on itself).

Applying Eq. (5.7), we have

$$\vec{F}_i + \sum_{j=1}^{N} \vec{f}_{ij} = \dot{\vec{p}}_i, \tag{5.10}$$

where $\vec{p}_i = m_i \vec{v}_i$ is the momentum of particle i. Equation (5.10) represents N equations, one for each particle. Let

$$\vec{F} = \sum_{i=1}^{N} \vec{F}_i \quad \text{and} \quad \vec{p} = \sum_{i=1}^{N} \vec{p}_i, \tag{5.11}$$

where $\vec{F}$ is the total external force acting on the system of particles and $\vec{p}$ is the total momentum of the system of particles. Recalling that Newton's third law requires that $\sum_{i=1}^{N} \left(\sum_{j=1}^{N} \vec{f}_{ij} \right) = \vec{0}$, we see that the sum for i going from 1 to N of Eq. (5.10), along with Eqs. (5.11), gives

$$\boxed{\vec{F} = \dot{\vec{p}}.} \tag{5.12}$$

We encountered this relationship in a slightly different form in Chapter 3 (see Eq. (3.73) on p. 237). Equation (5.12) states that internal forces play no role in changing the total momentum of a particle system. In addition, using the definition of center of mass of a closed system of particles, we have

$$\vec{p} = \sum_{i=1}^{N} \vec{p}_i = \sum_{i=1}^{N} m_i \vec{v}_i = m \vec{v}_G, \tag{5.13}$$

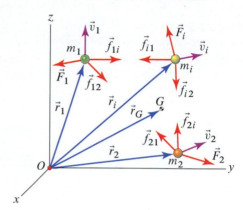

Figure 5.5
A system of particles under the action of internal and external forces. The mass center is at G.

Concept Alert

Internal forces do not change momentum. Internal forces cannot change the momentum of a system. This statement is a fundamental axiom of mechanics.

Helpful Information

Closed versus open systems. Equation (5.12) only applies to closed systems. For an *open* system, we *cannot* say that

$$\frac{d}{dt} \sum_{i=1}^{N} \vec{p}_i = \sum_{i=1}^{N} \dot{\vec{p}}_i$$

since N is not constant and its change would need to be accounted for when taking the time derivative.

where m is the total mass of the system and $\vec{v}_G$ is the velocity of the mass center of the system. Using Eq. (5.13), we can write Eq. (5.12) as

$$\vec{F} = \frac{d}{dt}\left(m\vec{v}_G\right) = m\vec{a}_G, \tag{5.14}$$

where we have used the fact that the mass of a closed system of particles is constant. Equation (5.14) states that we can view a system of particles as a single particle, whose mass is equal to the total mass of the system of particles, moving with the mass center of the system of particles.

Following the work of Euler, Eq. (5.12) is generally recognized to hold for any body whose mass is constant and for any internal force system. For this reason, Eq. (5.12) is generally considered to be a more fundamental axiom governing the motion of bodies than Newton's second law, and it is referred to as *Euler's first law*.

As argued earlier in this section, to directly investigate the relationships between changes in momentum and the applied forces, we can integrate both sides of Eq. (5.12) with respect to time, which yields

$$\int_{t_1}^{t_2} \vec{F}(t)\,dt = \vec{p}(t_2) - \vec{p}(t_1), \tag{5.15}$$

where the left-hand side of Eq. (5.15) is the *total external impulse exerted on the system between t_1 and t_2*. Even though Eqs. (5.6) and (5.15) are visually identical to one another, it is important to keep in mind that the definitions of $\vec{F}$ and $\vec{p}$ are different in the two equations.

Conservation of linear momentum

A number of engineering problems require the analysis of *isolated systems*, which are systems that are not acted upon by any external force. In other applications, including impact problems, it is common to have a situation in which there is some fixed direction in which the force is zero. In these cases, the impulse-momentum principle can take the form of a *conservation law* for which the conserved quantity is momentum (we saw conservation of energy in Section 4.1). To understand this, consider Fig. 5.6, the left side of which shows two billiard balls hitting one another. Now, let q be the line tangent to the impact surface and w be the line perpendicular to that surface. If we ignore friction between the two surfaces during impact (this is a reasonable approximation for billiard balls), then the only force in the qw plane acting on the 5 ball is directed along the w axis and there is no net external force on the two balls taken together (they form an isolated system). If the impact begins at time t_1 and ends at time t_2, and if we define our system to consist of both balls, then during the interval $t_1 \leq t \leq t_2$, the lower FBD in Fig. 5.6 indicates that we have

$$\left[\vec{F}(t)\right]_{\text{two-ball system}} = \vec{0}, \tag{5.16}$$

and therefore, by Eq. (5.15) we have

$$\left[\vec{p}(t_1)\right]_{\text{two-ball system}} = \left[\vec{p}(t_2)\right]_{\text{two-ball system}}, \tag{5.17}$$

that is, the total momentum of the two-ball system remains constant.

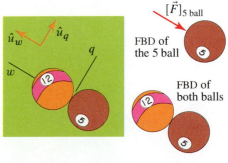

Figure 5.6
Two billiard balls hitting one another as well as the FBD of the 5 ball and the FBD of both balls during the impact.

If we now define the system to consist of only the 5 ball, for $t_1 \leq t \leq t_2$, the upper FBD in Fig. 5.6 indicates that

$$\left[\vec{F}(t)\right]_{5\text{ ball}} \cdot \hat{u}_q = 0. \tag{5.18}$$

Dotting Eq. (5.12) with $\hat{u}_q$, assuming that $\hat{u}_q$ remains constant for $t_1 \leq t \leq t_2$, and using Eq. (5.18), we obtain

$$\left[\dot{\vec{p}}\right]_{5\text{ ball}} \cdot \hat{u}_q = 0 \quad \Rightarrow \quad \left[\dot{p}_q\right]_{5\text{ ball}} = 0 \quad \Rightarrow \quad \left[p_q(t)\right]_{5\text{ ball}} = \text{const.} \tag{5.19}$$

during the impact. The last of Eqs. (5.19) states that, as long as the total external force acting on a system has a component that is zero throughout the time interval of interest, then the corresponding component of the system's momentum remains constant. For the billiard ball impact in Fig. 5.6, this means that the q component of momentum of the 5 ball remains constant (similarly, the q component of momentum of the 12 ball remains constant), and it means that the momentum vector of the system of two balls remains constant (i.e., its momentum is constant in any direction) since there is no external force acting on the two-ball system. Since the total resultant force acting on an isolated system is equal to zero, we know that *the total momentum of an isolated system never changes.* In addition, using Eq. (5.13) and noting that the mass of an isolated system does not change with time, we know that *the velocity of the center of mass of an isolated system is constant.*

End of Section Summary

We learned in Chapter 3 that forces lead to changes in velocities since forces cause accelerations. We began this section by learning that forces acting over time change momentum (not just velocity). By integrating Newton's second law, we obtained the *impulse-momentum principle*, which is given by

Eq. (5.6), p. 335

$$\vec{p}(t_1) + \int_{t_1}^{t_2} \vec{F}(t)\, dt = \vec{p}(t_2),$$

where the *linear momentum* (or *momentum*) was defined to be

Eq. (5.3), p. 334

$$\vec{p}(t) = m\vec{v}(t)$$

and a force acting over some time interval was called the *impulse* (or *linear impulse*) and is given by

Eq. (5.5), p. 335

$$\int_{t_1}^{t_2} \vec{F}(t)\, dt.$$

In addition, we found that without detailed knowledge of the force acting on a particle at every instant in time, we could not determine the change in momentum. On the other hand, knowing just the change in momentum allows us

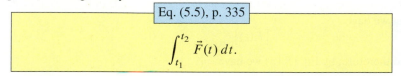

to determine the *average force* acting on a particle during the corresponding time interval, that is,

Eq. (5.9), p. 335

$$\vec{F}_{\text{avg}} = \frac{\vec{p}(t_2) - \vec{p}(t_1)}{t_2 - t_1}.$$

Impulse-momentum principle for systems of particles. When dealing with closed systems of particles, we discovered that we can write the impulse-momentum principle as

Eq. (5.12), p. 336, and Eq. (5.15), p. 337

$$\vec{F} = \dot{\vec{p}} \quad \text{and} \quad \int_{t_1}^{t_2} \vec{F}(t)\, dt = \vec{p}(t_2) - \vec{p}(t_1),$$

where $\vec{F}$ is the total external force on the particle system and $\vec{p} = \sum_{i=1}^{N} m_i \vec{v}_i$ is the total momentum of the system of particles. Using the definition of the mass center of a system of particles, the impulse-momentum principle can also be written as

Eq. (5.14), p. 337

$$\vec{F} = \frac{d}{dt}\left(m\vec{v}_G\right) = m\vec{a}_G,$$

where m is the total mass of the system of particles, $\vec{v}_G$ is the velocity of its mass center, and $\vec{a}_G$ is the acceleration of its mass center.

Conservation of linear momentum. When there is a direction in which the external force on a system of particles is zero, then the momentum in that direction is constant and is said to be conserved. If the total external force on a system of particles is zero, that is, $\vec{F} = 0$, then the momentum in every direction is constant and the mass center of the system of particles will move with constant velocity.

EXAMPLE 5.1 *Average Force on an Aircraft due to the Arresting Cable*

Figure 1
Military airplane landing on an aircraft carrier with a detailed view of the arresting cable.

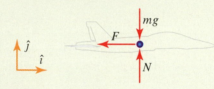

Figure 2
FBD of an airplane landing on a carrier. The force F is the force exerted by the tailhook.

To successfully land on an aircraft carrier, a pilot must use the airplane's tailhook to catch one of four steel arresting cables stretched across the landing deck (see Fig. 1). The hydraulic system activated by the motion of the arresting cable can stop a 54,000 lb aircraft traveling at a speed of 150 mph in little less than 2 s. Using this information, estimate the average braking force exerted on the plane by the arresting cable.

SOLUTION

Road Map & Modeling   Since we are given a *change of velocity* over a known *time interval* and we need to estimate the value of a force, we can apply the impulse-momentum principle, which provides a natural approach to relate force, velocity, and time information. We will model the entire plane as a *particle*, and we will focus on what happens between the time at which the tailhook first engages one of the arresting cables t_e and the time at which the plane comes to a stop t_s. While there are numerous forces acting on the plane during landing (e.g., engine thrust, lift, and drag), as can be seen in Fig. 2, we only include the force applied by the arresting cable, gravity, and the contact force between the plane and the landing deck. Since we want to determine the *average braking force*, which is the force that would act on the plane if this force were *constant*, our solution strategy will consist of writing the impulse-momentum principle under the assumption that the force exerted on the plane by the arresting cable is constant.

Governing Equations

Balance Principles   Applying Eq. (5.6) in component form to the FBD in Fig. (2), we obtain

$$x \text{ momentum:} \qquad \int_{t_e}^{t_s} F(t)\, dt = p_x(t_s) - p_x(t_e), \qquad (1)$$

$$y \text{ momentum:} \qquad \int_{t_e}^{t_s} (N - mg)\, dt = p_y(t_s) - p_y(t_e), \qquad (2)$$

where $p_x = mv_x$ and $p_y = mv_y$.

Force Laws   All forces that we are modeling are accounted for on the FBD.

Kinematic Equations   Since we are assuming that the plane only moves in a straight line parallel to the landing deck, Fig. 2 tells us that the motion is only along the x axis with $v_x(t_e) = 150$ mph. Since the plane comes to a stop at time t_s, we have

$$v_x(t_e) = 220\,\text{ft/s}, \quad v_x(t_s) = 0\,\text{ft/s}, \quad v_y(t_e) = 0\,\text{ft/s}, \quad v_y(t_s) = 0\,\text{ft/s}. \quad (3)$$

Computation   The last two of Eqs. (3) tell us there is no motion in the y direction, so Eq. (2) tells us that the force N and mg cancel each other throughout the time interval of interest. Hence, treating F as a constant, Eqs. (1) and (3) give

$$F(t_s - t_e) = mv_x(t_e) \quad \Rightarrow \quad \boxed{F = \frac{mv_x(t_e)}{t_s - t_e} = 184{,}000\,\text{lb.}} \qquad (4)$$

Discussion & Verification   The acceleration of the airplane is $a_x = F/m = 110\,\text{ft/s}^2 = 3.42g$. Carrier landings are usually considered to be maneuvers with roughly $3g$. Since we have computed an *average* force value, the plane will have accelerations in excess of $3.42g$. However, these higher accelerations would not be sustained for the entire duration of the maneuver (i.e., 2 s).

E X A M P L E   5.2   *Impulse-Momentum Applied to a Racquetball Hitting a Wall*

Figure 1 shows a 1.4 oz racquetball hitting a wall at 85 mph with an angle $\theta_1 = 65°$. The duration of the impact is 2.5 ms, and the rebound speed is 72.5 mph with the rebound angle $\theta_2 = 61.9°$. Determine the change in momentum, the impulse, and the average force applied to the ball during the impact.

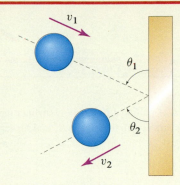

Figure 1
Racquetball approaching a wall at speed v_1 and then rebounding at speed v_2.

SOLUTION

Road Map & Modeling   We are given the mass of the ball and its pre- and postimpact velocities, so we can readily compute its pre- and postimpact momenta ($\vec{p}_1$ and $\vec{p}_2$, respectively). This will allow us to compute the change in momentum and impulse via Eq. (5.6) and the average force via Eq. (5.9). We will model the ball as a particle, and we will ignore the effects of gravity during the 2.5 ms impact. The FBD of the ball during the impact in Fig. 2 reflects this model. We have broken the contact force into a normal force N and a friction force F, both of which are a function of time.

Governing Equations

Balance Principles   Applying Eq. (5.6), the impulse-momentum principle, to the ball during impact, we obtain

$$\Delta\vec{p} = \int_{t_1}^{t_2} \vec{F}(t)\,dt = \vec{p}_2 - \vec{p}_1, \tag{1}$$

where $\Delta\vec{p}$ is the change in momentum, $\int_{t_1}^{t_2}\vec{F}(t)\,dt$ is the impulse applied to the racquetball, $\vec{p}_1 = m\vec{v}_1$ is the momentum just before impact, and $\vec{p}_2 = m\vec{v}_2$ is the momentum just after impact.

Force Laws   All forces are accounted for on the FBD.

Kinematic Equations   From Fig. 1, we can see the pre- and postimpact velocities of the racquetball are

$$\vec{v}_1 = v_1(-\sin\theta_1\,\hat{\imath} + \cos\theta_1\,\hat{\jmath}), \tag{2}$$
$$\vec{v}_2 = v_2(\sin\theta_2\,\hat{\imath} + \cos\theta_2\,\hat{\jmath}). \tag{3}$$

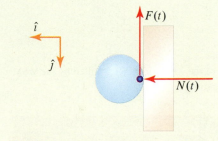

Figure 2
FBD of the racquetball during its impact with the wall.

Computation   Substituting Eqs. (2) and (3) into Eq. (1) and evaluating the result using the given quantities, which, when converted, are $m = 0.002717$ slug, $v_1 = 124.7$ ft/s, and $v_2 = 106.3$ ft/s, we obtain the impulse and change in momentum as

$$\boxed{\begin{aligned} \vec{p}_2 - \vec{p}_1 &= m\big[(v_2\sin\theta_2 + v_1\sin\theta_1)\,\hat{\imath} + (v_2\cos\theta_2 - v_1\cos\theta_1)\,\hat{\jmath}\big] \\ &= (0.5618\,\hat{\imath} - 0.0072\,\hat{\jmath})\ \text{lb·s.} \end{aligned}} \tag{4}$$

The average force is found by applying Eq. (5.9) to obtain

$$\boxed{\vec{F}_{\text{avg}} = \frac{\vec{p}_2 - \vec{p}_1}{t_2 - t_1} = \frac{(0.5618\,\hat{\imath} - 0.0072\,\hat{\jmath})\ \text{lb·s}}{(0.0025 - 0)\ \text{s}} = (225\,\hat{\imath} - 2.88\,\hat{\jmath})\ \text{lb.}} \tag{5}$$

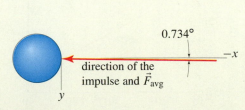

Figure 3
Vector showing the direction of the change in momentum, impulse, and average force acting on the racquetball during its impact with the wall. This vector deviates from the x axis by less than 1°.

Discussion & Verification   Figure 3 shows the direction of the change in momentum of the ball. This direction is also the direction of the impulse acting on the ball and of the average force on the ball during its impact with the wall. Equation (4) tells us that most of the change in momentum occurs in the x direction. Since the momentum changes very little in the direction parallel to the wall, the frictionless impact assumption we frequently use for impacts is often a good one. The magnitude of the average force is 225 lb, which seems about right if you have ever been hit by a racquetball.

EXAMPLE 5.3   *A Person Pulling a Commercial Airliner*

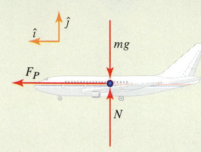

Figure 1

A man at A pulling a Boeing 737 airliner using both his arms (with the rope AB) and his legs.

Figure 2

FBD of the 737 as it is being pulled by the WSM competitor.

An event at the World's Strongest Man (WSM) competition involves the timed pull of a Boeing 737 airliner over a distance of 82 ft. Given that a Boeing 737 weighs about 75 tons, that the competitor starts pulling with a force of 850 lb, that his strength decreases linearly by 30% during the course of the pull, and that he completes the pull in 40 s, determine how fast he is moving at the end of the pull.

SOLUTION

Road Map & Modeling   We are given enough information to determine the force as a function of time with which the competitor pulls. We know that the system starts from rest and that we want to find its final speed after a given amount of time, so this problem lends itself to the application of the impulse-momentum principle. We neglect the rolling resistance of the airplane and the mass of the competitor, treat the plane as a particle, and assume that the force with which the person pulls the plane is parallel to the motion. Using these assumptions, the FBD of the plane is shown in Fig. 2.

Governing Equations

Balance Principles   Applying the impulse-momentum principle in the x direction to the FBD in Fig. 2 from the time that the person begins pulling until he pulls for 40 s gives

$$x \text{ momentum:} \quad p_1 + \int_{t_1}^{t_2} F_P \, dt = p_2, \tag{1}$$

where $t_1 = 0\,\text{s}$ is the time at which he starts pulling, $t_2 = 40\,\text{s}$ is the time at which he stops pulling, $p_1 = mv_1$ is the momentum of the plane at time t_1, and $p_2 = mv_2$ is the momentum of the plane at time t_2.

Force Laws   To derive the force law $F_P(t)$ for the WSM competitor, we note that he starts pulling with a force of 850 lb and it decreases linearly by 30% over the course of the pull. Therefore, the force versus time curve must be as shown in Fig. 3, and the equation for the curve is

$$F_P = -\frac{0.3(850)}{40}t + 850 = (-6.375t + 850)\,\text{lb}. \tag{2}$$

Kinematic Equations   The kinematic equation for this problem is that the competitor starts from rest, so

$$v_1 = 0. \tag{3}$$

Computation   Substituting Eqs. (2) and (3) into Eq. (1), integrating, and using $m = 4658\,\text{slug}$, we obtain the final speed v_2:

$$\int_0^{40} (-6.375t + 850)\, dt = mv_2 \quad \Rightarrow \quad \boxed{v_2 = 6.204\,\text{ft/s} = 4.23\,\text{mph}.} \tag{4}$$

Discussion & Verification   The competitor and the plane are both moving at a little over 4 mph at the end of the 82 ft pull. This is in the range of typical walking speeds, so it is not an unreasonable result. Thus, the value specified for F_P was appropriate. Note that a 75 ton Boeing 737 would have substantial rolling resistance, which we have ignored. It is also noted that a person *really* did pull a 75 ton Boeing 737 a distance of 82 ft in 40 s during a WSM competition. Therefore, the initial force generated by the WSM competitor must have actually been larger than the 850 lb that we have estimated.

Helpful Information

Why have we ignored the mass of the competitor? To include the mass of the competitor in our solution, we would have to realize that the force F_P with which the competitor is pulling the airplane is also pulling *him*. Of course, adding in a 300 lb strongman barely changes the result since he is only 0.2% of the weight of the airplane.

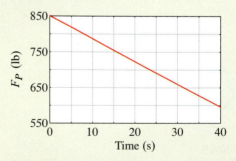

Figure 3

The force F_P versus time curve for the WSM competitor.

EXAMPLE 5.4 *Walking on a Floating Platform: Conservation of Momentum*

A person of mass m_p is at end A of a floating platform of mass m_{fp} and length L_{fp}. The person and the platform are initially at rest. The platform is touching the pier (as shown in Fig. 1) when the person starts moving toward B with a constant speed v_0, relative to the platform. Determine the distance between the platform and the pier when the person reaches the other end of the platform at B.

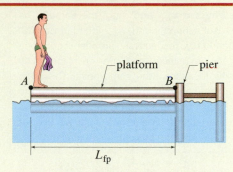

Figure 1
Person on a floating platform.

SOLUTION

Road Map & Modeling   We will model both the person and the platform as particles and will ignore the vertical motion of the platform as the person walks along its length. We will also ignore the drag force between the platform and the water. Since we do not know the forces acting between the person and the platform, our solution will be based on the analysis of the forces *external* to the system, where the system consists of the person + platform. These assumptions imply the FBD shown in Fig. 2, in which we notice that there are no forces in the horizontal direction, so momentum in that direction must be conserved.

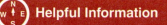

> **Helpful Information**
>
> **An alternate solution method.** This example has a lot in common with Example 3.9, which was solved without the use of the impulse-momentum principle. You should compare the two examples.

Governing Equations

Balance Principles   Applying the impulse-momentum principle for a system of particles, i.e., Eq. (5.15), in the x direction to the FBD in Fig. 2, we obtain

$$\int_{t_1}^{t_2} F_x(t)\,dt = p_x(t_2) - p_x(t_1), \tag{1}$$

where t_1 is when the person is at A, t_2 is any later time, F_x is the total force on the system in the x direction, and p_x is the momentum of the system in the x direction. Using Eq. (5.11), we can write $p_x(t_1)$ and $p_x(t_2)$ as

$$p_x(t_1) = m_p \dot{x}_{p1} + m_{fp} \dot{x}_{fp1} \quad \text{and} \quad p_x(t_2) = m_p \dot{x}_{p2} + m_{fp} \dot{x}_{fp2}, \tag{2}$$

where x_p and x_{fp} are defined in Fig. 3 (the subscripts p and fp stand for person and floating platform, respectively). The FBD in Fig. 2 tells us that $F_x = 0$, and so Eqs. (1) and (2) become

$$m_p \dot{x}_{p1} + m_{fp} \dot{x}_{fp1} = m_p \dot{x}_{p2} + m_{fp} \dot{x}_{fp2}, \tag{3}$$

which simply says that x momentum for the system is conserved.

Force Laws   All forces have been accounted for on the FBD.

Kinematic Equations   Both the person and platform start from rest, so

$$\dot{x}_{p1} = \dot{x}_{fp1} = 0. \tag{4}$$

In addition, referring to Fig. 3, applying the relative velocity equation $\vec{v}_p = \vec{v}_{p/fp} + \vec{v}_{fp}$, and neglecting the system's vertical motion, we have

$$\dot{x}_{p2} = -v_0 + \dot{x}_{fp2}. \tag{5}$$

The t_2 we are interested in is the time when the person gets to the far end of the platform, that is, $t_2 = t_f$. Since the person walks at a constant speed v_0 a distance L_{fp}, the time at which he reaches the far end of the platform must be

$$t_f = \frac{L_{fp}}{v_0}. \tag{6}$$

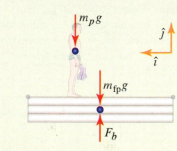

Figure 2
FBD of the system, i.e., the person + platform. The force F_b is the buoyancy force.

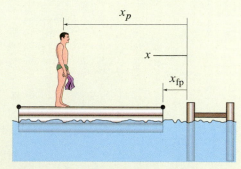

Figure 3
Kinematics diagram showing the origin of the x axis as well as the definition of the positions of the person and the platform.

Why are x_{pf} and $x_{\mathrm{fp}f}$ the same? The results for x_{pf} and $x_{\mathrm{fp}f}$ found in Eqs. (9) and (10), respectively, are identical to one another since they are each providing the location of the same point. This is so because when the person has walked the length of the platform, he is at point B and point B is also the point we use for measuring the position of the platform.

What is the effect of v_0? Equation (10) implies that v_0 plays no role in determining the position of the platform after the person has walked the length. What is the role played by the value of v_0? The value of v_0 determines the value of t_f; that is, the person can do nothing to control the position of the platform when he gets to the other end, but he can choose *when* this event takes place.

Computation   Substituting Eq. (4) into Eq. (3), we obtain

$$0 = m_p \dot{x}_{p2} + m_{\mathrm{fp}} \dot{x}_{\mathrm{fp}2}. \tag{7}$$

Equations (5) and (7) are two equations for the two unknowns $\dot{x}_{p2}$ and $\dot{x}_{\mathrm{fp}2}$. Solving those resulting two equations, we obtain

$$\dot{x}_{p2} = \left(\frac{-m_{\mathrm{fp}}}{m_p + m_{\mathrm{fp}}} \right) v_0 \quad \text{and} \quad \dot{x}_{\mathrm{fp}2} = \left(\frac{m_p}{m_p + m_{\mathrm{fp}}} \right) v_0. \tag{8}$$

Recalling that v_0 is constant, we see Eqs. (8) can be integrated to obtain

$$\int_{L_{\mathrm{fp}}}^{x_{pf}} dx_{p2} = \int_0^{t_f} \frac{-m_{\mathrm{fp}} v_0}{m_p + m_{\mathrm{fp}}} \, dt \quad \Rightarrow \quad x_{pf} = \left(\frac{m_p}{m_p + m_{\mathrm{fp}}} \right) L_{\mathrm{fp}}, \tag{9}$$

and

$$\int_0^{x_{\mathrm{fp}f}} dx_{\mathrm{fp}2} = \int_0^{t_f} \frac{m_p v_0}{m_p + m_{\mathrm{fp}}} \, dt \quad \Rightarrow \quad \boxed{x_{\mathrm{fp}f} = \left(\frac{m_p}{m_p + m_{\mathrm{fp}}} \right) L_{\mathrm{fp}},} \tag{10}$$

where we have substituted in Eq. (6) for t_f and where we have replaced the subscript 2 with f. We were seeking to find the distance between the platform and the pier after the person walks the length of the platform, and this answer is provided by Eq. (10) since $x_{\mathrm{fp}f}$ measures exactly that distance.

Discussion & Verification   The final results in Eqs. (9) and (10) both have the dimension of length, which is what they should be. In addition, notice that Eq. (10) tells us that the more massive the person (relative to the platform), the farther from the pier the platform will end up, with the limit being when m_p is infinitely more massive than m_{fp} and then $x_{\mathrm{fp}f} = L_{\mathrm{fp}}$. In addition, it says that if something with very little mass (e.g., a flea) walks across the platform, the platform will barely move. Both of these observations probably agree with your intuition.

A Closer Look   Notice that the answer is independent of v_0; that is, it does not matter how fast the person walks, the platform will always end up the same distance from the pier. In the problems section, we will solve this problem using an alternative, though completely equivalent, solution method.

EXAMPLE 5.5   *A Floating Platform Tied to the Pier: Modeling the Motion*

A person of mass m_p is at end A of a floating platform of mass m_{fp} and length L_{fp}. The person and the platform are initially at rest. The platform is touching the pier (as shown in Fig. 1) when the person starts moving toward B with a constant speed v_0, relative to the platform. Unlike in Example 5.4, the platform is moored to the pier via a massless and inextensible rope of length L_r. Determine

(a) the velocity of the platform relative to the water,

(b) the time at which the rope becomes taut, and

(c) the tension in the rope as a function of time if the rope becomes taut before the person arrives at B.

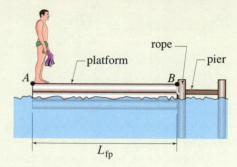

Figure 1
Floating platform moored at a fixed pier.

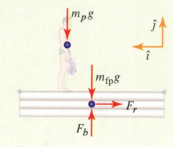

SOLUTION

Road Map & Modeling   As with Example 5.4, we will treat the person and platform as particles, we will ignore the drag force between the platform and the water, and we will neglect the vertical motion of the system. Since we do not know the forces acting between the person and the platform, our solution will again be based on the analysis of the forces *external* to the system consisting of the person + platform. This model implies the FBD shown in Fig. 2, where the force F_r, which is exerted by the rope on the platform, is nonzero only when the rope is taut. As with Example 5.4, the impulse-momentum principle will be used.

Figure 2
FBD of the system, i.e., the person + platform. The buoyancy force is F_b and the rope force is F_r.

Governing Equations

Balance Principles   Applying the impulse-momentum principle for a system of particles, i.e., Eq. (5.15), in the x direction to the FBD in Fig. 2, we obtain

$$\int_{t_1}^{t_2} -F_r \, dt = p_x(t_2) - p_x(t_1), \tag{1}$$

where $t_1 = 0$, t_2 is any time after the person begins walking, and p_x is the momentum of the system in the x direction. Using Eq. (5.11), we can write $p_x(t_1)$ and $p_x(t_2)$ as

$$p_x(t_1) = m_p \dot{x}_{p1} + m_{fp} \dot{x}_{fp1} \quad \text{and} \quad p_x(t_2) = m_p \dot{x}_{p2} + m_{fp} \dot{x}_{fp2}, \tag{2}$$

where x_p and x_{fp} are defined in Fig. 3.

Force Laws   To describe the tension provided by an *inextensible* rope, we need to distinguish between two cases. If the rope is not fully extended, i.e., if the distance between the rope's ends is smaller than its length, then $F_r = 0$. If the distance between the rope's ends is equal to its length, then F_r must take on whatever value is required to prevent the rope from stretching. Mathematically, we are saying that

$$F_r = 0 \text{ for } x_{fp} < L_r \quad \text{and} \quad F_r \geq 0 \text{ for } x_{fp} = L_r. \tag{3}$$

Kinematic Equations   Both the person and platform start from rest, so

$$\dot{x}_{p1} = \dot{x}_{fp1} = 0. \tag{4}$$

In addition, referring to Fig. 3, applying the relative velocity equation $\vec{v}_p = \vec{v}_{p/fp} + \vec{v}_{fp}$, and neglecting the system's vertical motion, we have

$$\dot{x}_{p2} = -v_0 + \dot{x}_{fp2}. \tag{5}$$

Computation   Now that we have the governing equations, we can start to obtain some answers.

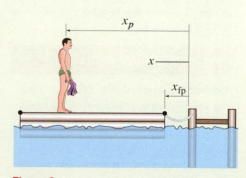

Figure 3
Diagram showing the origin of the x axis as well as the definition of the positions of the person and the platform.

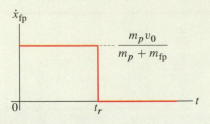

Figure 4
Plot of the platform's horizontal velocity versus time. The fact that $\dot{x}_{\text{fp}} > 0$ for $0 < t < t_r$ simply means that the platform moves to the left until the rope becomes taut. The time at which the rope becomes taut t_r is given by Eq. (8).

──────── **Part (a): Velocity of the Platform Relative to the Water** ────────

As we saw in Example 5.4, momentum is conserved when we substitute the first of Eqs. (3) into Eq. (1). In addition, if we also substitute Eqs. (2), (4), and (5) into Eq. (1) and then solve for $\dot{x}_{\text{fp2}}$, we obtain

$$\dot{x}_{\text{fp2}} = \frac{m_p v_0}{m_p + m_{\text{fp}}} \quad \text{for } x_{\text{fp}} < L_r \text{ or } 0 < t_2 < t_r, \tag{6}$$

where t_r is the time at which the rope becomes taut. This solution is only valid until the rope becomes taut, and for $t > t_r$ the velocity of the platform is equal to zero since the platform can no longer move due to the constraint enforced by the *inextensible* rope. The velocity of the platform as a function of time is described by the plot in Fig. 4.

──────── **Part (b): Time at Which the Rope Becomes Taut** ────────

To determine the time at which the rope becomes taut, we can integrate Eq. (6) to obtain

$$\int_0^{x_{\text{fp}}} dx_{\text{fp}} = \int_0^{t_r} \frac{m_p v_0}{m_p + m_{\text{fp}}} \, dt \quad \Rightarrow \quad x_{\text{fp}} = \left(\frac{m_p v_0}{m_p + m_{\text{fp}}} \right) t_r. \tag{7}$$

Letting $x_{\text{fp}} = L_r$ in Eq. (7), we find that the time at which the rope becomes taut is

$$t_r = \left(\frac{m_p + m_{\text{fp}}}{m_p v_0} \right) L_r. \tag{8}$$

──────── **Part (c): Tension in the Rope as a Function of Time** ────────

To calculate the force in the rope after it becomes taut, we can differentiate Eq. (1) with respect to time and then substitute Eq. (4) into the result to obtain

$$-F_r = \frac{d}{dt} \left(m_p \dot{x}_{p2} + m_{\text{fp}} \dot{x}_{\text{fp2}} \right). \tag{9}$$

We already know that for $x_{\text{fp}} < L_r$, the force in the rope is zero. For $x_{\text{fp}} = L_r$, the platform can no longer move, so $\dot{x}_{\text{fp2}} = 0$. Therefore we have

$$-F_r = \frac{d}{dt} \left(m_p \dot{x}_{p2} \right) = \frac{d}{dt} \left(-m_p v_0 \right) = 0 \quad \Rightarrow \quad \boxed{F_r = 0,} \tag{10}$$

where we have used Eq. (5) and the fact that m_p and v_0 are both constant.

Discussion & Verification   The dimensions on the right side of Eqs. (6), (8), and (10) are those of velocity, time, and force, respectively, as they should be.

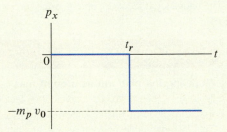

Figure 5
Plot of the system's horizontal momentum versus time. The time at which the rope becomes taut t_r is given by Eq. (8).

🖈 **A Closer Look**   The force in the rope is zero for $t < t_r$, and Eq. (10) tells us that the force in the rope is *also* zero for $t > t_r$. To understand this behavior, consider the plot of its horizontal momentum versus time given in Fig. 5. For $0 < t < t_r$, i.e., until the rope becomes taut, the system behaves as if it were *isolated* since no horizontal external forces are applied to it. The resulting momentum is therefore constant and equal to zero since the system was initially at rest. When the rope becomes taut, it applies a force to the system so as to provide the system with a finite momentum. However, since the cord is inextensible and the person maintains his or her *constant* relative velocity with respect to the platform, for $t > t_r$ the system's momentum is again constant, although not equal to zero. Recall that the time derivative of the momentum yields the force acting on the system. This is why F_r, the only horizontal external force present, is equal to zero even for $t > t_r$. The question that remains to be answered is, What happens at $t = t_r$? At $t = t_r$ the system undergoes an instantaneous *jump* in its horizontal momentum, which can be interpreted as implying that at $t = t_r$ the rope exerts an *infinite* force on the system. This behavior, clearly not possible in real physical systems, is the result of the model we have used that says that the rope is inextensible.

═══ PROBLEMS ═══

Problem 5.1

Use the definition of impulse given in Eq. (5.5) to compute the impulse of the forces shown during the interval $0 \le t \le 2\,\text{s}$.

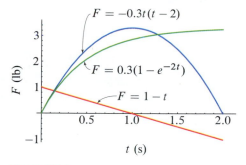

Figure P5.1

Problem 5.2

The total mass of the Earth is $m_e = 5.9736 \times 10^{24}\,\text{kg}$. Modeling the Earth (with everything in and on it) as an isolated system and assuming that the center of the Earth is also the center of mass of the Earth, determine the displacement of the center of the Earth due to

(a) a 2 m jump off the surface by a 85 kg person;

(b) the Space Shuttle, with a mass of 124,000 kg, reaching an orbit of 200 km;

(c) 170,000 km^3 of water being elevated 50 m (these numbers are estimates based on publicly available information about the Aswan Dam at the border between Egypt and Sudan). Use $1\,\text{g/cm}^3$ for the density of water.

Problem 5.3

Consider an elevator that moves with an operating speed of 2.5 m/s. Suppose that a person who boards the elevator on the ground floor gets off on the fifth floor. Assuming that the elevator has achieved operating speed by the time it reaches the second floor and that it is still moving at its operating speed as it passes the fourth floor, determine the momentum change of a person with a mass of 80 kg between the second and fourth floors if each floor is 4 m high. In addition, determine the impulse of the person's weight during the same time interval.

Figure P5.3

Problem 5.4

A 180 gr (7000 gr = 1 lb) bullet goes from rest to 3300 ft/s in 0.0011 s. Determine the magnitude of the impulse imparted to the bullet during the given time interval. In addition, determine the magnitude of the average force acting on the bullet.

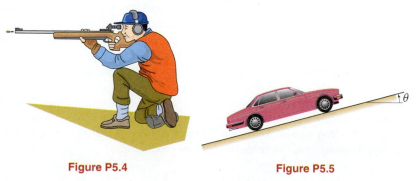

Figure P5.4   **Figure P5.5**

Problem 5.5

A 3400 lb car is parked as shown. Determine the impulse of the normal reaction force acting on the car during the span of an hour if $\theta = 15°$.

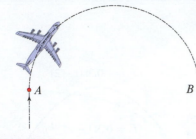

Figure P5.6

💡 Problem 5.6 💡

An airplane performs a turn at constant speed and elevation so as to change its course by 180°. Let *A* and *B* designate the beginning and end points of the turn. Assuming that the change in mass of the plane due to fuel consumption is negligible, is the airplane's momentum at *A* different from the airplane's momentum at *B*? In addition, again neglecting the change in mass between *A* and *B*, is the total work done on the plane between *A* and *B* positive, negative, or equal to zero?

Note: Concept problems are about *explanations*, not computations.

Problems 5.7 and 5.8

The takeoff runway on carriers is much too short for a modern jetplane to take off on its own. For this reason, the takeoff of carrier planes is assisted by *hydraulic catapults* (Fig. A). The catapult system is housed below the deck except for a relatively small *shuttle* that slides along a rail in the middle of the runway (Fig. B). The front landing gear of carrier planes is equipped with a *tow bar* that, at takeoff, is attached to the catapult shuttle (Fig. C). When the catapult is activated, the shuttle pulls the airplane along the runway and helps the plane reach its takeoff speed. The takeoff runway is approximately 300 ft long, and most modern carriers have three or four catapults.

A   B   C

Figure P5.7 and P5.8

Problem 5.7   In a catapult-assisted takeoff, assume that a 45,000 lb plane goes from 0 to 165 mph in 2 s while traveling along a rectilinear and horizontal trajectory. Also assume that throughout the takeoff the plane's engines are providing 32,000 lb of thrust.

(a) Determine the average force exerted by the catapult on the plane.

(b) Now suppose that the takeoff order is changed so that a small trainer aircraft must take off first. If the trainer's weight and thrust are 13,000 and 5850 lb, respectively, and if the catapult is not reset to match the takeoff specifications for the smaller aircraft, estimate the average acceleration to which the trainer's pilots would be subjected and express the answer in terms of *g*. What do you think would happen to the trainer's pilot?

Problem 5.8   If the carrier takeoff of a 45,000 lb plane subject to the 32,000 lb thrust of its engines were not assisted by a catapult, estimate how long it would take for a plane to safely take off, i.e., to reach a speed of 165 mph starting from rest. Also, how long a runway would be needed under these conditions?

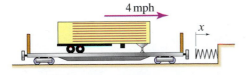

Figure P5.9

Problem 5.9

A 60 ton railcar and its cargo, a 27 ton trailer, are moving to the right at 4 mph when they come into contact with a bumper that is able to bring the system to a stop in 0.78 s. Determine the magnitude of the average force exerted on the railcar by the bumper.

Problem 5.10

In a simple force-controlled experiment, two curling stones A and B are made to slide over a sheet of ice. Initially, A and B are at rest on the start line. Then they are acted upon by identical and constant forces $\vec{F}$, which continually push A and B all the way to the finish line. Let $\vec{p}_{A_{\text{FL}}}$ and $\vec{p}_{B_{\text{FL}}}$ denote the momentum of A and B *at the finish line*, respectively, assuming that the forces $\vec{F}$ are the only nonnegligible forces acting in the plane of motion. If $m_A < m_B$, which of the following statements is true?

(a) $\left|\vec{p}_{A_{\text{FL}}}\right| < \left|\vec{p}_{B_{\text{FL}}}\right|$.

(b) $\left|\vec{p}_{A_{\text{FL}}}\right| = \left|\vec{p}_{B_{\text{FL}}}\right|$.

(c) $\left|\vec{p}_{A_{\text{FL}}}\right| > \left|\vec{p}_{B_{\text{FL}}}\right|$.

(d) There is not enough information given to make a comparison between $\left|\vec{p}_{A_{\text{FL}}}\right|$ and $\left|\vec{p}_{B_{\text{FL}}}\right|$.

Note: Concept problems are about *explanations*, not computations.

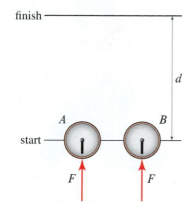

Figure P5.10

Problem 5.11

A 30,000 lb airplane is flying on a horizontal trajectory with a speed $v_0 = 650$ mph when, at point A, it maneuvers so that at point B it is set on a steady climb with $\theta = 40°$ and a speed of 600 mph. Assuming that the change in mass of the plane between A and B is negligible, determine the impulse that had to be exerted on the plane in going from A to B.

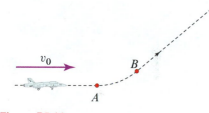

Figure P5.11

Problem 5.12

A 1600 kg car, when on a rectilinear and horizontal stretch of road, can go from rest to 100 km/h in 5.5 s.

(a) Assuming that the car travels on such a road, estimate the average value of the force acting on the car for the car to match the expected performance.

(b) Recalling that the force propelling a car is caused by the friction between the driving wheels and the road, and again assuming that the car travels on a rectilinear and horizontal stretch of road, estimate the average value of the friction force acting on the car for the car to match the expected performance. Also estimate the coefficient of friction required to generate such a force.

Figure P5.12

Problem 5.13

A 1600 kg car, when on a rectilinear and horizontal stretch of road and when the tires do not slip, can go from rest to 100 km/h in 5.5 s. Assuming that the car travels on a straight stretch of road with a 40% slope and that no slip occurs, determine how long it would take to attain a speed of 100 km/h if the car were propelled by the same maximum average force that can be generated on a horizontal road.

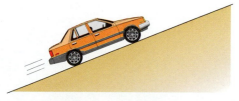

Figure P5.13

Problem 5.14

A $5\frac{1}{8}$ oz baseball traveling at 80 mph rebounds off a bat with a speed of 160 mph. The ball is in contact with the bat for roughly 10^{-3} s. The incoming velocity of the ball is horizontal, and the outgoing trajectory forms an angle $\alpha = 31°$ angle with respect to the incoming trajectory.

(a) Determine the impulse provided to the baseball by the bat.

(b) Determine the average force exerted by the bat on the ball.

(c) Determine how much the angle α would change (with respect to $31°$) if we were to neglect the effects of the force of gravity on the ball.

Figure P5.14

Problem 5.15

In an unfortunate incident, a 2.75 kg laptop computer is dropped onto the floor from a height of 1 m. Assuming that the laptop starts from rest, that it rebounds off the floor up to a height of 5 cm, and that the contact with the floor lasts 10^{-3} s, determine the impulse provided by the floor to the laptop and the average acceleration to which the laptop is subjected when in contact with the floor (express this result in terms of g, the acceleration of gravity).

Figure P5.15

Problem 5.16

A train is moving at a constant speed v_t relative to the ground, when a person who initially at rest (relative to the train) starts running and gains a speed v_0 (relative to the train) after a time interval Δt. Had the person started from rest on the ground (as opposed to the moving train), would the magnitude of the total impulse exerted on the person during Δt be smaller than, equal to, or larger than the impulse needed to cause the same change in relative velocity in the same amount of time on the moving train? Assume that the person always moves in the direction of motion of the train.
Note: Concept problems are about *explanations*, not computations.

Figure P5.16 and P5.17

Problem 5.17

A train is decelerating at a constant rate, when a person who initially at rest (relative to the train) starts running and gains a speed v_0 (again relative to the train) after a time interval Δt. Had the person started from rest on the ground (as opposed to the moving train), would the magnitude of the total impulse exerted on the person during Δt be smaller than, equal to, or larger than the impulse needed to cause the same change in velocity in the same amount of time on the moving train? Assume that the person always moves in the direction of motion of the train and that the train does not reverse its motion during the time interval Δt.
Note: Concept problems are about *explanations*, not computations.

Problem 5.18

A car of mass m collides head-on with a truck of mass $50m$. What is the ratio between the magnitude of the impulse provided by the car to the truck and the magnitude of the impulse provided by the truck to the car during the collision?
Note: Concept problems are about *explanations*, not computations.

Problems 5.19 through 5.21

These problems are an introduction to perfectly plastic impact (which we will cover in Section 5.2). In each problem, model the vehicles A and C as particles and treat the swarm of bugs B hitting the vehicles as a single particle. Also assume that the swarm of bugs sticks perfectly to each vehicle (this is what is meant by a *perfectly plastic impact*).

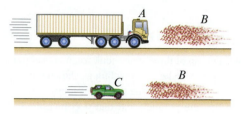

Figure P5.19–P5.21

Problem 5.19   An 80,000 lb semitruck A (the maximum weight allowed in many states) is traveling at 70 mph when it encounters a swarm of mosquitoes B. The swarm is traveling at 1 mph in the opposite direction of the truck. Assuming that the entire swarm sticks to the truck, the mass of each mosquito is 2 mg, and that all of these mosquitoes do not significantly damage the truck, how many mosquitoes must have hit the truck if it slows down by 2 mph on impact? If the same number of mosquitoes hit a small SUV C weighing 3000 lb, by how much will the SUV slow down?

Problem 5.20   An 80,000 lb semitruck A (the maximum weight allowed in many states) is traveling at 70 mph when it encounters a swarm of worker bees B. The swarm is traveling at 12 mph in the opposite direction of the truck. Assuming that the entire swarm sticks to the truck, the mass of each bee is 0.1 g, and that all of these bees do not significantly damage the truck, how many bees must have hit the truck if it slows down by 2 mph on impact? If the same number of bees hit a small SUV C weighing 3000 lb, by how much will the SUV slow down?

Problem 5.21   An 80,000 lb semitruck A (the maximum weight allowed in many states) is traveling at 70 mph when it encounters a swarm of dragonflies B. The swarm is traveling at 33 mph in the opposite direction of the truck. Assuming that the entire swarm sticks to the truck, the mass of each dragonfly is 0.25 g, and that all of these dragonflies do not significantly damage the truck, how many dragonflies must have hit the truck if it slows down by 2 mph on impact? If the same number of dragonflies hit a small SUV C weighing 3000 lb, by how much will the SUV slow down?

Problem 5.22

Solve Example 5.4 by directly applying Eq. (5.14), using the same assumptions made in that solution. Note that, unlike Example 5.4, the velocity of the person relative to the platform does not appear in the solution.

Problems 5.23 through 5.25

Two persons A and B weighing 140 and 180 lb, respectively, jump off a floating platform (in the same direction) with a velocity relative to the platform that is completely horizontal and with magnitude $v_0 = 6$ ft/s for both A and B. The floating platform weighs 800 lb. Assume that A, B, and the platform are initially at rest.

Problem 5.23   Neglecting the water resistance to the horizontal motion of the platform, determine the speed of the platform after A and B jump at the same time.

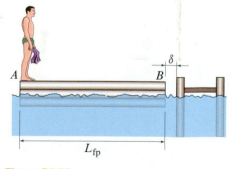

Figure P5.22

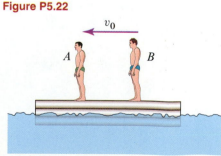

Figure P5.23–P5.25

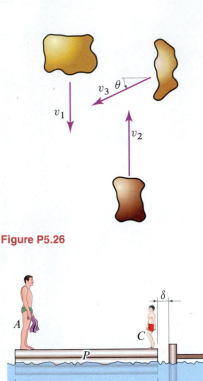

Figure P5.26

Figure P5.27 and P5.28

Problem 5.24 ¦ Neglecting the water resistance to the horizontal motion of the platform, and knowing that B jumps first, determine the speed of the platform after both A and B have jumped.

Problem 5.25 ¦ Neglecting the water resistance to the horizontal motion of the platform, and knowing that A jumps first, determine the speed of the platform after both A and B have jumped.

Problem 5.26 ¦

At the instant shown a group of three space-junk fragments with masses $m_1 = 7.45\,\text{kg}$, $m_2 = 3.22\,\text{kg}$, and $m_3 = 8.45\,\text{kg}$ are traveling as shown with $v_1 = 7701\,\text{m/s}$, $v_2 = 6996\,\text{m/s}$, and $v_3 = 6450\,\text{m/s}$. Assume that the velocity vectors of the fragments are coplanar, that $\vec{v}_1$ and $\vec{v}_2$ are parallel, and that θ is measured with respect to a line perpendicular to the direction of both $\vec{v}_1$ and $\vec{v}_2$. Furthermore, assume that the system is *isolated* and that, because of gravity, the fragments will eventually form a single body. Determine the common velocity of the fragments after they come together if $\theta = 25°$.

Problem 5.27 ¦

A 180 lb man A and a 40 lb child C are at the opposite ends of a 250 lb floating platform P with a length $L_{\text{fp}} = 15\,\text{ft}$. The man, child, and platform are initially at rest at a distance $\delta = 1\,\text{ft}$ from a mooring dock. The child and the man move toward each other with the same speed v_0 relative to the platform. Determine the distance d from the mooring dock where the child and man will meet. Assume that the resistance due to the water to the horizontal motion of the platform is negligible.

Problem 5.28 ¦

A man A, with a mass $m_A = 85\,\text{kg}$, and a child C, with a mass $m_C = 18\,\text{kg}$, are at the opposite ends of a floating platform P, with a mass $m_P = 150\,\text{kg}$ and a length $L_{\text{fp}} = 6\,\text{m}$. Assume that the man, child, and platform are initially at rest and that the resistance due to the water to the horizontal motion of the platform is negligible. Suppose that the man and child start moving toward each other in such a way that the platform does not move relative to the water. Determine the distance covered by the child until meeting the man.

🖥 Problem 5.29 🖥

The 28,000 lb A-10 Thunderbolt is flying at a constant speed of 375 mph when it fires a 4 s burst from its forward-facing seven-barrel Gatling gun. The gun fires 13.2 oz projectiles at a rate of 4200 rounds/min. The muzzle velocity of each projectile is 3250 ft/s. Assuming that each of the plane's two jet engines maintains a constant thrust of 9000 lb, that the plane is subject to a constant air resistance while the gun is firing (equal to that before the burst), and that the plane flies straight and level, determine the plane's change in velocity at the end of the 4 s burst.

Figure P5.29

Problem 5.30

A person P on a cart on rails is receiving packages from people standing on a stationary platform. Assume that person P and the cart have a combined weight of 350 lb and start from rest. In addition suppose that a person P_A throws a package A weighing 60 lb, which is received by person P with a horizontal speed $v_A = 4.5$ ft/s. After person P has received the package from person P_A, a second person P_B throws a package B weighing 80 lb, which is received by person P with a horizontal speed relative to P and in the same direction as the velocity of P of 5.25 ft/s. Determine the final velocity of the person P and the cart. Neglect any friction or air resistance acting on P and the cart.

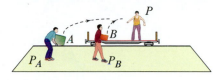

Figure P5.30

Problem 5.31

The spacecraft shown is out in space and is far enough from any other mass (e.g., planets, etc.) so as not to be affected by any gravitational influence (i.e., the net external force on the rocket is approximately zero). The system (i.e., the spacecraft *and all* its fuel) is at rest when it starts at A, and it thrusts all the way to B along the straight line shown using internal chemical rockets (which work by ejecting the fuel mass at very high speeds out the tail of the rocket). We are given that the mass of the system at A is m and that it has ejected half of its mass in thrusting from A to B. What will be the location of the system's mass center when the spacecraft reaches B?

Note: Concept problems are about *explanations*, not computations.

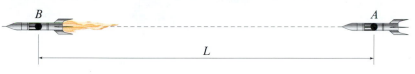

Figure P5.31

Problem 5.32

Energy storage devices that use spinning flywheels to store energy are starting to become available.* To store as much energy as possible, it is important that the flywheel spin as fast as possible. Unfortunately, if it spins too fast, internal stresses in the flywheel cause it to come apart catastrophically. Therefore, it is important to keep the speed at the edge of the flywheel below about 1000 m/s. In addition, it is critical that the flywheel be almost perfectly balanced to avoid the tremendous vibrations that would otherwise result. With this in mind, let the flywheel D, whose diameter is 0.3 m, rotate at $\omega = 60{,}000$ rpm. In addition, assume that the cart B is constrained to move rectilinearly along the guide tracks. Given that the flywheel is not perfectly balanced, that the unbalanced weight A has mass m_A, and that the total mass of the flywheel D, cart B, and electronics package E is m_B, determine the following as a function θ, the masses, the diameter, and the angular speed of the flywheel:

(a) the amplitude of the motion of the cart,

(b) the maximum speed achieved by the cart.

Neglect the mass of the wheels, assume that initially everything is at rest, and assume that the unbalanced mass is at the edge of the flywheel. Finally, evaluate your answers to Parts (a) and (b) for $m_A = 1$ g (about the mass of a paper clip) and $m_B = 70$ kg (the mass of the flywheel might be about 40 kg).

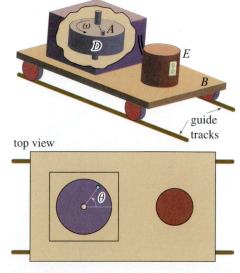

top view

guide tracks

Figure P5.32

---

* We will see the details in Chapter 8, but for now we can think of this as a bunch of particles moving in circles, each of which stores $\frac{1}{2}mv^2$ in kinetic energy.

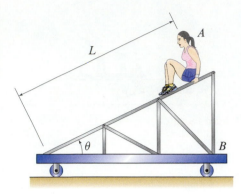

L

θ

A

B

Figure P5.33

Problem 5.33

The 135 lb woman A sits atop the 90 lb cart B, both of which are initially at rest. If the woman slides down the frictionless incline of length $L = 11$ ft, determine the velocity of both the woman and the cart when she reaches the bottom of the incline. Ignore the mass of the wheels on which the cart rolls and any friction in their bearings. The angle $\theta = 26°$.

Problem 5.34

An Apollo Lunar Module A and Command and Service Module B are moving through space far from any other bodies (so that their gravitational effects can be ignored). When $\theta = 30°$, the two craft are separated using an internal linear elastic spring whose constant is $k = 200{,}000$ N/m and is precompressed 0.5 m. Noting that the mass of the Command and Service Module is about 29,000 kg and that the mass of the Lunar Module is about 15,100 kg, determine their postseparation velocities if their common preseparation velocity is 11,000 m/s.

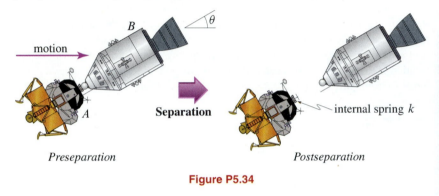

B θ

motion

A

Separation

internal spring k

Preseparation   *Postseparation*

Figure P5.34

Problems 5.35 through 5.38

In the ride shown, a person A sits in a seat that is attached via a cable of length L to a freely moving trolley B of mass m_B. The total mass of the person and the seat is m_A. The trolley is constrained by the beam to move only in the horizontal direction. The system is released from rest at the angle $\theta = \theta_0$ and it is allowed to swing in the vertical plane. Neglect the mass of the cable and treat the person and the seat as a single particle.

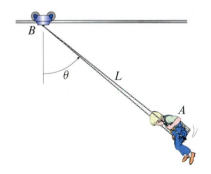

B

θ L

A

Figure P5.35–P5.38

Problem 5.35 Determine the velocities of the trolley and the rider the first time that $\theta = 0°$. Evaluate your solution for $W_A = 100$ lb, $W_B = 20$ lb, $L = 15$ ft, and $\theta_0 = 70°$.

Problem 5.36 As in Prob. 5.35, determine the velocities of the trolley and the rider the first time that $\theta = 0°$. After doing so, for given g, L, m_A, and θ_0, determine the maximum velocity achievable by the rider at $\theta = 0°$ and the corresponding value of m_B. Evaluate your solution for $W_A = 100$ lb, $L = 15$ ft, and $\theta_0 = 70°$. What would be the motion of B for this value of m_B?

Problem 5.37 Determine the velocity of the trolley and the speed of the rider for any arbitrary value of θ.

Problem 5.38 Determine the equations needed to find the velocity of the trolley and the rider for any arbitrary value of θ. Clearly label all equations and list the corresponding unknowns, showing that you have as many equations as you have unknowns. Solve the equations for the unknowns, and then plot the velocity of the trolley and the speed of the rider as a function of the angle θ for both halves of a full swing of the rider. Use $W_A = 100\,\text{lb}$, $W_B = 20\,\text{lb}$, $L = 15\,\text{ft}$, and $\theta_0 = 70°$.

Problem 5.39

A tower crane is lifting a 10,000 lb object B at a constant rate of 7 ft/s while rotating at a constant rate of $\dot{\theta} = 0.15\,\text{rad/s}$. In addition, B is moving outward with a radial velocity of 1.5 ft/s. Assume that the object B does not swing relative to the crane (i.e., it always hangs vertically) and that the crane is fixed to the ground at O.

(a) Determine the radial velocity required of the 20 ton counterweight A to prevent the horizontal motion of the system's center of mass.

(b) Find the total force acting on A and on B.

(c) Determine the velocity and acceleration of the mass center of the system when A moves as determined in Part (a).

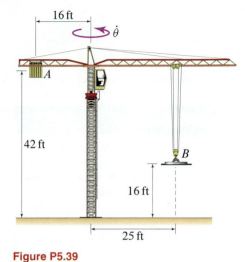

Figure P5.39

Figure 5.7
Two cars traveling on a horizontal road.

5.2   Impact

A collision between two cars

Figure 5.7 shows two cars A and B, of masses m_A and m_B and constant velocities $\vec{v}_A$ and $\vec{v}_B$, respectively, traveling in the same lane along a horizontal stretch of road. Car A catches up with car B and a collision occurs. The classical impact problem is to predict the postimpact velocities of the cars, knowing their preimpact velocities.*

We begin the modeling of this impact by assuming that the collision spans an *infinitesimal* amount of time around a specific instant t_i, the *time of impact*. The expressions *pre-* and *postimpact* will mean an *infinitesimal* amount of time before and after t_i, respectively. Pre- and postimpact quantities will be denoted by the superscripts − and +, respectively.

Next, consider the FBD of the two-car-system at t_i in Fig. 5.8(a). Since they turn out not to be important here, our FBD neglects any external forces acting in the horizontal direction. In addition, we will neglect the cars' vertical motion and we model the cars as particles. As shown in Fig. 5.8(b), we will also assume that the contact force P between the cars is purely horizontal.

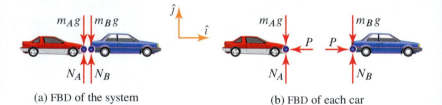

(a) FBD of the system   (b) FBD of each car

Figure 5.8. (a) Free body diagram of the overall system formed by cars A and B. (b) Free body diagram of each car during impact.

As there are no *external* forces acting in the x direction, we can conclude that the x component of the system's momentum is conserved during the impact, i.e.,

$$m_A v_{Ax}^- + m_B v_{Bx}^- = m_A v_{Ax}^+ + m_B v_{Bx}^+ \tag{5.20}$$

Since we are neglecting any vertical motion, the problem is one-dimensional with two unknowns v_{Ax}^+ and v_{Bx}^+, both of which appear in Eq. (5.20). Therefore, we need one additional equation in v_{Ax}^+ and v_{Bx}^+ to get a solution. Observing that Eq. (5.20) is a consequence of Newton's second law and that the kinematics for the problem are known (the motion is only in the x direction), the second equation we need must be a *force law* describing the role played by the material makeup of the colliding objects. While this equation can be very complicated, for now, assume that A and B interlock (stick together) after the collision. This is called a *perfectly plastic* impact and is described by the following equation:

$$v_{Ax}^+ = v_{Bx}^+. \tag{5.21}$$

Equations (5.20) and (5.21) can be solved to obtain

$$v_{Ax}^+ = v_{Bx}^+ = \frac{m_A v_{Ax}^- + m_B v_{Bx}^-}{m_A + m_B}. \tag{5.22}$$

---

* The problem of finding the preimpact velocities given the postimpact velocities is solved similarly and is essentially the problem we find in accident reconstruction.

What is the essence of an impact problem?

To understand how to extend our model for impact, consider Fig. 5.9, which displays the solution for the following choice of parameters: $m_A = 1200\,\text{kg}$, $m_B = 950\,\text{kg}$, $v_{Ax}^- = 18.1\,\text{m/s}$, and $v_{Bx}^- = 14.0\,\text{m/s}$. Notice how the velocity for each car *jumps* at $t = t_i$ from pre- to postimpact values. For this to happen, the cars must undergo a tremendous acceleration during the impact. In fact, our *mathematical* model can be viewed as implying that this acceleration is *infinite*! What kind of force can cause an "infinite" acceleration? Answering this question will help us deal with more general impact scenarios, including cases in which we do have external forces and in which a more general force law for impacts can be incorporated.

The car collision contains the two key elements of the solution to any impact problem:

1. Application of the impulse-momentum principle for the *system* formed by the colliding bodies.

2. A force law that tells us how the colliding objects rebound.

We will see that the basic solution strategy for any impact problem remains the same: combining conservation of momentum with an impact force law.

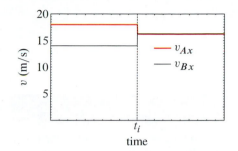

Figure 5.9
Plot of the pre- and postimpact velocities for cars A and B. This plot was obtained by using Eq. (5.22) and the following parameters: $m_A = 1200\,\text{kg}$, $m_B = 950\,\text{kg}$, $v_{Ax}^- = 18.1\,\text{m/s}$, and $v_{Bx}^- = 14.0\,\text{m/s}$.

Impulsive forces

Having discovered that forces during a collision can cause a jump in velocity, we now go back to Eq. (5.6) and focus instead on *changes in momentum*. In Fig. 5.10, consider a particle acted upon by a *constant* force $\vec{F}_1$ during the time

Figure 5.10. Particle moving along a rectilinear trajectory under the action of a constant force.

interval $(t, t + \Delta t_1)$. Using Eq. (5.6), the corresponding change of momentum $\Delta \vec{p}_1$ is

$$\Delta \vec{p}_1 = \int_t^{t+\Delta t_1} \vec{F}_1 \, dt = \vec{F}_1 \Delta t_1. \tag{5.23}$$

What kind of force can provide a *finite momentum change* when the time over which it acts Δt goes to zero? To answer this question, consider a time interval $\Delta t_n = \Delta t_1/n$, with $n = 2, 3, \ldots$, and let's determine the constant force $\vec{F}_n$ needed to produce $\Delta \vec{p}_n = \Delta \vec{p}_1$. Using $\Delta \vec{p}_n = \Delta \vec{p}_1$ and Eq. (5.23), we must have

$$\vec{F}_n \Delta t_n = \vec{F}_1 \Delta t_1 \quad \Rightarrow \quad \vec{F}_n = n \vec{F}_1. \tag{5.24}$$

As shown in Fig. 5.11, increasing n (decreasing Δt_n) means that $|\vec{F}_n|$ increases, and as $n \to \infty$ ($\Delta t_n \to 0$), we know that $|\vec{F}_n| \to \infty$. By letting $n \to \infty$ in Eq. (5.24), we learn that *to generate a finite change in momentum in an infinitesimal amount of time, one needs a force with infinite magnitude*. Such a force is called an *impulsive force*. Notice that *unless a force has an infinite magnitude, that force does not play a role during an impact* since it cannot cause a momentum change in an infinitesimal time interval. This is important

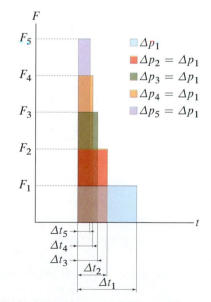

Figure 5.11
Relationship between a time interval and the corresponding force that preserves the change in momentum.

because when we sketch the FBD for an impact problem, we can neglect all nonimpulsive forces. For example, even if we had wanted to take into account a force such as air resistance in the car collision, it would not have played a role since its magnitude cannot become infinite.

Impulsive vs. nonimpulsive forces

Let's examine some common forces and classify each as either impulsive or nonimpulsive. Since impulsive forces must have an infinite magnitude, *any force with finite magnitude cannot be impulsive.* In addition, when distinguishing between impulsive and nonimpulsive forces, we keep in mind that our model of impact assumes that they occur in an infinitesimal time interval. This also means that objects do not move during the impact. Using these ideas, we can say the following:

- *Weight forces*, being finite, are *nonimpulsive.*

- A *spring force* (or of any elastic object), having a magnitude proportional to the spring's change in length, is *nonimpulsive* because, in general, springs do not stretch to infinite lengths.

- *Drag forces*, which are typically an increasing function of speed, are *nonimpulsive* because objects cannot have infinite speed.

- As was seen in Example 5.4, the *tension in an inextensible rope* can be *impulsive* because no limit is placed on the tension.

- The *tension or compression in a rigid bar* can be *impulsive* since no limit is placed on the load that the rigid bar can tolerate.

- The *contact forces* between two impacting objects can be *impulsive* if the impacting objects do not shatter.

Recognizing impulsive forces: two models of the same event

To gain some insight into our model of impact, let's look at the same event modeled in two very different ways, both of which give the same answer when energy is conserved.

Consider the two different models of the impact of a rubber ball with a solid surface, as shown in Fig. 5.12. In each case, the particle hits the wall with a speed v_0 that is perpendicular to the wall, and then it rebounds. The top half of the figure shows the particle impact model (PIM) we have been talking about, and the bottom half shows a model in which the particle only interacts with the wall through an elastic spring that is attached to the particle and a small bumper A that hits the wall (the particle-spring model or PSM).

In the PIM, the impact takes an *infinitesimal time interval*, the particle *does not move* during the impact, the normal force between the particle and the wall is infinite, and the particle momentum reverses direction instantaneously. If we assume that no energy is dissipated during the impact, then the particle will rebound with the same speed with which it hit the wall.

The PSM is very different (you may want to review Example 3.3). The particle and the bumper at A will both hit the wall with speed v_0, and the spring will deform as the particle approaches the wall. The speed of the particle will decrease as the spring compresses until the particle reaches zero velocity and

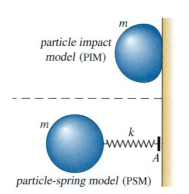

particle impact model (PIM)

m

m k

A

particle-spring model (PSM)

Figure 5.12
Two different models of particle impact that give the same result. The spring is used in the PSM to represent the elasticity of the ball. In the PIM, the ball is shown as deformed as a visual aid.

begins to rebound. The speed of the particle will then increase until it reaches speed v_0, the bumper A will then separate from the wall, and the impact will be complete. With the PSM, the impact takes a *finite time*, the particle *does move* during the impact (it moves twice the total compression of the spring), the normal force is equal to the force in the spring and is never infinite, and the momentum reverses direction, but not instantaneously. As with the PIM, the speed of the particle at the end of the impact will be v_0. The contrast between the two models is summarized in Table 5.1.

The message we should take away from the comparison of these two models is that the same phenomenon can almost always be modeled in different ways. Often the models give different results, but they can also give identical results (as they do in this case), even though, as can be seen in Table 5.1, the details of what happens to the particle between the instant just before the impact and the instant right after the impact are vastly different. So, why don't we use the PSM for impact? We could, but dealing with more complicated impacts would get *very* difficult with the PSM and, as we will see in this section, they are tractable using the PIM.

■ **Mini-Example.**   Now that we know a little about impulsive forces, let's re-sketch the FBD in Fig. 5.8(a), showing only those forces that are *relevant* to the impact problem under two different assumptions: (1) the ground is compliant and (2) the ground is rigid.

Solution. For both cases, forces such as the car weights and air resistance are nonimpulsive. For case 1, the ground's compliance requires that the reaction force between the ground and the car be proportional to the ground's deformation; i.e., the ground acts as a spring (Fig. 5.13). These reaction forces are nonimpulsive and therefore do not appear in the impact-relevant FBD. By contrast, for case 2, the ground's material behavior is similar to that of a rigid rod, and therefore the contact forces with the ground *might* need to be accounted for on the impact-relevant FBD (Fig. 5.14). We say "might" because it depends on whether or not the geometry of the collision pushes one of the cars into the ground. If neither of the cars is pushed into the ground by the collision, then the contact forces would not increase beyond their (finite) preimpact value and are thus nonimpulsive. In this case the impact-relevant FBD would be the one in Fig. 5.13. On the other hand, if we know that one of the cars is pushed into the ground, then we can exclude the contact force acting on the *other* car, as this force would be decreasing during the impact.

For the impact modeled by the FBD in Fig. 5.13, the system is isolated (there are no external impulsive forces) and its *total* momentum is conserved during the impact. By contrast, for the system in Fig. 5.14, only the component of the momentum *perpendicular* to the contact force is conserved. ────■

Constrained versus unconstrained impacts

As we saw in the above Mini-Example, two colliding objects for which there are no *external* impulsive forces form a system that is isolated. In this case, we have that the total linear momentum of the system is conserved, and the impact is said to be an *unconstrained impact*. By contrast, an impact that occurs in the presence of external impulsive forces is called a *constrained impact*. In general, unconstrained impact problems are easier to solve than constrained problems.

Table 5.1

Summary of the differences between the PIM and PSM impact models.

Physical characteristic	PIM	PSM
time	infinitesimal	finite
displacement	zero	finite[a]
normal force	infinite	finite[b]
Δmomentum[c]	instantaneous	finite time

[a] Equal to twice the full spring compression.
[b] Equal to the spring force.
[c] Reverses in both models.

Figure 5.13
FBD for the case with compliant ground.

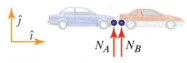

Figure 5.14
FBD for the case with rigid ground.

Coefficient of restitution

We will now formulate an impact force law to deal with rebounds. We begin by recalling that force laws are based on *experimental observations*. Therefore, consider the experiment shown in Fig. 5.15, in which we record the pre-

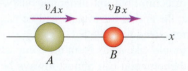

Figure 5.15. Geometry of an experiment to measure pre- and postimpact velocities.

and the postimpact velocities of two balls A and B. Before interpreting the experimental data, we make two observations from everyday experience:

1. The severity of an impact depends on the *relative* preimpact velocity, also called the *approach velocity*.

2. The "quality" of the *rebound* is described by the postimpact *relative* velocity, also called the *separation velocity*.

Therefore, when studying the experimental data, we will want to correlate the approach and separation velocities as is done in Fig. 5.16. This figure shows that the separation velocity depends on the approach velocity and that, for higher approach velocities, there is lack of proportionality between pre- and postimpact relative velocities. This is typically due to the fact that more permanent damage is caused by the collision for higher approach velocities. Figure 5.16 also shows that there is a regime (near the origin) in which the approach and separation velocities are *linearly proportional* to one another. Based on this experimental observation, we will only study impacts in their linear range, and therefore we will adopt the following *impact force law*:

$$e = \frac{\text{separation velocity}}{\text{approach velocity}} = \frac{v_{Bx}^+ - v_{Ax}^+}{v_{Ax}^- - v_{Bx}^-}. \qquad (5.25)$$

Equation (5.25) is called the *coefficient of restitution equation*, and e is the experimentally measured proportionality constant called the *coefficient of restitution* (COR).

Having obtained an impact force law, we make the following observations.

1. The COR e is *dimensionless*.

2. To ensure that $e \geq 0$ for any impact, the approach and separation velocities have been taken to be $v_{A/Bx}^-$ and $v_{B/Ax}^+$, respectively.

3. It is an experimental fact that the rebound speed is never greater than the approach speed for passive materials. Along with item 1, this tells us

$$0 \leq e \leq 1. \qquad (5.26)$$

4. The COR e depends on the material makeup of *both* colliding objects.

An impact is said to be *plastic* if $e = 0$, *elastic* if $0 < e < 1$, and *perfectly elastic* if $e = 1$ (Table 5.2).

Concept Alert

Relative velocities and impact. Intuition tells us that the head-on impact between two cars, one traveling at 35 mph and the other at 40 mph, is essentially equivalent to the collision of a car traveling 75 mph with one that is stationary. Therefore, what's really important is $\vec{v}_{A/B}^- = \vec{v}_A^- - \vec{v}_B^-$ and not $\vec{v}_A^-$ and $\vec{v}_B^-$ individually.

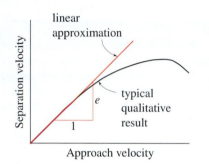

Figure 5.16
Qualitative trend usually found in collision experiments when plotting the post- versus preimpact *relative* velocity (black curve).

Helpful Information

Be careful when using Eq. (5.25). When you are given an e for use in Eq. (5.25), it can be assumed to be appropriate for the given conditions (e.g., approach velocities, material makeup of colliding bodies, etc.), but be careful when applying it under other conditions.

Table 5.2
Impact type as a function of COR value.

COR value	Impact type
$e = 0^a$	plastic impact
$0 < e < 1$	elastic impact
$e = 1$	perfectly elastic impact

[a] An impact is perfectly plastic if $\vec{v}_A^+ = \vec{v}_B^+$, that is, the objects stick together after impact.

■ **Mini-Example.** For the car collision example at the beginning of this section, compare the postimpact velocity obtained using Eq. (5.25) instead of Eq. (5.21). Use the same masses, same preimpact velocities, and $e = 0.5$. Plot the solution as was done in Fig. 5.9.

Solution. This impact problem is governed by Eqs. (5.20) and (5.25), which form a linear system of two equations in the two unknowns v_{Ax}^+ and v_{Bx}^+. The solution of this system yields the following result:

$$v_{Ax}^+ = \frac{v_{Ax}^-(m_A - m_B e) + m_B v_{Bx}^-(1 + e)}{m_A + m_B} = 15.4\,\text{m/s}, \qquad (5.27)$$

$$v_{Bx}^+ = \frac{m_A v_{Ax}^-(1 + e) + v_{Bx}^-(m_B - m_A e)}{m_A + m_B} = 17.4\,\text{m/s}. \qquad (5.28)$$

For $e = 0.5$ and for the parameters used to generate Fig. 5.9, Eqs. (5.27) and (5.28) give the plot shown in Fig. 5.17. ────────────────────■

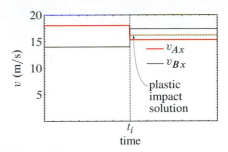

Figure 5.17
Pre- and postimpact velocities with $e = 0.5$ (compare with plot in Fig. 5.9).

Impulse-momentum principle and COR

In addition to the experimental approach used earlier, Eq. (5.25) can be obtained in a purely *theoretical* way; the key to this alternative derivation is to view colliding objects as *deformable bodies* instead of particles.

Referring to Fig. 5.18, at time t^- let two *deformable* objects S_A and S_B, with mass centers at points A and B, respectively, begin to impact one another. Since A and B are deformable bodies, after the impact begins they move closer to one another until the bodies reach maximum deformation at time t_C, at which point they share a common velocity v_C. After t_C, A and B push away from one another until time t^+, at which the bodies S_A and S_B separate. The process occurring between t^- and t_C is called *deformation*, and the process occurring between t_C and t^+ is called *restitution*.

If we measured the contact force between the colliding objects as a function of time, we would obtain a curve much like that seen in Fig. 5.19. Knowing that this curve exists and using the impulse-momentum principle, we can calculate the change in velocity of A and B. During the deformation process, for A we have

$$m_A v_A^- - \int_{t^-}^{t_C} D(t)\, dt = m_A v_C, \qquad (5.29)$$

where m_A is the mass of object S_A, v_A^- is the velocity of A at t^-, and D is the deformation force. Repeating the same operation for B, we obtain

$$m_B v_B^- + \int_{t^-}^{t_C} D(t)\, dt = m_B v_C, \qquad (5.30)$$

where m_B is the mass of object S_B and v_B^- is the velocity of B at t^-. Eliminating the common velocity v_C between Eqs. (5.29) and (5.30), we obtain the relative preimpact velocity

$$v_A^- - v_B^- = \left(\frac{1}{m_A} + \frac{1}{m_B}\right) \int_{t^-}^{t_C} D(t)\, dt. \qquad (5.31)$$

As far as the restitution process is concerned, by proceeding in a similar way, the impulse-momentum principle applied to A and B gives

$$m_A v_C - \int_{t_C}^{t^+} R(t)\, dt = m_A v_A^+, \qquad (5.32)$$

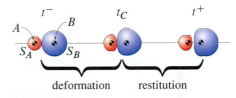

Figure 5.18
Collision between two deformable bodies with centers of mass at A and B, respectively. Note that at the end of the restitution process there may be some permanent deformation.

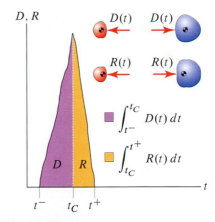

Figure 5.19
Internal deformation force $D(t)$ and restitution force $R(t)$ between two objects versus time.

$$m_B v_C + \int_{t_C}^{t^+} R(t)\, dt = m_B v_B^+, \qquad (5.33)$$

respectively. Again, eliminating the common velocity v_C, we derive the following expression for the postimpact relative velocity:

$$v_B^+ - v_A^+ = \left(\frac{1}{m_A} + \frac{1}{m_B}\right) \int_{t_C}^{t^+} R(t)\, dt. \qquad (5.34)$$

Constructing the ratio of postimpact relative velocity to the preimpact relative velocity, we obtain the following important result:

$$\frac{v_B^+ - v_A^+}{v_A^- - v_B^-} = \frac{\displaystyle\int_{t_C}^{t^+} R(t)\, dt}{\displaystyle\int_{t^-}^{t_C} D(t)\, dt} = e. \qquad (5.35)$$

This representation of the COR is important because it tells us that e can be viewed as the ratio of the restitution impulse to the deformation impulse. Equation (5.35) makes it easier to understand why the COR is constrained to values between 0 and 1 as these values define the full spectrum of possibilities ranging from no restitution to full restitution. Another remarkable aspect of Eq. (5.35) is that the masses of the colliding bodies do not play a role. When we first introduced the COR equation, we implicitly decided to neglect the effect of the masses involved in the impact. Equation (5.35) *demonstrates* that the masses of the colliding bodies do not play a role in controlling the ratio of separation to approach velocities.

Line of impact

We now extend our impact model to impacts in two dimensions. To begin, we introduce a component system describing the orientation of the contact forces in a collision. Referring to Fig. 5.20, consider the impact of two billiard balls. Experience tells us that one of the crucial elements in the execution of the shot is the *orientation of the plane tangent to the contact*. Knowing this plane is important in any type of collision and not just in billiards.

From geometry we know that the orientation of a plane is specified by the direction *perpendicular* to the plane in question. In the study of impacts, this is done by identifying *the line perpendicular to the contact surface at the point of contact*. This line is called the ***line of impact*** (LOI). Referring to Fig. 5.21, it is easy to see that the LOI, labeled by y in the figure, can be taken as one of the axes of a frame of reference (x and y in the figure), which is *intrinsic* to the colliding objects. Since we have assumed that an impact spans an infinitesimal time interval, we will also assume that this xy frame of reference defines the geometry of the impact *throughout the impact*.

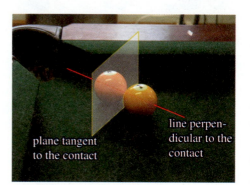

Figure 5.20
Photograph of two billiard balls in contact.

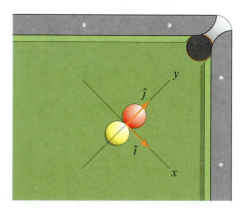

Figure 5.21
Frame of reference formed by lines perpendicular and tangent to the contact point between two colliding billiard balls.

Classification of impacts

Strategies for solving impact problems depend on how the preimpact velocities are oriented relative to the LOI. Hence, we will classify unconstrained impacts according to the orientations of the preimpact velocities, and then we will review a strategy to solve them. At the end of the section, we will discuss how to approach constrained impacts.

Referring to Fig. 5.22(a), a *direct impact* between two particles is one in which the preimpact velocities of both particles are *parallel* to the LOI. Referring to Fig. 5.22(b), if one of the colliding particles has a preimpact velocity that is not parallel to the LOI, the impact is called an *oblique impact*. An impact is said to be *central* if the LOI contains the mass centers of the impacting bodies; otherwise the impact is called *eccentric*. If we view a particle as something that coincides with its own center of mass, we come to the conclusion that particles can only have central impacts.

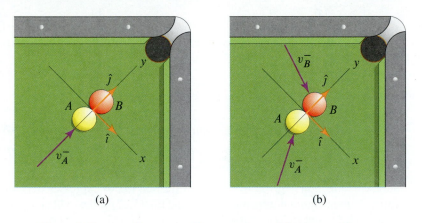

(a)   (b)

Figure 5.22. (a) A *direct central impact*. (b) An *oblique central impact*.

These classifications of impact (i.e., direct versus oblique and central versus eccentric) are then applied to every impact so that, for example, an impact between two particles in which at least one of the preimpact velocities is not parallel to the LOI is called an *oblique central impact*. These ideas are summarized in Table 5.3 and will be seen again in Chapter 8.

Table 5.3. Classification of impacts.

Impact geometry criteria		
Preimpact velocities	*Centers of mass*	**Impact type**
parallel to LOI	on LOI	direct central
parallel to LOI	not on LOI	direct eccentric
not parallel to LOI	on LOI	oblique central
not parallel to LOI	not on LOI	oblique eccentric

Direct central impact

Referring to Fig. 5.22(a), consider the impact of billiard balls A and B such that the preimpact velocities and the LOI coincide. This is a *direct central impact*, and this type of impact can always be viewed as one-dimensional along the LOI. Referring to Fig. 5.23(a), the impact-relevant FBD for particles A and B does not contain any external impulsive forces. Applying conservation of momentum and the COR equation along the LOI, we see that a direct central

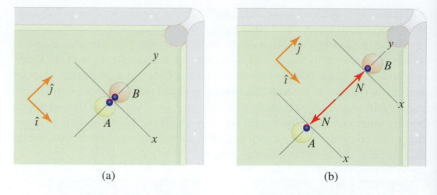

Figure 5.23. (a) FBD of the impact of two billiard balls modeled as particles. (b) FBD of billiard balls A and B individually.

impact is governed by the following two equations:

$$m_A v_{Ay}^- + m_B v_{By}^- = m_A v_{Ay}^+ + m_B v_{By}^+, \tag{5.36}$$

$$v_{By}^+ - v_{Ay}^+ = e\left(v_{Ay}^- - v_{By}^-\right), \tag{5.37}$$

where the y direction is given by the LOI. Equation (5.36) expresses the conservation of momentum during the impact, and Eq. (5.37) is the COR equation. Equations (5.36) and (5.37) form a linear system of equations that completely describes the direct central impact of two particles.

Oblique central impact

Figure 5.22(b) shows the impact of billiard balls A and B in which the LOI is not parallel to the preimpact velocities. This is an *oblique central impact*. Assuming we know the preimpact velocities, we will find the postimpact velocities, which means that we need four equations to solve for the two componenents of each postimpact velocity. The impact-relevant FBDs for this impact are shown in Fig. 5.23(a) and (b), which shows that we can conserve the y component of momentum for A and B together (Fig. 5.23(a)) and the x component of momentum for A and B individually (Fig. 5.23(b)), i.e.,

$$m_A v_{Ay}^- + m_B v_{By}^- = m_A v_{Ay}^+ + m_B v_{By}^+, \tag{5.38}$$

$$v_{Ax}^- = v_{Ax}^+, \tag{5.39}$$

$$v_{Bx}^- = v_{Bx}^+, \tag{5.40}$$

where we have canceled the mass on both sides of Eqs. (5.39) and (5.40) and have assumed that the impact is *frictionless*. As we have already seen, the final equation is the force law given by the COR equation applied along the LOI, which in this case is

$$v_{By}^+ - v_{Ay}^+ = e\left(v_{Ay}^- - v_{By}^-\right). \tag{5.41}$$

Equations (5.38)–(5.41) form a system of four equations that can be used to find the postimpact velocity components of A and B.

Perfectly plastic impact. This means that the two colliding particles stick together to become a single object, and the FBD in Fig. 5.23(a) applies. (Do you see why the FBD in Fig. 5.23(b) is no longer true?) This means that after the impact, we must have

$$v_{Ay}^+ = v_{By}^+, \tag{5.42}$$
$$v_{Ax}^+ = v_{Bx}^+, \tag{5.43}$$

and that during the impact, momentum is conserved in both the x and y directions, that is,

$$m_A v_{Ax}^- + m_B v_{Bx}^- = m_A v_{Ax}^+ + m_B v_{Bx}^+, \tag{5.44}$$
$$m_A v_{Ay}^- + m_B v_{By}^- = m_A v_{Ay}^+ + m_B v_{By}^+. \tag{5.45}$$

Equations (5.42)–(5.45) are the four equations that govern any 2D perfectly plastic impact.

Impact and energy

Since impacts occur in an infinitesimal amount of time and the impacting objects do not move during the impact, these objects cannot experience a change in potential energy during an impact. Similarly, there cannot be work done by any force that requires a finite displacement of the colliding objects. Hence, the work done during an impact, which is done by the impulsive forces, can only be measured by comparing the pre- and postimpact kinetic energies.

Experience tells us that the *total kinetic energy* of a system of colliding objects can only decrease. The total kinetic energy of a system of colliding objects stays constant only if the collisions are perfectly elastic, which is an idealization. For this reason, it is common to say that there is energy lost in a collision. In a collision between two particles A and B, the energy loss is usually expressed as a percentage of the preimpact total kinetic energy, i.e.,

$$\text{Percentage of energy loss} = \frac{T^- - T^+}{T^-} \times 100\%, \tag{5.46}$$

where

$$T^- = \tfrac{1}{2} m_A (v_A^-)^2 + \tfrac{1}{2} m_B (v_B^-)^2, \tag{5.47}$$
$$T^+ = \tfrac{1}{2} m_A (v_A^+)^2 + \tfrac{1}{2} m_B (v_B^+)^2, \tag{5.48}$$

and T^- and T^+ are the pre- and postimpact kinetic energies, respectively.

Constrained impact

Referring to Fig. 5.24, consider the collision between truck A and car B. Given the preimpact velocities, we want to compute the postimpact velocities, where the LOI is perpendicular to the contact surface between the truck and car and is as shown. We will model the impact as a two-dimensional oblique central impact between two particles, and we will assume that the contact between the truck and the car is frictionless (this assumption may not always be realistic). This means that the contact force between A and B is completely parallel to the LOI. We will also assume that the ground is *rigid* and frictionless. Due to

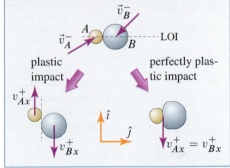

Concept Alert

Plastic impact versus perfectly plastic impact. A common misconception is that in an impact with $e = 0$ (i.e., a plastic impact), the colliding objects stick together after impact. If $e = 0$ in Eq. (5.41), it means that the *component* of the separation velocity along the LOI is equal to zero, but nothing can be said about the separation velocity perpendicular to the LOI. Therefore, an impact with $e = 0$ (a *plastic impact*) is *not*, in general, the same as a *perfectly plastic impact*.

Interesting Fact

Is energy really lost? The principles of thermodynamics tell us that there is no such thing as energy lost—energy is simply transformed into different forms. Specifically, in a collision, energy that appears to be lost is transformed into heat, sound energy, and energy locked into permanent deformation.

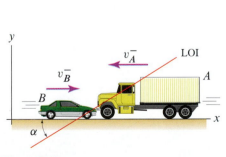

Figure 5.24
Head-on collision between a truck and a small passenger car. The orientation of the LOI is such that the car will be pushed into the ground and the truck will be lifted off the ground.

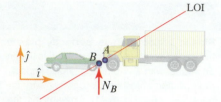

Figure 5.25
Impact-relevant FBD for the truck and car system, assuming that the ground is both *rigid* and *frictionless*. The reaction between the truck and the ground has been omitted because it is not impulsive since the truck will tend to be lifted from the ground.

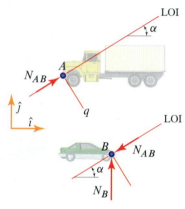

Figure 5.26
Impact-relevant FBDs of A and B.

the orientation of the LOI, B will be pushed into the ground during the impact, and due to the ground's rigidity, the reaction between B and the ground will be impulsive. Newton's third law then tells us that A will tend to be lifted off the ground by this impulsive force, and so the reaction between A and the ground will be nonimpulsive. The impact-relevant FBD of the system formed by A and B is shown in Fig. 5.25.

Given the FBD in Fig. 5.25, we conclude that only the x component of the total momentum is conserved, that is,

$$m_A v_{Ax}^- + m_B v_{Bx}^- = m_A v_{Ax}^+ + m_B v_{Bx}^+. \tag{5.49}$$

Since the impact is two-dimensional, we have four unknowns, namely, the components of the postimpact velocities, so we need three more equations. Another equation can be found by noting that the rigid road prevents B from moving vertically and so

$$v_{By}^+ = 0. \tag{5.50}$$

To find the final two equations, we need to consider the FBDs of A and B individually, which are shown in Fig. 5.26. Having assumed that the contact is frictionless, we see that there are no impulsive forces acting on A in the q direction, and therefore we can conclude that the linear momentum of A is conserved along that direction, that is, $m_A v_{Aq}^- = m_A v_{Aq}^+$, or

$$v_{Ax}^- \sin\alpha - v_{Ay}^- \cos\alpha = v_{Ax}^+ \sin\alpha - v_{Ay}^+ \cos\alpha. \tag{5.51}$$

The last equation comes from applying the COR equation along the LOI, which gives

$$\left(v_B^+\right)_{\text{LOI}} - \left(v_A^+\right)_{\text{LOI}} = e\left[\left(v_A^-\right)_{\text{LOI}} - \left(v_B^-\right)_{\text{LOI}}\right]. \tag{5.52}$$

The COR for a constrained impact cannot be expected to be equal to the one we would measure if the impact were unconstrained, since the impulse on A is not equal and opposite to that acting on B, and therefore the arguments we used to derive Eq. (5.35) no longer apply. We can still use the COR equation; it just needs to be measured under these circumstances. Writing Eq. (5.52) in terms of x and y components, we obtain

$$\begin{aligned}\left(v_{Bx}^+ - v_{Ax}^+\right)\cos\alpha + \left(v_{By}^+ - v_{Ay}^+\right)\sin\alpha \\ = e\left[\left(v_{Ax}^- - v_{Bx}^-\right)\cos\alpha + \left(v_{Ay}^- - v_{By}^-\right)\sin\alpha\right].\end{aligned} \tag{5.53}$$

We now have four equations, that is, Eqs. (5.49), (5.50), (5.51), and (5.53), in the four unknowns v_{Ax}^+, v_{Ay}^+, v_{Bx}^+, and v_{By}^+. In summary, our strategy had the following elements:

1. Identification of those directions along which the momentum is conserved through the impact for one and/or both of the bodies, as was done in Eqs. (5.49) and (5.51).

2. Kinematic description of the constraint, as was done in Eq. (5.50).

3. Application of the COR equation along the LOI, as was done in Eq. (5.53).

As we can see, identification of the constraint (the ground's rigidity) and the expression of its consequences are key to solving the problem.*

End of Section Summary

This section discussed impact between particles. We introduced a model based on the assumptions summarized in Table 5.4. We discovered that there are two key elements to *every* impact problem: (1) the application of the impulse-momentum principle and (2) a force law telling us how the colliding objects rebound.

Impulsive forces are forces that generate a finite change in momentum in an infinitesimal amount of time. When we apply the impulse-momentum principle during an impact, only impulsive forces play a role, so they are the only forces included on the FBD. Problems involving the impact between two particles generally involve four unknowns, so four equations are needed. The geometry of an *unconstrained impact* (for which there are no external impulsive forces) between two particles is shown in Fig. 5.27. When the impact is frictionless, the four equations come from

1. Application of the impulse-momentum principle to both particles along the LOI (the *y* direction), which gives

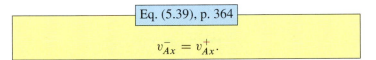

Eq. (5.38), p. 364

$$m_A v_{Ay}^- + m_B v_{By}^- = m_A v_{Ay}^+ + m_B v_{By}^+ .$$

2. Application of the impulse-momentum principle to particle *A* in the *x* direction, which gives

Eq. (5.39), p. 364

$$v_{Ax}^- = v_{Ax}^+ .$$

3. Application of the impulse-momentum principle to particle *B* in the *x* direction, which gives

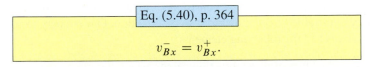

Eq. (5.40), p. 364

$$v_{Bx}^- = v_{Bx}^+ .$$

4. Application of the COR equation along the LOI, which is given by

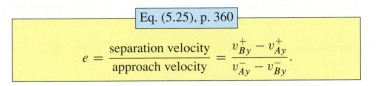

Eq. (5.25), p. 360

$$e = \frac{\text{separation velocity}}{\text{approach velocity}} = \frac{v_{By}^+ - v_{Ay}^+}{v_{Ay}^- - v_{By}^-} .$$

Table 5.4
Assumptions used in our impact model.

Physical characteristic	*Impact assumption*
duration of impact	infinitesimal
displacement of particle	zero
force on particle	infinite
change in momentum	instantaneous

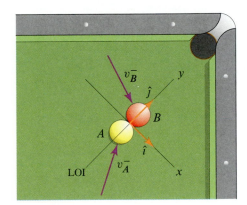

Figure 5.27
Geometry of two impacting particles.

---

* It is not difficult to conceive of problems in which this strategy would fail. For example, if the ground, in addition to being rigid, had friction, then momentum would not be conserved in the *x* direction. This should remind us that our theory, as any theory, has limitations.

The coefficient of restitution e determines the nature of the rebound between the two particles. When $e = 0$, the impact is called *plastic* (the objects do not necessarily stick together in a plastic impact); when $0 < e < 1$, the impact is called *elastic*; and when $e = 1$, it is called *perfectly elastic*.

In a *perfectly plastic impact* the colliding objects stick together after the impact. Therefore, in a perfectly plastic impact, the fact that the objects stick together is reflected in the equations

Eqs. (5.42) and (5.43), p. 365

$$v_{Ay}^+ = v_{By}^+ \quad \text{and} \quad v_{Ax}^+ = v_{Bx}^+.$$

In addition, momentum is conserved in all directions during the impact, which gives

Eqs. (5.44) and (5.45), p. 365

$$m_A v_{Ax}^- + m_B v_{Bx}^- = m_A v_{Ax}^+ + m_B v_{Bx}^+,$$
$$m_A v_{Ay}^- + m_B v_{By}^- = m_A v_{Ay}^+ + m_B v_{By}^+.$$

Impact and energy.   Unless an impact is perfectly elastic, mechanical energy must be lost during the impact. Generally, the energy loss is computed as a percentage of the preimpact total kinetic energy, or as

Eq. (5.46), p. 365

$$\text{Percentage of energy loss} = \frac{T^- - T^+}{T^-} \times 100\%,$$

where T^- is the preimpact total kinetic energy and T^+ is the postimpact total kinetic energy, and they are given by

Eqs. (5.47) and (5.48), p. 365

$$T^- = \tfrac{1}{2}m_A(v_A^-)^2 + \tfrac{1}{2}m_B(v_B^-)^2,$$
$$T^+ = \tfrac{1}{2}m_A(v_A^+)^2 + \tfrac{1}{2}m_B(v_B^+)^2.$$

Constrained impact.   In a constrained impact, one of the objects is physically constrained from moving in some direction. Our strategy to solve these problems had the following elements:

1. Identification of those directions along which the momentum is conserved through the impact for one and/or both of the bodies.

2. Kinematic description of the constraint, which means that the constrained body can only move normal to the constraint after the impact.

3. Application of the COR equation along the LOI.

EXAMPLE 5.6   *Direct Central Impact of Two Bowling Balls*

Bowling ball A, traveling at 6 ft/s, arrives at a return station and collides with ball B, which is at rest. Balls A and B have weights $W_A = 13$ lb and $W_B = 16$ lb, respectively, they have identical diameters, and the COR for the collision is $e = 0.98$. Determine the postimpact velocities of balls A and B.

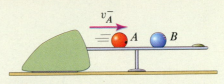

Figure 1
Ball arriving at a bowling ball return station.

SOLUTION

Road Map & Modeling   The motion of the two balls takes place along a horizontal line, and so the collision is one-dimensional. Since the diameters of the balls are identical, the LOI is also horizontal and so the contact force between the balls will be horizontal. This means that the vertical reactions between the balls and the ball return are nonimpulsive. Hence, the impact-relevant FBD for the two balls is shown in Fig. 2. Treating A and B as particles, we can model this impact as an unconstrained direct central impact.

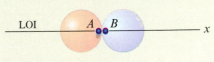

Figure 2
Impact-relevant FBD of the colliding particles.

Governing Equations

Balance Principles   Referring to Fig. 2, since there are no impulsive external forces acting on the system, we can conserve momentum in the x direction as

$$m_A v_{Ax}^- + m_B v_{Bx}^- = m_A v_{Ax}^+ + m_B v_{Bx}^+, \tag{1}$$

where m_A and m_B denote the masses of A and B, respectively.

Force Laws   As we have seen, the force law characterizing impacts is the COR equation, which for this problem is

$$v_{Bx}^+ - v_{Ax}^+ = e\left(v_{Ax}^- - v_{Bx}^-\right). \tag{2}$$

Kinematic Equations   Since the problem is one-dimensional, the only remaining kinematic information to formulate consists of listing the known velocities. This information is given by the following relations:

$$v_{Ax}^- = 6 \text{ ft/s} \quad \text{and} \quad v_{Bx}^- = 0 \text{ ft/s.} \tag{3}$$

Computation   After substituting Eq. (3) into Eqs. (1) and (2), we are left with a system of two equations in the two unknowns v_{Ax}^+ and v_{Bx}^+. Solving this system, we obtain

$$v_{Ax}^+ = \frac{m_A v_{Ax}^- + m_B\left[v_{Bx}^- + e\left(v_{Bx}^- - v_{Ax}^-\right)\right]}{m_A + m_B} = -0.554 \text{ ft/s,} \tag{4}$$

$$v_{Bx}^+ = \frac{m_B v_{Bx}^- + m_A\left[v_{Ax}^- + e\left(v_{Ax}^- - v_{Bx}^-\right)\right]}{m_A + m_B} = 5.33 \text{ ft/s.} \tag{5}$$

Discussion & Verification   The solution conforms to common experience in that ball A, being lighter than ball B, rebounds backward (to the left). Also, given that the value of the COR is less than 1, we expect the impact to imply a loss of kinetic energy. A quick calculation based on the given data and the final results tells us that the pre- and postimpact total kinetic energies for the system are $T^- = 7.27$ ft·lb and $T^+ = 7.11$ ft·lb, respectively. Since $T^+ < T^-$, we can say that our solution behaves as expected.

> ⊕ **Helpful Information**
>
> **What about conservation of y momentum?** We could, of course, conserve momentum in the y direction, but that would simply be an equation stating that $0 = 0$.

E X A M P L E   5.7   *Direct Central Impact of a Ball with the Ground*

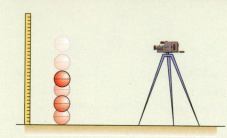

Figure 1
Video camera recording the motion of a ball falling on the ground and rebounding.

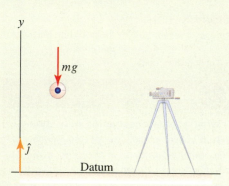

Figure 2
FBD of the ball before and after the impact with the ground.

Figure 3
FBD of the ball during the impact with the ground/Earth.

N W E S **Helpful Information**

The time of impact. We have defined four states, namely, ①–④, but it is important to remember that states ② and ③ are assumed to happen at essentially the same time; i.e., $t_2 \approx t_3 \approx t_i$, where t_i is the time of impact. This is so because we view impacts as events that take place in an infinitesimal amount of time and because the event separating ② and ③ is the impact between the ball and the ground.

An experiment is carried out to measure the COR for the impact of a ball with the ground (Fig. 1). The experiment consists of video recording the motion of a ball that is initially at rest and that is dropped onto the ground from a known height. By using the video recording it is possible to measure the maximum rebound height of the ball. Use the measurements of the release height and the maximum rebound height to measure the COR for the impact.

S O L U T I O N

Road Map & Modeling   Neglecting air resistance, the FBD before and after impact with the ground is shown in Fig. 2. The FBD of the ball and the ground during the impact is shown in Fig. 3. Referring to Fig. 2, we can use the work-energy principle to relate the release height of the ball to the speed with which it hits the ground. Similarly, we can relate the rebound height to the speed with which the ball rebounds off the ground. Referring to Fig. 3, we can apply the direct central impact equations (Eqs. (5.36) and (5.37)) to the impact between the ball and the ground.

──────────── **Application of the Work-Energy Principle** ────────────

Governing Equations

Balance Principles   We will let ① denote the state of the ball at the instant of release, ② the state of the ball right before impact with the ground, ③ the state of the ball right after impact, and ④ the state of the ball at its maximum rebound height. Therefore, since all forces are conservative, the application of the work-energy principle between ① and ② and between ③ and ④ takes on the form

$$T_1 + V_1 = T_2 + V_2, \tag{1}$$
$$T_3 + V_3 = T_4 + V_4. \tag{2}$$

The expressions for the kinetic energies are given by

$$T_1 = \tfrac{1}{2}mv_1^2, \quad T_2 = \tfrac{1}{2}mv_2^2, \quad T_3 = \tfrac{1}{2}mv_3^2, \quad T_4 = \tfrac{1}{2}mv_4^2, \tag{3}$$

where v_1 is the speed at release, v_2 is the speed with which the ball collides with the ground, v_3 is the speed with which the ball rebounds off the ground, and v_4 is the speed at the maximum rebound height.

Force Laws   Selecting the ground for our datum, we have

$$V_1 = mgh_i, \quad V_2 = 0, \quad V_3 = 0, \quad V_4 = mgh_f, \tag{4}$$

where h_i is its release height and h_f is its rebound height (the subscripts i and f stand for initial and final, respectively).

Kinematic Equations   The kinematic equations relate the speeds of the ball at ① and ④, which are

$$v_1 = 0 \quad \text{and} \quad v_4 = 0. \tag{5}$$

Combining Eqs. (1)–(5) will allow us to solve for v_2 and v_3.

──────────── **Application of the Impulse-Momentum Principle** ────────────

Governing Equations

Balance Principles   Referring to the impact-relevant FBD of the ball impacting the ground in Fig. 3, we see that there are no external impulsive forces in the y direction. Therefore, we will solve this problem exactly as we solved Example 5.6, except that now the Earth is one of the particles. Applying conservation of momentum in the y direction, we obtain

$$mv_2 + m_e v_e^- = mv_3 + m_e v_e^+, \tag{6}$$

where m_e is the mass of the Earth and v_e^- and v_e^+ are the pre- and postimpact velocities of the Earth, respectively.

Force Laws   The COR equation for the ball and the Earth is

$$v_3 - v_e^+ = e(v_e^- - v_2). \tag{7}$$

Kinematic Equations   The kinematics equations in this situation come from considering the fact that the mass of the Earth is *much* greater than the mass of the ball. Dividing both sides of Eq. (6) by m_e, we obtain

$$\frac{m}{m_e} v_2 + v_e^- = \frac{m}{m_e} v_3 + v_e^+ \quad \Rightarrow \quad v_e^- = v_e^+ = 0, \tag{8}$$

where we have used $m/m_e \approx 0$ and the fact that the Earth isn't moving before the impact to obtain $v_e^- = v_e^+ = 0$. If the Earth isn't moving before the impact, it certainly isn't moving much after!

Computation   Combining Eqs. (1)–(5), we have

$$mgh_i = \tfrac{1}{2}mv_2^2 \quad \text{and} \quad \tfrac{1}{2}mv_3^2 = mgh_f, \tag{9}$$

which can be solved for v_2 and v_3 to obtain

$$v_2 = -\sqrt{2gh_i} \quad \text{and} \quad v_3 = \sqrt{2gh_f}, \tag{10}$$

where, based on the choice of coordinate system, we have chosen the negative root for v_2 and the positive root for v_3.

Substituting Eq. (8) into Eq. (7), we obtain

$$v_3 = -ev_2 \quad \Rightarrow \quad \boxed{e = \sqrt{\frac{h_f}{h_i}},} \tag{11}$$

where Eqs. (10) have been used.

Discussion & Verification   To verify that our final COR formula is correct, we can compare its determination with our everyday experience. If we drop a ball from a height h_i, the maximum rebound height h_f is smaller than h_i, i.e., $h_f < h_i$. Furthermore, in general, $h_f \geq 0$. Hence, using Eq. (11), we can conclude that $0 \leq e \leq 1$, which is to say that e behaves as expected. Furthermore, it can be seen that for $e = 1$ we must have $h_f = h_i$, which is to say that the rebound height is equal to the release height only if the collision is perfectly elastic.

🔎 **A Closer Look**   Note that we need not write both Eqs. (6) and (7) every time we have to solve a problem in which a particle impacts a nonmoving surface. We can just write

$$v_{\text{particle}}^+ - v_{\text{surface}}^+ = e\left(v_{\text{surface}}^- - v_{\text{particle}}^-\right), \tag{12}$$

along the LOI and then let $v_{\text{surface}}^- = v_{\text{surface}}^+ = 0$, to obtain the result that

$$v_{\text{particle}}^+ = -ev_{\text{particle}}^-. \tag{13}$$

> **Interesting Fact**
>
> **Motion of the Earth due to impact (and other things).** The Earth does move a very small amount when an object impacts it — conservation of momentum tells us that this *must* be so. As Eq. (8) tells us, the mass ratios are so small that the motion of the Earth in a typical impact, such as the one solved here, cannot be measured. Some objects on the Earth that move can cause changes in the Earth's motion. For example, earthquakes, which involve the motion of large portions of the Earth's crust, do produce changes in the Earth's motion that can be detected.

EXAMPLE 5.8   *A Sequence of Particle Collisions*

Figure 1
Ball arriving at a bowling ball return station.

Bowling ball A, traveling at 6 ft/s, arrives at a return station and collides with ball B, which is in contact with ball C. Balls B and C are initially at rest. Let $W_A = 16$ lb, $W_B = 13$ lb, and $W_C = 15$ lb be the weights of A, B, and C, respectively. In addition, let $e_{AB} = 0.98$ and $e_{BC} = 0.94$ be the CORs measured for the individual impacts between balls A and B and balls B and C, respectively. Use the given information to determine the postimpact velocities of all three balls.

SOLUTION

Road Map & Modeling   The theory developed in this section applies only to the impact of two particles at a time. However, since A, B, and C are arranged in series, we can make the problem tractable if we assume that balls B and C are separated by a thin gap, instead of being in contact. Then the problem can be viewed as a *sequence of collisions*. This sequence consists of at least an impact between A and B followed by an impact between B and C, though additional impacts may occur. The final motion of the balls must satisfy the inequality

$$v_A^+ \leq v_B^+ \leq v_C^+, \tag{1}$$

where we are assuming all motion occurs in the x direction (see Fig. 2) and so that subscript is omitted.

The balls all have the same diameter, so the LOI is parallel to the guides of the ball return. Therefore, the reaction forces between the balls and the guides are not affected by the impact and can be considered *nonimpulsive*. The resulting impact-relevant FBDs for the A-B and B-C impacts are shown in Fig. 2. As the motion is in the direction parallel to the LOI, all the impacts are unconstrained direct central impacts.

--- **First (A-B) and Second (B-C) Collisions** ---

Governing Equations

Balance Principles   The A-B and B-C impacts are *separate but sequential* events. In view of Fig. 2, we can say that momentum is conserved in the x direction for these collisions, that is,

$$m_A v_{A1}^- + m_B v_{B1}^- = m_A v_{A1}^+ + m_B v_{B1}^+, \tag{2}$$

$$m_B v_{B2}^- + m_C v_{C2}^- = m_B v_{B2}^+ + m_C v_{C2}^+, \tag{3}$$

where v_{A1}^+ and v_{B1}^+ are the velocities of A and B right after the first collision; v_{B2}^- and v_{C2}^- are the velocities of B and C right before the second collision; and v_{B2}^+ and v_{C2}^+ are the velocities of B and C right after the second collision. Note that the velocity of B after the first collision must equal the velocity of B before the second, that is,

$$v_{B1}^+ = v_{B2}^-. \tag{4}$$

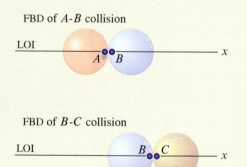

FBD of A-B collision

FBD of B-C collision

Figure 2
Impact-relevant FBDs for impacts between A and B and between B and C.

Force Laws   For the first impact (A-B) and second impact (B-C), we have two corresponding COR equations, i.e.,

$$v_{B1}^+ - v_{A1}^+ = e_{AB}\left(v_{A1}^- - v_{B1}^-\right), \tag{5}$$

$$v_{C2}^+ - v_{B2}^+ = e_{BC}\left(v_{B2}^- - v_{C2}^-\right). \tag{6}$$

Kinematic Equations   Listing the known values of velocity for the various balls, we have

$$v_{A1}^- = 6 \text{ ft/s}, \quad v_{B1}^- = 0 \text{ ft/s}, \quad \text{and} \quad v_{C2}^- = 0 \text{ ft/s}. \tag{7}$$

Computation After we insert Eqs. (4) and (7), Eqs. (2), (3), (5), and (6) form a system of four equations in the four unknowns v_{A1}^+, v_{B1}^+, v_{B2}^+, and v_{C2}^+. For the first collision, we solve Eqs. (2) and (5) for v_{A1}^+ and v_{B1}^+ and then substitute this solution into Eqs. (3) and (6), which can then be solved for v_{B2}^+ and v_{C2}^+. Solving, we obtain

$$v_{A1}^+ = 0.6745\,\text{ft/s}, \qquad v_{B1}^+ = 6.554\,\text{ft/s},$$
$$v_{B2}^+ = -0.2575\,\text{ft/s}, \qquad v_{C2}^+ = 5.903\,\text{ft/s}. \tag{8}$$

Discussion & Verification At this point, our candidate postimpact velocities are v_{A1}^+, v_{B2}^+, and v_{C2}^+. The values given in Eqs. (8) do not satisfy the inequalities in Eq. (1). While the result for C is acceptable, the values for v_{A1}^+ and v_{B2}^+ say that ball A is moving to the right and ball B is moving to the left — therefore A and B will collide again. Hence, we need to study this third collision and determine if it is the last.

——————————————— **Third Collision (A-B)** ———————————————

Governing Equations

Balance Principles Referring again to the top half of Fig. 2, we can write a conservation of momentum equation for A and B during the third collision as

$$m_A v_{A3}^- + m_B v_{B3}^- = m_A v_{A3}^+ + m_B v_{B3}^+. \tag{9}$$

We need to realize that A is not involved in the second collision, and so the velocity of A after the first collision is the same as the velocity of A before the third collision. In addition, the velocity of B right after the second collision is equal to the velocity of B right before the third collision. Mathematically, these two statements give

$$v_{A1}^+ = v_{A3}^- \quad \text{and} \quad v_{B2}^+ = v_{B3}^-. \tag{10}$$

Force Laws The COR equation for this third impact is given by

$$v_{B3}^+ - v_{A3}^+ = e_{AB}\left(v_{A3}^- - v_{B3}^-\right). \tag{11}$$

Kinematic Equations The known velocity components for this collision are

$$v_{A3}^- = v_{A1}^+ = 0.6745\,\text{ft/s} \quad \text{and} \quad v_{B3}^- = v_{B2}^+ = -0.2575\,\text{ft/s}. \tag{12}$$

Computation Substituting Eqs. (12) into Eqs. (9) and (11), we have a system of two equations in the two unknowns v_{A3}^+ and v_{B3}^+. The solution to these two equations yields the following results:

$$v_{A3}^+ = -0.153\,\text{ft/s} \quad \text{and} \quad v_{B3}^+ = 0.761\,\text{ft/s}. \tag{13}$$

With these results, we now satisfy Eq. (1) and so the final answers are

$$\boxed{v_A^+ = -0.153\,\text{ft/s}, \quad v_B^+ = 0.761\,\text{ft/s}, \quad v_C^+ = 5.90\,\text{ft/s}.} \tag{14}$$

Discussion & Verification At the end of third collision in the overall sequence, we know that ball A is traveling to the left while ball B and ball C are both traveling to the right. Furthermore, a fourth collision, this time between B and C, cannot occur because ball C is moving to the right faster than B. In addition, we should expect that the kinetic energy will decrease in a non-perfectly elastic collision. The final total kinetic energy of the system is 8.24 ft·lb, whereas the initial kinetic energy was 8.94 ft·lb.

EXAMPLE 5.9   *Oblique Central Impact in Air Hockey*

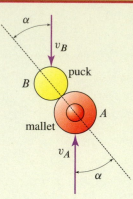

Figure 1

A hockey puck B (yellow) colliding with a mallet A (red) in an air hockey game.

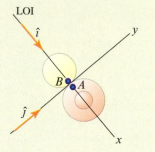

Figure 2

FBD for A and B combined. We have chosen the x axis to coincide with the LOI.

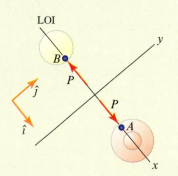

Figure 3

FBDs for A and B taken individually. Force P is the impulsive contact force between A and B during the impact.

In an air hockey game* a 6 oz mallet A is let go[†] and collides with a 1 oz puck B while the puck is in motion (Fig. 1). Assuming that the collision is perfectly elastic, that $\alpha = 40°$, and that the preimpact speeds of the mallet and puck are 9 and 20 ft/s, respectively, determine the postimpact velocities of the mallet and the puck. Finally, verify that the total pre- and postimpact kinetic energies are equal in a perfectly elastic collision.

SOLUTION

Road Map & Modeling   Modeling both the puck and the mallet as particles, we can easily identify the LOI as the line connecting their centers, which is the dashed line in Fig. 1. Since the preimpact velocities are not parallel to the LOI, we can see that it is an oblique central impact. The impact-relevant FBD for A and B taken together is shown in Fig. 2. The individual FBDs of A and B are shown in Fig. 3, in which we have assumed that the impact surface is frictionless and so internal impulsive forces are parallel to the LOI.

Governing Equations

Balance Principles   Following the discussion of the oblique central impact presented earlier in the section, whenever the contact between two impacting objects is frictionless, it is most convenient to (1) write a single conservation of momentum equation along the LOI and (2) write two additional conservation of momentum equations, one for each particle, in the direction orthogonal to the line of impact. Proceeding in this fashion, we have

$$m_A v_{Ax}^- + m_B v_{Bx}^- = m_A v_{Ax}^+ + m_B v_{Bx}^+, \tag{1}$$

$$m_A v_{Ay}^- = m_A v_{Ay}^+ \quad \Rightarrow \quad v_{Ay}^- = v_{Ay}^+, \tag{2}$$

$$m_B v_{By}^- = m_B v_{By}^+ \quad \Rightarrow \quad v_{By}^- = v_{By}^+. \tag{3}$$

Equations (2) and (3) reflect the fact that, for each individual particle, there are no impulsive forces acting in the direction perpendicular to the LOI (see Fig. 3).

Force Laws   Since the collision is perfectly elastic (i.e., $e = 1$), the COR equation for this impact is

$$v_{Bx}^+ - v_{Ax}^+ = v_{Ax}^- - v_{Bx}^-. \tag{4}$$

Kinematic Equations   The kinematic equations for this problem, as for most impact problems, consist of the list of the velocity components that are known. The preimpact velocities are known for both particles, and referring to Fig. 4, they are

$$v_{Ax}^- = -v_A^- \cos\alpha \quad \Rightarrow \quad v_{Ax}^- = -6.894\,\text{ft/s}, \tag{5}$$

$$v_{Ay}^- = v_A^- \sin\alpha \quad \Rightarrow \quad v_{Ay}^- = 5.785\,\text{ft/s}, \tag{6}$$

$$v_{Bx}^- = v_B^- \cos\alpha \quad \Rightarrow \quad v_{Bx}^- = 15.32\,\text{ft/s}, \tag{7}$$

$$v_{By}^- = -v_B^- \sin\alpha \quad \Rightarrow \quad v_{By}^- = -12.86\,\text{ft/s}. \tag{8}$$

---

* Air hockey is a game in which the players use hard plastic mallets to hit a plastic puck, which then slides over a horizontal table. The objective of the game is to hit the puck into the opponent's goal. To facilitate the sliding motion of the puck, the table used for the game has a surface with an array of small-diameter holes through which air is pushed by a small compressor. Hence, the puck is essentially sliding over a thin air cushion.

[†] "Loosing control of the mallet while the puck is in play" is considered a foul according to the official air hockey rules.

Computation   Substituting Eqs. (5)–(8) into Eqs. (1)–(4), gives a system of four equations in the four unknowns v_{Ax}^+, v_{Ay}^+, v_{Bx}^+, and v_{By}^+. Instead of dealing with four equations and four unknowns, it is helpful to notice that the equations in the x direction are decoupled from those in the y direction. This means that we can view Eqs. (1) and (4) as forming a system of two equations in the two unknowns v_{Ax}^+ and v_{Bx}^+. The solution of these two equations gives

$$v_{Ax}^+ = -0.547\,\text{ft/s} \quad \text{and} \quad v_{Bx}^+ = -22.8\,\text{ft/s}. \tag{9}$$

As far as the remaining two unknowns are concerned, these can be obtained directly by substituting Eqs. (6) and (8) into Eqs. (2) and (3), respectively, thus giving

$$v_{Ay}^+ = 5.79\,\text{ft/s} \quad \text{and} \quad v_{By}^+ = -12.9\,\text{ft/s}. \tag{10}$$

We can now verify the claim that, for an elastic collision, the pre- and postimpact kinetic energies have the same value. The preimpact total kinetic energy is

$$T^- = \tfrac{1}{2}m_A(v_A^-)^2 + \tfrac{1}{2}m_B(v_B^-)^2 = 0.504\,\text{lb}\cdot\text{ft}, \tag{11}$$

and the postimpact total kinetic energy is

$$T^+ = \tfrac{1}{2}m_A(v_A^+)^2 + \tfrac{1}{2}m_B(v_B^+)^2 = 0.504\,\text{lb}\cdot\text{ft}, \tag{12}$$

thus verifying that the energy is unchanged during a perfectly elastic collision.

Discussion & Verification   This problem offers an example of how to deal with typical oblique central impacts. The solution discussed here requires that the contact between the colliding objects be frictionless. In addition, in this particle impact, the angle between the incoming velocity of A and the LOI was equal to the angle between the incoming velocity of B and the LOI (both were α). This will not always be the case, but it is simple to modify Eqs. (5)–(8) to reflect different incoming angles — the basic strategy we have followed is general.

It is also helpful to look geometrically at the postimpact velocities. Referring to Fig. 5, we see that

$$\beta = 180° + \tan^{-1}\!\left[\frac{v_{Ay}^+}{v_{Ax}^+}\right] = 95.4° \quad \text{and} \quad \gamma = \tan^{-1}\!\left[\frac{v_{By}^+}{v_{Bx}^+}\right] = 29.5°. \tag{13}$$

These angles seem reasonable. If we had found, say, that $\beta = -80°$, we would have known that something was wrong.

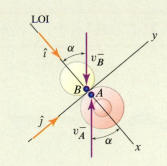

Figure 4
Geometry of the impact between the puck and mallet.

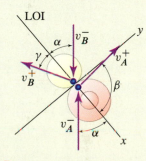

Figure 5
Postimpact geometry of the puck and the mallet.

E X A M P L E   5.10   *Constrained Impact Between a Car and a Truck*

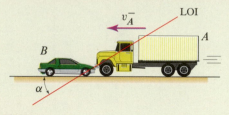

Figure 1
Low-velocity impact between a passenger car and a medium-size truck. The angle $\alpha = 25°$, the COR for the collision is $e = 0.1$, $m_A = 15{,}000\,\text{kg}$, $m_B = 1100\,\text{kg}$, and $v_A^- = 7.3\,\text{km/h}$.

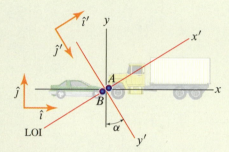

Figure 2
Impact-relevant FBD for the *A-B* collision under the assumption that the pavement is compliant.

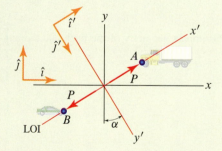

Figure 3
Impact-relevant FBDs for *A* and *B* taken individually, under the assumption that the pavement is compliant.

A truck *A* collides with a parked car *B* (Fig. 1). Assuming that the friction between *A* and *B* and between *B* and the ground is negligible, compare the postimpact velocities if (a) the pavement is compliant and (b) the pavement is rigid.

SOLUTION

Road Map   This example illustrates some of the differences between unconstrained and constrained impacts. In Part (a), the problem is essentially identical to that in Example 5.9. By contrast, in Part (b), the impact is *constrained* and must be dealt with using the ideas discussed beginning on p. 365.

--------------------------- **Part (a): Compliant Pavement** ---------------------------

Modeling   Modeling both *A* and *B* as particles, we will neglect friction, and we will neglect gravity as it is nonimpulsive. Due to the orientation of the LOI, *A* will push *B* into the pavement, which is *compliant* and thus has a springlike behavior. Hence, the reaction between *B* and the pavement must be treated as a nonimpulsive force (see discussion on p. 358). Therefore, the impact-relevant FBD for the *A-B* system under these assumptions is shown in Fig. 2, and the impact will be treated as an unconstrained oblique central impact (since v_A^- is not parallel to the LOI). The corresponding individual FBDs for *A* and *B* are those in Fig. 3.

Governing Equations

Balance Principles   As discussed in Example 5.9, the FBDs in Figs. 2 and 3 allow us to write the following statements of conservation of linear momentum:

$$m_A v_{Ax'}^- + m_B v_{Bx'}^- = m_A v_{Ax'}^+ + m_B v_{Bx'}^+, \tag{1}$$

$$m_A v_{Ay'}^- = m_A v_{Ay'}^+ \quad \Rightarrow \quad v_{Ay'}^- = v_{Ay'}^+, \tag{2}$$

$$m_B v_{By'}^- = m_B v_{By'}^+ \quad \Rightarrow \quad v_{By'}^- = v_{By'}^+. \tag{3}$$

Force Laws   The COR equation applied along the LOI gives

$$v_{Bx'}^+ - v_{Ax'}^+ = e\left(v_{Ax'}^- - v_{Bx'}^-\right). \tag{4}$$

Kinematic Equations   The preimpact velocities are known to be

$$v_{Ax}^- = -2.028\,\text{m/s}, \quad v_{Ay}^- = 0.0\,\text{m/s}, \quad v_{Bx}^- = 0.0\,\text{m/s}, \quad v_{By}^- = 0.0\,\text{m/s}. \tag{5}$$

The transformation equations from the xy to the $x'y'$ coordinate systems are given by

$$\hat{\imath}' = \cos\alpha\,\hat{\imath} + \sin\alpha\,\hat{\jmath} \quad \text{and} \quad \hat{\jmath}' = \sin\alpha\,\hat{\imath} - \cos\alpha\,\hat{\jmath}. \tag{6}$$

Computation   The solution of this problem is extremely similar to that discussed in Example 5.9. Therefore, we will present only the final answers, which we have expressed in the xy coordinate system:

$$\boxed{\begin{array}{ll} v_{Ax}^+ = -1.90\,\text{m/s}, & v_{Ay}^+ = 0.0584\,\text{m/s}, \\[4pt] v_{Bx}^+ = -1.71\,\text{m/s}, & v_{By}^+ = -0.796\,\text{m/s}. \end{array}} \tag{7} \tag{8}$$

--------------------------- **Part (b): Rigid Pavement** ---------------------------

Modeling   If the pavement is assumed to be *rigid*, then it is capable of providing impulsive reaction forces. As *B* is pushed into the pavement, there is an impulsive reaction force exerted on *B* by the pavement, and there is no impulsive reaction force on *A*. Hence, the impact-relevant FBD for the *A-B* system is that in Fig. 4, whereas the corresponding FBDs for *A* and *B* taken individually are given in Fig. 5. This is a *constrained oblique central impact*.

Governing Equations

Balance Principles   In the FBD in Fig. 4, there are no impulsive external forces in the x direction. Hence, the linear momentum for the A-B system is conserved in that direction. Using a similar argument, we know that A's linear momentum is conserved along the y' direction (see Fig. 5). However, the same cannot be said about B. Hence, the application of the impulse-momentum principle yields the following two equations:

$$m_A v_{Ax}^- + m_B v_{Bx}^- = m_A v_{Ax}^+ + m_B v_{Bx}^+, \tag{9}$$

$$m_A v_{Ay'}^- = m_A v_{Ay'}^+ \quad \Rightarrow \quad v_{Ay'}^- = v_{Ay'}^+. \tag{10}$$

Force Laws   As discussed on p. 366, the COR equation for this impact is taken to be the same one as in the unconstrained case. Therefore, Eq. (4) still applies:

$$v_{Bx'}^+ - v_{Ax'}^+ = e\left(v_{Ax'}^- - v_{Bx'}^-\right). \tag{11}$$

Kinematic Equations   Kinematically, the crucial difference between the current and the previous case is that B's movement occurs in the x direction, i.e.,

$$v_{By}^+ = 0. \tag{12}$$

This equation must be added to the rest of the kinematic information we already possess, as expressed by Eqs. (5).

Computation   Choosing to express our final answer in the xy coordinate system and since v_{By}^+ is already known, we have three unknowns for this problem: v_{Ax}^+, v_{Ay}^+, and v_{Bx}^+. Hence, we need three corresponding equations. One of these equations is Eq. (9). The other two equations will be obtained by rewriting Eqs. (10) and (11) in xy components. Using Eqs. (6) and (12), this change of components becomes

$$\left(v_{Bx}^+ - v_{Ax}^+\right)\cos\alpha - v_{Ay}^+ \sin\alpha = e v_{Ax}^- \cos\alpha, \tag{13}$$

$$v_{Ax}^- \sin\alpha = v_{Ax}^+ \sin\alpha - v_{Ay}^+ \cos\alpha. \tag{14}$$

Solving Eqs. (9), (13), and (14) for v_{Ax}^+, v_{Ay}^+, and v_{Bx}^+ gives

$$\boxed{v_{Ax}^+ = -1.88\,\text{m/s}, \quad v_{Ay}^+ = 0.0700\,\text{m/s}, \quad v_{Bx}^+ = -2.05\,\text{m/s}.} \tag{15}$$

Discussion & Verification   There are several differences between the two solutions. First, we see that if the car B is not allowed to sink into the pavement, then the postimpact velocity of truck A in the y direction for the constrained impact is higher than in the unconstrained impact case. For the constrained impact case, right after impact, the car B moves to the left faster than the truck A, contrary to what happens in the unconstrained impact. Therefore, the solution for the constrained case predicts that B actually moves away from A. By contrast, the unconstrained solution predicts that A is actually going over B. This is not to say that one solution is right and the other is wrong. Both solutions might be correct depending on the mechanical properties of the vehicles and pavement. Clearly, all of the considerations mentioned here are limited by the fact that we are modeling A and B as particles, and that therefore we are ignoring myriad additional effects concerning the objects' finite size (and therefore, for example, rotations).

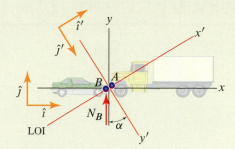

Figure 4

Impact-relevant FBD for the A-B collision under the assumption that the pavement is rigid. Force N_B denotes the impulsive reaction force exerted on B by the pavement.

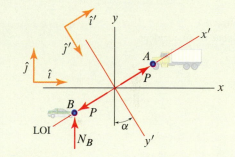

Figure 5

Impact-relevant FBDs for A and B taken individually, under the assumption that the pavement is rigid. Note that the impulsive force P for this case will have, in general, a different intensity than that in Fig. 3.

EXAMPLE 5.11 *Oblique Central Impact: Accident Reconstruction*

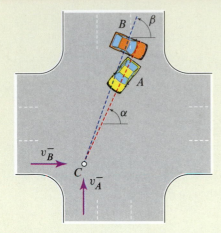

Figure 1
Scene of an accident.

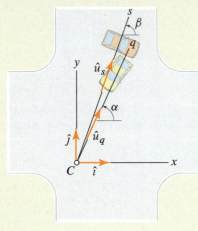

Figure 2
Coordinate axes for the accident reconstruction. Car A slides in the q direction after impact, and car B slides in the s direction.

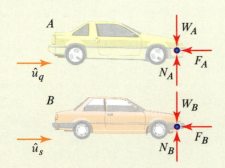

Figure 3
Postimpact FBDs of cars A and B.

The driver of car A has caused a collision by failing to stop at a red light. To better ascertain the culpability of driver A, the following data is recorded (Fig. 1): $W_A = 5400$ lb, $W_B = 4630$ lb, $\alpha = 67°$, $\beta = 70°$, $d_A = 60.7$ ft, $d_B = 69.9$ ft, and $\mu_k = 0.8$, where W_A and W_B are the weights of A and B, respectively, d_A and d_B are the distances covered by A and B after colliding at C, and μ_k is the coefficient of kinetic friction. The evidence collected also indicates that the collision did not cause any significant vertical motion for either car and that the velocities of the cars before the impact were perpendicular to one another and were oriented as shown in Fig. 1. Assuming the intersection is horizontal, determine the preimpact speeds of cars A and B, the coefficient of restitution for the impact, and the orientation of the LOI.

SOLUTION

Road Map We first need to find the postimpact velocities of the cars given their final position. That can be done by applying the work-energy principle. Once we know the postimpact velocities, we will solve the impact problem at C to find the desired four unknowns, i.e., the components of the preimpact velocities.

Post-Impact Problem: Application of the Work-Energy Principle

Modeling We will assume that, after the impact, A and B slide along two straight lines oriented by α and β, respectively, and with common origin at C (Fig. 2). Let ① be immediately after the impact, and let ② be the final position of A and B. The FBDs describing the forces acting on A and B in going from ① to ② are shown in Fig. 3.

Governing Equations

Balance Principles The work-energy principle between ① and ② for each car gives

$$T_{A1} + V_{A1} + (U_{1\text{-}2})^A_{np} = T_{A2} + V_{A2}, \tag{1}$$

$$T_{B1} + V_{B1} + (U_{1\text{-}2})^B_{np} = T_{B2} + V_{B2}. \tag{2}$$

The kinetic energies in ① correspond to the postimpact speeds, and the kinetic energies in ② correspond to the speeds of the cars after they have come to a stop, that is,

$$T_{A1} = \tfrac{1}{2}m_A(v_A^+)^2, \; T_{A2} = \tfrac{1}{2}m_A v_{As}^2, \; T_{B1} = \tfrac{1}{2}m_B(v_B^+)^2, \; T_{B2} = \tfrac{1}{2}m_B v_{Bs}^2. \tag{3}$$

Force Laws All potential energy terms are equal to zero since the motion from ① to ② occurs on a horizontal surface; that is,

$$V_{A1} = V_{A2} = 0 \quad \text{and} \quad V_{B1} = V_{B2} = 0. \tag{4}$$

The horizontal nature of the motion also implies that $N_A = W_A$ and $N_B = W_B$. Assuming Coulomb friction, we have $F_A = \mu_k W_A$ and $F_B = \mu_k W_B$. Since F_A and F_B are constant between ① and ②, the work done by friction is

$$(U_{1\text{-}2})^A_{np} = -\mu_k W_A d_A \quad \text{and} \quad (U_{1\text{-}2})^B_{np} = -\mu_k W_B d_B. \tag{5}$$

Kinematic Equations The final speed of the two cars is zero, so

$$v_{As} = 0 \quad \text{and} \quad v_{Bs} = 0. \tag{6}$$

Computation   Substituting Eqs. (3)–(6) into Eqs. (1) and (2), we obtain

$$\tfrac{1}{2}m_A(v_A^+)^2 - \mu_k m_A g d_A = 0 \quad \text{and} \quad \tfrac{1}{2}m_B(v_B^+)^2 - \mu_k m_B g d_B = 0, \qquad (7)$$

where we have used $W_A = m_A g$ and $W_B = m_B g$. Equations (7) can be solved to obtain

$$v_A^+ = \sqrt{2\mu_k g d_A} = 55.92\,\text{ft/s} \quad \text{and} \quad v_B^+ = \sqrt{2\mu_k g d_B} = 60.01\,\text{ft/s}. \qquad (8)$$

Equations (8) give us the cars' postimpact *speeds*. Since the postimpact paths of A and B are known (Fig. 2), we also know the postimpact *velocities* of the two cars, i.e.,

$$\vec{v}_A^+ = v_A^+ \, \hat{u}_q \;\Rightarrow\; \vec{v}_A^+ = v_A^+ (\cos\alpha\,\hat{\imath} + \sin\alpha\,\hat{\jmath}) = (21.85\,\hat{\imath} + 51.48\,\hat{\jmath})\,\text{ft/s}, \qquad (9)$$

$$\vec{v}_B^+ = v_B^+ \, \hat{u}_s \;\Rightarrow\; \vec{v}_B^+ = v_A^+ (\cos\beta\,\hat{\imath} + \sin\beta\,\hat{\jmath}) = (20.52\,\hat{\imath} + 56.39\,\hat{\jmath})\,\text{ft/s}. \qquad (10)$$

Now that the postimpact velocities are known, we can proceed with the calculation of the preimpact velocities by solving the impact problem.

—— **Impact Problem: Application of the Impulse-Momentum Principle** ——

Modeling   Referring to the impact-relevant FBD in Fig. 4, we see that the impact between A and B is unconstrained. Referring to Fig. 1, we see that this is an *oblique central impact*. Finally, we assume that the contact between A and B is frictionless.

Governing Equations

Balance Principles   Since this is an oblique central impact of an isolated system, we first write a conservation of momentum equation in the direction of the LOI, which is

$$m_A v_{Ay'}^- + m_B v_{By'}^- = m_A v_{Ay'}^+ + m_B v_{By'}^+. \qquad (11)$$

Since we are assuming the impact is frictionless, the FBDs of A and B individually are given by those found in Fig. 5. This implies that the momentum of A and B, taken individually, is conserved in the direction perpendicular to the LOI, which implies

$$v_{Ax'}^- = v_{Ax'}^+ \quad \text{and} \quad v_{Bx'}^- = v_{Bx'}^+. \qquad (12)$$

Force Laws   The COR equation, applied along the LOI, takes on the following form:

$$e\big(v_{Ay'}^- - v_{By'}^-\big) = v_{By'}^+ - v_{Ay'}^+. \qquad (13)$$

Kinematic Equations   Referring to Fig. 1, the preimpact velocities of A and B are

$$v_{Ax}^- = 0 \quad \text{and} \quad v_{By}^- = 0. \qquad (14)$$

Computation   Since our equations have been written in two different component system (i.e., the xy and $x'y'$ systems), we need to rewrite our equations using a single component system. Since we have already computed the postimpact velocities in the xy coordinate system (Eqs. (9) and (10)), we will rewrite Eqs. (11)–(13) in terms of the xy coordinate system. Doing this, while taking into account Eqs. (14), yields

$$m_A v_{Ay}^- \cos\gamma - m_B v_{Bx}^- \sin\gamma = m_A\big(v_{Ay}^+ \cos\gamma - v_{Ax}^+ \sin\gamma\big)$$
$$\qquad\qquad\qquad + m_B\big(v_{By}^+ \cos\gamma - v_{Bx}^+ \sin\gamma\big), \qquad (15)$$

$$v_{Ay}^- \sin\gamma = v_{Ax}^+ \cos\gamma + v_{Ay}^+ \sin\gamma, \qquad (16)$$

$$v_{Bx}^- \cos\gamma = v_{Bx}^+ \cos\gamma + v_{By}^+ \sin\gamma, \qquad (17)$$

$$e\big(v_{Ay}^- \cos\gamma + v_{Bx}^- \sin\gamma\big) = \big(v_{By}^+ - v_{Ay}^+\big)\cos\gamma - \big(v_{Bx}^+ - v_{Ax}^+\big)\sin\gamma. \qquad (18)$$

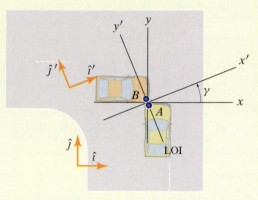

Figure 4
Geometry and FBD of the system at the moment of impact. The y' axis coincides with the LOI. The angle γ is one of the unknowns.

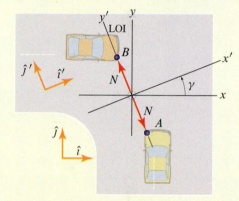

Figure 5
Individual FBDs for A and B. The frictionless contact between A and B implies that there are no impulsive forces in the x' direction.

> **Helpful Information**
>
> **Transforming from a component system to another.** In order for us to easily transform equations written in the $x'y'$ component system into equations written in the xy component system, we can make use of the following relations among the unit vectors of the two systems:
>
> $$\hat{\imath}' = \cos\gamma\,\hat{\imath} + \sin\gamma\,\hat{\jmath},$$
> $$\hat{\jmath}' = -\sin\gamma\,\hat{\imath} + \cos\gamma\,\hat{\jmath}.$$

We can now see that Eqs. (15)–(18) form a system of four equations in the four unknowns v_{Ay}^-, v_{Bx}^-, e, and γ (all the postimpact velocity components have been determined and the masses are both known). This system of equations can be solved to yield the following results:

$$v_{Ay}^- = 99.8\,\text{ft/s}, \quad v_{Bx}^- = 46.0\,\text{ft/s}, \quad e = 0.0457, \quad \gamma = 24.3°, \tag{19}$$

or, converting from ft/s to mph, we find that

$$v_{Ay}^- = 68.1\,\text{mph} \quad \text{and} \quad v_{Bx}^- = 31.4\,\text{mph}. \tag{20}$$

Discussion & Verification In view of Eqs. (14), the values we have found for the velocity components v_{Ay}^- and v_{Bx}^- coincide with the values of the cars' speeds. We can therefore conclude that the driver of car A not only failed to stop at a red light but also was probably speeding. The value we obtained for the COR indicates that the impact was very nearly plastic. In order to get a better sense of how plastic the impact was, we can calculate the percentage of energy lost during the collision. This calculation yields the following result:

$$\text{Percentage of energy lost} = \frac{T^- - T^+}{T^-} \times 100\% = 47.2\%, \tag{21}$$

where T denotes the kinetic energy of the system. Almost one-half of the preimpact kinetic energy has been lost in the permanent deformation of the cars.

PROBLEMS

Problem 5.40

An 8600 lb Ford Excursion A traveling with a speed $v_A = 55$ mph collides head-on with a 1990 lb Smart Fortwo B traveling in the opposite direction with a speed $v_B = 35$ mph. Determine the postimpact velocity of the two cars if the impact is perfectly plastic.

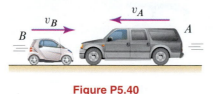

Figure P5.40

Problems 5.41 through 5.43

The ballistic pendulum used to be a common tool for the determination of the muzzle velocity of bullets as a measure of the performance of firearms and ammunition (nowadays, the ballistic pendulum has been replaced by the ballistic chronograph, an electronic device). The ballistic pendulum is a simple pendulum that allows one to record the maximum swing angle of the pendulum arm caused by the firing of a bullet into the pendulum bob.

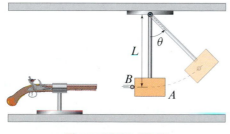

Figure P5.41–P5.43

Problem 5.41

Letting L be the length of the pendulum's arm (whose mass is assumed to be negligible), m_A be the bob's mass, and m_B be the mass of the bullet, and assuming that the pendulum is at rest when the weapon is fired, derive the formula that relates the pendulum's maximum swing angle to the impact velocity of the bullet.

Problem 5.42

Let $L = 1.5$ m and $m_A = 6$ kg. For George Washington's 0.58 caliber pistol, which fired a roundball of mass $m_B = 87$ g, it is found that the maximum swing angle of the pendulum is $\theta_{max} = 46°$. Determine the preimpact speed of the bullet B.

Problem 5.43

Suppose we want to build a ballistic pendulum to test rifles using standard NATO 7.62 mm ammunition, i.e., ammunition for which a (single) cartridge weighs roughly 147 gr (1 lb $= 7000$ gr) and the muzzle speed is typically 2750 ft/s. If the pendulum's length is taken to be 5 ft, and if we are to fire from a short distance so that there is a negligible decrease in speed before the bullet reaches the pendulum, what is the minimum weight we need to give to the pendulum bob to avoid having the pendulum swing to an angle greater than 90°?

Problem 5.44

A 323 gr bullet (1 lb = 7000 gr) hits a 2 kg block that is initially at rest. After the collision, the bullet becomes embedded in the block, and they slide a distance of 0.31 m. If the coefficient of friction between the block and the ground is $\mu_k = 0.7$, determine the preimpact speed of the bullet. Although the definition of the unit "grain" is given in terms of pounds, express the answer in SI units.

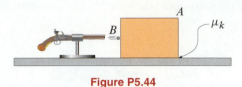

Figure P5.44

Problem 5.45

The official rules of tennis specify that

> "The ball shall have a [re]bound of more than 53 in. (134.62 cm) and less than 58 in. (147.32 cm) when dropped 100 in. (254.00 cm) upon a concrete base."

Understanding the expression "when dropped" as "when dropped from rest," determine the range of acceptable CORs for the collision of a tennis ball with concrete.

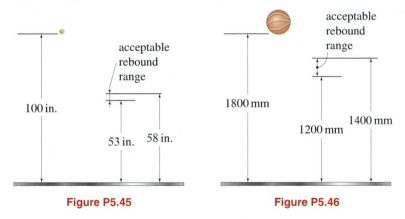

Figure P5.45   **Figure P5.46**

Problem 5.46

The official rules of basketball specify that a basketball is properly inflated

> "such that when it is dropped onto the playing surface from a height of about 1800 mm measured from the bottom of the ball, it will rebound to a height, measured to the top of the ball, of not less than about 1200 mm nor more than about 1400 mm."

Based on this rule, and understanding the expression "when it is dropped" as "when it is dropped from rest," determine the range of acceptable CORs for the collision between the ball and the court's surface.

Problem 5.47

Consider a direct central impact for two spheres. Let m_A, m_B, and e denote the mass of sphere A, the mass of sphere B, and the COR, respectively. If sphere B is at rest before the collision, determine the relation that m_A, m_B, and e need to satisfy in order for A to come to a complete stop right after impact.

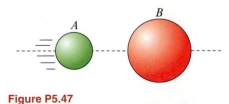

Figure P5.47

Problems 5.48 and 5.49

Car A, with $m_A = 1550$ kg, is stopped at a red light. Car B, with $m_B = 1865$ kg and a speed of 40 km/h, fails to stop before impacting car A. After impact, cars A and B slide over the pavement with a coefficient of friction $\mu_k = 0.65$.

Problem 5.48   How far will the cars slide if the cars become entangled?

Problem 5.49   How far will the cars slide if the COR for the impact is $e = 0.2$?

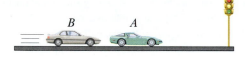

Figure P5.48 and P5.49

Problems 5.50 and 5.51

A platform bench scale consists of a 120 lb plate resting on linear elastic springs whose combined spring constant is $k = 5000$ lb/ft. Let $W = k(\delta - \delta_0)$ be the weight measurement actually provided by the scale (that is, it reads zero pounds when nothing is on the plate), where δ_0 is the spring's compression due to the weight of the scale's plate.

Problem 5.50   A 50 lb sack of portland cement is dropped (from rest) onto the scale from a height $h = 4$ ft measured from the scale's plate (there is no rebound of the sack). Determine the maximum weight displayed by the scale.

Problem 5.51   Repeat Prob. 5.50 with $h = 0$ ft.

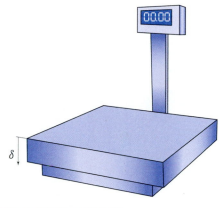

Figure P5.50 and P5.51

Problems 5.52 through 5.54

A 31,000 lb truck A and a 3970 lb sports car B collide at an intersection. Right before the collision the truck and the sports car are traveling at $v_A^- = 60$ mph and $v_B^- = 50$ mph. Assume that the entire intersection forms a horizontal surface.

Problem 5.52   Letting the line of impact be parallel to the ground and to the preimpact velocity of the truck, determine the postimpact velocities of A and B if A and B become entangled. Furthermore, assuming that the truck and the car slide after impact and that the coefficient of kinetic friction is $\mu_k = 0.7$, determine the position at which A and B come to a stop relative to the position they occupied at the instant of impact.

Problem 5.53   Letting the line of impact be parallel to the ground and to the preimpact velocity of the truck, determine the postimpact velocities of A and B if the contact between A and B is frictionless and the COR $e = 0$. Furthermore, assuming that the truck and the car slide after impact and that the coefficient of kinetic friction is $\mu_k = 0.7$, determine the position at which A and B come to a stop relative to the position they occupied at the instant of impact.

Problem 5.54   Letting the line of impact be parallel to the ground and to the preimpact velocity of the truck, determine the postimpact velocities of A and B if the contact between A and B is frictionless and the COR $e = 0.1$. Furthermore, assuming that the truck and the car slide after impact and that the coefficient of kinetic friction is $\mu_k = 0.7$, determine the position at which A and B come to a stop relative to the position they occupied at the instant of impact.

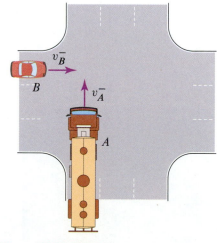

Figure P5.52–P5.54

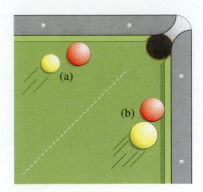

Figure P5.55

Problem 5.55

Although competition rules prohibit significant difference in size, typical coin-operated pool tables may present players with a significant difference in diameter between the typical object ball (i.e., a colored ball) and the cue ball (i.e., the white ball). In fact, once an object ball goes into a pocket, it is captured by the table whereas a cue ball must always be returned to the player; and it is not uncommon for the return mechanism to use the difference in ball diameter to separate the cue ball from the rest. Given this, suppose we want to hit a ball resting against the bumper in such a way that, after the collision, it moves along the bumper. Modeling the contact between balls as frictionless, establish whether or not it is possible to execute the shot in question with (a) an undersized cue ball and (b) an oversized cue ball.

Note: Concept problems are about *explanations*, not computations.

Problems 5.56 and 5.57

Competition billiard balls and tables need to adhere to strict standards (see the Billiard Congress of America for standards in the United States). Specifically, billiard balls must weigh between 5.5 and 6 oz, and they must be 2.25 ± 0.005 in. in diameter.

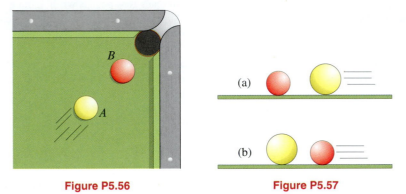

Figure P5.56   **Figure P5.57**

Problem 5.56 Using the theory presented in this section, establish whether or not it is possible to have a moving ball A hit a stationary ball B so that A stops right after the impact, if A and B have the same diameter but not the same weight (since it appears possible to have a weight difference of up to 0.5 oz while staying within regulations). Assume that the COR $e = 1$.

Problem 5.57 Professional billiard players can easily impart to a ball a speed of 20 mph. Assume the tolerance on the ball diameter to be 1/100 in. instead of 5/1000 in. and determine the outcome of the collision between (a) a 2.26 in. diameter ball traveling at 20 mph with a stationary 2.24 in. diameter ball (i.e., each ball is at the extreme limit of tolerance relative to the nominal diameter) and (b) a 2.24 in. diameter ball traveling at 20 mph with a stationary 2.26 in. diameter ball. Assume that the COR $e = 1$ and that the weights of the two balls are identical. Furthermore, assume that the contact between the balls and the table can be treated as essentially frictionless.

Problem 5.58

On a billiard table, the COR for the impact between a ball and any of the four bumpers should be the same. Assuming that this is the case, determine the angle β after two banks as a function of the initial incidence angle α.

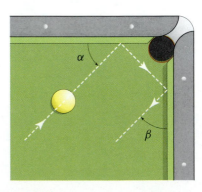

Figure P5.58

Problems 5.59 through 5.61

Newton's cradle is a common desk toy consisting of a number of identical pendulums with steel balls as bobs. These pendulums are arranged in a row in such a way that, when at rest, each ball is tangent to the next and the cords are all vertical. Assume that the COR for the impact of a ball with the next is $e = 1$.

Problem 5.59 Explain why if you release the ball to the far left from a certain angle, the ball in question comes to a stop after impact while all the other balls do not seem to move except for the ball to the far right, which swings upward and achieves a maximum swing angle equal to the initial release angle of the ball to the far left.

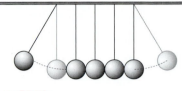

Figure P5.59

Problem 5.60 Explain why if you release the two balls to the far left from a certain angle, the balls in question come to a stop right after impact while all the other balls do not seem to move except for the two balls to the far right, which swing upward and achieve a maximum swing angle equal to the initial release angle of the two balls to the far left.

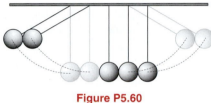
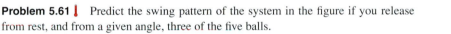

Figure P5.60

Problem 5.61 Predict the swing pattern of the system in the figure if you release from rest, and from a given angle, three of the five balls.

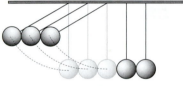

Figure P5.61

💡 Problem 5.62 💡

If an impact is an event spanning an infinitesimally small time interval, is the total potential energy of two colliding objects conserved through the impact? What about the potential energy of each individual object?
Note: Concept problems are about *explanations*, not computations.

💡 Problem 5.63 💡

If an impact is an event spanning an infinitesimally small time interval, is the total kinetic energy of two colliding objects conserved through an impact? What about the kinetic energy of each individual object?
Note: Concept problems are about *explanations*, not computations.

Problems 5.64 and 5.65

Two spheres, A and B, with masses $m_A = 1.35\,\text{kg}$ and $m_B = 2.72\,\text{kg}$, respectively, collide with $v_A^- = 26.2\,\text{m/s}$, and $v_B^- = 22.5\,\text{m/s}$.

Problem 5.64 Compute the postimpact velocities of A and B if $\alpha = 45°$, $\beta = 16°$, the COR is $e = 0.57$, and the contact between A and B is frictionless.

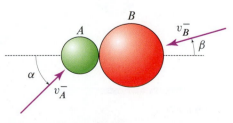

Problem 5.65 Compute the postimpact velocities of A and B if $\alpha = 45°$, $\beta = 16°$, the COR is $e = 0$, and the contact between A and B is frictionless.

Figure P5.64 and P5.65

Problems 5.66 and 5.67

A 1.34 lb ball is dropped on a 10 lb incline with $\alpha = 33°$. The ball's release height is $h_1 = 5$ ft, and the height of the impact point relative to the ground is $h_2 = 0.3$ ft. Assume that the contact between the ball and the incline is frictionless, and let the COR for the impact be $e = 0.88$.

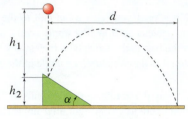

Figure P5.66 and P5.67

Problem 5.66 Compute the distance d at which the ball will hit ground for the first time if the incline cannot move relative to the floor.

Problem 5.67 Compute the distance d at which the ball will hit ground for the first time if the incline can slide without friction relative to the floor.

Problem 5.68

A 1.34 lb ball is dropped on a 10 lb incline with $\alpha = 33°$. The ball's release height is $h_1 = 5$ ft, and the height of the impact point relative to the ground is $h_2 = 0.3$ ft. Assume that the contact between the ball and the incline is frictionless, and let the COR for the impact be $e = 0.88$. Compute the distance d at which the ball will hit ground for the first time if the combined stiffness of the supporting springs is $k = 50$ lb/in. Assume that the incline can move only vertically.

Problem 5.69

Consider two balls A and B that are stacked one on top of the other and dropped from rest from a height h. Let $e_{AG} = 1$ be the COR for the collision of ball A with the ground, and let $e_{AB} = 1$ be the COR for the collision between balls A and B. Finally, assume that the balls can move only vertically and that $m_A \gg m_B$, that is, that $m_B/m_A \approx 0$. Model the combined collision as a sequence of impacts, and predict the rebound speed of ball B as a function of h and g, the acceleration due to gravity.

Problem 5.70

Consider a stack of N balls dropped from rest from a height h. Let all impacts be perfectly elastic, and assume that $m_i \gg m_{i+1}$, that is, that $m_{i+1}/m_i \approx 0$, with $i = 1, \ldots, N-1$ and m_i being the mass of the ith ball. Model the combined collision as a sequence of impacts, and predict the rebound speed of the topmost ball. Assume that the balls can move only vertically.

Problem 5.71

A ball is dropped from rest from a height $h_0 = 1.5$ m. The impact between the ball and the floor has a COR $e = 0.92$. Find the formula that allows you to compute the rebound height h_i of the ith rebound. Furthermore, find the formula that provides the total time required to complete i rebounds. Finally, compute the time t_{stop} that the ball will take to stop bouncing. *Hint:* A formula you may find useful in the solution of this problem is that of the limit value of a geometric series: $\sum_{i=0}^{N-1} e^i = (e^N - 1)/(e - 1)$, with $|e| < 1$.

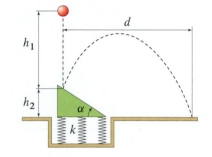

Figure P5.68

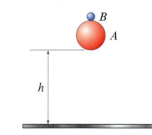

Figure P5.69

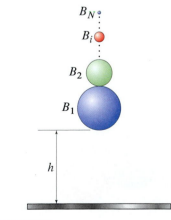

Figure P5.70

DESIGN PROBLEMS

Design Problem 5.1

In the manufacture of steel balls of the type used for ball bearings, it is important that their material properties be sufficiently uniform. One way to detect gross differences in their material properties is to observe how a ball rebounds when dropped on a hard strike plate. Assuming that each ball has a radius $R = 0.3$ in., design a sorting device to select the balls with $0.900 < \text{COR} < 0.925$. The device consists of an incline defined by the angle θ and length L. The strike plate is placed at the bottom of a well with depth h and width w. Finally, at a distance ℓ from the end of the incline, there is a trap with a diameter d. By releasing a ball from rest at the top of the incline and assuming that the ball slides with negligible friction, the ball will reach the bottom of the incline with a speed v_0, rebound off the strike plate, and then fall directly in the trap (you may want to add a design element that prevents balls from simply rolling into the trap). In your design, choose appropriate values of L, $\theta < 45°$, h, w, d, and ℓ to accomplish the desired task while ensuring that the overall dimensions of the device do not exceed 4 ft in both the horizontal and vertical directions.

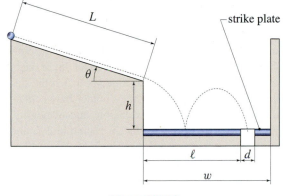

Figure DP5.1

5.3   Angular Momentum

The concept of *angular momentum* arises in the study of orbital motion such as that of a satellite about a planet. Angular momentum is also a key concept in generalizing the applicability of Newton's laws from particles to more complex models such as a rigid body. In this generalization, a *fundamental law* is formulated called the *angular impulse-momentum principle*.

A skater's spin

Spins are common in ice-skating performances. In a spin, a skater rotates about a vertical axis that runs through the skater's body. The skater controls his or her spin rate by extending or retracting the arms and/or legs. How is it possible to control the spin rate *without external help*?

Our intuition tells us that the spin rate is controlled by adjusting the distribution of the skater's mass in relation to the spin axis. Specifically, when arms and/or legs are brought closer to the spin axis, more of the body mass is placed closer to the spin axis and the spin rate increases. Conversely, when the mass is distributed away from the spin axis, e.g., by extending arms and/or legs outward, the spin rate decreases.

To verify our intuitive explanation we can use the simple model depicted in Fig. 5.29, in which a disk D of mass m orbits a fixed point O. Disk D is connected to O by a string that can be pulled in or let out of the hole in the table at O. This setup allows for a variable distance between the disk and the spin axis z. For simplicity, we assume that the contact between D and the table is frictionless, and that the cord is inextensible and of negligible mass. The system in Fig. 5.29 is very similar to that in Example 3.7, and this fact will allow us to immediately delve into the solution of the problem. Specifically, let's focus on Eq. (2) of Example 3.7, which we repeat here in the following form:

$$ma_\theta = m(r\ddot\theta + 2\dot r\dot\theta) = 0. \tag{5.54}$$

Figure 5.28
Three snapshots of a *forward spin*. From left to right, notice how first a leg is retracted and then both arms are retracted. By doing so the skater is dramatically increasing her spin rate.

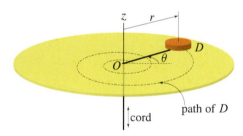

Figure 5.29
Disk moving in orbit around point O while constrained by a string. The contact between the disk and the horizontal surface is frictionless.

Referring to the FBD in Fig. 5.30, Eq. (5.54) states that $F_\theta = 0$, i.e., there is no force in the transverse direction promoting the disk's rotation about O. In our model, this corresponds to the fact that there is nothing helping the skater's rotation. By using the product rule of calculus, and for $r \neq 0$, Eq. (5.54) can be rewritten as

$$\frac{1}{r}\frac{d}{dt}\left[m(r^2\dot\theta)\right] = 0 \quad \Rightarrow \quad mr^2\dot\theta = \text{constant}. \tag{5.55}$$

The constant value taken on by $mr^2\dot\theta$ is determined by how D was put in motion at the initial time t_i. Hence, the spinning motion of D about O is subject to the following *conservation* requirement:

$$mr^2\dot\theta = mr^2(t_i)\dot\theta(t_i). \tag{5.56}$$

This result helps us understand the skater's spin. In fact, recalling that r measures the distance between D and the spin axis and solving Eq. (5.56) for the spin rate $\dot\theta$, we have

$$\dot\theta = \left(\frac{r_i}{r}\right)^2\dot\theta_i, \tag{5.57}$$

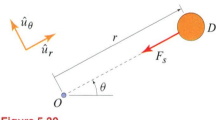

Figure 5.30
FBD of the disk D.

where $r_i = r(t_i)$ and $\theta_i = \theta(t_i)$. Now notice that if the distance r between D and O decreases, the ratio r_i/r increases and $\dot{\theta}$ increases as well. This is analogous to what happens to the skater when the arms and/or legs are brought closer to the spin axis. Similarly, if r increases, $\dot{\theta}$ decreases.

Our solution showed us that the skater's ability to control the spin rate relies on the fact that the quantity in Eq. (5.55) remains *constant*. Now we want to better understand what this constant quantity represents. Recall that in polar coordinates $r\dot{\theta} = v_\theta$, and therefore Eq. (5.55) can be rewritten in the following manner:

$$r(mv_\theta) = \text{constant.} \tag{5.58}$$

Equation (5.58) tells us that what is being *conserved* is a *momentlike quantity* since the distance $\overline{OD} = r$ can be interpreted as a moment arm and since mv_θ is the component of the momentum of D perpendicular to $\overline{OD}$. It is not difficult to see that $r(mv_\theta)$ is the moment with respect to O of the momentum of D (see Fig. 5.31). We will now formalize this discovery via the concept of angular momentum, which will be defined as the moment of the momentum vector.

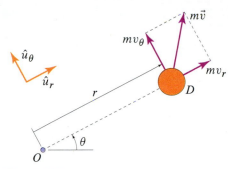

Figure 5.31
The quantity $r(mv_\theta)$ is the moment with respect to O of the momentum of D.

Definition of angular momentum of a particle

Referring to Fig. 5.32, given a particle Q of mass m and momentum $\vec{p}_Q = m\vec{v}_Q$, the particle's *angular momentum with respect to a point P* is denoted by $\vec{h}_P$ and is defined as the moment of $\vec{p}_Q$ about P, i.e.,

$$\boxed{\vec{h}_P = \vec{r}_{Q/P} \times \vec{p}_Q = \vec{r}_{Q/P} \times m\vec{v}_Q,} \tag{5.59}$$

where $\vec{r}_{Q/P}$ denotes the position of Q relative to P and where the reference point P is called the *moment center*. One important aspect of this definition is that there are no limitations to how the moment center P is chosen; i.e., P can be another particle or a convenient geometric reference point, and it can be moving or stationary. Angular momentum is typically expressed using units of kg·m^2/s and slug·ft^2/s in the SI and U.S. Customary systems, respectively.

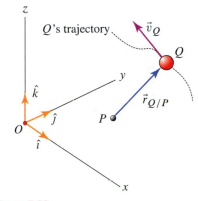

Figure 5.32
A particle Q in motion relative to a point P. Point P need *not* be a stationary point.

A final remark about the skater's spin problem. In the skater's spin problem we chose the point O in Fig. 5.29 as the moment center. Due to this choice, $\vec{r}_{D/O} = r\,\hat{u}_r$, and since the linear momentum in cylindrical coordinates is given by $\vec{p} = mv_r\,\hat{u}_r + mv_\theta\,\hat{u}_\theta + mv_z\,\hat{k}$, we have

$$\vec{h}_O = -rmv_z\,\hat{u}_\theta + rmv_\theta\,\hat{k}. \tag{5.60}$$

Therefore, recalling that the z axis is the spin axis, we now see that the conserved quantity $r(mv_\theta)$ is the component of the angular momentum $\vec{h}_O$ along the spin axis. For this observation to be fully appreciated we must now establish a connection between the concept of angular momentum of a particle and Newton's second law.

Concept Alert

Angular momentum is the *moment of momentum*. While the term *angular momentum* is most commonly used, conceptually and computationally, it is useful to remember it as the *moment of momentum* since that is what $\vec{h}_P$ in Eq. (5.59) is. As such, the angular momentum is meaningless unless we clearly indicate the moment center with respect to which the moment of the momentum is calculated.

Angular impulse-momentum principle for a particle

The relation between Newton's second law and angular momentum can be discovered by taking the time derivative of Eq. (5.59), which yields

$$\dot{\vec{h}}_P = \dot{\vec{r}}_{Q/P} \times m\vec{v}_Q + \vec{r}_{Q/P} \times m\dot{\vec{v}}_Q. \tag{5.61}$$

Recalling that

1. $\dot{\vec{r}}_{Q/P} = \vec{v}_Q - \vec{v}_P$, where $\vec{v}_Q$ and $\vec{v}_P$ are the velocities of Q and P, respectively,

2. by Newton's second law, $m\dot{\vec{v}}_Q = \vec{F}$, where $\vec{F}$ is the total force on Q,

we can rewrite Eq. (5.61) as

$$\dot{\vec{h}}_P = (\vec{v}_Q - \vec{v}_P) \times m\vec{v}_Q + \vec{r}_{Q/P} \times \vec{F}. \tag{5.62}$$

The term

$$\vec{r}_{Q/P} \times \vec{F} \tag{5.63}$$

is the *moment with respect to P of all the forces acting on Q*, which we will denote by $\vec{M}_P$. Making use of Eq. (5.63) in Eq. (5.62) and recognizing that $\vec{v}_Q \times m\vec{v}_Q = \vec{0}$, we obtain the following important result:

$$\boxed{\vec{M}_P = \dot{\vec{h}}_P + \vec{v}_P \times m\vec{v}_Q.} \tag{5.64}$$

Equation (5.64) is called the *angular impulse-momentum principle*, and it describes the relation between the moment of the total force acting on Q and the angular momentum of Q, both of which are with respect to the moment center P. Equation (5.64) can be simplified in the following cases:

1. When the reference point P is fixed, i.e., if $\vec{v}_P = \vec{0}$ or

2. When $\vec{v}_P$ is parallel to $\vec{v}_Q$, i.e., $\vec{v}_P \times m\vec{v}_Q = \vec{0}$.

In each of the above two cases, we have

$$\boxed{\vec{M}_P = \dot{\vec{h}}_P,} \tag{5.65}$$

i.e., the time rate of change of angular momentum (with respect to P) is *completely* equal to $\vec{M}_P$, the moment of the force acting on Q (with respect to P). Whenever Eq. (5.65) holds, integrating this equation with respect to time over a time interval $t_1 \le t \le t_2$, we have

$$\boxed{\vec{h}_{P1} + \int_{t_1}^{t_2} \vec{M}_P\, dt = \vec{h}_{P2},} \tag{5.66}$$

where $\vec{h}_{P1} = \vec{h}_P(t_1)$, $\vec{h}_{P2} = \vec{h}_P(t_2)$, and the integral term in Eq. (5.66) is called the *angular impulse with respect to P* of all forces acting on Q.

■ **Mini-Example.** Apply the angular impulse-momentum principle to the skater's problem discussed earlier to show that the skater's angular momentum remains constant.

Solution. Referring to Fig. 5.33 and recalling that we had chosen O as our moment center, we have $\vec{M}_O = \vec{0}$ since the line of action of $\vec{F}_s$ goes through O. Also, because O is a fixed point, the application of Eq. (5.66) tells us that $\vec{h}_O(t_2) = \vec{h}_O(t_1)$; i.e., the angular momentum with respect to O of the disk D remains constant. Equations (5.64) and (5.66) are very useful in the study of those problems in which a particle moves under the action of a central force, i.e., a force always pointing toward a fixed point. The most classical of these problems is the motion of planets about the Sun. ▬▬▬▬▬■

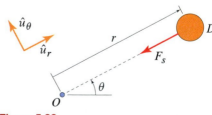

Figure 5.33
Figure 5.30 repeated. FBD of disk D.

Angular impulse-momentum for a system of particles

We now extend the angular impulse-momentum principle to systems of particles. As was done for the linear impulse-momentum principle for systems, we begin by considering a system of N particles subject to a system of external forces and interacting with one another in a pairwise fashion (see Fig. 5.34). In addition, we assume that the system under consideration is *closed*, i.e., does not exchange mass with its surroundings (see discussion on p. 336 of Section 5.1). We choose a point P with velocity $\vec{v}_P$ as moment center. Applying Eq. (5.64) to the ith particle in the system, we have

$$\vec{M}_{Pi} = \dot{\vec{h}}_{Pi} + \vec{v}_P \times m_i \, \vec{v}_i, \tag{5.67}$$

where

$$\vec{M}_{Pi} = \vec{r}_{i/P} \times \left(\vec{F}_i + \sum_{\substack{j=1 \\ j \neq i}}^{N} \vec{f}_{ij} \right), \tag{5.68}$$

in which

- $\vec{r}_{i/P}$ denotes the position of particle i relative to P,

- $\vec{F}_i$ is the *external* force acting on particle i, and

- $\vec{f}_{ij}$ is the *internal* force acting on particle i due to its interaction with particle j.

Thus, the term $\vec{F}_i + \sum_{\substack{j=1 \\ j \neq i}}^{N} \vec{f}_{ij}$, in the parentheses of Eq. (5.68), represents the *total* force acting on particle i. Summing the contributions in Eq. (5.67) over all the particles in the system, we obtain

$$\sum_{i=1}^{N} \vec{M}_{Pi} = \sum_{i=1}^{N} \dot{\vec{h}}_{Pi} + \sum_{i=1}^{N} \vec{v}_P \times m_i \, \vec{v}_i. \tag{5.69}$$

To simplify Eq. (5.69), we start with the sum on the left-hand side, which can be rewritten as

$$\sum_{i=1}^{N} \vec{M}_{Pi} = \sum_{i=1}^{N} \left(\vec{r}_{i/P} \times \vec{F}_i \right) + \sum_{i=1}^{N} \left(\vec{r}_{i/P} \times \sum_{\substack{j=1 \\ j \neq i}}^{N} \vec{f}_{ij} \right). \tag{5.70}$$

To rewrite the last term in Eq. (5.70), recall that Newton's third law is expressed by the following *two* relationships:*

$$\vec{f}_{ij} = -\vec{f}_{ji} \quad \text{and} \quad \vec{f}_{ij} \times (\vec{r}_i - \vec{r}_j) = \vec{0}. \tag{5.71}$$

Next we consider how a given pair of particles d and e contributes to the last term in Eq. (5.70). Referring to Fig. 5.35, consider the terms in the sum

$$\vec{r}_{e/P} \times \vec{f}_{ed} + \vec{r}_{d/P} \times \vec{f}_{de}, \tag{5.72}$$

---

*These two relationships were first given in Eqs. (1.4) and (1.5), respectively, and have been repeated here for convenience.

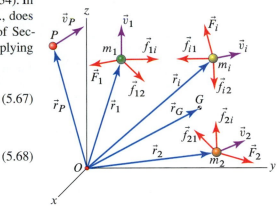

Figure 5.34
A system of particles under the action of internal and external forces. Point P, in general, is a moving point.

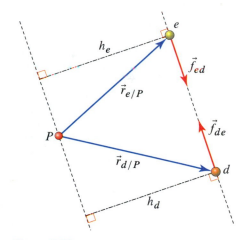

Figure 5.35
Moment of a pair of internal forces that are required to be equal and opposite as well as collinear by Newton's third law. Observe that the internal forces $\vec{f}_{ed}$ and $\vec{f}_{de}$ have equal moment arms relative to the line going through P and parallel to $\vec{f}_{ed}$ and $\vec{f}_{de}$, i.e., $h_e = h_d$. Therefore, the overall moment of $\vec{f}_{ed}$ and $\vec{f}_{de}$ with respect to P is equal to zero.

which, by using the first of Eqs. (5.71), can be rewritten as

$$\vec{r}_{e/P} \times \vec{f}_{ed} + \vec{r}_{d/P} \times (-\vec{f}_{ed}) = (\vec{r}_{e/P} - \vec{r}_{d/P}) \times \vec{f}_{ed}. \tag{5.73}$$

Using $\vec{r}_{e/P} - \vec{r}_{d/P} = (\vec{r}_e - \vec{r}_P) - (\vec{r}_d - \vec{r}_P) = \vec{r}_e - \vec{r}_d$ and the second of Eqs. (5.71), Eq. (5.73) becomes

$$(\vec{r}_e - \vec{r}_d) \times \vec{f}_{ed} = \vec{0}. \tag{5.74}$$

Equation (5.74) is true for every pair of particles, so the last term in Eq. (5.70) must be zero and Eq. (5.70) can be simplified to read

$$\vec{M}_P = \sum_{i=1}^{N} \vec{r}_{i/P} \times \vec{F}_i, \tag{5.75}$$

where $\vec{M}_P$ is the total moment with respect to P of *only* the *external* forces.

Now we consider the angular momentum terms in the last summation in Eq. (5.69). Observe that the term $\vec{v}_P$ is not characterized by the index i. This means that we can rewrite the summation containing the term in question as

$$\sum_{i=1}^{N} \vec{v}_P \times m_i \vec{v}_i = \vec{v}_P \times \sum_{i=1}^{N} m_i \vec{v}_i = \vec{v}_P \times m\vec{v}_G, \tag{5.76}$$

where we have used Eq. (5.13) to replace the sum of all momentum contributions with the term $m\vec{v}_G$ (see the Helpful Information note in the margin), where G is the system's center of mass. Next, we define the system's *total angular momentum computed with respect to the moment center P* to be the sum of the angular momenta with respect to P of all the particles in the system, i.e.,

$$\vec{h}_P = \sum_{i=1}^{N} \vec{h}_{Pi}. \tag{5.77}$$

Using Eqs. (5.75)–(5.77), we obtain the *angular impulse-momentum principle for a closed particle system* by rewriting Eq. (5.69) as

$$\vec{M}_P = \dot{\vec{h}}_P + \vec{v}_P \times m\vec{v}_G. \tag{5.78}$$

Angular momentum for systems of particles: important special cases

We now consider those special, but important, cases in which Eq. (5.78) can be simplified by making $\vec{v}_P \times m\vec{v}_G = \vec{0}$. This happens when

1. P is a fixed point, i.e., when $\vec{v}_P = \vec{0}$,

2. G is a fixed point, i.e., when $\vec{v}_G = \vec{0}$,

3. P coincides with the center of mass so that $\vec{v}_P = \vec{v}_G$, or

4. Vectors $\vec{v}_P$ and $\vec{v}_G$ are parallel to one another.

In all of these cases, Eq. (5.78) takes on the following simplified form:

$$\vec{M}_P = \dot{\vec{h}}_P. \tag{5.79}$$

Furthermore, if any of the conditions listed before hold over a time interval $t_1 \le t \le t_2$, we can integrate Eq. (5.79) with respect to time to obtain

$$\vec{h}_{P1} + \int_{t_1}^{t_2} \vec{M}_P \, dt = \vec{h}_{P2}, \tag{5.80}$$

where $\vec{h}_{P1} = \vec{h}_P(t_1)$ and $\vec{h}_{P2} = \vec{h}_P(t_2)$.

Conservation of angular momentum

Referring to Eq. (5.80), if $\vec{M}_P = \vec{0}$ for $t_1 \le t \le t_2$, then we have

$$\vec{h}_{P1} = \vec{h}_{P2}, \tag{5.81}$$

and we say that there is *conservation of angular momentum*.

In some cases it is possible to have conservation of a *component* of the angular momentum instead of the entire angular momentum. This happens when $\vec{M}_P$ is not zero, but a component of $\vec{M}_P$ *along a fixed axis is zero*. For example, consider Fig. 5.36 showing the FBD of a conical pendulum with bob Q swinging in a horizontal plane. The moment of all forces about point O is

$$\begin{aligned}
\vec{M}_O &= \vec{r}_{Q/O} \times \vec{F}_Q \\
&= L\left(\sin\phi \, \hat{u}_R - \cos\phi \, \hat{k}\right) \times \left[-mg \, \hat{k} + T(\cos\phi \, \hat{k} - \sin\phi \, \hat{u}_r)\right] \\
&= mgL \sin\phi \, \hat{u}_\theta,
\end{aligned} \tag{5.82}$$

where $\vec{F}_Q$ is the total force on Q. Now recall that $\vec{M}_O = \dot{\vec{h}}_O$ by Eq. (5.79). Since we can write $\vec{h}_O = h_{OR} \hat{u}_R + h_{O\theta} \hat{u}_\theta + h_{Oz} \hat{k}$, we have

$$\dot{\vec{h}}_O = \dot{h}_{Or} \hat{u}_R + h_{Or} \dot{\hat{u}}_R + \dot{h}_{O\theta} \hat{u}_\theta + h_{O\theta} \dot{\hat{u}}_\theta + \dot{h}_{Oz} \hat{k} + h_{Oz} \dot{\hat{k}}. \tag{5.83}$$

In cylindrical coordinates, we have $\dot{\hat{k}} = 0$ because the z axis is fixed, and we have $\dot{\hat{u}}_R = \dot{\theta} \hat{u}_\theta$ and $\dot{\hat{u}}_\theta = -\dot{\theta} \hat{u}_R$ because the R and θ directions rotate as the pendulum swings. Hence, Eq. (5.83) simplifies to

$$\dot{\vec{h}}_O = (\dot{h}_{Or} - \dot{\theta} h_{O\theta}) \hat{u}_R + (\dot{h}_{O\theta} + \dot{\theta} h_{OR}) \hat{u}_\theta + \dot{h}_{Oz} \hat{k}. \tag{5.84}$$

Equating Eqs. (5.82) and (5.84) component by component, we can then conclude that $M_{Oz} = 0$ implies $\dot{h}_{Oz} = 0$ and therefore $h_{Oz} = $ constant. By contrast, for example, the fact that $M_{OR} = 0$ only implies that $\dot{h}_{Or} - \dot{\theta} h_{O\theta} = 0$. Again, it is important to keep in mind that $M_{Oz} = 0$ implies $h_{Oz} = $ constant *because the z direction is fixed.*

Common Pitfall

Same appearance, different meaning. Equations (5.79) and (5.80) are identical in appearance to Eqs. (5.65) and (5.66), respectively. However, the quantity $\vec{h}_P$ in Eqs. (5.79) and (5.80) is the total angular momentum of a *system* and therefore is very different from the quantity $\vec{h}_P$ in Eqs. (5.65) and (5.66), which is the angular momentum of a single particle. The reason these equations appear the same is to reinforce the fact that there is a single physical principle at play in both cases.

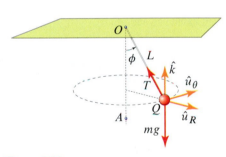

Figure 5.36
FBD of a conical pendulum swinging in a horizontal plane.

Euler's first and second laws of motion

Newton's laws of motion regulate the motion of a single mass point (or particle), where a mass point is an *abstract* entity with *finite* mass and *zero* (or undefined) volume. Using Newton's laws, we derived relations that must be satisfied by the motion of any *closed* system of mass points with pairwise internal forces. These relations are

$$\vec{F} = m\vec{a}_G \quad \text{and} \quad \vec{M}_P = \dot{\vec{h}}_P + \vec{v}_P \times m\vec{v}_G, \tag{5.85}$$

given in Eq. (5.14) on p. 337 and Eq. (5.78) on p. 392, respectively.

Many common models of the physical world do *not* view objects as consisting of mass points. For example, the modeling of beams, rods, cables, etc., is based on the notion of *continuous body*, which is not an assemblage of mass points. Even *rigid bodies*, studied later in this book, need not be viewed as assemblages of mass points. Therefore, we need to confront the following crucial question:

> What laws of motion do we use in models that are not based on the idea of mass point?

The answer to this question was first given by Euler* who recognized that Eqs. (5.85) were not just mere by-products of Newton's laws applicable only to special particle systems, but could be used as the foundation of the study of motion of *any* physical system. Therefore, Euler *postulated* that, in an inertial frame of reference, the motion of *any* physical object that does not exchange mass with its surroundings must satisfy two laws, which we state as follows:

$$\text{Euler's first law:} \quad \vec{F} = m\vec{a}_G, \tag{5.86}$$

$$\text{Euler's second law:} \quad \vec{M}_P = \dot{\vec{h}}_P + \vec{v}_P \times m\vec{v}_G, \tag{5.87}$$

where, for a physical system of constant mass m and center of mass G,

- $\vec{F}$ is the resultant of all *external* forces.

- $\vec{v}_G$ and $\vec{a}_G$ are the velocity and acceleration of G, respectively.

- P is an arbitrarily chosen moment center.

- $\vec{M}_P$ is the resultant of the moments of all *external* forces and torques computed with respect to P.

- $\vec{h}_P$ is the angular momentum computed with respect to P.

In postulating Eqs. (5.86) and (5.87), Euler felt no need to add any specific reference to Newton's third law, which can be shown to be embedded and generalized by Eqs. (5.86) and (5.87) (see C. Truesdell, *Essays in the History of Mechanics*, Springer-Verlag, Berlin, 1968).

In the remainder of this book, we will consider Eqs. (5.86) and (5.87) as the *fundamental laws of motion* of bodies. To recognize Euler's contribution to the formulation of fundamental laws of mechanics while retaining an awareness of the Newtonian origin of these laws, we will refer to these laws as the *Newton-Euler* equations.

---

* See the historical note about Euler on p. 6 in Chapter 1.

End of Section Summary

In this section we developed the concept of angular momentum for a single particle and for systems of particles. In addition, we presented the *angular impulse-momentum principle*.

Referring to Fig. 5.37, the *angular momentum of a particle Q with respect to the moment center P* is given by

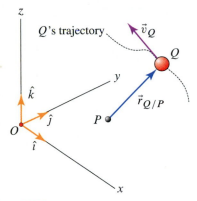

Eq. (5.59), p. 389

$$\vec{h}_P = \vec{r}_{Q/P} \times \vec{p}_Q = \vec{r}_{Q/P} \times m\vec{v}_Q,$$

where P can be fixed or moving; $\vec{r}_{Q/P}$ is the position of Q relative to P; m and $\vec{v}_Q$ are the mass and the velocity of Q, respectively; and $\vec{p}_Q = m\vec{v}_Q$ is the linear momentum of Q.

The *angular impulse-momentum principle* for a single particle is given by

Eq. (5.64), p. 390

$$\vec{M}_P = \dot{\vec{h}}_P + \vec{v}_P \times m\vec{v}_Q,$$

Figure 5.37
Figure 5.32 repeated. A particle Q in motion relative to a point P. Point P need *not* be a stationary point.

where $\vec{M}_P$ is the moment with respect to P of all the forces acting on Q. If one of the following conditions is satisfied:

1. The reference point P is fixed, i.e., if $\vec{v}_P = \vec{0}$.

2. $\vec{v}_P$ is parallel to $\vec{v}_Q$, i.e., $\vec{v}_P \times m\vec{v}_Q = \vec{0}$.

then the angular impulse-momentum principle for a single particle can be simplified to read

Eq. (5.65), p. 390

$$\vec{M}_P = \dot{\vec{h}}_P.$$

If conditions (1) or (2) are satisfied for $t_1 \leq t \leq t_2$, then the angular impulse-momentum principle for a single particle can be integrated with respect to time to read

Eq. (5.66), p. 390

$$\vec{h}_{P1} + \int_{t_1}^{t_2} \vec{M}_P \, dt = \vec{h}_{P2},$$

where $\vec{h}_{P1} = \vec{h}_P(t_1)$ and $\vec{h}_{P2} = \vec{h}_P(t_2)$.

For a *closed system of particles*, the angular impulse-momentum principle can be given the form

Eq. (5.78), p. 392

$$\vec{M}_P = \dot{\vec{h}}_P + \vec{v}_P \times m\vec{v}_G,$$

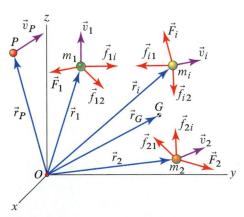

Figure 5.38
Figure 5.34 repeated. A system of particles under the action of internal and external forces. Point P, in general, is a moving point.

where, with reference to Fig. 5.38, $\vec{M}_P$ is the moment with respect to P of *only the external forces* acting on the system, $m = \sum_{i=1}^{N} m_i$ is the total mass of the system, G is the system's center of mass, $\vec{v}_G$ is the velocity of G, and $\vec{h}_P$ is the total angular momentum, which is defined as

Eq. (5.77), p. 392

$$\vec{h}_P = \sum_{i=1}^{N} \vec{h}_{Pi}.$$

As we saw earlier, the angular impulse-momentum principle takes on a simpler form if certain conditions are satisfied. For a closed system of particles, these conditions are any of the following:

1. When P is a fixed point, i.e., when $\vec{v}_P = \vec{0}$.

2. When G is a fixed point, i.e., when $\vec{v}_G = \vec{0}$.

3. When P coincides with the center of mass and therefore $\vec{v}_P = \vec{v}_G$.

4. The vectors $\vec{v}_P$ and $\vec{v}_G$ are parallel to one another.

Under any of these conditions, the angular impulse-momentum principle for a closed system of particles simplifies to

Eq. (5.79), p. 393

$$\vec{M}_P = \dot{\vec{h}}_P.$$

Furthermore, if any of the above conditions are satisfied over a time interval $t_1 \le t \le t_2$, then the angular impulse-momentum principle for a closed system of particles can be integrated with respect to time to read

Eq. (5.80), p. 393

$$\vec{h}_{P1} + \int_{t_1}^{t_2} \vec{M}_P \, dt = \vec{h}_{P2}.$$

EXAMPLE 5.12   *Angular Momentum to Find the Speed of a Satellite*

A satellite orbits the Earth as shown in Fig. 1. The minimum and maximum distances from the center of the Earth are $R_P = 4.5 \times 10^7$ m and $R_A = 6.163 \times 10^7$ m, respectively, where the subscripts P and A stand for *perigee* (the point on the orbit closest to Earth) and *apogee* (the point on the orbit farthest from Earth), respectively. If the satellite's speed at P is $v_P = 3.2 \times 10^3$ m/s, determine the satellite's speed at A.

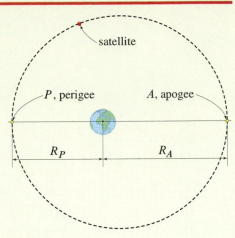

SOLUTION

Road Map & Modeling   Referring to Fig. 2, we model both the Earth and the satellite as *particles*, we assume the Earth to be *fixed*, and we consider the center of the Earth E as the origin of an inertial frame of reference. We also assume that the only force acting on the satellite is F_G, the gravitational attraction exerted by the Earth. Finally, we assume that the satellite's orbit is planar so that we can study the satellite's motion, using a polar coordinate system with origin at E. These assumptions imply that the force on the satellite is central (it always points toward the fixed point E) and its moment about E is always equal to zero. Therefore, the satellite's angular momentum with respect to E is *conserved* throughout the motion. This condition will be the key to the problem's solution.

Figure 1
Satellite in elliptical orbit around the Earth. The orbit is drawn to scale with respect to the Earth.

Governing Equations

Balance Principles   Choosing the fixed point E as our moment center, we can use the angular impulse-momentum principle in Eq. (5.64) on p. 390:

$$\vec{M}_E = \dot{\vec{h}}_E, \qquad (1)$$

where

$$\vec{M}_E = \vec{r} \times \vec{F}_G \quad \text{and} \quad \vec{h}_E = \vec{r} \times m\vec{v}, \qquad (2)$$

and where $\vec{r}$, m, and $\vec{v}$ are the position, mass, and velocity of the satellite, respectively.

Force Laws $\vec{F}_G$ is given by (see Eq. (1.6) on p. 5)

$$\vec{F}_G = -G \frac{m_e m}{r^2} \hat{u}_r, \qquad (3)$$

where G is the universal gravitational constant, m_e is the mass of the Earth, r is the distance between the center of the Earth and the satellite, and $\hat{u}_r$ is the unit vector pointing from the Earth to the satellite.

Kinematic Equations   To use Eq. (2), we need $\vec{r}$ and $\vec{v}$ in polar coordinates:

$$\vec{r} = r\,\hat{u}_r \quad \text{and} \quad \vec{v} = v_r\,\hat{u}_r + v_\theta\,\hat{u}_\theta. \qquad (4)$$

Since we need to relate the speed at A with the speed at P, using the second of Eqs. (4), recall that the velocity components and speed are related as follows:

$$v = \sqrt{v_r^2 + v_\theta^2}. \qquad (5)$$

We note that

$$v_{rP} = \dot{r}_P = 0 \quad \text{and} \quad v_{rA} = \dot{r}_A = 0, \qquad (6)$$

because, by definition, P and A are the points where the distances between the satellite and the Earth are minimum and maximum, respectively. Therefore, combining Eqs. (5) and (6), we have

$$v_P = \sqrt{v_{\theta P}^2} = |v_{\theta P}| \quad \text{and} \quad v_A = \sqrt{v_{\theta A}^2} = |v_{\theta A}|. \qquad (7)$$

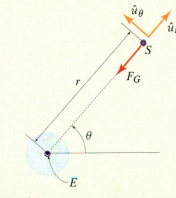

Figure 2
FBD of the satellite along with the chosen coordinate system.

Interesting Fact

Can the Earth be treated as a particle?
The answer is: yes, as long as the satellite and Earth are "sufficiently" far apart that the way in which the mass of the Earth (and possibly the mass of the satellite) is *distributed* in space does not affect the calculation of the force of gravity. To account for the effects of the Earth's mass *distribution*, Eq. (1.6) on p. 5 cannot be applied "as is" but would need to be reformulated as an integral over the regions across which the mass is distributed.

Computation   Substituting Eqs. (3) and (4) into the first of Eqs. (2), we have

$$\vec{M}_E = r\,\hat{u}_r \times \left(-\frac{Gm_e m}{r^2}\right)\hat{u}_r = \vec{0}, \tag{8}$$

since the line of action of $\vec{F}_G$ goes through the moment center E (we chose E as the moment center to obtain $\vec{M}_E = \vec{0}$). Therefore, from Eq. (1) we have

$$\dot{\vec{h}}_E = \vec{0} \quad \Rightarrow \quad \vec{h}_E = \text{constant throughout the motion}, \tag{9}$$

so that we have

$$\vec{h}_{EP} = \vec{h}_{EA} \quad \Rightarrow \quad mR_P v_{\theta P}\,\hat{k} = mR_A v_{\theta A}\,\hat{k}, \tag{10}$$

where $\vec{h}_{EP}$ and $\vec{h}_{EA}$ are the values of $\vec{h}_E$ at P and A, respectively, and where we have used Eqs. (4) and the second of Eqs. (2). Computing the magnitude of both sides of Eq. (10) and using Eqs. (7), we have

$$mR_P v_P = mR_A v_A, \tag{11}$$

which can be solved for v_A to obtain

$$\boxed{v_A = \frac{R_P}{R_A}\,v_P = 2340\,\text{m/s}.} \tag{12}$$

Discussion & Verification   Since the ratio R_P/R_A is nondimensional, the result in Eq. (12) is dimensionally correct. In addition, our solution confirms what we have learned in the analysis of the skater's spin problem presented at the beginning of the section. When the angular momentum is conserved, increasing the distance of mass from the spin axis causes a reduction in speed proportional to the ratio between the small and large values of distance from the spin axis. Therefore, the result in Eq. (12) is as expected given that $v_A < v_P$.

EXAMPLE 5.13   *Orbit of a Satellite: Angular Momentum and Work-Energy*

At perigee, a satellite in orbit around the Earth has a speed $v_P = 3.2 \times 10^3$ m/s and a distance from the center of the Earth $R_P = 4.5 \times 10^7$ m. Treating the Earth as stationary, (a) establish whether the satellite's trajectory is circular and (b) determine R_A and the satellite's speed at the *apogee* A, i.e., the point on the orbit farthest from Earth.

SOLUTION

Road Map & Modeling   We model both the Earth and the satellite as particles, with the Earth being fixed and its center E as the origin of an inertial reference frame (Fig. 2). The satellite is assumed to be subject only to the Earth's gravitational attraction $\vec{F}_G$. Finally, assuming that the satellite's orbit is planar, we study the satellite's motion using a polar coordinate system with origin at E. If the orbit were circular, we would have $R_A = R_P$ and $v_A = v_P$. Since this problem is a central force problem, we can obtain one relation concerning R_A and v_A via the conservation of angular momentum. We can then take advantage of the fact that the force of gravity is conservative to obtain a second relation between R_A and v_A using the work-energy principle. We will see that, in answering question (a), we will also answer question (b).

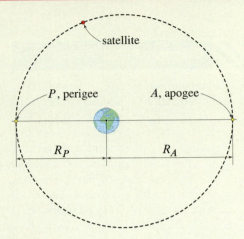
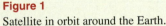

Figure 1
Satellite in orbit around the Earth.

Governing Equations

Balance Principles   Applying Eq. (5.66), we have

$$\vec{h}_{EP} = \text{constant} = \vec{h}_E, \tag{1}$$

where $\vec{h}_{EP}$ is the angular momentum of the satellite about E evaluated at P, where, at a generic position along the satellite's orbit,

$$\vec{h}_E = \vec{r} \times m\vec{v}, \tag{2}$$

and where m, $\vec{r}$, and $\vec{v}$ are the mass, position, and velocity of the satellite, respectively. The work-energy principle, applied between P and a generic point along the orbit, reads

$$T_P + V_P = T + V, \tag{3}$$

where

$$T_P = \tfrac{1}{2}mv_P^2 \quad \text{and} \quad T = \tfrac{1}{2}mv^2, \tag{4}$$

and where v_P and v are the satellite's speed at P and at a generic point on the orbit, respectively.

Force Laws   The potential energy of the force $\vec{F}_G$ is (see Eq. (4.35) on p. 280)

$$V = -G\frac{m_e m}{r}, \tag{5}$$

where we recall that $G = 6.674 \times 10^{-11}$ m^3/(kg·s^2) is the universal gravitational constant and $m_e = 5.9736 \times 10^{24}$ kg is the mass of the Earth.

Kinematic Equations   In polar coordinates we have

$$\vec{r} = r\,\hat{u}_r, \quad \vec{v} = v_r\,\hat{u}_r + v_\theta\,\hat{u}_\theta, \quad \text{and} \quad v = \sqrt{v_r^2 + v_\theta^2}. \tag{6}$$

Hence, Eq. (2) simplifies to

$$\vec{h}_E = mrv_\theta\,\hat{k}. \tag{7}$$

We also recall (see Example 5.12) that at perigee and apogee we must have

$$v = |v_\theta|. \tag{8}$$

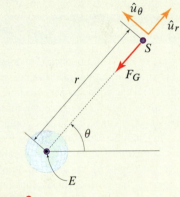

Figure 2
FBD of satellite along with a description of the chosen coordinate and component system.

Computation   Substituting Eq. (7) into Eq. (1) and recalling that $r_P = R_P$, then anywhere along the orbit we have

$$mR_P v_{\theta P} = mrv_\theta. \tag{9}$$

Computing the absolute value of Eq. (9) and evaluating it at those points along the orbit for which Eq. (8) is true (i.e., for the perigee or apogee), we obtain

$$R_P v_P = rv. \tag{10}$$

Next, substituting Eqs. (4) and (5) into Eq. (3), we obtain

$$\tfrac{1}{2}mv_P^2 - G\frac{m_e m}{R_P} = \tfrac{1}{2}mv^2 - G\frac{m_e m}{r}. \tag{11}$$

Equations (10) and (11) form a system of two equations in the two unknowns r and v, which has the following two solutions:

Solution 1: $v_1 = v_P,$ $r_1 = R_P,$ (12)

Solution 2: $v_2 = \dfrac{2Gm_e - R_P v_P^2}{R_P v_P},$ $r_2 = \dfrac{R_P^2 v_P^2}{2Gm_e - R_P v_P^2}.$ (13)

Solution 1 indicates that P is one of the places on the orbit where Eq. (8) is satisfied. Solution 2 tells us that there is another place on the orbit where Eq. (8) is satisfied.

──────────────── **Part (a): Is the Satellite's Trajectory Circular?** ────────────────

If the trajectory were circular, everywhere along the orbit we would have $r = R_P$ and $v = v_P$ and therefore we would have $v_1 = v_2$ and $r_1 = r_2$, so that we would have

$$v_1 = v_2 \quad \Rightarrow \quad v_P = \sqrt{Gm_e/R_P}. \tag{14}$$

Since the values of v_P, R_P, m_e, and G are given, by substituting these values in Eq. (14) we can verify whether or not the trajectory is circular. Doing so yields

$$
\boxed{
\begin{aligned}
(v_P)_{\text{circular orbit}} &= \sqrt{\frac{\left[6.6732\times10^{-11}\,\text{m}^3/(\text{kg}\cdot\text{s}^2)\right]5.9736\times10^{24}\,\text{kg}}{4.5\times10^7\,\text{m}}} \\
&= 2976\,\text{m/s} \neq (v_P)_{\text{given}} = 3200\,\text{m/s}.
\end{aligned}
}
\tag{15}
$$

Therefore, the answer to Part (a) is that the satellite's orbit is *not* circular.

──────────────── **Part (b): Conditions at Apogee.** ────────────────

Since the satellite's trajectory is not circular, there is only one apogee with corresponding values of v_A and R_A given by Solution 2. Substituting the given numerical data in Eqs. (13), we have

$$\boxed{v_2 = v_A = 2340\,\text{m/s} \quad \text{and} \quad r_2 = R_A = 6.160\times10^7\,\text{m}.} \tag{16}$$

Discussion & Verification   The solution in Eqs. (16) was obtained from Eqs. (13), which are dimensionally correct. Also, notice that $v_A < v_P$ as we expect due to the conservation of angular momentum. Hence, our overall solution appears to be correct.

🔎 **A Closer Look**   We could have ended this solution with Eqs. (12) and (13); that is, we could have substituted all known quantities into those equations, and we would have obtained the results given in Eqs. (16). Doing so would have also made it clear that the orbit is not circular. On the other hand, by obtaining the condition on the relationship between v_P and R_P for a circular orbit found in Eq. (14), we have created a useful result that can be used in other situations and we have gained some insight into what it means for an orbit to be circular.

Common Pitfall

The unknowns in Eqs. (10) and (11). We stated that the unknowns in Eqs. (10) and (11) are r and v. However, it is important to recall that in formulating Eq. (10) we used Eq. (8). This means that r and v are not the radial coordinate and the corresponding speed at any possible point; rather they are the radial coordinate and corresponding speed of any point for which Eq. (8) is satisfied!

Common Pitfall

Proper use of Eq. (14). Equation (14) does not say anything about the *given* values of v_P and R_P. This equation simply states that *if the satellite's orbit were circular*, then the values of v_P and R_P would not be independent — they would be related in the way specified by this equation. Conversely, if the given values of v_P and R_P were not to satisfy Eq. (14), then we would conclude that the satellite's orbit is not a circle.

E X A M P L E   5.14   *Controlling the Spin Rate of a Satellite*

Figure 1 shows a satellite that has been deployed with an angular speed $\omega_S(t_d)$ about the spin axis z, where t_d is the time of deployment. To control the satellite's attitude (its orientation) and overall spin rate, the satellite is equipped with internal motors that can spin up and spin down internal masses. In this problem, we consider a simple setup with just two internal spheres, each of mass m_{int} at a distance R_{int} from the spin axis (where the subscript "int" stands for internal). Assuming that, at deployment, the spheres and the satellite are spinning at the rate ω_S, find the rotational speed of the two internal masses to cause the satellite's body to spin down to one-half of $\omega_S(t_d)$.

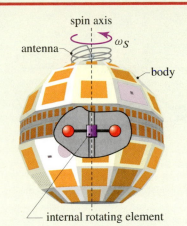

spin axis

antenna

ω_S

body

internal rotating element

Figure 1
Satellite spinning about an axis. Notice the presence of the *internal* moving element.

SOLUTION

Road Map & Modeling   We will model the satellite as a *system of mass points*, and we will then discover that by controlling the motion of some of these points, we can also control the motion of the rest of the system. Referring to Fig. 2, we will assume that

1. The satellite is not acted upon by external forces.

2. The satellite's only motion is a rotation about the z axis, which is assumed to be fixed. This assumption allows us to choose an inertial reference frame with origin at the satellite's center of mass G and with one of the axes coinciding with the spin axis z.

3. The satellite's body has a mass $m_B = m_3 + m_4 + m_5$ (the subscript B is for body). The mass m_B is *lumped* into three identical point masses m_3, m_4, and m_5 arranged as shown in Fig. 2, i.e., placed at a distance R_{ext} from the spin axis and separated from each other by equal angles of $120°$. The system's symmetry is such that the satellite's center of mass G is on the spin axis.

The position of the internal and external masses can be characterized by the angular coordinates θ_{int} and θ_B, respectively, shown in Fig. 2. Since the system is isolated, its angular momentum about the spin axis is *conserved*. This observation is the key to the problem's solution.

Governing Equations

Balance Principles   Choosing the center of mass G as the moment center and recalling that the system is isolated, we see the system's angular momentum is *constant* and therefore equal to what it was at deployment, i.e.,

$$\vec{h}_G = \text{constant} = \vec{h}_G(t_d), \tag{1}$$

where

$$\vec{h}_G = \sum_{i=1}^{5} \vec{r}_i \times m_i \vec{v}_i, \tag{2}$$

where the subscript i denotes the ith particle in the system.

Force Laws   All forces (in this case none) are accounted for on the FBD.

Kinematic Equations   Referring to Figs. 3 and 4, for the position vectors we have

$$\vec{r}_i = R_i \hat{u}_{ri}, \tag{3}$$

where $\hat{u}_{ri}$ is the unit vector pointing from G to particle i and where

$$R_i = \begin{cases} R_{\text{int}} & \text{for } i = 1, 2 \\ R_{\text{ext}} & \text{for } i = 3, 4, 5. \end{cases} \tag{4}$$

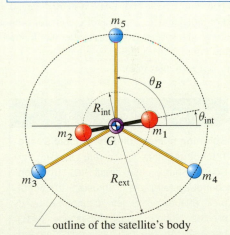

m_5

θ_B

R_{int}

θ_{int}

m_2 m_1

G

m_3 R_{ext} m_4

outline of the satellite's body

Figure 2
View down the z axis of *lumped-mass* model. Masses m_3, m_4, and m_5 rotate together; masses m_1 and m_2 rotate together, but they can spin at a different rate than the others. This diagram is also an FBD of the system. No external forces are present since the system is isolated.

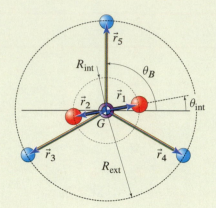

Figure 3

Position vectors for the chosen point masses.

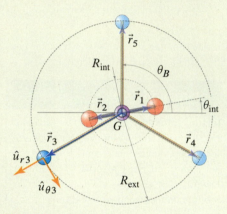

Figure 4

When using a polar coordinate system, there is a set of radial and transverse unit vectors for each point in the system. As an example, this figure shows the radial and transverse unit vectors for m_3 at a particular instant.

For the velocity vectors we have

$$\vec{v}_i = R_i \dot{\theta}_i \hat{u}_{\theta i}, \tag{5}$$

where $\hat{u}_{\theta i}$ is the unit vector perpendicular to $\hat{u}_{ri}$ pointing in the direction of growing θ. In addition, we have

$$\dot{\theta}_i = \begin{cases} \dot{\theta}_{\text{int}} & \text{for } i = 1, 2 \\ \dot{\theta}_{\text{ext}} & \text{for } i = 3, 4, 5. \end{cases} \tag{6}$$

Computation   Substituting the kinematic relations into Eqs. (1) and (2), we obtain the z component of the angular momentum as

$$3 m_{\text{ext}} R_{\text{ext}}^2 \dot{\theta}_{\text{ext}} + 2 m_{\text{int}} R_{\text{int}}^2 \dot{\theta}_{\text{int}} = \left(3 m_{\text{ext}} R_{\text{ext}}^2 + 2 m_{\text{int}} R_{\text{int}}^2 \right) \dot{\theta}(t_d), \tag{7}$$

where

$$m_{\text{int}} = m_1 = m_2, \quad \dot{\theta}(t_d) = \omega_S(t_d), \quad m_{\text{ext}} = m_3 = m_4 = m_5 = \frac{m_B}{3}. \tag{8}$$

In the end, we want to achieve the following result:

$$\dot{\theta}_{\text{ext}} = \tfrac{1}{2} \omega_S(t_d). \tag{9}$$

Hence, substituting the desired value of spin rate in Eq. (7) and solving for $\dot{\theta}_{\text{int}}$, we have

$$\boxed{ \dot{\theta}_{\text{int}} = \frac{2 m_{\text{int}} R_{\text{int}}^2 + m_B R_{\text{ext}}^2 / 2}{2 m_{\text{int}} R_{\text{int}}^2} \, \omega_S(t_d). } \tag{10}$$

Discussion & Verification   Since the fraction on the right-hand side of Eq. (10) is nondimensional, our final result is dimensionally correct. Notice that the result in Eq. (10) can be written as

$$\dot{\theta}_{\text{int}} = \left(1 + \frac{m_B R_{\text{ext}}^2}{4 m_{\text{int}} R_{\text{int}}^2} \right) \omega_S(t_d), \tag{11}$$

which shows that $\dot{\theta}_{\text{int}}$ has the same sign as $\omega_S(t_d)$ and must always be greater than $\omega_S(t_d)$. That is, the internal masses need to rotate in the same direction as the body's initial angular velocity at a rate higher than $\omega_S(t_d)$. This makes physical sense in that if we want to slow down the outer system while keeping the overall angular momentum constant, the internal masses must speed up. Hence, overall our solution appears to be correct.

EXAMPLE 5.15   *Conservation of Angular Momentum: An Advanced Example*

At time t_0 a pendulum bob B is put in motion along a horizontal circle so that the pendulum describes a right cone with vertex at O, an initial side length $L_0 = 1$ m, and initial opening angle $\phi_0 = 20°$. During a time interval $t_1 \leq t \leq t_2$, with $t_1 > t_0$, L is made to decrease and then is held constant for $t \geq t_2$. The decrease of L occurs so slowly that, for time $t \geq t_2$, the trajectory of B can be viewed as being a circle lying in a horizontal plane, i.e., a circle parallel to the initial trajectory of the pendulum. Use the angular impulse-momentum principle, and determine the pendulum's initial speed v_0. In addition, at $t = t_2$ determine the speed v_2 and length L_2 if $\phi_2 = 50°$.

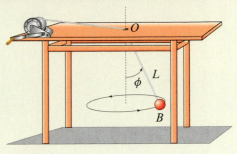

Figure 1
A pendulum with its bob B tracing a horizontal circle and being slowly reeled in.

SOLUTION

Road Map & Modeling   Considering the FBD in Fig. 2, we model B as a particle subject only to its own weight mg and the tension in the cord F_c. We model the cord as inextensible and describe the motion of B using a cylindrical coordinate system with origin at O. As requested by the problem statement, we will obtain the problem's solution via the angular impulse-momentum principle. In doing so we need to achieve the following objectives: (1) establish a relation between L, v, and ϕ when the trajectory of B is horizontal and (2) determine how the motions for $t < t_1$ and $t \geq t_2$ are related, i.e., determine how the conditions with which the pendulum is initially put in motion affect the motion after the pendulum cord length is decreased.

Governing Equations

Balance Principles   Choosing the fixed point O as moment center, we apply the angular impulse-momentum principle as given in Eq. (5.65):

$$\vec{M}_O = \dot{\vec{h}}_O, \tag{1}$$

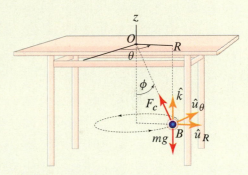

Figure 2
FBD of B with the chosen coordinate and component systems.

where

$$\vec{M}_O = \vec{r}_{B/O} \times (\vec{F}_c - mg\,\hat{k}) \quad \text{and} \quad \vec{h}_O = \vec{r}_{B/O} \times m\vec{v}, \tag{2}$$

and where $\vec{r}_{B/O}$ is the position of B relative to O and $\vec{v}$ is the velocity of B.

Force Laws   Since $\vec{F}_c$ is directed from B to O, we must have

$$\vec{F}_c = -F_c \frac{\vec{r}_{B/O}}{|\vec{r}_{B/O}|}. \tag{3}$$

Kinematic Equations   Referring to Fig. 3, for $\vec{r}_{B/O}$, we have

$$\vec{r}_{B/O} = L(\sin\phi\,\hat{u}_R - \cos\phi\,\hat{k}). \tag{4}$$

Leaving the discussion of the motion during $t_1 < t < t_2$ for later, for both $t \leq t_1$ and $t \geq t_2$, L is constant and the vertical component of the velocity of B is equal to zero, so that, duing these time intervals, we have

$$\vec{v} = v_\theta\,\hat{u}_\theta \quad \text{with} \quad v_\theta = L\dot{\theta}\sin\phi. \tag{5}$$

To compute $\dot{\vec{h}}_O$ in Eq. (1), we need the time derivatives of the unit vectors $\hat{u}_R$, $\hat{u}_\theta$, and $\hat{k}$, which are (see Sections 2.6 and 2.8)

$$\dot{\hat{u}}_R = \dot{\theta}\,\hat{u}_\theta, \quad \dot{\hat{u}}_\theta = -\dot{\theta}\,\hat{u}_R, \quad \dot{\hat{k}} = \vec{0}. \tag{6}$$

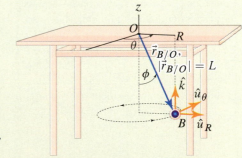

Figure 3
Depiction of the position vector $\vec{r}_{B/O}$. Notice that $\vec{r}_{B/O}$ has no component in the transverse direction and that $|\vec{r}_{B/O}| = L$.

Computation Since O is on the line of action of $\vec{F}_c$, by using the first of Eqs. (4), the first of Eqs. (2) yields

$$\vec{M}_O = mgL \sin\phi \, \hat{u}_\theta. \tag{7}$$

Substituting Eqs. (4) and (5) into the second of Eqs. (2), we have

$$\vec{h}_O = mv_\theta L \cos\phi \, \hat{u}_R + mv_\theta L \sin\phi \, \hat{k}. \tag{8}$$

Differentiating Eq. (8) with respect to time and using Eqs. (6), we have

$$\dot{\vec{h}}_O = \frac{d}{dt}(mv_\theta L \cos\phi)\hat{u}_R + (mv_\theta L \cos\phi)\dot{\theta}\,\hat{u}_\theta + \frac{d}{dt}(mv_\theta L \sin\phi)\hat{k}. \tag{9}$$

Enforcing Eq. (1), i.e., equating Eqs. (7) and (9) component by component, we have

$$0 = \frac{d}{dt}(mv_\theta L \cos\phi) \quad\Rightarrow\quad v_\theta L \cos\phi = K_R, \tag{10}$$

$$mgL \sin\phi = (mv_\theta L \cos\phi)\dot{\theta} \quad\Rightarrow\quad gL \sin\phi = v_\theta^2 \frac{\cos\phi}{\sin\phi}, \tag{11}$$

$$0 = \frac{d}{dt}(mv_\theta L \sin\phi) \quad\Rightarrow\quad v_\theta L \sin\phi = K_z, \tag{12}$$

where K_R and K_z are constants of integration and where, in Eq. (11), we have used the second of Eqs. (5) to express $\dot{\theta} = v_\theta/(L \sin\phi)$. Equations (10)–(12) describe the state of motion of the system for $t_0 \le t \le t_1$ and for $t \ge t_2$. Recalling that $L_0 = 1$ m and $\phi_0 = 20°$, by using Eq. (11) we can solve for the initial value of v_0

$$v_0 = |v_{\theta 0}| = \sqrt{gL_0 \frac{\sin^2\phi_0}{\cos\phi_0}} = 1.105 \,\text{m/s}, \tag{13}$$

where $|v_\theta| = v$ given that for $t_0 \le t \le t_1$ the motion of B is circular.

For $t \ge t_2$ we only know ϕ_2, and it would seem that there are not enough equations to solve for all the unknowns of the problem, which are L_2, v_2, K_{R2}, and K_{z2}. However, referring to Fig. 2, observe that, *for any time $t \ge t_0$, none of the forces on B provides a moment about the z axis. Since this axis is fixed, as discussed on p. 393, we must have conservation of angular momentum about the z axis. We now observe that Eq. (12) expresses such a conservation requirement and is valid not just for the time intervals discussed earlier, but for any possible time (this is not true for Eqs. (10) and (11)). This means that the value of the right-hand side of Eq. (12) must be the same at t_0 as it is at t_2! Therefore, at $t = t_2$, we can use Eqs. (11) and (12) to write

$$gL_2 \sin\phi_2 = v_2^2 \frac{\cos\phi_2}{\sin\phi_2} \quad\text{and}\quad v_2 L_2 \sin\phi_2 = v_0 L_0 \sin\phi_0, \tag{14}$$

where it is assumed that $v_{\theta 2} > 0$ so that $v_{\theta 2} = v_2$ (see note in the margin). Equations (14) form a system of two equations in the unknowns L_2 and v_2 with solution

$$v_2 = \sqrt[3]{g \tan\phi_2 v_0 L_0 \sin\phi_0} = 1.64 \,\text{m/s}, \tag{15}$$

$$L_2 = \sqrt[3]{(v_0 L_0 \sin\phi_0)^2 \cos\phi_2/(g \sin^4\phi_2)} = 0.301 \,\text{m}. \tag{16}$$

Helpful Information

From Eqs. (11) and (12) to Eqs. (14). Equations (14) are perfectly consistent with the discussion leading to them. This is so because the first of Eqs. (14) is given by Eq. (11) evaluated *just at time t_2*. By contrast, we see that the left-hand side of the second of Eqs. (14) is Eq. (12) evaluated at time t_2, whereas the right-hand side is Eq. (12) evaluated at time t_0.

Discussion & Verification We leave it to the reader to verify that Eqs. (15) and (16) are dimensionally correct. Observe that as the pendulum cord is decreased, the distance between B and the spin axis z is also decreased. Consequently, in compliance with the conservation of angular momentum requirement, we would expect $v_2 > v_0$, as is indeed the case. Hence, overall our solution appears to be correct.

A Closer Look This example was deemed advanced because we had to recognize that part of the solution, namely, Eq. (12), was applicable outside the assumptions underlying the rest of the solution. It was this realization that allowed us to determine how the initial conditions affected the motion at time t_2 without having to explicitly compute the motion that takes the system from t_1 to t_2.

PROBLEMS

Problems 5.72 through 5.74

At the instant shown, a truck A, of weight $W_A = 31{,}000\,\text{lb}$, and a car B, of weight $W_B = 3970\,\text{lb}$, are traveling with speeds $v_A = 35\,\text{mph}$ and $v_B = 34\,\text{mph}$, respectively.

Problem 5.72   Choosing point O as the moment center, determine the angular momentum (with respect to O) of A and B individually at this instant.

Problem 5.73   Choosing point O as the moment center, determine the angular momentum (with respect to O) of the *particle system* formed by A and B at this instant.

Problem 5.74   Choosing point Q as the moment center, determine the angular momentum (with respect to Q) of the *particle system* formed by A and B at this instant.

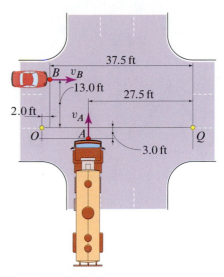

Figure P5.72–P5.74

💡 Problem 5.75 💡

Consider the situation depicted in the figure. At the instant shown, how are the angular momenta of particle P with respect to O and Q related?
Note: Concept problems are about *explanations*, not computations.

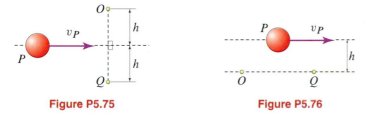

Figure P5.75   **Figure P5.76**

💡 Problem 5.76 💡

Consider the situation depicted in the figure. At the instant shown, how are the angular momenta of particle P with respect to O and Q related?
Note: Concept problems are about *explanations*, not computations.

Problem 5.77

A rotor consists of four horizontal blades each of length $L = 4\,\text{m}$ and mass $m = 90\,\text{kg}$ cantilevered off of a vertical shaft. Assume that each blade can be modeled as having its mass concentrated at its midpoint. The rotor is initially at rest when it is subjected to a moment $M = \beta t$, with $\beta = 60\,\text{N·m/s}$. Determine the angular speed of the rotor after 10 s.

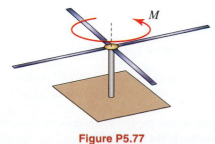

Figure P5.77

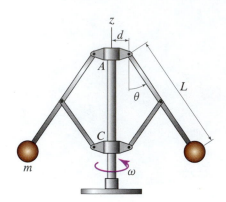

Problem 5.78

The object shown is called a *speed governor*, a mechanical device for the regulation and control of the speed of mechanisms. The system consists of two arms of negligible mass at the end of which are attached two spheres, each of mass m. The upper end of each arm is attached to a fixed collar A. The system is then made to spin with a given angular speed ω_0 at a set opening angle θ_0. Once it is in motion, the opening angle of the governor can be varied by adjusting the position of the collar C (by the application of some force). Let θ represent the generic value of the governor opening angle. If the arms are free to rotate, that is, if no moment is applied to the system about the spin axis after the system is placed in motion, determine the expression of the angular velocity ω of the system as a function of ω_0, θ_0, m, d, and L, where L is the length of each arm and d is the distance of the top hinge point of each arm from the spin axis. Neglect any friction at A and C.

Problems 5.79 through 5.81

Consider the motion of a projectile P of mass $m_P = 18.5\,\text{kg}$, which is shot with an initial speed $v_P = 1675\,\text{m/s}$ as shown in the figure. Ignore aerodynamic drag forces.

Problem 5.79
Compute the projectile's angular momentum with respect to the point O as a function of time from the time it exits the barrel until the time it hits the ground.

Problem 5.80
Choose point O as moment center. Then verify the validity of the angular impulse-momentum principle as given in Eq. (5.65) by showing that the time derivative of the angular momentum does, in fact, equal the moment.

Problem 5.81
Knowing that the helicopter E happens to have the same horizontal coordinate of the projectile at the instant the projectile leaves the gun and that it moves at a constant speed $v_E = 15\,\text{m/s}$ as shown, and treating E as a moving moment center, verify the angular impulse-momentum principle as given in Eq. (5.64).

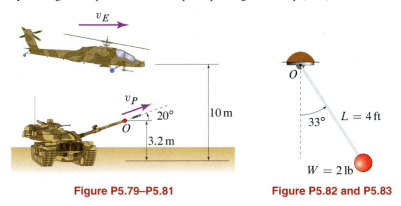

Figure P5.79–P5.81   **Figure P5.82 and P5.83**

Problems 5.82 and 5.83

The simple pendulum in the figure is released from rest as shown.

Problem 5.82
Knowing that the bob's weight is $W = 2\,\text{lb}$, determine its angular momentum computed with respect to O as a function of the angle θ.

Problem 5.83
Use the angular impulse-momentum principle in Eq. (5.65) to determine the equations of motion of the pendulum bob.

Problem 5.84

At the lowest and highest points on its trajectory, the pendulum cord, with a length $L = 2\,\text{ft}$, forms angles $\phi_1 = 15°$ and $\phi_2 = 50°$ with the vertical direction, respectively. Determine the speed of the pendulum bob corresponding to ϕ_1 and ϕ_2.

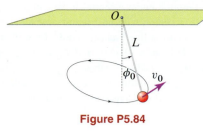

Figure P5.84

Problems 5.85 and 5.86

A collar with mass $m = 2\,\text{kg}$ is mounted on a rotating arm of negligible mass that is initially rotating with an angular velocity $\omega_0 = 1\,\text{rad/s}$. The collar's initial distance from the z axis is $r_0 = 0.5\,\text{m}$ and $d = 1\,\text{m}$. At some point, the restraint keeping the collar in place is removed so that the collar is allowed to slide. Assume that the friction between the arm and the collar is negligible.

Problem 5.85 If no external forces and moments are applied to the system, with what speed will the collar impact the end of the arm?

Problem 5.86 Compute the moment that must be applied to the arm, as a function of position along the arm, to keep the arm rotating at a constant angular velocity while the collar travels toward the end of the arm.

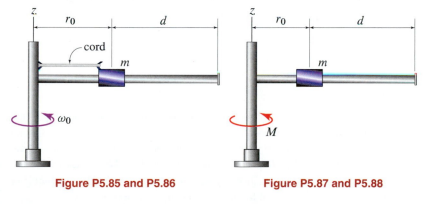

Figure P5.85 and P5.86   **Figure P5.87 and P5.88**

Problems 5.87 and 5.88

A collar of mass m is initially at rest on a horizontal arm when a constant moment M is applied to the system to make it rotate. Assume that the mass of the horizontal arm is negligible and that the collar is free to slide without friction.

Problem 5.87 Derive the equations of motion of the system, taking advantage of the angular impulse-momentum principle. *Hint:* Applying the angular impulse momentum principle yields only one of the needed equations of motion.

Problem 5.88 Continue Prob. 5.87 by integrating the collar's equations of motion and determine the time the collar takes to reach the end of the arm. Assume that

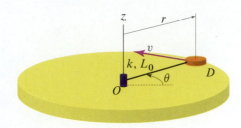

Figure P5.89

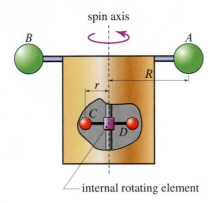

Figure P5.90

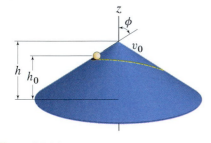

Figure P5.91

the collar weighs 1.2 lb and that $M = 20$ ft·lb. Also, at the initial time let $r_0 = 1$ ft and $d = 3$ ft.

Problem 5.89

A simple model of orbital motion under a central force can be constructed by considering the motion of a disk D sliding with no friction over a horizontal surface while connected to a fixed point O by a linear elastic cord of constant k and unstretched length L_0. Let the mass of D be $m = 0.45$ kg and $L_0 = 1$ m. Suppose that when D is at its maximum distance from O, this distance is $r_0 = 1.75$ m and the corresponding speed of D is $v_0 = 4$ m/s. Determine the elastic cord constant k such that the minimum distance between D and O is equal to the unstretched length L_0.

Problem 5.90

The body of the satellite shown has a weight that is negligible with respect to the two spheres A and B that are rigidly attached to it, which weigh 150 lb each. The distance between A and B from the spin axis of the satellite is $R = 3.5$ ft. Inside the satellite there are two spheres C and D weighing 4 lb mounted on a motor that allows them to spin about the axis of the cylinder at a distance $r = 0.75$ ft from the spin axis. Suppose that the satellite is released from rest and that the internal motor is made to spin up the internal masses at a constant time rate of 5.0 rad/s² for a total of 10 s. Treating the system as isolated, determine the angular speed of the satellite at the end of spin-up.

Problem 5.91

A sphere of mass m slides over the outer surface of a cone with angle ϕ and height h. The sphere was released at a height h_0 with a velocity of magnitude v_0 and a direction that was completely horizontal. Assume that the opening angle of the cone and the value of v_0 are such that the sphere does not separate from the surface of the cone once put in motion. In addition, assume that the friction between the sphere and the cone is negligible. Determine the vertical component of the sphere's velocity as a function of the vertical position z (measured from the base of the cone), v_0, h, h_0, and ϕ.

Problem 5.92

Consider a planet orbiting the Sun, and let P_1, P_2, P_3, and P_4 be the planet's position at four corresponding time instants t_1, t_2, t_3, and t_4 such that $t_2 - t_1 = t_4 - t_3$. Letting O denote the position of the Sun, determine the ratio between the areas of the orbital sectors $P_1 O P_2$ and $P_3 O P_4$. *Hint:* (1) The area of triangle OAB defined by the two planar vectors $\vec{c}$ and $\vec{d}$ as shown is given by Area(ABC) $= |\vec{c} \times \vec{d}|$; (2) the solution of this problem is a demonstration of Kepler's second law (see Section 1.1).

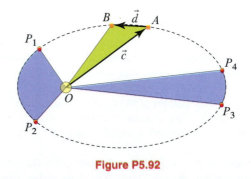

Figure P5.92

DESIGN PROBLEMS

Design Problem 5.2

The device shown is designed to remotely rotate a video camera C by counterrotating the two equal masses m via an internal motor that couples the masses and the camera. Assume that the system is mounted on bearings that allow the counterrotating masses and the camera to rotate freely on the vertical mounting post. Model the video camera as two particles, each weighing 3.1 lb and each at a distance of 4.1 in. from the rotation axis. Design the radius of the counterrotating masses, their mass, and their maximum angular velocity so that the angular velocity of the camera never exceeds 10 rpm and so that the camera rotates 90° when the masses have rotated 360°. Treat the counterrotating masses as particles, and neglect the mass of the rod on which they are mounted.

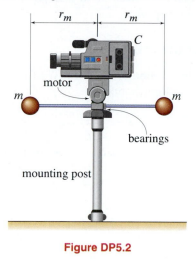

Figure DP5.2

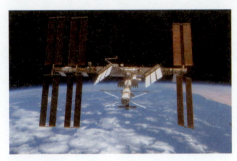

Figure 5.39
The International Space Station (ISS) as of June
2008. The equations for two-body orbital motion developed in this section form the basis for
accurately predicting the orbit of the ISS about
the Earth.

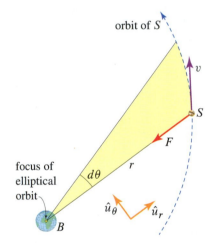

Figure 5.40
Satellite S of mass m orbiting a primary body
B of mass m_B under the action of Newton's
universal law of gravitation.

Helpful Information

Evaluating Gm_B for the Earth. If the
primary body B is the Earth, then from
Eq. (1.11) on p. 6 we can write

$$Gm_B = Gm_e = gr_e^2,$$

where m_e is the mass of the Earth, g is
the acceleration due to gravity, and r_e is
the radius of the Earth, which is 6371 km,
or 3959 mi.

5.4   Orbital Mechanics

Central forces have a special role in dynamics. In Example 3.7 on p. 218 we
saw that central force motion is related to an aspect of the motion that is constant, and in Section 5.3 we discovered that this constant is called *angular momentum*. In Section 4.2 on p. 277 we saw that the work done by a central force
such as a spring force or gravity depends only on the initial and final positions
of the point of application of the force, and so those forces are conservative.
We will now study one special central force, namely, the central force that
arises from Newton's universal law of gravitation. The solutions to the problem of one body orbiting another are provided by combining Newton's second
law and this central force. These solutions form the foundation that allows
us to predict the motion of planets, Earth satellites (Fig. 5.39), high-altitude
rockets, and interplanetary probes.

Determination of the orbit

We are interested in the motion of a satellite S of mass m that is subject to
Newton's universal law of gravitation,

$$F = \frac{Gm_B m}{r^2}, \tag{5.88}$$

where m_B is the mass of the primary, central, or attracting body, G is the
universal gravitational constant,* and r is the distance between the centers of
mass of the two bodies (see Fig. 5.40). We begin by making three important
assumptions:

1. The attracting bodies B and S are treated as particles. This assumption
   is exactly correct if each body has a spherically symmetric mass distribution. It is a good approximation if the distance between the two
   bodies is large compared with their dimensions, which is the case for
   many planets and artificial satellites in high orbits. Because the Earth is
   oblate and has a nonuniform mass distribution, this assumption starts to
   break down for artificial satellites in low Earth orbit.

2. The only force acting on the satellite S is the gravitational force F.

3. The primary body B is fixed in space. This assumption works well when
   m_B is very large compared with m.

Applying Newton's second law to S in the polar component system shown in
Fig. 5.40, we obtain

$$\sum F_r: \quad -\frac{Gm_B m}{r^2} = m(\ddot{r} - r\dot{\theta}^2), \tag{5.89}$$

$$\sum F_\theta: \quad 0 = m(r\ddot{\theta} + 2\dot{r}\dot{\theta}), \tag{5.90}$$

where we have substituted in the kinematic equations in polar components and
have also used Eq. (5.88). In Section 5.3 we saw that Eq. (5.90) is equivalent
to the conservation of angular momentum of S about B, that is,

$$h_B = mr^2\dot{\theta} = \text{constant.} \tag{5.91}$$

---

* Recall that the generally accepted value of this constant is $G = 6.674 \times 10^{-11} \text{ m}^3/(\text{kg} \cdot \text{s}^2) = 3.439 \times 10^{-8} \text{ ft}^3/(\text{slug} \cdot \text{s}^2)$.

Equation (5.91) is also a reflection of Kepler's second law of planetary motion. To see this, refer to Fig. 5.40 and note that the area dA swept by the radial line of length r during the time dt (i.e., the yellow area) is equal to $\frac{1}{2}(r\,d\theta)r$. Therefore, the *areal velocity*, which is the time rate at which area is swept by the radial line r, is constant according to Eq. (5.91) and is given by

$$\frac{dA}{dt} = \tfrac{1}{2}r^2\dot{\theta} = \text{constant.} \tag{5.92}$$

This is Kepler's second law (see p. 2 in Section 1.1)!

Solution of the governing equations

To find the trajectory of S, we need to solve Eqs. (5.89) and (5.90). This is traditionally done by eliminating t and then solving for $r(\theta)$.

To begin, we first rewrite Eqs. (5.89) and (5.90) as

$$\ddot{r} - r\dot{\theta}^2 = -\frac{Gm_B}{r^2}, \tag{5.93}$$

$$r^2\dot{\theta} = \kappa, \tag{5.94}$$

where κ (the Greek letter kappa) is the *angular momentum per unit mass* of the satellite S, which is equal to h_B/m from Eq. (5.91). The chain rule allows us to write $\dot{r}$ and $\ddot{r}$ as

$$\dot{r} = \frac{dr}{d\theta}\frac{d\theta}{dt} = \frac{dr}{d\theta}\frac{\kappa}{r^2} = -\kappa\frac{d}{d\theta}\left(\frac{1}{r}\right), \tag{5.95}$$

$$\ddot{r} = \frac{d\dot{r}}{d\theta}\frac{d\theta}{dt} = \frac{d\dot{r}}{d\theta}\frac{\kappa}{r^2} = \frac{\kappa}{r^2}\frac{d}{d\theta}\left[-\kappa\frac{d}{d\theta}\left(\frac{1}{r}\right)\right] = -\frac{\kappa^2}{r^2}\frac{d^2}{d\theta^2}\left(\frac{1}{r}\right). \tag{5.96}$$

Using Eqs. (5.94) and (5.96), Eq. (5.93) becomes

$$-\frac{\kappa^2}{r^2}\frac{d^2}{d\theta^2}\left(\frac{1}{r}\right) - r\left(\frac{\kappa}{r^2}\right)^2 = -\frac{Gm_B}{r^2}. \tag{5.97}$$

Letting $u = 1/r$, Eq. (5.97) becomes

$$-\kappa^2 u^2 \frac{d^2 u}{d\theta^2} - \kappa^2 u^3 = -Gm_B u^2, \tag{5.98}$$

which, upon canceling u^2 and rearranging, becomes

$$\frac{d^2 u}{d\theta^2} + u = \frac{Gm_B}{\kappa^2}. \tag{5.99}$$

As we will see again in Section 9.2, this is a second-order, constant-coefficient, nonhomogeneous, differential equation whose solution can be verified by direct substitution to be

$$u = \frac{1}{r} = C\cos(\theta - \beta) + \frac{Gm_B}{\kappa^2}, \tag{5.100}$$

where C and β are constants of integration.

Equation (5.100) determines the unpowered or free-flight trajectory of the satellite. It is the polar coordinate representation of a *conic section*, which is

---

Interesting Fact

The primary mass really moves. In reality, the satellite S and the primary body B orbit about their common mass center. If this motion of B is taken into account, Eqs (5.89) and (5.90) no longer apply and the governing equations become

$$-\frac{Gm_B m}{r^2} = \frac{m_B m}{m_B + m}(\ddot{r} - r\dot{\theta}^2),$$

$$0 = \frac{m_B m}{m_B + m}(r\ddot{\theta} + 2\dot{r}\dot{\theta}),$$

where r is the distance between mass centers of B and S. Notice that if $m_B \gg m$, then these two equations are approximated by Eqs. (5.89) and (5.90). See Prob. 2.63 and 2.64 for a one-dimensional version of the above two equations.

Helpful Information

Periapsis and apoapsis. An *apsis* is the point of maximum or minimum distance from the focus containing the center of attraction in an elliptical orbit. The point at which the orbital radius is a minimum is called the *periapsis*, and the point at which the orbital radius is a maximum is called the *apoapsis*. For specific celestial bodies, names that apply specifically to orbits about those bodies are generally used rather than periapsis and apoapsis. The following table lists some of these.

Body	Minimum r	Maximum r
Earth	perigee	apogee
Sun	perihelion	aphelion
Mars	periareion	apoareion
Jupiter	perijove	apojove

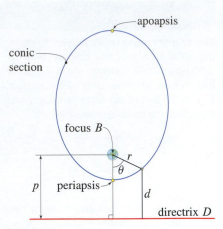

Figure 5.41
Geometry used to define a conic section. In this case, the conic section is an ellipse with eccentricity $e = 0.661$.

defined as follows: Given a point B, called the *focus*, and a line D, called the *directrix*, a conic section is defined as the locus of points for which the ratio (see Fig. 5.41)

$$e = \frac{\text{distance to } B}{\text{distance to } D} = \frac{r}{d} \qquad (5.101)$$

is a constant. As we shall soon see, circles, ellipses, parabolas, and hyperbolas are all conic sections. The ratio in Eq. (5.101) is called the *eccentricity* of the conic section. We now need to obtain the constants of integration in Eq. (5.100).

The constant β in Eq. (5.100) can be eliminated by letting the $\theta = 0$ axis be at periapsis, that is, at the point where the orbital radius r is a minimum. At periapsis, $\dot{r} = 0$, which implies that $du/d\theta = 0$ from Eq. (5.95). Applying this condition, we find that $\beta = 0$, which simplifies Eq. (5.100) to

$$\boxed{\frac{1}{r} = C \cos \theta + \frac{Gm_B}{\kappa^2}.} \qquad (5.102)$$

Referring to Eq. (5.101) and Fig. 5.41, we can use the definition of eccentricity e to write

$$r = ed = e(p - r \cos \theta), \qquad (5.103)$$

where p is the *focal parameter*. Rearranging this equation, we obtain

$$\frac{1}{r} = \frac{1}{p} \cos \theta + \frac{1}{ep}. \qquad (5.104)$$

Comparing Eqs. (5.102) and (5.104), we see that

$$p = \frac{1}{C} \qquad (5.105)$$

and that

$$\boxed{e = \frac{C\kappa^2}{Gm_B}.} \qquad (5.106)$$

Using Eq. (5.106), we can write Eq. (5.102) as

$$\boxed{\frac{1}{r} = \frac{Gm_B}{\kappa^2}(1 + e \cos \theta).} \qquad (5.107)$$

Equations (5.102) and Eq. (5.107) are equivalent to one another. In the former, we need to determine the constants C and κ, and in the latter we need to determine the constants e and κ (Eq. (5.106) provides the link between these three constants). In either case, these constants are determined by knowing the position and velocity of the satellite at some point on its trajectory.

Launch conditions at periapsis.   Consider a satellite in orbit about a primary body, as shown in Fig. 5.42. If instead of knowing the launch conditions at an arbitrary position S, we know the launch conditions at periapsis (point P), then $\phi_P = 90°$ and $\theta_P = 0°$. Under these conditions, Eq. (5.94) tells us that the angular momentum per unit mass κ becomes

$$\boxed{\kappa = r_P^2 \dot{\theta}_P = r_P v_P.} \qquad (5.108)$$

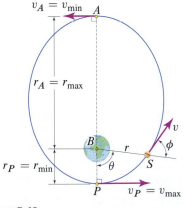

Figure 5.42
A satellite S in orbit about a primary body B. The launch conditions are at periapsis P and are r_P and v_P.

To determine C, we evaluate Eq. (5.102) at P to find that

$$C = \frac{1}{r_P}\left(1 - \frac{Gm_B}{r_P v_P^2}\right),$$

(5.109)

where we have used κ from Eq. (5.108). Substituting Eqs. (5.108) and (5.109) into Eq. (5.102), we obtain the following equation for the trajectory of the satellite:

$$\frac{1}{r} = \frac{1}{r_P}\left(1 - \frac{Gm_B}{r_P v_P^2}\right)\cos\theta + \frac{Gm_B}{r_P^2 v_P^2}.$$

(5.110)

Conic sections

We have already mentioned that the solution $r(\theta)$ to the governing orbital equations is a conic section. The type of conic section depends on the eccentricity of the trajectory e (see Fig. 5.43), which, in turn, depends on C and κ via Eq. (5.106). We will now look at each of the four possible cases: $e = 0$, $0 < e < 1$, $e = 1$, and $e > 1$.

Circular orbit ($e = 0$). If r_P and v_P are chosen so that $e = 0$, then Eq. (5.106) tells us that $C\kappa^2 = 0$, which implies that $C = 0$ since κ cannot be zero. If $C = 0$, then Eq. (5.109) tells us that the speed in a circular orbit of radius r_P is given by

$$\frac{1}{r_P}\left(1 - \frac{Gm_B}{r_P v_c^2}\right) = 0 \quad\Rightarrow\quad 1 = \frac{Gm_B}{r_P v_c^2} \quad\Rightarrow\quad \boxed{v_c = \sqrt{\frac{Gm_B}{r_P}}}.$$

(5.111)

Elliptical orbit ($0 < e < 1$). For an elliptical orbit, $0 < e < 1$, and when evaluated at periapsis ($\theta = 0°$), Eq. (5.110) gives $r = r_P$ as expected since r_P was chosen at $\theta = 0°$. If we instead evaluate Eq. (5.110) at $\theta = 180°$, we will find r at apoapsis, that is, r_A. Doing so gives

$$\frac{1}{r_A} = \frac{-1}{r_P}\left(1 - \frac{Gm_B}{r_P v_P^2}\right) + \frac{Gm_B}{r_P^2 v_P^2} = \frac{2Gm_B - r_P v_P^2}{r_P^2 v_P^2},$$

(5.112)

or, by simplifying and rearranging,

$$\boxed{r_A = \frac{r_P}{2Gm_B/(r_P v_P^2) - 1}},$$

(5.113)

where we recall that the launch conditions are at periapsis. It is left as an exercise (see Prob. 5.93) to show that r_A can be written in terms of e as

$$r_A = r_P\left(\frac{1+e}{1-e}\right).$$

(5.114)

An additional set of relations for elliptical orbits is obtained by evaluating Eq. (5.104) at periapsis ($\theta = 0°$) and apoapsis ($\theta = 180°$) to obtain

$$\frac{1}{r_P} = \frac{1}{p} + \frac{1}{ep} \quad\Rightarrow\quad r_P = \frac{pe}{1+e}$$

(5.115)

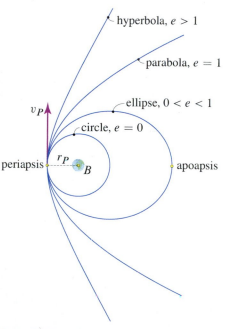

hyperbola, $e > 1$

parabola, $e = 1$

ellipse, $0 < e < 1$

v_P

circle, $e = 0$

periapsis r_P B apoapsis

Figure 5.43
The four conic sections, showing r_P and v_P.

and

$$\frac{1}{r_A} = \frac{-1}{p} + \frac{1}{ep} \quad \Rightarrow \quad r_A = \frac{pe}{1-e}, \tag{5.116}$$

respectively, where p is the focal parameter shown in Fig. 5.41. Referring to Fig. 5.44, and using Eqs. (5.115) and (5.116), we obtain

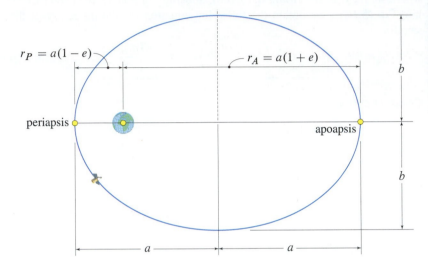

Figure 5.44. The semimajor axis a and semiminor axis b of an ellipse.

$$2a = r_P + r_A = pe\left(\frac{1}{1+e} + \frac{1}{1-e}\right) \quad \Rightarrow \quad a = \frac{pe}{1-e^2}, \tag{5.117}$$

where a is referred to as the *semimajor axis* of the ellipse. Solving Eq. (5.117) for p and substituting the result into Eq. (5.104) give

$$\frac{1}{r} = \frac{1 + e\cos\theta}{a(1-e^2)}, \tag{5.118}$$

which, when evaluated at periapsis and apoapsis, gives

$$r_P = a(1-e) \quad \text{and} \quad r_A = a(1+e). \tag{5.119}$$

Equation (5.118) is a statement of Kepler's first law, which states that orbits of planets are ellipses with the Sun at one focus (see p. 2).

To find the period of an elliptical orbit, we refer to the definition of areal velocity in Eq. (5.92) and integrate that equation over the entire ellipse to find that

$$\frac{dA}{dt} = \tfrac{1}{2}r^2\dot\theta = \frac{\kappa}{2} \quad \Rightarrow \quad \int_0^A dA = \int_0^\tau \frac{\kappa}{2}\,dt \quad \Rightarrow \quad A = \frac{\kappa}{2}\tau, \tag{5.120}$$

where τ is the *orbital period* and we have used Eq. (5.94) to write $\kappa = r^2\dot\theta$, which is constant. For an ellipse, analytical geometry tells us that its area is equal to $\pi a b$, where b is the semiminor axis of the ellipse (see Fig. 5.44), and so we can write Eq. (5.120) as

$$\pi a b = \frac{\kappa}{2}\tau \quad \Rightarrow \quad \boxed{\tau = \frac{2\pi a b}{\kappa}.} \tag{5.121}$$

Figure 5.45
A portrait of Johannes Kepler painted in 1610.

Now, it can be shown that (see Prob. 5.94)

$$b = \sqrt{r_P r_A}, \tag{5.122}$$

and since Eq. (5.117) tells us that $a = (r_P + r_A)/2$, we can write Eq. (5.121) as

$$\tau = \frac{\pi}{\kappa}(r_P + r_A)\sqrt{r_P r_A}, \tag{5.123}$$

where κ can be found by using Eq. (5.108).

Kepler's third law (see p. 2) states that the square of the orbital period is proportional to the cube of the semimajor axis of that orbit. To show this, we start with Eq. (5.123) and substitute in Eqs. (5.119) to obtain

$$\tau = \frac{\pi}{\kappa}\left[a(1-e)+a(1+e)\right]\sqrt{a(1-e)a(1+e)} = \frac{2\pi}{\kappa}a^2\sqrt{1-e^2}. \tag{5.124}$$

Squaring both sides and noting from Eq. (5.117) that $a(1-e^2) = pe$, we have

$$\tau^2 = \frac{4\pi^2}{\kappa^2}a^3 pe. \tag{5.125}$$

Using Eqs. (5.105) and (5.106), we can write the product pe as $\kappa^2/(Gm_B)$, which means that Eq. (5.125) becomes

$$\tau^2 = \frac{4\pi^2}{Gm_B}a^3, \tag{5.126}$$

which is Kepler's third law.

Parabolic trajectory ($e = 1$).   If r_P and v_P are chosen so that the trajectory is parabolic, then $e = 1$. Figure 5.43 indicates that the parabolic trajectory is that which divides those trajectories that are periodic from those that are not. That is, it divides trajectories that return to their initial starting point from those that do not. For a given r_P, the launch velocity v_{par} required to achieve a parabolic trajectory can be found by letting $e = 1$ in Eq. (5.106), which gives $Gm_B = C\kappa^2$, and then substituting C from Eq. (5.109) into that result to obtain

$$Gm_B = \frac{1}{r_P}\left(1 - \frac{Gm_B}{r_P v_{par}^2}\right)(r_P v_{par})^2, \tag{5.127}$$

which, when solved for v_{par}, becomes

$$v_{par} = v_{esc} = \sqrt{\frac{2Gm_B}{r_P}}. \tag{5.128}$$

The speed given in Eq. (5.128) is often referred to as *escape velocity* since it is the speed required to completely escape the influence of the primary body B.

Hyperbolic trajectory ($e > 1$).   For a hyperbolic trajectory, the governing equations given by Eqs. (5.106)–(5.110) are used for values of $e > 1$.

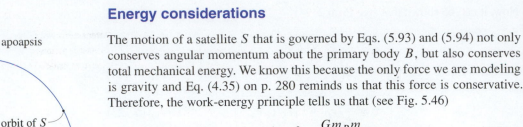

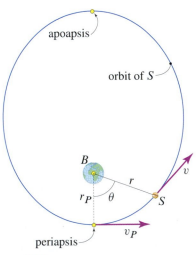

Figure 5.46
A satellite S in an elliptical orbit about a primary body B. The orbital parameters a, r_P, r_A, r, and θ are shown.

Figure 5.47
A satellite S of mass m orbiting a primary body B of mass m_B on a conic section, which in this case is an ellipse. The angle θ and the launch conditions r_P and v_P are defined from periapsis.

Energy considerations

The motion of a satellite S that is governed by Eqs. (5.93) and (5.94) not only conserves angular momentum about the primary body B, but also conserves total mechanical energy. We know this because the only force we are modeling is gravity and Eq. (4.35) on p. 280 reminds us that this force is conservative. Therefore, the work-energy principle tells us that (see Fig. 5.46)

$$T_P + V_P = \tfrac{1}{2}mv_P^2 - \frac{Gm_Bm}{r_P} = \text{constant}, \tag{5.129}$$

which, after dividing through by m, gives

$$\tfrac{1}{2}v_P^2 - \frac{Gm_B}{r_P} = E \quad \text{or} \quad \frac{\kappa^2}{2r_P^2} - \frac{Gm_B}{r_P} = E, \tag{5.130}$$

where E is the *mechanical energy per unit mass* (sometimes called the *specific energy*) of the satellite and we have used Eq. (5.108) to introduce κ. Next, we evaluate Eq. (5.107) at periapsis (i.e., $\theta = 0$) and solve for κ to obtain

$$\kappa^2 = Gm_B(1+e)r_P = Gm_B(1+e)a(1-e) = Gm_Ba(1-e^2), \tag{5.131}$$

where we have used the first of Eqs. (5.119) for r_P. Substituting Eq. (5.131) and the first of Eqs. (5.119) into Eq. (5.130), we obtain

$$E = \frac{Gm_Ba(1-e^2)}{2a^2(1-e)^2} - \frac{Gm_B}{a(1-e)} \quad \Rightarrow \quad \boxed{E = -\frac{Gm_B}{2a},} \tag{5.132}$$

where we see that the total energy of the satellite depends on *only* the semimajor axis a of the orbit.

Now that we have Eq. (5.132), we can apply the work-energy principle at an arbitrary position in the orbit to obtain

$$E = -\frac{Gm_B}{2a} = \tfrac{1}{2}v^2 - \frac{Gm_B}{r}. \tag{5.133}$$

Solving this equation for v, we have

$$\boxed{v = \sqrt{Gm_B\left(\frac{2}{r} - \frac{1}{a}\right)},} \tag{5.134}$$

which is very useful for solving orbital transfer problems.

End of Section Summary

In this section, we studied the motion of a satellite S of mass m that is subject to Newton's universal law of gravitation due to a body B of mass m_B, which is the primary or attracting body (see Fig. 5.47). We began with these important assumptions:

1. The primary body B and the satellite S are both treated as particles.

2. The only force acting on the satellite S is the force of mutual attraction between B and S.

3. The primary body B is fixed in space.

Determination of the orbit

Solving the governing equations in polar coordinates, we found that the trajectory of the satellite S under these assumptions is a conic section, whose equation can be written as

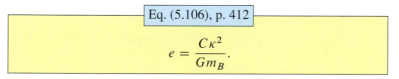

Eqs. (5.102) and (5.107), p. 412

$$\frac{1}{r} = C \cos \theta + \frac{Gm_B}{\kappa^2} \quad \text{or}$$

$$\frac{1}{r} = \frac{Gm_B}{\kappa^2}(1 + e \cos \theta),$$

where r is the distance between the centers of mass of S and B; θ is the orbital angle measured relative to periapsis; G is the universal gravitational constant; κ is the angular momentum per unit mass of the satellite S measured about B; C is a constant to be determined; and e is the *eccentricity* of the trajectory, which can be written as

Eq. (5.106), p. 412

$$e = \frac{C\kappa^2}{Gm_B}.$$

If, as is generally the case here, the orbital conditions are known at periapsis, then κ is

Eq. (5.108), p. 412

$$\kappa = r_P v_P,$$

the constant C is

Eq. (5.109), p. 413

$$C = \frac{1}{r_P}\left(1 - \frac{Gm_B}{r_P v_P^2}\right),$$

and the equation describing the trajectory becomes

Eq. (5.110), p. 413

$$\frac{1}{r} = \frac{1}{r_P}\left(1 - \frac{Gm_B}{r_P v_P^2}\right)\cos \theta + \frac{Gm_B}{r_P^2 v_P^2}.$$

Conic sections. Equations (5.102), (5.107), and (5.110) represent equivalent conic sections in polar coordinates. The type of conic section depends on the eccentricity of the trajectory e (see Fig. 5.48), which, in turn, depends on C and κ via Eq. (5.106). There are four types of conic section, which are determined by the value of the eccentricity e, that is, $e = 0, 0 < e < 1, e = 1$, and $e > 1$.

For a *circular orbit* ($e = 0$), the radius is r_P and the speed in the orbit is equal to

Eq. (5.111), p. 413

$$v_c = \sqrt{\frac{Gm_B}{r_P}}.$$

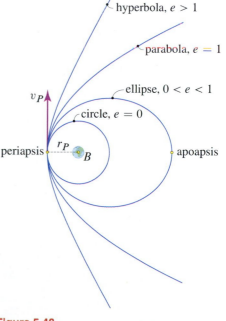

Figure 5.48
Figure 5.43 repeated. The four conic sections, showing r_P and v_P.

For an *elliptical orbit* $(0 < e < 1)$, as expected, the radius at periapsis is r_P. The radius at apoapsis is given by

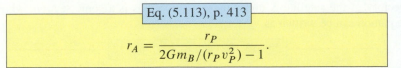

$$r_A = \frac{r_P}{2Gm_B/(r_P v_P^2) - 1}.$$

Referring to Fig. 5.49, additional relationships between the semimajor axis a

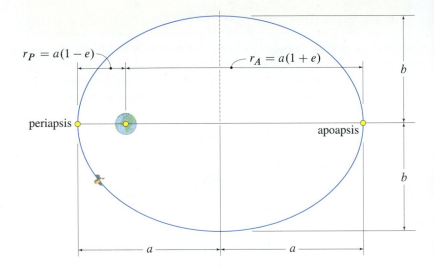

Figure 5.49. *Figure 5.44 repeated.* The semimajor axis a and semiminor axis b of an ellipse.

of an elliptical orbit, the eccentricity of the orbit, and the radii at periapsis and apoapsis are, respectively,

Eqs. (5.119), p. 414

$$r_P = a(1-e) \quad \text{and} \quad r_A = a(1+e).$$

The period of an elliptical orbit τ can be written in the following two ways

Eq. (5.121), p. 414, and Eq. (5.123), p. 415

$$\tau = \frac{2\pi a b}{\kappa} = \frac{\pi}{\kappa}(r_P + r_A)\sqrt{r_P r_A};$$

or, reflecting Kepler's third law, the orbital period can be written as

Eq. (5.126), p. 415

$$\tau^2 = \frac{4\pi^2}{Gm_B}a^3.$$

A *parabolic trajectory* $(e = 1)$ is that which divides periodic orbits (that return to their starting location) from trajectories that are not periodic. For a given r_P, the speed v_{par} required to achieve a parabolic trajectory is given by

Eq. (5.128), p. 415

$$v_{par} = v_{esc} = \sqrt{\frac{2Gm_B}{r_P}},$$

which is also referred to as the *escape velocity* since it is the speed required to completely escape the influence of the primary body B.

For a *hyperbolic trajectory* ($e > 1$), the governing equations given by Eqs. (5.106)–(5.110) are used for values of $e > 1$.

Energy considerations

Using the work-energy principle, we discovered that the total mechanical energy in an orbit depends on only the semimajor axis a of the orbit and is given by

Eq. (5.132), p. 416

$$E = -\frac{Gm_B}{2a},$$

where E is the *mechanical energy per unit mass* of the satellite. Applying the work-energy principle at an arbitrary location within the orbit, we found that the speed can be written as

Eq. (5.134), p. 416

$$v = \sqrt{Gm_B\left(\frac{2}{r} - \frac{1}{a}\right)}.$$

E X A M P L E 5.16 *Apogee and Perigee Speeds for a Given Orbit*

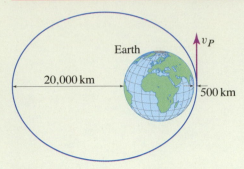

Figure 1
An elliptical orbit about Earth with a launch altitude of 500 km and an apogee altitude of 20,000 km. It can be shown that the orbit has an eccentricity e of 0.587. The figure is drawn to scale.

An artificial satellite is launched from an altitude of 500 km with a velocity that is parallel to the surface of the Earth (Fig. 1). Requiring that the altitude at apogee be 20,000 km and using 6371 km for the radius of the Earth, determine

(a) the required speed at perigee v_P,

(b) the speed at apogee v_A, and

(c) the period of the orbit.

SOLUTION

Road Map & Modeling This is an Earth orbit, and we are given r_P and the required r_A, so Eq. (5.113) will allow us to determine the speed at perigee v_P. Knowing r_P and having found v_P, we see that Eq. (5.108) allows us to find κ, which will then allow us to find v_A given that κ is conserved and allow us to find the orbital period τ using Eq. (5.123). Note that the orbital eccentricity mentioned in the caption of Fig. 1 can be found from either of Eqs. (5.119).

Computation Referring to Fig. 1, we see that an altitude of 500 km at perigee corresponds to

$$r_P = (500 + 6371)\,\text{km} = 6871 \times 10^3\,\text{m}. \tag{1}$$

Similarly, the altitude of 20,000 km at apogee means

$$r_A = (20{,}000 + 6371)\,\text{km} = 26{,}370 \times 10^3\,\text{m}. \tag{2}$$

Using Eq. (1.11) on p. 6 in Eq. (5.113) and solving for v_P, we find that

$$r_A = \frac{r_P^2 v_P^2}{2gr_e^2 - r_P v_P^2} \quad \Rightarrow \quad v_P = \sqrt{\frac{2gr_A r_e^2}{(r_A + r_P)r_P}} \tag{3}$$

$$\Rightarrow \quad \boxed{v_P = 9589\,\text{m/s} = 34{,}500\,\text{km/h},} \tag{4}$$

where r_e is the radius of the Earth.

To find the speed at apogee, we note from Eq. (5.108) that the angular momentum per unit mass κ is constant and so

$$\kappa = r_P v_P = r_A v_A \quad \Rightarrow \quad v_A = \frac{r_P}{r_A} v_P \tag{5}$$

$$\Rightarrow \quad \boxed{v_A = 2499\,\text{m/s} = 9000\,\text{km/h}.} \tag{6}$$

To determine the period of the orbit, we can use Eq. (5.123) after substituting in κ from Eq. (5) as follows:

$$\boxed{\tau = \frac{\pi}{r_P v_P}(r_P + r_A)\sqrt{r_P r_A} = 21{,}340\,\text{s} = 5.93\,\text{h}.} \tag{7}$$

Discussion & Verification It can be seen that both Eqs. (3) and (5) have dimensions of velocity, as they should. Equation (7) has the dimension of time, as it should. We also note that the speed at apogee is less than the speed at perigee, as it should be since κ is constant.

PROBLEMS

For the problems in this section, use 6371 km or 3959 mi for the radius of the Earth. *Not all orbits or objects are drawn to scale.*

Problem 5.93

Starting with Eq. (5.113) and using Eqs. (5.106), (5.108), and (5.109), show that the radius at apoapsis r_A can be written as shown in Eq. (5.114).

Problem 5.94

Using the lengths shown as well as the property of an ellipse that states that the sum of the distances from each of the foci (i.e., points O and B) to any point on the ellipse is a constant, prove Eq. (5.122), that is, that the length of the semiminor axis can be related to the periapsis and apoapsis radii via $b = \sqrt{r_P r_A}$.

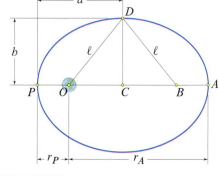

Figure P5.94

Problem 5.95

An artificial satellite is launched from an altitude of 500 km with a velocity v_P that is parallel to the surface of the Earth. Requiring that the altitude at apogee be 20,000 km, determine the velocity at B, that is, the position in the orbit when the velocity is first orthogonal to the launch velocity.

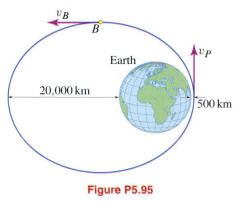

Figure P5.95

Problem 5.96

The S-IVB third stage of the Saturn V rocket, which was used for the Apollo missions, would burn for about 2.5 min to place the spacecraft into a "parking orbit"* and then, after several orbits, would burn for about 6 min to accelerate the spacecraft to escape velocity to send it to the Moon. Assuming a circular parking orbit with an altitude of 170 km, determine the change in speed needed at P to go from the parking orbit to escape velocity. Assume that the change in speed occurs instantaneously so that you need not worry about changes in orbital position during the engine thrust.

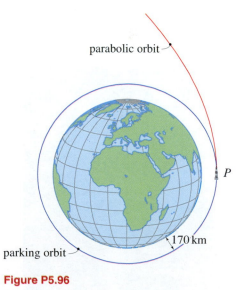

Figure P5.96

Problem 5.97

Using the last of Eqs. (5.131), along with Eq. (5.132), solve for the eccentricity e as a function of E, κ, and Gm_B.

---

* A *parking orbit* is a temporary orbit of an artificial satellite or spacecraft in preparation for thrusting to another orbit or trajectory.

(a) Using that result, along with fact that $e \geq 0$, show that $E < 0$ corresponds to an elliptical orbit, $E = 0$ corresponds to a parabolic trajectory, and $E > 0$ corresponds to a hyperbolic trajectory.

(b) Show that for $e = 0$, the expression you found for e leads to Eq. (5.111).

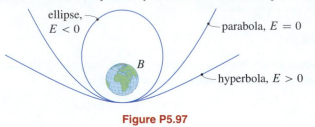

Figure P5.97

Problem 5.98

Assuming that the Sun is the only significant body in the solar system (the mass of the Sun accounts for 99.8% of the mass of the solar system), determine the escape velocity from the Sun as a function of the distance r from its center. What is the value of the escape velocity (expressed in km/h) when r is equal to the radius of Earth's orbit? Use 1.989×10^{30} kg for the mass of the Sun and 150×10^6 km for the radius of Earth's orbit.

Problem 5.99

In 1705, Edmund Halley (1656–1742), an English astronomer, claimed that the comet sightings of 1531, 1607, and 1682 were all the same comet and that, after some rough calculations accounting for the influence of the larger planets, this comet would return again in 1758. Halley did not live to see the comet's return, but it did return late in 1758 and reached perihelion in March 1759. In honor of his prediction, this comet was named "Halley."

Each elliptical orbit of Halley is slightly different, but the average value of the semimajor axis a is about 17.95 AU.* Using this value, along with the fact that its orbital eccentricity is 0.967 (the orbit is drawn to scale, but the Sun is shown to be 36 times bigger than it should be), determine

(a) the orbital period in years of Halley's comet, and

(b) its distance, in AU, from the Sun at perihelion P and at aphelion A. Look up the orbits of the planets of our solar system on the Web. What planetary orbits is Halley near to at perihelion and aphelion?

Use 1.989×10^{30} kg for the mass of the Sun.

orbit of Halley's comet

Figure P5.99

Problems 5.100 and 5.101

Explorer 7 was launched on October 13, 1959, with an apogee altitude above the Earth's surface of 1073 km and a perigee altitude of 573 km above the Earth's surface. Its orbital period was 101.4 min.

Problem 5.100   Using this information, calculate Gm_e for the Earth and compare it with $g r_e^2$.

Problem 5.101   Determine the eccentricity of the Explorer 7's orbit as well as its speeds at perigee and apogee.

Figure P5.100 and P5.101

---

* One *astronomical unit* (AU) is the distance between the center of mass of the Earth and that of the Sun and is approximately 1.496×10^8 km $= 9.296 \times 10^7$ mi.

Problem 5.102

A *geosynchronous equatorial orbit* is a circular orbit above the Earth's equator that has a period of 1 day (these are sometimes called *geostationary orbits*). These geostationary orbits are of great importance for telecommunications satellites because a satellite orbiting with the same angular rate as the rotation rate of the Earth will appear to hover in the same point in the sky as seen by a person standing on the surface of the Earth. Using this information, determine the altitude h_g and radius r_g of a geostationary orbit (in miles). In addition, determine the speed v_g of a satellite in such an orbit (in miles per hour).

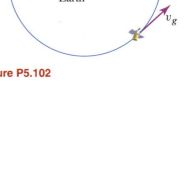

Figure P5.102

Problem 5.103

The mass of the planet Jupiter is 318 times that of Earth, and its equatorial radius is 71,500 km. If a space probe is in a circular orbit about Jupiter at the altitude of the Galilean moon Callisto (orbital altitude 1.812×10^6 km), determine the change in speed Δv needed in the outer orbit so that the probe reaches a minimum altitude at the orbital radius of the Galilean moon Io (orbital altitude 3.502×10^5 km). Assume that the probe is at the maximum altitude in the transfer orbit when the change in speed occurs and that change in speed is impulsive, that is, it occurs instantaneously.

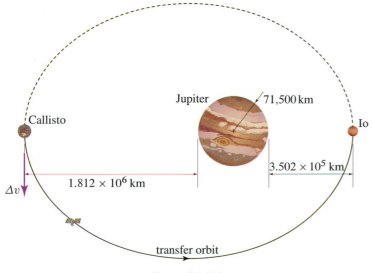

Figure P5.103

Problem 5.104

The on-orbit assembly of the International Space Station (ISS) began in 1998 and continues today. The ISS has an apogee altitude above the Earth's surface of 341.9 km and a perigee altitude of 331.0 km above the Earth's surface. Determine its maximum and minimum speeds in orbit, its orbital eccentricity, and its orbital period. Research its actual orbital period and compare it with your calculated value.

Figure P5.104

Problems 5.105 through 5.107

The optimal way (from an energy standpoint) to transfer from one circular orbit about a primary body B to another circular orbit is via the so-called *Hohmann transfer*, which involves transferring from one circular orbit to another using an elliptical orbit that is tangent to both at the periapsis and apoapsis of the ellipse. The ellipse is uniquely

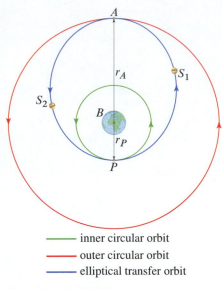

—— inner circular orbit
—— outer circular orbit
—— elliptical transfer orbit

Figure P5.105–P5.109

defined because we know r_P (the radius of the inner circular orbit) and r_A (the radius of the outer circular orbit), and therefore we know the semimajor axis a via Eq. (5.117) and the eccentricity e via Eq. (5.114) or Eqs. (5.119). Performing a Hohmann transfer requires two maneuvers, the first to leave the inner (outer) circular orbit and enter the transfer ellipse and the second to leave the transfer ellipse and enter the outer (inner) circular orbit.

Problem 5.105 A spacecraft S_1 needs to transfer from circular low Earth parking orbit with altitude 120 mi above the surface of the Earth to a circular geosynchronous orbit with altitude 22,240 mi. Determine the change in speed Δv_P required at perigee P of the elliptical transfer orbit and the change in speed Δv_A required at apogee A. In addition, compute the time required for the orbital transfer. Assume that the changes in speed are impulsive, that is, they occur instantaneously.

Problem 5.106 A spacecraft S_2 needs to transfer from a circular Earth orbit whose period is 12 h (i.e., it is overhead twice per day) to a low Earth circular orbit with an altitude of 110 mi. Determine the change in speed Δv_A required at apogee A of the elliptical transfer orbit and the change in speed Δv_P required at perigee P. In addition, compute the time required for the orbital transfer. Assume that the changes in speed are impulsive; that is, they occur instantaneously.

Problem 5.107 A spacecraft S_1 is transferring from circular low Earth parking orbit with altitude 100 mi to a circular orbit with radius r_A. Plot, as a function of r_A for $r_P \leq r_A \leq 100 r_P$, the change in speed Δv_P required at perigee of the elliptical transfer orbit as well as the change in speed Δv_A required at apogee. In addition, plot the time as a function of r_A, again for $r_P \leq r_A \leq 100 r_P$, required for the orbital transfer. Assume that the changes in speed are impulsive; that is, they occur instantaneously.

Problem 5.108

Referring to the description given for Probs. 5.105–5.107, for a Hohmann transfer from an inner circular orbit to an outer circular orbit, what would you expect to be the signs on the change in speed at periapsis and at apoapsis?
Note: Concept problems are about *explanations*, not computations.

Problem 5.109

Referring to the description given for Probs. 5.105–5.107, for a Hohmann transfer from an outer circular orbit to an inner circular orbit, what would you expect to be the signs on the change in speed at periapsis and at apoapsis?
Note: Concept problems are about *explanations*, not computations.

Problem 5.110

During the Apollo missions, while the astronauts were on the Moon with the lunar module (LM), the command module (CM) would fly in a circular orbit around the Moon at an altitude of 60 mi. After the astronauts were done exploring the Moon, the LM would launch from the Moon's surface (at L) and undergo powered flight until burnout at P, which occurred when the LM was approximately 15 mi above the surface of the Moon with its velocity v_{bo} parallel to the surface of the Moon (i.e., at periapsis). It would then fly under the influence of the Moon's gravity until reaching apoapsis A, at which point it would rendezvous with the CM. The radius of the Moon is 1079 mi, and its mass is 0.0123 times that of the Earth.

(a) Determine the required speed v_{bo} at burnout P.

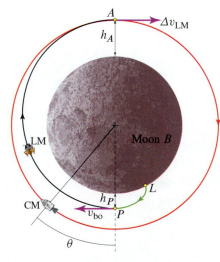

Figure P5.110

(b) What is the change in speed Δv_{LM} required of the LM at the rendezvous point A?

(c) Determine the time it takes the LM to travel from P to A.

(d) In terms of the angle θ, where should the CM be when the LM reaches P so that they can rendezvous at A?

Assume that the changes in speed are impulsive; that is, they occur instantaneously.

Problem 5.111

One option when traveling to Mars from the Earth is to use a Hohmann transfer orbit like that described in Probs. 5.105–5.109. Assuming that the Sun is the primary gravitational influence and ignoring the gravitational influence of Earth and Mars (since the Sun accounts for 99.8% of the mass of the solar system), determine the change in speed required at the Earth Δv_e (perihelion in the transfer orbit) and the required change in speed at Mars Δv_m (aphelion in the transfer orbit) to accomplish the mission to Mars using a Hohmann transfer. In addition, determine the amount of time τ it would take for orbital transfer. Use 1.989×10^{30} kg for the mass of the Sun, assume that the orbits of Earth and Mars are circular, and assume that the changes in speed are impulsive, that is, they occur instantaneously. In addition, use 150×10^6 km for the radius of Earth's orbit and 228×10^6 km for the radius of Mars' orbit.

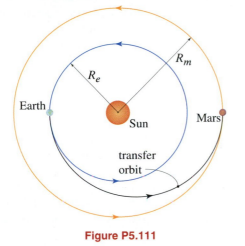

Figure P5.111

Problem 5.112

Use the work-energy principle applied between periapsis P and $r = \infty$, along with the potential energy of the force of gravity given in Eq. (4.35).

(a) Show that a satellite on a hyperbolic trajectory arrives at $r = \infty$ with speed

$$v_\infty = \sqrt{\frac{r_P v_P^2 - 2Gm_B}{r_P}}.$$

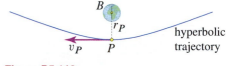

Figure P5.112

(b) In addition, using Eqs. (5.106) and (5.109), show that for a hyperbolic trajectory, $r_P v_P^2 > 2Gm_B$, which means that the square root in the above equation must always yield a real value.

5.5   Mass Flows

In this section we apply the impulse-momentum principles to physical systems that exchange mass with their surrounding environment. We will focus on problems involving the motion of fluids as can be found in oil refinement plants and rocket propulsion. The subject matter we cover is also applicable to other systems that, while not fluids, move in a fluidlike fashion. For example, in some cases the steady motion of bottles along the bottling line shown in Fig. 5.50 can be modeled as a *continuous* flow of mass. We will discover that the impulse-momentum principles offer a direct way to compute the relevant forces present in simple fluid motions.

Open and closed systems

Before we begin, it is important to recall the distinction between *open* and *closed* systems (which were introduced on p. 336), since it is at the core of the derivations presented in this section. A *closed system* does not exchange mass with its environment, whereas an *open system* does exchange mass with its environment. The mass of a closed system is constant. An open system can have constant or variable mass. If a physical system is referred to as a *variable mass system*, then such a system is necessarily open. *It is crucial to keep in mind that the impulse-momentum principles presented earlier in the chapter are only applicable to closed systems.*

Steady flows

Figure 5.51 shows part of a pipe filled with a fluid in motion. To simplify our analysis, we assume that the fluid flow is *steady*, where by steady we mean that the velocity of a fluid particle going through a certain location within the pipe depends only on that location but is otherwise independent of time. On the other hand, the velocity of a particle *can* change as the particle moves from a point to another point within the pipe. In our analysis we also assume that the velocity of the particles at a given cross section is the same for all of the particles in question.

If the velocity of the fluid particles changes, then these particles are accelerating, and by Newton's second law, we conclude that the fluid must be subject to a force, which we now proceed to compute. Because a pipe can be a large structure, we determine only the force exerted on the fluid contained in a given *portion* of the pipe. We will refer to the chosen pipe portion as a *control volume* CV. Referring to Fig. 5.52, we consider a CV defined by the two cross sections A and B. We assume that the fluid enters the CV at A with velocity $\vec{v}_A$ and exits at B with velocity $\vec{v}_B$. We assume that the flow is such that $\vec{v}_A$ and $\vec{v}_B$ are perpendicular to the cross sections A and B, respectively.

Since mass flows in and out of the CV, *a CV is an open system.* When the flow is steady, then the mass of the fluid in the CV remains constant. This implies that if Δm_f (where the subscript f stands for flux) is the mass that flows into the CV at A during any time interval Δt, then Δm_f is also the mass of the fluid that flows out of the CV at B during the same time interval.

The tools we have to relate forces and motion are the impulse-momentum principles, which only apply to closed systems. We will first apply the impulse-momentum principle to a closed system containing the open system of interest,

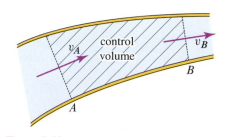

Figure 5.50
Bottles moving along a bottling line. The bottles can be modeled as individual particles/bodies or as mass elements in a continuous mass flow.

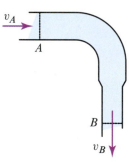

Figure 5.51
A fluid flowing through a curved pipe with variable cross section.

Figure 5.52
A CV corresponding to the portion of a pipe between cross sections A and B.

and then we will "shrink" this closed system to make it coincide with the open system contained therein on an instant-by-instant basis. First, referring to Fig. 5.53, we select a body of fluid which at time t fills the chosen CV and has a small element of mass Δm_f about to enter the CV at A. *This will be our closed system.* The fluid element that is about to enter the CV is chosen to be small enough to allow us to assume that, at time t, $\vec{v}_A$ is the velocity of all of its particles. Second, we let Δt be the time taken by the fluid element that is entering at A to flow into the CV. Because the flow is steady, at time $t + \Delta t$, the body will completely fill the CV between A and B and will also include an element of mass Δm_f, which has exited the CV at B (see Fig. 5.53). Since Δm_f is small, we can assume that all of the particles to the right of B have the same velocity $\vec{v}_B$. Since the selected body is a closed system, we can apply to it Eq. (5.15) on p. 337, which yields

$$\int_t^{t+\Delta t} \vec{F}\, dt = \vec{p}(t + \Delta t) - \vec{p}(t), \qquad (5.135)$$

where $\vec{F}$ is the total external force acting on the system and, using the stated assumptions, the linear momenta $\vec{p}(t)$ and $\vec{p}(t + \Delta t)$ can be written as

$$\vec{p}(t) = \Delta m_f\, \vec{v}_A + \vec{p}_{\text{cv}}(t), \qquad (5.136)$$
$$\vec{p}(t + \Delta t) = \vec{p}_{\text{cv}}(t + \Delta t) + \Delta m_f\, \vec{v}_B. \qquad (5.137)$$

The quantity $\vec{p}_{\text{cv}}$ denotes the momentum of the fluid within the CV. Because the flow is steady, we must have

$$\vec{p}_{\text{cv}}(t) = \vec{p}_{\text{cv}}(t + \Delta t). \qquad (5.138)$$

Substituting Eqs. (5.136)–(5.138) into Eq. (5.135), simplifying, and dividing all terms by Δt, we have

$$\frac{1}{\Delta t}\int_t^{t+\Delta t} \vec{F}\, dt = \frac{\Delta m_f}{\Delta t}(\vec{v}_B - \vec{v}_A). \qquad (5.139)$$

Equation (5.139) holds for a closed system that occupies a volume larger than the selected CV. Recalling that Δt is the time it takes Δm_f to flow into and out of the CV, we see that letting Δt go to zero implies that Δm_f must also go to zero and the fluid in the closed system at time t will fill the CV *exactly!* Therefore, by letting Δt go to zero, Eq. (5.139) yields the result (see the Helpful Information note in the margin)

$$\boxed{\vec{F} = \dot{m}_f(\vec{v}_B - \vec{v}_A),} \qquad (5.140)$$

where the quantity

$$\boxed{\dot{m}_f = \lim_{\Delta t \to 0} \frac{\Delta m_f}{\Delta t}} \qquad (5.141)$$

is called the *mass flow rate* or *mass flux* and measures the amount of mass flowing into and out of the chosen CV per unit time. The result in Eq. (5.140) applies to open systems and was possible because, in going from Eq. (5.139) to Eq. (5.140), we took a limit that forced the chosen closed system to coincide with the open system we wanted to characterize at time t.

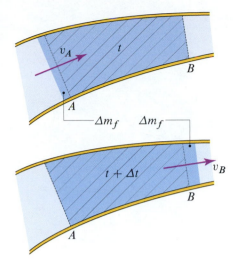

Figure 5.53
A fluid body flowing through a CV defined by the cross sections A and B and behaving as a closed system.

Helpful Information

The limit of the left-hand side of Eq. (5.139). To evaluate the limit as $\Delta t \to 0$ of the left-hand side of Eq. (5.139), we used the fundamental theorem of calculus which states that

$$\lim_{\Delta x \to 0} \frac{1}{\Delta x}\int_{x_0}^{x_0+\Delta x} f(x)\, dx = f(x_0).$$

Common Pitfall

The mass in the CV is constant. Often $\dot{m}_f$ is misinterpreted as the time rate of change of the mass contained in the CV. However, the mass of the fluid within the CV is constant because the flow is steady. The quantity $\dot{m}_f$ is simply a measure of the rate at which mass flows into and out of the CV.

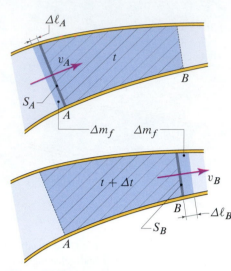

Figure 5.54

Volumes occupied by the fluid element with mass Δm_f upon entering (top) and exiting (bottom) a chosen CV.

Volumetric flow rate

In addition to the mass flux, there is another commonly used measure of the amount of fluid moving through a CV called the *volumetric flow rate*. Referring to Fig. 5.54, we see that, at time t, the volume occupied by the fluid element of mass Δm_f is approximately given by $\Delta \ell_A S_A$, where $\Delta \ell_A$ is the pipe length occupied by Δm_f and S_A is the area of the cross section at A. Similarly, at time $t + \Delta t$, the volume occupied by the fluid element of mass Δm_f is $\Delta \ell_B S_B$. Because the fluid motion is steady, the quantity Δm_f is the same at A and B. Therefore, letting ρ_A and ρ_B be the fluid density at A and B, respectively, we have

$$\Delta m_f = \rho_A \, \Delta \ell_A \, S_A = \rho_B \, \Delta \ell_B \, S_B. \qquad (5.142)$$

Since we have assumed that $\vec{v}_A$ and $\vec{v}_B$ are perpendicular to the cross sections A and B, respectively, we have

$$\lim_{\Delta t \to 0} \frac{\Delta \ell_A}{\Delta t} = v_A \quad \text{and} \quad \lim_{\Delta t \to 0} \frac{\Delta \ell_B}{\Delta t} = v_B, \qquad (5.143)$$

where v_A and v_B are the values of the speed of the fluid at A and B, respectively. If S is the area of a generic cross section along the pipe and if v is the fluid speed at that cross section, we define the *volumetric flow rate* as the quantity

$$\boxed{Q = vS.} \qquad (5.144)$$

Dividing Eq. (5.142) by Δt, letting $\Delta t \to 0$, and using the definition in Eq. (5.144), we have

$$\boxed{\dot{m}_f = \rho_A Q_A = \rho_B Q_B,} \qquad (5.145)$$

where Q_A and Q_B are the volume flow rates at A and B, respectively.

Moment acting on the fluid

Sometimes it is useful to relate the change in the fluid's angular momentum, computed with respect to a chosen moment center, to the corresponding moment acting on the fluid. Referring to Fig. 5.55, to compute this moment, we choose as moment center the *fixed* point P, we select a body of fluid in the same way as was done for the force calculation, and then we apply the angular impulse-momentum principle given in Eq. (5.80) on p. 393 between times t and $t + \Delta t$. Doing so gives

$$\vec{h}_P(t) + \int_t^{t+\Delta t} \vec{M}_P \, dt = \vec{h}_P(t + \Delta t), \qquad (5.146)$$

where $\vec{h}_P$ is the angular momentum of the selected fluid body and $\vec{M}_P$ is the moment we intend to compute. Because we have assumed that all of the particles in the volume elements of mass Δm_f are moving with velocity $\vec{v}_A$ at time t and $\vec{v}_B$ at time $t + \Delta t$, we have

$$\vec{h}_P(t) = \vec{r}_{C/P} \times \Delta m_f \, \vec{v}_A + \left(\vec{h}_P\right)_{\text{cv}}, \qquad (5.147)$$

$$\vec{h}_P(t + \Delta t) = \left(\vec{h}_P\right)_{\text{cv}} + \vec{r}_{D/P} \times \Delta m_f \, \vec{v}_B, \qquad (5.148)$$

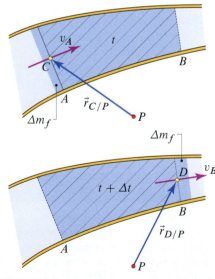

Figure 5.55

Choice of moment center P for the determination of the moment acting on the fluid contained in the CV (shaded area).

where C and D are the centers of the cross sections A and B, respectively, $\vec{r}_{C/P}$ and $\vec{r}_{D/P}$ are the positions of C and D with respect to P, respectively, and $\left(\vec{h}_P\right)_{cv}$ is the angular momentum with respect to P of the fluid contained in the CV. Since the flow is steady, $\left(\vec{h}_P\right)_{cv}$ is a constant. Substituting Eqs. (5.147) and (5.148) into Eq. (5.146), simplifying, and rearranging terms, we have

$$\frac{1}{\Delta t}\int_t^{t+\Delta t}\vec{M}_P\,dt = \frac{\Delta m_f}{\Delta t}(\vec{r}_{D/P}\times\vec{v}_B - \vec{r}_{C/P}\times\vec{v}_A), \qquad (5.149)$$

where we have divided all terms by Δt. By proceeding as in the case of the force calculation, i.e., letting $\Delta t \to 0$, Eq. (5.149) yields

$$\boxed{\vec{M}_P = \dot{m}_f(\vec{r}_{D/P}\times\vec{v}_B - \vec{r}_{C/P}\times\vec{v}_A).} \qquad (5.150)$$

Variable mass flows and propulsion

Referring to Fig. 5.56, consider a body A (the rocket) propelled via the continuous ejection of some material B (combustion gas). Since B used to be part of A before ejection, the mass of A changes with time so that A is a *variable mass system*.* We want to determine the force acting on A, and we will do as was done in the case of CVs. That is, first we will apply the impulse-momentum principle to a closed system containing the open system of interest, and then we will "shrink" this closed system to make it coincide with the open system contained therein on an instant-by-instant basis.

Referring to Fig. 5.57, we consider a body with mass $m(t)$ at time t such that almost all of its particles travel with a velocity $\vec{v}(t)$. Some particles, which at time t have a total mass that is negligible with respect to $m(t)$, are being ejected from the body. After an amount of time Δt, the body will have lost an amount of mass Δm_o (the subscript o stands for outflow), and we write

$$m(t + \Delta t) = m(t) - \Delta m_o. \qquad (5.151)$$

We assume that all the particles contributing to Δm_o have the same *inertial* velocity $\vec{v} + \vec{v}_o$, where $\vec{v}_o$ is the *relative* velocity of the particles in question with respect to the main body.

As long as the physical system we analyze consists of both the particles of mass Δm_o and the main body of mass m, our system is a closed system. Applying to this system the impulse-momentum principle given in Eq. (5.15) on p. 337, between times t and $t + \Delta t$, we have

$$\int_t^{t+\Delta t}\vec{F}\,dt = \vec{p}(t + \Delta t) - \vec{p}(t), \qquad (5.152)$$

where $F(t)$ is the total external force acting on the system and $\vec{p}(t)$ is the total momentum of the system. Using the stated assumptions, we have

$$\vec{p}(t) = m(t)\vec{v}(t), \qquad (5.153)$$

$$\vec{p}(t + \Delta t) = m(t + \Delta t)\vec{v}(t + \Delta t) \\ + \Delta m_o[\vec{v}(t + \Delta t) + \vec{v}_o(t + \Delta t)]. \qquad (5.154)$$

---

* As mentioned at the beginning of this section on p. 426, a variable mass system is *necessarily* open.

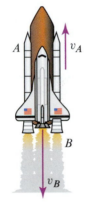

Figure 5.56
A rocket A being propelled by the ejection of combustion gases, which we have labeled B.

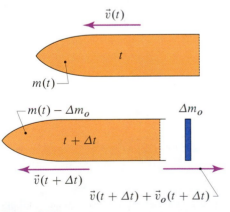

Figure 5.57
A variable mass system that ejects a material element of mass Δm_o over the time interval Δt.

Substituting Eq. (5.151) into Eq (5.154), we have

$$\vec{p}(t + \Delta t) = [m(t) - \Delta m_o]\vec{v}(t + \Delta t) + \Delta m_o[\vec{v}(t + \Delta t) + \vec{v}_o(t + \Delta t)]$$
$$= m(t)\vec{v}(t + \Delta t) + \Delta m_o\, \vec{v}_o(t + \Delta t). \qquad (5.155)$$

Substituting Eqs. (5.153) and (5.155) into Eq. (5.152) and collecting the term $m(t)$, we have

$$\int_t^{t+\Delta t} \vec{F}\, dt = m(t)[\vec{v}(t + \Delta t) - \vec{v}(t)] + \Delta m_o\, \vec{v}_o(t + \Delta t). \qquad (5.156)$$

Dividing Eq. (5.156) by Δt, we obtain

$$\frac{1}{\Delta t}\int_t^{t+\Delta t} \vec{F}\, dt = m(t)\frac{\vec{v}(t + \Delta t) - \vec{v}(t)}{\Delta t} + \frac{\Delta m_o}{\Delta t}\vec{v}_o(t + \Delta t). \qquad (5.157)$$

By the definition of time derivative, we have

$$\lim_{\Delta t \to 0}\frac{\vec{v}(t + \Delta t) - \vec{v}(t)}{\Delta t} = \vec{a}(t) \quad \text{and} \quad \lim_{\Delta t \to 0}\frac{\Delta m_o}{\Delta t} = \dot{m}_o(t), \qquad (5.158)$$

where $\vec{a}(t)$ is the acceleration of the main body at time t and $\dot{m}_o(t)$ (with $\dot{m}_o \geq 0$) is the rate at which mass flows out of the main body. In addition, by the fundamental theorem of calculus, we have

$$\lim_{\Delta t \to 0}\frac{1}{\Delta t}\int_t^{t+\Delta t} \vec{F}\, dt = \vec{F}. \qquad (5.159)$$

Therefore, taking the limit as $\Delta t \to 0$ of the terms in Eq. (5.157) and using Eqs. (5.158) and (5.159), we obtain

$$\vec{F} = m\vec{a} + \dot{m}_o\vec{v}_o, \qquad (5.160)$$

where all the terms in Eq. (5.160) are evaluated at time t. Equation (5.160) applies only to systems that lose mass. However, by following steps analogous to those that gave us Eq. (5.160) and referring to Fig. 5.58, it is not difficult to show that if the system also gains mass at the rate $\dot{m}_i$ (the subscript i stands for inflow), with $\dot{m}_i \geq 0$ and such that the inflowing mass has a velocity $\vec{v}_i$ *relative* to the main body, then Eq. (5.160) can be generalized to

$$\boxed{\vec{F} = m\vec{a} + \dot{m}_o\vec{v}_o - \dot{m}_i\vec{v}_i,} \qquad (5.161)$$

where we note that the contribution of the inflowing mass has a sign opposite to that of the outflowing mass.

Equation (5.161) is an important result that can be viewed as the generalization of Newton's second law to an open system with variable mass. We arrived at Eq. (5.161), starting from a balance principle applied to a *closed* system, whose mass can only be constant. This was possible because, in going from Eq. (5.157) to Eq. (5.160), we took a limit that forced the chosen closed system to coincide with the variable mass system we wanted to characterize at the time instant t.

In the field of rocket propulsion $\dot{m}_i = 0$, and it is often common to move the term $\dot{m}_o\vec{v}_o$ in Eq. (5.161) to the left-hand side of the equation and then to refer to the term $-\dot{m}\vec{v}_o$ as the *thrust force* provided by the propulsion system. In jet propulsion, we have both mass outflow and mass inflow so that the thrust is given by the term $\dot{m}_i\vec{v}_i - \dot{m}_o\vec{v}_o$.

Figure 5.58
A plane with a jet engine. The airplane is taking in air with a mass flow rate $\dot{m}_i$ while combustion gases are ejected with a mass flow rate $\dot{m}_o$. The vector $\vec{v}_i$ is the velocity of the inflowing air *relative* to the plane. The vector $\vec{v}_o$ is the velocity of the outflowing combustion gases *relative* to the plane.

End of Section Summary

In this section we have considered mass flows. Specifically, we have considered (1) *steady* mass flows, in which a fluid moves through a conduit with a velocity that depends only on the position within the conduit, and (2) *variable* mass flows, such as the flow of combustion gases out of a rocket.

Steady flows. Given the control volume (CV) shown in Fig. 5.59, where by *control volume* we mean a portion of a conduit delimited by two cross sections, we showed that, in the case of a steady flow, the total external force $\vec{F}$ acting on the fluid in the CV is

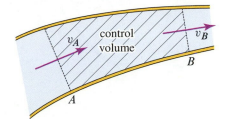

Figure 5.59
Figure 5.52 repeated. A CV corresponding to the portion of a pipe between cross sections A and B.

> **Eq. (5.140), p. 427**
> $$\vec{F} = \dot{m}_f (\vec{v}_B - \vec{v}_A),$$

where as long as the cross sections are perpendicular to the flow velocity, $\dot{m}_f$ is the *mass flow rate*, i.e., the amount of mass flowing through a cross section per unit time, and where $\vec{v}_A$ and $\vec{v}_B$ are the flow velocities at the cross sections A and B, respectively. In addition to the mass flow rate, we defined the *volumetric flow rate* as the quantity

> **Eq. (5.144), p. 428**
> $$Q = vS,$$

where v is the speed of the fluid at a given cross section and S is the area of the cross section in question. We showed that

> **Eq. (5.145), p. 428**
> $$\dot{m}_f = \rho_A Q_A = \rho_B Q_B,$$

where ρ_A and ρ_B are the values of the mass density of the fluid at A and B, respectively. Referring to Fig. 5.60, we also showed that, given a fixed point P, the total moment $\vec{M}_P$ acting on the fluid in the CV is

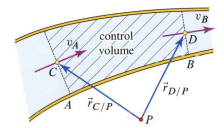

Figure 5.60
A fluid flowing through a CV along with a choice of moment center P for the calculation of angular momenta and moments.

> **Eq. (5.150), p. 429**
> $$\vec{M}_P = \dot{m}_f (\vec{r}_{D/P} \times \vec{v}_B - \vec{r}_{C/P} \times \vec{v}_A),$$

where C and D are the centers of the cross sections A and B, respectively.

Variable mass flows. With reference to Fig. 5.61, for a body with time varying mass $m(t)$ due to an inflow of mass with rate $\dot{m}_i$ and an outflow of mass with the rate $\dot{m}_o$, the total external force acting on the body is given by

Figure 5.61
Figure 5.58 repeated. A plane with a jet engine. The airplane is taking in air with a mass flow rate $\dot{m}_i$ while combustion gases are ejected with a mass flow rate $\dot{m}_o$. The vector $\vec{v}_i$ is the velocity of the inflowing air *relative* to the plane. The vector $\vec{v}_o$ is the velocity of the outflowing com-

> **Eq. (5.161), p. 430**
> $$\vec{F} = m\vec{a} + \dot{m}_o \vec{v}_o - \dot{m}_i \vec{v}_i,$$

where $\vec{a}$ is the acceleration of the main body, $\vec{v}_o$ is the *relative* velocity of the outflowing mass with respect to the main body, and $\vec{v}_i$ is the *relative* velocity of the inflowing mass, again relative to the main body.

bustion gases *relative* to the plane.

EXAMPLE 5.17   *Force of an Open Water Jet*

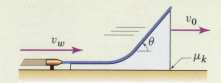

Figure 1

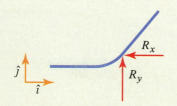

Figure 2
FBD of the incline. The force F is the friction force due to the sliding relative to the ground.

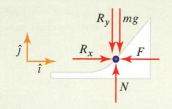

Figure 3
FBD of the water jet. The forces R_x and R_y are equal and opposite to those indicated in Fig. 2 to comply with Newton's third law.

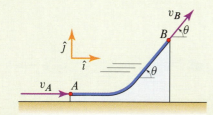

Figure 4
Velocities of the steady flow at A and B as perceived by an observer moving with the incline.

A water jet is let out of a nozzle attached to the ground. The jet has a constant mass flow rate and a speed $v_w = 65\,\text{ft/s}$ relative to the nozzle. The jet strikes a 25 lb incline and causes it to slide at a constant speed $v_0 = 5.5\,\text{ft/s}$. The kinetic coefficient of friction between the incline and the ground is $\mu_k = 0.43$. Neglecting the effect of gravity and air resistance on the water flow, as well as friction between the water jet and the incline, determine the mass flow rate of the water jet at the nozzle if $\theta = 50°$.

SOLUTION

Road Map & Modeling   Modeling the incline as a particle and referring to the FBD in Fig. 2, we represent the effect of the water jet on the incline via the reaction forces R_x and R_y. To solve the problem, we need to relate R_x and R_y to the mass flow rate out of the nozzle and then use these relations when applying Newton's second law to the incline. In this way we will be able to relate the mass flow rate to the friction force opposing the motion of the incline. The key to determining R_x and R_y is to realize that if the friction between the water jet and the incline is negligible, then there is no force that will slow down the flow of water over the incline. This fact, along with the fact that the incline moves at a constant velocity, allows us to model the flow of water over the incline as *steady* and with a constant speed relative to the incline.

────── **Determination of R_x and R_y in Terms of Mass Flow Rate** ──────
Governing Equations

Balance Principles   Here we apply the force balance relation for CVs in Eq. (5.140) on p. 427. The terms in this equation must be measured using an *intertial frame of reference*. Because the incline moves at a constant velocity, a reference frame attached to the incline is inertial.* Choosing such a frame and referring to the water jet FBD in Fig. 3, we choose our CV to be the volume occupied by the water flowing over the top surface of the incline. Although this CV is moving with respect to the ground, our choice is acceptable because such a CV is *stationary* with respect to the chosen inertial frame and, as discussed above, the water flow is *steady* over the incline. Then Eq. (5.140) in component form yields (see Fig. 4)

$$\sum F_x: \quad -R_x = \dot{m}_f(v_{Bx} - v_{Ax}), \tag{1}$$

$$\sum F_y: \quad R_y = \dot{m}_f(v_{By} - v_{Ay}), \tag{2}$$

where $\dot{m}_f$ is the mass flow rate that goes past the cross section at A and $\vec{v}_A$ and $\vec{v}_B$ are the velocities of the water flow at A and B, respectively. We note again that $\dot{m}_f$, $\vec{v}_A$, and $\vec{v}_B$ are measured by the inertial observer who moves with the incline.

Force Laws   All forces are accounted for on the FBD.

Kinematic Equations   Using relative kinematics, an observer moving with the incline measures

$$v_{Ax} = v_w - v_0 \quad \text{and} \quad v_{Ay} = 0, \tag{3}$$

where v_w and v_0 are the speeds of the water jet and of the incline measured relative to the ground. Since we are neglecting the friction between the water and the incline, we must have $|\vec{v}_A| = |\vec{v}_B|$. Hence, the components of the velocity of the water at B are

$$v_{Bx} = (v_w - v_0)\cos\theta \quad \text{and} \quad v_{By} = (v_w - v_0)\sin\theta. \tag{4}$$

─────────────────────────────

*This statement is predicated on the implicit assumption that the ground over which the incline slides can be chosen as an inertial reference frame (see discussion on p. 188).

Let $(\dot{m}_f)_{nz}$ be the mass flow rate measured at the nozzle. This quantity is the unknown we want to determine. We assume that the water jet has a constant cross section even when in contact with the incline. Then, letting S denote the flow cross-sectional area, we see from Eq. (5.144) the volumetric flow rate at the nozzle is $Q_{nz} = v_w S$, and therefore the mass flow rate at the nozzle is

$$(\dot{m}_f)_{nz} = \rho S v_w, \tag{5}$$

where ρ denotes the mass density of the water. Similarly, the mass flow rate $\dot{m}_f$ measured by an observer moving with the incline is

$$\dot{m}_f = \rho S v_A = \rho S (v_w - v_0) \quad \Rightarrow \quad \dot{m}_f = (\dot{m}_f)_{nz}(v_w - v_0)/v_w, \tag{6}$$

where we have used Eq. (5) in the first of Eqs. (6).

Computation   Substituting Eqs. (3), (4), and the last of Eqs. (6) into Eqs. (1) and (2) gives

$$R_x = \frac{(\dot{m}_f)_{nz}}{v_w}(1 - \cos\theta)(v_w - v_0)^2 \quad \text{and} \quad R_y = \frac{(\dot{m}_f)_{nz}}{v_w}\sin\theta\,(v_w - v_0)^2. \tag{7}$$

Discussion & Verification   Recalling that the mass flow rate has dimensions of mass over time, we know that Eqs. (7) are dimensionally correct. In addition, given that the right-hand sides of Eqs. (7) have a positive sign under all circumstances, we see that the directions of R_x and R_y are as expected.

────────── $\vec{F} = m\vec{a}$ **for the Incline and Determination of** $(\dot{m}_f)_{nz}$ ──────────

Governing Equations

Balance Principles   Using the FBD in Fig. 2, the application of Newton's second law to the incline yields

$$\sum F_x: \qquad R_x - F = ma_x, \tag{8}$$

$$\sum F_y: \quad -R_y - mg + N = ma_y. \tag{9}$$

Force Laws   Since the incline is sliding, we have

$$F = \mu_k N. \tag{10}$$

Kinematic Equations   Because the incline moves with constant velocity, we have

$$a_x = 0 \quad \text{and} \quad a_y = 0. \tag{11}$$

Computation   Substituting Eqs. (11) into Eqs. (8) and (9), solving for F and N, and then substituting the result into Eq. (10) yield

$$R_x = \mu_k(R_y + mg). \tag{12}$$

Substituting Eqs. (7) into Eq. (12) and solving for $(\dot{m}_f)_{nz}$, we have

$$\boxed{(\dot{m}_f)_{nz} = \frac{\mu_k mg\, v_w}{(v_w - v_0)^2(1 - \cos\theta - \mu_k\sin\theta)} = 7.10\,\text{slug/s.}} \tag{13}$$

Discussion & Verification   Since the terms μ_k and $1 - \cos\theta - \mu_k\sin\theta$ in Eq. (13) are nondimensional and since the term $g v_w/(v_w - v_0)^2$ has dimensions of 1 over time, our result is dimensionally correct. We also note that our result is directly proportional to the weight of the incline as well as to the friction coefficient. This is reasonable since, keeping v_w and v_0 fixed, we expect that more water mass per unit time is needed to move a heavier incline over a rougher surface. Hence, overall our solution appears to be correct.

EXAMPLE 5.18   *Geometry of Fluid Motion and Structural Loads*

Figure 1

An example of intricate pipeline geometry in a refinement plant.

Pipelines can be quite intricate (see Fig. 1). The fluid going through a bend and/or a change in cross section can exert significant structural loads on the line. Referring to Fig. 2, consider two straight pipes connected by a flanged diverter/reducer of length $\ell = 88$ in., height $h = 62$ in., and internal volume $V = 34\,\text{ft}^3$. Suppose that the flow is steady, the fluid specific weight is $\gamma = 42.5\,\text{lb/ft}^3$ (this is typical of gasoline), and the cross sections at A and B are circular with radii $R_A = 13$ in. and $R_B = 9$ in., respectively. Assume that the center of mass G of the fluid between A and B is located as shown with $d = \ell/2$ and $q = 33$ in. In addition, let $p_A = 1400$ psi and $p_B = 1390$ psi be known measurements of the static pressure at A and B, respectively. If the fluid speed at A is $v_A = 6\,\text{ft/s}$, determine the loads that the fluid exerts on the diverter/reducer. Finally, neglect the weight of the diverter/reducer and sketch its FBD, showing the internal forces at A and B.

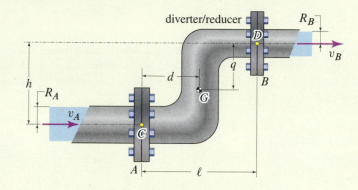

Figure 2. A diverter/reducer connecting two straight pipe segments in the vertical plane. Points C and D indicate the centers of the cross sections A and B, respectively.

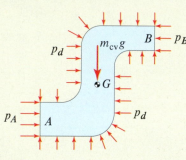

Figure 3

FBD of the fluid moving through the chosen CV, which was taken to be the interior volume of the diverter.

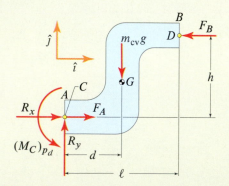

Figure 4

FBD of the fluid within the CV obtained using the concept of equivalent force system.

SOLUTION

Road Map & Modeling   The CV is the region occupied by the fluid between the cross sections A and B. Referring to Fig. 3, the fluid in question is subject to the pressure distributions p_A and p_B due to the fluid outside the CV. We assume that p_A and p_B are uniform over A and B, respectively. The fluid in the CV is also subject to the pressure distribution p_d due to the contact with the inner walls of the diverter. Finally, the fluid in the CV is subject to gravity, which is represented by the weight $m_{cv}g$, applied at the fluid's center of mass G. Although we do not have a detailed knowledge of p_d, we can describe the overall effect of p_d via the concept of equivalent force system.* Using this concept, the force system acting on the fluid in the CV can be represented as shown in Fig. 4. The forces F_A and F_B, applied as shown, are equivalent to the pressure distributions p_A and p_B, respectively. The force system equivalent to the pressure distribution p_d consists of the forces R_x and R_y, as well as the moment $(M_C)_{p_d}$, where C has been chosen as the moment center because its location is known (we could have chosen some other convenient reference point such as G or D). It is this force system that we need to compute, and we will do so by applying the force and moment balance for steady flows. Because R_x, R_y, and $(M_C)_{p_d}$ describe the action of the diverter on the fluid in the CV, in sketching the FBD of the diverter we need to include these forces and moment but with opposite sign to abide by Newton's third law.

_____
* See your statics textbook for the concept of equivalent force system.

Governing Equations

Balance Principles   Referring to the FBD in Fig. 4, choosing C as the moment center, and recalling that $m_{vc}g = \gamma V$, the force and impulse-momentum principles in Eqs. (5.140) and (5.150), in component form, give

$$\sum F_x: \qquad R_x + F_A - F_B = \dot{m}_f(v_{Bx} - v_{Ax}), \qquad (1)$$

$$\sum F_y: \qquad R_y - \gamma V = \dot{m}_f(v_{By} - v_{Ay}), \qquad (2)$$

$$\sum M_C: \quad (M_C)_{p_d} - \gamma V d + F_B h = \dot{m}_f(v_{By}\ell - v_{Bx}h). \qquad (3)$$

Force Laws   Since the cross sections at A and B are circular with radii R_A and R_B, respectively, and since we have assumed that the pressure distributions over A and B are uniform, we have

$$F_A = \pi R_A^2 p_A \quad \text{and} \quad F_B = \pi R_B^2 p_B. \qquad (4)$$

Kinematic Equations   Based on the flow depicted in Fig. 2, we have

$$v_{Ax} = v_A, \quad v_{Ay} = 0, \quad v_{Bx} = v_B, \quad v_{By} = 0. \qquad (5)$$

Furthermore, applying Eqs. (5.144) and (5.145), we must have

$$\dot{m}_f = (\gamma/g)\pi R_A^2 v_A = (\gamma/g)\pi R_B^2 v_B \quad \Rightarrow \quad v_B = v_A R_A^2/R_B^2. \qquad (6)$$

Computation   Substituting Eqs. (4)–(6) into Eqs. (1)–(3), we see the force system that is equivalent to the pressure distribution p_d is given by

$$R_x = \pi\left(p_B R_B^2 - p_A R_A^2\right) + \frac{\gamma\pi v_A^2 R_A^2(R_A^2 - R_B^2)}{g R_B^2} = -389\times10^3 \text{ lb}, \qquad (7)$$

$$R_y = gV = 1440 \text{ lb}, \qquad (8)$$

$$(M_C)_{p_d} = \gamma V d - \pi p_B R_B^2 h - \frac{\gamma\pi v_A^2 R_A^4 h}{g R_B^2} = -1.82\times10^6 \text{ ft·lb}. \qquad (9)$$

Now that we have R_x, R_y, and $(M_C)_{p_d}$, by applying Newton's third law, the FBD for the diverter is that given in Fig. 5, where the internal force system over the cross section A consists of the forces N_A (tension), V_A (shear force), and M_{Ci} (bending moment). Similarly, the internal force system at B is given by N_B, V_B, and M_{Di}.

Discussion & Verification   The dimensions of pressure and mass density are force per unit area and mass over length cubed, respectively. Hence our results are dimensionally correct. The result in Eq. (7) makes sense because the fact that R_x is negative is consistent with the idea that the fluid motion in the x direction is being hindered by the presence of the bend in the line. The result in Eq. (8) also makes sense since it confirms that the diverter is supporting the weight of the fluid in the CV. To explain the result in Eq. (9), recall that if there is no flow (i.e., $v_A = 0$), then the diverter must provide a positive moment to balance the weight. However, if $v_A \neq 0$ and we neglect the weight, then common experience tells us that the flow would cause a counterclockwise rotation of the system and the diverter must exert a moment in the clockwise direction to prevent the rotation in question. This means that both positive and negative moments are to be expected, and the sign of the result in Eq. (9) tells us that, in our case, the moment due to the fluid motion has the greater effect.

🔎 **A Closer Look**   Referring to Fig. 5, if end B of the diverter were free, then the internal actions at B would be equal to zero and an elementary equilibrium calculation would show that we must have $N_A = -R_x$, $V_A = R_y$, and $(M_{C_A})_i = -(M_{C_A})_{p_d}$. That is, if one end is free, we can compute the internal actions at the other end directly in terms of forces computed from the force balance for CVs.

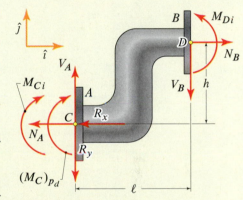
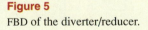

Figure 5
FBD of the diverter/reducer.

EXAMPLE 5.19   *Hovering Using a Jet Pack*

Figure 1
A pilot with a jet pack.

A jet pack is a rocket propulsion device worn on a person's back that allows the person to become airborne and fly (see Fig. 1). Suppose that the jet pack can hold 75 lb of fuel. Suppose further that when there is no fuel in the pack, the combined weight of the pilot and the jet pack is 180 lb. It is assumed that, once in operation, the outflow speed v_o of the ejected material relative to the pack is constant. Neglecting the amount of time it takes for the pilot to start hovering a few feet off the ground, and assuming that the jets are oriented in the direction of gravity, determine v_o so that the fuel will be completely spent after the pilot hovers for 45 s.

SOLUTION

Road Map & Modeling   The pilot and the pack form a simple variable mass system. We will therefore apply to this system the force balance given in Eq. (5.161) while enforcing the requirement that the mass outflow rate be equal to the time rate of decrease of the system's mass. By doing so we will be able to relate the speed v_o to the time it takes to exhaust all of the fuel. When we use Eq. (5.161), the thrust due to the ejection of matter from the pack is *not* considered an external force. Hence, given that the pilot is simply hovering, the system's FBD is that shown in Fig. 2, in which we have only included the system's weight. After solving the problem, we will discuss another approach to the solution of propulsion problems according to which the thrust acting on the system is shown on the FBD and, at the same time, the force balance law is made to take on the form $\vec{F} = m\vec{a}$.

Governing Equations

Balance Principles   Observing that there are no forces acting in the horizontal direction, we see that the only significant component of the force balance law is that in the y direction. Hence we have

$$-mg = ma_y + \dot{m}_o\vec{v}_o \cdot \hat{j}, \tag{1}$$

where m is the time-varying combined mass of the pilot and of the pack, $\dot{m}_o$ is the time rate at which mass is being ejected from the pack, $\vec{v}_o \cdot \hat{j}$ is the y component of the velocity of the ejected matter relative to the main system, and we have accounted for the fact that the system does not gain mass (i.e., $\dot{m}_i = 0$).

Force Laws   All forces are accounted for on the FBD.

Kinematic Equations   Since the pilot (with the pack) is hovering, the system is stationary with respect to the ground, which is chosen as our inertial frame. Therefore we must have

$$a_y = 0 \quad \text{and} \quad \vec{v}_o = -v_o\,\hat{j}. \tag{2}$$

In addition, as already observed, the mass of the system decreases at the same rate at which mass is ejected from the pack, so we have

$$\dot{m} = -\dot{m}_o. \tag{3}$$

Computation   Substituting Eqs. (2) and (3) into Eq. (1), we have

$$-mg = v_o\dot{m}. \tag{4}$$

Recalling that $\dot{m} = dm/dt$, Eq. (4) can be written as

$$-\frac{g}{v_o}\,dt = \frac{dm}{m}. \tag{5}$$

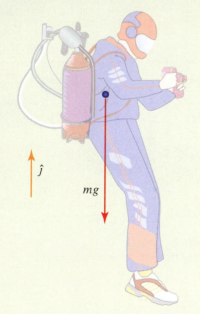

Figure 2
FBD of the system consisting of the pilot and the jet pack.

Integrating this equation from $t = 0$ to the final time $t_f = 45\,\text{s}$, we have

$$-\int_0^{t_f} \frac{g}{v_o}\,dt = \int_{m(0)}^{m(t_f)} \frac{dm}{m} \quad \Rightarrow \quad -\frac{g}{v_o}t_f = \ln\frac{m(t_f)}{m(0)}. \tag{6}$$

Recalling that $m(0) = (180 + 75)\,\text{lb}$ is the combined mass of the pilot, the pack, and 75 lb of fuel, and that $m(t_f) = 180\,\text{lb}$ is the combined mass of the pilot and the empty pack, we can solve Eq. (6) for v_o to obtain

$$\boxed{v_o = -\frac{gt_f}{\ln\left[m(t_f)/m(0)\right]} = \frac{gt_f}{\ln\left[m(0)/m(t_f)\right]} = 4160\,\text{ft/s}.} \tag{7}$$

Discussion & Verification   Since the argument of the natural logarithm in Eq. (7) is nondimensional, and since the product of acceleration and time has the dimensions of length over time, the result in Eq. (7) has the correct dimensions. As far as the numerical value obtained for v_o is concerned, this result is not far from what we would expect from simple monopropellant rocket engines whose exhaust speeds are typically of the order of 5600 ft/s (although they can get close to 10,000 ft/s).

✎ **A Closer Look**   The problem discussed in this example can be approached by writing the force balance law as $\vec{F} = m\vec{a}$, which is meant to resemble Newton's second law.* If we approach the force balance for a variable mass system using the expression $\vec{F} = m\vec{a}$, then the force $\vec{F}$ includes both those forces that would be considered external according to a strict interpretation of the impulse-momentum principle and those forcelike terms that result from the inflow and outflow of mass. Therefore, our FBD would have been that in Fig. 3, where $-\dot{m}_o\vec{v}_o$ is the thrust force provided by the rocket engine.

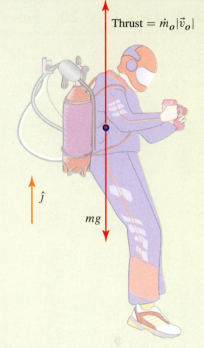

Thrust $= \dot{m}_o|\vec{v}_o|$

$\hat{j}$

mg

Figure 3
Alternate FBD of the system. The force balance law that must be used with this system is $\vec{F} = m\vec{a}$, where $\vec{F} = -mg\,\hat{j} - \dot{m}_o\vec{v}_o$.

---

* Writing $\vec{F} = m\vec{a}$ for variable mass systems *cannot* be considered to be the same as applying Newton's second law. This is so because *Newton's second law cannot be applied to variable mass systems.* If $\vec{F} = m\vec{a}$ is applied to a variable mass system, the only correct interpretation that can be given is that what is being applied is actually Eq. (5.161), with $\vec{F} = \vec{F}_{\text{ext}} - \dot{m}_o\vec{v}_o + \dot{m}_i\vec{v}_i$, where $\vec{F}_{\text{ext}}$ is the total external force applied to the system according to a strict interpretation of the impulse-momentum principle.

E X A M P L E 5.20   *Forces in a Falling String*

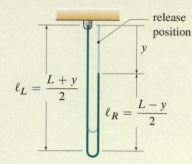

Figure 1
A string falling vertically down.

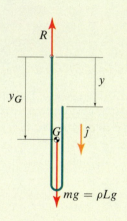

Figure 2
FBD of the string as a whole. The weight of the string has been placed at the string's center of mass, which has been denoted by G.

In Example 4.11 on p. 309, we discovered that the velocity of the free end of a falling inextensible string of length L, released from rest, is given by (see Fig. 1)

$$\dot{y} = \sqrt{gy\frac{2L-y}{L-y}}, \tag{1}$$

where g is the acceleration due to gravity and y is the position of the free end of the string. Letting ρ be the string's mass density per unit length, use Eq. (1) to determine the reaction force R at the ceiling as a function of y by modeling the whole string as a closed system and by modeling the two branches to the right and left of the bend as variable mass systems.

SOLUTION

Modeling the Whole String as a Closed System

Road Map & Modeling   If we model the whole string as a closed system, then the string's FBD is that shown in Fig. 2, where the only force other than R is the string's weight (this is consistent with the solution of Example 4.11). Since the system is closed, we can apply the impulse-momentum principle as given in Eq. (5.14) on p. 337.

Governing Equations

Balance Principles   Using the FBD in Fig. 2 and Eq. (5.14), we obtain

$$\sum F_y: \quad \rho L g - R = \rho L a_G, \tag{2}$$

where a_G is the acceleration of the string's mass center and ρL is the string's mass.

Force Laws   All forces are accounted for on the FBD.

Kinematic Equations   Since $a_G = \ddot{y}_G$, we first find y_G via Eq. (3.71) on p. 237. Recalling that the mass of the string is ρL and referring to Fig. 1, we have

$$\rho L y_G = \rho \ell_L(\ell_L/2) + \rho \ell_R(y + \ell_R/2) \quad \Rightarrow \quad y_G = \frac{1}{4L}(L^2 + 2Ly - y^2), \tag{3}$$

where $\rho \ell_L$ and $\rho \ell_R$ are the masses of the left and right branches of the string, respectively. Differentiating the final result in Eq. (3) twice with respect to time, we obtain

$$a_G = \frac{1}{2L}[\ddot{y}(L-y) - \dot{y}^2]. \tag{4}$$

Using Eq. (1) along with the chain rule, we have

$$\ddot{y} = \frac{d\dot{y}}{dy}\dot{y} = \frac{\sqrt{g}(2L^2 - 2Ly + y^2)}{2(L-y)^{3/2}\sqrt{y(2L-y)}}\dot{y} \quad \Rightarrow \quad \ddot{y} = \frac{g}{2}\left[1 + \frac{L^2}{(L-y)^2}\right]. \tag{5}$$

Substituting Eq. (1) and the final result in Eq. (5) into Eq. (4), after simplifying we obtain

$$a_G = \frac{g}{4}\left(3 - 3\frac{y}{L} - \frac{L}{L-y}\right). \tag{6}$$

Computation   Substituting Eqs. (6) into Eq. (2) and solving for R, we have

$$\boxed{R = \frac{\rho L g}{4}\left(1 + 3\frac{y}{L} + \frac{L}{L-y}\right).} \tag{7}$$

──────────── **Variable Mass Systems Modeling** ────────────

Road Map & Modeling   We can model the left and right branches of the string as variable mass systems that exchange mass with each other. Specifically, the left branch gains mass at the expense of the right branch. In this case, separating these systems with a cut at the bend, we have the FBDs in Fig. 3 (see the Helpful Information marginal note for further comments on these FBDs). Then we can apply to each branch the force balance for variable mass systems given in Eq. (5.161).

Governing Equations

Balance Principles   Using the FBDs in Fig. 3 along with the force balance for variable mass systems, for the left and right branches of the string we have, respectively,

$$\sum F_{yL}: \quad \rho\ell_L g - R = \rho\ell_L a_{yL} - \dot{m}_i \vec{v}_i \cdot \hat{j}, \tag{8}$$

$$\sum F_{yR}: \quad \rho\ell_R g = \rho\ell_R a_{yR} + \dot{m}_o \vec{v}_o \cdot \hat{j}, \tag{9}$$

where $\dot{m}_i$ is the time rate of mass gain of the left branch, $\vec{v}_i$ is the velocity of the mass joining the left branch relative to the velocity of the left branch itself, $\dot{m}_o$ is the time rate of mass loss of the right branch, and $\vec{v}_o$ is the velocity of the mass leaving the right branch relative to the right branch itself.

Force Laws   All forces are accounted for on the FBDs.

Kinematic Equations   Because of inextensibility, all the mass elements on the left branch must move with the same velocity. The same is true for the mass elements on the right branch. Observing that one point on the left branch is fixed to the ceiling and that the acceleration of the top end of the right branch is $\ddot{y}$, we must have

$$a_{yL} = 0 \quad \text{and} \quad a_{yR} = \ddot{y}. \tag{10}$$

Referring to Fig. 1, the time derivatives of the lengths of the two branches are $\dot{\ell}_L = \dot{y}/2$ and $\dot{\ell}_R = -\dot{y}/2$. Hence, since $m_L = \rho\ell_L$ and $m_R = \rho\ell_R$, we have

$$\dot{m}_i = \dot{m}_L = \rho\dot{\ell}_L = \rho\dot{y}/2 \quad \text{and} \quad \dot{m}_o = -\dot{m}_R = -\rho\dot{\ell}_R = \rho\dot{y}/2. \tag{11}$$

The velocity of the mass elements joining the left branch and leaving the right branch must match the time rate of lengthening and shortening of these branches, i.e.,

$$\vec{v}_i = \dot{\ell}_L \hat{j} = \tfrac{1}{2}\dot{y}\,\hat{j} \quad \text{and} \quad \vec{v}_o = \dot{\ell}_R \hat{j} = -\tfrac{1}{2}\dot{y}\,\hat{j}. \tag{12}$$

Computation   Substituting Eqs. (10)–(12) into Eq. (8) and solving for R, we have

$$R = \rho\ell_L g + \tfrac{1}{4}\rho\dot{y}^2, \tag{13}$$

Recalling that $\dot{y}$ is given in Eq. (1) and $\ell_L = (L+y)/2$, after simplification, we have

$$\boxed{R = \frac{\rho L g}{4}\left(1 + 3\frac{y}{L} + \frac{L}{L-y}\right).} \tag{14}$$

────────────────────────────────────

Discussion & Verification   Since we have obtained the same result via two very different methods, we can be confident that our final result is correct.

🔍 **A Closer Look**   We present a plot of R as a function of y in Fig. 4. Notice that as the string becomes vertical, i.e., as $y \to L$, R goes to infinity. This is so because the free end of the string moves with infinite speed when the string is *almost* completely vertical (i.e., $\dot{y} \to \infty$ as $y \to L$), and therefore R must become *impulsive* to bring the string to a complete stop as soon as the string becomes vertical.

Finally, we note that we did not take advantage of Eq. (9). The reason for this is that substituting Eqs. (10)–(12) into Eq. (9) yields an equation whose solution coincides with Eq. (1) (see Prob. 5.149). If we had not been given Eq. (1), we would have had to use Eq. (9) to obtain the velocity of the free end as a function of y.

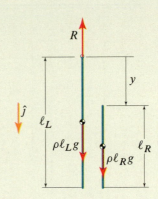

Figure 3
FBDs of the left and right branches of the falling string modeled as variable mass systems (see the Helpful Information marginal note for further comments on these FBDs).

Helpful Information

Is something missing from the FBDs in Fig. 3? When we cut some structure and we sketch the FBD of the cut structure, we place on the FBD those forces that act internally to the structure at the location of the cut. However, here we are modeling the two branches as *variable mass systems*, and in this case, we do not include the forces at the cut because of how we derived Eq. (5.161). Specifically, the external forces that appear in Eq. (5.161) do not include any effects due to the exchange of mass.

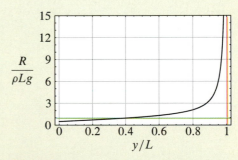

Figure 4
The reaction R (nondimensionalized with respect to the string weight $\rho L g$) at the top of the string as a function of the nondimensional fall distance (y/L). The vertical red line corresponds to the end of the fall, and the horizontal green line corresponds to the weight of the string.

PROBLEMS

Problem 5.113

A fluid is in steady motion in the conduit shown. The lines depicted are tangent to the velocity of the fluid particles in the conduit (these lines are called streamlines). Explain whether or not the control volume defined by the cross sections A and B in the figure is consistent with the assumptions laid out in this section.

Note: Concept problems are about *explanations*, not computations.

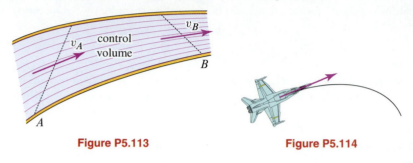

Figure P5.113   **Figure P5.114**

Problem 5.114

A hydraulic system is being used to actuate the control surfaces of a plane. Suppose that there is a time interval during which (a) the speed of the hydraulic fluid within a particular line is constant relative to the line itself and (b) the plane is performing a turn. Explain whether or not the force balance for control volumes presented in this section is applicable to the analysis of the hydraulic fluid in question.

Note: Concept problems are about *explanations*, not computations.

Problem 5.115

The cross sections labeled A and B in case (a) are identical to the corresponding cross sections in case (b). Assume that, in both (a) and (b), a fluid in steady motion flows through A with speed v_1 and exits the system at B with a speed v_2. If the pipe sections are to remain stationary and if the mass flow rate is identical in the two cases, determine whether the magnitude of the horizontal force acting on the pipes due to the water flow in case (a) is smaller than, equal to, or larger than that in case (b). In addition, for both (a) and (b), establish the direction of the force.

Note: Concept problems are about *explanations*, not computations.

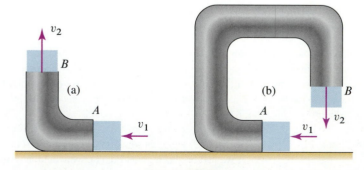

Figure P5.115

Problem 5.116

Experience tells us that when a steady water jet comes out of a nozzle, the line attached to the nozzle is in tension, that is, the nozzle exerts a force on the line that is in the direction of the flow. If the end of the nozzle at B were capped to stop the water flow, would the force exerted by the nozzle on the line decrease, stay the same, or increase? **Note:** Concept problems are about *explanations*, not computations.

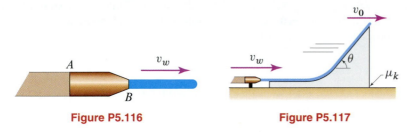

Figure P5.116   Figure P5.117

Problem 5.117

Revisit Example 5.17 and use the numerical result in Eq. (13) of the example, along with the fact that the specific weight of water is $62.4\,\text{lb/ft}^3$, to determine the volumetric flow rate at the nozzle and the nozzle diameter.

Problem 5.118

The tip B of a nozzle is 1.5 in. in diameter whereas the diameter at A where the line is attached is 3 in. If water is flowing through the nozzle at 95 gpm ("gpm" stands for gallons per minute, 1 U.S. gallon is defined as 231 in.^{3}) and the water static pressure in the line is 300 psi, determine the force necessary to hold the nozzle stationary. Recall that the specific weight of water is $\gamma = 62.4\,\text{lb/ft}^3$, and neglect the atmospheric pressure at B.

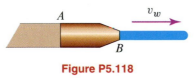

Figure P5.118

Problem 5.119

The rocket shown has 7 lb of propellant with a burnout time (time required to burn all the fuel) of 7 s. Assume that the mass flow rate is constant and that the speed of the exhaust relative to the rocket is also constant and equal to 6500 ft/s. If the rocket is fired from rest, determine the initial weight of the body for the rocket if the rocket is to experience an initial acceleration of $6g$.

Problem 5.120

An intubed fan is mounted on a cart connected to a fixed wall via a linear elastic spring with constant $k = 50\,\text{lb/ft}$. Assume that in a test the fan draws air at A with essentially zero speed and that the outgoing flow causes the cart to displace to the left so that the spring is stretched by 0.5 ft from its unstretched position. Assuming that the specific weight of the air $\gamma = 7.5\times10^{-2}\,\text{lb/ft}^3$ is constant, and letting the diameter of the tube at B be $d = 4\,\text{ft}$ (the cross section is assumed circular), determine the airspeed at B.

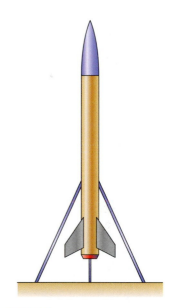

Figure P5.119

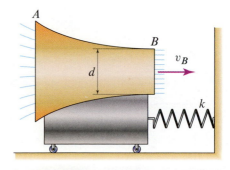

Figure P5.120

Problem 5.121

A test is conducted in which an 80 kg person sitting in a 15 kg cart is propelled by the jets emitted by two household fire extinguishers with a combined initial mass of 18 kg. The cross section of the exhaust nozzles is 3 cm in diameter, and the density of the exhaust is $\rho = 1.98\,\text{kg/m}^3$. The vehicle starts from rest, and it is determined that the initial acceleration of the "jet cart" is $1.8\,\text{m/s}^2$. Recalling that the mass flow rate out of the nozzle is given by $\dot{m}_o = \rho S v_o$, where S is the area of the nozzle cross section and v_o is the exhaust speed, determine v_o at the initial time. Ignore any resistance to the horizontal motion of the cart.

Problem 5.122

Consider a rocket in space so that it can be assumed that no external forces act on the rocket. Let v_o be the speed of the exhaust gases relative to the rocket. In addition, let $m_b + m_f$ and m_b be the total mass of the rocket and its fuel at the initial time and the mass of the body after all the fuel is burned, respectively. If the rocket is fired from rest, determine an expression for the maximum speed that the rocket can achieve.

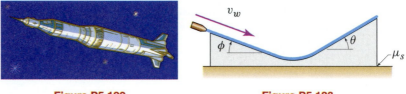

Figure P5.122   **Figure P5.123**

Problem 5.123

A stationary 4 cm diameter nozzle emits a water jet with a speed of 30 m/s. The water jet impinges on a vane with a mass of 15 kg. Recalling that water has a mass density of $1000\,\text{kg/m}^3$, determine the minimum static friction coefficient with the ground such that the vane does not move if $\phi = 20°$ and $\theta = 30°$. Neglect the weight of the water layer in contact with the vane as well as friction between the water and the vane.

Problem 5.124

A diffuser is attached to a structure whose rigidity in the horizontal direction can be modeled via a linear spring with constant k. The diffuser is hit by a water jet issued with a speed $v_w = 55\,\text{ft/s}$ from a 2 in. diameter nozzle. Assume that the friction between the jet and the diffuser is negligible and that the diffuser's motion in the vertical direction can be neglected. Recalling that the specific weight of water is $\gamma = 62.4\,\text{lb/ft}^3$, if the opening angle of the diffuser is $\theta = 40°$, determine k such that the horizontal displacement of the diffuser does not exceed 0.25 in. from the diffuser's rest position. Assume that the water jet splits symmetrically over the diffuser.

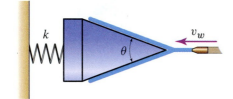

Figure P5.124

Problem 5.125

A water jet with a mass flow rate $\dot{m}_f$ at the nozzle impinges with a speed v_w on a fixed flat vane inclined at an angle θ with respect to the horizontal. Assuming that there is no friction between the water jet and the vane, the jet will split into two flows with mass flow rates $\dot{m}_{f1}$ and $\dot{m}_{f2}$. Neglecting the weight of the water, determine how $\dot{m}_{f1}$ and $\dot{m}_{f2}$ depend on $\dot{m}_f$, v_w, and θ. *Hint:* Due to the no-friction assumption, there is no force that slows down the water in the direction tangent to the vane, and this implies that the momentum in that direction is conserved.

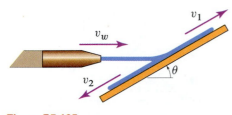

Figure P5.125

Problem 5.126

A person wearing a jet pack lifts off from rest and ascends along a straight vertical trajectory. Let M denote the initial combined mass of the pilot and the equipment, including the fuel in the pack. Assume that the mass flow rate m_o and exhaust gas speed v_o are known constants and that the pilot can take off as soon as the rocket engine is started. If the exhaust engine is completely directed in the direction of gravity, determine the expression of the pilot's speed as a function of time, M, m_o, v_o, and g (the acceleration due to gravity) while the pack is providing a thrust. Neglect air resistance and assume that gravity is constant.

Figure P5.126

Problem 5.127

A 28,000 lb A-10 Thunderbolt is flying at a constant speed of 375 mph when it fires a 4 s burst from its forward-facing seven-barrel Gatling gun. The gun fires 13.2 oz projectiles at a constant rate of 4200 rounds/min. The muzzle velocity of each projectile is 3250 ft/s. Assume that each of the plane's two jet engines maintains a constant thrust of 9000 lb, that the plane is subject to a constant air resistance while the gun is firing (equal to that before the burst), and that the plane flies straight and level during that time. Determine the plane's change in velocity at the end of the 4 s burst, modeling the airplane's change of mass due to firing as a continuous mass loss.

Figure P5.127

Problem 5.128

A faucet is letting out water at a rate of 15 L/min. Assume that the internal diameter d of the faucet is uniform and equal to 1.5 cm, the distance $\ell = 20$ cm, and the static water pressure at the wall is 0.30 MPa. Neglecting the weight of the water inside the faucet as well as the weight of the faucet itself, determine the forces and the moment that the wall exerts on the faucet. Recall that the density of water is $\rho = 1000\,\text{kg/m}^3$, and neglect the atmospheric pressure at the spout. *Hint:* Define your control volume using a section along the wall.

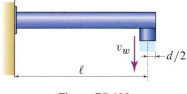

Figure P5.128

Problem 5.129

Consider a wind turbine with a diameter $d = 110$ m and the airflow streamlines shown, which are symmetric relative to the axis of the turbine. Since the airflow is tangent to the streamlines (by definition), these lines can be taken to define the top and bottom surfaces of a control volume. Suppose that pressure measurements indicate that the flow experiences atmospheric pressure at the cross sections A and B (as well as outside the control volume) where the wind speed is $v_A = 7\,\text{m/s}$ and $v_B = 2.5\,\text{m/s}$, respectively. Furthermore, assume that the average pressure along the streamlines defining the control volume is also atmospheric. Finally, assume that the diameter of the flow cross section at A is 85% of the rotor diameter and that the rotor hub is at a distance $h = 75$ m above the ground. If the density of air is constant and equal to $\rho = 1.25\,\text{kg/m}^3$, determine the force exerted by the air on the wind turbine and the reaction moment at the base of the support.

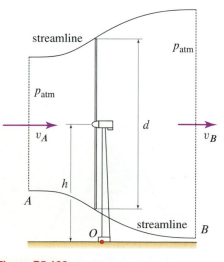

Figure P5.129

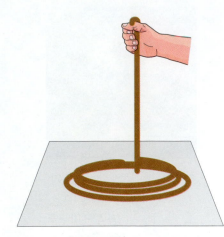

Figure P5.130–P5.132

Problem 5.130

A rope with weight per unit length of 0.1 lb/ft is lifted at a constant upward speed $v_0 = 8$ ft/s. Treating the rope as inextensible, determine the force applied to the top end of the rope after it is lifted 9 ft. Assume that the top end of the rope is initially at rest and on the floor. In addition, disregard the horizontal motion associated with the uncoiling of the rope.

Problem 5.131

A rope with mass per unit length of 0.05 kg/m is lifted at a constant upward acceleration $a_0 = 6$ m/s^2. Treating the rope as inextensible, determine the force that must be applied at the top end of the rope after it is lifted 3 m. Assume that the top end of the rope is initially at rest and on the floor. In addition, disregard the horizontal motion associated with the uncoiling of the rope.

Problem 5.132

A rope with mass per unit length of 0.05 kg/m is lifted by applying a constant vertical force $F = 10$ N. Treating the rope as inextensible, plot the velocity and position of the top end of the string as a function of time for $0 \leq t \leq 3$ s. Assume that the top end of the rope is initially at rest and 1 mm off the floor. In addition, disregard the horizontal motion associated with the uncoiling of the rope.

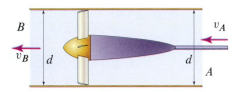

Figure P5.133

Problem 5.133

Let p_A and p_B be given static pressure measurements at the cross sections A and B in the air duct shown. Assume that any cross section between A and B is circular with diameter d. Assume that the flow is steady and that the mass density ρ_A at A is known along with v_A, the speed of the flow at A, and v_B, the speed of the flow at B. Determine the expression of mass density at B and the expression of the force F acting on the fan.

Problem 5.134

An amateur rocket with a body weight of 6.5 lb is equipped with a rocket engine holding 2.54 lb of solid propellant with a burnout time (time required to burn all the fuel) of 5.25 s (this is the typical data made available by amateur rocket engine manufacturers). The initial thrust is 68 lb. Assuming that the mass flow rate and the speed of the exhaust relative to the rocket remain constant, determine the exhaust mass flow rate m_o and the speed relative to the rocket v_o. In addition, determine the maximum speed achieved by the rocket v_{max} if the rocket is launched from rest and moves in the direction opposite to gravity. Neglect air resistance and assume that gravity does not change with elevation.

Problem 5.135

Continue Prob. 5.134 and determine the maximum height reached by the rocket, again neglecting air resistance and changes of gravity with elevation. *Hint:* For $0 < t < t_0$,

$$\int \ln\left(1 - \frac{t}{t_0}\right) dt = (t_0 - t)\left[1 - \ln\left(1 - \frac{t}{t_0}\right)\right] + C.$$

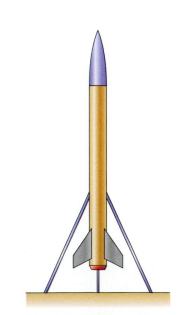

Figure P5.134 and P5.135

Problem 5.136

A Pelton impulse wheel, as shown in Fig. P5.136(a), typically found in hydroelectric power plants consists of a wheel at the periphery of which are attached a series of buckets. As shown in Fig. P5.136(b), water jets impinge on the buckets and cause the wheel to spin about its axis (labeled O). Let v_w and $(\dot{m}_f)_{nz}$ be the speed and the mass flow rate of the water jets at the nozzles (the nozzles are stationary), respectively. As the wheel spins, a given water jet will impinge on a given bucket only for a very small portion of the bucket's trajectory. This fact allows us to model the motion of a bucket relative to a given jet (during the time the bucket interacts with that jet) as essentially rectilinear and with constant relative speed, as was done in Example 5.17. Although each bucket moves away from the jet, the fact that they are arranged in a wheel is such that the effective mass flow rate experienced by the vanes is $(\dot{m}_f)_{nz}$ instead of the reduced mass flow rate computed in Eq. (6). With this in mind, consider a bucket, as shown in Fig. P5.136(c), that is moving with a speed v_0 horizontally away from a fixed nozzle but subject to a mass flow rate $(\dot{m}_f)_{nz}$. The inside of the bucket is shaped so as to redirect the water jet laterally out (away from the plane of the wheel). The angle θ describes the orientation of the velocity of the fluid relative to the (moving) bucket at B, the point at which the water leaves the bucket. Determine θ and v_0 such that the power transmitted by the water to the wheel is maximum. Express v_0 in terms of v_w.

(a)   (b)   (c)

Figure P5.136

5.6 Chapter Review

Momentum and impulse

We learned in Chapter 3 that forces lead to changes in velocities since forces cause accelerations. We began this section by learning that forces acting over time change momentum (not just velocity) by integrating Newton's second law to get the *impulse-momentum principle*, which is given by

> Eq. (5.6), p. 335
>
> $$\vec{p}(t_1) + \int_{t_1}^{t_2} \vec{F}(t)\,dt = \vec{p}(t_2),$$

where the *linear momentum* (or *momentum*) was defined to be

> Eq. (5.3), p. 334
>
> $$\vec{p}(t) = m\vec{v}(t)$$

and a force acting over some time interval was called the *impulse* (or *linear impulse*) and is given by

> Eq. (5.5), p. 335
>
> $$\int_{t_1}^{t_2} \vec{F}(t)\,dt.$$

In addition, we found that without detailed knowledge of the force acting on a particle at every instant in time, we could not determine the change in momentum. On the other hand, knowing just the change in momentum does allow us to determine the *average force* acting on a particle during the corresponding time interval, that is,

> Eq. (5.9), p. 335
>
> $$\vec{F}_{\text{avg}} = \frac{\vec{p}(t_2) - \vec{p}(t_1)}{t_2 - t_1}.$$

Impulse-momentum principle for systems of particles. When dealing with systems of particles, we discovered that we can write the impulse-momentum principle as

> Eq. (5.12), p. 336, and Eq. (5.15), p. 337
>
> $$\vec{F} = \dot{\vec{p}} \quad \text{and} \quad \int_{t_1}^{t_2} \vec{F}(t)\,dt = \vec{p}(t_2) - \vec{p}(t_1),$$

where $\vec{F}$ is the total external force on the particle system and $\vec{p} = \sum_{i=1}^{N} m_i \vec{v}_i$ is the total momentum of the system of particles. Using the definition of the mass center of a system of particles, the impulse-momentum principle can also be written as

Eq. (5.14), p. 337

$$\vec{F} = \frac{d}{dt}(m\vec{v}_G) = m\vec{a}_G,$$

where m is the total mass of the system of particles, $\vec{v}_G$ is the velocity of its mass center, and $\vec{a}_G$ is the acceleration of its mass center.

Conservation of linear momentum. When there is a direction in which the external force on a system of particles is zero, then the momentum in that direction is constant and is said to be conserved. If the total external force on a system of particles is zero, that is, $\vec{F} = 0$, then the momentum in every direction is constant and the mass center of the system of particles will move with constant velocity.

Impact

This section discussed impact between particles. We introduced a model based on the assumptions summarized in Table 5.5. We discovered that there are two key elements to *every* impact problem: (1) the application of the impulse-momentum principle and (2) a force law telling us how the colliding objects rebound.

Impulsive forces are forces that generate a finite change in momentum in an infinitesimal amount of time. When we apply the impulse-momentum principle during an impact, only impulsive forces play a role, so they are the only forces included on the FBD. Problems involving the impact between two particles generally involve four unknowns, so four equations are needed. The geometry of an *unconstrained impact* (for which there are no external impulsive forces) between two particles is shown in Fig. 5.62. When the impact is frictionless, the four equations come from

1. Application of the impulse-momentum principle to both particles along the LOI (the y direction), which gives

Eq. (5.38), p. 364

$$m_A v_{Ay}^- + m_B v_{By}^- = m_A v_{Ay}^+ + m_B v_{By}^+,$$

2. Application of the impulse-momentum principle to particle A in the x direction, which gives

Eq. (5.39), p. 364

$$v_{Ax}^- = v_{Ax}^+,$$

3. Application of the impulse-momentum principle to particle B in the x direction, which gives

Eq. (5.40), p. 364

$$v_{Bx}^- = v_{Bx}^+,$$

Table 5.5
Assumptions used in our impact model.

Physical characteristic	Impact assumption
duration of impact	infinitesimal
displacement of particle	zero
force on particle	infinite
change in momentum	instantaneous

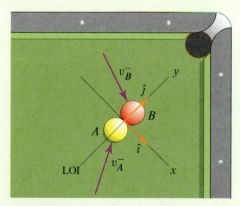

Figure 5.62
Geometry of two impacting particles.

4. Application of the COR equation along the LOI, which is given by

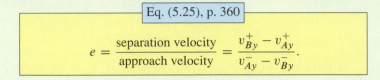

Eq. (5.25), p. 360

$$e = \frac{\text{separation velocity}}{\text{approach velocity}} = \frac{v_{By}^+ - v_{Ay}^+}{v_{Ay}^- - v_{By}^-}.$$

The coefficient of restitution e determines the nature of the rebound between the two particles. When $e = 0$, the impact is called *plastic* (the objects do not necessarily stick together in a plastic impact); when $0 < e < 1$, the impact is called *elastic*; and when $e = 1$, it is called *perfectly elastic*.

In a *perfectly plastic impact* the colliding objects stick together after the impact. Therefore, in a perfectly plastic impact, the fact the objects stick together is reflected in the equations

Eqs. (5.42) and (5.43), p. 365

$$v_{Ay}^+ = v_{By}^+ \quad \text{and} \quad v_{Ax}^+ = v_{Bx}^+.$$

In addition, momentum is conserved in all directions during the impact, which gives

Eqs. (5.44) and (5.45), p. 365

$$m_A v_{Ax}^- + m_B v_{Bx}^- = m_A v_{Ax}^+ + m_B v_{Bx}^+,$$
$$m_A v_{Ay}^- + m_B v_{By}^- = m_A v_{Ay}^+ + m_B v_{By}^+.$$

Impact and energy.   Unless an impact is perfectly elastic, mechanical energy must be lost during the impact. Generally, the energy loss is computed as a percentage of the preimpact total kinetic energy, or as

Eq. (5.46), p. 365

$$\text{Percentage of energy loss} = \frac{T^- - T^+}{T^-} \times 100\%,$$

where T^- is the preimpact total kinetic energy and T^+ is the postimpact total kinetic energy and they are given by

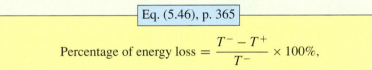

Eqs. (5.47) and (5.48), p. 365

$$T^- = \tfrac{1}{2} m_A (v_A^-)^2 + \tfrac{1}{2} m_B (v_B^-)^2,$$
$$T^+ = \tfrac{1}{2} m_A (v_A^+)^2 + \tfrac{1}{2} m_B (v_B^+)^2.$$

Constrained impact.   In a constrained impact, one of the objects is physically constrained from moving in some direction. Our strategy to solve these problems had the following elements:

1. Identification of those directions along which the momentum is conserved through the impact for one and/or both of the bodies.

2. Kinematic description of the constraint, which means that the constrained body can only move normal to the constraint after the impact.

3. Application of the COR equation along the LOI.

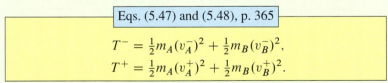

Angular momentum

In the section on angular momentum we developed the concept of angular momentum for a single particle and for systems of particles. In addition, we presented the *angular impulse-momentum principle*.

Definition of angular momentum of a particle.   Referring to Fig. 5.63, the *angular momentum of a particle Q with respect to the moment center P* is given by

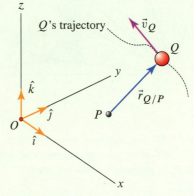

$$\boxed{\text{Eq. (5.59), p. 389}}$$

$$\vec{h}_P = \vec{r}_{Q/P} \times \vec{p}_Q = \vec{r}_{Q/P} \times m\vec{v}_Q,$$

where P can be fixed or moving; $\vec{r}_{Q/P}$ is the position of Q relative to P; m and $\vec{v}_Q$ are the mass and the velocity of Q, respectively; and $\vec{p}_Q = m\vec{v}_Q$ is the linear momentum of Q.

Figure 5.63
Figure 5.32 repeated. A particle Q in motion relative to a point P. Point P need *not* be a stationary point.

Angular impulse-momentum principle for a particle.   Newton's second law tells us that the time rate of change of the angular momentum of a particle, computed about a given moment center, is related to the moment of the total force acting on the particle about the same moment center. The *angular impulse-momentum principle* for a single particle is given by

$$\boxed{\text{Eq. (5.64), p. 390}}$$

$$\vec{M}_P = \dot{\vec{h}}_P + \vec{v}_P \times m\vec{v}_Q,$$

where $\vec{M}_P$ is the moment with respect to P of all the forces acting on Q. If one of the following conditions is satisfied:

1. The reference point P is fixed, i.e., if $\vec{v}_P = \vec{0}$.

2. $\vec{v}_P$ is parallel to $\vec{v}_Q$, i.e., $\vec{v}_P \times m\vec{v}_Q = \vec{0}$.

then the angular impulse-momentum principle for a single particle can be simplified to read

$$\boxed{\text{Eq. (5.65), p. 390}}$$

$$\vec{M}_P = \dot{\vec{h}}_P.$$

If conditions (1) or (2) are satisfied for $t_1 \leq t \leq t_2$, then the angular impulse-momentum principle for a single particle can be integrated with respect to time to read

$$\boxed{\text{Eq. (5.66), p. 390}}$$

$$\vec{h}_{P1} + \int_{t_1}^{t_2} \vec{M}_P \, dt = \vec{h}_{P2},$$

where $\vec{h}_{P1} = \vec{h}_P(t_1)$ and $\vec{h}_{P2} = \vec{h}_P(t_2)$.

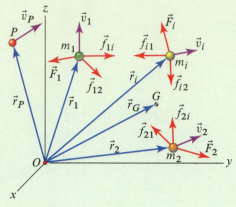

Figure 5.64

Figure 5.34 repeated. A system of particles under the action of internal and external forces. Point P, in general, is a moving point.

Angular impulse-momentum for a system of particles. When applied to a closed particle system, the angular impulse-momentum principle can be given the form

Eq. (5.78), p. 392

$$\vec{M}_P = \dot{\vec{h}}_P + \vec{v}_P \times m\vec{v}_G,$$

where, with reference to Fig. 5.64, $\vec{M}_P$ is the moment with respect to P of *only the external forces* acting on the system, $m = \sum_{i=1}^{N} m_i$ is the total mass of the system, G is the system's center of mass, $\vec{v}_G$ is the velocity of G, and $\vec{h}_P$ is the total angular momentum, which is defined as

Eq. (5.77), p. 392

$$\vec{h}_P = \sum_{i=1}^{N} \vec{h}_{Pi}.$$

As we saw earlier, the angular impulse-momentum principle takes on a simpler form if certain conditions are satisfied. For a closed system of particles, these conditions are any of the following:

1. When P is a fixed point, i.e., when $\vec{v}_P = \vec{0}$.

2. When G is a fixed point, i.e., when $\vec{v}_G = \vec{0}$.

3. When P coincides with the center of mass and therefore $\vec{v}_P = \vec{v}_G$.

4. When the point P and the center of mass travel parallel to one another, i.e., when the vectors $\vec{v}_P$ and $\vec{v}_G$ are parallel.

Under any of these conditions the angular impulse-momentum principle for a closed system of particles simplifies to

Eq. (5.79), p. 393

$$\vec{M}_P = \dot{\vec{h}}_P.$$

Furthermore, if any of the above conditions are satisfied over a time interval $t_1 \le t \le t_2$, then the angular impulse-momentum principle for a closed system of particles can be integrated with respect to time to read

Eq. (5.80), p. 393

$$\vec{h}_{P1} + \int_{t_1}^{t_2} \vec{M}_P \, dt = \vec{h}_{P2}.$$

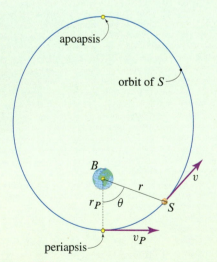

Figure 5.65

Figure 5.47 repeated. A satellite S of mass m orbiting a primary body B of mass m_B on a conic section, which in this case is an ellipse. The angle θ and the launch conditions r_P and v_P are defined from periapsis.

Orbital mechanics

In orbital mechanics, we studied the motion of a satellite S of mass m that is subject to Newton's universal law of gravitation due to a body B of mass m_B, which is the primary or attracting body (see Fig. 5.65). We began with these important assumptions:

1. The primary body B and the satellite S are both treated as particles.

2. The only force acting on the satellite S is the force of mutual attraction between B and S.

3. The primary body B is fixed in space.

Determination of the orbit.   Solving the governing equations in polar coordinates, we found that the trajectory of the satellite S under these assumptions is a conic section, whose equation can be written as

> Eqs. (5.102) and (5.107), p. 412
>
> $$\frac{1}{r} = C\cos\theta + \frac{Gm_B}{\kappa^2} \quad \text{or}$$
> $$\frac{1}{r} = \frac{Gm_B}{\kappa^2}(1 + e\cos\theta),$$

where r is the distance between the centers of mass of S and B; θ is the orbital angle measured relative to periapsis; G is the universal gravitational constant; κ is the angular momentum per unit mass of the satellite S measured about B; C is a constant to be determined; and e is the *eccentricity* of the trajectory, which can be written as

> Eq. (5.106), p. 412
>
> $$e = \frac{C\kappa^2}{Gm_B}.$$

If, as is generally the case here, the orbital conditions are known at periapsis, then κ is

> Eq. (5.108), p. 412
>
> $$\kappa = r_P v_P,$$

the constant C is

> Eq. (5.109), p. 413
>
> $$C = \frac{1}{r_P}\left(1 - \frac{Gm_B}{r_P v_P^2}\right),$$

and the equation describing the trajectory becomes

> Eq. (5.110), p. 413
>
> $$\frac{1}{r} = \frac{1}{r_P}\left(1 - \frac{Gm_B}{r_P v_P^2}\right)\cos\theta + \frac{Gm_B}{r_P^2 v_P^2}.$$

Conic sections.   Equations (5.102), (5.107), and (5.110) represent equivalent conic sections in polar coordinates. The type of conic section depends on the eccentricity of the trajectory e (see Fig. 5.66), which, in turn, depends on C and κ via Eq. (5.106). There are four types of conic section, which are determined by the value of the eccentricity e, that is, $e = 0$, $0 < e < 1$, $e = 1$, and $e > 1$.

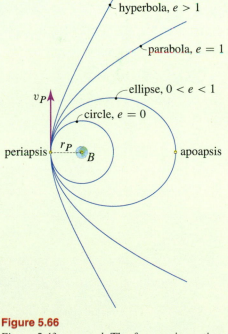

Figure 5.66

Figure 5.43 repeated. The four conic sections, showing r_P and v_P.

For a *circular orbit* ($e = 0$), the radius is r_P and the speed in the orbit is equal to

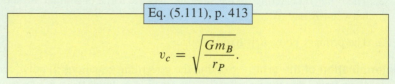

Eq. (5.111), p. 413

$$v_c = \sqrt{\frac{Gm_B}{r_P}}.$$

For an *elliptical orbit* ($0 < e < 1$), as expected, the radius at periapsis is r_P. The radius at apoapsis is given by

Eq. (5.113), p. 413

$$r_A = \frac{r_P}{2Gm_B/(r_P v_P^2) - 1}.$$

Referring to Fig. 5.67, additional relationships between the semimajor axis a

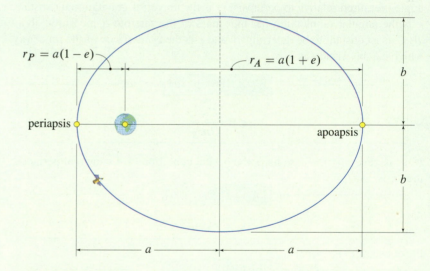

Figure 5.67. *Figure 5.44 repeated.* The semimajor axis a and semiminor axis b of an ellipse.

of an elliptical orbit, the eccentricity of the orbit, and the radii at periapsis and apoapsis are, respectively,

Eqs. (5.119), p. 414

$$r_P = a(1 - e) \quad \text{and} \quad r_A = a(1 + e).$$

The period of an elliptical orbit τ can be written in the following two ways

Eq. (5.121), p. 414, and Eq. (5.123), p. 415

$$\tau = \frac{2\pi ab}{\kappa} = \frac{\pi}{\kappa}(r_P + r_A)\sqrt{r_P r_A},$$

or, reflecting Kepler's third law, the orbital period can be written as

Eq. (5.126), p. 415

$$\tau^2 = \frac{4\pi^2}{Gm_B}a^3.$$

A *parabolic trajectory* ($e = 1$) is that which divides periodic orbits (that return to their starting location) from trajectories that are not periodic. For a given r_P, the speed v_{par} required to achieve a parabolic trajectory is given by

Eq. (5.128), p. 415

$$v_{\text{par}} = v_{\text{esc}} = \sqrt{\frac{2Gm_B}{r_P}},$$

which is also referred to as the *escape velocity* since it is the speed required to completely escape the influence of the primary body B.

For a *hyperbolic trajectory* ($e > 1$), the governing equations given by Eqs. (5.106)–(5.110) are used for values of $e > 1$.

Energy considerations.   Using the work-energy principle, we discovered that the total mechanical energy in an orbit depends on only the semimajor axis a of the orbit and is given by

Eq. (5.132), p. 416

$$E = -\frac{Gm_B}{2a},$$

where E is the *mechanical energy per unit mass* of the satellite. Applying the work-energy principle at an arbitrary location within the orbit, we found that the speed can be written as

Eq. (5.134), p. 416

$$v = \sqrt{Gm_B\left(\frac{2}{r} - \frac{1}{a}\right)}.$$

Mass flows

In Section 5.5 we have considered mass flows, i.e., the motion of fluids or systems whose motion can be modeled as being "fluidlike." Specifically, we have considered (1) *steady* mass flows, in which a fluid moves through a conduit with a velocity that depends only on the position within the conduit; and (2) *variable* mass flows, such as the flow of combustion gases out of a rocket.

Steady flows.   Given the control volume (CV) shown in Fig. 5.68, where by *control volume* we mean a portion of a conduit delimited by two cross sections,

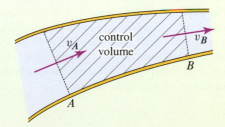

Figure 5.68
Figure 5.52 repeated. A CV corresponding to the portion of a pipe between cross sections A and B.

we showed that, in the case of a steady flow, the total external force $\vec{F}$ acting on the fluid in the CV is

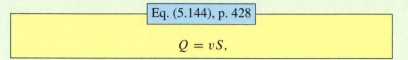

Eq. (5.140), p. 427

$$\vec{F} = \dot{m}_f(\vec{v}_B - \vec{v}_A),$$

where as long as the cross sections are perpendicular to the flow velocity, $\dot{m}_f$ is the *mass flow rate*, i.e., the amount of mass flowing through a cross section per unit time, and where $\vec{v}_A$ and $\vec{v}_B$ are the flow velocities at the cross sections A and B, respectively. In addition to the mass flow rate, we defined the *volumetric flow rate* as the quantity

Eq. (5.144), p. 428

$$Q = vS,$$

where v is the speed of the fluid at a given cross section and S is the area of the cross section in question. We showed that

Eq. (5.145), p. 428

$$\dot{m}_f = \rho_A Q_A = \rho_B Q_B,$$

where ρ_A and ρ_B are the values of the mass density of the fluid at A and B, respectively. Referring to Fig. 5.69, we also showed that, given a fixed point P, the total moment $\vec{M}_P$ acting on the fluid in the CV is

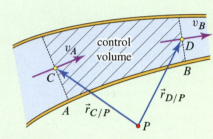

Figure 5.69

Figure 5.60 repeated. A fluid flowing through a CV along with a choice of moment center P for the calculation of angular momenta and moments.

Eq. (5.150), p. 429

$$\vec{M}_P = \dot{m}_f(\vec{r}_{D/P} \times \vec{v}_B - \vec{r}_{C/P} \times \vec{v}_A),$$

where C and D are the centers of the cross sections A and B, respectively.

Variable mass flows.   With reference to Fig. 5.70, for a body with time varying mass $m(t)$ due to an inflow of mass with rate $\dot{m}_i$ and an outflow of mass with the rate $\dot{m}_o$, the total external force acting on the body is given by

Figure 5.70

Figure 5.58 repeated. A plane with a jet engine. The airplane is taking in air with a mass flow rate $\dot{m}_i$ while combustion gases are ejected with a mass flow rate $\dot{m}_o$. The vector $\vec{v}_i$ is the velocity of the inflowing air *relative* to the plane. The vector $\vec{v}_o$ is the velocity of the outflowing combustion gases *relative* to the plane.

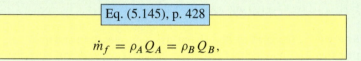

Eq. (5.161), p. 430

$$\vec{F} = m\vec{a} + \dot{m}_o\vec{v}_o - \dot{m}_i\vec{v}_i,$$

where $\vec{a}$ is the acceleration of the main body, $\vec{v}_o$ is the *relative* velocity of the outflowing mass with respect to the main body, and $\vec{v}_i$ is the *relative* velocity of the inflowing mass, again relative to the main body.

REVIEW PROBLEMS

Problem 5.137

In Major League Baseball, a pitched ball has been known to hit the head of the batter (sometimes unintentionally and sometimes not). Let the pitcher be, for example, Nolan Ryan who can throw a $5\frac{1}{8}$ oz baseball that crosses the plate at 100 mph.* Studies have shown that the impact of a baseball with a person's head has a duration of about 1 ms. So using Eq. (5.9) on p. 335 and assuming that the rebound speed of the ball after the collision is negligible, determine the magnitude of the average force exerted on the person's head during the impact.

Problem 5.138

A 0.6 kg ball that is initially at rest is dropped on the floor from a height of 1.8 m and has a rebound height of 1.25 m. If the ball spends a total of 0.01 s in contact with the ground, determine the average force applied to the ball by the ground during the rebound. In addition, determine the ratio between the magnitude of the impulse provided to the ball by the ground and the magnitude of the impulse provided to the ball by gravity during the time interval that the ball is in contact with the ground. Neglect air resistance.

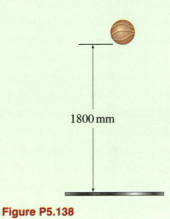

1800 mm

Figure P5.138

Problem 5.139

A person P is initially standing on a cart on rails, which is moving to the right with a speed $v_0 = 2$ m/s. The cart is not being propelled by any motor. The combined mass of person P, the cart, and all that is being carried on the cart is 270 kg. At some point a person P_A standing on a stationary platform throws to person P a package A to the right with a mass $m_A = 50$ kg. Package A is received by P with a horizontal speed $v_{A/P} = 1.5$ m/s. After receiving the package from A, person P throws a package B with a mass $m_B = 45$ kg toward a second person P_B. The package intended for P_B is thrown to the right, i.e., in the direction of the motion of P, and with a horizontal speed $v_{B/P} = 4$ m/s relative to P. Determine the final velocity of the person P. Neglect any friction or air resistance acting on P and the cart.

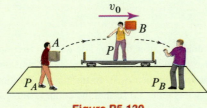

Figure P5.139

Problem 5.140

A Ford Excursion A, with a mass $m_A = 3900$ kg, traveling with a speed $v_A = 85$ km/h collides head-on with a Mini Cooper B, with a mass $m_B = 1200$ kg, traveling in the opposite direction with a speed $v_B = 40$ km/h. Determine the postimpact velocities of the two cars if the impact's coefficient of restitution is $e = 0.22$. In addition, determine the percentage of kinetic energy loss.

Figure P5.140

---

* This means that he must have thrown the ball at about 108 mph since the ball loses speed at the rate of 1 mph for every 7 ft it travels.

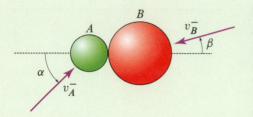

Figure P5.141

Problem 5.141

The two spheres, A and B, with masses $m_A = 1.35\,\text{kg}$ and $m_B = 2.72\,\text{kg}$, respectively, collide with $v_A^- = 26.2\,\text{m/s}$ and $v_B^- = 22.5\,\text{m/s}$. Let $\alpha = 45°$ and compute the value of β if the component of the postimpact velocity of B along the LOI is equal to zero and if the COR is $e = 0.63$.

Problem 5.142

A 31,000 lb truck A and a 3970 lb sports car B collide at an intersection. At the moment of the collision the truck and the sports car are traveling with speeds $v_A^- = 60\,\text{mph}$ and $v_B^- = 50\,\text{mph}$, respectively. Assume that the entire intersection forms a horizontal surface. Letting the line of impact be parallel to the ground and rotated counterclockwise by $\alpha = 20°$ with respect to the preimpact velocity of the truck, determine the postimpact velocities of A and B if the contact between A and B is frictionless and the COR $e = 0.1$. Furthermore, assuming that the truck and the car slide after impact and that the coefficient of kinetic friction is $\mu_k = 0.7$, determine the position at which A and B come to a stop relative to the position they occupied at the instant of impact.

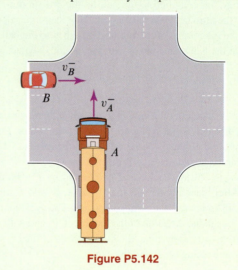

Figure P5.142

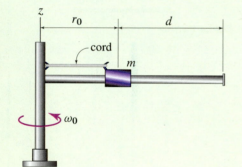

Figure P5.143

Problem 5.143

Consider a collar with mass m that is free to slide with no friction along a rotating arm of negligible mass. The system is initially rotating with a constant angular velocity ω_0 while the collar is kept at a distance r_0 from the z axis. At some point, the restraint keeping the collar in place is removed so that the collar is allowed to slide. Determine the expression for the moment that you need to apply to the arm, as a function of time, to keep the arm rotating at a constant angular velocity while the collar travels toward the end of the arm. *Hint:*

$$\int \frac{1}{\sqrt{x^2 - 1}}\, dx = \ln\left(x + \sqrt{x^2 - 1}\right) + C.$$

Problem 5.144

A satellite is launched parallel to the Earth's surface at an altitude of 450 mi with a speed of 17,500 mph. Determine the apogee altitude h_A above the Earth's surface as well as the period of the satellite.

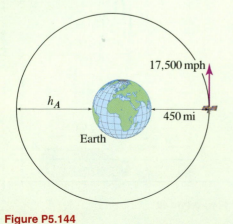

Figure P5.144

Problem 5.145 🌡

A spacecraft is traveling at 19,000 mph parallel to the surface of the Earth at an altitude of 250 mi, when it fires a retrorocket to transfer to a different orbit. Determine the change in speed Δv necessary for the spacecraft to reach a minimum altitude of 110 mi during the ensuing orbit. Assume that the change in speed is impulsive; that is, it occurs instantaneously.

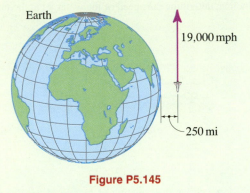

Earth

19,000 mph

250 mi

Figure P5.145

Problems 5.146 and 5.147

The optimal way (from an energy standpoint) to transfer from one circular orbit about a primary body (in this case, the Sun) to another circular orbit is via the *Hohmann transfer*, which involves transferring from one circular orbit to another using an elliptical orbit that is tangent to both at the periapsis and apoapsis of the ellipse. This ellipse is uniquely defined because we know the perihelion radius r_e (the radius of the inner circular orbit) and the aphelion radius r_j (the radius of the outer circular orbit), and therefore we know the semimajor axis a via Eq. (5.117) and the eccentricity e via Eq. (5.114) or Eqs. (5.119). Performing a Hohmann transfer requires two maneuvers, the first to leave the inner (outer) circular orbit and enter the transfer ellipse and the second to leave the transfer ellipse and enter the outer (inner) circular orbit. Assume that the orbits of Earth and Jupiter are circular, use 150×10^6 km for the radius of Earth's orbit, use 779×10^6 km for the radius of Jupiter's orbit, and note that the mass of the Sun is 333,000 times that of the Earth.

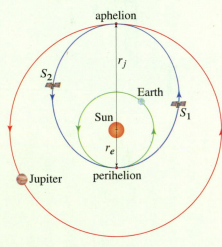

aphelion

r_j

S_2

Earth

Sun

S_1

r_e

Jupiter   perihelion

Figure P5.146 and P5.147

Problem 5.146 🌡 A space probe S_1 is launched from Earth to Jupiter via a Hohmann transfer orbit. Determine the change in speed Δv_e required at the radius of Earth's orbit of the elliptical transfer orbit (perihelion) and the change in speed Δv_j required at the radius of Jupiter's orbit (aphelion). In addition, compute the time required for the orbital transfer. Assume that the changes in speed are impulsive; that is, they occur instantaneously.

Problem 5.147 🌡 A space probe S_2 is at Jupiter and is required to return to the radius of Earth's orbit about the Sun so that it can return samples taken from one of Jupiter's moons. Assuming that the mass of the probe is 722 kg, determine the change in kinetic energy required at Jupiter ΔT_j for the maneuver at aphelion. In addition, determine the change in kinetic energy required at Earth ΔT_e for the perihelion maneuver. Finally, what is the change in potential energy ΔV of the spacecraft in going from Jupiter to the Earth?

Problem 5.148

A water jet is emitted from a nozzle attached to the ground. The jet has a constant mass flow rate $(\dot{m}_f)_{nz} = 15\,\text{kg/s}$ and a speed v_w relative to the nozzle. The jet strikes a 12 kg incline and causes it to slide at a constant speed $v_0 = 2\,\text{m/s}$. The kinetic coefficient of friction between the incline and the ground is $\mu_k = 0.25$. Neglecting the effect of gravity and air resistance on the water flow, as well as friction between the water jet and the incline, determine the speed of the water jet at the nozzle if $\theta = 47°$.

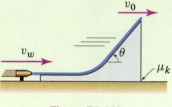

Figure P5.148

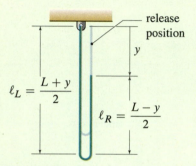

Figure P5.149

Problem 5.149

Revisit Example 5.20 and derive the equation of motion of the free end of the string starting from the force balance for the right branch of the string modeled as a variable mass system.

Problem 5.150

An intubed fan (a fan rotating within a tube or other conduit) is mounted on a cart that is connected to a fixed wall via a linear elastic spring with constant $k = 70\,\text{N/m}$. Assume that in a particular test the fan draws air that enters the tube at A with a speed v_A. The outgoing flow at B has a speed v_B. The flow of air through the tube causes the cart to displace to the left so that the spring is stretched by 0.25 m from its unstretched position. Assume that the density of air is constant throughout the tube and equal to $\rho = 1.25\,\text{kg/m}^3$. In addition let the tube's cross section be circular, and let the cross-sectional diameters at A and B be $d_A = 3\,\text{m}$ and $d = 1.5\,\text{m}$, respectively. Determine the velocities of the airflow at A and B.

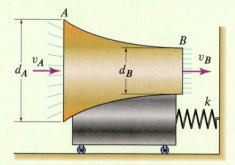

Figure P5.150

Planar Rigid Body Kinematics

This chapter begins the study of *rigid body motion* by developing the *planar kinematics of a rigid body*. As we did in Chapter 2, we will describe motion without addressing what causes the motion. We will assume that (1) the body is rigid and its mass is distributed over a *region* of space (see Section 1.2, p. 9) and (2) the velocity of each of the body's points is parallel to a common plane. This may appear daunting because the description of a body's motion requires that we know the motion of each point in the body. However, the rigidity assumption alleviates this difficulty, and we will discover that it is possible to completely characterize the planar motion of a rigid body using only three functions of time. To understand how *rigidity* helps us develop a kinematics that is so efficient, we begin by giving a qualitative description of rigid body motions, and then we proceed with a quantitative analysis. General 3D rigid body motions are examined in Chapter 10.

6.1 Fundamental Equations, Translation, and Rotation About a Fixed Axis

Crank, connecting rod, and piston motion

A typical car engine converts the chemical energy released by the combustion of air and fuel into mechanical energy, i.e., into motion. As shown in Fig. 6.1, in a typical internal combustion engine, pistons slide up and down inside cylinders. Each piston is connected to a shaft, called the crankshaft (or crank), by a connecting rod, which is pin-connected to both the piston and the crank. Figure 6.2 shows three sequential views of what we would see in one of the cylinders if we looked down the crankshaft (i.e., with the crankshaft axis perpendicular to the page). The point labeled A represents the axis of the crankshaft, whereas points labeled B and C represent the pin connections between the connecting rod and the crank and between the connecting rod and the piston, respectively. Because of these connections, the motion of the piston causes a rotation of the crankshaft. As the crankshaft goes through a complete rotation, the piston reverses its motion and slides back up the cylinder so that the overall motion is repeated. This is an example of a *slider-crank mechanism*.*

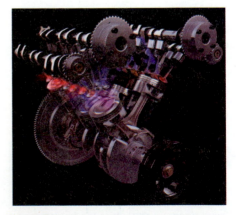

Figure 6.1
Interior view of the EcoBoost engine from Ford, revealing the pistons and the crankshaft.

---

* A mechanism is "a system of elements arranged to transmit motion in a predetermined fashion." From R. L. Norton *Design of Machinery,* 4th ed., McGraw-Hill, New York, 2008.

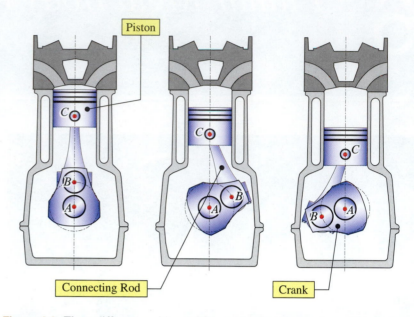

Figure 6.2. Three different positions of the mechanical system formed by a piston, connecting rod, and crank in a typical internal combustion engine. The axis of the crankshaft is perpendicular to the page, and it is represented by the points labeled A.

> ### Interesting Fact
>
> **What if a body is deformable?** In Chapter 3 we saw that, for planar motion, the equations of motion for a particle consist of two second-order ordinary differential equations (ODEs), i.e., as many ODEs as the number of degrees of freedom. In Chapter 7, we will see that, for planar motion, the equations of motion of a rigid body consist of three second-order ODEs, which is the number of degrees of freedom of a rigid body in planar motion. If the body is deformable, it has *infinitely many* degrees of freedom. In that case, the equations of motion are *partial differential equations* instead of infinitely many ODEs.

Figure 6.3
Left: a lift with an articulated boom and a platform holding a worker. Right: a schematic of the lift's range of motion (the gray area) showing that the platform remains parallel to the ground while moving along a curved path.

> ### Common Pitfall
>
> **Translations are not necessarily rectilinear.** In general, the fact that a body is translating does not imply that the trajectory of its points is a straight line.

The geometry of the system in Fig. 6.2 influences the motion of the piston, connecting rod, and crankshaft and is part of what determines the overall engine performance. Before we delve into the mathematics of how the geometry determines the motion, a qualitative overview of some specific motions is useful. Building a catalog of rigid body motions will help us build our physical intuition and understand the equations we will derive. We will use this example to illustrate most of the kinematics we will present in this chapter.

Qualitative description of rigid body motion

Translation

Referring to Fig. 6.2, assume that the cylinder within which the piston moves is stationary. Looking at the piston and modeling it as rigid, we see, for example, that the piston rings (appearing as horizontal bands at the top of the piston) always remain horizontal as the piston moves. Therefore, we conclude that the velocity of point C is the same as that of *any* other point on the piston. This type of motion is called a *translation*, and it is defined by saying that *any line segment connecting two points in the body maintains its original orientation throughout the motion*. It is important to realize that a translating body does not necessarily move in a straight line, as illustrated in Fig. 6.3. Therefore, it is customary to classify translations into one of two categories, *rectilinear translations* or *curvilinear translations*, based on whether the trajectory of each point is a straight line or not, respectively. The motion of the piston in Fig. 6.2 is a rectilinear translation, whereas the motion of the platform in Fig. 6.3 is a curvilinear translation.

Rotation about a fixed axis

Figure 6.4, which emphasizes the motion of the crank in Fig. 6.2, shows that point A on the axis of the shaft does not move. Any point on the crank lying

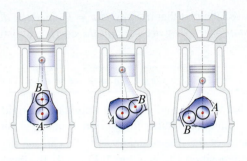

Figure 6.4. A modification of Fig. 6.2 emphasizing the motion of the crank.

on the axis perpendicular to the plane of the figure and going through point A does not move. Because the crank is rigid, off-axis points on the crank (e.g., point B) can only move in a circle centered at A. Any segment connecting points to the shaft axis rotates with the same angular velocity as any other segment. This type of motion, where there is a line of points with zero velocity functioning as an axis of rotation, is a *rotation about a fixed axis*.

General planar motion

Now looking at the connecting rod, which is shown in Fig. 6.5, we realize that this motion does not correspond to either of the two cases discussed above since (1) we cannot find an axis that is fixed throughout the motion and (2) no two points define a segment that does not change its orientation. The only distinguishing feature of the rod's motion is that none of its points has a component of velocity perpendicular to the plane of the page, which is called the *plane of motion*. This kind of motion is called *general plane motion* or *general planar motion*. This motion can be viewed as the composition of a translation with a rotation about an axis perpendicular to the plane of motion. Note that both the piston's translation and crankshaft's rotation are also planar motions.

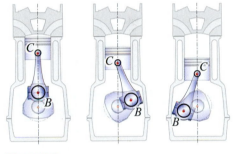

Figure 6.5
A modification of Fig. 6.2 emphasizing the motion of the connecting rod.

General motion of a rigid body

To describe a body's motion, we need to describe the motion of *every* point in the body. However, for a rigid body, we can describe the motion of all points by describing (1) the motion of a *single* point along with (2) *the rate of change in the body's orientation*. Let's find out why.

To express the velocity of a point B in relation to the velocity of another point A, we can use Eq. (2.106) on p. 136, which is

$$\vec{v}_B = \vec{v}_A + \vec{v}_{B/A}. \tag{6.1}$$

Next, noting that $\vec{v}_{B/A} = \dot{\vec{r}}_{B/A}$ and referring to Fig. 6.6 to write $\vec{r}_{B/A}$ as $|\vec{r}_{B/A}|\,\hat{u}_{B/A}$, we can use Eq. (2.62) on p. 92, which says that the time derivative of a vector is equal to its time rate of change in magnitude plus its change in direction, which is given by the angular velocity of the vector crossed with the vector itself, to obtain

$$\vec{v}_{B/A} = \frac{d\,|\vec{r}_{B/A}|}{dt}\,\hat{u}_{B/A} + \vec{\omega}_{AB} \times \vec{r}_{B/A} = \vec{\omega}_{AB} \times \vec{r}_{B/A}, \tag{6.2}$$

Figure 6.6
Aerial view of an aircraft carrier performing a maneuver. We model the carrier's motion as a planar rigid body motion.

where we have used the fact that $d|\vec{r}_{B/A}|/dt = 0$ since *the distance between A and B is constant when they are both on the carrier*. Substituting Eq. (6.2) into Eq. (6.1) gives

$$\vec{v}_B = \vec{v}_A + \vec{\omega}_{AB} \times \vec{r}_{B/A} = \vec{v}_A + \vec{v}_{B/A}, \qquad (6.3)$$

where we note that when A and B are two points on a rigid body, $\vec{v}_{B/A} = \vec{\omega}_{AB} \times \vec{r}_{B/A}$. This equation relating the motion of points A and B depends on the angular velocity of the line segment $\overline{AB}$. We can see that as the aircraft carrier moves from position 1 to position 2 (Fig. 6.7), the three lines segments $\overline{AB}$, $\overline{BC}$, and $\overline{CA}$ all rotate the same amount. This implies that *any* line seg-

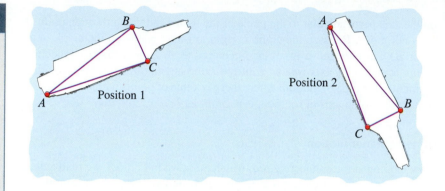

Figure 6.7. Aerial view of an aircraft carrier moving from position 1 to position 2 showing that *all* line segments rotate the same amount as a rigid body moves.

ment on the aircraft carrier will have the same rate of rotation as any other line segment, and so they must *all* rotate at the same rate. Thus, *the angular velocity of the body* $\vec{\omega}_{AB}$ is a property of the body as a whole and not any particular part of it. This means that Eq. (6.3) applies to any two points A and B on the *same* rigid body. The subscripts on angular velocities (and soon, angular accelerations) should be viewed as labels referring to a particular body. For example, in Eq. (6.3), the subscript AB on $\vec{\omega}$ tells us that it is the angular velocity of the body containing points A and B.

The equation relating the acceleration of two points on the same rigid body is found by differentiating Eq. (6.3) with respect to time to obtain

$$\vec{a}_B = \vec{a}_A + \dot{\vec{\omega}}_{AB} \times \vec{r}_{B/A} + \vec{\omega}_{AB} \times \dot{\vec{r}}_{B/A}. \qquad (6.4)$$

The quantity $\dot{\vec{\omega}}_{AB}$ is the *angular acceleration of the body* and is denoted by $\vec{\alpha}_{AB}$. Recalling that $\dot{\vec{r}}_{B/A} = \vec{v}_{B/A} = \vec{\omega}_{AB} \times \vec{r}_{B/A}$, Eq. (6.4) can be written as

$$\vec{a}_B = \vec{a}_A + \vec{\alpha}_{AB} \times \vec{r}_{B/A} + \vec{\omega}_{AB} \times (\vec{\omega}_{AB} \times \vec{r}_{B/A}) = \vec{a}_A + \vec{a}_{B/A}, \qquad (6.5)$$

where we note that when A and B are two points on a rigid body, $\vec{a}_{B/A} = \vec{\alpha}_{AB} \times \vec{r}_{B/A} + \vec{\omega}_{AB} \times (\vec{\omega}_{AB} \times \vec{r}_{B/A})$.

Applying Eqs. (6.3) and (6.5)

Equations (6.3) and (6.5) will be applied to analyze mechanisms consisting of several rigid bodies. In using these equations, points A and B *must* belong to the same rigid body, and $\vec{\omega}_{\text{body}}$ and $\vec{\alpha}_{\text{body}}$ must be the body's angular velocity and acceleration, respectively. For example, referring to Fig. 6.8, we can use Eqs. (6.3) and (6.5) to relate the velocities and accelerations, respectively, of points A and B because these points are both on the crank. We can do the same for points B and C because they are both on the connecting rod. Notice that B is a point on *both* the crank and the connecting rod, and so it allows us to relate the motion of the connecting rod to that of the crank. We will frequently make use of common points on connected bodies in the kinematic analysis of rigid bodies.

Notion of extended rigid body

Referring to Fig. 6.9, a hammer handle D is attached to a hammer head H. If both bodies are modeled as rigid, and if they are *rigidly connected*, then D

Figure 6.9. A claw hammer consisting of two rigid bodies, the head and the handle.

and H can be viewed as *parts* of a single composite rigid body—a hammer. In this case, there exists a single $\vec{\omega}_{\text{ham}}$ and a single $\vec{\alpha}_{\text{ham}}$ for both D and H, and we can use Eqs. (6.3) and (6.5) to relate the velocity and acceleration of any pair of points on the *composite body*. We can generalize this idea by noting that we can always conceive of a physical body D as being a part of a *fictitious* larger rigid body H, which can be thought of as occupying all the space surrounding D (see Fig. 6.10). This fictitious rigid body is referred to as the

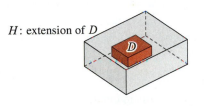

Figure 6.10. Notion of *extended rigid body*.

extended rigid body of the original body, and we can use Eqs. (6.3) and (6.5) to relate velocities and accelerations of pairs of points on the extended rigid body and/or the physical body.

Elementary rigid body motions: translations

Figure 6.11 shows the deployment of a basketball goal. The backboard AB and hoop are attached to an arm, which, in turn is hinged to two parallel bars CD and EF. The design is meant to ensure that the hoop remains parallel to the floor. Therefore, by definition, the backboard, hoop, and supporting arm are in *translation*, i.e., a motion in which the body never changes its orientation. This means that the angular velocity and angular acceleration of a translating rigid body are equal to zero, that is,

$$\vec{\omega}_{AB} = \vec{\omega}_{\text{body}} = \vec{0} \quad \text{and} \quad \vec{\alpha}_{AB} = \vec{\alpha}_{\text{body}} = \vec{0}. \tag{6.6}$$

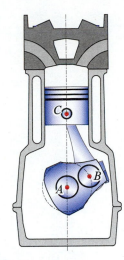

Figure 6.8
Slider-crank mechanism.

> ### 🔦 Concept Alert
>
> **Relating points A and C in Fig. 6.8.** We cannot use Eqs. (6.3) and (6.5) to relate the motion of points A and C because these points belong to two distinct bodies—the crank and the piston, respectively.

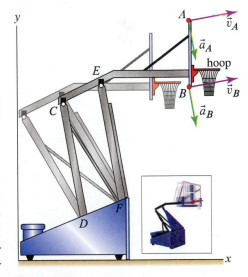

Figure 6.11
The deployment of a portable basketball goal (see inset photo).

Substituting Eqs. (6.6) into Eqs. (6.3) and (6.5) gives

$$\vec{v}_B = \vec{v}_A \quad \text{and} \quad \vec{a}_B = \vec{a}_A, \tag{6.7}$$

where A and B are any two arbitrarily chosen points on the body, as shown in Fig. 6.11. Equation (6.7) says that the motion of a translating rigid body is characterized by a single value of velocity and a single value of acceleration.

Elementary rigid body motions: rotation about a fixed axis

We now go back to a question we asked earlier: How does the geometry of a slider-crank mechanism affect its motion? We begin by analyzing the motion of point B on the crank (see Fig. 6.12). We will continue the analysis of other parts of the slider-crank mechanism in Section 6.2.

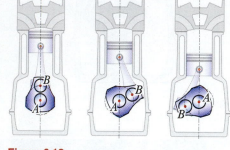

We have seen that the crank is *rotating about a fixed axis*. Referring to Fig. 6.13, the axis of rotation is perpendicular to the plane of motion and goes through point A, which is fixed, i.e., $\vec{v}_A = \vec{0}$ and $\vec{a}_A = \vec{0}$. Equations (6.3) and (6.5) then give

$$\vec{v}_B = \vec{\omega}_{AB} \times \vec{r}_{B/A}, \tag{6.8}$$

$$\vec{a}_B = \vec{\alpha}_{AB} \times \vec{r}_{B/A} + \vec{\omega}_{AB} \times (\vec{\omega}_{AB} \times \vec{r}_{B/A}). \tag{6.9}$$

Figure 6.12
Crank motion in a slider-crank mechanism.

Since the motion is planar, $\vec{\omega}_{AB}$ and $\vec{\alpha}_{AB}$ can be written in terms of the body's orientation. This is done by noting that points A and B are in the plane of motion and that the body's orientation θ can be defined using the line connecting A and B (Fig. 6.13). Therefore, the vectors $\vec{\omega}_{AB}$ and $\vec{\alpha}_{AB}$ can be written as:

$$\vec{\omega}_{AB} = \omega_{AB}\,\hat{k} = \dot{\theta}\,\hat{k} \quad \text{and} \quad \vec{\alpha}_{AB} = \alpha_{AB}\,\hat{k} = \ddot{\theta}\,\hat{k}, \tag{6.10}$$

where $\hat{k} = \hat{u}_r \times \hat{u}_\theta = \hat{\imath} \times \hat{\jmath}$ is perpendicular to the plane of motion, and ω_{AB} and α_{AB} denote the components of the angular velocity and acceleration, respectively, in the $\hat{k}$ direction. Using the polar component system shown in Fig. 6.13, letting $\vec{r}_{B/A} = R\,\hat{u}_r$, and substituting Eqs. (6.10) into Eqs. (6.8) and (6.9), $\vec{v}_B$ and $\vec{a}_B$ become

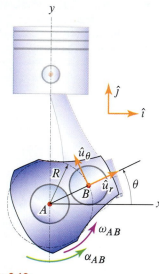

$$\vec{v}_B = R\dot{\theta}\,\hat{u}_\theta \quad \text{and} \quad \vec{a}_B = R\ddot{\theta}\,\hat{u}_\theta - R\dot{\theta}^2\,\hat{u}_r, \tag{6.11}$$

which isn't surprising since B is in circular motion about A.

The acceleration coming from the term $\vec{\omega}_{AB} \times (\vec{\omega}_{AB} \times \vec{r}_{B/A})$ is equal to $-R\dot{\theta}^2\,\hat{u}_r$. Noting that $\dot{\theta}^2 = \omega_{AB}^2$ and that $-R\,\hat{u}_R = -\vec{r}_{B/A}$, for planar motion, we can write

$$\vec{\omega}_{AB} \times (\vec{\omega}_{AB} \times \vec{r}_{B/A}) = -R\dot{\theta}^2\,\hat{u}_r = -\omega_{AB}^2\,\vec{r}_{B/A}, \tag{6.12}$$

Figure 6.13
Detailed view of the crank of a slider-crank mechanism.

We can see this geometrically if we note that $\vec{r}_{B/A}$ is always in the plane of motion and that $\vec{\omega}_{AB}$ is always perpendicular to it (see Fig. 6.14). Taking their cross product $\vec{\omega}_{AB} \times \vec{r}_{B/A}$ results in a vector that is in the plane of motion and perpendicular to both $\vec{\omega}_{AB}$ and $\vec{r}_{B/A}$. Finally, taking $\vec{\omega}_{AB} \times (\vec{\omega}_{AB} \times \vec{r}_{B/A})$ (the cross product of the perpendicular purple vectors in Fig. 6.14) results in the vector $-\omega_{AB}^2\,\vec{r}_{B/A}$ (the green vector in Fig. 6.14). Using Eq. (6.12), we can write Eq. (6.9) as

$$\vec{a}_B = \vec{\alpha}_{AB} \times \vec{r}_{B/A} - \omega_{AB}^2\,\vec{r}_{B/A}, \tag{6.13}$$

which is a form that can save substantial computation when we compute accelerations for planar problems.

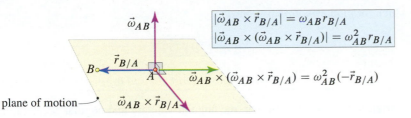

Figure 6.14. Geometric demonstration of the equivalence of $\vec{\omega}_{AB} \times (\vec{\omega}_{AB} \times \vec{r}_{B/A})$ and $-\omega_{AB}^2 \vec{r}_{B/A}$ for planar motion.

Graphical interpretation of Eq. (6.8).   Referring to the crank in Fig. 6.15, consider the velocity of points H, B, and Q lying on the radial line ℓ with origin at the center of rotation A. Equation (6.8), or the first of Eqs. (6.11), implies that $\vec{v}_H$, $\vec{v}_B$, and $\vec{v}_Q$ are all perpendicular to ℓ (and parallel to one another) and have a magnitude *proportional* to their distance from A. The constant of proportionality is ω_{AB}, which Fig. 6.15 shows is also $\tan \psi$. Therefore, the distribution of the velocities of points on radial lines can be represented graphically via a triangle as shown.

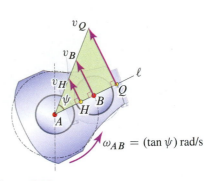

Figure 6.15
Graphical representation of the velocities of points on radial lines originating at the center of rotation.

Planar motion in practice

For planar motions, we just saw that it is *always* possible to express the term $\vec{\omega}_{AB} \times (\vec{\omega}_{AB} \times \vec{r}_{B/A})$ as $-\omega_{AB}^2 \vec{r}_{B/A}$. Equation (6.5) can then be written as

$$\vec{a}_B = \vec{a}_A + \vec{\alpha}_{AB} \times \vec{r}_{B/A} - \omega_{AB}^2 \vec{r}_{B/A}. \qquad (6.14)$$

When relating the motion of two points A and B on the *same* rigid body, we will generally use this version of the acceleration equation.

End of Section Summary

This section began our study of the dynamics of rigid bodies. As with particles, we start with the study of kinematics, which is the focus of this chapter. The key kinematic idea is that a rigid body has only one angular velocity and one angular acceleration; i.e., each is a property of the body as a whole. This allowed us to use the relative velocity equation (Eq. (2.106) on p. 136), and the relation for the time derivative of a vector (Eq. (2.62) on p. 92) to relate the velocities of *two points A and B on the same rigid body* using (see Fig. 6.16)

Eq. (6.3), p. 462

$$\vec{v}_B = \vec{v}_A + \vec{\omega}_{AB} \times \vec{r}_{B/A} = \vec{v}_A + \vec{v}_{B/A},$$

and the accelerations using

Eq. (6.5), p. 462

$$\vec{a}_B = \vec{a}_A + \vec{\alpha}_{AB} \times \vec{r}_{B/A} + \vec{\omega}_{AB} \times (\vec{\omega}_{AB} \times \vec{r}_{B/A}) = \vec{a}_A + \vec{a}_{B/A},$$

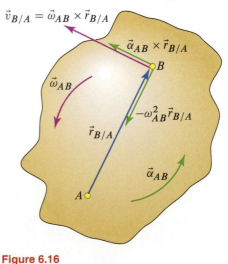

Figure 6.16
A rigid body showing the quantities used in Eqs. (6.3), (6.5), and (6.14).

which for *planar motion* becomes

Eq. (6.14), p. 465

$$\vec{a}_B = \vec{a}_A + \vec{\alpha}_{AB} \times \vec{r}_{B/A} - \omega_{AB}^2 \vec{r}_{B/A}.$$

Translation.   For this motion, the angular velocity and angular acceleration of the body are equal to zero, i.e.,

Eq. (6.6), p. 463

$$\vec{\omega}_{\text{body}} = \vec{0} \quad \text{and} \quad \vec{\alpha}_{\text{body}} = \vec{0},$$

so that the velocity and acceleration relations for the body reduce to

Eq. (6.7), p. 464

$$\vec{v}_B = \vec{v}_A \quad \text{and} \quad \vec{a}_B = \vec{a}_A,$$

where A and B are any two points on the body.

Rotation about a fixed axis.   For this special motion, there is an axis of rotation perpendicular to the plane of motion that does not move. All points not on the axis of rotation can only move in a circle about that axis. If the axis of rotation is at point A, then the velocity of B is given by

Eq. (6.8), p. 464

$$\vec{v}_B = \vec{\omega}_{AB} \times \vec{r}_{B/A},$$

and its acceleration is

Eqs. (6.9) and (6.13), p. 464

$$\vec{a}_B = \vec{\alpha}_{AB} \times \vec{r}_{B/A} + \vec{\omega}_{AB} \times (\vec{\omega}_{AB} \times \vec{r}_{B/A}),$$
$$\vec{a}_B = \vec{\alpha}_{AB} \times \vec{r}_{B/A} - \omega_{AB}^2 \vec{r}_{B/A},$$

where the vectors $\vec{\omega}_{AB}$ and $\vec{\alpha}_{AB}$ are the angular velocity and angular acceleration of the body, respectively, and $\vec{r}_{B/A}$ is the vector that describes the position of B relative to A.

EXAMPLE 6.1   *Engine Pulleys: Fixed Axis Rotation*

Most car engines have a number of belts connecting pulleys on the engine (Fig. 1). The belts are not supposed to slip relative to the pulleys they connect and are used to transmit as well as synchronize motion between engine parts. For the belt connecting pulley A, which rotates with the crankshaft, with pulley B, which drives the alternator, determine the angular speed of pulley B if the crankshaft is spinning at 2550 rpm and the radii of pulleys A and B are $R_A = 4.25$ in. and $R_B = 2.5$ in., respectively.

SOLUTION

Road Map Referring to Fig. 2, the no-slip condition between the belt and pulleys means that any two points on the belt and pulley that are in contact at a given instant, e.g., C and D or P and Q, must have the same velocity (i.e., $\vec{v}_{C/D} = \vec{v}_{P/Q} = \vec{0}$). Combining this observation with the fact that pulleys A and B rotate about fixed axes and the assumption that the belt is inextensible (all points on the belt must have the same *speed*) will allow us to solve the problem.

Computation From Fig. 2, the inextensibility of the belt implies that

$$|\vec{v}_C| = |\vec{v}_P|. \tag{1}$$

Using this result and recalling the no slip condition, we find

$$\vec{v}_C = \vec{v}_D \quad \text{and} \quad \vec{v}_P = \vec{v}_Q \quad \Rightarrow \quad |\vec{v}_D| = |\vec{v}_Q|. \tag{2}$$

Referring to Fig. 2, pulleys A and B undergo fixed axis rotation about their respective centers. Applying Eq. (6.8) to each of the pulleys to obtain $\vec{v}_D$ and $\vec{v}_Q$, we find

$$\vec{v}_D = \vec{\omega}_A \times \vec{r}_{D/A} \quad \text{and} \quad \vec{v}_Q = \vec{\omega}_B \times \vec{r}_{Q/B}, \tag{3}$$

and since $\vec{\omega}$ and $\vec{r}$ are perpendicular to each other in each case, the corresponding speeds are

$$|\vec{v}_D| = |\vec{\omega}_A| R_A \quad \text{and} \quad |\vec{v}_Q| = |\vec{\omega}_B| R_B. \tag{4}$$

Substituting Eq. (4) into the last of Eqs. (2), we obtain

$$|\vec{\omega}_A| R_A = |\vec{\omega}_B| R_B \quad \Rightarrow \quad |\vec{\omega}_B| = \frac{R_A}{R_B} |\vec{\omega}_A| = 4340 \text{ rpm}, \tag{5}$$

where we have plugged in $|\vec{\omega}_A| = 2550$ rpm, R_A, and R_B.

Discussion & Verification To verify the result in Eq. (5), Fig. 3 shows that the no-slip condition between the belt and pulley A implies that if pulley A rotates through an angle θ_A, then the belt must move an amount $\Delta L = \theta_A R_A$ around pulley A. Since the belt does not slip with respect to pulley B, we must also have $\Delta L = \theta_B R_B$, where θ_B is the angle through which pulley B rotates. Therefore, we must have $\theta_B = (R_A/R_B)\theta_A$. This relation implies that the rates of change of the angle θ_B will be proportional to θ_A via the ratio of the pulley radii R_A/R_B —this is what we obtained in Eq. (5).

Figure 1
Typical car engine with several belts.

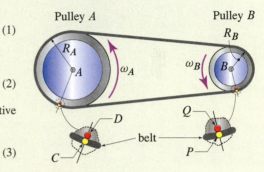

Figure 2
Schematic of the system consisting of pulleys A and B and the belt connecting them. The pulleys are attached to the engine, which is assumed to be stationary.

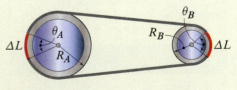

Figure 3
Measure of the linear length of belt going around the pulleys under the no-slip condition.

E X A M P L E   6.2   *Gears: Fixed Axis Rotation*

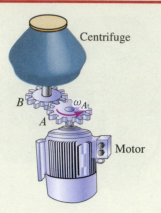

Centrifuge

B

ω_A

A

Motor

Figure 1
A centrifuge driven by an electric motor via a gear system.

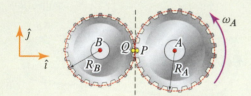

$\hat{j}$

$\hat{\imath}$

B

Q P

A

R_B R_A

ω_A

Figure 2
Two meshing gears. Note that the gears' radii are the radii of the circumferences indicated by the dashed lines.

An electric motor with a top angular speed of 3450 rpm is used to spin a centrifuge. The motor's motion is transmitted to the centrifuge via two gears A and B, with radii R_A and R_B, respectively (see Fig. 1). Determine the ratio R_A/R_B if the centrifuge is to achieve 6000 rpm as its top angular speed. In addition, determine the ratio between the magnitude of the angular acceleration of A to that of B during spin-up.

SOLUTION

Road Map   Gear teeth are designed so that the gears behave as two wheels rolling without slip on each other. The radius of a gear is the radius of the wheel that the gear is designed to represent (see the dashed red line in Fig. 2), as opposed to the inner or outer radius of the teeth. Therefore, referring to Fig. 2, the no-slip condition between the gears tells us that the velocity of points P and Q must be the same when they are in contact. Furthermore, gears A and B are rotating about the fixed axes of the motor shaft and centrifuge shaft, respectively.

Computation   The velocity of points P and Q, which are moving in circular motion about A and B, respectively, can be found by applying Eq. (6.8), which gives

$$\vec{v}_P = \vec{\omega}_A \times \vec{r}_{P/A} = \omega_A\,\hat{k} \times (-R_A)\,\hat{\imath} = -\omega_A R_A\,\hat{\jmath}, \tag{1}$$

$$\vec{v}_Q = \vec{\omega}_B \times \vec{r}_{Q/B} = \omega_B\,\hat{k} \times R_B\,\hat{\imath} = \omega_B R_B\,\hat{\jmath}, \tag{2}$$

where we have assumed that the angular velocities of A and B are both in the positive $\hat{k}$ direction. Enforcing the no-slip condition between the gears, we have

$$\vec{v}_P = \vec{v}_Q \quad \Rightarrow \quad -R_A\omega_A\,\hat{\jmath} = R_B\omega_B\,\hat{\jmath} \quad \Rightarrow \quad -R_A\omega_A = R_B\omega_B, \tag{3}$$

where the minus sign tells us that the two gears spin in opposite directions since R_A and R_B are both positive. Since $|\omega_A| = 3450$ rpm and $|\omega_B| = 6000$ rpm, we can solve Eq. (3) for R_A/R_B to obtain

$$\boxed{\frac{R_A}{R_B} = \left|\frac{\omega_B}{\omega_A}\right| = \frac{6000\text{ rpm}}{3450\text{ rpm}} = 1.739.} \tag{4}$$

To find the ratio of the angular acceleration of gear A to that of gear B, we can obtain the desired information by taking the time derivative of the result in Eq. (3) to obtain

$$-R_A\alpha_A = R_B\alpha_B \quad \Rightarrow \quad \boxed{\left|-\frac{\alpha_A}{\alpha_B}\right| = \frac{R_B}{R_A} = \frac{1}{1.739} = 0.575.} \tag{5}$$

Discussion & Verification   These results are dimensionally correct and are consistent with the result in Example 6.1; that is, Eq. (5) indicates that the angular speed and the magnitude of the angular accelerations of two gears are proportional to each other via the ratio of the gears' radii.

🔍 **A Closer Look**   The result in Eq. (4) could be found using the ideas developed in Chapter 2 (see, for example, Eq. (2.46) on p. 59 or Eq. (2.93) on p. 120), along with the knowledge of how gears rotate. From Chapter 2, we know that for circular motion, the speed is equal to the radius of the path times the angular velocity and the direction is determined by the tangent to the path at the point of interest. Therefore, from Fig. 2 we have $\vec{v}_P = R_A|\omega_A|(-\hat{\jmath}) = -R_A|\omega_A|\,\hat{\jmath}$ and $\vec{v}_Q = R_B|\omega_B|\,\hat{\jmath}$ (assuming Q moves in the positive $\hat{\jmath}$ direction). Determining the gear ratio R_A/R_B is then a matter of once again setting $\vec{v}_P = \vec{v}_Q$ as was done in Eqs. (3) and (4).

EXAMPLE 6.3   *A Carnival Ride: Translation*

Motion platforms, which are used in many of today's amusement parks, are a type of carnival ride in which a platform with seats is made to move always parallel to the ground. Figure 1 shows an elementary type of motion platform, more often found in traveling carnivals than in big amusement parks. Given that the motion platform in the figure (see also Fig. 2) is designed so that the rotating arms AB and CD are of equal length $L = 10$ ft and remain parallel to each other, determine the velocity and acceleration of a person P onboard the ride when ω_{AB} is constant and equal to 1.25 rad/s.

Figure 1
A carnival ride consisting of a motion platform

SOLUTION

Road Map   To simplify the problem, we will assume that the platform and the persons on it form a single rigid body. This means that the two rotating arms and the platform form a *four-bar linkage*.* Because the arms AB and CD are identical in size and are always parallel to each other, the four-bar linkage $ABCD$ always forms a parallelogram, and so the platform BC does not change its orientation and has zero angular velocity. Knowing this and applying the kinematics of fixed axis rotation will allow us to determine $\vec{v}_P$ and $\vec{a}_P$.

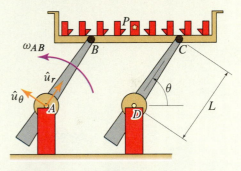

Figure 2
Coordinate definition and geometry for the carnival ride.

Computation   Again, from Fig. 2, we see that arms AB and CD are always parallel to each other, and so platform BC does not change its orientation. From the discussion on p. 463, this means the angular velocity of the platform is zero, that is,

$$\omega_{BC} = 0. \tag{1}$$

Also, observe that points B and C move along circles centered at A and D, respectively, where A and D are fixed points. Given the fact that the trajectories of points on BC are not a straight line, BC's motion is a *curvilinear translation*. Consequently, all points on BC, or any rigid extension of it, i.e., any of the passengers, share the same value of velocity as well as acceleration. Therefore we have

$$\boxed{\vec{v}_P = \vec{v}_B = \omega_{AB} L\, \hat{u}_\theta = 12.5 \,\text{ft/s}\, \hat{u}_\theta} \tag{2}$$

and

$$\boxed{\vec{a}_P = \vec{a}_B = -\omega_{AB}^2 L\, \hat{u}_r = -15.6 \,\text{ft/s}^2\, \hat{u}_r} \tag{3}$$

where the velocity and acceleration of B were computed using circular motion formulas in Eqs. (6.11).

Discussion & Verification   The dimensions and units in Eqs. (2) and (3) are correct, and the magnitudes of the velocity and acceleration of P are reasonable. In particular, the acceleration is not far from those found in general public (as opposed to extreme) carnival rides, and it amounts to a little less than 50% of the acceleration of gravity.

🔎 **A Closer Look**   It is important to remember that the direction of $\vec{v}_P$ in Eq. (2) and the direction of $\vec{a}_P$ in Eq. (3) are those shown at point A in Fig. 2. That is, since P is *not* in fixed axis rotation about A, the $\hat{u}_r$ and $\hat{u}_\theta$ in Eqs. (2) and (3) are not those of P relative to A.

---

* A *four-bar linkage* is a mechanism with four members or links in which one of the links is fixed (link AD in this example) and the other three move in a predetermined fashion (links AB, BC, and CD in this example).

PROBLEMS

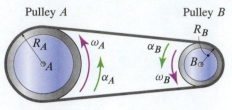

Pulley A Pulley B

Figure P6.1

Problem 6.1

Letting $R_A = 7.2$ in. and $R_B = 4.6$ in., and assuming that the belt does not slip relative to pulleys A and B, determine the angular velocity and angular acceleration of pulley B when pulley A rotates at 340 rad/s while accelerating at 120 rad/s^2.

Problem 6.2

Letting $R_A = 8$ in., $R_B = 4.2$ in., and $R_C = 6.5$ in., determine the angular velocity of gears B and C when gear A has an angular speed $|\omega_A| = 945$ rpm in the direction shown.

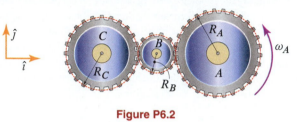

Figure P6.2

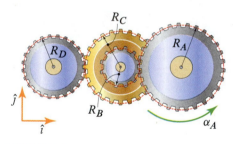

Figure P6.3

Problem 6.3

Letting $R_A = 203$ mm, $R_B = 107$ mm, $R_C = 165$ mm, and $R_D = 140$ mm, determine the angular acceleration of gears B, C, and D when gear A has an angular acceleration with magnitude $|\alpha_A| = 47$ rad/s^2 in the direction shown. Note that gears B and C are mounted on the same shaft and they rotate together as a unit.

Problem 6.4

The bevel gears A and B have nominal radii $R_A = 20$ mm and $R_B = 5$ mm, respectively, and their axes of rotation are mutually perpendicular. If the angular speed of gear A is $\omega_A = 150$ rad/s, determine the angular speed of gear B.

Figure P6.4

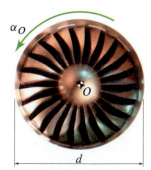

Figure P6.5

Problem 6.5

A rotor with a fixed spin axis identified by point O is accelerated from rest with an angular acceleration $\alpha_O = 0.5$ rad/s^2. If the rotor's diameter is $d = 15$ ft, determine the time it takes for the points at the outer edge of the rotor to reach a speed $v_0 = 300$ ft/s. Finally, determine the magnitude of the acceleration of these points when the speed v_0 is achieved.

Problem 6.6

Do points A and B on the surface of the bevel gear (bg), which rotates with angular velocity ω_{bg}, move relative to each other? At what rate does the distance between A and B change?
Note: Concept problems are about *explanations*, not computations.

Figure P6.6

Problem 6.7

A Pelton turbine (a type of turbine used in hydroelectric power generation) is spinning at 1100 rpm when the water jets acting on it are shut off, thus causing the turbine to slow down. Assuming that the angular deceleration rate is constant and equal to 1.31 rad/s², determine the time it takes for the turbine to stop. In addition, determine the number of revolutions of the turbine during the spin-down.

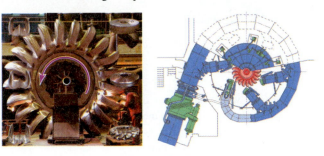

Figure P6.7

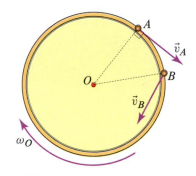

Figure P6.8

Problem 6.8

The velocities of points A and B on a disk, which is undergoing planar motion, are such that $|\vec{v}_A| = |\vec{v}_B|$. Is it possible for the disk to be rotating about a fixed axis going through the center of the disk at O? Explain.
Note: Concept problems are about *explanations*, not computations.

Problem 6.9

The velocity of a point A and the acceleration of a point B on a disk undergoing planar motion are shown. Is it possible for the disk to be rotating about a fixed axis going through the center of the disk at O? Explain.
Note: Concept problems are about *explanations*, not computations.

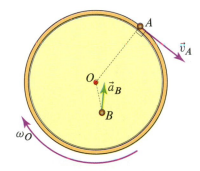

Figure P6.9 and P6.10

Problem 6.10

Assuming that the disk shown is rotating about a fixed axis going through its center at O, determine whether the disk's angular velocity is constant, increasing, or decreasing.
Note: Concept problems are about *explanations*, not computations.

Problem 6.11

The sprinkler shown consists of a pipe AB mounted on a hollow vertical shaft. The water comes in the horizontal pipe at O and goes out the nozzles at A and B, causing the pipe to rotate. Letting $d = 7$ in., determine the angular velocity of the sprinkler ω_s, and $|\vec{a}_B|$, the magnitude of the acceleration of B, if B is moving with a constant speed $v_B = 20$ ft/s. Assume that the sprinkler does not roll on the ground.

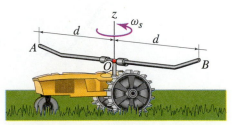

Figure P6.11

Problem 6.12

In a carnival ride, two gondolas spin in opposite directions about a fixed axis. If $\ell = 4\,\text{m}$, determine the maximum constant angular speed of the gondolas if the magnitude of the acceleration of point A is not to exceed $2.5g$.

Figure P6.12

Problem 6.13

The tractor shown is stuck with its right track off the ground, and therefore the track is able to move without causing the tractor to move. Letting the radius of sprocket A be $d = 2.5\,\text{ft}$ and the radius of sprocket B be $\ell = 2\,\text{ft}$, determine the angular speed of wheel B if the sprocket A is rotating at 1 rpm.

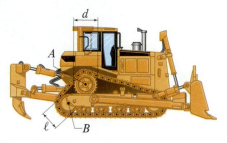

Figure P6.13

Problem 6.14

A battering ram is suspended in its frame via bars AD and BC, which are identical and pinned at their endpoints. At the instant shown, point E on the ram moves with a speed $v_0 = 15\,\text{m/s}$. Letting $H = 1.75\,\text{m}$ and $\theta = 20°$, determine the magnitude of the angular velocity of the ram at this instant.

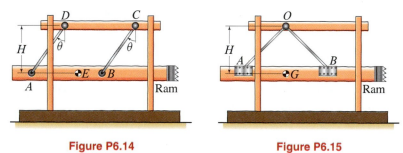

Figure P6.14   **Figure P6.15**

Problem 6.15

A battering ram is suspended in its frame via bars OA and OB, which are pinned at O. At the instant shown, point G on the ram is moving to the right with a speed $v_0 = 15\,\text{m/s}$. Letting $H = 1.75\,\text{m}$, determine the angular velocity of the ram at this instant.

Problems 6.16 and 6.17

The hammer of a Charpy impact toughness test machine has the geometry shown, where G, is the mass center of the hammer head. Use Eqs. (6.8) and (6.13) and write your answers in terms of the component system shown.

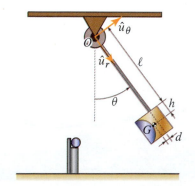

Figure P6.16 and P6.17

Problem 6.16 Determine the velocity and acceleration of G, assuming $\ell = 500$ mm, $h = 65$ mm, $d = 25$ mm, $\dot{\theta} = -5.98$ rad/s, and $\ddot{\theta} = -8.06$ rad/s^2.

Problem 6.17 Determine the velocity and acceleration of G as a function of the geometric parameters shown, $\dot{\theta}$, and $\ddot{\theta}$.

Problem 6.18

At the instant shown the paper is being unrolled with a speed $v_p = 7.5$ m/s and an acceleration $a_p = 1$ m/s^2. If at this instant the outer radius of the roll is $r = 0.75$ m, determine the angular velocity ω_s and acceleration α_s of the roll.

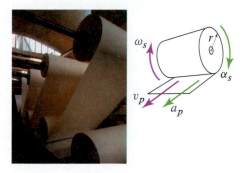

Figure P6.18

Problem 6.19

At the instant shown, the propeller is rotating with an angular velocity $\omega_p = 400$ rpm in the positive z direction and an angular acceleration $\alpha_p = -2$ rad/s^2 in the negative z direction, where the z axis is also the spin axis of the propeller. Consider the cylindrical coordinate system shown, with origin O on the z axis and unit vector $\hat{u}_R$ pointing toward point Q on the propeller which is 14 ft away from the spin axis. Compute the velocity and acceleration of Q. Express your answer using the component system shown.

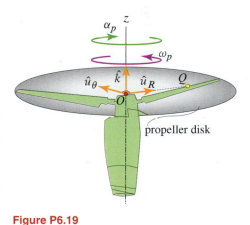

Figure P6.19

Problem 6.20

The wheel A, with diameter $d = 5$ cm, is mounted on the shaft of the motor shown and is rotating with a constant angular speed $\omega_A = 250$ rpm. The wheel B, with center at the fixed point O, is connected to A with a belt, which does not slip relative to A or B. The radius of B is $R = 12.5$ cm. At point C the wheel B is connected to a saw. If point C is at distance $\ell = 10$ cm from O, determine the velocity and acceleration of C when $\theta = 20°$. Express your answers using the component system shown.

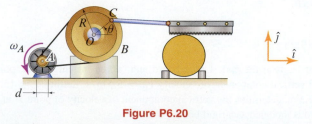

Figure P6.20

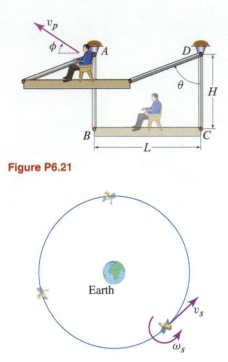

Figure P6.21

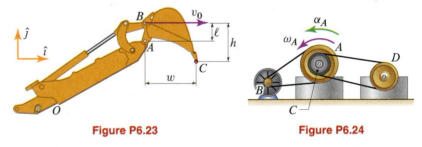

Figure P6.22

Problem 6.21

In a contraption built by a fraternity, a person is sitting at the center of a swinging platform with length $L = 12$ ft that is suspended via two identical arms each of length $H = 10$ ft. Determine the angle θ and the angular speed of the arms if the person is moving upward and to the left with a speed $v_p = 25$ ft/s at the angle $\phi = 33°$.

Problem 6.22

A *geosynchronous equatorial orbit* is a circular orbit above the Earth's equator that has a period of 1 day (these are sometimes called *geostationary orbits*). These geostationary orbits are of great importance for telecommunications satellites because a satellite orbiting with the same angular rate as the rotation rate of the Earth will appear to hover in the same point in the sky as seen by a person standing on the surface of the Earth. Using this information, modeling a geosynchronous satellite as a rigid body, and noting that the satellite has been stabilized so that the same side always faces the Earth, determine the angular speed ω_s of the satellite.

Problem 6.23

The bucket of a backhoe is being operated while holding the arm OA fixed. At the instant shown, point B has a horizontal component of velocity $v_0 = 0.25$ ft/s and is vertically aligned with point A. Letting $\ell = 0.9$ ft, $w = 2.65$ ft, and $h = 1.95$ ft, determine the velocity of point C. In addition, assuming that, at the instant shown, point B is not accelerating in the horizontal direction, compute the acceleration of point C. Express your answers using the component system shown.

Figure P6.23   **Figure P6.24**

Problem 6.24

Wheels A and C are mounted on the same shaft and rotate together. Wheels A and B are connected via a belt and so are wheels C and D. The axes of rotation of all the wheels are fixed, and the belts do not slip relative to the wheels they connect. If, at the instant shown, wheel A has an angular velocity $\omega_A = 2$ rad/s and an angular acceleration $\alpha_A = 0.5$ rad/s^2, determine the angular velocity and acceleration of wheels B and D. The radii of the wheels are $R_A = 1$ ft, $R_B = 0.25$ ft, $R_C = 0.6$ ft, and $R_D = 0.75$ ft.

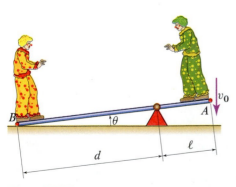

Figure P6.25

Problem 6.25

An acrobat lands at the end A of a board and, at the instant shown, point A has a downward vertical component of velocity $v_0 = 5.5$ m/s. Letting $\theta = 15°$, $\ell = 1$ m, and $d = 2.5$ m, determine the vertical component of velocity of point B at this instant if the board is modeled as a rigid body.

Problem 6.26

At the instant shown, A is moving upward with a speed $v_0 = 5\,\text{ft/s}$ and acceleration $a_0 = 0.65\,\text{ft/s}^2$. Assuming that the rope that connects the pulleys does not slip relative to the pulleys and letting $\ell = 6$ in. and $d = 4$ in., determine the angular velocity and angular acceleration of pulley C.

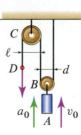

Figure P6.26

Problem 6.27

At the instant shown, the angle $\phi = 30°$, $|\vec{v}_A| = 292\,\text{ft/s}$, and the turbine is rotating clockwise. Letting $\overline{OA} = R$, $\overline{OB} = R/2$, $R = 182$ ft, and treating the blades as being equally spaced, determine the velocity of point B at the given instant and express it using the component system shown.

Problem 6.28

At the instant shown, the angle $\phi = 30°$, the turbine is rotating clockwise, and $\vec{a}_B = (70.8\,\hat{\imath} - 12.8\,\hat{\jmath})\,\text{m/s}^2$. Letting $\overline{OA} = R$, $\overline{OB} = R/2$, $R = 55.5$ m, and assuming the blades are equally spaced, determine the angular velocity and angular acceleration of the turbine blades as well as the acceleration of point A at the given instant.

Figure P6.27 and P6.28

Problems 6.29 through 6.31

A bicycle has wheels 700 mm in diameter and a gear set with the dimensions given in the table below.

Figure P6.29–P6.31

Crank			
Sprocket	C1	C2	C3
No. of Cogs	26	36	48
Radius (mm)	52.6	72.8	97.0

Cassette (9 speeds)									
Sprocket	S1	S2	S3	S4	S5	S6	S7	S8	S9
No. of Cogs	11	12	14	16	18	21	24	28	34
Radius (mm)	22.2	24.3	28.3	32.3	36.4	42.4	48.5	56.6	68.7

Problem 6.29   If a cyclist has a cadence of 1 Hz, determine the angular speed of the rear wheel in rpm when using the combination of C3 and S2. In addition, knowing that the speed of the cyclist is equal to the speed of a point on the tire relative to the wheel's center, determine the cyclist's speed in m/s.

Problem 6.30 If a cyclist has a cadence of 68 rpm, determine which combination of chain ring (a sprocket mounted on the crank) and (rear) sprocket would *most closely* make the rear wheel rotate with an angular speed of 127 rpm. Having found a chain ring/sprocket combination, determine the wheel's exact angular speed corresponding to the chosen chain ring/sprocket combination and the given cadence.

Problem 6.31 If a cyclist is pedaling so that the rear wheel rotates with an angular speed of 16 rad/s, determine all possible (rear) sprocket/chain ring (crank-mounted sprocket) combinations that would allow him or her to pedal with a frequency within the range 1.00–1.25 Hz.

6.2 Planar Motion: Velocity Analysis

In this section we continue the analysis of the slider-crank mechanism (see Fig. 6.17) begun in Section 6.1 on p. 459. Referring to Figs. 6.17 and 6.18, we want to determine $\vec{\omega}_{BC}$, the angular velocity of the connecting rod, and $\vec{v}_C$, the velocity of the piston, given the crank angle θ and the crank's angular velocity $\dot{\theta}$. We will develop three different approaches to the problem that are applicable in the velocity analysis of any planar rigid body motion: the vector approach, differentiation of constraints, and instantaneous center of rotation. In the main body of the section we present the basic ideas underlying the three methods, and then we will demonstrate them in the examples.

Vector approach

The connecting rod in the slider-crank mechanism (Fig. 6.17) is in general planar motion, which means that we can describe the velocity of any of its points by knowing the velocity of one point on the rod and the angular velocity of the rod (see the marginal note on p. 462). Since we are interested in velocities, recall that on p. 464 we found the velocity of point B, which is in fixed axis rotation about the centerline A of the crank (see Fig. 6.18). Points B and C are on the same rigid body, so we can relate the velocity of C to that of B using Eq. (6.3) on p. 462 in Section 6.1, which gives

$$\vec{v}_C = \vec{v}_B + \vec{\omega}_{BC} \times \vec{r}_{C/B}. \qquad (6.15)$$

In planar motion, Eq. (6.15) is a vector equation that represents *two* scalar equations. These two scalar equations are the key to determining $\vec{v}_C$ and $\vec{\omega}_{BC}$. This is so because $\vec{v}_B$ is known, the direction of $\vec{v}_C$ is known (the motion of the piston C is rectilinear along the y axis), $\vec{r}_{C/B}$ can be found in terms of the crank angle θ (see Fig. 6.18), and the axis of rotation for $\vec{\omega}_{BC}$ is known (it is perpendicular to the plane of motion). Therefore, the components v_C and ω_{BC}, which are the answers we seek, are the only unknowns in these two scalar equations. This will be shown in Example 6.5.

Rolling without slip: velocity analysis

Many applications in dynamics involve the motion of disks or wheels rolling over a surface, for example, car wheels rolling on a road, train wheels rolling on tracks, or meshed gears rolling over each other. An important special case of rolling motion is called *rolling without slip* (also called *rolling without slipping* or *rolling without sliding*).

Consider a wheel W rolling over a surface S (S can be moving), as shown in Fig. 6.19. If the wheel and the surface *remain in contact* during the motion and if the contact points P and Q (on W and S, respectively) were to move relative to each other, then the only direction in which relative motion can occur at any instant is along the line ℓ that is tangent to both W and S (assuming the wheel does not separate from the surface). To say that W is *rolling without slip* over S means that points P and Q *do not move relative to each other*. In terms of velocity, this definition implies that

$$\vec{v}_{P/Q} \cdot \hat{u}_t = 0 \quad \text{and} \quad \vec{v}_{P/Q} \cdot \hat{u}_n = 0, \qquad (6.16)$$

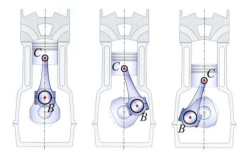

Figure 6.17
Schematic of a slider-crank mechanism emphasizing the motion of the connecting rod.

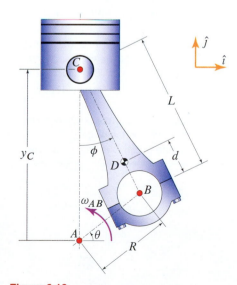

Figure 6.18
Definitions of the geometric parameters used in the slider-crank analysis.

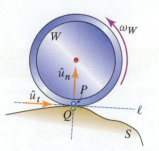

Figure 6.19
A wheel W rolling over a surface S. At the instant shown, the line ℓ is tangent to the path of points P and Q.

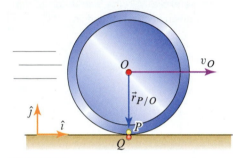

Figure 6.20
A wheel rolling without slip on a horizontal fixed surface.

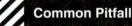

Common Pitfall

***P* is not a fixed point.** Do not interpret Eq. (6.19) as saying that P is fixed because $\vec{v}_P$ is equal to zero. Equation (6.19) *only* holds at the instant when P is touching the ground. In Section 6.3 we will discover that although $\vec{v}_P = \vec{0}$ when P is touching the ground, $\vec{a}_P \neq \vec{0}$ at that instant. That is, P only stops for an instant while it is accelerating away from its current position, so that some other point on the wheel can become the part of the wheel touching the ground.

where $\hat{u}_t$ is a unit vector parallel to ℓ and $\hat{u}_n$ is perpendicular to ℓ, or that

$$\vec{v}_{P/Q} = \vec{0} \quad \Rightarrow \quad \vec{v}_P = \vec{v}_Q. \tag{6.17}$$

As an application of Eq. (6.17), consider a wheel of radius R rolling without slipping on a flat stationary surface, and suppose we want to find the wheel's angular velocity when the center of the wheel is moving with a given speed v_O (Fig. 6.20). Since P and O are two points on the same rigid body, we must have

$$\vec{v}_P = \vec{v}_O + \vec{\omega}_O \times \vec{r}_{P/O}, \tag{6.18}$$

where, at the instant shown, P is the point on the wheel in contact with the ground, $\vec{v}_O = v_O\,\hat{\imath}$, $\vec{\omega}_O = \omega_O\,\hat{k}$ is the angular velocity of the wheel, and $\vec{r}_{P/O} = -R\,\hat{\jmath}$. Enforcing the no slip condition given in Eq. (6.17), we have

$$\vec{v}_P = \vec{v}_Q = \vec{0}, \tag{6.19}$$

where Q is the point on the ground in contact with P at this instant and $\vec{v}_Q = \vec{0}$ because the ground is *stationary*. Substituting Eq. (6.19) into Eq. (6.18) and simplifying, we have

$$\vec{0} = v_O\,\hat{\imath} + \omega_O\,\hat{k} \times (-R\,\hat{\jmath}) = v_O\,\hat{\imath} + \omega_O R\,\hat{\imath} \quad \Rightarrow \quad \boxed{\omega_O = -\frac{v_O}{R}.} \tag{6.20}$$

Notice that the wheel's center O is not the wheel's instantaneous center of rotation (because $\vec{v}_O \neq \vec{0}$). In fact, the point on the wheel that is in contact with the ground at each instant is the point that has zero velocity and is thus the IC.

Differentiation of constraints

Referring to Fig. 6.18, we can also determine $\vec{\omega}_{BC}$ and $\vec{v}_C$ as functions of θ by writing the appropriate constraint equations and then differentiating them with respect to time. We introduced this idea in Section 2.7 on p. 137, but it applies to rigid bodies as well as to particles.

To determine the motion of the piston C, we can write the constraint equation for the y coordinate of C as

$$y_C = R\sin\theta + L\cos\phi, \tag{6.21}$$

which can be differentiated with respect to time to find the velocity of C as

$$\dot{y}_C = v_C = R\dot{\theta}\cos\theta - L\dot{\phi}\sin\phi. \tag{6.22}$$

The quantities R, L, and θ are assumed to be known, though we see that v_C is also a function of ϕ, $\dot{\theta}$, and $\dot{\phi}$. Since the crank angle θ describes the orientation of the line AB, we can write

$$\dot{\theta} = \omega_{AB}, \tag{6.23}$$

where ω_{AB} is the known angular velocity of the crank. To determine ϕ and $\dot{\phi}$ in terms of known quantities, we note that the orientation ϕ of the connecting rod and the crank's orientation θ are related by $R\cos\theta = L\sin\phi$, that is,

$$\sin\phi = \frac{R}{L}\cos\theta. \tag{6.24}$$

Equation (6.24) can then be differentiated with respect to time to find $\dot{\phi}$, where $\dot{\phi}\,\hat{k} = \vec{\omega}_{BC}$, as a function of θ and $\dot{\theta}$ (see Prob. 6.73), thus concluding our analysis. We will demonstrate this method further in Example 6.6.

Instantaneous center of rotation

When a rigid body rotates about a fixed point Q (where Q is on the body or a rigid extension of the body; see p. 463), the velocity of any point C on the body is given by

$$\vec{v}_C = \vec{\omega}_{\text{body}} \times \vec{r}_{C/Q}, \tag{6.25}$$

because $\vec{v}_Q = \vec{0}$. If $\vec{v}_Q = \vec{0}$ for all times, we call Q the center of rotation. If $\vec{v}_Q = \vec{0}$ *only at a particular time instant*, then we call Q the *instantaneous center of rotation* or *instantaneous center* (IC).* If the motion is planar, then $\vec{\omega}_{\text{body}}$ and $\vec{r}_{C/Q}$ are mutually perpendicular, and we can write Eq. (6.25) as

$$|\vec{v}_C| = |\vec{\omega}_{\text{body}}||\vec{r}_{C/Q}| = |\vec{\omega}_{\text{body}}||\vec{r}_{C/\text{IC}}|, \tag{6.26}$$

that is, the speed of C is proportional to the distance from the IC. This formula provides a convenient tool in the study of planar motion if we can find the IC. It turns out that the IC can always be found if the motion is planar. We now describe the three different possibilities.

Given two nonparallel velocities on a body

When we apply the idea of IC to the connecting rod in Fig. 6.21, Eq. (6.25)

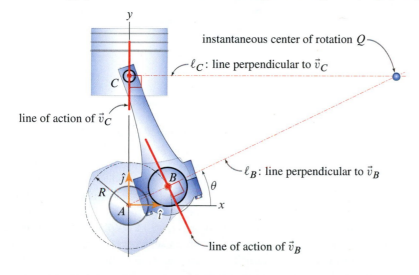

Figure 6.21. Graphical construction for the determination of the IC.

implies that $\vec{v}_C$ is *perpendicular* to both $\vec{\omega}_{BC}$ and $\vec{r}_{C/Q}$. That means that the IC for the connecting rod must lie in the plane of motion and at the intersection of the line that passes through point C and is perpendicular to $\vec{v}_C$ (line ℓ_C) with the line that passes through B and is perpendicular to $\vec{v}_B$ (line ℓ_B). Thus, at the instant shown, the IC of the connecting rod must be at Q, which is the intersection of ℓ_C and ℓ_B.

Given two parallel velocities on a body

The graphical procedure described above does not work if lines ℓ_B and ℓ_C are parallel. Consider two cases: (1) ℓ_B and ℓ_C are parallel and distinct and (2) ℓ_B and ℓ_C are parallel and coincide.

---

* The *instantaneous center* is also sometimes called the *instantaneous center of zero velocity*.

Concept Alert

The IC and general planar motion. In the velocity analysis of the planar motion of a rigid body, we can always view the body's motion as a rotation about an IC, though this IC may change on an instant-by-instant basis. For translation the IC is at infinity.

Common Pitfall

Do not use the IC for acceleration analysis. The IC may have zero velocity, but it does not, in general, have zero acceleration. Therefore, you should not use the IC concept for doing the analysis of accelerations.

Helpful Information

What if the IC is not on the body? Even though the IC of the connecting rod found in Fig. 6.21 is not on the connecting rod itself, we can still make use of Eq. (6.25) by invoking the notion of *extended rigid body* (see the discussion on p. 463).

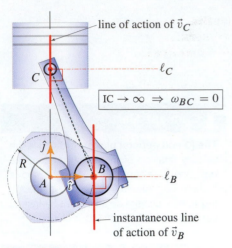

Figure 6.22

At this instant, $\vec{v}_B$ and $\vec{v}_C$ are parallel and ℓ_B and ℓ_C are distinct. The IC of BC is at infinity and $\omega_{BC} = 0$.

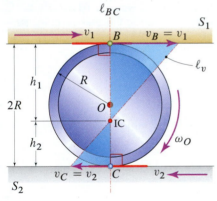

Figure 6.24

Determination of the IC for a wheel rolling without slipping between two parallel surfaces. The top surface is moving to the right, and the bottom is moving to the left.

Case 1: ℓ_B and ℓ_C are parallel and distinct. Referring to Fig. 6.22, at the instant when point B is on a line perpendicular to the line AC, then $\vec{v}_B$ and $\vec{v}_C$ are parallel and ℓ_B and ℓ_C are parallel and distinct. In this case, ℓ_B and ℓ_C intersect at infinity and the IC is infinitely far away. From Eq. (6.26), the only way that B and C can have finite velocities while being infinitely far from the IC is for the angular velocity of the body to equal zero! That is, if our geometric construction tells us that lines ℓ_B and ℓ_C are *parallel and distinct*, then we can conclude that $\vec{\omega}_{BC} = \vec{0}$ at that instant and thus $\vec{v}_B = \vec{v}_C$.

Case 2: ℓ_B and ℓ_C coincide. Consider the case of a wheel that is rolling without slip while in contact with two horizontal surfaces S_1 and S_2 moving with speeds v_1 and v_2 in opposite directions (Fig. 6.23). The no-slip condi-

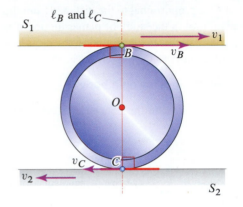

Figure 6.23. Wheel rolling without slipping over two surfaces simultaneously.

tion at B and C causes $\vec{v}_B$ and $\vec{v}_C$ to be parallel because they must match the velocities of the surfaces S_1 and S_2. The geometrical procedure for the determination of the IC tells us that ℓ_B and ℓ_C coincide. If all we know are the directions of $\vec{v}_B$ and $\vec{v}_C$, then the IC cannot be determined because every point on ℓ_B and ℓ_C is an intersection point for these lines. However, if the values of v_1 and v_2 are known, then we *can* find the IC as well as the body's angular velocity. Referring to Fig. 6.24, recall from Fig. 6.15 on p. 465 that the speed of points on line ℓ_{BC} can be graphically represented via right triangles with a vertex at the center of rotation. Therefore, the IC must be at the intersection of ℓ_{BC}, the (radial) line containing B and C, and ℓ_v, the line representing the velocity profile of points on ℓ_{BC}. Since the velocity of points on ℓ_{BC} is proportional to the distance from the IC via the angular speed, we can calculate the angular speed using Eq. (6.11) on p. 464 along with similar triangles as (see Fig. 6.24)

$$\frac{v_B}{h_1} = \frac{v_C}{h_2} = \omega_O. \tag{6.27}$$

Since $h_1 + h_2 = 2R$, Eq. (6.27) becomes

$$\frac{v_B}{\omega_O} + \frac{v_C}{\omega_O} = 2R \quad \Rightarrow \quad \omega_O = \frac{v_B + v_C}{2R}. \tag{6.28}$$

Since the motion is planar, we can assign a direction to ω_O (in Fig. 6.24, ω_O is clockwise since the top of the wheel is moving to the right and the bottom to the left) so that our geometrical calculation based on similar triangles allows us to compute the angular velocity of the body.

When both surfaces move in the same direction with known speeds, $\vec{v}_B$ and $\vec{v}_C$ are again parallel and the lines perpendicular to them that run through points B and C coincide in the single line ℓ_{BC} (Fig. 6.25 has both surfaces moving to the right with the top moving faster than the bottom). The same similar-triangles argument used for the case in Fig. 6.24 tells us that

$$\frac{v_B}{2R+h} = \frac{v_C}{h} = \omega_O. \qquad (6.29)$$

Solving the first equality for h and then substituting that into the second, we get

$$h = \frac{2Rv_C}{v_B - v_C} \quad \Rightarrow \quad \omega_O = \frac{v_B - v_C}{2R}. \qquad (6.30)$$

Given a velocity on a body and the body's angular velocity

This case is depicted in Fig. 6.26 for the crank in an internal combustion engine, for which we know the velocity of point B as well as the angular velocity ω_O of the crank. In this case the IC is located on line ℓ_B such that the distance from B to the IC is $R = v_B/\omega_O$. We can determine on which side of v_B the IC lies by considering the direction of rotation of the rigid body. In this case, it lies to the left of v_B since the rotation is counterclockwise.

End of Section Summary

This section presents three different ways to analyze the velocities of a rigid body in planar motion: the vector approach, differentiation of constraints, and instantaneous center of rotation.

Vector approach. We saw in Section 6.1 that the equation

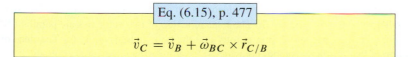

Eq. (6.15), p. 477

$$\vec{v}_C = \vec{v}_B + \vec{\omega}_{BC} \times \vec{r}_{C/B}$$

relates the velocity of two points on a rigid body, $\vec{v}_B$ and $\vec{v}_C$, via their relative position $\vec{r}_{C/B}$ and the angular velocity of the body $\vec{\omega}_{BC}$ (see Fig. 6.27).

Rolling without slip. When a body rolls without slip over another body (see, for example, the wheel W rolling over the surface S in Fig. 6.28), then the two points on the bodies that are in contact at any instant, points P and Q, must have the same velocity. Mathematically, this means that

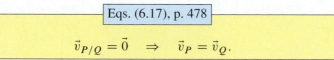

Eqs. (6.17), p. 478

$$\vec{v}_{P/Q} = \vec{0} \quad \Rightarrow \quad \vec{v}_P = \vec{v}_Q.$$

If a wheel of radius R is rolling without slipping over a flat, stationary surface, then the point P on the wheel in contact with the surface, must have zero velocity (see Fig. 6.29). The consequence is that the angular velocity of the

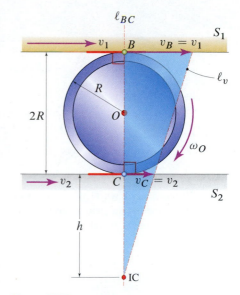

Figure 6.25
Determination of the IC for a wheel rolling without slipping over two parallel surfaces. Both surfaces are moving to the right, but the top is moving faster than the bottom.

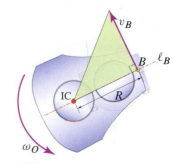

Figure 6.26
Determination of the IC for the case in which the velocity of a point on a body and the body's angular velocity are known.

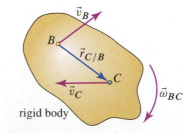

Figure 6.27
A rigid body on which we are relating the velocity of two points B and C.

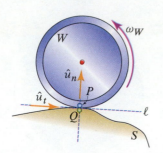

Figure 6.28
Figure 6.19 repeated. A wheel *W* rolling over a surface *S*. At the instant shown, the line ℓ is tangent to path of points *P* and *Q*.

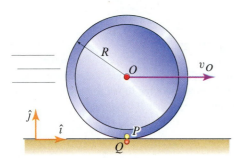

Figure 6.29
A wheel rolling without slip on a horizontal fixed surface.

wheel ω_O is related to the velocity of the center v_O and the radius of the wheel *R* according to

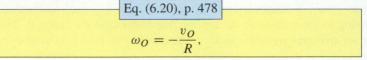

Eq. (6.20), p. 478

$$\omega_O = -\frac{v_O}{R},$$

in which the positive direction for ω_O is taken to be the positive *z* direction using the right-hand rule.

Differentiation of constraints.   As we discovered in Section 2.7, it is often convenient to write an equation describing the position of a point of interest, which can then be differentiated with respect to time to find the velocity of that point. For planar motion of rigid bodies, this idea can also apply for describing the position and velocity of a point on a rigid body as well as for describing the orientation of a rigid body, for which the time derivative provides its angular velocity.

Instantaneous center of rotation.   The point on a body (or imaginary extension of the body) whose velocity is zero at a particular instant is called the *instantaneous center of rotation* or *instantaneous center* (IC). The IC can be found geometrically if the velocity is known for two distinct points on a body or if a velocity on the body and the body's angular velocity are known. The three possible geometric constructions are shown in Fig. 6.30.

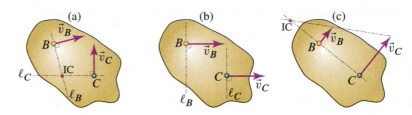

Figure 6.30. The three different possible motions for determining the IC. (a) Two non-parallel velocities are known. (b) The lines of action of two velocities are parallel and distinct; in this case, the IC is at infinity and the body is translating. (c) The lines of action of two parallel velocities coincide.

EXAMPLE 6.4   *Planetary Gears Rolling Without Slip: Vector Approach*

Planetary gear systems (Fig. 1) are used to transmit power between two shafts (a common application is in car transmissions). The center gear is called the *sun*, the outer gear is called the *ring*, and the inner gears are called *planets*. The planets are mounted on a component called the *planet carrier* (which is not shown in Fig. 1). Referring to Fig. 2, let $R_S = 2$ in., $R_P = 0.67$ in., the ring be fixed, and the angular velocity of the sun gear be $\omega_S = 1500$ rpm. Determine $\vec{\omega}_P$ and $\vec{\omega}_{PC}$, the angular velocities of the planet P and of the planet carrier (PC), respectively.

SOLUTION

Road Map   To compute $\vec{\omega}_P$, we need to determine the velocity of two points on P. Two promising candidates are points A' and Q' because their velocities are completely determined by the rolling without slip condition at the A-A' and Q-Q' contacts and because $\vec{v}_A$ and $\vec{v}_Q$ are easily computed. Once $\vec{\omega}_P$ is known, $\vec{\omega}_{PC}$ can be found by finding the velocity of two points on the planet carrier. We will choose O, because its velocity is zero, and C, because it is shared with the planet gear P.

Computation   Enforcing the rolling without slip condition at the A-A' and Q-Q' contacts yields

$$\vec{v}_{Q'} = \vec{v}_Q \quad \text{and} \quad \vec{v}_{A'} = \vec{v}_A = \vec{0}, \tag{1}$$

where $\vec{v}_A = \vec{0}$ because the ring is fixed. Next, because the sun gear rotates about the fixed point O, we can write

$$\vec{v}_Q = \vec{\omega}_S \times \vec{r}_{Q/O} = -\omega_S R_S \,\hat{\imath}, \tag{2}$$

where we have used $\vec{\omega}_S = \omega_S \,\hat{k}$ and $\vec{r}_{Q/O} = R_S \,\hat{\jmath}$. In addition, writing $\vec{v}_{Q'}$ using A' as a reference point yields

$$\vec{v}_{Q'} = \vec{v}_{A'} + \vec{\omega}_P \times \vec{r}_{Q'/A'} = 2\omega_P R_P \,\hat{\imath}, \tag{3}$$

where we have used $\vec{\omega}_P = \omega_P \,\hat{k}$ and $\vec{r}_{Q'/A'} = -2R_P \,\hat{\jmath}$, and we have let $v_{A'} = 0$ from Eqs. (1). Substituting Eqs. (2) and (3) into the first of Eqs. (1) gives

$$-\omega_S R_S = 2\omega_P R_P \quad \Rightarrow \quad \omega_P = -\frac{R_S}{2R_P}\omega_S = -2240 \text{ rpm.} \tag{4}$$

Now observe that C is shared by both planet gear P and the planet carrier. This means

$$\vec{v}_C = \vec{v}_{A'} + \omega_P \,\hat{k} \times \vec{r}_{C/A'} = \omega_P R_P \,\hat{\imath}, \tag{5}$$

and

$$\vec{v}_C = \vec{\omega}_{PC} \times \vec{r}_{C/O} = -\omega_{PC} R_{PC} \,\hat{\imath}, \tag{6}$$

where $R_{PC} = R_P + R_S$, $\vec{\omega}_{PC} = \omega_{PC} \,\hat{k}$, and we have enforced the second of Eqs. (1). The two expressions for $\vec{v}_C$ must be equal to each other, that is,

$$\omega_{PC} = -\frac{R_P}{R_{PC}}\omega_P = \frac{R_S}{2R_{PC}}\omega_S = 562 \text{ rpm,} \tag{7}$$

where we have taken advantage of the solution for ω_P in Eq. (4).

Discussion & Verification   To verify the correctness of our results, observe that since $\vec{v}_{A'} = \vec{0}$, point A' is the IC for the planet gear P. Hence, the speeds of point C and Q' are $|\omega_P|R_P$ and $|\omega_P|2R_P$, respectively, which is confirmed by Eqs. (4) and (7).

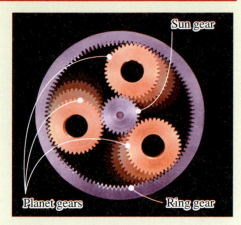

Figure 1
Photo of a planetary gear system.

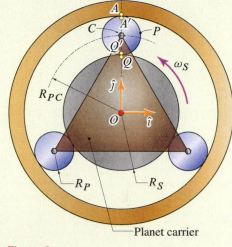

Figure 2
Schematic of a planetary gear system with three planets and a fixed ring.

EXAMPLE 6.5   *Completing the Velocity Analysis of the Connecting Rod*

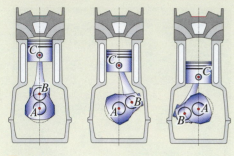

Figure 1

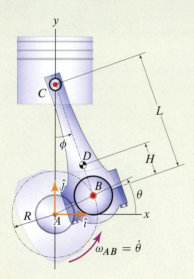

Figure 2
Detailed view of the connecting rod component of a slider-crank mechanism.

On p. 477 we outlined the vector approach for the velocity analysis of the connecting rod (CR) and piston in the slider-crank mechanism shown in Fig. 1. We will complete that analysis here.

Referring to Fig. 2, we are given the radius of the crank R, the length of the CR L, the position of the mass center of the CR H, the angular velocity of the crank ω_{AB}, and the crank angle θ. Determine the angular velocity of the CR $\vec{\omega}_{BC}$, the velocity of the piston $\vec{v}_C$, and the velocity of the mass center of the CR $\vec{v}_D$.

SOLUTION

Road Map   The road map for this problem was presented on p. 477.

Computation   We begin by recalling that, in Section 6.1, we found $\vec{v}_B$ to be (see Eq. (6.11) on p. 464 and Fig. 2)

$$\vec{v}_B = R\dot{\theta}\,\hat{u}_\theta = R\omega_{AB}(-\sin\theta\,\hat{\imath} + \cos\theta\,\hat{\jmath}), \tag{1}$$

where we have used $\dot{\theta} = \omega_{AB}$ and $\hat{u}_\theta = -\sin\theta\,\hat{\imath} + \cos\theta\,\hat{\jmath}$. Since B and C are both points on the CR, we can relate their velocities using Eq. (6.3) on p. 462, which gives

$$\vec{v}_C = \vec{v}_B + \vec{\omega}_{BC} \times \vec{r}_{C/B}. \tag{2}$$

As discussed on p. 477, Eq. (2) represents two scalar equations in the two unknowns v_C and ω_{BC}. Let's now work out the details to see how.

As for $\vec{\omega}_{BC}$, we can write it as

$$\vec{\omega}_{BC} = \omega_{BC}\,\hat{k}. \tag{3}$$

To complete the right-hand side of Eq. (2), we can write $\vec{r}_{C/B}$ as

$$\vec{r}_{C/B} = L(-\sin\phi\,\hat{\imath} + \cos\phi\,\hat{\jmath}), \tag{4}$$

where we note that the orientation ϕ of the CR and the crank's orientation θ are related by $R\cos\theta = L\sin\phi$, that is,

$$\sin\phi = \frac{R}{L}\cos\theta \quad \text{and} \quad \cos\phi = \frac{\sqrt{L^2 - R^2\cos^2\theta}}{L}. \tag{5}$$

Finally, enforcing the condition $v_{Cx} = 0$, we can write $\vec{v}_C$ as

$$\vec{v}_C = v_{Cy}\,\hat{\jmath}. \tag{6}$$

Substituting Eq. (1) and Eqs. (3)–(6) into Eq. (2) and carrying out the cross product give

$$v_{Cy}\,\hat{\jmath} = -\left(R\omega_{AB}\sin\theta + \omega_{BC}\sqrt{L^2 - R^2\cos^2\theta}\right)\hat{\imath} + R\left(\omega_{AB} - \omega_{BC}\right)\cos\theta\,\hat{\jmath}. \tag{7}$$

Equation (7) represents the two scalar equations

$$R\omega_{AB}\sin\theta + \omega_{BC}\sqrt{L^2 - R^2\cos^2\theta} = 0, \tag{8}$$

$$R\left(\omega_{AB} - \omega_{BC}\right)\cos\theta = v_{Cy}, \tag{9}$$

in the unknowns v_{Cy} and ω_{BC}. Solving, we obtain

$$\omega_{BC} = -\frac{\omega_{AB} \sin \theta}{\sqrt{(L/R)^2 - \cos^2 \theta}}, \tag{10}$$

$$v_{Cy} = R\omega_{AB} \cos \theta \left[1 + \frac{\sin \theta}{\sqrt{(L/R)^2 - \cos^2 \theta}} \right], \tag{11}$$

where the solutions have been written to emphasize that, at least for ω_{BC}, the geometry of the mechanism matters only through the ratio L/R, and the vectors $\vec{\omega}_{BC}$ and $\vec{v}_C$ are then given by Eqs. (3) and (6), respectively.

To find $\vec{v}_D$, we note that since $\vec{v}_B$ and $\vec{\omega}_{BC}$ are now both known, $\vec{v}_D$ is readily found by using Eq. (6.3) on p. 462 to relate the motion of D to that of B as

$$\vec{v}_D = \vec{v}_B + \vec{\omega}_{BC} \times \vec{r}_{D/B}. \tag{12}$$

Writing $\vec{r}_{D/B} = H(-\sin \phi \, \hat{i} + \cos \phi \, \hat{j})$; substituting Eqs. (1), (3), (5), and (10) into Eq. (12); carrying out the cross product; and simplifying; we obtain

$$\vec{v}_D = R\omega_{AB} \left\{ \sin \theta \left(\frac{H}{L} - 1 \right) \hat{i} \right.$$
$$\left. + \cos \theta \left[1 + \frac{H \sin \theta}{L\sqrt{(L/R)^2 - \cos^2 \theta}} \right] \hat{j} \right\}. \tag{13}$$

Discussion & Verification   The answers in Eqs. (10), (11), and (13) are somewhat complicated, but we can see that they have some expected behavior. For example, we expect the piston's velocity to be zero when $\theta = 90°$ and $\theta = 270°$ (when it reaches extreme positions along the y axis), and Eq. (11) tells us that it is. In addition, we expect the angular velocity of the CR to be zero at $\theta = 0°$ and $\theta = 180°$ since its rotation changes direction at those points — Eq. (10) verifies that this is true. Finally, inspection of our three final results tells us that they are all dimensionally correct.

🔍 **A Closer Look**   Our results are *general* because they apply for any value of θ and ω_{AB} as well as for any possible values of R, L, and H, i.e., the geometry of the mechanism. General relations such as these are useful because they allow us to know ω_{BC}, v_{Cy}, and $\vec{v}_D$ for *all* values for the parameters θ, ω_{AB}, R, L, and H when designing these machine components. This ability to see how one or more quantities change as parameters are changed is called *parametric analysis*. We now present plots of ω_{BC} and v_{Cy} for the operating conditions that are typical in car engines.

🖥️ ➡ Observe that ω_{BC} and v_{Cy} are periodic functions of θ so that we only need to plot them for one full crank rotation, that is, for $0 \leq \theta \leq 360°$. In addition, since ω_{BC} and v_{Cy} are directly proportional to ω_{AB}, the plots obtained for one value of ω_{AB} can be rescaled to obtain plots for other ω_{AB} values. Finally, we see that the geometry of the mechanism appears in the equations primarily through the ratio L/R. It is this ratio that is usually found in the analysis of car engine performance. The plots of the functions in Eqs. (10) and (11) are presented in Figs. 3 and 4, respectively, for $\omega_{AB} = 3500$ rpm (e.g., highway cruising), $L = 150$ mm (values for small block engines typically range between 140 and 155 mm), and three commonly found values of L/R. We see that the smaller L/R, the larger are ω_{BC} and v_{Cy}. In addition, as we discussed above, the piston's velocity is zero when $\theta = 90°$ and $\theta = 270°$. Finally, notice that v_{Cy} looks different for $0° \leq \theta < 180°$ when compared with $180° \leq \theta \leq 360°$. This difference is even more pronounced in the behavior of the piston's acceleration, discussed in Section 6.3. This lack of symmetry tends to disappear for larger values of L/R. ⬅ 🖥️

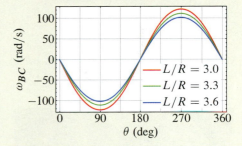

Figure 3
Plot of the angular velocity of the connecting rod for $\omega_{AB} = 3500$ rpm, $L = 150$ mm, and three values of L/R commonly found in practice.

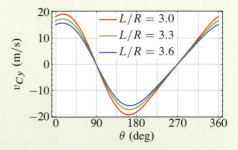

Figure 4
Plot of the piston's velocity for $\omega_{AB} = 3500$ rpm, $L = 150$ mm, and three values of L/R commonly found in practice.

EXAMPLE 6.6   *A Mechanism with a Slider: Differentiation of Constraints*

Figure 1
Model illustrating the components of a *swinging block* slider-crank mechanism.

Figure 2
The pneumatic door closers found on typical storm or screen doors are equivalent to the mechanism shown in Fig. 1.

Figure 1 shows a variant of the slider-crank mechanism called a *swinging block* slider crank. First used in various steam locomotive engines in the 1800s, this mechanism is often found in door closing systems (see Fig. 2). Referring to Fig. 3, note that the slider S is directly connected to the crank, and it slides within a swinging block that is free to swing about the pivot at O. For this mechanism we want to derive the relation between the angular velocity of the slider and that of the crank. Also, for $R = 8$ in., $H = 25$ in., $\theta = 20°$, and $\dot{\theta} = 265$ rad/s, we want to determine the velocity of the point P on the slider that is underneath O at this instant.

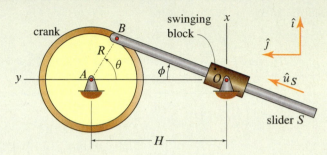

Figure 3. Representation of a *swinging block* slider-crank mechanism with a sliding contact at O.

SOLUTION

Road Map   Because θ describes the crank's orientation, the crank's angular velocity is given by $\vec{\omega}_{AB} = \dot{\theta}\,\hat{k}$. Similarly, the slider's angular velocity is $\vec{\omega}_S = -\dot{\phi}\,\hat{k}$, where the minus sign accounts for the fact that if $\dot{\phi} > 0$, the slider rotates clockwise. We can find $\dot{\phi}$ by first relating ϕ to θ and then differentiating the resulting equation with respect to time, which is the differentiation of constraint method of solution. Once $\vec{\omega}_S$ is known, the velocity of any point P on the slider can be found via the relation $\vec{v}_P = \vec{v}_B + \vec{\omega}_S \times \vec{r}_{P/B}$, where point B is on both the crank and the slider.

Computation   Focusing on the triangle AOB, throughout the motion we must have

$$R \sin\theta = (H - R\cos\theta)\tan\phi. \tag{1}$$

Differentiating Eq. (1) with respect to time, we obtain

$$R\dot{\theta}\cos\theta = R\dot{\theta}\sin\theta\tan\phi + (H - R\cos\theta)\dot{\phi}\sec^2\phi. \tag{2}$$

Solving Eq. (2) for $\dot{\phi}$ yields

$$\dot{\phi} = \frac{\cos\theta - \sin\theta\tan\phi}{(H - R\cos\theta)\sec^2\phi}R\dot{\theta}, \tag{3}$$

which can be simplified to read

$$\dot{\phi} = \frac{R(H\cos\theta - R)\dot{\theta}}{H^2 + R^2 - 2HR\cos\theta} \quad \Rightarrow \quad \vec{\omega}_S = \frac{R(R - H\cos\theta)\dot{\theta}}{H^2 + R^2 - 2HR\cos\theta}\,\hat{k}, \tag{4}$$

where we used Eq. (1) to write $\tan\phi = R\sin\theta/(H - R\cos\theta)$, as well as the identity $\sec^2\phi = 1 + \tan^2\phi$.

For the calculation of $\vec{v}_P$, let t_0 be the instant when $\theta = 20°$. At this instant we can write

$$\vec{v}_P(t_0) = \vec{v}_B(t_0) + \vec{\omega}_S(t_0) \times \vec{r}_{P/B}(t_0), \tag{5}$$

where P is the point on the slider that, at $t = t_0$, coincides with point O. Since B rotates about the fixed point A, we must have

$$\vec{v}_B = \vec{\omega}_{AB} \times \vec{r}_{B/A} = \dot{\theta}\,\hat{k} \times R(\sin\theta\,\hat{\imath} - \cos\theta\,\hat{\jmath}) = R\dot{\theta}(\cos\theta\,\hat{\imath} + \sin\theta\,\hat{\jmath})$$

$$\Rightarrow \quad \vec{v}_B(t_0) = (166.0\,\hat{\imath} + 60.42\,\hat{\jmath})\,\text{ft/s}, \tag{6}$$

where we have substituted in the given data for θ, $\dot{\theta}$, and R. As for $\vec{r}_{P/B}(t_0)$, since P coincides with the origin, we must have $\vec{r}_P(t_0) = \vec{0}$, so that $\vec{r}_{P/B}(t_0) = \vec{r}_P(t_0) - \vec{r}_B(t_0) = -\vec{r}_B(t_0)$. Hence, since $\vec{r}_B = R\sin\theta\,\hat{\imath} + (H - R\cos\theta)\,\hat{\jmath}$, we have

$$\vec{r}_{P/B}(t_0) = -\vec{r}_B(t_0) = (-0.2280\,\hat{\imath} - 1.457\,\hat{\jmath})\,\text{ft}. \tag{7}$$

Next, using Eq. (4), we have

$$\vec{\omega}_S(t_0) = (-104.9\,\hat{k})\,\text{rad/s}. \tag{8}$$

Finally, substituting the results in Eqs. (6), (7), and (8) into Eq. (5), we obtain

$$\boxed{\vec{v}_P(t_0) = (13.2\,\hat{\imath} + 84.3\,\hat{\jmath})\,\text{ft/s}.} \tag{9}$$

Discussion & Verification   Intuitively we would expect $|\omega_S|$ to be smaller than $|\dot{\theta}|$ for all θ. 🖥️ ➡ This is the case for the result in Eq. (8). By plotting the function $\omega_S/\dot{\theta}$ (see Fig. 4), we see that $|\omega_S/\dot{\theta}| < 1$ for any θ, that is, $|\omega_S|$ behaves as expected. ⬅🖥️

🔍 **A Closer Look**   Referring to Fig. 3, notice that the axes of the slider and of the swinging block must always coincide; otherwise the mechanism would jam. We can express this condition by saying that, on an instant-by-instant basis, given a point Q on the slider having the same x and y coordinates of a corresponding point Q' on the swinging block, we must have

$$\vec{v}_{Q/Q'} = v_{Q/Q'}\,\hat{u}_S \quad \text{with} \quad \hat{u}_S = \sin\phi\,\hat{\imath} + \cos\phi\,\hat{\jmath}, \tag{10}$$

where $\hat{u}_S$ is a unit vector identifying the orientation of the slider's axis (see Fig. 3). Recall that at time t_0, P has the same x and y coordinates as point O, which is a fixed point. Therefore, rewriting Eq. (10) for points P and O, we have

$$\vec{v}_{P/O}(t_0) = \vec{v}_P(t_0) - \vec{0} = v_P(t_0)\,\hat{u}_S(t_0). \tag{11}$$

Equation (11) says that the vectors $\hat{u}_S(t_0)$ and $\vec{v}_P(t_0)$ must be parallel. This gives the opportunity to check our calculations by comparing the direction of these two vectors. From the second of Eqs. (10), the direction of $\hat{u}_S$ can be expressed as

$$\frac{(\hat{u}_S(t_0))_x}{(\hat{u}_S(t_0))_y} = \frac{\sin\phi(t_0)}{\cos\phi(t_0)} = \tan\phi_0 = 0.157, \tag{12}$$

where we used Eq. (1) to compute $\tan\phi$ and evaluate it at $t = t_0$. Repeating the calculation for $\vec{v}_P(t_0)$, using Eq. (9), we have

$$\frac{(v_P(t_0))_x}{(v_P(t_0))_y} = \frac{13.2}{84.3} = 0.157, \tag{13}$$

which implies that $\vec{v}_P$ has the direction we expected.

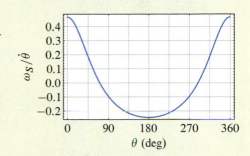

Figure 4
Plot of the function $\omega_S/\dot{\theta}$ as a function of θ over an entire cycle. The expected behavior of ω_S is to always be smaller than $\dot{\theta}$ in absolute value.

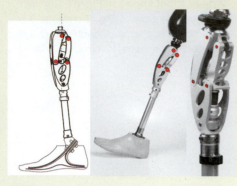

Figure 1

Figure 1 shows three views of a prosthetic leg with an artificial knee joint. The primary kinematic component of this artificial knee joint is a four-bar linkage (see the right panel in Fig. 1 and the system $ABCD$ in Fig. 2). The four-bar linkage consists of the four segments AB, BC, CD, and DA, which are pin-connected and can therefore rotate relative to one another. The mechanism is built in such a way that, given the motion of two of its segments, the motion of the other two is uniquely determined. By varying the relative proportions of its elements, a four-bar linkage system can provide an extremely large variety of controlled motions, and for this reason, four-bar linkages have myriad applications, including engines, sport machines, prosthesis components, drafting tools, and carnival rides. For the mechanism in Fig. 2, assume that the segment AD is fixed, and using the information given in Table 1 for the instant shown, determine the angular velocity of segment BC (the lower leg) if segment AB rotates counterclockwise with a rate $|\vec{\omega}_{AB}| = 1.5\,\text{rad/s}$, i.e., as if walking forward (negative x direction). The acceleration analysis is presented in Example 6.12 on p. 508.

Table 1. Approximate values of the coordinates of the pin centers A, B, C, and D for the system shown in Fig. 2 at the time instant considered.

Points	A	B	C	D
Coordinates (mm)	(0.0, 0.0)	(−27.0, 120)	(26.0, 124)	(30.0, 15.0)

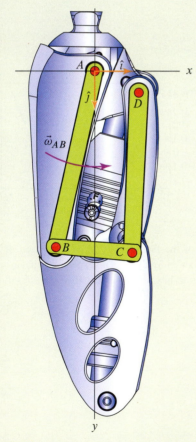

Figure 2

Geometry of the four-bar linkage in the prosthetic leg.

SOLUTION

Road Map   This linkage system is a *kinematic chain*, i.e., a system in which motion information is passed from one component to the next along the chain. This means we will start from a point whose motion is known, say, A because it is fixed, and then compute the velocity of point B, which is the next point along the chain $ABCD$, using the equation $\vec{v}_B = \vec{\omega}_{AB} \times \vec{r}_{B/A}$. We repeat this process for segments BC and CD. In doing this, we will generate enough equations to determine the angular velocity of each element along the chain. It is important to keep in mind that the calculations performed in this example hold only at a given instant in time.

Computation   For the velocity of B, we have

$$\vec{v}_B = \vec{v}_A + \omega_{AB}\,\hat{k} \times \vec{r}_{B/A}. \tag{1}$$

Letting $\omega_{AB} = -1.5\,\text{rad/s}$, observing that $\vec{r}_{B/A} = \vec{r}_B = (-27\,\hat{i} + 120\,\hat{j})\,\text{mm}$, and recalling that $\vec{v}_A = \vec{0}$, Eq. (1) yields

$$\vec{v}_B = (180\,\hat{i} + 40.5\,\hat{j})\,\text{mm/s}. \tag{2}$$

Since point C is shared by both segments BC and CD, we can express the velocity of C in two independent ways as follows:

$$\vec{v}_C = \vec{v}_B + \omega_{BC}\,\hat{k} \times \vec{r}_{C/B}, \tag{3}$$

and

$$\vec{v}_C = \vec{v}_D + \omega_{CD}\,\hat{k} \times \vec{r}_{C/D}. \tag{4}$$

Observing that

$$\vec{r}_{C/B} = \vec{r}_C - \vec{r}_B = (53\,\hat{i} + 4\,\hat{j})\,\text{mm} \tag{5}$$

$$\vec{r}_{C/D} = \vec{r}_C - \vec{r}_D = (-4\,\hat{\imath} + 109\,\hat{\jmath})\,\text{mm}, \tag{6}$$

recalling that $\vec{v}_D = \vec{0}$, and noting that the $\vec{v}_C$ obtained from Eq. (3) must be the same as that obtained from Eq. (4), we obtain

$$[(180\,\tfrac{\text{mm}}{\text{s}}) - (4\,\text{mm})\omega_{BC}]\,\hat{\imath} + [(40.5\,\tfrac{\text{mm}}{\text{s}}) + (53\,\text{mm})\omega_{BC}]\,\hat{\jmath}$$
$$= -(109\,\text{mm})\omega_{CD}\,\hat{\imath} - (4\,\text{mm})\omega_{CD}\,\hat{\jmath}. \tag{7}$$

Equation (7) is a vector equation equivalent to a linear system of two scalar equations in the unknowns ω_{BC} and ω_{CD}. These equations are

$$(180\,\tfrac{\text{mm}}{\text{s}}) - (4\,\text{mm})\omega_{BC} = -(109\,\text{mm})\omega_{CD}, \tag{8}$$
$$(40.5\,\tfrac{\text{mm}}{\text{s}}) + (53\,\text{mm})\omega_{BC} = -(4\,\text{mm})\omega_{CD}, \tag{9}$$

which can be solved to obtain

$$\boxed{\omega_{BC} = -0.638\,\text{rad/s} \quad \text{and} \quad \omega_{CD} = -1.67\,\text{rad/s}.} \tag{10}$$

Discussion & Verification   The solution we have obtained seems reasonable in that both segment BC and segment CD rotate counterclockwise, as one would expect, when attempting to walk forward. What is interesting is that the proportions of the segments in the mechanism are such that, in the configuration shown, point B moves down and to the right, that is, in a direction that would cause the foot to move into the ground, which, again, is consistent with what happens when we begin to walk forward from a standing position.

As far as the solution technique that we have used is concerned, what needs to be observed is that, in the analysis of kinematic chains, one always ends up expressing the velocity of one point in two independent ways that must be made to be consistent with each other. The enforcement of this consistency requirement produces useful equations in terms of the problem's unknowns.

EXAMPLE 6.8   *A Carnival Ride: Instantaneous Center Analysis*

Figure 1

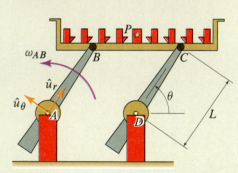

Figure 2
Coordinate definition and geometry for the carnival ride.

Motion platforms, which are used in many of today's amusement parks, are a type of carnival ride in which a platform with seats is made to move always parallel to the ground. Figure 1 shows an elementary type of motion platform, found more often in traveling carnivals than in big amusement parks. Given that the motion platform in the figure (see also Fig. 2) is designed so that the rotating arms AB and CD are of equal length $L = 10$ ft and remain parallel to each other, determine the velocity and acceleration of a person P onboard the ride when ω_{AB} is constant and equal to 1.25 rad/s.

SOLUTION

Road Map   To simplify the problem, we will assume that the platform and the persons on it form a single rigid body. This means that the two rotating arms and the platform form a *four-bar linkage*. Because the arms AB and CD are identical in size and are always parallel to each other, the four-bar linkage $ABCD$ always forms a parallelogram. We will use this fact, along with the concept of instantaneous center of rotation, to determine the angular velocity of the platform. Knowing this and applying the kinematics of fixed axis rotation will allow us to determine $\vec{v}_P$ and $\vec{a}_P$.

Computation   Referring to Fig. 2, we see that the arms AB and CD are always parallel to each other. Also observe that points B and C move along circles centered at A and D, respectively, where A and D are fixed points. Hence, going through the geometrical procedure to identify the IC of the element BC, we see that lines ℓ_{AB} and ℓ_{CD}, which are perpendicular to $\vec{v}_B$ and $\vec{v}_C$, respectively, are parallel and distinct. This means that the IC of BC is at infinity, and therefore

$$\omega_{BC} = 0. \tag{1}$$

Since the result in Eq. (1) is independent of the value of the angle θ of the arms AB and CD, the motion of element BC is translation. Given the fact that the trajectories of points on BC are not straight lines, BC's motion is a *curvilinear translation*. Consequently, all points on BC, or any rigid extension of it, i.e., any of the passengers, share the same value of velocity as well as acceleration. In view of this fact we have

$$\boxed{\vec{v}_P = \vec{v}_B = \vec{v}_C = \omega_{AB}L\,\hat{u}_\theta = (12.5\text{ ft/s})\,\hat{u}_\theta,} \tag{2}$$

and

$$\boxed{\vec{a}_P = \vec{a}_B = \vec{a}_C = -\omega_{AB}^2 L\,\hat{u}_r = (-15.6\text{ ft/s}^2)\,\hat{u}_r,} \tag{3}$$

where the velocity and acceleration of B were computed using circular motion formulas (see Eqs. (6.11) on p. 464).

Discussion & Verification   The dimensions and units in Eqs. (2) and (3) are correct. The solution of this problem is very elementary and performed in a conceptual manner to illustrate the concept of curvilinear translation and its relation to the concept of the IC. As far as the acceleration values computed are concerned, these are not far from those found in actual general public (as opposed to extreme) carnival rides, and they amount to a little less than 50% of the acceleration of gravity.

PROBLEMS

Problem 6.32

A carrier is maneuvering so that, at the instant shown, $|\vec{v}_A| = 25$ knots (1 kn is *exactly* equal to 1.852 km/h) and $\phi = 33°$. Letting the distance between A and B be 220 m and $\theta = 22°$, determine $\vec{v}_B$ at the given instant if the ship's turning rate at this instant is $\dot{\theta} = 2°/s$ clockwise.

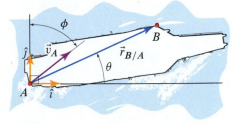

Figure P6.32

Problem 6.33

At the instant shown, the pinion is rotating between two racks with an angular velocity $\omega_P = 55$ rad/s. If the nominal radius of the pinion is $R = 4$ cm and if the lower rack is moving to the right with a speed $v_L = 1.2$ m/s, determine the velocity of the upper rack.

Problem 6.34

At the instant shown the lower rack is moving to the right with a speed of $v_L = 4$ ft/s, while the upper rack is fixed. If the nominal radius of the pinion is $R = 2.5$ in., determine ω_P, the angular velocity of the pinion, as well as the velocity of point O, i.e., the center of the pinion.

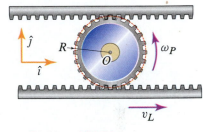

Figure P6.33 and P6.34

Problem 6.35

A bar of length $L = 2.5$ m is pin-connected to a roller at A. The roller is moving along a horizontal rail as shown with $v_A = 5$ m/s. If at a certain instant $\theta = 33°$ and $\dot{\theta} = 0.4$ rad/s, compute the velocity of the bar's midpoint C.

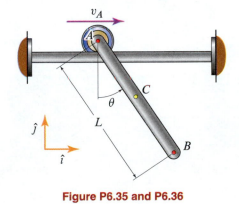

Figure P6.35 and P6.36

Problem 6.36

If the motion of the bar is planar, what would the speed of A need to be for $\vec{v}_C$ to be perpendicular to the bar AB? Why?
Note: Concept problems are about *explanations*, not computations.

Problem 6.37

Points A and B are both on the trailer part of the truck. If the relative velocity of point B with respect to A is as shown, is the body undergoing a planar rigid body motion?
Note: Concept problems are about *explanations*, not computations.

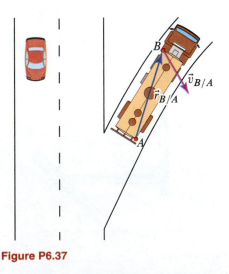

Figure P6.37

Problem 6.38

A truck is moving to the right with a speed $v_0 = 12\,\text{km/h}$ while the pipe section with radius $R = 1.25\,\text{m}$ and center at C rolls without slipping over the truck's bed. The center of the pipe section C is moving to the right at $2\,\text{m/s}$ relative to the truck. Determine the angular velocity of the pipe section and the absolute velocity of C.

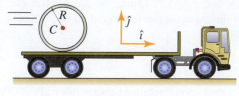

Figure P6.38

Problem 6.39

A wheel W of radius $R_W = 7\,\text{mm}$ is connected to point O via the rotating arm OC, and it rolls without slip over the stationary cylinder S of radius $R_S = 15\,\text{mm}$. If, at the instant shown, $\theta = 47°$ and $\omega_{OC} = 3.5\,\text{rad/s}$, determine the angular velocity of the wheel and the velocity of point Q, where point Q lies on the edge of W and along the extension of the line OC.

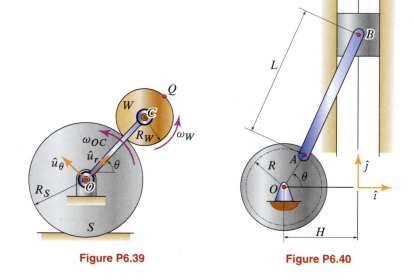

Figure P6.39   **Figure P6.40**

Problem 6.40

For the slider-crank mechanism shown, let $R = 20\,\text{mm}$, $L = 80\,\text{mm}$, and $H = 38\,\text{mm}$. Use the concept of instantaneous center of rotation to determine the values of θ, with $0° \leq \theta \leq 360°$, for which $v_B = 0$. Also, determine the angular velocity of the connecting rod at these values of θ.

Problem 6.41

At the instant shown bars AB and BC are perpendicular to each other, and bar BC is rotating counterclockwise at $20\,\text{rad/s}$. Letting $L = 2.5\,\text{ft}$ and $\theta = 45°$, determine the angular velocity of bar AB as well as the velocity of the slider C.

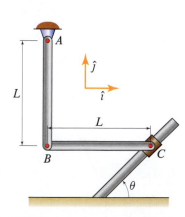

Figure P6.41

Problems 6.42 and 6.43

A ball of radius $R_A = 3$ in. is rolling without slip in a stationary spherical bowl of radius $R_B = 8$ in. Assume that the ball's motion is planar.

Problem 6.42 If the speed of the center of the ball is $v_A = 1.75$ ft/s and if the ball is moving down and to the right, determine the angular velocity of the ball.

Problem 6.43 If the angular speed of the ball $|\omega_A| = 4$ rad/s is counterclockwise, determine the velocity of the center of the ball.

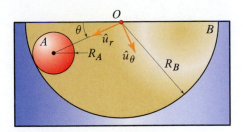

Figure P6.42 and P6.43

Problem 6.44

One way to convert rotational motion into linear motion and vice versa is via the use of a mechanism called a Scotch yoke, which consists of a crank C that is connected to a slider B via a pin A. The pin rotates with the crank while sliding within the yoke, which, in turn, rigidly translates with the slider. This mechanism has been used, for example, to control the opening and closing valves in pipelines. Letting the radius of the crank be $R = 1.5$ ft, determine the angular velocity ω_C of the crank so that the maximum speed of the slider is $v_B = 90$ ft/s.

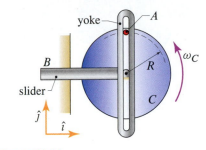

Figure P6.44

Problems 6.45 through 6.47

The system shown consists of a wheel of radius $R = 14$ in. rolling on a horizontal surface. A bar AB of length $L = 40$ in. is pin-connected to the center of the wheel and to a slider A that is constrained to move along a vertical guide. Point C is the bar's midpoint.

Problem 6.45 If, when $\theta = 72°$, the wheel is moving to the right so that $v_B = 7$ ft/s, determine the angular velocity of the bar as well as the velocity of the slider A.

Problem 6.46 If, when $\theta = 53°$, the slider is moving downward with a speed $v_A = 8$ ft/s, determine the velocity of points B and C.

Problem 6.47 If the wheel rolls without slip with a constant counterclockwise angular velocity of 10 rad/s, determine the velocity of the slider A when $\theta = 45°$.

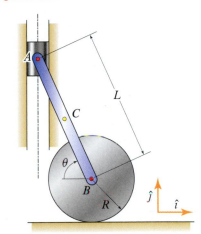

Figure P6.45–P6.47

Problem 6.48

At the instant shown, the lower rack is moving to the right with a speed of 2.7 m/s while the upper rack is moving to the left with a speed of 1.7 m/s. If the nominal radius of the pinion O is $R = 0.25$ m, determine the angular velocity of the pinion, as well as the position of the pinion's instantaneous center of rotation relative to point O.

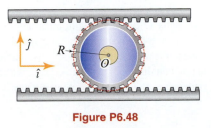

Figure P6.48

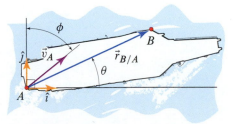

Figure P6.49

Problem 6.49

A carrier is maneuvering so that, at the instant shown, $|\vec{v}_A| = 22$ kn, $\phi = 35°$, $|\vec{v}_B| = 24$ kn (1 kn is equal to 1 nautical mile (nml) per hour or 6076 ft/h). Letting $\theta = 19°$

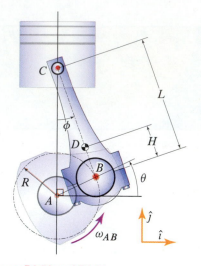

Figure P6.50 and P6.51

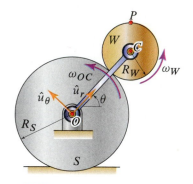

Figure P6.52

and the distance between A and B be 720 ft, determine the ship's turning rate at the given instant if the ship is rotating clockwise.

Problems 6.50 and 6.51

For the slider-crank mechanism shown, let $R = 1.9$ in., $L = 6.1$ in., and $H = 1.2$ in. Also, at the instant shown, let $\theta = 27°$ and $\omega_{AB} = 4850$ rpm.

Problem 6.50   Determine the velocity of the piston at the instant shown.

Problem 6.51   Determine $\dot{\phi}$ and the velocity of point D at the instant shown.

Problem 6.52

A wheel W of radius $R_W = 7$ mm is connected to point O via the rotating arm OC, and it rolls without slip over the stationary cylinder S of radius $R_S = 15$ mm. If, at the instant shown, $\theta = 63°$ and $\omega_W = 9$ rad/s, determine the angular velocity of the arm OC and the velocity of point P, where point P lies on the edge of W and is vertically aligned with point C.

Problem 6.53

At the instant shown, the center O of a spool with inner and outer radii $r = 1$ m and $R = 2.2$ m, respectively, is moving up the incline at speed $v_O = 3$ m/s. If the spool does not slip relative to the ground or relative to the cable C, determine the rate at which the cable is wound or unwound, that is, the length of rope being wound or unwound per unit time.

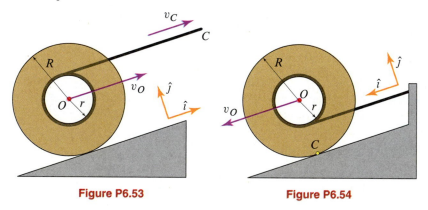

Figure P6.53   **Figure P6.54**

Problem 6.54

At the instant shown, the center O of a spool with inner and outer radii $r = 3$ ft and $R = 7$ ft, respectively, is moving down the incline at a speed $v_O = 12.2$ ft/s. If the spool does not slip relative to the rope and if the rope is fixed at one end, determine the velocity of point C (the point on the spool that is in contact with the incline) as well as the rope's unwinding rate, that is, the length of rope being unwound per unit time.

Problem 6.55

The bucket of a backhoe is the element AB of the four-bar linkage system $ABCD$. Assume that the points A and D are fixed and that, at the instant shown, point B is vertically aligned with point A, point C is horizontally aligned with point B, and point B is moving to the right with a speed $v_B = 1.2$ ft/s. Determine the velocity of point C

at the instant shown, along with the angular velocities of elements BC and CD. Let $h = 0.66\,\text{ft}$, $e = 0.46\,\text{ft}$, $l = 0.9\,\text{ft}$, and $w = 1.0\,\text{ft}$.

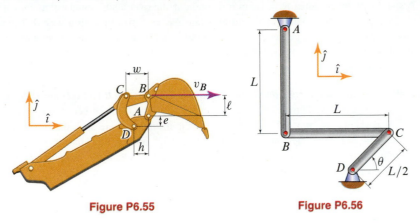

Figure P6.55   **Figure P6.56**

Problem 6.56

Bar AB is rotating counterclockwise with an angular velocity of $15\,\text{rad/s}$. Letting $L = 1.25\,\text{m}$, determine the angular velocity of bar CD when $\theta = 45°$.

Problem 6.57

At the instant shown, bars AB and CD are vertical and point C is moving to the left with a speed of $35\,\text{ft/s}$. Letting $L = 1.5\,\text{ft}$ and $H = 0.6\,\text{ft}$, determine the velocity of point B.

Problem 6.58

Collars A and B are constrained to slide along the guides shown and are connected by a bar with length $L = 0.75\,\text{m}$. Letting $\theta = 45°$, determine the angular velocity of the bar AB at the instant shown if, at this instant, $v_B = 2.7\,\text{m/s}$.

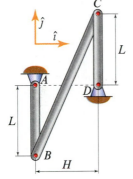

Figure P6.57

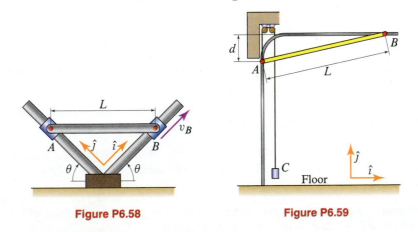

Figure P6.58   **Figure P6.59**

Problem 6.59

At the instant shown, an overhead garage door is being shut with point B moving to the left within the horizontal part of the door guide at a speed of $5\,\text{ft/s}$, while point A is moving vertically downward. Determine the angular velocity of the door and the velocity of the counterweight C at this instant if $L = 6\,\text{ft}$ and $d = 1.5\,\text{ft}$.

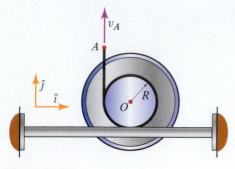

Figure P6.60

Problem 6.60

A spool with inner radius $R = 1.5\,\text{m}$ rolls without slip over a horizontal rail as shown. If the cable on the spool is unwound at a rate $v_A = 5\,\text{m/s}$ in such a way that the unwound cable remains perpendicular to the rail, determine the angular velocity of the spool and the velocity of the spool's center O.

Problem 6.61

In the four-bar linkage system shown, the lengths of the bars AB and CD are $L_{AB} = 46\,\text{mm}$ and $L_{CD} = 25\,\text{mm}$, respectively. In addition, the distance between points A and D is $d_{AD} = 43\,\text{mm}$. The dimensions of the mechanism are such that when the angle $\theta = 132°$, the angle $\phi = 69°$. For $\theta = 132°$ and $\dot\theta = 27\,\text{rad/s}$, determine the angular velocity of bars BC and CD as well as the velocity of the point E, the midpoint of bar BC. Note that the figure is drawn to scale and that bars BC and CD are not collinear.

Problem 6.62

A person is closing a heavy gate with rusty hinges by pushing the gate with car A. If $w = 24\,\text{m}$ and $v_A = 1.2\,\text{m/s}$, determine the angular velocity of the gate when $\theta = 15°$.

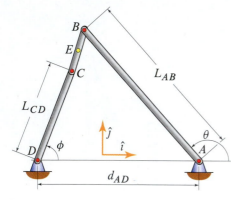

Figure P6.61

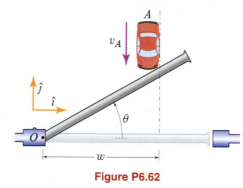

Figure P6.62

Problems 6.63 through 6.65

In the four-bar linkage system shown, let the circular guide with center at O be fixed and such that, when $\theta = 0°$, the bars AB and BC are vertical and horizontal, respectively. In addition, let $R = 2\,\text{ft}$, $L = 3\,\text{ft}$, and $H = 3.5\,\text{ft}$.

Problem 6.63   When $\theta = 0°$, the collar at C is sliding downward with a speed of $23\,\text{ft/s}$. Determine the angular velocities of the bars AB and BC at this instant.

Problem 6.64   When $\theta = 37°$, $\beta = 25.07°$, $\gamma = 78.71°$, and the collar is sliding clockwise with a speed $v_C = 23\,\text{ft/s}$. Determine the angular velocities of the bars AB and BC.

Problem 6.65   Determine the general expression for the angular velocities of bars AB and BC as a function of θ, β, γ, R, L, H, and $\dot\theta$.

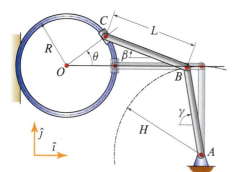

Figure P6.63–P6.65

Problem 6.66

At the instant shown, the arm OC rotates counterclockwise with an angular velocity of $35\,\text{rpm}$ about the fixed sun gear S of radius $R_S = 3.5\,\text{in}$. The planet gear P with radius $R_P = 1.2\,\text{in}$. rolls without slip over both the fixed sun gear and the outer ring gear. Finally, notice that the ring gear is not fixed and it rolls without slip over the sun gear. Determine the angular velocity of the ring gear and the velocity of the center of the ring gear at the instant shown.

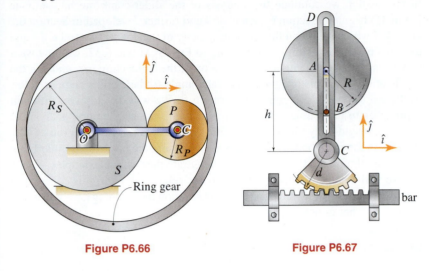

Figure P6.66

Figure P6.67

Problem 6.67

The crank AB is rotating counterclockwise at a constant angular velocity of $12\,\text{rad/s}$ while the pin B slides within the slot in the bar CD, which is pinned at C. Letting $R = 0.5\,\text{m}$, $h = 1\,\text{m}$, and $d = 0.25\,\text{m}$, determine the angular velocity of CD at the instant shown (with points A, B, and C vertically aligned) as well as the velocity of the horizontal bar to which bar CD is connected.

Problems 6.68 through 6.72

For the slider-crank mechanism shown, let $R = 20\,\text{mm}$, $L = 80\,\text{mm}$, and $H = 38\,\text{mm}$.

Problem 6.68   If $\dot\theta = 1700\,\text{rpm}$, determine the angular velocity of the connecting rod AB and the speed of the slider B for $\theta = 90°$.

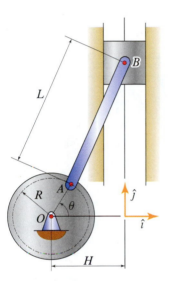

Figure P6.68–P6.72

Problem 6.69   Determine the angular velocity of the crank OA when $\theta = 20°$ and the slider is moving downward at $15\,\text{m/s}$.

Problem 6.70   Determine the general expression for the velocity of the slider B as a function of θ, $\dot\theta$, and the geometrical parameters R, H, and L, using the *vector approach*.

Problem 6.71   Determine the general expression for the velocity of the slider B as a function of θ, $\dot\theta$, and the geometrical parameters R, H, and L, using *differentiation of constraints*.

Problem 6.72   Plot the velocity of the slider C as a function of θ, for $0 \le \theta \le 360°$, and for $\dot\theta = 1000\,\text{rpm}$, $\dot\theta = 3000\,\text{rpm}$, and $\dot\theta = 5000\,\text{rpm}$.

Problem 6.73

Complete the velocity analysis of the slider-crank mechanism, using differentiation of constraints that was outlined beginning on p. 478. That is, determine the velocity of the piston C and the angular velocity of the connecting rod as a function of the given quantities θ, ω_{AB}, R, and L. Use the component system shown for your answers.

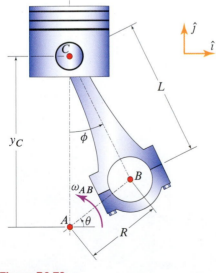

Figure P6.73

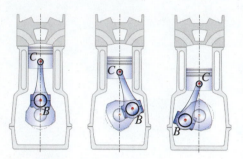

Figure 6.31

A slider-crank mechanism emphasizing the motion of the connecting rod.

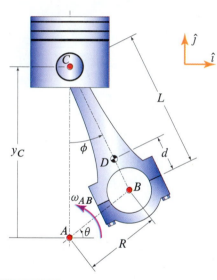

Figure 6.32

The geometric parameters used in the slider-crank analysis.

6.3   Planar Motion: Acceleration Analysis

In this section we continue the analysis of the slider-crank mechanism (see Fig. 6.31) begun in Section 6.1 on p. 459 and further developed in Section 6.2 on p. 477 where we found the angular velocity of the connecting rod $\vec{\omega}_{BC}$ and the velocity of the piston $\vec{v}_C$. Referring to Figs. 6.31 and 6.32, we now want to determine $\vec{\alpha}_{BC}$, the angular acceleration of the connecting rod, and $\vec{a}_C$, the acceleration of the piston, given the crank angle θ, the crank's angular velocity $\dot{\theta}$, and the crank's angular acceleration $\ddot{\theta}$. We will develop two different approaches to the problem that are applicable to the acceleration analysis of any planar rigid body motion: the vector approach and differentiation of constraints. In the main body of the section we present the basic ideas underlying the two methods, and then we fully demonstrate them in the examples.

We will assume that the crank's angular velocity $\dot{\theta} = \omega_{AB}$ is *constant* in the discussion below. More generally, the solution methodology we will apply is valid whether or not ω_{AB} is constant.

Vector approach

The connecting rod in the slider-crank mechanism (Fig. 6.31) is in general planar motion, which means that we can describe the acceleration of any of its points by knowing the acceleration of one point on the rod as well as the rotational motion of the rod (i.e., its angular velocity and angular acceleration).* The velocity analysis was done in Example 6.5 on p. 484, and so we know the angular velocity of the rod $\vec{\omega}_{BC}$. The acceleration analysis proceeds in much the same way.

We found the acceleration of point B, which is in fixed axis rotation about the centerline of the crank A (see Fig. 6.32), on p. 464. Points B and C are on the same rigid body, so we can relate the acceleration of C to that of B using either Eq. (6.5) on p. 462 or Eq. (6.14) on p. 465 in Section 6.1, which gives,

$$\vec{a}_C = \vec{a}_B + \vec{\alpha}_{BC} \times \vec{r}_{C/B} + \vec{\omega}_{BC} \times (\vec{\omega}_{BC} \times \vec{r}_{C/B}), \qquad (6.31)$$

or

$$\vec{a}_C = \vec{a}_B + \vec{\alpha}_{BC} \times \vec{r}_{C/B} - \omega_{BC}^2 \vec{r}_{C/B}, \qquad (6.32)$$

respectively, where $\vec{\alpha}_{BC}$ is the angular acceleration of bar BC and $\vec{\omega}_{BC}$ is known. In planar motion, either of the above equations is a vector equation that represents *two* scalar equations. These two scalar equations are the key to determining $\vec{a}_C$ and $\vec{\alpha}_{BC}$. This is so because $\vec{a}_B$ is known, the direction of $\vec{a}_C$ is known (the motion of the piston C is rectilinear along the y axis), $\vec{r}_{C/B}$ can be found in terms of the crank angle θ (see Fig. 6.32), and the axis of rotation for $\vec{\alpha}_{BC}$ is known (it is perpendicular to the plane of motion). Therefore, the components a_C and α_{BC} are the only unknowns in these two scalar equations. This will be shown in Example 6.10.

Differentiation of constraints

Referring to Fig. 6.32, we can also determine $\vec{\alpha}_{BC}$ and $\vec{a}_C$ as functions of θ by writing the appropriate constraint equations and then differentiating them with

---

* See the marginal note on p. 462.

respect to time. We applied this idea to the analysis of slider-crank velocities in Section 6.2 on p. 477 — now we apply it to accelerations.

As was done in Eq. (6.21) on p. 478, we can write the constraint equation for the y coordinate of C as

$$y_C = R \sin \theta + L \cos \phi, \tag{6.33}$$

which can be differentiated twice with respect to time to find the acceleration of C as

$$\ddot{y}_C = a_C = R\ddot{\theta} \cos \theta - R\dot{\theta}^2 \sin \theta - L\ddot{\phi} \sin \phi - L\dot{\phi}^2 \cos \phi. \tag{6.34}$$

As with the velocity analysis, quantities R, L, and $\theta(t)$ are assumed to be known, which implies that we also know $\dot{\theta} = \omega_{AB}$ and $\ddot{\theta} = \alpha_{AB}$, though we see that a_C is also a function of ϕ, $\dot{\phi}$, and $\ddot{\phi}$. In the velocity analysis using differentiation of constraints, we said that we can determine ϕ and $\dot{\phi}$ in terms of known quantities by relating the orientation ϕ of the connecting rod to the crank's orientation θ, using the second constraint equation $\sin \phi = (R/L) \cos \theta$. This equation was differentiated once with respect to time to find $\dot{\phi}$ as a function of θ and $\dot{\theta}$. It can be differentiated twice to obtain $\ddot{\phi}$, where $\ddot{\phi}\hat{k} = \vec{\alpha}_{BC}$, as a function of θ, $\dot{\theta}$, and $\ddot{\theta}$ (see Prob. 6.109), which then completes the analysis. We will demonstrate this method further in Example 6.11.

Rolling without slip: acceleration analysis

We now consider the acceleration of a point on a body in contact with a rolling surface. Referring to Fig. 6.33, the definition of rolling without slip given on p. 477 stated that if the body W is *rolling without slip* over the surface S (S can be moving), then the contact points P and Q (on W and S, respectively) have zero velocity relative to each other (see Eq. (6.17)). In terms of the relative acceleration of P and Q, this condition implies that the component of $\vec{a}_{P/Q}$ tangent to the contact must be equal to zero, that is,

$$\vec{a}_{P/Q} \cdot \hat{u}_t = 0, \tag{6.35}$$

where $\hat{u}_t$ is a unit vector parallel to ℓ, the line tangent to both W and S at their contact. In component form, Eq. (6.35) takes on the form

$$\boxed{(a_{P/Q})_t = 0 \quad \Rightarrow \quad a_{Pt} = a_{Qt}.} \tag{6.36}$$

As an application of Eq. (6.36), suppose we want to determine the acceleration of the point in contact with the ground for a wheel rolling without slip on a flat stationary surface. If the wheel's center O moves as shown in Fig. 6.34, then

$$\vec{v}_O = v_O \hat{\imath} \quad \text{and} \quad \vec{a}_O = a_O \hat{\imath}. \tag{6.37}$$

We have already discovered that $\vec{v}_P = \vec{0}$ and that $\vec{\omega}_O = -(v_O/R)\hat{k}$, so the acceleration of point P is given by

$$\vec{a}_P = \vec{a}_O + \alpha_O \hat{k} \times \vec{r}_{P/O} - \omega_O^2 \vec{r}_{P/O}, \tag{6.38}$$

which, recalling that $\vec{r}_{P/O} = -R \hat{\jmath}$, can be written as

$$\vec{a}_P = a_O \hat{\imath} + \alpha_O \hat{k} \times (-R) \hat{\jmath} - \left(-\frac{v_O}{R}\right)^2 (-R \hat{\jmath})$$

$$= (a_O + \alpha_O R) \hat{\imath} + \left(\frac{v_O^2}{R}\right) \hat{\jmath}. \tag{6.39}$$

Common Pitfall

Why are we not using the instantaneous center (IC) for acceleration analysis? As we stated on p. 479, the IC may have zero velocity, but it does not, in general, have zero acceleration. Therefore, we should *not* use the IC concept for doing the analysis of accelerations.

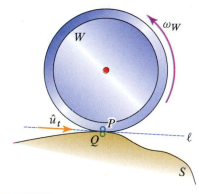

Figure 6.33
A wheel W rolling over a surface S. At the instant shown, the line ℓ is tangent to the path of P and Q.

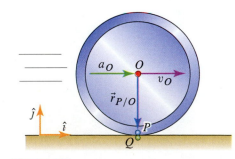

Figure 6.34
A wheel rolling without slip on a horizontal fixed surface.

Observe that the tangent to the wheel at the contact point P is in the x direction. Hence, the application of Eq. (6.36) reads

$$a_{Px} = a_{Qx} = 0 \quad \Rightarrow \quad \boxed{\alpha_O = -\frac{a_O}{R},} \tag{6.40}$$

since the point Q on the ground is stationary. This implies that the point P on the wheel is accelerating along the y axis according to

$$\boxed{\vec{a}_P = \frac{v_O^2}{R}\,\hat{j}.} \tag{6.41}$$

This result tells us that although P has zero velocity at the instant considered, it is accelerating, i.e., it is in the process of gaining velocity, which, in turn, will allow P to move away from its current position and thus allow some other point to become the next contact point.

End of Section Summary

This section presents two different ways to analyze the accelerations of a rigid body in planar motion: the vector approach and differentiation of constraints.

Vector approach.   We saw in Section 6.1 that, for planar motion, either of the equations

Eqs. (6.31) and (6.32), p. 498

$$\vec{a}_C = \vec{a}_B + \vec{\alpha}_{BC} \times \vec{r}_{C/B} + \vec{\omega}_{BC} \times (\vec{\omega}_{BC} \times \vec{r}_{C/B}),$$
$$\vec{a}_C = \vec{a}_B + \vec{\alpha}_{BC} \times \vec{r}_{C/B} - \omega_{BC}^2 \vec{r}_{C/B},$$

relates the acceleration of two points on a rigid body, $\vec{a}_B$ and $\vec{a}_C$, via their relative position $\vec{r}_{C/B}$, the angular acceleration of the body $\vec{\alpha}_{BC}$, and the angular velocity of the body $\vec{\omega}_{BC}$ (see Fig. 6.35).

Differentiation of constraints.   As we discovered in Section 2.7, it is often convenient to write an equation describing the position of a point of interest, which can then be differentiated once with respect to time to find the velocity of that point and twice with respect to time to find the acceleration. For planar motion of rigid bodies, this idea can also apply for describing the position, velocity, and acceleration of a point on a rigid body, as well as for describing the orientation of a rigid body, for which the first and second time derivatives provide its angular velocity and angular acceleration, respectively (we saw this again in Section 6.2 for velocities).

Rolling without slip.   When a body rolls without slip over another body (see Fig. 6.36), then the two points on the bodies that are in contact at any instant, points P and Q, must have the same velocity. For accelerations, this means that the component of $\vec{a}_{P/Q}$ tangent to the contact must be equal to zero, that is,

Eqs. (6.36), p. 499

$$(a_{P/Q})_t = 0 \quad \Rightarrow \quad a_{Pt} = a_{Qt}.$$

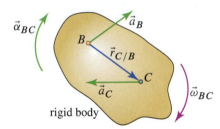

Figure 6.35
A rigid body on which we are relating the acceleration of two points B and C.

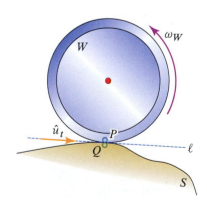

Figure 6.36
Figure 6.33 repeated. A wheel W rolling over a surface S. At the instant shown, the line ℓ is tangent to the path of P and Q.

If a wheel of radius R is rolling without slip over a flat, stationary surface, then the point P on the wheel in contact with the surface must have zero velocity (see Fig. 6.37). The consequence is that the angular acceleration of the wheel α_O is related to the acceleration of the center a_O and the radius of the wheel R according to

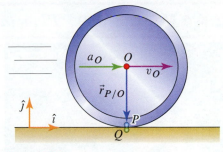

Figure 6.37

Figure 6.34 repeated. A wheel rolling without slip on a horizontal fixed surface.

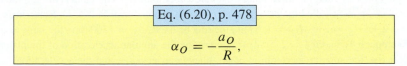

Eq. (6.20), p. 478

$$\alpha_O = -\frac{a_O}{R},$$

in which the positive direction for α_O is taken to be the positive z direction. The point P on the wheel that is in contact with the ground at this instant is accelerating along the y axis according to

Eq. (6.41), p. 500

$$\vec{a}_P = \frac{v_O^2}{R}\,\hat{j}.$$

Even though P has zero velocity at the instant considered, it is accelerating.

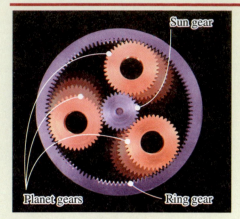

Figure 1
Photo of a planetary gear system.

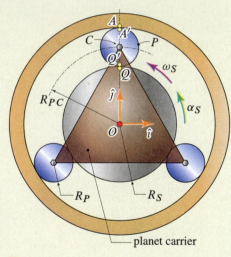

Figure 2
Schematic of a planetary gear system with three planets and a fixed ring.

EXAMPLE 6.9 *Planetary Gears Rolling Without Slip: Vector Approach*

Recall from Example 6.4 on p. 483 that planetary gear systems (see Fig. 1) are used to transmit power between two shafts. The center gear is called the *sun*, the outer gear is called the *ring*, and the three inner gears are called *planets*. The planets are mounted on a component called the *planet carrier* (which is not shown in Fig. 1). Referring to Fig. 2, let $R_S = 2$ in., $R_P = 0.67$ in., the ring be fixed, $\omega_S = 1500$ rpm (the same as in Example 6.4), and $\alpha_S = 2.7$ rad/s². Determine the angular acceleration of the planet gear $\vec{\alpha}_P$, the angular acceleration of the planet carrier $\vec{\alpha}_{PC}$, and the accelerations of points A' and Q'.

SOLUTION

Road Map The solution of this problem has two essential elements: (1) the enforcement of the rolling without slip conditions for the A-A' and the Q-Q' contacts and (2) the enforcement of the constraint that point C on the planet gear P must move along a circle of radius $R_{PC} = R_S + R_P$. We carry out element 1 by first writing the accelerations of points A, A', Q, and Q'. To enforce 2, we write $\vec{a}_C$ twice, the first time viewing C as part of the planet carrier and the second time viewing C as part of P. We then force the two resulting expressions to be equal to each other. In general, before doing the acceleration analysis, we first need to find the angular velocities of all components in the system. This was done in Example 6.4 on p. 483, in which we discovered that $\omega_P = -(R_S/2R_P)\omega_S = -2240$ rpm and $\omega_{PC} = (R_S/2R_{PC})\omega_S = 562$ rpm.

Computation The accelerations of A (a fixed point) and A' can be expressed as

$$\vec{a}_A = \vec{0} \quad \text{and} \quad \vec{a}_{A'} = a_{A'x}\,\hat{\imath} + a_{A'y}\,\hat{\jmath}. \tag{1}$$

Observe that the line tangent to the A-A' contact is parallel to the x direction, so that the rolling without slip condition between P and the ring gear implies

$$a_{Ax} = a_{A'x} \quad \Rightarrow \quad a_{A'x} = 0. \tag{2}$$

To find the accelerations of Q and Q', since the sun gear is rotating about the (fixed) z axis, we can express $\vec{a}_Q$ as follows:

$$\vec{a}_Q = \vec{\alpha}_S \times \vec{r}_{Q/O} - \omega_S^2 \vec{r}_{Q/O} = -\alpha_S R_S\,\hat{\imath} - \omega_S^2 R_S\,\hat{\jmath}, \tag{3}$$

where we have set $\vec{\alpha}_S = \alpha_S\,\hat{k}$ and $\vec{r}_{Q/O} = R_S\,\hat{\jmath}$. For Q', using A' as a reference point, we can write

$$\begin{aligned}
\vec{a}_{Q'} &= \vec{a}_{A'} + \vec{\alpha}_P \times \vec{r}_{Q'/A'} - \omega_P^2 \vec{r}_{Q'/A'} \\
&= a_{A'y}\,\hat{\jmath} + \alpha_P\,\hat{k} \times (-2R_P)\,\hat{\jmath} - \omega_P^2 (-2R_P)\,\hat{\jmath} \\
&= 2R_P\alpha_P\,\hat{\imath} + \left(a_{A'y} + 2R_P\omega_P^2\right)\hat{\jmath}.
\end{aligned} \tag{4}$$

We now enforce the rolling without slip condition at the Q-Q' contact by observing that the tangent line to the contact is parallel to the x axis. Therefore, from Eqs. (3) and (4), we have

$$a_{Qx} = a_{Q'x} \quad \Rightarrow \quad -\alpha_S R_S = 2R_P\alpha_P, \tag{5}$$

or

$$\alpha_P = -\frac{R_S}{2R_P}\alpha_S = -4.03 \text{ rad/s}^2 \quad \Rightarrow \quad \boxed{\vec{\alpha}_P = (-4.03 \text{ rad/s}^2)\,\hat{k}.} \tag{6}$$

As for the acceleration of point C, when viewed as part of the planet carrier, point C is rotating about point O, so that $\vec{a}_C$ can be given the form

$$\vec{a}_C = \vec{\alpha}_{PC} \times \vec{r}_{C/O} - \omega_{PC}^2 \vec{r}_{C/O}$$

$$= -\alpha_{PC} R_{PC}\, \hat{\imath} - \omega_{PC}^2 R_{PC}\, \hat{\jmath}, \qquad (7)$$

where we have set $\vec{r}_{C/O} = R_{PC}\, \hat{\jmath}$ and $\vec{\alpha}_{PC} = \alpha_{PC}\, \hat{k}$, and we note that $R_{PC} = R_S + R_P$. Viewing point C as part of the planet gear P and relating its acceleration to A', we have

$$\vec{a}_C = \vec{a}_{A'} + \vec{\alpha}_P \times \vec{r}_{C/A'} - \omega_P^2 \vec{r}_{C/A'}$$

$$= a_{A'y}\, \hat{\jmath} + \alpha_P\, \hat{k} \times \vec{r}_{C/A'} - \omega_P^2 \vec{r}_{C/A'}$$

$$= \alpha_P R_P\, \hat{\imath} + \left(a_{A'y} + \omega_P^2 R_P\right) \hat{\jmath}. \qquad (8)$$

The expressions for $\vec{a}_C$ in Eqs. (7) and (8) must be equal to one another. Therefore,

$$-\alpha_{PC} R_{PC} = \alpha_P R_P \quad \text{and} \quad -\omega_{PC}^2 R_{PC} = a_{A'y} + \omega_P^2 R_P. \qquad (9)$$

Equations (9) can be solved for the unknowns α_{PC} and $a_{A'y}$ to obtain

$$\alpha_{PC} = -\frac{R_P}{R_{PC}} \alpha_P \quad \text{and} \quad a_{A'y} = -\omega_P^2 R_P - \omega_{PC}^2 R_{PC}. \qquad (10)$$

Using Eq. (6) for α_P, the expressions for ω_P and ω_{PC} from the Road Map, and Eq. (2) for $a_{A'x}$, Eqs. (10) become

$$\boxed{\vec{\alpha}_{PC} = \frac{R_S}{2R_{PC}} \alpha_S\, \hat{k} = \left(1.01\ \text{rad/s}^2\right) \hat{k},} \qquad (11)$$

$$\boxed{\vec{a}_{A'} = -\frac{R_S^2}{4R_P}\left(1 + \frac{R_P}{R_{PC}}\right) \omega_S^2\, \hat{\jmath} = \left(-3840\ \text{ft/s}^2\right) \hat{\jmath}.} \qquad (12)$$

Finally, substituting the results in Eqs. (6) and (12) and ω_P from the Road Map into Eq. (4), we obtain

$$\boxed{\vec{a}_{Q'} = -\alpha_S R_S\, \hat{\imath} + \frac{R_S^2}{4R_P}\left(1 - \frac{R_P}{R_{PC}}\right) \omega_S^2\, \hat{\jmath} = \left(-0.45\, \hat{\imath} + 2300\, \hat{\jmath}\right) \text{ft/s}^2.} \qquad (13)$$

Discussion & Verification   The dimensions of the symbolic answers are all as they should be, and so the units of the numerical answers are also correct.

The results in Eqs. (6) and (11) are not hard to verify by differentiating the corresponding angular velocity equations (we will see this in Example 6.11 on p. 506). While the procedure we have used in our solution is applicable in general, obtaining the component of an angular acceleration by simply differentiating the corresponding component of the angular velocity can be done only under special circumstances, such as when the components in question are with respect to a fixed axis (in our case the z axis). As far as the acceleration of point A' is concerned, based on our discussion of the rolling without slip condition earlier in this section, because A' was in contact with a stationary surface we should have expected $\vec{a}_{A'}$ to be completely in the negative y direction and proportional to ω_P^2. This is exactly what we obtained, given that ω_P is proportional to ω_S. As far as $\vec{a}_{Q'}$ is concerned, our expectation was that the x component had to match the motion of the sun gear (and therefore be in the negative x direction) while the y component had to be in the positive y direction and, again, be proportional to ω_P^2, i.e., ω_S^2. Again, these expectations match the obtained results.

EXAMPLE 6.10   *Completing the Acceleration Analysis of the Connecting Rod*

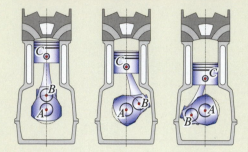

Figure 1

On p. 498 we outlined the vector approach for the acceleration analysis of the connecting rod (CR) and piston in the slider-crank mechanism shown in Fig. 1. We will complete that analysis here.

Referring to Fig. 2, we are given the radius of the crank R, the length of the CR L, the position of the mass center of the CR H, the *constant* angular velocity of the crank $\omega_{AB} = \dot{\theta}$, and the crank angle θ. Determine the angular acceleration of the CR $\vec{\alpha}_{BC}$ and the acceleration of the piston $\vec{a}_C$.

SOLUTION

Road Map   The road map for this problem was laid out on p. 498.

Computation   As with the velocity analysis, we begin by determining the motion of point C, which is constrained to move along the y axis and so $a_{Cx} = 0$. Applying Eq. (6.14) on p. 465 to the CR and using point B as a reference point for the body, we have

$$\vec{a}_C = \vec{a}_B + \vec{\alpha}_{BC} \times \vec{r}_{C/B} - \omega_{BC}^2 \vec{r}_{C/B}, \tag{1}$$

where $\vec{\alpha}_{BC} = \alpha_{BC}\,\hat{k}$ is the angular acceleration of the CR. Recall that in Example 6.5 on p. 484 we found that the angular velocity of the CR is

$$\omega_{BC} = -\frac{\omega_{AB}\sin\theta}{\sqrt{(L/R)^2 - \cos^2\theta}}, \tag{2}$$

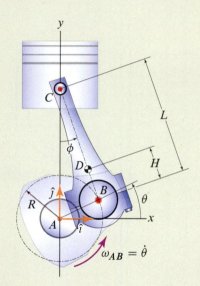

Figure 2
A slider-crank mechanism showing the relevant dimensions.

and the position of C relative to B is given by

$$\vec{r}_{C/B} = -L\sin\phi\,\hat{\imath} + L\cos\phi\,\hat{\jmath}, \tag{3}$$

where $\sin\phi$ and $\cos\phi$ are found from the equations

$$\sin\phi = \frac{R}{L}\cos\theta \quad \text{and} \quad \cos\phi = \frac{\sqrt{L^2 - R^2\cos^2\theta}}{L}. \tag{4}$$

Since B is in fixed axis rotation about A with $\omega_{AB} = \dot{\theta} = $ constant, its acceleration is given by (see p. 464 of Section 6.1)

$$\vec{a}_B = \vec{\alpha}_{AB} \times \vec{r}_{B/A} - \omega_{AB}^2 \vec{r}_{B/A} = -R\omega_{AB}^2(\cos\theta\,\hat{\imath} + \sin\theta\,\hat{\jmath}), \tag{5}$$

where we have used $\alpha_{AB} = \ddot{\theta} = 0$. Substituting Eqs. (3)–(5) into Eq. (1) and simplifying, we obtain

$$\vec{a}_C = R\left[\cos\theta\,(\omega_{BC}^2 - \omega_{AB}^2) - \alpha_{BC}\sqrt{(L/R)^2 - \cos^2\theta}\,\right]\hat{\imath}$$
$$- R\left[\alpha_{BC}\cos\theta + \omega_{BC}^2\sqrt{(L/R)^2 - \cos^2\theta} + \omega_{AB}^2\sin\theta\right]\hat{\jmath}. \tag{6}$$

Enforcing the condition $a_{Cx} = 0$ in Eq. (6) and solving for α_{BC} yields

$$\alpha_{BC} = \frac{\cos\theta\,(\omega_{BC}^2 - \omega_{AB}^2)}{\sqrt{(L/R)^2 - \cos^2\theta}}. \tag{7}$$

Substituting Eq. (2) into Eq. (7) and simplifying give

$$\alpha_{BC} = \frac{\left[1 - (L/R)^2\right]\cos\theta}{\left[(L/R)^2 - \cos^2\theta\right]^{3/2}}\omega_{AB}^2 \Rightarrow \boxed{\vec{\alpha}_{BC} = \frac{\left[1 - (L/R)^2\right]\cos\theta}{\left[(L/R)^2 - \cos^2\theta\right]^{3/2}}\omega_{AB}^2\,\hat{k},}$$

$$\tag{8}$$

Helpful Information

Another way to compute α_{BC}. The method used to obtain Eqs. (8) was laborious, but it only requires a series of algebraic steps. We could have obtained α_{BC} by differentiating ω_{BC} in Eq. (2) with respect to time (this is true here since the motion is planar), provided that we correctly apply the product and the chain rules of calculus.

where we used $\vec{\alpha}_{BC} = \alpha_{BC}\,\hat{k}$. Substituting Eqs. (2) and (8) into Eq. (6), we obtain

$$\vec{a}_C = -R\omega_{AB}^2 \left\{ \frac{\left[1 - (L/R)^2\right]\cos^2\theta}{\left[(L/R)^2 - \cos^2\theta\right]^{3/2}} + \frac{\sin^2\theta}{\sqrt{(L/R)^2 - \cos^2\theta}} + \sin\theta \right\}\hat{j}, \quad (9)$$

where we have used $a_{Cx} = 0$.

Discussion & Verification   The method we illustrated here, i.e., based on the systematic application of the (general) equation $\vec{a}_B = \vec{a}_A + \vec{\alpha}_{AB} \times \vec{r}_{B/A} + \vec{\omega}_{AB} \times (\vec{\omega}_{AB} \times \vec{r}_{B/A})$, while at times laborious, is relatively straightforward in that it only involves a series of algebraic steps, as opposed to computing accelerations directly through differentiation with respect to time. There are many situations in which it is indeed simpler to differentiate with respect to time than it is to apply the acceleration formula for a rigid body. This strategy for the calculation of accelerations will be demonstrated in Example 6.11 on p. 506.

🔍 **A Closer Look**   We can now complete the parametric analysis begun in Example 6.5 on p. 484.

💻 ➡ We plot the angular acceleration of the CR α_{BC} as well as the acceleration of the piston a_{Cy}, for the same conditions considered in Example 6.11. What is remarkable is the magnitude of the accelerations undergone by the CR. We refer to Figs. 3 and 4 and recall that, for a value of $\dot{\theta} = 3500$ rpm (a car's engine is easily capable

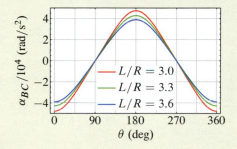

Figure 3. The angular acceleration of the CR for $\dot{\theta} = 3500$ rpm, $L = 150$ mm, and three values of L/R.

Finding velocities and accelerations by time differentiation. We mentioned that we could have computed Eqs. (8) by time differentiating Eq. (2). In doing this, a common mistake is to take the derivative at a particular instant rather than the general function of time. For example, suppose that we had computed ω_{BC} at a *particular instant* t_0 to be $\omega_{BC}(t_0) = 1234$ rad/s. We shouldn't say that $\alpha_{BC}(t_0) = 0$ because the time derivative of the number 1234 rad/s equals zero. While it is true that the time derivative of a constant is equal to zero, we must first take a time derivative of the *function* $\omega_{BC}(t)$ and then evaluate the result at the time of interest t_0. This concept is illustrated in Example 6.11.

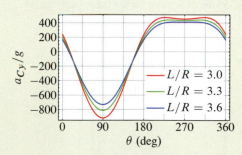

Figure 4
The piston's acceleration for $\dot{\theta} = 3500$ rpm, $L = 150$ mm, and three values of L/R.

of 7000 rpm, though it is most efficient operating at 2200–2500 rpm), we see that the angular acceleration reaches values in excess of $40{,}000$ rad/s^2 and the acceleration of the piston reaches values that are 900 times the acceleration of gravity g! Since the accelerations in question are proportional to $\dot{\theta}^2$, if the engine's angular velocity is increased by a factor of, for example, 2, the accelerations we plotted increase by a factor of 4! Thus, for an engine running at 7000 rpm, the piston's accelerations can easily reach $3600g$. In Chapter 7 we will learn how to translate this information into a computation of the forces and moments that a mechanism such as the slider-crank must be able to sustain. This will allow us to understand why components such as a CR in an engine are typically made of high-grade steel. As a final remark on Fig. 4, notice that the piston's acceleration has two different behaviors: one for $0° \le \theta < 180°$ and another for $180° \le \theta \le 360°$. We already observed this lack of symmetry during the velocity analysis, and we see now that it is even more pronounced in the acceleration behavior. It is the geometry of the mechanism that generates this lack of symmetry, and the behavior becomes more symmetric as the ratio L/R is increased. ⬅ 💻

EXAMPLE 6.11   *Motion of a Propped Ladder: Differentiation of Constraints*

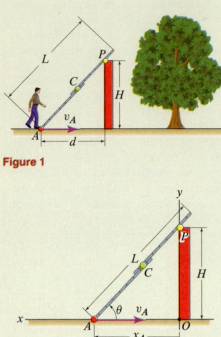

Figure 1

Figure 2
Coordinate system for the ladder kinematics.

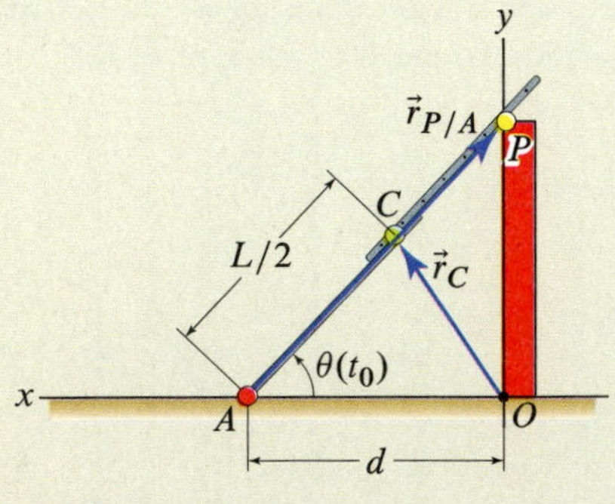

Figure 3
Definition of $\vec{r}_C$ and $\vec{r}_{P/A}$.

A person is propping up a ladder against a wall by pushing the end A to the right along the ground (see Fig. 1). If, at the instant shown, the speed of A is constant and equal to $v_A = 0.8\,\text{m/s}$, the length $L = 6\,\text{m}$, the height $H = 4\,\text{m}$, and the distance $d = 1.57\,\text{m}$, determine the angular velocity and angular acceleration of the ladder. In addition, determine the acceleration of the midpoint of the ladder C and the acceleration of the point P on the ladder that, at the given instant, is in contact with the wall.

SOLUTION

Road Map   Referring to Fig. 2, we can use θ to describe the ladder's orientation so that we can write $\vec{\omega}_L = -\dot{\theta}\,\hat{k}$ and $\vec{\alpha}_L = -\ddot{\theta}\,\hat{k}$ (the minus signs are needed to reconcile the positive direction of θ with the component system used). Thus, we will need to relate $\dot{\theta}$ and $\ddot{\theta}$ to the given data (v_A, H, and d). This can be accomplished by differentiating with respect to time the constraint relating θ to the position of A. We will compute $\vec{a}_C$ using differentiation of constraints and $\vec{a}_P$ using the vector approach to relate $\vec{a}_P$ to the acceleration of a known point on the ladder.

Computation   Referring to Fig. 2 and focusing on the triangle AOP, at any time during the ladder's motion we must have

$$x_A \tan\theta = H \quad\Rightarrow\quad \dot{x}_A \tan\theta + x_A \dot{\theta} \sec^2\theta = 0, \tag{1}$$

where the second equation was obtained by differentiating the first with respect to time. By using $\sec^2\theta = 1 + \tan^2\theta$, $\tan\theta = H/x_A$, and $\dot{x}_A = -v_A$, the second of Eqs. (1) can be solved for $\dot{\theta}$ to obtain

$$\dot{\theta} = \frac{v_A H}{x_A^2 + H^2} \quad\Rightarrow\quad \dot{\theta}(t_0) = 0.1733\,\text{rad/s}, \tag{2}$$

where t_0 is the time at which $x_A = d = 1.57\,\text{m}$. Having computed $\dot{\theta}$, we can now obtain $\ddot{\theta}$ by time differentiating the first of Eqs. (2) with respect to time to obtain

$$\ddot{\theta} = \frac{2 x_A v_A^2 H}{\left(x_A^2 + H^2\right)^2} \quad\Rightarrow\quad \ddot{\theta}(t_0) = 0.02358\,\text{rad/s}^2, \tag{3}$$

where we used the fact that v_A is constant. Since $\vec{\omega}_L = -\dot{\theta}\,\hat{k}$ and $\vec{\alpha}_L = -\ddot{\theta}\,\hat{k}$, Eqs. (2) and (3) imply that

$$\boxed{\vec{\omega}_L(t_0) = -0.173\,\hat{k}\,\text{rad/s},} \tag{4}$$

$$\boxed{\vec{\alpha}_L(t_0) = -0.0236\,\hat{k}\,\text{rad/s}^2.} \tag{5}$$

We now compute $\vec{a}_C(t_0)$ by differentiation of constraints. This means that we need to take two time derivatives of the general constraint equations for the position of C. Referring to Fig. 3, we can write

$$\vec{r}_C = \left(x_A - \frac{L}{2}\cos\theta\right)\hat{\imath} + \frac{L}{2}\sin\theta\,\hat{\jmath}. \tag{6}$$

Taking one and then two time derivatives of Eq. (6), we obtain

$$\vec{v}_C = \left(-v_A + \frac{L}{2}\dot{\theta}\sin\theta\right)\hat{\imath} + \frac{L}{2}\dot{\theta}\cos\theta\,\hat{\jmath}, \tag{7}$$

and

$$\vec{a}_C = \frac{L}{2}\left[\left(\ddot{\theta}\sin\theta + \dot{\theta}^2\cos\theta\right)\hat{\imath} + \left(\ddot{\theta}\cos\theta - \dot{\theta}^2\sin\theta\right)\hat{\jmath}\right], \tag{8}$$

where we used the fact that v_A is constant. We now find $\theta(t_0)$ by substituting $x_A(t_0) = d = 1.57\,\text{m}$ and $H = 4\,\text{m}$ into the first of Eqs. (1) to obtain

$$\theta(t_0) = \tan^{-1}(4/1.57) = 68.57°. \tag{9}$$

Substituting the results in Eqs. (2), (3), and (9) into Eq. (8), we obtain

$$\boxed{\vec{a}_C(t_0) = \left(98.8\times10^{-3}\,\hat{\imath} - 58.0\times10^{-3}\,\hat{\jmath}\right)\text{m/s}^2.} \tag{10}$$

Now we compute $\vec{a}_P$ by using the vector method to relate the acceleration of P to that of A. Choosing A as a reference point is convenient because $\vec{a}_A = \vec{0}$ (A is moving at a constant velocity). Since we need to compute $\vec{a}_P$ *at the instant shown*, we can write

$$\vec{a}_P(t_0) = \vec{\alpha}_L(t_0) \times \vec{r}_{P/A}(t_0) - \omega_L^2(t_0)\vec{r}_{P/A}(t_0). \tag{11}$$

Referring to Fig. 3, at time t_0 we have

$$\vec{r}_{P/A}(t_0) = -d\,\hat{\imath} + d\tan\theta(t_0)\,\hat{\jmath}. \tag{12}$$

Substituting Eq. (12), along with the results in Eqs. (4), (5), and (9), into Eq. (11), we have

$$\boxed{\vec{a}_P(t_0) = \left(141\times10^{-3}\,\hat{\imath} - 83.1\times10^{-3}\,\hat{\jmath}\right)\text{m/s}^2.} \tag{13}$$

Discussion & Verification   The dimensions and therefore the units of all of our results are as they should be. The signs for the angular velocity and acceleration of the ladder match our expectation given that the ladder is rotating counterclockwise and the $\hat{k}$ direction is into the page. As far as the values of acceleration, these are harder to verify without using an alternative solution strategy.

A Closer Look   Without actually performing the calculations, if we used the formula $\vec{a}_C = \vec{\alpha}_L \times \vec{r}_{C/A} - \omega_L^2\vec{r}_{C/A}$, we would readily see that the vectors $\vec{\alpha}_L \times \vec{r}_{C/A}$ and $-\omega_L^2\vec{r}_{C/A}$ have positive x components, thus matching the fact that we obtained a positive value for $a_{Cx}(t_0)$. As far as their y components are concerned, we would find that the terms $\vec{\alpha}_L \times \vec{r}_{C/A}$ and $-\omega_L^2\vec{r}_{C/A}$ have positive and negative y components, respectively. However, given that ω_L is larger than α_L in absolute value, we expect that the term with ω_L^2 would dominate with respect to the α_L term. Therefore, overall we expect $a_{Cy}(t_0)$ to be negative, which is exactly what we found. A similar logic can be applied to the discussion of the $\vec{a}_P$ result.

Again, this example is meant to show that the techniques we learned in Chapter 2 are still relevant to the study of rigid bodies and can be used together with the vector method discussed in this chapter.

EXAMPLE 6.12   *Acceleration Analysis of a Four-Bar Linkage: Vector Approach*

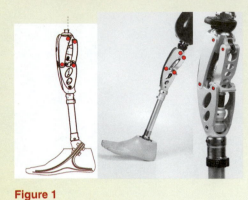

Figure 1

We now continue the kinematic analysis of a prosthetic leg with an artificial knee joint presented in Example 6.7 on p. 488 (see Fig. 1). The primary kinematic component of the artificial knee joint shown is the four-bar linkage system highlighted in Fig. 2. Since the determination of accelerations is crucial in the determination of the forces and moments acting on a mechanism, we will determine the angular accelerations of each link in the system in Fig. 2 at the instant shown, assuming that the AD segment is fixed. We will also determine the acceleration of point P, which is the midpoint of the link BC. As was done in Example 6.7, we will use the coordinates of points A, B, C, and D given in Table 1. The segment AB is rotating counterclockwise at $1.5\,\text{rad/s}$ and that rate is decreasing at $0.8\,\text{rad/s}^2$.

Table 1. Approximate values of the coordinates of the pin centers A, B, C, and D for the system shown in Fig. 2 at the time instant considered.

Points	A	B	C	D
Coordinates (mm)	(0.0, 0.0)	(−27.0,120)	(26.0,124)	(30.0, 15.0)

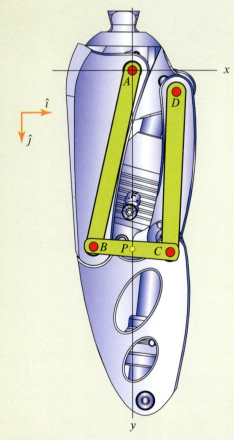

Figure 2
Geometry of the four-bar linkage in the prosthetic leg.

SOLUTION

Road Map As we have seen, an acceleration analysis is generally preceded by the corresponding velocity analysis. This was done for this linkage in Example 6.7 on p. 488, and we will use the angular velocities found there, which were $\omega_{BC} = -637.8 \times 10^{-3}\,\text{rad/s}$ and $\omega_{CD} = -1.675\,\text{rad/s}$.* The calculation of accelerations is done similarly to the velocity analysis; that is, we will compute the acceleration of B and then the acceleration of C. Because point C is shared by both links BC and CD, we will obtain two independent statements for $\vec{a}_C$, which we will then *require* to be equal. This will allow us to obtain two equations for the angular accelerations of the links BC and CD. Once the angular accelerations are known, then we will compute the acceleration of point P.

Computation Recalling that A is fixed, $\vec{a}_B$ is given by

$$\vec{a}_B = \alpha_{AB}\,\hat{k} \times \vec{r}_{B/A} - \omega_{AB}^2 \vec{r}_{B/A} = -(35.25\,\hat{\imath} + 291.6\,\hat{\jmath})\,\text{mm/s}^2, \tag{1}$$

where we set $\alpha_{AB} = +0.8\,\text{rad/s}^2$ and used $\vec{r}_{B/A} = (-27\,\hat{\imath} + 120\,\hat{\jmath})\,\text{mm}$ from Table 1. Next, since C is shared by both segments BC and CD, we can express $\vec{a}_C$ in the following two independent ways:

$$\vec{a}_C = \vec{a}_B + \alpha_{BC}\,\hat{k} \times \vec{r}_{C/B} - \omega_{BC}^2 \vec{r}_{C/B}, \tag{2}$$

and

$$\vec{a}_C = \vec{a}_D + \alpha_{CD}\,\hat{k} \times \vec{r}_{C/D} - \omega_{CD}^2 \vec{r}_{C/D}, \tag{3}$$

where

$$\vec{r}_{C/B} = \vec{r}_C - \vec{r}_B = (53\,\hat{\imath} + 4\,\hat{\jmath})\,\text{mm}, \tag{4}$$

$$\vec{r}_{C/D} = \vec{r}_C - \vec{r}_D = (-4\,\hat{\imath} + 109\,\hat{\jmath})\,\text{mm}, \tag{5}$$

---

* We have reported the angular velocities here to 4 significant figures since they are intermediate results in this context.

and $\vec{a}_D = \vec{0}$. Substituting Eqs. (1), (4), (5), and the known angular velocities into Eqs. (2) and (3) and setting two expressions for $\vec{a}_C$ equal to one another, we obtain the following vector equation:

$$-\left[56.81\,\tfrac{\text{mm}}{\text{s}^2} + (4\,\text{mm})\alpha_{BC}\right]\hat{\imath} + \left[-293.2\,\tfrac{\text{mm}}{\text{s}^2} + (53\,\text{mm})\alpha_{BC}\right]\hat{\jmath}$$

$$= \left[11.22\,\tfrac{\text{mm}}{\text{s}^2} - (109\,\text{mm})\alpha_{CD}\right]\hat{\imath} - \left[305.7\,\tfrac{\text{mm}}{\text{s}^2} + (4\,\text{mm})\alpha_{CD}\right]\hat{\jmath}. \quad (6)$$

Equating $\hat{\imath}$ components and equating $\hat{\jmath}$ components yield the linear system of two equations

$$-56.81\,\tfrac{\text{mm}}{\text{s}^2} - (4\,\text{mm})\alpha_{BC} = 11.22\,\tfrac{\text{mm}}{\text{s}^2} - (109\,\text{mm})\alpha_{CD}, \quad (7)$$

$$-293.2\,\tfrac{\text{mm}}{\text{s}^2} + (53\,\text{mm})\alpha_{BC} = 305.7\,\tfrac{\text{mm}}{\text{s}^2} + (4\,\text{mm})\alpha_{CD}, \quad (8)$$

in the two unknowns α_{BC} and α_{CD}, whose solution is

$$\boxed{\alpha_{BC} = -0.282\,\text{rad/s}^2 \quad \text{and} \quad \alpha_{CD} = 0.614\,\text{rad/s}^2.} \quad (9)$$

Now that the angular accelerations are known, we can find $\vec{a}_P$ by using

$$\vec{a}_P = \vec{a}_B + \alpha_{BC}\,\hat{k} \times \vec{r}_{P/B} - \omega_{BC}^2\,\vec{r}_{P/B}. \quad (10)$$

Since P is the midpoint between B and C, we have

$$\vec{r}_P = \frac{\vec{r}_B + \vec{r}_C}{2} \quad \Rightarrow \quad \vec{r}_{P/B} = \vec{r}_P - \vec{r}_B = \frac{\vec{r}_C - \vec{r}_B}{2} = (26.5\,\hat{\imath} + 2\,\hat{\jmath})\,\text{mm}. \quad (11)$$

Using Eqs. (1), (9), (11), and the previously computed value of ω_{BC}, from Eq. (10) we obtain

$$\boxed{\vec{a}_P = -(45.5\,\hat{\imath} + 300\,\hat{\jmath})\,\text{mm/s}^2.} \quad (12)$$

Discussion & Verification   As a first check, we see that the dimensions and thus the units of all results are as they should be.

Another way to argue that the results we obtained are reasonable is to observe that, *in the position shown*, this four-bar linkage is such that links AB and BC are nearly parallel to each other, while link BC is oriented such that point C has a larger y coordinate than point B. Thus, *in the position shown*, the behavior of this four-bar linkage should not be that different from a similarly sized parallelogram. Therefore, in the position shown, we would expect the angular velocity of CD to have the same sign as α_{AB}. By the same token, we would expect the angular acceleration of BC to have a sign opposite to that of AB. This is exactly what we obtained. However, it should also be said that obtaining an intuitive understanding of the signs and/or magnitudes of accelerations is not as easy as for velocities. Therefore, when it comes to accelerations, double-checking our calculations is more important than having an intuitive understanding of the mechanism's motion.

PROBLEMS

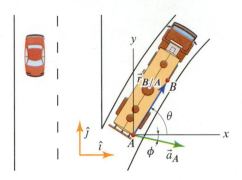

Figure P6.74

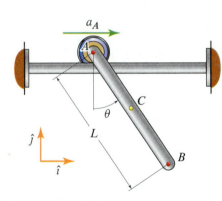

Figure P6.75–P6.77

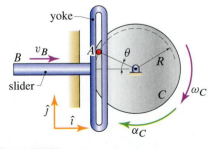

Figure P6.79

Problem 6.74

A truck on an exit ramp is moving in such a way that, at the instant shown, $|\vec{a}_A| = 17\,\text{ft/s}^2$, $\dot\theta = -0.3\,\text{rad/s}$, and $\ddot\theta = -0.1\,\text{rad/s}^2$. If the distance between points A and B is $d_{AB} = 12\,\text{ft}$, $\theta = 57°$, and $\phi = 13°$, determine $\vec{a}_B$.

Problems 6.75 through 6.77

Let $L = 4\,\text{ft}$, let point A travel parallel to the guide shown, and let C be the midpoint of the bar.

Problem 6.75   If point A is accelerating to the right with $a_A = 27\,\text{ft/s}^2$ and $\dot\theta = 7\,\text{rad/s} = \text{constant}$, determine the acceleration of point C when $\theta = 24°$.

Problem 6.76   If point A is accelerating to the right with $a_A = 27\,\text{ft/s}^2$, $\dot\theta = 7\,\text{rad/s}$, and $\ddot\theta = -0.45\,\text{rad/s}^2$, determine the acceleration of point C when $\theta = 26°$.

Problem 6.77   If, when $\theta = 0°$, A is accelerating to the right with $a_A = 27\,\text{ft/s}^2$ and $\vec{a}_C = \vec{0}$, determine $\dot\theta$ and $\ddot\theta$.

Problem 6.78

A wheel W of radius $R_W = 5\,\text{cm}$ rolls without slip over the stationary cylinder S of radius $R_S = 12\,\text{cm}$, and the wheel is connected to point O via the arm OC. If $\omega_{OC} = \text{constant} = 3.5\,\text{rad/s}$, determine the acceleration of point Q, which lies on the edge of W and along the extension of the line OC.

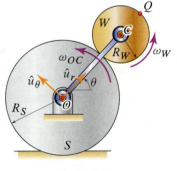

Figure P6.78

Problem 6.79

One way to convert rotational motion into linear motion and vice versa is via the use of a mechanism called the Scotch yoke, which consists of a crank C that is connected to a slider B via a pin A. The pin rotates with the crank while sliding within the yoke, which, in turn, rigidly translates with the slider. This mechanism has been used, for example, to control the opening and closing of valves in pipelines. Letting the radius of the crank be $R = 25\,\text{cm}$, determine the angular velocity ω_C and the angular acceleration α_C of the crank at the instant shown if $\theta = 25°$ and the slider is moving to the right with a constant speed $v_B = 40\,\text{m/s}$.

Problem 6.80

Collar C moves along a circular guide with radius $R = 2$ ft with a constant speed $v_C = 18$ ft/s. At the instant shown, the bars AB and BC are vertical and horizontal, respectively. Letting $L = 4$ ft and $H = 5$ m, determine the angular accelerations of the bars AB and BC at this instant.

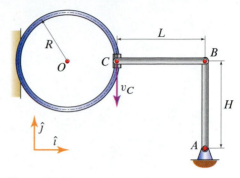

Figure P6.80

Problems 6.81 and 6.82

A ball of radius $R_A = 5$ in. is rolling without slip inside a stationary spherical bowl of radius $R_B = 17$ in. Assume that the motion of the ball is planar.

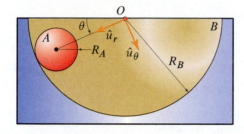

Figure P6.81 and P6.82

Problem 6.81 If, at the instant shown, the center of the ball is traveling counterclockwise with a speed $v_A = 32$ ft/s and such that $\dot{v}_A = 0$, determine the acceleration of the center of the ball as well as the acceleration of the point on the ball that is contact with the bowl.

Problem 6.82 If, at the instant shown, the center of the ball is traveling counterclockwise with a speed $v_A = 32$ ft/s and such that $\dot{v}_A = 24$ ft/s^2, determine the acceleration of the center of the ball as well as the acceleration of the point on the ball that is contact with the bowl.

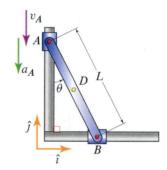

Problem 6.83

A bar of length $L = 2.5$ m is falling so that, when $\theta = 34°$, $v_A = 3$ m/s and $a_A = 8.7$ m/s^2. At this instant, determine the angular acceleration of the bar AB and the acceleration of point D, where D is the midpoint of the bar.

Figure P6.83–P6.85

Problem 6.84

A bar of length $L = 8$ ft and midpoint D is falling so that when $\theta = 27°$, $|\vec{v}_D| = 18$ ft/s, and the vertical acceleration of point D is 23 ft/s^2 downward. At this instant, compute the angular acceleration of the bar and the acceleration of point B.

Problem 6.85

Assuming that, for $0° \le \theta \le 90°$, v_A is constant, compute the expression for the acceleration of point D, the midpoint of the bar, as a function of θ and v_A.

Problem 6.86

A truck on an exit ramp is moving in such a way that, at the instant shown, $|\vec{a}_A| = 6$ m/s^2 and $\phi = 13°$. Let the distance between points A and B be $d_{AB} = 4$ m. If, at this instant, the truck is turning clockwise, $\theta = 59°$, $a_{Bx} = 6.3$ m/s^2, and $a_{By} = -2.6$ m/s^2, determine the angular velocity and angular acceleration of the truck.

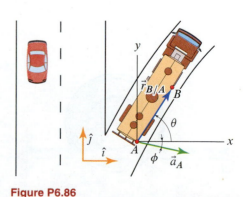

Figure P6.86

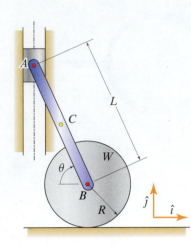

Figure P6.87–P6.89

Problems 6.87 through 6.89

The system shown consists of a wheel of radius $R = 1.4$ m rolling without slip on a horizontal surface. A bar AB of length $L = 3.7$ m is pin-connected to the center of the wheel and to a slider A constrained to move along a vertical guide. Point C is the bar's midpoint.

Problem 6.87 If the wheel is rolling clockwise with a constant angular speed of 2 rad/s, determine the angular acceleration of the bar when $\theta = 72°$.

Problem 6.88 If the slider A is moving downward with a constant speed 3 m/s, determine the angular acceleration of the wheel when $\theta = 53°$.

Problem 6.89 Determine the general relation expressing the acceleration of the slider A as a function of θ, L, R, the angular velocity of the wheel ω_W, and the angular acceleration of the wheel α_W.

Problems 6.90 through 6.93

For the slider-crank mechanism shown, let $R = 0.75$ m and $H = 2$ m, and let the length of bar BC be $L_{BC} = 3.25$ m.

Problem 6.90 Assume that $\dot{\theta} = 50$ rad/s = constant and compute the angular acceleration of the slider for $\theta = 27°$.

Problem 6.91 Assume that, at the instant shown, $\theta = 27°$, $\dot{\theta} = 50$ rad/s, and $\ddot{\theta} = 15$ rad/s^2. Compute the angular acceleration of the slider at this instant as well as the acceleration of point C.

Problem 6.92 Assuming that $\dot{\theta}$ is constant, determine the expression of the angular acceleration of the slider as a function of θ and $\dot{\theta}$ (and the accompanying geometrical parameters).

Problem 6.93 Letting $\dot{\theta} = 300$ rpm = constant, plot the angular acceleration of the slider as a function of θ for $0° \leq \theta \leq 360°$. In addition, plot the speed of point C for the same range of θ.

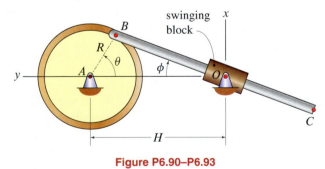

Figure P6.90–P6.93

Problem 6.94

A spool with inner radius $R = 5$ ft is made to roll without slip over a horizontal rail as shown. If the cable on the spool is unwound in such a way that the free or vertical portion of cable remains perpendicular to the rail, determine the angular acceleration of the spool and the acceleration of the spool's center O. The vertical component of the velocity of point A is $v_A = 12$ ft/s, and the vertical component of its acceleration is $a_A = 2$ ft/s^2.

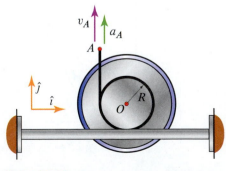

Figure P6.94

Problem 6.95

At the instant shown, bars AB and BC are perpendicular to each other while the slider C has a velocity $v_C = 24$ m/s and an acceleration $a_C = 2.5$ m/s^2 in the directions shown. Letting $L = 1.75$ m and $\theta = 45°$, determine the angular acceleration of bars AB and BC.

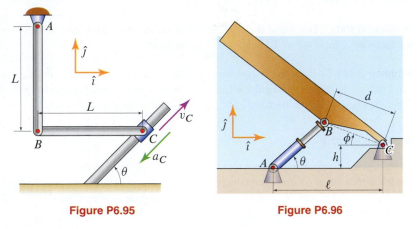

Figure P6.95   Figure P6.96

Problem 6.96

A flood gate is controlled via the hydraulic cylinder AB. If the length of the cylinder is increased with a constant time rate of 2.5 ft/s, determine the angular acceleration of the gate when $\phi = 0°$. Let $\ell = 10$ ft, $h = 2.5$ ft, and $d = 5$ ft.

Problem 6.97

The bucket of a backhoe is the element AB of the four-bar linkage system $ABCD$. Assume that the points A and D are fixed and that the bucket rotates with a constant angular velocity $\omega_{AB} = 0.25$ rad/s. In addition, suppose that, at the instant shown, point B is aligned vertically with point A, and C is aligned horizontally with B. Determine the acceleration of point C at the instant shown along with the angular accelerations of the elements BC and CD. Let $h = 0.66$ ft, $e = 0.46$ ft, $l = 0.9$ ft, and $w = 1.0$ ft.

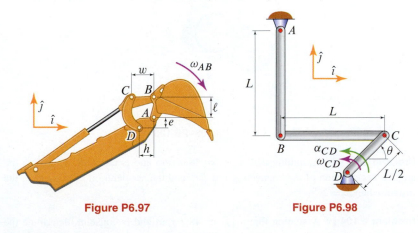

Figure P6.97   Figure P6.98

Problem 6.98

At the instant shown, bar CD is rotating with an angular velocity 20 rad/s and with angular acceleration 2 rad/s^2 in the directions shown. Furthermore, at this instant $\theta = 45°$. Letting $L = 2.25$ ft, determine the angular accelerations of bars AB and BC.

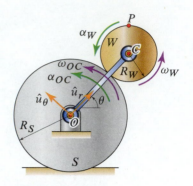

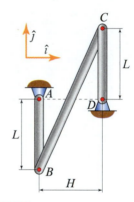

Figure P6.99–P6.101

Figure P6.102

Problems 6.99 through 6.101

A wheel W of radius $R_W = 2$ in. rolls without slip over the stationary cylinder S of radius $R_S = 5$ in., and the wheel is connected to point O via the arm OC.

Problem 6.99 Determine the acceleration of the point on the wheel W that is in contact with S for $\omega_{OC} = 7.5$ rad/s = constant.

Problem 6.100 Determine the acceleration of the point on the wheel W that is in contact with S for $\omega_{OC} = 7.5$ rad/s and $\alpha_{OC} = 2$ rad/s^2.

Problem 6.101 If, at the instant shown, $\theta = 63°$, $\omega_W = 9$ rad/s, and $\alpha_W = -1.3$ rad/s^2, determine the angular acceleration of the arm OC and the acceleration of point P, where P lies on the edge of W and is aligned vertically with point C.

Problem 6.102

At the instant shown, bars AB and CD are vertical. In addition, point C is moving to the left with a speed of 4 m/s, and the magnitude of the acceleration of C is 55 m/s^2. Letting $L = 0.5$ m and $H = 0.2$ m, determine the angular accelerations of bars AB and BC.

Problems 6.103 through 6.106

For the slider-crank mechanism shown, let $R = 1.9$ in., $L = 6.1$ in., and $H = 1.2$ in.

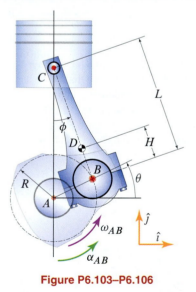

Figure P6.103–P6.106

Problem 6.103 Assuming that $\omega_{AB} = 4850$ rpm and is constant, determine the angular acceleration of the connecting rod BC and the acceleration of point C at the instant when $\theta = 27°$.

Problem 6.104 Assuming that $\omega_{AB} = 4850$ rpm and is constant, determine the acceleration of point D at the instant when $\phi = 10°$.

Problem 6.105 Assuming that, at the instant shown, $\theta = 31°$, $\omega_{AB} = 4850$ rpm, and $\alpha_{AB} = \dot{\omega}_{AB} = -280$ rad/s^2, determine the angular acceleration of the connecting rod and the acceleration of point C.

Problem 6.106 ❘ Determine the general expression of the acceleration of the piston C as a function of L, R, θ, $\omega_{AB} = \dot{\theta}$, and $\alpha_{AB} = \ddot{\theta}$.

Problems 6.107 and 6.108

In the four-bar linkage system shown, let the circular guide with center at O be fixed and such that, for $\theta = 0°$, the bars AB and BC are vertical and horizontal, respectively. In addition, let $R = 0.6\,\text{m}$, $L = 1\,\text{m}$, and $H = 1.25\,\text{m}$.

Problem 6.107 ❘ When $\theta = 37°$, $\beta = 25.07°$, and $\gamma = 78.71°$, assume collar C is sliding clockwise with a speed $7\,\text{m/s}$. Assuming that, at the instant in question, the speed is increasing and that $|\vec{a}_C| = 93\,\text{m/s}^2$, determine the angular accelerations of the bars AB and BC.

Problem 6.108 🖥 Use the method of differentiation of constraints to derive expressions for the angular accelerations of bars AB and BC as a function of θ, $\dot{\theta}$, and $\ddot{\theta}$. Finally, Let $\theta = (0.3\,\text{rad/s}^2)t^2$ and plot the angular accelerations of AB and BC for $0 \le t \le 1\,\text{s}$.

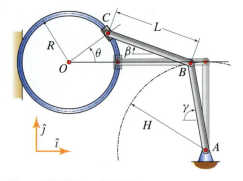

Figure P6.107 and P6.108

Problem 6.109 ❘

Complete the acceleration analysis of the slider-crank mechanism using differentiation of constraints that was outlined beginning on p. 498. That is, determine the accleration of the piston C and the angular acceleration of the connecting rod as a function of the given quantities θ, ω_{AB}, R, and L. Assume that ω_{AB} is constant, and use the component system shown for your answers.

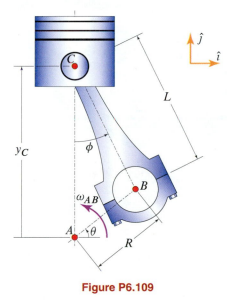

Figure P6.109

DESIGN PROBLEMS

Design Problem 6.1

In some high-performance mountain bikes, one element of the frame is attached to the rest via a four-bar linkage system. Referring to Fig. DP6.1, the four-bar linkage system is defined by the points A, B, C, and D. Notice that this system is also connected to the shock absorber pinned at points E and F. Research the available commercial literature (this information is readily available on the Web), and select a bicycle frame containing a four-bar linkage system such as that shown below. For the frame you select, obtain the necessary geometric information (again, this information is often made available on the Web by manufacturers), and assume that the points attached to the front part of the frame, which, in Fig. DP6.1 are points A and D, are *fixed*. Then calling ℓ the length of the shock absorber, which in Fig. DP6.1 is the distance between E and F, analyze the kinematics of the linkage system and determine $\dot{\ell}$ and $\ddot{\ell}$ as a function of both the angular velocity and the angular acceleration of the part of the frame to which the wheel is attached, which, in Fig. DP6.1, is the part to the left of points B and C.

Figure DP6.1

6.4 Rotating Reference Frames

Up until now, we have applied rigid body kinematics by either relating pairs of points on the same rigid body using Eqs. (6.3) and (6.5) or by differentiating constraint equations (see Example 6.6). Both of these ideas rely on the fact that the reference frame used to describe the motions is not rotating (this was assumed every time we took a derivative with respect to time). It turns out that we can solve Example 6.6, and problems like it, with a vector-based approach that utilizes both a primary reference frame (this will often be inertial) *and* a secondary reference frame that rotates (and translates). This approach will have far-reaching applications since it will allow us to describe the motion of any point using a reference frame that is translating and rotating relative to some inertial reference frame.

Motion of an airplane propeller blade

Say that the airplane P shown in Fig. 6.38 is rolling, pitching, and yawing as it flies along some path, and say that we are interested in the motion of point Q, which is at the mass center of one of the propeller blades, relative to the XYZ reference frame (the primary frame).* With this goal in mind, we attach the xyz reference frame shown in Fig. 6.38 to the airplane at its center C. This reference frame *must* move as the airplane moves. Now, we have frequently described the motion of points like C, so it isn't hard to imagine that we could do that here too. If we could then describe the motion of Q relative to C, we could write

$$\vec{a}_Q = \vec{a}_C + \vec{a}_{Q/C}, \tag{6.42}$$

and we would have $\vec{a}_Q$. This isn't easy to do since Q's motion relative to C, as seen by the primary XYZ frame, involves the airplane's three rotations ω_x, ω_y, and ω_z as well as the propeller's rotation ω_{prop}, and so it is *very* complicated. On the other hand, the motion of Q as seen by someone sitting in the airplane, i.e., as seen by the xyz frame, is *very* simple — it is just circular motion about the x axis! It seems that our goal is to find a way to be able to write $\vec{a}_{Q/C}$ by taking advantage of the fact that the motion relative to the frame attached to the airplane is very simple. While the equations we will derive are applicable to three-dimensional motion, we won't analyze the airplane until we cover three-dimensional rigid body motion in Chapter 10, though we will apply them to a planar mechanism we solved earlier in this chapter.

Sliding contacts

Sliding contacts occur frequently in linkage systems, and we have already analyzed the kinematics of a mechanism with a sliding contact in Example 6.6 on p. 486. Figure 6.39 shows a photo of that swinging block slider-crank mechanism, and Fig. 6.40 shows a schematic of the mechanism defining all relevant dimensions. In Example 6.6 we related the angular velocity of the slider to that of the crank — we are going to do that again here.

---

* We will see in Chapters 7 and 10 that to find the forces and moments required to keep the propeller attached to the shaft, one needs to know the acceleration of the mass center of the propeller blade.

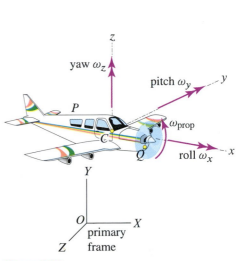

Figure 6.38

A secondary or moving xyz reference frame attached to an airplane, along with a primary or inertial XYZ reference frame.

Figure 6.39

Photo of the swinging block slider-crank. From Cornell University's Kinematic Models for Design Digital Library (KMODDL). http://kmoddl.library.cornell.edu/index.php

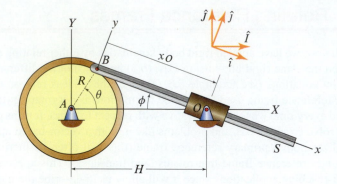

Figure 6.40. Schematic of the swinging block slider-crank showing the primary and secondary/moving reference frames. This mechanism was used in some steam engines (including locomotive engines) in the 19th century. One modern application is in linkages for door opening mechanisms in which the cylinder at O is a passive damper.

In Example 6.6 we differentiated a constraint equation to find $\dot{\phi}(\theta, \dot{\theta})$. Here, we are going to make use of the relation

$$\vec{r}_O = \vec{r}_B + \vec{r}_{O/B} = \vec{r}_B + x_O \hat{\imath}, \tag{6.43}$$

where point O is at the pin connection on the swinging block, and $\hat{\imath}$ and $\hat{\jmath}$ are the unit vectors corresponding to the xyz reference frame that is *attached to* the slider S. Differentiating this relation, we get

$$\vec{v}_O = \vec{0} = \vec{v}_B + \dot{x}_O \hat{\imath} + x_O \dot{\hat{\imath}}, \tag{6.44}$$

where we have used the fact that the pin at O is fixed and $\dot{x}_O \hat{\imath}$ is the velocity of point O as seen by someone in the xyz frame. In Section 2.4, Eq. (2.60) on p. 92 tells us that $\dot{\hat{\imath}}$ is given by

$$\dot{\hat{\imath}} = \vec{\omega}_{\hat{\imath}} \times \hat{\imath} = \omega_S \, \hat{k} \times \hat{\imath} = \omega_S \, \hat{\jmath}, \tag{6.45}$$

since the angular velocity of the xyz frame is the same as that of the slider. We can write $\vec{v}_B$ as

$$\vec{v}_B = \vec{\omega}_{AB} \times \vec{r}_{B/A} = \dot{\theta} \, \hat{K} \times R \big(\cos\theta \, \hat{I} + \sin\theta \, \hat{J} \big)$$
$$= R\dot{\theta} \big(-\sin\theta \, \hat{I} + \cos\theta \, \hat{J} \big), \tag{6.46}$$

where $\hat{I}$, $\hat{J}$, and $\hat{K}$ are the unit vectors corresponding to the primary XYZ frame. Substituting Eqs. (6.45) and (6.46) into Eq. (6.44), we obtain

$$\vec{0} = R\dot{\theta} \big(-\sin\theta \, \hat{I} + \cos\theta \, \hat{J} \big) + \dot{x}_O \hat{\imath} + x_O \omega_S \, \hat{\jmath}. \tag{6.47}$$

Here we have unit vectors associated with both reference frames. Transforming the primary frame to the secondary one, we get

$$\hat{I} = \cos\phi \, \hat{\imath} + \sin\phi \, \hat{\jmath} \qquad \text{and} \qquad \hat{J} = -\sin\phi \, \hat{\imath} + \cos\phi \, \hat{\jmath}. \tag{6.48}$$

Substituting Eq. (6.48) into Eq. (6.47), we now get

$$\vec{0} = -R\dot{\theta}\sin\theta(\cos\phi\,\hat{\imath} + \sin\phi\,\hat{\jmath}) + R\dot{\theta}\cos\theta(-\sin\phi\,\hat{\imath} + \cos\phi\,\hat{\jmath})$$
$$+ \dot{x}_O\,\hat{\imath} + x_O\omega_S\,\hat{\jmath}. \quad (6.49)$$

Since we can find $\phi(\theta)$, Eq. (6.49) yields two scalar equations for two unknowns: $\dot{x}_O$ and ω_S. First, let's find $\sin\phi$ and $\cos\phi$ as functions of θ.

Referring to Fig. 6.40, the law of sines tells us that

$$\frac{x_O}{\sin\theta} = \frac{R}{\sin\phi} \quad \text{so that} \quad \sin\phi = \frac{R}{x_O}\sin\theta, \quad (6.50)$$

and

$$\cos\phi = \sqrt{1 - \left(\frac{R}{x_O}\sin\theta\right)^2} = \frac{H - R\cos\theta}{x_O}, \quad (6.51)$$

where x_O is found by using the law of cosines to be

$$x_O^2 = R^2 + H^2 - 2HR\cos\theta \quad \Rightarrow \quad x_O = \sqrt{R^2 + H^2 - 2HR\cos\theta}. \quad (6.52)$$

Substituting Eqs. (6.52) into Eq. (6.51), substituting that result into Eq. (6.49), and then solving the corresponding two scalar equations for $\dot{x}_O$ and ω_S, we get

$$\omega_S = \frac{R\dot{\theta}(R - H\cos\theta)}{H^2 + R^2 - 2HR\cos\theta}, \quad (6.53)$$

$$\dot{x}_O = \frac{HR\dot{\theta}\sin\theta}{\sqrt{H^2 + R^2 - 2HR\cos\theta}}. \quad (6.54)$$

The result for ω_S in Eq. (6.53) is identical to that obtained in Eq. (4) of Example 6.6 on p. 486. In addition, Example 6.6 found the velocity of the point on the slider S that is directly underneath O to be $\vec{v}_P = (13.2\,\hat{\imath} + 84.3\,\hat{\jmath})$ ft/s (for the configuration of Example 6.6). Evaluating the magnitude of $\vec{v}_P$ gives $|\vec{v}_P| = 85.4$ ft/s, and evaluating $\dot{x}_O$ in Eq. (6.54) for the parameters in Example 6.6 gives $\dot{x}_O = 85.4$ ft/s—they agree as they should (also see marginal note).

Helpful Information

Can we get the velocity vector $\vec{v}_P$ in Example 6.6 from $\dot{x}_O$? We know that $|\vec{v}_P| = \dot{x}_O$, but what about the velocity *vectors*? The direction of $\dot{x}_O$ is along the slider S, so we can write the velocity of the point on the slider directly underneath point O, call it $\vec{v}_Q$, as

$$\vec{v}_Q = \dot{x}_O(\cos\phi\,\hat{\imath} - \sin\phi\,\hat{\jmath}),$$

which, when evaluated at the parameters given in Example 6.6, is equal to $\vec{v}_P$.

The general kinematic equations for the motion of a point relative to a rotating reference frame

Referring to Fig. 6.41, to determine the motion of point P, we will make use of two different reference frames: (1) a reference frame xyz that translates *and* rotates, which we will call the *rotating reference frame*,* and (2) a reference frame XYZ with respect to which we are measuring the motion of P and that we will call the *primary reference frame*. For us, the primary frame will usually be inertial. In addition, since the rotating reference frame is almost always attached to a moving rigid body (as it is in Fig. 6.41), we will often call it the *body-fixed frame*. With this as background, we will describe the motion of point P in terms of

1. The motion of the origin (point A) of the rotating reference frame.

2. The angular velocity $\vec{\Omega}$ and angular acceleration $\dot{\vec{\Omega}}$ of the rotating reference frame.†

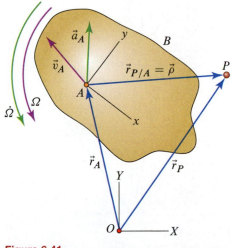

Figure 6.41
Reference frame, position vector, velocity, and acceleration definitions used in describing the motion of a point P relative to a rigid body B.

---

* This is also called the *secondary reference frame*. or *moving reference frame*.
† The symbol Ω is the uppercase Greek letter omega.

3. How P is seen to be moving by an observer *attached to* the rotating reference frame.

To begin, we note from Fig. 6.41 that the position of P can be written as

$$\vec{r}_P = \vec{r}_A + \vec{r}_{P/A} = \vec{r}_A + \vec{\rho}, \tag{6.55}$$

where $\vec{r}_A$ is the position of the origin of the rotating frame and $\vec{r}_{P/A} = \vec{\rho}$ is the position of P relative to the rotating frame. Writing $\vec{\rho}$ in terms of the xyz frame as $\vec{\rho} = \rho_x\,\hat{\imath} + \rho_y\,\hat{\jmath}$, Eq. (6.55) becomes*

$$\vec{r}_P = \vec{r}_A + \rho_x\,\hat{\imath} + \rho_y\,\hat{\jmath}. \tag{6.56}$$

To find velocities and accelerations, we need to differentiate Eq. (6.56) with respect to time. As we do so, we need to realize that the xyz frame is attached to the body B, and so $\hat{\imath}$ and $\hat{\jmath}$ rotate with B and, therefore, are not constant.

Velocity using a rotating frame

Differentiating Eq. (6.56) with respect to time, we get

$$\vec{v}_P = \vec{v}_A + \dot{\rho}_x\,\hat{\imath} + \rho_x\dot{\hat{\imath}} + \dot{\rho}_y\,\hat{\jmath} + \rho_y\dot{\hat{\jmath}}, \tag{6.57}$$

where $\vec{v}_P$ is the velocity of P and $\vec{v}_A$ is the velocity of the origin A of the rotating frame. We can now rewrite this expression, using our knowledge of the time derivative of a unit vector from Eq. (2.60) on p. 92, that is,

$$\dot{\hat{\imath}} = \vec{\omega}_{\hat{\imath}} \times \hat{\imath} \quad \text{and} \quad \dot{\hat{\jmath}} = \vec{\omega}_{\hat{\jmath}} \times \hat{\jmath}, \tag{6.58}$$

where $\vec{\omega}_{\hat{\imath}}$ and $\vec{\omega}_{\hat{\jmath}}$ are the angular velocities of $\hat{\imath}$ and $\hat{\jmath}$, respectively. Since the xyz frame is body-fixed, we know that all the unit vectors rotate with the body at angular velocity $\vec{\Omega}$ and so

$$\begin{aligned}
\vec{v}_P &= \vec{v}_A + \dot{\rho}_x\,\hat{\imath} + \dot{\rho}_y\,\hat{\jmath} + \rho_x\vec{\Omega} \times \hat{\imath} + \rho_y\vec{\Omega} \times \hat{\jmath} \\
&= \vec{v}_A + \dot{\rho}_x\,\hat{\imath} + \dot{\rho}_y\,\hat{\jmath} + \vec{\Omega} \times (\rho_x\,\hat{\imath} + \rho_y\,\hat{\jmath}) \\
&= \vec{v}_A + \underbrace{\dot{\rho}_x\,\hat{\imath} + \dot{\rho}_y\,\hat{\jmath}}_{\vec{v}_{P\text{rel}}} + \vec{\Omega} \times \vec{\rho},
\end{aligned} \tag{6.59}$$

$$\underbrace{\phantom{\dot{\rho}_x\,\hat{\imath} + \dot{\rho}_y\,\hat{\jmath} + \vec{\Omega} \times \vec{\rho}}}_{\dot{\vec{\rho}}}$$

or

$$\boxed{\vec{v}_P = \vec{v}_A + \vec{v}_{P\text{rel}} + \vec{\Omega} \times \vec{r}_{P/A},} \tag{6.60}$$

where

$$\vec{v}_{P\text{rel}} = \dot{\rho}_x\,\hat{\imath} + \dot{\rho}_y\,\hat{\jmath} \tag{6.61}$$

is the velocity of P relative to the rotating or body-fixed frame (i.e., *as seen by an observer moving with the body B*) and we have replaced $\vec{\rho}$ with $\vec{r}_{P/A}$. Equation (6.60) is an important development since it allows us to relate the velocities of two points that are *not* on the same rigid body — in this case P and A. In words, and referring to Fig. 6.42, Eq. (6.60) tells us that the velocity

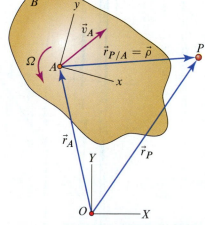

Figure 6.42

Simplified version of Fig. 6.41 showing the essential ingredients for finding velocities in rotating reference frames.

---

*For planar motion, we can, without loss of generality, consider the plane of motion to be the $z = 0$ plane.

of P can be found by using

$\vec{v}_A$ = velocity of the origin of the rotating or body-fixed reference frame (point A in Fig. 6.42)

$\vec{v}_{P\text{rel}}$ = velocity of P as seen by an observer moving with the rotating reference frame xyz

$\vec{\Omega}$ = angular velocity of the rotating reference frame xyz

$\vec{r}_{P/A}$ = vector from the origin of the rotating reference frame to point P

Before looking at accelerations, let's see how to apply Eq. (6.60) by looking at a short example.

■ **Mini-Example.**   To illustrate the use of Eq. (6.60), let's find the velocity of the point P that is moving in the slot in the disk shown in Fig. 6.43. Assume the disk is rotating at a constant angular velocity ω_0 and that $s(t)$ is known.
Solution. We can't apply the kinematic equations developed in Section 6.1 (P is *not* a point fixed on the disk), so we will apply Eq. (6.60) as

$$\vec{v}_P = \vec{v}_O + \vec{v}_{P\text{rel}} + \vec{\Omega} \times \vec{r}_{P/O}, \qquad (6.62)$$

where we have attached the rotating xy frame to the disk with its origin at the center O, as shown in Fig. 6.43. Let's now interpret each of these terms. First, $\vec{v}_O = \vec{0}$ since it is the velocity of the origin of the rotating xy frame, which is not moving. Second, $\vec{v}_{P\text{rel}}$ is the velocity of P as seen by an observer rotating with the disk. If we are sitting on the disk, we see P moving in just the x direction and so $\vec{v}_{P\text{rel}} = \dot{s}\,\hat{\imath}$. Finally, $\vec{\Omega}$ is the angular velocity of the rotating frame, which is $\omega_0\,\hat{k}$, and $\vec{r}_{P/O}$ is the vector from the origin of the rotating frame to P, which is $\vec{\rho} = s\,\hat{\imath}$. Putting this all in Eq. (6.62), we get

$$\vec{v}_P = \dot{s}\,\hat{\imath} + \omega_0\,\hat{k} \times s\,\hat{\imath} = \dot{s}\,\hat{\imath} + s\omega_0\,\hat{\jmath}. \qquad (6.63)$$

Thinking back to Chapter 2, we see this problem could be handled using polar coordinates. Let's see how.

Referring to Fig. 6.44, we can use polar coordinates to write $\vec{v}_P$ as

$$\vec{v}_P = \dot{r}\,\hat{u}_r + r\dot{\theta}\,\hat{u}_\theta. \qquad (6.64)$$

Notice that $r = s$, $\dot{r} = \dot{s}$, $\dot{\theta} = \omega_0$, and the $\hat{\imath}$ and $\hat{\jmath}$ are in the same direction as $\hat{u}_r$ and $\hat{u}_\theta$, respectively. That means that Eqs. (6.63) and (6.64) are giving identical results (as they should!). ────────────■

Acceleration using a rotating frame

Now that we have the velocity of P, we would like to find its acceleration. We start by differentiating Eq. (6.59) with respect to time (we use Eq. (6.59) rather than Eq. (6.60) so that we have the component form of $\vec{v}_{P\text{rel}}$) to get

$$\vec{a}_P = \vec{a}_A + \ddot{\rho}_x\,\hat{\imath} + \dot{\rho}_x\,\dot{\hat{\imath}} + \ddot{\rho}_y\,\hat{\jmath} + \dot{\rho}_y\,\dot{\hat{\jmath}} + \dot{\vec{\Omega}} \times \vec{\rho} + \vec{\Omega} \times \dot{\vec{\rho}} \qquad (6.65)$$

$$= \vec{a}_A + \ddot{\rho}_x\,\hat{\imath} + \ddot{\rho}_y\,\hat{\jmath} + \dot{\rho}_x\,\vec{\Omega} \times \hat{\imath} + \dot{\rho}_y\,\vec{\Omega} \times \hat{\jmath}$$

$$+ \dot{\vec{\Omega}} \times \vec{\rho} + \vec{\Omega} \times \left(\vec{v}_{P\text{rel}} + \vec{\Omega} \times \vec{\rho}\right) \qquad (6.66)$$

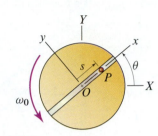

Figure 6.43
A particle moving in the radial slot of a rotating rigid disk.

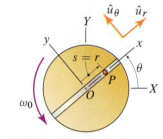

Figure 6.44
The spinning disk of Fig. 6.43 with polar coordinates defined.

Common Pitfall

The velocity and acceleration can be expressed in either the primary or rotating reference frame. We have expressed our final answers in terms of components in the rotating or body-fixed reference frame. Any vector can be expressed in terms of the components of either the rotating frame or the primary frame. For example, in terms of the XY frame, we can write the velocity of P in the mini-example as

$$\vec{v}_P = \dot{s}\left(\cos\theta\,\hat{I} + \sin\theta\,\hat{J}\right)$$
$$+ s\omega_0\left(-\sin\theta\,\hat{I} + \cos\theta\,\hat{J}\right)$$
$$= (\dot{s}\cos\theta - s\omega_0\sin\theta)\,\hat{I}$$
$$+ (\dot{s}\sin\theta + s\omega_0\cos\theta)\,\hat{J}$$

where $\hat{I}$ and $\hat{J}$ are the unit vectors associated with the XY frame.

What if P is a point *on* the rigid body B?
If P is attached to the body B as shown below, then $\vec{v}_{P\text{rel}} = \vec{0}$ in Equation (6.60).

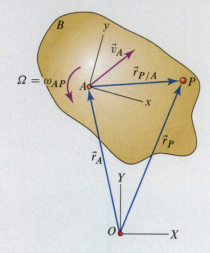

Then we have

$$\vec{v}_P = \vec{v}_A + \vec{\Omega} \times \vec{r}_{P/A} = \vec{v}_A + \vec{\omega}_{AP} \times \vec{r}_{P/A},$$

which is just Eq. (6.3) on p. 462. A corresponding simplification applies to the acceleration in Eq. (6.68) leading to Eq. (6.5).

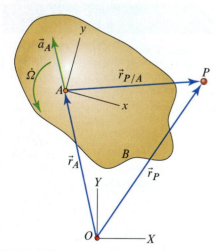

Figure 6.45
Simplified version of Fig. 6.41 showing the essential quantities needed for finding accelerations in rotating reference frames.

$$= \vec{a}_A + \underbrace{\ddot{\rho}_x\,\hat{\imath} + \ddot{\rho}_y\,\hat{\jmath}}_{\vec{a}_{P\text{rel}}} + \vec{\Omega} \times (\underbrace{\dot{\rho}_x\,\hat{\imath} + \dot{\rho}_y\,\hat{\jmath}}_{\vec{v}_{P\text{rel}}})$$

$$+ \dot{\vec{\Omega}} \times \vec{\rho} + \vec{\Omega} \times (\vec{v}_{P\text{rel}} + \vec{\Omega} \times \vec{\rho})$$

$$= \vec{a}_A + \vec{a}_{P\text{rel}} + 2\vec{\Omega} \times \vec{v}_{P\text{rel}} + \dot{\vec{\Omega}} \times \vec{\rho} + \vec{\Omega} \times (\vec{\Omega} \times \vec{\rho}), \qquad (6.67)$$

where we have used the time derivative of the unit vectors from Eq. (6.58) and $\dot{\vec{\rho}} = \vec{v}_{P\text{rel}} + \vec{\Omega} \times \vec{\rho}$ from Eq. (6.59) to go from Eq. (6.65) to Eq. (6.66). Noticing that there are two $\vec{\Omega} \times \vec{v}_{P\text{rel}}$ terms, we obtain

$$\boxed{\vec{a}_P = \vec{a}_A + \vec{a}_{P\text{rel}} + 2\vec{\Omega} \times \vec{v}_{P\text{rel}} + \dot{\vec{\Omega}} \times \vec{r}_{P/A} + \vec{\Omega} \times (\vec{\Omega} \times \vec{r}_{P/A}),}$$

$$(6.68)$$

where we have replaced $\vec{\rho}$ with $\vec{r}_{P/A}$ and

$$\vec{a}_{P\text{rel}} = \ddot{\rho}_x\,\hat{\imath} + \ddot{\rho}_y\,\hat{\jmath}, \qquad (6.69)$$

is the *acceleration of P relative to the rotating frame* (i.e., *as seen by an observer moving with the body B*). Taken as a whole, Eq. (6.68) looks complicated, but it is readily applied as long as each term is considered individually. Let's go through each term in Eq. (6.68) and define it in words, and then we will look again at the spinning disk mini-example.

Referring to Fig. 6.45, the acceleration of a point P as given by Eq. (6.68) consists of the following terms:

$\vec{a}_A =$ acceleration of the origin of the rotating or body-fixed reference frame

$\vec{a}_{P\text{rel}} =$ acceleration of P as seen by an observer moving with the rotating reference frame

$2\vec{\Omega} \times \vec{v}_{P\text{rel}} =$ Coriolis acceleration of P; this term results from two equal, but different effects: (1) the change in direction of $\vec{v}_{P\text{rel}}$ due to $\vec{\Omega}$ and (2) the effect of $\vec{\Omega}$ on the change in magnitude of $\vec{r}_{P/A}$ relative to the rotating reference frame

$\dot{\vec{\Omega}} \times \vec{r}_{P/A} =$ tangential acceleration of a point attached to the moving reference frame and coincident with P at any given time t as the point moves on a sphere of radius $|\vec{r}_{P/A}|$ about point A, where $\dot{\vec{\Omega}}$ is the angular acceleration of the rotating reference frame

$\vec{\Omega} \times (\vec{\Omega} \times \vec{r}_{P/A}) =$ normal acceleration of a point attached to the moving reference frame and coincident with P at any given time t as the point moves on a sphere of radius $|\vec{r}_{P/A}|$ about point A

Given the descriptions of the last two terms, we can see that the sum $\vec{a}_A + \dot{\vec{\Omega}} \times \vec{r}_{P/A} + \vec{\Omega} \times (\vec{\Omega} \times \vec{r}_{P/A})$ represents the acceleration of a point attached to the moving reference frame and coincident with P at any given time t. Now, let's revisit the mini-example on p. 521 and find the acceleration of P.

■ **Mini-Example.**   To illustrate the use of Eq. (6.68), let's find the acceleration of the point P that is moving in the slot in the disk shown in Fig. 6.46.
Assume the disk is rotating with angular velocity ω_0 and angular acceleration
α_0 and $s(t)$ is known.

Solution. We can't apply the kinematic equations developed in Section 6.1
since P is moving *relative to* the disk, so we will apply Eq. (6.68) as

$$\vec{a}_P = \vec{a}_O + \vec{a}_{P\text{rel}} + 2\vec{\Omega} \times \vec{v}_{P\text{rel}} + \dot{\vec{\Omega}} \times \vec{r}_{P/O} + \vec{\Omega} \times (\vec{\Omega} \times \vec{r}_{P/O}), \quad (6.70)$$

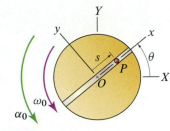

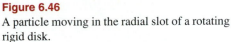

Figure 6.46
A particle moving in the radial slot of a rotating
rigid disk.

where we have attached the rotating xy frame to the disk with its origin at the
center O as shown in Fig. 6.46. First, $\vec{a}_O = \vec{0}$ since the origin of the rotating
xy frame is not moving. Second, $\vec{a}_{P\text{rel}}$ is the acceleration of P as seen by an
observer rotating with the disk, which means that $\vec{a}_{P\text{rel}} = \ddot{s}\,\hat{\imath}$. Using similar
reasoning, $\vec{v}_{P\text{rel}} = \dot{s}\,\hat{\imath}$. Finally, $\vec{\Omega}$ is the angular velocity of the rotating frame,
which is $\omega_0\,\hat{k}$; $\dot{\vec{\Omega}}$ is the angular acceleration of the rotating frame, which is
$\alpha_0\,\hat{k}$; and $\vec{r}_{P/O}$ is the vector from the origin of the rotating frame to P, which
is $s\,\hat{\imath}$. Putting this all in Eq. (6.70), we get

$$\begin{aligned}
\vec{a}_P &= \ddot{s}\,\hat{\imath} + 2\omega_0\,\hat{k} \times \dot{s}\,\hat{\imath} + \alpha_0\,\hat{k} \times s\,\hat{\imath} + \omega_0\,\hat{k} \times (\omega_0\,\hat{k} \times s\,\hat{\imath}) \\
&= (\ddot{s} - s\omega_0^2)\,\hat{\imath} + (s\alpha_0 + 2\dot{s}\omega_0)\,\hat{\jmath}. \quad (6.71)
\end{aligned}$$

Using polar coordinates gives us the same result (see Prob. 6.110). ────■

Before we end this section, there are some things to note.

- For planar motion, which is the type of motion we are concentrating on
  in this section, the acceleration given in Eq. (6.68) can be written as

$$\vec{a}_P = \vec{a}_A + \vec{a}_{P\text{rel}} + 2\vec{\Omega} \times \vec{v}_{P\text{rel}} + \dot{\vec{\Omega}} \times \vec{r}_{P/A} - \Omega^2 \vec{r}_{P/A}, \quad (6.72)$$

  where $\Omega = |\vec{\Omega}|$.

- The terms $\vec{v}_{P\text{rel}}$ and $\vec{a}_{P\text{rel}}$ in Eqs. (6.60) and (6.68) give us a new per-
  spective on the motion of points that we haven't seen until now, that
  is, they tell us the *apparent* motion of something as seen by a moving
  observer. For example, referring to Fig. 6.47, if the person at A is just
  standing on the merry-go-round (i.e., not moving relative to it), which is
  rotating at the constant rate ω_m, and the person at B is walking at a constant speed along the sidewalk in the direction shown, then the velocity
  and acceleration of B as seen by A are *not* $\vec{v}_{B/A}$ and $\vec{a}_{B/A}$, respectively;
  they are instead

$$\vec{v}_{B\text{rel}} = \vec{v}_B - \vec{v}_A - \vec{\omega}_m \times \vec{r}_{B/A}, \quad (6.73)$$

$$\vec{a}_{B\text{rel}} = \vec{a}_B - \vec{a}_A - 2\vec{\omega}_m \times \vec{v}_{P\text{rel}} - \vec{\alpha}_m \times \vec{r}_{B/A} + \omega_m^2 \vec{r}_{B/A}, \quad (6.74)$$

respectively. In the given situation, $\vec{\alpha}_m$ and $\vec{a}_B$ would both be zero. In
Example 6.15, we will attach the rotating reference frame to the merry-
go-round with its origin at the person at A.

- When deciding where to attach the rotating frame, it is important that
  we choose a frame that makes it easy to determine each of the terms in
  Eqs. (6.60) and (6.68). For example, when relating the angular velocity

> 🏛 **Concept Alert**
>
> ***Rotating reference frame, body-fixed ref-
> erence frame, rigid body B, moving ref-
> erence frame,*** and ***secondary reference
> frame.*** We will see all these terms used
> when referring to what we are calling a *ro-
> tating* or *body-fixed reference frame*, and
> we need to be aware that all these terms
> are talking about the same thing—that is,
> they are all referring to a reference frame
> that is translating and rotating relative to
> some primary or inertial reference frame.

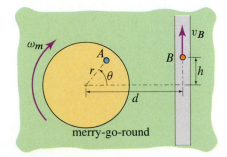

Figure 6.47
A person A standing on a spinning merry-go-
round as the person B walks on a sidewalk in
the direction shown.

of the slider to that of the crank for the mechanism in Fig. 6.40, we attached the rotating frame to the slider S since it was then very easy to describe the motion of point O relative to the rotating frame (the term $\vec{v}_{B\text{rel}}$ in Eq. (6.60)). Instead, we could have attached the rotating frame to the disk centered at A, but in that case the motion of point O would have been very hard to describe relative to the rotating frame.

End of Section Summary

In this section, we developed the kinematic equations that allow us to relate the motion of two points that *are not* on the same rigid body. Referring to Fig. 6.48, this means we can relate the velocity of point P to that of A using the relation

Eq. (6.60), p. 520

$$\vec{v}_P = \vec{v}_A + \vec{v}_{P\text{rel}} + \vec{\Omega} \times \vec{r}_{P/A},$$

where

$\vec{v}_A$ = velocity of the origin of the rotating or body-fixed reference frame (point A in Fig. 6.48)

$\vec{v}_{P\text{rel}}$ = velocity of P as seen by an observer moving with the rotating reference frame

$\vec{\Omega}$ = angular velocity of the rotating reference frame

$\vec{r}_{P/A}$ = position of point P relative to A

We also found that we can relate the acceleration of point P to that of A by using

Eq. (6.68), p. 522

$$\vec{a}_P = \vec{a}_A + \vec{a}_{P\text{rel}} + 2\vec{\Omega} \times \vec{v}_{P\text{rel}} + \dot{\vec{\Omega}} \times \vec{r}_{P/A} + \vec{\Omega} \times \left(\vec{\Omega} \times \vec{r}_{P/A}\right),$$

where, in addition to the terms defined for $\vec{v}_P$, we have

$\vec{a}_A$ = acceleration of the origin of the rotating or body-fixed reference frame (point A in Fig. 6.48)

$\vec{a}_{P\text{rel}}$ = acceleration of P as seen by an observer moving with the rotating reference frame

$2\vec{\Omega} \times \vec{v}_{P\text{rel}}$ = Coriolis acceleration of P

$\dot{\vec{\Omega}}$ = angular acceleration of the rotating reference frame

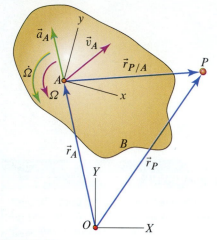

Figure 6.48
The essential ingredients for finding velocities and accelerations using a rotating reference frame.

EXAMPLE 6.13 *A Particle in a Rotating Slot: Equation of Motion and Forces*

In Fig. 1, we have one object *sliding* relative to another object, which is rotating. Assume that the disk is rotating in the horizontal plane with angular velocity ω_0 and angular acceleration α_0. The particle P is constrained to move in the slot, which is a distance d from the center of the disk. In addition, a linear elastic spring of constant k is attached to the particle such that the spring is undeformed when the particle is at $s = 0$. For this system, determine the equation(s) of motion of the particle and the normal force between the particle and the slot.

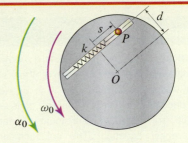

Figure 1

SOLUTION

Road Map & Modeling   We will apply Newton's second law to find the equation of motion for the particle and the normal force between the particle and the slot. In doing so, we will need to find the acceleration of the particle. Since the particle is sliding *relative to* the disk, we will attach a rotating reference frame to the disk since the motion of the particle relative to an observer who rotates with the disk is easy to describe.

The FBD of the particle at an arbitrary position s is shown in Fig. 2, where F_s is the spring force acting on the particle and N is the normal force acting on the particle due to the slot. We neglect friction between the particle and the slot. In addition, we attach a rotating xy frame to the disk with its origin at point O, which is the center of the disk. Since this system has one degree of freedom, we expect to obtain one equation of motion.

> **⊕ Helpful Information**
>
> **Why is there just one degree of free-dom?** Since motion of the disk is completely specified, we will always know the orientation of the slot in the disk. Therefore, the only degree of freedom is the motion of the particle in the slot, and so the system shown in Fig. 1 has only one degree of freedom.

Governing Equations

Balance Principles   Referring to Fig. 2, Newton's second law applied to the particle gives

$$\sum F_x: \quad -F_s = ma_{Px}, \qquad (1)$$

$$\sum F_y: \quad N = ma_{Py}, \qquad (2)$$

where a_{Px} and a_{Py} are the x and y components, respectively, of the acceleration $\vec{a}_P$ of the particle.

Force Laws   Since the spring is undeformed at $s = 0$, the law for the spring force is given by

$$F_s = ks. \qquad (3)$$

Kinematic Equations   To determine a_{Px} and a_{Py}, we use Eq. (6.72), which is

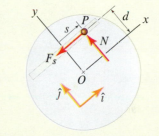

Figure 2
The FBD of the particle as well as the definition of the rotating reference frame. The unit vectors $\hat{\imath}$ and $\hat{\jmath}$ rotate with the disk.

$$\vec{a}_P = a_{Px}\hat{\imath} + a_{Py}\hat{\jmath} = \vec{a}_O + \vec{a}_{P\text{rel}} + 2\vec{\Omega} \times \vec{v}_{P\text{rel}} + \dot{\vec{\Omega}} \times \vec{r}_{P/O} - \Omega^2 \vec{r}_{P/O}, \qquad (4)$$

in which, referring to Fig. 2, $\vec{a}_O = \vec{0}$ since point O is not moving, $\vec{a}_{P\text{rel}} = \ddot{s}\,\hat{\imath}$, $\vec{v}_{P\text{rel}} = \dot{s}\,\hat{\imath}$, $\vec{\Omega} = \omega_0\,\hat{k}$, $\dot{\vec{\Omega}} = \alpha_0\,\hat{k}$, and $\vec{r}_{P/O} = s\,\hat{\imath} + d\,\hat{\jmath}$. Substituting each of these terms into Eq. (4), we obtain

$$\vec{a}_P = \ddot{s}\,\hat{\imath} + 2\omega_0\,\hat{k} \times \dot{s}\,\hat{\imath} + \alpha_0\,\hat{k} \times (s\,\hat{\imath} + d\,\hat{\jmath}) - \omega_0^2(s\,\hat{\imath} + d\,\hat{\jmath})$$
$$= (\ddot{s} - d\alpha_0 - s\omega_0^2)\,\hat{\imath} + (2\dot{s}\omega_0 + s\alpha_0 - d\omega_0^2)\,\hat{\jmath}, \qquad (5)$$

so that

$$a_{Px} = \ddot{s} - d\alpha_0 - s\omega_0^2 \quad \text{and} \quad a_{Py} = 2\dot{s}\omega_0 + s\alpha_0 - d\omega_0^2. \qquad (6)$$

Computation Substituting Eqs. (3) and (6) into Eqs. (1) and (2), we obtain

$$-ks = m\left(\ddot{s} - d\alpha_0 - s\omega_0^2\right) \quad \text{and} \quad N = m\left(2\dot{s}\omega_0 + s\alpha_0 - d\omega_0^2\right). \tag{7}$$

Rearranging the first of Eqs. (7), we obtain the equation of motion of the particle as

$$\ddot{s} + \left(\frac{k}{m} - \omega_0^2\right)s = d\alpha_0. \tag{8}$$

The second of Eqs. (7) tells us that once we know $s(t)$ from the solution of Eq. (8), then we can find the normal force between the particle and the slot from

$$N = m\left(2\dot{s}\omega_0 + s\alpha_0 - d\omega_0^2\right). \tag{9}$$

Discussion & Verification Note that the dimension of each term in Eq. (8) is that of acceleration, and the dimension of each term in Eq. (9) is that of force, so both results are dimensionally consistent.

🔎 **A Closer Look** While the only rigid body in this problem (i.e., the disk) has a known motion, it is an example of the power of the kinematics we have developed in this section. The motion of the particle P is actually quite complex, and yet we are easily able to write down its acceleration. The power of the kinematics we have developed lies in the idea that the motion relative to a frame that *rotates with the disk* is very simple, that is, it is rectilinear. Using that motion, along with a knowledge of the motion of the rotating frame, we are easily able to obtain the acceleration. We will see this in the remaining examples as well as in the homework problems.

EXAMPLE 6.14   *Analysis of a Reciprocating Rectilinear Motion Mechanism*

The *reciprocating rectilinear motion* mechanism shown in Fig. 1 consists of a disk pinned at its center at A that rotates with a constant angular velocity ω_{AB}, a slotted arm CD that is pinned at C, and a bar that can oscillate within the guides at E and F. As the disk rotates, the peg at B moves within the slotted arm, causing it to rock back and forth. As the arm rocks, it provides a slow advance and a quick return to the reciprocating bar due to the change in distance between C and B. For the position shown ($\theta = 30°$), determine the

(a) Angular velocity and angular acceleration of the slotted arm CD

(b) Velocity and acceleration of the bar

We will evaluate our results for $\omega_{AB} = 60\,\text{rpm}$, $R = 0.1\,\text{m}$, $h = 0.2\,\text{m}$, and $d = 0.12\,\text{m}$.

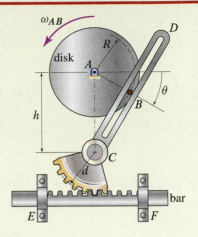

Figure 1

SOLUTION

Road Map   This is a mechanism with a sliding contact, so we cannot use Eqs. (6.3) and (6.5) to relate the motion of the disk to that of the slotted arm CD. Therefore, we will use a rotating reference frame to relate the motion of the pin at B to the pivot point at C, and we will attach the frame to arm CD as shown in the schematic of the mechanism in Fig. 2.

Computation   Since we are relating the motion of B to C, for velocities we can write

$$\vec{v}_B = \vec{v}_C + \vec{v}_{B\text{rel}} + \vec{\Omega}_{CD} \times \vec{r}_{B/C}, \qquad (1)$$

where $\vec{v}_{B\text{rel}}$ is the velocity of B as seen by an observer in the rotating frame and $\vec{\Omega}_{CD} = \vec{\omega}_{CD}$ is *both* the angular velocity of the rotating frame as well as the angular velocity of the arm CD. Now, since B is rotating in a circle about A, $\vec{v}_B$ is found from using Eq. (6.8) to be

$$\vec{v}_B = \vec{\omega}_{AB} \times \vec{r}_{B/A} = \omega_{AB}\,\hat{k} \times R\,\hat{\imath} = R\omega_{AB}\,\hat{\jmath}. \qquad (2)$$

In addition, we can see that

$$\vec{v}_C = \vec{0}, \quad \vec{v}_{B\text{rel}} = \dot{r}_{B/C}\,\hat{\jmath}, \quad \vec{\Omega}_{CD} = \Omega_{CD}\,\hat{k}, \quad \vec{r}_{B/C} = r_{B/C}\,\hat{\jmath}, \qquad (3)$$

where $r_{B/C} = \sqrt{h^2 - R^2} = 0.1732\,\text{m}$ at this instant. Substituting Eqs. (2) and (3) into Eq. (1), we obtain

$$R\omega_{AB}\,\hat{\jmath} = \dot{r}_{B/C}\,\hat{\jmath} + \Omega_{CD}\,\hat{k} \times r_{B/C}\,\hat{\jmath}, \qquad (4)$$

which is equivalent to two scalar equations for Ω_{CD} and $\dot{r}_{B/C}$. Solving, we find that $\dot{r}_{B/C} = R\omega_{AB} = 0.628\,\text{m/s}$ and

$$\Omega_{CD} = 0\,\text{rad/s} \quad \Rightarrow \quad \boxed{\vec{\Omega}_{CD} = \vec{\omega}_{CD} = \vec{0}\,\text{rad/s.}} \qquad (5)$$

For the acceleration analysis, we apply Eq. (6.68),

$$\vec{a}_B = \vec{a}_C + \vec{a}_{B\text{rel}} + 2\vec{\Omega}_{CD} \times \vec{v}_{B\text{rel}} + \dot{\vec{\Omega}}_{CD} \times \vec{r}_{B/C} - \Omega_{CD}^2\,\vec{r}_{B/C} \qquad (6)$$

$$= \vec{a}_C + \vec{a}_{B\text{rel}} + \dot{\vec{\Omega}}_{CD} \times \vec{r}_{B/C}, \qquad (7)$$

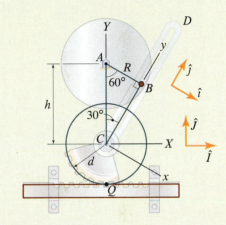

Figure 2
The primary and rotating reference frame for the reciprocating rectilinear motion mechanism.

where to go from Eq. (6) to Eq. (7) we have used Eq. (5). Since B is moving in a circle centered at A, we can use Eq. (6.13) to find $\vec{a}_B$ as

$$\vec{a}_B = -\omega_{AB}^2 \vec{r}_{B/A} = -\omega_{AB}^2 R \hat{\imath}. \tag{8}$$

In addition, $\vec{a}_{B\text{rel}}$ and $\dot{\vec{\Omega}}_{CD}$ in Eq. (7) are given by

$$\vec{a}_{B\text{rel}} = \ddot{r}_{B/C}\,\hat{\jmath} \quad \text{and} \quad \dot{\vec{\Omega}}_{CD} = \dot{\Omega}_{CD}\,\hat{k}. \tag{9}$$

Finally, noting that $\vec{a}_C = \vec{0}$ and substituting the last of Eqs. (3) and Eqs. (8) and (9) into Eq. (7), we obtain

$$-\omega_{AB}^2 R \hat{\imath} = \ddot{r}_{B/C}\,\hat{\jmath} + \dot{\Omega}_{CD}\,\hat{k} \times r_{B/C}\,\hat{\jmath}, \tag{10}$$

which is equivalent to two scalar equations for $\ddot{r}_{B/C}$ and $\dot{\Omega}_{CD}$. Solving, we find that $\ddot{r}_{B/C} = 0$ and

$$\boxed{\dot{\Omega}_{CD} = \frac{R}{r_{B/C}}\omega_{AB}^2 = 22.8\,\text{rad/s}^2 \quad \Rightarrow \quad \dot{\vec{\Omega}}_{CD} = \vec{\alpha}_{CD} = \left(22.8\,\text{rad/s}^2\right)\hat{k}.} \tag{11}$$

Referring to Fig. 2, now that we have the angular velocity and angular acceleration of the slotted arm CD, we can find the velocity of the bar using

$$\boxed{\vec{v}_{\text{bar}} = \vec{v}_Q = \vec{\omega}_{CD} \times \vec{r}_{Q/C} = \vec{0},} \tag{12}$$

since $\vec{\omega}_{CD} = \vec{0}$ at this instant. In addition, the acceleration of the bar is given by

$$\vec{a}_{\text{bar}} = \vec{a}_Q \cdot \hat{I} = \vec{\alpha}_{CD} \times \vec{r}_{Q/C} = \alpha_{CD}\,\hat{K} \times (-d\hat{J}) = d\alpha_{CD}\,\hat{I}, \tag{13}$$

that is, $\vec{a}_{\text{bar}}$ is just the tangential component of acceleration of Q. Therefore, the acceleration of the bar at this instant is

$$\boxed{\vec{a}_{\text{bar}} = d\alpha_{CD}\,\hat{I} = \left(2.74\,\text{m/s}^2\right)\hat{I}.} \tag{14}$$

Discussion & Verification The dimensions and thus the units of the final results are as they should be.

Looking more deeply, we note that the angular velocity of the slotted arm CD is zero at this instant. This makes sense since the line AB on the rotating disk is perpendicular to the slot in the arm at this instant. Therefore, *at this instant*, the velocity of B is parallel to the slot and so it is not inducing any rotation of the arm containing the slot. Consequently, the velocity of the bar must be zero at this instant since it is the angular velocity of the slotted arm that imparts a velocity to the bar.

On the other hand, we note that the slotted arm does have an angular acceleration. This also makes sense since an instant before the arm is in this position, and an instant after, the slotted arm must have an angular velocity and therefore there must be an angular acceleration causing this change in angular velocity.

In Problems 6.115 and 6.116, we will see why this is mechanism is sometimes referred to as a *quick-return mechanism*.

E X A M P L E 6.15   *Actual Versus Perceived Motion: Finding $\vec{v}_{Brel}$ and $\vec{a}_{Brel}$*

The person A is standing on the merry-go-round (i.e., not moving relative to it), which is rotating at the constant rate ω_m, and the person B is walking in a straight line at constant speed v_B along the sidewalk in a stationary frame (see Fig. 1). Determine the velocity and acceleration of B as seen by A.[*] Evaluate the results for $v_B = 4$ mph, $\omega_m = 3$ rad/s, $\theta = 45°$, $d = 30$ ft, and $h = 10$ ft.

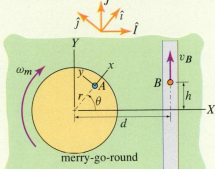

Figure 1

SOLUTION

Road Map   Since A is not moving relative to the merry-go-round, the motion of B as seen by A is equivalent to the motion of B as seen by an observer rotating with the merry-go-round. As we mention on p. 523, the velocity of B as seen by A is *not* equal to $\vec{v}_{B/A}$ since that quantity gives the velocity of B as seen by A only if A is *not rotating*. Therefore, recall that the term $\vec{v}_{Brel}$ in Eq. (6.60) is *the velocity of P as seen by an observer moving with the rotating frame* — this is exactly what we want (along with $\vec{a}_{Brel}$ for the acceleration of B as seen by A).

Computation   Referring to Fig. 2, the primary XY frame is as shown, and the rotating xy frame is attached to the merry-go-round with its origin at A. We want to compute $\vec{v}_{Brel}$, which is given by Eq. (6.60) to be

$$\vec{v}_{Brel} = \vec{v}_B - \vec{v}_A - \vec{\omega}_m \times \vec{r}_{B/A}, \qquad (1)$$

where $\vec{\omega}_m$ is also the angular velocity of the rotating xy frame since the frame is attached to the merry-go-round. Evaluating the three terms above, we find

$$\vec{v}_B = v_B\,\hat{J} = v_B(\sin\theta\,\hat{\imath} + \cos\theta\,\hat{\jmath}), \qquad (2)$$

$$\vec{v}_A = -r\omega_m\,\hat{\jmath}, \qquad (3)$$

$$\vec{\omega}_m \times \vec{r}_{B/A} = -\omega_m\,\hat{k} \times (-r\,\hat{\imath} + d\,\hat{I} + h\,\hat{J})$$
$$= -\omega_m\,\hat{k} \times [-r\,\hat{\imath} + d(\cos\theta\,\hat{\imath} - \sin\theta\,\hat{\jmath}) + h(\sin\theta\,\hat{\imath} + \cos\theta\,\hat{\jmath})]$$
$$= \omega_m(h\cos\theta - d\sin\theta)\hat{\imath} - \omega_m(d\cos\theta + h\sin\theta - r)\hat{\jmath}. \qquad (4)$$

Substituting Eqs. (2)–(4) into Eq. (1), we get

$$\boxed{\begin{aligned} \vec{v}_{Brel} &= (v_B\sin\theta - h\omega_m\cos\theta + d\omega_m\sin\theta)\,\hat{\imath} \\ &\quad + (v_B\cos\theta + h\omega_m\sin\theta + d\omega_m\cos\theta)\,\hat{\jmath} \\ &= (46.6\,\hat{\imath} + 89.0\,\hat{\jmath})\,\text{ft/s}, \end{aligned}} \qquad \begin{aligned}(5)\\[2mm](6)\end{aligned}$$

where the numerical result was obtained by evaluating the answer using the given values. Figure 3 shows the actual velocity of B, i.e., $\vec{v}_B$, as well as the velocity B as seen by A, i.e., $\vec{v}_{Brel}$.

Now computing $\vec{a}_{Brel}$ using Eq. (6.72), we have

$$\vec{a}_{Brel} = \vec{a}_B - \vec{a}_A - 2\vec{\omega}_m \times \vec{v}_{Brel} - \vec{\alpha}_m \times \vec{r}_{B/A} + \omega_m^2\vec{r}_{B/A}, \qquad (7)$$

where, in the given situation, $\vec{\alpha}_m$ and $\vec{a}_B$ are both zero. Computing the other terms in Eq. (7), we have

$$\vec{a}_A = -r\omega_m^2\,\hat{\imath}, \qquad (8)$$

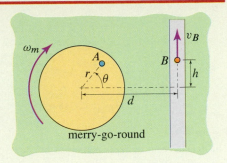

Figure 2

Definition of the primary and rotating frames of reference.

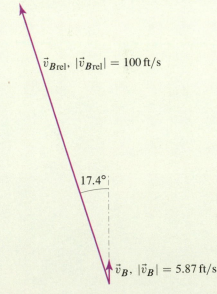

$\vec{v}_{Brel}$, $|\vec{v}_{Brel}| = 100$ ft/s

$17.4°$

$\vec{v}_B$, $|\vec{v}_B| = 5.87$ ft/s

Figure 3

A comparison of the velocity of B in the stationary frame, i.e., $\vec{v}_B$, with the velocity B as seen by A, i.e., $\vec{v}_{Brel}$.

---

[*] Person A cannot turn his or her head to follow B as the merry-go-round rotates, or else we will have introduced yet another rotation.

$$2\vec{\omega}_m \times \vec{v}_{B\text{rel}} = 2\omega_m \left[h\omega_m \sin\theta + (v_B + d\omega_m)\cos\theta \right] \hat{\imath}$$
$$+ 2\omega_m \left[h\omega_m \cos\theta - (v_B + d\omega_m)\sin\theta \right] \hat{\jmath}, \tag{9}$$

$$\omega_m^2 \vec{r}_{B/A} = \omega_m^2 \left[(h\sin\theta + d\cos\theta - r)\hat{\imath} + (h\cos\theta - d\sin\theta)\hat{\jmath} \right]. \tag{10}$$

Substituting Eqs. (8)–(10) as well as $\vec{\alpha}_m = \vec{0}$ and $\vec{a}_B = \vec{0}$ into Eq. (7), we obtain

$$\vec{a}_{B\text{rel}} = -\omega_m \left[(2v_B + d\omega_m)\cos\theta + h\omega_m \sin\theta \right] \hat{\imath}$$
$$+ \omega_m \left[(2v_B + d\omega_m)\sin\theta - h\omega_m \cos\theta \right] \hat{\jmath} \tag{11}$$
$$= (-279\,\hat{\imath} + 152\,\hat{\jmath})\,\text{ft/s}^2, \tag{12}$$

where the numerical result was obtained by evaluating the answer using the given values. Figure 4 shows the acceleration of B as perceived by A, i.e., $\vec{a}_{B\text{rel}}$, as well as the actual acceleration of B, which is zero, so only a dot is shown.

Discussion & Verification The dimensions of both $\vec{v}_{B\text{rel}}$ and $\vec{a}_{B\text{rel}}$ are as they should be, that is, velocity and acceleration, respectively. More importantly, this example illustrates an important idea — the motion of an object (i.e., its velocity and acceleration) that one perceives when on a rotating frame is *very* different from the actual motion of the object. Figure 3 shows the vast difference between the actual velocity vector of B and the velocity of B as seen by A who is rotating with the merry-go-round. Person B is walking at 4 mph but A sees B moving at 68.5 mph. As far as acceleration is concerned, B is not accelerating at all, yet A sees B moving at $318\,\text{ft/s}^2 = 9.87g$!

🔍 **A Closer Look** Notice that the relative velocity $\vec{v}_{B\text{rel}}$ and acceleration $\vec{a}_{B\text{rel}}$ of B as seen by A do not depend on r, which is the radius of the circular path of A. This happens in this particular example because the person A, who is observing the relative quantities $\vec{v}_{B\text{rel}}$ and $\vec{a}_{B\text{rel}}$, is not moving relative to the moving reference frame. That is, A and the moving reference frame are a single rigid body. Because of this, the motion of A always cancels with part of the motion of B as seen by A. For example, referring to the velocities, we see that $\vec{v}_A = -r\omega_m\,\hat{\jmath}$ and that part of $\vec{\omega}_m \times \vec{r}_{B/A}$ is given by $r\omega_m\,\hat{\jmath}$. Therefore, when these terms are combined in Eq. (1), they cancel. A similar cancellation occurs for the relative acceleration.

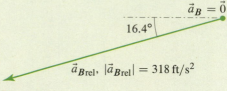

Figure 4
The acceleration of B as perceived by A, i.e., $\vec{a}_{B\text{rel}}$, as well as the actual acceleration of B, which is zero so only a dot is shown.

PROBLEMS

Problem 6.110

Obtain the acceleration of point P using the information from the mini-example on p. 522; that is, obtain Eq. (6.71), using polar coordinates.

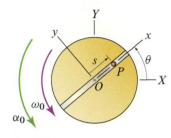

Figure P6.110

Problems 6.111 through 6.115

The *reciprocating rectilinear motion* mechanism consists of a disk pinned at its center at A that rotates with a constant angular velocity ω_{AB}, a slotted arm CD that is pinned at C, and a bar that can oscillate within the guides at E and F. As the disk rotates, the peg at B moves within the slotted arm, causing it to rock back and forth. As the arm rocks, it provides a slow advance and a quick return to the reciprocating bar due to the change in distance between C and B.

Problem 6.111   For the position shown, determine

(a) The angular velocity of the slotted arm CD and the velocity of the bar

(b) The angular acceleration of the slotted arm CD and the acceleration of the bar

Evaluate your results for $\omega_{AB} = 120\,\text{rpm}$, $R/h = 0.5$, and $d = 0.12\,\text{m}$.

Problem 6.112   For the position shown, determine

(a) The angular velocity of the slotted arm CD and the velocity of the bar

(b) The angular acceleration of the slotted arm CD and the acceleration of the bar

Evaluate your results for $\omega_{AB} = 90\,\text{rpm}$, $R/h = 0.5$, and $d = 0.12\,\text{m}$.

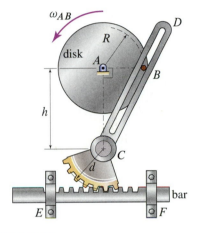

Figure P6.111

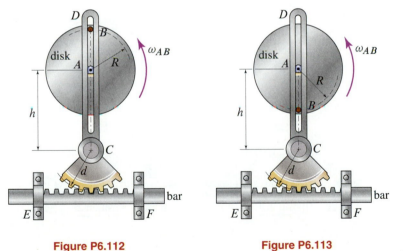

Figure P6.112   **Figure P6.113**

Problem 6.113   For the position shown, determine

(a) The angular velocity of the slotted arm CD and the velocity of the bar

(b) The angular acceleration of the slotted arm CD and the acceleration of the bar

Evaluate your results for $\omega_{AB} = 60\,\text{rpm}$, $R/h = 0.5$, and $d = 0.12\,\text{m}$.

Problem 6.114   For the arbitrary position shown, determine

(a) The angular velocity of the slotted arm CD and the velocity of the bar

(b) The angular acceleration of the slotted arm CD and the acceleration of the bar

as functions of θ, $\delta = R/h$, d, and ω_{AB}

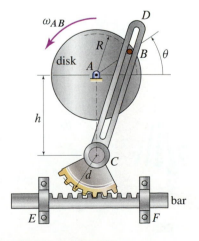

Figure P6.114 and P6.115

Problem 6.115 Determine the angular velocity and angular acceleration of the slotted arm CD as functions of θ, $\delta = R/h$, d, and ω_{AB}. After doing so, plot the velocity and acceleration of the bar as a function of the disk angle θ for one full cycle of the disk's motion and for $\omega_{AB} = 90$ rpm, $d = 0.12$ m, and

(a) $\delta = R/h = 0.1$

(b) $\delta = R/h = 0.3$

(c) $\delta = R/h = 0.6$

(d) $\delta = R/h = 0.9$

Explain why this mechanism is often referred to as a *quick return mechanism.*

Problem 6.116

The *reciprocating rectilinear motion* mechanism of Probs. 6.111–6.115 is often referred to as a *quick return mechanism* since it can move much more quickly in one direction than the other. To see this, we will determine the velocity of the bar in each of the two positions shown (position 1 when B is farthest from C and position 2 when B is closest to C) under the assumption that the disk, whose center is at A, rotates with a constant angular velocity $\omega_{AB} = 120$ rpm. Find the velocity of the bar in positions 1 and 2 for $d = 0.12$ m and for

(a) $R/h = 0.3$

(b) $R/h = 0.8$

Comment on which of the two R/h values would provide a better quick return and why.

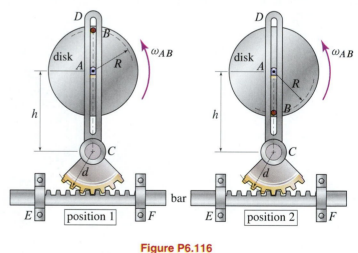

Figure P6.116

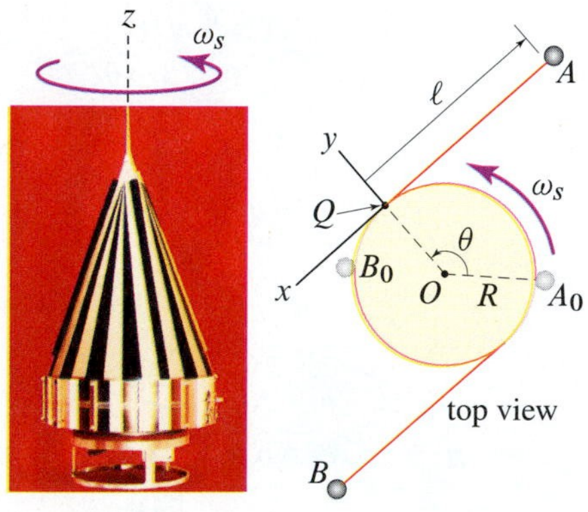

Problem 6.117

The Pioneer 3 spacecraft was a spin-stabilized spacecraft launched on December 6, 1958, by the U.S. Army Ballistic Missile agency in conjunction with NASA. It was designed with a despin mechanism consisting of two equal masses A and B that could be spooled out to the end of two wires of variable length $\ell(t)$ when triggered by a hydraulic timer.[*] As a prelude to Probs. 7.64, 7.65, and 8.77, we will find the velocity

Figure P6.117

---

[*] The spacecraft was intended as a lunar probe, but failed to go past the Moon and into a heliocentric (sun-centered) orbit as planned, but it did reach an altitude of $107,400$ km before falling back to Earth. It was a cone-shaped probe 58 cm high and 25 cm diameter at its base and was

and acceleration of each the two masses. To do this, assume that masses A and B are initially at positions A_0 and B_0, respectively. After the masses are released, they begin to unwind symmetrically, and the length of the cord attaching each mass to the spacecraft of radius R is $\ell(t)$. Given that the angular velocity of the spacecraft at each instant is $\omega_s(t)$, determine

(a) The velocity of mass A

(b) The acceleration of mass A

in components expressed in the rotating reference frame whose origin is at Q, as well as R, $\ell(t)$, and $\omega_s(t)$. Note that the rotating frame is always aligned with the unwinding cord, and Q is the point on the cord that is about to unwind at time t.

Problem 6.118

Let frames A and B be the frames with origins at points A and B, respectively. Point B does not move relative to point A. The velocity and acceleration of point P relative to frame B are

$$\vec{v}_{P\mathrm{rel}} = (-6.14\,\hat{\imath}_B + 23.7\,\hat{\jmath}_B)\,\text{ft/s} \quad\text{and}\quad \vec{a}_{P\mathrm{rel}} = (3.97\,\hat{\imath}_B + 4.79\,\hat{\jmath}_B)\,\text{ft/s}^2.$$

Knowing that, at the instant shown, frame B rotates relative to frame A at a constant angular velocity $\omega_B = 1.2\,\text{rad/s}$, that the position of point P relative to frame B is $\vec{r}_{P/B} = (8\,\hat{\imath}_B + 4.5\,\hat{\jmath}_B)\,\text{ft}$, and that frame A is fixed, determine the velocity and acceleration of P at the instant shown and express the results using the frame A component system.

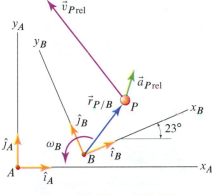

Figure P6.118 and P6.119

Problem 6.119

Repeat Prob. 6.118, but express the results using the component system of frame B.

Problem 6.120

A vertical shaft has a base B that is stationary relative to an inertial reference frame with vertical axis Z. Arm OA is attached to the vertical shaft and rotates about the Z axis with an angular velocity $\omega_{OA} = 5\,\text{rad/s}$ and an angular acceleration $\alpha_{OA} = 1.5\,\text{rad/s}^2$. The z axis is coincident with the Z axis but is part of a reference frame that

designed with a despin mechanism consisting of two 7 g weights that could be spooled out to the end of two 150 cm wires when triggered by a hydraulic timer 10 h after launch. The weights would slow the spacecraft spin from 400 to 6 rpm, and then the weights and wires would be released.

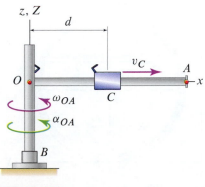

Figure P6.120

rotates with the arm OA. The x axis of the rotating reference frame coincides with the axis of the arm OA. At the instant shown, the y axis is perpendicular to the page and directed into the page. At this instant, the collar C is sliding along OA with a constant speed $v_C = 5$ ft/s and is at a distance $d = 1.2$ ft from the z axis. Compute the inertial acceleration of the collar, and express it relative to the rotating coordinate system.

Problem 6.121

A vertical shaft has a base B that is stationary relative to an inertial reference frame with vertical axis Z. Arm OA is attached to the vertical shaft and rotates about the Z axis with an angular velocity $\omega_{OA} = 5$ rad/s and an angular acceleration $\alpha_{OA} = 1.5$ rad/s^2. The z axis is coincident with the Z axis but is part of a reference frame that rotates with the arm OA. The x axis of the rotating reference frame coincides with the axis of the arm OA. At the instant shown, the y axis is perpendicular to the page and directed into the page. At this instant, the collar C is sliding along OA with a constant speed $v_C = 3.32$ m/s and is rotating with a constant angular velocity $\omega_C = 2.3$ rad/s relative to the arm OA. At the instant shown, point D happens to be in the xz plane and is at a distance $\ell = 0.05$ m from the x axis and at a distance $d = 0.75$ m from the z axis. Compute the inertial acceleration of point D, and express it relative to the rotating reference frame.

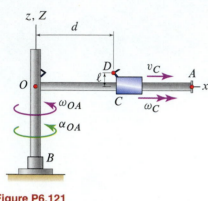

Figure P6.121

Problem 6.122

The wheel D rotates with a constant angular velocity $\omega_D = 14$ rad/s about the fixed point O, which is assumed to be stationary relative to an inertial frame of reference. The xy frame rotates with the wheel. Collar C slides along the bar AB with a constant velocity $v_C = 4$ ft/s relative to the xy frame. Letting $\ell = 0.25$ ft, determine the inertial velocity and acceleration of C when $\theta = 25°$. Express the result with respect to the xy frame.

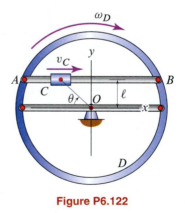

Figure P6.122

Problem 6.123

The wheel D rotates without slipping over a flat surface. The XY frame shown is inertial, whereas the xy frame is attached to D at O and rotates with it at a constant angular velocity $\omega_D = 14$ rad/s. Collar C slides along the bar AB with a constant velocity $v_C = 4$ ft/s relative to the xy frame. Letting $\ell = 0.25$ ft and $R = 1$ ft, determine the inertial velcity and acceleration of C when $\theta = 25°$ and the xy frame is parallel to the XY frame as shown. Express your result in both the xy and XY frames.

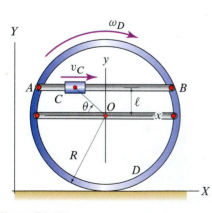

Figure P6.123

Problem 6.124

The wheel D rotates without slipping over a flat surface. The XY frame shown is inertial whereas the xy frame is attached to D and rotates with it at a constant angular velocity ω_D. Collar C slides along the bar AB with a velocity v_C relative to the xy frame. Suppose that ℓ, θ, and R are given and that we want to determine the inertial acceleration of C when the xy frame is parallel to the XY frame as shown. Would the expression of the inertial acceleration of the collar in the two frames be different or the same?

Note: Concept problems are about *explanations*, not computations.

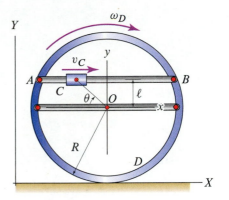

Figure P6.124

Problem 6.125

At the instant shown, the wheel D rotates without slipping over a flat surface with an angular velocity $\omega_D = 14\,\text{rad/s}$ and an angular acceleration $\alpha_D = 1.1\,\text{rad/s}^2$. The XY frame shown is inertial whereas the xy frame is attached to D. At the instant shown, the collar C is sliding along the bar AB with a velocity $v_C = 4\,\text{ft/s}$ and acceleration $a_C = 7\,\text{ft/s}^2$, both relative to the xy frame. Letting $\ell = 0.25\,\text{ft}$ and $R = 1\,\text{ft}$, determine the inertial velocity and acceleration of C when $\theta = 25°$ and the xy frame is parallel to the XY frame as shown. Express your result in both the xy and the XY frames.

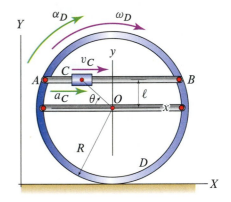

Figure P6.125

Problem 6.126

A floodgate is controlled by the motion of the hydraulic cylinder AB. If the gate BC is to be lifted with a constant angular velocity $\omega_{BC} = 0.5\,\text{rad/s}$, determine $\dot{d}_{AB}$ and $\ddot{d}_{AB}$, where d_{AB} is the distance between points A and B when $\phi = 0$. Let $\ell = 10\,\text{ft}$, $h = 2.5\,\text{ft}$, and $d = 5\,\text{ft}$.

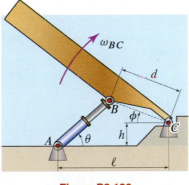

Figure P6.126

6.5   Chapter Review

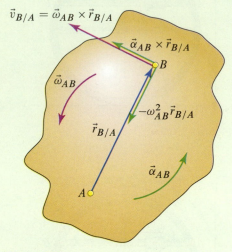

Figure 6.49
Figure 6.16 repeated. A rigid body showing the quantities used in Eqs. (6.3), (6.5), and (6.14).

Fundamental equations, translation, and rotation about a fixed axis

This section began our study of the dynamics of rigid bodies. As with particles, we started with the study of kinematics, which is the focus of this chapter. The key kinematic idea is that a rigid body has only one angular velocity and one angular acceleration; i.e., each is a property of the body as a whole. This allowed us to use the relative velocity equation (Eq. (2.106) on p. 136) and the relation for the time derivative of a vector (Eq. (2.62) on p. 92) to relate the velocities of *two points A and B on the same rigid body* using (see Fig. 6.49)

Eq. (6.3), p. 462

$$\vec{v}_B = \vec{v}_A + \vec{\omega}_{AB} \times \vec{r}_{B/A} = \vec{v}_A + \vec{v}_{B/A},$$

and the accelerations using

Eq. (6.5), p. 462

$$\vec{a}_B = \vec{a}_A + \vec{\alpha}_{AB} \times \vec{r}_{B/A} + \vec{\omega}_{AB} \times (\vec{\omega}_{AB} \times \vec{r}_{B/A}) = \vec{a}_A + \vec{a}_{B/A},$$

which for *planar motion* becomes

Eq. (6.14), p. 465

$$\vec{a}_B = \vec{a}_A + \vec{\alpha}_{AB} \times \vec{r}_{B/A} - \omega_{AB}^2 \vec{r}_{B/A}.$$

Translation.   For this motion, the angular velocity and angular acceleration of the body are equal to zero, i.e.,

Eqs. (6.6), p. 463

$$\vec{\omega}_{\text{body}} = \vec{0} \quad \text{and} \quad \vec{\alpha}_{\text{body}} = \vec{0},$$

so that the velocity and acceleration relations for the body reduce to

Eqs. (6.7), p. 464

$$\vec{v}_B = \vec{v}_A \quad \text{and} \quad \vec{a}_B = \vec{a}_A,$$

where A and B are any two points on the body.

Rotation about a fixed axis.   For this special motion, there is an axis of rotation perpendicular to the plane of motion that does not move. All points not on the axis of rotation can only move in a circle about that axis. If the axis of rotation is at point A, then the velocity of B is given by

Eq. (6.8), p. 464

$$\vec{v}_B = \vec{\omega}_{AB} \times \vec{r}_{B/A},$$

and its acceleration is

Eqs. (6.9) and (6.13), p. 464

$$\vec{a}_B = \vec{\alpha}_{AB} \times \vec{r}_{B/A} + \vec{\omega}_{AB} \times (\vec{\omega}_{AB} \times \vec{r}_{B/A}),$$

$$\vec{a}_B = \vec{\alpha}_{AB} \times \vec{r}_{B/A} - \omega_{AB}^2 \vec{r}_{B/A},$$

where the vectors $\vec{\omega}_{AB}$ and $\vec{\alpha}_{AB}$ are the angular velocity and angular acceleration of the body, respectively, and $\vec{r}_{B/A}$ is the vector that describes the position of B relative to A.

Planar motion: velocity analysis

This section presented three different ways to analyze the velocities of a rigid body in planar motion: the vector approach, differentiation of constraints, and instantaneous center of rotation.

Vector approach.   We saw in Section 6.1 that the equation

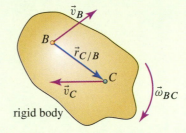

Figure 6.50
Figure 6.27 repeated. A rigid body on which we are relating velocity of two points B and C.

Eq. (6.15), p. 477

$$\vec{v}_C = \vec{v}_B + \vec{\omega}_{BC} \times \vec{r}_{C/B}$$

relates the velocity of two points on a rigid body $\vec{v}_B$ and $\vec{v}_C$ via their relative position $\vec{r}_{C/B}$ and the angular velocity of the body $\vec{\omega}_{BC}$ (see Fig. 6.50).

Rolling without slip.   When a body rolls without slip over another body (see, for example, the wheel W rolling over the surface S in Fig. 6.51), then the two points on the bodies that are in contact at any instant, points P and Q, must have the same velocity. Mathematically, this means that

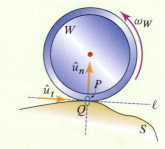

Figure 6.51
Figure 6.19 repeated. A wheel W rolling over a surface S. At the instant shown, the line ℓ is tangent to the path of points P and Q.

Eqs. (6.17), p. 478

$$\vec{v}_{P/Q} = \vec{0} \quad \Rightarrow \quad \vec{v}_P = \vec{v}_Q.$$

If a wheel of radius R is rolling without slip over a flat, stationary surface, then the point P on the wheel in contact with the surface must have zero velocity (see Fig. 6.52). The consequence is that the angular velocity of the wheel ω_O is related to the velocity of the center v_O and the radius of the wheel R according to

Eq. (6.20), p. 478

$$\omega_O = -\frac{v_O}{R},$$

in which the positive direction for ω_O is taken to be the positive z direction using the right-hand rule.

Differentiation of constraints.   As we discovered in Section 2.7, it is often convenient to write an equation describing the position of a point of interest, which can then be differentiated with respect to time to find the velocity of that point. For planar motion of rigid bodies, this idea can also apply for describing the position and velocity of a point on a rigid body as well as for describing the orientation of a rigid body, for which the time derivative provides its angular velocity.

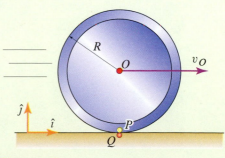

Figure 6.52
Figure 6.29 repeated. A wheel rolling without slip on a horizontal fixed surface.

Instantaneous center of rotation.   The point on a body (or imaginary extension of the body) whose velocity is zero at a particular instant is called the *instantaneous center of rotation* or *instantaneous center* (IC). The IC can be found geometrically if the velocity is known for two distinct points on a body or if a velocity on the body and the body's angular velocity are known. The three possible geometric constructions are shown in Fig. 6.53.

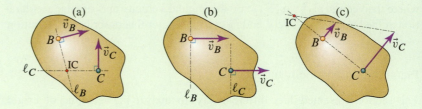

Figure 6.53. *Figure 6.30 repeated.* The three different possible motions for determining the IC. (a) Two nonparallel velocities are known. (b) The lines of action of two velocities are parallel and distinct; in this case, the IC is at infinity and the body is translating. (c) The lines of action of two parallel velocities coincide.

Planar motion: acceleration analysis

This section presented two different ways to analyze the accelerations of a rigid body in planar motion: the vector approach and differentiation of constraints.

Vector approach.   We saw in Section 6.1 that, for planar motion, either of the equations

> **Eqs. (6.31) and (6.32), p. 498**
>
> $$\vec{a}_C = \vec{a}_B + \vec{\alpha}_{BC} \times \vec{r}_{C/B} + \vec{\omega}_{BC} \times (\vec{\omega}_{BC} \times \vec{r}_{C/B}),$$
> $$\vec{a}_C = \vec{a}_B + \vec{\alpha}_{BC} \times \vec{r}_{C/B} - \omega_{BC}^2 \vec{r}_{C/B},$$

relates the acceleration of two points on a rigid body $\vec{a}_B$ and $\vec{a}_C$ via their relative position $\vec{r}_{C/B}$, the angular acceleration of the body $\vec{\alpha}_{BC}$, and the angular velocity of the body $\vec{\omega}_{BC}$ (see Fig. 6.54).

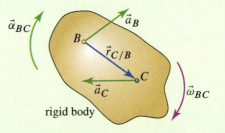

Figure 6.54. *Figure 6.35 repeated.* A rigid body on which we are relating the acceleration of two points B and C.

Differentiation of constraints.   As we discovered in Section 2.7, it is often convenient to write an equation describing the position of a point of interest, which can then be differentiated once with respect to time to find the velocity of that point and twice with respect to time to find the acceleration. For planar motion of rigid bodies, this idea can also apply for describing the position, velocity, and acceleration of a point on a rigid body, as well as for describing the orientation of a rigid body, for which the first and second time derivatives provide its angular velocity and angular acceleration, respectively (we saw this again in Section 6.2 for velocities).

Rolling without slip.   When a body rolls without slip over another body (see Fig. 6.55), then the two points on the bodies that are in contact at any instant, points P and Q, must have the same velocity. For accelerations, this means that the component of $\vec{a}_{P/Q}$ tangent to the contact must be equal to zero, that is,

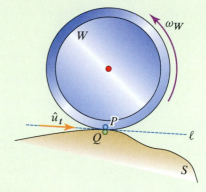

Figure 6.55
Figure 6.33 repeated. A wheel W rolling over a surface S. At the instant shown, the line ℓ is tangent to the path of P and Q.

Eqs. (6.36), p. 499

$$(a_{P/Q})_t = 0 \quad \Rightarrow \quad a_{Pt} = a_{Qt}.$$

If a wheel of radius R is rolling without slip over a flat, stationary surface, then the point P on the wheel in contact with the surface, must have zero velocity (see Fig. 6.56). The consequence is that the angular acceleration of the wheel α_O is related to the acceleration of the center a_O and the radius of the wheel R according to

Eq. (6.20), p. 478

$$\alpha_O = -\frac{a_O}{R},$$

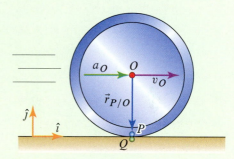

Figure 6.56
Figure 6.34 repeated. A wheel rolling without slip on a horizontal fixed surface.

in which the positive direction for α_O is taken to be the positive z direction. The point P on the wheel that is in contact with the ground at this instant is accelerating along the y axis according to

Eq. (6.41), p. 500

$$\vec{a}_P = \frac{v_O^2}{R}\,\hat{j}.$$

Even though P has zero velocity at the instant considered, it is accelerating.

Rotating reference frames

In this section, we developed the kinematic equations that allow us to relate the motion of two points that *are not* on the same rigid body. Referring to Fig. 6.57, this means we can relate the velocity of point P to that of A using the relation

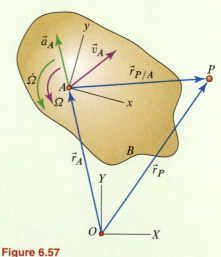

Figure 6.57
Figure 6.48 repeated. The essential ingredients for finding velocities and accelerations using a rotating reference frame.

Eq. (6.60), p. 520

$$\vec{v}_P = \vec{v}_A + \vec{v}_{P\text{rel}} + \vec{\Omega} \times \vec{r}_{P/A},$$

where

$\vec{v}_A =$ velocity of the origin of the rotating or body-fixed reference frame (point A in Fig. 6.57)

$\vec{v}_{P\text{rel}} =$ velocity of P as seen by an observer moving with the rotating reference frame

$\vec{\Omega}$ = angular velocity of the rotating reference frame

$\vec{r}_{P/A}$ = position of point P relative to A

We also found that we can relate the acceleration of point P to that of A using

Eq. (6.68), p. 522

$$\vec{a}_P = \vec{a}_A + \vec{a}_{P\text{rel}} + 2\vec{\Omega} \times \vec{v}_{P\text{rel}} + \dot{\vec{\Omega}} \times \vec{r}_{P/A} + \vec{\Omega} \times (\vec{\Omega} \times \vec{r}_{P/A}),$$

where, in addition to the terms defined for $\vec{v}_P$, we have

$\vec{a}_A$ = acceleration of the origin of the rotating or body-fixed reference frame (point A in Fig. 6.48)

$\vec{a}_{P\text{rel}}$ = acceleration of P as seen by an observer moving with the rotating reference frame

$2\vec{\Omega} \times \vec{v}_{P\text{rel}}$ = Coriolis acceleration of P

$\dot{\vec{\Omega}}$ = angular acceleration of the rotating reference frame

REVIEW PROBLEMS

Problem 6.127

The manually operated road barrier shown has a total length $l = 15.7$ ft and has a counterweight C whose position is identified by the distances $d = 2.58$ ft and $\delta = 1.4$ ft. The barrier is pinned at O and can move in the vertical plane. Suppose that the barrier is raised and then released so that the end B of the barrier hits the support with a speed $v_B = 1.5$ ft/s. Determine the speed of the counterweight C when B hits A.

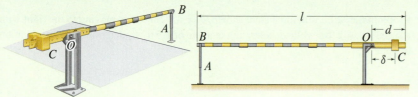

Figure P6.127

Problem 6.128

A slender bar AB of length $L = 1.45$ ft is mounted on two identical disks D and E pinned at A and B, respectively, and of radius $r = 1.5$ in. The bar is allowed to move within a cylindrical bowl with center at O and diameter $d = 2$ ft. At the instant shown, the center G of the bar is moving with a speed $v = 7$ ft/s. Determine the angular velocity of the bar at the instant shown.

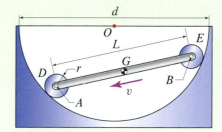

Figure P6.128

Problem 6.129

Assuming that the rope does not slip relative to any of the pulleys in the system, determine the velocity and acceleration of A and D, knowing that the angular velocity and acceleration of pulley B are $\omega_B = 7$ rad/s and $\alpha_B = 3$ rad/s^2, respectively. The diameters of pulleys B and C are $d = 25$ cm and $\ell = 34$ cm, respectively.

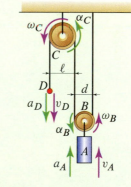

Figure P6.129

Problem 6.130

At the instant shown, the hammer head H is moving to the right with a speed $v_H = 45$ ft/s and the angle $\theta = 20°$. Assuming that the belt does not slip relative to wheels A and B and assuming that wheel A is mounted on the shaft of the motor shown, determine the angular velocity of the motor at the instant shown. The diameter of wheel A is $d = 0.25$ ft, the radius of wheel B is $R = 0.75$ ft, and point C is at a distance $\ell = 0.72$ ft from O, which is the center of the wheel A. Finally, let CD have a length $L = 2$ ft, and assume that, at the instant shown, $\phi = 25°$.

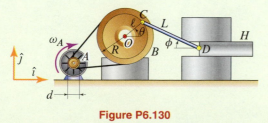

Figure P6.130

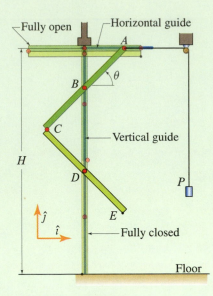

Figure P6.131 and P6.132

Problems 6.131 and 6.132

An overhead fold-up door with height $H = 30$ ft consists of two identical sections hinged at C. The roller at A moves along a horizontal guide, whereas the rollers at B and D, which are the midpoints of sections AC and CE, move along a vertical guide. The door's operation is assisted by a counterweight P. Express your answers using the component system shown.

Problem 6.131 If at the instant shown, the angle $\theta = 55°$ and P is moving upward with a speed $v_P = 15$ ft/s, determine the velocity of point E as well as the angular velocities of sections AC and CE.

Problem 6.132 If at the instant shown, $\theta = 45°$, and A is moving to the right with a speed $v_A = 2$ ft/s while decelerating at a rate of 1.5 ft/s^2, determine the acceleration of point E.

Problem 6.133

At the instant shown, bar AB rotates with a constant angular velocity $\omega_{AB} = 24$ rad/s. Letting $L = 0.75$ m and $H = 0.85$ m, determine the angular acceleration of bar BC when bars AB and CD are as shown, i.e., parallel and horizontal.

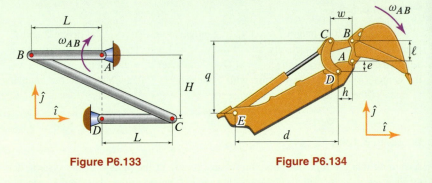

Figure P6.133   **Figure P6.134**

Problem 6.134

The bucket of a backhoe is the element AB of the four-bar linkage system $ABCD$. The bucket's motion is controlled by extending or retracting the hydraulic arm EC. Assume that the points A, D, and E are fixed and that the bucket is made to rotate with a constant angular velocity $\omega_{AB} = 0.25$ rad/s. In addition, suppose that, at the instant shown, point B is vertically aligned with point A and point C is horizontally aligned with B. Letting d_{EC} denote the distance between points E and C, determine $\dot{d}_{EC}$ and $\ddot{d}_{EC}$ at the instant shown. Let $h = 0.66$ ft, $e = 0.46$ ft, $l = 0.9$ ft, $w = 1.0$ ft, $d = 4.6$ ft, and $q = 3.2$ ft.

Problem 6.135

Let frames A and B have their origins at points A and B, respectively. Point B does not move relative to point A. The velocity and acceleration of point P relative to frame A, which is fixed, are

$$\vec{v}_P = (-14.9\,\hat{\imath}_A + 19.4\,\hat{\jmath}_A)\,\text{ft/s} \quad \text{and} \quad \vec{a}_P = (1.78\,\hat{\imath}_A + 5.96\,\hat{\jmath}_A)\,\text{ft/s}^2.$$

Knowing that frame B rotates relative to frame A at a constant angular velocity $\omega_B = 1.2\,\text{rad/s}$ and that the position of point P relative to frame B is $\vec{r}_{P/B} = (8\,\hat{\imath}_B + 4.5\,\hat{\jmath}_B)\,\text{ft}$, determine the velocity and acceleration of P relative to frame B at the instant shown and express the results using the frame A component system.

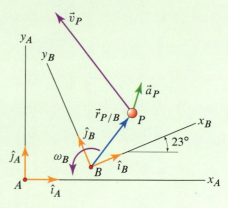

Figure P6.135

Problem 6.136

A vertical shaft has a base B that is stationary relative to an inertial reference frame with vertical axis Z. Arm OA is attached to the vertical shaft and rotates about the Z axis with an angular velocity ω_{OA} and an angular acceleration α_{OA}. The z axis is coincident with the Z axis but is part of a reference frame that rotates with the arm OA. The x axis of the rotating reference frame coincides with the axis of the arm OA. At the instant shown, the y axis is perpendicular to the page and directed into the page. At this instant, the collar C is at a distance d from the Z axis, is sliding along OA with a constant speed v_C (relative to the arm OA), and is rotating with a constant angular velocity ω_C (relative to the xyz frame). Point D is attached to the collar and is at a distance ℓ from the x axis. At the instant shown, point D happens to be in the plane that is rotated by an angle ϕ from the xz plane. Compute the expression of the inertial velocity and acceleration of point D at the instant shown in terms of the parameters given, and express it relative to the rotating coordinate system.

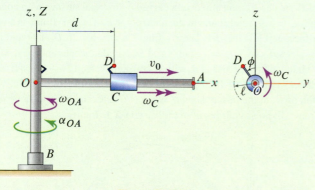

Figure P6.136

Problem 6.137 ▎

The wheel D rotates without slip over a curved cylindrical surface with constant radius
of curvature $L = 1.6\,\text{ft}$ and center at the fixed point E. The XY frame is attached to
the rolling surface, and it is inertial. The xy frame is attached to D at O and rotates
with it at a constant angular velocity $\omega_D = 14\,\text{rad/s}$. Collar C slides along the bar AB
with a constant velocity $v_C = 4\,\text{ft/s}$ relative to the xy frame. At the instant shown,
points O and E are vertically aligned. Letting $\ell = 0.25\,\text{ft}$ and $R = 1\,\text{ft}$, determine the
inertial velocity and acceleration of C at the instant shown when $\theta = 25°$ and the xy
frame is parallel to the XY frame. Express your result in the xy and XY frames.

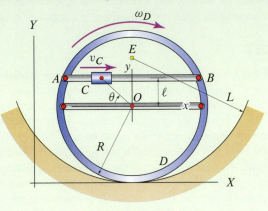

Figure P6.137

Problem 6.138 ▎

The wheel D rotates without slip over a curved cylindrical surface with constant radius
of curvature $L = 1.6\,\text{ft}$ and center at the fixed point E. The XY frame is attached to the
rolling surface, and it is inertial. The xy frame is attached to D at O and rotates with
it at an angular velocity $\omega_D = 14\,\text{rad/s}$ and an angular acceleration $\alpha_D = 1.3\,\text{rad/s}^2$.
Collar C slides along the bar AB with a constant velocity $v_C = 4\,\text{ft/s}$ relative to
the xy frame. Letting $\ell = 0.25\,\text{ft}$ and $R = 1\,\text{ft}$, determine the inertial velocity and
acceleration of C, at the instant shown, when $\theta = 25°$ and the xy frame is parallel to
the XY frame. Express your result in the xy and XY frames.

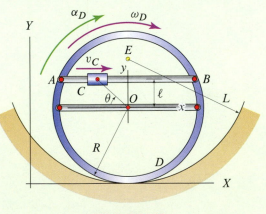

Figure P6.138

Newton-Euler Equations for Planar Rigid Body Motion

We now begin the study of rigid body kinetics. In this chapter, we develop the balance principles for forces and moments, this time for a rigid body. In Chapter 8 we will study energy and momentum methods for rigid bodies, and in Chapter 9 we will study the vibration of both particles and rigid bodies. By the time we complete this chapter, we will know how to combine the Newton-Euler equations for a rigid body with the force laws and the kinematic equations to obtain the equations governing the body's motion.

In rigid body kinetics we will be writing force *and* moment equations. Therefore, in the spirit of the discussion in Section 5.3, when referring to the force and moment equations for a rigid body, we refer to these equations as *Newton-Euler equations* instead of Newton's second law.

Note to the reader. This chapter consists of a single section designed to unify the subject of rigid body kinetics and to help you solve dynamics problems on your own. This unified treatment of the governing equations for rigid bodies still reflects the traditional categories of rigid body motion (i.e., translation, rotation about a fixed axis, and general plane motion), but does so via the titles of examples and the organization of the homework problems rather than artificial divisions in the exposition. This makes it a section that takes more time to cover than other sections in this book. Because this chapter consists of a single section, no separate chapter review is included.

7.1 Bodies Symmetric with Respect to the Plane of Motion

A motorcycle popping a wheelie

Figure 7.1 shows a Kawasaki Ninja ZX-14 sport bike, which, at the time of this writing, is the most powerful and the fastest production motorcycle in history. This motorcycle can go from a standstill to 60 mph in under 2.5 s, and its top

Figure 7.1
A Kawasaki Ninja® ZX™-14 motorcycle.

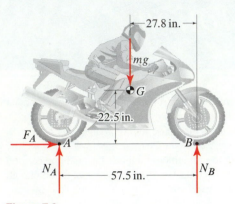

Figure 7.2
The FBD of the motorcycle and rider of Fig. 7.1.

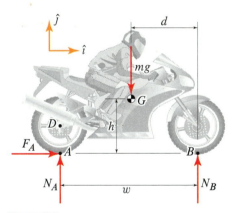

Figure 7.3
FBD of the Kawasaki Ninja with the addition of a component system. In addition, we have let the dimensions be $w = 57.5$ in., $d = 27.8$ in., and $h = 22.5$ in.

speed is 186 mph (300 km/h).* With this performance, it isn't hard to imagine that one could easily lose control of this bike by having the front tire come off the ground when accelerating. The question is, How much acceleration is too much, that is, what is the maximum acceleration that the bike and rider can experience and still keep the front wheel on the ground? To answer this question, we can no longer model the system as a particle since the rotation (or tendency for rotation) of the motorcycle plays an important role.

Let's consider a motorcycle and a rider with a combined weight of 650 lb and for which μ_s and μ_k between the tires and the road are 0.9 and 0.75, respectively. Figure 7.2 shows the FBD of the motorcycle and rider of Fig. 7.1 based on the following modeling assumptions:

1. The rider and the motorcycle are both modeled as rigid bodies.

2. The rider does not move relative to the motorcycle, and a common mass center is used.

3. The rotational inertia of the front wheel is ignored (the validity of this assumption is explored in Prob. 7.38).

4. The motorcycle is driven by its rear wheel (notice that the friction force exerted *on* the rear wheel by the ground acts in the direction of motion — this is so because the motorcycle wants to "push" the ground backward, so Newton's third law says that the ground must "push" the motorcycle forward).

Assumption 1 underlies everything we will do in rigid body dynamics. Assumption 2 will allow us to treat the rider and motorcycle as a single body, and we won't have to worry about their separate accelerations or separate weights. Assumption 3 is the reason there is no friction force on the front tire, and its validity depends on the inertia properties of the wheels. Assumption 4 is what provides the friction force on the rear wheel. Making these modeling assumptions and the associated FBD requires some experience. For example, the reason for the absence of a friction force on the front tire may not be obvious at this point; it will soon become clear.

Now that we have a model and an FBD, we can write the governing equations starting, as usual, with the balance principles, which in this case will be the Newton-Euler equations.[†] In Chapter 6 we saw that, in general, a rigid body in planar motion has three degrees of freedom (translation of the center of mass in two orthogonal directions and rotation). Therefore we need to write three Newton-Euler equations. We have already seen these equations in Section 5.3, namely, Eq. (5.86), which governs the motion of the body's center of mass, and Eq. (5.87), which governs the body's rotational motion. Referring to Fig. 7.3, we can sum forces in the x and y directions to obtain

$$\sum F_x: \qquad\qquad F_A = ma_{Gx}, \qquad\qquad (7.1)$$

$$\sum F_y: \quad N_A + N_B - mg = ma_{Gy}, \qquad\qquad (7.2)$$

---

* It can likely go much faster; the quoted top speed is due to electronic restriction imposed by the manufacturer.

† The rationale for using the expression *Newton-Euler equations* instead of *Newton's laws of motion* was presented on p. 394.

where $m = 650\,\text{lb}/(32.2\,\text{lb/slug}) = 20.2\,\text{slug}$ is the combined mass of the rider and motorcycle and $\vec{a}_G$ is the acceleration of the mass center. Choosing the mass center G as moment center, for the moment equation we obtain

$$\sum \vec{M}_G: \quad [N_B d - N_A(w-d) + F_A h]\,\hat{k} = \dot{\vec{h}}_G. \qquad (7.3)$$

Notice that the one quantity in Eq. (7.3) that we don't know how to handle is $\dot{\vec{h}}_G$, so the question now becomes, What is $\dot{\vec{h}}_G$ for a rigid body? We will soon learn that $\dot{\vec{h}}_G = I_G \alpha_{\text{mot}}\,\hat{k}$, where I_G is the *mass moment of inertia of the rider and motorcycle about an axis perpendicular to the page through G* (see App. A) and α_{mot} is the angular acceleration of the rider and motorcycle. Thus, Eq. (7.3), in scalar form, becomes

$$\sum M_G: \quad N_B d - N_A(w-d) + F_A h = I_G \alpha_{\text{mot}}. \qquad (7.4)$$

We have the Newton-Euler equations, so the force laws are next. Recall that we want to find the maximum acceleration for which the front wheel will not lift off the ground. Once we have determined F_A and N_A, we will use

$$|F_A| \leq \mu_s |N_A|, \qquad (7.5)$$

to check whether or not the rear wheel slips under the imposed conditions. The fact that we want the front wheel to be imminently lifting off the ground implies that

$$N_B = 0. \qquad (7.6)$$

Once we have the force laws, the last step in assembling the problem's governing equations consists of writing the kinematic equations. Again, we want the *largest acceleration* such that the front wheel *does not lift off the ground.* Therefore, the motorcycle does *not* rotate, and so $\omega_{\text{mot}} = 0$ and $\alpha_{\text{mot}} = 0$. In addition, we know that

$$\vec{a}_G = \vec{a}_D + \vec{\alpha}_{\text{mot}} \times \vec{r}_{G/D} - \omega_{\text{mot}}^2 \vec{r}_{G/D}, \qquad (7.7)$$

since G and D are two points on the same rigid body, i.e., the rigid body that is the motorcycle plus rider. Since ω_{mot} and α_{mot} are both zero, this becomes $\vec{a}_G = \vec{a}_D$. Noting that D *must* move horizontally, G must also do so and therefore $a_{Gy} = 0$. Summarizing these kinematic conditions, we have

$$\omega_{\text{mot}} = 0, \qquad \alpha_{\text{mot}} = 0, \qquad a_{Gy} = 0. \qquad (7.8)$$

Now that we have written our usual three sets of equations, that is, the Newton-Euler equations, the force laws, and the kinematic equations, it is time to solve them. Substituting Eqs. (7.6) and (7.8) into Eqs. (7.1), (7.2), and (7.4), we obtain

$$F_A = m a_{Gx}, \qquad (7.9)$$
$$N_A - mg = 0, \qquad (7.10)$$
$$-N_A(w-d) + F_A h = 0. \qquad (7.11)$$

Equations (7.9)–(7.11) are three equations for the three unknowns N_A, F_A, and a_{Gx}, which can be solved to obtain

$$N_A = mg = 650\,\text{lb}, \qquad (7.12)$$

Mass moment of inertia. The mass moment of inertia of a body can be thought of in an analogous way as the mass of a body. That is, the mass of a body is a measure of its resistance to accelerations under the action of forces. Similarly, the mass moment of inertia of a body is a measure of its resistance to angular acceleration under the action of moments.

$$F_A = \left(\frac{w-d}{h}\right)mg = 858\,\text{lb}, \tag{7.13}$$

$$a_{Gx} = \left(\frac{w-d}{h}\right)g = 42.5\,\text{ft/s}^2 = 1.32g, \tag{7.14}$$

where the acceleration in Eq. (7.14) is the answer we desired, that is, the largest acceleration such that the motorcycle doesn't pop a wheelie. Finally, let's check the friction inequality in Eq. (7.5). We were given that $\mu_s = 0.9$, and so the maximum available friction is $\mu_s N_A = 585\,\text{lb}$, which is *less than* the needed friction, which is $F_A = 858\,\text{lb}$. Therefore, we can conclude that the rear tire will slip and that the acceleration needed to pop a wheelie, i.e., $42.5\,\text{ft/s}^2$, is not achievable without increased friction. The acceleration needed to pop a wheelie is 1.32 times the acceleration due to gravity — under these conditions, holding onto the handlebars would require some effort!

Let's now formally develop the governing equations for a rigid body and then see how to apply them in the example problems.

Linear momentum: translational equations

To obtain the translational equations of motion, we only need Euler's first law, which is given by Eq. (5.86) on p. 394 and applies to any body or system of bodies with constant mass:

$$\boxed{\vec{F} = m\vec{a}_G,} \tag{7.15}$$

where $\vec{F}$ is the resultant of all *external* forces, m is the total mass of the system, and $\vec{a}_G$ is the inertial acceleration of its mass center. The acceleration of the mass center is given by

$$\vec{a}_G = \frac{d^2\vec{r}_G}{dt^2}, \tag{7.16}$$

where, referring to Fig. 7.4, the *mass center* is the point with position

$$\vec{r}_G = \frac{1}{m}\int_B \vec{r}_{dm}\,dm, \tag{7.17}$$

where the integral is performed over the body B and $\vec{r}_{dm}$ is the position of the infinitesimal element of mass dm.

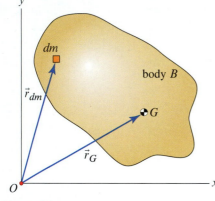

Figure 7.4
Vectors needed for the determination of the position of the mass center $\vec{r}_G$ of a rigid body.

Angular momentum: rotational equations

Since Euler's second law relates moments to angular momentum, it will give us the rotational equations of motion. We begin with the moment–angular momentum relation given by Eq. (5.87) on p. 394, which is

$$\vec{M}_P = \dot{\vec{h}}_P + \vec{v}_P \times m\vec{v}_G, \tag{7.18}$$

where

- $\vec{M}_P$ is the moment of all external forces about point P, *which is an arbitrary point in space*

- $\vec{h}_P$ is the angular momentum (or moment of momentum) of the entire system with respect to point P

- $\vec{v}_P$ is the velocity of the arbitrary point P

- $\vec{v}_G$ is the velocity of the mass center of the system

- m is the total mass of the system

When we apply Eq. (7.18) to the rigid body shown in Fig. 7.4, except for $\dot{\vec{h}}_P$, all terms are easily interpreted in terms of their original system of particle definitions. For a system of particles, $\vec{h}_P$ is computed via the definitions in Eqs. (5.59) and (5.77) (on pp. 389 and 392, respectively), which yield

$$\vec{h}_P = \sum_{i=1}^{N} \vec{r}_{i/P} \times m_i \vec{v}_i, \qquad (7.19)$$

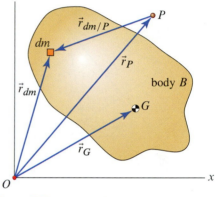

where N is the number of particles in the system, $\vec{r}_{i/P}$ is the position of particle i relative to the moment center P, and $m_i \vec{v}_i$ is the linear momentum of particle i. For a rigid body whose mass is distributed in space, we can generalize Eq. (7.19) by replacing a summation over the number of particles with an integral over the body. In doing so, the particle of mass m_i becomes the infinitesimal element of mass dm and Eq. (7.19) becomes

$$\vec{h}_P = \int_B \vec{r}_{dm/P} \times \vec{v}_{dm} \, dm, \qquad (7.20)$$

Figure 7.5
Points and corresponding position vectors needed to develop the moment–angular momentum relationship for a rigid body.

where $\vec{v}_{dm}$ is the velocity of the infinitesimal element of mass dm (see Fig. 7.5). What we really need is $\dot{\vec{h}}_P$, so we differentiate Eq. (7.20) with respect to time to obtain

$$\dot{\vec{h}}_P = \int_B \vec{v}_{dm/P} \times \vec{v}_{dm} \, dm + \int_B \vec{r}_{dm/P} \times \vec{a}_{dm} \, dm. \qquad (7.21)$$

Using $\vec{v}_{dm/P} = \vec{v}_{dm} - \vec{v}_P$, Eq. (7.21) becomes

$$\dot{\vec{h}}_P = \int_B (\vec{v}_{dm} - \vec{v}_P) \times \vec{v}_{dm} \, dm + \int_B \vec{r}_{dm/P} \times \vec{a}_{dm} \, dm$$

$$= \underbrace{-\int_B \vec{v}_P \times \vec{v}_{dm} \, dm}_{\text{integral } A} + \underbrace{\int_B \vec{r}_{dm/P} \times \vec{a}_{dm} \, dm.}_{\text{integral } B} \qquad (7.22)$$

Since $\vec{v}_P$ does not depend on dm, integral A can be written as

$$-\int_B \vec{v}_P \times \vec{v}_{dm} \, dm = -\vec{v}_P \times \int_B \vec{v}_{dm} \, dm = -\vec{v}_P \times \frac{d}{dt} \int_B \vec{r}_{dm} \, dm$$

$$= -\vec{v}_P \times \frac{d}{dt}(m\vec{r}_G) = -\vec{v}_P \times m\vec{v}_G, \qquad (7.23)$$

where we have used the definition of the mass center of a rigid body given in Eq. (7.17). Substituting Eq. (7.23) into Eq. (7.22) gives

$$\dot{\vec{h}}_P = -\vec{v}_P \times m\vec{v}_G + \underbrace{\int_B \vec{r}_{dm/P} \times \vec{a}_{dm} \, dm.}_{\text{integral } B} \qquad (7.24)$$

Helpful Information

Which terms can be pulled outside the integral signs? In going from Eq. (7.22) to Eq. (7.23), $\vec{v}_P$ is pulled out of the integral. We can pull $\vec{v}_P$ out since it is a property of a single point and does not vary with dm. We will see this again in Eq. (7.35) where, for example, the same is true for a_{Gy}, which is also a property of a particular point, and for α_B, which is a property of the body as a whole.

Substituting Eq. (7.24) into Eq. (7.18), we are left with just integral B, that is,

$$\vec{M}_P = \int_B \vec{r}_{dm/P} \times \vec{a}_{dm}\, dm, \tag{7.25}$$

since $-\vec{v}_P \times m\vec{v}_G$ in Eq. (7.24) cancels with $\vec{v}_P \times m\vec{v}_G$ in Eq. (7.18). This leaves us with the final task of interpreting the integral in Eq. (7.25). We will do this for the case in which the rigid body is symmetric with respect to the plane of motion (Chapter 10 covers the case when it is not).

Bodies symmetric with respect to the plane of motion

Referring to Fig. 7.6, since the body B is rigid, we can relate $\vec{a}_{dm}$ to $\vec{a}_G$ via Eq. (6.13) on p. 464 as

$$\vec{a}_{dm} = \vec{a}_G + \vec{\alpha}_B \times \vec{q} - \omega_B^2 \vec{q}, \tag{7.26}$$

where $\vec{q}$ is the position of dm relative to G. To continue, it is convenient to write all vectors in Cartesian components as follows:

$$\vec{M}_P = M_{Px}\,\hat{\imath} + M_{Py}\,\hat{\jmath} + M_{Pz}\,\hat{k}, \tag{7.27}$$

$$\vec{a}_G = a_{Gx}\,\hat{\imath} + a_{Gy}\,\hat{\jmath}, \tag{7.28}$$

$$\vec{q} = q_x\,\hat{\imath} + q_y\,\hat{\jmath}, \tag{7.29}$$

$$\vec{r}_{dm/P} = \vec{r}_{G/P} + \vec{q} = (x_{G/P} + q_x)\,\hat{\imath} + (y_{G/P} + q_y)\,\hat{\jmath}, \tag{7.30}$$

$$\vec{\omega}_B = \omega_B\,\hat{k} \quad \text{and} \quad \vec{\alpha}_B = \alpha_B\,\hat{k}, \tag{7.31}$$

where Eqs. (7.28) and (7.31) reflect the constraint that the body's motion is planar and Eqs. (7.29) and (7.30) reflect our assumption that the body is symmetric with respect to the plane of motion. Substituting Eqs. (7.28), (7.29), and (7.31) into Eq. (7.26), we obtain $\vec{a}_{dm}$ in component form

$$\vec{a}_{dm} = \left(a_{Gx} - \alpha_B q_y - \omega_B^2 q_x\right)\hat{\imath} + \left(a_{Gy} + \alpha_B q_x - \omega_B^2 q_y\right)\hat{\jmath}. \tag{7.32}$$

We can now substitute Eqs. (7.27), (7.30), and (7.32) into Eq. (7.25); expand the cross product; and equate components; to obtain the following three expressions for the rotational equations of motion:

$$M_{Px} = 0, \tag{7.33}$$

$$M_{Py} = 0, \tag{7.34}$$

$$
\begin{aligned}
M_{Pz} = {}& \alpha_B \int_B \left(q_x^2 + q_y^2\right) dm + \left(x_{G/P}\, a_{Gy} - y_{G/P}\, a_{Gx}\right) \int_B dm \\
& + \left(a_{Gy} + \alpha_B x_{G/P} + \omega_B^2 y_{G/P}\right) \int_B q_x\, dm \\
& + \left(-a_{Gx} + \alpha_B y_{G/P} - \omega_B^2 x_{G/P}\right) \int_B q_y\, dm,
\end{aligned} \tag{7.35}
$$

where we have taken all terms that do not depend on the element of mass dm outside of the integrals. Observe that *there are no moments about the x or y axes when the body is symmetric with respect to the plane of motion*. We can now interpret all the integrals in Eq. (7.35).

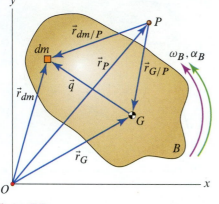

Figure 7.6
The definitions of all needed kinematic and kinetic quantities for a rigid body moving in planar motion.

The easiest integral to interpret in Eq. (7.35) is the second one, which is

$$\int_B dm = m, \qquad (7.36)$$

that is, it is just the total mass of the body. Recalling that Eq. (7.17) defines the position of the mass center relative to point O (see Fig. 7.4), in component form, Eq. (7.17) becomes

$$\vec{r}_G = x_{G/O}\,\hat{\imath} + y_{G/O}\,\hat{\jmath} = \frac{1}{m}\int_B (x_{dm/O}\,\hat{\imath} + y_{dm/O}\,\hat{\jmath})\,dm, \qquad (7.37)$$

or

$$m x_{G/O} = \int_B x_{dm/O}\,dm \quad \text{and} \quad m y_{G/O} = \int_B y_{dm/O}\,dm. \qquad (7.38)$$

Since $\vec{q} = q_x\,\hat{\imath} + q_y\,\hat{\jmath}$ defines the position of each element of mass dm relative to the mass center G, the component form of the definition of the mass center of a rigid body in Eq. (7.38) tells us that (see Fig. 7.6)

$$\int_B q_x\,dm = \int_B x_{dm/G}\,dm = m x_{G/G} = 0, \qquad (7.39)$$

$$\int_B q_y\,dm = \int_B y_{dm/G}\,dm = m y_{G/G} = 0, \qquad (7.40)$$

where we see that these integrals just define the position of the mass center relative to the mass center. There is only one integral left to interpret, which is

$$\int_B (q_x^2 + q_y^2)\,dm = I_G. \qquad (7.41)$$

This integral defines the *mass moment of inertia of the rigid body about an axis perpendicular to the plane of motion passing through the mass center G*, and as can be seen in Eq. (7.41), we will label this quantity I_G. Because the motion is planar, we will usually refer to the term in question simply as the *mass moment of inertia*, and the subscript on I will be used to uniquely identify the axis with respect to which I is calculated (see App. A).

Using Eqs. (7.36)–(7.41), we can see that Eq. (7.35) becomes

$$\boxed{M_P = I_G \alpha_B + m(x_{G/P} a_{Gy} - y_{G/P} a_{Gx}),} \qquad (7.42)$$

where $M_{Pz} = M_P$ to simplify the notation for the planar, symmetric case. Equation (7.42) is the most general rotational equation for the planar motion of a rigid body. Equation (7.42) and the two equations represented by Eq. (7.15) give the three equations needed to describe the planar motion of a rigid body that is symmetric with respect to the plane of motion.

Now that we have the rotational equation we need, note that

- If the body is *not* symmetric with respect to the xy plane of motion, then moments in the x and/or y directions will be required to maintain planar motion.

- Equation (7.42) can be written using vector notation as

$$\boxed{\vec{M}_P = I_G \vec{\alpha}_B + \vec{r}_{G/P} \times m\vec{a}_G,} \qquad (7.43)$$

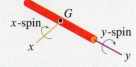
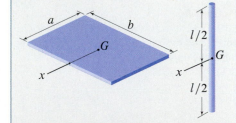

where, since the motion is planar, $\vec{M}_P = M_P\,\hat{k}$, $\vec{\alpha}_B = \alpha_B\,\hat{k}$, $\vec{r}_{G/P} = x_{G/P}\,\hat{i} + y_{G/P}\,\hat{j}$, and $\vec{a}_G = a_{Gx}\,\hat{i} + a_{Gy}\,\hat{j}$. We want to emphasize again that *Eqs. (7.42) and (7.43) apply when P is an arbitrary point in space*, that is, it *does not* have to be a point on the rigid body B.

- If *any* one of the following conditions is true:

  1. Point P is the mass center G, so that $\vec{r}_{G/P} = \vec{0}$.

  2. $\vec{a}_G = \vec{0}$ (i.e., G is fixed or moves with constant velocity).

  3. $\vec{r}_{G/P}$ is parallel to $\vec{a}_G$.

  then Eq. (7.42) reduces to

$$\boxed{M_P = I_G\alpha_B.} \tag{7.44}$$

■ **Mini-Example.**   Anyone who has played baseball or softball has experienced that feeling when the ball is hit just right; that is, you can barely feel the bat in your hands, but the ball goes *a long way*. This may happen when you hit the ball on the "sweet spot" of the bat. There is some debate as to physically what point on a bat defines the sweet spot, but it is thought to involve the vibrational modes of the bat as well as a point called the *center of percussion.**
Figure 7.7 shows a ball hitting a bat at a distance d from the knob A when the batter has "choked up" a distance δ. We will assume that the batter swings at his or her wrists (i.e., the pivot point is O) and will then determine the distance d at which the ball should be hit so that, no matter how large the force applied at P to the bat by the ball, the lateral force (i.e., perpendicular to the bat) at O is zero. That point P defines the *center of percussion,* and its location depends on the pivot point O. We will assume the bat has mass m, its mass center is at G, and I_G is its mass moment of inertia.

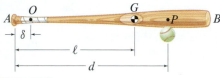

Figure 7.7
A baseball bat showing the location of its mass center G, pivot point O, and the point of impact with a baseball.

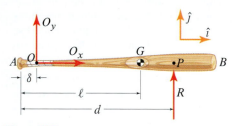

Figure 7.8
FBD of a bat as it is striking a ball. Point O is considered to be fixed since the batter is swinging at the wrists.

Solution. The FBD of the bat as it is striking the ball is shown in Fig. 7.8. We have neglected the moment that the batter applies to the bat because the effect of this moment is negligible next to that of the forces due to the collision of the ball against the bat. We have neglected the weight of the bat for the same reason. Applying Eqs. (7.15) and (7.44) to this FBD, we obtain the Newton-Euler equations

$$\sum F_x: \qquad\qquad O_x = ma_{Gx}, \tag{7.45}$$

$$\sum F_y: \qquad\qquad O_y + R = ma_{Gy}, \tag{7.46}$$

$$\sum M_G: \quad R(d - \ell) - O_y(\ell - \delta) = I_G\alpha_{\text{bat}}. \tag{7.47}$$

Since all forces are accounted for on the FBD, there are no force laws to write. Assuming the bat is swung with angular velocity ω_{bat} and angular acceleration α_{bat}, and recalling that the bat has been assumed to rotate in the xy plane about the fixed point O, we obtain the following kinematic equations:

$$a_{Gx} = -(\ell - \delta)\omega_{\text{bat}}^2 \qquad \text{and} \qquad a_{Gy} = (\ell - \delta)\alpha_{\text{bat}}. \tag{7.48}$$

---

* See R. Cross, "The Sweet Spot of a Baseball Bat," *American Journal of Physics,* **66**(9), 1998, pp. 772–779.

Plugging Eqs. (7.48) into Eqs. (7.45)–(7.47), we get

$$O_x = -m(\ell - \delta)\omega_{\text{bat}}^2,$$ (7.49)

$$O_y + R = m(\ell - \delta)\alpha_{\text{bat}},$$ (7.50)

$$R(d - \ell) - O_y(\ell - \delta) = I_G\alpha_{\text{bat}}.$$ (7.51)

Eliminating α_{bat} from Eqs. (7.50) and (7.51) and solving for O_y, we get

$$O_y = \frac{mR(d - \ell)(\ell - \delta) - RI_G}{I_G + m(\ell - \delta)^2} = 0.$$ (7.52)

Setting this result equal to zero and then solving for the distance d, we find

$$d = \frac{I_G + m\ell(\ell - \delta)}{m(\ell - \delta)} = 2.26\,\text{ft} = 27.1\,\text{in.},$$ (7.53)

where we have used $\delta = 2\,\text{in.}$, and for a typical bat used in Major League Baseball whose stated weight is 32 oz and whose length is 34 in., $\ell = 22.5\,\text{in.}$, $m = 0.0630\,\text{slug}$, and $I_G = 0.0413\,\text{slug·ft}^2$. As one can see in Prob. 7.13, the location of the center of percussion depends on the pivot point for the bat and is not an inherent property of the bat. ■

What if the moment center is on the rigid body?   Equation (7.43) is valid for any possible choice of moment center P. We now end the development of the rotational equations by deriving a version of Eq. (7.43) that is applicable when point P is a point *on* the rigid body or on an arbitrary extension of the rigid body (see the discussion on extended rigid body on p. 463).

Referring to Fig. 7.9, if points P and G are both on the rigid body, we can write the acceleration of G as

$$\vec{a}_G = \vec{a}_P + \vec{\alpha}_B \times \vec{r}_{G/P} - \omega_B^2 \vec{r}_{G/P}$$ (7.54)

and use the parallel axis theorem (see App. A) to write I_G as

$$I_G = I_P - mr_{G/P}^2,$$ (7.55)

where I_P is the mass moment of inertia of the body about an axis perpendicular to the plane of motion passing through point P. Substituting Eqs. (7.54) and (7.55) into Eq. (7.43), we obtain

$$\vec{M}_P = (I_P - mr_{G/P}^2)\vec{\alpha}_B + \vec{r}_{G/P} \times m(\vec{a}_P + \vec{\alpha}_B \times \vec{r}_{G/P} - \omega_B^2 \vec{r}_{G/P})$$ (7.56)

$$= (I_P - mr_{G/P}^2)\vec{\alpha}_B + \vec{r}_{G/P} \times m\vec{a}_P + m(\vec{r}_{G/P} \cdot \vec{r}_{G/P})\vec{\alpha}_B$$
$$\quad - m(\vec{r}_{G/P} \cdot \vec{\alpha}_B)\vec{r}_{G/P}$$ (7.57)

$$= (I_P - mr_{G/P}^2)\vec{\alpha}_B + \vec{r}_{G/P} \times m\vec{a}_P + mr_{G/P}^2\vec{\alpha}_B,$$ (7.58)

where we have used the vector identity $\vec{a} \times (\vec{b} \times \vec{c}) = (\vec{a} \cdot \vec{c})\vec{b} - (\vec{a} \cdot \vec{b})\vec{c}$, $\vec{r}_{G/P} \times \vec{r}_{G/P} = \vec{0}$, $\vec{r}_{G/P} \cdot \vec{r}_{G/P} = r_{G/P}^2$, and the fact that $\vec{r}_{G/P}$ is orthogonal to $\vec{\alpha}_B$ (so that $\vec{r}_{G/P} \cdot \vec{\alpha}_B = 0$). By making the final simplification by canceling the two terms involving $mr_{G/P}^2\vec{\alpha}_B$, Eq. (7.58) becomes

$$\boxed{\vec{M}_P = I_P\vec{\alpha}_B + \vec{r}_{G/P} \times m\vec{a}_P.}$$ (7.59)

Because of the assumptions in going from Eq. (7.43) to Eq. (7.59), Eq. (7.59) is subject to the restriction that point P *must be* a point on the rigid body B or the extended rigid body. Finally, we note that if *any* of the following is true:

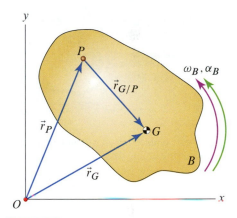

Figure 7.9
An arbitrary rigid body with the point P being a point *on* the rigid body.

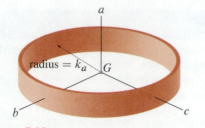

Figure 7.10

A thin ring whose mass is all a distance k_a from the a axis. The radius of gyration of this object with respect to the a axis would be k_a since its mass moment of inertia would be $I_a = mk_a^2$.

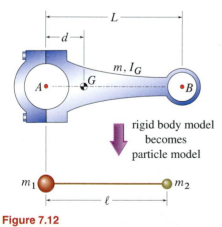

Figure 7.11

Photo of two connecting rods manufactured by Metaldyne (Plymouth, MI). Metaldyne uses sintering to make powder metal connecting rods for gasoline and diesel engines.

1. Point P is the mass center G so that $\vec{r}_{G/P} = \vec{0}$.

2. $\vec{a}_P = \vec{0}$ (i.e., P is fixed or moves with constant velocity).

3. $\vec{r}_{G/P}$ is parallel to $\vec{a}_P$.

then Eq. (7.59) becomes

$$M_P = I_P \alpha_B, \tag{7.60}$$

where we have used the scalar form to reflect the fact that the motion is planar.

Radius of gyration

In some handbooks and other references, the property of mass moment of inertia is described in terms of the *radius of gyration*. The radius of gyration k_a with respect to an axis a is defined in terms of the mass moment of inertia about that axis as

$$k_a = \sqrt{\frac{I_a}{m}}, \tag{7.61}$$

where m is the mass of the body in question. Therefore, if a body of mass m had all its mass concentrated at a distance k_a from the axis a, its mass moment of inertia would be I_a (see Fig. 7.10).*

Dynamically equivalent mass system

In mechanism and machine dynamics, engineers often want to replace a rigid body model of a component with some *equivalent* particle system. This concept of *dynamical equivalency* or *dynamically equivalent mass system* is important to many engineering disciplines. For two systems to be dynamically equivalent, they must have

1. The same total mass.

2. The same center of mass location.

3. The same mass moment of inertia I_G.

This is typically done by replacing the rigid body model with a model consisting of two particles connected by a massless rod. Let's apply this idea to the connecting rod shown in Fig. 7.11.

Referring to Fig. 7.12, the goal is to replace the connecting rod with mass m, mass moment of inertia I_G, and whose mass center is a distance d from the crank end (shown in the top part of the figure) with the two particles of mass m_1 and m_2, which are separated by a distance ℓ (shown in the bottom part of the figure). Since the three criteria listed above will generate three equations, we will only be able to solve for three unknowns; these will be the mass m_1, the mass m_2, and the distance ℓ between them. Choosing the location of m_1 to be at point A and writing these three equations, we get

$$m_1 + m_2 = m \qquad \text{(same mass)} \tag{7.62}$$

Figure 7.12

A rigid body model of a connecting rod being replaced by a dynamically equivalent two-particle model. For calculations, we use $d = 36.4$ mm, $m = 0.439$ kg, and $I_G = 0.00144$ kg·m². For comparison, $L = 141$ mm.

---

* The radius of gyration can also be interpreted in terms of the statistical measure known as *standard deviation*. In fact, the radius of gyration can be thought of as the "standard deviation of the mass distribution."

$$m_2 \ell = md \qquad \text{(same location of } G) \qquad (7.63)$$

$$m_1 d^2 + m_2 (\ell - d)^2 = I_G \qquad \text{(same } I_G). \qquad (7.64)$$

Solving these equations for m_1, m_2, and ℓ, we obtain

$$m_1 = \frac{m}{1 + md^2/I_G} = 0.313\,\text{kg}, \qquad (7.65)$$

$$m_2 = \frac{m}{1 + I_G/(md^2)} = 0.126\,\text{kg}, \qquad (7.66)$$

$$\ell = d + \frac{I_G}{md} = 127\,\text{mm}. \qquad (7.67)$$

Referring to Eq. (7.53), notice that the location of mass m_2 is the center of percussion for the connecting rod if the connecting rod were in fixed axis rotation about the crank (point A).

We should note that the way we calculated a dynamically equivalent system using Eqs. (7.62)–(7.64) is not the only way to do it. For example, some people place the two masses at the two connecting points of the link (points A and B for the connecting rod in Fig. 7.12).* Unfortunately, to achieve the same mass moment of inertia I_G, this sometimes requires that we attach an object with no mass and *negative mass moment of inertia* to the equivalent two-particle system (that would be the case with the connecting rod shown here).

Graphical interpretation of the equations of motion

Equation (7.43) has a graphical interpretation that can help us remember the equation and understand physically what it is saying. Unfortunately, this interpretation is also *very* easy to misapply and therefore we need to be careful. We present it here since it does provide a convenient way of applying and remembering Eq. (7.43).

We begin by referring to Fig. 7.13. The left side of the figure shows a rigid body that is being acted upon by a number of external forces and moments — it is just the FBD of the rigid body. The right side of the figure introduces a new diagram called a *kinetic diagram* (KD). The kinetic diagram *always* contains the $I_G \vec{\alpha}_B$ vector and the $m\vec{a}_G$ vector, which, although it has units of force, we will color green (as we do all accelerations) since these vectors originate from accelerations. Now, if we always draw the KD in this way, then by setting the FBD *equal to* the KD, we always recover Eqs. (7.15) and (7.43), which are the Newton-Euler equations for a rigid body. We can see this by noting that the left-hand side of Eq. (7.15) (i.e., $\vec{F}_R$) and the left-hand side of Eq. (7.43) (i.e., $\vec{M}_P$) are readily obtained from the FBD in Fig. 7.13. The right-hand side of Eq. (7.15) is simply $m\vec{a}_G$, which is what one obtains from the KD. The right-hand side of Eq. (7.43) comes from taking moments about P on the KD in Fig. 7.13. Thus, if we always draw the KD by including the $m\vec{a}_G$ vector (written in a convenient component system) and the $I_G \alpha_B$ vector and we equate forces and moments on the FBD with forces and moments on the KD, we will always end up with the correct Newton-Euler equations for that rigid body.

* See B. Paul, *Kinematics and Dynamics of Planar Machinery*, Prentice-Hall, Englewood Cliffs, N.J., 1979, pp. 439–442.

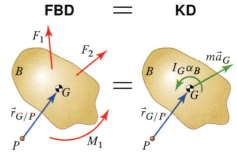

FBD $=$ **KD**

Figure 7.13
The free-body diagram and kinetic diagram of a general rigid body. Equating them and writing the associated equations always give the correct equations of motion, i.e., Eqs. (7.15) and (7.43).

Common Pitfall

The KD must be consistent with the kinematics. The positive directions of $I_G \alpha_B$ and $m\vec{a}_G$ on the KD must be consistent with the positive directions for α_B and $\vec{a}_G$ in the kinematic equations. If they are not, sign errors will end up polluting the problem solution.

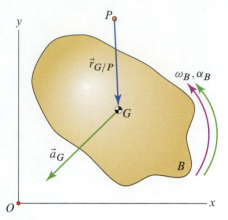

Figure 7.14
The relevant kinematic quantities for the Newton-Euler equations for a rigid body.

End of Section Summary

In this section, we have developed the Newton-Euler equations (equations of motion) for a rigid body. We began by showing that the translational equations are given by *Euler's first law*, which is

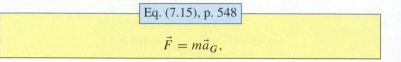

$$\vec{F} = m\vec{a}_G,$$

Eq. (7.15), p. 548

where $\vec{F}$ is the resultant of all *external* forces, m is the mass of the rigid body, and $\vec{a}_G$ is the inertial acceleration of its mass center (see Fig. 7.14).

Bodies symmetric with respect to the plane of motion.   For rigid bodies, we also need rotational equations of motion. Applying *Euler's second law*, i.e., the moment-angular momentum relationship for a system of particles, we were able to show that, *for a rigid body that is symmetric with respect to the plane of motion*, the most general form of the rotational equations of motion is given by (see Fig. 7.14)

Eqs. (7.42) and (7.43), p. 551

$$M_P = I_G \alpha_B + m\left(x_{G/P} a_{Gy} - y_{G/P} a_{Gx}\right),$$
$$\vec{M}_P = I_G \vec{\alpha}_B + \vec{r}_{G/P} \times m\vec{a}_G,$$

where

- M_P is the total moment about P in the z direction

- I_G is the mass moment of inertia of the body about its mass center G

- α_B is the angular acceleration of the body

- m is the total mass of the body

- $\vec{a}_G = a_{Gx}\,\hat{\imath} + a_{Gy}\,\hat{\jmath}$ is the acceleration of the mass center

- $\vec{r}_{G/P} = x_{G/P}\,\hat{\imath} + y_{G/P}\,\hat{\jmath}$ is the position of the mass center G relative to the moment center P

- The second equation is simply the vector form of the first

Now, in addition, if *any* one of the following conditions is true:

1. Point P is the mass center G, so that $\vec{r}_{G/P} = \vec{0}$.

2. $\vec{a}_G = \vec{0}$ (i.e., G moves with constant velocity).

3. $\vec{r}_{G/P}$ is parallel to $\vec{a}_G$.

then Eqs. (7.42) and (7.43) reduce to

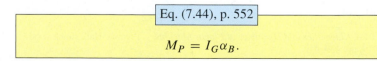

$$M_P = I_G \alpha_B.$$

Eq. (7.44), p. 552

Finally, if point P is on the rigid body or an arbitrary extension of the rigid body, then an alternate form of Eq. (7.43) is (see Fig. 7.15)

Eq. (7.59), p. 553

$$\vec{M}_P = I_P \vec{\alpha}_B + \vec{r}_{G/P} \times m\vec{a}_P,$$

where I_P is the mass moment of inertia of the body about an axis perpendicular to the plane of motion passing through point P and $\vec{a}_P$ is the acceleration of point P.

If *any* one of the following is true:

1. Point P is the mass center G so that $\vec{r}_{G/P} = \vec{0}$.

2. $\vec{a}_P = \vec{0}$ (i.e., P moves with constant velocity).

3. $\vec{r}_{G/P}$ is parallel to $\vec{a}_P$.

then Eq. (7.59) becomes

Eq. (7.60), p. 554

$$M_P = I_P \alpha_B.$$

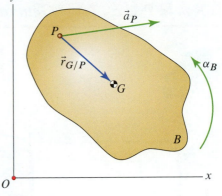

Figure 7.15
The relevant kinematic quantities for the rotational equations of motion of a rigid body when the moment center P is a point on the rigid body.

Graphical interpretation of the equations of motion.   There is a graphical/visual way of obtaining Eqs. (7.15) and (7.43). Referring to Fig. 7.16, we begin by drawing the FBD of the rigid body, including all forces and moments, and then we draw the KD (kinetic diagram) of the rigid body, which includes the vectors $I_G \alpha_B$ and $m\vec{a}_G$. As shown in Fig. 7.16, we graphically equate these two diagrams, and we write the equations generated by that equation. In doing so, we automatically obtain Eqs. (7.15) and (7.43).

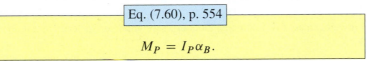

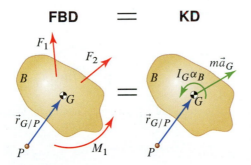

Figure 7.16. *Figure 7.13 repeated.* The free-body diagram and kinetic diagram of a general rigid body. Equating them and writing the associated equations always give the correct equations of motion, i.e., Eqs. (7.15) and (7.43).

EXAMPLE 7.1   *General Plane Motion: Analysis of a Falling Rigid Body*

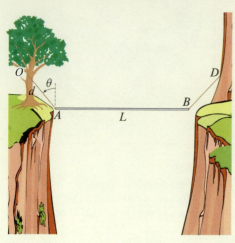

Figure 1

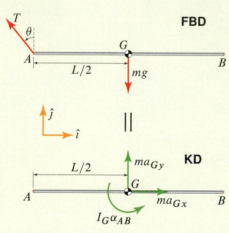

Figure 2
FBD (top) and KD (bottom) of the platform in Fig. 1.

As part of a movie stunt, a long, thin 388 lb platform whose length is $L = 39$ ft has been jury-rigged across a ravine using two ropes OA and BD. Rope OA of length $d = 13.4$ ft is securely tied to the tree, but rope BD has been tied to a carabiner at D that has not been adequately fastened to the rock face. After everything is set up as shown in Fig. 1, the carabiner at D breaks free and the platform starts to fall. Determine the angular acceleration of the platform and the tension in the rope OA immediately after the rope BD breaks free. The initial value of θ is $39°$.

SOLUTION

Road Map & Modeling   The FBD of the platform immediately after the carabiner breaks is shown in Fig. 2. We model the platform as a slender rod, and our key assumption is that, immediately after the rope BD breaks, all velocities are zero. In addition, we will neglect the mass of each rope and assume that they are inextensible. If the ropes are massless, then neither gravity nor acceleration will affect the behavior of the ropes. Therefore, assuming that rope OA is in tension, it will behave as a straight-line segment with constant length. That is, at the time instant considered, OA can be treated as if it were a massless rigid body. Of course, we will need to verify that it doesn't go slack.

Using this model, applying the Newton-Euler equations to the platform will allow us to determine the forces and accelerations once we have determined the kinematics of the platform immediately after rope BD breaks.

Governing Equations

Balance Principles   By equating the FBD and KD of the platform shown in Fig. 2 (this is equivalent to applying Eqs. (7.15) and (7.43) to the FBD in Fig. 2), the Newton-Euler equations for the platform are

$$\sum F_x: \qquad -T \sin \theta = ma_{Gx}, \tag{1}$$

$$\sum F_y: \qquad T \cos \theta - mg = ma_{Gy}, \tag{2}$$

$$\sum M_G: \qquad -\frac{L}{2} T \cos \theta = I_G \alpha_{AB}, \tag{3}$$

where T is the tension in the rope and I_G is the mass moment of inertia of the platform, which is given by

$$I_G = \tfrac{1}{12} m L^2. \tag{4}$$

Force Laws   All forces are accounted for on the FBD, although we must verify that $T > 0$ to make sure that the rope doesn't go slack. If the rope does go slack, we will solve the problem with the knowledge that $T = 0$.

Kinematic Equations   We can see from Eqs. (1)–(3) that we need to relate $\vec{a}_G$ to the angular acceleration of the platform, subject to the constraint that point A moves in a circle about O and all velocities are zero immediately after release. Relating A to O, we get

$$\vec{a}_A = \vec{a}_O + \vec{\alpha}_{OA} \times \vec{r}_{A/O} - \omega_{OA}^2 \vec{r}_{A/O}$$
$$= \alpha_{OA} \hat{k} \times d(\sin \theta \, \hat{\imath} - \cos \theta \, \hat{\jmath}) = d\alpha_{OA} \cos \theta \, \hat{\imath} + d\alpha_{OA} \sin \theta \, \hat{\jmath}, \tag{5}$$

Helpful Information

Equation (5) and our rope model. Consistent with our model of the ropes, Eq. (5) implies that we have viewed the rope OA as a rigid body.

since $\vec{a}_O = \vec{0}$ and all velocities are zero. Relating G to A, we get

$$\vec{a}_G = \vec{a}_A + \vec{\alpha}_{AB} \times \vec{r}_{G/A} - \omega_{AB}^2 \vec{r}_{G/A}$$
$$= d\alpha_{OA}(\cos\theta\,\hat{\imath} + \sin\theta\,\hat{\jmath}) + \alpha_{AB}\,\hat{k} \times (L/2)\hat{\imath}$$
$$= d\alpha_{OA}\cos\theta\,\hat{\imath} + (d\alpha_{OA}\sin\theta + L\alpha_{AB}/2)\,\hat{\jmath}, \tag{6}$$

where we have used the expression for $\vec{a}_A$ from Eq. (5) and set all velocities to zero.

Computation   We can substitute Eqs. (4) and (6) into Eqs. (1)–(3) to obtain the following three equations in the three unknowns T, α_{AB}, and α_{OA}:

$$-T\sin\theta = md\alpha_{OA}\cos\theta, \tag{7}$$

$$T\cos\theta - mg = m\left(d\alpha_{OA}\sin\theta + \frac{L}{2}\alpha_{AB}\right), \tag{8}$$

$$-\frac{L}{2}T\cos\theta = \tfrac{1}{12}mL^2\alpha_{AB}. \tag{9}$$

When solved, these give

$$T = \frac{2mg\cos\theta}{5 + 3\cos(2\theta)} = 107\,\text{lb}, \tag{10}$$

$$\alpha_{AB} = -\frac{12g\cos^2\theta}{L[5 + 3\cos(2\theta)]} = -1.06\,\text{rad/s}^2, \tag{11}$$

$$\alpha_{OA} = -\frac{2g\sin\theta}{d[5 + 3\cos(2\theta)]} = -0.538\,\text{rad/s}^2, \tag{12}$$

where we have used all known values to get the numerical answers.

Discussion & Verification

- The dimensions and the units of the final results in Eqs. (10)–(12) are all correct.

- The initial angular accelerations of both the rope OA and the bar AB are negative, as expected, since they should both initially rotate clockwise.

- It is difficult to have a sense of the magnitude of the tension in the rope OA, but it should certainly be positive, which it is. The fact that $T > 0$ confirms that our use of Eq. (5) was correct. In Eq. (5) we treated the rope as a massless rigid body, and this is acceptable only as long as the rope does not go slack.

- Referring to Fig. 2, right after the carabiner fails, we expect point G to accelerate downward, i.e., $a_{Gy} < 0$. Since from Eq. (2) we have $T = m(g + a_{Gy})/\cos\theta$, the expectation that $a_{Gy} < 0$ implies the expectation that $T < mg/\cos\theta = 499.3\,\text{lb}$. The result in Eq. (10) is consistent with this expectation.

EXAMPLE 7.2   *Rotation About a Fixed Axis: Moments About a Fixed Point*

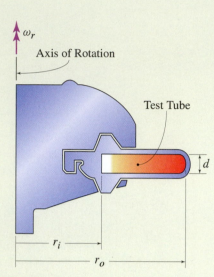

Figure 2

Cross-section of one-half of the centrifuge rotor shown in Fig. 1(b). r_i = 63.1 mm, r_o = 120.5 mm, d = 11.0 mm, and ω_r = 60,000 rpm. Recall that the double-headed arrow for the angular velocity indicates its direction via the right-hand rule.

Centrifuges like the one shown in Fig. 1(a) can generate accelerations exceeding 1 million g. With the *swinging-bucket* rotor shown in Fig. 1(b), the centrifuge can spin at 60,000 rpm and achieve an acceleration of $485,000g$ at the ends of the buckets. As the rotor spins up, the buckets that hang from the bottom of the rotor swing up and eventually assume the horizontal position shown in Fig. 2.

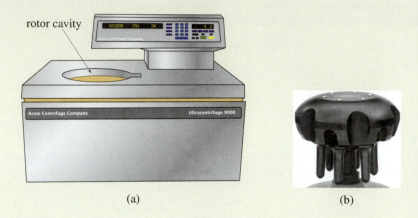

(a)   (b)

Figure 1. (a) A tabletop ultracentrifuge. (b) Swinging-bucket centrifuge rotor that holds six sample tubes.

(a) Determine the radial force and the moment parallel to the axis of rotation required to hold the test tube in place when the rotor is spinning at its maximum rated speed of 60,000 rpm.

(b) What are the implications of these loads on the rotor bearings?

Assume that the test tube and its contents (e.g., blood) can be modeled as a uniform circular cylinder with mass 10 g and ignore gravity.

SOLUTION

Road Map & Modeling   The forces shown on the FBD of the test tube in Fig. 3 form the *equivalent force-couple system* to the system of forces that are actually acting on the tube. The FBD should also show a force in the z direction, R_z, as well as moments in both the x and y directions, M_{Ox} and M_{Oy}, respectively. Summing forces in the z direction would simply tell us that $R_z = mg$. Since the body is symmetric with respect to the plane of motion, it follows that $M_{Ox} = M_{Oy} = 0$, and so those moment equations become part of a statics problem. In this case, we are only interested in R_x and M_t (the subscripts t stands for tube), so we have chosen the FBD given in Fig. 3. We will approximate the test tube as a uniform circular cylinder, thus ignoring any motion of the fluid within the tube and the nonuniform shape and mass distribution of the tube. Since we *know* the motion of the test tube, we can find all the velocities and accelerations needed to use Eq. (7.15) as well as all the moment equations we have developed. Therefore, the forces and moments will fall right out of the Newton-Euler equations once the kinematics is done.

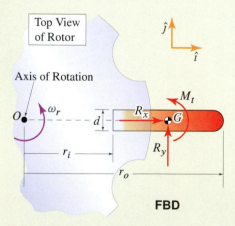

FBD

Figure 3

Top view of the FBD of the test tube in Fig. 2. The mass center of the test tube is at G.

Governing Equations

Balance Principles   Equating the FBD and KD of the test tube shown in Figs. 3 and 4, respectively (this is equivalent to applying Eqs. (7.15) and (7.43) to the FBD in Fig. 3), the Newton-Euler equations for the test tube are

$$\sum F_x: \qquad\qquad R_x = ma_{Gx}, \qquad\qquad (1)$$

$$\sum F_y: \qquad\qquad R_y = ma_{Gy}, \qquad\qquad (2)$$

$$\sum M_O: \quad M_t + R_y\left(\frac{r_o + r_i}{2}\right) = I_G\alpha_r + m[x_{G/O}a_{Gy} - y_{G/O}a_{Gx}], \quad (3)$$

where I_G is the mass moment of inertia of the test tube and a_{Gx} and a_{Gy} are the x and y components of the acceleration of the mass center G, respectively. Modeling the test tube as a uniform circular cylinder, we calculate its mass moment of inertia as

$$I_G = \tfrac{1}{12}m\left(3r^2 + h^2\right) = \tfrac{1}{12}(0.01\,\text{kg})\left[3(0.0055\,\text{m})^2 + (0.0574\,\text{m})^2\right]$$

$$= 2.821 \times 10^{-6}\,\text{kg} \cdot \text{m}^2, \qquad\qquad (4)$$

where we have used $m = 0.01\,\text{kg}$, $r = d/2 = 0.0055\,\text{m}$, and $h = r_o - r_i = 0.0574\,\text{m}$.

Force Laws   All forces are accounted for on the FBD.

Kinematic Equations   As for the accelerations on the right-hand side of Eqs. (1)–(3), since we know the motion of the rotor, these are readily found by relating $\vec{a}_G$ to $\vec{a}_O$, where point O is on an arbitrary rigid body extension of the test tube. Doing this, we obtain

$$\vec{a}_G = \vec{a}_O + \vec{\alpha}_r \times \vec{r}_{G/O} - \omega_r^2\vec{r}_{G/O}, \qquad\qquad (5)$$

in which we note that $\vec{a}_O = \vec{0}$ since it is on the axis of rotation and $\vec{\alpha}_r = \vec{0}$ since the rotor has reached its constant final speed. Substituting in $\omega_r = 60,000\,\text{rpm} = 6283\,\text{rad/s}$ and $\vec{r}_{G/O} = (r_o + r_i)/2\,\hat{i} = 0.0918\,\text{m}\,\hat{i}$, we have

$$\vec{a}_G = -(6283\,\text{rad/s})^2(0.0918\,\text{m}\,\hat{i}) = (-3.624 \times 10^6\,\text{m/s}^2)\,\hat{i}. \qquad (6)$$

Finally, note that we computed $x_{G/O}$ and $y_{G/O}$ when we found $\vec{r}_{G/O}$, that is

$$x_{G/O} = 0.0918\,\text{m} \quad \text{and} \quad y_{G/O} = 0.0\,\text{m}. \qquad\qquad (7)$$

Computation   Substituting Eqs. (4), (6), (7), and $\alpha_r = 0$ into Eqs. (1)–(3), we obtain

$$R_x = -(0.01\,\text{kg})(3.624 \times 10^6\,\text{m/s}^2) = -36,200\,\text{N}, \qquad (8)$$

$$R_y = 0\,\text{N}, \qquad\qquad (9)$$

$$M_t + R_y\left(\frac{0.1205\,\text{m} + 0.0631\,\text{m}}{2}\right) = 0 \quad \Rightarrow \quad M_t = 0\,\text{N} \cdot \text{m}. \qquad (10)$$

where we have substituted $R_y = 0$ from Eq. (9) to obtain the final result in Eq. (10).

Discussion & Verification   The force needed to keep the test tube in place is 36,200 N (8147 lb), even though the test tube only has a mass of 10 g (an average paperclip has a mass of about 1 g). Note that an equal and opposite force is acting on the rotor bearing and that force is rotating around the bearing 60,000 times per minute or 1000 times per second! Therefore, not only is the rotor subject to failure due to a huge load imbalance, but also it is subject to failure due to fatigue loading (see marginal note). This means that balancing the rotor is essential to safely operate the centrifuge.

  🔎 **A Closer Look**   Instead of using Eq. (3), we could have applied any of Eqs. (7.44), (7.59), or (7.60) to get the moment equation for the test tube. This will almost always be the case; that is, we will usually have more than one choice as to which moment equation we apply. We will explore this more in the examples and exercises that follow.

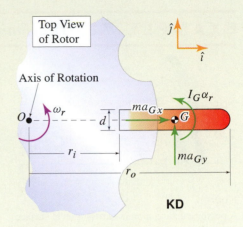

Figure 4
Top view of the KD of the test tube in Fig. 2. Equating this KD to the FBD in Fig. 3, we obtain the test tube's Newton-Euler equations.

Interesting Fact

Cyclic loading and fatigue. The fact that, under the given conditions, the rotor bearing experiences a cyclic load 1000 times per second means that it will quickly experience a large number of load cycles. It turns out that even a rather low stress can cause an object to break after millions of load cycles. The higher the stress, the smaller the number of cycles required. This mechanism of failure is called *fatigue*. Since the number of load cycles on the rotor bearing of a centrifuge grows quickly, even a small imbalance can cause failure due to fatigue. To learn more about fatigue, see W. D. Callister, Jr., *Materials Science and Engineering: An Introduction*, 7th ed., John Wiley & Sons, 2006.

EXAMPLE 7.3 *Translation: Slipping Versus Tipping with Friction*

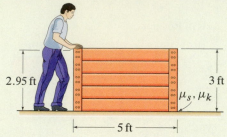

Figure 1
A person pushing a crate across a level surface.

A uniform flat crate is pushed with a constant horizontal force of 95 lb across a rough surface (Fig. 1). The force is applied 2.95 ft above the floor, and the crate is 5 ft long, 3 ft high, and weighs 120 lb. The coefficients of static and kinetic friction between the crate and the surface are $\mu_s = 0.4$ and $\mu_k = 0.35$, respectively. Verify that the crate slips and does not tip, and determine its acceleration.

SOLUTION

Road Map & Modeling   The FBD of the crate is shown in Fig. 2, where P is the pushing force applied at a height d above the floor, F is the friction force, and N is the equivalent normal force on the crate due to the ground. In addition, w, h, and mg are the crate's width, height, and weight, respectively. We know how hard and where the person is pushing on the crate, and the other forces acting on the crate will be determined from the governing equations. We are told to verify that the crate does not tip, so we will first solve the problem by assuming just that. During the verification, we will discuss what to look for if the crate were to tip. In Fig. 2 the normal force N has been placed at an arbitrary location on the bottom of the crate. We will determine its exact location by determining ℓ.*

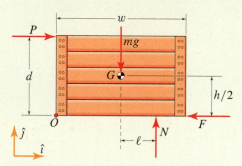

Figure 2
FBD of the crate shown in Fig. 1. The mass center is at G.

Governing Equations

Balance Principles   Applying Eqs. (7.15) and (7.44), the Newton-Euler equations for the FBD in Fig. 2 are

$$\sum F_x: \qquad\qquad P - F = ma_{Gx}, \qquad (1)$$

$$\sum F_y: \qquad\qquad N - mg = ma_{Gy}, \qquad (2)$$

$$\sum M_G: \quad N\ell - P(d - h/2) - Fh/2 = I_G\alpha_c, \qquad (3)$$

where a_{Gx} and a_{Gy} are the x and y components of the acceleration of the mass center G, respectively, α_c is the crate's angular acceleration, and I_G is the crate's mass moment of inertia, which is given by

$$I_G = \tfrac{1}{12}m(w^2 + h^2). \qquad (4)$$

Force Laws   The friction force can be related to N using the Coulomb law for sliding friction, which is

$$F = \mu_k N. \qquad (5)$$

Kinematic Equations   Letting ω_c denote the crate's angular velocity, we can relate the acceleration of G to O using $\vec{a}_G = \vec{a}_O + \vec{\alpha}_c \times \vec{r}_{G/O} - \omega_c^2 \vec{r}_{G/O}$, and since we are assuming that the crate slips and does not tip, we have $\vec{a}_O$ is only in the x direction. Therefore, we can write

$$\omega_c = 0, \quad \alpha_c = 0, \quad \text{and} \quad a_{Gy} = 0. \qquad (6)$$

Computation   Plugging Eqs. (4)–(6) into Eqs. (1)–(3), we obtain the following three equations for the unknowns N, a_{Gx}, and ℓ:

$$P - \mu_k N = ma_{Gx}, \qquad (7)$$

$$N - mg = 0, \qquad (8)$$

---

* To review this idea, see Sections 5.2 and 9.1 of M. E. Plesha, G. L. Gray, and F. Costanzo, *Engineering Mechanics: Statics*, McGraw-Hill Publishing, Chicago, 2010.

$$N\ell - P(d - h/2) - h\mu_k N/2 = 0. \tag{9}$$

Solving Eqs. (7)–(9), we obtain

$$N = mg = 120\,\text{lb}, \tag{10}$$

$$a_{Gx} = P/m - \mu_k g = 14.2\,\text{ft/s}^2, \tag{11}$$

$$\ell = \tfrac{1}{2}h\mu_k + \frac{P}{mg}(d - h/2) = 1.67\,\text{ft}, \tag{12}$$

where we have used the given data to obtain the numerical results.

   We have the acceleration of the crate, but we need to verify that it doesn't tip. The key is that the normal force needs to be located *within* the crate; that is, it can't be located outside the right or left edges of the crate. The idea behind this criterion is that we have assumed that the crate *does not* tip, and so our solution must be compatible with that assumption. A normal force outside the boundaries of the crate would mean that a wider base (all other parameters being the same) would be required to prevent tipping. With this said, since $\ell \le w/2$ (i.e., 1.67 ft $\le$ 2.5 ft), the crate does not tip.

Discussion & Verification

- The dimensions of the solutions in Eqs. (10)–(12) are all as they should be.

- It is reasonable that a crate with the given dimensions would not tip under the given circumstances.

- Although it is hard to know whether $0.44g$ is reasonable for a_{Gx}, it is certainly true that its direction is as expected.

A Closer Look

- The plot in Fig. 3 shows how ℓ (i.e., the distance of N to the right of the center-line of the crate) varies as P (the applied load) is increased from 0 to 200 lb. As the statics solution tells us, before P reaches $\mu_s N$, the crate does not even move and so we get the *dark red* preslip curve. Once the crate starts to slip, there is a small sudden drop in ℓ since there is a small sudden drop in the friction force as it goes from $\mu_s N$ to $\mu_k N$, and we get the *dark green* postslip curve. The *red* lines indicate those values of ℓ and P at which the normal force reaches the edge of the crate; at this point, the crate would also start to tip.

- We assumed that the crate slipped, but did not tip. We were then able to verify that assumption. It is important to realize that we can assume any motion we like and that the correctness of our assumption can always be verified using our solution. For example, under the given conditions, if we assumed that the crate tips and slips, we would discover that an impossible motion is obtained and we would then be able to rule out that possibility. We will explore this possibility and others in the exercises.

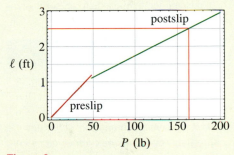

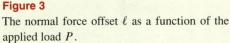

Figure 3
The normal force offset ℓ as a function of the applied load P.

EXAMPLE 7.4   *General Plane Motion: Rolling Without Slip*

Figure 1
Mazda Miata © 2006 Mazda Motor of America, Inc. Used by permission.

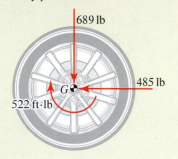

Figure 2
The loads applied by the axle to one of the rear wheels of the roadster shown in Fig. 1.

> **Interesting Fact**
>
> **Where do the forces shown in Fig. 2 come from?** In Prob. 7.50, one can find the forces on the rear wheels due to the axle by first performing an analysis of the entire car to get the friction and normal forces on the rear wheels, and then isolating one of the rear wheels and analyzing it.

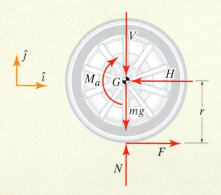

Figure 3
The FBD of one of the rear wheels of the car shown in Fig. 1.

When accelerating from 0 to 60 mph, the rear-wheel drive roadster shown in Fig. 1 has the loads shown in Fig. 2 applied to *each* of the two rear wheels from the rear axle of the 2570 lb car. Given that each wheel weighs 47 lb, has a mass moment of inertia I_G of 0.989 slug·ft^2, and has a diameter of 24.3 in., determine

(a) The normal and frictional forces between the wheel and the ground

(b) The minimum coefficient of static friction required for the wheel to roll without slip

(c) The time it takes for the car to reach 60 mph

Assume that the car accelerates uniformly while moving over a flat and level surface.

SOLUTION

Road Map & Modeling   The FBD of the wheel is shown in Fig. 3, where, from Fig. 2, the vertical force is V, the horizontal force is H, and the moment is M_a. Since we know all the loads causing the wheel to move and the inertia properties of the wheel, we should be able to determine how the wheel moves via the solution of the Newton-Euler equations we will write. We will use the normal and friction forces that we find to determine the minimum friction coefficient required for rolling without slip.

Governing Equations

Balance Principles   Based on the FBD in Fig. 3, the Newton-Euler equations are

$$\sum F_x: \qquad F - H = ma_{Gx}, \tag{1}$$

$$\sum F_y: \quad N - V - mg = ma_{Gy}, \tag{2}$$

$$\sum M_G: \qquad Fr - M_a = I_G\alpha_w, \tag{3}$$

where a_{Gx} and a_{Gy} are the x and y components of the acceleration of the mass center G, the friction force acting at the bottom of the wheel is F, the normal force between the ground and the wheel is N, the radius of the wheel is $r = (24.3/2)$ in. $= 1.012$ ft, and α_w is the angular acceleration of the wheel. Also, the inertia properties of the wheel are its mass m and its mass moment of inertia I_G, which in this case are

$$m = \frac{47\,\text{lb}}{32.2\,\text{ft/s}^2} = 1.460\,\text{slug} \quad \text{and} \quad I_G = 0.9890\,\text{slug·ft}^2. \tag{4}$$

Force Laws   The inequality that must be satisfied for the wheel to roll without slip is

$$|F| \le \mu_s|N|, \tag{5}$$

where μ_s is the coefficient of static friction between the wheel and the ground.

Kinematic Equations   Since the car is on a flat surface, the center of the wheel cannot undergo any vertical motion, and since we are assuming that the wheel is rolling without slip, we have the following two kinematic constraints

$$a_{Gy} = 0 \quad \text{and} \quad a_{Gx} = -r\alpha_w, \tag{6}$$

where the minus sign comes from the fact that α_w has been assumed to be positive in the positive z direction.

Computation   Substituting Eqs. (6) into Eqs. (1)–(3), we obtain the following three equations:

$$F - H = -mr\alpha_w, \tag{7}$$

$$N - V - mg = 0, \tag{8}$$

$$Fr - M_a = I_G\alpha_w, \tag{9}$$

for the three unknowns N, F, and α_w. Solving, we obtain

$$N = V + mg = 736\,\text{lb}, \tag{10}$$

$$F = \frac{I_G H + mrM_a}{I_G + mr^2} = 504\,\text{lb}, \tag{11}$$

$$\alpha_w = \frac{rH - M_a}{I_G + mr^2} = -12.6\,\text{rad/s}^2, \tag{12}$$

where we have used the given parameters to obtain the final numerical answers. Now that we know F and N, we can find the minimum value of μ_s that is compatible with the no slip assumption by simply using the equality in Eq. (5), that is,

$$\mu_s \geq \left|\frac{F}{N}\right| \quad \Rightarrow \quad \boxed{(\mu_s)_{\text{min}} = 0.685.} \tag{13}$$

Finally, to determine the time it takes for the car to reach 60 mph, we first find a_{Gx} from the second of Eqs. (6) as

$$a_{Gx} = -(1.012\,\text{ft})(-12.6\,\text{rad/s}^2) = 12.75\,\text{ft/s}^2, \tag{14}$$

and then we apply Eq. (2.41) on p. 59 since we are assuming the acceleration is uniform, that is,

$$v = v_0 + a_{Gx}t \quad \Rightarrow \quad 88\,\text{ft/s} = (12.75\,\text{ft/s}^2)t \quad \Rightarrow \quad \boxed{t = 6.90\,\text{s.}} \tag{15}$$

Discussion & Verification

- The dimensions of each of the results in Eqs. (10)–(12) are correct.

- The value of static friction found in Eq. (13) is reasonable for a tire on asphalt.

- The "0 to 60" time we found in Eq. (15) is consistent with times found in the product literature for a roadster like the one analyzed here.

A Closer Look   We are given all the nonconstraint forces acting on a wheel, which allows us to then determine the motion of the wheel and whether it slips. In this case, since one of the things we were looking for was the minimum μ_s for rolling without slip, we could assume no slip and then find the μ_s compatible with that assumption. Had we not been told whether the wheel slips, then we could have assumed no slip and verified that assumption in the usual way by comparing the needed friction with the available friction (we would need to be given the static friction coefficient). If we discovered that the wheel slips, then Eq. (5) would become $F = \mu_k N$ and the second of Eqs. (6) would no longer be valid.

EXAMPLE 7.5 *General Plane Motion: Example 3.6 Revisited*

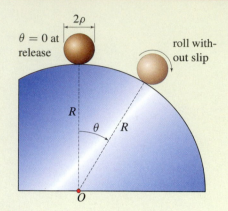

$\theta = 0$ at release

2ρ

roll without slip

R

θ R

O

Figure 1

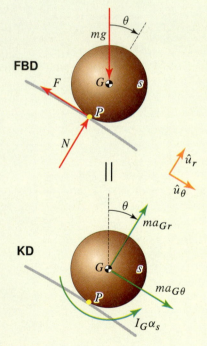

mg θ

FBD

F G s

P

N

$||$

$\hat{u}_r$

$\hat{u}_\theta$

θ ma_{Gr}

KD

G s

P $ma_{G\theta}$

$I_G\alpha_s$

Figure 2
FBD of the sphere s and the polar component system drawn at an arbitrary angle θ.

In Example 3.6, we released a small sphere from the top of a semicylinder and, by modeling it as a particle, we determined that it separated from the semicylinder at $\theta = 48.2°$ (see Fig. 1). Here we wish to determine the value of θ at which the small sphere separates if we treat it as a uniform sphere of radius ρ and mass m. We release the sphere from the top of the semicylinder by giving it a *slight* nudge to the right, and we assume that there is sufficient friction between the semicylinder and the sphere for the sphere to roll without slip.

SOLUTION

Road Map & Modeling   As with Example 3.6, the key is to find the normal force between the sphere and the semicylinder as a function of θ and then say that the sphere separates at the location where this force becomes zero.

The FBD of the sphere as it slides down the semicylinder is shown in Fig. 2. The FBD has been drawn at an arbitrary angle θ since we need to find that angle at which N becomes zero and so we need to find N for *any* θ. Since the motion of the mass center of the sphere is along a circular path until it separates from the surface, we will use polar coordinates for the solution. Note that the friction force F has been drawn in the indicated direction since the sphere is passively rolling down the semicylinder, i.e., it is not driven.

Governing Equations

Balance Principles   Equating the FBD and KD of the sphere shown in Fig. 2 (this is equivalent to applying Eqs. (7.15) and (7.43) to the FBD in Fig. 2), the Newton-Euler equations for the sphere are

$$\sum F_r: \qquad N - mg\cos\theta = ma_{Gr}, \tag{1}$$

$$\sum F_\theta: \qquad -F + mg\sin\theta = ma_{G\theta}, \tag{2}$$

$$\sum M_P: \qquad mg\rho\sin\theta = I_G\alpha_s + \rho ma_{G\theta}, \tag{3}$$

where m is the mass of the sphere, F is the friction force between the sphere and semicylinder, and the mass moment of inertia I_G of the sphere is

$$I_G = \tfrac{2}{5}m\rho^2. \tag{4}$$

Force Laws   To ensure that the sphere rolls without slip, we must have

$$|F| \le \mu_s|N|. \tag{5}$$

Kinematic Equations   If we want N as a function of θ, then the accelerations need to be expressed as functions of θ. We begin by writing $\vec{a}_G$ using polar coordinates as

$$\vec{a}_G = -\dot{\theta}^2(R+\rho)\,\hat{u}_r + \ddot{\theta}(R+\rho)\,\hat{u}_\theta, \tag{6}$$

since G is moving in a circle centered at O. Equation (6) implies that

$$a_{Gr} = -\dot{\theta}^2(R+\rho), \tag{7}$$

$$a_{G\theta} = \ddot{\theta}(R+\rho). \tag{8}$$

Now that we have a_{Gr} and $a_{G\theta}$ as a function of θ, we need $\alpha_s(\theta)$. We can get this by finding $\omega_s(\theta)$, the sphere's angular velocity, and then differentiating with respect to time. Relating $\vec{v}_G$ to $\vec{v}_P$, we obtain

$$\vec{v}_G = \vec{v}_P + \vec{\omega}_s \times \vec{r}_{G/P} \quad \Rightarrow \quad \dot{\theta}(R+\rho)\,\hat{u}_\theta = \vec{v}_P + \vec{\omega}_s \times \vec{r}_{G/P}, \tag{9}$$

where v_G has been written using polar coordinates. Noting that $\vec{v}_P = \vec{0}$ and using components, Eq. (9) becomes

$$\dot{\theta}(R + \rho)\,\hat{u}_\theta = \omega_s\,\hat{u}_z \times \rho\,\hat{u}_r \quad \Rightarrow \quad \omega_s = \left(\frac{R + \rho}{\rho}\right)\dot{\theta}, \tag{10}$$

where we have used $\hat{u}_z \times \hat{u}_r = \hat{u}_\theta$. Differentiating Eq. (10), we obtain

$$\alpha_s = \left(\frac{R + \rho}{\rho}\right)\ddot{\theta}. \tag{11}$$

Computation We get the equations of motion for the sphere by substituting Eqs. (4), (7), (8), and (11) into Eqs. (1)–(3), which gives

$$N - mg\cos\theta = -m\dot{\theta}^2(R + \rho), \tag{12}$$

$$-F + mg\sin\theta = m\ddot{\theta}(R + \rho), \tag{13}$$

$$g\sin\theta = \tfrac{7}{5}(R + \rho)\ddot{\theta}, \tag{14}$$

which are three equations to solve for N, F, and θ (a differential equation must be solved to get θ). However, all we really want is $N(\theta)$. To get $N(\theta)$, we can see from Eq. (12) that we will need to get $\dot{\theta}$ as a function of θ—we can do this by using the chain rule, i.e., $\ddot{\theta} = \dot{\theta}\,d\dot{\theta}/d\theta$, and then integrating Eq. (14) as follows:

$$\int_0^{\dot{\theta}} \dot{\theta}\,d\dot{\theta} = \frac{5g}{7(R + \rho)} \int_0^{\theta} \sin\theta\,d\theta \quad \Rightarrow \quad \dot{\theta}^2 = \frac{10g}{7(R + \rho)}(1 - \cos\theta). \tag{15}$$

Substituting Eqs. (15) into Eq. (12), N as a function of θ is

$$N = \tfrac{1}{7}mg(17\cos\theta - 10). \tag{16}$$

Therefore, calling θ_{sep} the separation angle, the sphere separates from the surface, i.e., N becomes zero, when

$$17\cos\theta_{\text{sep}} - 10 = 0 \quad \Rightarrow \quad \theta_{\text{sep}} = \pm 54.0° + n360°, \; n = 0, \pm 1, \ldots, \pm\infty. \tag{17}$$

Since we are only interested in $0° \le \theta \le 90°$, the only acceptable answer is

$$\boxed{\theta_{\text{sep}} = 54.0°.} \tag{18}$$

Discussion & Verification When we compare this example with Example 3.6, we see that when rotatory inertia plays a role in the dynamics, as it does in this example, the object separates from the surface almost 6° farther down the cylinder, independent of R, ρ, g, and m. Given that this result is "in the same ballpark" as the 48.2° separation angle for a particle, it helps build some confidence that the result for a sphere is correct.

🔎 **A Closer Look** In Example 3.6, we treated the object sliding down the semicylinder as a particle. A finite-sized sphere would act as a particle if the contact interface were frictionless. Therefore, we should be able to recover the result of Example 3.6 if, in this example, (1) we let the friction go to zero and (2) we account for the fact that the mass center is $R + \rho$ from the center of the semicylinder at O. We will see this in Prob. 7.36.

Common Pitfall

Can we *really* satisfy Eq. (5)? We stated at the beginning that there is sufficient friction between the surface and the sphere such that the sphere will not slip on the surface. Is this possible? It isn't and let's quickly see why. To obtain $F(\theta)$, we can solve Eq. (14) for $\ddot{\theta}$, substitute the result into Eq. (13), and then solve for F. Now $N(\theta)$ is found in Eq. (16). Taking the ratio of the two as given by Eq. (5), we obtain

$$\mu_s = \left|\frac{F}{N}\right| = \left|\frac{2\sin\theta}{17\cos\theta - 10}\right|.$$

Notice that as the sphere rolls down the semicylinder and θ approaches the separation position, the denominator $17\cos\theta - 10$ goes to zero (see Eq. (17)) and so μ_s goes to ∞. This actually tells us that *it isn't possible for the sphere to roll without slipping until it separates from the semicylinder since infinite friction would be required.*

Helpful Information

Why does the sphere go 6° farther than the particle? The answer lies in the speed of the objects as they fall down the cylinder. We haven't covered the work-energy principle for rigid bodies, but we know that in conservative systems, a decrease in potential energy leads to a corresponding increase in kinetic energy. As the particle and sphere move down the cylinder, for a given height change, they each experience the same increase in kinetic energy. For the particle, all the kinetic energy goes into its speed. For the sphere, some goes into its translational speed, but some also goes into the energy associated with its rotation. Either the sphere or the particle separates from the cylinder when they are moving fast enough so that their v^2/ρ acceleration overcomes the normal component of mg. It takes the sphere a little longer to get up to that speed since some of its energy goes into rotation, so it separates at a larger angle.

EXAMPLE 7.6   *General Plane Motion: A System with Multiple Rigid Bodies*

Figure 1

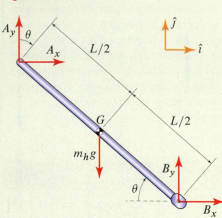

Figure 2
FBD of the handle of the lawn roller shown in Fig. 1.

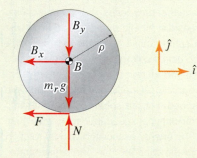

Figure 3
FBD of the roller of the lawn roller shown in Fig. 1.

A man starts pushing the lawn roller shown in Fig. 1 such that the angle θ remains at a constant $40°$ and the center of the lawn roller at B accelerates to the right at a constant $0.45\,\text{m/s}^2$. Given that the mass of the roller is $100\,\text{kg}$, the mass of the handle is $4\,\text{kg}$, $\rho = 25\,\text{cm}$, and $L = 1.1\,\text{m}$, determine the force at A that the man must apply to the handle to achieve this motion and the minimum necessary coefficient of static friction between the roller and ground if the roller is to roll without slip. Treat the roller as a uniform circular cylinder and the handle as a thin rod.

SOLUTION

Road Map & Modeling   The FBD of the handle is shown in Fig. 2, and the FBD of the roller is shown in Fig. 3. We have let m_r be the mass of the roller, m_h be the mass of the handle, and F and N be the friction and normal forces, respectively, between the roller and the ground. Since we are treating the bar and roller as uniform, we have placed their mass centers at their geometric centers. These FBDs tell us that this is a problem in which we must analyze *two* rigid bodies. To do so, we will write a set of Newton-Euler equations for each, which will result in a set of *six* equations. Since the kinematics are entirely known, the unknowns will then turn out to be six forces in the system.

Governing Equations

Balance Principles   The Newton-Euler equations corresponding to the FBD of the handle in Fig. 2 are

$$\sum F_x: \qquad\qquad\qquad\qquad A_x + B_x = m_h a_{Gx}, \quad (1)$$

$$\sum F_y: \qquad\qquad\qquad A_y + B_y - m_h g = m_h a_{Gy}, \quad (2)$$

$$\sum M_G: \quad B_x \frac{L}{2}\sin\theta + B_y \frac{L}{2}\cos\theta - A_x \frac{L}{2}\sin\theta - A_y \frac{L}{2}\cos\theta = I_G \alpha_{AB}, \quad (3)$$

where $I_G = \frac{1}{12}m_h L^2$. The Newton-Euler equations corresponding to the FBD of the roller in Fig. 3 are

$$\sum F_x: \qquad -B_x - F = m_r a_{Bx}, \quad (4)$$

$$\sum F_y: \quad N - B_y - m_r g = m_r a_{By}, \quad (5)$$

$$\sum M_B: \qquad -F\rho = I_B \alpha_r, \quad (6)$$

where $I_B = \frac{1}{2}m_r \rho^2$ and α_r is the angular acceleration of the roller.

Force Laws   The force law for this system is the friction inequality that must be satisfied for the roller to roll without slip, that is,

$$|F| \leq \mu_s |N|. \quad (7)$$

All other forces are accounted for on the FBD.

Kinematic Equations   Kinematically, we know θ is a constant, and so the bar AB is in pure translation. This implies that

$$\omega_{AB} = \alpha_{AB} = 0 \quad \Rightarrow \quad \vec{a}_G = \vec{a}_B. \quad (8)$$

In addition, since the roller is rolling without slip over a flat surface, we can say that a_{Bx} is known and is equal to the given acceleration of $0.45\,\text{m/s}^2$ and that

$$a_{By} = 0 \quad \text{and} \quad \alpha_r = -\frac{a_{Bx}}{\rho}. \quad (9)$$

Computation Substituting Eqs. (7)–(9) into Eqs. (1)–(6), we obtain the six equations

$$A_x + B_x = m_h a_{Bx}, \tag{10}$$

$$A_y + B_y - m_h g = 0, \tag{11}$$

$$\frac{L}{2}\left[(B_x - A_x)\sin\theta + (B_y - A_y)\cos\theta\right] = 0, \tag{12}$$

$$-B_x - F = m_r a_{Bx}, \tag{13}$$

$$N - B_y - m_r g = 0, \tag{14}$$

$$F\rho = \tfrac{1}{2} m_r \rho a_{Bx}, \tag{15}$$

which we can solve for the six unknowns A_x, A_y, B_x, B_y, F, and N. Solving this system of equations, we obtain

$$A_x = \left(m_h + \tfrac{3}{2} m_r\right) a_{Bx} = 69.3\,\text{N}, \tag{16}$$

$$A_y = \tfrac{1}{2}\left[m_h g - (m_h + 3m_r) a_{Bx} \tan\theta\right] = -37.8\,\text{N}, \tag{17}$$

$$B_x = -\tfrac{3}{2} m_r a_{Bx} = -67.5\,\text{N}, \tag{18}$$

$$B_y = \tfrac{1}{2}\left[m_h g + (m_h + 3m_r) a_{Bx} \tan\theta\right] = 77.0\,\text{N}, \tag{19}$$

$$F = \tfrac{1}{2} m_r a_{Bx} = 22.5\,\text{N}, \tag{20}$$

$$N = \tfrac{1}{2}\left[(m_h + 2m_r) g + (m_h + 3m_r) a_{Bx} \tan\theta\right] = 1060\,\text{N}, \tag{21}$$

so that the force that must be applied at A is given by

$$\boxed{\vec{A} = (69.3\,\hat{\imath} - 37.8\,\hat{\jmath})\,\text{N},} \tag{22}$$

or $|\vec{A}| = \sqrt{A_x^2 + A_y^2} = 78.9\,\text{N}$ at the angle shown in Fig. 4.

Now that we have the friction and normal forces, we can use the friction inequality in Eq. (7) to determine how much friction is needed to ensure that the roller rolls without slipping, that is,

$$\boxed{\mu_s \geq \left|\frac{F}{N}\right| = 0.0213.} \tag{23}$$

Discussion & Verification The dimension of each final result in Eqs. (16)–(21) is correct, and the magnitude of the required force at A is reasonable. Notice that Eq. (23) tells us that not much friction is needed for the roller to roll without slip. This is so because of the (considerable) weight of the roller in relation to the small value of acceleration that is being imparted to the roller by the person pushing it.

A Closer Look The force the man must apply to the handle of the lawn roller is given either by Eqs. (16) and (17) or by Eq. (22). Notice that the force that must be applied at A is not parallel to the handle. The reason is that the handle has mass—if we were to let m_h be zero in our model, we would find that the angle in Eq. (22) would be $-40°$ and the force at A would be directed along the handle.

If we now compute the magnitude of the force at B, we obtain

$$|\vec{B}| = \sqrt{B_x^2 + B_y^2} = 102\,\text{N}, \tag{24}$$

where we have also shown this force on the handle at B in Fig. 4. Notice that even though the bar is pin-connected at each end, it is not a two-force member. This is so because of two distinct reasons. The first is that the weight of the bar has not been neglected, and the second is that the center of mass of the bar is accelerating.

Helpful Information

Solving six simultaneous equations. While these equations are not difficult to solve by hand, it is certainly tedious. A system of equations like this is the perfect "excuse" to learn a mathematical software package such as Mathematica, Maple, MATHCAD, or MATLAB.

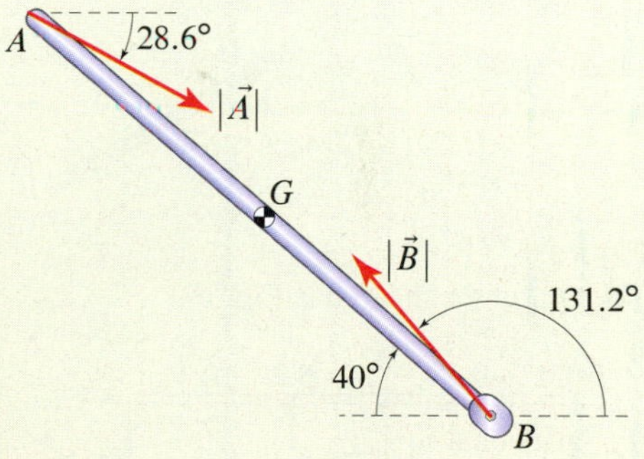

Figure 4
The forces at the ends of the lawn roller handle.

EXAMPLE 7.7 *General Plane Motion: Derivation of Equations of Motion*

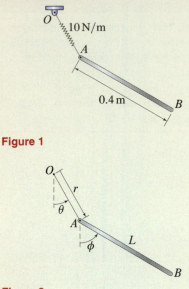

Figure 1

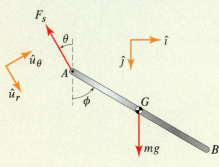

Figure 2
Definition of the coordinates r, θ, and ϕ that will be used to define the position of the rod.

Figure 3
The FBD of the thin rod in Fig. 1.

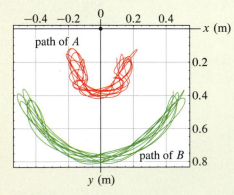

Figure 4
Paths of ends A and B for the first set of initial conditions.

If one places a thin rod at the end of a spring or an elastic band, suspends the elastic band from the ceiling, and lets the rod swing, the motion looks very complicated. Write the equations of motion for a system like this, and then study the motion of the rod for two different sets of initial conditions using computer simulations. Referring to Fig. 1, use a rod 0.4 m long with a mass of 0.1 kg. Assume the spring is linear elastic with constant 10 N/m and with an unstretched length of 0.2 m. In the first simulation, use $\theta(0) = 30°$, $\phi(0) = 60°$, $r(0) = 0.2$ m, and $\dot{\theta}(0) = \dot{\phi}(0) = \dot{r}(0) = 0$; and in the second use the same conditions except let $r(0) = 0.4$ m.

SOLUTION

Road Map & Modeling   The coordinates r, θ, and ϕ that we will use to define the position of the rod are shown in Fig. 2, where r is the length of the spring, L is the length of the rod, and θ and ϕ define the angle of the spring and of the rod with respect to the vertical, respectively. If we ignore the mass of the spring, the hanging rod has three degrees of freedom (we could have also used two coordinates to locate point A and then one to give the orientation of the bar). Therefore, we will need to derive three equations of motion. We will obtain these by writing the Newton-Euler equations for the rod using the FBD shown in Fig. 3, where F_s is the force on the rod due to the spring. Notice that we are using two coordinate systems in the FBD — a global Cartesian system and a polar coordinate system aligned with the spring (and thus the spring force) that we will use to describe the motion of point A.

Governing Equations

Balance Principles   The Newton-Euler equations corresponding to the FBD in Fig. 3 are given by

$$\sum F_x: \qquad -F_s \sin\theta = ma_{Gx}, \qquad (1)$$

$$\sum F_y: \qquad mg - F_s \cos\theta = ma_{Gy}, \qquad (2)$$

$$\sum M_G: \qquad \vec{r}_{A/G} \times \vec{F}_s = I_G \alpha_{AB}\,\hat{k}, \qquad (3)$$

where $I_G = \frac{1}{12}mL^2$,

$$\vec{r}_{A/G} = \tfrac{L}{2}(-\sin\phi\,\hat{\imath} - \cos\phi\,\hat{\jmath}) \quad \text{and} \quad \vec{F}_s = F_s(-\sin\theta\,\hat{\imath} - \cos\theta\,\hat{\jmath}), \qquad (4)$$

and α_{AB} is the angular acceleration of the rod. Substituting Eqs. (4) into Eq. (3), Eq. (3) becomes

$$\tfrac{1}{2}LF_s \sin(\phi - \theta) = I_G \alpha_{AB}, \qquad (5)$$

where we have used the trigonometric identity $\sin\phi\cos\theta - \sin\theta\cos\phi = \sin(\phi - \theta)$.

Force Laws   The one force that has not been accounted for in Fig. 3 is that of the linear elastic spring, which is

$$F_s = k(r - r_0), \qquad (6)$$

where r_0 is the unstretched length of the spring.

Kinematic Equations   Since we are using r, θ, and ϕ as the three coordinates to define the position of the rod, we need to write $\vec{a}_G$ and α_{AB} in terms of those coordinates and their derivatives. We can do this by relating $\vec{a}_G$ to $\vec{a}_A$ as follows:

$$\vec{a}_G = \vec{a}_A + \vec{\alpha}_{AB} \times \vec{r}_{G/A} - \omega_{AB}^2 \vec{r}_{G/A}, \qquad (7)$$

where ω_{AB} is the angular velocity of the rod and we can write

$$\vec{\alpha}_{AB} = -\ddot{\phi}\,\hat{k}, \quad \omega_{AB} = -\dot{\phi}, \quad \text{and} \quad \vec{r}_{G/A} = -\vec{r}_{A/G} = \frac{L}{2}(\sin\phi\,\hat{\imath} + \cos\phi\,\hat{\jmath}). \quad (8)$$

Using the polar coordinate system defined in Fig. 3, we can write $\vec{a}_A$ as

$$\vec{a}_A = (\ddot{r} - r\dot{\theta}^2)\,\hat{u}_r + (r\ddot{\theta} + 2\dot{r}\dot{\theta})\,\hat{u}_\theta, \quad (9)$$

in which

$$\hat{u}_r = \sin\theta\,\hat{\imath} + \cos\theta\,\hat{\jmath} \quad \text{and} \quad \hat{u}_\theta = \cos\theta\,\hat{\imath} - \sin\theta\,\hat{\jmath}. \quad (10)$$

Substituting Eqs. (8)–(10) into Eq. (7), the components of $\vec{a}_G$ become

$$a_{Gx} = (\ddot{r} - r\dot{\theta}^2)\sin\theta + (r\ddot{\theta} + 2\dot{r}\dot{\theta})\cos\theta + \frac{L}{2}\ddot{\phi}\cos\phi - \frac{L}{2}\dot{\phi}^2\sin\phi, \quad (11)$$

$$a_{Gy} = (\ddot{r} - r\dot{\theta}^2)\cos\theta - (r\ddot{\theta} + 2\dot{r}\dot{\theta})\sin\theta - \frac{L}{2}\ddot{\phi}\sin\phi - \frac{L}{2}\dot{\phi}^2\cos\phi. \quad (12)$$

Computation   Now that we have assembled all the pieces, the equations of motion are obtained by substituting Eqs. (6), (8), (11), and (12) into Eqs. (1), (2), and (5) to obtain the three equations of motion:

$$(\ddot{r} - r\dot{\theta}^2)\sin\theta + (r\ddot{\theta} + 2\dot{r}\dot{\theta})\cos\theta + \frac{L}{2}\ddot{\phi}\cos\phi - \frac{L}{2}\dot{\phi}^2\sin\phi$$
$$+ \frac{k}{m}(r - r_0)\sin\theta = 0, \quad (13)$$

$$(\ddot{r} - r\dot{\theta}^2)\cos\theta - (r\ddot{\theta} + 2\dot{r}\dot{\theta})\sin\theta - \frac{L}{2}\ddot{\phi}\sin\phi - \frac{L}{2}\dot{\phi}^2\cos\phi$$
$$+ \frac{k}{m}(r - r_0)\cos\theta = g, \quad (14)$$

$$\frac{L}{6}\ddot{\phi} + \frac{k}{m}(r - r_0)\sin(\phi - \theta) = 0. \quad (15)$$

💻 ➡ Computer simulations of the motion of the rod are shown in Figs. 4–7. ⬅ 💻

Discussion & Verification

- Each term in Eqs. (13) and (14) has been divided by m, so each term should have the units of acceleration, which it does.

- Each term in Eq. (15) has been divided by mL, so each term should have the units of acceleration, which it does.

🔎 **A Closer Look**   Figures 4 and 5 show the trajectories of the ends of bar for the first 10 s after release for the first and second sets of initial conditions, respectively. The first set of initial conditions releases the bar when the spring is unstretched, and the second set has the bar stretched to twice its unstretched length at release (everything else is equal). In the second case, adding that additional initial energy to the system dramatically changes the ensuing motion; that is, it goes from a fairly regular pattern to one in which the bar is moving very irregularly.

Figures 6 and 7 are stroboscopic images of the movement of the thin rod for the first 10 s of motion. In each case, the earliest image is the one in light gray (labeled "initial"), and each successive image becomes a darker purple (with the last one labeled "final"). The time between successive images is 0.5 s. These figures demonstrate the regular motion associated with the first set of initial conditions and the irregular motion associated with the second set. Systems, such as this one, whose motion is described by a set of nonlinear differential equations can be *very* sensitive to how the system is put in motion. This is one of the subjects of *chaos theory*.*

---

* See, for example, S. H. Strogatz, *Nonlinear Dynamics and Chaos: With Applications to Physics, Biology, Chemistry and Engineering*, Perseus Books, Reading, Mass., 1994.

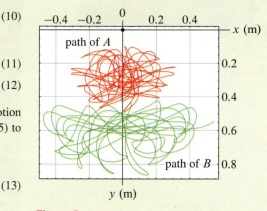

Figure 5
Paths of ends A and B for the second set of initial conditions.

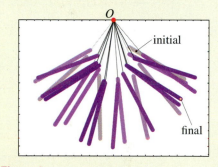

Figure 6
Stroboscopic image sequence of the rod for the first set of initial conditions.

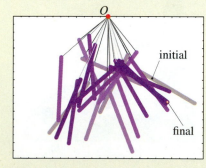

Figure 7
Stroboscopic image sequence of the rod for the second set of initial conditions.

PROBLEMS

Figure P7.1

Translation Problems

Problem 7.1

The roadster weighs 2750 lb, its mass is evenly distributed between its front and rear wheels, and it accelerates from 0 to 60 mph in 7.0 s. If the acceleration is uniform and if the rear wheels do not slip, determine the forces on each of the front and rear wheels due to the pavement. Also determine the minimum coefficient of static friction compatible with this motion. Assume that the mass is evenly distributed between the right and left sides of the car, neglect the rotational inertia of the front wheels, and assume that the front wheels roll freely.

Problem 7.2

The conveyor is moving the cans at a constant speed $v_0 = 18$ ft/s when, to proceed to the next step in packaging, the cans are transferred onto a stationary surface at A. If each can weighs 0.95 lb, $w = 2.71$ in., $h = 5$ in., and $\mu_k = 0.3$ between the cans and the stationary surface, determine the time and distance it takes for each can to stop. In addition, show that the cans don't tip. Treat each can as a uniform circular cylinder.

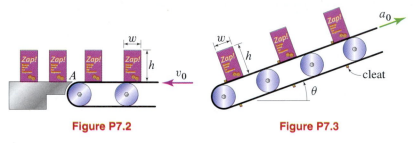

Figure P7.2   **Figure P7.3**

Problem 7.3

Determine the maximum acceleration a_0 of the conveyor so that the cans do not tip over the cleats. The cleats completely prevent slipping, but are not tall enough to dynamically influence tipping. Treat each can as a uniform circular cylinder of mass m.

Problem 7.4

A person is pushing a lawn mower of mass $m = 38$ kg and with $h = 0.75$ m, $d = 0.25$ m, $\ell_A = 0.28$ m, and $\ell_B = 0.36$ m. Assuming that the force exerted on the lawn mower by the person is completely horizontal, the mass center of the lawn mower is at G, and neglecting the rotational inertia of the wheels, determine the minimum value of this force that causes the rear wheels (labeled A) to lift off the ground. In addition, determine the corresponding acceleration of the mower.

Problem 7.5

Suppose that the mower shown is self-propelled; i.e., that the rear wheels cause the mower to move forward due to the friction between them and the ground. If a person were to apply a purely horizontal force, would this force help or hinder the rear wheels' contribution to the forward motion of the mower? That is, would the rear wheels slip less easily or more easily?

Note: Concept problems are about *explanations*, not computations.

Figure P7.4 and P7.5

Problems 7.6 and 7.7

The uniform slender bar AB has a weight $W_{AB} = 150\,\text{lb}$ while the crate's weight is $W_C = 500\,\text{lb}$. The bar AB is rigidly attached to the cage containing the crate. Neglect the mass of the cage, and assume that the mass of the crate is uniformly distributed. Furthermore, let $L = 8.5\,\text{ft}$, $d = 2.5\,\text{ft}$, $h = 4\,\text{ft}$, and $w = 6\,\text{ft}$.

Problem 7.6   If the trolley is accelerating with $a_0 = 11\,\text{ft/s}^2$, determine θ so that the bar-crate system translates with the trolley.

Problem 7.7   If the bar-crate system is translating with the trolley so that $\theta = 26°$, determine the acceleration a_0 of the trolley.

Problem 7.8

A conveyor belt must accelerate the cans from rest to $v = 18\,\text{ft/s}$ as quickly as possible. Treating each can as a uniform circular cylinder weighing 1.1 lb, find the minimum possible time to reach v so that the cans do not tip or slip on the conveyer. Assume that the acceleration is uniform and use $w = 2.71\,\text{in.}$, $h = 5\,\text{in.}$, and $\mu_s = 0.5$.

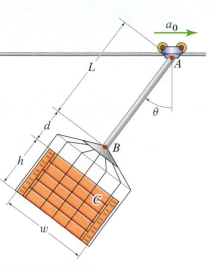

Figure P7.6 and P7.7

Problem 7.9

The uniform slender bar AB, with mass $m_{AB} = 75\,\text{kg}$ and length $L = 4.5\,\text{m}$, is pin-connected at A to a trolley accelerating with $a_0 = 3\,\text{m/s}^2$ along a horizontal rail. A crate with uniformly distributed mass $m_C = 250\,\text{kg}$, height $h = 1.5\,\text{m}$, and width $w = 2\,\text{m}$ is contained in a cage with negligible mass that is pin-connected to AB at B. The distance between B and the top of the crate us $d = 0.75\,\text{m}$. Determine the angles ϕ and θ so that the bar and the crate translate with the trolley.

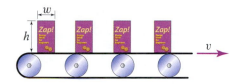

Figure P7.8

Problem 7.10

The 3300 lb front-wheel-drive car whose mass center is at A is pulling a 4300 lb trailer whose mass center is at B. The car and trailer start from rest and accelerate uniformly to 60 mph in 18 s. Determine the forces on all tires as well as the total force acting on the car due to the trailer. In addition, determine the friction required so that the wheels of the car do not slip. Assume that the car and trailer are laterally symmetric and that the rotational inertia of the wheels is negligible. Note that the mass center of the trailer is directly above the axle of the rear wheel.

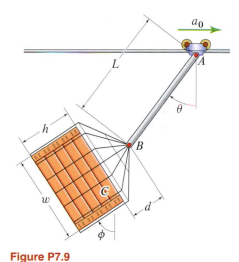

Figure P7.9

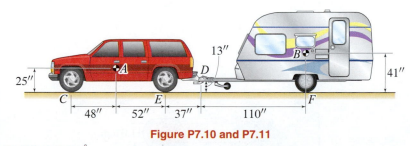

Figure P7.10 and P7.11

Problem 7.11

The 3300 lb front-wheel drive car, which is pulling a 4300 lb trailer, is traveling 60 mph and applies its brakes to come to a stop. Assuming that all four wheels of the car assist in the braking and that $\mu_s = 0.85$, determine the minimum possible stopping distance and find the forces on all tires as well as the total force acting on the car due to the trailer. Assume that the car and trailer are laterally symmetric. Note that the mass center of the trailer is directly above the axle of the rear wheel.

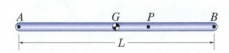

Figure P7.12

Figure P7.13

Rotation About a Fixed Axis

Problem 7.12

For the uniform thin bar of mass m and length L that is pinned at each end, determine the dynamically equivalent mass system, using the three requirements for a dynamically equivalent system given on p. 554.

(a) Place the point mass m_1 at A and determine the size of m_1 as well as the size and location P of the point mass m_2.

(b) Place the point mass m_1 at A and the point mass m_2 at B. Determine the size of m_1, the size of m_2, and the additional mass moment of inertia $(I_G)_{\text{extra}}$ needed. *Hint:* you will need to attach an object with no mass and a negative mass moment of inertia $(I_G)_{\text{extra}}$ to the dynamically equivalent system.

Problem 7.13

Following up on the center of percussion mini-example on p. 552, now assume that the pivot point O is close to the center of mass of the batter so that it is a distance $\delta = 3$ in. from the knob of the bat at A. Determine the location of the center of percussion P relative to the knob at A and show that the position of P is independent of the location of point C at which the batter grips the bat. Recall that P is the point at which the ball should be hit so that, no matter how large the force applied at P to the bat by the ball, the lateral force (i.e., perpendicular to the bat) felt by the batter at the grip C is zero. Assume the bat has mass m, its mass center is at G, and I_G is its mass moment of inertia. Evaluate your answer for a typical bat used in Major League Baseball whose weight is 32 oz and whose length is 34 in., $\ell = 22.5$ in., $m = 0.0630$ slug, and $I_G = 0.0413$ slug·ft². Ignore the weight force on the bat.

Problem 7.14

For the centrifuge rotor and test tube given in Example 7.2 on p. 560, assume that all the test tubes are locked into their horizontal position and that the rotor is uniformly accelerated from rest to 60,000 rpm in 9.5 min. Determine, as a function of time, the forces and moments on one of the test tubes during this spin-up phase of motion. Assume that each test tube and its contents can be modeled as a uniform circular cylinder with a mass of 10 g, and ignore gravity.

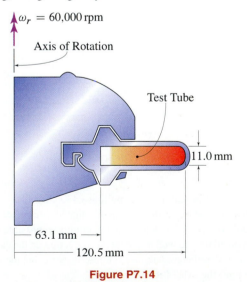

Figure P7.14

Problem 7.15

The driveway gate is hinged at its right end and is to be pushed open with a force P. As the gate is pushed open, where should the force P be applied (i.e., where should A be located) so that the force acting on the hinge due to the gate always acts along a line parallel to the gate during the entire time the gate is opening? Neglect the weight force acting on the gate, and model the gate as a uniform thin bar as shown below the photo. **Note:** Concept problems are about *explanations*, not computations.

Figure P7.15–P7.17

Problems 7.16 and 7.17

The driveway gate is hinged at its right end and can swing freely in the horizontal plane. The gate is pushed open by the force P that always acts perpendicular to the plane of the gate at point A, which is a horizontal distance d from the gate hinge. The weight of the gate is $W = 215\,\text{lb}$, and its mass center is at G, which is a distance $w/2$ from each end of the gate, where $w = 16\,\text{ft}$. Assume that the gate is initially at rest and model the gate as a uniform thin bar as shown below the photo.

Problem 7.16   Given that a force of $P = 20\,\text{lb}$ is applied at the center of mass of the gate (i.e., $d = w/2$), determine the reactions at the hinge O after the force P has been continuously applied for 2 s.

Problem 7.17   Given that a force of $P = 20\,\text{lb}$ is applied at the center of percussion of the gate, determine the reactions at the hinge O after the force P has been continuously applied for 2 s.

Problem 7.18

The uniform thin bar of length L and mass m is released from rest in the horizontal position shown. Determine the distance d at which the pin should be located from the end of the bar so that it has the maximum possible angular acceleration α_{max}. In addition, determine the value of α_{max}.

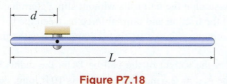

Figure P7.18

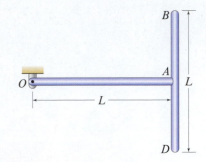

Figure P7.19 and P7.20

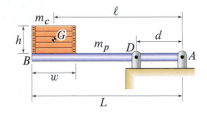

Figure P7.21

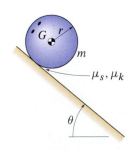

Figure P7.23 and P7.24

Problems 7.19 and 7.20

The T bar consists of two thin rods, OA and BD, each of length $L = 1.5\,\text{m}$ and mass $m = 12\,\text{kg}$, that are connected to the frictionless pin at O. The rods are welded together at A and lie in the vertical plane.

Problem 7.19 If the rods are released from rest in the position shown, determine the force on the pin at O as well as the angular acceleration of the rods immediately after release.

Problem 7.20 If, at the instant shown, the system is rotating clockwise with angular velocity $\omega_0 = 7\,\text{rad/s}$, determine the force on the pin at O as well as the angular acceleration of the rods.

Problem 7.21

The uniform thin platform AB of length L and mass m_p is pinned both at A and at D. A uniform crate of height h, width w, and mass m_c is placed at the end of the platform a distance ℓ from the pin at A. The system is at rest when the pin at A breaks. Determine the angular acceleration of the platform and crate, as well as the force on the platform due to the pin at D, immediately after the pin at A breaks. Assume that the crate and the platform do not separate immediately after the pin fails.

General Plane Motion Problems

Problem 7.22

The sphere, cylinder, and thin ring each have mass m and radius r. Each is released from rest on identical inclines. Assuming they all roll without slipping, which will have the largest initial angular acceleration? In addition, which will reach the bottom of the incline first?
Note: Concept problems are about *explanations*, not computations.

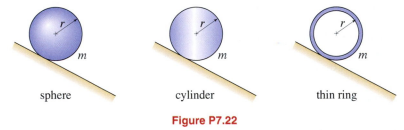

sphere   cylinder   thin ring

Figure P7.22

Problems 7.23 and 7.24

A bowling ball of radius r, mass m, and radius of gyration k_G is released from rest on a rough surface that is inclined at the angle θ with respect to the horizontal. The coefficients of static and kinetic friction between the ball and the incline are μ_s and μ_k, respectively. Assume that the mass center G is at the geometric center.

Problem 7.23 Assuming the ball rolls without slip, determine the angular acceleration of the ball and the friction and normal force between the ball and the incline. In addition, find the minimum value of μ_s that is compatible with this motion.

Problem 7.24 Let the weight of the ball be $14\,\text{lb}$, the radius be $4.25\,\text{in.}$, and the radius of gyration be $k_G = 2.6\,\text{in.}$ If the incline is $10\,\text{ft}$ long, determine the time it takes the ball to reach the bottom of the incline and the speed of G when it reaches the bottom. Use $\theta = 40°$, $\mu_s = 0.2$, and $\mu_k = 0.15$.

Problems 7.25 and 7.26

A bowling ball is thrown onto a lane with a backspin ω_0 and forward velocity v_0. The mass of the ball is m, its radius is r, its radius of gyration is k_G, and the coefficient of kinetic friction between the ball and the lane is μ_k. Assume the mass center G is at the geometric center.

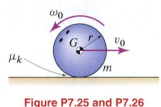

Figure P7.25 and P7.26

Problem 7.25 ⚠️ Find the acceleration of G and the ball's angular acceleration while the ball is slipping.

Problem 7.26 ⚠️ For a 14 lb ball with $r = 4.25$ in., $k_G = 2.6$ in, $\omega_0 = 10\,\text{rad/s}$, and $v_0 = 17\,\text{mph}$, determine the time it takes for the ball to start rolling without slip and its speed when it does so. In addition, determine the distance it travels before it starts rolling without slip. Use $\mu_k = 0.10$.

Problems 7.27 and 7.28

A bowling ball is thrown onto a lane with a forward spin ω_0 and forward velocity v_0. The mass of the ball is m, its radius is r, its radius of gyration is given by k_G, and the coefficient of kinetic friction between the ball and the lane is μ_k. Assume the mass center G is at the geometric center.

Problem 7.27 ⚠️ Assuming that $v_0 > r\omega_0$, determine the acceleration of the center of the ball and the angular acceleration of the ball until it starts rolling without slip.

Problem 7.28 ⚠️ Assuming that $v_0 < r\omega_0$, determine the acceleration of the center of the ball and the angular acceleration of the ball until it starts rolling without slip.

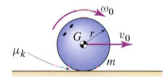

Figure P7.27 and P7.28

Problem 7.29 ⚠️

Solve Example 7.3 on p. 562 by assuming that the crate slips *and* tips. In doing so, show that this motion is not possible for the given conditions since part of your solution will not be physically admissible.

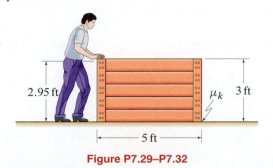

Figure P7.29–P7.32

Figure P7.30 and P7.31

Problems 7.30 and 7.31

A shop sign, with a uniformly distributed mass $m = 30\,\text{kg}$, $h = 1.5\,\text{m}$, $w = 2\,\text{m}$, and $d = 0.6\,\text{m}$, is at rest when cord AB suddenly breaks.

Problem 7.30 🌡 Modeling AB and CD as inextensible and with negligible mass, determine the tension in cord CD and the acceleration of the sign's center of mass immediately after AB breaks.

Problem 7.31 🌡 Modeling AB and CD as elastic cords with negligible mass and stiffness $k = 8000\,\text{N/m}$, determine the tension in cord CD and the acceleration of the sign's center of mass immediately after AB breaks.

Problem 7.32 🌡

Solve Example 7.3 on p. 562 by assuming that the crate *just* tips. In doing so, show that this motion is not possible for the given conditions since part of your solution will not be physically admissible.

Problem 7.33 🌡

Referring to the system in Example 7.5 on p. 566 (and conveniently ignoring the Pitfall on p. 567 so that we can assume that the object rolls without slip), how would the results change if one were to release a uniform cylinder, of mass m and radius ρ, instead of a sphere?

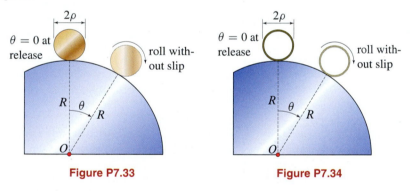

Figure P7.33   **Figure P7.34**

Problem 7.34 🌡

Referring to the system in Example 7.5 on p. 566 (and conveniently ignoring the Pitfall on p. 567 so that we can assume that the object rolls without slip), how would the results change if one were to release a uniform thin ring, of mass m and radius ρ, instead of a sphere?

Problem 7.35 🌡

Refer to the systems in Example 3.6 on p. 216 (particle separating from semicylinder) and Example 7.5 on p. 566 (sphere separating from semicylinder).

(a) Determine the speed of the particle and that of the sphere when each separates from the semicylinder.

(b) Compare their speeds of separation and explain the sources of any difference.

(c) Determine the value of ρ such that the sphere and the particle separate at the same speed.

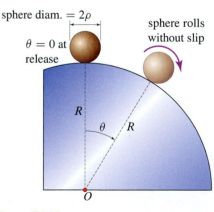

Figure P7.35

Problem 7.36 🌡

Referring to the systems in Example 3.6 on p. 216 and Example 7.5 on p. 566, show that the sphere dynamically behaves just as a particle if the interface between the sphere and the semicylinder is frictionless. In this case, that will mean that the sphere separates from the semicylinder at the same location as the particle.

Problem 7.37 🌡

The cord, which is wrapped around the inner radius of the spool of mass m, is pulled vertically at A by a constant force P, causing the spool to roll over the horizontal bar BD. Assuming that the cord is inextensible and of negligible mass, that the spool rolls without slip, and that its radius of gyration is k_G, determine the angular acceleration of the spool and the total force between the spool and the bar.

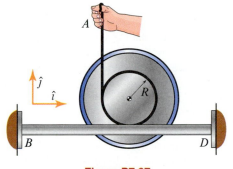

Figure P7.37

Problem 7.38 🌡

In Prob. 7.1 you were told to neglect the rotational inertia of the front wheels — would including it really make a difference? Let's see.

A certain roadster can go from 0 to 60 mph in 7.0 s, the weight of the car (including the two front wheels) is 2750 lb, the weight of each of its front wheels is 47 lb, and they each have a mass moment of inertia I_G of 0.989 slug·ft². To determine the effect of the rotational inertia of the front wheels, perform the following analysis:

(a) Isolate one of the front wheels and determine the friction force that must be acting on the wheel for it to accelerate as given. *Hint:* The weight of the car on the front wheel is not known, but it is not needed to find the friction force since we are assuming that friction is sufficient to prevent slipping of the front wheels.

(b) Next, note that it is the friction force that makes the rotational motion of each front wheel possible. In addition, note that if the mass moment of inertia I_G of the front wheels were zero, then the friction force would be zero. Therefore, by neglecting the rotational inertia of the front wheels, the car would not be slowed by the friction forces found in (a). In other words, when we *do* account for the rotational inertia of the front wheels, we can then conclude that there is a force equal to twice the friction force that is "retarding" the motion of the car. Use this fact, and your result from (a), to determine the 0 to 60 mph time of this same roadster with front wheels that have no rotational inertia.

24.3 in.

Figure P7.38

Problems 7.39 through 7.42

The uniform thin rod AB couples the slider A, which moves along a frictionless guide, to the wheel B, which rolls without slip over a horizontal surface.

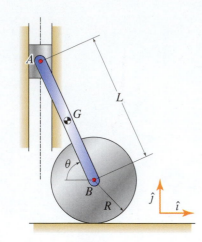

Figure P7.39–P7.42

Problem 7.39 Assuming that A and B have negligible mass, that the mass of AB is m_{AB}, and that the system is released from rest at the angle θ, determine, immediately after release, the angular acceleration of the rod AB, the acceleration of the center of the wheel at B, and the angular acceleration of the wheel.

Problem 7.40 Assuming that A has negligible mass, B is a uniform disk of mass m_B, the mass of AB is m_{AB}, and the system is released from rest at the angle θ, determine, immediately after release, the angular acceleration of the rod AB, the acceleration of the center of the wheel at B, and the angular acceleration of the wheel.

Problem 7.41 Assuming that AB has negligible mass, the mass of A is m_A, B is a uniform disk of mass m_B, and the system is released from rest at the angle θ, determine, immediately after release, the angular acceleration of the rod AB, the acceleration of the center of the wheel at B, and the angular acceleration of the wheel.

Problem 7.42 Assuming that the mass of A is m_A, B is a uniform disk of mass m_B, the mass of AB is m_{AB}, and that the system is released from rest at the angle θ, determine, immediately after release, the angular acceleration of the rod AB, the acceleration of the center of the wheel at B, and the angular acceleration of the wheel.

Problem 7.43

A spool of mass $m = 220\,\text{kg}$, inner and outer radii $\rho = 1.75\,\text{m}$ and $R = 2.25\,\text{m}$, respectively, and radius of gyration $k_G = 1.9\,\text{m}$, is being lowered down an incline with $\theta = 29°$. If the static and kinetic friction coefficients between the incline and the spool are $\mu_s = 0.4$ and $\mu_k = 0.35$, respectively, determine the acceleration of G, the angular acceleration of the spool, and the tension in the cable.

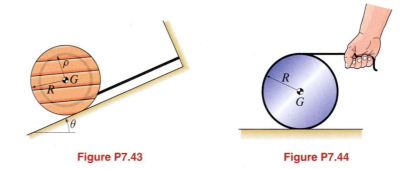

Figure P7.43   **Figure P7.44**

Problem 7.44

An inextensible cord of negligible mass is wound around a homogeneous circular object. Assume that the cord is pulled to the right while remaining horizontal, and determine the value of the object's mass moment of inertia I_G such that the object rolls without slip no matter how large the tension in the cord. What is the shape of such an object?

Problem 7.45

Figure P7.45

A spool of mass $m = 300\,\text{kg}$, inner and outer radii $\rho = 1.5\,\text{m}$ and $R = 2\,\text{m}$, respectively, and radius of gyration $k_G = 1.8\,\text{m}$, is placed on an incline with $\theta = 43°$. The cable that is wrapped around the spool and attached to the wall is initially taut. If the static and kinetic friction coefficients between the incline and the spool are $\mu_s = 0.35$

and $\mu_k = 0.3$, respectively, determine the acceleration of G, the angular acceleration of the spool, and the tension in the cable once the spool is released from rest.

Problem 7.46

The spool of mass m, radius of gyration k_G, inner radius r_i, and outer radius r_o is placed on a horizontal conveyer belt. The cable that is wrapped around the spool and attached to the wall is initially taut. Both the spool and the conveyer belt are initially at rest when the conveyer belt starts moving with acceleration a_c. If the coefficient of static friction between the conveyer belt and spool is μ_s, determine

(a) The maximum acceleration of the conveyer belt so that the spool rolls without slipping on the belt

(b) The initial tension in the cable that attaches the spool to the wall

(c) The angular acceleration of the spool

Evaluate your answers for $m = 500\,\text{kg}$, $k_G = 1.3\,\text{m}$, $\mu_s = 0.5$, $r_i = 0.8\,\text{m}$, and $r_o = 1.6\,\text{m}$.

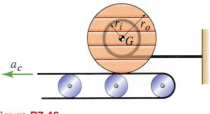

Figure P7.46

Problems 7.47 through 7.49

The thin uniform bar AB of mass m and length L hangs from a wheel at A, which rolls freely on the horizontal bar DE. In the following problems, neglect the mass of the wheel and assume that the wheel never separates from the horizontal bar.

Problem 7.47 If the bar is released from rest at the angle θ, determine, immediately after release, the angular acceleration of the bar, the force on the bar at A, and the acceleration of end A.

Problem 7.48 Find the equation(s) of motion of the bar, using the coordinates x and θ shown on the figure as the dependent variables.

 Problem 7.49 Find the equation(s) of motion of the bar using the coordinates x and θ shown on the figure as the dependent variables and then simulate the system's behavior by numerically solving the equations of motion for 5 s, using $m = 2\,\text{kg}$, $L = 0.6\,\text{m}$, $x(0) = 0\,\text{m}$, $\dot{x}(0) = 0\,\text{m/s}$, $\theta(0) = 60°$, and $\dot{\theta}(0) = 0\,\text{rad/s}$. Plot x and θ for $0 \le t \le 5\,\text{s}$.

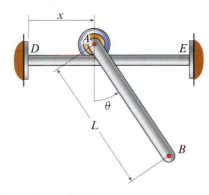

Figure P7.47–P7.49

Problem 7.50

The roadster weighs 2570 lb, and its mass is evenly distributed between its front and rear wheels. It can accelerate from 0 to 60 mph in 6.98 s. The rear wheel, shown in the blowup above the roadster, weighs 47 lb, its mass center is at its geometric center, and its mass moment of inertia I_B is $0.989\,\text{slug·ft}^2$. With this in mind, we want to determine the forces on the rear wheel shown in Fig. 2 of Example 7.4.

(a) Assuming that its acceleration is uniform, determine the forces on the front and rear wheels due to the pavement.

(b) Now that you have the normal and friction forces between the rear wheels and the pavement, isolate one of the rear wheels and determine the forces and moments exerted by the axle on that rear wheel.

Assume that the mass is evenly distributed between the right and left sides of the car and that friction is sufficient to prevent slipping of the wheels.

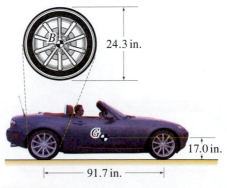

Figure P7.50

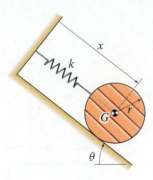

Figure P7.51 and P7.52

Problem 7.51

A spool of mass m, radius r, and radius of gyration k_G rolls without slipping on the incline, whose angle with respect to the horizontal is θ. A linear elastic spring with constant k and unstretched length L_0 connects the center of the spool to a fixed wall. Determine the equation(s) of motion of the spool, using the x coordinate shown.

Problem 7.52

A spool of mass $m = 200\,\text{kg}$, radius $r = 0.8\,\text{m}$, and radius of gyration $k_G = 0.65\,\text{m}$ rolls without slipping on the incline, whose angle with respect to the horizontal is $\theta = 38°$. A linear elastic spring with constant $k = 500\,\text{N/m}$ and unstretched length $L_0 = 1.5\,\text{m}$ connects the center of the spool to a fixed wall. Determine the equation(s) of motion of the spool, using the x coordinate shown; solve them for 15 s, using the initial conditions $x(0) = 2.5\,\text{m}$ and $\dot{x}(0) = 0\,\text{m/s}$; and then plot x versus t. What is the approximate period of oscillation of the spool?

Problem 7.53

The uniform bar AB of mass m and length L is leaning against the corner with $\theta \approx 0$ when end B is given a slight nudge so that end A starts sliding down the wall as B slides along the floor. Assuming that friction is negligible between the bar and the two surfaces against which it is sliding, determine the angle θ at which end A will lose contact with the vertical wall.

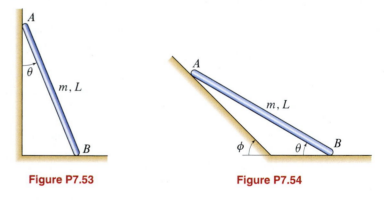

Figure P7.53   **Figure P7.54**

Problem 7.54

The uniform thin bar, which is leaning on the incline, is released from rest in the position shown and slides in the vertical plane. The contacts between the bar and the surface at ends A and B have negligible friction. Determine the angular acceleration of the bar immediately after it is released. Evaluate your answer for $m = 3\,\text{kg}$, $L = 0.75\,\text{m}$, $\phi = 45°$, and $\theta = 30°$.

Problem 7.55

An important problem in billiards or pool is the determination of the height at which you should hit the cue ball to give it backspin, topspin, or no spin. With this in mind, at what height h should the cue hit the ball so that the ball always rolls without slip, regardless of how hard the ball is hit and how much friction is available? Assume a uniform ball of mass m and radius r. You can determine this position without having to worry about the impact between the cue and the ball by studying an arbitrary horizontal force applied to a ball at height h.

Figure P7.55

Problems 7.56 and 7.57

The disk A rolls without slipping on a horizontal surface. End B of bar BC is pinned to the edge of the disk A, and end C of the bar BC can slide freely along the horizontal surface. In addition, bar BC is pushed by the force $P = mg$ at its left end. The mass of bar BC is m_{BC}, and the mass of the disk A is m_A. The system is initially at rest.

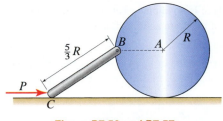

Figure P7.56 and P7.57

Problem 7.56   Determine the acceleration of the center of the disk A and the angular acceleration of bar BC immediately after the force P is applied if $m_A = m_{BC} = m$.

Problem 7.57   Determine the acceleration of the center of disk A and the angular acceleration of bar BC immediately after force P is applied if $m_{BC} = m$ and the mass of the disk m_A is negligible.

Problems 7.58 through 7.61

The uniform ball of radius ρ and mass m is gently placed in the bowl B with inner radius R and is released. The angle ϕ measures the position of the center of the ball at G with respect to a vertical line, and the angle θ measures the rotation of the ball with respect to a vertical line. Assume that the system lies in the vertical plane. *Hint:* In working the following problems, we recommend using the $r\phi$ coordinate system shown.

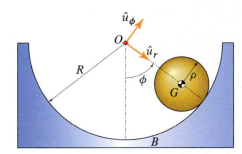

Figure P7.58–P7.61

Problem 7.58   Assuming that the ball rolls without slip, determine the acceleration of the center of the ball at G, the angular acceleration of the ball, and the force on the ball due to the bowl immediately after the ball is released.

Problem 7.59   Assuming that the ball rolls without slip, that it weighs 3 lb, is at the position $\phi = 40°$, and is moving clockwise at 10 ft/s, determine the acceleration of the center of the ball at G and the normal and friction force between the ball and the bowl. Use $R = 4$ ft and $\rho = 1.2$ ft.

Problem 7.60   Assuming that friction is sufficient to prevent slipping, derive the equation(s) of motion of the ball in terms of the angle ϕ.

Problem 7.61   Assume that friction is sufficient to prevent slipping.

(a) Derive the equation(s) of motion of the ball in terms of the angle ϕ.

(b) Determine the friction force as a function of ϕ.

(c) Letting $\phi(0) = \phi_0$ and $\dot{\phi}(0) = 0$, with $0° < \phi_0 < 90°$ integrate the equation(s) of motion to determine the normal force as a function of ϕ.

(d) Using the results of Parts (b) and (c), given a value for μ_s, determine the maximum value of $\phi(0) = \phi_0$ so that the ball does not slip.

Problem 7.62

An SUV is pushing a large drum to the right with force P, using its front bumper. The drum has mass m and radius of gyration k_G. The static and kinetic friction coefficients between the drum and the ground and between the drum and the SUV are μ_s and μ_k, respectively.

(a) Assuming that there is no slipping between the drum and the ground, determine the acceleration of the drum and the minimum value of μ_s that is consistent with this motion.

(b) Determine the acceleration of point G and the angular acceleration of the drum if P is increased so that the drum slips relative to the ground.

Figure P7.62

Problem 7.63

The crank AB in the slider-crank mechanism is rotating counterclockwise with constant angular velocity ω_{AB}. The crankshaft radius is R, the length of the connecting rod BC is L, and the distance from the mass center of the connecting rod D to the end of the crank at B is d. The mass of the connecting rod is m_D, the mass moment of inertia of the connecting rod is I_D, and the mass of the piston is m_C.

(a) Using the component system shown, determine the x and y components of the forces on the connecting rod at B and C as functions of the crank angle θ.

(b) Using, $\omega_{AB} = 5700$ rpm, $R = 48.5$ mm, $L = 141$ mm, $d = 36.4$ mm, $m_D = 0.439$ kg, $I_D = 0.00144$ kg·m², and $m_C = 0.434$ kg, plot each of the four force components, the magnitude of the forces at B and C, and the moment acting on the connecting rod about point D, all as a function of θ, for one full rotation of the crank.

Hint: The kinematics of this problem have been considered in Example 6.10 on p. 504.

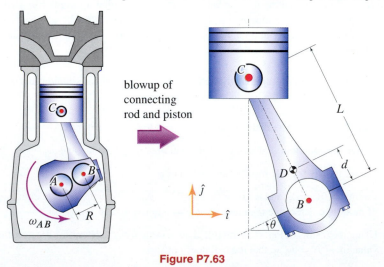

Figure P7.63

Problems 7.64 and 7.65

The Pioneer 3 spacecraft was a spin-stabilized spacecraft launched on December 6, 1958, by the U.S. Army Ballistic Missile agency in conjunction with NASA. It was designed with a despin mechanism consisting of two equal masses A and B, each of mass m, that could be spooled out to the end of two wires of variable length $\ell(t)$ when triggered by a hydraulic timer. As the masses unwound, they would slow the spacecraft's spin from an initial angular velocity $\omega_s(0)$ to the final angular velocity $(\omega_s)_{\text{final}}$, and then the weights and wires would be released. Assume that masses A and B are initially at positions A_0 and B_0, respectively, before the wire begins to unwind, that the mass moment of inertia of the spacecraft is I_O (this does not include the two masses A and B), and that gravity and the mass of each wire are negligible. *Hint:* Refer to Prob. 6.117 if you need help with the kinematics.

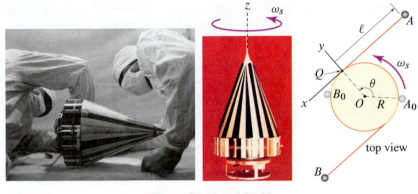

Figure P7.64 and P7.65

Problem 7.64   Derive the equation(s) of motion of the system in terms of the dependent variables $\ell(t)$ and $\omega_s(t)$.

Problem 7.65   Derive the equation(s) of motion of the system in terms of the dependent variables $\ell(t)$ and $\omega_s(t)$. After doing so:

(a) Use a computer to solve the equations of motion for $0 \leq t \leq 4\,\text{s}$, using $R = 12.5\,\text{cm}$, $m = 7\,\text{g}$, $I_O = 0.0277\,\text{kg·m}^2$, and the initial conditions $\omega_s(0) = 400\,\text{rpm}$, $\ell(0) = 0.01\,\text{m}$, and $\dot{\ell}(0) = 0\,\text{m/s}$.

(b) Determine the time at which the angular velocity of the spacecraft becomes zero (this can be done by plotting the solution for ω_s and then estimating the time or by using numerical root finding to determine when $\omega_s = 0$).

(c) Determine the length ℓ of each of the wires at the instant that the angular velocity of the spacecraft becomes zero.

Review Problems

Problem 7.66

Two identical uniform bars are pinned together at one end, and each bar has a roller at its other end. The rollers can roll freely along the horizontal surface as shown. Each bar has mass m and length L, and a horizontal force P is applied to bar BC at B. Although the bars can rotate relative to each other, for a given value of P there exists a corresponding value of θ such that the system moves with θ constant.

(a) Find the forces on the bars at A and B and show that they are independent of P.

(b) Determine θ as a function of P and find θ_0 as $P \to 0$ and θ_∞ as $P \to \infty$.

Neglect the mass of the rollers and any friction in their bearings.

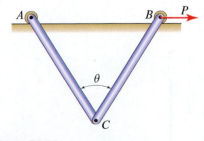

Figure P7.66

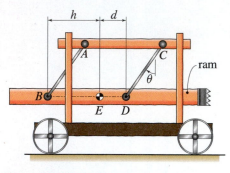

Figure P7.67–P7.72

Problems 7.67 through 7.72

The figure shows a weapon called a battering ram (modern large battering rams are typically mounted on armored vehicles). The ram has a weight $W_r = 2500$ lb, center of mass at E, and radius of gyration $k_E = 6.5$ ft. Also let the distance between points A and B and between points C and D be 6 ft. In addition, let $h = 4.5$ ft and $d = 3$ ft. Finally, let the connections at points A, B, C, and D be pin connections, and assume that the cart does not move while the ram swings.

Problem 7.67   Assuming that the ram is suspended via inextensible cords of negligible mass, determine the tension in the cords and the acceleration of E immediately after the ram is released from rest at $\theta = 75°$.

Problem 7.68   Let the ram be at rest with $\theta = 0°$. Assume that the cords AB and CD are inextensible and of negligible mass. Also assume that cord AB breaks suddenly. Determine the tension in cord CD and the acceleration of E immediately after AB breaks.

Problem 7.69   Assume that AB and BC are inextensible cords with negligible mass. In addition, assume that, at the instant shown, $\theta = 10°$ and the ram is swinging forward with $|\vec{v}_E| = 7$ ft/s. At this instant, determine the acceleration of E as well as the reaction forces at points A and C.

Problem 7.70   Assume that AB and BC are uniform thin rods weighing 100 lb each. If the ram is released from rest when $\theta = 63°$, determine the acceleration of E as well as the reaction forces at points A and C immediately after release.

Problem 7.71   Assume that AB and BC are uniform thin rods weighing 100 lb each and that the ram is at rest with $\theta = 0°$. Assume that bar CD breaks suddenly, and determine the acceleration of E and the forces at A immediately after CD breaks.

Problem 7.72   Assume that AB and BC are uniform thin rods weighing 100 lb each. Assume that, at the instant shown, $\theta = 10°$ and the ram is swinging forward with $|\vec{v}_E| = 7$ ft/s. At this instant, determine the acceleration of the ram as well as the reaction forces at points A and C.

Problems 7.73 through 7.75

Figure P7.73–P7.75

A spool has a weight $W = 450$ lb, outer and inner radii $R = 6$ ft and $\rho = 4.5$ ft, respectively, radius of gyration $k_G = 4.0$ ft, and mass center at G. The spool is being pulled to the right as shown, and the cable wrapped around the spool is inextensible and of negligible mass.

Problem 7.73   Assume that the spool rolls without slipping with respect to both the cable and the ground. If the pickup truck pulls the cable with a force $P = 125$ lb, determine the acceleration of the center of the spool and the minimum value of the static friction coefficient between the spool and the ground.

Problem 7.74   Assume that the static friction coefficient between the spool and the ground is $\mu_s = 0.75$, and determine the maximum value of the force that the truck could exert on the cable without causing the spool to slip relative to the ground.

Problem 7.75   Assume that the static and kinetic friction coefficients between the spool and the ground are $\mu_s = 0.25$ and $\mu_k = 0.2$. Furthermore, assume that the spool is initially at rest and that the pickup truck pulls the spool with a force of 550 lb, and determine the acceleration of the center of the spool at the initial time.

Problems 7.76 and 7.77

The car, as seen from the front, is traveling at a constant speed v_c on a turn of constant radius R that is banked at an angle θ with respect to the horizontal. The coefficient of static friction between the tires and the road is μ_s. The car's center of mass is at G.

Problem 7.76 Determine the bank angle θ so that there is no tendency to slip or tip, i.e., so that no friction is required to keep the car on the road.

Problem 7.77 For a given bank angle θ, and assuming that the car does not tip, find the maximum speed v_m that the car can achieve without slipping.

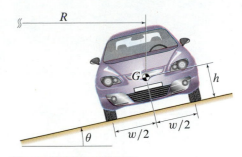

Figure P7.76 and P7.77

Problems 7.78 through 7.80

An overhead fold-up door, with height h and mass m, consists of two identical sections hinged at C. The roller at A moves along a horizontal guide, whereas the rollers at B and D, which are the midpoints of sections AC and CE, move along a vertical guide. The door's operation is assisted by two identical springs attached to the horizontally moving rollers (only one of the two springs is shown). The springs are stretched an amount δ_0 when the door is fully open.

Problem 7.78 Let $h = 10$ m and $m = 380$ kg. In addition, let $k = 2400$ N/m and $\delta_0 = 0.15$ m. Assuming that the door is released from rest when $\theta = 10°$ and that all sources of friction are negligible, determine the angular acceleration of each section of the door right after release.

Problem 7.79 Assuming that friction between the rollers and the guide can be neglected, determine the equation(s) of motion of the system.

Problem 7.80 Let $h = 10$ m and $m = 320$ kg. In addition, let $k = 2400$ N/m and $\delta_0 = 0.15$ m. Assuming that the door is released from rest when $\theta = 5°$, determine the time the door will take to close and the speed of E at closing.

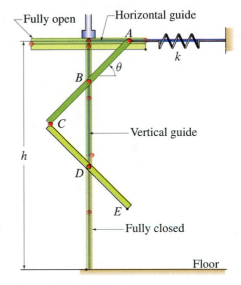

Figure P7.78–P7.80

Problems 7.81 through 7.83

A wheel with center O, radius R, weight W, radius of gyration k_G, and center of mass G at a distance ρ from O is released from rest on a rough incline. The angle ϕ is the angle between the segment OG (which rotates with the wheel) and the horizontal direction.

Problem 7.81 Let $R = 1.5$ ft, $\rho = 0.8$ ft, $k_G = 0.6$ ft, $W = 4$ lb, and $\theta = 25°$. In addition, let $\phi = 35°$ at the instant of release. Determine the minimum coefficient of static friction so that the wheel starts moving while rolling without slip. In addition, determine the corresponding angular acceleration right after release.

Problem 7.82 Assuming that there is enough friction for the wheel to roll without slip, determine the equation(s) of motion of the wheel as well as the constraint force equations, that is, those equations that would allow you to compute the reaction forces at the contact point with the incline if the motion were known.

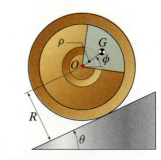

Figure P7.81–P7.83

Problem 7.83 Let $R = 1.5$ ft, $\rho = 0.8$ ft, $k_G = 0.6$ ft, $W = 4$ lb, and $\theta = 25°$, and let $\phi = 60°$ at the instant of release. Assuming that there is sufficient friction for the wheel to roll without slip and that the incline is sufficiently long that we need not worry about the wheel reaching the end of the incline, determine the equation(s) of motion of the wheel and expressions for the friction and normal forces at the point of contact between the wheel and the incline. Then, integrate the equation(s) of motion as a function of time for $0 \leq t \leq 2$ s. Plot the normal force as a function of time over the given time interval and determine if and when the wheel loses contact with the incline.

Problems 7.84 and 7.85

The uniform slender bar AB has mass m_{AB} and length L. The crate has a uniformly distributed mass m_C and dimensions h and w. Bar AB is pin-connected to the trolley at A and to the crate at B. The trolley is constrained to move along the horizontal guide shown. Point O on the trolley's guide is a fixed reference point.

Problem 7.84 Neglecting the mass of the trolley and friction, derive the equations of motion of the system and express them in terms of the variables x_A, θ, and ϕ along with their time derivatives.

Problem 7.85 Let $m_{AB} = 75$ kg, $L = 4.5$ m, $m_C = 250$ kg, $d = 0.5$ m, $h = 1.5$ m, and $w = 2$ m. Neglect the mass of the trolley and friction. Finally, assume that the system is released from rest when $x_A = 0$, $\theta = 30°$, and $\phi = 45°$. Plot x_A, θ, and ϕ as functions time for $0 \leq t \leq 15$ s.

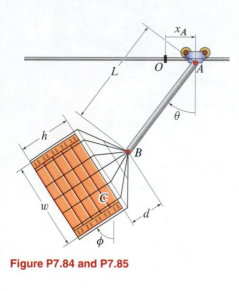

Figure P7.84 and P7.85

Problems 7.86 and 7.87

A uniform thin rod is slightly nudged at B from the $\theta = 0$ position so that it falls to the right. The coefficient of static friction between the rod and the floor is μ_s.

Problem 7.86

(a) Determine as a function of θ the normal force (N) and the frictional force (F) exerted by the ground on the rod as the rod falls over.

(b) Knowing that the rod will slip when $|F/N|$ exceeds μ_s, determine whether the rod will slip as it falls.

Problem 7.87

(a) Determine as a function of θ the normal force (N) and the frictional force (F) exerted by the ground on the rod as the rod falls over.

(b) Knowing that the rod will slip when $|F/N|$ exceeds μ_s, determine whether the rod will slip as it falls.

(c) Plot $F/(mg)$, $N/(mg)$, and $|F/N|$ as a function of θ for $0 \leq \theta \leq \pi/2$ rad. Use those plots to show that for smaller values of μ_s end A of the rod slips to the left, and for larger values of μ_s it slips to the right.

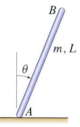

Figure P7.86 and P7.87

Problem 7.88

A drum of mass m_d, radius R, radius of gyration k_G, and with center of mass at G is placed on a cart of mass m_c for transport. The system is initially at rest, and the cart is pushed to the right with the force P. The coefficient of static friction between the cart and the drum is μ_s. Neglecting the mass of the wheels, determine the maximum force P that can be applied to the cart so that the drum does not slip on the cart, and find the corresponding acceleration of the cart and of point G.

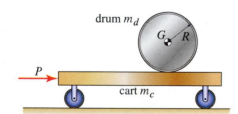

Figure P7.88

Problem 7.89

The uniform bar AB of mass m and length L is leaning against the corner with $\theta \approx 0$. A small box is placed on top of the bar at A. End B of the bar is given a slight nudge

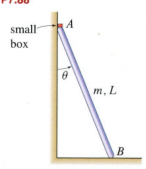

Figure P7.89

so that end A starts sliding down the wall as B slides along the floor. Assuming that friction is negligible between the bar and the two surfaces against which it is sliding and neglecting the weight of the box, determine the angle θ at which the box will lose contact with the bar.

Problem 7.90

Bars AB and BC are uniform with masses $m_{AB} = 2\,\text{kg}$ and $m_{BC} = 1\,\text{kg}$, respectively. Their lengths are $L = 1.25\,\text{m}$ and $H = 0.75\,\text{m}$. Bar BC is pin-connected to a fixed support at C that is a distance $\delta = 0.2\,\text{m}$ from the ground. Bar AB is pin-connected at A to a uniform wheel with radius $R = 0.60\,\text{m}$ and mass $m_{OA} = 5\,\text{kg}$. Note that A is at a distance ρ from the center of the wheel. At the instant shown, A is vertically aligned with O; in addition, AB and BC are parallel and perpendicular to the ground, respectively. At this instant, bar BC is rotating clockwise with an angular velocity of $2\,\text{rad/s}$ and an angular acceleration of $1.2\,\text{rad/s}^2$. Assuming that the wheel rolls without slip, determine the force P that is applied to the wheel at the instant shown. In addition, determine the minimum static coefficient of friction necessary for the wheel not to slip.

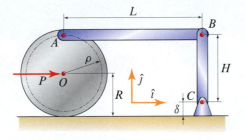

Figure P7.90

DESIGN PROBLEMS

Design Problem 7.1

Revisit the calculations done at the beginning of the chapter concerning the determination of the maximum acceleration that can be achieved by a motorcycle without causing the front wheel to lift off the ground. Specifically, construct a new model of the motorcycle by selecting a real-life motorcycle and researching its geometry and inertia properties, including the inertia properties of the wheels. Then analyze your model to determine how the maximum acceleration in question depends on the horizontal and vertical positions of the center of mass with respect to the points of contact between the ground and the wheels. Include in your analysis a comparison of results that account for the inertia of the front wheel with results that neglect the inertia of the front wheel.

Figure DP7.1

Design Problem 7.2

One end of a seat belt on passenger cars is wound around a ratchet wheel that can be locked when the deceleration of the car exceeds a set value. In the sketch shown, consider a ratchet wheel that can rotate about the fixed point A. The ratchet wheel has cogs and is positioned above a weight. The weight can pivot about point B and is rigidly connected to a pawl that, for a sufficiently large rotation, will lock the motion of the ratchet wheel. Research realistic dimensions for the ratchet wheel, and design a locking mechanism such that the belt's motion will be stopped for horizontal decelerations greater than $0.5g$, where g is the acceleration due to gravity. In your design you may use a spring to limit the motion of the weight.

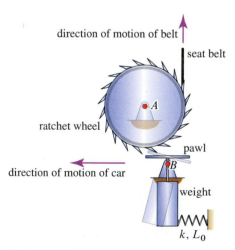

direction of motion of belt

seat belt

A

ratchet wheel

pawl

B

direction of motion of car

weight

k, L_0

Figure DP7.2

8

Energy and Momentum Methods for Rigid Bodies

In this chapter we apply three fundamental balance laws to rigid bodies: the work-energy principle, the linear impulse-momentum principle, and the angular impulse-momentum principle. Each of these balance principles stems from the Newton-Euler equations we studied in Chapter 7, and they make it easier for us to tackle the solution of many problems that would be far more challenging if solved using the Newton-Euler approach directly, as presented in Chapter 7.

8.1 Work-Energy Principle for Rigid Bodies

Flywheels as energy storage devices

When we turn on an electrical appliance, we tap into the local electric grid to get the electricity we need. This electricity was produced only a few milliseconds prior to its use. While we might marvel at the technology that delivers electric power on demand, it is easy to see a fundamental weakness in our current energy management: as soon as the demand exceeds the production capability, some of us suffer a brownout or even a blackout! To avoid such a situation, new technologies are being developed to store large amounts of energy and thus manage the production and consumption of electric power.

One of the devices being considered for electric energy storage is the *flywheel*, a schematic of which is shown in Fig. 8.1 and an advanced model of which is shown in Fig. 8.2. Conceptually, a flywheel is a very simple object consisting of a rotor that spins about a fixed axis. The energy stored in the flywheel is the kinetic energy of the rotor, which is often supported using magnetic levitation bearings and is housed in a vacuum chamber to limit the amount of energy dissipated by friction and air resistance. To see how much energy is stored in a flywheel, consider the rotor B in Fig. 8.1, which is modeled as a rigid body rotating with an angular speed ω_B about the (fixed) z axis. If dm is a mass element a distance r away from the z axis, the speed of dm is $v_{dm} = \omega_B r$ and the *kinetic energy* of dm is

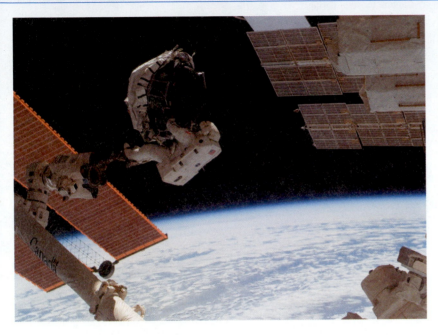

Figure 8.1
Sketch of a rotor B spinning about a fixed z axis.

$$dT = \tfrac{1}{2}(dm)v_{dm}^2 = \tfrac{1}{2}(dm)\omega_B^2 r^2. \qquad (8.1)$$

591

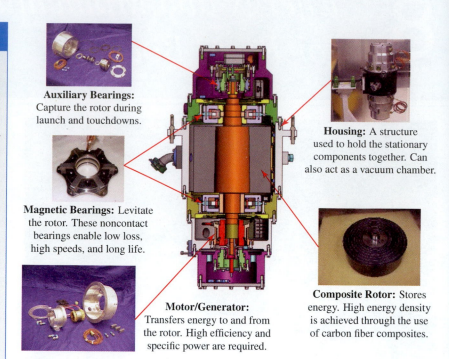

Auxiliary Bearings: Capture the rotor during launch and touchdowns.

Magnetic Bearings: Levitate the rotor. These noncontact bearings enable low loss, high speeds, and long life.

Housing: A structure used to hold the stationary components together. Can also act as a vacuum chamber.

Motor/Generator: Transfers energy to and from the rotor. High efficiency and specific power are required.

Composite Rotor: Stores energy. High energy density is achieved through the use of carbon fiber composites.

Figure 8.2. A flywheel for energy storage designed by NASA.

The kinetic energy of the rotor for fixed axis rotation is then

$$T = \int_B dT = \int_B \tfrac{1}{2}\omega_B^2 r^2 \, dm = \tfrac{1}{2}\omega_B^2 \int_B r^2 \, dm = \tfrac{1}{2}\omega_B^2 I_z, \qquad (8.2)$$

where I_z is the rotor's mass moment of inertia with respect to the z axis (see App. A for a review of mass moments of inertia).* Equation (8.2) tells us that T is directly proportional to the mass moment of inertia with respect to the spin axis and to the square of the spin rate. The dependence on I_z indicates that the flywheel energy depends on how the rotor's mass is distributed relative to the spin axis and not just the rotor's mass. Equation (8.2) is also interesting because it says that $T = \tfrac{1}{2}(\text{mass property})(\text{speed})^2$ and therefore is similar to the kinetic energy of a particle, i.e., $\tfrac{1}{2}mv^2$. However, the expression for the kinetic energy in Eq. (8.2) only applies to fixed axis rotation. We now need to find the general expression of the kinetic energy of a rigid body, which requires some additional development.

Kinetic energy of rigid bodies in planar motion

Referring to Fig. 8.3, consider an infinitesimal mass element dm with velocity $\vec{v}_{dm}$ and kinetic energy $dT_{dm} = \tfrac{1}{2}(dm)v_{dm}^2$. The kinetic energy of the body B containing dm is therefore

$$T = \int_B \tfrac{1}{2}v_{dm}^2 \, dm. \qquad (8.3)$$

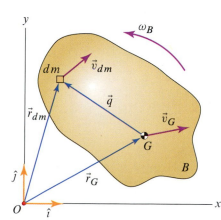

Figure 8.3
A rigid body in planar motion.

---

* As stated in Chapter 7, the subscript on I will be used to uniquely identify the axis with respect to which I is calculated. Hence, I_A denotes the mass moment of inertia of the rigid body about an axis perpendicular to the plane of motion passing through point A.

Because G and dm are two points on the same rigid body, we have

$$\vec{v}_{dm} = \vec{v}_G + \vec{\omega}_B \times \vec{q}, \qquad (8.4)$$

where G and $\vec{\omega}_B$ are the mass center and angular velocity of B, respectively, and $\vec{q}$ is the position of dm relative to G. In planar motion, $\vec{\omega}_B = \omega_B \hat{k}$ and $\vec{q} = q_x \hat{\imath} + q_y \hat{\jmath}$, so $\vec{\omega}_B \times \vec{q} = \omega_B(-q_y \hat{\imath} + q_x \hat{\jmath})$. Using Eq. (8.4), we can then write

$$
\begin{aligned}
v_{dm}^2 &= \vec{v}_{dm} \cdot \vec{v}_{dm} \\
&= \vec{v}_G \cdot \vec{v}_G + 2\vec{v}_G \cdot \omega_B(-q_y \hat{\imath} + q_x \hat{\jmath}) + \omega_B^2(q_x^2 + q_y^2) \\
&= v_G^2 + 2\omega_B(-v_{Gx} q_y + v_{Gy} q_x) + \omega_B^2(q_x^2 + q_y^2), \quad (8.5)
\end{aligned}
$$

where we have used $\vec{v}_G = v_{Gx}\hat{\imath} + v_{Gy}\hat{\jmath}$. Substituting Eq. (8.5) in Eq. (8.3),

$$T = \int_B \left[\tfrac{1}{2} v_G^2 + \omega_B(-v_{Gx} q_y + v_{Gy} q_x) + \tfrac{1}{2}\omega_B^2(q_x^2 + q_y^2) \right] dm. \qquad (8.6)$$

We can simplify Eq. (8.6), starting with the integral of the term $\tfrac{1}{2} v_G^2$, as

$$\int_B \tfrac{1}{2} v_G^2 \, dm = \tfrac{1}{2} v_G^2 \int_B dm = \tfrac{1}{2} m v_G^2, \qquad (8.7)$$

where $m = \int_B dm$ is the total mass of B. The last term in the integral of Eq. (8.6) can be written as

$$\int_B \tfrac{1}{2}\omega_B^2(q_x^2 + q_y^2)\, dm = \tfrac{1}{2}\omega_B^2 \int_B (q_x^2 + q_y^2)\, dm = \tfrac{1}{2} I_{Gz}\omega_B^2, \qquad (8.8)$$

where, recalling Eq. (7.41) on p. 551, $\int_B (q_x^2 + q_y^2)\, dm = I_{Gz}$. The second term in the integral of Eq. (8.6) can be simplified as

$$
\begin{aligned}
\int_B \omega_B(-v_{Gx} q_y + v_{Gy} q_x)\, dm &= -\omega_B v_{Gx} \int_B q_y \, dm \\
&\quad + \omega_B v_{Gy} \int_B q_x \, dm = 0, \quad (8.9)
\end{aligned}
$$

where we have used the fact that both $\int_B q_x \, dm$ and $\int_B q_y \, dm$ are zero because they measure the position of G with respect to G (see Eqs. (7.39) and (7.40) on p. 551). Substituting Eqs. (8.7)–(8.9) into Eq. (8.6) yields

$$\boxed{ T = \tfrac{1}{2} m v_G^2 + \tfrac{1}{2} I_G \omega_B^2, } \qquad (8.10)$$

where we have written the shorthand I_G for I_{Gz} (as done in Chapter 7). From Eq. (8.10), the kinetic energy of a rigid body in planar motion consists of

1. The term $\tfrac{1}{2} m v_G^2$, often called the *translational* kinetic energy, which is the kinetic energy of the body in translation

2. The term $\tfrac{1}{2} I_G \omega_B^2$, often called the *rotational* kinetic energy, which is the kinetic energy of the body in fixed axis rotation about the mass center G and with the axis of rotation perpendicular to the plane of motion

Equation (8.10) shows that the form of the kinetic energy of a rigid body is not that different from that of a particle because this energy consists of terms of the form $\tfrac{1}{2}(\text{mass property})(\text{speed})^2$.

Helpful Information

Planar motion of 3D bodies. In planar motion, when we write

$$\vec{\omega}_B \times \vec{q} = \omega_B(-q_y \hat{\imath} + q_x \hat{\jmath})$$

we are *not* assuming that the body is 2D. For a 3D body we have $\vec{q} = q_x \hat{\imath} + q_y \hat{\jmath} + q_z \hat{k}$ and so

$$\omega_B \hat{k} \times (q_x \hat{\imath} + q_y \hat{\jmath} + q_z \hat{k})$$

is *still* equal to $\omega_B(-q_y \hat{\imath} + q_x \hat{\jmath})$. The z component of $\vec{q}$ does not contribute to the final result because the motion is planar.

Concept Alert

Kinetic energy of a rigid body. The kinetic energy of a rigid body consists of two parts. The first part, $\tfrac{1}{2} m v_G^2$, is due to motion of the center of mass (translational kinetic energy), and the second, $\tfrac{1}{2} I_G \omega_B^2$, is due to the rotational motion of points *relative* to the center of mass (rotational kinetic energy). A rigid body can have kinetic energy even when the center of mass does not move.

Helpful Information

Kinetic energy of a rigid body vs. kinetic energy of a particle system. In Section 4.3, we found that the kinetic energy of a particle system is (Eq. (4.74) on p. 306)

$$T = \tfrac{1}{2} m v_G^2 + \tfrac{1}{2} \sum_{i=1}^{n} m_i v_{i/G}^2.$$

The first term is identical to the first term in Eq. (8.10). The term $\tfrac{1}{2} \sum_{i=1}^{n} m_i v_{i/G}^2$ serves the same purpose as the term $\tfrac{1}{2} I_G \omega_B^2$ in Eq. (8.10)—it accounts for the motion of points in the body *relative* to the mass center.

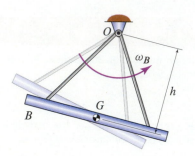

Figure 8.4
A platform B swinging about the pin at O.

Kinetic energy: rotation about a fixed axis

Referring to Fig. 8.4, consider the platform B supported by two bars that are both pinned at O and rigidly connected to B. The platform B rotates about the fixed axis perpendicular to the plane of the figure and going through O. If ω_B is the angular speed of B, rigid body kinematics tells us that the center of mass G of B has speed $v_G = \omega_B h$, where h is the distance between G and O. Equation (8.10) then tells us that the kinetic energy of B is

$$T = \tfrac{1}{2}mv_G^2 + \tfrac{1}{2}I_G\omega_B^2 = \tfrac{1}{2}m\omega_B^2 h^2 + \tfrac{1}{2}I_G\omega_B^2, \qquad (8.11)$$

which simplifies to

$$T = \tfrac{1}{2}(mh^2 + I_G)\omega_B^2 = \tfrac{1}{2}I_O\omega_B^2, \qquad (8.12)$$

where, using the parallel axis theorem, $I_O = mh^2 + I_G$ is the mass moment of inertia of B relative to the axis of rotation (see App. A). This calculation shows that the kinetic energy of a rigid body B in a fixed axis rotation can always be written as

$$\boxed{T = \tfrac{1}{2}I_O\omega_B^2,} \qquad (8.13)$$

where I_O is the mass moment of inertia relative to the axis of rotation. This result is useful because fixed axis rotations are common in applications.

Using the IC to find kinetic energy.   Since the kinetic energy only depends on the velocities, we can also use Eq. (8.13) to determine the kinetic energy of a rigid body using the IC as point O. Referring to Fig. 8.5, which shows a uniform thin bar AB of mass m and length L sliding down a corner, we see that the IC is easy to locate and that it is a distance $\ell = L/2$ from the mass center G of the bar. If we know the bar's orientation θ and either v_A or v_B, then we can find ω_{AB} using either of the following expressions

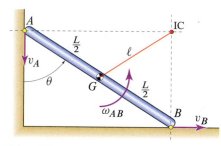

Figure 8.5
A bar sliding down a corner, identifying the kinematic parameters needed to find its kinetic energy using its IC.

$$\omega_{AB} = \frac{v_A}{L\sin\theta} \quad \text{or} \quad \omega_{AB} = \frac{v_B}{L\cos\theta}. \qquad (8.14)$$

Knowing ω_{AB}, we can then compute the kinetic energy of bar AB as

$$T = \tfrac{1}{2}I_{IC}\omega_{AB}^2 = \tfrac{1}{2}(I_G + m\ell^2)\omega_{AB}^2, \qquad (8.15)$$

where I_G is the mass moment of inertia of the bar. Example 8.1 uses the IC to compute the kinetic energy of a bicycle.

Work-energy principle for a rigid body

A rigid body can be viewed as a rigid *system* of mass elements. Thus, the form of the work-energy principle for a rigid body is the same as that obtained for a *system* of particles, which is given in Eq. (4.69) on p. 305 as

$$T_1 + (U_{1\text{-}2})_{\text{ext}} + (U_{1\text{-}2})_{\text{int}} = T_2, \qquad (8.16)$$

where T is the system's kinetic energy and $(U_{1\text{-}2})_{\text{ext}}$ and $(U_{1\text{-}2})_{\text{int}}$ are the work done in going from ① to ② by the forces that are external and internal to the system, respectively. Since we have already discussed the kinetic energy T, here we focus on the term $(U_{1\text{-}2})_{\text{int}}$, leaving the discussion about $(U_{1\text{-}2})_{\text{ext}}$ for later. As discussed in Section 4.3, the internal forces do work only when the system *deforms*. In the case of a *rigid* body, regardless of the internal forces, the *rigidity* of the body prevents deformation, and therefore the internal forces do no work, i.e.,

$$(U_{1\text{-}2})_{\text{int}} = 0. \tag{8.17}$$

Substituting Eq. (8.17) into Eq. (8.16), we get the work-energy principle for a rigid body

$$\boxed{T_1 + U_{1\text{-}2} = T_2,} \tag{8.18}$$

where $U_{1\text{-}2}$ is *only the work of the external forces and couples*. Equation (8.18) shows that the work-energy principle for a rigid body has the same form as that for a single particle (the kinetic energy for a rigid body is *not* computed in the same way as that of a particle). Next we discuss how to compute term $U_{1\text{-}2}$ in Eq. (8.18).

Work done on rigid bodies

In statics we learned that a general force system consists of both forces and couples. Here we will briefly review how to compute the work of forces, and then we will learn how to compute the work of a couple.

Referring to Fig. 8.6, consider a rigid body acted upon by n forces, labeled $\vec{F}_i$ ($i = 1, \ldots, n$). The work of this force system is given by

$$U_{1\text{-}2} = \sum_{i=1}^{n} \int_{(\mathcal{L}_{1\text{-}2})_i} \vec{F}_i \cdot d\vec{r}_i, \tag{8.19}$$

where $(\mathcal{L}_{1\text{-}2})_i$ is the *path of the point of application of force* $\vec{F}_i$ and $\vec{r}_i$ is the position of the *point of application* force $\vec{F}_i$. Equation (8.19) states that the work done by a force system is the sum of work contributions due to each force, where each contribution is computed by applying what we learned in Chapter 4.

Referring to Fig. 8.7, consider a rigid body subject to a couple consisting of two equal and opposite forces $\vec{F}_A$ and $\vec{F}_B$ with parallel lines of action separated by the distance h. The moment of this couple is

$$\vec{M} = \vec{r}_{A/B} \times \vec{F}_A = \vec{r}_{B/A} \times \vec{F}_B, \tag{8.20}$$

where A and B are any two arbitrarily chosen points on the lines of action of $\vec{F}_A$ and $\vec{F}_B$, respectively. Applying Eq. (8.19) to the case of $\vec{F}_A$ and $\vec{F}_B$, we

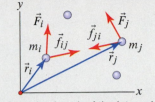

Helpful Information

Work of internal forces in rigid bodies.

The infinitesimal work of the internal forces between particles i and j in a physical system is $dU = \vec{f}_{ij} \cdot d\vec{r}_{i/j}$, where $\vec{f}_{ij}$ is parallel to $\vec{r}_{i/j}$. In general, $dU \neq 0$ whenever $d\vec{r}_{i/j} \neq \vec{0}$, i.e., in the presence of relative motion. If the motion is that of a rigid body, then even though $d\vec{r}_{i/j} \neq \vec{0}$, we have that $d\vec{r}_{i/j}$ is *always* perpendicular to $\vec{r}_{i/j}$ and therefore to $\vec{f}_{ij}$, thus implying $dU = 0$. The relative motion of two points in a rigid body motion does not allow internal forces to do work.

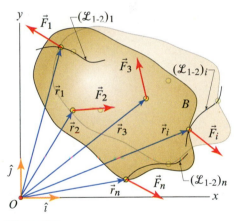

Figure 8.6
A rigid body under the action of a force system.

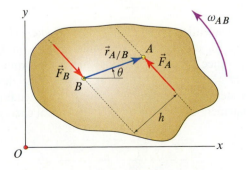

Figure 8.7. Rigid body subject to a couple.

have

$$U_{1\text{-}2} = \int_{(\mathcal{L}_{1\text{-}2})_A} \vec{F}_A \cdot d\vec{r}_A + \int_{(\mathcal{L}_{1\text{-}2})_B} \vec{F}_B \cdot d\vec{r}_B. \tag{8.21}$$

Since $d\vec{r}_A = \vec{v}_A\, dt$ and $d\vec{r}_B = \vec{v}_B\, dt$, Eq. (8.21) can be rewritten as

$$U_{1\text{-}2} = \int_{t_1}^{t_2} \left(\vec{F}_A \cdot \vec{v}_A + \vec{F}_B \cdot \vec{v}_B\right) dt = \int_{t_1}^{t_2} \vec{F}_A \cdot (\vec{v}_A - \vec{v}_B)\, dt, \tag{8.22}$$

where we have used the fact that $\vec{F}_B = -\vec{F}_A$. Because the body is rigid, $\vec{v}_A - \vec{v}_B = \vec{\omega}_{AB} \times \vec{r}_{A/B}$, so that Eq. (8.22) becomes

$$U_{1\text{-}2} = \int_{t_1}^{t_2} \vec{F}_A \cdot (\vec{\omega}_{AB} \times \vec{r}_{A/B})\, dt = \int_{t_1}^{t_2} \vec{\omega}_{AB} \cdot (\vec{r}_{A/B} \times \vec{F}_A)\, dt, \tag{8.23}$$

where the identity $\vec{a} \cdot (\vec{b} \times \vec{c}) = \vec{b} \cdot (\vec{c} \times \vec{a})$ has been used. The term $\vec{r}_{A/B} \times \vec{F}_A$ is equal to $\vec{M}$ from Eq. (8.20), and since the motion is planar, we can represent the vectors $\vec{\omega}_{AB}$ and $\vec{M}$ as $\vec{\omega}_{AB} = \omega_{AB}\,\hat{k}$ and $\vec{M} = M\,\hat{k}$, respectively. Equation (8.23) can then be rewritten as

$$U_{1\text{-}2} = \int_{t_1}^{t_2} M\omega_{AB}\, dt = \int_{\theta_1}^{\theta_2} M\, d\theta, \tag{8.24}$$

since in planar rigid body motions $d\theta = \omega_{AB}\, dt$, where $d\theta$ is the infinitesimal angular displacement of the body. If M is constant, Eq. (8.24) simplifies to $U_{1\text{-}2} = M(\theta_2 - \theta_1)$, where $\theta_2 - \theta_1$ is the angular displacement of the body between ① and ②.

Potential energy and conservation of energy

If some of the forces acting on a rigid body are conservative, we can account for their work using their associated potential energy.* Therefore, the work-energy principle for a rigid body moving between ① and ② under a general force system can be given the form

$$T_1 + V_1 + (U_{1\text{-}2})_{\text{nc}} = T_2 + V_2, \tag{8.25}$$

where V is the total potential energy of the body and where $(U_{1\text{-}2})_{\text{nc}}$ is the work of nonconservative forces, i.e, forces for which we do not have a potential energy function. Notice that Eq. (8.25) has the same form as Eq. (4.50) on p. 282, which is the general form of the work-energy principle for a particle. If the entire force system consists of conservative forces, then Eq. (8.25) reduces to the familiar statement of conservation of mechanical energy, which is

$$T_1 + V_1 = T_2 + V_2. \tag{8.26}$$

---

* See Section 4.2 on p. 276 to review the concepts of conservative forces and potential energy.

Potential energy of a torsional spring

Figure 8.8 displays a variety of *torsional* springs. These springs are designed to provide a moment in response to an angular displacement. A typical application of a torsional spring is shown in Fig. 8.9(a) where we see a bar pinned at one end with a spring. If the bar is subject to an angular displacement θ, then the deformation in the torsional spring causes a moment to be applied to the bar in the direction opposite to the angular displacement, as shown in Fig. 8.9(b). The magnitude of the moment generated is a function of the

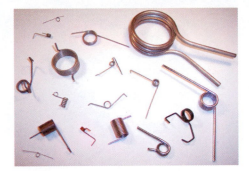

Figure 8.8
A collection of torsional springs.

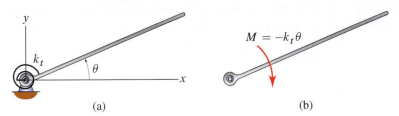

(a)   (b)

Figure 8.9. Bar with linear torsional spring (a) and spring's reaction moment (b).

angular displacement. For example, for a *linear* torsional spring, the moment–angular displacement relation is

$$M = -k_t\theta, \tag{8.27}$$

where k_t is the *torsional spring constant*, which has dimensions of force $\times$ length/angle. Therefore, the SI units of k_t are N·m/rad and in U.S. Customary units they are ft·lb/rad.

If we apply Eq. (8.24) to compute the work done by a torsional spring, we obtain

$$(U_{1\text{-}2})_{\text{torsional spring}} = -\int_{\theta_1}^{\theta_2} k_t\theta\, d\theta = -\left(\tfrac{1}{2}k_t\theta_2^2 - \tfrac{1}{2}k_t\theta_1^2\right). \tag{8.28}$$

Equation (8.28) tells us that the work of a torsional spring depends only on the initial and final positions of the spring. This means that we can account for the work of a torsional spring using a potential energy function. Since the relation between work and potential energy is $U_{1\text{-}2} = -(V_2 - V_1)$, for a linear torsional spring the potential energy is

$$\boxed{V_{\text{torsional spring}} = \tfrac{1}{2}k_t\theta^2.} \tag{8.29}$$

Potential energy of a constant gravitational force

A particle occupies a single point, so finding its change in height to compute the work done on it by gravity is straightforward. For rigid bodies, it may not be immediately obvious what point should be used to compute the change in height. Referring to the left side of Fig. 8.10, we see that, for example, point A moves to A' when the rigid body moves from ① to ②, but the net force of gravity does not act at A, it acts at G. Therefore, when computing the change in height of a rigid body for purposes of computing its gravitational potential energy, we still use Eq. (4.33) on p. 280, that is,

$$\boxed{V_g = mgy,} \tag{8.30}$$

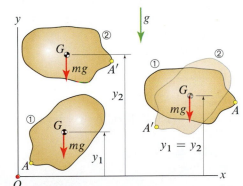

Figure 8.10
Left: A rigid body that rotates and whose mass center G has changed its height. Right: A rigid body that rotates and whose mass center G has *not* changed its height.

where y now *measures the height of the mass center* with respect to the arbitrarily chosen datum line. Referring to the right side of Fig. 8.10, we see that a rigid body that simply rotates about its mass center G has no change in gravitational potential, even though *infinitely many* points on the body change their height (e.g., A to A').

Work-energy principle for systems

Whether we view a physical system as consisting of particles, rigid bodies, or a mixture of the two, the statement of the work-energy principle takes on a form that is always the same. Therefore, avoiding unnecessary rederivations, we simply state the work-energy principle for any physical system as

$$T_1 + V_1 + (U_{1\text{-}2})_{\text{nc}}^{\text{ext}} + (U_{1\text{-}2})_{\text{nc}}^{\text{int}} = T_2 + V_2, \tag{8.31}$$

where

- T is the total kinetic energy, given by the sum of the kinetic energy of each individual part;

- V is the total potential energy, consisting of contributions from all conservative forces whether external or internal to the system;

- $(U_{1\text{-}2})_{\text{nc}}^{\text{ext}}$ is the work of all *external* forces without a potential energy; and

- $(U_{1\text{-}2})_{\text{nc}}^{\text{int}}$ is the work of all *internal* forces without a potential energy.

Equation (8.31) has the same form as Eq. (4.70) on p. 305.

> **Common Pitfall**
>
> **Work of internal forces for rigid bodies.** The internal work term $(U_{1\text{-}2})_{\text{nc}}^{\text{int}}$ in Eq. (8.31) does *not* refer to work internal to a single rigid body — it refers to the work done between two or more rigid bodies in a system of rigid bodies.

Power

We first discussed the power developed by a force in Section 4.4 on p. 320. Whether we model an object as a particle or as a rigid body, the power developed by the force $\vec{F}$ as its point of application moves with a velocity $\vec{v}$ is the work done by the force per unit time and is

$$\text{Power developed by a force } \vec{F} = \frac{dU}{dt} = \vec{F} \cdot \vec{v}. \tag{8.32}$$

As far as the power of a couple is concerned, referring to Fig 8.11, we can derive its form as a direct application of the fundamental theorem of calculus to the expression of the work of a couple given in Eq. (8.23), i.e.,

$$\begin{array}{l}\text{Power developed}\\\text{by a couple}\end{array} = \frac{dU}{dt} = \vec{\omega}_{AB} \cdot \left(\vec{r}_{A/B} \times \vec{F}_A\right) = \vec{M} \cdot \vec{\omega}_{AB}, \tag{8.33}$$

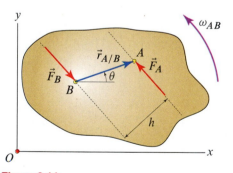

Figure 8.11
Figure 8.7 repeated. Rigid body subject to a couple.

where $\vec{M}$ is the moment of the couple.

End of Section Summary

The kinetic energy T of a rigid body B in planar motion is given by

Eq. (8.10), p. 593

$$T = \tfrac{1}{2}mv_G^2 + \tfrac{1}{2}I_G\omega_B^2,$$

or by

Eq. (8.13), p. 594

$$T = \tfrac{1}{2} I_O \omega_B^2,$$

where m, G, and I_G are the body's mass, the center of mass, and mass moment of inertia, respectively, and the second equation applies to fixed axis rotation, where I_O is the mass moment of inertia relative to the axis of rotation. The work-energy principle for a rigid body is

Eq. (8.18), p. 595

$$T_1 + U_{1\text{-}2} = T_2,$$

where $U_{1\text{-}2}$ is the work done on the body in going from ① to ② by only the external forces. If we make use of the potential energy V of conservative forces, the work-energy principle can also be written as

Eqs. (8.25) and (8.26), p. 596

$$T_1 + V_1 + (U_{1\text{-}2})_{nc} = T_2 + V_2 \quad \text{(general systems)},$$
$$T_1 + V_1 = T_2 + V_2 \quad \text{(conservative systems)},$$

where the second expression applies only to a conservative system and states that the total mechanical energy of the system is conserved. For systems of rigid bodies or mixed systems of rigid bodies and particles, the work-energy principle reads

Eq. (8.31), p. 598

$$T_1 + V_1 + (U_{1\text{-}2})_{nc}^{ext} + (U_{1\text{-}2})_{nc}^{int} = T_2 + V_2,$$

where V is the total potential energy (of both external and internal conservative forces) and $(U_{1\text{-}2})_{nc}^{ext}$ and $(U_{1\text{-}2})_{nc}^{int}$ are the work contributions due to external and internal forces for which we do not have a potential energy, respectively.

Referring to Fig. 8.12, in planar rigid body motion, the work of a couple is

Eq. (8.24), p. 596

$$U_{1\text{-}2} = \int_{t_1}^{t_2} M \omega_{AB}\, dt = \int_{\theta_1}^{\theta_2} M\, d\theta,$$

where $d\theta$ is the body's infinitesimal angular displacement and M is the component of the moment of the couple in the direction perpendicular to the plane of motion, taken to be positive in the direction of positive θ. In addition, the power developed by a couple with moment $\vec{M}$ is computed as

Eq. (8.33), p. 598

$$\text{Power developed by a couple} = \frac{dU}{dt} = \vec{M} \cdot \vec{\omega}_{AB},$$

where $\vec{\omega}_{AB}$ is the body's angular velocity.

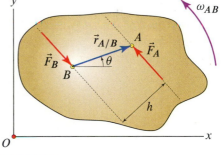

Figure 8.12
Figure 8.7 repeated. Rigid body subject to a couple.

EXAMPLE 8.1   *Kinetic Energy Computation with Rolling without Slip*

Figure 1
A cyclist riding on a horizontal road.

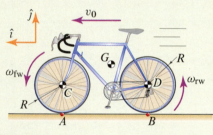

Figure 2
Bicycle moving at constant speed v_0. The points G, C, and D are the centers of mass of the frame, front wheel, and rear wheel, respectively.

A cyclist rides on a horizontal road at a constant speed $v_0 = 35\,\text{km/h}$. The mass of the frame (the bicycle without the wheels) is $m_f = 2.25\,\text{kg}$. The front and the rear wheels have the same diameter $d = 0.7\,\text{m}$. The masses of the front and rear wheel are $m_{\text{fw}} = 0.737\,\text{kg}$ and $m_{\text{rw}} = 0.933\,\text{kg}$, respectively, and the mass moments of inertia of the front and rear wheel are $I_C = 0.0510\,\text{kg·m}^2$ and $I_D = 0.0501\,\text{kg·m}^2$, respectively (see Fig. 2). Assuming the wheels roll without slip, compute the kinetic energy of just the bicycle. Neglect the kinetic energy of the pedals, chain, and other components not specifically mentioned.

SOLUTION

Road Map   We are given a system's motion, and we are asked to compute the system's kinetic energy, where the system consists of a bicycle frame and two wheels. The kinetic energy of the bicycle is found by computing the kinetic energy of each part and then adding the individual contributions. The solution does not require any knowledge of the forces acting on the system. Therefore, as in kinematics problems, our solution will consist of only the computation step.

Computation   Letting T denote the kinetic energy of the system, we have

$$T = T_f + T_{\text{fw}} + T_{\text{rw}}, \tag{1}$$

where T_f, T_{fw}, and T_{rw} are the kinetic energies of the frame, front wheel, and rear wheel, respectively. Because the system is traveling along a horizontal road, the frame is translating in the x direction with speed v_0 (see Fig. 2). Therefore, since the angular velocity of the frame is equal to zero, applying Eq. (8.10) on p. 593, we have

$$T_f = \tfrac{1}{2}m_f v_0^2 = 106.3\,\text{J}. \tag{2}$$

Applying Eq. (8.10) to the front wheel, we have

$$T_{\text{fw}} = \tfrac{1}{2}m_{\text{fw}}v_C^2 + \tfrac{1}{2}I_C \omega_{\text{fw}}^2, \tag{3}$$

where v_C and ω_{fw} are the speed of the center of mass and the angular speed of the front wheel, respectively. Because the wheel rolls without slip, we have

$$v_C = v_0 = R\omega_{\text{fw}} \quad \Rightarrow \quad \omega_{\text{fw}} = \omega_0 = \frac{v_0}{R} = 27.78\,\text{rad/s}, \tag{4}$$

where ω_0 is the angular speed of both wheels since they have the same radius $R = d/2 = 0.35\,\text{m}$. Substituting Eq. (4) into Eq. (3) gives

$$T_{\text{fw}} = \tfrac{1}{2}m_{\text{fw}}R^2\omega_0^2 + \tfrac{1}{2}I_C\omega_0^2 = \tfrac{1}{2}\left(I_C + m_{\text{fw}}R^2\right)\omega_0^2 = 54.52\,\text{J}. \tag{5}$$

Because the rear wheel is undergoing the same motion as the front wheel, the expression for T_{rw} has the same form as that in Eq. (5). Therefore, we have

$$T_{\text{rw}} = \tfrac{1}{2}\left(I_D + m_{\text{rw}}R^2\right)\omega_0^2 = 63.43\,\text{J}. \tag{6}$$

Substituting the results in Eqs. (2), (5), and (6) into Eq. (1), the total kinetic energy of the bicycle is

$$\boxed{T = 224\,\text{J}.} \tag{7}$$

Discussion & Verification   Our answer was obtained by combining contributions from Eqs. (2), (5), and (6), each of which is dimensionally correct. The units used in the final answer are appropriate since the problem's data were given in SI units.

One way to check whether or not our result is reasonable is to verify that the computed kinetic energy is larger than what we would compute if the system were just translating. If the wheels did not rotate, their rotational kinetic energy would not contribute to the system's total kinetic energy and therefore the corresponding system's kinetic energy would have to be smaller than that in Eq. (7). Computing the kinetic energy as if the system were just translating gives $T_{\text{translation}} = \frac{1}{2}(m_f + m_{\text{fw}} + m_{\text{rw}})v_0^2 = 185\,\text{J}$, which is less than our computed value, as expected.

🔍 **A Closer Look**   An important observation about our calculation is that we solved the problem by applying Eq. (8.10) on p. 593, i.e., the general formula for the kinetic energy of a rigid body in planar motion. In so doing, for the wheels, we ended up with the following two expressions

$$T_{\text{fw}} = \frac{1}{2}\left(I_C + m_{\text{fw}}R^2\right)\omega_0^2 \quad \text{and} \quad T_{\text{rw}} = \frac{1}{2}\left(I_D + m_{\text{rw}}R^2\right)\omega_0^2. \tag{8}$$

The first term in parentheses is equivalent to applying the parallel axis theorem to find mass moment of inertia of the front wheel about A, that is,

$$I_A = I_C + m_{\text{fw}}R^2, \tag{9}$$

where, referring to Fig. 2, point A is the contact point between the wheel and the ground. What is special about point A is that *it is the instantaneous center of rotation of the front wheel*. Similarly, for the rear wheel we have $I_D + m_{\text{rw}}R^2 = I_B$, which is the mass moment of inertia of the rear wheel about B, where point B is the IC of the rear wheel. These observations point to the fact that we could have computed the kinetic energy of the wheels using Eq. (8.13) on p. 594 as

$$T_{\text{fw}} = \frac{1}{2}I_A\omega_0^2 \quad \text{and} \quad T_{\text{rw}} = \frac{1}{2}I_B\omega_0^2. \tag{10}$$

E X A M P L E   8.2   *The Work-Energy Principle for Rotation about a Fixed Axis*

The manually operated road barrier shown in Fig. 1 is easily opened and closed by hand due to the counterweight. Referring to Fig. 2, model the arm of the barrier as a *uniform*

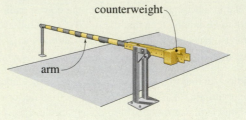

Figure 1. A manually operated road barrier. Note the counterweight on the right end.

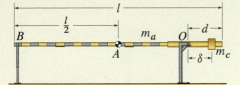

Figure 2
The relevant dimensions of the road barrier. The arm has mass m_a and the counterweight has mass m_c.

thin bar of mass m_a that is pinned at O and with mass center at A, and model the counterweight as a particle of mass m_c. Determine the angular velocity of the arm and the speed of end B as the arm reaches the horizontal position if it is nudged from rest when it is vertical and falls freely. Evaluate your answers for $l = 15.7$ ft, $d = 2.58$ ft, $\delta = 1.4$ ft, a 45 lb arm, and a 160 lb counterweight.

SOLUTION

Road Map & Modeling   Since we want to relate the change in speed of the arm to its displacement, we will apply the work-energy principle. As can be seen in the FBD of the arm in Fig. 3, only weight forces do work, so this is a conservative system. We will assume that there are no losses due to friction or drag.

Governing Equations

Balance Principles   Since the system is conservative, the work-energy principle gives

$$T_1 + V_1 = T_2 + V_2, \tag{1}$$

where ① is when the arm is vertical ($\theta = \pi/2$ rad) and ② is when it is horizontal ($\theta = 0$). At ①, the kinetic energy is zero, and at ②, the kinetic energy must account for the arm and the counterweight, and so

$$T_1 = 0 \quad \text{and} \quad T_2 = \underbrace{\tfrac{1}{2}m_c v_{c2}^2}_{\text{counterweight}} + \underbrace{\tfrac{1}{2}I_O \omega_{a2}^2}_{\text{arm}}, \tag{2}$$

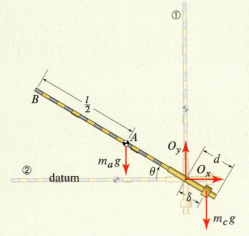

Figure 3
The FBD of the road barrier as it is falling from the vertical position.

where v_{c2} is the speed of the counterweight at ②, I_O is the mass moment of inertia of the arm with respect to point O, ω_{a2} is the angular velocity of the arm at ②, and we have used Eq. (8.13) to compute the kinetic energy of the arm. Using the parallel axis theorem, the mass moment of inertia of the arm with respect to point O is given by

$$I_O = \tfrac{1}{12}m_a l^2 + m_a\left(\tfrac{1}{2}l - d\right)^2. \tag{3}$$

Force Laws   The potential energies of the system at ① and ② are

$$V_1 = m_a g\left(\tfrac{1}{2}l - d\right) - m_c g\delta \quad \text{and} \quad V_2 = 0. \tag{4}$$

Kinematic Equations   Since we want to solve for ω_{a2}, we will need to write v_{c2} in terms of ω_{a2}, which is readily done as

$$v_{c2} = \delta\omega_{a2}. \tag{5}$$

Computation   Substituting Eqs. (2)–(5) into Eq. (1), we obtain

$$m_a g\left(\tfrac{1}{2}l - d\right) - m_c g\delta = \tfrac{1}{2}m_c(\delta\omega_{a2})^2 + \tfrac{1}{2}\left[\tfrac{1}{12}m_a l^2 + m_a\left(\tfrac{1}{2}l - d\right)^2\right]\omega_{a2}^2, \quad (6)$$

which, solving for ω_{a2}, gives

$$\boxed{\omega_{a2} = \sqrt{\frac{2g\left[m_a(l/2 - d) - m_c\delta\right]}{m_c\delta^2 + m_a\left[l^2/12 + (l/2 - d)^2\right]}} = 0.583\,\text{rad/s},} \quad (7)$$

and so the speed of the end of the arm at B is

$$\boxed{v_{B2} = (l - d)\omega_{a2} = 7.65\,\text{ft/s}.} \quad (8)$$

Discussion & Verification   The dimensions in Eqs. (7) and (8) are correct. If $m_c\delta$ is increased, the argument of the square root in Eq. (7) becomes negative and the solution is no longer meaningful. This makes sense because we expect that if $m_c\delta$ exceeds a critical value, the barrier will never reach the horizontal position. Different design configurations for the arm can be explored in the exercises and design problems.

🔍 **A Closer Look**   Problems 8.33 and 8.34 take a closer look at the problem examined in this example by taking into account the rotational inertia (and thus rotational kinetic energy) of the counterweight. Even without solving those problems, we can predict what the inclusion of rotational inertia will do to our results in Eqs. (7) and (8).

It won't change the potential energy of the counterweight, but it will add a kinetic energy term to the right side of Eq. (6). Once this kinetic energy term is written in terms of ω_{a2}, we see that the denominator in Eq. (7) will get larger and so, all other things being equal, ω_{a2} will decrease.

EXAMPLE 8.3   *Work-Energy Principle and Rolling without Slip*

Figure 1
A roadster on a horizontal stretch of road. Mazda Miata © 2006 Mazda Motor of America, Inc. Used by permission.

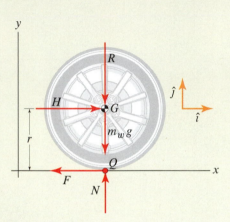

Figure 2
FBD of one of the front wheels. The point G identifies both the geometric center of the wheel as well as the wheel's center of mass.

A rear-wheel-drive car accelerates from rest to a final speed v_f over a distance L along a horizontal stretch. If the front wheels do not slip, we neglect rolling resistance, bearing friction, and air drag; and if the geometric center of each wheel is also the wheel's mass center, we determine those forces on each front wheel that do work and express the work done by those forces in terms of the given speed v_f, the wheel's radius r, mass m_w, and mass moment of inertia I_G, where G denotes the mass center of the wheel. Can we use this expression for work to find the average friction force acting on the wheel?

SOLUTION

Road Map & Modeling   The work-energy principle tells us that the work done on a wheel is the difference between the wheel's final and initial kinetic energy. Since we have enough information to determine these kinetic energies, the problem is solved by computing their difference. As for calculating the average friction force at the ground, we will need to see if and how the friction force appears in the expression for the work done on the wheel. Since the work-energy principle accounts for the work of all forces acting on the wheel, it is important to sketch the wheel's FBD, as shown in Fig. 2. The only friction we considered is that at the ground, and no couple is assumed to be acting on the wheel because the front wheels are *not* the driving wheels. The force H comes from the fact that the front axle of the car is pushing the wheel forward and the force R is due to the weight of the car pushing down on the tire.

Governing Equations

Balance Principles   The work-energy principle for each front wheel is

$$T_1 + U_{1\text{-}2} = T_2, \tag{1}$$

where the roadster's speed is zero at ① and v_0 at ②. The kinetic energy T can then be written as

$$T_1 = 0 \quad \text{and} \quad T_2 = \frac{1}{2} m_w v_{G2}^2 + \frac{1}{2} I_G \omega_{w2}^2, \tag{2}$$

where v_{G2} is the speed of G and ω_{w2} is the angular speed of the wheel, both at ②.

Force Laws   To express the work of each force appearing on the FBD, recall that if the point of application of a force $\vec{P}$ displaces by $d\vec{r}$, the corresponding work dU is

$$dU = \vec{P} \cdot d\vec{r} \quad \text{with} \quad d\vec{r} = \vec{v}\, dt, \tag{3}$$

where $\vec{v}$ is the velocity of the point of application of $\vec{P}$. Referring to Fig. 2,

1. Forces R and $m_w g$ are oriented vertically and do no work because their points of application move horizontally.

2. F and N do no work because the rolling without slip condition demands that $\vec{v}_Q = \vec{0}$, where Q is the point of application of F and N.

Therefore, the only force doing work is the force H and we can write

$$U_{1\text{-}2} = \int_{(\mathcal{L}_{1\text{-}2})_G} H\,\hat{\imath} \cdot d\vec{r}_G = \int_{x_{G1}}^{x_{G2}} H\, dx_G, \tag{4}$$

where G is the point of application of H and x_G is the horizontal position of G.

Kinematic Equations   At ② we have

$$v_{G2} = v_f. \tag{5}$$

To find ω_{w2}, we apply the kinematics of rolling without slip using

$$\vec{v}_G = \vec{v}_Q + \omega_w \, \hat{k} \times r \, \hat{j} \quad \text{with} \quad \vec{v}_Q = \vec{0}, \tag{6}$$

so that at ② we have

$$\vec{v}_{G2} = v_f \, \hat{i} = \omega_{w2} \, \hat{k} \times r \, \hat{j} \quad \Rightarrow \quad \omega_{w2} = -\frac{v_f}{r}. \tag{7}$$

Computation   Combining Eqs. (5) and (7) with Eq. (2), we have

$$T_1 = 0 \quad \text{and} \quad T_2 = \frac{1}{2} m_w v_f^2 + \frac{1}{2} I_G \left(-\frac{v_f}{r} \right)^2. \tag{8}$$

Substituting Eqs. (8) into Eq. (1) and solving for $U_{1\text{-}2}$, we have

$$U_{1\text{-}2} = \int_{x_{G1}}^{x_{G2}} H \, dx_G = \frac{1}{2} m_w v_f^2 + \frac{1}{2} I_G \left(-\frac{v_f}{r} \right)^2, \tag{9}$$

where we have also made use of Eq. (4). Equation (9) can be simplified to

$$\boxed{U_{1\text{-}2} = \int_{x_{G1}}^{x_{G2}} H \, dx_G = \frac{1}{2} \left(I_G + m_w r^2 \right) \left(\frac{v_f}{r} \right)^2.} \tag{10}$$

Now that we have found the expression for $U_{1\text{-}2}$, we see that the friction force does not appear in it. Therefore, we *cannot* calculate the average value of the friction force acting on the wheel directly from Eq. (10).

Discussion & Verification   The term v_f/r in Eq. (10) has dimensions of $1/\text{time}$. Since I_G is a mass moment of inertia, it has dimensions of mass × length². Therefore, overall the result in Eq. (10) has dimensions of $\left(\text{mass} \times \text{length/time}^2\right)(\text{length})$, i.e., dimensions of work, as it should.

✎ **A Closer Look**   In general, rolling without slip requires a friction force to act on the rolling body at the point of contact between the body in question and the rolling surface. However, this friction force does not dissipate any energy because the absence of slip prevents the friction force from doing work (i.e., the friction is being applied at a point that is not moving at that instant). Therefore, if a *rigid* body is rolling without slip on a *rigid* surface, the mechanical energy of the body will be conserved during the motion. This is an important result because in many problems involving rolling without slip we can apply conservation of energy even if a friction force does appear on the body's FBD.

Interesting Fact

If friction does no work, then what is "rolling resistance"? Rolling resistance occurs when one object rolls without slip over another object *and* one or both of the objects deform. Not only does the deformation itself dissipate energy, but also the equivalent normal force between the object and the ground impedes the motion.

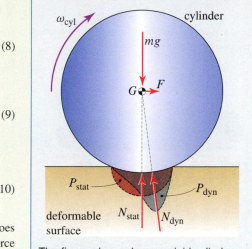

The figure above shows a rigid cylinder on a deformable surface. When $F = 0$ and the object is at rest, the pressure distribution on the object due to the deformation of the surface is given by P_{stat} and the equivalent normal force is given by the vertical force N_{stat}. On the other hand, when a force F is applied to the cylinder so that it rolls with a constant angular velocity ω_{cyl}, then the pressure distribution is that given by P_{dyn} and the equivalent normal force is N_{dyn}. Notice that there is a horizontal component of N_{dyn} that is impeding the forward motion of the cylinder. Since the cylinder is moving with a constant velocity, the horizontal component of N_{dyn} must be equal and opposite to F.

EXAMPLE 8.4   *Computing Engine Power Output and Wheel Torque*

Figure 1
A roadster on a horizontal stretch of road. Mazda Miata © 2006 Mazda Motor of America, Inc. Used by permission.

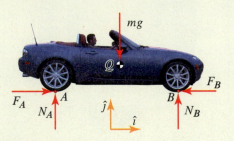

Figure 2
FBD of the car as a whole. Q is the center of mass of the whole car, and A and B are the contact points between the ground and the rear and front wheels, respectively.

The 2600 lb* rear-wheel-drive roadster shown in Fig. 1 accelerates from rest to 60 mph in 7 s. Each of the four wheels has a diameter $d = 24.3$ in., weighs 47 lb, and has a mass moment of inertia $I_G = 0.989\,\text{slug·ft}^2$ with respect to its mass center. Assuming that the car accelerates uniformly, that we neglect rolling resistance, bearing friction, and air drag, and that the wheels do not slip relative to the ground, estimate the average engine power output as well as the torque provided to the rear wheels.

SOLUTION

Road Map & Modeling   We can use the work-energy principle to compute the total work done on the car as the difference between the final and initial kinetic energies. Referring to the FBD in Fig. 2, we see that none of the external forces does work since the wheels roll without slip and the car's trajectory is horizontal. All the work done on the car is due to internal forces, namely, the torque provided by the engine. The corresponding average power output is then computed by dividing this work by the given time interval. To measure the average torque provided by the engine, we need to consider the forces internal to the car, depicted in Fig. 3. The work done by the forces

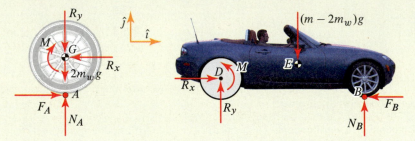

Figure 3. FBDs of the rear wheels and of the rest of car. Point E is the mass center of the car without the rear wheels. Points G and D are coincident: G is part of the wheel whereas D is part of the axle on which the wheel in mounted.

R_x and R_y applied at G on a rear wheel is equal and opposite to the work done by R_x and R_y applied at D because G and D do not move relative to one another. By contrast, the torque M does work because the rear wheels rotate relative to the rest of the car. Therefore, we can estimate M by dividing the work of the torque by the angle swept by the wheels during the car's motion.

Governing Equations

Balance Principles   The general form of the work-energy principle for a system is

$$T_1 + V_1 + (U_{1\text{-}2})_{\text{nc}}^{\text{ext}} + (U_{1\text{-}2})_{\text{nc}}^{\text{int}} = T_2 + V_2, \tag{1}$$

where the car's speed is equal to zero at ① and is equal to 60 mph = 88.00 ft/s at ②. The system's kinetic energy at ① and ② can be written as

$$T_1 = 0 \quad \text{and} \quad T_2 = \underbrace{\tfrac{1}{2}mv_2^2}_{\text{tr KE}} + \underbrace{4\left(\tfrac{1}{2}I_G\omega_{w2}^2\right)}_{\text{rot KE}} = \tfrac{1}{2}mv_2^2 + 2I_G\omega_{w2}^2, \tag{2}$$

where tr KE is the translational kinetic energy of the car, rot KE is the rotational kinetic energy of all four wheels, v_2 is the translational speed of the car and the wheels, and ω_{w2} is the angular velocity of the wheels.

---

* This weight includes the wheels, the fuel, and the passenger.

Force Laws   The external forces do no work, so we can write

$$(U_{1\text{-}2})_{nc}^{ext} = 0, \quad V_1 = 0, \quad \text{and} \quad V_2 = 0. \tag{3}$$

As for $(U_{1\text{-}2})_{nc}^{int}$, referring to Fig. 3, because the points G and D do not move relative to one another, R_x and R_y do no work. Consequently, all the internal work is done by the torque M, which can be written as

$$(U_{1\text{-}2})_{nc}^{int} = \int_{\theta_1}^{\theta_2} M\, d\theta = M\,(\theta_2 - \theta_1), \tag{4}$$

where θ measures the rotation of the rear wheels (see Fig. 4) and we have assumed that M is constant between ① and ②.

Kinematic Equations   At ②, the translational speed v_2 is given, and the corresponding ω_{w2} can be found by enforcing the rolling without slip condition, which gives

$$v_2 = 88.00\,\text{ft/s} \quad \text{and} \quad \omega_{w2} = \frac{v_2}{d/2} = 86.91\,\text{rad/s}. \tag{5}$$

Note that to compute M, we will need the value of $\theta_2 - \theta_1$, which can be calculated if the distance L traveled by the car between ① and ② is known. The distance L can be computed since we have assumed that the car is uniformly accelerating. Letting $\Delta t = 7\,\text{s}$, then the constant acceleration of the car is $a = v_2/\Delta t = 12.57\,\text{ft/s}^2$. Therefore, using constant acceleration equations, we have

$$L = \frac{1}{2}\,a\,\Delta t^2 = 308\,\text{ft} \quad \Rightarrow \quad \theta_2 - \theta_1 = \frac{L}{d/2} = 304.2\,\text{rad}. \tag{6}$$

Computation   Combining Eqs. (5) with Eqs. (2) and substituting in known values, we have

$$T_1 = 0 \quad \text{and} \quad T_2 = 327{,}600\,\text{ft·lb}. \tag{7}$$

Substituting Eqs. (7) into Eq. (1) and using Eqs. (3), for the average power P_{avg} we have

$$\boxed{(U_{1\text{-}2})_{nc}^{int} = 327{,}600\,\text{ft·lb} \quad \Rightarrow \quad P_{avg} = \frac{(U_{1\text{-}2})_{nc}^{int}}{\Delta t} = 46{,}800\,\frac{\text{ft·lb}}{\text{s}} = 85.1\,\text{hp}.} \tag{8}$$

Using Eq. (4), along with the first of Eqs. (8) and the result in Eq. (6), we have

$$\boxed{M\,(\theta_2 - \theta_1) = (U_{1\text{-}2})_{nc}^{int} \quad \Rightarrow \quad M = \frac{(U_{1\text{-}2})_{nc}^{int}}{\theta_2 - \theta_1} = 1080\,\text{ft·lb}.} \tag{9}$$

Discussion & Verification   The answers are all dimensionally correct. As far as the acceptability of each value is concerned, the calculated power may seem low with respect to the power of typical sports car engines (e.g., 170 hp at an engine speed of 6000 rpm). However, the power indicated in ads and product literature is a peak value corresponding to a specific angular velocity of the crankshaft. In our case, we have calculated an *average* power value, and, as such, it is not surprising that it is smaller than the peak values seen in advertisements. As far as the torque is concerned, the value we have obtained is not that unusual if we keep in mind that it measures the torque provided to the driving wheels. That is, the computed torque value should not be confused with the typically much smaller value of torque reported in product literature, which is the torque output at the crankshaft, i.e., before the transmission gets it to the wheels.

Work of internal forces. The work done by R_x and R_y on the rear wheels is equal and opposite to the work done by these forces on the rest of the car because the relative displacement of points G and D (see Fig. 3) is equal to zero. Therefore, the overall work of the internal forces R_x and R_y is equal to zero. If the displacement of G relative to D were not equal to zero, the car would be broken into two parts. By contrast, the rear wheels rotate relative to the car so that the *relative angular displacement* between these wheels and the rest of the car is not equal to zero. This allows the torque M to do (internal) work.

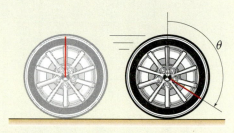

Figure 4

Definition of the angle θ measuring the rotation of the rear wheels. The red line is an arbitrarily chosen reference line that rotates with the wheels.

EXAMPLE 8.5   *An Overhead Fold-Up Door: Conservation of Energy*

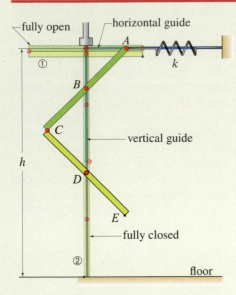

Figure 1
Side view of an overhead door at various positions during its operation.

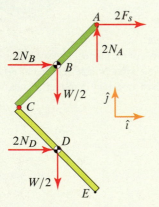

Figure 2
Side view of the system's FBD for a generic position between ① and ②. The factor of 2 in front of F_s, N_A, N_B, and N_D is necessary because there are two springs and two sets of rollers on the door (in the view shown only one set is visible). Note that the centers of mass of sections AC and CE coincide with points B and D, respectively.

An overhead door, with height $h = 30$ ft and weight $W = 800$ lb, consists of two identical sections hinged at C. The roller at A moves along a horizontal guide, whereas the rollers at B and D, which are the midpoints of sections AC and CE, move along a vertical guide. The door's operation is assisted by two identical springs attached to the horizontally moving rollers (only one of the two springs is shown). The springs are stretched 0.25 ft when the door is fully open. Determine the minimum value of the spring constant k if the roller at A is to strike the left end of the horizontal guide with a maximum speed of 1.5 ft/s after the door is released from rest in the fully open position.

SOLUTION

Road Map & Modeling   Since the desired value of k is found by relating changes in speed to changes in position, we will use the work-energy principle as a solution method. We will treat the two identical sections AC and CE as uniform thin rigid bodies, neglect the inertia of the rollers, neglect the mass of the springs, and neglect all friction. Letting ① and ② be the fully open and fully closed positions, respectively, the system's FBD for a generic position between ① and ② is that shown in Fig. 2. The force in each of the two springs is F_s, and N_A, N_B, and N_D are the reactions at each of the rollers A, B, and D, respectively. Note that the reactions at the rollers do no work because each roller moves in a direction perpendicular to the reaction force acting on it. The remaining forces W and F_s are conservative so that we have conservation of mechanical energy. Our solution strategy will be to find the value of k for which the speed of A is $v_{max} = 1.5$ ft/s and then show that larger values of k result in values of the speed of A at ② that are less than v_{max}.

Governing Equations

Balance Principles   Since mechanical energy is conserved, we can write

$$T_1 + V_1 = T_2 + V_2, \tag{1}$$

where, since the system is at rest at ①, T_1 and T_2 can be expressed as

$$T_1 = 0 \quad \text{and} \quad T_2 = \tfrac{1}{2}m_{AC}v_{B2}^2 + \tfrac{1}{2}I_B\omega_{AC2}^2 + \tfrac{1}{2}m_{CE}v_{D2}^2 + \tfrac{1}{2}I_D\omega_{CE2}^2, \tag{2}$$

where m_{AC} and I_B are the mass and mass moment of inertia of AC, respectively, and m_{CE} and I_D are the mass and mass moment of inertia of CE, respectively. Since AC and CE are identical, we have

$$m_{AC} = m_{CE} = \frac{W/2}{g} \quad \text{and} \quad I_B = I_D = \frac{1}{12}\left(\frac{W}{2g}\right)\left(\frac{h}{2}\right)^2 = \frac{Wh^2}{96g}. \tag{3}$$

Force Laws   Choosing the datum for gravitational potential energy as shown in Fig. 3 and recalling that there are two springs, we have

$$V_1 = 2\left(\tfrac{1}{2}k\delta_1^2\right) \quad \text{and} \quad V_2 = \frac{W}{2}y_{B2} + \frac{W}{2}y_{D2} + 2\left(\tfrac{1}{2}k\delta_2^2\right), \tag{4}$$

where $\delta_1 = 0.25$ ft, and where

$$y_{B2} = -h/4, \quad y_{D2} = -3h/4, \quad \text{and} \quad \delta_2 = \delta_1 + h/4. \tag{5}$$

Kinematic Equations   When at ②, B and D are at the lower limit of their respective motion ranges. In addition, given that B and D cannot move in the x direction, it must therefore be true that $\vec{v}_{B2} = \vec{0} = \vec{v}_{D2}$, and so

$$v_{B2} = 0 \quad \text{and} \quad v_{D2} = 0. \tag{6}$$

Referring to Fig. 3, observe that AC and CE rotate in opposite directions while remaining mirror images of one another relative to the line bisecting the angle ACE. Therefore, we must have

$$\omega_{AC} = -\omega_{CE} \quad \Rightarrow \quad \omega_{AC2} = -\omega_{CE2}. \tag{7}$$

In addition, since $\vec{v}_B = \vec{v}_A + \omega_{AC}\,\hat{k} \times \vec{r}_{B/A}$ and since $\vec{v}_{B2} = \vec{0}$, at ② we can write

$$\vec{v}_{B2} = \vec{0} = (v_{Ax})_2\,\hat{\imath} + \omega_{AC2}\,\hat{k} \times (-h/4)\,\hat{\jmath} = \left[(v_{Ax})_2 + h\omega_{AC2}/4\right]\hat{\imath}, \tag{8}$$

where we have used the fact that A can only move in the horizontal direction. Solving Eq. (8) for ω_{AC2} and using Eq. (7), we obtain

$$\omega_{AC2} = -4(v_{Ax})_2/h \quad \text{and} \quad \omega_{CE2} = 4(v_{Ax})_2/h. \tag{9}$$

Computation   Using Eq. (2), along with Eqs. (3), (6), and (9), we have

$$T_1 = 0 \quad \text{and} \quad T_2 = \frac{Wh^2}{96g}\frac{16(v_{Ax})_2^2}{h^2} = \frac{W(v_{Ax})_2^2}{6g}. \tag{10}$$

Substituting Eqs. (5) into Eqs. (4) and simplifying, we have

$$V_1 = k\delta_1^2 \quad \text{and} \quad V_2 = k\left(\delta_1^2 + \frac{\delta_1 h}{2} + \frac{h^2}{16}\right) - \frac{Wh}{2}. \tag{11}$$

Substituting Eqs. (10) and (11) into Eq. (1) and simplifying, we obtain

$$0 = \frac{W(v_{Ax})_2^2}{6g} + k\left(\frac{\delta_1 h}{2} + \frac{h^2}{16}\right) - \frac{Wh}{2}, \tag{12}$$

which can be solved for k to obtain

$$\boxed{k = 8\frac{W}{g}\frac{3hg - (v_{Ax})_2^2}{24\delta_1 h + 3h^2} = 200\,\text{lb/ft},} \tag{13}$$

where we have let $(v_{Ax})_2 = v_{\text{max}} = 1.5\,\text{ft/s}$.

Discussion & Verification   The result in Eq. (13) is consistent with the fact that k has dimensions of force over length and the units used in expressing the numerical result are therefore appropriate. In addition, the result matches our expectation that the slower $(v_{Ax})_2$ is, the stiffer the spring must be. Therefore the value of $k = 200\,\text{lb/ft}$ is the value of k we were looking for.

✎ A Closer Look   If we were to use the value of k in Eq. (13), each of the springs would be subject to $F_{s2} = k\delta_2 = 1550\,\text{lb}$ when at ②. If this value of force is judged to be too high, we can design a door with a smaller value of the spring constant (and therefore of force) using a system of counterweights that are lowered when the door opens and that are lifted when the door closes. A properly designed counterweight system can make the use of springs unnecessary.

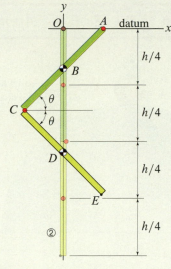

Figure 3
Coordinate system used with the indication of the datum choice. Note that, for convenience, only ② is shown. When the system is at ①, sections AC and CE lie on the x axis since their thickness has been neglected.

EXAMPLE 8.6   *Application of the Work-Energy Principle and $\vec{F} = m\vec{a}$*

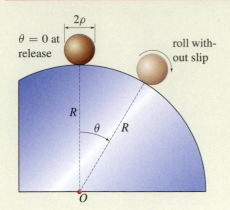

2ρ

$\theta = 0$ at release

roll without slip

R

θ R

O

Figure 1

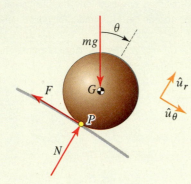

mg θ

F G $\hat{u}_r$

P $\hat{u}_\theta$

N

Figure 2
FBD of the sphere drawn at an arbitrary angle $\theta = \theta_2$, which corresponds to ②.

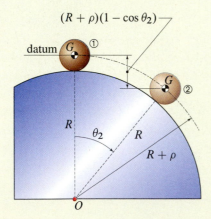

$(R + \rho)(1 - \cos\theta_2)$

datum G ①

G ②

R θ_2 R

$R + \rho$

O

Figure 3
Placement of the datum line coincides with the location of the center of the sphere at ①.

Let's revisit Example 7.5 (on p. 566), in which a solid uniform sphere was released from the top of a semicylinder and we determined where it separated from the semicylinder as it rolled *without slipping*. As was done in Example 7.5, we assume that the sphere is uniform; has mass m and radius ρ; and is released from rest by giving it a *slight* nudge to the right so that it begins to roll along the surface (Fig. 1). Assuming that there is sufficient friction between the surface and the sphere such that the sphere will not slip on the surface, we want to determine the angle θ at which the sphere separates from the surface. The purpose of this example is to show how the work-energy principle can be combined with the methods studied in Chapter 7 to obtain a solution without direct integration of the system's equations of motion.

SOLUTION

Road Map & Modeling   The FBD of the sphere is shown in Fig. 2. The sphere will begin to separate from the cylinder when the reaction N becomes zero. Therefore we need to write $\vec{F} = m\vec{a}_G$ in the direction of N (which is the radial direction) and solve for the value of θ at which $N = 0$. This equation will involve N, the sphere's weight, and a_{Gr}. As long as the sphere does not separate, G moves in a circle so that a_{Gr} depends on $\dot{\theta}$ but not $\ddot{\theta}$. This observation is important because it tells us that to find θ such that $N = 0$, we must relate $\dot{\theta}$ to θ, i.e., position to velocity, and this problem is ideal for the work-energy principle. Therefore our solution strategy will be to combine $F_r = ma_{Gr}$ with the work-energy principle.

Governing Equations

Balance Principles   Letting ① be when $\theta = 0$ and ② be at an arbitrary angle $\theta = \theta_2$, the Newton-Euler equations in the radial direction at ② and the work-energy principle applied to the sphere between ① and ② are, respectively,

$$\sum F_r: \quad N_2 - mg\cos\theta_2 = m(a_{Gr})_2, \tag{1}$$

$$T_1 + V_1 + (U_{1\text{-}2})_{\text{nc}} = T_2 + V_2, \tag{2}$$

where T_1 and T_2 can be written as

$$T_1 = 0 \quad \text{and} \quad T_2 = \tfrac{1}{2}mv_{G2}^2 + \tfrac{1}{2}I_G\omega_{s2}^2, \tag{3}$$

where v_{G2} and ω_{s2} are the velocity of the mass center of the sphere and the angular velocity of the sphere, respectively, at ②. The sphere's mass moment of inertia is given by

$$I_G = \tfrac{2}{5}m\rho^2. \tag{4}$$

Force Laws   Referring to Fig. 2, since the sphere rolls without slip, both F and N do no work so that the only force doing work is the sphere's weight. Referring to the datum line in Fig. 3, we have

$$(U_{1\text{-}2})_{\text{nc}} = 0, \quad V_1 = 0, \quad \text{and} \quad V_2 = -mg(R + \rho)(1 - \cos\theta_2). \tag{5}$$

Kinematic Equations   Recalling that we are using polar coordinates and that G moves along a circle of radius $R + \rho$, we have

$$(a_{Gr})_2 = -(R + \rho)\dot{\theta}_2^2. \tag{6}$$

Next we focus on the information needed to compute the kinetic energy. We know the kinetic energy at ①, and at ②, since G moves in the θ direction along a circle of radius $R + \rho$, we have

$$v_{G2} = (R + \rho)\dot{\theta}_2. \tag{7}$$

To compute ω_{s2} we need to apply the kinematics of rolling without slip. Rolling without slip implies that $\vec{v}_{P2} = \vec{0}$ (see Fig. 4), and so we can write

$$\vec{v}_{P2} = \vec{0} = \vec{v}_{G2} + \omega_{s2}\,\hat{k} \times (-\rho\,\hat{u}_r). \tag{8}$$

From Eq. (7), we know that $\vec{v}_{G2} = (R + \rho)\dot{\theta}_2\,\hat{u}_\theta$, so that Eq. (8) yields

$$(R + \rho)\dot{\theta}_2\,\hat{u}_\theta - \rho\omega_{s2}\,\hat{u}_\theta = \vec{0} \quad \Rightarrow \quad \omega_{s2} = \frac{R + \rho}{\rho}\dot{\theta}_2, \tag{9}$$

where we recall that $\hat{k}$ is defined such that $\hat{u}_r \times \hat{u}_\theta = \hat{k}$.

Computation   Combining Eqs. (7) and (9) with Eq. (3) yields

$$T_1 = 0 \quad \text{and} \quad T_2 = \frac{1}{2}m(R + \rho)^2\dot{\theta}_2^2 + \frac{1}{2}I_G\left(\frac{R + \rho}{\rho}\right)^2\dot{\theta}_2^2, \tag{10}$$

which, using Eq. (4), can be simplified to

$$T_1 = 0 \quad \text{and} \quad T_2 = \tfrac{7}{10}m(R + \rho)^2\dot{\theta}_2^2. \tag{11}$$

Substituting Eqs. (5) and (11) into Eq. (2) gives

$$0 = -mg(R + \rho)(1 - \cos\theta_2) + \tfrac{7}{10}m(R + \rho)^2\dot{\theta}_2^2, \tag{12}$$

which can be solved for $\dot{\theta}_2^2$ to obtain

$$\dot{\theta}_2^2 = \frac{10g(1 - \cos\theta_2)}{7(R + \rho)}. \tag{13}$$

Substituting Eq. (6) into Eq. (1) and solving for N_2, we obtain

$$N_2 = mg\cos\theta_2 - m(R + \rho)\dot{\theta}_2^2, \tag{14}$$

which, using Eq. (13), yields

$$N_2 = \tfrac{1}{7}mg(17\cos\theta_2 - 10). \tag{15}$$

As discussed above, we want to solve for the value of θ corresponding to $N = 0$. Therefore, letting $N_2 = 0$ gives

$$\boxed{\theta_2 = \cos^{-1}\left(\tfrac{10}{17}\right) = 54.0°,} \tag{16}$$

where, recalling that $\cos^{-1}(10/17) = \pm 54.0° + n360°, n = 0, \pm 1, \ldots, \pm\infty$, we have selected the only physically meaningful solution for θ_2.

Discussion & Verification   As expected, our solution matches that found in Example 7.5 on p. 566. What is important here is to notice that we were able to solve the same problem given in Example 7.5 by using a mixed solution strategy that combined both the use of $\vec{F} = m\vec{a}_G$ and the work-energy principle. The use of this principle allowed us to completely bypass the derivation and subsequent integration of the equation of motion in the θ direction.

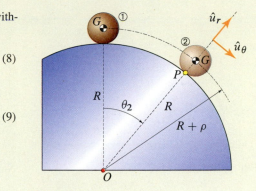

Figure 4
Component system at ②.

PROBLEMS

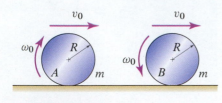

Figure P8.1 and P8.2

💡 Problem 8.1 💡

At the instant shown, the centers of the two identical uniform disks A and B are moving to the right with the same speed v_0. In addition, disk A is rolling clockwise with an angular speed ω_0 while disk B has a backspin with angular speed equal to ω_0. Letting T_A and T_B be the kinetic energies of A and B, respectively, state which of the following statements is true and why: (a) $T_A < T_B$; (b) $T_A = T_B$; (c) $T_A > T_B$.
Note: Concept problems are about *explanations*, not computations.

Problem 8.2

At the instant shown, the centers of the two identical uniform disks A and B, each with mass m and radius R, are moving to the right with the same speed $v_0 = 4\,\text{m/s}$. In addition, disk A is rolling clockwise with an angular speed $\omega_0 = 5\,\text{rad/s}$, while disk B has a backspin with angular speed $\omega_0 = 5\,\text{rad/s}$. Letting $m = 45\,\text{kg}$ and $R = 0.75\,\text{m}$, determine the kinetic energy of each disk.

💡 Problem 8.3 💡

Two identical battering rams are mounted on their respective frames as shown. Bars BC and AD are identical and pinned at B and C and at A and D, respectively. Bars FO and HO are rigidly attached to the ram and are pinned at O. At the instant shown, the mass centers of rams 1 and 2, at E and G, respectively, are moving horizontally with speed v_0. Letting T_1 and T_2 be the kinetic energies of rams 1 and 2, respectively, state which of the following statements is true and why: (a) $T_1 < T_2$; (b) $T_1 = T_2$; (c) $T_1 > T_2$.
Note: Concept problems are about *explanations*, not computations.

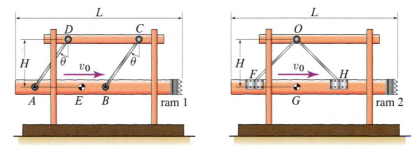

Figure P8.3 and P8.4

Problem 8.4

Two identical battering rams are mounted on their respective frames as shown. Bars BC and AD are identical and pinned at B and C and at A and D, respectively. Bars FO and HO are rigidly attached to the ram and are pinned at O. At the instant shown, the centers of mass of rams 1 and 2, at E and G, respectively, are moving horizontally with a speed $v_0 = 20\,\text{ft/s}$. Treating the rams as slender bars with length $L = 10\,\text{ft}$ and weight $W = 1250\,\text{lb}$, and letting $H = 3\,\text{ft}$, compute the kinetic energy of the two rams.

Problem 8.5

A pendulum consists of a uniform disk A of diameter $d = 0.15\,\text{m}$ and mass $m_A = 0.35\,\text{kg}$ attached at the end of a uniform bar B of length $L = 0.75\,\text{m}$ and mass $m_B = 0.8\,\text{kg}$. At the instant shown, the pendulum is swinging with an angular velocity $\omega = 0.24\,\text{rad/s}$ clockwise. Determine the kinetic energy of the pendulum at this instant, using Eq. (8.10) on p. 593.

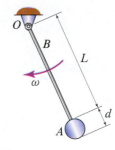

Figure P8.5

Problem 8.6

A 2570 lb car (this includes the weight of the wheels) is traveling on a horizontal flat road at 60 mph. If each wheel has a diameter $d = 24.3\,\text{in.}$ and a mass moment of inertia with respect to its mass center equal to $0.989\,\text{slug·ft}^2$, determine the kinetic energy of the car. Neglect the rotational energy of all parts of the car except for the wheels.

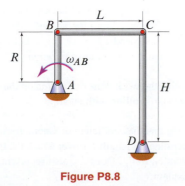

Figure P8.6

Problem 8.7

In Example 7.2 on p. 560 we analyzed the forces acting on a test tube in an ultra-centrifuge. Recalling that the center of mass G of the test tube was assumed to be at a distance $r = 0.0918\,\text{m}$ from the centrifuge's spin axis, that the test tube had a mass $m = 0.01\,\text{kg}$ and a mass moment of inertia $I_G = 2.821 \times 10^{-6}\,\text{kg·m}^2$, determine the kinetic energy of the test tube when it is spun at $\omega = 60{,}000\,\text{rpm}$. In addition, if you were to convert the computed kinetic energy to gravitational potential energy, at what height (expressed in meters) relative to the ground could you lift a 10 kg weight?

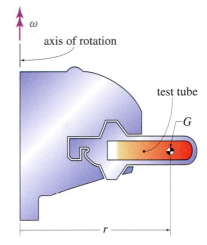

Figure P8.7

Problem 8.8

The uniform thin bars AB, BC, and CD have masses $m_{AB} = 2.3\,\text{kg}$, $m_{BC} = 3.2\,\text{kg}$, and $m_{CD} = 5.0\,\text{kg}$, respectively. The connections at A, B, C, and D are pinned joints. Letting $R = 0.75\,\text{m}$, $L = 1.2\,\text{m}$, and $H = 1.55\,\text{m}$, and knowing that bar AB rotates at an angular velocity $\omega_{AB} = 4\,\text{rad/s}$, compute the kinetic energy T of the system at the instant shown.

Figure P8.8

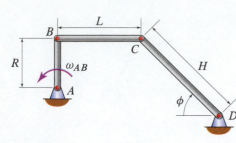

Figure P8.9

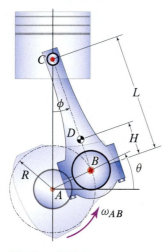

Figure P8.11 and P8.12

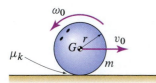

Figure P8.13–P8.15

Problem 8.9

The weights of the uniform thin pin-connected bars AB, BC, and CD are $W_{AB} = 4$ lb, $W_{BC} = 6.5$ lb, and $W_{CD} = 10$ lb, respectively. Letting $\phi = 47°$, $R = 2$ ft, $L = 3.5$ ft, and $H = 4.5$ ft, and knowing that bar AB rotates at an angular velocity $\omega_{AB} = 4$ rad/s, compute the kinetic energy T of the system at the instant shown.

Problem 8.10

The uniform slender bar AB has length $L = 1.45$ ft and weight $W_{AB} = 20$ lb. Rollers D and E, which are pinned at A and B, respectively, can be modeled as two identical uniform disks, each with radius $r = 1.5$ in. and weight $W_r = 0.35$ lb. Rollers D and E roll without slip on the surface of a cylindrical bowl with center at O and radius $R = 1$ ft. Determine the system's kinetic energy when G (the center of mass of bar AB) moves with a speed $v = 7$ ft/s.

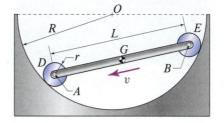

Figure P8.10

Problems 8.11 and 8.12

For the slider-crank mechanism shown, let $L = 141$ mm, $R = 48.5$ mm, and $H = 36.4$ mm. In addition, observing that D is the center of mass of the connecting rod, let the mass moment of inertia of the connecting rod be $I_D = 0.00144$ kg·m² and the mass of the connecting rod be $m = 0.439$ kg.

Problem 8.11 Letting $\omega_{AB} = 2500$ rpm, compute the kinetic energy of the connecting rod for $\theta = 90°$ and for $\theta = 180°$.

Problem 8.12 Plot the kinetic energy of the connecting rod as a function of the crank angle θ over one full cycle of the crank for $\omega_{AB} = 2500$ rpm, 5000 rpm, and 7500 rpm.

Problems 8.13 and 8.14

A 14 lb bowling ball is thrown onto a lane with a backspin angular speed $\omega_0 = 10$ rad/s and forward velocity $v_0 = 17$ mph. After a few seconds, the ball starts rolling without slip and moving forward with a speed $v_f = 17.2$ ft/s. Let $r = 4.25$ in. be the radius of the ball, and let $k_G = 2.6$ in. be its radius of gyration.

Problem 8.13 Determine the work done by friction on the ball from the initial time until the time that the ball starts rolling without slip.

Problem 8.14 Knowing that the coefficient of kinetic friction between the lane and the ball is $\mu_k = 0.1$, determine the length L_f over which the friction force acts in order to slow down the ball from v_0 to v_f. Does L_f also represent the distance traveled by the center of the ball? Explain.

Problem 8.15

A bowling ball is thrown onto a lane with a forward velocity v_0 and no angular velocity ($\omega_0 = 0$). Because of friction between the lane and the ball, after a short time, the ball starts rolling without slip and moving forward with speed v_f. Let L_G be the distance traveled by the center of the ball while slowing down from v_0 to v_f. In addition, let L_f be the length over which the friction force had to act in order to slow down the ball from v_0 to v_f. State which of the following relations is true and why: (a) $L_G < L_f$; (b) $L_G = L_f$; (c) $L_G > L_f$.
Note: Concept problems are about *explanations*, not computations.

Problem 8.16

A conveyor is moving cans at a constant speed v_0 when, to proceed to the next step in packaging, the cans are transferred onto a stationary surface at A. The cans each have mass m, width w, and height h. Assuming that there is friction between each can and the stationary surface, under what conditions would we be able to compute the stopping distance of the cans, using the work-energy principle for a particle?
Note: Concept problems are about *explanations*, not computations.

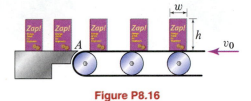

Figure P8.16

Problem 8.17

One of the basement doors is left open in the vertical position when it is given a nudge and allowed to freely fall to the closed position. Given that the door has mass m and that it is modeled as a uniform thin plate of width w and length d, determine its angular velocity when it reaches the closed position. *Hint:* Treat the problem as symmetric with respect to a plane of motion in which the acceleration due to gravity is $g \cos \theta$ rather than g.

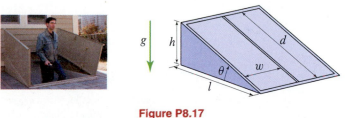

Figure P8.17

Problem 8.18

The disk D, which has mass m, center of mass G, and radius of gyration k_G, is at rest on a flat horizontal surface when the constant moment M is applied to it. The disk is attached at its center to a vertical wall by a linear elastic spring of constant k. The spring is unstretched when the system is at rest. Assuming that the disk rolls without slipping and that it has not yet come to a stop, determine the angular velocity of the disk after its center G has moved a distance d. After doing so, determine the distance d_s that the disk moves before it comes to a stop.

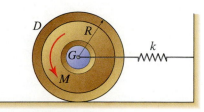

Figure P8.18

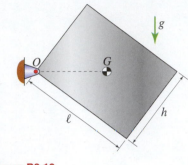

Figure P8.19

Problem 8.19

The uniform rectangular plate of length ℓ, height h, and mass m lies in the vertical plane and is pinned at one corner. If the plate is released from rest in the position shown, determine its angular velocity when the center of mass G is directly below the pivot O. Neglect any friction at the pin at O.

Problem 8.20

A turbine rotor with weight $W = 3000$ lb, center of mass at the fixed point G, and radius of gyration $k_G = 15$ ft is brought from rest to an angular velocity $\omega = 1500$ rpm in 20 revolutions by applying a constant torque M. Neglecting friction, determine the value of M needed to spin up the rotor as described.

Figure P8.20

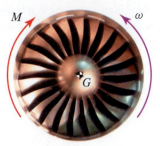

Figure P8.21

Problem 8.21

A turbine rotor with weight $W = 3000$ lb, center of mass at the fixed point G, and radius of gyration $k_G = 15$ ft is spinning with an angular speed $\omega = 1200$ rpm when a braking system is engaged that applies a constant torque $M = 3000$ ft·lb. Determine the number of revolutions needed to bring the rotor to a stop.

Problems 8.22 and 8.23

In a contraption built by a fraternity, a person sits at the center of a swinging platform with mass $m = 400$ kg and length $L = 4$ m suspended via two identical arms of length $H = 3$ m.

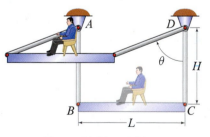

Figure P8.22 and P8.23

Problem 8.22

Neglecting the mass of the arms and of the person, neglecting friction, and assuming that the platform is released from rest when $\theta = 180°$, compute the speed of the person as a function θ for $0° \le \theta \le 180°$. In addition, find the speed of the person for $\theta = 0°$.

Problem 8.23

Neglecting the mass of the person, neglecting friction, letting the mass of each arm be $m_A = 150$ kg, and assuming that the platform is released from rest when $\theta = 180°$, compute the speed of the person as a function θ for $0° \leq \theta \leq 180°$. In addition, find the speed of the person for $\theta = 0°$.

Problem 8.24

An eccentric wheel with weight $W = 250\,\text{lb}$, mass center G, and radius of gyration $k_G = 1.32\,\text{ft}$ is initially at rest in the position shown. Letting $R = 1.75\,\text{ft}$ and $h = 0.8\,\text{ft}$, and assuming that the wheel is gently nudged to the right and rolls without slip, determine the speed of O when G is closest to the ground.

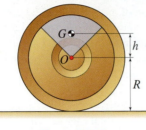

Figure P8.24

Problem 8.25

An eccentric wheel with mass $m = 150\,\text{kg}$, mass center G, and radius of gyration $k_G = 0.4\,\text{m}$ is placed on the incline shown such that the wheel's center of mass G is vertically aligned with P, which is the point of contact with the incline. If the wheel rolls without slip once it is gently nudged away from its initial placement, letting $R = 0.55\,\text{m}$, $h = 0.25\,\text{m}$, $\theta = 25°$, and $d = 0.5\,\text{m}$, determine whether the wheel arrives at B and, if yes, determine the corresponding speed of the center O. Note that the angle POG is not equal to $90°$ at release.

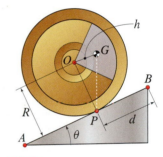

Figure P8.25

Problems 8.26 and 8.27

In a contraption built by a fraternity, a person sits at the center of a swinging platform with weight $W_p = 800\,\text{lb}$ and length $L = 12\,\text{ft}$ suspended via two identical arms each of length $H = 10\,\text{ft}$ and weight $W_a = 200\,\text{lb}$. The platform, which is at rest when $\theta = 0$, is put in motion by a motor that pumps the ride by exerting a constant moment M in the direction shown whenever $0 \leq \theta \leq \theta_p$ while exerting zero moment for any other value of θ.

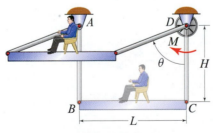

Figure P8.26 and P8.27

Problem 8.26   Neglecting the mass of the person, neglecting friction, letting $M = 900\,\text{ft·lb}$, and letting $\theta_p = 25°$, find the minimum number of swings necessary to achieve $\theta > 90°$ and the ensuing speed achieved by the person at the lowest point in the swing. Model the arms AB and CD as uniform thin bars.

Problem 8.27   Neglecting the mass of the person, neglecting friction, and letting $\theta_p = 20°$, determine the value of M required to achieve a maximum value of θ equal to $90°$ in 6 full swings. Model the arms AB and CD as uniform thin bars.

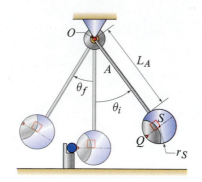

Figure P8.28 and P8.29

Problems 8.28 and 8.29

The *Charpy impact test* is one test that measures the resistance of a material to fracture. In this test, the fracture toughness is assessed by measuring the energy required to break a specimen of a given geometry. This is done by releasing a heavy pendulum from rest at an angle θ_i and then measuring the maximum swing angle θ_f reached by the pendulum after the specimen is broken.

Problem 8.28 Consider a test rig in which the striker S (the pendulum's bob) can be modeled as a uniform disk of mass $m_S = 19.5\,\text{kg}$ and radius $r_S = 150\,\text{mm}$, and the arm can be modeled as a thin rod of mass $m_A = 2.5\,\text{kg}$ and length $L_A = 0.8\,\text{m}$. Neglecting friction and noting that the striker and the arm are rigidly connected, determine the fracture energy (i.e., the kinetic energy lost in breaking the specimen) in an experiment where $\theta_i = 158°$ and $\theta_f = 43°$.

Problem 8.29 Consider a test rig in which the striker S (the pendulum's bob) can be modeled as a uniform disk of weight $W_S = 40\,\text{lb}$ and radius $r_S = 6\,\text{in.}$, and the arm can be modeled as a thin rod of weight $W_A = 5.5\,\text{lb}$ and length $L_A = 2.75\,\text{ft}$. If the release angle of the striker is $\theta_i = 158°$ and if the striker impacts the specimen when the pendulum's arm is vertical, determine the speed of the point Q on the striker immediately before the striker impacts with the specimen. Neglect friction and observe that the striker and the arm are rigidly connected.

Problem 8.30

A crate, with weight $W = 155\,\text{lb}$ and mass center G, is placed on a slide and released from rest as shown. The lower part of the slide is circular with radius $R = 6\,\text{ft}$. Model the crate as a uniform body with $b = 3.6\,\text{ft}$ and $h = 2\,\text{ft}$, take into account the gap between the crate and the slide when the crate is in its lowest position, and assume that when the crate is in its lowest position on the slide, the crate's center of mass is moving to the left with a speed $v_G = 12\,\text{ft/s}$. Determine the work done by friction on the crate as the crate moves from the release point to the lowest point on the slide.

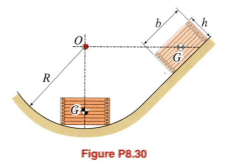

Figure P8.30

Problem 8.31

The disk D, which has weight W, mass center G coinciding with the disk's geometric center, and radius of gyration k_G, is at rest on an incline when the constant moment M is applied to it. The disk is attached at its center to a wall by a linear elastic spring of constant k. The spring is unstretched when the system is at rest. Assuming that the disk rolls without slipping and that it has not yet come to a stop, determine the angular velocity of the disk after its center G has moved a distance d down the incline. After doing so, using $k = 5\,\text{lb/ft}$, $R = 1.5\,\text{ft}$, $W = 10\,\text{lb}$, and $\theta = 30°$, determine the value of the moment M for the disk to stop after rolling $d_s = 5\,\text{ft}$ down the incline.

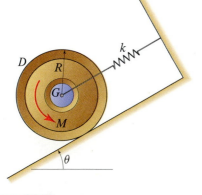

Figure P8.31

Problem 8.32

The figure shows the cross section of a garage door with length $L = 9$ ft and weight $W = 175$ lb. At A and B there are rollers of negligible mass constrained to move in the guide whose horizontal portion is at a distance $H = 11$ ft from the floor. The door's motion is assisted by two springs, each with constant k (only one spring is shown). The door is released from rest when $d = 26$ in. and the spring is stretched 4 in. Neglecting friction, knowing that, when A touches the floor, B is in the vertical portion of the guide, and modeling the door as a uniform thin plate, determine the minimum value of k so that A will strike the ground with a speed no greater than 1 ft/s.

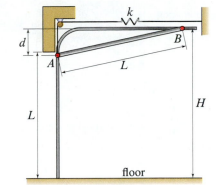

Figure P8.32

Problem 8.33

In Example 8.2 on p. 602 we ignored the rotational inertia of the counterweight. Let's revisit that example and remove that simplifying assumption. Assume that the arm AD is still a uniform thin bar of length $L = 15.7$ ft and weight 45 lb. The hinge O is still $d = 2.58$ ft from the right end of arm, and the 160 lb counterweight C is still $\delta = 1.4$ ft from the hinge. Now model the counterweight as a uniform block of height $h = 14$ in. and width $w = 9$ in. With this new assumption, solve for the angular velocity of the arm as it reaches the horizontal position after being nudged from the vertical position. Determine the percent change in angular velocity compared with that found in Example 8.2.

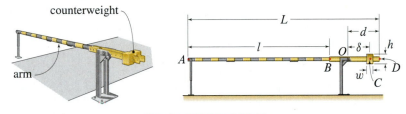

Figure P8.33 and P8.34

Problem 8.34

For the barrier gate shown, assume that the arm consists of a section of aluminum tubing from A to B of length $l = 11.6$ ft and weight 20 lb and a steel support section from B to D of weight 40 lb. The overall length of the arm is $L = 15.7$ ft. In addition, the 120 lb counterweight C is placed a distance δ from the hinge at O, and the hinge is $d = 2.58$ ft from the right end of section BD. Model the two sections AB and BD as uniform thin bars, and model the counterweight as a uniform block of height $h = 14$ in. and width $w = 9$ in. Using these new assumptions, determine the distance δ so that the angular velocity of the arm is 0.25 rad/s as it reaches the horizontal position after being nudged from the vertical position.

Problem 8.35

The figure shows the cross section of a garage door with length $L = 2.5$ m and mass $m = 90$ kg. At the ends A and B there are rollers of negligible mass constrained to move in the guide whose horizontal portion is at a distance $H = 3$ m from the floor. The door's motion is assisted by two counterweights C, each of mass m_C (only one counterweight is shown). If the door is released from rest when $d = 53$ cm, neglecting friction and modeling the door as a uniform thin plate, determine the minimum value of m_C so that A will strike the ground with a speed no greater than 0.25 m/s.

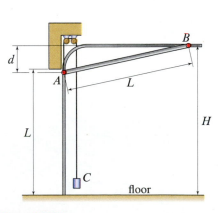

Figure P8.35

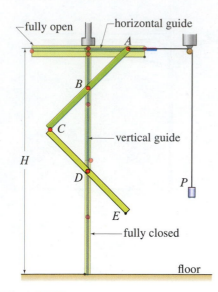

Figure P8.36

Figure P8.37 and P8.38

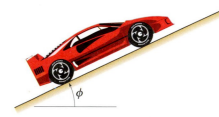

Figure P8.39

Problem 8.36

Revisit Example 8.5 on p. 608 and replace the two springs with a system of two counterweights P (only one counterweight is shown) each of weight W_P. Recalling that the door's weight is $W = 800$ lb and that the total height of the door is $H = 30$ ft, if the door is released from rest in the fully open position and friction is negligible, determine the minimum value of W_P so that A will strike the left end of the horizontal guide with a speed no greater than 0.5 ft/s.

Problems 8.37 through 8.39

Torsional springs provide a simple propulsion mechanism for toy cars. When the rear wheels are rotated as if the car were moving backward, they cause a torsional spring (with one end attached to the axle and the other to the body of the car) to wind up and store energy. Therefore, a simple way to charge the spring is to place the car onto a surface and to pull it backward, making sure that the wheels roll without slipping. Note that the torsional spring can only be wound by pulling the car backward; that is, *the forward motion of the car unwinds the spring*.

Problem 8.37   Let the weight of the car (body and wheels) be $W = 5$ oz, the weight of each of the wheels be $W_w = 0.15$ oz, and the radius of the wheels be $r = 0.25$ in., where the wheels roll without slip and can be treated as uniform disks. Neglecting friction internal to the car and letting the car's torsional spring be linear with constant $k_t = 0.0002$ ft·lb/rad, determine the maximum speed achieved by the car if it is released from rest after pulling it back a distance $L = 0.75$ ft from a position in which the spring is unwound.

Problem 8.38   Let the weight of the car (body and wheels) be $W = 5$ oz, the weight of each of the wheels be $W_w = 0.15$ oz, and the radius of the wheels be $r = 0.25$ in., where the wheels roll without slip and can be treated as uniform disks. In addition, let the torque M provided by the nonlinear torsional spring be given by $M = -\beta\theta^3$, where $\beta = 0.5\times10^{-6}$ ft·lb/rad³, θ is the angular displacement of the rear axle, and the minus sign in front of β indicates that M acts opposite to the direction of θ. Neglecting any friction internal to the car, determine the maximum speed achieved by the car if it is released from rest after pulling it back a distance $L = 0.75$ ft from a position in which the spring is unwound.

Problem 8.39   Let the mass of the car (body and wheels) be $m = 120$ g, the mass of each of the wheels be $m_w = 5$ g, and the radius of the wheels be $r = 6$ mm, where the wheels roll without slip and can be treated as uniform disks. In addition, let the car's torsional spring be linear with constant $k_t = 0.00025$ N·m/rad. Neglecting any friction internal to the car, if the angle of the incline is $\phi = 25°$ and the car is released from rest after pulling it back a distance $L = 25$ cm from a position in which the spring is unwound, determine the maximum distance d_{max} that the car will travel up the incline (from its release point), the maximum speed v_{max} achieved by the car, and the distance $d_{v\text{max}}$ (from the release point) at which v_{max} is achieved.

Problems 8.40 and 8.41

The double pulley D has mass of 15 kg, center of mass G coinciding with its geometric center, radius of gyration $k_G = 10$ cm, outer radius $r_o = 15$ cm, and inner radius $r_i = 7.5$ cm. It is connected to the pulley P with radius R via a cord of negligible mass that unwinds from the inner and outer spools of the double pulley D. The crate C, which has a mass of 20 kg, is released from rest.

Problem 8.40 Neglecting the mass of the pulley P, determine the speed of the crate C and the angular velocity of the pulley D after the crate has dropped a distance $h = 2\,\text{m}$.

Problem 8.41 Assuming that the pulley P has a mass of 1.5 kg and a radius of gyration $k_A = 3.5\,\text{cm}$, determine the speed of the crate C and the angular velocity of the pulley D after the crate has dropped a distance $h = 2\,\text{m}$.

Problems 8.42 through 8.44

The uniform thin rod AB is pin-connected to the slider S, which moves along the frictionless guide, and to the disk D, which rolls without slip over the horizontal surface. The pins at A and B are frictionless, and the system is released from rest. Neglect the vertical dimension of S.

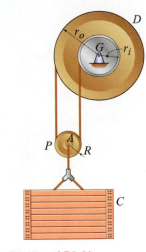

Figure P8.40 and P8.41

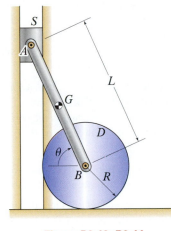

Figure P8.42–P8.44

Problem 8.42 Letting $L = 1.75\,\text{m}$ and $R = 0.6\,\text{m}$, assuming that S and D are of negligible mass, that the mass of rod AB is $m_{AB} = 7\,\text{kg}$, and that the system is released from the angle $\theta_0 = 65°$, determine the speed of the slider S when it strikes the ground.

Problem 8.43 Letting $L = 4.5\,\text{ft}$ and $R = 1.2\,\text{ft}$, assuming that AB is of negligible mass, the weight of S is $W_S = 3\,\text{lb}$, D is a uniform disk of weight $W_D = 9\,\text{lb}$, and the system is released from the angle $\theta_0 = 67°$, determine the speed of the slider S when it strikes the ground.

Problem 8.44 Letting $L = 1.75\,\text{m}$ and $R = 0.6\,\text{m}$, assuming that the mass of S is $m_S = 4.2\,\text{kg}$, D is a uniform disk of mass $m_D = 12\,\text{kg}$, the mass of AB is $m_{AB} = 7\,\text{kg}$, and that the system is released from the angle $\theta = 69°$, determine the speed and the direction of motion of point B when the slider S strikes the ground.

Problem 8.45

The figure shows the cross section of a garage door with length $L = 2.5\,\text{m}$ and mass $m = 90\,\text{kg}$. At the ends A and B there are rollers of negligible mass constrained to move in a vertical and a horizontal guide, respectively. The door's motion is assisted by two counterweights (only one counterweight is shown), each of mass $m_C = 42.5\,\text{kg}$. If the door is released from rest when horizontal, neglecting friction and modeling the door as a uniform thin plate, determine the speed with which B strikes the left end of the horizontal guide.

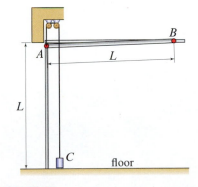

Figure P8.45

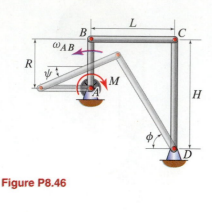

Figure P8.46

Problem 8.46

The uniform thin pin-connected bars AB, BC, and CD have masses $m_{AB} = 2.3$ kg, $m_{BC} = 3.2$ kg, and $m_{CD} = 5.0$ kg, respectively. In addition, $R = 0.75$ m, $L = 1.2$ m, and $H = 1.55$ m. When bars AB and CD are vertical, AB is rotating with angular speed $\omega_{AB} = 4$ rad/s in the direction shown. At this instant, the motor connected to AB starts to exert a constant torque M in the direction opposite to ω_{AB}. If the motor stops AB after AB has rotated 90° counterclockwise, determine M and the maximum power output of the motor during the stopping phase. In the final position, $\phi = 64.36°$ and $\psi = 29.85°$.

Problem 8.47

A stick of length L and mass m is in equilibrium while standing on its end A when end B is gently nudged to the right, causing the stick to fall. Model the stick as a uniform slender bar, and assume that there is friction between the stick and the ground. Under these assumptions, there is a value of θ, let's call it $\theta_{\max}$, such that the stick *must* start slipping before reaching $\theta_{\max}$ for *any* value of the coefficient of static friction μ_s. To find the value of $\theta_{\max}$, follow the steps below.

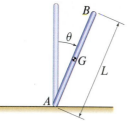

Figure P8.47

(a) Letting F and N be the friction and normal forces, respectively, between the stick and the ground, draw the FBD of the stick as it falls. Then set the sum of forces in the horizontal and vertical directions equal to the corresponding components of $m\vec{a}_G$. Express the components of $\vec{a}_G$ in terms of θ, $\dot{\theta}$, and $\ddot{\theta}$. Finally, express F and N as functions of θ, $\dot{\theta}$, and $\ddot{\theta}$.

(b) Use the work-energy principle to find an expression for $\dot{\theta}^2(\theta)$. Differentiate the expression for $\dot{\theta}^2(\theta)$ with respect to time, and find an expression for $\ddot{\theta}(\theta)$.

(c) Substitute the expressions for $\dot{\theta}^2(\theta)$ and $\ddot{\theta}(\theta)$ into the expressions for F and N to obtain F and N as functions of θ. For impending slip, $|F/N|$ must be equal to the coefficient of static friction. Use this fact to determine $\theta_{\max}$.

Problem 8.48

A stick of length L and mass m is in equilibrium while standing on its end A when the end B is gently nudged to the right, causing the stick to fall. Letting μ_s be the coefficient of static friction between the stick and the ground and modeling the stick as a uniform slender bar, find the largest value of μ_s for which the stick slides to the left as well as the corresponding value of θ at which sliding begins. To solve this problem, follow the steps below.

(a) Let F and N be the friction and normal forces, respectively, between the stick and the ground, and let F be positive to the right and N positive upward. Draw the FBD of the stick as it falls. Then set the sum of forces in the horizontal and vertical directions equal to the corresponding components of $m\vec{a}_G$. Express the components of $\vec{a}_G$ in terms of θ, $\dot{\theta}$, and $\ddot{\theta}$. Finally, express F and N as functions of θ, $\dot{\theta}$, and $\ddot{\theta}$.

(b) Use the work-energy principle to find an expression for $\dot{\theta}^2(\theta)$. Differentiate the expression for $\dot{\theta}^2(\theta)$ with respect to time, and find an expression for $\ddot{\theta}(\theta)$.

(c) Substitute the expressions for $\dot{\theta}^2(\theta)$ and $\ddot{\theta}(\theta)$ into the expressions for F and N to obtain F and N as functions of θ. When slip is impending (i.e., when $|F| = \mu_s|N|$), $|F/N|$ must be equal to the static coefficient of friction. Therefore, compute the maximum value of $|F/N|$ by differentiating it with respect to θ and setting the resulting derivative equal to zero.

Problem 8.49

A stick of length L and mass m is in equilibrium while standing on its end A when end B is gently nudged to the right, causing the stick to fall. Letting the coefficient of static friction between the stick and the ground be $\mu_s = 0.7$ and modeling the stick as a uniform slender bar, find the value of θ at which end A of the stick starts slipping and determine the corresponding direction of slip. As part of the solution, plot the absolute value of the ratio between the friction and normal force as a function of θ. To solve this problem, follow the steps below.

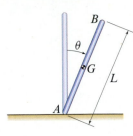

Figure P8.49

(a) Letting F and N be the friction and normal forces, respectively, between the stick and the ground, draw the FBD of the stick as it falls. Then set the sum of forces in the horizontal and vertical directions equal to the corresponding components of $m\vec{a}_G$. Express the components of $\vec{a}_G$ in terms of θ, $\dot{\theta}$, and $\ddot{\theta}$. Finally, express F and N as functions of θ, $\dot{\theta}$, and $\ddot{\theta}$.

(b) Use the work-energy principle to find an expression for $\dot{\theta}^2(\theta)$. Differentiate the expression for $\dot{\theta}^2(\theta)$ with respect to time, and find an expression for $\ddot{\theta}(\theta)$.

(c) After substituting the expressions for $\dot{\theta}^2(\theta)$ and $\ddot{\theta}(\theta)$ into the expressions for F and N, plot $|F/N|$ as a function of θ. For impending slip, $|F/N|$ must be equal to μ_s. Therefore, the desired value of θ corresponds to the intersection of the plot of $|F/N|$ with the horizontal line intercepting the vertical axis at the value 0.7. After determining the desired value of θ, the direction of slip can be found by determining the sign of F evaluated at the θ computed.

DESIGN PROBLEMS

Design Problem 8.1

The opening and closing of the manually operated road barrier is assisted by the counterweight C and linear elastic torsional spring with constant k_t that is mounted at O. Assume that the length of the arm is $L = 15.7\,\text{ft}$ and that it consists of a section of aluminum tubing from A to B of length $l = 11.6\,\text{ft}$ and weight 20 lb and a steel support section from B to D of weight 40 lb. Model both sections of the arm as uniform thin rods. Model the counterweight as a uniform rectangular rigid body of weight W_c, height h, and width w, and let the hinge O be a distance $d = 2.58\,\text{ft}$ from the right end of section BD.

Using these assumptions, design the unspecified parameters δ, h, w, W_C, and the torsional spring (its stiffness k_t and the position at which it is undeformed) so that a small nudge will close the barrier from the vertical position and so that the arm will reach the closed position with an angular velocity that is less than $0.25\,\text{rad/s}$. In addition, make sure that the barrier is still easy to open.

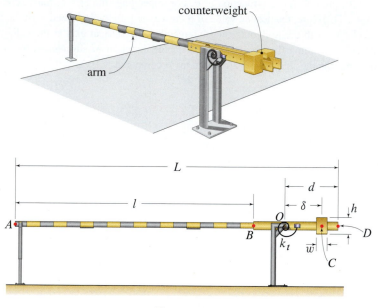

Figure DP8.1

8.2   Momentum Methods for Rigid Bodies

In this section, we develop both the linear and the angular impulse-momentum principles for rigid bodies. This is a departure from what was done in Chapter 5, where we devoted individual sections to each of these principles. The reason for this different approach is that, for rigid bodies, the linear and angular impulse-momentum principles must often be applied together to get a complete picture of a body's motion, as shown in the following example.

Rigid body modeling in crash reconstructions

Consider the impending collision shown in Fig. 8.13. Before the collision each vehicle is moving along a straight line. In Fig. 8.14, we see that after the collision both vehicles are displaced *and* rotated relative to their positions at the time of impact. Is it possible to predict the postimpact velocities, including angular velocities, if we know the preimpact motion? In Section 5.2 (on p. 356) we learned how to predict postimpact velocities from the preimpact velocities, but we did not deal with angular velocities (the concept of angular velocity does not pertain to particles). To determine the postimpact motion, observe that if we model the impact as two-dimensional and model the vehicles as rigid bodies, *we need six equations to solve the problem!* We see this by referring to Fig. 8.15 and calling C and D the mass centers of vehicles A and B, respectively. We need four equations to find v_{Cx}^+, v_{Cy}^+, v_{Dx}^+, and v_{Dy}^+ (recall that $+$ stands for postimpact) and we need two equations to find ω_A^+ and ω_B^+, which are the postimpact angular velocities of A and B, respectively. The impact theory we learned in Section 5.2 (on p. 356) gives us only four of the equations we need:

1. Conservation of the system's momentum along the LOI (line of impact);

2. Conservation of the momentum of A in the direction normal to the LOI;

3. Conservation of the momentum of B in the direction normal to the LOI;

4. COR (coefficient of restitution) equation along the LOI.

What physical principle can we use to write the remaining two equations? The answer is found by observing that the equations listed in points 1–4 are based on the *linear* impulse-momentum principle and say nothing about rotational motion. Rigid body motion is governed by both a force equation and a *moment* equation. Therefore, we need to take advantage of the moment equations of bodies A and B to obtain the two equations we are seeking. Specifically, we need to supplement the linear impulse-momentum principle with the *angular* impulse-momentum principle, which is derived from the moment equation. This is what the current section is about. After we derive these two balance principles, we will come back to the rigid body impact problem in Section 8.3 (on p. 646).

Impulse-momentum principle for a rigid body

A rigid body's mass center moves according to Eq. (7.15) on p. 548 (first given for general systems in Eq. (5.86) on p. 394), i.e.,

$$\vec{F} = m\vec{a}_G, \qquad (8.34)$$

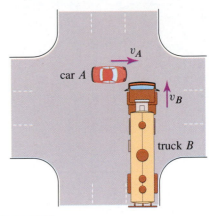

Figure 8.13
A car and a truck that are about to collide.

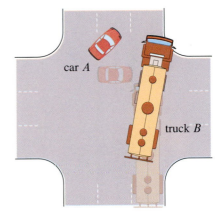

Figure 8.14
Vehicles before (in transparency) and after the collision.

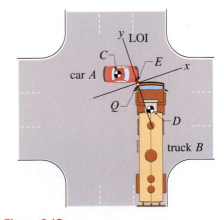

Figure 8.15
Vehicles at the instant of collision and in relation to the line of impact.

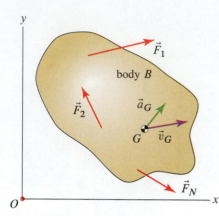

Figure 8.16
A rigid body under the action of a system of forces.

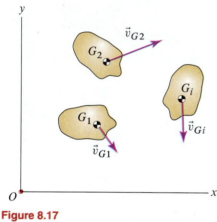

Figure 8.17
A system of rigid bodies.

where, referring to Fig. 8.16, $\vec{F} = \vec{F}_1 + \vec{F}_2 + \cdots + \vec{F}_N$ is the total force acting on the body and m and $\vec{a}_G$ are the body's mass and acceleration of the mass center, respectively. Integrating Eq. (8.34) over time for $t_1 \leq t \leq t_2$, we obtain

$$\int_{t_1}^{t_2} \vec{F}\, dt = \int_{t_1}^{t_2} m\vec{a}_G\, dt = m\vec{v}_G(t_2) - m\vec{v}_G(t_1), \qquad (8.35)$$

where, using concepts introduced in Section 5.1, the first term in Eq. (8.35) is the total *linear impulse* acting on the body and $m\vec{v}_G(t)$ is the body's *linear momentum*, which we denote by $\vec{p}$. As we have done for a particle (see Eq. (5.6) on p. 335), we can rewrite Eq. (8.35) as

$$m\vec{v}_{G1} + \int_{t_1}^{t_2} \vec{F}\, dt = m\vec{v}_{G2} \quad \text{or} \quad \vec{p}_1 + \int_{t_1}^{t_2} \vec{F}\, dt = \vec{p}_2, \qquad (8.36)$$

where the subscripts 1 and 2 indicate the values of a quantity at t_1 and t_2, respectively. Equations (8.36) express the linear impulse-momentum principle for a rigid body.

Extension of Eq. (8.36) for a system

Recalling that $\vec{F}$ is the sum of only the *external forces*, Eqs. (8.36) can be applied even to a system of rigid bodies if we properly compute $\vec{p}$. For a system of N rigid bodies (see Fig. 8.17), the system's total momentum is

$$\vec{p} = \sum_{i=1}^{N} m_i \vec{v}_{Gi}(t), \qquad (8.37)$$

where m_i and $\vec{v}_{Gi}$ $(i = 1, \ldots, N)$ are the mass and the velocity, respectively, of the mass center of rigid body i.

Conservation of linear momentum

If $\vec{F} = \vec{0}$ for $t_1 \leq t \leq t_2$, Eqs. (8.36) reduce to

$$m\vec{v}_{G1} = m\vec{v}_{G2} \quad \text{or} \quad \vec{p}_1 = \vec{p}_2, \qquad (8.38)$$

which states that the system's momentum is conserved for $t_1 \leq t \leq t_2$. In many applications the total external force $\vec{F}$ is not equal to zero over the time interval considered, but there is a direction, say q, along which the *component* $F_q = 0$ for $t_1 \leq t \leq t_2$. In this case we can write

$$m(v_{Gq})_1 = m(v_{Gq})_2 \quad \text{or} \quad p_{q1} = p_{q2}, \qquad (8.39)$$

that is, the momentum is conserved in the q direction. Equations (8.39) are useful in many situations, especially in the study of impacts.

Angular impulse-momentum principle for a rigid body

The moment equation governing the motion of a rigid body is Eq. (7.18) on p. 548 (first given for general systems in Eq. (5.87) on p. 394), i.e.,

$$\vec{M}_P = \dot{\vec{h}}_P + \vec{v}_P \times m\vec{v}_G, \qquad (8.40)$$

where P is an arbitrarily chosen moment center (see Fig. 8.18), $\vec{v}_P$ is the velocity of P, and $\vec{M}_P$ is the total moment relative to P due to the external force system acting on the body. The quantity $\vec{h}_P$ is the body's angular momentum relative to P and was defined in Eq. (7.20) on p. 549 as

$$\vec{h}_P = \int_B \vec{r}_{dm/P} \times \vec{v}_{dm}\, dm. \tag{8.41}$$

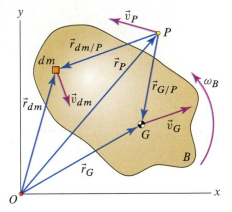

Figure 8.18
The quantities needed to obtain the angular momentum relationships for a rigid body.

While Eqs. (8.40) and (8.41) are valid for any type of body and for any motion, the applications we focus on in this chapter concern the planar motion of rigid bodies that are symmetric with respect to the plane of motion. For this kind of application, as shown in Eq. (B.25) of App. B, $\vec{h}_P$ can be given the following compact form

$$\boxed{\vec{h}_P = I_G\vec{\omega}_B + \vec{r}_{G/P} \times m\vec{v}_G,} \tag{8.42}$$

where I_G and $\vec{\omega}_B$ are the body's mass moment of inertia and angular velocity, respectively.* In addition, by choosing the point P in the same plane as G, and such that

1. P is a fixed point (i.e., $\vec{v}_P = \vec{0}$); or

2. P coincides with G (i.e., $\vec{v}_P = \vec{v}_G \Rightarrow \vec{v}_P \times \vec{v}_G = \vec{0}$); or

3. P and G move parallel to one another (i.e., $\vec{v}_P \times \vec{v}_G = \vec{0}$);

Eq. (8.40) simplifies to

$$\boxed{\vec{M}_P = \dot{\vec{h}}_P.} \tag{8.43}$$

The derivation of Eq. (8.42) from Eq. (8.41) using the stated assumptions is shown in App. B on p. 741.

If the assumptions underlying Eq. (8.43) hold throughout a time interval $t_1 \le t \le t_2$, then integrating Eq. (8.43) over this time interval, we obtain the traditional form of the angular impulse-momentum principle, i.e.,

$$\boxed{\vec{h}_{P1} + \int_{t_1}^{t_2} \vec{M}_P\, dt = \vec{h}_{P2},} \tag{8.44}$$

where $\vec{h}_{P1}$ and $\vec{h}_{P2}$ are the values of $\vec{h}_P$ at times t_1 and t_2, respectively. Equation (8.44) was first obtained in Eq. (5.80) on p. 393 when studying particle systems. If the moment center is taken to be the mass center G, then $\vec{r}_{G/P} = \vec{0}$ since P and G coincide. In this case, by combining Eqs. (8.44) and (8.42), the angular impulse-momentum principle takes on the form

$$\boxed{I_{G1}\omega_{B1} + \int_{t_1}^{t_2} M_{Gz}\, dt = I_{G2}\omega_{B2},} \tag{8.45}$$

where M_{Gz} is the z component of the moment about G, subscripts 1 and 2 indicate the value of a quantity at t_1 and t_2, respectively, and Eq. (8.45)

---

* As stated in Chapter 7, the subscript on I will be used to uniquely identify the axis with respect to which I is calculated. Hence, I_A denotes the mass moment of inertia of the rigid body about an axis perpendicular to the plane of motion passing through A.

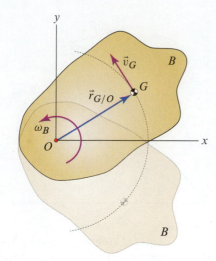

Figure 8.19
A rigid body in a fixed axis rotation.

has been written in scalar form because, under the current assumptions, the only nonzero component of Eq. (8.44) is perpendicular to the plane of motion. Equation (8.45) is valid even if I_G changes with time (see Example 8.9).

A variant of Eq. (8.44) is obtained in the case depicted in Fig. 8.19, in which a body is in fixed axis rotation about a point O. In this case, $\vec{v}_G = \vec{\omega}_B \times \vec{r}_{G/O}$ and by choosing the center of rotation O as our moment center, Eq. (8.42) becomes

$$
\begin{aligned}
\vec{h}_O &= I_G\vec{\omega}_B + \vec{r}_{G/O} \times m(\vec{\omega}_B \times \vec{r}_{G/O}) \\
&= (I_G + m|\vec{r}_{G/O}|^2)\vec{\omega}_B = I_O\vec{\omega}_B,
\end{aligned} \tag{8.46}
$$

where, by the parallel axis theorem, $I_O = I_G + m|\vec{r}_{G/O}|^2$ is the body's mass moment of inertia about the axis of rotation. Using Eq. (8.46), the angular impulse-momentum principle for a body that is symmetric with respect to the plane of motion and under fixed axis rotation takes on the form

$$
I_{O1}\omega_{B1} + \int_{t_1}^{t_2} M_{Oz}\, dt = I_{O2}\omega_{B2}, \tag{8.47}
$$

where O is the center of rotation and M_{Oz} is the z component of $\vec{M}_O$. We have written Eq. (8.47) in scalar form because, under the current assumptions, the only nonzero component of Eq. (8.44) is normal to the plane of motion.

Angular impulse-momentum principle for a system

Equations (8.44) and (8.45) apply to systems of rigid bodies if the assumptions underlying these equations are satisfied by each element of the system. Referring to Fig. 8.20, for a system of N rigid bodies, in which body i has angular velocity $\vec{\omega}_i$, mass center G_i, mass m_i, and mass moment of inertia I_{Gi}, $\vec{h}_P$ is

$$
\vec{h}_P = \sum_{i=1}^{N}(I_{Gi}\vec{\omega}_i + \vec{r}_{Gi/P} \times m_i\vec{v}_{Gi}), \tag{8.48}
$$

where $\vec{v}_{Gi}$ is the velocity of G_i and $\vec{r}_{Gi/P}$ is the position of G_i relative to P.

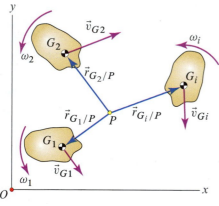

Figure 8.20
A system of rigid bodies.

Conservation of angular momentum

If $\vec{M}_P = \vec{0}$ for $t_1 \le t \le t_2$, Eq. (8.44) implies that

$$
\vec{h}_{P1} = \vec{h}_{P2} = \text{constant}, \tag{8.49}
$$

which states that the body's angular momentum relative to P is conserved. Another useful result is obtained when $\vec{M}_P \ne \vec{0}$ but there is a *fixed* direction, say q, along which $M_{Pq} = 0$. In this case, we can write

$$
(h_{Pq})_1 = (h_{Pq})_2 = \text{constant} \tag{8.50}
$$

and apply conservation of angular momentum in the q direction.

Common Pitfall

Assumptions for conservation of angular momentum. The fact that $\vec{M}_P = \vec{0}$ by itself does not allow us to say that $\vec{h}_P$ is conserved! In order to use Eq. (8.49), we must first verify that one of the three assumptions listed above Eq. (8.43) is satisfied.

End of Section Summary

The linear impulse-momentum principle for a rigid body reads (see Fig. 8.21)

Eqs. (8.36), p. 626

$$m\vec{v}_{G1} + \int_{t_1}^{t_2} \vec{F}\, dt = m\vec{v}_{G2} \quad \text{or} \quad \vec{p}_1 + \int_{t_1}^{t_2} \vec{F}\, dt = \vec{p}_2,$$

where $\vec{F}$ is the total external force on B, $\vec{p} = m\vec{v}_G$ is B's linear momentum, and $\vec{v}_G$ is the velocity of B's center of mass. If $\vec{F} = \vec{0}$, we have

Eqs. (8.38), p. 626

$$m\vec{v}_{G1} = m\vec{v}_{G2} \quad \text{or} \quad \vec{p}_1 = \vec{p}_2,$$

and we say that the body's momentum is conserved. If P in Fig. 8.22 is a moment center coplanar with G and if B is symmetric relative to the plane of motion, the angular momentum of B relative to P is

Eq. (8.42), p. 627

$$\vec{h}_P = I_G\vec{\omega}_B + \vec{r}_{G/P} \times m\vec{v}_G,$$

where I_G is the mass moment of inertia of B, $\vec{\omega}_B$ is the angular velocity of B, and $\vec{r}_{G/P}$ is the position of G relative to P. If P is chosen so that (1) P is fixed or (2) P coincides with G or (3) P and G move parallel to one another, then $\vec{M}_P = \dot{\vec{h}}_P$, where $\vec{M}_P$ is the moment relative to P of the *external* force system acting on B. When this equation holds, by integrating with respect to time over a time interval $t_1 \le t \le t_2$, we have

Eq. (8.44), p. 627

$$\vec{h}_{P1} + \int_{t_1}^{t_2} \vec{M}_P\, dt = \vec{h}_{P2}.$$

When P coincides with G or if the body undergoes a fixed axis rotation about a point O as shown in Fig. 8.23, then the above equation becomes

Eq. (8.45), p. 627, and Eq. (8.47), p. 628

$$I_{G1}\omega_{B1} + \int_{t_1}^{t_2} M_{Gz}\, dt = I_{G2}\omega_{B2},$$

$$I_{O1}\omega_{B1} + \int_{t_1}^{t_2} M_{Oz}\, dt = I_{O2}\omega_{B2},$$

respectively, where I_O is the mass moment of inertia about the fixed axis of rotation.

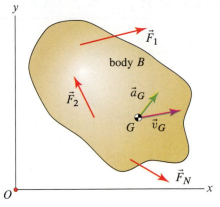

Figure 8.21
Figure 8.16 repeated. A rigid body under the action of a system of force.

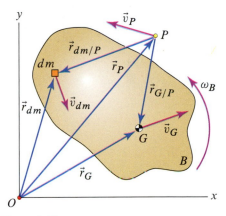

Figure 8.22
Figure 8.18 repeated. The quantities needed to obtain the angular momentum relationships for a rigid body.

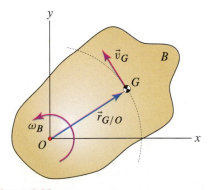

Figure 8.23
A rigid body in a fixed axis rotation.

EXAMPLE 8.7   *A Rolling Wheel: Computing Angular Momentum*

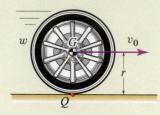

Figure 1

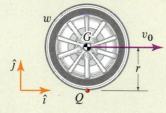

Figure 2
Wheel rolling without slip.

The wheel w shown in Fig. 1 has radius r and mass m. Point G is both the wheel's center of mass and the wheel's geometric center. The radius of gyration is k_G. The wheel is rolling without slip with G moving to the right with a speed v_0. Compute the wheel's angular momentum relative to both G and Q, which is the point on the wheel in contact with the ground.

SOLUTION

Road Map   This problem is solved by a direct application of the expression for the angular momentum of a rigid body given in Eq. (8.42) on p. 627.

Computation   Referring to Fig. 2 and applying Eq. (8.42), the angular momentum of the wheel relative to its center of mass is

$$\vec{h}_G = I_G \vec{\omega}_w + \vec{r}_{G/G} \times m\vec{v}_G, \tag{1}$$

where $\vec{r}_{G/G} = \vec{0}$ and $I_G = mk_G^2$. Recalling that because of rolling without slip we must have $\vec{\omega}_w = -(v_0/r)\,\hat{k}$, Eq. (1) can be rewritten as

$$\boxed{\vec{h}_G = -mk_G^2 \left(\frac{v_0}{r}\right)\hat{k}.} \tag{2}$$

Again referring to Fig. 2 and applying Eq. (8.42), the angular momentum of the wheel relative to Q is

$$\vec{h}_Q = I_G \vec{\omega}_w + \vec{r}_{G/Q} \times m\vec{v}_G. \tag{3}$$

In this case we have

$$\vec{r}_{G/Q} = r\,\hat{j} \quad \text{and} \quad \vec{v}_G = v_0\,\hat{i} \quad \Rightarrow \quad \vec{r}_{G/Q} \times m\vec{v}_G = -mrv_0\,\hat{k}. \tag{4}$$

Substituting the last of Eqs. (4) into Eq. (3) and recalling that $I_G \vec{\omega}_w = -mk_G^2 (v_0/r)\,\hat{k}$, we have

$$\boxed{\vec{h}_Q = -mrv_0\,\hat{k} - mk_G^2 \frac{v_0}{r}\,\hat{k} = -mv_0 \left(r + \frac{k_G^2}{r}\right)\hat{k}.} \tag{5}$$

Discussion & Verification   To verify that the results in Eqs. (2) and (5) are dimensionally correct, recall that angular momentum has the dimensions of *moment of the momentum*, i.e., of mass × velocity × length, which is what we see in Eqs. (2) and (5).

✎ **A Closer Look**   Using the parallel axis theorem, we can give the expression in Eq. (5) a much more compact form. Let I_Q denote the wheel's mass moment of inertia relative to Q. Using the parallel axis theorem, we have

$$I_Q = I_G + mr^2 = m(k_G^2 + r^2). \tag{6}$$

Going back to Eq. (5) and factoring $1/r$ out of the term in parentheses, we can rewrite this equation as

$$\vec{h}_Q = -m\frac{v_0}{r}\left(r^2 + k_G^2\right)\hat{k} = I_Q \vec{\omega}_w, \tag{7}$$

where we have taken advantage of Eq. (6) and the fact that $\vec{\omega}_w = -(v_0/r)\,\hat{k}$. Comparing Eq. (7) with Eq. (8.46) on p. 628, we could interpret the result by saying that the wheel appears as though it were in a fixed axis rotation about Q. This interpretation is appropriate because Q is the wheel's instantaneous center of rotation.

EXAMPLE 8.8   *A Rolling Pipe: Application of Impulse and Momentum*

A pipe section A of radius r, center G, and mass m is gently placed (i.e., with zero velocity) on a conveyor belt moving with a constant speed v_0 to the right as shown in Fig. 1. The friction between the belt and pipe will cause the pipe to move to the right as well as to rotate and, eventually, to roll without slip. Determine the velocity of G and the angular velocity of the pipe when it rolls without slip.

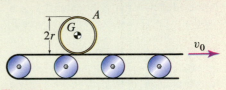

Figure 1
A pipe section of radius r lowered very gently over a conveyor belt.

SOLUTION

Road Map & Modeling   Because the pipe A is stationary when it is placed on the conveyor belt, A must slip relative to the belt until rolling without slip begins. Modeling A as a uniform rigid body, until rolling without slip begins, the FBD of A is that shown in Fig. 2. Assuming that G does not move in the vertical direction, the motion of the body is determined by the impulse provided by the friction force, which is the only force acting in the horizontal direction. We can then solve the problem by applying the linear and angular impulse-momentum principles along with the kinematic relations that describe rolling without slip over a moving surface.

Governing Equations

Balance Principles   Let t_1 denote the time at which A is placed on the conveyor and t_2 denote the time at which the pipe starts rolling without slip. The impulse-momentum principle in the x direction reads

$$m(v_{Gx})_1 + \int_{t_1}^{t_2} F\, dt = m(v_{Gx})_2. \tag{1}$$

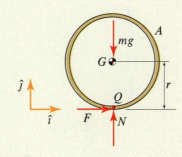

Figure 2
FBD of the pipe section at an instant between the time at which the pipe is placed over the conveyor and the time at which the pipe starts rolling without slip.

Choosing the mass center G as moment center, the angular impulse-momentum principle applied between t_1 and t_2 reads

$$\vec{h}_{G1} + \int_{t_1}^{t_2} (\vec{r}_{Q/G} \times F\,\hat{\imath})\, dt = \vec{h}_{G2}, \tag{2}$$

where, using Eq. (8.42) on p. 627 and modeling the pipe section as a thin ring,

$$\vec{h}_{G1} = I_G \vec{\omega}_{A1} = m r^2 \omega_{A1}\, \hat{k} \quad \text{and} \quad \vec{h}_{G2} = I_G \vec{\omega}_{A2} = m r^2 \omega_{A2}\, \hat{k}. \tag{3}$$

Force Laws   All forces are accounted for on the FBD.

Kinematic Equations   At time t_1, A is stationary so we have

$$(v_{Gx})_1 = 0 \quad \text{and} \quad \omega_{A1} = 0. \tag{4}$$

At time t_2, A rolls without slip over the moving belt, which means that $\vec{v}_{Q2} = v_0\,\hat{\imath}$. For G we have

$$\vec{v}_{G2} = \vec{v}_{Q2} + \omega_{A2}\,\hat{k} \times \vec{r}_{G/Q} \quad \Rightarrow \quad \vec{v}_{G2} = (v_0 - r\omega_{A2})\,\hat{\imath}. \tag{5}$$

Computation   After expanding the cross-product, the integrand in the second term of Eq. (2) can be written as follows:

$$\vec{r}_{Q/G} \times F\,\hat{\imath} = -r\,\hat{\jmath} \times F\,\hat{\imath} = Fr\,\hat{k}. \tag{6}$$

Substituting Eqs. (3), the second of Eqs. (4), and Eq. (6) into Eq. (2), we have

$$r \int_{t_1}^{t_2} F\, dt = m r^2 \omega_{A2}, \tag{7}$$

Helpful Information

Is the friction force F constant? In the integral in Eq. (7) we left the friction force F inside the integral because we did not know F as a function of time. The important point to understand here is that the final solution to this problem does not require that we know F as a function of time. With this said, starting from the system's FBD and what we learned in Chapter 7, we can show that in this problem F is constant for $t_1 < t < t_2$ and then becomes equal to zero as soon as the pipe section starts rolling without slip.

where we have pulled r outside the integral because it is constant.

Substituting the first of Eqs. (4) and the last of Eqs. (5) into Eq. (1), we have

$$\int_{t_1}^{t_2} F\, dt = m(v_0 - r\omega_{A2}). \tag{8}$$

Substituting Eq. (8) into Eq. (7), we obtain

$$mr(v_0 - r\omega_{A2}) = mr^2\omega_{A2} \quad \Rightarrow \quad \boxed{\omega_{A2} = \frac{v_0}{2r}.} \tag{9}$$

Substituting ω_{A2} from Eq. (9) into the last of Eqs. (5), we have

$$\boxed{\vec{v}_{G2} = \tfrac{1}{2} v_0\,\hat{\imath}.} \tag{10}$$

Discussion & Verification   The result we obtained is dimensionally correct and consistent with the FBD in that friction will cause the pipe section to move to the right and rotate counterclockwise.

🔍 **A Closer Look**   The problem's solution does not depend on the mass of the object, only on its shape. That is, we would have obtained a different result had we modeled the pipe section as, say, a cylinder.

Note that we could have obtained the solution by enforcing the conservation of angular momentum about point Q without invoking the linear impulse-momentum principle. To see this, referring to the system FBD in Fig. 2, observe that the moment of the external forces about Q is equal to zero. Normally this observation does not help much since Q is neither a fixed point nor the system's center of mass. However, referring to the list preceding Eq. (8.43) on p. 627, point Q does move parallel to the center of mass G. Therefore, since $\vec{M}_Q = \vec{0}$, Eq. (8.43) implies that $\vec{h}_{Q1} = \vec{h}_{Q2}$. Furthermore, since the pipe section was stationary when it was placed on the conveyor belt, we must have $\vec{h}_{Q1} = \vec{0}$. This fact, along with Eq. (8.42) on p. 627, yields

$$I_G \omega_{A2}\,\hat{k} + \vec{r}_{G/Q} \times m\vec{v}_{G2} = \vec{0} \quad \Rightarrow \quad mr^2\omega_{A2} - r(v_{Gx})_2 = 0. \tag{11}$$

The result in Eq. (11), along with the rolling without slip condition in Eq. (5), yields the same solution we derived in Eqs. (9) and (10).

EXAMPLE 8.9   *A Spinning Skater: Conservation of Angular Momentum*

The skater in Fig. 1 begins to spin with her arms completely stretched out and then brings her arms close to her body to increase her spin rate. In Section 5.3 on p. 388, we modeled the skater using a single particle. Here we revisit the problem by modeling the skater as a system of rigid bodies, as shown in Fig. 2. Except for her arms,* her body is modeled as a cylinder of radius $r_b = 0.55$ ft, weight $W_b = 102$ lb, and radius of gyration $k_G = 0.3$ ft, where G is her body's center of mass. Each arm has weight $W_a = 7.4$ lb and length $\ell = 2.2$ ft and is divided into an upper arm and a forearm. The upper arm and forearm are each modeled as a uniform thin rod weighing $W_a/2$ and with length $\ell/2$. Assuming that the skater starts spinning with a rate $\omega_0 = 60$ rpm as shown and that her arms are stretched out, determine her spin rate if (a) her upper arms are kept stretched out and her forearms are folded so as to overlap with her upper arms; and (b) if her entire arms are placed vertically downward next to her body.

Figure 1
Three snapshots of a *forward spin*. By extending and retracting her arms and leg, the skater controls her spin rate.

SOLUTION

Road Map & Modeling   Neglecting friction between the skater and ice, the skater's FBD for $t_1 \leq t \leq t_2$ is shown in Fig. 3, where t_1 is the time at which the spin begins and t_2 is the time at which one of the positions corresponding to (a) or (b) is achieved. None of the external forces in the FBD contributes to a moment about the z axis. Assuming that the spin axis coincides with the z axis for $t_1 \leq t \leq t_2$, the condition $M_z = 0$ causes the angular momentum about this axis to be conserved. Since we need to find only one scalar unknown, namely, the angular velocity at t_2, satisfying this conservation statement will lead us to the solution.

Figure 2
Model of the skater as a system of rigid bodies.

Governing Equations

Balance Principles   Referring to Fig. 3, since the skater's body is in a fixed axis rotation about the z axis with $M_{Oz} = 0$, then Eq. (8.47) on p. 628 implies

$$I_{O1}\omega_{s1} = I_{O2}\omega_{s2}, \tag{1}$$

where ω_s is the angular velocity of the skater and I_O is the mass moment of inertia of the skater about O (or any other point along the z axis).

In this problem *the mass moment of inertia changes as the skater moves her arms!* Referring to Fig. 2, when completely outstretched, the upper arm and forearm can be viewed as forming a single uniform thin rod of mass m_a and length ℓ with mass center $r_b + \ell/2$ away from the z axis. Therefore, applying the parallel axis theorem, we have

$$I_{O1} = \underbrace{m_b k_G^2}_{\text{body}} + 2\underbrace{\left[\tfrac{1}{12}m_a\ell^2 + m_a\left(r_b + \tfrac{1}{2}\ell\right)^2\right]}_{\text{each arm}}, \tag{2}$$

where m_b is the mass of her body. Equation (2) can be simplified as

$$I_{O1} = m_b k_G^2 + 2m_a\left(r_b^2 + r_b\ell + \tfrac{1}{3}\ell^2\right). \tag{3}$$

Referring to Fig. 4(a), for case (a), when the skater folds her forearms horizontally, using the parallel axis theorem again, at time t_2 we have

$$(I_{O2})_{\text{out}} = \underbrace{m_b k_G^2}_{\text{body}} + 4\underbrace{\left[\frac{1}{12}\frac{m_a}{2}\left(\frac{\ell}{2}\right)^2 + \frac{m_a}{2}\left(r_b + \frac{\ell}{4}\right)^2\right]}_{\text{each upper arm and each forearm}}, \tag{4}$$

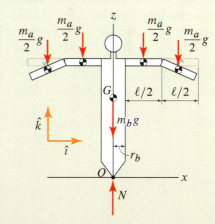

Figure 3
FBD of a spinning skater.

---

* We use *arm* according to its common meaning, i.e., everything from the shoulder to the tip of the fingers. However, in medical anatomy, an arm is only what lies between shoulder and elbow.

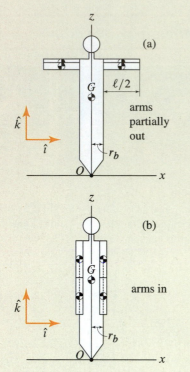

Figure 4
Configuration of the skater's arms for the two cases we are considering.

where the subscript *out* indicates that the arms are partially stretched out. Equation (4) can be simplified to

$$(I_{O2})_{\text{out}} = m_b k_G^2 + m_a\left(2r_b^2 + r_b\ell + \frac{\ell^2}{6}\right). \tag{5}$$

When her arms are folded completely downward, as in Fig. 4(b), we have

$$(I_{O2})_{\text{in}} = \underbrace{m_b k_G^2}_{\text{body}} + \underbrace{2m_a r_b^2}_{\text{each arm}}, \tag{6}$$

where the subscript *in* indicates that the arms are completely downward.

Force Laws   All forces are accounted for on the FBD.

Kinematic Equations   We know that the skater is initially spinning at ω_0 and so

$$\omega_{s1} = \omega_0. \tag{7}$$

Computation   Substituting Eq. (7) into Eq. (1) and solving for ω_{s2}, we have $\omega_{s2} = (I_{O1}/I_{O2})\omega_0$, and therefore for the two cases considered we have

$$(\omega_{s2})_{\text{out}} = \frac{m_b k_G^2 + 2m_a\left(r_b^2 + r_b\ell + \frac{\ell^2}{3}\right)}{m_b k_G^2 + m_a\left(2r_b^2 + r_b\ell + \frac{\ell^2}{6}\right)}\omega_0 = 116\,\text{rpm} \tag{8}$$

and

$$(\omega_{s2})_{\text{in}} = \frac{m_b k_G^2 + 2m_a\left(r_b^2 + r_b\ell + \frac{\ell^2}{3}\right)}{m_b k_G^2 + 2m_a r_b^2}\omega_0 = 244\,\text{rpm}, \tag{9}$$

where we have used Eqs. (3), (5), and (6) and where we have plugged in the given data to obtain the numerical results.

Discussion & Verification   In each of Eqs. (8) and (9) the final spin rate is greater than the initial spin rate ω_0, as expected. In addition, the result in Eq. (9) is larger than that in Eq. (8); i.e., the increase in spin rate for the case where the skater's arms are completely against her body is larger than that when only the forearms are folded, again as expected. Finally, the spin rates that we have obtained are certainly within reach of professional skaters (see the Interesting Fact in the margin).

A Closer Look   The particle solution to this problem was given in Eq. (5.57) on p. 388, which, in terms of the current variables, is

$$\omega_{s2} = (r_1^2/r_2^2)\omega_0, \tag{10}$$

where r_1 and r_2 are the distances between the arms and the spin axis at times t_1 and t_2, respectively. The simplicity of Eq. (10) is due to the fact that we ignored the dimensions of the body and its parts and only considered the mass of the arms. However, it is important to understand that the particle and rigid body solutions are not that different in spirit. We can rewrite Eq. (10) as

$$\omega_{s2} = \frac{2m_a}{2m_a}\frac{r_1^2}{r_2^2}\omega_0 = \frac{2m_a r_1^2}{2m_a r_2^2}\omega_0 = \frac{(I_{O1})_p}{(I_{O2})_p}\omega_0 \Rightarrow (I_{O1})_p\omega_0 = (I_{O2})_p\omega_{s2}, \tag{11}$$

where $(I_{O1})_p$ and $(I_{O2})_p$ represent the mass moments of inertia relative to the spin axis for the particle model. Recalling that $\omega_0 = \omega_{s1}$ and comparing the last of Eqs. (11) with Eq. (1), we see that the only difference between the two models is how the mass moments of inertia are calculated.

EXAMPLE 8.10   *Space Shuttle Docking with ISS: Conservation of Momentum*

Figure 1 shows the Space Shuttle docked with the International Space Station (ISS). To explore what docking entails, we consider the simplified 2D scenario in Fig. 2, in which the Shuttle A docks to the ISS B with a speed $v_0 = 0.03$ m/s. We wish to determine the velocities of A and B *immediately after* they dock, assuming that no spacecraft attitude controls are exerted on A or B and assuming that, after docking, A and B *form a single rigid body*. Referring to Fig. 2, we will use the following data: the mass and mass moment of inertia of A are $m_A = 120 \times 10^3$ kg and $I_C = 14 \times 10^6$ kg·m^2, respectively; the mass and mass moment of inertia of B are $m_B = 180 \times 10^3$ kg and $I_D = 34 \times 10^6$ kg·m^2, respectively; the dimensions are $\ell = 24$ m and $h = 8$ m. Note that we are *not* assuming that A and B are rectangles in Fig. 2. Since a body's mass and mass moment of inertia completely describe it, these rectangles are used only to describe the relative position of points C and D.

SOLUTION

Road Map & Modeling   Since we are assuming that A and B join to form a single rigid body, we can use rigid body kinematics to describe the motion of the A-B-body via the motion of only two points, namely, C and D, provided we know their relative position, which is given in Fig. 2. We will neglect all gravitational effects and assume that B is initially at rest relative to an inertial frame. Since we want the motion immediately after docking, we can assume that the positions of A and B are still the same as those at the time of docking. This allows us to make no distinction between the positions of the system immediately before and after docking. Finally, recalling that no attitude controls are used, the FBD of the system right before *and* right after docking is that in Fig. 3, so that the system's linear and angular momenta are conserved. These conservation statements give us three scalar equations which, when combined with the assumption that A and B form a single rigid body, are sufficient to solve the problem.

Governing Equations

Balance Principles   In components, the conservation of total linear momentum reads

$$m_A(v_{Cx})_1 + m_B(v_{Dx})_1 = m_A(v_{Cx})_2 + m_B(v_{Dx})_2, \qquad (1)$$
$$m_A(v_{Cy})_1 + m_B(v_{Dy})_1 = m_A(v_{Cy})_2 + m_B(v_{Dy})_2, \qquad (2)$$

where the subscripts 1 and 2 indicate right before and right after docking, respectively.

Choosing the fixed point O as the moment center, the conservation of total angular momentum reads

$$(\vec{h}_O)_{A1} + (\vec{h}_O)_{B1} = (\vec{h}_O)_{A2} + (\vec{h}_O)_{B2}, \qquad (3)$$

where, because A and B do not move significantly between times t_1 and t_2, we have

$$(\vec{h}_O)_{A1} = I_C\vec{\omega}_{A1} + \vec{r}_{C/O} \times m_A\vec{v}_{C1}, \quad (\vec{h}_O)_{B1} = I_D\vec{\omega}_{B1} + \vec{r}_{D/O} \times m_B\vec{v}_{D1}, \quad (4)$$
$$(\vec{h}_O)_{A2} = I_C\vec{\omega}_{A2} + \vec{r}_{C/O} \times m_A\vec{v}_{C2}, \quad (\vec{h}_O)_{B2} = I_D\vec{\omega}_{B2} + \vec{r}_{D/O} \times m_B\vec{v}_{D2}. \quad (5)$$

Force Laws   All forces are accounted for on the FBD.

Kinematic Equations   Before docking,

$$(v_{Cx})_1 = -v_0, \quad (v_{Cy})_1 = 0, \quad \omega_{A1} = 0, \qquad (6)$$
$$(v_{Dx})_1 = 0, \quad (v_{Dy})_1 = 0, \quad \omega_{B1} = 0. \qquad (7)$$

Figure 1
Artist's rendition of the Space Shuttle Discovery docked to the International Space Station.

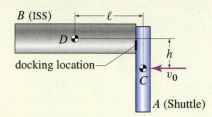

Figure 2
Relative positions of the mass centers C and D of A and B, respectively, at docking. The rectangles shown are not physical models, they are used only to describe the relative position of C and D.

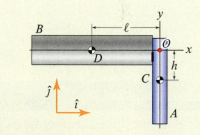

Figure 3
FBD of A and B right before *and* after docking. The coordinate system shown is *fixed in space*, and A and B can move relative to it.

After docking A and B form a single rigid body, so that we have

$$\omega_{A2} = \omega_{B2} = \omega_{AB} \quad \text{and} \quad \vec{v}_{C2} = \vec{v}_{D2} + \omega_{AB}\, \hat{k} \times \vec{r}_{C/D}, \tag{8}$$

where ω_{AB} is the common angular velocity of A and B immediately after docking. The relative position vectors in Eqs. (4), (5), and (8) are given by

$$\vec{r}_{C/O} = -h\,\hat{j}, \quad \vec{r}_{D/O} = -\ell\,\hat{i}, \quad \text{and} \quad \vec{r}_{C/D} = \ell\,\hat{i} - h\,\hat{j}. \tag{9}$$

Computation   Substituting the first two of Eqs. (6) and (7) into Eqs. (1) and (2), we have

$$-m_A v_0 = m_A (v_{Cx})_2 + m_B (v_{Dx})_2, \tag{10}$$

$$0 = m_A (v_{Cy})_2 + m_B (v_{Dy})_2, \tag{11}$$

Referring to Fig. 4 and substituting Eqs. (6), (7), the first of Eqs. (8), and the first two of Eqs. (9) into Eqs. (4) and (5), we have

$$(\vec{h}_O)_{A1} = -m_A h v_0\,\hat{k}, \quad (\vec{h}_O)_{A2} = \left[I_C \omega_{AB} + m_A h (v_{Cx})_2\right]\hat{k}, \tag{12}$$

$$(\vec{h}_O)_{B1} = \vec{0}, \quad (\vec{h}_O)_{B2} = \left[I_D \omega_{AB} - m_A \ell (v_{Dy})_2\right]\hat{k}. \tag{13}$$

Substituting Eqs. (12) and (13) into Eq. (3), we obtain

$$-m_A h v_0 = (I_C + I_D)\omega_{AB} + m_A h (v_{Cx})_2 - m_B \ell (v_{Dy})_2. \tag{14}$$

Equation (14) is in scalar form because the only nonzero component of Eq. (3) is in the z direction. Finally, substituting the last of Eqs. (9) into the second of Eqs. (8), expanding the cross-product, and expressing the result in components, we have

$$(v_{Cx})_2 = (v_{Dx})_2 + \omega_{AB} h \quad \text{and} \quad (v_{Cy})_2 = (v_{Dy})_2 + \omega_{AB}\ell. \tag{15}$$

Equations (10), (11), (14), and (15) form a system of five equations in the five unknowns $(v_{Cx})_2$, $(v_{Cy})_2$, $(v_{Dx})_2$, $(v_{Dy})_2$, and ω_{AB}. The solution to these five equations is found to be

$$(v_{Cx})_2 = \frac{-m_A \left(I + m_B h^2 + \frac{m_A m_B}{m}\ell^2\right)v_0}{m_A m_B d^2 + mI} = -0.0129\,\text{m/s}, \tag{16}$$

$$(v_{Cy})_2 = \frac{-m_B \frac{m_A m_B}{m} h\ell v_0}{m_A m_B d^2 + mI} = -0.00264\,\text{m/s}, \tag{17}$$

$$(v_{Dx})_2 = \frac{-m_A \left(I + \frac{m_A m_B}{m}\ell^2\right)v_0}{m_A m_B d^2 + mI} = -0.0114\,\text{m/s}, \tag{18}$$

$$(v_{Dy})_2 = \frac{m_A \frac{m_A m_B}{m} h\ell v_0}{m_A m_B d^2 + mI} = 0.00176\,\text{m/s}, \tag{19}$$

$$\omega_{AB} = \frac{-m_A m_B h v_0}{m_A m_B d^2 + mI} = -0.000184\,\text{rad/s}, \tag{20}$$

where $m = m_A + m_B$, $d = \sqrt{h^2 + \ell^2}$, and $I = I_C + I_D$.

Discussion & Verification   The results appear reasonable since the computed velocities are comparable to v_0. In addition, the signs appear correct in that, after docking, we expect both A and B to move to the left and the AB-body to rotate clockwise. This rotation then causes C and D to move slightly downward and upward, respectively.

🔎 **A Closer Look**   We assumed that A and B form a rigid body after docking because we did not know the exact position of the docking location. A better assumption is that A and B become pinned to each other after docking. In this way we would better capture the effect of the local flexibility of the docking location. This possibility is considered in Prob. 8.94 on p. 665.

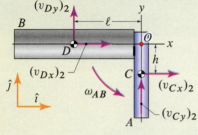

Figure 4
Sketch of the velocity components of the system right after docking.

Interesting Fact

Mass of the ISS. The assembly of the ISS began in 1998. Eventually the ISS mass will be about 472000 kg. The mass and moment of inertia given in the problem are rough estimates based on stage 9A.1 of the ISS assembly process (see J. A. Wojtowicz, "Dynamic Properties of the International Space Station throughout the Assembly Process," Report N. A282843, Air Force Institute of Technology, Wright-Patterson AFB, Ohio, 1998).

PROBLEMS

Problem 8.50

Disks A and B have identical masses and mass moments of inertia about their respective mass centers. Point C is both the geometric center and center of mass of disk A. Points O and D are the geometric center and center of mass of disk B, respectively. If at the instant shown, the two disks are rotating about their centers with the same angular velocity ω_0, determine which of the following statements is true and why? (a) $|(\vec{h}_C)_A| < |(\vec{h}_O)_B|$, (b) $|(\vec{h}_C)_A| = |(\vec{h}_O)_B|$, (c) $|(\vec{h}_C)_A| > |(\vec{h}_O)_B|$.
Note: Concept problems are about *explanations*, not computations.

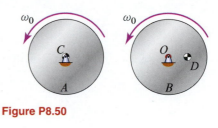

Figure P8.50

Problem 8.51

Body B has mass m and mass moment of inertia I_G, where G is the mass center of B. If B is translating as shown, determine which of the following statements is true and why: (a) $|(\vec{h}_E)_B| < |(\vec{h}_P)_B|$, (b) $|(\vec{h}_E)_B| = |(\vec{h}_P)_B|$, (c) $|(\vec{h}_E)_B| > |(\vec{h}_P)_B|$.
Note: Concept problems are about *explanations*, not computations.

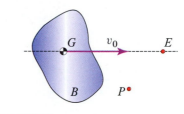

Figure P8.51

Problem 8.52

The uniform thin pin-connected bars AB, BC, and CD have masses $m_{AB} = 2.3\,\text{kg}$, $m_{BC} = 3.2\,\text{kg}$, and $m_{CD} = 5.0\,\text{kg}$, respectively. Letting $R = 0.75\,\text{m}$, $L = 1.2\,\text{m}$, and $H = 1.55\,\text{m}$, and knowing that bar AB rotates at a constant angular velocity $\omega_{AB} = 4\,\text{rad/s}$, compute the angular momentum of bar AB about A, of bar BC about A, and bar CD about D at the instant shown.

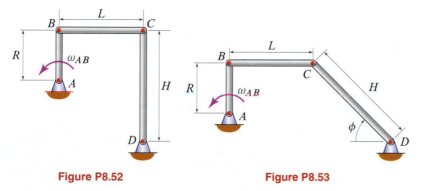

Figure P8.52   **Figure P8.53**

Problem 8.53

The weights of the uniform thin pin-connected bars AB, BC, and CD are $W_{AB} = 4\,\text{lb}$, $W_{BC} = 6.5\,\text{lb}$, and $W_{CD} = 10\,\text{lb}$, respectively. Letting $\phi = 47°$, $R = 2\,\text{ft}$, $L = 3.5\,\text{ft}$, and $H = 4.5\,\text{ft}$, and knowing that bar AB rotates at a constant angular velocity $\omega_{AB} = 4\,\text{rad/s}$, compute the magnitude of the linear momentum of the system at the instant shown.

Problem 8.54

A uniform disk W of radius $R_W = 7\,\text{mm}$ and mass $m_W = 0.15\,\text{kg}$ is connected to point O via the rotating arm OC. Disk W also rolls without slip over the stationary cylinder S of radius $R_S = 15\,\text{mm}$. Assuming that $\omega_W = 25\,\text{rad/s}$, determine the angular momentum of W about its own center of mass C as well as about point O.

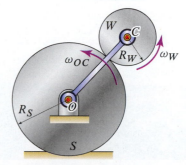

Figure P8.54

Problem 8.55

An eccentric wheel B weighing 150 lb has its mass center G at a distance $d = 4$ in. from the wheel's center O. The wheel is in the horizontal plane and is spun from rest by applying a constant torque $M = 32$ ft·lb. Determine the wheel's radius of gyration k_G if it takes 2 s to spin up the wheel to 140 rpm. Neglect all possible sources of friction.

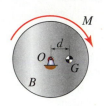

Figure P8.55

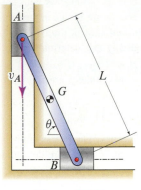

Figure P8.56

Problem 8.56

The uniform bar AB has length $L = 4.5$ ft and weight $W_{AB} = 14$ lb. At the instant shown, $\theta = 67°$ and $v_A = 5.8$ ft/s. Determine the magnitude of the linear momentum of AB as well as the angular momentum of AB about its mass center G at the instant shown.

Figure P8.57

Problem 8.57

The top of the Space Needle in Seattle, Washington, hosts a revolving restaurant that goes through one full revolution every 47 min under the action of a motor with a power output of 1.5 hp. The portion of the restaurant that rotates is a ring-shaped turntable with internal and external radii $r_i = 33.3$ ft and $r_o = 47.3$ ft, respectively, and approximate weight $W = 125$ tons (1 ton = 2000 lb). Use the given values of power output and angular speed to estimate the torque M that the engine provides. Then, assuming that the motor can provide a constant torque equal to M, neglecting all friction, and modeling the turntable as a uniform body, determine the time t_s that it takes to spin up the revolving restaurant from rest to its working angular speed.

Problem 8.58

Moving on a straight and horizontal stretch of road, the rear-wheel-drive car shown can go from rest to 60 mph in $\Delta t = 8$ s. The car weighs 2570 lb (the weight includes the wheels). Each wheel has diameter $d = 24.3$ in., mass moment of inertia relative to its own center of mass $I_G = 0.989$ slug·ft^2, and the center of mass of each wheel coincides with its geometric center. Determine the average friction force F_{avg} acting on the car during Δt. In addition, if the wheels roll without slip, for each wheel, determine the average moment M_{avg}, computed relative to the wheel's center, that is applied to the wheel during Δt.

Figure P8.58

Problem 8.59

The rear-wheel-drive car can go from rest to 60 mph in $\Delta t = 8\,\text{s}$. Assume that the wheels are all identical and that their geometric centers coincide with their mass centers. Let M_{rear} be the average moment applied to one of the rear wheels during Δt and computed relative to the wheel's center. Finally, let M_{front} be the average moment applied to one of the front wheels during Δt and computed relative to the wheel's center. Modeling the wheels as rigid bodies, determine which of the following statements is true and why. (a) $\left|M_{\text{rear}}\right| < \left|M_{\text{front}}\right|$, (b) $\left|M_{\text{rear}}\right| = \left|M_{\text{front}}\right|$, (c) $\left|M_{\text{rear}}\right| > \left|M_{\text{front}}\right|$. **Note:** Concept problems are about *explanations*, not computations.

Figure P8.59 and P8.60

Problem 8.60

The rear-wheel-drive car can go from rest to 60 mph in $\Delta t = 8\,\text{s}$. Assume that its wheels are identical, with their geometric centers coinciding with their mass centers. Let F_{avg} be the average friction force acting on the system during Δt due to contact with the ground. Modeling the car and the wheels as rigid bodies, does the value of F_{avg} change whether or not we account for the rotational inertia of the wheels? Why? **Note:** Concept problems are about *explanations*, not computations.

Problem 8.61

A rotor B with center of mass G, weight $W = 3000\,\text{lb}$, and radius of gyration $k_G = 15\,\text{ft}$ is spinning with an angular speed $\omega_B = 1200\,\text{rpm}$ when a braking system is applied to it, providing a time-dependent torque $M = M_0(1+ct)$, with $M_0 = 3000\,\text{ft·lb}$ and $c = 0.01\,\text{s}^{-1}$. If G is also the geometric center of the rotor and is a fixed point, determine the time t_s that it takes to bring the rotor to a stop.

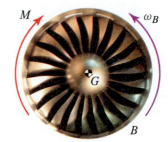

Figure P8.61

Problem 8.62

A uniform pipe section A of radius r, mass center G, and mass m is gently placed (i.e., with zero velocity) on a conveyor belt moving with a constant speed v_0 to the right. Friction between the belt and pipe causes the pipe to move to the right and eventually to roll without slip. If μ_k is the coefficient of kinetic friction between the pipe and the conveyor belt, find an expression for t_r, the time it takes for A to start rolling without slip. *Hint:* Using the methods of Chapter 7, we can show that the force between the pipe section and the belt is constant.

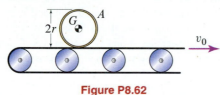

Figure P8.62

Problem 8.63

A 14 lb bowling ball is thrown onto a lane with a backspin $\omega_0 = 9\,\text{rad/s}$ and forward velocity $v_0 = 18\,\text{mph}$. Point G is both the geometric center and the mass center of the ball. After a few seconds, the ball starts rolling without slip. Let $r = 4.25\,\text{in.}$ and let the radius of gyration of the ball be $k_G = 2.6\,\text{in.}$ If the coefficient of kinetic friction between the ball and the floor is $\mu_k = 0.1$, determine the speed v_f that the ball will achieve when it starts rolling without slip. In addition, determine the time t_r the ball takes to achieve v_f. *Hint:* Using the methods of Chapter 7, we can show that the force between the ball and the floor is constant.

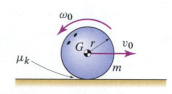

Figure P8.63

Problem 8.64

The uniform disk A, of mass $m_A = 1.2\,\text{kg}$ and radius $r_A = 0.25\,\text{m}$, is mounted on a vertical shaft that can translate along the horizontal guide C. The uniform disk B, of mass $m_B = 0.85\,\text{kg}$ and radius $r_B = 0.38\,\text{m}$, is mounted on a fixed vertical shaft. Both disks A and B can rotate about their own axes, namely, ℓ_A and ℓ_B, respectively. Disk A is initially spun with $\omega_A = 1000\,\text{rpm}$ and then brought into contact with B, which is initially stationary. The contact is maintained via a spring, and due to friction between A and B, disk B starts spinning and eventually A and B will stop slipping *relative to one another*. Neglecting any friction except at the contact between the two disks, determine the angular velocities of A and B when slipping stops.

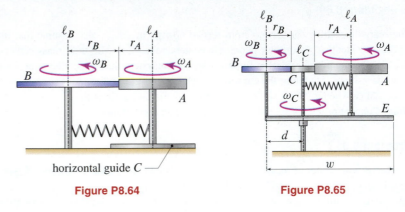

Figure P8.64   Figure P8.65

Problem 8.65

The uniform disk A, of mass $m_A = 1.2\,\text{kg}$ and radius $r_A = 0.25\,\text{m}$, is mounted on a vertical shaft that can translate along the horizontal arm E. The uniform disk B, of mass $m_B = 0.85\,\text{kg}$ and radius $r_B = 0.38\,\text{m}$, is mounted on a vertical shaft that is rigidly attached to arm E. Disk A can rotate about axis ℓ_A, disk B can rotate about axis ℓ_B, and the arm E, along with disk C, can rotate about the fixed axis ℓ_C. Disk C has negligible mass and is rigidly attached to E so that they rotate together. While keeping both B and C stationary, disk A is spun to $\omega_A = 1200\,\text{rpm}$. Disk A is then brought in contact with disk C (contact is maintained via a spring), and B and C (and the arm E) are then allowed to freely rotate. Due to friction between A and C, disks C (and arm E) and B start spinning. Eventually A and C stop slipping relative to one another. Disk B always rotates without slip over C. Let $d = 0.27\,\text{m}$ and $w = 0.95\,\text{m}$. If the only elements of the system that have mass are A and B, and if all friction in the system can be neglected except for that between A and C and between C and B, determine the angular speeds of A and C when they stop slipping relative to one another.

Problem 8.66

An 0.8 lb collar with center of mass at G and a uniform cylindrical horizontal arm A of length $L = 1\,\text{ft}$, radius $r_i = 0.022\,\text{ft}$, and weight $W_A = 1.5\,\text{lb}$ are rotating as shown with $\omega_0 = 1.5\,\text{rad/s}$ while the collar's mass center is at a distance $d = 0.44\,\text{ft}$ from the z axis. The vertical shaft has radius $e = 0.03\,\text{ft}$ and negligible mass. After the cord restraining the collar is cut, the collar slides with no friction relative to the arm. Assuming that no external forces and moments are applied to the system, determine the collar's impact speed with the end of A if (a) the collar is modeled as a particle coinciding with its own mass center (in this case neglect the collar's dimensions), and (b) the collar is modeled as a uniform hollow cylinder with length $\ell = 0.15\,\text{ft}$, inner radius r_i, and outer radius $r_o = 0.048\,\text{ft}$.

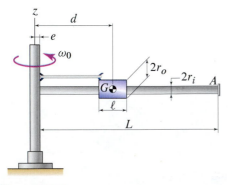

Figure P8.66

Problem 8.67

A crate A with weight $W_A = 250$ lb is hanging from a rope wound around a uniform drum D of radius $r = 1.2$ ft, weight $W_D = 125$ lb, and center C. The system is initially at rest when the restraining system holding the drum stationary fails, thus causing the drum to rotate, the rope to unwind, and, consequently, the crate to fall. Assuming that the rope does not stretch or slip relative to the drum and neglecting the inertia of the rope, determine the speed of the crate 1.5 s after the system starts to move.

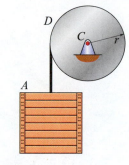

Figure P8.67

Problem 8.68

Some pipe sections of radius r and mass m are being unloaded and placed in a row against a wall. The first of these pipe sections, A, is made to roll without slipping into a corner with an angular velocity ω_0 as shown. Upon touching the wall, A does not rebound but slips against the ground and against the wall. Modeling A as a uniform thin ring with center at G and letting μ_g and μ_w be the coefficients of kinetic friction of the contacts between A and the ground and between A and the wall, respectively, determine an expression for the angular velocity of A as a function of time from the moment A touches the wall until it stops. *Hint:* Using the methods learned in Chapter 7, we can show that the friction forces at the ground and at the wall are constant.

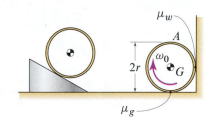

Figure P8.68

Problem 8.69

A cord, which is wrapped around the inner radius of the spool of mass $m = 35$ kg, is pulled vertically at A by a constant force $P = 120$ N (the cord is pulled in such a way that it remains vertical), causing the spool to roll over the horizontal bar BD. The inner radius of the spool is $R = 0.3$ m, and the center of mass of the spool is at G, which also coincides with the geometric center of the spool. The spool's radius of gyration is $k_G = 0.18$ m. Assuming that the spool starts from rest, that the cord's inertia and extensibility can be neglected, and that the spool rolls without slip, determine the speed of the spool's center 3 s after the application of the force. In addition, determine the minimum static friction coefficient for rolling without slip to be maintained during the time interval in question.

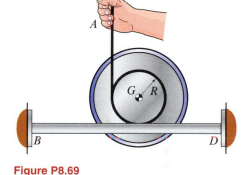

Figure P8.69

Problem 8.70

A spool has weight $W = 450$ lb, outer and inner radii $R = 6$ ft and $\rho = 4.5$ ft, respectively, center of mass G coinciding with its geometric center, and radius of gyration $k_G = 4.0$ ft. The spool is being pulled to the right as shown, and the cable wrapped around the spool can be modeled as being inextensible and of negligible mass. Assume that the spool rolls without slip relative to both the cable and the ground. If the cable is pulled with a force $P = 125$ lb, determine the speed of the center of spool after 2 s and the minimum value of the static friction coefficient between the spool and the ground necessary to guarantee rolling without slip.

Figure P8.70

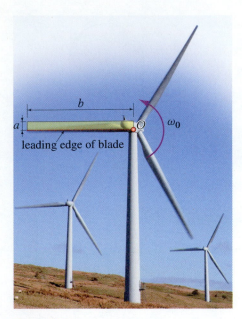

Problem 8.71

The wind turbine in the figure consists of three equally spaced blades that are rotating as shown about the fixed point O with an angular velocity $\omega_0 = 30$ rpm. Suppose that each 38,000 lb blade can be modeled as a narrow uniform rectangle of length $b = 182$ ft, width $a = 12$ ft, and negligible thickness, with one of its corners coinciding with the center of rotation O. The orientation of each blade can be controlled by rotating the blade about an axis going through the center O and coinciding with the blade's leading edge. Neglecting aerodynamic forces and any source of friction, and assuming that turbine is freely rotating, determine the turbine's angular velocity ω_f after each blade has been rotated 90° about its own leading edge.

Problem 8.72

A toy helicopter consists of a rotor A, a body B, and a small ballast C. The axis of rotation of the rotor goes through G, which is the center of mass of the body B and ballast C. While holding the body (and ballast) fixed, the rotor is spun as shown with a given angular velocity ω_0. If there is *no friction* between the helicopter's body and the rotor's shaft, will the body of the helicopter start spinning once the toy is released? **Note:** Concept problems are about *explanations*, not computations.

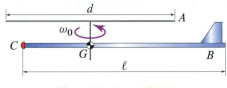

Figure P8.72 and P8.73

Problem 8.73

A toy helicopter consists of a rotor A with diameter $d = 10$ in. and weight $W_r = 0.09 \times 10^{-3}$ oz, a thin body B of length $\ell = 12$ in. and weight $W_B = 0.144 \times 10^{-3}$ oz, and a small ballast C placed at the front end of the body and with weight $W_C = 0.0723 \times 10^{-3}$ oz. The ballast's weight is such that the axis of the rotation of the rotor goes through G, which is the center of mass of the body B and ballast C. While holding the body (and ballast) fixed, the rotor is spun as shown with $\omega_0 = 150$ rpm. Neglecting aerodynamic effects, the weights of the rotor's shaft and the body's tail, and assuming there is friction between the helicopter's body and the rotor's shaft, determine the angular velocity of the body once the toy is released and the angular velocity of the rotor decreases to 120 rpm. Model the body as a uniform thin rod and the ballast as a particle. Assume that the rotor and the body remain horizontal after release.

Problem 8.74

A crane has a boom A of mass m_A and length ℓ that can rotate in the horizontal plane about a fixed point O. A trolley B of mass m_B is mounted on one side of A such that the mass center of B is always at a distance e from the longitudinal axis of A. The position of B is controlled via a cable and a system of pulleys. Both A and B are initially at rest in the position shown, where d is the initial distance of B from O measured along the longitudinal axis of A. The boom A is free to rotate about O and, for a short time interval $0 \leq t \leq t_f$, B moves with constant acceleration a_0 without reaching the end of A. Letting I_O be the mass moment of inertia of A, modeling B as a particle, and accounting only for the inertia of A and B, determine the direction of rotation of A and the angle θ swept by A from $t = 0$ to $t = t_f$. Neglect the mass of the cable and of the pulleys.

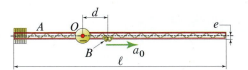

Figure P8.74

Problem 8.75

Cars A and B collide as shown. Neglecting the effect of friction, what would be the angular velocity of A and B immediately after impact if A and B were to form a single rigid body as a result of the collision? In solving the problem, let C and D be the mass centers of A and B, respectively, and use the following data: the weight of A is $W_A = 3130\,\text{lb}$, the radius of gyration of A is $k_C = 34.5\,\text{in.}$, the speed of A right before impact is $v_A = 12\,\text{mph}$, the weight of B is $W_B = 3520\,\text{lb}$, the radius of gyration of B is $k_D = 39.3\,\text{in.}$, the speed of B right before impact is $v_B = 15\,\text{mph}$, $d = 19\,\text{in.}$, and $\ell = 144\,\text{in.}$ Finally, assume that while A and B form a single rigid body right after impact, the mass center of the rigid body formed by A and B coincides with the mass center of the A-B system right before impact.

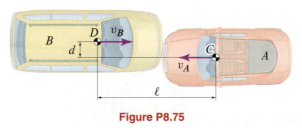

Figure P8.75

Problem 8.76

Some pipe sections are gently nudged from rest down an incline and roll without slipping all the way to a step of height b. Assume that each pipe section does not slide or rebound against the step, so that the pipes move as if hinged at the corner of the step. Modeling a pipe as a uniform thin ring of mass m and radius r, and letting d be the height from which the pipes are released, determine the minimum value of d so that the pipes can roll over the step. *Hint:* When a pipe hits the corner of the step, its motion changes almost instantaneously from rolling without slip on the ground to a fixed axis rotation about the corner of the step. Model this transition, using the ideas presented in Section 5.2 on p. 356. That is, assume that there is an infinitesimal time interval right after the impact between a pipe and the corner of the step, in which the pipe does not change its position significantly, the pipe loses contact with the ground, and its weight is negligible relative to the contact forces between the pipe and the step.

Figure P8.76

Problem 8.77

Following up on parts (b) and (c) of the Pioneer 3 despin in Prob. 7.65, it turns out that it is possible to analytically determine the length of the unwound wire needed to achieve *any* value of ω_s by making use of conservation of energy and conservation of angular momentum. In doing so, let the masses of A and B each be m, and the mass moment of inertia of the spacecraft body be I_O. Let the initial conditions of the system be $\omega_s(0) = \omega_0$, $\ell(0) = 0$, and $\dot{\ell}(0) = 0$, and neglect gravity and the mass of each wire.

(a) Find the velocity of each of the masses A and B as a function of the wire length $\ell(t)$ and the angular velocity of the spacecraft body $\omega_s(t)$ (and the radius of the spacecraft R). *Hint:* This part of the problem involves just kinematics — refer to Prob. 6.117 if you need help with the kinematics.

(b) Apply the work-energy principle to the spacecraft system between the time just before the masses start to unwind and any arbitrary later time. You should obtain an expression relating ℓ, $\dot{\ell}$, ω_s, and constants. *Hint:* No external work is done on the system.

(c) Since no external forces act on the system, its total angular momentum must be conserved about point O. Relate the angular momentum for this system between the time just before the masses start unwinding and any arbitrary later time. As with part (b), you should obtain an expression relating ℓ, $\dot{\ell}$, ω_s, and constants.

(d) Solve the energy and angular momentum equations obtained in parts (b) and (c), respectively, for $\dot{\ell}$ and ω_s. Now, letting $\omega_s = 0$, show that the length of the unwound wire when the angular velocity of the spacecraft body is zero is given by $\ell_{\omega_s=0} = \sqrt{(I_O + 2mR^2)/(2m)}$.

(e) From your solutions for $\dot{\ell}$ and ω_s in part (d), find the equations for $\ell(t)$ and $\omega_s(t)$. These are the general solutions to the nonlinear equations of motion found in Prob. 7.64.

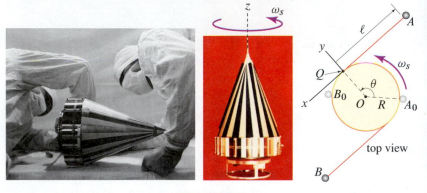

Figure P8.77

DESIGN PROBLEMS

Design Problem 8.2

The Pioneer 3 spacecraft was a spin-stabilized spacecraft launched on December 6, 1958, by the U.S. Army Ballistic Missile agency in conjunction with NASA. It was designed with a despin mechanism consisting of two equal masses A and B, each of mass m, that could be spooled out to the end of two wires of variable length $\ell(t)$ when triggered by a hydraulic timer. As the masses unwound, they would slow the spacecraft's spin from an initial angular velocity $\omega_s(0)$ to the final angular velocity $(\omega_s)_{\text{final}}$, and then the weights and wires would be released. Assume that masses A and B are initially at positions A_0 and B_0, respectively, before the wire begins to unwind, that the mass moment of inertia of the spacecraft is I_O (this does not include the two masses A and B), and neglect gravity and the mass of each wire. With this in mind, you are given the task of despinning the spacecraft body from an angular velocity of 400 rpm to *any* angular velocity in the range $-400\,\text{rpm} < \omega_s < 400\,\text{rpm}$. To do this, a transducer is used that senses the tension in one of the wires at every instant. Design the total length of the wires to achieve the desired range of angular velocities, and determine the tension in the wires as a function of the angular velocity of the spacecraft so that the sensor can know when the spacecraft has achieved the desired velocity and release the wires and masses.

Use $R = 12.5\,\text{cm}$, $m = 7\,\text{g}$, $I_O = 0.0277\,\text{kg·m}^2$, and the initial conditions $\ell(0) = 0.01\,\text{m}$ and $\dot{\ell}(0) = 0\,\text{m/s}$, with the despin mechanism shown. *Hint:* Refer to Prob. 6.117 on p. 532 if you need help with the kinematics.

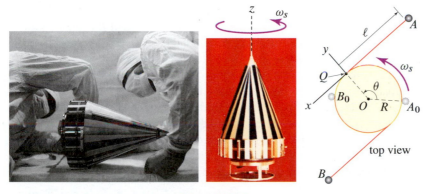

Figure DP8.2

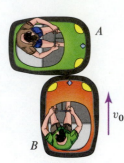

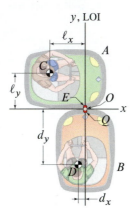

Figure 8.24
A collision between two bumper cars.

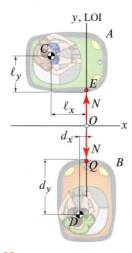

Figure 8.25
Impact-relevant FBD of A and B as a system. The contact points E and Q belong to A and B, respectively, and coincide with O.

Figure 8.26
Individual FBDs of A and B.

8.3 Impact of Rigid Bodies

In this section we continue the study of impacts, begun in Section 5.2 on p. 356, by considering impacts between rigid bodies. We will use the notation and concepts introduced in Section 5.2. As a reminder,

- We denote the value of a quantity right before and right after a collision by the superscripts $-$ and $+$, respectively (for example, v_A^+ denotes the speed of a point A right after impact).

- Our modeling of impacts is based on the concept of *impulsive force* (see p. 357). As in Section 5.2, this modeling assumption is reflected in our *impact-relevant* FBDs, which include *only* impulsive forces.

Collision of two bumper cars

Bumper car A in Fig. 8.24, while stopped for a moment, is rammed by car B, which is moving with a speed $v_0 = 9$ ft/s as shown. What will the postimpact velocities of A and B be if we model A and B as rigid bodies? We will answer this question by combining what we learned in Section 5.2 with the momentum principles in Section 8.2.

To begin, we count the problem's unknowns. Assume that we know the COR (coefficient of restitution) e for the collision, the masses of A and B, the location of their mass centers, their preimpact velocities, and their mass moments of inertia about their respective mass centers. Since the motion is planar, the postimpact velocities of A and B are described by six pieces of information: the velocity components of the mass centers of A and B (four unknowns) and the angular velocities of A and B (two unknowns). Consequently, we will need six scalar equations to solve for the six unknowns.

The system's FBD is shown in Fig. 8.25. Consistent with our modeling approach, all nonimpulsive forces have been neglected. Figure 8.25 also defines the LOI (line of impact) and a coordinate system whose y axis coincides with the LOI. Since there are no external impulsive forces, the system's total linear and angular momenta are conserved. Guided by what we learned in Section 5.2, instead of writing all of these conservation statements for the system, we only need to consider conservation of linear momentum along the LOI

$$m_A v_{Cy}^- + m_B v_{Dy}^- = m_A v_{Cy}^+ + m_B v_{Dy}^+, \tag{8.51}$$

in which $v_{Cy}^- = 0$ since A is stationary before impact.

Again following Section 5.2, we assume that the contact between A and B is frictionless. As a result, there are no impulsive forces perpendicular to the LOI, as shown in Fig. 8.26. Therefore, the individual linear momenta of A and B in the x direction are conserved

$$m_A v_{Cx}^- = m_A v_{Cx}^+ \quad \text{and} \quad m_B v_{Dx}^- = m_B v_{Dx}^+. \tag{8.52}$$

We now write the COR equation, which relates the component of the separation velocity along the LOI to the corresponding component of the approach velocity. The COR depends on the materials of the colliding objects *at the points of contact*. Therefore, referring to Fig. 8.25, we write the COR equation using the y velocity components of points E and Q

$$v_{Ey}^+ - v_{Qy}^+ = e\left(v_{Qy}^- - v_{Ey}^-\right). \tag{8.53}$$

Although Eq. (8.53) is not written in terms of the problem's unknowns, we can use rigid body kinematics to write

$$\vec{v}_E = \vec{v}_C + \omega_A \hat{k} \times \vec{r}_{E/C} \quad \text{and} \quad \vec{v}_Q = \vec{v}_D + \omega_B \hat{k} \times \vec{r}_{Q/D}, \tag{8.54}$$

where since the positions of A and B *do not change during the impact,*

$$\vec{r}_{E/C} = \vec{r}_{E/C}^{\,-} = \vec{r}_{E/C}^{\,+} \quad \text{and} \quad \vec{r}_{Q/D} = \vec{r}_{Q/D}^{\,-} = \vec{r}_{Q/D}^{\,+}. \tag{8.55}$$

Writing $\vec{v}_E = v_{Ex}\,\hat{i} + v_{Ey}\,\hat{j}$, $\vec{v}_C = v_{Cx}\,\hat{i} + v_{Cy}\,\hat{j}$, and $\vec{r}_{E/C} = \ell_x\,\hat{i} - \ell_y\,\hat{j}$ and looking at just the y component, we get

$$v_{Ey}^{\pm} = v_{Cy}^{\pm} + \omega_A^{\pm}\ell_x, \tag{8.56}$$

where the superscript $\pm$ is meant to indicate that Eq. (8.56) holds when evaluated both right before and right after impact. Similarly, writing $\vec{v}_Q = v_{Qx}\,\hat{i} + v_{Qy}\,\hat{j}$, $\vec{v}_D = v_{Dx}\,\hat{i} + v_{Dy}\,\hat{j}$, and $\vec{r}_{Q/D} = d_x\,\hat{i} + d_y\,\hat{j}$ and looking at just the y component, we get

$$v_{Qy}^{\pm} = v_{Dy}^{\pm} + \omega_B^{\pm}d_x. \tag{8.57}$$

Noting that $v_{Cy}^{-} = 0$, $\omega_A^{-} = 0$, and $\omega_B^{-} = 0$, Eqs. (8.56) and (8.57) become

$$v_{Ey}^{-} = 0, \qquad v_{Ey}^{+} = v_{Cy}^{+} + \omega_A^{+}\ell_x, \tag{8.58}$$

$$v_{Qy}^{-} = v_{Dy}^{-}, \qquad v_{Qy}^{+} = v_{Dy}^{+} + \omega_B^{+}d_x. \tag{8.59}$$

Substituting Eqs. (8.58) and (8.59) into Eq. (8.53), we obtain

$$v_{Cy}^{+} + \omega_A^{+}\ell_x - v_{Dy}^{+} - \omega_B^{+}d_x = ev_{Dy}^{-}. \tag{8.60}$$

We derived Eq. (8.51), Eqs. (8.52), and Eq. (8.60), which are four equations in six unknowns, using the *particle* impact theory. The remaining two equations must address the fact that we are modeling the bumper cars as *rigid bodies*. Note that the four equations in question are unrelated to the moment equation governing the rotational motion of a rigid body. Thus, we have yet to take advantage of the angular impulse-momentum principle.

Referring to Fig. 8.26, we choose point O as a *convenient* moment center for the application of the angular impulse-momentum principle because, at the time of impact, O coincides with E and Q and so the impulsive force N has no moment about O. In addition, unlike points E and Q, O is a *fixed* point. Hence, the individual angular momenta of A and B relative to O are conserved

$$\left(\vec{h}_O^{-}\right)_A = \left(\vec{h}_O^{+}\right)_A \quad \text{and} \quad \left(\vec{h}_O^{-}\right)_B = \left(\vec{h}_O^{+}\right)_B, \tag{8.61}$$

where, using Eq. (8.42) on p. 627,

$$\left(\vec{h}_O^{\pm}\right)_A = I_C\vec{\omega}_A^{\pm} + \vec{r}_{C/O} \times m_A\vec{v}_C^{\pm} \tag{8.62}$$

$$\left(\vec{h}_O^{\pm}\right)_B = I_D\vec{\omega}_B^{\pm} + \vec{r}_{D/O} \times m_B\vec{v}_D^{\pm}, \tag{8.63}$$

and where I_C and I_D are the mass moments of inertia of A and B, respectively.* Note that, consistent with our modeling assumptions, the relative

---

* As stated in Chapter 7, the subscript on I will be used to uniquely identify the axis with respect to which I is calculated. Hence, I_A denotes the mass moment of inertia of the rigid body about an axis perpendicular to the plane of motion passing through A.

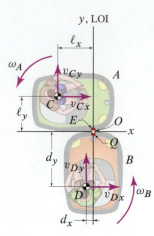

Figure 8.27
Velocity components of C and D together with the angular velocities of A and B.

position vectors $\vec{r}_{C/O}$ and $\vec{r}_{D/O}$ in Eqs. (8.62) and (8.63) remain constant during the impact and do not need the superscript $\pm$. Substituting Eqs. (8.62) and (8.63) into Eqs. (8.61), carrying out the cross-products, and recalling that the only nonzero components of Eqs. (8.61) are those in the z direction, we obtain the following two scalar equations (see Fig. 8.27)

$$0 = I_C \omega_A^+ - m_A v_{Cx}^+ \ell_y - m_A v_{Cy}^+ \ell_x, \tag{8.64}$$

$$-m_B v_0 d_x = I_D \omega_B^+ + m_B v_{Dx}^+ d_y - m_B v_{Dy}^+ d_x, \tag{8.65}$$

in which we have accounted for the preimpact motion of A and B. Equations (8.51), (8.52), (8.60), (8.64), and (8.65) are the six equations we were looking for. Solving them yields

$$v_{Cx}^+ = 0 \text{ ft/s}, \tag{8.66}$$

$$v_{Cy}^+ = \frac{I_C I_D m_B v_0 (1+e)}{I_C I_D (m_A + m_B) + m_A m_B (d_x^2 I_C + \ell_x^2 I_D)} = 1.57 \frac{\text{ft}}{\text{s}}, \tag{8.67}$$

$$\omega_A^+ = \frac{\ell_x I_D m_A m_B v_0 (1+e)}{I_C I_D (m_A + m_B) + m_A m_B (d_x^2 I_C + \ell_x^2 I_D)} = 5.08 \frac{\text{rad}}{\text{s}}, \tag{8.68}$$

$$v_{Dx}^+ = 0 \text{ ft/s}, \tag{8.69}$$

$$v_{Dy}^+ = \frac{\left[m_A m_B (I_C d_x^2 + I_D \ell_x^2) + I_C I_D (m_B - e m_A) \right] v_0}{I_C I_D (m_A + m_B) + m_A m_B (d_x^2 I_C + \ell_x^2 I_D)} = 7.47 \frac{\text{ft}}{\text{s}}, \tag{8.70}$$

$$\omega_B^+ = \frac{-d_x I_C m_A m_B v_0 (1+e)}{I_C I_D (m_A + m_B) + m_A m_B (d_x^2 I_C + \ell_x^2 I_D)} = -0.908 \frac{\text{rad}}{\text{s}}, \tag{8.71}$$

where, to obtain the numerical results, we used the following data: $W_A = 630$ lb (weight of A including the driver), $W_B = 645$ lb (weight of B including the driver), $I_C = 15.1$ slug·ft^2, $I_D = 15.2$ slug·ft^2, $e = 0.8$, $\ell_x = 2.5$ ft, $\ell_y = 2.5$ ft, $d_x = 0.45$ ft, and $d_y = 3.9$ ft. The results in Eqs. (8.66)–(8.71) match our intuition since it indicates that car A will rotate counterclockwise, whereas car B will rotate clockwise. Furthermore, letting T be the system's kinetic energy, we have $T^+ = 784$ ft·lb $< T^- = 811$ ft·lb, as it should since the impact was not perfectly elastic.

We solved this rigid body impact problem without needing any new theory. Hence, this section will not offer any new theoretical developments; rather we will carefully discuss the assumptions needed to solve rigid body impact problems that are within the bounds of the theory we already possess. This discussion will also emphasize what makes a rigid body impact different from a particle impact.

Rigid body impact: basic nomenclature and assumptions

As in the case of particle impacts (see Section 5.2 on p. 356), a rigid body impact is called *plastic* if the COR $e = 0$. We call a rigid body impact *perfectly plastic* if the colliding bodies form a single rigid body after impact. An impact is *elastic* if $0 < e < 1$, and the ideal case with $e = 1$ is called *perfectly elastic*. In addition, an impact is *unconstrained* if the impacting objects are not acted upon by any external impulsive forces; otherwise the impact is called *constrained*.

Rigid body impacts can be very complex, and we will consider only those cases, such as the bumper car collision problem, in which the following assumptions hold:

1. The impact involves only two bodies in planar motion where no impulsive force has a component perpendicular to the plane of motion.

2. Contact between any two rigid bodies occurs only at one point, and at this point we can clearly define the LOI.

3. The contact between the bodies is frictionless.

Classification of impacts

In Section 5.2 (p. 356), we classified impacts based on (1) the position of the mass centers of the colliding objects in relation to the LOI and (2) the orientation of the preimpact velocities in relation to the LOI. For convenience, we repeat Table 5.3 (p. 363) in Table 8.1, in which the expression *preimpact*

Table 8.1. *Table 5.3 repeated.* Classification of impacts.

Impact Geometry Criteria		Impact Type
Preimpact velocities	*Centers of mass*	
parallel to LOI	on LOI	direct central
parallel to LOI	not on LOI	direct eccentric
not parallel to LOI	on LOI	oblique central
not parallel to LOI	not on LOI	oblique eccentric

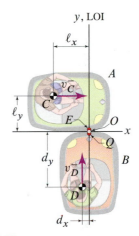

Figure 8.28
Preimpact system configuration of an oblique eccentric impact.

velocities refers to the velocities of the mass centers of the colliding objects and not to their angular velocities (which are perpendicular to the plane of motion). Based on Table 8.1, the bumper car collision problem was a *direct eccentric* impact because the cars' mass centers were not on the LOI and because the preimpact velocity of the mass center of B was parallel to the LOI while the velocity of A was equal to zero. If, right before impact, car A had been moving as shown in Fig. 8.28, then one of the mass centers would have had a preimpact velocity not parallel to the LOI and the impact would have been oblique and eccentric.

Classifying impacts helps us assess how involved the solution process might be. Generally speaking, direct central impacts are the simplest to analyze, and the oblique eccentric impacts are the most complex.

Central impact

In a *central impact*, the centers of mass of the two colliding objects lie on the LOI at the time of impact. There are two types of central impacts: direct and oblique (see Table 8.1). No matter the type, under assumptions 1–3 introduced above, rigid body central impacts have two important characteristics:

1. The angular velocities are *conserved* through the impact.

2. The COR equation can be written directly in terms of the velocity components (along the LOI) of the mass centers.

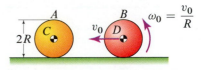

Figure 8.29
Example of a direct central impact.

We will now demonstrate properties 1 and 2 by considering the collision of two identical billiard balls A and B, as shown in Fig. 8.29. Assume that A

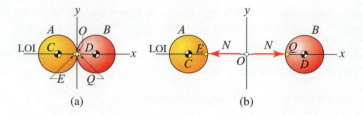

Figure 8.30. Impact-relevant FBDs (a) of the system as a whole and (b) of the two balls individually.

is initially stationary, while B is rolling without slip to the left at a speed v_0. Modeling the collision as perfectly elastic, we want to determine the postimpact velocities of the two balls. As always, we begin by sketching the system's (impact-relevant) FBD, shown in Fig. 8.30(a), and the (impact-relevant) FBDs of the colliding bodies separately, shown in Fig. 8.30(b). As with all impact problems, the FBDs should show the LOI and a convenient coordinate system (typically aligned with the LOI). Points E and Q belong to A and B, respectively. The origin O of the chosen coordinate system coincides with E and Q at the time of impact but is otherwise understood to be a *fixed* point. Notice that the FBDs in Fig. 8.30 do not show any reaction force between the balls and the table. This is so because we have regarded the reaction forces in question as *nonimpulsive*. The reason for this choice is rooted in assumption 3 on p. 649. The assumption that the contact between the two balls is frictionless implies that no impulsive friction force can be generated along the y direction in response to the sliding of point Q relative to point E (before impact, ball B rolls without slip so that point Q has a vertical component of velocity at the time of impact). In turn, the absence of a vertical impulsive force at the point of contact implies that no corresponding impulsive *reaction* force is generated in the vertical direction at the supporting surface. If such a force were present, then it would be an *external* impulsive force and the impact would be a *constrained* impact.

Because the impact is central, the impulsive force N has no moment with respect to C and D. Summing moments about the mass center of each ball using Eq. (8.45) on p. 627, we can then say that $I_C \omega_A^- = I_C \omega_A^+$ and $I_D \omega_B^- = I_D \omega_B^+$, where I_C and I_D are the mass moments of inertia of A and B, respectively. Therefore, we conclude that

$$\omega_A^- = \omega_A^+ \quad \text{and} \quad \omega_B^- = \omega_B^+, \qquad (8.72)$$

that is, the angular velocities of A and B are conserved through the impact.

Because there are no impulsive forces in the direction perpendicular to the LOI, the components of velocity in that direction do not change through the impact. Thus we have $v_{Cy}^- = v_{Cy}^+ = 0$ and $v_{Dy}^- = v_{Dy}^+ = 0$ and so the only remaining unknowns are v_{Cx}^+ and v_{Dx}^+. To find these unknowns, we enforce the conservation of the system's linear momentum along the LOI, i.e.,

$$m_A v_{Cx}^- + m_B v_{Dx}^- = m_A v_{Cx}^+ + m_B v_{Dx}^+, \qquad (8.73)$$

and the COR equation for points E and Q, that is,

$$v_{Ex}^+ - v_{Qx}^+ = e\big(v_{Qx}^- - v_{Ex}^-\big). \qquad (8.74)$$

📢 **Concept Alert**

Central impacts and angular velocities.
If the contact between the colliding bodies is assumed to be frictionless (see assumption 3 on p. 649) and if the impact is central, the angular velocities of the two impacting bodies are completely unaffected by the impact.

Using rigid body kinematics, we have

$$\vec{v}_E^{\pm} = \vec{v}_C^{\pm} + \omega_A^{\pm}\,\hat{k} \times \vec{r}_{E/C} \quad \text{and} \quad \vec{v}_Q^{\pm} = \vec{v}_D^{\pm} + \omega_B^{\pm}\,\hat{k} \times \vec{r}_{Q/D}, \qquad (8.75)$$

where we have used the fact that $\vec{r}_{E/C}$ and $\vec{r}_{Q/D}$ do not change during the impact. Referring to Fig. 8.31, $\vec{r}_{E/C} = R\,\hat{\imath}$ and $\vec{r}_{Q/D} = -R\,\hat{\imath}$ and so

$$\omega_A^{\pm}\,\hat{k} \times \vec{r}_{E/C} = \omega_A^{\pm} R\,\hat{\jmath} \quad \text{and} \quad \omega_B^{\pm}\,\hat{k} \times \vec{r}_{Q/D} = -\omega_B^{\pm} R\,\hat{\jmath}. \qquad (8.76)$$

Equations (8.75) and (8.76) tell us that the x components (i.e., along the LOI) of $\vec{v}_E^{\pm}$ and $\vec{v}_Q^{\pm}$ must be identical to the x components of $\vec{v}_C^{\pm}$ and $\vec{v}_D^{\pm}$, respectively, that is,

$$v_{Ex}^{\pm} = v_{Cx}^{\pm} \quad \text{and} \quad v_{Qx}^{\pm} = v_{Dx}^{\pm}. \qquad (8.77)$$

Equations (8.77) allow us to write the COR equation directly in terms of the velocity of the mass centers, i.e.,

$$\boxed{v_{Cx}^{+} - v_{Dx}^{+} = e\left(v_{Dx}^{-} - v_{Cx}^{-}\right).} \qquad (8.78)$$

Going back to the billiard ball impact problem, recalling the given preimpact conditions and that $e = 1$, we obtain the following solution:

$$v_{Cx}^{+} = -v_0, \quad \omega_A^{+} = 0, \quad v_{Dx}^{+} = 0, \quad \text{and} \quad \omega_B^{+} = v_0/R. \qquad (8.79)$$

Interestingly, even though $v_{Dx}^{+} = 0$, because $\omega_B^{+} = v_0/R$, ball B does *not* stop after impact. The equations $v_{Dx}^{+} = 0$ and $\omega_B^{+} = v_0/R$ taken together imply that right after impact the mass center of ball B has zero velocity for an instant while ball B slips against the pool table. Assuming that there is friction between the table and balls, the friction force due to sliding will cause the center of mass of B to start moving again to the left, as if to chase ball A. This result is important because we would never have reached it had we modeled the balls as particles (we would have concluded that B simply stopped after impact). In turn, this shows that while a central impact of rigid bodies might be as simple to solve as a particle impact problem, the two models yield different predictions.

Eccentric impact

Referring to Fig. 8.32, for an impact to be *eccentric* at least one of the mass centers of the colliding bodies cannot lie on the LOI. The key feature of an eccentric impact is that not only does the collision affect the velocities of the mass centers, but also it affects the bodies' angular velocities.

The bumper car collision analyzed at the beginning of the section is an example of eccentric impact. Therefore, to avoid unnecessary repetition, here we simply list the equations required to solve unconstrained rigid body eccentric impact problems and review the physical justification for these equations.

Referring to Fig. 8.32, consider two impacting rigid bodies moving in the plane as shown. As usual, we start the problem solution with the FBD of the system and the FBDs of the individual bodies at the time of impact, shown in Fig. 8.33(a) and (b), respectively. On these diagrams we clearly indicate the LOI and the chosen coordinate system. In addition, we choose the origin O so as to coincide with the points of contact E and Q between the two rigid

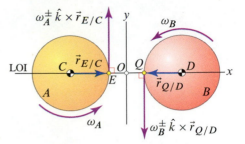

Figure 8.31
Kinematics of a central impact.

🗼 **Concept Alert**

Central impacts and COR equation. Under the stated assumptions, in a central impact the COR equation can be written directly in terms of the velocity components of the mass centers.

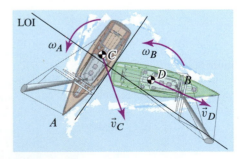

Figure 8.32
Collision between two boats.

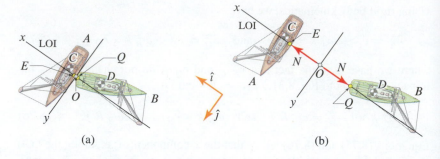

Figure 8.33. FBDs of the colliding bodies (a) as a system and (b) individually.

bodies at the time of impact, and we keep in mind that O is a *fixed* point. The reason for choosing O as stated is that it is a convenient point to use as the moment center when enforcing the angular impulse-momentum principle.

The solution of an unconstrained rigid body eccentric impact problem is governed by six scalar equations. The first of these equations enforces the conservation of linear momentum for the system along the LOI, that is,

$$m_A v_{Cx}^- + m_B v_{Dx}^- = m_A v_{Cx}^+ + m_B v_{Dx}^+, \tag{8.80}$$

The second and third of these equations are

$$v_{Cy}^- = v_{Cy}^+ \quad \text{and} \quad v_{Dy}^- = v_{Dy}^+, \tag{8.81}$$

which follow from the frictionless contact assumption (assumption 3 on p. 649). This assumption implies that the individual linear momentum of the colliding bodies is conserved in the direction perpendicular to the LOI.

The fourth equation is the COR equation, which is first written in terms of the components of the velocities of the contact points along the LOI, i.e.,

$$v_{Ex}^+ - v_{Qx}^+ = e\left(v_{Qx}^- - v_{Ex}^-\right), \tag{8.82}$$

and then rewritten in terms of the velocity components of the mass centers, making sure to satisfy rigid body kinematics, which requires that

$$\vec{v}_E^{\pm} = \vec{v}_C^{\pm} + \omega_A^{\pm}\hat{k} \times \vec{r}_{E/C} \quad \text{and} \quad \vec{v}_Q^{\pm} = \vec{v}_D^{\pm} + \omega_B^{\pm}\hat{k} \times \vec{r}_{Q/D}. \tag{8.83}$$

Referring to Fig. 8.34, because E, Q, and O coincide at the time of impact,

$$\vec{r}_{E/C} = -\vec{r}_C = -x_C\,\hat{\imath} - y_C\,\hat{\jmath} \quad \text{and} \quad \vec{r}_{Q/D} = -\vec{r}_D = -x_D\,\hat{\imath} - y_D\,\hat{\jmath}. \tag{8.84}$$

Then, substituting Eqs. (8.84) into Eqs. (8.83), carrying out the cross-products, and rearranging terms, we obtain

$$\vec{v}_E^{\pm} = \left(v_{Cx}^{\pm} + \omega_A^{\pm} y_C\right)\hat{\imath} + \left(v_{Cy}^{\pm} - \omega_A^{\pm} x_C\right)\hat{\jmath}, \tag{8.85}$$

$$\vec{v}_Q^{\pm} = \left(v_{Dx}^{\pm} + \omega_B^{\pm} y_D\right)\hat{\imath} + \left(v_{Dy}^{\pm} - \omega_B^{\pm} x_D\right)\hat{\jmath}. \tag{8.86}$$

Using Eqs. (8.85) and (8.86), we can give Eq. (8.82) the following final form:

$$v_{Cx}^+ + \omega_A^+ y_C - v_{Dx}^+ - \omega_B^+ y_D$$
$$= e\left(v_{Dx}^- + \omega_B^- y_D - v_{Cx}^- - \omega_A^- y_C\right). \tag{8.87}$$

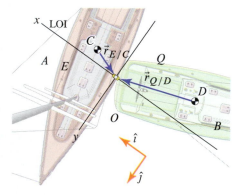

Figure 8.34

Relative position vectors of E and Q with respect to C and D, respectively.

The remaining two equations say that each of the angular momenta of A and B relative to O is conserved, that is,

$$\left(\vec{h}_O^-\right)_A = \left(\vec{h}_O^+\right)_A \quad \text{and} \quad \left(\vec{h}_O^-\right)_B = \left(\vec{h}_O^+\right)_B, \qquad (8.88)$$

where $\left(\vec{h}_O\right)_A$ and $\left(\vec{h}_O\right)_B$ are the angular momenta of A and B relative to O, respectively, and, applying Eq. (8.42) on p. 627, can be written as

$$\left(\vec{h}_O^\pm\right)_A = I_C \vec{\omega}_A^\pm + \vec{r}_{C/O} \times m_A \vec{v}_C^\pm, \qquad (8.89)$$

$$\left(\vec{h}_O^\pm\right)_B = I_D \vec{\omega}_B^\pm + \vec{r}_{D/O} \times m_B \vec{v}_D^\pm, \qquad (8.90)$$

where I_C and I_D are the mass moments of inertia of A and B, respectively. Again we note that, in Eqs. (8.89) and (8.90), the relative position vectors $\vec{r}_{C/O}$ and $\vec{r}_{D/O}$ are assumed to be constant through the impact and therefore do not need the superscript $\pm$. Referring to Fig. 8.33(b), the physical justification for Eqs. (8.88) is that the impulsive forces acting on A and B provide no moment about the fixed point O. Observe that Eqs. (8.88) yield only two scalar equations because the only nonzero component of these equations is perpendicular to the plane of motion.

Equations (8.80), (8.81), (8.87), and (8.88) are all that is needed to solve the most general case of unconstrained oblique eccentric impact of two rigid bodies (under the assumptions stated on p. 648).

Constrained eccentric impact

In a *constrained impact* one or both of the colliding bodies are subject to external impulsive forces (see p. 365). Modeling these impacts can be challenging, and we consider only a simple case in which one of the impacting bodies is constrained to move in fixed axis rotation.

Referring to Fig. 8.35, consider a ballistic pendulum consisting of a uniform thin rod of mass m_r and length L pinned at O and a target block of mass m_t, width w, and height h. Suppose that a bullet of mass m_b is fired with a speed v_0 as shown. Assuming that the bullet becomes embedded in the block, what will the postimpact velocity of the pendulum-bullet system be?

The key to solving the problem is that the only admissible motion of the pendulum is a fixed axis rotation about O. Therefore, the only piece of information we need to describe the system's postimpact behavior is ω_p^+, the postimpact angular velocity of the pendulum. As usual, we sketch the system's FBD at the time of impact, given in Fig. 8.36, making sure to include only impulsive forces. The pin reactions R_x and R_y appear on the FBD because they will take on whatever value is required of them to keep O from moving, and, as such, they are impulsive. The presence of R_x and R_y makes the impact a *constrained* impact. Observe that R_x and R_y, while impulsive, provide no moment about the *fixed* point O so that the system's total angular momentum relative to O must be conserved through the impact. Calling $\left(\vec{h}_O\right)_b$ and $\left(\vec{h}_O\right)_p$ the angular momenta relative to O of the bullet and the pendulum, respectively, we have

$$\left(\vec{h}_O^-\right)_b + \left(\vec{h}_O^-\right)_p = \left(\vec{h}_O^+\right)_b + \left(\vec{h}_O^+\right)_p, \qquad (8.91)$$

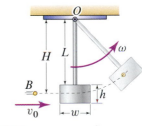

Figure 8.35
A ballistic pendulum.

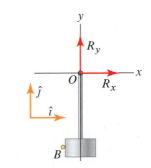

Figure 8.36
FBD of the ballistic pendulum and bullet system at the time of impact.

where, modeling the bullet as a particle and recalling that the pendulum can move only in a fixed axis rotation about O (see Eq. (8.46) on p. 628),

$$(\vec{h}_O^{\pm})_b = \vec{r}_{B/O} \times m_b \vec{v}_B^{\pm} \quad \text{and} \quad (\vec{h}_O^{\pm})_p = (I_O)_p \vec{\omega}_p^{\pm}, \tag{8.92}$$

where $(I_O)_p$ is the mass moment of inertia of the pendulum relative to O and where the relative position vector $\vec{r}_{B/O}$ does not have the superscript $\pm$ because it is treated as a constant through the impact. Since the pendulum is initially stationary and the bullet becomes embedded in the block, substituting Eqs. (8.92) into Eq. (8.91), carrying out the cross-products, and simplifying yield

$$m_b v_0 H = \left[(I_O)_p + m_b H^2 \right] \omega_p^+. \tag{8.93}$$

Solving Eq. (8.93) for the postimpact angular velocity of the pendulum-bullet system, we have

$$\omega_p^+ = \frac{m_b v_0 H}{(I_O)_p + m_b H^2}. \tag{8.94}$$

The problem just discussed illustrates a key element of the solution of most constrained impact problems, namely, the identification of a fixed point about which the system's angular momentum is conserved. Examples 8.12 and 8.13 demonstrate the use of this strategy in the case of more involved physical situations.

End of Section Summary

In this section we studied planar rigid body impacts. We learned that, contrary to particle impacts, rigid bodies can experience *eccentric* impacts. These are collisions in which at least one of the mass centers of the impacting bodies does not lie on the LOI. We also learned that the basic concepts used in particle impacts are applicable to rigid body impacts, and we reviewed solution strategies for a variety of situations.

As with particle impacts, we say that a rigid body impact is *plastic* if the COR $e = 0$. A rigid body impact will be called *perfectly plastic* if the colliding bodies form a single rigid body after impact. An impact is *elastic* if the COR e is such that $0 < e < 1$. Finally, the ideal case with $e = 1$ is referred to as a *perfectly elastic* impact. In addition, an impact is *unconstrained* if the system consisting of the two impacting objects is not subject to external impulsive forces; otherwise, the impact is called *constrained*.

We have considered only impacts satisfying the following assumptions:

1. The impact involves only two bodies in planar motion where no impulsive force has a component perpendicular to the plane of motion.

2. Contact between any two rigid bodies occurs at only one point, and at this point we can clearly define the LOI.

3. The contact between the bodies is frictionless.

We recommend organizing the solution of any impact problem as follows:

- Begin with an FBD of the impacting bodies as a system and FBDs for each of the colliding bodies. Neglect nonimpulsive forces.

- Choose a coordinate system with the origin coincident with the points that come into contact at the time of impact. Recall that the origin of such a coordinate system is a fixed point.

- Enforce the linear and/or the angular impulse-momentum principles for the system and/or for the individual bodies. In applying the angular impulse-momentum principle for the whole system, the moment center should be a fixed point whereas for an individual body the moment center should be a fixed point or the body's mass center.

- For plastic, elastic, and perfectly elastic impacts, the COR equation is first written using the velocity components along the LOI of the points that actually come into contact. For example, for the impact shown in Fig. 8.37, the COR equation is first written as

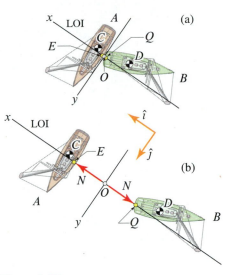

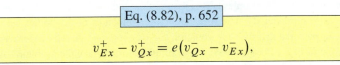

$$v_{Ex}^{+} - v_{Qx}^{+} = e\left(v_{Qx}^{-} - v_{Ex}^{-}\right),$$

Eq. (8.82), p. 652

where the COR e is such that $0 \le e \le 1$. The COR equation must then be rewritten in terms of the colliding bodies' angular velocities and velocities of the mass centers using rigid body kinematics. For the situation in Fig. 8.37, this means rewriting the COR equation using the relations $\vec{v}_E^{\pm} = \vec{v}_C^{\pm} + \omega_A^{\pm}\hat{k} \times \vec{r}_{E/C}$ and $\vec{v}_Q^{\pm} = \vec{v}_D^{\pm} + \omega_B^{\pm}\hat{k} \times \vec{r}_{Q/D}$. Notice that the relative position vectors $\vec{r}_{E/C}$ and $\vec{r}_{Q/D}$ do not have the $\pm$ superscript because they are treated as constants during the impact.

- In perfectly plastic impacts, kinematic constraint equations must be enforced that express the fact that two bodies form a single rigid body after impact.

Figure 8.37
FBDs of the colliding bodies (a) as a system and (b) individually.

EXAMPLE 8.11 *Rigid Body Central Impact*

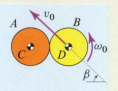

Figure 1
Two colliding hockey pucks.

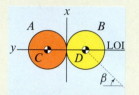

Figure 2
Impact-relevant FBD of the two pucks as a system.

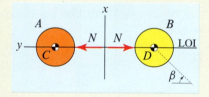

Figure 3
Impact-relevant FBDs of the each of the two pucks.

Two identical hockey pucks sliding over ice collide as shown in Fig. 1. If $\beta = 43°$, the COR is $e = 0.95$, A is initially at rest, and puck B is moving with speed $v_0 = 30\,\text{m/s}$ and angular speed $\omega_0 = 4\,\text{rad/s}$ as shown, determine the postimpact velocities of A and B.

SOLUTION

Road Map & Modeling We start with the FBD of the system and the FBDs of the colliding bodies separately, shown in Figs. 2 and 3, respectively, where we also indicate the LOI and a convenient coordinate system. Because C and D lie on the LOI, the impact is *central*. As discussed earlier in the section, we assume that the contact between the pucks is frictionless. This assumption and the fact that the impact is central allow us to immediately state that the angular velocities of the bodies are unaffected by the impact and that the COR equation can be expressed directly in terms of velocity components of the mass centers. These simplifications make the problem solution very similar to that of a particle impact problem.

Governing Equations

Balance Principles There are no external impulsive forces acting on the system (Fig. 2). Therefore, the system's linear momentum is conserved along the LOI, which gives

$$m_A v_{Cy}^- + m_B v_{Dy}^- = m_A v_{Cy}^+ + m_B v_{Dy}^+. \tag{1}$$

No impulsive force acts on A or B in the direction perpendicular to the LOI (Fig. 3). Therefore we can write

$$v_{Cx}^- = v_{Cx}^+ \quad \text{and} \quad v_{Dx}^- = v_{Dx}^+. \tag{2}$$

Since the impact is central (and the contact between the pucks is frictionless), the angular impulse-momentum principle for A and B individually yields the following two equations

$$\omega_A^+ = \omega_A^- \quad \text{and} \quad \omega_B^+ = \omega_B^-. \tag{3}$$

Force Laws The COR equation can be expressed directly in terms of the velocity components along the LOI of the mass centers, so that we have

$$v_{Cy}^+ - v_{Dy}^+ = e\left(v_{Dy}^- - v_{Cy}^-\right). \tag{4}$$

Kinematic Equations The preimpact velocities of A and B are

$$\omega_A^- = 0, \qquad v_{Cx}^- = 0, \qquad v_{Cy}^- = 0, \tag{5}$$

$$\omega_B^- = \omega_0, \qquad v_{Dx}^- = v_0 \sin\beta, \qquad v_{Dy}^- = v_0 \cos\beta. \tag{6}$$

Computation Equations (1)–(4) form a system of six equations in the six unknowns $\omega_A^+, v_{Cx}^+, v_{Cy}^+, \omega_B^+, v_{Dx}^+,$ and v_{Dy}^+. Recalling that $m_A = m_B$, this system of equations can be solved to obtain

$$\omega_A^+ = 0, \qquad v_{Cx}^+ = 0, \qquad v_{Cy}^+ = \frac{1+e}{2} v_0 \cos\beta, \tag{7}$$

$$\omega_B^+ = \omega_0, \qquad v_{Dx}^+ = v_0 \sin\beta, \qquad v_{Dy}^+ = \frac{1-e}{2} v_0 \cos\beta, \tag{8}$$

which, upon substitution of the given data, yields the following numerical answer:

$$\omega_A^+ = 0\,\text{rad/s}, \qquad v_{Cx}^+ = 0\,\text{m/s}, \qquad v_{Cy}^+ = 21.4\,\text{m/s}, \tag{9}$$

$$\omega_B^+ = 4\,\text{rad/s}, \qquad v_{Dx}^+ = 20.5\,\text{m/s}, \qquad v_{Dy}^+ = 0.549\,\text{m/s}. \tag{10}$$

Discussion & Verification The solution appears to be reasonable in that, as expected, the center of mass of puck A moves only along the LOI: this is the expected behavior of any impact in which there is no friction between the colliding bodies. Since the impact is central (and the contact between the pucks is frictionless), the angular velocities of the colliding bodies are conserved. In addition to these considerations, we should check that the postimpact kinetic energy of the system is smaller than the preimpact kinetic energy since the COR used was less than 1. Going through this verification is a bit involved in this problem because neither the masses of A and B nor their respective mass moments of inertia, I_C and I_D, were given. Using Eqs. (5) and (6), the preimpact kinetic energy T^- is

$$T^- = \tfrac{1}{2}m_A(v_C^-)^2 + \tfrac{1}{2}I_C\omega_A^- + \tfrac{1}{2}m_B(v_D^-)^2 + \tfrac{1}{2}I_D\omega_B^-$$
$$= \tfrac{1}{2}m_B v_0^2 + \tfrac{1}{2}I_D\omega_0^2. \tag{11}$$

Using Eqs. (7) and (8), the postimpact kinetic energy T^+ is

$$T^+ = \tfrac{1}{2}m_A(v_C^+)^2 + \tfrac{1}{2}I_C\omega_A^+ + \tfrac{1}{2}m_B(v_D^+)^2 + \tfrac{1}{2}I_D\omega_B^+$$
$$= \tfrac{1}{8}m_A(1+e)^2 v_0^2 \cos^2\beta$$
$$+ \tfrac{1}{8}m_B[4\sin^2\beta + (1-e)^2\cos^2\beta]v_0^2 + \tfrac{1}{2}I_D\omega_0^2. \tag{12}$$

Since the two pucks are identical, $I_C = I_D$. Letting $m = m_A = m_B$, subtracting Eq. (12) from Eq. (11), and simplifying, we have

$$T^- - T^+ = \tfrac{1}{4}(1-e^2)mv_0^2\cos^2\beta, \tag{13}$$

where we have used the fact that $\tfrac{1}{2}m_B v_0^2 - \tfrac{1}{2}m_B v_0^2 \sin^2\beta = \tfrac{1}{2}mv_0^2\cos^2\beta$. Finally, since $e < 1$, we have $1 - e^2 > 0$ and consequently the right-hand side of Eq. (13) is positive, i.e.,

$$T^- - T^+ > 0 \quad \Rightarrow \quad T^+ < T^-, \tag{14}$$

as expected.

EXAMPLE 8.12   *Constrained Impact of Two Rigid Bodies*

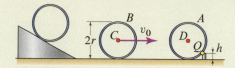

Figure 1
Pipe sections being placed horizontally. Points C and D are the centers of B and A, respectively. Point Q is the point on A in contact with the block of height h.

Identical pipe sections of radius $r = 1.5\,\text{ft}$ and weight $W = 200\,\text{lb}$ are being unloaded and aligned horizontally. A block of height $h = 6\,\text{in}$. is fixed to the ground and used to hold in place A, the first pipe in the row (see Fig. 1). If the next pipe section B rolls without slip to the right with a speed v_0 and bumps into A, the COR for the A-B collision is $e = 0.85$, and A does not rebound off the block or slide relative to it, then determine the smallest value of v_0 that would make A roll over the block.

SOLUTION

Road Map   The solution can be organized in two parts. In the first part, we will study the collision of A and B and determine the postimpact motion of A and B. Once the motion of A is described as a function of v_0, in the second part we will find the *smallest* value of v_0 that makes A roll over the block by relating the postimpact kinetic energy of A to the amount of work needed to move A by a vertical distance h against the direction of gravity.

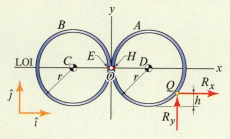

Figure 2
Impact-relevant FBD of A and B combined at the time of impact.

--- **The Impact of A and B** ---

Modeling   The FBD of the system is shown in Fig. 2, and the FBDs of each colliding body are shown in Fig. 3. At the time of impact, the origin of the coordinate system O, which is a fixed point, coincides with the contact points E and H, belonging to B and A, respectively. Notice that we are dealing with a constrained impact due to the presence of external impulsive forces at Q. These forces exist because A is initially in contact with a fixed block and does not slide relative to it. This also means that A moves as if in a fixed axis rotation about Q.

Governing Equations

Balance Principles   The external impulsive forces in Fig. 2 provide no moment about the fixed point Q, and therefore the system's angular momentum about Q is conserved,

$$(\vec{h}_Q^-)_A + (\vec{h}_Q^-)_B = (\vec{h}_Q^+)_A + (\vec{h}_Q^+)_B, \tag{1}$$

where, viewing both pipe sections as uniform thin rings of radius r,

$$(\vec{h}_Q^\pm)_A = I_Q \vec{\omega}_A^\pm, \quad \text{with} \quad I_Q = mr^2 + mr^2 = 2mr^2, \tag{2}$$

$$(\vec{h}_Q^\pm)_B = I_C \vec{\omega}_B^\pm + \vec{r}_{C/Q} \times m\vec{v}_C^\pm, \quad \text{with} \quad I_C = mr^2, \tag{3}$$

where I_Q and I_C are the mass moments of inertia of A relative to Q and of B relative to C, respectively. Note that, in calculating I_Q, we used the parallel axis theorem. The first of Eqs. (2) holds because A moves as if pinned at Q.

The impulsive force N, which acts on B, points toward C. Therefore, since m (the mass of B) and I_C are constant, the y component of the linear momentum of B as well as the angular momentum of B relative to C are conserved, that is,

$$v_{Cy}^- = v_{Cy}^+ \quad \text{and} \quad \omega_B^- = \omega_B^+. \tag{4}$$

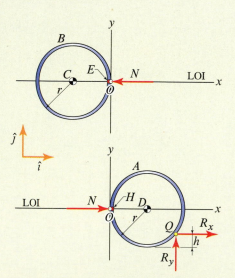

Figure 3
Impact-relevant FBD of A and B individually at the time of impact.

Note that the impulsive force system acting on A does *not* allow us to write relations for A analogous to those in Eqs. (4) for B.

Force Laws   Since the contact points between A and B are E and H, the COR equation reads

$$v_{Hx}^+ - v_{Ex}^+ = e(v_{Ex}^- - v_{Hx}^-). \tag{5}$$

Kinematic Equations Before impact, A is stationary and B rolls without slip, so that

$$v_{Hx}^- = 0, \quad \omega_A^- = 0, \quad v_{Cy}^- = 0, \quad \text{and} \quad \omega_B^- = -v_0/r. \tag{6}$$

To express $v_{Ex}^{\pm}$ and v_{Hx}^+ in Eq. (5) in terms of $\vec{v}_C^+$, ω_B^+, and ω_A^+, recall that $\vec{v}_E = \vec{v}_C + \vec{\omega}_B \times \vec{r}_{E/C}$ and $\vec{v}_H = \vec{\omega}_A \times \vec{r}_{H/Q}$ (A rotates about Q). Consequently, we have

$$v_{Ex}^- = v_0, \quad v_{Ex}^+ = v_{Cx}^+, \quad \text{and} \quad v_{Hx}^+ = -\omega_A^+(r - h). \tag{7}$$

Computation Substituting Eqs. (2) and (3) into Eq. (1), taking advantage of Eqs. (4) and the last two of Eqs. (6), expanding the cross-products, and simplifying, we have

$$-v_0(r - h) = -v_{Cx}^+(r - h) + 2r^2\omega_A^+, \tag{8}$$

where we have written this equation in scalar form because the only nonzero component of Eq. (1) is in the z direction. Substituting the first of Eqs. (6) and all of Eqs. (7) into Eq. (5), we have

$$-\omega_A^+(r - h) - v_{Cx}^+ = ev_0. \tag{9}$$

Equations (8) and (9) form a system of two equations for v_{Cx}^+ and ω_A^+. Solving these equations, we get the postimpact motion of A and B in terms of v_0:

$$\omega_A^+ = -\frac{(1 + e)(r - h)v_0}{h^2 - 2hr + 3r^2} \quad \text{and} \quad v_{Cx}^+ = \frac{h^2 - 2hr + (1 - 2e)r^2}{h^2 - 2hr + 3r^2}v_0. \tag{10}$$

—————————— **The Work-Energy Principle Applied to A** ——————————

Modeling After the impact, A rolls over the block as if pinned at Q, and A's FBD prior to reaching the top of the block is that in Fig. 4. Since Q is fixed, the reactions at Q do no work, so that energy is conserved during the postimpact motion of A.

Governing Equations

Balance Principles Choosing ① to be immediately after impact and ② when A gets to the top of the block with zero velocity (because we want to compute the *minimum* v_0), we have

$$T_{A1} + V_{A1} = T_{A2} + V_{A2}, \quad \text{where} \quad T_{A1} = \tfrac{1}{2}I_Q\omega_{A1}^2 \quad \text{and} \quad T_{A2} = \tfrac{1}{2}I_Q\omega_{A2}^2, \tag{11}$$

and where $I_Q = 2(W/g)r^2$.

Force Laws Choosing the datum line as shown in Fig. 4, we have

$$V_{A1} = 0 \quad \text{and} \quad V_{A2} = Wh. \tag{12}$$

Kinematic Equations Based on our modeling assumptions, we have

$$\omega_{A1} = \omega_A^+ \quad \text{and} \quad \omega_{A2} = 0. \tag{13}$$

Computation Combining Eqs. (11)–(13), recalling that ω_A^+ is given by the first of Eqs. (10), and solving for v_0, we have

$$\boxed{v_0 = \frac{\sqrt{gh}\left(h^2 - 2hr + 3r^2\right)}{r(1 + e)(r - h)} = 7.95\,\text{ft/s.}} \tag{14}$$

Discussion & Verification The solution seems reasonable since the computed v_0, along with Eqs. (10), gives $\omega_A^+ = -2.67\,\text{rad/s}$ and $v_{Cx}^+ = -4.08\,\text{ft/s}$, i.e., after impact A rotates clockwise and B moves to the left, as one would expect. Furthermore, the value of v_0 was expected to be larger than $\sqrt{gh} = 4.01\,\text{ft/s}$, which is the v_0 needed by B to get to a height equal to h off the ground by simply rolling without slip along any path and without any collisions. Finally, using the computed value of v_0 to compute the pre- and postimpact values of the system's kinetic energy, we have $T^- = 393\,\text{ft·lb} > T^+ = 348\,\text{ft·lb}$, as it should.

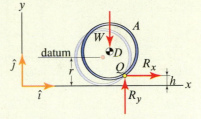

Figure 4
FBD of A following the impact with B.

EXAMPLE 8.13   *Modeling a Catch as a Rigid Body Impact*

Figure 1
A midair catch. The acrobat hanging at the knee from the trapeze is called the *catcher* whereas the other acrobat is called the *flyer*.

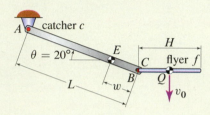

Figure 2
Model of the catch between two trapeze artists. The situation shown is at the time of the catch.

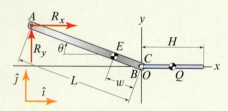

Figure 3
Impact-relevant FBD of the catcher-flyer system. Points B, C, and O are coincident.

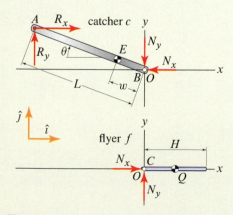

Figure 4
Impact-relevant FBD of c and f individually.

In a perfect catch (see the acrobats in Fig. 1), catcher and flyer grasp each other when they both have zero velocity. What if this ideal situation is not met? For example, what is the average force with which the catcher needs to catch the flyer if the catcher is in an ideal position (i.e., has zero velocity) and the flyer has a free-fall speed $v_0 = 2\,\text{m/s}$ (after dropping about 20 cm)? We will use the model* shown in Fig. 2, where the catcher/trapeze system is viewed as the nonuniform slender bar of mass $m_c = 90\,\text{kg}$, length $L = 4\,\text{m}$, mass center at E, mass moment of inertia $I_E = 30\,\text{kg·m}^2$, and $w = 1\,\text{m}$. The flyer is modeled as a uniform slender bar of length $H = 2\,\text{m}$ and mass $m_f = 70\,\text{kg}$. We model the grasp between catcher and flyer as a pin connection and assume that, at the instant shown, catcher and flyer both have zero angular velocity. Finally, we assume that catcher and flyer can establish a firm grasp in 0.15 s (the best athletes' reaction times are 120–160 ms), and since the catch happens quickly, we model it as an impact.

SOLUTION

Road Map & Modeling   Modeling the catch as an impact is useful because (1) by relating the pre- and postcatch velocities via the linear impulse-momentum principle we can compute the impulse exerted by c on f, and (2) by dividing this impulse by the reaction time we can compute the desired average force. Note that the impact we are modeling is not plastic or elastic since c and f do become joined. At the same time, the impact is not perfectly plastic because c and f can rotate relative to one another. Hence, we will need to adapt our knowledge of rigid body impact and create an ad hoc solution strategy. As is typical of impact problems, we consider the system's FBD and the FBDs of c and f individually, shown in Figs. 3 and 4, respectively. At the time of catch, the origin O of the chosen coordinate system, which is a fixed point, coincides with points B and C at which c and f grasp one another. Note that c can only move in a fixed axis rotation about A. Therefore, the postcatch velocities consist of four quantities: the angular velocities of c and f and the two components of $\vec{v}_Q$.

Governing Equations

Balance Principles   Referring to Fig. 4 and to Eq. (8.36) on p. 626, applying the linear impulse-momentum principle to f gives

$$m_f \vec{v}_Q^- + \int_{t^-}^{t^+} (N_x \hat{\imath} + N_y \hat{\jmath})\, dt = m_f \vec{v}_Q^+. \tag{1}$$

The forces R_x and R_y have no moment about the fixed point A (see Fig. 3), and so the system's angular momentum about A is conserved, that is,

$$(\vec{h}_A^-)_c + (\vec{h}_A^-)_f = (\vec{h}_A^+)_c + (\vec{h}_A^+)_f, \tag{2}$$

where

$$(\vec{h}_A^\pm)_c = \left[I_E + m_c(L-w)^2\right]\vec{\omega}_c^\pm \quad \text{and} \quad (\vec{h}_A^\pm)_f = I_Q \vec{\omega}_f^\pm + \vec{r}_{Q/A} \times m_f \vec{v}_Q^\pm. \tag{3}$$

Note that the first of Eqs. (3) accounts for the fixed axis rotation of c about A. Next, since C and O coincide at the time of catch (see Fig. 4), N_x and N_y provide no moment on f about O. Thus, the angular momentum of f relative to O, $(\vec{h}_O)_f$, must be conserved, that is,

$$(\vec{h}_O^-)_f = (\vec{h}_O^+)_f \quad \text{where} \quad (\vec{h}_O^\pm)_f = I_Q \vec{\omega}_f^\pm + \vec{r}_{Q/O} \times m_f \vec{v}_Q^\pm. \tag{4}$$

* A better model would account for the trapeze and the flexibility of the human body.

Force Laws   All forces are accounted for on the FBD. Also, in this impact we do not have a COR equation.

Kinematic Equations   After the catch $\vec{v}_B^+ = \vec{v}_C^+$ so that we must have

$$\vec{\omega}_c^+ \times \vec{r}_{B/A} = \vec{v}_Q^+ + \vec{\omega}_f^+ \times \vec{r}_{C/Q}, \tag{5}$$

where Eq. (5) enforces both rigid body kinematics and the fact that A is a fixed point.

Computation   Substituting Eqs. (3) into Eq. (2), we have

$$- m_f v_0 [L\cos\theta + (H/2)] = \Big[I_E + m_c(L-w)^2 \Big] \omega_c^+ + I_Q \omega_f^+$$
$$+ m_f v_{Qx}^+ L\sin\theta + m_f v_{Qy}^+ [L\cos\theta + (H/2)], \tag{6}$$

where Eq. (6) is in scalar form since the only nonzero component of Eq. (2) is in the z direction. Proceeding similarly with Eqs. (4), we have

$$-m_f v_0 (H/2) = I_Q \omega_f^+ + m_f v_{Qy}^+ (H/2). \tag{7}$$

Expanding the products in Eq. (5) and expressing the result in components, we have

$$\omega_c^+ L\sin\theta = v_{Qx}^+ \quad \text{and} \quad \omega_c^+ L\cos\theta = v_{Qy}^+ - \omega_f^+ (H/2). \tag{8}$$

Equations (6)–(8) form a system of four equations in the four unknowns $\omega_c^+, \omega_f^+, v_{Qx}^+$, and v_{Qy}^+. Recalling that $I_Q = m_f H^2/12$, the solution of this system of equations is

$$\omega_c^+ = -\frac{L m_f v_0 \cos\theta}{4[I_E + m_c(L-w)^2] + L^2 m_f(1 + 3\sin^2\theta)} = -0.1080 \, \text{rad/s}, \tag{9}$$

$$\omega_f^+ = -\frac{3v_0}{H} \frac{2I_E + 2m_f L^2 \sin^2\theta + 2m_c(L-w)^2}{4[I_E + m_c(L-w)^2] + L^2 m_f(1 + 3\sin^2\theta)} = -1.196 \, \text{rad/s}, \tag{10}$$

$$v_{Qx}^+ = -\frac{L^2 m_f v_0 \cos\theta \sin\theta}{4[I_E + m_c(L-w)^2] + L^2 m_f(1 + 3\sin^2\theta)} = -0.1477 \, \text{m/s}, \tag{11}$$

$$v_{Qy}^+ = -v_0 \frac{3I_E + m_f L^2(1 + 2\sin^2\theta) + 3m_c(L-w)^2}{4[I_E + m_c(L-w)^2] + L^2 m_f(1 + 3\sin^2\theta)} = -1.601 \, \text{m/s}. \tag{12}$$

Recalling that $t^+ - t^- = 0.15$ s, using the definition of average force over the time interval $t^+ - t^-$, and employing Eq. (1), we have $\vec{N}_{\text{avg}} = \frac{1}{t^+ - t^-} \int_{t^-}^{t^+} (N_x\,\hat{\imath} + N_y\,\hat{\jmath})\, dt = \frac{m_f}{t^+ - t^-}(\vec{v}_Q^+ - \vec{v}_Q^-)$, which gives

$$\boxed{(N_x)_{\text{avg}} = -68.9\,\text{N} \quad \text{and} \quad (N_y)_{\text{avg}} = 186\,\text{N} \quad \Rightarrow \quad |\vec{N}_{\text{avg}}| = 198\,\text{N}.} \tag{13}$$

Discussion & Verification   The signs of the velocities we computed are all as expected. In addition, as expected, $v_{Qy}^+ < v_0$, i.e., the catcher slows down the flyer's drop. As far as the forces are concerned, the signs are also as expected, and their value is consistent with the velocity result and the given reaction time.

✎ **A Closer Look**   Our model predicts that the catcher is required to exert a force roughly equal to 200 N to catch a flyer weighing 687 N (this is the weight of person with a mass of 70 kg) who had been in free fall over a distance of roughly 20 cm. Such a value of force is certainly well within the capabilities of a fit person, and therefore, using this value of force as a criterion, we could conclude that missing the ideal catch conditions by a few centimeters is not that bad. However, to more completely assess the consequences of missing the ideal catch position, we need to consider other important factors in the maneuver, such as the overall maximum force that needs to be exerted over the entire swing following the catch.

PROBLEMS

Figure P8.78

💡 Problem 8.78 💡

A *stop shot* is a pool shot in which the cue ball (white) stops upon striking the object ball (red). Modeling the collision between the two balls as a perfectly elastic collision of two rigid bodies with frictionless contact, determine which condition must be true for the preimpact angular velocity of the cue ball in order to properly execute a stop shot: (a) $\omega_0 < 0$; (b) $\omega_0 = 0$; (c) $\omega_0 > 0$.
Note: Concept problems are about *explanations*, not computations.

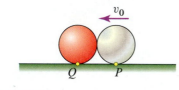

Figure P8.79

Problem 8.79

The cue ball (white) is rolling without slip to the left, and its center is moving with a speed $v_0 = 6\,\text{ft}$ while the object ball (red) is stationary. The diameter d of the two balls is the same and is equal to 2.25 in. The coefficient of restitution of the impact is $e = 0.98$. Let $W_c = 6\,\text{oz}$ and $W_o = 5.5\,\text{oz}$ be the weights of the cue ball and object ball, respectively. Let P and Q be the points on the cue ball and on the object ball, respectively, that are in contact with the table at the time of impact. Assuming that the contact between the two balls is frictionless and modeling the balls as uniform spheres, determine the postimpact velocities of P and Q.

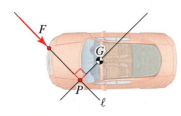

Figure P8.80

💡 Problem 8.80 💡

Consider the impact-relevant FBD of a car involved in a collision. Assume that, at the time of impact, the car was stationary. In addition, assume that the impulsive force F, with line of action ℓ, is the only impulsive force acting on the car at the time of impact. The point P at the intersection of ℓ and the line perpendicular to ℓ and passing through G, the center of mass of the car, is sometimes referred to as the *center of percussion* (for an alternative definition of center of percussion see the Mini-Example on p. 552). Is it true that, at the time of impact, the instantaneous center of rotation of the car lies on the same line as P and G?
Note: Concept problems are about *explanations*, not computations.

Problem 8.81

A basketball with mass $m = 0.6\,\text{kg}$ is rolling without slipping as shown when it hits a small step with $\ell = 7\,\text{cm}$. Letting the ball's diameter be $r = 12.0\,\text{cm}$, modeling the ball as a thin spherical shell (the mass moment of inertia of a spherical shell about its mass center is $\frac{2}{3}mr^2$), and assuming that the ball does not rebound off the step or slip relative to it, determine v_0 such that the ball barely makes it over the step.

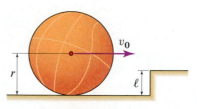

Figure P8.81

Problem 8.82

A bullet B of mass m_b is fired with a speed v_0 as shown against a uniform thin rod A of length ℓ, mass m_r, and that is pinned at O. Determine the distance d such that no horizontal reaction is felt at the pin when the bullet strikes the rod.

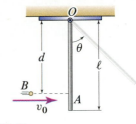

Figure P8.82

Problem 8.83

Solve the problem in the Mini-Example on p. 552 using momentum methods and the concept of impulsive force. Specifically, consider a ball hitting a bat at a distance d from the handle when the batter has "choked up" a distance δ. Find the "sweet spot" P (more properly called the center of percussion*) of the bat B by determining the distance d at which the ball should be hit so that the lateral force (i.e., perpendicular to the bat) at O is zero. Assume that the bat is pinned at O, it has mass m, the mass center is at G, and the mass moment of inertia is I_G.

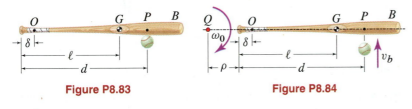

Figure P8.83   **Figure P8.84**

Problem 8.84

A batter is swinging a 34 in. long bat with weight $W_B = 32\,\text{oz}$, mass center G, and mass moment of inertia $I_G = 0.0413\,\text{slug·ft}^2$. The center of rotation of the bat is point Q. Compute the distance d identifying the position of point P, the bat's "sweet spot" or center of percussion, such that the batter will not feel any impulsive forces at O where he is grasping the bat. In addition, knowing that the ball, weighing 5 oz, is traveling at a speed $v_b = 90\,\text{mph}$ and that the batter is swinging the bat with an angular velocity $\omega_0 = 45\,\text{rad/s}$, determine the speed of the ball and the angular velocity of the bat immediately after impact. To solve the problem, use the following data: $\delta = 6\,\text{in.}$, $\rho = 14\,\text{in.}$, $\ell = 22.5\,\text{in.}$, and COR $e = 0.5$.

Problem 8.85

A bullet B weighing 147 gr (1 lb = 7000 gr) is fired with a speed v_0 as shown and becomes embedded in the center of a rubber block of dimensions $h = 4.5\,\text{in.}$ and $w = 6\,\text{in.}$ weighing $W_{rb} = 2\,\text{lb}$. The rubber block is attached to the end of a uniform thin rod A of length $L = 1.5\,\text{ft}$, weight $W_r = 5\,\text{lb}$, and that is pinned at O. After the impact, the rod (with the block and the bullet embedded in it) swings upward to an angle of $60°$. Determine the speed of the bullet right before impact.

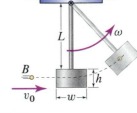

Figure P8.85

Problem 8.86

A thin homogeneous bar A of length $\ell = 1.75\,\text{m}$ and mass $m = 23\,\text{kg}$ is translating as shown with a speed $v_0 = 12\,\text{m/s}$ when it collides with the fixed obstacle B. Modeling the contact between the bar and obstacle as frictionless, letting $\beta = 32°$, and letting the distance $d = 0.46\,\text{m}$, determine the angular velocity of the bar immediately after the collision, knowing that the COR for the impact is $e = 0.74$.

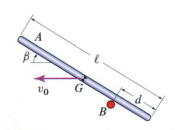

Figure P8.86

---

* See R. Cross, "The Sweet Spot of a Baseball Bat," *American Journal of Physics*, **66**(9), 1998, pp. 772–779.

Problem 8.87

A drawbridge of length $\ell = 30\,\text{ft}$ and weight $W = 600\,\text{lb}$ is released in the position shown and freely pivots clockwise until it strikes the right end of the moat. If the COR for the collision between the bridge and the ground is $e = 0.45$ and if the contact point between the bridge and the ground is effectively ℓ away from the bridge's pivot point, determine the angle to which the bridge rebounds after the collision. Neglect any possible source of friction.

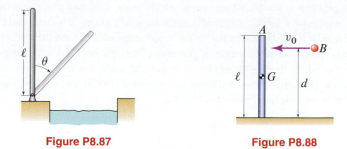

Figure P8.87   Figure P8.88

Problem 8.88

A stick A with length $\ell = 1.55\,\text{m}$ and mass $m_A = 6\,\text{kg}$ is in static equilibrium as shown when a ball B with mass $m_B = 0.15\,\text{kg}$ traveling at a speed $v_0 = 30\,\text{m/s}$ strikes the stick at distance $d = 1.3\,\text{m}$ from the lower end of the stick. If the COR for the impact is $e = 0.85$, determine the velocity of the mass center G of the stick as well as the stick's angular velocity right after the impact.

Problem 8.89

A uniform bar A with a hook H at the end is dropped as shown from a height $d = 3\,\text{ft}$ over a fixed pin B. Letting the weight and length of A be $W = 100\,\text{lb}$ and $\ell = 7\,\text{ft}$, respectively, determine the angle θ that the bar will sweep through if the bar becomes hooked with B and does not rebound. Although bar A becomes hooked with B, assume that there is no friction between the hook and the pin.

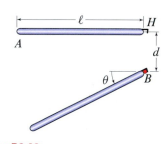

Figure P8.89

Problem 8.90

A gymnast on the uneven parallel bars has a vertical speed v_0 and no angular speed when she grasps the upper bar. Model the gymnast as a single uniform rigid bar A of weight $W = 92\,\text{lb}$ and length $\ell = 6\,\text{ft}$. Neglecting all friction, letting $\beta = 12°$, and assuming that the upper bar B does not move after the gymnast grasps it, determine v_0 if the gymnast is to swing (counterclockwise) so as to become horizontal. Assume that, during the motion, the friction between the gymnast's hands and the upper bar is negligible.

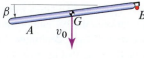

Figure P8.90

Problem 8.91

A uniform thin ring A of mass $m = 7\,\text{kg}$ and radius $r = 0.5\,\text{m}$ is released from rest as shown and rolls without slip until it meets a step of height $\ell = 0.45\,\text{m}$. Letting $\beta = 12°$ and assuming that the ring does not rebound off the step or slip relative to it, determine the distance d such that the ring barely makes it over the step.

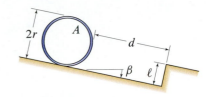

Problem 8.92

Two identical uniform bars AB and BD are pin-connected at B, and bar BD has a hook at the free end. The two bars are dropped as shown from a height $d = 3\,\text{ft}$ over a fixed pin E (shown in cross section). Letting the weight and length of each bar be $W = 100\,\text{lb}$ and $\ell = 7\,\text{ft}$, respectively, determine the angular velocities of AB and BD immediately after bar BD becomes hooked on E and does not rebound. *Hint:* The angular momentum of bar AB is conserved about B during impact.

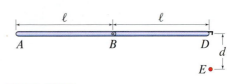

Figure P8.92

Problem 8.93

Cars A and B collide as shown. Determine the angular velocities of A and B immediately after impact if the COR is $e = 0.35$. In solving the problem, let C and D be the mass centers of A and B, respectively. In addition, enforce assumption 3 on p. 649 and use the following data: $W_A = 3130\,\text{lb}$ (weight of A), $k_C = 34.5\,\text{in.}$ (radius of gyration of A), $v_C = 12\,\text{mph}$ (speed of the mass center of A), $W_B = 3520\,\text{lb}$ (weight of B), $k_D = 39.3\,\text{in.}$ (radius of gyration of B), $v_D = 15\,\text{mph}$ (speed of the mass center of B), $d = 19\,\text{in.}$, $\ell = 79\,\text{in.}$, $\delta = 7.1\,\text{in.}$, $\rho = 65\,\text{in.}$, and $\beta = 12°$.

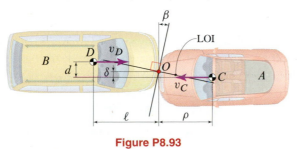

Figure P8.93

Problem 8.94

Consider the collision of two rigid bodies A and B, which, referring to Example 8.10 on p. 635, models the docking of the Space Shuttle (body A) to the International Space Station (body B). As in Example 8.10, we assume that B is stationary relative to an inertial frame of reference and that A translates as shown. In contrast to Example 8.10, here we assume that A and B join at point Q but, due to the flexibility of the docking system, can rotate relative to one another. Determine the angular velocities of A and B right after docking if $v_0 = 0.03\,\text{m/s}$. In solving the problem, let C and D be the centers of mass of A and B, respectively. In addition, let the mass and mass moment of inertia of A be $m_A = 120 \times 10^3\,\text{kg}$ and $I_C = 14 \times 10^6\,\text{kg·m}^2$, respectively, and the mass and mass moment of inertia of B be $m_B = 180 \times 10^3\,\text{kg}$ and $I_D = 34 \times 10^6\,\text{kg·m}^2$, respectively. Finally, use the following dimensions: $\ell = 24\,\text{m}$, $d = 8\,\text{m}$, $\rho = 2.6\,\text{m}$, and $\delta = 2.4\,\text{m}$.

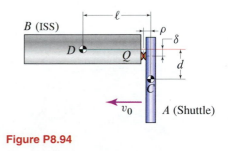

Figure P8.94

8.4   Chapter Review

Work-energy principle for rigid bodies

In this section, we discovered that the kinetic energy T of a rigid body B in planar motion is given by

Eq. (8.10), p. 593

$$T = \tfrac{1}{2}mv_G^2 + \tfrac{1}{2}I_G\omega_B^2,$$

or by

Eq. (8.13), p. 594

$$T = \tfrac{1}{2}I_O\omega_B^2,$$

where m, G, and I_G are the body's mass, the center of mass, and mass moment of inertia, respectively, and the second equation applies to fixed axis rotation, where I_O is the mass moment of inertia relative to the axis of rotation. The work-energy principle for a rigid body is

Eq. (8.18), p. 595

$$T_1 + U_{1\text{-}2} = T_2,$$

where $U_{1\text{-}2}$ is the work done on the body in going from ① to ② by only the external forces. If we make use of the potential energy V of conservative forces, the work-energy principle can also be written as

Eqs. (8.25) and (8.26), p. 596

$$T_1 + V_1 + (U_{1\text{-}2})_{\text{nc}} = T_2 + V_2 \qquad \text{(general systems)},$$
$$T_1 + V_1 = T_2 + V_2 \qquad \text{(conservative systems)},$$

where the second expression applies only to a conservative system and states that the total mechanical energy of the system is conserved. For systems of rigid bodies or mixed systems of rigid bodies and particles, the work-energy principle reads

Eq. (8.31), p. 598

$$T_1 + V_1 + (U_{1\text{-}2})_{\text{nc}}^{\text{ext}} + (U_{1\text{-}2})_{\text{nc}}^{\text{int}} = T_2 + V_2,$$

where V is the total potential energy (of both external and internal conservative forces) and $(U_{1\text{-}2})_{\text{nc}}^{\text{ext}}$ and $(U_{1\text{-}2})_{\text{nc}}^{\text{int}}$ are the work contributions due to external and internal forces for which we do not have a potential energy, respectively.

Referring to Fig. 8.38, in planar rigid body motion, the work of a couple is

Eq. (8.24), p. 596

$$U_{1\text{-}2} = \int_{t_1}^{t_2} M\omega_{AB}\, dt = \int_{\theta_1}^{\theta_2} M\, d\theta,$$

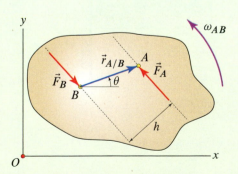

Figure 8.38
Figure 8.7 repeated. Rigid body subject to a couple.

where $d\theta$ is the body's infinitesimal angular displacement and M is the component of the moment of the couple in the direction perpendicular to the plane of motion, taken to be positive in the direction of positive θ. In addition, the power developed by a couple with moment $\vec{M}$ is computed as

Eq. (8.33), p. 598

$$\text{Power developed by a couple} = \frac{dU}{dt} = \vec{M} \cdot \vec{\omega}_{AB},$$

where $\vec{\omega}_{AB}$ is the body's angular velocity.

Momentum methods for rigid bodies

In this section we derived the linear and the angular impulse-momentum principles for rigid bodies. Referring to Fig. 8.39, the linear impulse-momentum principle for a rigid body reads

Eqs. (8.36), p. 626

$$m\vec{v}_{G1} + \int_{t_1}^{t_2} \vec{F}\, dt = m\vec{v}_{G2} \quad \text{or} \quad \vec{p}_1 + \int_{t_1}^{t_2} \vec{F}\, dt = \vec{p}_2,$$

where $\vec{F}$ is the total external force on B, $\vec{p} = m\vec{v}_G$ is B's linear momentum, and $\vec{v}_G$ is the velocity of B's center of mass. If $\vec{F} = \vec{0}$, we have

Eqs. (8.38), p. 626

$$m\vec{v}_{G1} = m\vec{v}_{G2} \quad \text{or} \quad \vec{p}_1 = \vec{p}_2,$$

and we say that the body's momentum is conserved. If P in Fig. 8.40 is a moment center coplanar with G and if B is symmetric relative to the plane of motion, the angular momentum of B relative to P is

Eq. (8.42), p. 627

$$\vec{h}_P = I_G \vec{\omega}_B + \vec{r}_{G/P} \times m\vec{v}_G,$$

where I_G is the mass moment of inertia of B, $\vec{\omega}_B$ is the angular velocity of B, and $\vec{r}_{G/P}$ is the position of G relative to P. If P is chosen so that (1) P is fixed or (2) P coincides with G or (3) P and G move parallel to one another, then $\vec{M}_P = \dot{\vec{h}}_P$, where $\vec{M}_P$ is the moment relative to P of the *external* force system acting on B. When this equation holds, by integrating with respect to time over a time interval $t_1 \le t \le t_2$, we have

Eq. (8.44), p. 627

$$\vec{h}_{P1} + \int_{t_1}^{t_2} \vec{M}_P\, dt = \vec{h}_{P2}.$$

When P coincides with G or if the body undergoes a fixed axis rotation about a point O as shown in Fig. 8.41, then the above equation becomes

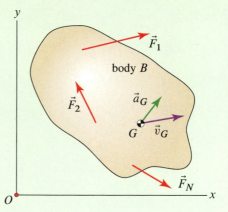

Figure 8.39
Figure 8.16 repeated. A rigid body under the action of a system of forces.

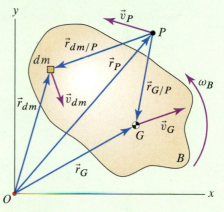

Figure 8.40
Figure 8.18 repeated. The quantities needed to obtain the angular momentum relationships for a rigid body.

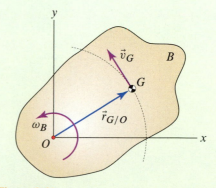

Figure 8.41
Figure 8.23 repeated. A rigid body in a fixed axis rotation.

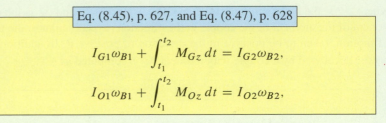

Eq. (8.45), p. 627, and Eq. (8.47), p. 628

$$I_{G1}\omega_{B1} + \int_{t_1}^{t_2} M_{Gz}\, dt = I_{G2}\omega_{B2},$$

$$I_{O1}\omega_{B1} + \int_{t_1}^{t_2} M_{Oz}\, dt = I_{O2}\omega_{B2},$$

respectively, where I_O is the mass moment of inertia about the fixed axis of rotation.

Impact of rigid bodies

In this section we studied planar rigid body impacts. We learned that, contrary to particle impacts, rigid bodies can experience *eccentric* impacts. These are collisions in which at least one of the mass centers of the impacting bodies does not lie on the LOI. We also learned that the basic concepts used in particle impacts are applicable to rigid body impacts, and we reviewed solution strategies for a variety of situations.

As with particle impacts, we say that a rigid body impact is *plastic* if the COR $e = 0$. A rigid body impact will be called *perfectly plastic* if the colliding bodies form a single rigid body after impact. An impact is *elastic* if the COR e is such that $0 < e < 1$. Finally, the ideal case with $e = 1$ is referred to as a *perfectly elastic* impact. In addition, an impact is *unconstrained* if the system consisting of the two impacting objects is not subject to external impulsive forces; otherwise, the impact is called *constrained*.

We have considered only impacts satisfying the following assumptions:

1. The impact involves only two bodies in planar motion where no impulsive force has a component perpendicular to the plane of motion.

2. Contact between any two rigid bodies occurs at only one point, and at this point we can clearly define the LOI.

3. The contact between the bodies is frictionless.

We recommend organizing the solution of any impact problem as follows:

- Begin with an FBD of the impacting bodies as a system and FBDs for each of the colliding bodies. Neglect nonimpulsive forces.

- Choose a coordinate system with the origin coincident with the points that come into contact at the time of impact. Recall that the origin of such a coordinate system is a fixed point.

- Enforce the linear and/or the angular impulse-momentum principles for the system and/or for the individual bodies. In applying the angular impulse-momentum principle for the whole system, the moment center should be a fixed point whereas for an individual body the moment center should be a fixed point or the body's mass center.

- For plastic, elastic, and perfectly elastic impacts, the COR equation is first written using the velocity components along the LOI of the points that actually come into contact. For example, for the impact shown in Fig. 8.42, the COR equation is first written as

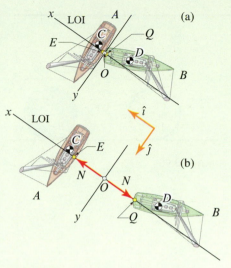

Eq. (8.82), p. 652

$$v_{Ex}^{+} - v_{Qx}^{+} = e\left(v_{Qx}^{-} - v_{Ex}^{-}\right),$$

where the COR e is such that $0 \leq e \leq 1$. The COR equation must then be rewritten in terms of the colliding bodies' angular velocities and velocities of the mass centers using rigid body kinematics. For the situation in Fig. 8.37, this means rewriting the COR equation using the relations $\vec{v}_{E}^{\pm} = \vec{v}_{C}^{\pm} + \omega_{A}^{\pm}\hat{k} \times \vec{r}_{E/C}$ and $\vec{v}_{Q}^{\pm} = \vec{v}_{D}^{\pm} + \omega_{B}^{\pm}\hat{k} \times \vec{r}_{Q/D}$. Notice that the relative position vectors $\vec{r}_{E/C}$ and $\vec{r}_{Q/D}$ do not have the $\pm$ superscript because they are treated as constants during the impact.

• In perfectly plastic impacts, kinematic constraint equations must be enforced that express the fact that two bodies form a single rigid body after impact.

Figure 8.42
Figure 8.37 repeated. FBDs of the colliding bodies (a) as a system and (b) individually.

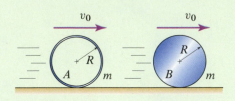

Figure P8.95

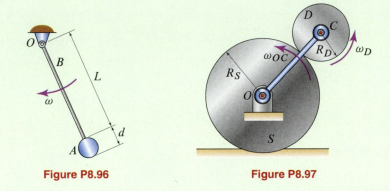

Figure P8.96   **Figure P8.97**

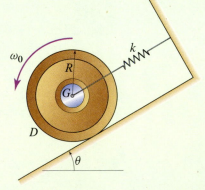

Figure P8.98

💡 **Problem 8.95** 💡

A uniform thin ring A and a uniform disk B roll without slip as shown. Letting T_A and T_B be the kinetic energies of A and B, respectively, if the two objects have the same mass and radius and if their centers are moving with the same speed v_0, state which of the following statements is true and why: (a) $T_A < T_B$; (b) $T_A = T_B$; (c) $T_A > T_B$. **Note:** Concept problems are about *explanations*, not computations.

Problem 8.96 🌡

A pendulum consists of a uniform disk A of diameter $d = 5$ in. and weight $W_A = 0.25$ lb attached at the end of a uniform bar B of length $L = 2.75$ ft and weight $W_B = 1.3$ lb. At the instant shown, the pendulum is swinging with an angular velocity $\omega = 0.55$ rad/s clockwise. Determine the kinetic energy of the pendulum at this instant, using Eq. (8.13) on p. 594.

Problem 8.97 🌡 .

A uniform disk D of radius $R_D = 7$ mm and mass $m_D = 0.15$ kg is connected to point O via the rotating arm OC and rolls without slip over the stationary cylinder S of radius $R_S = 15$ mm. Assuming that $\omega_D = 25$ rad/s and treating the arm OC as a uniform slender bar of length $L = R_D + R_S$ and mass $m_{OC} = 0.08$ kg, determine the kinetic energy of the system.

Problem 8.98 🌡

At the instant shown, the disk D, which has mass m and radius of gyration k_G, is rolling without slip down the flat incline with angular velocity ω_0. The disk is attached at its center to a wall by a linear elastic spring of constant k. If, at the instant shown, the spring is unstretched, determine the distance d down the incline that the disk rolls before coming to a stop. Use $k = 65$ N/m, $R = 0.3$ m, $m = 10$ kg, $k_G = 0.25$ m, $\omega_0 = 60$ rpm, and $\theta = 30°$.

Problem 8.99 🌡

The figure shows the cross section of a garage door with length $L = 9$ ft and weight $W = 175$ lb. At the ends A and B there are rollers of negligible mass constrained to move in a vertical and a horizontal guide, respectively. The door's motion is assisted by two springs (only one is shown), each with constant $k = 9.05$ lb/ft. If the door is released from rest when horizontal and the spring is stretched 4 in., neglecting friction, and modeling the door as a uniform thin plate, determine the speed with which B strikes the left end of the horizontal guide.

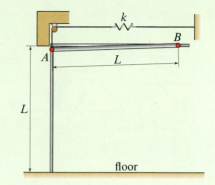

Figure P8.99

🔆 Problem 8.100 🔆

Body B has mass m and mass moment of inertia I_G, where G is the mass center of B. If B is in fixed axis rotation about its center of mass G, determine which of the following statements is true and why: (a) $\left|(\vec{h}_E)_B\right| < \left|(\vec{h}_P)_B\right|$, (b) $\left|(\vec{h}_E)_B\right| = \left|(\vec{h}_P)_B\right|$, (c) $\left|(\vec{h}_E)_B\right| > \left|(\vec{h}_P)_B\right|$.

Note: Concept problems are about *explanations*, not computations.

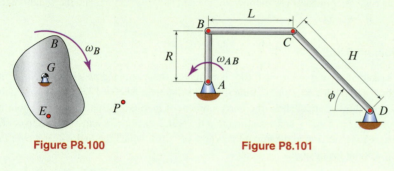

Figure P8.100	**Figure P8.101**

Problem 8.101 🌡

The weights of the uniform thin pin-connected bars AB, BC, and CD are $W_{AB} = 4\,\text{lb}$, $W_{BC} = 6.5\,\text{lb}$, and $W_{CD} = 10\,\text{lb}$, respectively. Letting $\phi = 47°$, $R = 2\,\text{ft}$, $L = 3.5\,\text{ft}$, and $H = 4.5\,\text{ft}$, and knowing that bar AB rotates at a constant angular velocity $\omega_{AB} = 4\,\text{rad/s}$, compute the angular momentum of the system about D at the instant shown.

🔆 Problem 8.102 🔆

Consider Prob. 8.55 on p. 637 in which an eccentric wheel B is spun from rest under the action of a known torque M. In that problem, it was said that the wheel was in the *horizontal* plane. Is it possible to solve Prob. 8.55 by just applying Eq. (8.44) on p. 627 if the wheel is in the *vertical* plane? Why?

Note: Concept problems are about *explanations*, not computations.

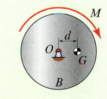

Figure P8.102

Problem 8.103 🌡

The uniform disk A, of mass $m_A = 1.2\,\text{kg}$ and radius $r_A = 0.25\,\text{m}$, is mounted on a vertical shaft that can translate along the horizontal rod E. The uniform disk B, of mass $m_B = 0.85\,\text{kg}$ and radius $r_B = 0.38\,\text{m}$, is mounted on a vertical shaft that is rigidly attached to E. Disk C has a negligible mass and is rigidly attached to E; i.e., C and E form a single rigid body. Disk A can rotate about the axis ℓ_A, disk B can rotate about the axis ℓ_B, and the arm E along with C can rotate about the fixed axis ℓ_C. While keeping both B and C stationary, disk A is initially spun with $\omega_A = 1200\,\text{rpm}$. Disk A is then brought in contact with C (contact is maintained via a spring), and at the same time, both B and C (and the arm E) are free to rotate. Due to friction between A and C, C along with E and disk B start spinning. Eventually A and C will stop slipping relative to one another. Disk B always rotates without slip over C. Let $d = 0.27\,\text{m}$ and $w = 0.95\,\text{m}$. Assuming that the only elements of the system that have mass are A, B, and E and that $m_E = 0.3\,\text{kg}$, and assuming that all friction in the system can be neglected except for that between A and C and between C and B, determine the

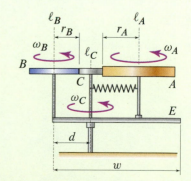

Figure P8.103

angular speeds of A, B, and C (the angular velocity of C is the same as that of E since they form a single rigid body), when A and C stop slipping relative to one another.

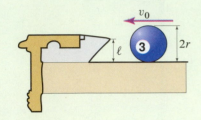

Figure P8.104

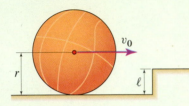

Figure P8.105

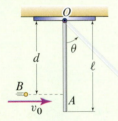

Figure P8.106

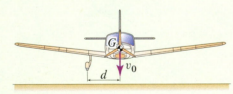

Figure P8.107

Problem 8.104

A billiard ball is rolling without slipping with a speed $v_0 = 6$ ft/s as shown when it hits the rail. According to regulations, the nose of the rail is at a height from the table bed of 63.5% of the ball's diameter (i.e., $\ell/(2r) = 0.635$). Model the impact with the rail as perfectly elastic, neglect friction between the ball and the rail as well as between the ball and the table, and neglect any vertical motion of the ball. Based on the stated assumptions, determine the velocity of the point of contact between the ball and the table right after impact. The diameter of the ball is $2r = 2.25$ in., and the weight of the ball is $W = 5.5$ oz.

Problem 8.105

A basketball with mass $m = 0.6$ kg is rolling without slipping as shown when it hits a small step with $\ell = 7$ cm. Letting the ball's diameter be $r = 12.0$ cm, modeling the ball as a thin spherical shell (the mass moment of inertia of a spherical shell about its mass center is $\frac{2}{3}mr^2$), and assuming that the ball does not rebound off the step or slip relative to it, determine the maximum value of v_0 for which the ball will roll over the step without losing contact with it.

Problem 8.106

A bullet B weighing 147 gr (1 lb = 7000 gr) is fired with a speed $v_0 = 2750$ ft/s as shown against a thin uniform rod A of length $\ell = 3$ ft, weight $W_r = 35$ lb, and pinned at O. If $d = 1.5$ ft and the COR for the impact is $e = 0.25$, determine the bar's angular velocity immediately after the impact. In addition, determine the maximum value of the angle θ to which the bar swings after impact.

Problem 8.107

An airplane is about to crash-land with a vertical component of speed $v_0 = 2$ ft/s and zero roll, pitch, and yaw. Determine the vertical component of velocity of the center of mass of the airplane G as well as the airplane's angular velocity immediately after touching down, assuming that (1) the only available landing gear is rigid and rigidly attached to the airplane, (2) the coefficient of restitution between the landing gear and the ground is $e = 0.1$, (3) the airplane can be modeled as a rigid body, (4) the mass center G and the point of first contact between the landing gear and the ground are in the same plane perpendicular to the longitudinal axis of the airplane, and (5) friction between the landing gear and the ground is negligible. In solving the problem use the following data: $W = 2500$ lb (weight of the airplane), G is the mass center of the airplane, $k_G = 3$ ft is the radius of gyration of the airplane, and $d = 5.08$ ft.

9 Mechanical Vibrations

A *vibration* is a type of dynamic behavior in which a system or part of a system oscillates about an equilibrium position. Vibrations occur in mechanical, electrical, thermal, and fluid systems, but we will consider only those occurring in mechanical systems. Vibrations in mechanical systems can be undesirable (e.g., vibration in structures can lead to failure and can create unwanted noise), or they can be desirable (e.g., vibration in fluid systems can dissipate unwanted energy, and music would be impossible without vibrations).

9.1 Undamped Free Vibration

Oscillation of a railcar after coupling

Recall Example 3.3 on p. 200 in which a railcar ran into a large spring that was designed to stop it (see Fig. 9.1). In that example, we were interested in the maximum compression of the spring and the time it took to stop the railcar. After maximum compression was reached, the spring would push the railcar back and if the railcar were to couple to the spring, the railcar would overshoot the equilibrium position of the spring and the railcar would start oscillating back and forth on the tracks. Let's look at this motion.

In that example, we found the equation of motion of the railcar to be

$$\ddot{x} + \frac{k}{m}x = 0, \qquad (9.1)$$

where k is the spring constant, m is the mass of the railcar and its load, and x is measured from the equilibrium position of the spring (see Fig. 9.1). Using $x(0) = x_i$ and $\dot{x}(0) = v_i$ for initial conditions, we were able to integrate this equation of motion to obtain time as a function of position (see Eq. (11) on

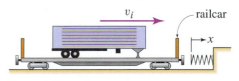

Figure 9.1
A railcar hitting a spring. The coordinate x measures the displacement of the spring from its equilibrium position. Recall from Example 3.3 that the railcar and its load weigh 87 tons and are moving at 4 mph at impact. Also recall that we found $k = 22{,}300$ lb/ft. We assume that the trailer does not move relative to the railcar.

673

p. 201), which can be inverted to obtain

$$x(t) = \frac{\sqrt{v_i^2 + \frac{k}{m}x_i^2}}{\sqrt{k/m}} \sin\left[\sqrt{\frac{k}{m}}\, t + \tan^{-1}\left(\frac{x_i\sqrt{k/m}}{v_i}\right)\right],\qquad(9.2)$$

where we have treated both x_i and v_i as positive quantities and then used the trigonometric identity $\sin^{-1}\left(1/\sqrt{z^2+1}\right) = \tan^{-1}(1/z)$. Equation (9.2) looks complicated, but it is really of the form

$$x(t) = C\sin(\omega_n t + \phi),\qquad(9.3)$$

where

$$\omega_n = \sqrt{\frac{k}{m}},\qquad(9.4)$$

$$C = \sqrt{\frac{v_i^2}{\omega_n^2} + x_i^2},\qquad(9.5)$$

$$\tan\phi = \frac{x_i\omega_n}{v_i}.\qquad(9.6)$$

The quantity ω_n is a constant called the *natural frequency*[*] of vibration, and it is expressed in rad/s in both the SI and U.S. Customary unit systems. The quantities C and ϕ, called the *amplitude* and *phase angle* of vibration, respectively, are constants that depend on the initial conditions and ω_n. Consistent with Eq. (9.5), the *amplitude* of vibration C is understood to be a positive quantity. The angle ϕ can be determined via Eq. (9.6) for $v_i \neq 0$. When $v_i = 0$, ϕ can be chosen to be equal to $-\pi/2$ or $\pi/2$ rad depending on whether $x_i < 0$ or $x_i > 0$, respectively.

Equation (9.3), which is a solution of Eq. (9.1), describes a *harmonic motion* with natural frequency ω_n. For this reason, the physical system modeled by Eq. (9.1) is called a *harmonic oscillator*. Equation (9.1) represents an *undamped* vibration because there are no terms that depend on $\dot{x}$, which would occur with viscous damping or in some models of aerodynamic drag. Equation (9.1) also represents a *free* vibration since it is homogeneous; that is, there are only terms containing the dependent variable x and no terms that are functions of time or are constant.[†] The function in Eq. (9.3) is plotted in Fig. 9.2.

The oscillator represented by Eq. (9.3) completes one cycle in the time

$$\tau = \frac{2\pi}{\omega_n}.\qquad(9.7)$$

The quantity τ (the Greek letter tau) is called the *period* of the vibration (see Fig. 9.2). Finally, the number of cycles of vibration per unit of time is called the *frequency*, and it is defined as

$$f = \frac{1}{\tau} = \frac{\omega_n}{2\pi}.\qquad(9.8)$$

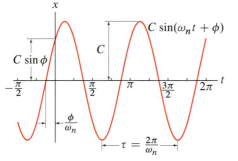

Figure 9.2
Plot of $x(t)$ in Eq. (9.3) showing the amplitude C, phase angle ϕ, and period τ of a harmonic oscillator.

---

[*] The natural frequency is also sometimes called the *circular frequency*.
[†] We will see in Section 9.2 that this is equivalent to saying that there is no forcing function in Eq. (9.1).

Generally the frequency f is expressed in cycles per second or *hertz* (Hz).

Applying these ideas to the railcar, we see that if it couples to the spring, its natural frequency is $\omega_n = 2.03\,\text{rad/s}$, its period is $\tau = (2\pi\,\text{rad})/(2.03\,\text{rad/s}) = 3.10\,\text{s}$, and its frequency is $f = 0.323\,\text{Hz}$, where we have used $k = 22{,}300\,\text{lb/ft}$ and $m = 5404\,\text{slug}$. The amplitude of vibration is $C = 2.89\,\text{ft}$ and the phase angle is $\phi = 0$, where we have used $x_i = 0\,\text{ft}$ and $v_i = 5.867\,\text{ft/s}$.

■ **Mini-Example.**   A jumper whose mass is $m = 65\,\text{kg}$ is hanging in equilibrium from a linear elastic bungee cord with constant $k = 200\,\text{N/m}$. The jumper is then pulled down 5 m and released from rest (see Fig. 9.3). Determine the equation governing the ensuing vibration, the period, the amplitude, and the phase angle of the vibration. Treat the jumper as a particle.

Solution. The FBD of the jumper after being pulled a distance y below the $y = 0$ static equilibrium position is shown in Fig. 9.4. Summing forces in the y direction gives

$$\sum F_y: \quad mg - F_s = ma_y, \tag{9.9}$$

where F_s is the force in the bungee cord and $a_y = \ddot{y}$. Since y is measured from the static equilibrium position, the force in the bungee cord must be (see Fig. 9.5)

$$F_s = mg + ky. \tag{9.10}$$

Substituting Eq. (9.10) into Eq. (9.9), we obtain

$$mg - (mg + ky) = m\ddot{y} \quad \Rightarrow \quad \ddot{y} + \frac{k}{m}y = 0. \tag{9.11}$$

Equation (9.11) is of exactly the same form as Eq. (9.1), and so we can immediately say that the jumper's natural frequency of vibration is $\omega_n = \sqrt{k/m} = 1.75\,\text{rad/s}$ and the corresponding period of vibration is $\tau = 2\pi/\omega_n = 3.58\,\text{s}$. Since $v_i = 0$ and $x_i = 5\,\text{m}$; Eq. (9.5) tells us that the jumper's amplitude of vibration is 5 m; i.e., as expected, the jumper oscillates about the static equilibrium position. Finally, since $v_i = 0$ and $x_i > 0$, we can choose the phase angle to be $\phi = \pi/2\,\text{rad}$. ■

One of the lessons of this mini-example is that *it is convenient to choose the origin of the displacement variable to be at the equilibrium position of the system* rather than at the position of zero spring deflection. Doing this allows us to ignore the equal and opposite forces associated with equilibrium.

Standard form of the harmonic oscillator

Based on the preceding development, we define a *harmonic oscillator* to be any one degree of freedom (DOF, see definition on p. 190) system whose equation of motion can be given the form

$$\boxed{\ddot{x} + \omega_n^2 x = 0,} \tag{9.12}$$

where ω_n is the natural frequency* of the oscillator and x is the coordinate for the one DOF system. Equation (9.12) is called the *standard form* of the harmonic oscillator equation.

---
* In the case of a spring-mass system, ω_n takes the form in Eq. (9.4).

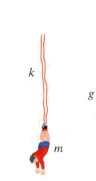

Figure 9.3
A bungee jumper of mass m hanging from a bungee cord of stiffness k.

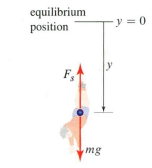

Figure 9.4
FBD of the bungee jumper, along with the coordinate system used in the mini-example.

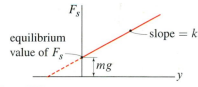

Figure 9.5
The force in the bungee cord as a function of y. The position $y = 0$ corresponds to static equilibrium.

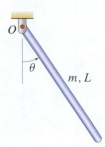

Figure 9.6
A uniform swinging bar.

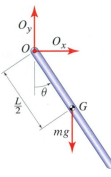

Figure 9.7
FBD of the bar in Fig. 9.6.

Mathematical justification for small-angle approximations. Figure 9.8 graphically demonstrates that $\sin x$ behaves as x for small x, but we can also see this with the Taylor series. The Taylor series expansion of $\sin x$ about $x = 0$ is

$$\sin x = x - \frac{x^3}{3!} + \frac{x^5}{5!} - \cdots,$$

where we see that if x is small, then x^3, x^5, and higher-order terms will be so small that $\sin x \approx x$. Similarly, the Taylor series expansion of $\cos x$ about $x = 0$ is

$$\cos x = 1 - \frac{x^2}{2!} + \frac{x^4}{4!} - \cdots,$$

where we see that if x is small, then x^2, x^4, and higher-order terms will be so small that $\cos x \approx 1$.

As we have seen in Eq. (9.3), *we know the complete vibrational solution for any system whose equation of motion can be put in the form of Eq. (9.12).* We also note that an alternative form of the solution to Eq. (9.12) is

$$x(t) = A \cos \omega_n t + B \sin \omega_n t, \tag{9.13}$$

where the solution in Eq. (9.3) is recovered from Eq. (9.13) if we let

$$C = \sqrt{A^2 + B^2} \quad \text{and} \quad \tan \phi = \frac{A}{B}. \tag{9.14}$$

Noting again that $x(0) = x_i$ and $\dot{x}(0) = v_i$, we find $A = x_i$ and $B = v_i / \omega_n$, and so Eq. (9.13) becomes

$$x(t) = x_i \cos \omega_n t + \frac{v_i}{\omega_n} \sin \omega_n t. \tag{9.15}$$

Linearizing nonlinear systems

Not all vibrating one DOF systems are harmonic oscillators like Eq. (9.12). However, many systems can be approximated as harmonic oscillators. As an example, consider the uniform thin bar that is pinned and suspended at one end (Fig. 9.6). Using the FBD in Fig. 9.7 and the methods of Chapter 7, the equation of motion for this bar is

$$\ddot{\theta} + \frac{3g}{2L} \sin \theta = 0. \tag{9.16}$$

This equation is *nonlinear* in θ because of the presence of $\sin \theta$, which is a *nonlinear* function of θ (in general, nonlinear equations are more challenging to solve than linear equations). Equation (9.16) can be written in the standard form $\ddot{\theta} + \omega_n^2 \theta = 0$ if we consider only vibrations for small values of θ, although this creates an approximate version of the original equation. As shown in Fig. 9.8, when θ is small, $\sin \theta$ behaves as θ and Eq. (9.16) becomes

$$\ddot{\theta} + \frac{3g}{2L} \theta = 0, \tag{9.17}$$

which is in standard form with $\omega_n = \sqrt{3g/(2L)}$. This process, in which a nonlinear ordinary differential equation is approximated to be linear, is called *linearization*. We will explore linearization further in the example problems.

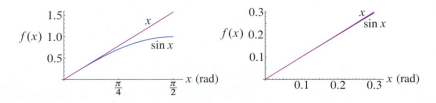

Figure 9.8. Plots of $f(x) = \sin x$ and $f(x) = x$ for two different ranges of x. The plot on the left shows that for large x, the two curves diverge. The plot on the right shows that for $x \lesssim 0.3$ rad the two curves are almost indistinguishable.

Energy method

The equations of motion for all three of the systems considered in this section were derived by applying the Newton-Euler equations to the FBD of the particle or body of interest. In addition, all three of these systems are conservative; that is, all forces doing work are conservative (the spring on the railcar, the bungee cord and gravity on the bungee jumper, and gravity on the swinging bar). For systems like these, the fact that energy is conserved can provide a way to derive the equation of motion.

Finding the equation of motion

To see how to find the equation of motion using conservation of mechanical energy, consider the bungee jumper mini-example on p. 675. Since the bungee jumper is a conservative system, we know that the work-energy principle gives

$$T_1 + V_1 = T_2 + V_2, \tag{9.18}$$

where ① is at release and ② is any subsequent position. This implies that

$$\boxed{T + V = \text{constant} \quad \Rightarrow \quad \frac{d}{dt}(T + V) = 0,} \tag{9.19}$$

where we have dropped the use of the subscript 2, to reinforce the idea that ② is *any* position following ①. If we now compute T and V at an arbitrary position for the bungee jumper, we find that the potential energy is given by (see Fig. 9.9)

$$V = V_e + V_g = \tfrac{1}{2}k(y + \delta_{st})^2 - mgy, \tag{9.20}$$

where y is measured from the static equilibrium position of the jumper and δ_{st} is the amount of stretch in the bungee cord at the static equilibrium position. The kinetic energy is given by

$$T = \tfrac{1}{2}m\dot{y}^2. \tag{9.21}$$

Substituting Eqs. (9.20) and (9.21) into Eq. (9.19) and taking the time derivative, we find that

$$\frac{d}{dt}(T + V) = k(y + \delta_{st})\dot{y} - mg\dot{y} + m\dot{y}\ddot{y} = 0. \tag{9.22}$$

Rewriting Eq. (9.22) as $[k(y + \delta_{st}) - mg + m\ddot{y}]\dot{y} = 0$, we see that this equation is satisfied for any value of $\dot{y}$ if and only if

$$k(y + \delta_{st}) - mg + m\ddot{y} = 0. \tag{9.23}$$

Since $k\delta_{st} = mg$, we recover the harmonic oscillator equation

$$m\ddot{y} + ky = 0, \tag{9.24}$$

which is equivalent to Eq. (9.11) for the same bungee jumper.

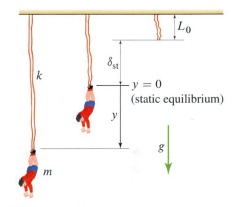

Figure 9.9
The unstretched length of the bungee cord L_0, the static equilibrium position of the jumper ($y = 0$), and the jumper at an arbitrary y position.

Helpful Information

Notation for the work-energy principle. Here we use the same notation introduced in Chapter 4. When we write ① (or ②), we mean "position 1" (or "position 2").

The energy method and linearization

When using the energy method to find the linearized equations of motion of a mechanical system, we can begin by approximating the kinetic and potential energies as quadratic functions of position and velocity before taking their time derivative. Then the time derivative of the quadratic approximation of the energy yields equations of motion that are *automatically* linear. The rationale for starting with the quadratic approximation of the energy is that, in some cases, linearizing the equations of motion obtained from the nonapproximated form of the energy is more involved than developing the quadratic approximation of the energy.

To see what we mean by "approximating the kinetic and potential energies as quadratic functions of position and velocity," consider again the uniform thin bar in Fig. 9.6. Writing the kinetic and potential energies at an arbitrary angle θ, we find that

$$T = \tfrac{1}{2} I_O \dot{\theta}^2 = \tfrac{1}{6} m L^2 \dot{\theta}^2, \tag{9.25}$$

$$V = -\tfrac{1}{2} m g L \cos \theta \approx -\tfrac{1}{2} m g L \left(1 - \tfrac{1}{2} \theta^2 \right). \tag{9.26}$$

Notice that the kinetic energy T in Eq. (9.25) is a quadratic function of the angular velocity $\dot{\theta}$ and therefore does not need to be approximated in any way. By contrast, notice that the potential energy $V = -\tfrac{1}{2} m g L \cos \theta$ is not a quadratic function of θ, but we were able to approximate it as such by using the first *two* terms in the Taylor series expansion of $\cos \theta$ (see the Helpful Information marginal note on p. 676). Using Eqs. (9.25) and (9.26) in $\frac{d}{dt}(T + V) = 0$, we obtain

$$\tfrac{1}{3} m L^2 \dot{\theta} \ddot{\theta} + \tfrac{1}{2} m g L \theta \dot{\theta} = 0 \quad \Rightarrow \quad \ddot{\theta} + \frac{3g}{2L} \theta = 0, \tag{9.27}$$

which is exactly what we obtained in Eq. (9.17). The simple message here is that when we use the energy method to derive the equation of motion, it is important to approximate the kinetic and potential energies as quadratic functions of position and velocities *before* taking their time derivatives.

For future reference, we note that in linearizing the sine function, and in approximating the cosine function as a quadratic function of its argument, we use the following relations

$$\sin \theta \approx \theta \quad \text{and} \quad \cos \theta \approx 1 - \theta^2/2, \tag{9.28}$$

respectively. In addition, we note that since the quadratic term in the power series expansion of the sine function is identically equal to zero (see the Helpful Information marginal note on p. 676), the linearized form of the sine function can also be viewed as the quadratic approximation of the sine function.

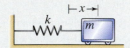

End of Section Summary

Any one DOF system whose equation of motion is of the form

> Eq. (9.12), p. 675

$$\ddot{x} + \omega_n^2 x = 0$$

is called a *harmonic oscillator*, and the above expression is referred to as the *standard form* of the harmonic oscillator equation. The solution of this equation can be written as (see Fig. 9.10)

> Eq. (9.3), p. 674

$$x(t) = C \sin(\omega_n t + \phi),$$

where ω_n is the *natural frequency*, C is the *amplitude*, and ϕ is the *phase angle* of vibration.

A simple example of a harmonic oscillator is a system consisting of a mass m attached at the free end of a spring with constant k and with the other end fixed (see Fig. 9.11). The natural frequency of such a system is given by

> Eq. (9.4), p. 674

$$\omega_n = \sqrt{\frac{k}{m}}.$$

In addition, the amplitude C and the phase angle ϕ are given by, respectively,

> Eqs. (9.5) and (9.6), p. 674

$$C = \sqrt{\frac{v_i^2}{\omega_n^2} + x_i^2} \quad \text{and} \quad \tan\phi = \frac{x_i \omega_n}{v_i},$$

where, by letting $t = 0$ be the initial time, $x_i = x(0)$ (i.e., x_i is the initial position) and $v_i = \dot{x}(0)$ (i.e., v_i is the initial velocity). If $v_i = 0$, then ϕ can be chosen equal to $-\pi/2$ or $\pi/2$ rad for $x_i < 0$ and $x_i > 0$, respectively. An alternative form of the solution to Eq. (9.12) is given by

> Eq. (9.15), p. 676

$$x(t) = x_i \cos\omega_n t + \frac{v_i}{\omega_n} \sin\omega_n t.$$

The *period* of the oscillation is given by

> Eq. (9.7), p. 674

$$\text{Period} = \tau = \frac{2\pi}{\omega_n},$$

and the *frequency* of vibration is

> Eq. (9.8), p. 674

$$\text{Frequency} = f = \frac{1}{\tau} = \frac{\omega_n}{2\pi}.$$

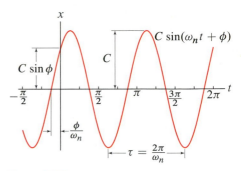

Figure 9.10

Figure 9.2 repeated. Plot of Eq. (9.3) showing the amplitude C, phase angle ϕ, and period τ of a harmonic oscillator.

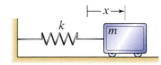

Figure 9.11

A simple spring-mass harmonic oscillator whose equation of motion is given by $m\ddot{x} + kx = 0$ and for which $\omega_n = \sqrt{k/m}$. The position x is measured from the location of m when the spring is undeformed.

Concept Alert

Natural frequency is proportional to square root of the stiffness over mass. The natural frequency of a harmonic oscillator is proportional to the square root of the stiffness of the system divided by its mass. Keep in mind that as we state in the Helpful Information on p. 678, the mass isn't always just m and the stiffness isn't always just k. Of course, we should also keep in mind that the dimensions of natural frequency must always be 1 over time.

Energy method.   For conservative systems, the work-energy principle tells us that the quantity $T + V$ is constant, and so its time derivative must be zero. This provides a convenient way to obtain the equations of motion via

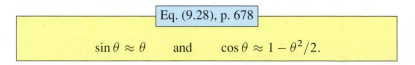

Eq. (9.19), p. 677

$$\frac{d}{dt}(T + V) = 0 \quad \Rightarrow \quad \text{equations of motion.}$$

When we apply the energy method to determine the linearized equations of motion, it is often convenient to first approximate the kinetic and potential energies as quadratic functions of position and velocity and then take derivatives with respect to time. This process yields equations of motion that are linear.

In approximating the sine and cosine functions as quadratic functions of their arguments, we use the relations

Eq. (9.28), p. 678

$$\sin \theta \approx \theta \qquad \text{and} \qquad \cos \theta \approx 1 - \theta^2/2.$$

Since the quadratic term in the power series expansion of the sine function is identically equal to zero, the linearized form of the sine function can also be viewed as the quadratic approximation of the sine function.

EXAMPLE 9.1   *Finding the Moment of Inertia of a Rigid Body*

When the connecting rod shown in Fig. 1 is suspended from the knife-edge at point O and displaced slightly so that it oscillates like a pendulum, its period of oscillation is 0.77 s. In addition, it is known that the mass center G is located a distance $L = 110$ mm from O and that the mass of the connecting rod is 661 g. Using this information, determine the mass moment of inertia of the connecting rod about G.

SOLUTION

Road Map & Modeling   If we write the equation of motion of the connecting rod for small angles, then, as with a pendulum, we should be able to write it in standard form and extract the natural frequency of vibration. The natural frequency will depend on the mass moment of inertia of the connecting rod, which should then allow us to solve for the moment of inertia. The FBD of the connecting rod is shown in Fig. 2.

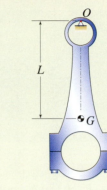

Figure 1
A connecting rod hinged on a knife-edge at O and allowed to oscillate freely in the plane of the page. Point G is the rod's mass center.

Governing Equations

Balance Principles   Summing moments about the fixed point O (see Fig. 2), we obtain

$$\sum M_O: \quad -mgL\sin\theta = I_O\alpha_{\text{cr}}, \tag{1}$$

where I_O is the moment of inertia of the connecting rod with respect to point O and α_{cr} is the angular acceleration of the connecting rod.

Force Laws   All forces are accounted for on the FBD.

Kinematic Equations   The only kinematic equation is $\alpha_{\text{cr}} = \ddot{\theta}$.

Computation   Substituting the kinematic equation into Eq. (1) and rearranging, we obtain

$$\ddot{\theta} + \frac{mgL}{I_O}\sin\theta = 0. \tag{2}$$

For small θ, $\sin\theta \approx \theta$ and so Eq. (2) becomes

$$\ddot{\theta} + \frac{mgL}{I_O}\theta = 0. \tag{3}$$

Equation (3) is in standard form so we know that

$$\omega_n^2 = \frac{mgL}{I_O} \quad \Rightarrow \quad I_O = \frac{mgL}{\omega_n^2} \quad \Rightarrow \quad I_O = \frac{mgL\tau^2}{4\pi^2}, \tag{4}$$

where we have used $\omega_n^2 = 4\pi^2/\tau^2$ from Eq. (9.7). Noting that the parallel axis theorem states that $I_G = I_O - mL^2$, we find

$$I_G = I_O - mL^2 = \frac{mgL\tau^2}{4\pi^2} - mL^2 \quad \Rightarrow \quad I_G = mL^2\left(\frac{g\tau^2}{4\pi^2 L} - 1\right), \tag{5}$$

which, when evaluated for the given quantities, yields

$$\boxed{I_G = 0.00271 \text{ kg·m}^2.} \tag{6}$$

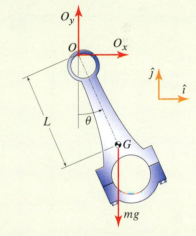

Figure 2
FBD of the connecting rod during oscillation.

Discussion & Verification   The dimensions of Eq. (5) are mass times length squared, as they should be. In addition, the first of Eqs. (4) tells us that the natural frequency of vibration is inversely proportional to the square root of the mass moment of inertia, which agrees with our intuition.

E X A M P L E   9.2   *Vibration of a Silicon Nanowire Modeled as a Rigid Bar*

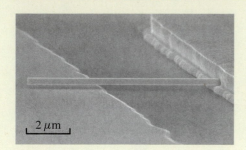

Figure 1
A field-emission scanning electronic micro-scope image of a silicon (Si) nanowire. From Mingwei Li et al., "Bottom-up Assembly of Large-Area Nanowire Resonator Arrays," *Nature Nanotechnology*, **3**(2), 2008, pp. 88–92.

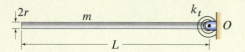

Figure 2
A rigid bar and torsional spring model of a flexible nanowire.

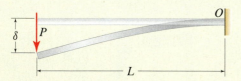

Figure 3
A flexible cantilever beam subject to a load P at its end. The deflection is given by Eq. (1).

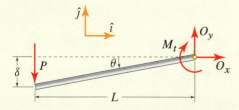

Figure 4
FBD of the rigid bar for finding the equivalent torsional spring constant k_t.

The natural frequency of a vibrating spring-mass system is a function of the equivalent mass and stiffness according to the relation $\omega_n = \sqrt{k_{eq}/m_{eq}}$. Therefore, if the cantilevered silicon nanowire (SiNW) shown in Fig. 1 were to vibrate, it would have a different natural frequency as shown than if additional mass were added to the end of the wire. Vibrating nanoelectromechanical systems (NEMS), such as this SiNW, have been proposed for use in chip-based sensor arrays as ultrasensitive mass detectors to detect masses in the zeptogram (zg) range. Such a wire would be capable of detecting small numbers of viruses!

Given a uniform Si nanowire with a circular cross section that is 9.8 μm long and 330 nm in diameter, compute its natural frequency, using a rigid bar model with all the flexibility lumped in a linear torsional spring at the base of the wire (see Fig. 2). Use $\rho = 2330\,kg/m^3$ for the density of silicon and $E = 152\,GPa$ for its modulus of elasticity.

SOLUTION

Road Map & Modeling   To obtain the linear torsional spring constant k_t, we will employ a result from mechanics of materials, which states that the deflection of a cantilevered bar subjected to a load P at its end is (see Fig. 3)

$$P = \frac{3EI_{cs}}{L^3}\delta, \tag{1}$$

where E is its modulus of elasticity and $I_{cs} = \frac{1}{4}\pi r^4$ is the centroidal *area* moment of inertia of the beam's cross section. Using Eq. (1), we will find the value of the torsional spring constant in Fig. 2 that gives this same deflection for a given load P. Once we have k_t, we can apply the Newton-Euler equations to then obtain the equation of motion in the form of Eq. (9.12).

──────── **Finding the Torsional Spring Constant** ────────

Governing Equations

Balance Principles   Referring to the FBD in Fig. 4, taking moments about point O, and noting that this is a *statics* problem for the purpose of finding k_t, we obain

$$\sum M_O: \quad PL - M_t = 0, \tag{2}$$

where M_t is the moment due to the torsional spring and we note that the moment arm for the load P is the distance L for small θ.

Force Laws   For a torsional spring, we have $M_t = k_t\theta$.

Kinematic Equations   For small θ, we relate δ and θ using $\delta = L\theta$.

Computation   Substituting the force law and the kinematic relation into Eq. (2) and solving the resulting equation for k_t, we obtain

$$k_t = \frac{PL^2}{\delta} \quad \Rightarrow \quad k_t = \frac{3EI_{cs}}{L}, \tag{3}$$

where we have used Eq. (1) in going from the first to the second expression for k_t.

—————————— **The Equation of Motion for the Rigid Bar** ——————————

Governing Equations

Balance Principles   Now that we have k_t, for the bar in Fig. 2, we draw the FBD shown in Fig. 5, and we sum moments about point O to obtain

$$\sum M_O: \quad -M_t = I_O \alpha_{\text{bar}}, \tag{4}$$

where the mass moment of inertia of the bar with respect to O is $I_O = \frac{1}{3}mL^2$.

Force Laws   The expression for M_t is unchanged and is given by $M_t = k_t\theta$.

Kinematic Equations   The kinematic equation relating α_{bar} to θ is

$$\alpha_{\text{bar}} = \ddot{\theta}. \tag{5}$$

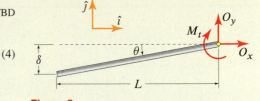

Figure 5
FBD of the bar while vibrating.

Computation   Substituting Eq. (3), the force law, and Eq. (5) into Eq. (4), we obtain the equation of motion as

$$-\frac{3EI_{\text{cs}}}{L}\theta = \frac{1}{3}mL^2\ddot{\theta} \quad\Rightarrow\quad \ddot{\theta} + \frac{9EI_{\text{cs}}}{mL^3}\theta = 0, \tag{6}$$

where we recall that $I_{\text{cs}} = \frac{1}{4}\pi r^4$. Comparing Eq. (6) to Eq. (9.12), we see that the natural frequency is

$$\omega_n = 3\sqrt{\frac{EI_{\text{cs}}}{mL^3}}. \tag{7}$$

To obtain a numerical value for ω_n, we find that the volume of the nanowire is $\pi r^2 L$ and that the mass is then $m = \rho\pi r^2 L = 1.953\times10^{-15}$ kg. The area moment of inertia is $I_{\text{cs}} = \frac{1}{4}\pi r^4 = 5.821\times10^{-28}$ m^4. Using these results, along with $L = 9.8\times10^{-6}$ m and $E = 152\times10^9$ N/m^2,* we find that

$$\omega_n = 2.08\times10^7 \text{ rad/s} \quad \text{and} \quad f = 3.31 \text{ MHz}. \tag{8}$$

Discussion & Verification   Closely examining Eq. (7), we find that the quantity under the square root has dimensions of 1 over time squared. Hence, the dimensions of ω_n are 1 over time as they should be. While it is hard to know what the frequency of vibration of a cantilevered bar of this size should be, in the Closer Look below, we will see that our model actually is quite good.

🔍 **A Closer Look**   From the theory of vibration of continuous systems, one can show that a continuous system like this nanowire vibrates with infinitely many natural frequencies. The first (i.e., the smallest) of these frequencies is given by

$$(\omega_n)_{\text{exact}} = 3.516\sqrt{\frac{EI_{\text{cs}}}{mL^3}}, \tag{9}$$

which, when compared with Eq. (7), tells us that our model is only off by about 15%. Evaluating Eq. (9) numerically, we find that

$$(\omega_n)_{\text{exact}} = 2.44\times10^7 \text{ rad/s} \quad \text{and} \quad f_{\text{exact}} = 3.88 \text{ MHz}. \tag{10}$$

In Prob. 9.17, we have the opportunity to examine another model for a cantilevered wire such as this and see how the natural frequency changes with the addition of a few zeptograms of virus to the end of the wire.

> **Interesting Fact**
>
> **Natural frequency of a wooden yardstick.** As a comparison, it is interesting to compute the natural frequency of a wooden yardstick using Eq. (9). Using properties typical of a wooden yardstick, that is, $E = 12$ GPa, $m = 0.0614$ kg, a 28 mm × 4 mm cross section (which gives $I_{\text{cs}} = 1.49\times10^{-10}$ m^4), and $L = 0.914$ m, we find that $\omega_n = 21.7$ rad/s, which corresponds to $f = 3.45$ Hz. The nanowire's natural frequency is 1.1 million times higher!

————————————
* Recall that 1 Pa = 1 N/m^2.

E X A M P L E 9.3 *Energy Method: Equation of Motion of a Diving Board*

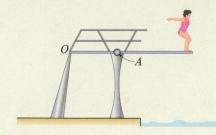

Figure 1

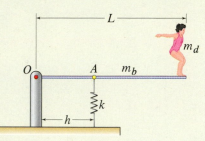

Figure 2

The model used to analyze the oscillation of the diving board and diver.

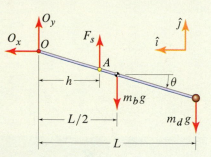

Figure 3

FBD of the board and diver as they oscillate. Note that the dimensions shown apply only when θ is small so that $\cos\theta \approx 1$.

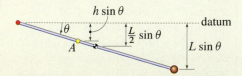

Figure 4

The vertical displacements of the relevant points on the diving board as it rotates.

A diver is causing the end of the diving board shown in Fig. 1 to oscillate. Referring to Fig. 2, we will model the board as a thin, uniform, *rigid* plate of mass m_b and length L and assume that the board is pinned at O. To model the elastic response of the board, we assume that the board oscillates due to a spring of stiffness k attached to the board at what was the fulcrum at A. In addition, we will model the diver as a point mass of mass m_d standing at the end of the board. Use the energy method to find the equation of motion for small rotations of the board.

SOLUTION

Road Map & Modeling The system is conservative since only the spring and the two weight forces do work as the system oscillates. This allows us to write the sum of the kinetic and potential energies of the system at an arbitrary position and then differentiate that sum with respect to time to obtain the equation of motion.

Governing Equations

Balance Principles Since energy is conserved, we can write that the sum of the kinetic and potential energies is constant so that

$$T + V = \text{constant} \quad \Rightarrow \quad \frac{d}{dt}(T + V) = 0. \tag{1}$$

The kinetic energy is given by

$$T = \tfrac{1}{2}I_O\dot\theta^2 + \tfrac{1}{2}m_d v_d^2, \tag{2}$$

where $I_O = \tfrac{1}{3}m_b L^2$ is the mass moment of inertia of the diving board with respect to point O and v_d is the speed of the diver.

Force Laws The potential energy of the system at an arbitrary angle θ is (see Fig. 4)

$$V = \tfrac{1}{2}k(\delta_{\text{st}} + h\sin\theta)^2 - m_b g\tfrac{L}{2}\sin\theta - m_d gL\sin\theta \tag{3}$$

$$\approx \tfrac{1}{2}k(\delta_{\text{st}} + h\theta)^2 - m_b g\tfrac{L}{2}\theta - m_d gL\theta, \tag{4}$$

where δ_{st} is the compression of the spring when the diving board is in static equilibrium with the diver on it, i.e., when $\theta = 0$, and where we have approximated V as a quadratic function of θ (i.e., position) in approximating Eq. (3) as Eq. (4).

Kinematic Equations Since the diver is rotating about the fixed point at O, the diver's speed is $v_d = L\dot\theta$.

Computation Substituting the potential energy in Eq. (4), the kinetic energy in Eq. (2), and the kinematic equation into Eq. (1), and then taking the time derivative, we obtain

$$(I_O + m_d L^2)\dot\theta\ddot\theta + k(\delta_{\text{st}} + h\theta)h\dot\theta - \tfrac{1}{2}m_b gL\dot\theta - m_d gL\dot\theta = 0. \tag{5}$$

Canceling $\dot\theta$, we note that the term $kh\delta_{\text{st}}$ is the moment about O due to the spring force needed to hold the system in equilibrium (i.e., at $\theta = 0$). This moment is equal and opposite to the moment about O created by the two weight forces, that is, $-\tfrac{1}{2}m_b gL - m_d gL$. Therefore, Eq. (5) reduces to the final equation of motion (we have used $I_O = \tfrac{1}{3}m_b L^2$)

$$\boxed{\ddot\theta + \frac{kh^2}{(\tfrac{1}{3}m_b + m_d)L^2}\theta = 0.} \tag{6}$$

Discussion & Verification The coefficient of θ in Eq. (6) has dimensions of 1 over time squared, as it should.

PROBLEMS

Problem 9.1

Show that Eq. (9.15) is equivalent to Eq. (9.3) if $C = \sqrt{A^2 + B^2}$ and $\tan \phi = A/B$.

Problem 9.2

Derive the formula for the mass moment of inertia of an arbitrarily shaped rigid body about its mass center based on the body's period of oscillation τ when suspended as a pendulum. Assume that the mass of the body m is known and that the location of the mass center G is known relative to the pivot point O.

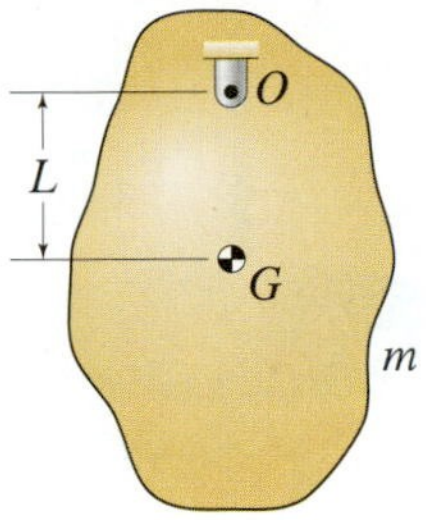

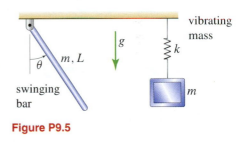

Figure P9.2

Problem 9.3

The thin ring of radius R and mass m is suspended by the pin at O. Determine its period of vibration if it is displaced a small amount and released.

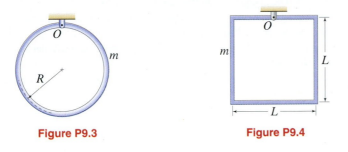

Figure P9.3   **Figure P9.4**

Problem 9.4

The thin square hoop has mass m and is suspended by the pin at O. Determine its period of vibration if it is displaced a small amount and released.

💡 Problem 9.5 💡

The swinging bar and the vibrating mass are made to vibrate on Earth, and their respective natural frequencies are measured. The two systems are then taken to the Moon and are again allowed to vibrate at their respective natural frequencies. How will the natural frequency of each system change when compared with that on the Earth, and which of the two systems will experience the larger change in natural frequency?
Note: Concept problems are about *explanations*, not computations.

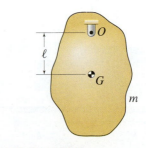

Figure P9.5

Problems 9.6 and 9.7

A rigid body of mass m, mass center at G, and mass moment of inertia I_G is pinned at an arbitrary point O and allowed to oscillate as a pendulum.

Problem 9.6   By writing the Newton-Euler equations, determine the distance ℓ from G to the pivot point O so that the pendulum has the highest possible natural frequency of oscillation.

Problem 9.7   Using the energy method, determine the distance ℓ from G to the pivot point O so that the pendulum has the highest possible natural frequency of oscillation.

Figure P9.6 and P9.7

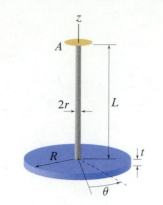

Figure P9.8

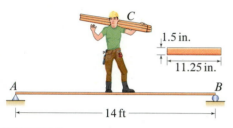

Figure P9.9

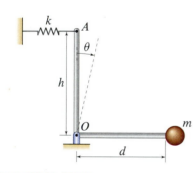

Figure P9.10–P9.13

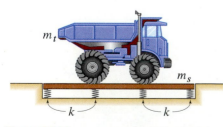

Figure P9.14

Problem 9.8

The uniform disk of radius R and thickness t is attached to the thin shaft of radius r, length L, and negligible mass. The end A of the shaft is fixed. From mechanics of materials, it can be shown that if a torque M_z is applied to the free end of the shaft, then it can be related to the twist angle θ via

$$\theta = \frac{M_z L}{GJ},$$

where G is the shear modulus of elasticity of the shaft and $J = \frac{\pi}{2} r^4$ is the polar moment of inertia of the cross-sectional area of the shaft. Letting ρ be the mass density of the disk and using the given relationship between M_z and θ, determine the natural frequency of vibration of the disk in terms of the given dimensions and material properties when it is given a small angular displacement θ in the plane of the disk.

Problem 9.9

A construction worker C is standing at the midpoint of a 14 ft long pine board that is simply supported. The board is a standard 2×12, so its cross-sectional dimensions are as shown. Assuming the worker weighs 180 lb and he flexes his knees once to get the board oscillating, determine his vibration frequency. Neglect the weight of the beam and use the fact that a load P applied to a simply supported beam will deflect the center of the beam $PL^3/(48EI_{cs})$, where L is the length of the beam, E is its modulus of elasticity, and I_{cs} is the area moment of inertia of the cross section of the beam. The elastic modulus of pine is 1.8×10^6 psi.

Problems 9.10 through 9.13

The L-shaped bar lies in the vertical plane and is pinned at O. One end of the bar has a linear elastic spring with constant k attached to it, and attached at the other end is a mass m of negligible size. The angle θ is measured from the equilibrium position of the system and it is assumed to be small.

Problem 9.10   Assuming that the L-shaped bar has negligible mass, determine the natural period of vibration of the system by writing the Newton-Euler equations.

Problem 9.11   Assuming that the L-shaped bar has negligible mass, determine the natural period of vibration of the system via the energy method.

Problem 9.12   Assuming that the L-shaped bar has mass per unit length ρ, determine the natural period of vibration of the system by writing the Newton-Euler equations.

Problem 9.13   Assuming that the L-shaped bar has mass per unit length ρ, determine the natural period of vibration of the system via the energy method.

Problem 9.14

An off-highway truck drives onto a concrete deck scale to be weighed, thus causing the truck and scale to vibrate vertically at the natural frequency of the system. The empty truck weighs 74,000 lb, the scale platform weighs 51,000 lb, and the platform is supported by eight identical springs (four of which are shown), each with constant $k = 3.6 \times 10^5$ lb/ft. Modeling the truck, its contents, and the concrete deck as a single particle, if a vibration frequency of 3.3 Hz is measured, what is the weight of the payload being carried by the truck?

Problem 9.15

A mass m of 3 kg is in equilibrium when a hammer hits it, imparting a velocity v_0 of 2 m/s to it. If k is 120 N/m, determine the amplitude of the ensuing vibration and find the maximum acceleration experienced by the mass.

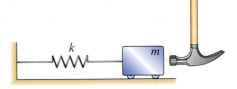

Figure P9.15

Problem 9.16

The buoy in the photograph can be modeled as a circular cylinder of diameter d and mass m. If the buoy is pushed down in the water, which has density ρ, it will oscillate vertically. Determine the frequency of oscillation. Evaluate your result for $d = 1.2$ m, $m = 900$ kg, and surface seawater, which has a density of $\rho = 1027$ kg/m^3. *Hint:* Use Archimedes' principle, which states that a body wholly or partially submerged in a fluid is buoyed up by a force equal to the weight of the displaced fluid.

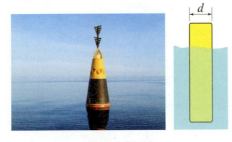

Figure P9.16

Problem 9.17

For the silicon nanowire in Example 9.2, use the lumped mass model shown, in which a point mass m is connected to a rod of negligible mass and length L that is pinned at O, to determine the natural frequency ω_n and frequency f of the nanowire. Use the values given in Example 9.2 for the mass of the lumped mass, the length of the massless rod, and the parameters used to determine the spring constant $k = 3EI_{\text{cs}}/L^3$. You may use either δ or θ as the position variable in your solution. Assume that the displacement of m is small so that it moves vertically.

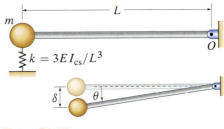

Figure P9.17

Problem 9.18

The small sphere A has mass m and is fixed at the end of the arm OA of negligible mass, which is pinned at O. If the linear elastic spring has stiffness k, determine the equation of motion for small oscillations, using

(a) the vertical position of the mass A as the position coordinate,

(b) the angle formed by the arm OA with the horizontal as the position coordinate.

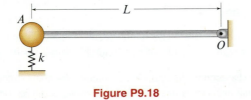

Figure P9.18

Problems 9.19 and 9.20

Grandfather clocks keep time by advancing the hands a set amount per oscillation of the pendulum. Therefore, the pendulum needs to have a very accurate period for the clock to keep time accurately. As a fine adjustment of the pendulum's period, many grandfather clocks have an adjustment nut on a bolt at the bottom of the pendulum disk. By screwing this nut inward or outward, the mass distribution of the pendulum can be changed and its period adjusted. In the following problems, model the pendulum as a uniform disk of radius r and mass m_p, which is at the end of a rod of negligible mass and length $L - r$. Model the adjustment nut as a particle of mass m_n, and let the distance between the bottom of the pendulum disk and the nut be d.

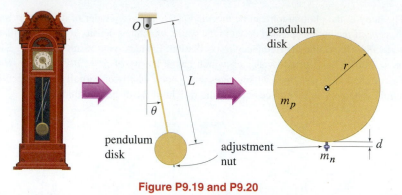

Figure P9.19 and P9.20

Problem 9.19 If the adjustment is initially at a distance $d = 9$ mm from the bottom of the pendulum disk, how much would the period of the pendulum change if the nut were screwed 4 mm closer to the disk? In addition, how much time would the clock gain or lose in a 24 h period if this were done? Let $m_p = 0.7$ kg, $r = 0.1$ m, $m_n = 8$ g, and $L = 0.85$ m.

Problem 9.20 The clock is running slow so that it is losing 1 minute every 24 hours (i.e., the clock takes 1441 minutes to complete a 1440 minute day). If the adjustment nut is at $d = 2$ cm, what would its mass need to have to correct the pendulum's period if the nut is moved to $d = 0$ cm? Let $m_p = 0.7$ kg, $r = 0.1$ m, and $L = 0.85$ m.

Problems 9.21 and 9.22

The uniform cylinder rolls without slipping on a flat surface. Let $k_1 = k_2 = k$ and $r = R/2$. Assume that the horizontal motion of G is small.

Problem 9.21 Determine the equation of motion for the cylinder by writing its Newton-Euler equations. Use the horizontal position of the mass center G as the degree of freedom.

Problem 9.22 Determine the equation of motion for the cylinder using the energy method. Use the horizontal position of the mass center G as the degree of freedom.

Problems 9.23 and 9.24

The uniform cylinder of mass m and radius R rolls without slipping on the inclined surface. The spring with constant k wraps around the cylinder as it rolls.

Problem 9.23 Determine the equation of motion for the cylinder by writing its Newton-Euler equations. Determine the numerical value of the period of oscillation of the cylinder using $k = 30$ N/m, $m = 10$ kg, $R = 30$ cm, and $\theta = 20°$.

Problem 9.24 Determine the equation of motion for the cylinder using the energy method. Determine the numerical value of the period of oscillation of the cylinder using $k = 30$ N/m, $m = 10$ kg, $R = 30$ cm, and $\theta = 20°$.

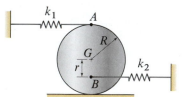

Figure P9.21 and P9.22

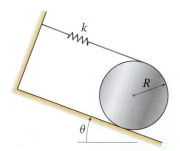

Figure P9.23 and P9.24

Problem 9.25

A uniform bar of mass m is placed off-center on two counter-rotating drums A and B. Each drum is driven with constant angular speed ω_0, and the coefficient of kinetic friction between the drums and the bar is μ_k. Determine the natural frequency of oscillation of the bar on the rollers. *Hint:* Measure the horizontal position of G relative to the midpoint between the two drums, and assume that the drums rotate sufficiently fast so that the drums are always slipping relative to the bar.

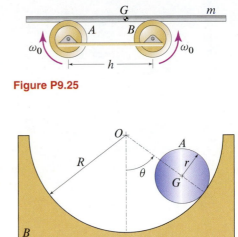

Figure P9.25

Problem 9.26

The uniform cylinder A of radius r and mass m is released from a small angle θ inside the large cylinder of radius R. Assuming that it rolls without slipping, determine the natural frequency and period of oscillation of A.

Problem 9.27

The uniform sphere A of radius r and mass m is released from a small angle θ inside the large cylinder of radius R. Assuming that it rolls without slipping, determine the natural frequency and period of oscillation of the sphere.

Figure P9.26

Figure P9.27

Problem 9.28

The U-tube manometer lies in the vertical plane and contains a fluid of density ρ that has been displaced a distance y and oscillates in the tube. If the cross-sectional area of the tube is A and the total length of the fluid in the tube is L, determine the natural period of oscillation of the fluid, using the energy method. *Hint:* As long as the curved portion of the tube is always filled with liquid (i.e., the oscillations don't get large enough to empty part of it), the contribution of the liquid in the curved portion to the potential energy is *constant*.

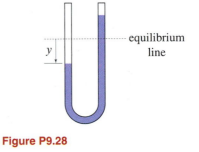

Figure P9.28

Problem 9.29

The uniform semicylinder of radius R and mass m rolls without slip on the horizontal surface. Using the energy method, determine the period of oscillation for small θ.

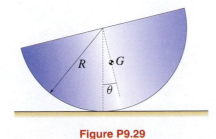

Figure P9.29

Problem 9.30

The thin shell semicylinder of radius R and mass m rolls without slip on the horizontal surface. Using the energy method, determine the period of oscillation for small θ.

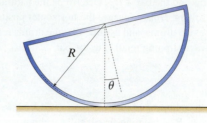

Figure P9.30

DESIGN PROBLEMS

Design Problem 9.1

As part of a manufacturing process, a uniform bar is placed on a centering device consisting of two counter-rotating drums A and B and a frictional slider C that provides light damping. When the bar is placed off-center on the two counter-rotating drums, it oscillates back and forth due to the sliding friction between each drum and the bar, and it eventually settles into the centered position due to the damper C. Once the bar has been centered, the next step in the manufacturing process can begin.

Assume that the steel bar has length $L = 1.2\,\text{m}$, radius $r = 22.5\,\text{mm}$, and density $\rho = 7.85\,\text{g/cm}^3$. Assuming that the initial misalignment (i.e., the initial distance between G and C) of each bar can be up to $0.25\,\text{m}$ and that sliding between the drums and the bar never ceases, design the drums (i.e., their radius and what they are made of) and determine their placement so that sliding is maintained for the entire range of motion. Since the damping is light, it can be neglected. In addition, assume that you want to drive each drum with a constant angular speed ω_0.

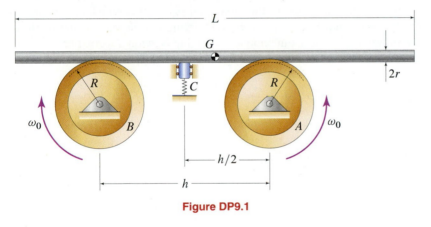

Figure DP9.1

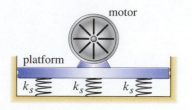

Figure 9.12

An unbalanced motor mounted on a platform that is elastically suspended on six springs, three of which are shown.

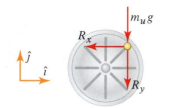

Figure 9.13

FBD of the eccentric mass m_u. The forces R_x and R_y are the forces exerted by the motor on the particle.

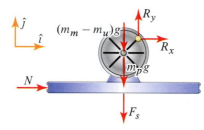

Figure 9.14

FBD of the motor and platform. The forces R_x and R_y are the forces exerted by the particle on the motor and, by Newton's third law, are equal and opposite to the corresponding forces in Fig. 9.13.

9.2 Undamped Forced Vibration

Not all vibration is of the type studied in the previous section — many systems vibrate due to external excitation that *forces* the system to vibrate. This section is devoted to the forced vibration of mechanical systems.

Vibration of an unbalanced motor

A motor is not perfectly balanced when the mass center of the rotor is not on its spin axis. When this happens, as the rotor spins, it transmits time-varying forces to the housing. In turn, these forces cause the motor and the table or platform on which it is mounted to vibrate. Figure 9.12 shows a motor of mass m_m mounted on a platform whose mass is m_p. The platform is supported on six linear elastic springs, each with constant k_s, whose equivalent spring constant is $k_{eq} = 6k_s$. The rotor spins inside the motor at a constant angular velocity ω_r, and the effect of the unbalance is equivalent to an *eccentric mass* m_u located a distance ε from the axis of rotation. Using this information, the FBDs of the eccentric mass and of the motor and platform combination are shown in Figs. 9.13 and 9.14, respectively. Observing that the motor and platform can move only in the y direction and applying Newton's second law in the y direction to the eccentric mass in Fig. 9.13, we find that

$$\left(\sum F_y \right)_{\text{ecc. mass}} : \quad -R_y - m_u g = m_u a_{uy}, \tag{9.29}$$

and doing the same for the motor and platform in Fig. 9.14, we obtain

$$\left(\sum F_y \right)_{\text{motor}} : \quad R_y - (m_m - m_u)g - m_p g - F_s$$
$$= (m_m - m_u + m_p)a_{my}, \tag{9.30}$$

where a_{uy} and a_{my} are the y components of acceleration of the eccentric mass and the motor/platform, respectively, and F_s is the force on the motor/platform due to the springs. Referring to Fig. 9.15, since y_m is measured from the static

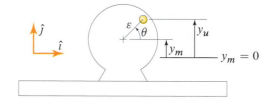

Figure 9.15. Kinematics of the motor/platform and the eccentric mass.

equilibrium position of the system, we have

$$F_s = k_{eq}(y_m - \delta_s), \tag{9.31}$$

where δ_s is the deflection of the spring when the system is in static equilibrium. Referring to Fig. 9.15, the kinematic equation for the eccentric mass is

$$y_u = y_m + \varepsilon \sin \theta \quad \Rightarrow \quad \dot{y}_u = \dot{y}_m + \varepsilon \dot{\theta} \cos \theta$$
$$\Rightarrow \quad \ddot{y}_u = \ddot{y}_m - \varepsilon \dot{\theta}^2 \sin \theta \quad \Rightarrow \quad a_{uy} = \ddot{y}_m - \varepsilon \omega_r^2 \sin \theta, \tag{9.32}$$

where $\dot{\theta} = \omega_r$, $\ddot{y}_u = a_{uy}$, and $\ddot{\theta} = 0$ since ω_r is constant. For the motor/platform, we have

$$a_{my} = \ddot{y}_m. \tag{9.33}$$

Substituting Eqs. (9.31)–(9.33) into Eqs. (9.29) and (9.30) and then eliminating R_y, we obtain

$$
\begin{aligned}
-m_u g - m_u(\ddot{y}_m - \varepsilon\omega_r^2 \sin\theta) - (m_m - m_u)g - m_p g \\
- k_{eq}(y_m - \delta_s) = (m_m - m_u + m_p)\ddot{y}_m. \tag{9.34}
\end{aligned}
$$

Canceling terms and rearranging, we obtain

$$(m_m + m_p)\ddot{y}_m + k_{eq}y_m + (m_m + m_p)g - k_{eq}\delta_s = m_u\varepsilon\omega_r^2 \sin\theta, \tag{9.35}$$

which, upon noting that $(m_m + m_p)g = k_{eq}\delta_s$ and that $\theta = \omega_r t$,[*] becomes

$$\ddot{y}_m + \frac{k_{eq}}{m_m + m_p}y_m = \frac{m_u\varepsilon\omega_r^2}{m_m + m_p}\sin\omega_r t. \tag{9.36}$$

The left-hand side of this equation has the same form as that of the harmonic oscillator equation we studied in Section 9.1. However, we now have a *time-dependent* term on the right-hand side due to the eccentric mass that *forces* or *drives* the oscillator. In this section we will learn how to solve equations like Eq. (9.36). We will complete the analysis of this system in Example 9.4.

Standard form of the forced harmonic oscillator

Equation (9.36) is of the form

$$\boxed{\ddot{x} + \omega_n^2 x = \frac{F_0}{m}\sin\omega_0 t.} \tag{9.37}$$

Equation (9.37) is the standard form of the forced harmonic oscillator equation and is a *nonhomogeneous* version of Eq. (9.12) on p. 675 because of the term $\frac{F_0}{m}\sin\omega_0 t$. A simple system whose equation of motion is given by Eq. (9.37) is shown in Fig. 9.16. The term on the right-hand side of Eq. (9.37) is a function of *only* the independent variable t and is often called a *forcing function* because it forces the system to vibrate. This particular type of forcing is harmonic because it is a harmonic function of time.

The theory of differential equations tells us that the *general solution* of Eq. (9.37) is the sum of the *complementary solution* and a *particular solution*. The *complementary solution*[†] is the solution of the associated homogeneous equation (i.e., Eq. (9.12)) given in Eq. (9.3) (or in Eq. (9.13)). The *particular solution* is *any* solution of Eq. (9.37). A way to obtain a particular solution is to guess its form and then verify whether or not the guess is correct. Since it seems reasonable that the response of a harmonically forced harmonic oscillator should resemble the forcing, we conjecture that the particular solution x_p is of the form

$$x_p = D\sin\omega_0 t, \tag{9.38}$$

---

[*] We have assumed that $\theta(0) = 0$.
[†] The complementary solution is sometimes called the *homogeneous solution*.

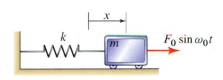

Figure 9.16
A forced harmonic oscillator whose equation of motion is given by Eq. (9.37) with $\omega_n = \sqrt{k/m}$. The position x is measured from the equilibrium position of the system when $F_0 = 0$.

where D is a constant to be determined. We can verify whether our guess is correct by substituting Eq. (9.38) into Eq. (9.37). Doing so yields

$$-D\omega_0^2 \sin \omega_0 t + \omega_n^2 D \sin \omega_0 t = \frac{F_0}{m} \sin \omega_0 t. \tag{9.39}$$

Canceling $\sin \omega_0 t$ and solving for D, we obtain

$$D = \frac{F_0/m}{\omega_n^2 - \omega_0^2} = \frac{F_0/k}{1 - (\omega_0/\omega_n)^2}, \tag{9.40}$$

where we have assumed that $\omega_0 \neq \omega_n$ and have used the fact that $\omega_n^2 = k/m$. Choosing D as in Eq. (9.40), we see that the guess in Eq. (9.38) is correct and the corresponding particular solution is

$$x_p = \frac{F_0/k}{1 - (\omega_0/\omega_n)^2} \sin \omega_0 t. \tag{9.41}$$

Combining the complementary solution in Eq. (9.13), which we label as x_c, with the particular solution in Eq. (9.41), the general solution to Eq. (9.37) is

$$x = x_c + x_p = A \sin \omega_n t + B \cos \omega_n t + \frac{F_0/k}{1 - (\omega_0/\omega_n)^2} \sin \omega_0 t, \tag{9.42}$$

where, as usual, A and B are constants determined by enforcing the initial conditions.

Equation (9.42) tells us that the vibration of a forced harmonic oscillator is composed of two parts: the complementary solution x_c that describes the *free vibration* of the system and the particular solution x_p that describes the *forced vibration* due to $F_0 \sin \omega_0 t$. As we will see in Section 9.3, the free vibration corresponding to x_c will die out with any amount of damping or energy dissipation, which is always present in real physical systems. For this reason, free vibration is often referred to as *transient vibration*. On the other hand, the forced vibration corresponding to x_p will be there as long as the forcing is there, and so it is often called *steady-state vibration.*[*]

When a vibration is forced, it is important to know the amplitude of the motion since that will determine the deformation and the deformation-related forces that the system has to endure. Equation (9.41) tells us that the amplitude of the steady-state vibration, which is given by

$$x_{\text{amp}} = \frac{F_0/k}{1 - (\omega_0/\omega_n)^2}, \tag{9.43}$$

depends on the *frequency ratio* ω_0/ω_n. If we now define the *magnification factor* MF for this case to be the ratio of the amplitude x_{amp} of steady-state vibration to the static deflection F_0/k caused by the force F_0, we find it to be

$$\text{MF} = \frac{x_{\text{amp}}}{F_0/k} = \frac{1}{1 - (\omega_0/\omega_n)^2}, \tag{9.44}$$

a plot of which is shown in Fig. 9.17. Notice that when ω_0 is small, MF $\to 1$.

---

[*] We do not discuss the solution of Eq. (9.37) for $\omega_0 = \omega_n$ because such a solution does not describe a steady-state vibration.

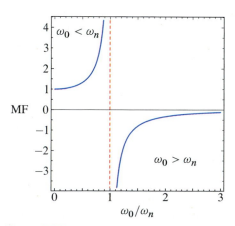

Figure 9.17
MF as a function of the frequency ratio ω_0/ω_n.

That is, for low-frequency forcing, the motion of the block is in the direction of the forcing (they are said to be *in phase* with one another). As the forcing frequency ω_0 approaches the natural frequency of the system ω_n, MF increases dramatically and goes to infinity as $\omega_0/\omega_n \to 1$. The situation in which $\omega_0 \approx \omega_n$ is called *resonance* and results in very large vibration amplitudes. Resonances in engineering systems are generally undesirable as they result in large displacements and deformations, often leading to premature failure. However, there do exist beneficial resonances in engineering systems, such as in amplifiers or in devices designed to aid sputum clearance of the airways of respiratory patients.[*] As we will see in Section 9.3, in a system with even a small amount of energy dissipation or damping, resonance does not result in infinite amplitudes, but the amplitudes can still grow to be *very* large.

Looking back at the system in Fig. 9.16, when $\omega_0 > \omega_n$, the MF is negative and the forcing is out of phase with the motion of the block. Finally, when $\omega_0 \gg \omega_n$, the force changes direction so rapidly compared to the natural frequency of the system that the system remains almost stationary and the MF $\to 0$.

Harmonic excitation of the support

Devices that measure vibration generally depend on the motion of the support structure of the device as forcing. For example, MacBook[®] series notebook computers sold by Apple[®] Inc. contain a three-axis accelerometer designed to detect large accelerations of the computer (e.g., when it is dropped or when the surface on which it is resting undergoes severe vibration).[†] Should such accelerations occur, the computers are designed to instantly park the hard drive heads to help reduce the risk of damage. Such a scenario might occur if a MacBook were on a desk in the same room as a large, severely unbalanced motor (see Fig. 9.18). In this case, the floor and the desk would transmit the vibration of the motor to the computer.

We will assume that the motor is causing large lateral vibrations of the floor and desk.[‡] We already know that an unbalanced motor vibrates harmonically, so we will let the lateral displacement of the floor be given by $x_f = X \sin \omega_f t$, where X is the amplitude of its lateral motion. In addition, we will model the coupling between the floor and the notebook computer as a linear spring of constant k, which is shown in Fig. 9.19. Treating the computer

Figure 9.18
A notebook computer with motion sensors being vibrated through its support by an unbalanced motor.

Figure 9.19. A simple model for the lateral vibration of the notebook computer on a desk.

---

[*] See L. C. de Lima et al., "Mechanical Evaluation of a Respiratory Device," *Medical Engineering & Physics*, **27**, 2005, pp. 181–187.

[†] This was true at the time this was written.

[‡] If we were to study the vertical vibrations, we would obtain equations of the same form as those obtained for the lateral vibrations—only the effective mass and effective stiffness coefficients would be different.

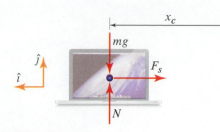

Figure 9.20
FBD of the notebook computer on the desk.

as a particle, its FBD is as shown in Fig. 9.20, where F_s is the force in the spring. Summing forces in the x direction gives

$$\sum F_x: \quad -F_s = m\ddot{x}_c, \tag{9.45}$$

where since both ends of the spring are moving, the force in it is given by $F_s = k(x_c - x_f) = k(x_c - X\sin\omega_f t)$, so that Eq. (9.45) becomes

$$-k(x_c - X\sin\omega_f t) = m\ddot{x}_c, \tag{9.46}$$

or, upon rearranging,

$$\ddot{x}_c + \omega_n^2 x_c = \frac{kX}{m}\sin\omega_f t, \tag{9.47}$$

where $\omega_n^2 = k/m$. Notice that Eq. (9.47) is of the same form as that Eq. (9.37) with the term F_0 replaced by the term kX. The results given in Eqs. (9.41)–(9.44) are valid with that same replacement.

End of Section Summary

When a harmonic oscillator is subject to harmonic forcing, the standard form of the equation of motion is

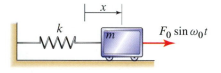

Figure 9.21
Figure 9.16 repeated. A forced harmonic oscillator whose equation of motion is given by Eq. (9.37) with $\omega_n = \sqrt{k/m}$. The position x is measured from the equilibrium position of the mass.

Eq. (9.37), p. 693

$$\ddot{x} + \omega_n^2 x = \frac{F_0}{m}\sin\omega_0 t,$$

where F_0 is the amplitude of the forcing and ω_0 is its frequency (see Fig. 9.21). The general solution to this equation consists of the sum of the complementary solution and a particular solution. The *complementary solution* x_c is the solution of the associated homogeneous equation, which is given by, for example, Eq. (9.13). For $\omega_0 \neq \omega_n$, a particular solution was found to be

Eq. (9.41), p. 694

$$x_p = \frac{F_0/k}{1 - (\omega_0/\omega_n)^2}\sin\omega_0 t,$$

and so the *general solution* is given by

Eq. (9.42), p. 694

$$x = x_c + x_p = A\sin\omega_n t + B\cos\omega_n t + \frac{F_0/k}{1 - (\omega_0/\omega_n)^2}\sin\omega_0 t,$$

where A and B are constants determined by the initial conditions. The amplitude of the steady-state vibration is

Eq. (9.43), p. 694

$$x_{\text{amp}} = \frac{F_0/k}{1 - (\omega_0/\omega_n)^2},$$

which means that the corresponding *magnification factor* MF is

Eq. (9.44), p. 694

$$MF = \frac{x_{\text{amp}}}{F_0/k} = \frac{1}{1 - (\omega_0/\omega_n)^2}.$$

The plot of the MF in Fig. 9.22 illustrates the phenomenon of *resonance*, which occurs when $\omega_0 \approx \omega_n$ and results in very large vibration amplitudes.

Harmonic excitation of the support.   If the support of a structure is excited harmonically rather than the structure itself (see Fig. 9.23), then Eq. (9.37) is still the governing equation, except that F_0 is replaced by the spring constant k times the amplitude of the support vibration X. All solutions described above are then valid with that same replacement.

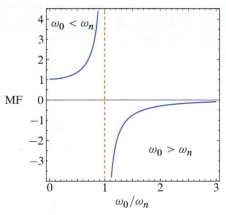

Figure 9.22
Figure 9.17 repeated. MF as a function of the frequency ratio ω_0/ω_n.

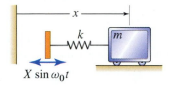

Figure 9.23. A harmonic oscillator whose support is being excited harmonically.

EXAMPLE 9.4 *Response and MF for the Unbalanced Motor*

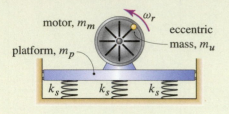

Figure 1

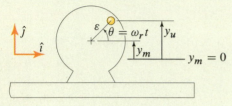

Figure 2
Kinematic definitions for the unbalanced motor and platform.

Find the general solution for the displacement, i.e., $y_m(t)$, of the unbalanced motor (Fig. 1) subject to the initial conditions $y_m(0) = 0$ and $\dot{y}_m(0) = v_{m0}$. After doing so, plot the solution for $0 < t < 1$ s, using $m_m = 40$ kg, $m_p = 15$ kg, $k_{eq} = 6k_s = 420,000$ N/m, $\varepsilon = 15$ cm, $\omega_r = 1200$ rpm, $y_m(0) = 0$ m, $\dot{y}_m(0) = 0.4$ m/s, and three different values of m_u: 10 g, 100 g, and 1000 g. From the plot, find the approximate maximum amplitude of the vibration. Finally, determine and plot the MF for the unbalanced rotor.

SOLUTION

Road Map & Modeling The equation of motion for our model of the unbalanced motor and platform was derived in Eq. (9.36), so we need only apply the general solution in Eq. (9.42) to determine and plot the response. For the MF, as with the MF found in Eq. (9.44), we will need to find a function of ω_r/ω_n where ω_r is the angular velocity of the unbalanced rotor inside the motor.

Governing Equations For convenience, we repeat the equation of motion for the unbalanced motor and platform, which was found in Eq. (9.36) to be

$$\ddot{y}_m + \frac{k_{eq}}{m_m + m_p} y_m = \frac{m_u \varepsilon \omega_r^2}{m_m + m_p} \sin \omega_r t, \tag{1}$$

where y_m is measured from the static equilibrium position of the system, as shown in Fig. 2.

Computation Equation (1) is of the same form as Eq. (9.37), which is repeated below for convenience

$$\ddot{x} + \omega_n^2 x = \frac{F_0}{m} \sin \omega_0 t, \tag{2}$$

where, comparing Eqs. (1) and (2), we have

$$\omega_n^2 = \frac{k_{eq}}{m_m + m_p}, \qquad \frac{F_0}{m} = \frac{m_u \varepsilon \omega_r^2}{m_m + m_p}, \qquad \text{and} \qquad \omega_0 = \omega_r. \tag{3}$$

Therefore, the general solution to Eq. (1) can be found using Eq. (9.42), which gives

$$y_m = A \sin \omega_n t + B \cos \omega_n t + \frac{m_u \varepsilon \omega_r^2 / k_{eq}}{1 - (\omega_r/\omega_n)^2} \sin \omega_r t. \tag{4}$$

For $t = 0$, Eq. (4) gives $y_m(0) = B$. Therefore, recalling that we must have $y_m(0) = 0$, we have

$$B = 0. \tag{5}$$

Differentiating y_m in Eq. (4) with respect to time, we obtain

$$\dot{y}_m = A\omega_n \cos \omega_n t - B\omega_n \sin \omega_n t + \frac{m_u \varepsilon \omega_r^3 / k_{eq}}{1 - (\omega_r/\omega_n)^2} \cos \omega_r t, \tag{6}$$

and then applying the initial condition $\dot{y}_m(0) = v_{m0}$, we get

$$A\omega_n + \frac{m_u \varepsilon \omega_r^3 / k_{eq}}{1 - (\omega_r/\omega_n)^2} = v_{m0} \quad \Rightarrow \quad A = \frac{v_{m0}}{\omega_n} - \frac{\omega_r}{\omega_n} \frac{m_u \varepsilon \omega_r^2 / k_{eq}}{1 - (\omega_r/\omega_n)^2}. \tag{7}$$

Combining Eqs. (4), (5), and (7), the general solution becomes

$$y_m = \left[\frac{v_{m0}}{\omega_n} - \frac{\omega_r}{\omega_n} \frac{m_u \varepsilon \omega_r^2 / k_{eq}}{1 - (\omega_r/\omega_n)^2} \right] \sin \omega_n t + \frac{m_u \varepsilon \omega_r^2 / k_{eq}}{1 - (\omega_r/\omega_n)^2} \sin \omega_r t, \tag{8}$$

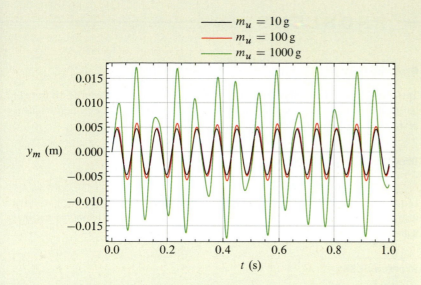

Figure 3. The response y_m of the motor and platform for increasing values of the eccentric mass m_u.

a plot of which is shown in Fig. 3. From this figure, we can see that the maximum amplitude of vibration for $m_u = 10\,\mathrm{g}$ is about 5 mm, for $m_u = 100\,\mathrm{g}$ it is about 6 mm, and for $m_u = 1\,\mathrm{kg}$ it is about 17 mm.

To compute MF, we take the amplitude of the particular solution and rearrange it so that ω_r and ω_n always appear as their ratio. Doing this gives

$$|y_{mp}| = \frac{m_u \varepsilon \omega_r^2 / k_{\text{eq}}}{1 - (\omega_r/\omega_n)^2} \quad \Rightarrow \quad |y_{mp}| = \frac{\frac{m_u \varepsilon}{m_m + m_p}(\omega_r/\omega_n)^2}{1 - (\omega_r/\omega_n)^2}. \tag{9}$$

Therefore, MF is given by

$$\boxed{\text{MF} = \frac{|y_{mp}|(m_m + m_p)}{m_u \varepsilon} = \frac{(\omega_r/\omega_n)^2}{1 - (\omega_r/\omega_n)^2},} \tag{10}$$

a plot of which is shown in Fig. 4. In our case, since $\omega_r = 1200\,\mathrm{rpm} = 125.7\,\mathrm{rad/s}$ and $\omega_n = 87.39\,\mathrm{rad/s}$, we have

$$\boxed{\text{MF} = -1.94.} \tag{11}$$

Discussion & Verification   The dimensions of all the terms in the general solution in Eq. (8) are length, as they should be. The amplitude of the vibration found by examining Fig. 3 is a few millimeters in all cases, which seems reasonable given the masses and stiffnesses involved. Finally, MF is dimensionless, again as it should be.

🔎 **A Closer Look**   It is interesting to compare the MF of the unbalanced motor plotted in Fig. 4, with the MF in Fig. 9.17, which applies to a mass that is forced directly (see Fig. 9.16). Notice that for small ω_r, the MF in Fig. 4 approaches 0 rather than 1, as it does in Fig. 9.17. This means that when the unbalanced rotor within the motor is spinning *very* slowly, it does not shake the motor and platform by any appreciable amount, which agrees with our intuition. On the other hand, when a mass is harmonically forced directly as in Fig. 9.16 and the forcing frequency is *very* small, the mass moves with the forcing and so the MF is 1.

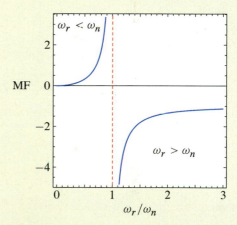

Figure 4
The MF for response of the unbalanced motor as given by Eq. (10).

PROBLEMS

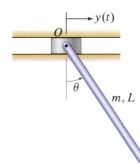

Figure P9.32

Problem 9.31

The magnification factor for a forced (undamped) harmonic oscillator is measured to be equal to 5. Determine the driving frequency of the forcing if the natural frequency of the system is 100 rad/s.

Problem 9.32

Suppose that equation of motion of a forced harmonic oscillator is given by $\ddot{x} + \omega_n^2 x = (F_0/m) \cos \omega_0 t$. Obtain the expression for the response of the oscillator, and compare it to the response presented in Eq. (9.42) (which is for a forced harmonic oscillator with the equation of motion given in Eq. (9.37)).

Problem 9.33

Derive the equations of motion for the unbalanced motor introduced in this section by applying Newton's second law to the center of mass of the system shown in Fig. 9.12.

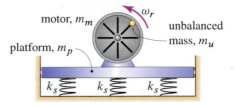

Figure P9.33 and P9.34

Problem 9.34

Determine the amplitude of vibration of the unbalanced motor we studied in Example 9.4 if the forcing frequency of the motor is $0.95\omega_n$.

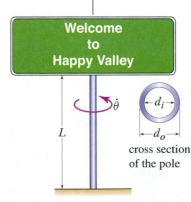

Figure P9.35

Problem 9.35

A uniform bar of mass m and length L is pinned to a slider at O. The slider is forced to oscillate horizontally according to $y(t) = Y \sin \omega_s t$. The system lies in the vertical plane.

(a) Derive the equation of motion of the bar for small angles θ.

(b) Determine the amplitude of steady-state vibration of the bar.

Problem 9.36

Figure P9.36

Consider a sign mounted on a circular hollow steel pole of length $L = 5$ m, outer diameter $d_o = 5$ cm, and inner diameter $d_i = 4$ cm. Aerodynamic forces due to wind provide a harmonic torsional excitation with frequency $f_0 = 3$ Hz and amplitude $M_0 = 10$ N·m about the z axis. The mass center of the sign lies on the central axis z of the pole. The mass moment of inertia of the sign is $I_z = 0.1$ kg·m^2. The torsional stiffness of the pole can be estimated as $k_t = \pi G_{st}(d_o^4 - d_i^4)/(32L)$, where G_{st} is the shear modulus of steel, which is 79 GPa. Neglecting the inertia of the pole, calculate the amplitude of vibration of the sign.

Problems 9.37 and 9.38

One of the propellers on the Beech King Air 200 is unbalanced such that the eccentric mass m_u is a distance R from the spin axis of the propeller. The propellers spin at a constant rate ω_r, and the mass of each engine is m_e (this includes the mass of the propeller). Assume that the wing is a uniform beam that is cantilevered at A, has mass m_w and bending stiffness EI, and whose mass center is at G. For each problem, evaluate your answers for $m_u = 3\,\text{oz}$, $m_e = 450\,\text{lb}$, $R = 5.1\,\text{ft}$, $\omega_r = 2000\,\text{rpm}$, $EI = 1.13 \times 10^{11}\,\text{lb·in.}^2$, $d = 8.7\,\text{ft}$, and $h = 10.9\,\text{ft}$.

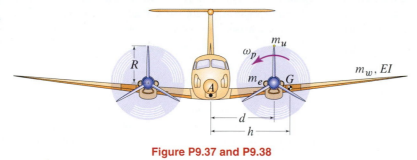

Figure P9.37 and P9.38

Problem 9.37   Neglect the mass of the wing and model the wing as done in Example 9.2. Determine the resonance frequency of the system, and find the MF for the given parameters.

Problem 9.38   Let the mass of the wing be $m_w = 350\,\text{lb}$, and model the wing as done in Example 9.2. Determine the resonance frequency of the system and find the MF for the given parameters.

Problem 9.39

An unbalanced motor is mounted at the tip of a rigid beam of mass m_b and length L. The beam is restrained by a torsional spring of stiffness k_t and an additional support of stiffness k located at the half length of the beam. In the static equilibrium position, the beam is horizontal and the torsional spring does not exert any moment on the beam. The mass of the motor is m_m, and the unbalance results in a harmonic excitation $F(t) = F \sin \omega_0 t$ in the vertical direction. Derive the equation of motion for the system assuming that θ is small.

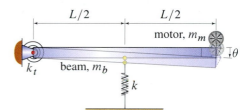

Figure P9.39

Problem 9.40

Revisit Example 9.4 and discuss whether it is possible to obtain the equation of motion of the system via the energy method.

Note: Concept problems are about *explanations*, not computations.

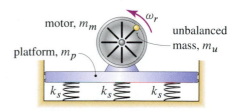

Figure P9.40

Problem 9.41

A fatigue-testing machine for electronic components consists of a platform with an unbalanced motor. Assume that the rotor in the motor spins at $\omega_0 = 3000\,\text{rpm}$, the mass of the platform is $m_p = 20\,\text{kg}$, the mass of the motor is $m_m = 15\,\text{kg}$, the eccentric mass is $m_u = 0.5\,\text{kg}$, and the equivalent stiffness of the platform suspension is $k = 5 \times 10^6\,\text{N/m}$. For the testing machine, the distance ε between the spin axis of the rotor and the location at which m_u is placed can be varied to obtain the desired vibration level. Calculate the range of values of ε that would provide amplitudes of the particular solution ranging from 0.1 mm to 2 mm.

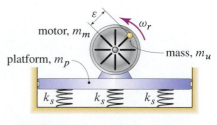

Figure P9.41

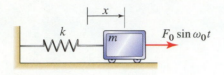

Figure P9.42

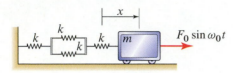

Figure P9.43

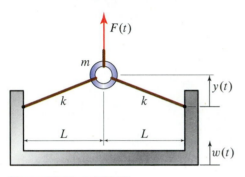

Figure P9.44 and P9.45

Problem 9.42

At time $t = 0$, a forced harmonic oscillator occupies position $x(0) = 0.1$ m and has a velocity $\dot{x}(0) = 0$. The mass of the oscillator is $m = 10$ kg, and the stiffness of the spring is $k = 1000$ N/m. Calculate the motion of the system if the forcing function is $F(t) = F_0 \sin \omega_0 t$, with $F_0 = 10$ N and $\omega_0 = 200$ rad/s.

Problem 9.43

The forced harmonic oscillator shown has a mass $m = 10$ kg. In addition the harmonic excitation is such that $F_0 = 150$ N and $\omega_0 = 200$ rad/s. If all sources of friction can be neglected, determine the spring constant k such that the magnification factor MF $= 5$.

Problem 9.44

A ring of mass m is attached by two linear elastic cords with elastic constant k and unstretched length $L_0 < L$ to a support, as shown. Assuming that the pretension in the cords is large, so that the cords' deflection due to the ring's weight can be neglected, find the linearized equation of motion for the case where $F(t) = F_0 \sin \omega_0 t$ and $w(t) = 0$ (i.e., the support is stationary). In addition, find the response of the system for $y(0) = 0$ and $\dot{y} = 0$.

Problem 9.45

A ring of mass m is attached by two linear elastic cords with elastic constant k and unstretched length $L_0 < L$ to a support, as shown. Assuming that the pretension in the cords is large, so that the cords' deflection due to the ring's weight can be neglected, find the linearized equation of motion for the case where $F(t) = 0$ and $w(t) = w_0 \sin \omega t$. In addition, find the response of the system for $y(0) = 0$ and $\dot{y}(0) = 0$.

Problem 9.46

Modeling the beam as a rigid uniform thin bar, ignoring the inertia of the pulleys, assuming that the system is in static equilibrium when the bar is horizontal, and assuming that the cord is inextensible and does not go slack, determine the linearized equation of motion of the system in terms of x, which is the position of A. Finally, determine the amplitude of the steady-state vibration of block A.

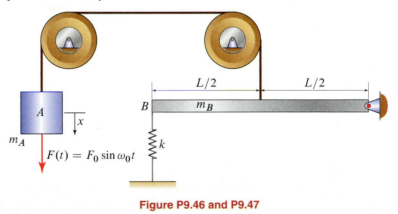

Figure P9.46 and P9.47

Problem 9.47

For the system in Prob. 9.46 determine the maximum forcing frequency ω_0 for steady state motion such that the cord does not go slack.

DESIGN PROBLEMS

Design Problem 9.2

The device shown can detect when the angular velocity and angular acceleration of a rigid body B achieve a combination of specified values. The device works using the principle that the vibration amplitude of the mass P depends on both the angular velocity and angular acceleration of the rigid body. When the angular velocity ω_B and angular acceleration α_B reach the appropriate combination, the mass P will contact the touch sensor, thus signaling that the specified values have been reached. With this as background, assume that the rigid body rotates in the horizontal plane with angular velocity ω_B and angular acceleration α_B, and that the mass P is constrained to move in the slot, which is at a distance d from the center of the disk. In addition, a linear elastic spring of constant k is attached to the mass such that the spring is undeformed when the mass is at $s = 0$.

(a) Derive the equation of motion for the mass P with s as the dependent variable.

(b) Assuming that the mass P is released from rest at $s = 0$, find the solution to the equation of motion found in (a), knowing that solution to ordinary differential equations of the type

$$\ddot{s} + \omega_n^2 s = D$$

is given by

$$s(t) = \frac{D}{\omega_n^2} + C_1 \cos \omega_n t + C_2 \sin \omega_n t,$$

where C_1 and C_2 are constants determined from the initial conditions and D is a known constant.

(c) Using the solution for $s(t)$ found above, for given values of d, k, m, ω_B, and α_B, determine the maximum distance from $s = 0$ that the mass P achieves in one cycle.

(d) For a disk-shaped rigid body B whose diameter is $1.5\,\mathrm{m}$, specify the mass of P (treat it as a particle), the spring constant k, the length h, and the distance d so that the touch sensor can detect when the rigid body reaches an angular velocity $\omega_{\mathrm{crit}} = 100\,\mathrm{rpm}$ for a constant angular acceleration $\alpha_B = 1\,\mathrm{rad/s}^2$.

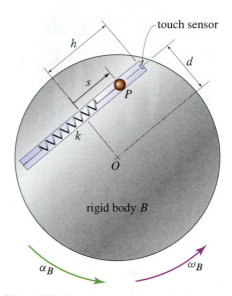

Figure DP9.2

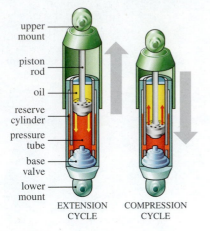

upper mount
piston rod
oil
reserve cylinder
pressure tube
base valve
lower mount

EXTENSION CYCLE   COMPRESSION CYCLE

Figure 9.24
A cutaway view of a typical shock absorber, which is used in many suspension systems.

9.3   Viscously Damped Vibration

All mechanical systems exhibit some energy dissipation or damping due to air drag, viscous fluids, friction, and other effects. If the damping is small enough, the undamped solutions obtained in Sections 9.1 and 9.2 will be in close agreement with the damped solution for a short period of time. On the other hand, if we need a solution for a longer period of time or if there is more damping, we need to resort to the solution of the equations that model damped mechanical systems.

In this section, we will consider *linear viscous damping*. This is damping in which the damping force is directly proportional and opposite in sign to the velocity of a body. This type of damping tends to occur when the energy dissipation is due to a fluid (e.g., oil, water, or air), as seen in the shock absorber in Fig. 9.24. In addition, even when the damping is due to other physical mechanisms, linear viscous damping can still be an effective model.

Viscously damped free vibration

The effect of viscous damping is usually modeled by an element called a *dashpot*, a schematic of which is shown in Fig. 9.25. Damping in a dashpot occurs

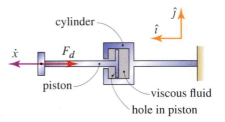

cylinder
$\hat{j}$
$\hat{i}$
$\dot{x}$ F_d
piston
viscous fluid
hole in piston

Figure 9.25. Schematic diagram of a dashpot illustrating its basic operation.

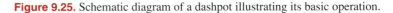

when the piston moves within the fluid-filled cylinder and forces the fluid to flow either around the piston or through one or more holes in it. This fluid motion results in energy dissipation. Referring to Fig. 9.25, if the piston is moving with velocity $\dot{x}$, then the damping force F_d on the piston is equal to

$$F_d = c\dot{x}, \qquad (9.48)$$

where c is a constant called the *coefficient of viscous damping*. This coefficient depends on the physical properties of the fluid and the geometry of the dashpot. The coefficient of viscous damping is expressed in lb·s/ft in U.S. Customary units and N·s/m in SI units.

Revisiting Example 3.3 on p. 200 in which a railcar runs into a large spring that was designed to stop it, a dashpot has now been added in parallel with the spring (see Fig. 9.26). We will assume that the railcar couples with the spring and dashpot after it hits them, which implies the FBD of the railcar shown in Fig. 9.27. Summing forces in the x direction, we find

$$\sum F_x: \quad -F_d - F_s = m\ddot{x}, \qquad (9.49)$$

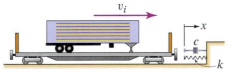

v_i
x
c
k

Figure 9.26
Railcar hitting a spring and dashpot. Recall from Example 3.3 that the railcar weighs 87 tons and is moving at 4 mph at impact, and $k = 22{,}300$ lb/ft.

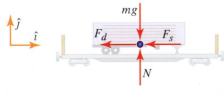

mg
$\hat{j}$
$\hat{i}$
F_d F_s
N

Figure 9.27
FBD of the railcar after it has hit and coupled with the spring and dashpot.

where F_d is the force due to the dashpot, F_s is the force due to the spring, and x measures the displacement of the spring from its equilibrium position. The forces F_d and F_s can be written as

$$F_d = c\dot{x} \quad \text{and} \quad F_s = kx, \tag{9.50}$$

which allows us to write Eq. (9.49) as

$$m\ddot{x} + c\dot{x} + kx = 0, \tag{9.51}$$

which is the *standard form of the viscously damped harmonic oscillator*. The theory of differential equations tells us that Eq. (9.51) is a linear, second-order, homogeneous, constant coefficient differential equation, and as such, it has solutions of the form

$$x = e^{\lambda t}, \tag{9.52}$$

where λ (the Greek letter lambda) is a constant to be determined. Substituting Eq. (9.52) into Eq. (9.51), we find

$$m\lambda^2 e^{\lambda t} + c\lambda e^{\lambda t} + k e^{\lambda t} = 0. \tag{9.53}$$

Factoring out $e^{\lambda t}$ in Eq. (9.53), we have

$$e^{\lambda t}(m\lambda^2 + c\lambda + k) = 0. \tag{9.54}$$

Since $e^{\lambda t}$ never vanishes, to have a solution to Eq. (9.51) we must have

$$m\lambda^2 + c\lambda + k = 0. \tag{9.55}$$

If λ is a root of this quadratic equation, called the *characteristic equation*, then $e^{\lambda t}$ is a solution to Eq. (9.51). The two roots of Eq. (9.55) are given by

$$\lambda_1 = -\frac{c}{2m} + \sqrt{\left(\frac{c}{2m}\right)^2 - \frac{k}{m}}, \qquad \lambda_2 = -\frac{c}{2m} - \sqrt{\left(\frac{c}{2m}\right)^2 - \frac{k}{m}}. \tag{9.56}$$

Referring to Eqs. (9.56), the theory of differential equations tells us that the general solution of Eq. (9.51) takes on one of three possible forms determined by the values of λ_1 and λ_2. Observe that the character of λ_1 and λ_2 depends on whether the term $(c/2m)^2 - k/m$ is positive, zero, or negative. Therefore, we introduce a special value of the damping coefficient called the *critical damping coefficient*, which we denote by c_c and define as the value of c that makes the term $(c/2m)^2 - k/m$ equal to zero, that is,

$$\left(\frac{c_c}{2m}\right)^2 - \frac{k}{m} = 0 \quad \Rightarrow \quad c_c = 2m\sqrt{\frac{k}{m}} = 2m\omega_n, \tag{9.57}$$

where $\omega_n = \sqrt{k/m}$. We now distinguish three cases based on whether $c > c_c$, $c = c_c$, or $c < c_c$.

Overdamped system ($c > c_c$)

When $c > c_c$, the term $(c/2m)^2 - k/m$ is positive. Thus, λ_1 and λ_2 are real, distinct, and negative. In this case, the general solution of Eq. (9.51) is

$$x = e^{-(c/2m)t}\left(A e^{t\sqrt{(c/2m)^2 - k/m}} + B e^{-t\sqrt{(c/2m)^2 - k/m}}\right), \tag{9.58}$$

where A and B are constants that are determined from the initial conditions of the system. The motion represented by Eq. (9.58) is characterized by decaying exponentials, the system does not vibrate, and there is no period associated with the motion (see Fig. 9.28). This type of system is said to be *overdamped*.

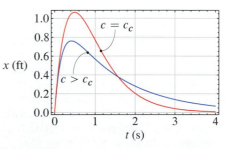

Figure 9.28
The position of the railcar as a function of time after it impacts the spring and dashpot. The blue curve is overdamped with $c = 35,000\ \text{lb·s/ft}$, and the red curve is critically damped with $c = c_c \approx 21,950\ \text{lb·s/ft}$.

Critically damped system ($c = c_c$)

When $c = c_c$, the term $(c/2m)^2 - k/m$ is zero. Thus $\lambda_1 = \lambda_2 = -c_c/2m$. In this case, the general solution of Eq. (9.51) is

$$x = (A + Bt)e^{-\omega_n t}, \qquad (9.59)$$

where, again, A and B are constants that are determined from the initial conditions. When $c = c_c$, then c has the smallest value for which no vibration occurs and the system is said to be *critically damped*. Referring to Fig. 9.28, notice that a critically damped system approaches equilibrium faster than an overdamped system. Critically damped systems are of great interest in engineering applications since they approach equilibrium in the minimum possible time.

Underdamped system ($c < c_c$)

When $c < c_c$, the term $(c/2m)^2 - k/m$ is negative. Thus, λ_1 and λ_2 are complex since they involve the square root of a negative quantity. In this case, the general solution of Eq. (9.51) is

$$x = e^{-(c/2m)t}(A \sin \omega_d t + B \cos \omega_d t), \qquad (9.60)$$

where A and B are determined from the initial conditions and ω_d is the *damped natural frequency*, which is given by

$$\omega_d = \sqrt{\frac{k}{m} - \left(\frac{c}{2m}\right)^2} = \omega_n \sqrt{1 - (c/c_c)^2}, \qquad (9.61)$$

and we recall that $\omega_n = \sqrt{k/m}$, $c_c = 2m\omega_n$. A system for which $c < c_c$ is said to be *underdamped*, and for any such system, Eq. (9.61) implies that ω_d is *always* less than ω_n. Using the definition of ω_d given in Eq. (9.61), the *period of damped vibration* is given by

$$\tau_d = 2\pi/\omega_d. \qquad (9.62)$$

Note that the solution for x given in Eq. (9.60) can also be written as

$$x = De^{-(c/2m)t} \sin(\omega_d t + \phi), \qquad (9.63)$$

where D and ϕ are constants determined by the initial conditions. The solution in Eq. (9.63) has been plotted in Fig. 9.29 using the data specified in Fig. 9.26 and a value of c that makes the system underdamped.

Damping ratio

In practice, the three cases just discussed are often classified in terms of a nondimensional parameter called the *damping ratio* or *damping factor*, which is usually denoted by the symbol ζ (the Greek letter zeta) and is defined as

$$\zeta = c/c_c. \qquad (9.64)$$

Using ζ, the standard form of the viscously damped harmonic oscillator in Eq. (9.51) is rewritten as

$$\ddot{x} + 2\zeta\omega_n x + \omega_n^2 x = 0. \qquad (9.65)$$

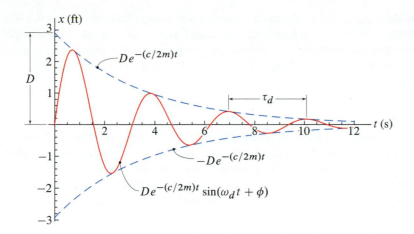

Figure 9.29. The position of the railcar as a function of time after it impacts the spring and dashpot. The red curve is the underdamped solution for $c = 3000 \, \text{lb·s/ft}$.

Overdamped system ($\zeta > 1$). In terms of ζ, an overdamped system is characterized by $\zeta > 1$, and the general solution in Eq. (9.58) is rewritten as

$$x = e^{-\zeta \omega_n t} \left(A e^{\sqrt{\zeta^2 - 1}\,\omega_n t} + B e^{-\sqrt{\zeta^2 - 1}\,\omega_n t} \right). \qquad (9.66)$$

Critically damped system ($\zeta = 1$). In terms of ζ, a critically damped system is characterized by $\zeta = 1$, and the general solution in Eq. (9.59) is unchanged.

Underdamped system ($\zeta < 1$). In terms of ζ, an underdamped system is characterized by $\zeta < 1$, and the solutions in Eqs. (9.60) and (9.63) are rewritten as

$$x = e^{-\zeta \omega_n t} (A \sin \omega_d t + B \cos \omega_d t) = D e^{-\zeta \omega_n t} \sin(\omega_d t + \phi), \qquad (9.67)$$

respectively, where, referring to Eq. (9.61), ω_d is expressed in terms of ζ as

$$\boxed{\omega_d = \omega_n \sqrt{1 - \zeta^2}.} \qquad (9.68)$$

Viscously damped forced vibration

Here we consider the case of the vibration of a one degree of freedom system that is both damped *and* forced. Consider, for example, the simple system shown in Fig. 9.30, in which we have simply added a dashpot to the system in Fig. 9.16 on p. 693. The equation of motion for this system is given by

$$\boxed{m\ddot{x} + c\dot{x} + kx = F_0 \sin \omega_0 t,} \qquad (9.69)$$

where x is measured from the equilibrium position of the mass. Using the damping ratio ζ, Eq. (9.69) can be written as

$$\boxed{\ddot{x} + 2\zeta \omega_n \dot{x} + \omega_n^2 x = \frac{F_0}{m} \sin \omega_0 t.} \qquad (9.70)$$

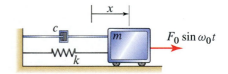

Figure 9.30
A simple damped harmonic oscillator that is harmonically forced.

If, rather than the harmonic forcing being applied to m, the support in Fig. 9.30 is displaced harmonically according to $Y \sin \omega_0 t$, we cannot simply

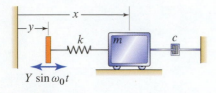

$Y \sin \omega_0 t$

Figure 9.31
Harmonic support displacement of a viscously damped harmonic oscillator.

replace F_0 with kY, similar to what we did in Eq. (9.47) in Section 9.2 (see Prob. 9.55 for way to handle this situation). On the other hand, if the dashpot remains attached to a fixed support and we harmonically displace the support to which the spring is attached (see Fig. 9.31), then the equation of motion is of the form

$$m\ddot{x} + c\dot{x} + kx = kY \sin \omega_0 t, \qquad (9.71)$$

so that we *can* replace F_0 by kY in all the corresponding solutions.

As was the case with undamped forced vibration in Section 9.2, the solution to Eq. (9.69) or Eq. (9.70) is the sum of the complementary solution and a particular solution. As a reminder, the complementary solution x_c is the solution to the associated homogeneous equation, which is Eq. (9.51) and for which we found the solution to depend on the level of damping (i.e., whether c is greater than, equal to, or less than the critical damping c_c). Regardless of the level of damping, this complementary solution will die out in time, and thus its contribution is *transient* (see p. 694). As discussed in Section 9.2 on p. 694, the forced vibration associated with a particular solution x_p will exist as long as the forcing does and is thus a *steady-state vibration*. Since we have already found the complementary solution, it will be the particular solution that we focus on here.

As with undamped forced vibration, the response of a damped forced system should also resemble the forcing. Therefore, we will assume a particular solution of either of the following forms

$$\boxed{x_p = A \sin \omega_0 t + B \cos \omega_0 t = D \sin(\omega_0 t - \phi),} \qquad (9.72)$$

where, in the first expression, A and B are constants to be determined and, in the second expression, D and ϕ are constants to be determined and D is assumed to be a positive quantity.* While either of these expressions can be used, the latter gives a more easily interpreted result since the amplitude D and phase ϕ are immediately apparent, and so we will use it and proceed to determine D and ϕ. Substituting the second expression for x_p from Eq. (9.72) into Eq. (9.69), we obtain

$$-Dm\omega_0^2 \sin(\omega_0 t - \phi) + Dc\omega_0 \cos(\omega_0 t - \phi) + Dk \sin(\omega_0 t - \phi)$$
$$= F_0 \sin \omega_0 t. \quad (9.73)$$

Using the trigonometric identities $\sin(\alpha - \beta) = \sin \alpha \cos \beta - \cos \alpha \sin \beta$ and $\cos(\alpha - \beta) = \cos \alpha \cos \beta + \sin \alpha \cos \beta$ and then collecting terms, we obtain

$$D(-m\omega_0^2 \cos \phi + c\omega_0 \sin \phi + k \cos \phi) \sin \omega_0 t$$
$$+ D(m\omega_0^2 \sin \phi + c\omega_0 \cos \phi - k \sin \phi) \cos \omega_0 t = F_0 \sin \omega_0 t. \quad (9.74)$$

Since this equation must be true for all time, we can equate the coefficients of $\sin \omega_0 t$ and $\cos \omega_0 t$ to obtain two equations for the unknowns D and ϕ. Doing this for $\cos \omega_0 t$ allows us to solve for $\tan \phi$ as

$$\boxed{\tan \phi = \frac{c\omega_0}{k - m\omega_0^2} = \frac{2(c/c_c)(\omega_0/\omega_n)}{1 - (\omega_0/\omega_n)^2} = \frac{2\zeta\omega_0/\omega_n}{1 - (\omega_0/\omega_n)^2},} \qquad (9.75)$$

---

* In Eq. (9.72) we have used a sine function of the form $\sin(\omega_0 t - \phi)$ instead of $\sin(\omega_0 t + \phi)$ because it results in a more convenient expression for $\tan \phi$.

where the definitions of ω_n, c_c, and ζ, in Eqs. (9.4), Eq. (9.57), and Eq. (9.64), respectively, have been used. The phase angle ϕ, which is plotted in Fig. 9.32 as a function of the frequency ratio ω_0/ω_n for different values of ζ, represents the amount of a cycle by which the response of the system lags the forcing applied to it. Equating the coefficients of $\sin \omega_0 t$, we obtain

$$D = \frac{F_0/\cos \phi}{k - m\omega_0^2 + c\omega_0 \tan \phi} = \frac{F_0/\cos \phi}{c\omega_0/\tan \phi + c\omega_0 \tan \phi}, \tag{9.76}$$

where, to get the second expression for D, we have used Eq. (9.75). Now multiplying the numerator and denominator by $\tan \phi$ and noting that $1 + \tan^2 \phi = 1/\cos^2 \phi$, we obtain

$$D = \frac{F_0 \sin \phi}{c\omega_0} = \frac{F_0}{c\omega_0} \left\{ \frac{2\zeta\omega_0/\omega_n}{\sqrt{[1 - (\omega_0/\omega_n)^2]^2 + (2\zeta\omega_0/\omega_n)^2}} \right\}, \tag{9.77}$$

where we have used the trigonometric identity that if $\alpha = \tan^{-1} x$, then $\sin \alpha = x/\sqrt{1 + x^2}$. Again using the definitions of ω_n, c_c, and ζ as above, D simplifies to

$$D = \frac{F_0/k}{\sqrt{[1 - (\omega_0/\omega_n)^2]^2 + (2\zeta\omega_0/\omega_n)^2}}. \tag{9.78}$$

Now that we have the particular solution, we will, as was done in Section 9.2, define a *magnification factor* for a damped forced harmonic oscillator as the ratio of the amplitude of steady-state vibration D to the static deflection F_0/k, thus obtaining

$$MF = \frac{D}{F_0/k} = \frac{1}{\sqrt{[1 - (\omega_0/\omega_n)^2]^2 + (2\zeta\omega_0/\omega_n)^2}}. \tag{9.79}$$

A plot of the MF is shown in Fig. 9.33 for various values of the damping ratio ζ. Figure 9.33 illustrates some important features of the behavior of a damped forced harmonic oscillator:

* The magnitude of the oscillation can be small in two ways: by keeping the forcing frequency ω_0 away from the natural frequency and/or by increasing the amount of damping, i.e., increasing ζ.

* As the amount of damping is increased, the peak in the MF moves farther to the left away from $\omega_0/\omega_n = 1$. The peak for any given value of ζ can be found by using the usual technique from calculus to find the maximum value of a function (see Prob. 9.52).

* Comparing Fig. 9.33 with Fig. 9.17, observe that the effect of damping, while quite noticeable near the resonance frequency, becomes very small for values of ω_0/ω_n away from 1.

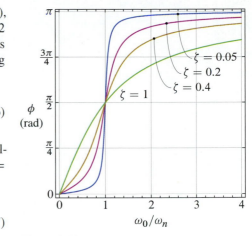

Figure 9.32
The phase angle ϕ as a function of the frequency ratio ω_0/ω_n.

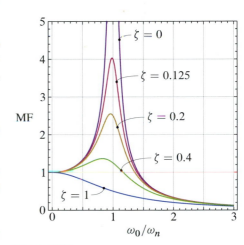

Figure 9.33
The MF as a function of the frequency ratio ω_0/ω_n for various values of the damping ratio ζ.

End of Section Summary

Viscously damped free vibration. The standard form of the equation of motion for a one degree of freedom viscously damped harmonic oscillator is

Eq. (9.51), p. 705

$$m\ddot{x} + c\dot{x} + kx = 0,$$

where m is the mass, c is the *coefficient of viscous damping*, and k is the linear spring constant. The character of the solution to this equation depends on the amount of damping relative to a specific amount of damping called the *critical damping coefficient*, which is defined as

Eq. (9.57), p. 705

$$c_c = 2m\sqrt{\frac{k}{m}} = 2m\omega_n,$$

where $\omega_n = \sqrt{k/m}$. In particular, if $c \geq c_c$, then the motion is nonoscillatory, whereas if $c < c_c$, then the motion is oscillatory. When $c > c_c$, the system is said to be *overdamped*, and the solution is given by

Eq. (9.58), p. 705

$$x = e^{-(c/2m)t}\left(Ae^{t\sqrt{(c/2m)^2-k/m}} + Be^{-t\sqrt{(c/2m)^2-k/m}}\right),$$

where A and B are constants to be determined from the initial conditions. When $c = c_c$, the system is said to be *critically damped*, and the solution is given by

Eq. (9.59), p. 706

$$x = (A + Bt)e^{-\omega_n t},$$

where, again, A and B are constants to be determined from the initial conditions. Finally, when $c < c_c$, the system is said to be *underdamped*, and the solution is given by

Eq. (9.60), p. 706

$$x = e^{-(c/2m)t}(A\sin\omega_d t + B\cos\omega_d t),$$

or, equivalently, by

Eq. (9.63), p. 706

$$x = De^{-(c/2m)t}\sin(\omega_d t + \phi),$$

where A and B are constants to be determined from the initial conditions in the first solution, D and ϕ are analogous constants in the second solution, and ω_d is the *damped natural frequency*, which is given by

Eq. (9.61), p. 706, and Eq. (9.68), p. 707

$$\omega_d = \sqrt{\frac{k}{m} - \left(\frac{c}{2m}\right)^2} = \omega_n\sqrt{1 - (c/c_c)^2} = \omega_n\sqrt{1 - \zeta^2},$$

where $\zeta = c/c_c$ is the *damping ratio*.

Viscously damped forced vibration. The standard form of a viscously damped forced harmonic oscillator is

Eq. (9.69), p. 707

$$m\ddot{x} + c\dot{x} + kx = F_0 \sin \omega_0 t,$$

where F_0 is the amplitude of the forcing function and ω_0 is the frequency of the forcing function. When expressed using the damping ratio ζ, the equation above takes on the form

Eq. (9.70), p. 707

$$\ddot{x} + 2\zeta\omega_n\dot{x} + \omega_n^2 x = \frac{F_0}{m}\sin\omega_0 t.$$

The general solution to either of these equations is the sum of the complementary solution and a particular solution. The complementary solution is transient; i.e., it vanishes as time increases. The particular or steady-state solution is of the form

Eq. (9.72), p. 708

$$x_p = D\sin(\omega_0 t - \phi),$$

where ϕ and D are given by

Eq. (9.75), p. 708, and Eq. (9.78), p. 709

$$\tan\phi = \frac{c\omega_0}{k - m\omega_0^2} = \frac{2(c/c_c)(\omega_0/\omega_n)}{1 - (\omega_0/\omega_n)^2} = \frac{2\zeta\omega_0/\omega_n}{1 - (\omega_0/\omega_n)^2},$$

$$D = \frac{F_0/k}{\sqrt{[1 - (\omega_0/\omega_n)^2]^2 + (2\zeta\omega_0/\omega_n)^2}}.$$

The *magnification factor* for a damped, forced harmonic oscillator is

Eq. (9.79), p. 709

$$\mathrm{MF} = \frac{D}{F_0/k} = \frac{1}{\sqrt{[1 - (\omega_0/\omega_n)^2]^2 + (2\zeta\omega_0/\omega_n)^2}},$$

a plot of which is shown in Fig. 9.34 for various values of the damping ratio ζ.

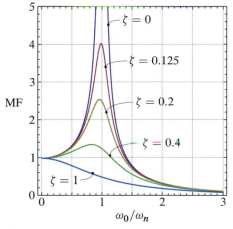

Figure 9.34
Figure 9.33 repeated. The MF as a function of the frequency ratio ω_0/ω_n for various values of the damping ratio ζ.

EXAMPLE 9.5   *Critically Damped Free Vibration of a Gate*

Figure 1
A carnival ride with a gate that counts riders.

The carnival ride in Fig. 1 has a gate, which is shown in Fig. 2, that is used to count people entering the ride. The gate has a linear elastic torsional spring of stiffness k and a torsional damper with constant c at the pin O that control how it returns to the closed position after being opened. Determine k and c so that the gate is critically damped and returns to within $\theta = 4°$ of the closed position less than 2.5 s after starting from rest at $\theta = 80°$. Model the gate as a thin bar of mass m_b and length L with a point mass m at its end. Neglect friction at the pin O as well as air resistance.

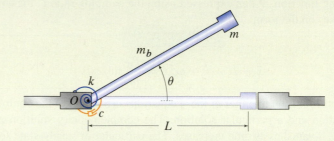

Figure 2. A carnival ride gate for which $L = 42$ in., the weight of the thin bar is 4 lb, and the weight of the point mass is 3 lb. The gate lies in the horizontal plane.

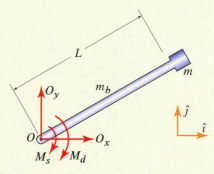

Figure 3
FBD of the carnival ride gate in Fig. 2.

SOLUTION

Road Map & Modeling   We will first need to derive the equation of motion for the gate in the standard form of Eq. (9.51), using the FBD of the arm shown in Fig. 3. Once we have the equation of motion, since we want the system to be critically damped and we know its initial conditions, we can find its response using Eq. (9.59). From the response, we can determine the value of k needed to get the arm closed in the required time and then use that to find the c needed to critically damp the arm.

Governing Equations

Balance Principles   Summing moments about point O, we obtain

$$\sum M_O: \quad -M_s - M_d = I_O \alpha_{\text{gate}}, \tag{1}$$

where M_s is the restoring moment due to the spring, M_d is the damping moment due to the dashpot, and α_{gate} is the angular acceleration of the gate. The mass moment of inertia of the gate with respect to O is computed as

$$I_O = mL^2 + \tfrac{1}{12}m_b L^2 + m_b(L/2)^2 = \left(\tfrac{1}{3}m_b + m\right)L^2. \tag{2}$$

Force Laws   The moment laws for the spring and dashpot are

$$M_s = k\theta \quad \text{and} \quad M_d = c\dot{\theta}. \tag{3}$$

Kinematic Equations   The angular acceleration of the gate can be written in terms of θ as $\alpha_{\text{gate}} = \ddot{\theta}$.

Computation   Substituting the kinematic relation as well as Eqs. (2) and (3) into Eq. (1) and rearranging, we obtain the equation of motion of the gate as

$$\left(\tfrac{1}{3}m_b + m\right)L^2\ddot{\theta} + c\dot{\theta} + k\theta = 0. \tag{4}$$

This implies that the natural frequency ω_n is given by

$$\omega_n = \sqrt{\frac{k}{\left(\frac{1}{3}m_b + m\right)L^2}} = 0.7788\sqrt{k} \text{ rad/s.} \qquad (5)$$

To determine k, we will enforce the condition requiring the gate to be within $4°$ of the closed position in 2.5 s or less to the solution of Eq. (4), which is given by $\theta = (A + Bt)e^{-\omega_n t}$. But first we need to find A and B. Enforcing the condition that $\theta(0) = 80° = 1.396$ rad, we obtain

$$\theta(0) = A = 1.396 \text{ rad.} \qquad (6)$$

We now enforce the condition that $\dot{\theta}(0) = 0$ rad/s, which gives

$$\dot{\theta} = Be^{-\omega_n t} + (A + Bt)\left(-\omega_n e^{-\omega_n t}\right) \quad \Rightarrow \quad \dot{\theta}(0) = B - A\omega_n = 0$$

$$\Rightarrow \quad B = 1.396\omega_n. \quad (7)$$

Substituting Eqs. (5)–(7) into the critically damped solution for θ, we obtain

$$\theta = \left(1.396 + 1.087t\sqrt{k}\right)e^{-0.7788t\sqrt{k}}. \qquad (8)$$

To obtain k, we will say that we want $\theta(2.5) = 4° = 0.06981$ rad, which gives

$$0.06981 = \left(1.396 + 2.718\sqrt{k}\right)e^{-1.947\sqrt{k}}, \qquad (9)$$

which is a transcendental equation for k. This equation can be solved using almost any mathematical software package. Solving Eq. (9) gives

$$\boxed{k = 5.94 \text{ ft·lb/rad.}} \qquad (10)$$

Now that we have k, the condition for critical damping in Eq. (9.57) tells us that the damping coefficient must be given by

$$c_c = 2\left(\frac{1}{3}m_b + m\right)L^2\omega_n \quad \Rightarrow \quad \boxed{c_c = 6.26 \text{ ft·lb·s.}} \qquad (11)$$

A plot of the solution in Eq. (8) (using k from Eq. (10)) can be found in Fig. 4.

Discussion & Verification The dimensions of each of the two torsional constants found in Eqs. (10) and (11) are as they should be.

🔍 **A Closer Look** Note that the gate could be returned to the closed position even more quickly by increasing the value of k (compare the curves for two different values of k in Fig. 4). Unfortunately, this has potentially undesirable consequences. Referring to Fig. 5, we see that larger values of k result in the angular velocity of the gate being larger for *every* value of θ. This means that someone going through the gate before it closes will get hit harder by the gate (compare $\dot{\theta}$ at points A and B in Fig. 5). In addition, referring to Eq. (11), we see that larger values of k mean that the damping coefficient must be larger to achieve critical damping. This would likely imply that the damping mechanism must be more substantial and more expensive.

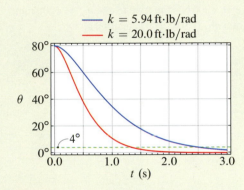

Figure 4
Plot of response of the gate for two different values of the torsional spring constant.

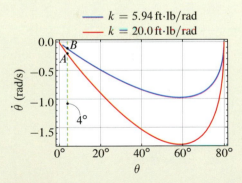

Figure 5
A plot of the angular velocity of the gate $\dot{\theta}$ as a function of its position θ for two different values of the torsional spring constant.

EXAMPLE 9.6   *Response and MF for a Rotating Unbalanced Mass*

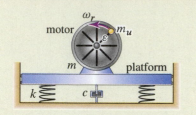

Figure 1

An unbalanced motor on a platform that is elastically supported, and whose vertical motion is damped by dashpots. The values of the system parameters are $m = 55$ kg, $k = 420{,}000$ N/m, $c = 4000$ N·s/m, $\varepsilon = 15$ cm, $\omega_r = 1200$ rpm, and m_u is 10 g, 100 g, or 1000 g.

If we add dashpots in parallel with the springs that are supporting the platform to the unbalanced motor we studied in Section 9.2, we obtain the system shown in Fig. 1. We will assume that all the springs supply a total spring constant k, the dashpots provide a total damping coefficient c, and the combined mass of the motor and of the platform is m. Using these definitions and referring to Section 9.2, we can show that the equation of motion for the unbalanced motor is (see Prob. 9.54)

$$\ddot{y} + 2\zeta\omega_n\dot{y} + \omega_n^2 y = \frac{m_u\varepsilon\omega_r^2}{m}\sin\omega_r t, \tag{1}$$

where y is the vertical position of the motor measured from its static equilibrium position, m_u is the eccentric mass (m includes m_u), ε is the distance from the unbalanced mass to the rotor axis, and ω_r is the angular velocity of the rotor. Using $c = 4000$ N·s/m, determine and plot the steady-state solution, using the parameters given in Example 9.4. In addition, determine and plot the MF for the unbalanced motor and compare it with the MF shown in Fig. 9.33, which applies to Eq. (9.70).

SOLUTION

Road Map & Modeling   We know that the steady-state solution to an equation of the form in Eq. (1) is given by Eqs. (9.72), (9.75), and (9.78). Therefore, we need to interpret the forcing amplitude on the right-hand side of Eq. (1) in that context to obtain the steady-state solution. The MF is found from the amplitude of the steady-state solution, so we will look at the expression for D that we obtain from Eq. (9.78) after interpreting the right-hand side of Eq. (1). Once we write the amplitude as a function of ω_r/ω_n and ζ, we will have the desired MF.

Governing Equations   As discussed above, the steady-state solution is given by Eq. (9.72), i.e.,

$$y_{ss} = D\sin(\omega_r t - \phi), \tag{2}$$

where D is given by Eq. (9.78), i.e.,

$$D = \frac{F_0/k}{\sqrt{[1 - (\omega_0/\omega_n)^2]^2 + (2\zeta\omega_0/\omega_n)^2}}, \tag{3}$$

and ϕ is given by Eq. (9.75), i.e.,

$$\tan\phi = \frac{c\omega_0}{k - m\omega_0^2} = \frac{2\zeta\omega_0/\omega_n}{1 - (\omega_0/\omega_n)^2}, \tag{4}$$

in which $\omega_n = \sqrt{k/m} = 87.39$ rad/s and $\zeta = c/(2m\omega_n) = 0.4161$. Comparing Eq. (1) with Eq. (9.70) on p. 707, we see that

$$\omega_0 = \omega_r \quad \text{and} \quad F_0 = m_u\varepsilon\omega_r^2. \tag{5}$$

Computation   The steady-state solution is now found by substituting Eqs. (3)–(5) into Eq. (2). Doing this and then substituting in all given parameters, we find

$$\boxed{y_{ss} = 0.00352 m_u \sin(125.7t + 0.842) \text{ m},} \tag{6}$$

which is plotted in Fig. 2 for the three given values of m_u.

To find the MF, we substitute Eqs. (5) into Eq. (3) to obtain

$$D = \frac{m_u\varepsilon\omega_r^2/k}{\sqrt{[1 - (\omega_r/\omega_n)^2]^2 + (2\zeta\omega_r/\omega_n)^2}}. \tag{7}$$

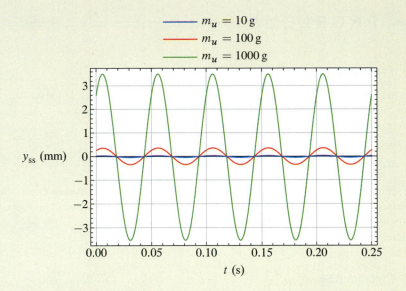

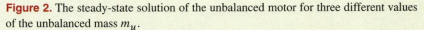

Figure 2. The steady-state solution of the unbalanced motor for three different values of the unbalanced mass m_u.

Focusing on the numerator, we see that we can write it as

$$\frac{m_u \varepsilon \omega_r^2}{k} = \frac{m_u \varepsilon \omega_r^2}{k} \frac{m}{m} = \frac{m_u \varepsilon}{m} \frac{m}{k} \omega_r^2 = \frac{m_u \varepsilon}{m} \frac{\omega_r^2}{\omega_n^2}, \tag{8}$$

where, to obtain the last equality, we have used the fact that $m/k = 1/\omega_n^2$. Substituting Eq. (8) into Eq. (7) and moving $m_u \varepsilon / m$ to the left-hand side, we obtain the MF as

$$\boxed{\text{MF} = \frac{mD}{m_u \varepsilon} = \frac{(\omega_r/\omega_n)^2}{\sqrt{[1 - (\omega_r/\omega_n)^2]^2 + (2\zeta\omega_r/\omega_n)^2}},} \tag{9}$$

a plot of which is shown in Fig. 3 for various values of ζ.

Discussion & Verification Careful examination of the steady-state response given in Eq. (2), with Eqs. (3)–(5) substituted in, reveals that it has the dimension of length as should be expected. In addition, we see in Fig. 2 that as we increase the amount of the unbalanced mass m_u, the oscillation amplitude increases as expected. The MF in Eq. (9) is dimensionless, as it should be.

🔍 **A Closer Look** Comparing Fig. 3 with Fig. 9.33 on p. 709, we see that for low forcing frequencies:

- When a harmonic oscillator is forced by applying the forcing directly to the mass, the mass follows the forcing (MF → 1 in Fig. 9.33).
- When a harmonic oscillator is forced by an internal rotating unbalanced mass, the mass barely moves (MF → 0 in Fig. 3).

For high forcing frequencies:

- When a harmonic oscillator is forced by applying the forcing directly to the mass, the mass barely moves (MF → 0 in Fig. 9.33).
- When a harmonic oscillator is forced by an internal rotating unbalanced mass, the mass moves with the forcing (MF → 1 in Fig. 3).

These observations are consistent with our intuition regarding the behavior of these systems.

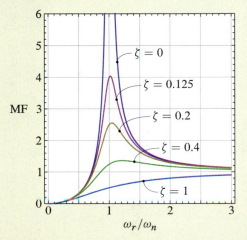

Figure 3
The unbalanced motor MF as a function of the frequency ratio ω_r/ω_n for various values of the damping ratio ζ.

PROBLEMS

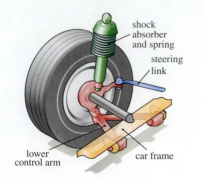

Figure P9.48

Problem 9.48

In the design of a MacPherson strut suspension, what would you choose for the damping ratio ζ? Explain your answer in terms of automotive ride and comfort.
Note: Concept problems are about *explanations*, not computations.

Problem 9.49

For identical systems, one with damping and the other without, would you expect the period of damped vibration to be greater, less than, or equal to the period of undamped vibration? Explain your answer.
Note: Concept problems are about *explanations*, not computations.

Problem 9.50

A vibration test is performed on a structure, in which both the magnification factor MF and the phase angle ϕ are recorded as a function of excitation frequency ω_0. After the test, it is discovered that, for some unfortunate reason, the recording of the magnification factor data is corrupted so that only the phase angle data is available for analysis. Is it possible to determine the resonant frequency from the available data? What can be inferred about the amount of damping in the system from the phase data?
Note: Concept problems are about *explanations*, not computations.

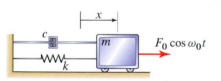

Figure P9.51

Problem 9.51

Suppose that equation of motion of a damped forced harmonic oscillator is given by $\ddot{x} + 2\zeta\omega_n\dot{x} + \omega_n^2 x = (F_0/m)\cos\omega_0 t$, where x is measured from the equilibrium position of the system. Obtain the expression for the amplitude of the steady-state response of the oscillator, and compare it with the expression presented in Eq. (9.78) (which is for a system with equation of motion $\ddot{x} + 2\zeta\omega_n\dot{x} + \omega_n^2 x = (F_0/m)\sin\omega_0 t$).

Problem 9.52

Differentiate Eq. (9.79) with respect to ω_0/ω_n and set the result equal to zero to determine the frequency ω_0 at which peaks in the MF curve occur as a function of ζ and ω_n. Use this result to show that the peak always occurs at $\omega_0/\omega_n \leq 1$. Finally, determine the value of ζ for which the MF has no peak.

Problem 9.53

Calculate the response described by the equations listed below, in which x is measured in feet and time is measured in seconds.

(a) $5\ddot{x} + 10\dot{x} + 100x = 0$, with $x(0) = 0.1$ and $\dot{x}(0) = -0.1$

(b) $3\ddot{x} + 15\dot{x} + 12x = 0$, with $x(0) = 0$ and $\dot{x}(0) = 0.5$

(c) $\ddot{x} + 10\dot{x} + 25x = 0$, with $x(0) = 0.15$ and $\dot{x}(0) = 0$

(d) $25\ddot{x} + 200\dot{x} + 1500x = 0$, with $x(0) = 0.01$ and $\dot{x}(0) = 0$

Problem 9.54

Derive the equation of motion given in Eq. (1) of Example 9.6 for the system in that example. The independent variable y is measured from the equilibrium position of the system, m is the mass of the motor and platform, c is the total damping coefficient of the dashpots, k is the total constant of the linear elastic springs, ω_r is the angular velocity of the unbalanced rotor, ε is the distance of the eccentric mass from the rotor axis, and m_u is the eccentric mass. Note that m *includes* the eccentric mass so that the nonrotating mass is equal to $m - m_u$.

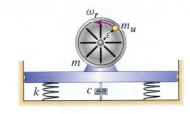

Figure P9.54

Problem 9.55

The mass m is coupled to the support A, which is displacing harmonically according to $y = Y \sin \omega_0 t$, by the linear elastic spring of constant k and the dashpot with constant c.

(a) Derive its equation of motion, using x as the independent variable, and explain in what way the resulting equation of motion is not in the form of Eq. (9.71).

(b) Next, let $z = x - y$ and substitute it into the equation of motion found in part (a). After doing so, show that you obtain an equation of motion in z that is of the same form as Eq. (9.71).

(c) Find the steady-state solution to the equation of motion found in part (b) and then using that, determine the steady-state solution for x.

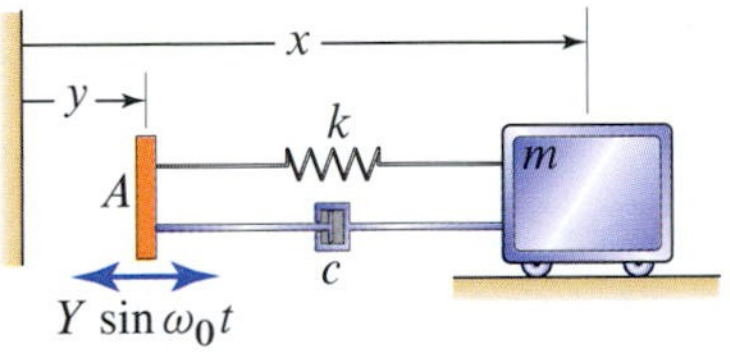

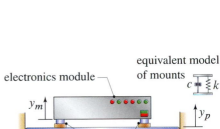

Figure P9.55

Problem 9.56

A module with sensitive electronics is mounted on a panel that vibrates due to excitation from a nearby diesel generator. To prevent fatigue failure, the module is placed on vibration-absorbing mounts. The displacement of the panel is measured to be $y_p(t) = y_0 \sin \omega_0 t$, where $y_0 = 0.001$ m, $\omega_0 = 300$ rad/s, and the time t is measured in seconds. Letting the mass of the electronic module be $m = 0.5$ kg, calculate the amplitude of the vibration of the module if the equivalent stiffness and damping coefficients for all the mounts combined are $k = 10{,}000$ N/m and $c = 40$ N·s/m, respectively.

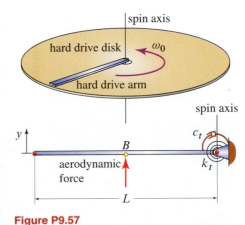

Figure P9.56

Problem 9.57

A hard drive arm undergoes flow-induced vibration caused by the vortices of air produced by a platter that rotates at $\omega_0 = 10{,}000$ rpm. The arm has length $L = 0.037$ m and mass $m = 0.00075$ kg, and it is made from aluminum with a modulus of elasticity $E = 70$ GPa. In addition, assume that the cross section of the arm has an area moment of inertia $I_{cs} = 8.5 \times 10^{-14}$ m^4. Following the steps in Example 9.2 on p. 682, the arm can be modeled as a rigid rod that is pinned at one end and is restrained by a torsional spring with equivalent spring constant $k_t = 3EI_{cs}/L$. In addition to the torsional spring, assume that the arm's motion is affected by a torsional damper with torsional damping coefficient c_t. Assuming that the damping ratio is $\zeta = 0.02$ and that the vortices produce an aerodynamic force with the same frequency as the rotation of the platter, determine the amplitude of the aerodynamic force needed to cause a steady-state vibration amplitude of 0.0001 m at the tip of the arm. Assume that the aerodynamic force is applied at the midpoint B of the hard drive. What vibration amplitude will result if the same excitation is applied to a hard drive arm assembly with the damping ratio of 0.05?

Figure P9.57

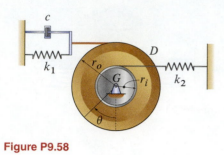

Figure P9.58

Problem 9.58

The mechanism consists of a disk D pinned at G, which is both the geometric center of the disk and its mass center. The outer circumference of the disk has radius $r_o = 0.1$ m and is connected to an element consisting of a linear spring with stiffness $k_1 = 100$ N/m in parallel with a dashpot with damping coefficient $c = 50$ N·s/m. The disk has a hub of radius $r_i = 0.05$ m that is connected to a linear spring with constant $k_2 = 350$ N/m. Knowing that for $\theta = 0$ the disk is in static equilibrium and that the mass moment of inertia of the disk is $I_G = 0.001$ kg·m^2, derive the linearized equation of motion of the disk in terms of θ. In addition, calculate the resulting vibrational motion if the system is released from rest with an initial angular displacement $\theta_i = 0.05$ rad.

Problem 9.59

A box of mass 0.75 kg is thrown on a scale, causing both the scale and the box to move vertically downward with an initial speed of 0.5 m/s. Before the box lands on the scale, the scale is in equilibrium. The total mass of the scale's moving platform and the box is $m = 1.25$ kg. Modeling the platform's support as a spring and dashpot with stiffness $k = 1000$ N/m and damping coefficient $c = 70.7$ N·s/m, find the response of the scale. *Hint:* Place the origin of the y axis at the position of the platform corresponding to the equilibrium configuration of the platform and box together.

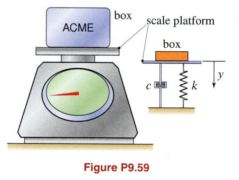

Figure P9.59

Problem 9.60

Consider a simple viscously damped harmonic oscillator governed by Eq. (9.51), and analyze the case in which the damping coefficient c is negative. Calculate the general expression for the response (without taking into account specific initial conditions), using $m = 1$ kg, $c = -1$ N·s/m, and $k = 10$ N/m. Comment on the system's response.

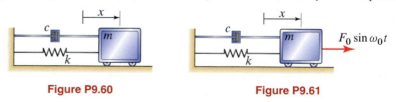

Figure P9.60   **Figure P9.61**

Problem 9.61

The MF for a harmonically excited spring-mass-damper system at $\omega_0/\omega_n \approx 1$ is equal to 5. Calculate the damping ratio of the system. What would the damping ratio be if the MF were equal to 10? Sketch the magnification factor at $\omega_0/\omega_n \approx 1$ as a function of the damping ratio.

Problem 9.62

A slider moves in the horizontal plane under the action of the harmonic forcing $F(t) = F_0 \sin \omega_0 t$. The slider is connected to two identical linear springs, each of which has constant k. When $t = 0$, $x(0) = 0$, the springs are unstretched, $\theta = 45°$, and $L = L_0$. The slider is also connected to a damper with damping coefficient c. Treating F_0, k, c, and L_0 as known quantities, neglecting friction, and letting $\dot{x}(0) = v_i$, (a) derive the equations of motion of the system, (b) derive the linearized equations of motion about the initial position, and (c) determine the amplitude of the steady-state vibrations for the linearized equations of motion.

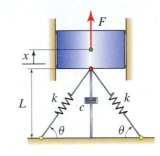

Figure P9.62

Problem 9.63

The mechanism shown is a pendulum consisting of a pendulum bob B with mass m and a T bar, which is pinned at O and has negligible mass. The horizontal portion of the T bar is connected to two supports, each of which has an identical spring and dashpot system, each with spring constant k and damping coefficient c. The springs are unstretched when B is vertically aligned with the pin at O. Modeling B as a particle, derive the linearized equations of motion of the system. In addition, assuming that the system is underdamped, derive the expression for the damped natural frequency of vibration of the system.

Figure P9.63

Problem 9.64

The engine in the rocket shown is supposed to provide a constant thrust of 5000 kN. The turbopump unit in the engine nominally operates at 7000 rpm, and as a result of a design issue, the actual thrust provided by the engine oscillates harmonically with an amplitude of 10 kN at the same rotational frequency of the turbopump unit. The mass of the engine is $m = 5000$ kg. The rest of the rocket is much heavier than the engine and can be treated as being fixed. The engine is mounted to the rocket via two structural members, each of which can be modeled as consisting of a linear spring of stiffness k in parallel with a dashpot with linear viscous damping coefficient c. Determine the smallest values of k and c such that the static deflection due to the constant component of the thrust is less than 0.01 m and so that the vibration amplitude in the nominal operating regime is less than 0.001 m. Ignore the stiffness and damping due to the piping. *Hint:* If x is measured from the equilibrium position of the engine that results from the combined effect of the thrust and gravity, then the engine is subject to an externally applied forcing equal to $(10 \text{ kN}) \sin \omega_0 t$, where ω_0 is the rotational frequency of the turbopump.

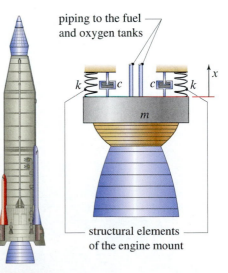

Figure P9.64

piping to the fuel and oxygen tanks

structural elements of the engine mount

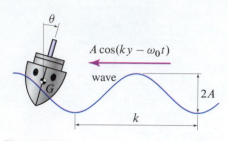

Figure P9.65

Problem 9.65

A simple model for a ship rolling on waves[*] treats the waves as *sinusoids*. Using this model, it can be shown that a linear model for the roll angle θ is given by

$$I_G\ddot{\theta} + c\dot{\theta} + mg\theta = -I_G k\omega_0^2 \sin\omega_0 t,$$

where G denotes the mass center of the ship, I_G is the ship's mass moment of inertia, c is a rotational viscous damping constant coefficient, m is the mass of the ship, A is the wave amplitude, and k is the wavelength of the waves.

(a) What is the natural frequency of the system?

(b) Find the magnification factor for the system.

(c) Assuming that the damping is negligible (i.e., $c \approx 0$), if the maximum amplitude of oscillation that the ship can undergo without capsizing is $\theta_{max} = 1$ rad, find the maximum A so that the crew remains safe.

Problem 9.66

A delicate instrument of mass m must be isolated from excessive vibration of the ground, which is described by the function $u(t) = A \sin\omega_0 t$. To do so, we need to design a *vibration isolating mount*, modeled by the spring and dashpot system shown.

(a) Find the equation of motion of the instrument and reduce it to standard form.

(b) Find the steady-state response $y(t)$.

(c) Find the *displacement transmissibility*, i.e., the response amplitude D divided by A, where A is the amplitude of the ground's vibration.

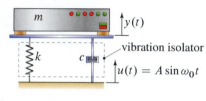

Figure P9.66

[*] See J. M. T. Thompson, R. C. T. Rainey, and M. S. Soliman, "Mechanics of Ship Capsize under Direct and Parametric Wave Excitation," *Philosophical Transactions of the Royal Society of London A*, **338**(1651), 1992, pp. 471–490.

DESIGN PROBLEMS

Design Problem 9.3

As a result of firing a projectile, a 300 kg naval gun assembly gains momentum in the x direction. We can consider the motion of the assembly to start from the equilibrium position $x(0) = 0$ with an initial velocity $\dot{x}(0) = 50$ m/s. Choose values of spring stiffness k and damping coefficient c to provide the fastest return of the assembly to its equilibrium position without oscillation. In addition, make sure that the maximum displacement of the assembly does not exceed 0.1 m. Finally, estimate the time it takes for the gun assembly to return to within 1% of its maximum displacement.

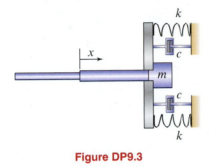

Figure DP9.3

9.4 Chapter Review

In this chapter, we studied the vibration or oscillation of mechanical systems about their equilibrium position. We considered only harmonic oscillators, although we did study the effects of viscous damping and harmonic forcing on the response of a harmonic oscillator.

Undamped free vibration

Any one DOF system whose equation of motion is of the form

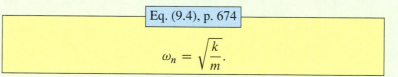

Eq. (9.12), p. 675

$$\ddot{x} + \omega_n^2 x = 0$$

is called a *harmonic oscillator*, and the above expression is referred to as the *standard form* of the harmonic oscillator equation. The solution of this equation can be written as (see Fig. 9.35)

Eq. (9.3), p. 674

$$x(t) = C \sin(\omega_n t + \phi),$$

where ω_n is the *natural frequency*, C is the *amplitude*, and ϕ is the *phase angle* of vibration.

A simple example of a harmonic oscillator is a system consisting of a mass m attached at the free end of a spring with constant k and with the other end fixed (see Fig. 9.36). The natural frequency of such a system is given by

Eq. (9.4), p. 674

$$\omega_n = \sqrt{\frac{k}{m}}.$$

In addition, the amplitude C and the phase angle ϕ are given by, respectively,

Eqs. (9.5) and (9.6), p. 674

$$C = \sqrt{\frac{v_i^2}{\omega_n^2} + x_i^2} \qquad \text{and} \qquad \tan\phi = \frac{x_i \omega_n}{v_i},$$

where we let $t = 0$ be the initial time, $x_i = x(0)$ (i.e., x_i is the initial position), and $v_i = \dot{x}(0)$ (i.e., v_i is the initial velocity). If $v_i = 0$, then ϕ can be chosen equal to $-\pi/2$ or $\pi/2$ rad for $x_i < 0$ and $x_i > 0$, respectively. An alternative form of the solution to Eq. (9.12) is given by

Eq. (9.15), p. 676

$$x(t) = x_i \cos\omega_n t + \frac{v_i}{\omega_n} \sin\omega_n t.$$

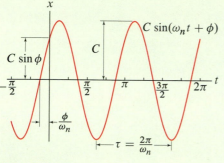

Figure 9.35

Figure 9.2 repeated. Plot of Eq. (9.3) showing the amplitude C, phase angle ϕ, and period τ of a harmonic oscillator.

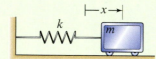

Figure 9.36

Figure 9.11 repeated. A simple spring-mass harmonic oscillator whose equation of motion is given by $m\ddot{x} + kx = 0$ and for which $\omega_n = \sqrt{k/m}$.

The *period* of the oscillation is given by

Eq. (9.7), p. 674

$$\text{Period} = \tau = \frac{2\pi}{\omega_n},$$

and the *frequency* of vibration is

Eq. (9.8), p. 674

$$\text{Frequency} = f = \frac{1}{\tau} = \frac{\omega_n}{2\pi}.$$

Energy method.   For conservative systems, the work-energy principle tells us that the quantity $T + V$ is constant, and so its time derivative must be zero. This provides a convenient way to obtain the equations of motion via

Eq. (9.19), p. 677

$$\frac{d}{dt}(T + V) = 0 \quad \Rightarrow \quad \text{equations of motion.}$$

When we apply the energy method to determine the linearized equations of motion, it is often convenient to first approximate the kinetic and potential energies as quadratic functions of position and velocity and then take derivatives with respect to time. This process yields equations of motion that are linear.

In approximating the sine and cosine functions as quadratic functions of their arguments, we use the relations:

Eqs. (9.28), p. 678

$$\sin \theta \approx \theta \quad \text{and} \quad \cos \theta \approx 1 - \theta^2/2.$$

Since the quadratic term in the power series expansion of the sine function is identically equal to zero, the linearized form of the sine function can also be viewed as the quadratic approximation of the sine function.

Undamped forced vibration

When a harmonic oscillator is subject to harmonic forcing, the standard form of the equation of motion is

Eq. (9.37), p. 693

$$\ddot{x} + \omega_n^2 x = \frac{F_0}{m} \sin \omega_0 t,$$

where F_0 is the amplitude of the forcing and ω_0 is its frequency (see Fig. 9.37). The general solution to this equation consists of the sum of the complementary solution and a particular solution. The *complementary solution* x_c is the solution of the associated homogeneous equation, which is given by, for example,

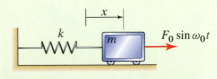

Figure 9.37
Figure 9.16 repeated. A forced harmonic oscillator whose equation of motion is given by Eq. (9.37) with $\omega_n = \sqrt{k/m}$. The position x is measured from the equilibrium position of the mass.

Eq. (9.13). For $\omega_0 \neq \omega_n$, a particular solution is

Eq. (9.41), p. 694

$$x_p = \frac{F_0/k}{1 - (\omega_0/\omega_n)^2} \sin \omega_0 t,$$

and so the *general solution* is given by

Eq. (9.42), p. 694

$$x = x_c + x_p = A \sin \omega_n t + B \cos \omega_n t + \frac{F_0/k}{1 - (\omega_0/\omega_n)^2} \sin \omega_0 t,$$

where A and B are constants determined by the initial conditions. The amplitude of the steady-state vibration is

Eq. (9.43), p. 694

$$x_{\text{amp}} = \frac{F_0/k}{1 - (\omega_0/\omega_n)^2},$$

which means that the corresponding *magnification factor* MF is

Eq. (9.44), p. 694

$$\text{MF} = \frac{x_{\text{amp}}}{F_0/k} = \frac{1}{1 - (\omega_0/\omega_n)^2}.$$

The plot of the MF in Fig. 9.38 illustrates the phenomenon of *resonance*, which occurs when $\omega_0 \approx \omega_n$ and results in very large vibration amplitudes.

Harmonic excitation of the support.   If the support of a structure is excited harmonically rather than the structure itself (see Fig. 9.39), then Eq. (9.37) is still the governing equation, except that F_0 is replaced by the spring constant k times the amplitude of the support vibration X. All solutions described above are then valid with that same replacement.

Viscously damped vibration

Viscously damped free vibration.   The standard form of the equation of motion for a one degree of freedom viscously damped harmonic oscillator is

Eq. (9.51), p. 705

$$m\ddot{x} + c\dot{x} + kx = 0,$$

where m is the mass, c is the *coefficient of viscous damping*, and k is the linear spring constant. The character of the solution to this equation depends on the amount of damping relative to a specific amount of damping called the *critical damping coefficient*, which is defined as

Eq. (9.57), p. 705

$$c_c = 2m\sqrt{\frac{k}{m}} = 2m\omega_n,$$

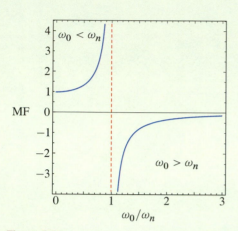

Figure 9.38
Figure 9.17 repeated. MF as a function of the frequency ratio ω_0/ω_n.

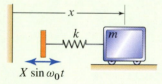

Figure 9.39
Figure 9.23 repeated. A harmonic oscillator whose support is being excited harmonically.

where $\omega_n = \sqrt{k/m}$. In particular, if $c \geq c_c$, then the motion is nonoscillatory, whereas if $c < c_c$, then the motion is oscillatory. When $c > c_c$, the system is said to be *overdamped*, and the solution is given by

Eq. (9.58), p. 705

$$x = e^{-(c/2m)t} \left(A e^{t\sqrt{(c/2m)^2 - k/m}} + B e^{-t\sqrt{(c/2m)^2 - k/m}} \right),$$

where A and B are constants to be determined from the initial conditions. When $c = c_c$, the system is said to be *critically damped*, and the solution is given by

Eq. (9.59), p. 706

$$x = (A + Bt)e^{-\omega_n t},$$

where, again, A and B are constants to be determined from the initial conditions. Finally, when $c < c_c$, the system is said to be *underdamped*, and the solution is given by

Eq. (9.60), p. 706

$$x = e^{-(c/2m)t}(A \sin \omega_d t + B \cos \omega_d t),$$

or, equivalently, by

Eq. (9.63), p. 706

$$x = D e^{-(c/2m)t} \sin(\omega_d t + \phi),$$

where A and B are constants to be determined from the initial conditions in the first solution, D and ϕ are analogous constants in the second solution, and ω_d is the *damped natural frequency*, which is given by

Eq. (9.61), p. 706, and Eq. (9.68), p. 707

$$\omega_d = \sqrt{\frac{k}{m} - \left(\frac{c}{2m}\right)^2} = \omega_n \sqrt{1 - (c/c_c)^2} = \omega_n \sqrt{1 - \zeta^2},$$

where $\zeta = c/c_c$ is the *damping ratio*.

Viscously damped forced vibration. The standard form of a viscously damped forced harmonic oscillator is

Eq. (9.69), p. 707

$$m\ddot{x} + c\dot{x} + kx = F_0 \sin \omega_0 t,$$

where F_0 is the amplitude of the forcing function and ω_0 is the frequency of the forcing function. When expressed using the damping ratio ζ, the equation above takes on the form

Eq. (9.70), p. 707

$$\ddot{x} + 2\zeta\omega_n \dot{x} + \omega_n^2 x = \frac{F_0}{m} \sin \omega_0 t.$$

The general solution to either of these equations is the sum of the complementary solution and a particular solution. The complementary solution is transient; i.e., it vanishes as time increases. The particular or steady-state solution is of the form

Eq. (9.72), p. 708

$$x_p = D \sin(\omega_0 t - \phi),$$

where ϕ and D are given by

Eq. (9.75), p. 708, and Eq. (9.78), p. 709

$$\tan \phi = \frac{c\omega_0}{k - m\omega_0^2} = \frac{2(c/c_c)(\omega_0/\omega_n)}{1 - (\omega_0/\omega_n)^2} = \frac{2\zeta\omega_0/\omega_n}{1 - (\omega_0/\omega_n)^2},$$

$$D = \frac{F_0/k}{\sqrt{[1 - (\omega_0/\omega_n)^2]^2 + (2\zeta\omega_0/\omega_n)^2}}.$$

The *magnification factor* for a damped, forced harmonic oscillator is

Eq. (9.79), p. 709

$$\mathrm{MF} = \frac{D}{F_0/k} = \frac{1}{\sqrt{[1 - (\omega_0/\omega_n)^2]^2 + (2\zeta\omega_0/\omega_n)^2}},$$

a plot of which is shown in Fig. 9.40 for various values of the damping ratio ζ.

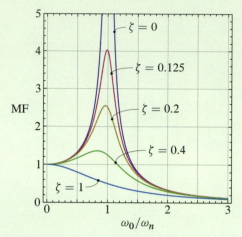

Figure 9.40

Figure 9.33 repeated. The MF as a function of the frequency ratio ω_0/ω_n for various values of the damping ratio ζ.

REVIEW PROBLEMS

Problem 9.67

When the connecting rod shown is suspended from the knife-edge at point O and displaced slightly so that it oscillates as a pendulum, its period of oscillation is 0.77 s. In addition, it is known that the mass center G is located a distance $L = 110$ mm from O and that the mass of the connecting rod is 661 g. Using the energy method, determine the mass moment of inertia of the connecting rod I_G.

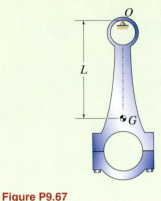

Figure P9.67

Problem 9.68

Derive the equation of motion for the system, in which the springs with constants k_1 and k_2 connecting m to the wall are joined in series. Neglect the mass of the small wheels, and assume that the attachment point A between the two springs has negligible mass. *Hint:* The force in the two springs must be the same; use this fact, along with the fact that the total deflection of the mass must equal the sum of the deflections of the springs to find an equivalent spring constant k_{eq}.

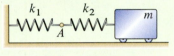

Figure P9.68

Problem 9.69

Revisit Example 9.2 and compute the natural frequency of the silicon nanowire, using the energy method. Use a uniform Si nanowire with a circular cross section that is 9.8 μm long and 330 nm in diameter and with all its flexibility lumped in a torsional spring at the base of the wire. In addition, use $\rho = 2330\,\text{kg/m}^3$ for the density of silicon and $E = 152\,\text{GPa}$ for its modulus of elasticity.

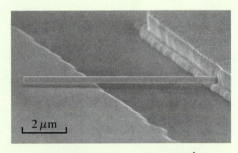

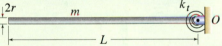

Figure P9.69

Problem 9.70

Structural health monitoring technology detects damage in civil, aerospace, and other structures. Structural damage is usually comprised of cracking, delaminations, or loose fasteners, which result in the reduction of stiffness. Many structural health monitoring methods are based on tracking changes in natural frequencies. Modeling a structure as a one DOF harmonic oscillator, calculate the change in stiffness needed to cause a 3% reduction in the natural frequency of the structure being monitored.

Problem 9.71

The harmonic oscillator shown has a mass $m = 5$ kg, a spring with constant $k = 4000$ N/m, and a dashpot with a damping coefficient $c = 20$ N·s/m. Calculate the amplitude F_0 of the sinusoidal excitation force that is necessary to produce a steady-state vibration with a velocity amplitude of 10 m/s at resonance. What is the corresponding amplitude of the acceleration?

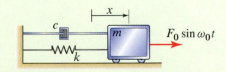

Figure P9.71

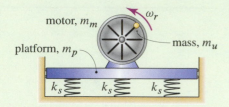

Figure P9.72

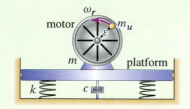

Figure P9.74

Problem 9.72

Revisit Example 9.4 and derive the equations of motion of the motor, using the following equation, called *Lagrange's equation*,

$$\frac{d}{dt}\left(\frac{\partial T}{\partial \dot{y}_m}\right) - \frac{\partial T}{\partial y_m} + \frac{\partial V}{\partial y_m} = 0,$$

where T and V are the kinetic and potential energies of the system, respectively.

Problem 9.73

Modeling the beam as a uniform thin bar, ignoring the inertia of the pulleys, assuming that the system is in static equilibrium when the bar is horizontal, and assuming that the cord is inextensible and does not go slack, determine the linearized equation of motion of the system. In addition, determine the system's natural frequency of vibration. Treat the parameters shown in the figure as known.

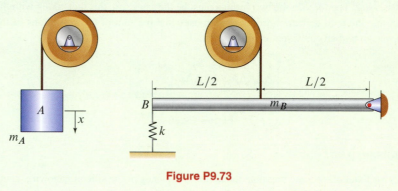

Figure P9.73

Problem 9.74

Revisit Example 9.6 and obtain the expression for the force transmitted to the floor, using the expression for the steady-state response of the unbalanced motor.

Problem 9.75

The system shown is released from rest when both springs are unstretched and $x = 0$. Neglecting the inertia of the pulley P and assuming that the disk rolls without slip, derive the equation of motion of the system in terms of x. Assume that point G is both the mass center of the disk and its geometric center. Treat the quantities k_1, k_2, c, m_1, m_2, and I_G as known, where I_G is the mass moment of inertia of the disk. Finally, assuming that the system is underdamped, derive an expression for the damped natural frequency of the system.

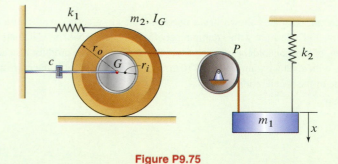

Figure P9.75

Problem 9.76

A ring of mass m is attached by two linear elastic cords to the vertical supports as shown. The cords have elastic constant k and unstretched length $L_0 < L$. Assuming that the pretension in the cords is large enough that the deflection of the cords due to the ring's weight can be neglected, find the nonlinear equation of motion for the mass m.

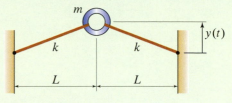

Problem 9.77

A ring of mass m is attached by two linear elastic cords to the vertical supports as shown. The cords have elastic constant k and unstretched length $L_0 < L$. Assuming that the pretension in the cords is large enough that the deflection of the cords due to the ring's weight can be neglected, use Newton's second law to find the linearized equation of motion about $y = 0$ for the mass m. In addition, determine the natural frequency of the ring's vibration.

Figure P9.76–P9.78

Problem 9.78

Solve Prob. 9.77 by finding the linearized equations of motion via the energy method.

A | *Mass Moments of Inertia*

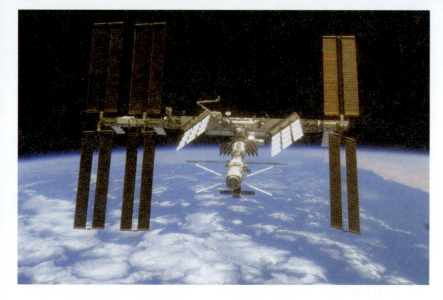

Mass moments and products of inertia are measures of how the mass is distributed within a body. Mass moments and products of inertia of a body depend on its geometry (size and shape), the density of the material at each point in the body, and the axes selected for measuring them.

Mass moments of inertia and *mass products of inertia* are measures of how the mass is distributed within a body. Mass moments and products of inertia arise in the rotational equations of motion for a rigid body (Chapter 7), the kinetic energy of a rigid body (Chapter 8), the angular momentum of a rigid body (Chapter 8 and Appendix B), and the three-dimensional dynamics of rigid bodies.

Definition of mass moments and products of inertia

The *mass moments of inertia* for the body B shown in Fig. A.1 are defined as

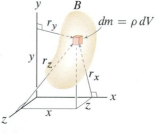

$$I_x = \int_B r_x^2 \, dm = \int_B (y^2 + z^2) \, dm, \tag{A.1}$$

$$I_y = \int_B r_y^2 \, dm = \int_B (x^2 + z^2) \, dm, \tag{A.2}$$

$$I_z = \int_B r_z^2 \, dm = \int_B (x^2 + y^2) \, dm, \tag{A.3}$$

Figure A.1
An object with mass m, density ρ, and volume V. The scalar quantities r_x, r_y, and r_z are radial distances from the x, y, and z axes, respectively, to the center of mass of mass element dm.

where:

r_x, r_y, and r_z are shown in Fig. A.1 and are the radial distances (i.e., moment arms) from the x, y, and z axes, respectively, to the center of mass of the mass element dm.

x, y, and z are shown in Fig. A.1 and are the coordinates of the center of mass of the mass element dm.

731

I_x, I_y, and I_z are the *mass moments of inertia of the body B about the x, y, and z axes*, respectively.

The *mass products of inertia* for the body B shown in Fig. A.1 are defined as

$$I_{xy} = I_{yx} = \int_B xy \, dm, \tag{A.4}$$

$$I_{yz} = I_{zy} = \int_B yz \, dm, \tag{A.5}$$

$$I_{xz} = I_{zx} = \int_B xz \, dm, \tag{A.6}$$

where:

I_{xy}, I_{yz}, and I_{xz} are the *products of inertia of the mass about the xy, yz, and xz axes*, respectively.

Remarks

- When referring to mass moments of inertia, we often omit the word "mass" when it is obvious from the context that we are dealing with mass moments of inertia as opposed to area moments of inertia.

- In each of Eqs. (A.1)–(A.3), two equivalent integral expressions are provided, and each is useful depending on the geometry of the object under consideration.

- The moments of inertia in Eqs. (A.1)–(A.6) measure the *second moment* of the mass distribution. That is, to determine I_x, I_y, and I_z in Eqs. (A.1)–(A.3), the moment arms r_x, r_y, and r_z are *squared*. The second integral in each of these expressions is obtained by noting that $r_x^2 = y^2 + z^2$, and similarly for r_y^2 and r_z^2. For the products of inertia I_{xy}, I_{yz}, and I_{xz} in Eqs. (A.4)–(A.6), the product of two different moment arms is used.

- In Eqs. (A.1)–(A.6), x, y, and z have dimensions of length, and *dm* has the dimension of mass. Hence, all mass moments of inertia have dimensions of $(mass)(length)^2$ and are expressed in slug·ft^2 and kg·m^2 in the U.S. Customary and SI unit systems, respectively.

- When the x, y, and z axes pass through the center of mass of an object, we denote these axes as x', y', and z', and we refer to the moments of inertia associated with these axes as *mass center moments of inertia* with the designations $I_{x'}$, $I_{y'}$, etc.

- The quantities I_x, I_y, and I_z are never negative.* The products of inertia I_{xy}, I_{yz}, and I_{xz} may be positive, zero, or negative, as discussed below.

- Evaluation of moments of inertia using composite shapes is possible using the parallel axis theorem, as discussed later in this appendix.

---

*For the thin rod shown in the Table of Properties of Solids on the inside back cover, I_x is positive and nonzero, but since it is much smaller than I_y and I_z, it is usually taken to be zero.

How are mass moments of inertia used?

It is useful to discuss why there are six mass moments of inertia, how they differ from each other, and how they are used.

Moments of inertia I_x, I_y, and I_z. In Fig. A.2, the International Space Station with a docked Space Shuttle is shown. Imagine that a moment M_x about the x axis is applied to the Space Station. Assuming the Space Station is rigid,* it will begin to undergo an angular acceleration about the x axis. The value of the angular acceleration is directly proportional to the mass moment of inertia about the x axis I_x. Furthermore, the larger I_x is, the lower the angular acceleration will be for a given value of M_x. Similar remarks apply to moments applied about the y and z axes and the influence that moments of inertia I_y and I_z have on angular accelerations about these axes.

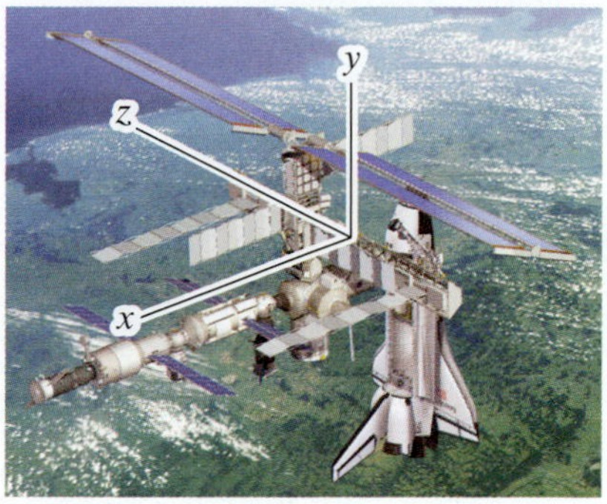

Products of inertia I_{xy}, I_{yz}, and I_{xz}. Products of inertia measure the asymmetry of a body's mass distribution with respect to the xy, yz, and xz planes. Products of inertia can have a positive, zero, or negative value, depending on the shape and mass distribution of an object, the selection of the x, y, and z directions, and the location of the origin of the xyz coordinate system. Figure A.3 shows the cross section of a uniform body at some arbitrary z coordinate. The orange shaded region shows that part of the cross section that is

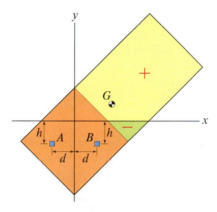

Figure A.2
The International Space Station with a Space Shuttle docked to it.

Figure A.3. A uniform body and a coordinate system for calculating the products of inertia. The cross section of the body is assumed to be the same at every z coordinate.

symmetric about the y axis. The two mass elements A and B (shown in blue) are an equal distance d from the yz plane and both have the same y coordinate, i.e., $y = -h$. Referring to the integrals for I_{xy} and I_{xz} in Eqs. (A.4) and (A.6), that means that $xy \, dm$ and $xz \, dm$ for the left element A have the opposite sign of the analogous quantities for the right element B. Therefore, the mass in the orange region does not contribute to the products of inertia I_{xy} and I_{xz}. On the other hand, the mass in the yellow shaded region has no corresponding region that is symmetric with respect to the yz plane and since the product xy for all points in that region is positive, it contributes positively to

---

* The International Space Station is very flexible, as are most space structures. If a moment were applied about the x axis shown in Fig. A.2 then in addition to the rotations discussed above, the structure would vibrate. Control of vibrations in space structures is very important and receives considerable attention.

I_{xy}. Using a similar argument, the green shaded region contributes negatively to I_{xy}. If the body in Fig. A.3 had a uniform mass distribution, it would have $I_{xy} > 0$ since the yellow region is larger than the green. Note that nothing can be said about the sign of I_{xz} since we don't know if the cross section in Fig. A.3 is at a positive or negative z coordinate.

The preceding arguments lead us to the conclusion that if the body in Fig. A.3 consisted of only the area shaded in orange, then both I_{xy} and I_{xz} would be zero. This allows us to state the following:

> *All products of inertia containing a coordinate that is perpendicular to a plane of symmetry for a body must be zero as long as the origin of the coordinate system lies in that plane of symmetry.*

For example, if an object is symmetric about the xy plane, such as the mallet shown in Fig. A.4, then $I_{xz} = I_{yz} = 0$ and I_{xy} may be positive, zero or negative. If an object is symmetric about at least two of the xy, yz, and xz planes, then all of the products of inertia are zero. Thus, for example, the products of inertia are zero for a uniform solid of revolution as long as one of the coordinate directions coincides with the axis of revolution.

Objects that have one or more nonzero products of inertia may display complicated dynamics in three dimensional motions. In addition, forcing such a body to undergo planar motion generally requires the application of moments along directions in that plane of motion. For example, the Space Station in Fig. A.2 is nonsymmetric about the xyz axes shown and thus it has nonzero products of inertia. As a consequence, if, for example, a moment M_x about the x axis is applied, the Space Station, in addition to rotating about the x axis, will also rotate about the y and/or z axes (or moments about those axes will be required to prevent it from doing so).

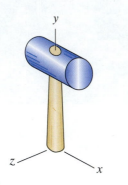

Figure A.4
If the mallet's geometry and mass distribution are both symmetric about the xy plane, then the mallet is said to be a *symmetric object*. If the mallet is also symmetric about the yz plane, it may be called a *doubly symmetric object*.

Radius of gyration

Rather than using mass moments of inertia to quantify the distribution of mass within a body, the *radii of gyration* are often used. The *radii of gyration* are directly related to the mass moments of inertia, and are defined as

$$k_x = \sqrt{\frac{I_x}{m}}, \qquad k_y = \sqrt{\frac{I_y}{m}}, \qquad k_z = \sqrt{\frac{I_z}{m}}, \tag{A.7}$$

where

> k_x, k_y, and k_z are the *radii of gyration of the body about the x, y, and z axes, respectively.*

m is the mass of the object.

The radii of gyration have units of *length*.

Parallel axis theorem

The parallel axis theorem relates mass moments and products of inertia I_x, I_y, I_z, I_{xy}, I_{yz}, and I_{xz} to the mass center moments and products of inertia $I_{x'}$, $I_{y'}$, $I_{z'}$, $I_{x'y'}$, $I_{y'z'}$, and $I_{x'z'}$. We will assume that the x and x' axes are parallel, the y and y' axes are parallel, the z and z' axes are parallel, the origin of the xyz axes is at an arbitrary point P, and the origin of the $x'y'z'$

system is at the mass center G. As can be seen in Chapters 7–9 and the three-dimensional dynamics of rigid bodies, the parallel axis theorem is important for the dynamics of rigid bodies.

Parallel axis theorem for moments of inertia

Consider the object with mass m and center of mass G shown in Fig. A.5. An $x'y'z'$ coordinate system is defined, with origin at the center of mass G

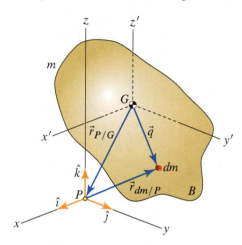

Figure A.5. An object with mass m and center of mass at point G. The x and x' axes, the y and y' axes, and the z and z' axes are parallel to one another, respectively.

of the object and the x, y, and z axes are parallel to the x', y', and z' axes, respectively. We begin by noting that the mass moment of inertia of the body B about the x axis is given by Eq. (A.1), which is repeated here for convenience as

$$I_x = \int_B \left(y^2 + z^2\right) dm. \tag{A.8}$$

In addition, we can write the position of a mass element dm relative to G as

$$\vec{q} = \vec{r}_{P/G} + \vec{r}_{dm/P}. \tag{A.9}$$

Writing Eq. (A.9) in component form, we obtain

$$x'\hat{\imath} + y'\hat{\jmath} + z'\hat{k} = \left[(r_{P/G})_x\,\hat{\imath} + (r_{P/G})_y\,\hat{\jmath} + (r_{P/G})_z\,\hat{k}\right] \\ + \left(x\,\hat{\imath} + y\,\hat{\jmath} + z\,\hat{k}\right). \tag{A.10}$$

Substituting Eq. (A.10) into Eq. (A.8) results in

$$I_x = \int_B \left[(y' - (r_{P/G})_y)^2 + (z' - (r_{P/G})_z)^2\right] dm, \tag{A.11}$$

$$= \int_B (y'^2 + z'^2)\,dm + \left[(r_{P/G})_y^2 + (r_{P/G})_z^2\right]\int_B dm$$

$$- 2(r_{P/G})_y \int_B y'\,dm - 2(r_{P/G})_z \int_B z'\,dm. \tag{A.12}$$

The first term in Eq. (A.12) is the mass center moment of inertia about the x' axis $I_{x'}$ (see Eq. (A.1)). The second integral in Eq. (A.12) is the mass m of B.

Finally, since x', y', and z' measure the position of dm relative to G, the last two terms in Eq. (A.12) measure the position of the mass center of B relative to G and so they must both be zero. Therefore, Eq. (A.12) becomes

$$I_x = I_{x'} + m\big[(r_{P/G})_y^2 + (r_{P/G})_z^2\big]. \tag{A.13}$$

Referring to Fig. A.6, we see that $(r_{P/G})_y^2 + (r_{P/G})_z^2$ is the square of the

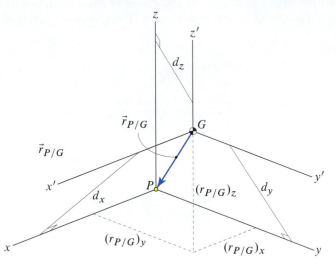

Figure A.6. The x and x' axes are parallel with separation distance d_x, the y and y' axes are parallel with separation distance d_y, and the z and z' axes are parallel with separation distance d_z.

perpendicular distance d_x between the x and x' axes, i.e.,

$$d_x^2 = (r_{P/G})_y^2 + (r_{P/G})_z^2, \tag{A.14}$$

so that Eq. (A.13) becomes

$$I_x = I_{x'} + md_x^2. \tag{A.15}$$

Similarly, substituting Eq. (A.10) into Eqs. (A.2) and (A.3), we obtain the following two equations

$$I_y = I_{y'} + m\big[(r_{P/G})_x^2 + (r_{P/G})_z^2\big] = I_{y'} + md_y^2, \tag{A.16}$$

$$I_z = I_{z'} + m\big[(r_{P/G})_x^2 + (r_{P/G})_y^2\big] = I_{z'} + md_z^2, \tag{A.17}$$

where, referring to Fig. A.6, we have used the fact that

$$d_y^2 = (r_{P/G})_x^2 + (r_{P/G})_z^2, \tag{A.18}$$

and

$$d_z^2 = (r_{P/G})_x^2 + (r_{P/G})_y^2. \tag{A.19}$$

Summarizing these results, we have the *parallel axis theorem for moments of inertia*, which relates the mass moments of inertia with respect to the x, y, and z axes to the mass center moments of inertia as follows

$$I_x = I_{x'} + md_x^2 = I_{x'} + m\big[(r_{P/G})_y^2 + (r_{P/G})_z^2\big], \tag{A.20}$$

$$I_y = I_{y'} + md_y^2 = I_{y'} + m\big[(r_{P/G})_x^2 + (r_{P/G})_z^2\big], \tag{A.21}$$

$$I_z = I_{z'} + md_z^2 = I_{z'} + m\big[(r_{P/G})_x^2 + (r_{P/G})_y^2\big]. \tag{A.22}$$

Common Pitfall

Parallel axis theorem. Consider the body whose center of mass is at point G, as well as the parallel axes x_1, x_2, and mass center axis x'.

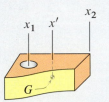

If I_{x_1} is known, a common error is to use the parallel axis theorem to *directly* determine I_{x_2}. In the parallel axis theorem, one of the axes is always a mass center axis. Thus, with I_{x_1} known, the parallel axis theorem must be used *twice*, the first time to determine $I_{x'}$, and the second time to use $I_{x'}$ to determine I_{x_2}.

Parallel axis theorem for products of inertia

Again consider the object with mass m and center of mass G shown in Fig. A.7. An $x'y'z'$ coordinate system is defined, with origin at the center of mass G of the object and the x, y, and z axes are parallel to the x', y', and z' axes, respectively. We begin by noting that the product of inertia of the body B about the x and y axes is given by Eq. (A.4), which is repeated here for convenience as

$$I_{xy} = \int_B xy \, dm. \tag{A.23}$$

Substituting in Eq. (A.10) for x and y, Eq. (A.23) becomes

$$I_{xy} = \int_B \left[x' - (r_{P/G})_x \right] \left[y' - (r_{P/G})_y \right] dm, \tag{A.24}$$

$$= \int_B x'y' \, dm + (r_{P/G})_x (r_{P/G})_y \int_B dm$$

$$\quad - (r_{P/G})_y \int_B x' \, dm - (r_{P/G})_x \int_B y' \, dm. \tag{A.25}$$

Again, the last two integrals are zero by definition of center of mass. The first integral is the mass center product of inertia about the $x'y'$ axes and the second integral defines the mass of the body B. Therefore, Eq. (A.25) becomes

$$I_{xy} = I_{x'y'} + m(r_{P/G})_x (r_{P/G})_y. \tag{A.26}$$

Using a similar approach, we can readily find I_{yz} and I_{xz} in terms of the mass center products of inertia. In summary, the *parallel axis theorem for products of inertia* is

$$I_{xy} = I_{x'y'} + m(r_{P/G})_x (r_{P/G})_y, \tag{A.27}$$
$$I_{xz} = I_{x'z'} + m(r_{P/G})_x (r_{P/G})_z, \tag{A.28}$$
$$I_{yz} = I_{y'z'} + m(r_{P/G})_y (r_{P/G})_z, \tag{A.29}$$

where we recall that $I_{x'y'}$, $I_{x'z'}$, and $I_{y'z'}$ are the mass center products of inertia and $(r_{P/G})_x$, $(r_{P/G})_y$, and $(r_{P/G})_z$ are defined in Eq. (A.10) and can be seen in Fig. A.6.

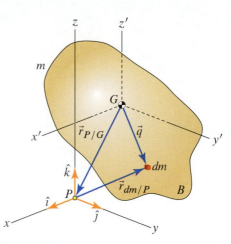

Figure A.7
Figure A.5 repeated. An object with mass m and center of mass at point G. The x and x' axes, the y and y' axes, and the z and z' axes are parallel to one another, respectively.

Principal moments of inertia

We have seen that the moments and products of inertia for a rigid body B depend on both the origin and orientation of the coordinate axes used to compute them. For a given origin, it turns out that we can find a unique orientation of the coordinate axes such that all the products of inertia are zero when computed with respect to those axes. These special axes are called the *principal axes of inertia* and the corresponding moments of inertia are called the *principal moments of inertia* $\bar{I}_x$, $\bar{I}_y$, and $\bar{I}_z$. Let's see how we find these axes and the corresponding moments of inertia.

Equation (B.16) in Appendix B states that the angular momentum of a rigid body with respect to its mass center G, that is $\vec{h}_G$, can be written as

$$\vec{h}_G = \left(I_x \omega_{Bx} - I_{xy}\omega_{By} - I_{xz}\omega_{Bz} \right) \hat{\imath}$$
$$\quad + \left(-I_{xy}\omega_{Bx} + I_y \omega_{By} - I_{yz}\omega_{Bz} \right) \hat{\jmath}$$

$$+ \left(-I_{xz}\omega_{Bx} - I_{yz}\omega_{By} + I_z\omega_{Bz} \right)\hat{k}, \qquad (A.30)$$

$$= \{h_G\} = [I_G]\{\omega_B\}, \qquad (A.31)$$

where $\vec{\omega}_B = \{\omega_B\}$ is the angular velocity of the rigid body B and where $[I_G]$ is the *inertia tensor*, which is written as

$$[I_G] = \begin{bmatrix} I_x & -I_{xy} & -I_{xz} \\ -I_{xy} & I_y & -I_{yz} \\ -I_{xz} & -I_{yz} & I_z \end{bmatrix}. \qquad (A.32)$$

We now assume that the rigid body is spinning about one of its principal axes of inertia whose moment of inertia is $\bar{I}$. If this is the case, then the angular momentum can be written as

$$\vec{h}_G = \bar{I}\vec{\omega}_B = \bar{I}\omega_{Bx}\,\hat{\imath} + \bar{I}\omega_{By}\,\hat{\jmath} + \bar{I}\omega_{Bz}\,\hat{k}. \qquad (A.33)$$

The expressions for $\vec{h}_G$ given in Eq. (A.30) and Eq. (A.33) must be equal to one another. Since the Cartesian basis vectors are linearly independent, we can equate the coefficients of $\hat{\imath}$, $\hat{\jmath}$, and $\hat{k}$ in these two equations to obtain a linear system of equations in the unknowns ω_{Bx}, ω_{By}, and ω_{Bz}, which can be expressed as

$$\begin{bmatrix} I_x - \bar{I} & -I_{xy} & -I_{xz} \\ -I_{xy} & I_y - \bar{I} & -I_{yz} \\ -I_{xz} & -I_{yz} & I_z - \bar{I} \end{bmatrix} \begin{Bmatrix} \omega_{Bx} \\ \omega_{By} \\ \omega_{Bz} \end{Bmatrix} = \begin{Bmatrix} 0 \\ 0 \\ 0 \end{Bmatrix}. \qquad (A.34)$$

This system of equations has a solution if and only if the determinant of the coefficient matrix is zero, which implies that

$$\bar{I}^3 - (I_x + I_y + I_z)\bar{I}^2 - (I_{xy}^2 + I_{xz}^2 + I_{yz}^2 - I_xI_y - I_xI_z - I_yI_z)\bar{I}$$
$$+ (I_xI_{yz}^2 + I_yI_{xz}^2 + I_zI_{xy}^2 - I_xI_yI_z + 2I_{xy}I_{yz}I_{xz}) = 0. \quad (A.35)$$

The three roots of this equation represent the three principal moments of inertia $\bar{I}_x$, $\bar{I}_y$, and $\bar{I}_z$ and their corresponding directions are given by the vector $\vec{\omega}_B$ when each root in substituted into Eq. (A.34).

Moment of inertia about an arbitrary axis

Referring to Fig. A.8, suppose we know all the components of the inertia tensor relative to the xyz axes shown and that we wish to find the moment of inertia of the body about the arbitrarily oriented axis ℓ. The direction of ℓ is defined by the unit vector $\hat{u}_\ell$. We know that the definition of mass moment of inertia states the moment of inertia about the ℓ axis is given by

$$I_\ell = \int_B |\vec{r}_{dm/P}\sin\theta|^2\,dm, \qquad (A.36)$$

where θ is the angle between the positive directions of the $\vec{r}_{dm/P}$ and $\hat{u}_\ell$ vectors and $|\vec{r}_{dm/P}\sin\theta|$ is the perpendicular distance from ℓ to dm. If we now notice that

$$|\vec{r}_{dm/P} \times \hat{u}_\ell| = |\vec{r}_{dm/P}||\hat{u}_\ell|\sin\theta = |\vec{r}_{dm/P}|\sin\theta, \qquad (A.37)$$

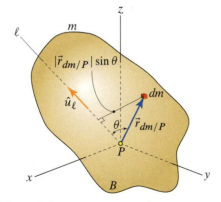

Figure A.8
The quantities needed to find the moment of inertia of a body B about an arbitrarily oriented axis ℓ.

and therefore that

$$|\vec{r}_{dm/P} \sin\theta|^2 = (\vec{r}_{dm/P} \times \hat{u}_\ell) \cdot (\vec{r}_{dm/P} \times \hat{u}_\ell), \qquad (A.38)$$

then we can write the integral in Eq. (A.36) as

$$I_\ell = \int_B (\vec{r}_{dm/P} \times \hat{u}_\ell) \cdot (\vec{r}_{dm/P} \times \hat{u}_\ell)\,dm, \qquad (A.39)$$

If we now write $\vec{r}_{dm/P}$ and $\hat{u}_\ell$ in component form as

$$\vec{r}_{dm/P} = x\,\hat{\imath} + y\,\hat{\jmath} + z\,\hat{k}, \qquad (A.40)$$

$$\hat{u}_\ell = u_{\ell x}\,\hat{\imath} + u_{\ell y}\,\hat{\jmath} + u_{\ell z}\,\hat{k}, \qquad (A.41)$$

so that

$$\vec{r}_{dm/P} \times \hat{u}_\ell = (u_{\ell z}y - u_{\ell y}z)\,\hat{\imath} + (u_{\ell x}z - u_{\ell z}x)\,\hat{\jmath} + (u_{\ell y}x - u_{\ell x}y)\,\hat{k}, \quad (A.42)$$

then Eq. (A.39) becomes

$$I_\ell = \int_B \Big[(u_{\ell z}y - u_{\ell y}z)^2 + (u_{\ell x}z - u_{\ell z}x)^2$$
$$+ (u_{\ell y}x - u_{\ell x}y)^2 \Big]\,dm \qquad (A.43)$$

$$= u_{\ell x}^2 \int_B (y^2 + z^2)\,dm + u_{\ell y}^2 \int_B (x^2 + z^2)\,dm$$

$$+ u_{\ell z}^2 \int_B (x^2 + y^2)\,dm - 2u_{\ell x}u_{\ell y} \int_B xy\,dm$$

$$- 2u_{\ell x}u_{\ell z} \int_B xz\,dm - 2u_{\ell y}u_{\ell z} \int_B yz\,dm. \qquad (A.44)$$

The integrals in Eq. (A.44) are the *known* components of the inertia tensor relative to the xyz axes so that we can write Eq. (A.44) as

$$\boxed{\begin{aligned} I_\ell = &\ u_{\ell x}^2 I_x + u_{\ell y}^2 I_y + u_{\ell z}^2 I_z \\ &- 2u_{\ell x}u_{\ell y} I_{xy} - 2u_{\ell x}u_{\ell z} I_{xz} - 2u_{\ell y}u_{\ell z} I_{yz}. \end{aligned}} \qquad (A.45)$$

Thus, if we know the moments and products of inertia for a body B about a given set of axes xyz (i.e., the inertia tensor relative to these axes), we can find the moment of inertia of that body about an arbitrarily oriented axis ℓ using Eq. (A.45), where we note that $u_{\ell x}$, $u_{\ell y}$, and $u_{\ell z}$ are the direction cosines between the positive $\hat{u}_\ell$ direction and the positive x, y, and z axes, respectively.

Evaluation of moments of inertia using composite shapes

The parallel axis theorem written for composite shapes is

$$\boxed{I_x = \sum_{i=1}^{n} (I_{x'} + d_x^2 m)_i,} \qquad (A.46)$$

where n is the number of shapes, $I_{x'}$ is the mass moment of inertia for shape i about its mass center x' axis, d_x is the *shift distance* for shape i (i.e., the distance between the x axis and the x' axis for shape i), and m is the mass for shape i. Similar expressions may be written for I_y and I_z. To use Eq. (A.46), it is necessary to know the mass moment of inertia for each of the composite shapes about its mass center axis, and this generally must be obtained by integration or, when possible, by consulting a table of moments of inertia for common shapes, such as the Table of Properties of Solids on the inside back cover. A common error is to use the parallel axis theorem to relate moments of inertia between two parallel axes where neither of them is a mass center axis.

Angular Momentum of a Rigid Body

In Section 8.2, we used the angular momentum of a rigid body to derive the angular impulse-momentum principle for a rigid body. We will now show how we obtained Eq. (8.42) on p. 627 for the angular momentum of a rigid body. Along the way, we will derive the angular momentum for a rigid body in three-dimensional motion since that result will be useful in Chapter 10.

We will now show how we obtained the angular momentum of a rigid body $\vec{h}_P$ as given by Eq. (8.42) on p. 627, which we repeat here as

$$\vec{h}_P = I_G \vec{\omega}_B + \vec{r}_{G/P} \times m\vec{v}_G, \tag{B.1}$$

where I_G and $\vec{\omega}_B$ are the body's mass moment of inertia and angular velocity, respectively (see Fig. B.1). At the time, we said that this equation is applicable to rigid bodies in planar motion that are symmetric with respect to the plane of motion. We could derive Eq. (B.1) directly, but it will be useful to derive the general three-dimensional form of the angular momentum of a rigid body and then simplify our result for rigid bodies in planar motion that are symmetric with respect to the plane of motion. We will find that the expression for $\vec{h}_P$ that we obtain for the three-dimensional motion of a rigid body will also be useful in Chapter 10.

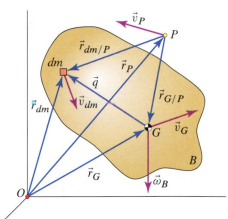

Figure B.1
A rigid body in motion with the definitions of the vectors needed to characterize the body's angular momentum.

Angular momentum of a rigid body undergoing three-dimensional motion

Referring to Fig. B.1, recall that the angular momentum of a rigid body B about the arbitrary point P was defined in Eq. (7.20) on p. 549 to be

$$\vec{h}_P = \int_B \vec{r}_{dm/P} \times \vec{v}_{dm}\, dm. \tag{B.2}$$

The vectors $\vec{r}_{dm/P}$ and $\vec{v}_{dm}$ indicate the position of the infinitesimal mass element dm relative to P and the velocity of dm, respectively. Letting $\vec{q}$ indicate the position of dm relative to the mass center G, and noting that G and dm are two points on a rigid body, we can write

$$\vec{r}_{dm/P} = \vec{q} + \vec{r}_{G/P} \quad \text{and} \quad \vec{v}_{dm} = \vec{v}_G + \vec{\omega}_B \times \vec{q}. \tag{B.3}$$

Substituting Eqs. (B.3) into Eq. (B.2) and expanding the cross products, we have

$$\vec{h}_P = \int_B \vec{q} \times (\vec{\omega}_B \times \vec{q})\, dm + \int_B \vec{q} \times \vec{v}_G\, dm$$
$$+ \int_B \vec{r}_{G/P} \times (\vec{\omega}_B \times \vec{q})\, dm + \int_B \vec{r}_{G/P} \times \vec{v}_G\, dm. \tag{B.4}$$

To simplify the right-hand side of Eq. (B.4), recall that $\vec{q}$ measures the position of each mass element dm relative to the mass center G of the body. Therefore, the definition of center of mass requires that

$$\int_B \vec{q}\, dm = m\vec{r}_{G/G} = \vec{0}. \tag{B.5}$$

Since the quantities $\vec{v}_G, \vec{r}_{G/P}$, and $\vec{\omega}_B$ are not a function of position within B, the last three terms on the right-hand side of Eq. (B.4) become

$$\int_B \vec{q} \times \vec{v}_G\, dm = \left(\int_B \vec{q}\, dm \right) \times \vec{v}_G = \vec{0}, \tag{B.6}$$

$$\int_B \vec{r}_{G/P} \times (\vec{\omega}_B \times \vec{q})\, dm = \vec{r}_{G/P} \times \left[\vec{\omega}_B \times \left(\int_B \vec{q}\, dm \right) \right] = \vec{0}, \tag{B.7}$$

and

$$\int_B \vec{r}_{G/P} \times \vec{v}_G\, dm = (\vec{r}_{G/P} \times \vec{v}_G) \int_B dm = \vec{r}_{G/P} \times m\vec{v}_G. \tag{B.8}$$

Substituting Eqs. (B.6)–(B.8) into Eq. (B.4), it becomes

$$\vec{h}_P = \int_B \vec{q} \times (\vec{\omega}_B \times \vec{q})\, dm + \vec{r}_{G/P} \times m\vec{v}_G. \tag{B.9}$$

The second term on the right side of Eq. (B.9) needs no further interpretation since $\vec{r}_{G/P}$ is the position of the mass center of the rigid body relative to the reference point P, and $\vec{v}_G$ is the velocity of the mass center of the rigid body. As for the integral on the right side of Eq. (B.9), writing $\vec{q}$ and $\vec{\omega}_B$ in Cartesian components as (see Fig. B.2)

$$\vec{q} = q_x \hat{\imath} + q_y \hat{\jmath} + q_z \hat{k} \quad \text{and} \quad \vec{\omega}_B = \omega_{Bx} \hat{\imath} + \omega_{By} \hat{\jmath} + \omega_{Bz} \hat{k}, \tag{B.10}$$

and then substituting them into the integral on the right side of Eq. (B.9), we obtain

$$\int_B \vec{q} \times (\vec{\omega}_B \times \vec{q})\, dm$$
$$= \left\{ \int_B \left[(q_y^2 + q_z^2)\, \omega_{Bx} - q_x q_y \omega_{By} - q_x q_z \omega_{Bz} \right] dm \right\} \hat{\imath}$$
$$+ \left\{ \int_B \left[-q_x q_y \omega_{Bx} + (q_x^2 + q_z^2)\, \omega_{By} - q_y q_z \omega_{Bz} \right] dm \right\} \hat{\jmath}$$

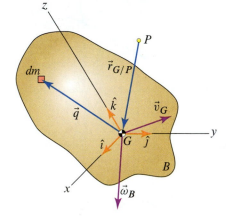

Figure B.2

$$+ \left\{ \int_B \left[-q_x q_z \omega_{Bx} - q_y q_z \omega_{By} + \left(q_x^2 + q_y^2 \right) \omega_{Bz} \right] dm \right\} \hat{k}. \quad \text{(B.11)}$$

Since the components of $\vec{\omega}_B$ are not a function of position within B, they can be brought outside of the integrals. Bringing these angular velocity components outside the integrals and then using the definitions of mass moment and mass product of inertia given in Eqs. (A.1)–(A.6) on p. 731, we obtain

$$\int_B \vec{q} \times \left(\vec{\omega}_B \times \vec{q} \right) dm = \left(I_x \omega_{Bx} - I_{xy} \omega_{By} - I_{xz} \omega_{Bz} \right) \hat{i}$$
$$+ \left(-I_{xy} \omega_{Bx} + I_y \omega_{By} - I_{yz} \omega_{Bz} \right) \hat{j}$$
$$+ \left(-I_{xz} \omega_{Bx} - I_{yz} \omega_{By} + I_z \omega_{Bz} \right) \hat{k}, \quad \text{(B.12)}$$

where it is understood that the moments and products of inertia *are with respect to the mass center* G since $\vec{q}$ measures the position of dm with respect to G. Writing $\vec{r}_{G/P}$ and $\vec{v}_G$ in Cartesian components as

$$\vec{r}_{G/P} = x_{G/P}\, \hat{i} + y_{G/P}\, \hat{j} + z_{G/P}\, \hat{k}, \quad \text{(B.13)}$$

$$\vec{v}_G = v_{Gx}\, \hat{i} + v_{Gy}\, \hat{j} + v_{Gz}\, \hat{k}, \quad \text{(B.14)}$$

respectively, then, substituting Eqs. (B.12)–(B.14) into Eq. (B.9), we find that the *angular momentum of a rigid body* is given by

$$\vec{h}_P = \left[I_x \omega_{Bx} - I_{xy} \omega_{By} - I_{xz} \omega_{Bz} \right.$$
$$\left. + m \left(y_{G/P} v_{Gz} - z_{G/P} v_{Gy} \right) \right] \hat{i}$$
$$+ \left[-I_{xy} \omega_{Bx} + I_y \omega_{By} - I_{yz} \omega_{Bz} \right.$$
$$\left. + m \left(z_{G/P} v_{Gx} - x_{G/P} v_{Gz} \right) \right] \hat{j}$$
$$+ \left[-I_{xz} \omega_{Bx} - I_{yz} \omega_{By} + I_z \omega_{Bz} \right.$$
$$\left. + m \left(x_{G/P} v_{Gy} - y_{G/P} v_{Gx} \right) \right] \hat{k}. \quad \text{(B.15)}$$

If any of the following conditions is satisfied:

1. The reference point P is the mass center G of the rigid body so that $\vec{r}_{G/P} = \vec{0}$.

2. The mass center G of the rigid body is a fixed point so that $\vec{v}_G = \vec{0}$.

3. The velocity of the mass center G, $\vec{v}_G$, is parallel to the position of G relative to P, $\vec{r}_{G/P}$.

then Eq. (B.15) simplifies to

$$\vec{h}_P = \left(I_x \omega_{Bx} - I_{xy} \omega_{By} - I_{xz} \omega_{Bz} \right) \hat{i}$$
$$+ \left(-I_{xy} \omega_{Bx} + I_y \omega_{By} - I_{yz} \omega_{Bz} \right) \hat{j}$$
$$+ \left(-I_{xz} \omega_{Bx} - I_{yz} \omega_{By} + I_z \omega_{Bz} \right) \hat{k}, \quad \text{(B.16)}$$
$$= h_{Px}\, \hat{i} + h_{Py}\, \hat{j} + h_{Pz}\, \hat{k}. \quad \text{(B.17)}$$

Conditions 1 and 2 are very common in practice, so Eq. (B.16) is frequently used.

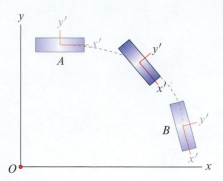

Figure B.3
A rectangular rigid body rotating in planar motion relative to the *fixed xy* frame. The *x'y'* is *attached* to the body.

Practical use of Eqs. (B.15) and (B.16)

The xyz reference frame referred to in Eqs. (B.15) and (B.16) can be any frame (inertial or not) as long as velocity and angular velocity components are computed with respect to an inertial frame. With that in mind, if the body B rotates relative to the xyz frame, the moments and products of inertia in Eqs. (B.15) and (B.16) will be *time dependent*. For example, referring to Fig. B.3, notice that as the rectangular body moves in the xy plane, its distribution of mass changes relative to the non-moving xyz reference frame and therefore its moments and products of inertia will change (i.e., be time dependent) relative to the frame. On the other hand, the moments and products of inertia relative to the $x'y'z'$ frame that is attached to the body will be *constant*. We will see in applications, especially when dealing with the dynamics of rigid bodies in three dimensions, that attaching a reference frame to the body will be the preferred strategy for solving problems. We will refer to these frames as *body fixed*.

A compact way to write Eqs. (B.15) and (B.16)

Using a combination of matrix and vector notation, Eq. (B.15) can be written as

$$\{h_P\} = [I_G]\{\omega_B\} + \vec{r}_{G/P} \times m\vec{v}_G, \tag{B.18}$$

where $\{h_P\}$ is the angular momentum vector of the body B with respect to P, $\{\omega_B\}$ is the angular velocity vector of B, and $[I_G]$ is the *inertia matrix* or *inertia tensor* of B with respect the xyz axes. In matrix form *and* vector form, these quantities are given by

$$\{h_P\} = \begin{Bmatrix} h_{Px} \\ h_{Py} \\ h_{Pz} \end{Bmatrix} = \vec{h}_P = h_{Px}\,\hat{\imath} + h_{Py}\,\hat{\jmath} + h_{Pz}\,\hat{k}, \tag{B.19}$$

$$\{\omega_B\} = \begin{Bmatrix} \omega_{Bx} \\ \omega_{By} \\ \omega_{Bz} \end{Bmatrix} = \vec{\omega}_B = \omega_{Bx}\,\hat{\imath} + \omega_{By}\,\hat{\jmath} + \omega_{Bz}\,\hat{k}, \tag{B.20}$$

and

$$[I_G] = \begin{bmatrix} I_x & -I_{xy} & -I_{xz} \\ -I_{xy} & I_y & -I_{yz} \\ -I_{xz} & -I_{yz} & I_z \end{bmatrix}.$$

Standard matrix multiplication of $[I_G]$ and $\{\omega_B\}$, with the addition of the vector cross product $\vec{r}_{G/P} \times m\vec{v}_G$, then gives $\{h_P\}$ as given in either Eq. (B.15) or (B.18). With this as background, we see that Eq. (B.16) can be written much more compactly as

$$\{h_P\} = [I_G]\{\omega_B\}. \tag{B.21}$$

Angular momentum of a rigid body in planar motion

We are now in a position to develop Eq. (B.1), which we recall is the angular momentum of a rigid body in planar motion that is symmetric with respect to the plane of motion.

If the rigid body B is in planar motion and the motion is occurring in the xy plane, then

$$\omega_{Bx} = \omega_{By} = 0, \quad v_{Gz} = 0, \quad \text{and} \quad z_{G/P} = 0, \qquad \text{(B.22)}$$

which means that Eq. (B.15) becomes

$$\vec{h}_P = -I_{xz}\omega_{Bz}\,\hat{\imath} - I_{yz}\omega_{Bz}\,\hat{\jmath}$$
$$+ \left[I_z\omega_{Bz} + m\left(x_{G/P}v_{Gy} - y_{G/P}v_{Gx}\right)\right]\hat{k}, \qquad \text{(B.23)}$$

where, again, the mass moments and products of inertia are computed with respect to the mass center G of the rigid body.

Finally, if the xy plane is also a plane of symmetry for the body, then $I_{xz} = 0$ and $I_{yz} = 0$ and so Eq. (B.23) can be written as

$$\vec{h}_P = \left[I_z\omega_{Bz} + m\left(x_{G/P}v_{Gy} - y_{G/P}v_{Gx}\right)\right]\hat{k}. \qquad \text{(B.24)}$$

Noting that for planar motion, $I_z = I_G$, $\omega_{Bz}\,\hat{k} = \vec{\omega}_B$, and that $\vec{r}_{G/P} \times m\vec{v}_G = m\left(x_{G/P}v_{Gy} - y_{G/P}v_{Gx}\right)$, we can write Eq. (B.24) as

$$\vec{h}_P = I_G\vec{\omega}_B + \vec{r}_{G/P} \times m\vec{v}_G, \qquad \text{(B.25)}$$

which is the form of $\vec{h}_P$ used in Eq. (8.42).

Answers to Selected Problems

The answers to most even-numbered problems are provided in a freely down-loadable PDF file at:

www.mhhe.com/pgc/

Providing answers in this manner allows for more complex information than would otherwise be possible. In addition to final numerical and/or symbolic answers, selected problems have more extensive information such as free body diagrams and/or plots for Computer Problems. For example, Prob. 7.52 and the corresponding answer (as given in the online answers) are shown below.

Problem 7.52

A spool of mass $m = 200\,\text{kg}$, radius $r = 0.8\,\text{m}$, and radius of gyration $k_G = 0.65\,\text{m}$ rolls without slipping on the incline, whose angle with respect to the horizontal is $\theta = 38°$. A linear elastic spring with constant $k = 500\,\text{N/m}$ and unstretched length $L_0 = 1.5\,\text{m}$ connects the center of the spool to a fixed wall. Determine the equation(s) of motion of the spool using the x coordinate shown; solve them for 15 s, using the initial conditions $x(0) = 2.5\,\text{m}$ and $\dot{x}(0) = 0\,\text{m/s}$; and then plot x versus t. What is the approximate period of oscillation of the spool?

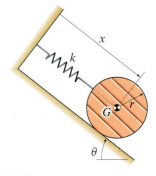

Figure P7.52

Problem 7.52 The equation of motion is

$$m\left(1 + \frac{k_G^2}{r^2}\right)\ddot{x} + kx = mg\sin\theta + kL_0.$$

The FBD used to obtain the equation of motion and the required plot are shown below.

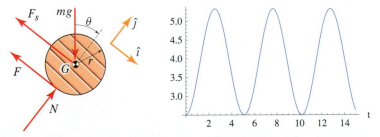

The approximate period of oscillation is seen to be 5.1–5.2 s.

This feature not only provides more complete answers in selected circumstances, but also provides the FBDs that are sometimes needed to provide the modeling kick-start to get you started on the homework problems. Furthermore, the FBDs provided in some answers will give you the opportunity to work extra problems when preparing for quizzes and exams.

CREDITS

PHOTO CREDITS

Chapter 1

Page 1, Opener: © SuperStock/Super-Stock, Inc.; p. 3, **Figure 1.1:** Portrait of Newton by Sr. Godfrey Kneller, 1689, © Photo by Jeremy Whitaker; p. 5, **Figure 1.3(both):** NASA; p. 6, **Fig 1.4:** © Index Stock Imagery/Jupiter Images; p. 25, **Figure 1.19:** © M.S.C.U.A., University of Washington, Farquharson, 12.

Chapter 2

Page 29, Opener: © Paul Slaughter/www.slaughterphoto.com; p. 29, **Figure 2.1:** © Adam Pretty/Getty Images; p. 43, **Ex. 2.6, Figure 1:** © Colin Anderson/Blend Images/CORBIS; p. 45, **Ex. 2.7, Figure 1:** © Tim de Waele/CORBIS; p. 53, **Figure P2.34 and P2.35:** NASA; p. 63, **Ex. 2.9, Figure 1:** © Terrance Klassen/Alamy; p. 65, **Ex.2.10, Figure 1:** U.S. Navy photo by Seaman Daniel A. Barker; p. 72, **Figure P2.71:** © Universal Studios; p. 72, **Figure P2.73-75:** © David Lees/CORBIS; p. 73, **Figure 2.18:** © Geoff Dann/Getty Images; p. 83, **Figure P2.85:** US Army Photo; p. 101, **Figure P2.119:** "Design and Analysis of a Surface Micromachined Spiral-Channel Viscous Pump", by M. I. Kilani, P.C. Galambos, Y.S. Haik, C.H. Chen, *Journal of Fluids Engineering,* Vol. 125, pp. 339–344, 2003; p. 104, **Figure 2.32 and Figure 2.33:** Courtesy of Watkins Glenn International Raceway; p. 104, **Figure 2.34:** Courtesy of FIA; p. 109, **Ex. 2.17, Figs. 1-3:** Courtesy of FIA; p. 111, **Ex. 2.19, Figure 1:** NASA; p. 114, **Figure P2.131:** Courtesy of FIA; p. 115, **Figure P2.137:** Department of Energy; p. 123, **Ex. 2.21, Figure 1:** © Charles O'Rear/CORBIS; p. 132, **Figure P2.173-174:** "Design and Analysis of a Surface Micromachined Spiral-Channel Viscous Pump", by M.I. Kilani, P.C. Galambos, Y.S. Haik, C.H. Chen, *Journal of Fluids Engineering* Vol. 125, pp. 339–344, 2003.; p. 154, **Figure 2.54:** © Joe Jennings/Jennings Productions; p. 160, **Ex. 2.29, Figures 1a-b,** p. 166 **Figure P2.229,** p. 181, **Figure P2.260:** author photos.

Chapter 3

Page 183, Opener: © Robert King: p. 183, **Figure 3.1:** © David A. Northcott/CORBIS;

p. 187, **Figure 3.6:** author photo; p. 212, **Ex. 3.4, Figure 1:** © Scott Halleran/Getty Images; p. 232, **Figure 3.23:** U.S. Navy.

Chapter 4

Page 259, Opener: © Eric Gaillard/Reuters/CORBIS; p. 265, **Ex. 4.1, Figure 1:** U.S. Navy photo by Mass Communication Specialist 3rd Class Torrey W. Lee; p. 270, **Figure P4.1:** NASA; p. 273, **Figure P4.16:** U.S. Navy photo by Mass Communication Specialist 3rd Class Torrey W. Lee; p. 276, **Figure 4.8:** © Michael Dahms/Lift-World.info; p. 276, **Figure 4.9:** © Michael Busselle/Robert Harding World Imagery/CORBIS; p. 286, **Ex. 4.5, Figure 1:** © GFC Collection/Alamy; p. 301, **Figure DP4.1:** © tbkmedia.de/Alamy; p. 309, **Ex. 4.11, Figure 1:** author photo; p. 320, **Figure 4.23:** © AP Photo/Ferrari Press Office, HO; p. 322, **Ex. 4.13, Figure 1:** U.S. Navy photo by Mass Communication Specialist 3rd Class Torrey W. Lee; p. 326, **Figure P4.77:** Mazda Miata © 2006 Mazda Motor of America, Inc. Used by permission.

Chapter 5

Page 333, Opener: NASA; p. 333, **Figure 5.1:** © Steve Cole/Getty Images; p. 334, **Figure 5.2:** © Loren M. Winters, Durham, NC; p. 340, **Ex. 5.1, Figure 1:** PH3 Christopher Mobley/U.S. Navy; p. 347, **Figure P5.3** © Rooney, Irving & Associates, Ltd.; p. 348, **Figure P5.7a:** U.S. Navy photo by Photographer's Mate 2nd Class H. Dwain Willis; p. 348, **Figure P5.7b:** PHAN James Farrally II, US Navy; p. 348, **Figure P5.7c:** U. S. Navy photo by Photographer's Mate 3rd Class (AW) J. Scott Campbell; p. 350, **Figure P5.15:** © Chip Simons/Jupiter Images; p. 362, **Figure 5.20:** author photo; p. 388, **Figure 5.28:** © Lucinda Dowell; p. 410, **Figure 5.39:** NASA; p. 414, **Figure 5.45:** NASA; p. 422, **Figure P5.100-101:** NASA; p. 423, **Figure P5.104:** NASA; p. 426, **Figure 5.50:** © Sean Gallup/Getty Images; p. 434, **Ex. 5.18, Figure 1:** © Royalty-Free/CORBIS; p. 436, **Ex. 5.19, Figure 1,** p. 443 **Figure P5.126a:** Courtesy of JetPack International, LLC; p. 445, **Figure P5.136a:** Courtesy of Andritz Hydro, Austria.

Chapter 6

Page 459, CO6: © John Peter Photography/Alamy; p. 459, **Figure 6.1:** © Ford Motor

Company; p. 463, **Fig 6.11:** Courtesy of First Team Sports, www.firstteaminc.com; p. 467, **Ex. 6.1, Figure 1:** © Ford Motor Company; p. 469, **Ex. 6.3, Figure 1:** © AP Photo/Majdi Mohammed; p. 470, **Figure P6.4:** © Arrow Gear Company; p. 470, **Figure P6.5:** NASA; p. 471, **Figure P6.6:** Value RF| © Lawrence Manning/CORBIS; p. 471, **Figure P6.7a:** © Andritz Hydro, Austria; p. 471, **Figure P6.6:** Value RF| © Lawrence Manning/CORBIS; p. 471, **Figure P6.7b:** Courtesy of Voith Hydro, Germany; p. 472, **Figure P6.12:** author photo; p. 473, **Figure P6.18:** © David Lees/CORBIS; p. 475, **Figure P6.27-28:** © Martin Child/Getty Images RF; p. 475, **Figure P6.29-31:** author photos; p. 481 **Ex. 7.4 Figure 1:** Mazda Miata © 2006 Mazda Motor of America, Inc. Used by permission; p. 483, **Ex. 6.4, Figure 1:** © Lester Lefkowitz/CORBIS; p. 486, **Ex. 6.6, Figure 1:** © Jon Reis; p. 486 **Ex 6.6, Figure 2:** © The McGraw-Hill Companies, Inc./Photo by Lucinda Dowell; p. 488, **Ex. 6.7, Figure 1:** Courtesy of Otto Bock HealthCare, Germany; p. 490, **Ex. 6.8, Figure 1:** © AP Photo/Majdi Mohammed; p. 502, **Ex. 6.9, Figure 1:** © Lester Lefkowitz/CORBIS; p. 508, **Ex. 6.12, Figure 1:** Courtesy of Otto Bock HealthCare, Germany; p. 516, **Figure DP6.1:** Courtesy of Specialized Bicycles; p. 517, **Figure 6.39:** © Jon Reis; p. 532, **Figure P6.117:** NASA.

Chapter 7

Page 545, Opener: © Quinn Rooney/Getty Images; p. 545, **Figure 7.1:** Courtesy of Kawasaki Motors Corp., U.S.A.; p. 554, **Figure 7.11:** Courtesy of Metaldyne Corporation; pp. 560-561, **Ex. 7.2, Figs. 1-4:** Courtesy of Beckman Coulter, Inc.; p. 564, **Ex. 7.4, Figure 1 and** p. 572, **Figure P7.1:** Mazda Miata © 2006 Mazda Motor of America, Inc. Used by permission; p. 575, **Figure P7.15-17:** Courtesy of Amazing Gates of America; p. 579, **Figure P7.38 and** p. 581, **Figure P. 7.50:** Mazda Miata © 2006 Mazda Motor of America, Inc. Used by permission, p. 585, **P7.64-65a,b (both):** NASA; p. 590, **Figure DP7.1:** Courtesy of Ducati Motor Holding.

Chapter 8

P. 591, Opener: NASA; p. 592, **Figure 8.2:** NASA; p. 597, **Figure 8.8:** © Joe Kilmer/Penninsula Spring Corp.; p. 600, **Ex. 8.1,**

INDEX

Plesha et al., Dynamics

Properties of lines and areas. Length L, area A, centroid C, and area moments of inertia.

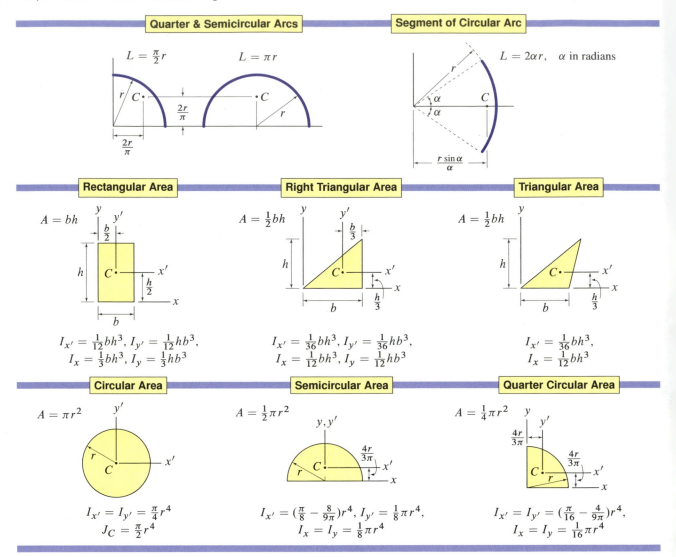

Quarter & Semicircular Arcs

$L = \frac{\pi}{2}r$ $L = \pi r$

Segment of Circular Arc

$L = 2\alpha r$, α in radians

$\frac{r \sin \alpha}{\alpha}$

Rectangular Area

$A = bh$

$I_{x'} = \frac{1}{12}bh^3, I_{y'} = \frac{1}{12}hb^3,$
$I_x = \frac{1}{3}bh^3, I_y = \frac{1}{3}hb^3$

Right Triangular Area

$A = \frac{1}{2}bh$

$I_{x'} = \frac{1}{36}bh^3, I_{y'} = \frac{1}{36}hb^3,$
$I_x = \frac{1}{12}bh^3, I_y = \frac{1}{12}hb^3$

Triangular Area

$A = \frac{1}{2}bh$

$I_{x'} = \frac{1}{36}bh^3,$
$I_x = \frac{1}{12}bh^3$

Circular Area

$A = \pi r^2$

$I_{x'} = I_{y'} = \frac{\pi}{4}r^4$
$J_C = \frac{\pi}{2}r^4$

Semicircular Area

$A = \frac{1}{2}\pi r^2$

$I_{x'} = (\frac{\pi}{8} - \frac{8}{9\pi})r^4, I_{y'} = \frac{1}{8}\pi r^4,$
$I_x = I_y = \frac{1}{8}\pi r^4$

Quarter Circular Area

$A = \frac{1}{4}\pi r^2$

$I_{x'} = I_{y'} = (\frac{\pi}{16} - \frac{4}{9\pi})r^4,$
$I_x = I_y = \frac{1}{16}\pi r^4$